AF843903

Nineteenth International Seaweed Symposium

Developments in Applied Phycology 2

Series Editor:
Michael A. Borowitzka
School of Biological Sciences & Biotechnology
Murdoch University, Murdoch, Western Australia

Nineteenth International Seaweed Symposium

Proceedings of the 19th International Seaweed Symposium, held in Kobe, Japan, 26–31 March, 2007

Edited By

Michael A. Borowitzka, Alan T. Critchley, Stefan Kraan, Akira Peters, Kjersti Sjøtun, Masahiro Notoya

Meeting coordinated by the INTERNATIONAL SEAWEED ASSOCIATION

Hosted by:
The Japan Seaweed Association
The Japanese Society of Phycology
The Japanese Society of Marine Biotechnology

Sponsors:
Portopia '81 Foundation
The Tsutomu Nakauchi Foundation
Japan Plant Science Foundation
The late Mr. Kiyoshi Kadoya
Japan Seaweed Association
Japanese Society of Applied Phycology

Previously published in the Journal of Applied Phycology Volume 20, Issue 5, 2008

Editors
Dr. Michael A. Borowitzka
School of Biological Sciences
Algal Biotechnology
Murdoch University
Murdoch WA 6150, Australia
m.borowitzka@murdoch.edu.au

Akira Peters
Bezhin Rosco
28 Rue de Perharidy
29680 Roscoff, France

Alan T. Critchley
Acadian Seaplants Ltd.
30 Brown Avenue
Dartmouth NS B3B 1X8, Canada

Kjersti Sjøtun
University of Bergen
Department of Biology
5020 Bergen, Norway

Stefan Kraan
National University of Ireland, Galway
Martin Ryan Institute
Irish Seaweed Centre
University Road Galway, Ireland

Masahiro Notoya
Tokyo University of Marine Science & Technology
Department of Marine Bioscience
4 Konan, Tokyo
Minato-ku, 108-8477 Japan

Cover Picture
Machine to manufacture sheets of the red alga nori (*Porphyra*) for sushi on Shikoku Island, Japan
(Photo: Thierry Chopin)

ISBN: 978-1-4020-6918-1 eISBN: 978-1-4020-9619-8

DOI: 10.1007/978-1-4020-9619-8

Library of Congress Control Number: 2009920519

Printed on acid-free paper.

springer.com

Contents

International Seaweed Association Executive Council and Officers
Harris "Pete" Bixler (President and Treasurer), USA
Guillermo Garcia Reyna, Spain
Masao Ohno, Japan
Dimitri Stancioff, USA
Juan Correa, Chile
Alan Critchley (Editor-in-Chief), Canada
In Kyu Lee, Korea
Michael Friedlander, Israel
Peter Salling, Spain
Thierry Chopin (Vice-President), Canada
Robert Anderson, South Africa
Rhodora Azanza, The Philippines
Eurico C. Oliveira (Secretary), Brazil
Michael Guiry (Webmaster), Ireland
Honorary members
Jack McLachlan, Canada
Mark Ragan, Australia

Local Organizing Committee
Chair: Yusho Aruga, Professor Emeritus, Tokyo University of Fisheries;
Secretary: Masao Ohno, Professor Emeritus, Kochi University;
Treasurer: Masahiro Notoya, Professor, Tokyo University of Marine Science and Technology;
Officers: Hideomi Amano, Professor, Mie University;
Yoshiaki Hara, Professor, Yamagata University;
Hiroshi Kawai, Professor, Kobe University;
Tadashi Matsunaga, Professor, Tokyo University of Agriculture and Technology;
Naotsune Saga, Professor, Hokkaido University;
Katsuaki Iwamoto, President, Marine Science Co., Ltd.;
Fumiyoshi Kasahara, President, KIMICA Corportation, Co., Ltd.;
Hiroshi Tsukakoshi, Chairman, Ina Food Industry, Co., Ltd.

Reviewers:
The Editors of these Proceedings are grateful to the following people for reviewing manuscripts (in some cases more than one): Chris Amsler, Robert Anderson, Rhodora Verzo Azanza, David Ballantyne, R Barreiro, Inka Bartsch, Kai Bischof, Harris (Pete) Bixler, John Bolton, John H Bothwell, Matthew Bracken, TL Bray, Anders Broberg, Juliet Brodie, Alejandro Buschmann, P Chigbu, Han Gil Choi, Thierry Chopin, Hartvig Christie, J Mark Cock, Gloria Collantes, Solene Connan, Peter Cook, Jeffrey Cosgrove, TE Cox, James A Coyer, James S Craigie, D Kapraun, Jose Darias, Bellwood R David, Mariana C de Oliveira, Olivier Declerck, De Lin Duan, Nils Ekelund, R G Elkin, William Farnham, W Fenical, F Fernandez, Helen Fitton, Sofia Fredriksson, Yolanda Freile Pelegrin, Michael Friedlander, Meenakshisundaram Ganesan, Thierry Givernaud, F J Gordillo, Michael Graham, Bruce Grindley, Martin Gullstrøm, F Gurgel, Jeff Hafting, Gustaaf Hallegraeff, Chris Halling, Leila Hayashi, Chris Hepburn, Ignacio Hernandez Carrero, John Huisman, Catriona Hurd, Anicia Q Hurtado, Vivian Husa, Alvaro Israel, Joanna Kain, S Kaehler, Lena Kautsky, Yukihiro Kitade, Samuli Korpinen, A Kraberg, John Kuo, Tiit Kutser, AWD Larkum, G Le Gall, D Les, Klaus Luening, Jim Thinh Luong-Van, JC McComb, Michael Melkonian, Subhash Minocha, Hiroyuki Mizuta, Steve Nelson, Wendy Nelson, Amir Neori, A D N'Yeurt, Eurico Oliveira, ShaoJun Pang, GF Pauli, Leonel Pereira, Stephan Pflugmacher, Siew-Moi Phang, Alastair Poore, Benoit Queguineur, CRK Reddy, Jeffrey Reimers, Eli Rinde, Deborah Robertson-Andersson, Daniel Robledo, Andrew Rodger, Hector Romo, V Roussis, Jan Rueness, Lisa Russell, Ari Ruuskanen, Dinabandhu Sahoo, Dalia Sanchez-Machado, Bernabe Santelices, Britta Schaffelke, Songdong Shen, Knut Sivertsen, Christophe Six, AJ Smit, Anna Soler, A Strezov, D Suggett, P Szefer, Jane Teas, Eduard Titlyanov, Giuseppe Torzillo, Raul Ugarte, Charles Santhanaraju Vairappan, Juan Vergara, Inger Wallentinus, Susse Wegeberg, Florian Weinberger, Christian Wiencke, Robert Wilkes, CW Wright, JT Wu, Tom Wydrzynski, Nair Yokoya, Norishige Yotsukura, GC Zuccarello.

Welcoming Remarks by Dr. H.J. Bixler, President of the International Seaweed Association (delivered by Dr. Thierry Chopin, Vice-President of the International Seaweed Association)

By tradition the outgoing president of ISA makes a few welcoming remarks.

Unfortunately in my case some delayed surgery on my ankle prevents me from being in attendance—much to my loss. You are about to participate in a marvelous program of Plenary Addresses, Oral Presentations, Poster Sessions and Workshops. The breadth of topics being covered convinces me that the field of algal science and engineering is alive and well, and can look forward to a healthy future as we use algae to tackle challenges in protecting the environment and making better use of renewable resources for food, energy and bioactive chemicals.

I would be remiss not to mention the Exhibition of seaweed products taking place in an adjoining hall. We Westerners have much to learn from the diversity of Asian uses of seaweeds.

My wife, Ann, will particularly miss the events for accompanying persons. Fortunately for me, I had a chance to visit most of the venues for these events during an ISA site visit in 2005.

As many of you know, I came from the hydrocolloids industry and not even as a card-carrying algologist but as a chemical engineer. Our industry lost a giant in the supply of temperate water seaweeds from Chile late last year when Hugo Pino lost his battle with cancer. Hugo, as managing director of Multiexports, added a touch of class to a much maligned segment of our industry—the seaweed traders. Hugo was not only a business acquaintance but, as a frequent attendee to ISSs, a true friend to many of us here. May we have learned from his high ethical standards and may his soul rest in peace.

I now declare the 19th International Seaweed Symposium open for its activities.

Marinalg Awards
From ISS XVIII Proceedings
(Presented in absentia by Dr. H.J. Bixler, Marinalg President)

Marinalg International is pleased once more to award $US1,000 each to the best papers presented at the 18th Symposium in Bergen that were of most interest to producers of carrageenan, agar and alginates. It should be pointed out that these awards at their inception were specifically for papers on these three hydrocolloids. However, as time has gone on, the interests of the scientists presenting papers at these symposia have veered away from this limited aspect of seaweed science. Nevertheless, Marinalg members maintain a strong interest in such fields as seaweed cultivation, resource management and environmental issues relating to seaweed harvesting, seaweed diseases and the like. In fact, the awards being announced today are all concerned with one or more of these issues.

Let me turn now to the specific awards.

For *carrageenan* the papers of most interest ranged from characterizing the cell wall carrageenans in *Chondrus pinnulatus* to analysis of an epiphyte infestation at a particular seaweed farming site in the Philippines. The winning paper, however, was:

Romo H, Avila M, Núñez M, Pérez R, Candia A, Aroca G.
Culture of *Gigartina skottsbergii* (Rhodophyta) in Southern Chile.
J Appl Phycol. (2006) 18: 307–314.

This paper confirms that effort expended on cultivating of this industrially important carrageenophyte is progressing and may in the not too distant future relieve stress on the natural harvest.

For *agar*, papers of most interest ranged from analysis of the reduction in *Gelidium* beds in Toyama Bay, Japan, to comparing factors affecting growth of two *Gracilaria* species in Korea. Marinalg's reviewer chose as the winning paper:

Bouzon ZL, Ouriques LC, Oliveira EC.
Spore adhesion in cell wall formation in *Gelidium floridanum* (Rhodophyta, Gelidiales).
J Appl Phycol. (2006) 18: 287–294.

This paper appealed to Marinalg's reviewer for its progress in further understanding the chemistry of *mucillages* in commercially important seaweeds.

While there were no papers on *alginate* per se presented at Bergen, there were several papers on resource management of alginate-bearing seaweed species. The reviewer for this award also noted that a few papers focused on the growing importance of these brown seaweeds as sources of human food and plant and animal nutrients. These value-added applications may soon replace alginates as the driving force for resource management and cultivation of brown seaweeds. The best paper from Marinalg's member's perspective was:

Kirihara S, Nakamura T, Kon N, Fujita D, Notoya M.
Recent fluctuations in distribution and biomass of cold and warm water species of Laminarialean algae at Cape Ohma, northern Honshu, Japan.
J Appl Phycol. (2006) 18: 521–527.

The particular appeal of this paper was its observation that increased water temperature over a 20-year period had significantly changed the distribution of seaweed species and reduced biomass. Whether or not this is a consequence of global warming it is an environmental trend that should raise concerns.

JSA Awards

Critchley AT, Hurtado AQ, Bleicher-Lhonneur Gm Leflamand A and Mazoyer J. Observations on the duration of culture of *Kapaphycus alvarezii* var. *tambalang*.

Hwang EK, Baeck JM and Park CS. The green alga, *Codium fragile*, cultivation by artificial seed production in Korea.

Hayashi Y and Notoya M. Morphological studies of a non vegetative propagation strain of *Ecklonia stolonifera* Okamura and *E. kurome* Okamura (Laminariales, Phaeophyta) from Oki Islands, Shimane Prefecture, Japan.

University of British Columbia Graduate Student Paper Awards

Parente M.I., Rousseau F., Fletcher R.L., Neto A.I., Vidal N. and de Reviers B. Re-evaluating orders within the Phaeophyceae: Nemodermatales Ord. Nov. Petrodermatales Ord. Nov. and Ralfsiales Ord. Nov., three new orders of crustose brown algae.

Robertson-Andersson D.V., Bolton J.J., Troell M., Anderson R.J., Maneveldt G., Halling C., Smit A.J., Probyn T. and Peall S. The evolution of integrated seaweed cultivation in temperate Southern Africa.

Nerio Y., Karato M., Koyama T., Yamaguchi K., Nemoto M., Shigemori H., Kamitani T., Toda T. and Yazawa K. Effect of *Laminaria japonica* extract on blood glucose levels in mice.

The International Seaweed Association would like to thank Dr. Juliet Brodie for taking the lead in organizing the selection panel comprised of her and Drs. Dennis Hanisak, Sandra Lindstrom, Wendy Nelson, Amir Neori, Mariana Oliveira, Akira Peters, Isabel Sousa-Pinto, John West, Nair Yokoya and Jose Zertuche.

Closing Remarks of the New President of the International Seaweed Association, Professor Thierry Chopin (Canada)

The 19th International Seaweed Symposium: a turning point in the seaweed industry sector?

Dear friends and colleagues, let me first carry out my official duties as the new President of the International Seaweed Association (ISA) and pronounce a few words for this Closing Ceremony.

I would like, first, to thank Dr Yusho Aruga (Chair), Dr Masao Ohno (Secretary), the Local Organizing Committee, the sponsors and the numerous volunteers for a wonderful 19th International Seaweed Symposium (ISS) this week in Kobe. The organization was superb, the venue beautiful and the hospitality measured up to all expectations.

We had 561 participants [including a healthy number (94) of students] from 47 countries, 4 plenary lectures, 102 presentations distributed among 22 mini-symposium themes, 67 presentations in 12 contributed paper sessions, 158 posters in 4 sessions, 1 special lecture and 4 workshops. We also had, for the first time in the history of the ISSs, a very interesting concurrent International Seaweed Exhibition of 61 seaweed product companies (57 from Japan and 4 from other countries): what an eye opener for participants from around the world, especially Westerners! I hope that such an initiative will be continued in the future ISSs.

On Thursday, the five mid-symposium excursions took us to different places according to our tastes: visits to nori and wakame farms, visits to seaweed processing factories, a seaweed collection trip, a visit to the Himeji Castle and the Suma Aqualife Museum, and a visit to Kyoto. I visited the impressive Himeji Castle on Sunday and imperial Kyoto last year, so I opted for the visits to nori and wakame farms. I can tell you that the cameras were making a lot of noise when people saw these nori harvesting boats going under the nets to "mow" the precious purple material: unbelievable pictures for all our phycology courses and presentations!

The Welcome Reception was an opportunity to get re-acquainted in a nice setting. The Sakura Festival introduces us to Japanese culture and gastronomy, even if our hosts were not expecting that an unleashed group of phycologists would behave like a bunch of grasshoppers! The Symposium Banquet was again delectable and in a very warm atmosphere.

I heard that the 61 accompanying persons had a wonderful time discovering Kobe and were culturally immersed in the interesting Ikebana, Chanoyu, Origami and Shodo programmes to give them a flavor of Japanese traditions.

On Monday, the International Seaweed Association Council (ISAC) had its triennial face to face meeting. The ISA governing body conducts its business electronically in-between the ISSs to be sure that we meet again every three years, each time in very interesting and varied places. Four members of the ISAC have their mandate expiring with this ISS: Guillermo Garcia Reina, In Kyu Lee, Masao Ohno, who gave so much in the preparation of this ISS, and Dimitri Stancioff, a pillar of the ISA for so many years. To all four we express our gratitude for their commitment towards the ISA. Four new members have joined the ISAC: Iain Neish, who has been elected Vice-President, Jose Zertuche, who will be the host of the next ISS in Ensenada, Mexico, Masahiro Notoya and Jean-Paul Deveau. To all four we extend our welcome and forewarn you to be ready to roll up your sleeves!

Now, I would like to talk about a subject close to my heart: whether we are witnessing a turning point in the seaweed industry sector with the 19th ISS?

The International Seaweed Exhibition in the adjoining building showed us that the seaweed industry sector is alive and in full mutation. Interestingly, the best known component of the seaweed-derived industry is that of the phycocolloids, the gelling, thickening, emulsifying, binding, stabilizing, clarifying and protecting agents known as carrageenans, alginates and agars. The phycocolloid industry is also historically at the origin of the ISA. For many years, it dominated the content of the ISSs and their Proceedings; it was also reflected in the composition of the ISAC. We owe a lot to the phycocolloid industry for having put seaweeds on the map for the production, harvesting, processing, and management of marine resources. Many of us here owe, directly or indirectly, our careers to the phycocolloid industry.

However, this component now represents only a minor volume (1.26 million wet tonnes or 11.2%) and value (US$650 million or 10.8%) of the entire seaweed-derived industry (Chopin and Sawhney 2008). The use of seaweeds as sea-vegetables for direct human consumption is much more significant in tonnage (8.59 million wet tonnes or 76.1%) and value (US$5.29 billion or 88.3%). Three genera, *Laminaria* (or kombu), *Porphyra* (or nori) and *Undaria* (or wakame) dominate the edible seaweed market. The phycosupplement industry is an emerging component estimated at 1.22 million wet tonnes, with a value of US$53 million. Most of the tonnage is used for the manufacturing of soil additives; however, the agrichemical and animal feed markets are comparatively much more lucrative if one considers the much smaller volume of seaweeds they require. The use of seaweeds in the development of pharmaceuticals, nutraceuticals, botanicals, cosmeceuticals, and as a source of pigments, bioactive compounds, and antiviral agents, is in full expansion. Presently, that market is difficult to evaluate accurately; the use of 3,000 wet tonnes of raw material to obtain 600 tonnes of products valued at US$3 million could be an underestimation.

The seaweed aquaculture production (92% of the world seaweed supply) doubled between 1996 and 2004, and is estimated at 11.3 million wet tonnes, with 99.7% of the biomass being cultivated in Asia [Food and Agriculture Organization of the United Nations (FAO) 2004; 2006a, b]. Brown seaweeds represent 63.8% of the production, while red seaweeds represent 36.0%, and the green seaweeds 0.2%. The seaweed aquaculture production is valued at US$5.7 billion (again with 99.7% of the value being provided by Asian countries). Brown seaweeds dominate with 66.8% of the value, while red seaweeds contribute 33.0%, and the green seaweeds 0.2%. Approximately 220 species of seaweeds are cultivated worldwide; however, 6 genera [*Laminaria* (kombu; 40.1%), *Undaria* (wakame; 22.3%), *Porphyra* (nori; 12.4%), *Eucheuma/Kappaphycus* (11.6%) and *Gracilaria* (8.4%)] are providing 94.8% of the seaweed aquaculture production, and 4 genera [*Laminaria* (47.9%), *Porphyra* (23.3%), *Undaria* (17.7%) and *Gracilaria* (6.7%)] are providing 95.6% of its value.

Published world statistics (FAO 2004, 2006a, b), which regularly include the notation "data exclude aquatic plants" in their tables, indicate that in 2004 the top four individual species produced by the global aquaculture (50.9% mariculture, 43.4% freshwater aquaculture and 5.7% brackishwater aquaculture) were the Pacific cupped oyster (*Crassostrea gigas* – 4.4 million wet tonnes), followed by three species of carp, the silver carp (*Hypophthalmichthys molitrix* – 4.0 million wet tonnes), the grass carp (*Ctenopharyngodon idellus* – 3.9 million wet tonnes) and the common carp (*Cyprinus carpio* – 3.4 million wet tonnes). However, in fact, the kelp, *Laminaria japonica*, was the first top species, with a production of 4.5 million wet tonnes.

Because the mariculture of aquatic plants (11.3 million wet tonnes of seaweeds and 2.6 million wet tonnes of unspecified "aquatic plants" reported by the FAO) has developed essentially in Asia, it remains mostly unknown in the West, and is often neglected or ignored in world statistics—a situation we can only explain as being due to a deeply rooted zoological bias in marine academics, resource managers, bureaucrats and policy advisors! However, in 2004, the seaweed aquaculture sector represented 45.9% of the biomass and 24.2% of the value of the world mariculture production, estimated in 2004 at 30.2 million wet tonnes, and worth US$28.1 billion (FAO 2006a, b). Mollusc aquaculture comes second at 43.0%, and finfish aquaculture, the subject of so many debates, actually only represents 8.9% of the world mariculture production.

The fact that seaweeds represent close to half of the biomass of the maricultured products is often not known outside phycological circles and often comes as a surprise to many. Moreover, the mitigation potential of seaweeds is often not recognized by the rest of the aquaculture world. Phycomitigation, through the development of integrated multi-trophic aquaculture (IMTA) systems, has been rediscovered in Western countries over the last 30 years, despite the fact that this sustainable aquaculture practice has existed for centuries, especially in Asian countries, through trial and error and experimentation (Chopin et al. 2001; Troell et al. 2003; Neori et al. 2004, 2007). It is difficult to give a value to the phycomitigation industry, inasmuch as presently no country has yet implemented guidelines and regulations regarding nutrient discharge into coastal waters. As the "user pays" concept is expected to gain momentum as a tool in integrated coastal management, one should soon be able to put a value to the phycomitigation services and benefits of IMTA systems for improving water quality and coastal health. Moreover, the conversion of fed aquaculture by-products into the production of salable biomass and biochemicals used in the sea-vegetable, phycocolloid and phycosupplement industries should increase the revenues generated by the phycomitigation component.

If the environmental, economic and societal services and benefits of IMTA are properly estimated and internalized in aquaculture operations, they will represent significant incentives for the cultivation of extractive species such as seaweeds (Chopin 2008). We have to make sure that seaweeds are considered in the global economy as nutrient (nitrogen, phosphorus, etc.) credits, in a similar way as they should be for carbon credits, as the aquaculture sector moves to becoming more efficient and sustainable. We also have to make sure that seaweeds figure prominently in the present debates about energy generation and biofuel production. We now have an opportunity to reposition the value and roles seaweeds have in coastal ecosystems, in the economy and in our society. As the experts on these organisms, let us not miss that opportunity for the development of marine agronomy. It is essential for the vitality of our discipline and for the future of our students.

Just a month ago, I attended the Aquaculture 2007 Conference of the World Aquaculture Society in San Antonio, Texas. Going through the book of 1,198 abstracts, I realized that:

- 66.3% of the abstracts were dedicated to fish (8.9% of the 2004 world mariculture production),
- 15.9% to mollusks (43.0% of the 2004 world mariculture production),
- 14.2% to crustaceans (1.8% of the 2004 world mariculture production), and
- 1.6% to other aquatic animals (0.4% of the 2004 world mariculture production).

By regrouping abstracts on microalgae (0.8%), freshwater plants (0.6%), and seaweeds (0.6%), I was able painfully to reach a combined 2.0% for all these organisms, whereas seaweeds represent, as indicated above, 45.9% of the world mariculture production!

We, obviously, have a huge educational role to play to help bring a balanced ecosystem approach to aquaculture development and to "convert" the animal-dominated aquaculture world that it needs to recognize and take advantage of the services and benefits of seaweeds for their extractive functions. For that to happen, we need to attend not only phycological conferences, the comfort of which we appreciate, but also non-phycological conferences to stop "preaching" only to the "converted". If, before the next ISS, we can all attend at least one non-phycological conference to disseminate our message of relevance in many everyday activities and global issues, we will already have achieved a lot.

In the meantime, I wish you all a safe trip back home and "Hasta la proxima" in Ensenada, Mexico, in 2010 for the 20th ISS!

References

Chopin T (2008) Integrated multi-trophic aquaculture (IMTA) will also have its place when aquaculture moves to the open ocean. Fish Farmer 31 (2):40–41

Chopin T, Sawhney M (2008) Seaweeds and their mariculture. In: Steele JH, Thorpe SA, Turekian KK (Eds.) The Encyclopedia of Ocean Sciences. Elsevier, Oxford (in press)

Chopin T, Buschmann AH, Halling C, Troell M, Kautsky N, Neori A et al (2001) Integrating seaweeds into marine aquaculture systems: a key towards sustainability. J Phycol 37:975–986 doi:10.1046/j.1529-8817.2001.01137.x

FAO (2004) The state of world fisheries and aquaculture 2004. Food and Agriculture Organization of the United Nations, Rome; ftp://ftp.fao.org/docrep/fao/007/y5600e/y5600e01.pdf

FAO (2006a) The state of world fisheries and aquaculture 2006. Food and Agriculture Organization of the United Nations, Rome; http://www.fao.org/docrep/009/a0699e/A0699E00.htm

FAO (2006b) State of world aquaculture 2006. Food and Agriculture Organization of the United Nations, FAO Fisheries Technical Paper 500, Rome; http://www.fao.org/docrep/009/a0874e/a0874e00.htm

Neori A, Chopin T, Troell M, Buschmann AH, Kraemer GP, Halling C et al (2004) Integrated aquaculture: rationale, evolution and state of the art emphasizing seaweed biofiltration in modern mariculture. Aquaculture 231:361–391 doi:10.1016/j.aquaculture.2003. ??11.015

Neori A, Troell M, Chopin T, Yarish C, Critchley A, Buschmann AH (2007) The need for a balanced ecosystem approach to blue revolution aquaculture. Environment 49(3):36–43 doi:10.3200/ENVT.49.3.36-43

Troell M, Halling C, Neori A, Chopin T, Buschmann AH, Kautsky N et al (2003) Integrated mariculture: asking the right questions. Aquaculture 226:69–90 doi:10.1016/S0044-8486(03)00469-1

List of registrants

Abdul Rokhmah
Ministry of Marine Affairs and Fisheries,
Jl.Medan Merdeka Timur16, Jakarta 10110
INDONESIA
abyana@yahoo.com

Abe Mahiko
National Research Institute of Fisheries Science
Fisheries Research Agency
2-12-4, Fukuura, Kanazawa-ku, Yokohama,
Kanagawa 236-8648
JAPAN
abemahi@affrc.go.jp

Ajisaka Tetsuro
Hall of Global Environmental Research
Kyoto University, Kitashirakawa-Oiwake-cho, Sakyo-ku,
Kyoto, 606-8502
JAPAN
ajisaka@kais.kyoto-u.ac.jp

Akakabe Yoshihiko
Faculty of Agriculture, Yamaguchi University
1677-1 Yoshida, Yamaguchi, 753-8515
JAPAN
akakabe@yamaguchi-u.ac.jp

Allard Marc R.
Nunavik Biosciences Inc.
1111 Dr. Frederik-Philips Blvd. (3rd Floor)
St. Hourent, Quebec H4M 2X6
CANADA
mallard@makiuik.org

Alquati Monica
ALIMENTOS MULTIEXPORT S.A.
Isidora Goyenechea 3621 17th Floor
Las Condes, Santiago, 6760412
CHILE
malquati@multiexport.cl

Altamira J. Benito
Productos Del Pacifico Sa De Cv
Calle Mexico 100-5, Mexicali BC 21100
MEXICO
alga@propasa.com.mx

Amano Hideomi
Lab. of Marine Biochem.,
Graduate School of Bioresources,
Mie University,
1577 Kurimamachiya-cho, Tsu, Mie, 514-8507
JAPAN
amano@bio.mie-u.ac.jp

Ando Wataru
Fisheries Infrastructure Development Center
Saiesuta Bldg., Tsukiji 2-14-5, Chuo-ku,
Tokyo, 104-0045
JAPAN
w-ando@fidec.or.jp

Ang Put O., Jr.
Department of Biology
The Chinese University of Hong Kong
Shatin, N. T., Hong Kong SAR
CHINA
put-ang@cuhk.edu.hk

Anggadiredja Jana Tjahjana
The Agency for the Assessment
and Application of Technology,
Republic of Indonesia
Gedung 2 BPPT, 19th Fl., JL. M.H. Thamrin 8
Jakarta 10340
INDONESIA
jana@bppt.go.id

Aoki Masakazu
Shimoda Marine Research Center
University of Tsukuba, 5-10-1 Shimoda, Shizuoka, 415-0025
JAPAN
m-aoki@kurofune.shimoda.tsukuba.ac.jp

Aragaki Tsunehide
Harmony Green, Inc., 240 East 47th St.,
Ste#2-F, New York, NY 10017
USA
harmonygusa@aragaki.com

Aruga Yusho
Nishikamata 2-4-21, Ota-ku, Tokyo, 144-0051
JAPAN
arugay@mx4.ttcn.ne.jp

Arvizu-Higuera Dora Luz
CICIMAR-IPN
Av. Instituto Politecnico Nacional s/n,
Col. Playa Palo de Sta Rita, La Paz,
Baja California Sur, 23096
MEXICO
darvizu@ipn.mx

Ask Erick Ingvald
FMC Corp.
1735 Market St., Philadelphia, PA 19103
USA
ERICK_ASK@FMC.COM

Atsuta Chikayoshi
Nogami 2-chome 4-16-206, Takarazuka-shi,
Hyogo, 665-0022
JAPAN
c-atsuta@jttk.zaq.ne.jp

Azanza Rhodora V.
The Marine Science Institute
University of the Philippines, Diliman
Velasquez Street, Quezon City, 1101
PHILIPPINES
rhod@upmsi.ph

Baba Masasuke
Demonstration Laboratory,
Marine Ecology Research Institute
4-7-17 Arahama, Kashiwazaki,
Niigata Pref., 945-0017
JAPAN
baba@kaiseiken.or.jp

Baek Jaemin
Seaweed Research Centre, National Fisheries
Research & Development Institute
Mokpo, 530-83
KOREA
jmbaek@momaf.go.kr

Baricuatro Farley L.
FMC Biopolymer Marine Colldips Phils
Ouano Compound, Looc,
Mandaue City, Cebu 6014
PHILIPPINES
farley-baricuatro@fmc.com

Barrett Anthony James
Arramara Teoranta
Kilkieran, Connemara, Co,
Galway
IRELAND
tbarrett@arramara.ie

Batista Vega Gloria Edelmira
University of Panama,
Vicepresidency of Research and Post Grade
Calle 59, Edificio Via Pacífica #23 B, San Francisco
PANAMA
gloriabv@cwpanama.net

Bernal Maria Beatriz de Barros Barreto
Universidade Federal Rural,
Rio de Janeiro
Antiga Rodovia, Rio-S.Paulo km 47,
Seropedica, RJ, 23851-970
BRAZIL
beatrizb@ufrrj.br

Bixler Harris "Pete" J.
Seaweed Industry Association of the Philippines
62 Bayside Rd., Northport, Maine 04849
USA
pbixler@isinc.to

Bleicher-Lhonneur Genevieve
Cargill Texturizing Solutions (ex-Degussa)
Baupte, 50500
FRANCE
Geneview_Bleicher-Lhonneur@cargill.com

Boo Sung Min
Department of Biology
Chungnam National University
Gungdong 220, Youseonggu, 305-76
KOREA
smboo@cnu.ac.kr

Boonprab Kangsadan
Department of Fishery Product,
Faculty of Fisheries Kasetsart University
50 Phaholyothin Rd., Chatuchak, Bangkok, 10900
THAILAND
bffisksb@yahoo.com

Borowitzka Michael A.
School of Biological Sciences & Biotechnology
Murdoch University
Murdoch, WA, 6150
AUSTRALIA
m.borowitzka@murdoch.edu.au

Bragg Aline J.
Exportaciones Pampamar S. A.
Bocaccio 9066-Las Condes, Santiago
CHILE
alinebragg@tie.cl

Brawley Susan Howard
School of Marine Sciences, University of Maine

5735 Hitchner Hall, Crono, ME 04469
USA
Brawley@maine.edu

Brodie Juliet Ann
Department of Botany, Natural History Museum
Cromwell Road, London, SW7 5BD
UK
J.Brodie@nhm.ac.uk

Broom Judy
Department of Biochemistry, Te Whare Matuora
University of Otago,
710 Cumberland Street,
Dunedin 9054, Aotearoa
NEW ZEALAND
judy.broom@stonebow.otago.ac.nz

Bujang Japar Sidik
Faculty of Agriculture and Food Sciences
University Putra Malaysia Bintulu Campus
Jalan Nyabau, P.O. Box 396, Sarawak 97008
Bintulu
MALAYSIA
japar@science.upm.edu.my

Buschmann Alejandro II
Universidad de Los Lagos
Camino Chinquihue km 6, Puerto Montt
CHILE
abuschma@ulagos.cl

Carnachan Susan Mary
Industrial Research Ltd
PO Box 31-310,
Lower Hutt 5040
NEW ZEALAND
s.carnachan@irl.cri.nz

Chapman Anthony
Botanica Marina
285 Duncans Cove Road, Ketch Harbour,
Nova Scotia, B3V 1K4
CANADA
ACHAPMAN@DAL.CA

Chen Cankun
Fuqing Hai Xing Health Protection Food Co., Ltd.
No.3 Rong Cheng Town, Yi Fu Lu, Fuqing,
Fujian, 350300
CHINA
haizao001@sohu.com

Chirapart Anong
Department of Fishery Biology,
Faculty of Fisheries, Kasetsart University
50 Phaholyothin Road, Chatuchak,
Bangkok 10900
THAILAND
ffisanc@ku.ac.th

Cho Ji Young
Department of Marine Biotechnology,
Soonchunghyang Univeristy,
646 Shinchang Myun,
Asan City 3360745
KOREA
hwshin@sch.ac.kr

Choi Chang Geun
Department of Aquaculture, Fisheries Science
Pukyong National University
608-737
KOREA
changgeuni@hanmail.net

Choi Han Gil
Faculty of Biological Science
Wonkwang University
344-2, Shinyong-dong, Iksan 570-749
KOREA
hgchoi@wonkwang.ac.kr

Chong Kwan Wai
JY Resources SDN BHD
No. 10, Lorong, Raja Udang 9,
Taman Kingfisher IT,
Kota Kinabalu, Sabah 88400
MALAYSIA
jyrccs@tm.net.my

Chopin Thierry
Department of Biology
University of New Brunswick
P. O. Box 5050,
Saint John, New Brunswick E2L 4L5
CANADA
tchopin@unbsj.ca

Chow Fungyi
Departamento de Botanica, Instituto de Biociencias,
University of São Paulo
Rua do Matão, Travessa 14, No.321, 05508-900
BRAZIL
fchow@ib.usp.br

Chung Ik Kyo
Pusan National University
San #30, Jangjung-dong, Geumjung-gu
Busan Metro City, 609-735
KOREA
ikchung@pusan.ac.kr

Ciancia Marina
Cátedra de Química Orgánica-Departamento de Biología Aplicada y Alimentos,
Facultad de Agronomía,
Universidad de Buenos Aires
Av. San Martín 4453,
Buenos Aires C1417DSE
ARGENTINA
ciancia@agro.uba.ar

Connan Solene
Dept. of Botany, Martin Ryan Institute,
National University of Ireland, Galway
IRELAND
solene.connan@nuigalway.ie

Copertino Margareth da Silva
Departamento de Oceanografia,
Fundação Universidade Federal do Rio Grande
Av. Itália, km 8, c.p. 474, Rio Grande, RS,
Cep 96201-900
BRAZIL
doccoper@furg.br

Craigie James S.
Acadian Seaplants Ltd
Institute for Marine Biosciences
1411 Oxford Street,
Halifax, NS, B3H 3Z1
CANADA
James.Craigie@nrc-cnrc.gc.ca

Critchley Alan
Acadian Seaplants Ltd
30 Brown Avenue,
Dartmouth, NS, B3B 1X8
CANADA
Alan.Critchley@acadian.ca

Dakay Benson UY
Shemberg Marketing Corporation
Seaweed Industry Assn. of the Phills.
S. Jayme St. Pakna-An, Maridaue City,
Cebu City 6014
PHILIPPINES

Dan Akinori
Fisheries Division, Tokushima Prefecture
Mandaicho 1, Tokushima, 770-8570
JAPAN
dan_akinori_1@pref.tokushima.lg.jp

Dang Diem Hong
Institute of Biotechnology,
Vietnamese Academy of Science and Technology
18 Hoang Quoc Viet, Cau Giay,
Hanoi 1000
VIETNAM
ddhong@hn.vnn.vn

de la Piedra Pablo
Alimentos Multiexport, S.A.
Isidora Goyenechea 3621 17th Floor,
Las Condes 6760412
CHILE
pdelapiedra@multiexport.cl

Debonsi Navickiene Hosana Maria
FCFRP-USP
Avenida Caramuru, numero 1280, Apto 22,
Bairro Republica-Ribeirão Preto-SP, 14030000
BRAZIL
hosana@fcfrp.usp.br

Deslandes Eric
LEBHAM, European Institute for Marine Studies
Place Nicolas Copernic,
Plouzane, 29280
FRANCE
eric.deslandes@univ-brest.fr

Deveau Jean-Paul
Acadian Seaplants Limited
30 Brown Avenue, Dartmouth,
Nova Scotia B3B 1X8
CANADA
jpdeveau@acadian.ca

Deveau Louis E.
Acadian Seaplants Limited
30 Brown Avenue, Dartmouth,
Nova Scotia B3B 1X8
CANADA
ledeveau@acadian.ca

Deysher Larry E.
Ocean Imaging
201 Lomas Santa Fe Drive, Suite 370,

Solana Beach, CA 92075
USA
deysher@oceani.com

Di Tommaso Donata
VALAGRO SPA
Zona Industriale Atessa, 66040
ITALY
d.ditommaso@valagro.com

Dion Patrick
CEVA-Seaweed Manufacturing Technology Center
Presqu'lle de Pen-Lan BP 3, F-22610, Pleubian
FRANCE
patrick.dion@ceva.fr

Duan Delin
Institute of Oceanology
The Chinese Academy of Sciences
No.7 Nanhai Road,
Qingdao, 266071
CHINA
dlduan@ms.qdio.ac.cn

Dunton Kenneth
University of Texas at Austin,
Marine Science Institute
750 Channel View Drive,
Port Aransas, TE 78373-5015
USA
dunton@utmsi.utexas.edu

Dy Arturo UY
Shemberg Marketing Corporation
Seaweed Industry Assn. of the Phills.
S. Jayme St. Pakna-An, Maridaue City,
Cebu City 6014
PHILIPPINES

Edding Mario E.
Departamento de Biologia Marina
Facultad de Ciencias del Mar
Universidad Catolica del Norte
Larrondo 1281, Coquimbo, 178-1421
CHILE
medding@ucn.cl

Ekelund Nils Gustav
Department of Natural Sciences
Mid Sweden University
85170
SWEDEN
nils.ekelund@miun.se

El-Shora Hamed M.
Botany Department, Faculty of Science
Mansoura University,
Mansoura 050
EGYPT
shora@mum.mans.edu.eg

Endo Hirotoshi
Faculty of Fisheries,
Hokkaido University
3-1-1, Minatocho, Hakodate, Hokkaido, 041-8611
JAPAN
endo@fish.hokudai.ac.jp

Engelen Aschwin Hillebrand
CCMAR, University of Algarve, FCMA
Campus de Gambelas, Faro, 8005-139
PORTUGAL
aengelen@ualq.pt

Eriksen Mikael Rubjerg
EUROGUM A/S, Rumancevej 27,
DK-2730 Herlev
DENMARK
ME@EUROGUM.DK

Falshaw Ruth
Industrial Research Limited
69 Gracefield Rd., PO Box 31-310,
Lower Hutt 5040
NEW ZEALAND
R.Falshaw@irl.cri.nz

Fazal Murtaza
C-Weeds Corp.
Box 9712, Dar-Es-Salam
TANZANIA
fazal@raha.com

Faye Etienne Jean
Port and Airport Research Institute (PARI)
3-1-1 Nagase, Yokosuka,
Kanagawa. 239-0826
JAPAN
Etienne@pari.go.jp

Fei Xiugeng
Institute of Oceanology
Chinese Academy of Sciences
7 Nanhai Road,
Quigdao, 266071
CHINA
fei@public.qd.sd.cn

Fernandez Javier
Hispanagar S.A.
Lopez Bravo 98,
Poligono Industrial Villalonquejar
Burgos, 09001
SPAIN
javierfdez@hispanagar.es

Fernandez Luis
Inca, Islas Canarias S.A.
Francisco Giralte 2, Madrid 28035
SPAIN
Mamen.fando@ogicenter.es

Fitch Robert Scott
Biology Department,
Wenatchee Valley College
1300 Fifth Street, Wenatchee, WA 98801
USA
rfitch@wvc.edu

Fitton J. Helen
Marinova Pty Ltd.
Level 7, 39 Murray Street,
Hobart, Tasmania 7000
AUSTRALIA
helen.fitton@marinova.com.au

Fortes Edna Ganzon
Marine Science Institute, C. S.
University of the Philippines
Diliman, Quezon City, 1101
PHILIPPINES
ednaf@upmsi.ph

Fotedar Ravi
Curtin University of Technology
Brodie Hall Building, 1 Turner Avenue,
Technology Park,
Bentley, WA 6102
AUSTRALIA
r.fotedar@curtin.edu.au

Freile-Pelegrin Yolanda
Cinvestav Departomento de Recursos del Mar Km 6
Antigua, Carretera a Progreso,
A.P. 73 Cordemex
Merida Yucatãn, 97310
MEXICO
freile@mda.cinvestav.mx

Friedlander Michael
Israel Oceanographic and Limnological Research
8 Mordechai St., Ramat, Hasharon, 47441
ISRAEL
michael@ocean.org.il

Fujii Mutue Toyota
Instituto de Botanica, Secretaria do Meio Ambiente
Av. Miguel Estefano, 368, Agua Funda,
São Paulo 04301-012
BRAZIL
mtfujii@terra.com.br

Fujimura Taichiro
Shiono Koryo Kaisha, Ltd.
5-17-75 Niitaka, Yodogawa-Ku, Osaka 532-0033
JAPAN
t-fujimura@shiono-koryo.co.jp

Fujita Daisuke
Tokyo University of Marine Science and Technology
4-5-7 Konan, Minato-ku, Tokyo, 108-8477
JAPAN
d-fujita@s.kaiyodai.ac.jp

Fujita Yuji
Motohara-machi 34-1-601,
Nagasaki, 852-8133
JAPAN
yfujita@net.nagasaki-u.ac.jp

Fujiyoshi Eiji
Seikai Natl. Fish. Res. Inst.
1551-8 Taira-machi, Nagasaki, 851-2213
JAPAN
fujieiji@affrc.go.jp

Galanin Dmitry Alexandrovich
Sakhalin Research Institute of Fisheries & Oceanography
196, Komsomolskaya st.,
Yuzhno-Sakhalinsk, 693023
RUSSIA
dgalanin@sakhniro.ru

Galatas Cristina
Hispanagar S.A.
Nunez de Balboa No.120, 2 Madrid, 28006
SPAIN
cristinag@afina.es

Gan KianTee
7 Taman Stephens,
Kota Kinabalu, Sabah, 88200
MALAYSIA
gankiantee@gmail.com

Gan Sook Yee
School of Pharmacy and Health Sciences,
International Medical University
Plaza Komanwel, Bukit Jalil, Kuala Lumpur, 57000
MALAYSIA
sookyee_gan@imu.edu.my

Gan Timothy Poi Chung
7 Taman Stephens, Kota Kinabalu, Sabah, 88200
MALAYSIA
tpcgan@gmail.com

Gao Kunshan
Marine Biology Institute, Science Center
Shantou University
Shantou, Guangdong, 515063
CHINA
ksgao@stu.edu.cn

Garrott Paul Robert
Marinova Pty Ltd.
Level 7, 39 Murray Street,
Hobart, Tasmania 7000
AUSTRALIA
paul.garrott@marinova.com.au

Gerung Grevo Soleman
Fac. of Fisheries and Marine Science
Sam Ratulangi University
Jl. Kampus Bahu, Manado, 95115
INDONESIA
ggrevo@hotmail.com

Givernaud Thierry
SETEXAM
Km 7 route de Tanger BP 210 Kenitra 1400
MOROCCO
gthierry@setexam.com

Gorostiaga Jose Maria
Dept. Plant Sciences & Ecology,
Faculty of Science & Technology
University of Basque Country
UPV/EHU, Apdo, 644 Bilbao, 48080
SPAIN
jm.gorostiaga@ehu.es

Graham Michael H.
Moss Landing Marine Laboratories
8272 Moss Landing Road,
Moss Landing, CA 95039-9647
USA
mgraham@mlml.calstate.edu

Greig James A.
International Specialty Products
Girvan, Ayrshire, KA26 9JN
UK
JGreig@ispcorp.com

Grindley Bruce
Department of Chemistry,
Dalhousie University,
Halifax, NS, B3H 4J3
CANADA
Bruce.Grindley@Dal.CA

Guinto Dara L.
FMC Biopolymer
Marine Colloids Phils,
Ouano Compound, Looc,
Mandaue City, Cebu, 6014
PHILIPPINES
dara-guinto@fmc.com

Guiry Michael D. R.
Martin Ryan Institute, NUI Galway,
University Rd,
Galway
IRELAND
michael.guiry@nuigalway.ie

Guo Jie
GEL JEL International Ltd.
Room, Merchants Bank Tower,
No. 36 Hong Kong Middle Road,
Qingdao, 266071
CHINA
guo-jie@vip.sina.com

Gurgel Carlos Frederico
Smithsonian Marine Station
701 Seaway Drive, Fort Pierce, FL 34949
USA
f_gurgel@yahoo.com

Hafting Jeff
Acadian Seaplants
30 Brown Avenue, Dartmouth,
Nova Scotia, B3B 1X8
CANADA
alohahafting@hawaii.rr.com

Hagen Nils T.
Department of Fisheries and Natural Science
Bodo University College
Bodo, N-8049

NORWAY
Nils.Hagen@hibo.no

Hagino Hiroshi
Shirako Co., Ltd.
6-1-17 Nakakasai, Edogawa-ku,
Tokyo, 134-0083
JAPAN
hagino.hiroshi@shirako-nori.co.jp

Hamada Minoru
Hamadaya-honten Co. Ltd.
1864-8 Heianura, Aiga-cho, Sumoto-shi,
Hyogo 656-2121
JAPAN
hamahont@mocha.ocn.ne.jp

Hamano Kaoru
Japan International Research Center
for Agricultural Sciences
c/o JIRCAS SE Asia Office, Phaholyothin Rd.,
Ladyao, Chatuchak, Bangkok, 10900
THAILAND
hamano@affrc.go.jp

Hanisak M. Dennis
Harbor Branch Oceanographic Institution
Fort Pierce, FL 34946
USA
hanisak@hboi.edu

Hanyuda Takeaki
Kobe University Research Center for Inland Seas
Rokkodai 1-1, Nada, Kobe 657-8501
JAPAN
hanyut@kobe-u.ac.jp

Hara Yoshiaki
Faculty of Science
Yamagata University
Kojirakawa-machi 1-4-12, Yamagata 990-8560
JAPAN
hara@sci.kj.yamagata-u.ac.jp

Hatakeda Hirohisa
Yanmar
Oita, Kunisaki
JAPAN

Hayashi Yuichi
Okabe Kaiyo (Ocean) Engineering Company Limited
3-14-4-3F, Narihira, Sumida-ku,
Tokyo, 130-0002
JAPAN
y-hayasi@okabe.co.jp

Hayashida Fumio
292-14 Miho, Shimizu, Shizuoka, 424-0901
JAPAN
hayashida@festa.ocn.ne.jp

Hayashizaki Kenichi
Kitasato University
Utoh 160-4, Okirai, Sanriku, Iwate, 022-0101
JAPAN
ken-ichi@kitasato-u.ac.jp

He Peimin
Shanghai Fisheries University, Aqua-life College
334 Jungong Road, Shanghai, 200090
CHINA
pmhe@shfu.edu.cn

Hennequart Franck
Molecular Glycobiotechnology Research Group
Dept. of Biochemistry, NUI, Galway
Galway City
IRELAND
fhennequart@eircom.net

Henry Eric Charles
Reed Mariculture Inc.
181 Laguna Corralitos, Watsonville, CA 95076
USA
eric@reedmariculture.com

Hernández Carmona Gustavo
Centro Interdisciplinario de Ciencias
Marinas-IPN
Apdo. Postal 592,
La Paz, Baja California Sur, 23000
MEXICO
gcarmona@prodigy.net.mx

Hertz Ole
Nordic Seaweed Project
Gudhjemuej 50, 3760
DENMARK
OLEHERTZ@POST.TDCADSL.DK

Hiraoka Masanori
Usa Marine Biological Institute,
Kochi University
194 Inoshiri, Usa, Tosa, Kochi, 781-1164
JAPAN
mhiraoka@cc.kochi-u.ac.jp

Hirose Norikazu
Tetra Co., Ltd.
3-11-34 Mita, Minato-ku, Tokyo 108-0073
JAPAN
hirose@tetra.co.jp

Ho Chai-Ling
Department of Cell & Molecular Biology,
Faculty of Biotechnology & Biomolecular Sciences
University of Putra Malaysia
Serdang, Selangor, 43400
MALAYSIA
clho@putra.upm.edu.my

Hoang Thi Lan Anh
Research Institute for Innovation in Sustainable Chemistry,
National Institute of Advanced Industrial Science
and Technology (AIST)
1-8-31 Midorigaoka, Ikeda, Osaka, 563-8577
JAPAN
hoanglananhcnt@yahoo.com

Honda Masaki
Central Research Institute of Electric Power Industry
1646 Abiko, Abiko-shi, Chiba-ken, 299-5111
JAPAN
m-honda@criepi.denken.or.jp

Hong Yong-Ki
Department of Biotechnology
Pukyong National University
Deyondong, Namku, Busan, 608-737
KOREA
ykhong@pknu.ac.kr

Hori Kanji
Graduate School of Biosphere Science
Hiroshima University
Kagamiyama 1-4-4, Higashi-Hiroshima,
Hiroshima 739-8528
JAPAN
kanhori@hiroshima-u.ac.jp

Hosooka Yosuke
KDI Trading Co., Ltd.
Tahima Bldg 5F, 4-2-8 Isobe-dori,
Chuo-ku, Kobe
JAPAN

Hotta Kenji
College of Science and Technology
Nihon University
7-24-1 Narashino-dai, Funabashi-shi
Chiba 274-8501
JAPAN
hotta@ocean.cst.nihon-u.ac.jp

Hreinsson Grettir
Efstasund 68, 104 Reykjavik,
ICELAND
grettirh@simnet.is

Huang Esther
Hong Kong Sheli Ltd.
Flat B, 37/F, Tower 3, Queen's Terrace 1,
Queen St., Hong Kong
CHINA
esther@sheli.com.cn

Huang Su-Fang
Research Department,
National Taiwan Museum
6F, No.71, Guan-Qian Rd.,
Taipei, 10047
TAIWAN
sfhuang@ntm.gov.tw

Hurtado Anicia Q.
Aquaculture Department,
SEAFDEC/ISDA
Iloilo City, 5000
PHILIPPINES
aqhurtado@gmail.com

Husain Muna M.
Environment Public Authority EPA-Kuwait
Al Surrah P.O. Box 221, 45703
KUWAIT
Mhusain@epa.org.kw

Huynh Quang Nang
Institute of Material Science Nha Trang Branch
2-Hung Vuong- Nha Trang City, 84
VIETNAM
hqnang@dng.vnn.vn

Hwang Eun Kyoung
Seaweed Research Center
National Fisheries Research & Development Institute
Mokpo, 530-831
KOREA
ekhwang@momaf.go.kr

Hwang Mi Sook
Seaweed Research Center
National Fisheries Research & Development Institute

Mokpo, Chunnam, 530-831
KOREA
mishwang@nfrdi.re.kr

Inoue Akira
Hokkaido University
3-1-1 Minato-cho, Hakodate, Hokkaido, 041-8611
JAPAN
inouea21@fish.hokudai.ac.jp

Inoue Osamu
Ina Food Industry Co., Ltd.
5074 Nishi-Haruchika, Ina-City,
Nagano, 399-4498
JAPAN
o-ino@kantenpp.co.jp

Iri Tadao
Imperial 690, Dept 201, Puerto Varas
CHILE
tiri@surnet.cl

Ishibashi Kiyohide
Nabocul Cosmetics Co., Ltd.
5-29-7 Sendagaya, Shibuya-ku,
Tokyo, 151-0051
JAPAN
nabocul@lake.ocn.ne.jp

Ishikawa Yoshio
Institute for Environmental Sciences
1-7 Ienomae, Obuchi, Rokkasho, Kamikita, Aomori
039-3212
JAPAN
ishisim@ies.or.jp

Iwamoto Katsuaki
Marine Science Co., Ltd.
Daiichi Sakamoto Building 7F,
2-1-11 Higashi-Kanda, Chiyoda-ku,
Tokyo, 101-0031
JAPAN
iwamoto@marine-science.co.jp

Iyoda Noriko
Penta Ocean Construction Co., Ltd.
Kamihacchoubori, Naka-ku, Hiroshima, 730-8542
JAPAN
Noriko.Iyoda@mail.penta-ocean.co.jp

Jones Joanna M.
Australian National University
3 Esk Place
Lyons ACT, 2606
AUSTRALIA
jo@jokain.net

Kaewsrithong Janthira
Faculty of Science and Technology
Prince of Songkla University, Pattani, 94000
THAILAND
kianthir@bunga.pn.psu.ac.th

Kajimura Mitsuo
Shimane University
2589 Tsuma, Okinoshima-cho,
Oki-gun, 685-0104
JAPAN
hkaji@lily.ocn.ne.jp

Kajiwara Tadahiko
Faculty of Agriculture,
Yamaguchi University
Yoshida 1677-1, Yamaguchi City, 753-8515
JAPAN
kajiwara@yamaguchi-u.ac.jp

Kakinuma Makoto
Laboratory of Marine Biochemistry
Faculty of Bioresources,
Mie University
1577 Kurima-machiya, Tsu, Mie, 514-8507
JAPAN
kakinuma@bio.mie-u.ac.jp

Kakita Hirotaka
AIST Shikoku, National Institute of Advanced Industrial
Science and Technology
2217-14 Hayashi, Takamatsu, Kagawa, 761-0395
JAPAN
h-kakita@aist.go.jp

Kamei Yuto
Coastal Bioenvironment Center,
Saga University
152-1 Shonan-cho, Karatsu, Saga, 847-0021
JAPAN
kameiy@cc.saga-u.ac.jp

Kamitani Toshiaki
Research & Development Laboratory
Fujicco Co., Ltd.
6-13-4 Minatojima-Nakamachi, Chuo-ku,
Kobe 650-8558
JAPAN
t-kamitani@fujicco.co.jp

Kamiya Mitsunobu
Faculty of Biotechnology
Fukui Prefectural University
1-1 Gakuencho, Obama, Fukui, 917-0003
JAPAN
mkamiya@fpu.ac.jp

Kato Ikunoshin
Takara Bio Inc.
Seta 3-4-1, Otsu, Shiga, 520-2193
JAPAN
katoi@takara-bio.co.jp

Kawaguchi Shigeo
Faculty of Agriculture,
Kyushu University
6-10-1 Hakozaki-cho, Higashi-ku,
Fukuoka 812-8581
JAPAN
skawagu@agr.kyushu-u.ac.jp

Kawai Hiroshi
Kobe University Research Center for Inland Seas
1-1 Rokkodai, Nada-ku, Kobe, 657-8501
JAPAN
kawai@kobe-u.ac.jp

Kawai Masanobu
3-32-15 Mori-dai, Atsugi, Kanagawa, 243-0037
JAPAN
kawai.masa3159@ayu.ne.jp

Kawata Yoshikazu
Research Institute for Innovation in Sustainable Chemistry,
National Institute of Advanced Industrial Science
and Technology (AIST)
1-8-31 Midorigaoka, Ikeda, Osaka, 563-8577
JAPAN
y-kawata@aist.go.jp

Khinouche Karim-Franck
ETIA
258 Jia Shan Lu, Bldg 1, Rm 1301,
Shanghai, 200020
CHINA
kf.khinouche@etia.fr

Kikuchi Norio
Coastal Branch of Natural History Museum
and Institute, Chiba
123 Yoshio, Katsuura, Chiba, 299-5242
JAPAN
kikuchin@chiba-muse.or.jp

Kim Gwang Hoon
Department of Biology,
Kongju National University
Kongju, 314-701
KOREA
ghkim@kongju.ac.kr

Kim Hyung Geun
Faculty of Marine Bioscience and Technology,
Kangnung National University
Gangneung, Gangwondo, 210-702
KOREA
kimhg@kangnung.ac.kr

Kim Jeong-Bae
South Sea Fisheries Research Institute, NFRDI
347, Anpo-Ri, Hwayang-Mynu, Yeosu, 556-823
KOREA
jbkim@momaf.go.kr

Kim Jeong Ha
Dept. of Biological Science
Sungkyunkwan University
Suwon, 440-746
KOREA
jhkimbio@skku.ac.kr

Kim Jong Seok
MSC Co., Ltd.
439-13, Soju-Ri, Ungsang-Up, Yangsan-City,
Kyeongnam
KOREA
mscjp@msckorea.com

Kim Kil Jae
MSC Co., Ltd.
4 39-13, Soju-Ri, Ungsang-Up, Yangsan-City,
Kyeongnam-do
KOREA
mscjp@msckorea.com

Kim Myung Sook
Pusan National University
Research Institute for Basic Sciences
Changjeon-dong, Keumjeong-gu,
Busan, 609-735
KOREA
myungskim@pusan.ac.kr

Kim Nam-Gil
Department of Marine Bioscience,
College of Marine Science,
Gyeongsang National University

445 Inpyeong Tongyoung, Kyoungnam, 650-160
KOREA
ngkim@gaechuk.gsnu.ac.kr

Kim Yeon-Kye
Biotechnology Research Center
National Fisheries R & D Institute
408-1, Gijang-up, Gijang-gun, Busan, 619-902
KOREA
yeonkyekim@momaf.go.kr

Kim Young Don
Dongsung Marine Development, Ltd.
1842-3 Gyo-Dong, Kangnung-City, Kangwon-Do,
KOREA
dss0068@hanmail.net

Kim Young Sik
School of Marine Life Science
Kunsan National University
San 68 Miroyong-dong, Kunsan, 573-701
KOREA
kimys@kunsan.ac.kr

Kimura Hajime
Fisheries Experimental Station,
Wakayama Prefectural Research Center of Agriculture, Forestry & Fisheries,
Kushimoto 1557-20, Kushimoto-cho,
Higashimuro-gun, Wakayama, 659-3503
JAPAN
kimura_h0005@pref.wakayama.lg.jp

Kirihara Shinji
Aomori Prefectural Fisheries Research Center,
Aquaculture Institute
Hiranai, Aomori, 039-3381
JAPAN
shinji_kirihara@pref.aomori.lg.jp

Kitade Yukihiro
Faculty of Fisheries Sciences, Hokkaido University
3-1-1 Minato-cho, Hakodate,
Hokkaido, 041-8611
JAPAN
ykitade@fish.hokudai.ac.jp

Kitaya Yoshiaki
Graduate School of Life and Environmental Sciences,
Osaka Prefecture University
Gakuen-cho, Naka-ku, Sakai, Osaka, 599-8531
JAPAN
kitaya@envi.osakafu-u.ac.jp

Kobayashi Masami
Institute of Materials Science,
University of Tsukuba
Tennoudai 1-1-1, Tsukuba, Ibaraki, 305-8573
JAPAN
masami@ims.tsukuba.ac.jp

Kogame Kazuhiro
Department of Natural History Sciences
Faculty of Science, Hokkaido University
Sapporo, 060-0810
JAPAN
kogame@sci.hokudai.ac.jp

Komatsu Teruhisa
Ocean Research Institute,
The University of Tokyo
1-15-1, Minamidai, Nakano-ku, Tokyo, 164-8639
JAPAN
komatsu@ori.u-tokyo.ac.jp

Kraan Stefan
Irish Seaweed Centre, Martin Ryan Institute,
National University of Ireland,
University Road, Galway
IRELAND
Stefan.Kraan@nuigalway.ie

Kradolfer Paul François
Peruvian Seaweeds S.R.L.
Av. Paul Poblet Parcela D-19, Lote 19 Lurin,
Lima 16
PERU
paul@peruvianseaweeds.com

Küpper Frithj Christian
Culture Collection of Algae and Protozoa (CCAP),
Scottish Association for Marine Science
Dunstaffnage Marine Laboratory, Oban, Argyll,
Scotland, PA37 1QA
UK
fck@sams.ac.uk

Kurashima Akira
Lab. of Phycology, Faculty of Bioresources
Mie University
1577 Kurimamachiya-cho, Tsu, Mie, 514-8507
JAPAN
kurasima@bio.mie-u.ac.jp

Kurihara Hideyuki
Faculty of Fisheries Sciences, Hokkaido University
3-3-1 Minato, Hakodate, Hokkaido, 041-8611

JAPAN
kuri@fish.hokudai.ac.jp

Kusnowirjono Soerianto
Agarindo Bogatama PT
Jl. Jelambar Selatan II No.4,
Graha Sakura Jakarta Barat, 11460
INDONESIA
soerianto@agarindo-bogatama.co.id

Kustantiny Anny
Ministry of Marine Affairs and Fisheries,
Republic of Indonesia
Jl.Medan Merdeka Timur16,
Jakarta 10110
INDONESIA
anny4yafame@yahoo.com

LaPointe Brian Edward
Center for Coastal Research,
Harbor Branch Oceanographic Institution,
5600 US 1 North, Fort Pierce,
FL 34946
USA
lapointe@hboi.edu

Largo Danilo Basnillo
Department of Biology,
University of San Carlos
Talamban, Cebu City, 6000
PHILIPPINES
largodb@yahoo.com

Lebbar Rachid
SETEXAM
Km 7 route de TANGER BP 210
MOROCCO
rlebbar@setexam.com

Lee Antonio
Solomon Seaweed
P.O. Box 138, Honiara
SOLOMON ISLANDS
lkk@business.com.sb

Lee Barry Wayne
Rural Industries Research & Development Corporation
P.O. Box 294, Gymea, NSW 2227
AUSTRALIA
barrywlee@ozemail.com.au

Lee Ji Hee
Lab. of Marine Biology,
Faculty of Biological Science
Wonkwang University
344-2, Iksan, Jeonbuk, 570-749
KOREA
easy567@nate.com

Lee Jin Ae
School of Environmental Science & Engineering
Inje University
#607 Obang-dong, Gimbae,
Gyeongnam, 621-749
KOREA
envjal@inje.ac.kr

Lee Joon-Baek
Department of Oceanography,
College of Ocean Sciences,
Cheju National University, Jeju 690-756
KOREA
jblee@cheju.ac.jp

Lee Michael
98-1858 Hapaki St.,
Aiea, Hawaii, 96701-1636
USA
kilokilopoepoe@hotmail.com

Leon Armando
Sinaloa Seafields International, Inc.
2365 Biola Avenue,
San Diego, CA 92154
USA
ArmandoALeon@aol.com

Lerat Yannick
CEVA Centre d'Etude et de Valorisation des Algues
BP3-22610 Pleubian
FRANCE
yannick.lerat@ceva.fr

Lesser Hans Jorge
Algas Marinas S.A.
Ou Yang Road, Building 12, Apartment 1201
Shanghai
CHINA
HANSLESSER@GMAIL.COM

Levine Ira Arthur
University of Southern Maine
51 Westminster Street,
Lewiston, Maine 04240
USA
ilevine@usm.maine.edu

Levy Israel
Noritech Ltd.
Zurit, M. P. Misgav 20104
ISRAEL
israel@noritech.co.il

Lewmanomont Khanjanapai
Faculty of Fisheries, Kasetsart University
Bangkok, 10900
THAILAND
ffiskpl@ku.ac.th

Lin Showe-Mei
Department of Natural Science Education
National Taitung University
684 Chunghua Rd., Sec. 1, 95002
TAIWAN
SMLIN@nttu.edu.tw

Lindstrom Sandra Christine
Department of Botany
University of British Columbia
#3529-6270 University Blvd.,
Vancouver, B.C. V6T 1Z4
CANADA
sandracl@interchange.ubc.ca

Litzia Paul-Chavez
Programa de Investigacion en Botanica Marina
Departamento de Biologia Marina UABCS
Apartado postal 19-B, km. 5.5. Carretera al sur
La Paz B.C.S. 23080
MEXICO
lchavez@uabcs.mx

Loftus Leon
15 Bennett Street,
Stawell, Victoria 3380
AUSTRALIA
yzh62878@bigpond.net.au

Loughnane Claran John
Dept. of Botany, Martin Ryan Institute,
National University of Ireland, Galway
IRELAND
ciaran.loughnane@nuigalway.ie

Ma Jia Hai
Aqua-life College,
Shanghai Fisheries University
334 Jungong Road, Shanghai, 200090
CHINA
jhma@shfu.edu.cn

Maneveldt Gavin William
Department of Biodiversity & Conservation Biology,
University of the Western Cape
P. Bag X17, Bellville, 7535
SOUTH AFRICA
gmaneveldt@uwc.ac.za

Mansilla Andrés
Magallanes University
Avenida Bulnes 01855, Punta Arenas
CHILE
andres.mansilla@umag.cl

Mantri Vaibhav A.
Central Salt & Marine Chemicals Research Institute
Gijubhai Badheka Marg, Bhavnagar, 364 002
INDIA
vaibhav@csmcri.org

Marsano Carlos
Peruvian Seaweeds
Los Castanos, SGO, Apt. 902, Lima 27
PERU
carlos@peruvianseaweeds.com

Marcos Roberto
Consultora Los Lagos Ltd.
Bahia Torngas 834, Lomas San Marino,
Ensenada, B.C. 22870.
MEXICO
algasbobe@yahoo.com.mx

Marec Erick
Danisco
La Grande Palud, La Forest Landerneau 29200
FRANCE
Erick.marec@danisco.com

Martin Georg
Estonian Marine Institute
Maealuge 10A, Tallinn, 12618
ESTONIA
georg.martin@vt.ee

Martin Kai
Estonian Marine Institute
Maealuge 10A, Tallinn, 12618
ESTONIA
kaimartin@hot.ee

Martinez-Goss Milagrosa R.
Institute of Biological Sciences
University of the Philippines

Los Banos College, Laguna, 4031
PHILIPPINES
mgoss2003@yahoo.com

Maruyama Hiroko
Department of Pathology,
School of Allied Health Sciences,
Kitasato University
Kitasato 1-15-1, Sagamihara, Kanagawa, 228-8555
JAPAN
maruyama@ahs.kitasato-u.ac.jp

Matsubara Kiminori
Department of Nutritional Science
Okayama Prefectural University
Kuboki 111, Soja, Okayama, 719-1197
JAPAN
kmatsuba@fhw.oka-pu.ac.jp

Matsuda Yoshiaki
5-72 Senshu-Kitanomaru, Akita, 010-0872
JAPAN
matsuday2006@yahoo.co.jp

Matsuoka Masayoshi
Aizumi, Itano,Tokushima
JAPAN
c/o dan_akinori_1@pref.tokushima.lg.jp

McGillivary Philip A.
Science Liaison, US Coast Guard
P.O. Box 5082,
Alameda, CA 94501
USA
Philip.a.mcgillivary@uscg.mil

Mekideche Nicol
Biotech Marine
BS65 Zone Industrielle, Pontrieux, 22260
FRANCE
nmekideche@biotechmarine.com

Mekking Suzanne
Springer
Van Godewijckstraat 30
P.O. Box 17,
3300 AA Dordrecht
NETHERLANDS
suzanne.mekking@springer.com

Mine Ichiro
Graduate School of Kuroshio Science
Kochi University
2-5-1 Akebono-cho, Kochi
JAPAN
mine@cc.kochi-u.ac.jp

Minocha Subhash
University of New Hampshire
Dept. of Plant Biology, Rudman Hall,
Durham, NH 03824
USA
sminocha@unh.edu

Minoura Katsuteru
Nabocul Cosmetics Co., Ltd.
5-29-7 Sendagaya, Shibuya-ku, Tokyo, 151-0051
JAPAN
nabocul@lake.ocn.ne.jp

Misbakhun Mukhamad
PT. Agar Sehat Makmur Lestari
JL. Raya Purwosari-Pasuruan KM. 5, Desa Bakalan
Kecamatan Purwosari-Pasuruan-Jawa Timur, 67162
INDONESIA
bakhunpmb@yahoo.com

Misonou Taku
Graduate School of Medicine and Engineering,
University of Yamanashi
4-3-11 Takeda, Kofu, Yamanashi, 404-8511
JAPAN
mist@ccn.yamanashi.ac.jp

Miyaji Eisuke
Seitaro Arai & Co., Ltd.
1-8, Onoe-cho, Naka-ku, Yokohama, 231-0015
JAPAN
kairikusan@seitaroarai.co.jp

Miyamura Shinichi
Graduate School of Life and Environmental Sciences,
University of Tsukuba
Tsukuba, Ibaraki, 305-8572
JAPAN
miyamura@sakura.cc.tsukuba.ac.jp

Miyata Masahiko
Natural History Museum & Institute, Chiba
955-2 Aoba-cho, Chuo-ku, Chiba City,
Chiba Prefecture, 260-8682
JAPAN
miyata@chiba-muse.or.jp

Mohamed Soud
Ministry of Agriculture, Livestock and Environemnt

P.O. Box 774, Zan Zibar,
TANZANIA
macemp@zanline.com

Montaño Marco Nemesio Estampador
Marine Science Institute, College of Science,
University of the Philippines
Diliman, Quezon City, 1101
PHILIPPINES
coke@upmsi.ph

Mori Toshiyuki
Biomedical Research Laboratories
Pharmaceutical Research Division
Takeda Pharmaceutical Company, Ltd.
2-17-85 Jusohonmachi, Yodogawa-ku,
Osaka 532-8686
JAPAN
Mori_Toshiyuki2@takeda.co.jp

Morita Naohiro
Morita Shouten Corporation
Gouchi 58-59, Aza-Gouminami, Ooaza-Fuki,
Taketoyo-cho, Chita-gun,
Aichi, 470-2531
JAPAN
moritan@kxb.biglobe.ne.jp

Morre Roger
C-Weeds Corp.
P.O. Box 333, Penba, Zanzibar
TANZANIA
roger@cweed.com

Motomura Taizo
Field Science Center for Northern Biosphere
Hokkaido University
1-13 Bokoi minami-machi, Muroran, 051-0003
JAPAN
motomura@bio.sci.hokudai.ac.jp

Mouradi Aziza
LBBE (Laboratoire de Biochimie,
Biotechnologie et Environment)
LLBE Faculte des Sciences Universite
IBN TOFAIL BP 210 Kenitra 14000
MOROCCO
Mouradi14@caramail.com

Munda Ivka Maria
Scientific Research Centre of the Slovenian
Academy of Sciences and Arts
Novi trg 2, SI-1000 Ljubljana
SLOVENIA
spela@zrc-sazu.si

Muraoka Daisuke
Tohoku National Fisheries Research Institute
Fisheries Research Agency
3-27-5, Shinhama-cho, Shiogama,
Miyagi, 985-0001
JAPAN
dmuraoka@affrc.go.jp

Murase Noboru
Department of Applied Aquabiology
National Fisheries University
Nagatahonmachi 2-7-1, Shimonoseki,
Yamaguchi 759-6533
JAPAN
murasen@fish-u.ac.jp

Muslim Ermalena
Government
JL Kesehatan 8/3, Jakarta 10160
INDONESIA
ermalena@telcom.net

Nagai Kouhei
Wakayama Industry Promotion Foundation
60 Ogura, Wakayama-shi, Wakayama, 649-6421
JAPAN
knagai@waka.kindai.ac.jp

Nagakubo Masao
DENSO
JAPAN

Nagasato Chikako
Field Science Center for Northern Biosphere
Hokkaido University
1-13 Bokoi minami-machi, Muroran, 051-0003
JAPAN
nagasato@bio.sci.hokudai.ac.jp

Nagumo Tamotsu
Department of Biology,
The Nippon Dental University
Fujimi 1-9-20, Chiyoda-ku, Tokyo, 102-8159
JAPAN
t-nagumo@tky.ndu.ac.jp

Nakamura Akira
17-16-2 Yachiyodai-Kita,
Yachiyo-shi, Chiba
JAPAN

Nakamura Tatsuo
Faculty of Environment and Information Sciences
Yokohama National University
79-7, Tokiwadai, Hodogaya-ku,
Yokohama, 240-8501
JAPAN
t-nakamu@ynu.ac.jp

Nakayama Yoshiro
Kibunfood Chemifa Co., Ltd.
2-1-1 Irihune, Chuo-ku,
Tokyo, 104-8553
JAPAN
ynakayama@kibunfc.co.jp

Nam Ki Wan
Department of Marine Biology,
Pukyong National University
Nam-Gu, Busan
KOREA
kwnam@pknu.ac.jp

Nanba Nobuyoshi
Faculty of Fisheries Sciences,
Kitasato University
Sanriku, Ofunato, Iwate, 022-0101
JAPAN
nanba@kitasato-u.ac.jp

Neefus Christopher D.
Department of Plant Biology
University of New Hampshire
G32 Spaulding Hall,
38 College Rd, NH 03824
USA
chris.neefus@unh.edu

Neish Iain Charles
Sea Plant. Net
Jalan Pettarani 45, Indosat Building,
Makassar 90222
INDONESIA
iain@seaplant.net

Nelson Wendy A.
National Institute of Water and Atmospheric Research
Private Bag 14-901, Wellington, 6241
NEW ZEALAND
w.nelson@niwa.co.nz

Neori Amir
Israel Oceanographic & Limnological Research Ltd.
National Center for Mariculture
P.O. Box 1212, Eilat, 88112
ISRAEL
aneori@gmail.com

Niasser Sacing Saio
Ministry of Agric., Livestock & Environment
P.O. Box 801, Zan Zibar,
TANZANIA
zapoco@zanzinet.com

Nishide Eiichi
11-10-508 Nibancho, Chiyoda-ku, Tokyo, 102-0084
JAPAN

Nishihara Gregory Naoki
Faculty of Fisheries, Kagoshima University
Shimoarata 4-50-20, Kagoshima City
Kagoshima Pref. 890-0056
JAPAN
gnishihara@gmail.com

Nishimura Mitsumasa
Daichu Foods Co, Ltd.
288 Higashi-Oizu-Cho, Ise-City, Mie, 515-0509
JAPAN
mitsu@daichu-foods.co.jp

Nishino Yasuto
Faculty of Bioindustry
Tokyo University of Agriculture
196 Yasaka, Abashiri, Hokkaido 099-2493
JAPAN
y3nishin@bioindustry.nodai.ac.jp

Niwa Kyosuke
Fisheries Technology Institute,
Hyogo Prefectural Technology Center for Agriculture, Forestry and Fisheries
Minami-futami 22-2, Akashi, Hyogo, 674-0093
JAPAN
kyousuke_niwa@pref.hyogo.jp

Noiraksar Thidarat
Institute of Marine Science
Burapha University, Bangsaen, Chon Buri, 20131
THAILAND
sargassum2005@yahoo.com

Norman Mark Philip
Taighde Mara Teoranta
Carna, Galway
IRELAND
mnorman@taighde.ie

Noro Tadahide
Faculty of Fisheries, Kagoshima University
4-50-20 Shimoarata, Kagoshima, 890-0056
JAPAN
noro@fish.kagoshima-u.ac.jp

Notoya Masahiro
Laboratory of Applied Phycology
Tokyo University of Marine Science & Technology
Konan-4, Minato-ku, Tokyo, 108-8477
JAPAN
notoya@kaiyodai.ac.jp

Nukaya Michio
1-5-5 Kusanagi-Sugimichi, Shimizu-ku,
Shizuoka-shi, 424-0885
JAPAN

Ogawa Hiroo
Tokyo University of Marine Science & Technology
4-5-7 Konan, Minato-ku, Tokyo, 108-8477
JAPAN
ogawahi@s.kaiyodai.ac.jp

Ogawa Hisao
Kitasato University
Okirai, Sanriku-cho, Ohfunato-shi,
Iwate-ken 022-0101
JAPAN
seaweeds@kitasato-u.ac.jp

Oh Jung Hoon
MSC Co., Ltd.
439-13, Soju-Ri, Ungsang-Up, Yangsan-City,
Kyeongnam-do
KOREA
mscjp@msckorea.com

Ohba Tatsuyuki
8-21-2 Okusawa, Setagaya-ku, Tokyo 158-0083
JAPAN
ohba@ec.catv.ne.jp

Ohno Masao
Usa Marine Biological Institute,
Kochi University
194 Inoshiri, Usa, Kochi, 781-1164
JAPAN
mohno@cc.kochi-u.ac.jp

Ojima Takao
Hokkaido University
3-1-1 Minato-cho, Hakodate, Hokkaido, 041-8611
JAPAN
ojima@fish.hokudai.ac.jp

Okuda Kazuo
Graduate School of Kuroshio Science
Kochi University
2-5-1 Akebono-cho, Kochi, 780-8520
JAPAN
okuda@cc.kochi-u.ac.jp

Oliveira Eurico C.
Universidade de São Paulo
Rua Dos Escultores, 389,
São Paulo, SP, 05469-010
BRAZIL
euricodo@usp.br

Oliveira Mariana C.
Botany Dept., Biosciences Institute
University of São Paulo
Rua do Matão 277,
São Paulo, SP, 05508-900
BRAZIL
mcdolive@usp.br

Ono Akio
Yamagataya Noriten Co., Ltd.
384-60 Aza-Onoue, Ashitaka, Numazu,
Shizuoka 410-0001
JAPAN
ono@yamagataya.co.jp

Onorio Komeri
Atoll Seaweed Company Limited
KIRIBATI
komeri@maurinet.ki

Orav-Kotta Helen
Estonian Marine Institute,
University of Tarty
Mäealuse 10a, Tallinn, 12618
ESTONIA
helen.orav@sea.ee

Orbi Abldellatif
INRH
2 Rue de Tiznit Casablanca
MOROCCO
orbi@inrh.org.ma

Orosco Christine Abellana
Dept. of Marine Science,
Faculty of Marirtime Studies and Marine Science,

Kolej Universiti Sains dan Teknologi Malaysia (KUSTEM)
Mengabang Telipot, Kuala Terengganu,
Terengganu, D.I., 21030
MALAYSIA
aishah@kustem.edu.my

Oyamada Kumi
Aqua-, Bio-technology and Chemical Dept.
JFE R&D Corporation
1-1 Minamiwatarida-cho, Kawasaki-ku,
Kawasaki 210-0855
JAPAN
k-oyamada@jfe-rd.co.jp

Paalme Tiina
Estonian Marine Institute,
University of Tartu
Maecaluse 10a, Tallinn, 12618
ESTONIA
tiina.paalme@ut.ee

Pablico Grace Tolentino
World Fish Center–Philippines,
Khush Hall, College, Los Baños,
Laguna 4031,
PHILIPPINES
g.pablico@cgiar.org

Pang Kyun Fung
TACARA
2nd Floor, Block D, Bandaran Baru II,, Tawau,
Sabah, 91000
MALAYSIA
tacara@tm.net.my

Pang Shao Jun
Institute of Oceanology
Chinese Academy of Sciences
7 Nanhai Road, Qingdao, 266071
CHINA
sjpang@ms.qdio.ac.cn

Park Chan Sun
Department of Marine Resources,
Mokpo National University,
Muan-gun, Jeonnam, 534-729
KOREA
cspark85@mokpo.ac.kr

Park Ki Eui
Biotechnology Research Center
National Fisheries R & D Institute
408-1, Gijang-up, Gigang-gun, Busan, 619-902
KOREA
cwlim@nfrda.re.kr

Payri Claude Elisabeth
Institut de Recherche pour le Developpement
1RD BP A5 98848, Noumea
NEW CALEDONIA
claude.payri@noumea.ird.nc

Pedersen Are
NIVA, Gaustadalléen 21, NO-0349 Oslo
NORWAY
are.pedersen@niva.no

Pérez-Gil Fernando
National Institute of Nutrition
Vasco de Quiroga 15, Delegacion Tlalpan,
Mexico City, 14000
MEXICO
silvicarrillo3@hotmail.com

Peters Akira F.
UMR 7139 CNRS-UPMC
Station Biologique, Place Georges Teissier, CP 74,
Town ROSCOFF, 29682
FRANCE
apeters@sb-roscoff.fr

Phang Siew-Moi
Institute of Biological Sciences, Faculty of Science,
University of Malaya, Kuala Lumpur, 50603
MALAYSIA
phang@um.edu.my

Phillips Julie Ann
University of Queensland
Centre for Microscopy & Microanalysis
St. Lucia, QLD, 4072
AUSTRALIA
j.Phillips@uq.edu.au

Piaggesi Alberto
VALAGRO SPA
Zona Industriale Atessa, 66040
ITALY
a.piaggesi@valagro.com

Piantini Rene
General Velasquez No.890, Office No.602,
Antofagasta
CHILE
rpiantini@prodalmar.cl

Podkorytova Antonina Vladimirovna
Department of Biochemistry & Technology of Fishes, Invertebrates and Seaweeds
Russian Federal Research Institute
of Fisheries & Oceanography
17, V. Krasnoselskaya, Moscow, 107140
RUSSIA
podkor@vniro.ru

Prathep Anchana
Seaweed and Seagrass Research Unit
Centre for Biodiversity,
Department of Biology,
Faculty of Science, PSU,
Hat Yai, Songkhla 90112
THAILAND
anchana.p@psu.ac.th

Qin Song
Institute of Oceanology
Chinese Academy of Sciences
7 Nanhai Road, Qingdao, 266071
CHINA
sqin@ms.qdio.ac.cn

Radi Naseem Ismail
Department of Biological Sciences,
Faculty of Applied Sciences,
Umm Al Qurah University
Makkah, PO Box 100569, Jeddah 21311
SAUDI ARABIA
nassemradi@hotmail.com

Radier Richard
Carqill Texturizing Solutions Phils., Inc.
28-C/F Ayala Life FGU Center, 6811 Ayala Avenue,
Makati City, Metro Manila, 1229
PHILIPPINES
pbi@pldtdsl.net

Rafael Riosmena Rodriguez
Programa de Investigacion en Botanica Marina
Departamento de Biologia Marina UABCS
Apartado postal 19-B, km. 5.5. Carretera al sur
La Paz B.C.S. 23080
MEXICO
riosmena@uabcs.mx

Raid Abdelwahab
SETEXAM
Km 7 route de Tanger BP 210, 14000
MOROCCO
awriad@setexam.com

Ramirez Alberto
Fundación Chile
Av. Parque Antonio Rabat Sur 6165,
Vitacura, Santiago
CHILE
aramirez@fundacionchile.cl

Rebours Celine
Bioforsk-Norwegian Institute for Agricultural
& Environmental Research
Artic Agriculture and Land Use Division
Biofork Nord Vagones, N-8076
NORWAY
celine.rebours@bioforsk.no

Reddy C R K
Central Salt and Marine Chemicals Research Institute
Gijubhai, Badheka Marg,
Bhavnagar, 364 002
INDIA
crk@csmcri.org/crkreddy@sancharnet.in

Reis Renata Perpetuo
Instituto de Pesquisas Jardim Botanico
do Rio de Janeiro/MMA
Rua Tonelero, 380/1007, Copacabana
Rio de Janeiro, RJ 22030-000
BRAZIL
rreis@jbrj.gov.br

Renoux Aline H.
307 Terrasses du Golf, 33 rue de Hourticq
64600 Anglet
FRANCE

Rhatigan Prannie
Streedagh House, Streedagh,
Grange, County Sligo
IRELAND
rhatigandoc@eircom.net

Robledo Daniel
Cinvestav
Departamento de Recursos del Mar,
Km 6 Antigua Carretera a Progreso,
A.P. 73 Cordemex
Merida Yucatãn, 97310
MEXICO
robledo@mda.cinvestav.mx

Rodriguez-Montesions Yoloxochitl Elizabeth
CICIMAR-IPN
Av. Instituto Politecnico Nacional s/n. Col. Plava

Palo de Sta Rita, La Paz Baja California Sur, 23096
MEXICO
yrodriguez@ipn.mx

Romo Hector
Departamento de Oceanografia
Universidad de Concepcion
Casilla 160-C
CHILE
hromo@udec.cl

Rothann Klaus
Taurus Products
Box 5534, Rivonia, 2128
SOUTH AFRICA
k.rothann@taurus-products.co.za

Rothman Mark Dino
Seaweed Unit, Marine and Coastal Management,
Pvt Bag X2, Roggebaai, 8012
SOUTH AFRICA
mrothman@botzoo.uct.ac.za

Ruangchuay Rapeeporn
Division of Fishery Technology,
Faculty of Science and Technology
Prince of Songkla University
181 Charoen- Pardit Rd.,
Rusamilae District
Amphur Muang, Pattani, 94000
THAILAND
rrapee@bunga.pn.psu.ac.th

Sade Ahemad
Department of Fisheries, Sabah,
Malaysia Fisheries Research Centre
Likas, Kota Kinabalu, Sabah, 89400
MALAYSIA
Ahemad.Sadi@sabah.gov.my

Saga Naotsune
Faculty of Fisheries Sciences,
Hokkaido University
3-1-1 Minato-cho, Hakodate, Hokkaido, 041-8611
JAPAN
nsaga@fish.hokudai.ac.jp

Sahoo Dinabandhu
Department of Botany,
University of Delhi,
Delhi, 110007
INDIA
dbsahoo@hotmail.com

Saigusa Miho
Miyagi Prefectural Fisheries Experimental Station
Kesennuma Branch
3-9-318 Higashi Shinjou, Kesennuma-shi, Miyagi
JAPAN
saigusa-mi762@pref.miyagi.jp

Saitoh Kenji
Tohoku National Fisheries Research Institute
Shinhama 3-27-5, Shiogama, Miyagi, 985-0001
JAPAN
ksaitoh@affrc.go.jp

Salling Peter
Sallingredients
Torre Bermeja 1801, Urb. Guadalmansa
Estepona, E29680
SPAIN
psalling@msn.com

San Segundo Victor
Hispanagar S.A.Nunez de Balboa No.120,
2 Madrid, 28006
SPAIN
victorssd@yahoo.es

Sánchez Ignacio
Centro Interdisciplinario de Ciencias Marinas-I.P.N
Av. Instituto Politecnico Nacional s/n Col.
Palo de Santa Rita.
La Paz Baja California Sur, 23096
MEXICO
isrnacho@hotmail.com; isanchez@ipn.mx

Santelices Bernabe
Pontificia Universidad Catolica de Chile
Facultad de Ciencias Biologicas,
Departamento de Ecologia,
Av. Lib. Bernardo O'Higgins 340, Santiago
CHILE
bsantelices@bio.puc.cl

Santos Rui Orlando Pimenta
Marine Plant Ecology Research Group
Center of Marine Sciences of Algarve
Universidade do Algarve
Gambelas, Faro, 8005-139
PORTUGAL
rosantos@ualg.pt

Sato Minoru
Tohoku University
Graduate School of Agricultural Science

Aoba-ku, Sendai, 981-8555
JAPAN
msato@bios.tohoku.ac.jp

Sato Riki
Institute for Environmental Sciences
1-7 Ienomae, Obuchi,
Rokkasho, Kamikita, Aomori
039-3212
JAPAN
rsato@ies.or.jp

Sato Yoichi
RIKEN FOOD Co., Ltd.
5-60 Miyauchi 2-chome, Tagajo-shi, Miyagi-ken
JAPAN
yoi_sato@rikenfood.co.jp

Satoh Hiroo
Tokyo University of Marine Science & Technology
4-5-7 Konan, Minato-ku,
Tokyo, 108-8477
JAPAN
h-satoh@kaiyodai.ac.jp

Sayegh Fotoon Abdullah
Department of Biological Sciences
Faculty of Science,
King Abdul Aziz University
P. O. Box 100569, Jeddah, 21311
SAUDI ARABIA
fsaygh@kau.edu.sa

Segovia Danilo Yuri
Fundación Chile
Av. Parque Antonio Rabat Sur 6165,
Vitacura, Santiago, 7660118
CHILE
dsegovia@fundacionchile.cl

Sewuster John M.
Acadian Seaplants Limited
30 Brown Avenue, Dartmouth,
Nova Scotia B3B1X8
CANADA
jsewuster@acadian.ca

Sharp Glyn John
Fisheries and Oceans Canada1 Challenger Drive,
Dartmouth,
Nova Scotia B2Y4A2
CANADA
sharpg@mar.dfo-mpo.gc.ca

Shibata Toshiyuku
Department of Life,
Environment and Materials Science,
Faculty of Engineering,
Fukuoka Institute of Technology
3-30-1 Wajiro-Higashi, Higashi-ku,
Fukuoka 811-0295
JAPAN
t-shibata@fit.ac.jp

Shibuya Masanobu
Shibuya Diving Industry, Co.
821-11 Nezakama, Hiratsuka City,
Kanagawa Pref. 254-0904
JAPAN
info@shibuya-diving.co.jp

Shihira-Ishikawa Ikuko
3-60-2-606 Kami-ishihara,
Chofu, Tokyo
JAPAN
shihira-ishikawa@dab.hi-ho.ne.jp

Shimada Satoshi
Creative Research Initiative "Sousei",
Hokkaido University
Kitaku, Kita 10, Nishi 8, Sapporo, 001-0021
JAPAN
sshimada@sci.hokudai.ac.jp

Shimizu Takeshi
Hokkaido Industrial Technology Center,
379 Kikyo-cho, Hakodate,
Hokkaido, 041-0801
JAPAN
shimizu@techakodate.or.jp

Sivakumar Kathirasen
Department of Botany,
Annamalai University
Annamalai Nagar, Tamil, Nadu, 608002
INDIA
kshivam69@yahoo.com

Sjøtun Kjersti
Department of Biology,
University of Bergen
P. O. Box 7800, N-5020 Bergen
NORWAY
Kjersti.Sjotun@bio.uib.no

Soriano Gonzalo
Soriano SA

J.C. Evans 40,
Gaiman, Chubut, 9105
ARGENTINA
gonza@soriano-sa.com.AR

Sousa Pinto Isabel
Centre for Marine and Environmental Research (CIMAR),
Department of Botany FCUP,
University of Porto
R. do Bragas, 289, 4050-123 Porto
PORTUGAL
ispinto@cimar.org

Stefanova Roumiana
National Research Council of Canada
Institute for Marine Biosciences
1411 Oxford Street, Rm 105,
Halifax, Nova Scotia B3H 3Z1
CANADA
Roumiana.Stefanova@nrc.gc.ca

Stekoll Michael S.
Chemistry and Biochemistry
University of Alaska Southeast
11120 Glacier Highway, Juneau, AK 99801
USA
ffmss@uaf.edu

Stiger-Pouvreau Valerie
LEBHAM-IUEM-UBO
Technopole Brest-Iroise, Place Nicolas Copernic,
29280 Plouzane
FRANCE
stiger@univ-brest.fr

Su Huei Meei
Biotechnology Division,
Fisheries Research Institute, Council of Agriculture
No.67 Fongyu St., Tungkang, Pingtung 92845
TAIWAN
hmsu@mail.tfrin.gov.tw

Sudo Yusuke
Okinawa Prefectural Fisheries & Ocean Research Center
1-4-1 Nishizaki, Itoman, Okinawa, 901-0305
JAPAN
sd3_80@yahoo.co.jp

Suzuki Hidekazu
Tokyo University of Marine Science & Technology
Konan 4-5-7, Minato-ku, Tokyo, 108-8477
JAPAN
hsuzuki@s.kaiyodai.ac.jp

Suzuki Minoru
Philagar Corporation
1-8-6-305 Kiyosumi, Koto-ku,
Tokyo, 135-0024
JAPAN
suzualga@xd5.so-net.ne.jp

Taino Seiya
Kochi Prefectural Fisheries Experimental Station
1153-23 Haigata, Uranouchi, Susaki-shi,
Kochi 785-0617
JAPAN
seiya_taino@ken4.pref.kochi.jp

Taira Toshiaki
Ina Food Industry Co., Ltd.
5074 Nishiharuchika, Ina City,
Nagano, 399-4498
JAPAN
t-taira@kantenpp.co.jp

Takagishi Seiichiro
Andes Trading Co., Ltd
Ohtaka Building 7F, 2-14, Kanda-Tsukasa-cho,
Chiyoda-ku, Tokyo
JAPAN
takagishi@andes-trading. co.jp

Takahashi Yoshiaki
Fisheries Experimental Station,
Wakayama Prefectural Research Center of Agriculture,
Forestry & Fisheries
Kushimoto 1557-20, Kushimoto-cho,
Higashimuro-gun, Wakayama, 649-3503
JAPAN
takahashi_y0016@pref.wakayama.lg.jp

Takeuchi Ichiro
Faculty of Agriculture,
Ehime University
3-5-7 Tarumi, Matsuyama, Ehime, 790-8566
JAPAN
takeuchi@agr.ehime-u.ac.jp

Takio Susumu
Kumamoto University
Kurokami 2-39-1, Kumamoto 860-8555
JAPAN
stakio@gpo.kumamoto-u.ac.jp

Tamaki Hitoshi
Ishinomaki Senshu University
1 Shinmito Minamisakai, Ishinomaki-shi

Miyagi 986-8580
JAPAN
thitoshi@isenshu-u.ac.jp

Tamura Shinya
Research & Development Laboratory
Fujicco Co., Ltd.
6-13-4 Minatojima-Nakamachi, Chuo-ku,
Kobe 650-8558
JAPAN
s-tamura@fujicco.co.jp

Tanaka Hiroshi
8-31-14 Yamamoto, Asa-minami-ku,
Hiroshima 731-0137
JAPAN
kaisou-hiroshi@giga.ocn.ne.jp

Tanaka Jiro
Tokyo University of Marine Science & Technology
Konan 4-5-7, Minato-ku,
Tokyo, 108-8477
JAPAN
jtanaka@s.kaiyodai.ac.jp

Tanaka Ken
Shirako Co., Ltd.
6-1-17 Nakakasai, Edogawa-ku, Tokyo, 134-0083
JAPAN

Tanaka Toshimitsu
Wakayama Industry Promotion Foundation,
Technology Promotion Division,
Office of Regional Intensive Research Project
Kushimoto 1557-20, Kushimoto-cho
Higashimuro-gun, Wakayama, 649-3503
JAPAN
tanaka_t0018@pref.wakayama.lg.jp

Tanios Hengky
JL. Gedong, Panjang, No 2 B/D, Jakarta 11240
INDONESIA
scoobyhk@hotmail.com

Tashiro Yuri
Tokyo University of Marine Science & Technology
4-5-7 Konan, Minato-ku, Tokyo, 108-8477
JAPAN
tyurip@s.kaiyodai.ac.jp

Tatsukawa Kenichi
General Study Group for River Areas
6-1 Yotsuba-1, Itabashi-ku, Tokyo 175-0084
JAPAN
ryuiki_jcom.home.ne.jp

Taylor Colin Gordon
International Finance Corporation Pef-Philippines
Unit 90 & 92, Landco Corporate Center,
JP Laurel Avenue, Davao City 8000
PHILIPPINES
CTaylor1@ifc.org

Teas Jane
University of South Carolina
2221 Devine Street, Room 230,
Columbia, SC 29205
USA
teas@gwm.sc.edu

Terada Ryuta
Faculty of Fisheries, Kagoshima University
Shimoarata 4-50-20, Kagoshima City,
Kagoshima Prefecture, 890-0056
JAPAN
terada@fish.kagoshima-u.ac.jp

Terawaki Toshinobu
Fisheries Research Agency
Queen's Tower-B 15F,
Minato-Mirai, Nishi-ku,
Yokohama, 220-6115
JAPAN
terawaki@affrc.go.jp

Toda Toshiya
Research & Development Laboratory
Fujicco Co., Ltd.
6-13-4 Minatojima-Nakamachi, Chuo-ku,
Kobe 650-8558
JAPAN
ttoda@fujicco.co.jp

Tsairi Boaz
SAKURA
PO BOX 8559, Jerusalem, 91083
ISRAEL
skura@netvision.net.il

Tsuji Keisuke
School of Human Science and Environment,
University of Hyogo
1-1-12 Shinzaike-Honcho, Himeji,
Hyogo, 670-0092
JAPAN
ktsuji@shse.u-hyogo.ac.jp

Tsukakoshi Hidehiro
Ina Food Industry Co., Ltd.
5074 Nishiharuchika,
Ina City, Nagano, 399-4498
JAPAN
thide@kantenpp.co.jp

Tsukakoshi Hiroshi
Ina Food Industry, Co., Ltd.
5074 Nishi-Haruchika, Ina-City, Nagano, 399-4498
JAPAN
htsuka@kantenpp.co.jp

Tsutsui Isao
Japan International Research Center for Agricultural Sciences
c/o JIRCAS SE Asia Office, Phaholyothin Rd.,
Ladvao, Chatuchak, Bangkok, 10900
THAILAND
cong@affrc.go.jp

Uchimura Masayuki
Port and Airport Research Institute (PARI)
3-1-1 Nagase, Yokosuka, Kanagawa, 239-0826
JAPAN
uchimura@pari.go.jp

Uede Takashi
Fisheries Experimental Station of Wakayama Prefecture
1557-20, Kushimoto, Kushimoto-cho,
Higashimuro district, Wakayama Pref., 649-3503
JAPAN
uede_t0001@pref.wakayama.lg.jp

Ugarte Raul Alejandro
90 First St. Rothesay, New Brunswick, E2H 1L9
CANADA
rugarte@acadian.ca

Uzuhashi Yuji
Ina Food Industry, Co., Ltd.
5074 Nishi-Haruchika,
Ina-City, Nagano, 399-4498
JAPAN
uzuhashi@kantenpp.co.jp

Vairappan Charles S.
Lab. Nat. Product. Chem.,
Inst. Tropical Biol. and Conservation,
Univ. Misia Sabah, Kota Kinabalu
Sabah 88999
MALAYSIA
csvairappan@yahoo.com/csv@ums.edu.my

Vásquez Julio Alberto
Departamento de Biología Marina.
Universidad Católica del Norte.
Center for Advanced Studies in Arid Zones (CEAZA)
Casilla 117, Coquimbo
CHILE
jvasquez@ucn.cl

Vea Jostein
FMC Biopolymer
Veamyr 43, Vedavagen, 4276
NORWAY
jostein_vea@fmc.com

Vilkova Olga
Coastal Research Laboratory
Russian Federal Research Institute of Fisheries & Oceanography,
17, V. Krasnoselskaya,
Moscow, 107140
RUSSIA
prelest_moya@mail.ru

Villena Gunter Alberto
Peruvian Seaweeds S.R.L.
Av. Paul Poblet Parcela D-19, Lote 19
Lurin, Lima 16
PERU
gunter@peruvianseaweeds.com

Vladimirescu Alexandru Filip
NIRDMI "Cantacuzino" Romania
Splaiul Independentei, No. 103, Sect. 5,
Bucharest, CP 1-525, 70100
ROMANIA
alexandruvl@yahoo.com

Vu Huu Thiem
Hailong Co., Ltd.
109 Truong Chinh Road, Kien An Dist,
Haiphong City
VIETNAM
hailongrobikahp@vnn.vn

Wang Xiang-jiang
Chenyu Seaweed Culture Company Ltd
Shinqiao Village, Gangyu County, Jiangsu, 222111
CHINA
c/o feixiugeng@ms.qdio.ac.cn

Wang Xiuliang
Institute of Oceanology
The Chinese Academy of Sciences

No.7 Nanhai Road, Qingdao, 266071
CHINA
xliangw@yahoo.com

Watanuki Akira
Alpha Hydraulic Engineering Consultants Co., Ltd.
4-15-35 Mita, Minato-ku, Tokyo, 108-0073
JAPAN
watanuki@ahec.jp

Waters John
Streedagh House, Streedagh,
Grange, Country Sligo
IRELAND
rhatigandoc@eircom.net

Watson Duika Burges
Institute of Health and Society
Newcastle University
21 Claremont Place,
Newcastle upon Tyne, NE2 4HH
UK
Duika.Burges-Watson@newcastle.ac.uk

Wegeberg Susse
Department of Phycology, Biological Institute,
University of Copenhagen
Øster Farimagsgade 2D,
Copenhagen K, DK-1353
DENMARK
sussew@bi.ku.dk

Werlinger Camilo
Departamento de Oceanografia
Universidad de Concepcion
Barrio Universitario, Concepcion
CHILE
cwerling@udec.cl

Wernberg Thomas
Centre for Ecosystem Management
Edith Cowan University
Joondalup, 6024 WA
AUSTRALIA
t.wernberg@ecu.edu.au

Werner Astrid
Queen's University BelfastQueen's University Marine Laboratory Portaferry,
12 The Strand,
Portaferry, Northern Ireland, BT22 1PF
UK
a.werner@qub.ac.uk

West John A.
School of Botany,
University of Melborne
Melborne
AUSTRALIA
jwest@unimelb.edu.au

Whyte Karen
West Coast Abalone
P.O. Box 185, Stompneusbaai, 7382
SOUTH AFRICA
kwhyte@wca.co.za

Wolf William
Thorvin Inc.
P.O. Box 458, New Castle, VA 24127
USA
bwolf@thorvin.com

Woo Seonock
Korea Ocean Research & Development Institute
Jangmok 391, Jangmok-myon, Geoje-si,
Kyungsangnam-do, 656-830
KOREA
cwoo@kordi.re.kr

Xu Pu
Changshu Institute of Technology
98 Yuanhe Road, Changshu,
Jiangsu, 215500
CHINA
xupu66@sina.com.cn

Yamagishi Yukimasa
Department of Marine Biotechnology,
Faculty of Life and Biotechnology,
Fukuyama University
985 Sanzo, Higashimura-cho, Fukuyama,
Hiroshima Prefecture, 729-0292
JAPAN
yamag@ma.fuma.fukuyama-u.ac.jp

Yamamoto Takashi
Kibunfood Chemifa Co., Ltd.
1600 Kaisuka, Kamogawa-shi, Chiba, 296-0004
JAPAN
t-yamamoto@kibunfc.co.jp

Yamauchi Makoto
Fisheries Experimental Station,
Wakayama Prefectural Research Center of Agriculture,
Forestry & Fisheries
Kushimoto 1557-20, Kushimoto-cho

Higashimuro-gun, Wakayama, 649-3503
JAPAN
yamauchi_m0002@pref.wakayama.lg.jp

Yan Xing-hong
Shanghai Fisheries University
College of Aqua-life Science and Technology
334 Jungong Road, Shanghai, 200090
CHINA
xhyan@shfu.edu.cn

Yang Yu-Feng
Institute of Hydrobiology,
Jinan University
Guangzhou, 510632
CHINA
tyyf@jnu.edu.cn

Yangthog Monsuang
King Monkut's Institute of Technology
Ladkrabang Chumphon Campus
17/1 T. Chumko, A. Pathiw, 86160
THAILAND
Kymonsru@kmitl.ac.th

Yap Roseller Propios
Shemberg Marketing Corporation
Seaweed Industry Assn. of the Phills.
S. Jayme St. Pakna-An, Maridaue City,
Cebu City 6014
PHILIPPINES
roseller.yap@shemberg.cp.ph

Yarish Charles
Dept. of Ecol. & Evol. Biology
University of Connecticut
1 University Place,
Stamford, CT 06901-2315
USA
charles.yarish@uconn.edu

Yashiro Takuya
NABOCUL COSMETICS Co., Ltd.
5-29-7 Sendagaya, Shibuya-ku,
Tokyo, 151-0051
JAPAN
nabocul@lake.ocn.ne.jp

Yatsuya Kousuke
Kyoto Institute of Oceanic and Fishery Science
1029-3 Odashukuno, Miyazu, Kyoto, 626-0052
JAPAN
k-yatsuya85@pref.kyoto.lg.jp

Yazawa Kazunaga
Lab. of Nutraceutical & Functional Foods Science
Graduate School of Marine Science & Technology,
Tokyo University of Marine Science & Technology
4-5-7 Konan, Minato-ku,
Tokyo, 108-8477
JAPAN
yazawa@s.kaiyodai.ac.jp

Yeong Huiyin
Algae Research Laboratory (B207),
Institute of Postgraduate Studies,
University of Malaya
Kuala Lumpur, 50603
MALAYSIA
yhui_yin@yahoo.com

Yokoya Nair Sumie
Instituto de Botanica, Seção de Ficologia
Av. Miguel Estefano, 3687. Agua Funda,
São Paulo 04301-012
BRAZIL
nyokoya@hotmail.com/casamarinha@terra.com.br

Yokoyama Takehiko
School of Fisheries Sciences,
Kitasato University
Ofunato, Iwate, 022-0101
JAPAN
yokotake@kitasato-u.ac.jp

Yoon Hwan Su
Department of Biogical Sciences,
University of Iowa
468 Biology Building,
Iowa City, IOWA, 52242
USA
hwansu@gmail.com; hwan-yoon@uiowa.edu

Yoon Yun Koo
Pusan National University
San #30, Tangjun-dong,Geumjung-gu,
Busan Metro City 609-735
KOREA
yunkyoon@yahoo.co.kr

Yoshida Goro
National Research Institute of Fisheries & Environment
of Inland Sea
2-17-5 Maruishi, Hatsukaichi,
Hiroshima, 739-0452
JAPAN
gorogoro@affrc.go.jp

Yoshizaki Makoto
Department of Biology,
Toho University
Miyama 2-2-1, Funabashi,
Chiba Pref. 274-8510
JAPAN
makoto@bio.sci.toho-u.ac.jp

Yotsukura Norishige
Muroran Marine Station,
Field Science Center for Northern Biophere,
Hokkaido University
Muroran, Hokkaido, 051-0003
JAPAN
yotsukur@bio.sci.hokudai.ac.jp

Zafar Mohammad
Institute of Marine Sciences
University of Chittagong
Chittagong, 4331
BANGLADESH
zafarims@yahoo.com,/zafar@imscu.ac.bd

Zamorano Jaime
Gelymar S.A.
Casilla 997-Puerto Montt, 5489002
CHILE
JZAMORANO@GELYMAR.COM

Zemke-White Lindsey
Division of Applied Sciences,
AUT University
Mail No.C-47, Private Bag 92006, Auckland,
NEW ZEALAND
lindsey.zemke-white@aut.ac.nz

Zertuche-Gonzalez Jose A.
Universidad Autonoma de Baja California
KM 106 Carr, Tijuana-Ensenada, Ensenada,
Baja California 22860
MEXICO
zertuche@uabc.mx

Zhang Shouyu
Shanghai Fisheries University
No.334, Jungong Road, Shanghai, 200090
CHINA
syzhang@shfu.edu.cn

Zhu Jianyi
Changshu Institute of Technology
98 Yuanhe Road, Changshu,
Jiangsu, 215500
CHINA
xupu66@sina.com.cn

Zingmark Richard G.
Marine Science Program,
University of South Carolina
Columbia, SC 29208
USA
zingmark@sc.edu

Zubia Arrieta Mayalen Lydie
Departamento de Recursos del Mar,
Cinvestav Km 6 Carretera Antiqua a Progreso
A.P. 73 Cordemex, 97310 Merida, Yucatan
MEXICO
mayalen_zubia@yahoo.fr

Student delegates

Alamsjah Mochammad Amin
Morimoto Apt 2F, Nishiki 1-2-32,
Nagasaki, 852-8045
JAPAN
alamsjah_70@yahoo.com

Alcantara-Creencia Lota
Marine Center, Faculty of Fisheries
Kagoshima University4-50-20 Shimoarata, Kagoshima-shi, 890-0054
JAPAN
lota082968@yahoo.com

Ang May Yen
Lab. Nat. Product Chem.,
Inst. Tropical Biol. & Conservation,
Univ. Misia Sabah,
Kota Kinabalu, Sabah 88999
MALAYSIA
angmayyen@yahoo.com

Barradas Ana
INIAP-IPIMAR CRIPSul Aquaculture Research Station
Praceta de S. Tome e Principe, Lote 11, 3 Esq,
Faro 8000-190
PORTUGAL
arbarradas@ualg.pt

Bast Felix
Usa Marine Biological Institute,
Kochi University
194 Inoshiri, Usa, Kochi 781-1161
JAPAN
felix.bast@gmail.com

Bi Yanhui
Key Laboratory of Aquatic Genetic Resources
& Aquaculture Ecosystem
Shanghai Fisheries University
Rome 809, 334 Jungong Road, Shanghai, 200090
CHINA
biomarine_bi@yahoo.com.cn

Bittner Lucie
MNHN - Systematique Evolution
Casepostale 39-57, rue Cuviers,
75231 Paris
FRANCE
lucie@canarochouf.com

Boavida Joana Rlela
University of Algarve
Rua Sao Francisco Xavier, N-66 Lisboa, 1400-331
PORTUGAL
JOANARBOAVIDA@GMAIL.COM

Boizard Sophie
University of British Columbia
20F Balaclava St.,
Vancouver British Columbia, V6K 4C3
CANADA
boizard@interchange.ubc.ca

Bondu Stéphanie
LEBHAM, European Institute for Marine Studies
Place Nicolas Copernic, Plouzane, 29280
FRANCE
stephanie.bondu@univ-brest.fr

Campbell Claire
Queen's University Belfast, QUB Marine Station
12 The Strand,
Portaferry, Northern Ireland, BT22 1PF
UK
clairecampbell99@hotmail.com

Camus Carolina
Departamento de Ecologia
Pontificia Universidad Catolica de Chile
Alameda 310, Santiago 6513677
CHILE
ccamust@bio.puc.cl

Carton Rhea Joy Barcelona
Tokyo University of Marine Science & Technology
Rm 710, Shinagawa International House
4-5-7 Konan, Minato-ku, Tokyo 108-8477
JAPAN
pitsel@yahoo.com

Carvalho Matheus Carvalho de
Kitasato University
Room 13, Mioujindou-16, Okirai, Sanriku-cho,
Ofunato-shi, Iwate-ken, 022-0101
JAPAN
matheus@samerica.com

Fujie Asuka
5-57-21-B-1 Fujimidai, Chikusa, Nagoya, 464-0015
JAPAN
506M318@m.mie-u.ac.jp

Furuya Megumi
A-107 Ferrys Mitsuzawa, 80-27 Tokiwa-dai
Hodogaya-ku, Yokohama, Kanagawa 240-0067
JAPAN
d06na041@ynu.ac.jp

Gabriel Daniela
University of the Azores
Apartado 322, Ponta Delgada, 9501-904
PORTUGAL
dgabriel@notes.uac.pt

Gabriela Andrade-Sorcia
Programa de Investigacion en Botanica Marina
Departamento de Biologia Marina UABCS
Apartado postal 19-B, km. 5.5. Carretera al sur
La Paz B.C.S. 23080
MEXICO
gandrade@uabcs.mx

Han Myoung Shog
Department of Aquaculture, Fisheries Science,
Pukyung National University
Busan 608-737
KOREA
sodo14@korea.com

Haraguchi Hiroko
Graduate School of Kuroshio Science
Kochi University,
194 Inoshiri, Usa, Tosa, Kochi, 781-1164
JAPAN
hirokoh@cc.kochi-u.ac.jp

Hasegawa Kazukiyo
Tokyo University of Marine Science & Technology
4-5-7 Konan, Minato-ku, Tokyo, 108-8477

JAPAN
Hasegawa_kazukiyi@yahoo.co.jp

Hayashi Leila
Hayashi Shintaro/Nishimura Teruyo
Av. Marari, 356-Jo Dupece, São Paulo, 04402-000
BRAZIL
hayashi@usp.br

Henkel Sarah K.
Department of EEMB
University of California Santa Barbara
Mail Code 9610, Santa Barbara, CA 93106-9610
USA
henkel@lifesci.ucsb.edu

Herkül Kristjan
Estonian Marine Institute, University of Tarty
Mäealuse 10a, Tallin, 12618
ESTONIA
kristjan.herkyl@sea.ee

Ikoma Tomoyuki
1-1-53-501 Ikura, Yokkaichi, Mie, 510-0832
JAPAN
505M318@m.mie-u.ac.jp

Ito Tomoko
Shimoda Marine Research Center
University of Tsukuba
5-10-1 Shimoda, Shizuoka, 415-0025
JAPAN
tomoko198365@hotmail.com

Jacquin Anne Gaelle
LEBHAM, European Institute for Marine Studies
Place Nicolas Copernic, Plouzane, 29280
FRANCE
Anne-Gaelle.jacquin@univ-brest.fr

Jazmin de Jesus Hernandez-Kantun
Programa de Investigacion en Botanica Marina
Departamento de Biologia Marina UABCS
Apartado postal 19-B, km. 5.5. Carretera al sur
La Paz B.C.S. 23080
MEXICO
jazmin@uabcs.mx

Kamiya Akira
Fukuroyama 200-15, Koshigaya, Saitama 343-0032
JAPAN
a_kamiya77@hotmail.com

Kang Yun Hee
Pusan National University
San #30, Jangjung-dong, Geumjung-gu
Busan Metro City, 609-735
KOREA
unirang@pusan.ac.kr

Kanno Mariko
4-7-2-208 Amakubo, Tsukuba-shi
JAPAN
S0310757@ipe.tsukuba.ac.jp

Kawagoshi Masaru
Tokyo University of Marine Science & Technology
4-5-7 Konan, Minato-ku, Tokyo, 108-8477
JAPAN
gosi1227bk@hotmail.com

Kimura Kei
Division of Biosphere Science,
Graduate School of Environmental Science,
Hokkaido University
Muroran, Hokkaido, 051-0003
JAPAN
kkim727@fsc.hokudai.ac.jp

Kitagawa Nobuhito
Marine Plants Laboratory,
Department of Natural Environmental Science,
Faculty of Science, Kochi University
Inoshiri 194, Usa-cho, Tosa, Kochi, 781-1164
JAPAN
dorcus-maro@hotmail.co.jp

Kumar Vivek
Curtin University of Technology,
Brodie Hall Building 1, Turner Avenue,
Technology Park, Bentley WA 6102
AUSTRALIA
V.Kumar@exchange.curtin.edu.au

Kushii Ayumi
Joyfull 208, 2-74 Edobashi, Tsu, Mie, 514-0001
JAPAN
503535@m.mie-u.ac.jp

Lee Jin Hwan
Sang Myung University
#7 Hongji-dong, Jongro-gu,
Seoul, 110-743
KOREA
jhlee@smu.ac.kr

Lee Ki-Hoon
Lab. of Marine Biology
Faculty of Biological Science
Wonkwang University,
344-2, Iksan, Jeonbuk, 570-749
KOREA
red1882@daum.net

Leon Karla Cisneros
Universidade dos Açores
Rua da Mãe de Deus, 8y"A" RC, Ponta Delgada
9500-321 Azores
PORTUGAL
cisneros@notes.uac.pt

Leon Deniz Lorena Violeta
Centro de Investigacion y de Estudios Avanzados del IPN,
Departamento de Recursos del Mar Laboratorio
de Ficoquimica
Antigua Carretera a Progeso Km 6 A. P. 73 Cordemex,
Merida Yucatan 97310
MEXICO
jorena@mda.cinvestav.mx

Mahon Anne Marie
Marine Laboratory, The Queens University Belfast
The Strand,
Portaferry, Northern Ireland, BT22 1PF
UK
amahon01@qub.ac.uk

Mata Leonardo
CCMAR-Centre of Marine Sciences
Universidade do Algarve, Campus de Gambelas
8005-139 Faro
PORTUGAL
lmata@ualg.pt

Mattio Lydiane
IRD-BPAS, 98848 Noumea
NEW CALEDONIA
lydiane.mattio@noumea.ird.nc

Metti D. Yola
Royal Botanic Gardens, Sydney
Mrs Macquaries Road,
Sydney, NSW, 2000
AUSTRALIA
yola.metti@rbgsyd.nsw.gov.au

Miki Ayako
Faculty of Environment & Information Sciences,
Yokohama National University
79-7, Tokiwadai, Hodogaya-ku,
Yokohama 240-8501
JAPAN
d05ha059@ynu.ac.jp

Miura Yosuke
Sunny Life Kurima A-105, 1741-3 Matiyacho,
Kurima, Tsu, Mie, 514-0001
JAPAN
503584@m.mie-u.ac.jp

Miyata Mitsuyoshi
Laboratory of Nutraceutical & Functional Foods Science,
Graduate School of Marine Science & Technology,
Tokyo University of Marine Science & Technology,
4-5-7 Konan, Minato-ku, Tokyo, 108-8477
JAPAN
m052030@edu.s.kaiyodai.ac.jp

Montolalu Roike Iwan
Tokyo University of Marine Science & Technology
4-5-7 Konan, Minato-ku, Tokyo, 108-8477
JAPAN
rmontolalu@yahoo.co.jp

Mortensen Agnes Mols
Department of Phycology, Biological Institute,
University of Copenhagen
Øster Farimagsgade 2D,
Copenhagen K, DK-1353
DENMARK
ammortensen@bi.ku.dk

Muñoz Julieta
Department of Botany,
University of Delhi
Delhi, 110007
INDIA
kappaphycus@hotmail.com

Nakamoto Chika
House Camphor Tree 301, 1415-2 Machiyacho,
Kurima, Tsu, Mie, 514-0102
JAPAN
503571@m.mie-u.ac.jp

Nakamura Cassio Kiyonori
Kobe University
Rokkodai 1-1, Nada-ku, Kobe-shi
JAPAN

Nerio Yuka
Laboratory of Nutraceutical & Functional Foods Science,

Tokyo University of Marine Science & Technology
4-5-7 Konan, Minato-ku, Tokyo, 108-8477
JAPAN
m062042@edu.s.kaiyodai.ac.jp

Nishio Hiroto
Graduate School of Medicine and Engineering,
University of Yamanashi
4-3-11 Takeda, Kofu, Yamanashi, 404-8511
JAPAN
g06mf011@ccn.yamanashi.ac.jp

O'Connell Elaine
Molecular Glycobiotechnology Research Group
Department of Biochemistry, NUI, Galway City
IRELAND
elaineoc22@gmail.com

Ohashi Shunsuke
Institute of Material Science,
University of Tsukuba
1-1-1 Tennodai, Tsukuba, Ibaraki, 305-8573
JAPAN
s-084@ims.tsukuba.ac.jp

Okuyama Satomi
Shitami 5-8-35-205, Higashi-Hiroshima,
Hiroshima 739-0047
JAPAN
sokuyama@hiroshima-u.ac.jp

Ooi Jillian
Department of Geography,
Faculty of Arts & Social Sciences,
University of Malaya
Kuala Lumpur 50603
MALAYSIA
jillian_03@um.edu.my

Ozalp Hasan Baris
Esenler Mahalles, Dikk Yopi, Sites I Block,
No. 2215, Canakkale, 17100
TURKEY
Jacenzo@yahoo.com

Parente Manuela Isabel
Departamento de Biologia
Universidade dos Açores
Rua da Mãe de Deus Apartado 1422
9501 - 801 Ponta Delgada (Açores) Codex,
9500-656
PORTUGAL
mparente@notes.uac.pt

Park Jung-Im
Department of Biology,
Pusan National University
30 Changiun-dong, Geumjeong Gu,
Busan, 609-735
KOREA
jip@pusan.ac.kr

Park Sang Rul
Pusan National University
San #30, Jangjung-dong, Geumjung-gu
Busan Metro City, 609-735
KOREA
srv99park@pusan.ac.kr

Paulo Diogo Soares
University of Algarve
Avenida Marques de Pombal N-3,
Sub Cave Direita Amadora, 2700
PORTUGAL
diogosoarespaulo@gmail.com

Phooprong Sukanya
Laboratory of Primary Productivity
School of Fisheries Science,
Kitasato University
160-4 Uto, Okirai, Sanriku. Ofunato,
Iwate, 022-0101
JAPAN
spooprong@yahoo.com

Queguineur Benoit
Irish Seaweed Centre, NUIG, MRI
University Road, Galway
IRELAND
benoitqueguineur@yahoo.fr

Rioux Laurie-Eve
Institut des nutraceutiques et des aliments fonctionnels,
Departement des sciences des aliments et de nutrition,
Pavillon comtois, Universite Laval.
Quebec, G1K 7P4
CANADA
laurie-eve.rioux.1@ulaval.ca

Robertson-Andersson Deborah Vivienne
University of Cape Town
P.O. Box 26218 Hout BAT, 7872
SOUTH AFRICA
droberts@botzoo.uct.ac.za

Rodger Andrew N. S.
Scottish Association for Marine Sciences

Oban, Argyll, Scotland, PA37 1QA
UK
andrew.rodger@sams.ac.uk

Rodriguez-Figueroa Griselda Margarita
CICIMAR-IPN
Av. IPN. Col. Ploya Palo de Santa-Rita S/N 23060
MEXICO
grodriguezf@ipn.mx

Sato Shinya
Alfred Wegener Institute for Polar & Marine Research
Am Handelschafen 12,
Bremerhaven, D-27570
GERMANY
ssato@awi-bremerhaven.de

Sato Tomomi
Graduate School of Biosphere Science
Hiroshima University
Kagamiyama 1-4-4, Higashi-Hiroshima,
Hiroshima 739-8528
JAPAN
stomomi@hiroshima-u.ac.jp

Schmid Daniela
MPI for Chemical Ecology
Department of Bioorganic Chemistry
Hans-Knoll-Str.8, Jena 07745
GERMANY
dschmid@ice.mpg.de

Sekimoto Satoshi
Konan University
8-9-1 Okamoto, Higashi-nada, Kobe, 658-8501
JAPAN
dn422001@center.konan-u.ac.jp

Soler Vila Anna
Irish Seaweed Centre, Martin Ryan Institute
National University of Ireland, Galway
IRELAND
anna.soler@nuigalway.ie

Son Moonho
Pusan National University
San #30, Jangjung-dong, Geumjung-gu,
Busan 609-735
KOREA
mhsohn77@yahoo.co.kr

Suárez Castillo Alvin Noe
Instituto Politecnico National Centro
Interdisciplinario de Ciencias Marinas
Calle Camino Del Norte Numero 150, 23083
MEXICO
protomono@hotmail.com

Sun Zhongmin
Tokyo University of Marine Science & Technology
Konan 4-5-7, Minato-ku, Tokyo, 108-8477
JAPAN
zhmsun@hotmail.com

Suzuki Masahiro
Graduate School of Science,
Toho University
Miyama 2-2-1, Funabashi, Chiba Pref. 274-8510
JAPAN
masshark@bio.sci.toho-u.ac.jp

Suzuki Satoshi
The United Graduate School of Agricultural Sciences,
Ehime University
3-5-7 Tarumi, Matsuyama, Ehime, 790-8566
JAPAN
sasuzuki@agr.ehime-u.ac.jp

Takahashi Monica Miyuki
Takahashi Hiroshi/Takahashi Fujie,
R. Do Pombo Correio, 1S4, Itaim, Paulista
São Paulo SP 08160-142
BRAZIL
momita@uol.com.br

Tan Kai Lee
Lab. Nat. Product. Chem.,
Inst. Tropical Biol. And Conservation,
Univ. Misia Sabah,
Kota Kinabalu, Sabah 88999
MALAYSIA
kailee622@yahoo.com

Tokuyama Akihide
Department of Natural Environmental Science
Marine Plants Laboratory, Faculty of Science
Kochi University
Inoshiri 194, Usa-cho, Tosa, Kochi, 781-1164
JAPAN
akiki_hon@hotmail.co.jp

Tominaga Hiroshi
2-301 Diamond Hills, 179-1,
Kurima-nakayama,
Tsu, Mie
JAPAN

Tuvikene Rando
Department of Natural Sciences,
Tallinn University
Narva Road 25, Tallinn, 10120
ESTONIA
rtu@akvaarium.com

Ueki Chika
1-3-13 Konan-cho, Muroran, Hokkaido 051-0032
JAPAN

Wada Shigeki
Life and Environmental Sciences
University of Tsukuba
1-1-1 Tennoudai, Tsukuba City, Ibaraki, 305-8572
JAPAN
shige_sbm@ies.life.tsukuba.ac.jp

Wallenstein Francisco Maledo
Heriot-Watt University
Rua do Negrao 28, Sao Pedro/Ponta Delgada/
Azores, 9500-334
PORTUGAL
fmacedo@notes.uac.pt

Wang Lei
Shanghai Fisheries University
No. 334, Jungong Road, Shanghai, 200090
CHINA
L_wang@stmail.shfu.edu.cn

Watanabe Tsuyoshi
Tokyo University of Marine Science
& TechnologyKonan 4-5-7,
Minato-ku, Tokyo, 108-8477
JAPAN
tsu4shi_w@yahoo.co.jp

Yoo Hyun-Il
Lab. of Marine Biology
Faculty of Biological Science
Wonkwang University
344-2, Iksan, Jeonbuk, 570-749
KOREA
non19@hanmail.net

Youn Seok Jea
Sang Myung University
#7 Hongji-dong, Jongro-gu,
Seoul, 110-743
KOREA
cookel45@smu.ac.kr

Yuen Yeong Shyan
Faculty of Science,
University of the Ryukyus
1 Senbaru, Nishihara,
Okinawa, 903-0213
JAPAN
yeongshyan@yahoo.com

Zhou Xiao Li
Lab. of Marine Biology
Faculty of Biological Science,
Wonkwang University,
344-2, Iksan, Jeonbuk, 570-749
KOREA
zhoulily82485100@163.com

Advances in cultivation of Gelidiales

Michael Friedlander

Originally published in the Journal of Applied Phycology, Vol 20, No 5, 1–6.
DOI: 10.1007/s10811-007-9285-1

Abstract Currently, *Gelidium* and *Pterocladia* (*Gelidiales*) are collected or harvested only from the sea. Despite several attempts to develop a cultivation technology for *Gelidium*, no successful methodology has yet been developed. Initial steps towards developmental efforts in Portugal, Spain, South Africa and Israel have been published. More developments have probably been performed but have not been published. Two different technological concepts have been tested for *Gelidium* cultivation: (1) the attachment of *Gelidium* fragments to concrete cylinders floating in the sea, and (2) free-floating pond cultivation technology. These vegetative cultivation technologies might be partially optimized by controlling physical, chemical and biological growth factors. The pond cultivation technology is the much more controllable option. The effects of all factors are discussed in detail in this review. It seems that the main difficulty with cultivation of *Gelidium* is its low growth rate. The claimed yields of the two technologies are far from being economically attractive at this stage of their development. It seems that in order to introduce *Gelidium* into commercial cultivation, major efforts in genetic improvement through selection or genetic engineering will be required. Only high yield strains will have the potential to compete economically with the present harvesting tradition. However, accumulated experience with genetic improvement of other useful seaweed species suggests that this is possible.

Keywords *Gelidium* · *Pterocladia* · *Gelidiales* · Cultivation

M. Friedlander (✉)
Israel Oceanographic and Limnological Research,
P.O.B. 8030, Haifa 31080, Israel
e-mail: michael@ocean.org.il

Introduction

As far as I know there is no current commercial cultivation of *Gelidiales*. Despite several attempts to develop a cultivation technology for *Gelidium* and *Pterocladia*, so far no successful methodology has been developed. Because of the proprietary nature of commercial cultivation, a successful technology may have been developed but has remained unpublished. *Gelidium* and *Pterocladia* (or *Pterocladiella*) are currently only collected or harvested, as opposed to other useful seaweeds for which cultivation technology has been developed. The reasons for this situation are discussed in this review, including all important variables affecting *Gelidium* and *Pterocladia* growth. This review will rely mostly on *Gelidium* studies since most of the relevant information in the literature deals with *Gelidium*.

Initial steps

Gelidium has been cultured in Korea since 1989 (Melo 1998) and in China (Fei and Huang 1991). Early cultivation systems for *Gelidium* species have also been described in Spain and Portugal (Salinas 1991; Seoane-Camba 1997; Melo 1998). Cultivation of intertidal *G. pristoides* (Turner) Kützing in the sea was reported from South Africa (Aken et al. 1993), and growth experiments were performed in tanks with *G. crinale* (Turner) Gaillon and *Pterocladiella capillacea* (Gmel.) Santelices and Hommersand in Israel (Gal-Or and Israel 2004; Boulus et al. 2007). The ecological studies carried out with *Gelidium* and *Pterocladia* can be reviewed by following the major factors controlling the growth of these seaweeds. Evaluation of these studies may answer the major question of "why is there no mass cultivation of *Gelidium* and *Pterocladia*"?

Technological factors

Two major technologies have been suggested for *Gelidium* cultivation: sea farming and tank culture. Both use *Gelidium* fragments for initiation of the cultures.

Sea culture

In one culture experiment in Galicia (Spain), fragments of *G. sesquipedale* (Clemente) Born. et Thur. were attached to limestone cylinders placed in the sea (Seoane-Camba 1997). This technology has technical, biological and ecological advantages. However, it became clear that non-contaminated seawater and a high level of hydrodynamics are major prerequisites for a *Gelidium* farm because of epiphytes and competitors (Seoane-Camba 1997). Cultivation of intertidal *G. pristoides* in the sea in South Africa was attempted by suspending plants in net bags, or growing on limpet shells on ropes at different depths. However, after significant growth for 2 weeks, the thalli degenerated for unknown reasons (Aken et al. 1993). Systems with *G. rex* Santelices et Abbott hanging in open sea were developed in Chile after reattachment to shells (Rojas et al. 1996).

Tank culture

In Israel, *Gelidium* and *Pterocladia* were grown in outdoor tank culture systems adopted from other seaweed cultivation methods, included the use of 40 L tanks with continuous water flow and aeration (Gal-Or and Israel 2004; Boulus et al. 2007). The advantages are that tank culture allows control of most variables, and is relatively protected from grazers and epiphytes. The disadvantage is its high cost. Other experiments with *Gelidium* free-floating tank cultures in Spain claimed to generate healthy plants in 2–3 months, with a nice deep-red color, but very low growth rates (Salinas 1991a, b; Sosa et al. 2006). Looking at the effects of the major growth factors and at the maximal reported growth rates might improve our understanding of the obstacles to developing mass cultivation.

Physical factors

The major physical factors involved in culture of *Gelidium* and *Pterocladia* are temperature, irradiance and relative seaweed velocity. The optimal combination of these factors might lead to maximal yields. It is possible to study the requirements for these factors from limited experiments conducted in the sea, in tanks or in the laboratory.

Irradiance

G. pulchellum (Turner) Kurtz was cultured under controlled laboratory conditions. Maximum growth rate (10.0% day $^{-1}$) was obtained at 130–240 µmol photons m $^{-2}$ s $^{-1}$ under continuous light and aeration (Sousa-Pinto et al. 1999). The photosynthetic capacity of *G. sesquipedale* decreased with depth, revealing a 'sun' and 'shade' acclimation pattern (Silva et al. 1998). Photosynthesis is probably limited in *G. canariensis* (Grunow) Seoane-Camba, *G. arbuscula* Bory ex Boergesen and *Pterocladiella capillacea* by incident light in their natural habitats (Mercado et al. 2001). *Gelidium sesquipedale* in chemostats showed higher growth rates under blue and red light than under white light of the same photon-fluence rate (40 µmol m $^{-2}$ s $^{-1}$; Carmona et al. 1996). Reattachment of apical portions of *G. sesquipedale* increased under longer days compared to under shorter days (Juanes and Puente 1993). All these studies indicated that irradiance is a significant growth factor, and indicated a positive response to irradiance range. However, summer and winter outdoor experiments conducted in small aerated seawater tanks with *G. crinale* in Israel, showed that light intensity changes were less or non effective, with ambient conditions being almost optimal (Boulus et al. 2007).

Temperature

Temperature was defined as the major seasonal limiting factor for *G. crinale*, and positively affected weekly growth rate. The average daily growth rate was about 6.5% day^{-1} for the months of active growth (Boulus et al. 2007). The relative abundance of *G. robustum* (Gardn.) Hollenberg et Abbott in Mexico under normal conditions and during La Niña showed a positive correlation with sea surface temperature (Hernandez-Guerrero et al. 2000). Reattachment of apical portions of *G. sesquipedale* in the laboratory increased under a limited range of 16–18°C as compared to 20–22°C (Juanes and Puente 1993). These studies support a strong effect of temperature change on *Gelidium* cultures.

Water movement

A lag period of 3 months showed a positive correlation in situ between relative abundance of *G. robustum* and upwelling index / wind speed (Hernandez-Guerrero et al. 2000). Zones with greater water movement in situ (1.4 m s $^{-1}$) stimulated the growth of *G. robustum* (1.0% day^{-1}). Growth was directly proportional to water movement under semi-controlled cultivation conditions in Mexico, and treatment with the greatest dynamics demonstrated growth of 3.6% day^{-1} (Pacheco-Ruiz and Zertuche-Gonzalez 1995).

Thus, all three physical factors—irradiance, temperature and water movement—seem to be major effective growth factors both in the sea and under controlled conditions.

Chemical factors

The major chemical factors are the concentration of inorganic nitrogen and phosphate in seawater, and also the concentration of dissolved inorganic carbon (DIC) in the form of CO_2 or HCO_3^-. Several laboratory and outdoor experiments can supply a basic understanding of these factors.

Nitrogen

Ammonium was a major limiting factor for tank-cultured *G. crinale*, and positively affected growth (Boulus et al. 2007). In outdoor 40 L tank experiments of 4 weeks duration, the growth rate of *Pterocladiella capillacea* reached a weekly growth rate of 28.3% in winter and 25.3% in summer with different ammonium regimes (Gal-Or and Israel 2004). Growth of vegetative fragments of *G. sclerophyllum* Taylor was stimulated by high (2–3 mM) nitrogen concentrations in the laboratory (Rodriguez 1996).

Phosphate

The growth rate of *G. robustum* grown in the laboratory generally increased with increasing phosphate concentration up to 21% day^{-1}, obtained at 150 μM (Sousa-Pinto et al. 1996). Branch and rhizoid production of vegetative fragments of *G. sclerophyllum* were stimulated by high P concentrations (100–150 μM; Rodriguez 1996).

Inorganic carbon

Contradictory evidence has been presented regarding the carbon source of *Gelidiales*. In a pH-stat cultivation system, growth rates of *P. capillacea* were largely unaffected by pH in the range of 6.5–8.5. However, the growth rate decreased significantly above pH 8.5. This suggests that *P. capillacea* can generally take up both the CO_2 and HCO_3^- forms of inorganic carbon (Ci), but has a low ability to use HCO_3^- at alkaline seawater pH (Gal-Or and Israel 2004). Photosynthesis by *G. sesquipedale* was enhanced when the CO_2 concentration was increased by a decrease in pH. Similarly, an increase in pH from 8.1 to 8.7 produced a significant reduction in O_2 evolution rates, indicating that *G. sesquipedale* has a very low capacity to use HCO_3^- (Mercado et al. 1998). No detectable external carbonic anhydrase activity, and high sensitivity to alkaline pH indicated that *G. canariensis*, *G. arbuscula* and *P. capillacea* have a low capacity for using the external pool of HCO_3^-; therefore photosynthetic rates in a controlled environment room were dependent on CO_2 availability (Mercado et al. 2001). On the other hand, the weekly outdoor growth rate of *G. crinale* was not significantly affected by addition of a carbon source (CO_2 or $NaHCO_3$) or by the controlled maintenance of pH 8. This suggests that a daily pH increase up to pH 9 does not decrease the growth rate under these conditions as this species consumes bicarbonate, which is more abundant at high pH (Boulus et al. 2007).

The positive growth effect of nitrogen and phosphate additions to *Gelidium* and *Pterocladia* culture is fully supported; however, the consumption of bicarbonate is not consistent in *Gelidium* and *Pterocladia* cultures.

Biological factors

The major biological factors include strain selection, reproduction methods, seaweed density, epiphytes, and interaction with fish effluents. Only limited information can be found about these factors.

Strain selection

In a single genetic study, backcross-5, red, male-sterile plants of *G. vagum* Okamura grew at a significantly higher rate than plants from other lines, suggesting that male-sterile gametophytes would probably be better candidates for aquaculture than normal wild-type plants of this species (Patwary and Van der Meer 1997).

Reproduction method

Culture of vegetative fragments of different sizes and from different parts of the thallus of *G. sclerophyllum* showed that growth in length and branch production was favored by medial fragments, and rhizoid production was favored by apices (Rodriguez 1996). Research on the culture of *G. rex* was approached from two points of view: growth of thalli from spores, and growth from re-attachment of fragments. Mollusc shells, which are very easy to handle, were used in both systems. The results achieved by these methods showed that the re-attachment technique obtains larger plants in a shorter time (Rojas et al. 1996). Seedlings of *Gelidium* were cultured indoors for 20–30 days in China, and then transferred to the sea, where their recorded growth rate was 3.84% day^{-1} (Pei et al. 1996). Reattachment of apical portions of *G. sesquipedale* was induced in the laboratory under spray cultivation conditions, and occurred mainly at the first stage of development when in contact with the substratum. A probable advantage of sporophytic fronds was defined as compared to gametophytic fragments

(Juanes and Puente 1993). Apices of vegetative *G. sesquipedale* developed rhizoidal clusters during re-attachment to artificial substrate followed by germlings under spray technology in the laboratory, showing higher growth rate than development from sporelings. The germlings were transferred with their substrate to the sea in Spain (Salinas 1991). A new method of plantlet production of *Gelidium* sp. was developed using cell aggregates of apical fragments after a freezing–thawing procedure. It showed a daily growth rate of 13% day^{-1}, and might be an efficient method for generation of mass planting material for cultivation (Titlyanov et al. 2006). In pond culture, only vegetatively propagated, free-floating fragments are utilized.

Density

G. amansii (Lamouroux) Lamouroux plants are able to grow at densities up to 1 kg m^{-2} at Izu Peninsula (Japan), this being the most favorable bed. At harvesting, a standing crop of up to 0.2 kg m^{-2} is usually left behind, because these plants grow back to 1 kg m^{-2} within 2 months of harvesting. Harvesting is carried out by divers three times in a season, reaching an annual yield per square meter of well over 2 kg. *Gelidium* is a perennial with a slow rate of growth. The most reliable technique for propagation is to develop more habitats, but only in favorable areas, by throwing heavy stones into the sea (Shunzo Suto 1971). *Gelidium* species with a density of erect seedlings of 3–19 seedlings cm^{-2} were studied in the sea in China. Growth rate of regenerated fragments in the sea was 3.84% day^{-1} (Pei et al. 1996). In free-floating tank culture, *Gelidium crinale* optimal densities can reach 4 kg m^{-2} (M. Friedlander, unpublished observations).

Epiphytes

Although displaying seasonal cycles in abundance, *G. latifolium* (Greville) Bornet et Thuret was colonized by epiphytes throughout the annual study period in Northern Spain. Distribution of epiphytes on hosts followed a trend of large epiphytes on basal portions of the host, and ephemeral, small-sized epiphytes appearing on the tips of the host thallus (Arrontes 1990). Juvenile plants of *G. rex* in the open sea in Chile suffered from problems of fouling, epiphytes and epibionts caused by bacteria, algae and sessile invertebrates, which resulted in growth inhibition and decay (Rojas et al. 1996). The development of epiphytes was a serious problem in tank culture experiments in Portugal with *G. sesquipedale*, in which fish pond effluents were used (R. Santos, personal communication). Most of the cylinders with attached *G. sesquipedale* in the sea suffered from continuous attacks by epiphytes and epibionts on the cylinders and had to be withdrawn from the culture sites (Seoane-Camba 1997). For *G. amansii* cultured in the sea in Japan, the elimination of harmful weeds such as *Ecklonia* and *Eisenia* is effectively inhibiting epiphytes for 1 or 2 years (Shunzo Suto 1971).

These studies show that sensitivity to epiphytes is a major problem in *Gelidium* culture. This might be partially alleviated by clean seawater resources in tank culture and low nutrient areas in open sea farming.

Interaction with fish effluents

While adapting *G. sesquipedale* to free-floating conditions, fishpond effluents rich in NH_4^+ has been tried (Salinas 1991). Nitrogen removal experiments using *G. amansii* revealed that it is possible to use macroalgae as biofilters (Liu et al. 2004). In tank culture experiments with *G. sesquipedale*, fish pond effluents were used (R. Santos, personal communication). These studies show that the integration of *Gelidium* culture with fish culture is feasible.

Summarizing the major biological factors shows clearly that reproduction by reattached or free floating fragments is more effective than spore reproduction, and that epiphytes are a major obstacle to *Gelidium* cultivation.

Table 1 Daily growth rate (DGR) of various *Gelidiales*

Species	DGR (%)	Growth conditions	Experiment duration	Reference
G. pulchellum	10.0	Laboratory		Sousa-Pinto et al. 1999
G. robustum	21.0	Laboratory		Sousa-Pinto et al. 1996
G. sclerophyllum	7.0	Laboratory	1 month	Rodriguez 1996
Gelidium sp.	13.2	Laboratory	2 months	Titlyanov et al. 2006
G. sesquipedale	1.9	Chemostats	14 days	Carmona et al. 1996
G. rex	33.3	Laboratory and in situ	2 months	Rojas et al. 1996
Gelidium	3.8	Laboratory and in situ	1 year	Pei et al. 1996
G. amansii	6.7	In situ	2 months	Shunzo Suto 1971
G. sesquipedale	3.0	In situ	6 months	Seoane-Camba 1997
G. robustum	1.0, 3.6	In situ and semi controlled	35 days	Pacheco-Ruiz and Zertuche-Gonzalez 1995
G. crinale	6.5	Outdoor tanks	1 year	Boulus et al. 2007
P. capillacea	28.3	Outdoor tanks	1 month	Gal-Or and Israel 2004

Economic aspects

The following limited information illustrates the conflicting data available regarding economic aspects of *Gelidium* culture. Cultivation of *G. amansii* from spores is not an efficient way of increasing production because the spores take 2 years to grow and reach harvesting size. Branches of plants attached to ropes hung into the sea grow well; however, the costs of plants for seed and the labor force leaves little room for profit. On the other hand, as reported by Shunzo Suto (1971), the expenditure of setting stones for *Gelidium* culture in favorable areas can be recovered within 4 years. Because of the current status of natural coastal stocks of this resource, future applications of this type of culture appears to be more related to the restoration of damaged seaweed beds in Spain rather than to attempts at commercial cultivation, as suggested by Sousa-Pinto et al. (1996). As *G. sesquipedale* is a slow-growing species, it seems impossible to obtain more than one harvest per year in culture under strong water agitation. From these data, it would seem impossible to make *Gelidium* aquaculture profitable in Morocco as suggested by Givernaud and Mouradi (2006).

Table 1 summarizes the growth potential of *Gelidiales* as presented by various authors to date. In short-term experiments, the maximal daily growth rate (DGR) in the laboratory can reach 21%, in the laboratory and in situ 33%, and in tank culture 28%, while an annual outdoor tank experiment reached a DGR of only 6%, and an annual growth experiment in the sea reached only 3.8%.

There is no published economic analysis of *Gelidium* cultivation. However, the above mentioned data lead to a maximal calculated annual yield of *Gelidium* of about 25 kg FW m^{-2}, which is still far from the break-even point for commercial pond cultivation of agarophytes, which is about 60 kg FW $m^{-2}year^{-1}$(17% m^{-2} day^{-1}) or about 6 kg DW $m^{-2}year^{-1}$. On the other hand the assumed production costs of *Gelidium* with such a yield would be 2–3 US \$ kg DW^{-1}. The production costs were calculated for other seaweed cultures like *Gracilaria*, and might be affected by prices of infrastructure, electricity and labor in each country. The major conclusion is that the annual yield has to be dramatically improved.

Future prospects

Clearly, the major reason for the current lack of commercial cultivation technology for *Gelidium* is the slow growth rate and low yields of the existing species. Therefore, major effort towards the introduction of *Gelidium* into commercial cultivation have to involve genetic improvement through selection, breeding and possibly genetic engineering. Genetically engineered commercial products are meeting with increased resistance in the marketplace, limiting this technology's utilization. Genetic engineering may not be necessary, considering that many successful high yield seaweed strains that are now commercially produced were selected through more traditional methods. Only high yield strains will have the potential to compete economically with the present harvesting tradition.

References

Aken ME, Griffin NJ, Robertson BL (1993) Cultivation of the agarophyte *Gelidium pristoides* in Algoa bay South-Africa. Hydrobiologia 268:169–178

Arrontes J (1990) Composition, distribution on host, and seasonality of epiphytes on three intertidal algae. Bot Mar 33:205–211

Boulus A, Spaneir E, Friedlander M (2007) Effect of outdoor conditions on growth rate and chemical composition of *Gelidium crinale* in culture. J Appl Phycol (in press)

Carmona R, Vergara JJ, Perez-Llorens JL, Figueroa FL, Niell FX (1996) Photosynthetic acclimation and biochemical responses of *Gelidium sesquipedale* cultured in chemostats under different qualities of light. Mar Biol 127:25–34

Fei XG, Huang LJ (1991) Artificial sporeling and field cultivation of *Gelidium* in China. Hydrobiologia 221:119–124

Gal-Or S, Israel A (2004) Growth responses of *Pterocladiella capillacea* (Rhodophyta) in laboratory and outdoor cultivation. J Appl Phycol 16:195–202

Givernaud T, Mouradi A (2006) Seaweed resources of Morocco. In: Critchley AT, Ohno M, Largo DB (eds) World seaweed resources. ETI Bioinformatics, Amsterdam

Hernandez-Guerrero CJ, Casas-Valdez M, Ortega-Garcia S, Hernandez-Vazquez S (2000) Effect of climatic variation on the relative abundance of the red alga *Gelidium robustum* in Baja California Sur, Mexico. J Appl Phycol 12:177–183

Juanes JA, Puente A (1993) Differential reattachment capacity of isomorphic life history phases of *Gelidium sesquipedale*. Hydrobiologia 260–261:139–144

Liu D, Pickering A, Sun J (2004) Preliminary study on the responses of three marine algae, *Ulva pertusa* (Chlorophyta), *Gelidium amansii* (Rhodophyta) and *Sargassum enerve* (Phaeophyta), to nitrogen source and its availability. J Ocean Univ China 3:75–79

Melo RA (1998) *Gelidium* commercial exploitation: Natural resources and cultivation. J Appl Phycol 10:303–314

Mercado JM, Carmona R, Niell FX (1998) Bryozoans increase available O_2 for photosynthesis in *Gelidium sesquipedale*. J Phycol 34:925–927

Mercado JM, Niell FX, Gil-Rodriguez MC (2001) Photosynthesis might be limited by light, not inorganic carbon availability, in three intertidal *Gelidiales* species. New Phytol 149:431–439

Pacheco-Ruiz I, Zertuche-Gonzalez JA (1995) Effect of water movement on the growth of *Gelidium robustum* (Gardn.) Hollenb. and Abb. (Rhodophyta). Cienc Mar 21:59–70

Patwary MU, Van der Meer JP (1997) Construction of backcrossed *Gelidium* male- sterile and male-fertile lines and their growth comparison. J Appl Phycol 8:483–486

Pei L, Luo Q, Fei Z, Ma B (1996) Study on tissue culture for *Gelidium* seedling. Chin J Oceanol Limnol 14:175–182

Rodriguez D (1996) Vegetative propagation by fragmentation of *Gelidium sclerophyllum* (*Gelidiales*, *Rhodophyta*). Hydrobiologia 326–327:361–365

Rojas R, Leon NM, Rojas RO (1996) Practical and descriptive techniques for *Gelidium rex* (*Gelidiales*, *Rhodophyta*) culture. Hydrobiologia 326–327:367–370

Salinas JM (1991) Spray system for re-attachment of *Gelidium sesquipedale* (Clem.) Born. et Thur. (*Gelidiales*, Rhodophyta). Hydrobiologia 221:107–117

Salinas JM (1991a) El proceso de refijación en *Gelidium sesquipedale* (Clem.) Born. et. Thur. (*Gelidiales*: Rhodophyta). Bul Inst Esp Oceanogr 7:3–58

Seoane-Camba JA (1997) *Gelidium sesquipedale* cultivation in Galicia (Spain). Lagascalia 19:179–186

Shunzo S (1971) Mariculture of seaweeds and its problems in Japan. Proceedings of the First US–Japan Meeting on Aquaculture at Tokyo, Japan

Silva J, Santos R, Serodio J, Melo A (1998) Light response curves for *Gelidium sesquipedale* from different depths, determined by two methods: O_2 evolution and chlorophyll fluorescence. J Appl Phycol 10:295–301

Sosa PA, Gomez- Pinchetti JL, Juanes JA (2006) The seaweed resources of Spain. In: Critchley AT, Ohno M, Largo DB (eds) DVD ROM. World Seaweed Resources. An authoritative reference system. ETI Information, Amsterdam, ISBN: 90 75000 80 4

Sousa-Pinto I, Lewis R, Polne-Fuller M (1996) The effect of phosphate concentration on growth and agar content of *Gelidium robustum* (*Gelidiaceae*, Rhodophyta) in culture. Hydrobiologia 326–327:437–443

Sousa-Pinto I, Murano E, Coelho S, Felga A, Pereira R (1999) The effect of light on growth and agar content of *Gelidium pulchellum* (*Gelidiaceae*, Rhodophyta) in culture. Hydrobiologia 398/399:329–338

Titlyanov EA, Titlyanova TV, Kadel P, Luning K (2006) Obtaining plantlets from apical meristem of the red alga *Gelidium* sp. J Appl Phycol 18:167–174

Production, use and fate of Chilean brown seaweeds: re-sources for a sustainable fishery

Julio A. Vásquez

Originally published in the Journal of Applied Phycology, Vol 20, No 5, 7–17.
DOI: 10.1007/s10811-007-9308-y

Abstract Chile is an important producer of brown seaweeds representing 10% of world supply. Landings of Chilean kelp fluctuated between 40,000 $t.year^{-1}$ in the early '80s to 250,000 $t.year^{-1}$ more recently. Commercialized algae come from natural populations and no mass-cultures of involved species have been established. Four species of brown algae are commercially collected in the country: *Lessonia trabeculata*, *L. nigrescens*, *Macrocystis pyrifera* and *M. integrifolia*. Since 2000, the demand of alginate sources and food for cultivated abalones dramatically raised the harvesting of these species. Direct evaluations of *Lessonia* spp. and *Macrocystis* standing-stocks were made along 700 km of coast in northern Chile. *Lessonia* spp. estimated populations exceed 900,000 t, whereas *M. integrifolia* does not exceed 300 t. The insights provided indicate that sustainability of Chilean kelp subjected to intense harvesting would require management programs including the following bio-ecological recommendations: (1) to harvest the entire plant including the holdfast; (2) to harvest plants larger than 20 cm in diameter; (3) to harvest plants sparsely, selecting mayor specimens; (4) rotation of harvesting areas; and (5) for *Macrocystis*, to cut the canopy 1–2 m from the surface. They must be implemented in a National Program of Kelp Management, elaborated by government, scientists, fisherman, and industry.

Keywords Kelp · Chile · *Lessonia* · *Macrocystis* · Abalone

J. A. Vásquez (✉)
Departamento de Biología Marina,
Universidad Católica del Norte,
Center for Advanced Studies in Arid Zones (CEAZA),
Casilla 117, Coquimbo, Chile
e-mail: jvasquez@ucn.cl

Introduction

Giant kelps dominate shallow, subtidal rocky-bottom areas in temperate and cold seas to a depth of ca. 40 m (Dayton et al. 1984; Harrold and Pearse 1987; Vásquez 1992; Graham et al. 2007). Many kelp species have economic and social impacts since many local human populations base their economic support on these resources (Shiel and Nelson 1990; Vásquez and Fonck 1993; Stekoll et al. 2006; Rothman et al. 2006; Ugarte et al. 2006). Geographic and local distribution of diverse kelp species in the southeast Pacific is variable: *Lessonia nigrescens* Bory forms intertidal belts along exposed rocky coasts; *Lessonia trabeculata* Villouta & Santelices dominates rocky subtidal environments to a depth of ca. 40 m; *Macrocystis integrifolia* Bory forms shallow kelp beds from the intertidal zone to ca. 15 m in northern latitudes, and is gradually replaced by two species towards southern areas: *Durvillaea antarctica*, which dominates the intertidal zone in wave-exposed areas (Hoffmann and Santelices 1997); and *Macrocystis pyrifera* (Linnaeus) Agardh in both exposed and protected subtidal habitats (Buschmann et al. 2004). While the two mentioned *Lessonia* species have almost continuous distribution along the entire Chilean continental coast, the distribution of *M. integrifolia* is fragmented into patchy populations in northern Chile (Vega et al. 2005), while *M. pyrifera* forms coastal belts from 42°S to Cape Horn, including the fjord zone and the Atlantic coast up to Chubut area, in Argentina. The extension of local populations of *M. pyrifera* from north to south, may vary from 100 m to 100 km. The observed distribution patterns are the result of complex life history strategies and interactions with environmental factors such as spatial and temporal variations in water movement, nutrient availability, and temperature (Muñoz et al. 2004; Vega et al. 2005; Graham et al. 2007).

Lessonia species are dominant organisms in coastal communities, both in cover and biomass, being the foundation species (sensu Dayton 1972) in the intertidal and shallow subtidal rocky bottom. Several invertebrate and fishes, some of which are of economic interest, use *Lessonia* kelps as habitat, food supply, and area for recruitment (Vásquez 1992; Vásquez et al. 2001), so management of these brown seaweeds is also relevant to conservation of many other coastal fisheries.

In Chile, brown algae, especially those collected on the beaches as result of natural mortality, have been used since the 1960s as raw material for alginic acid extraction, with 90% of this productive activity, which occurs exclusively in Northern Chile between 26° and 32°S (Anuario SERNAPESCA 1985–2006), supported by both *Lessonia* species. Plants which have been cast ashore are collected by fishermen, dried, bundled and then sold to intermediaries who sell them to the grinding industry. This fishery is currently not regulated since landings are mainly composed of dead plants. Since 1998, extensive harvesting of intertidal and subtidal populations of *Lessonia* spp. in northern Chile, has put fishery authorities, as well as fishermen, industry and scientists, on alert.

This article presents the most relevant information regarding distribution, abundance, productivity, destination and predicted use of commercially important brown algae from Northern Chile. This information brings together all elements needed for planning their sustainable management.

Materials and methods

Since 2004, bio-ecological studies of production and analyses of the socioeconomic aspects of brown algae fisheries in northern Chile have been developed. Their common goals were to elaborate recommendations for organize, and make more efficient, the administration and management of this activity. This study analyzes the availability of biomass from natural populations (2004–2005), the chain of production (2005–2006), and some socioeconomic aspects of this fishery (2004–2006). The study area includes the entire geographical region of this fishery, between 18° and 32°S. Considering that the majority of collection and harvesting, as well as use and destination, occur between 26°–32°S, most of the information comes from this sub-sector of the Chilean coast (Fig. 1).

Collection and harvesting statistics

All data regarding brown algae landings (wet and dry) were obtained from the National Fishery Service, which depends on the Ministry of Economy http://www.sernapesca.cl); data related to exportation of brown algae (US$ and tons) as algae meal, for alginate extraction, were obtained from the Central Bank of Chile http://www.bancocentralchile.cl).

Distribution and abundance of brown algae

From 2004 to 2005, the distribution (latitudinal and bathymetric) and abundance of *Lessonia trabeculata*, *L. nigrescens* and *Macrocystis integrifolia* were determined along approximately 700 km of coast between 26° and 32°S. In this area, 140 sampling stations were established (Fig. 1), separated by 4.5 linear km, excluding a priori sandy beaches and soft bottom (subtidal) areas.

The basal diameters of the plants' holdfasts were measured in situ, since a significant correlation exists between this dimension and the weight and size of the plants in all three species (Cancino and Santelices 1984; Vásquez and Santelices 1984; Villouta and Santelices 1984, 1986; Vásquez 1992, 1993). Density of plants and type of available substrate were also determined. All evaluations were non-destructive.

Lessonia trabeculata: distribution and abundance of this species was evaluated using 160-m transects, perpendicular to the coastline and located between the intertidal zone and 20-m depth, according the methodology described by Vásquez et al. (2006).

Lessonia nigrescens: distribution and abundance of this species was determined using 10 m^2 (10×1 m) quadrants. Morphological attributes of adult sporophytes (holdfast diameter, length, number of stipes and presence of reproductive structures), were evaluated using 1-m^2 quadrants.

Macrocystis integrifolia: since the fragmented distribution of *M. integrifolia*, 0.25 m^2 quadrants were randomly thrown within each algae bed (see Vega et al. 2005). When *Macrocystis* extended into the subtidal zone, sampling was done in layers by positioning four 0.25 m^2 quadrants in different depth ranges (0–2, 2–4, 4–6 >6 m). The limits of each population were georeferenced (GPS-GARMIN 12 Channels) to estimate the total abundance of each bed.

In each sampling area (10 m^2 for *L. trabeculata*, 1 m^2 for *L. nigrescens*, and 0.25 m^2 for *M. integrifolia*) three aspects were evaluated: (1) total number of plants, (2) basal diameter of holdfast, and (3) substrate type. Each sampling area was georeferenced.

Biomass estimation using GIS

The data obtained in the field for each population, and regarding their abundance and distribution, were incorporated into the Geographical Information System (GIS,

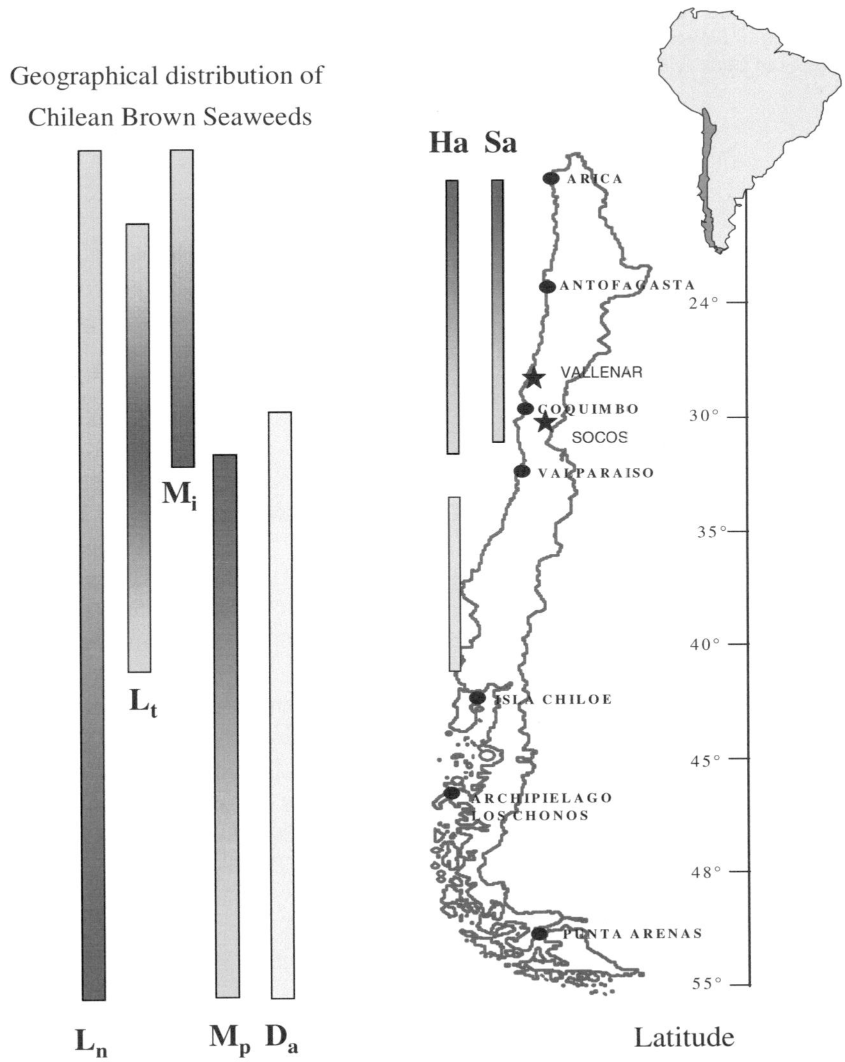

Fig. 1 Geographical distribution of brown seaweed of economic interest in Chile, including the study area of its fishery. L_t *Lessonia trabeculata*, L_n *Lessonia nigrescens*, M_I *Macrocystis integrifolia*, M_p *Macrocystis pyrifera*, D_a *Durvillaea antárctica*, H_a Study area, S_a Sampling area

ArcView and ArcGIS®). The coordinates of each sampling station and transect were acquired in sextadecimal format (degrees, minutes and seconds) from the Provisional South American spatial reference Datum from 1956 (Psad56).

The geostatistical analysis in the ArcGIS program generated variograms within and between sampled transects (Burrough and McDonnell 1998), and allowed construction of abundance polygons between transects (Klopatek and Francis 1999). An average distribution 160-m width was determined for *L. trabeculata* in agreement with Vásquez (1992) and Vásquez et al. (2006). A 2.5-m-wide belt in the intertidal zone was estimated for *L. nigrescens* (Santelices 1989). The abundance of *M. integrifolia* in each section of the coast was obtained by extrapolating the abundance values determined with quadrants along the depth gradient within the kelp forest (Vega et al. 2005). Standing stock of the three brown algae species was calculated as the sum of the abundance polygons from the entire study area.

Use and destination

The destination of extracted and collected brown algae from the study area was determined directly by compiling information from brown seaweed processing factories (milling centers) and abalone cultivation centers between 2005 and 2006. In each of these production activities, surveys were taken to evaluate: (1) species processed, (2) their origin, (3) method of collection/harvesting, (4) landed biomass. These same surveys were taken to the fishermen and commercial intermediaries who supply the processing and abalone cultivation centers.

Socioeconomic aspects

Evaluation of some socio-economic aspects of fishermen involved in the kelp collection and harvesting were made with special attention and care. They included determina-

Fig. 2 Annual production (wet biomass) of brown seaweed fishery in Chile (1984–2006)

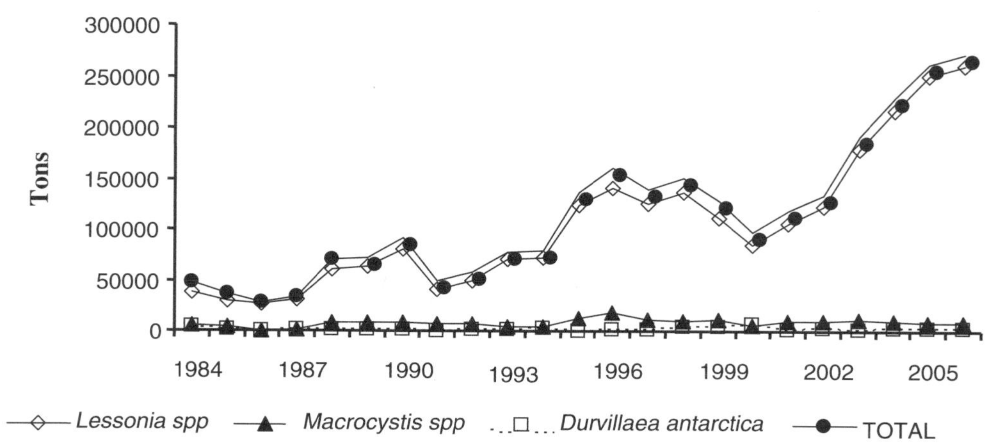

tion of age, sex, education level, and association to workers unions or consortia.

Results

The brown algae fishery in northern Chile is concentrated between 18° and 32°S. However, the majority of landings are collected or harvested between 26° and 32°S (Fig. 1). In the study area, most of the collection and milling centers are in Vallenar (28°32′S), Ovalle (30°36′) and Socos (30°40′). Abalone cultivation centers are located in Caldera (27°06′S) and Coquimbo (29°58′) (Fig. 1).

Collection, harvesting and destination of brown algae

During the last 25 years, brown algae landings have fluctuated between 40,000 and 280,000 t.year^{-1}, showing a constant increase since 1999 (Fig. 2). In Chile, *L. nigrescens* and *L. trabeculata* comprise more than 90% of the total production of brown algae, whereas *Macrocystis* sp. and local consumption of *Durvillaea antarctica* contribute only marginally to total landings (Fig. 2). Even if brown algae have an extensive distribution along the southeast Pacific coast (see Fig. 1), their fishery is concentrated between 20° and 32°S (Fig. 3). This is due to the arid climate of the wide Atacama Desert located immediately inland of this region, which allows for rapid drying and consequently a reduction in the productive fees.

Destination of brown algae

The increase in kelp extraction during the last 7 years is explained principally by the significant rise of their utilization as raw material for extraction of alginic acid, both in Chile and outside the country (Fig. 4). A smaller fraction of this increment can be explained by the yield reduction of kelp used for milling. This was consequence of higher humidity of recently processed plants compared to previous years (Fig. 4). From a different perspective, exports of brown algae during 1997–98, as in other "El Niño" phenomenon, could be related to higher mortalities generated by this large scale oceanographic event. The warming of the ocean surface and decreasing of nutrients concentrations, both associated with "El Niño", cause high mortalities in kelp populations which are then cast ashore by waves and finally collected by fishermen.

The new and impressively growing abalone cultivation industry in Northern Chile, consumed more than 4,800 t of

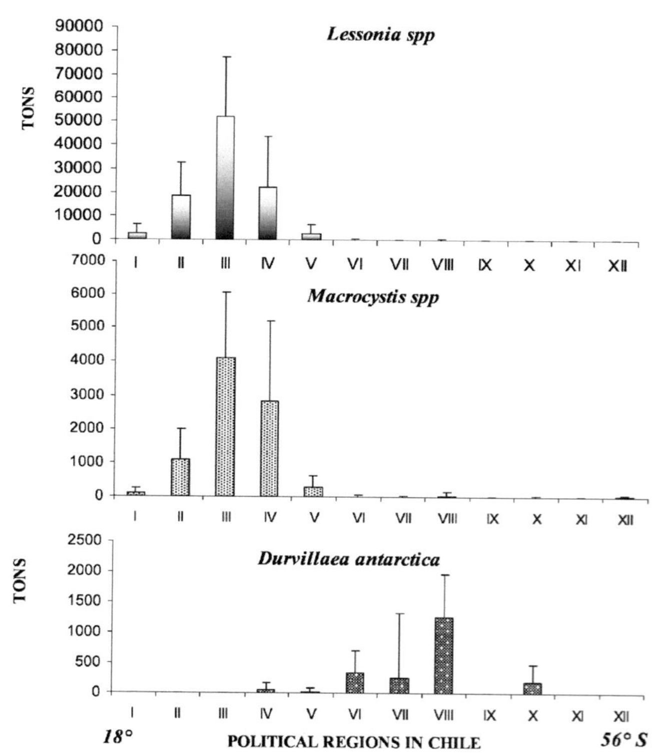

Fig. 3 Average landings (wet biomass) of brown seaweed (1984–2006), along the Chilean coastline: North (Regions *I to IV*), Central (*V to VIII*), and South (*IX to XII*)

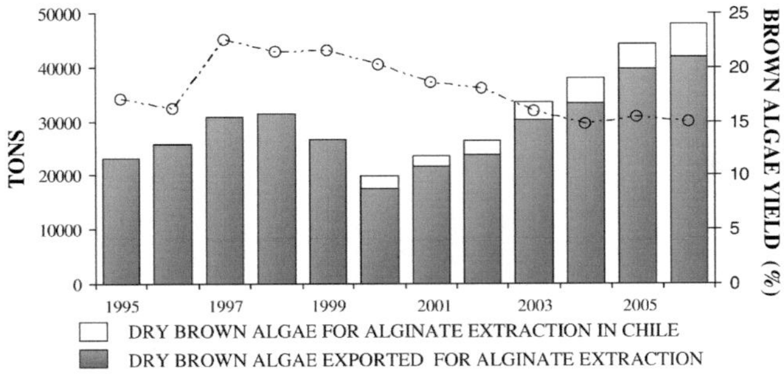

Fig. 4 Annual production (1995–2006) of raw material (dry biomass) for alginate extraction in Chile and overseas. Temporal variability of the percentage of brown seaweed yield (-o-)

fresh algae between 2005 and 2006, principally *Macrocystis* (Fig. 5). However, neither the utilization nor the economic yield of this activity is comparable, until now, to supplying material for alginate extraction (Fig. 6).

Distribution and abundance of brown algae

During 2004–2005, between 26° and 32°S, distribution and abundances of *L. trabeculata, L. nigrescens* and *M. integrifolia* were evaluated along more than 700 km of coast (Fig. 7). 140 sampling stations were established separated by 4.5 linear km, excluding a priori sandy beaches and soft bottom subtidal areas because of their unsuitability as a substratum for kelp spore settlement. *Lessonia trabeculata* is found in rocky subtidal habitats between 0 and 30 m depths in the entire study area (Fig. 7). Depending on the extension of the rocky platform into the subtidal zone, local biomasses up to 50,000 wet t were recorded. Abundance polygons indicate a standing stock of approximately 800,000 wet t of *L. trabeculata* in the study area. *Lessonia nigrescens* is distributed in the rocky intertidal zone of the entire study area with local biomass (sampling station) between 50 and 4,000 wet t (Fig. 7). The sum of the abundance polygons constructed with GIS shows a standing stock of more than 100,000 wet t of *L. nigrescens* in the study area. In contrast, *M. integrifolia* shows fragmented distribution within the study area (Fig. 7). Local populations are small, with biomasses between 2 and 12 wet t. The sum of local biomass determines a standing stock that does not exceed 200 wet t for the entire study area (Fig. 7).

Social indicators

The list of fishermen http://www.sernapesca.cl) shows 7,801 artisanal algae collectors and harvesters in the study area (Table 1). However, only 20% of these individuals showed evidence of permanent collection and harvesting activities during 2005–2006. More than the half of active fishermen (57%) show a preference for working independently, unassociated with consortia or unions; gender analysis shows that 83.4% of active participants are men. The marginality of this activity is reflected in the education of the active fishermen. Almost 70% of participants in this activity have only primary school education, which is generally incomplete. Among the elderly fishermen (>70 years), 4.6% have no formal education, being mostly illiterate. Only 26.2% have secondary education, which in most of cases is incomplete (Table 1).

Discusion and conclusion

Governments have attempted to introduce co-management in coastal waters granting territorial user rights to fishers (TURFs). The rationale behind TURFs is based on a common property approach, which proposes that a well-established rights-based system provides access, withdrawal, and management security for individuals and groups of individuals (Ostrom and Schlager 1996). With such assurances, fishermen would make commitments amonst each other and with the administration, and would develop long-term plans for investing in, and harvesting from, a common-pool resource in a sustainable manner. Accordingly, TURFs change the nature of resource extraction. In this sense, hunting becomes harvesting, a certain degree of predictability is introduced since extraction programs

Fig. 5 Wet biomass of brown seaweeds (*Macrocystis* and *Lessonia*), used as food for abalone culture in northern Chile

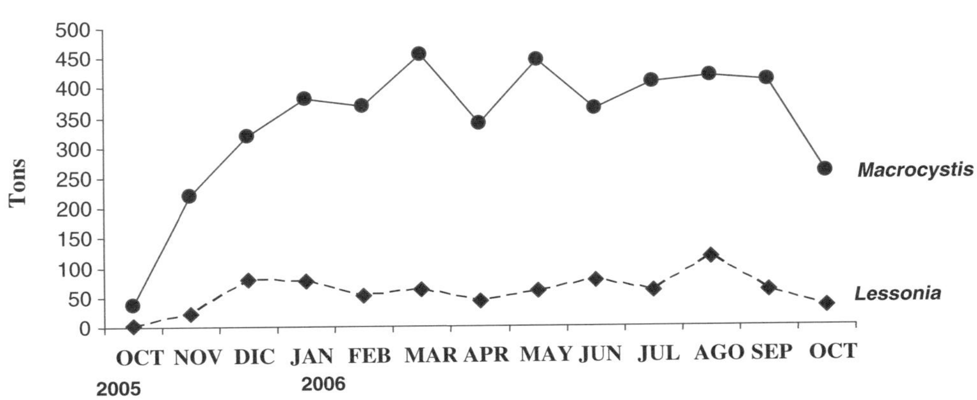

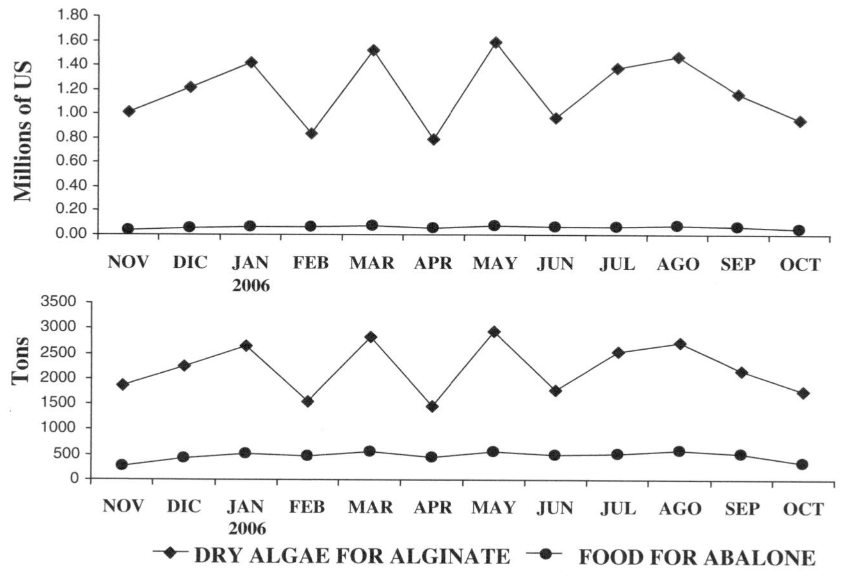

Fig. 6 Monthly variation of brown seaweed biomass used for alginate extraction, as well as fresh food for abalone culture, and respective incomes (US $) of both commercial activities

requires at least previous evaluation of target resources, and the skills of the fishermen (as hunters) become less important (Gelcich et al. 2005a).

Since 1997, Chilean Government has adopted co-management policies for the coastline, granting territorial user rights to fishermen (Bernal et al. 1999). Territorial rights related to users into the Chilean Fishery Law, adopted the legal title of Management and Exploitation Areas for Benthic (bottom dwelling) Resources (MEABR). Through this, the Governmental Entity that regulates fisheries assigns defined rights and norms of utilization to syndicates located in defined geographical coastal areas (Gelcich et al. 2005b). By August 2003, 188 exploitation areas (MEABR) had management plans in progress, and 649 were at diverse stages of application procedures (Gelcich et al. 2005b).

To date, between 18°–32°S (see Fig. 1), more than 50% of the coastline is under MEABRs regime, and this area represents the main geographical scenario for kelp fishery in Chile. In this context, fishermen syndicates would apply many of the proposed strategies for brown seaweed conservation in MEABRs. Currently, MEABRs that include brown seaweeds as target species have to harvest according biological and ecological recommendations. In addition, the results of harvesting procedure and the recovery of kelp beds after pruning or removal of individuals must be inserted into each management program, which is updated annually.

In open areas of Northern Chile without an MEABR regime, collection of plants cast ashore is the main activity of fisherman along the coast. Harvesting of *L. nigrescens* is the most difficult to control in terms of regulation enforcement along Chile's extensive coast. In contrast, due to the location of *Macrocystis*, harvesting is closely related with the proximity of abalone culture centers; the few and small beds of *Macrocystis* near them have been harvested under a co-management plan developed by fisherman jointly with the direct user, the abalone industry.

During the last 50 years, kelp fisheries have been supported by the natural mortality of coastal populations caused by the drag of bottom currents and wave impact on old and/or debilitated plants. The detached biomass is cast ashore by currents and hand-collected by fishermen. In this case, there was no need for administration and management of *L. trabeculata*, *L. nigrescens* and *M. integrifolia*. This feature of brown seaweed fishery was relevant in the past, not only for the conservation of *Lessonia* and *Macrocystis* populations but also for many other associated species with economic interest, such as keyhole limpets, sea urchins, and several fish and mollusc species (Vásquez and Santelices 1984; Vásquez and Vega 2005). Since 2000, extractive pressures from international markets for alginic acid extraction, as well as the emergent need of fresh food for abalone aquaculture, has dramatically raised the harvesting of brown algae. In this context, and considering the foundation species status (sensu Dayton 1972) of Chilean kelp species, any management program must include the whole coastal marine habitat.

The Kelp fishery is also important in other cold and temperate areas around the world, such as Alaska (Stekoll et al. 2006), Southern California, USA, and Baja California, Mexico (Barilotti and Zertuche-González 1990), Norway (Sjotun et al. 1993), South Africa (Anderson et al. 2006; Rothman et al. 2006), New Zealand (Shiel and Nelson 1990), Canada (Ugarte and Sharp 2001; Ugarte et al. 2006), France (Guillaumont et al. 1993), Japan (Terawaki et al.

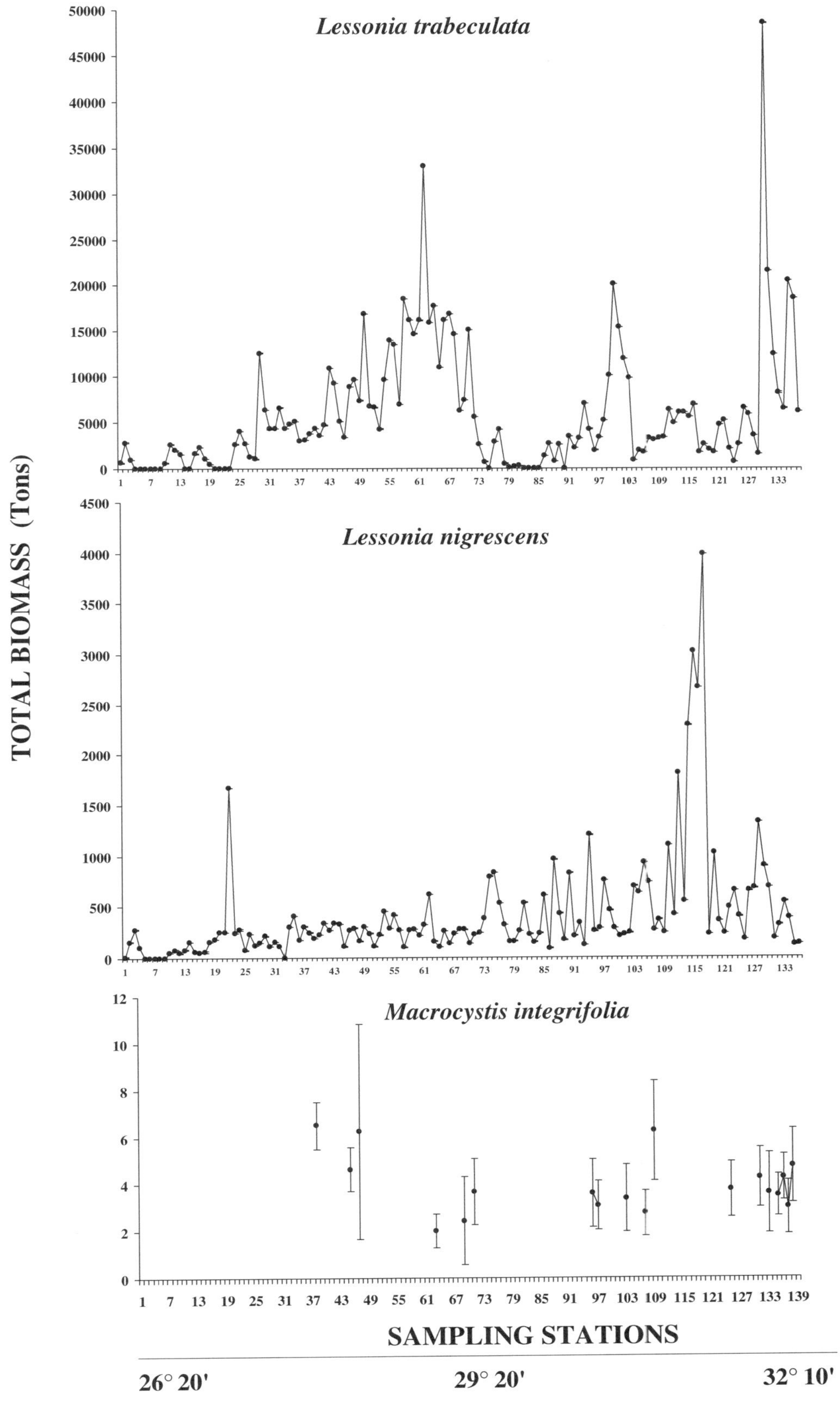

Fig. 7 Distribution and standing stock (wet biomass) of tree species of brown seaweed of economic interest along 700 km of coastline in northern Chile

Table 1 Socio-economic indicators of fisherman of brown seaweeds in Chile

No. of registered fisherman	No. active fisherman	Not-associated fisherman (%)	Female (%)	Male (%)	Education level (%)		
					Uneducated	Elementary	High school
7,801	1,549	57	16.6	83.4	4.6	69.2	26.2

2001), and Ireland (Werner and Kraan 2004). The production and sustainability of involved populations is mostly focused on biological and ecological knowledge, being the management based on a single species resource. In contrast, kelp management based on the ecosystem has been poorly documented.

As in this study, in addition to the bio-ecological features of target species, other management considerations such as maximum exploitation rate, cutting height, gear restriction, effects on associated species, and protected areas have been considered in the sustainability of *Ascophylum nodosum* in Canada (Ugarte and Sharp 2001) and *Laminaria* spp. in France and Norway (Werner and Kraan 2004). Similarly to *Ascophyllum* studies, universities, government authorities, and industry were involved in the production of a research and monitoring program for the sustainability of economically important kelp including also, social impact. In this context, the most important difference between the kelp fishery in Chile and other countries is the relevance we assigned to the social aspect of this fishery, since it provides the main economic resource for several families under high social risk in Northern Chile.

Location of growth meristems and reproductive structures in different kelp species (Edding et al. 1994; Hoffmann and Santelices 1997) determines that the way of harvesting has different consequences for different species: *Lessonia* has sori on apical fronds and along stipes (Edding et al. 1994; Venegas et al. 1992), whereas sporophylls are basally distributed in *Macrocystis*. The meristems of both genera are apically distrubuted, and only *Macrocystis* continually produces new stipes from the holdfast. Pruning *Lessonia* inhibits its growth, and regeneration is nonexistent (Vásquez and Santelices 1990; Vásquez 1992; Mendieta 1997), while harvesting of fronds and stipes eliminates gametangia and thus prevents sexual reproduction. In contrast, the regeneration of biomass after harvesting is a rapid and frequent process in *Macrocystis* (Graham et al. 2007); additionally, pruning only removes the canopy of individuals, and does not affect basal sporophylls involved in sexual reproduction.

Brown algae landings have fluctuated between 40,000 and 280,000 t during the last 25 years, with escalating rates since 1999. At present, approximately 300,000 wet t of brown algae are collected or extracted from the coast of Northern Chile, and *Lessonia* spp. represents more than 90% of this total. This productive activity creates yearly revenues of approximately US$22 million http://www.bancocentralchile.cl), which comes mainly from supplying raw sources for the alginic acid industry. Chile provides 10% of the world wide sources for alginate extraction, which is used primarily by pharmaceutical and food industries (Avila and Seguel 1993; Vásquez and Fonck 1993; Zemcke-White and Ohno 1999), though high quality alginate production within Chile is a growing activity.

In terms of fisheries statistics, the standardization of humidity content and yield of kelp for industrial processes are required at the national level. This would allow more precise predictions for the industry as functions of the requirements and availability of biomass stock from natural beds.

In Chile at present, the availability of fresh food is the most important constraint for the emerging aquaculture of abalone. In the study area, the estimated requirement of this industry reaches an average of 500 t per month of fresh algae, mainly *Macrocystis*. However, significant consumption of brown algae by cultivated abalone is foreseen. Preliminary studies of conversion rates (Pérez, personal communication) show that the production of 1 t of abalone requires 100 t of fresh algae. In consequence, the projected annual production of cultivated abalone for 2010 is 1,000 t, which would require 100,000 t of brown algae to reach this goal. Considering the low abundance and fragmented distribution of *M. integrifolia* in Northern Chile, this requirement greatly exceeds the production potential of natural populations. Consequently, the sustainability of abalone cultivation depends on the future culture of *Macrocystis*. In this context, successful efforts have been made in Chile to produce *Macrocystis* spp sporophytes using long-line culture systems (Gutierrez et al. 2006; Westermeier et al. 2006). From a different perspective and given the susceptibility of brown algae to interannual oceanographic events like El Niño (Vega et al. 2005; Vásquez et al. 2006), kelp culture would allow independence from unpredictable environmental catastrophic fluctuations and would help the natural recovery of coastal rocky areas after mass mortality events.

In Chile, a second epicenter for the abalone cultivation industry is located at ca. 40° S. At high latitudes, reaching the Pacific coast of Cape Horn and on the Atlantic coast near Chubut Argentina, *M. pyrifera* forms continuous belts that comprise the most important brown algae reservation in the world (Graham et al. 2007). In spite of this, and contrary to the perennial *M. integrifolia*, *M. pyrifera* populations are annual and therefore minimally available in the winter (Buschmann et al. 2006). For this reason, development of cultures should not be centred exclusively in Northern Chile.

Fishers, of whom only 17% are women, carry out collection and harvesting of brown algae in Chile. In the most active areas located between 18° and 32°S, 7,801 collectors are registered, but only 20% of them permanently participate in collection and harvesting activities. The majority of these fishers are independent, not belonging to any workers union or consortium. The level of education attained by active fishermen reflects the marginality of this work. Almost 70% of the workforce has only elementary education and in most cases incomplete.

The estimated standing stock of *Lessonia* (both species analyzed, more than 1 million t) is enough to sustain the actual brown algae landings Considering the morphologic and reproductive attributes of *Lessonia*, "how to harvest" seems to be more important than "how much to harvest". In terms of how to harvest, a few important considerations are relevant: (1) plants should be completely extracted including the holdfast; (2) large plants, with holdfast larger than 20 cm in diameter, should be selected for extraction; (3) harvesting would be sparse leaving a substratum space not larger than 1–2 m between plants. All these recommendations are relatively easy to implement and control in areas with an MEABR regime, even if. in areas with free access, they would be certainly more difficult to carry out. Management Plans would then consider these simple recommendations: (1) to liberate primary substratum for the settlement of juvenile sporophytes, (2) to allow for juvenile recruitment by diminishing competitive interaction between adults and sporelings, (3) to maintain a reproductive stock of adults, while allowing population regeneration, and (4) to prevent overgrazing of plants by benthic herbivores by maintaining a whiplash effect (see Dayton et al. 1984; Vásquez 1992).

In contrast to to the situation with *Lessonia*, where how to manage the stock is the most relevant issue, the sustainable fishery of *Macrocystis* should take into account its distribution and abundance, before authorizing the harvesting of natural populations in Northern Chile. This species has a unique genetic heritage due to its restricted and fragmented distribution. It also plays an important ecological role as an engineering (sensu Jones et al. 1994) and foundation species (Graham et al. 2007), because numerous invertebrates and fish depend on its existence (Vásquez and Vega 2005; Graham et al. 2007). Thus, the small beds of *Macrocystis* in Northern Chile would be under a harvesting program, which must be carried out by local fishermen, funded by the local abalone industry and supervised by fishery authorities.

Currently, some management strategies are being applied in Chile by means of a regionalized system that prohibits the extraction and collection of kelps in open areas (not under MEABR regime) during a certain period of time. Under these restrictions, users are allowed to obtain the resource, as long as they deliver the information required to determine: (1) harvesting and collection effort, (2) locations of collection and harvesting areas, (3) number of fishermen, (4) number intermediate buyers, (5) number and location of milling and abalone cultivation centers, and (6) amount of biomass used by each activity. Using the above information, we collaborated with governmental fishing authorities that organized technical work groups. In these collaborative instances, fishermen, industry representatives, technicians, scientists and administrators work together towards establishing sustainable management strategies that account for the distribution and abundance of brown algae and the requirements of the industries related to these resources.

Biological and ecological knowledge by itself does not guarantee the success of a harvesting program. In practice, social, political and economics aspects (see Vásquez and Westermeier 1993) must be considered for a correct strategy of management of the brown seaweed fishery. Among these, the most relevant are: (1) pressure of national and international markets for raw material, (2) unemployment indexes of coastal workers, which increase seasonal harvesters, (3) control of management recommendation along the extensive coast of northern Chile, and (4) low level of education and incomes of seaweed collectors.

Usually, most management strategies of natural resources start immediately before the imminent collapse or after evident decline of populations subjected to harvesting (e.g., *Gracilaria chilensis*; see Santelices and Doty 1989). This study is an example that illustrates the application of biological and ecological knowledge of target species having regard to the present and future uses of kelp species of economic interest. In this context, concerted actions of government, scientists, fishermen and industry have been taken and should be implemented in time. In the near future, management recommendation will be tested in order to evaluate the stocks of natural populations along Northern Chilean coasts, the production of brown seaweed under culture, and the fate and use of these important benthic marine resources.

Acknowledgements O. Castillo, R. Rojas, G. Vásquez, G. Aguirre, N. Kusmicic, E. Montanares, J. Johnson for their valuable work in the field. COPRAM Marine Seaweed Producers-Chile and Abalone Association-Chile.

References

Anderson RJ, Rothman MD, Share A, Drummond H (2006) Harvesting of the kelp *Ecklonia maxima* in South Africa affects its three obligate, red algal epiphytes. J Appl Phycol 18:343–349

Anuario Estadístico de Pesca (1985–2006) Servicio Nacional de Pesca, Ministerio de Economía Fomento y Reconstrucción, Gobierno de Chile

Avila M, Seguel M (1993) An overview of seaweed resources in Chile. J Appl Phycol 5:133–139

Barilotti DC, Zertuche-González JA (1990) Ecological effects of seaweed harvesting in the Gulf of California and Pacific Ocean off Baja California and California. Hydrobiologia 204/205:35–40

Bernal P, Oliva D, Aliaga B, Morales C (1999) New regulations in Chilean fisheries and aquaculture: ITQ's and territorial user rights. Ocean Coast Manag 42:119–142

Burrough PA, Mcdonell RA (1998) Principles of geoghraphic information systems, spatial information and geostatistics. Oxford University Press, Oxford, p 333

Buschmann AH, Vásquez, Osorio P, Reyes E, Filún L, Hernandez-González M, Vega AML (2004) The effect of water movement, temperatura and salinity on abundante and reproductive patterns of *Macrocystis* spp (Phaeophyta) at different latitudes. Mar Biol 145:849–862

Buschmann AH, Moreno C, Vásquez JA, Hernandez-Carmona MC (2006) Reproduction strategies of *Macrocystis pyrifera* (Phaeophyta) in southern Chile: the importance of population dynamics. J Appl Phycol 18:575–582

Cancino J, Santelices B (1984) Importancia ecológica de los discos adhesivos de *Lessonia nigrescens* Bory (Phaeophyta) en Chile central. Rev Chil Hist Nat 56:23–33

Dayton PK (1972) Toward an understanding of community resilience and the potential effects of enrichments to the benthos at McMurdo Sound, Antarctica. In: Parker BC (ed) Proceedings of the Colloquium on Conservation Problems in Antarctica. Allen Press, Virginia, pp 81–95

Dayton PK, Currie V, Gerrodette T, Keller BD, Rosenthal R, Vantresca D (1984) Patch dynamic and stability of some Californian kelp communities. Ecol Monogr 54:253–289

Edding M, Fonck E, Macchiavello J (1994) *Lessonia*. In: Akatsuka I (ed) Biology of economic algae. SPB Publishing, The Hague, pp 407–446

Gelcich S, Edwards-Jones G, Kaiser MJ, Watson E (2005a) Using discourses for policy evaluation: the case of marine common property rights in Chile. Soc Nat Res 18:377–391

Gelcich S, Edwards-Jones G, Kaiser MJ (2005b) Importance of attitudinal differences among artisanal fishers towards co-management and conservation of marine resources. Conserv Biol 19:865–875

Graham MH, Vásquez JA, Buschmann AH (2007) Global ecology of the giant kelp *Macrocystis*: from ecotypes to ecosystems. Ocean Mar Biol Ann Rev 45:39–88

Guillaumont B, Callens L, Dion P (1993) Spatial distribution of *Fucus* species and *Ascophyllum nodosum* beds in intertidal zones using spot imagery. Hydrobiologia 260/261:297–305

Gutierrez A, Correa T, Muñoz V, Santibañez A, Marcos R, Cáceres C, Buschmann AH (2006) Farming of the giant kelp *Macrocystis pyrifera* in southern Chile for development of novel food products. J Appl Phycol 18:259–267

Harrold C, Pearse JS (1987) The ecological role of echinoderms in kelp forests. In: Jangoux M, Lawrence JM (eds) Echinoderm studies, vol 2. Balkema, Rotterdam, pp 137–233

Hoffmann AJ, Santelices B (1997) Marine Flora of Central Chile. Ediciones Universidad Católica de Chile, Santiago, 434 pp

Jones CG, Lawton H, Shachak M (1994) Organisms as ecosystem engineers. Oikos 69, 373–386

Klopatek JM, Francis JM (1999) Spatial patterns analysis techniques. In: S Morain (ed) GIS Solutions in natural resource management: Balancing the technical-political equation. OnWord, Independence, KY, pp 17–40

Mendieta JJ (1997) Consideraciones ecológicas para el manejo de poblaciones de *Lessonia trabeculata* Villouta & Santelices (1986) en el norte de Chile. Tesis para optar al grado académico de Magister en Ciencias del Mar. Fac. Cs. del Mar, Universidad Católica del Norte, Coquimbo, Chile, 80 pp

Muñoz V, Hernández-González MC, Buschmann AH, Graham MH, Vásquez JA (2004) Variability in *per capita* oogonia and sporophyte production from giant kelp gametophyte (*Macrocystis pyrifera*, Phaeophyceae). Rev Chil Hist Nat 77: 639–647

Ostrom E, Schlager E (1996) The formation of property rights. In: Hanna S, Folke C, Maler K (eds) Rights to nature: ecological, economic, cultural and political principals of institutions for the environment. Island Press, Washington, pp 127–157

Santelices B (1989) Algas Marinas de Chile. Distribución, Ecología Utilización y Diversidad. Ediciones Universidad Católica de Chile, Santiago de Chile, 399 pp

Santelices B, Doty MS (1989) A review of *Gracilaria* farming. Aquaculture 78:95–133

Shiel DR, Nelson W (1990) The harvesting of macroalgae in New Zealand. Hydrobiologia 204/205:25–33

Sjotun K, Fredriksen S, Lein TE, Rueness J, Sivertsen K (1993) Population studies of *Laminaria hyperborea* in northern range of distribution in Norway. Hydrobiologia 260/261:215–221

Stekoll MS, Deysher LE, Hess M (2006) A remote sensing approach to estimating harvestable kelp biomass. J Appl Phycol 18:323–334

Rothman MD, Anderson RJ, Smit AJ (2006) The effects of harvesting of the South African kelp (*Ecklonia maxima*) on the kelp population structure, growth rate and recruitment. J Appl Phycol 18:335–341

Terawaki T, Hasegawa H, Arai S, Ohno M (2001) Management-free techniques for restoration of *Eisenia* and *Ecklonia* beds along the central Pacific coast of Japan. J Appl Phycol 13:13–17

Ugarte RA, Sharp G (2001) A new approach to seaweed management in Easter Canada: the case of *Ascophyllum nodosum*. Cah Biol Mar 42:63–70

Ugarte RA, Sharp G, Moore B (2006) Changes in the brown seaweed *Ascophyllum nodosum* (L.) Le Jol. Plant morphology and biomass produced by cutter rake harvest in southern New Brunswick, Canada. J Appl Phycol 18:351–359

Vásquez JA (1992) *Lessonia trabeculata*, a subtidal bottom kelp in northern Chile: a case of study for a structural and geographical comparison. In: Seeliger U (ed) Coastal plant communities of latin america. Academic Press, San Diego, pp 77–89

Vásquez JA (1993) Patrones de distribución de poblaciones submareales de *Lessonia trabeculata* (Laminariales, Phaeophyta) en el norte de Chile. SERIE OCASIONAL Facultad de Ciencias del Mar, Universidad Católica del Norte 2, pp 187–211

Vásquez JA, Fonck E (1993) Algas productoras de ácido algínico en Sudamérica: diagnóstico y proyecciones. In: Documento de Campo No 13 Situación actual de la industria de macroalgas productoras de ficocoloides en América Latina y el Caribe. FAO-ITALIA. Programa Cooperativo Gubernamental, pp 17–26

Vásquez JA, Santelices B (1984) Comunidades de macroinvertebrados en discos de adhesión de *Lessonia nigrescens* en Chile central. Rev Chil Hist Nat 57:131–154

Vásquez JA, Santelices B (1990) Ecological effects of harvesting *Lessonia* (Laminariales, Phaeophyta) in central Chile. Hydrobiologia 204/205:41–47

Vásquez JA, Vega JMA (2005). Macroinvertebrados asociados a discos de adhesión de algas pardas: Biodiversidad de comunidades discretas como indicadoras de perturbaciones locales y de gran escala. In: Figueroa E (ed) Biodiversidad Marina: Valoración, usos y perspectivas ¿Hacia donde va Chile? Editorial Universitaria Santiago-Chile, pp 429–450

Vásquez JA, Westermeier R (1993) Limiting factors in optimizing seaweed yield in Chile. Hydrobiologia 260/261:180–187

Vásquez JA, Véliz D, Pardo LM (2001) Biodiversidad bajo las grandes algas. In: Alveal K, Antezana T (eds) Sustentabilidad de la biosiversidad. Un problema actual, bases científico-técnicas, teorizaciones y perspectivas. Universidad de Concepción-Concepción, pp 293–308

Vásquez JA, Vega JMA, Buschmann AH (2006) Long term studies on El Niño-La Niña in northern Chile: effects on the structure and organization of subtidal kelp assemblages. J Appl Phycol 18:505–519

Vega JMA, Vásquez JA, Buschmann AH (2005) Biology of the subtidal kelps *Macrocystis integrifolia* and *Lessonia trabeculata* (Laminariales, Phaeophyceae) in an upwelling ecosystem of northern Chile: interannual variability and El Niño 1997–98. Rev Chil Hist Nat 78:33–50

Venegas M, Tala F, Fonck E, Vásquez JA (1992) Sporangial sori on stipes of *Lessonia nigrescens* Bory (Laminariaceae, Phaeophyta): a high frecuency phenomenom in intertidal populations of northern Chile. Bot Mar 35:573–578

Villouta E, Santelices B (1984) Estructura de la comunidad submareal de *Lessonia* (Phaeophyta, Laminariales) en Chile norte y central. Rev Chil Hist Nat 57:111–122

Villouta E, Santelices B (1986) *Lessonia trabeculata* sp. Nov. (Laminariales: Phaeophyta), a new kelp from Chile. Phycologia 25:81–86

Werner A, Kraan S (2004) Review of the potential mechanization of kelp harvesting in Ireland. Mar Envir Health Ser 17: 1–52

Westermeier R, Patiño D, Piel MI, Maier I, Mueller DG (2006) A new approach to kelp mariculture in Chile: production of free-floating sporophyte seedlings from gametophyte cultures of *Lessonia trabeculata* and *Macrocystis pyrifera*. Aquacult Res 37:164–171

Zemcke-White WL, Ohno M (1999) World seaweed utilization: An end-of-century summary. J Appl Phycol 11:369–376

Cultivation of the green alga, *Codium fragile* (Suringar) Hariot, by artificial seed production in Korea

Eun Kyoung Hwang · Jae Min Baek · Chan Sun Park

Originally published in the Journal of Applied Phycology, Vol 20, No 5, 19–25.
DOI: 10.1007/s10811-007-9265-5

Abstract *Codium fragile* (Suringar) Hariot is an edible green alga farmed in Korea using seed stock produced from regeneration of isolated utricles and medullary filaments. Experiments were conducted to reveal the optimal conditions for nursery culture and out-growing of *C. fragile*. Sampling and measurement of underwater irradiance were carried out at farms cultivating *C. fragile* at Wando, on the southwestern coast of Korea, from October 2004 to August 2005. Growth of erect thalli and underwater irradiance were measured over a range of depths for three culture stages. During the nursery cultivation stage (Stage I), growth rate was greatest at 0.5 m depth (0.055±0.032 mm day^{-1}), where the average midday irradiance over 60 days was 924± 32 μmol photons m^{-2} s^{-1}. During the pre-main cultivation stage (Stage II), the greatest growth rate occurred at a depth of 2 m (0.113±0.003 mm day^{-1}) with an average irradiance of 248±116 μmol photons m^{-2} s^{-1}. For the main cultivation stage (Stage III) of the alga, thalli achieved the greatest increase in biomass at 1 m depth (7.2±1.0 kg fresh wt m^{-1}). These results suggest that optimal growth at each cultivation stages of *C. fragile* could be controlled by depth of cultivation rope.

Keywords Artificial seed production · Cultivation · *Codium fragile*

E. K. Hwang · J. M. Baek
Seaweed Research Center,
National Fisheries Research and Development Institute,
Mokpo 530-831, South Korea

C. S. Park (✉)
Department of Marine Resources, Mokpo National University,
Jeonnam 534-729, South Korea
e-mail: cspark85@mokpo.ac.kr

Introduction

The marine green alga, *Codium fragile* (Suringar) Hariot, is widely distributed in temperature areas throughout the world (Silva and Womersley 1956; Yoshida 1998; Trowbridge 1998), and is eaten in Korea, China and Japan (Abbott 1988). In Korea *C. fragile* is an attractive seaweed species for cultivation because it commands a high market value compared to species such as *Porphyra* and *Undaria* (Sohn 1998). The alga is an additive of Kimchi, a traditional fermented vegetable. Small-scale cultivation of *C. fragile* from natural blooming zygotes has been conducted on the southwestern coast of Korea since 1987, but the success of production by natural seeding has varied due to annual variations in environmental conditions. Hwang et al. (2005a) reported that the regeneration capability of *C. fragile* could be a more reliable method for producing seed stock than reliance on wild zygote seeding, and published conditions for artificial seed production and nursery culture (Hwang et al. 2005b).

Codium species have two morphologically different thalli: spongy and filamentous. Spongy thalli contain numerous chloroplasts and an inner structure of intertwined colorless medullary filaments (Arasaki et al. 1956; Borden and Stein 1969; Yoshida 1998). The filamentous thallus consists of fine-branched filaments. Filamentous thalli have been observed in the field (Arasaki et al. 1956; Fletcher et al. 1989; Borden and Stein 1969), and are formed from isolated utricles, medullary filaments, zygotes and parthenogenetic female gametes of the spongy thalli (Borden and Stein 1969; Ramus 1972; Park and Sohn 1992; Yang et al. 1997).

A number of inter-related factors impact on the formation and growth of the spongy thalli, (i.e. irradiance and water movement). Water movement plays an important role in regulating formation of spongy thalli under laboratory culture condition (Ramus 1972; Park and Sohn 1992). Park and Sohn (1992) reported that the critical irradiance for formation of spongy thalli was over 3,000 lx (ca. 60 μmol photons m^{-2} s^{-1}). Nanba et al. (2005) reported that growth of the spongy thalli accelerated under a combination of high irradiance and high water velocity.

While other workers have demonstrated the importance of irradiance levels on controlling life cycle of *C. fragile* in the laboratory, there is no information on optimal irradiance, and consequently optimal culture depth, for the key stages of *C. fragile* cultivation in the field. A full understanding of the relationship between light availability and *C. fragile* light requirements requires site-and species-specific information. We report here the relationships between depth, irradiance, and growth of *C. fragile* during a cultivation cycle at one site.

Materials and methods

The *C. fragile* culture experiments were conducted at Wando (34°17′N, 126°42′E) on the southwestern coast of Korea. This is one of the most active seaweed culture areas around the southwestern sea of Korea. Environmental factors such as water temperature, salinity and conductivity, were measured by a logging multi-parameter probe (YSI-85, YSI Co, Yellow Springs, USA) at the experimental site. The light profile was measured every 7 days at the surface, and at depths of 0.5, 1, 2 and 3 m. Measurements were always taken within 2 h of solar noon, and five replicate measurements were taken at each depth on each occasion. Photosynthetically active radiation (PAR) was measured with a spherical quantum sensor (Li-193 SA; Li-Cor, Lincoln, USA) from October 2004 to August 2005 at the experimental culture ground. Values were recorded on a Li-1400 Data Logger (Li-Cor). The probe was held at each depth for 5 min each and an average reading was taken over that time. Surface PAR measurements were taken between each depth reading at the start and end of the depth profiles at each site. Attenuation coefficients were calculated on each sampling occasion by regression equation of natural logs of light values according to Carruthers et al. (2001).

Artificial seeding and tank culture methods of *C. fragile* were as described by Hwang et al. (2005a, b). Artificial seeding of isolated utricles and medullary filaments were conducted in July 2004. Vegetative thalli (Fig. 1a) were chopped using a hand blender (MR 5550CA, Braun, Barcelona, Spain) until the fragments were less than 5 mm in length. Isolated utricles and medullary filaments seeded onto a seed frame coiled with 100 m seed fiber (Daerim Fiber, Seoul, South Korea) (Fig. 1b). Most of the spongy fragments detached immediately after seeding onto the fibers, but some utricles and medullary filament became

Fig. 1a–h Procedure for artificial seeding and nursery culture of *Codium fragile*. **a** Vegetative thalli. **b** Seeding of the utricles and medullary filaments on seed frame with 100 m coiled seed fiber. **c** Seed frames are held in tank culture for 1 month. **d** Nursery culture of seed frames bearing medullary filaments on the strings in the sea at the different water depths. **e** Regenerated medullary filaments and young erect thalli after 40 days culture in situ. **f** After nursery culture, the seed fiber with young erect attached thalli is spooled around the main culture rope. **g** Young thalli of *C. fragile* after 5 months of culture in May 2005. **h** Fully grown *C. fragile* after 7 months of culture in August 2005. *Bars* **a** 10 cm, **e** 1 cm

attached to the fibers. The coils of fiber were then moved into a culture tank (Fig. 1c). The cultivation cycle for the alga can be divided into threestages: Stage I, nursery cultivation (Fig. 1d,e; 0–3 mm erect thallus length); Stage II, pre-main cultivation (Fig. 1f, stationary growth period, 0.3–4 cm thallus length); and Stage III, main cultivation (Fig. 1g,h; fast growth period, more than 4 cm thallus length and thalli become buoyant). Out-growing used a horizontal cultivation system modified from Sohn and Kain (1989), where horizontal ropes 100 m long were held at each of the experimental culture depths. For each experiment, a fresh batch of juvenile plants taken from replicate fiber coils from the original medullary filament settlement that had been held under the same environmental conditions (1 m depth) were used.

Seed frames were held in culture tanks for 60 days from July to September 2004 before the outdoor cultivation experiment commenced. From October to December 2004, seed frames of *C. fragile* were moved into the sea for Stage I experiment. For each experimental condition, plants were assessed every 30 days by measuring the maximum length of the medullary filaments or erect thalli using Venier Calipers, and measuring fresh weight on an electrical balance. The Stage II culture experiment ran from December 2004 to May 2005, and Stage III ran from May to August 2005. Water depths used for the Stage I, II and III cultures were 0.5, 1, 2 and 3 m, respectively. For in Stage II and III growth of *C. fragile*, plants were assessed every month by measuring the maximum length of erect thalli, fresh weight and mean number of erect thalli per 1 m of culture rope. One-way ANOVAs were used to test for significant differences in the mean growth rate among depths within each stage, and the mean biomass among depths. When significant differences were found ($P<0.01$), Fisher's LSD test was applied *a posteriori* to determine whether treatments varied significantly from each other.

Results

Water temperature varied from 8.0 to 24.8°C during the experimental period (Fig. 2). Maximum water temperature was recorded in September 2004, and the minimum in February 2005. Salinity ranged between 33.2 and 34.6 ppt. Mean water depth was 15–20 m in the culturing area. The mean current in the area was 0.5 ms^{-1}. Underwater irradiance varied according to month and water depth (Fig. 2). Maximum surface irradiance was recorded in August 2005, and the minimum in May 2005.

Growth

During Stage I (nursery cultivation), medullary filaments of *C. fragile* grew densely and developed into erect thalli with a spongy structure (Fig. 3a). Water temperature decreased during the culture period from 20.0 to 13.3°C. After 60 days culture, the mean length of the young erect thalli was greatest for plants held at 0.5 m depth (3.3±1.9 mm, Table 1). There was a significant difference in the mean growth rate among depths in Stage I (ANOVA, $P<0.01$). Subsequent *a posteriori* testing showed that these plants were significantly larger than at 2 m or 3 m depth (Table 1).

At the beginning of the Stage II cultivation trial, the mean length of *C. fragile* was 2.0±1.1 mm (Fig. 3b, Table 1). During the culture period, the water temperature increased from 8.0 to 13.9°C. At depths of 0.5 m and 1 m,

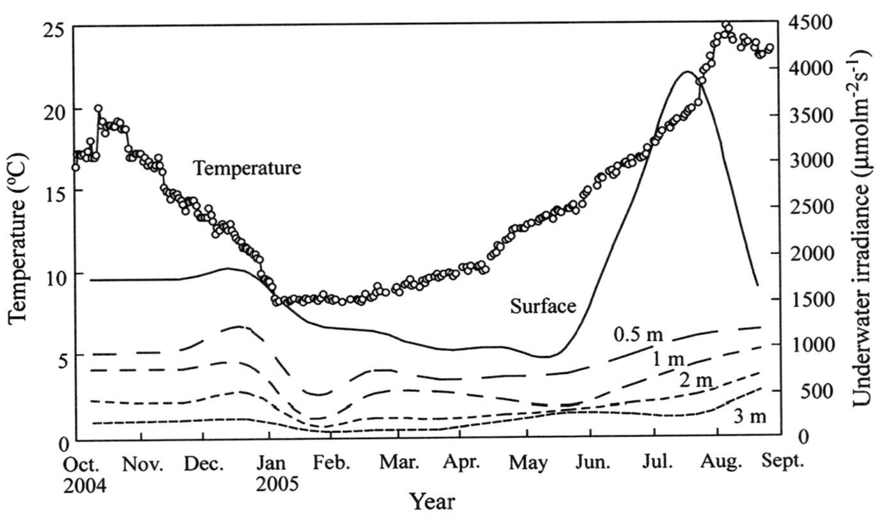

Fig. 2 Fluctuations in water temperature and underwater irradiance during the nursery and main culture periods of *C. fragile*. Irradiance was measured every month between October 2004 and August 2005 in the culture ground at Wando, Korea. Water temperature was recorded at 1 m depth

Fig. 3a–c Growth and development of *C. fragile* at the three stages with different water depth conditions. **a** Stage I (nursery cultivation) conducted from October to December 2004. **b** Stage II (pre-main cultivation) conducted from December to May 2005. **c** Stage III (main cultivation) conducted from May to August 2005

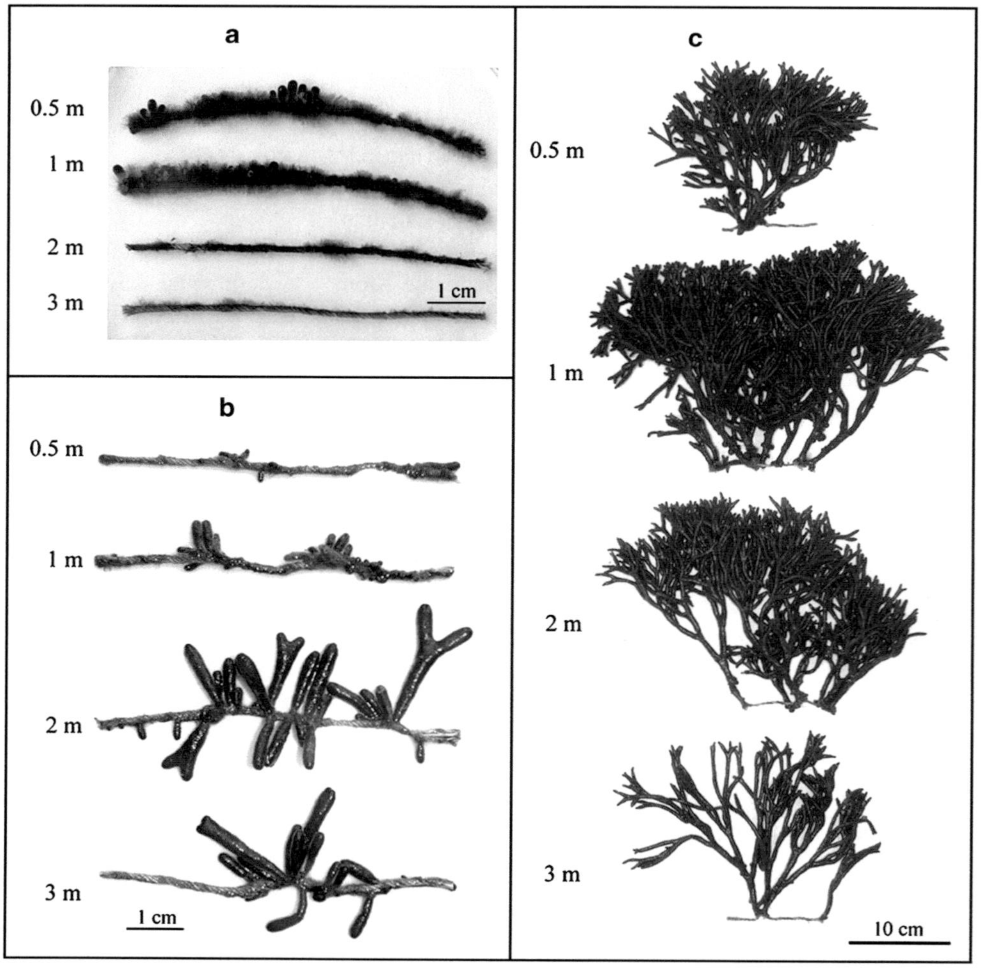

Table 1 Relation between growth rate and depth in *Codium fragile* during the different cultivation periods

Stage	Length growth of erect thalli					Regression of depth vs ln light
	Depth (m)	Initial length (mm)	Final length (mm)	Culture duration (days)	Growth rate (mm day^{-1})[d]	
I	0.5	0	3.3±1.9	60	0.055±0.032[a]	Y=−0.88x+7.48 (r^2=0.96)
	1	0	3.1±1.1	60	0.052±0.018 [a]	
	2	0	1.0±0.3	60	0.017±0.005 [b]	
	3	0	–[e]	60	–	
II	0.5	2.0±1.1	4.7±2.1	150	0.018±0.007 [a]	Y=−0.62x+6.86 (r^2=0.98)
	1	2.0±1.2	5.2±1.1	150	0.021±0.001 [a]	
	2	2.0±1.2	19.0±1.6	150	0.113±0.003 [b]	
	3	2.0±1.1	14.2±0.7	150	0.081±0.003 [c]	
III	0.5	30.2±1.5	142.1±15.3	90	1.24±0.15 [a]	Y=−0.54x+7.27 (r^2=0.91)
	1	30.2±1.3	233.2±18.1	90	2.26±0.19 [b]	
	2	30.2±1.4	198.5±21.6	90	1.87±0.22 [c]	
	3	30.2±1.5	168.4±20.1	90	1.54±0.21 [a]	

[d] Values (mean ± s.d. of triplicate groups) in same column having different superscripts are significantly different ($P<0.01$) according to *a posteriori* Fisher's LSD test
[e] No spongy thallus formation

erect thalli were seen to be bleached. A significant difference was found in the mean growth among depths in Stage II (ANOVA, $P<0.01$). *A posteriori* test revealed that after 150 days culture, the mean length of plants held at 2 m depth was significantly higher (19.0±1.6 mm) than at the other three depths, which did not vary significantly from each other (Table 1).

The largest *C. fragile* plants during the Stage III cultivation trial were observed after 90 days culture in August 2005 amongst plants held at 1 m depth (233.2± 18.1 mm, Fig. 3c, Table 1). There was a significant difference in the mean growth rate among depths in Stage III. *A posteriori* showed that the growth rate was significantly higher in 1 m depth than at other depths (Table 1). Water temperature increased during the culture period from 14.5 to 24.8°C. ANOVA showed that significant differences were found in the mean biomass among depths ($P<0.01$). Subsequently, Fisher's LSD test revealed that there was a significant difference in the mean biomass between all pairwise depths, except for between 0.5 m depth and 3 m depth (Table 2). This result indicates that, during the culture period, the greatest biomass of *C. fragile* was observed at 1 m depth after 90 days. This provided a maximum potential production estimate of 36,110 kg dry wt ha^{-1}.

Relation between underwater irradiance and depth

Underwater irradiance was found to vary with time of day, season and depth. In the first 1 m of the water column, PAR was reduced by around 70% compared to PAR at the surface (Fig. 2). Over the course of the three experiments, the attenuation coefficient varied from 0.54 in the summer to 0.88 in the autumn at the experimental culture ground. The seasonal changes of regression relation between depth vs ln light were reflected in the seasonal nature of each culture stage (Table 2). Growth of *C. fragile* was greater at 0.5–1 m than other depths during Stage I. During this culture stage the mean irradiance was 829±135 μmol photons m^{-2} s^{-1}. The lowest underwater irradiance measurements were observed during the Stage II cultivation trials. During this trial, mean irradiance at 2 m water depth was 248±116 μmol photons m^{-2} s^{-1} and the growth of *C. fragile* was greater at 2 m depth than at other depths. Stage III cultivation occurred during the summer months and, consequently, the highest underwater irradiance values were observed during this stage of the experiment as summarized in Fig. 4.

Discussion

This study has shown that numerous isolated utricles begin to form medullary filaments without a septum immediately after isolation, and forming of filamentous thallus is attributed to the transfer of chloroplasts from the utricles. The utricles rapidly lost their ability to form filamentous thalli after transferring their chloroplasts to the thalli, while the transfer leads the utricles to begin to attach to the substratum by rapidly producing elongated filamentous thalli (Fig. 1d). This study has also shown that the filamentous thalli and utricles of regenerated spongy thalli contain numerous fusiform chloroplasts (Fig. 1e–h). These results suggest that, as mentioned by Nanba et al. (2002), the protoplasm of *Codium* species, which contains numerous chloroplasts, plays an important role in the vegetative reproduction of the isolated utricles and filamentous thalli.

A number of experimental culture studies have revealed that environmental conditions such as temperature, irradiance, water flow, nitrogen source, and salinity are important in controlling the development of *C. fragile* under indoor culture conditions (Ramus 1972; Hanisak 1979; Park and Sohn 1992; Yang et al. 1997). *Codium* thalli are capable of survival and growth over a wide temperature range. In culture, only stationary growth occurs below 12°C. Growth increases as temperature increases up to 24°C. Reduced growth at 30°C is probably due to a sharp increase in respiration at higher temperatures (Hanisak 1979). While *Codium* grows more at warmer temperatures, it can survive

Table 2 Comparison of biomass, production and income of *C. fragile* cultivation in four different water depth conditions

Depth (m)	Biomass (kg fresh wt m^{-1})[a]	Biomass (kg dry wt 100 m^{-1})[b]	Production (kg dry wt ha^{-1})[c]	Estimate income (US$/ha)[d]	
				Codium	*Undaria*
0.5	1.58±0.51[a]	7.9±2.6[a]	790	9,480	790
1	7.22±1.03[b]	36.1±5.2[b]	36,110	433,320	36,110
2	3.43±0.62[c]	17.2±3.1[c]	1,720	20,640	1,720
3	1.03±0.31[a]	5.15±1.6[a]	515	6,180	515

[a] Values (mean ± s.d. of triplicate groups) in same column having different superscripts are significantly different ($P<0.01$) as a result of *a posteriori*

[b] Dry yield of *C. fragile* was 5% under 80°C after 24 h

[c] Main ropes of cultivation facilities assumed 100 horizontal ropes per 1 m interval from each culture ropes in a 1 ha area

[d] Local market price of *C. fragile* and *Undaria pinnatifida* is ca. 12 US$ and 1 US$ per kg dry weight, respectively

Fig. 4 Schedule of outdoor cultivation of *C. fragile* showing optimal conditions for different culture stages

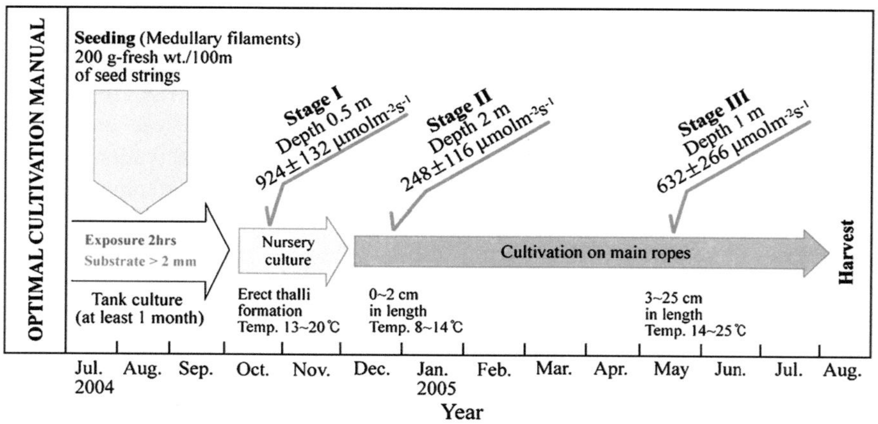

temperatures approaching the freezing point of seawater. In this study, the temperature range for the growth of isolated utricles and differentiated thalli was 8.0–24.8°C (Fig. 2). This tolerance to lower temperatures enables *Codium* to survive through the winter in temperate environments. Thus, while the response of *Codium* to temperature is somewhat dependent upon the stage in its life cycle, these stages are adapted to the range of temperatures they experience under natural conditions.

The filamentous thalli of several *Codium* species tolerate a wide range of environmental conditions in their life histories (Borden and Stein 1969; Fletcher et al. 1989). This study showed that isolated utricles have the ability to regenerate young spongy thalli with branched axes, and that their separated axes also have the ability to grow into spongy thalli. The utricles with medullary filaments isolated from various parts of spongy thalli have the ability to form filamentous thalli and attach to the substratum by themselves. Fragments of filamentous thalli grow into full-grown spongy thalli after transfer to the field (Fig. 3). These results suggest that utricles of this species detached by strong waves and grazers in the culture farm can grow into full-grown thalli via filamentous thalli. In an indoor culture experiment, Hwang et al. (2005a) reported that zygotes of *C. fragile* survived at 10–15°C and 10–20 μmol photons m^{-2} s^{-1}; however, medullary filaments of *C. fragile* could tolerate a wider range of temperatures (10–25°C) and irradiances (10–100 μmol photons m^{-2} s^{-1}). Yang et al. (1997) also observed that filamentous thalli of several *Codium* species grew over a broad range of environmental conditions, while Hanisak (1979) reported that filamentous thalli of *C. fragile* subspp. *tomentosoides* survived and grew in a narrower range of environmental conditions than spongy thalli. During this cultivation experiment, filamentous and spongy thalli of *C. fragile* co-existed on the culture ropes (Fig. 3). The filamentous thalli of *C. fragile* spread and wrapped around culture ropes; thereafter the spongy thalli grew densely. These findings indicate that the formation of filamentous thalli is an essential prerequisite for the formation of spongy thalli in the life history of *C. fragile*.

Hanisak (1979) reported the submarine irradiance levels for three populations of *Codium* in Rhode Island. Comparing the irradiance required for saturation of growth at various temperatures in culture with that measured in situ, irradiance levels were found to be above saturation during the summer (April–August or September). With one exception, irradiance levels suggested that radiant energy limited growth during the fall months when temperature was still adequate for growth. In this study, the indoor and field investigations showed that the optimum growth light doses of *C. fragile* differ according to growth stage (Table 1). Differences in the response to environmental factors for juvenile and mature stages suggest that this species is capable of altering its physiological activities to meet different environmental demands.

In this study we found that *C. fragile* shows acceptable commercial growth rates when cultured at depths receiving between 22% and 55% of surface irradiance (Table 1). Variation of the optimum light requirement suggests that the different life stages may be sensitive not only to total irradiance but also to spectral composition. These factors may vary between cultivation sites due to the variation in the scattering and light absorption characteristics of the particles and dissolved compounds in the water column (Kenworthy and Fonseca 1996).

Until 2004, fishermen collected naturally blooming zygotes of *C. fragile* and seeded them on cultivation ropes. The success of these natural algal seedings varied annually due to inter-annual variation in environmental conditions. In the case of zygote seeding, the mean biomass has been reported as being normally less than 1 kg fresh wt and the biomass by medullary filament seeding was ca. 7 kg fresh wt per 1 m culture rope under 1 m depth at harvest season, suggesting that the production of *C. fragile* by medullary filament seeding could be higher than that by zygote seeding.

For stable cultivation of this alga, artificial seeding of isolated medullary filament is therefore considered to be the most economic and effective method. The potential production of *C. fragile* has been estimated as 36,110 kg dry wt ha^{-1} (Table 2). However, to achieve the maximum yield it will be necessary to control the depth at which the plants are cultured. Understanding the appropriate light requirements will assist with site selection and help to expand commercial cultivation of *C. fragile.*

Acknowledgments This research was supported by a grant (F10400706A210000100) from the Fisheries Technology Development Program funded by Ministry of Maritime Affairs & Fisheries of Korean government and National Fisheries Research & Development Institute (RP-2007-AQ-025). We thank Dr. Phillip Heath (NIWA, New Zealand) for reviewing the English.

References

Abbott IA (1988) Food and food products from seaweeds. In: Lembi CA, Waaland JR (eds) Algae and human affairs. Cambridge Univ Press, New York, pp 135–147

Arasaki S, Tokuda H, Fujiyama K (1956) The reproduction and morphogeny in *Codium fragile*. Bot Mag Tokyo 69:39–44

Borden CA, Stein JR (1969) Reproduction and early development in *Codium fragile* (Suringar) Hariot: Chlorophyceae. Phycologia 8:91–99

Carruthers TJB, Longstaff BJ, Dennison WC, Abal EG, Aioi K (2001) Measurement of light penetration in relation to seagrass. In: Short FT, Coles RG (eds) Global seagrass research methods. Elsevier, Amsterdam, pp 369–392

Fletcher RL, Blunden G, Smith BE, Rogers DJ, Fish BC (1989) Occurrence of fouling, juvenile stage of *Codium fragile* ssp. *tomentosoides* (Goor) Silva (Chlorophyceae, Codiales). J Appl Phycol 1:227–237

Hanisak MD (1979) Physiological ecology of *Codium fragile* subsp. *tomentosoides*. PhD Thesis, University of Rhode Island

Hwang EK, Baek JM, Park CS (2005a) Artificial seed production using the reproduction methods in *Codium fragile* (Chlorophyta). J Korean Fish Soc 38:164–171

Hwang EK, Baek JM, Park CS (2005b) Artificial seed production and nursery culture conditions using regeneration of isolated utricles and medullary filaments of *Codium fragile* (Suringar) Hariot. J Korean Fish Soc 38:393–398

Kenworthy WJ, Fonseca MS (1996) Light requirements of seagrasses *Halodule wrightii* and *Syringodium filiforme* derived from the relationship between diffuse light attenuation and maximum depth distribution. Estuaries 19:740–750

Nanba N, Kado R, Ogawa H, Nakagawa, Komuro (2002) Formation and growth of filamentous thalli from isolated utricles with medullary filaments of *Codium fragile* spongy thalli. Aquat Bot 73:255–264

Nanba N, Kado R, Ogawa H, Nakagawa T, Sugiura Y (2005) Effects of irradiance and water flow on formation and growth of spongy and filamentous thalli of *Codium fragile*. Aquat Bot 81:315–325

Park CS, Sohn CH (1992) Effects of light and temperature on morphogenesis of *Codium fragile* (Suringar) Hariot in laboratory culture. Korean J Phycol 7:213–223

Ramus J (1972) Differentiation of the green alga *Codium fragile*. Am J Bot 59:478–482

Silva PC, Womersley HBS (1956) The genus *Codium* (Chlorophyta) in southern Australia. Aust J Bot 4:261–289

Sohn CH (1998) The seaweed resources of Korea. In: Critchley AT, Ohno M (eds), Seaweed resources of the world. Jap Int Coop Agency, pp 15–33

Sohn CH, Kain (Jones) JM (1989) *Undaria*, *Laminaria* and *Enteromorpha* cultivation in Korea. In: Kain (Jones) JM, Andrews JW, McGregor BJ (eds) Proceedings of the 2nd Workshop of COST 48 Subgroup 1. Port Erin, Isle of Man, British Isles, 1989

Trowbridge CD (1998) Ecology of the green macroalga *Codium fragile* (Suringar) Hariot 1989: invasive and non-invasive subspecies. Annu Rev Oceanogr Mar Biol 36:1–64

Yang MH, Blunden G, Huang FL, Fletcher RL (1997) Growth of dissociated, filamentous stage of *Codium* species in laboratory culture. J Appl Phycol 9:1–3

Yoshida T (1998) Marine algae of Japan. Uchida Rokakuho, Tokyo

Distribution and symptoms of epiphyte infection in major carrageenophyte-producing farms

Charles S. Vairappan · Chong Sim Chung · A. Q. Hurtado · Flower E. Soya · Genevieve Bleicher Lhonneur · Alan Critchley

Originally published in the Journal of Applied Phycology, Vol 20, No 5, 27–33.
DOI: 10.1007/s10811-007-9299-8

Abstract High density commercial farming of carrageenophyte *Kappaphycus alvarezii* is often plagued with "ice-ice" disease and epiphyte infection, which eventually leads to reduced production and in some cases collapse of crop. Epiphyte outbreak has been occurring regularly in major carrageenophyte farms in the Philippines, Indonesia, Malaysia and Tanzania. Infected materials from these countries were studied to establish baseline information on the epiphyte's identity, density, symptoms and secondary infection on the host seaweed. The causative organism was identified as *Neosiphonia apiculata* (Hollenberg) Masuda *et* Kogame, based on its morphological features. Epiphyte density on host seaweed materials decreased in the following order: the Philippines (88.5 epi cm^{-2}), Tanzania (69.0 epi cm^{-2}), Indonesia (56.5 epi cm^{-2}) and Malaysia (42.0 epi cm^{-2}). Initial symptoms were the presence of tiny black spots, indicating the embedded tetrasporeling in seaweed cortex layer. Vegetative form emerged after 2 weeks measuring less than 0.5 mm in length with a density of less than 25.0 epi cm^{-2}. Upon maturation, infected seaweed takes on a "hairy" appearance with "goose-bumps" like cortical swellings. The epiphyte appears as a solitary plant with multiple secondary rhizoids or as multiple epiphytes appearing from a single cortical opening. At the end of infection, the epiphytes left dark pits on the cortical swelling, and the carrageenophytes are infected by opportunistic bacteria. Bacterial enumeration of healthy and infected seaweed materials showed an increase of more than 300% in total bacterial count on infected materials dominated by *Alteromonas* sp., *Flavobacterium* sp. and *Vibrio* sp.

Keywords Carrageenophytes · Epiphyte · *Neosiphonia apiculata* · Secondary bacterial infection

C. S. Vairappan (✉) · C. S. Chung
Laboratory of Natural Products Chemistry, Institute for Tropical Biology and Conservation, University Malaysia Sabah,
88999 Kota Kinabalu, Sabah, Malaysia
e-mail: csvairappan@yahoo.com

A. Q. Hurtado
Aquaculture Department,
Southeast Asian Fisheries Development Center,
Tigbauan, Iloilo 5021, Philippines

A. Q. Hurtado
Integrated Services for the Development of Aquaculture and Fisheries (ISDA),
McArthur Highway Tabuc Suba,
Jaro, Iloilo 5000, Philippines

F. E. Soya
Institute of Marine Sciences, University of Dar es Salaam,
P.O. Box 668, Zanzibar, Tanzania

G. B. Lhonneur · A. Critchley
Raw Materials Procurement, Cargill Texturizing Solutions
(Ex-Degussa Food Ingredients/BL Texturant Systems),
50500 Baupte, France

Introduction

Red algae genus *Kappaphycus* are farmed as raw material for the extraction of carrageenan, and it is considered an important commercial commodity in the Philippines, Indonesia, Malaysia and Tanzania. More than 90% of the major carrageenophyte farms are located in the coastal waters of these countries (Vairappan et al. 2003; Werner et al. 2004), with the Philippines producing more than 70%

of the needed raw material. *Kappaphycus* farming is a labour intensive activity and the production is often dependent on culture conditions and disease outbreak. Two significant disease outbreaks often encountered in high density commercial farming are (1) ice-ice disease, and (2) epiphyte infection.

Emergence of "ice-ice" disease was first reported in 1974 in the Philippines by Trono (1974). The involvement of pathogenic microbes and the role of culture conditions were described in detailed by Uyenco et al. (1981) and Largo et al. (1995a, b). Their findings lead to the understanding of how stress induced by culture conditions could trigger the pathogenesis. The causative microbes were identified as *Vibrio-Aeromonas* complex and *Cytophaga-Flavobacterium* complex, and their mode of action was suggested to be based on lytic enzymatic activities of bacterial exudates (Largo et al. 1995a).

On the other hand, epiphytic infection of red filamentous microalgae is not a new phenomenon; it has been known to exist for the last 30 years. An epiphytic outbreak in farmed *Kappaphycus* was reported as early as 1975 by Doty and Alvarez (1975). However, not much interest has been focused on this problem until recently when there has been an increase in epiphyte outbreaks and their impact on production (Critchley et al. 2004; Hurtado et al. 2006; Vairappan 2006). Recent investigation of this problem indicated *Neosiphonia savateri* (Hariot) M.S. Kim *et* I.K. Lee as the causative filamentous red algae, and revealed some insights of its symptoms and seasonality at culture farms in Sabah, Malaysia (Vairappan 2006). Hurtado et al. (2006) described *Polysiphonia* sp. as the causative organism in the Philippines farms, with a different type of symptoms. The occurrence of epiphyte outbreaks in Malaysia and the Philippines has resulted in a tremendous reduction of biomass production and a decline in carrageenan quality (Vairappan et al. 2007).

Due to the regular outbreaks and economical loss caused by epiphytes, research collaboration was initiated among researchers from the Philippines, Indonesia, Malaysia and Tanzania to gather substantial information on the causative agent in these major seaweed producing countries. Infected seaweed specimens from cultivation farms in the respective countries were preserved, transported and studied at the Laboratory of Natural Products Chemistry, Institute for Tropical Biology and Conservation, Universiti Malaysia Sabah. This investigation involved identification of the causative organism, infection density, stages of infection and occurrence of secondary bacterial infection after the epiphytes had dropped off. Information on the symptoms of the infected seaweed and its secondary bacterial infection was documented using stereo microscope and scanning electron microscope (SEM) images and microbiological techniques.

Materials and methods

Infected seaweed samples of *Kappaphycus alvarezii*, at various stages of infection, were collected from cultivation farms at Davao, Philippines (6°45′24″N, 125°20′35″E), Tawi-Tawi, Indonesia (4°55′10″N, 120°00′31″E), Semporna, Malaysia (6°02′10″N, 116°01′02″E), and Zanzibar, Tanzania (6°00′08″S, 39°18′21″E). Diseased and healthy seaweeds were collected and analyzed between October 2005 and March 2006. Specimens were cleaned of organic detritus and contaminants, preserved in 10% formalin in seawater and transported to the laboratory in Kota Kinabalu, Sabah, Malaysia. At the laboratory, specimens were further cleaned using a soft tip brush and preserved in freshly prepared formalin (5%) in seawater buffer solution (pH 7.8, salinity 30 ppt).

Epiphyte identification and enumeration

Epiphyte infected specimens were cleaned with a soft brush to remove dirt and other contaminants. Sections of the seaweed were prepared manually using a surgical blade, immersed in seawater and viewed under a stereo microscope (Stemi-2000 CS; Carl Zeiss, Germany). Images were taken using an attached Nikon digital camera. Epiphyte densities were enumerated visually on 1×1 cm^2 sections of seaweed at 5× magnification. A total of 5 specimens for each location and 10 sections were enumerated for each specimen.

Epiphytes were removed from the host using pith stick and forceps under stereo microscope, and stained with 0.5% (w/v) cotton blue in lactic acid/phenol/glycerol/water [1:1:1:1 (v/v)] solution and mounted in 50% glycerol/seawater on microscope slides. Slides were viewed at 10× and 40× magnifications under a Carl Zeiss compound microscope (Axioskop 40; Carl Zeiss, Germany).

Scanning electron microscope study

Sections of epiphyte infected seaweed were viewed under a stereo microscope, and sections identified for detailed viewing were cropped and fixed in 4% glutaraldehyde in 0.1 M cacodylate buffer solution (salinity: 30 ppt) solution prior to processing. During processing, specimens fixed in 4% glutaraldehyde were taken out of the solution and rinsed twice in distilled water for 10 min per rinse. Sections were then trimmed to dimensions of 5×10 mm. Specimens were post fixed in 1% OsO_4 at 4°C for 2 h, followed by dehydration with graded acetone series. The sections were finally immersed in *t*-butyl alcohol and left for 24 h in a freezer until the solution had completely frozen. The samples were then placed in a freeze dryer (JFD-310 Freeze Drying Device) and the *t*-butyl alcohol was allowed to sublime completely. Dehydrated specimens were mounted on stubs and coated with 10- to 30-nm layer of gold (JEOL JFC-1600 auto fine coater) before observations were made with a Leica Cambridge S360 electron microscope.

Bacterial enumeration

An area measuring 1 cm^2 was cut aseptically from both healthy and infected thallus (Sabah specimen), homogenized separately and placed into respective vials containing 10 ml sterile seawater. Serial dilution was carried out to obtain a range of 1×10^{-1} to 1×10^{-5}. Triplicates subsamples of the suspension (100 μL) were inoculated onto modified Zobell Marine Agar (added 5% healthy seaweed aqueous extract) and spread with a glass rod. Inoculated plates were allowed to dry in a transfer chamber, sealed with parafilm to prevent desiccation, and maintained inverted during incubation at 28°C for 14 days. After this period, bacterial colonies with different morphological characters were counted, isolated, purified and maintained on modified Zobell Marine Agar. Quantitative enumerations of total number of culturable bacteria and bacterial species were recorded. Isolated pure cultures were identified using conventional biochemical and physiological methods as described by Baumann et al. (1972); Kong and Chan (1979); Sawabe et al. (1992, 1998, 2000); Largo et al. (1995a, b) and Vairappan et al. (2001).

Results

Epiphytes were removed from the infected specimens and viewed under a compound microscope at 10× and 40× magnifications. Filamentous red algae isolated from infected seaweeds from all four countries were similar in morphological features. Detailed observations revealed its thalli as brownish-red, rigid and with percurrent main axes. Its basal attachment system was composed of a primary rhizoid, and later it formed a tuft of rhizoids due to the production of many secondary rhizoids. Primary rhizoids penetrated into basiphyte tissue and well developed thalli also attached themselves to the basiphyte via secondary rhizoids, which were formed from pericentral cells of main axis lower segments and from proximal first-order branches. The epiphyte's main axis became thicker upwards and abruptly tapered at the apices. Each of its vegetative segments consisted of 4 pericentral cells and lacked cortical cells. The most prominent features were its rhizoids, which were cut off from the pericentral cells, the production of lateral-branch initials from successive segments in a spiral arrangement, and a 3-celled carpogonial branches. Its axis also bore vegetative trichoblasts from each segment in a spiral manner. Relative densities of the epiphytes are shown in Fig. 1, and were in the following order; the Philippines > Tanzania > Indonesia > Malaysia (88.5>69.0>56.5>42.0 epi cm^{-2}).

Infection phases

Detailed observation of the epiphyte's infection phases were possible in specimens collected from Semporna, Sabah. Corresponding phases were also noticed in samples collected from the Philippines, Indonesia and Tanzania. Details of this epiphyte's infection phases on *K. alvarezii* are shown in Fig. 2a–f. The first sign of epiphyte infection was observed with the appearance of tiny black spots on the cell layers of the cuticle surface (Fig. 2a), indicating the presence of embedded tetrasporeling between the outer cortex cells. Epiphytes emerge after 2–4 weeks depending on the seawater temperature and salinity conditions, and form a rough surface as shown in Fig. 2b-c. After another 4–6 weeks, the epiphytes mature and are seen as shown in Fig. 2d. The sites where epiphyte entered the basiphyte had protruded to feature a typical "goose-bumps"-like symptom. This seems to be a prominent feature in all the specimens investigated. After about 10 weeks, the epiphytes reached the end of their life cycle and start to drop off from their attachment as shown in Fig. 2e. Their initial sites of attachment turned dark and started to disintegrate as shown in Fig. 2f.

Tissue degradation and secondary infection

Scanning electron microscope images of the tissue degradation and bacterial infection are observed in Fig. 3A–F. Locations where the epiphytes penetrated the basiphyte is a characteristic feature of its infection, featuring a "goose

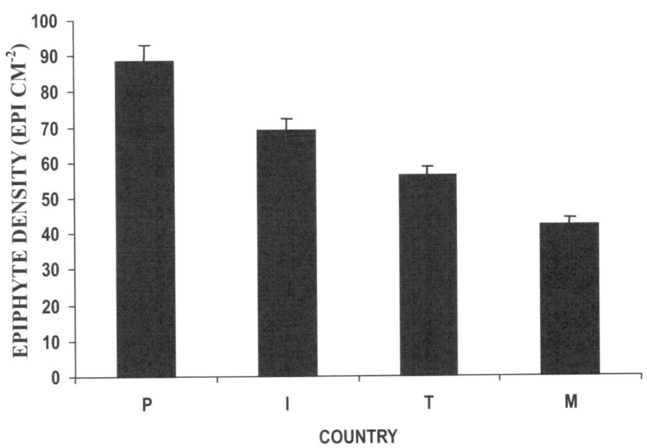

Fig. 1 Epiphyte (*Neosiphonia apiculata*) density (number/cm^2) on *Kappaphycus alvarezii* specimens collected from Philippines (*P*), Indonesia (*I*), Tanzania (*T*) and Malaysia (*M*). Values are means ± SD for 10 replicates for each specimen

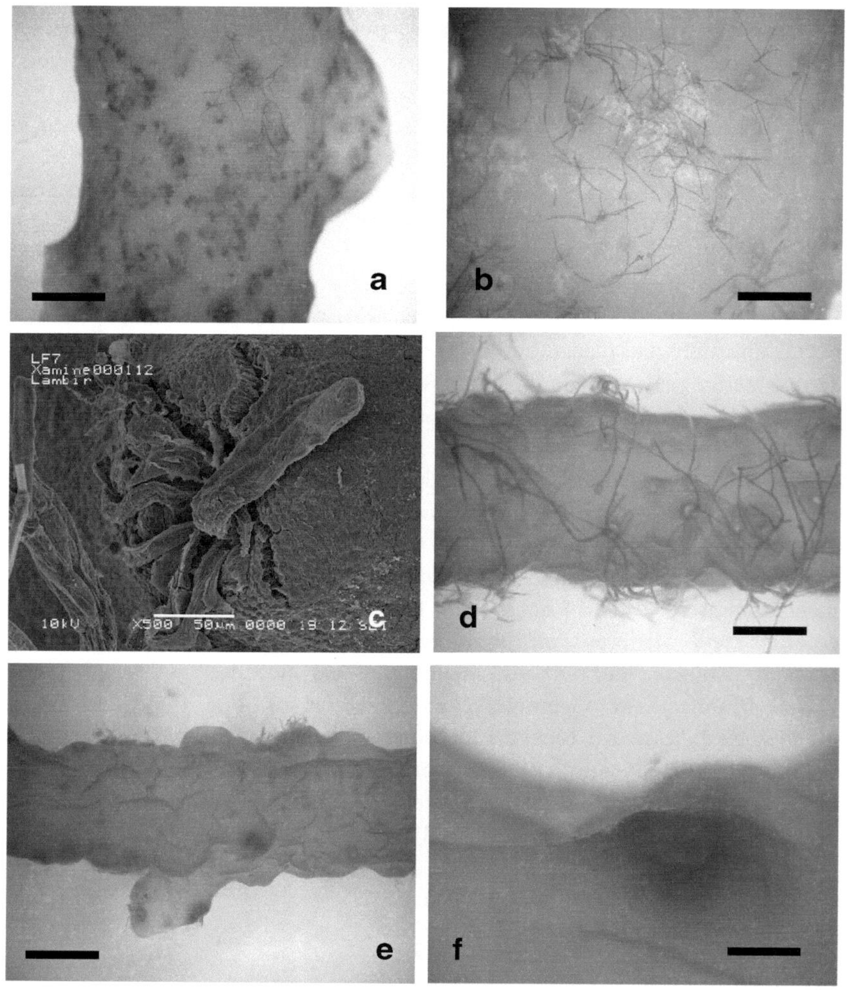

Fig. 2 Phases of epiphyte infection on *Kappaphycus alvarezii* thallus. **a** Host plant with early stage of epiphytes imbedded as tiny black spots in outer cell layers (*scale bar*=300 μm), **b** Host plant with epiphyte germlings (*scale bar*=300 μm). **c** Scanning electron microscope micrographs showing emergence of epiphyte germling from the cortex cell layers. **d** Host plant with mature epiphyte (*scale bar*=300 μm). **e** Host thallus with "goose-bump"-like symptoms at the end of the epiphyte infection phase (*scale bar*= 300 μm). **f** Epiphyte infected "mounts" turn dark with the onset of the secondary bacterial infection (*scale bar*=300 μm)

bump"-like swelling and formation of a pit in the middle where its basal primary rhizoid is still loosely attached. Secondary rhizoids and their upper main thalli drop off, leaving multiple "bumps" with a pit in the middle throughout the seaweed's surface (Fig. 3a). Early tissue degradation could be noticed with the appearance of tiny pores around the tip and sides of the swelling. Then, the remaining rhizoid of the epiphyte started to drop off at this point as shown in Fig. 3b–c. Further progression leads to the disintegration of the "goose bump" walls as shown in Fig. 3d–e. Secondary bacterial infection was observed to set in during this phase as shown by the presence of rod-shaped bacteria in Fig. 3f. The characteristic mount at the location where the epiphyte enters the tissue was completely disintegrated, exposing the host thallus to further attack from grazers and other microbes. The total numbers of culturable bacteria and culturable bacteria species for the healthy and infected seaweed are presented in Table 1. The total number of culturable bacteria from diseased seaweed thalli was more than 300% greater than in the healthy seaweed. A total of five bacterial species (*Alteromonas* sp., *Aeromonas* sp., *Escherichia coli*, *Flavobacterium* sp. and *Vibrio* sp.) were isolated, identified and their relative abundance determined from healthy seaweed. But only three species (*Alteromonas* sp., *Flavobacterium* sp. and *Vibrio* sp.) with much higher abundance were isolated and identified in diseased seaweed thallus as shown in Table 1.

Discussion

Epiphyte outbreaks in farms in all four countries were reported to occur during drastic fluctuations in seawater salinity, temperature and nutrient levels (Hurtado et al. 2006; Vairappan 2006). Newly infected seaweed materials had young epiphytes with multiple turts like secondary rhizoids and vegetative trichoblasts from each segment arranged in a spiral manner. The identity was established based on characteristic features such as rhizoids separated from the parental pericentral cells by a cross wall, its procarp with a 3-celled carpogonial branch, its spermatangial branch arising from the first fork of a fertile trichoblast.

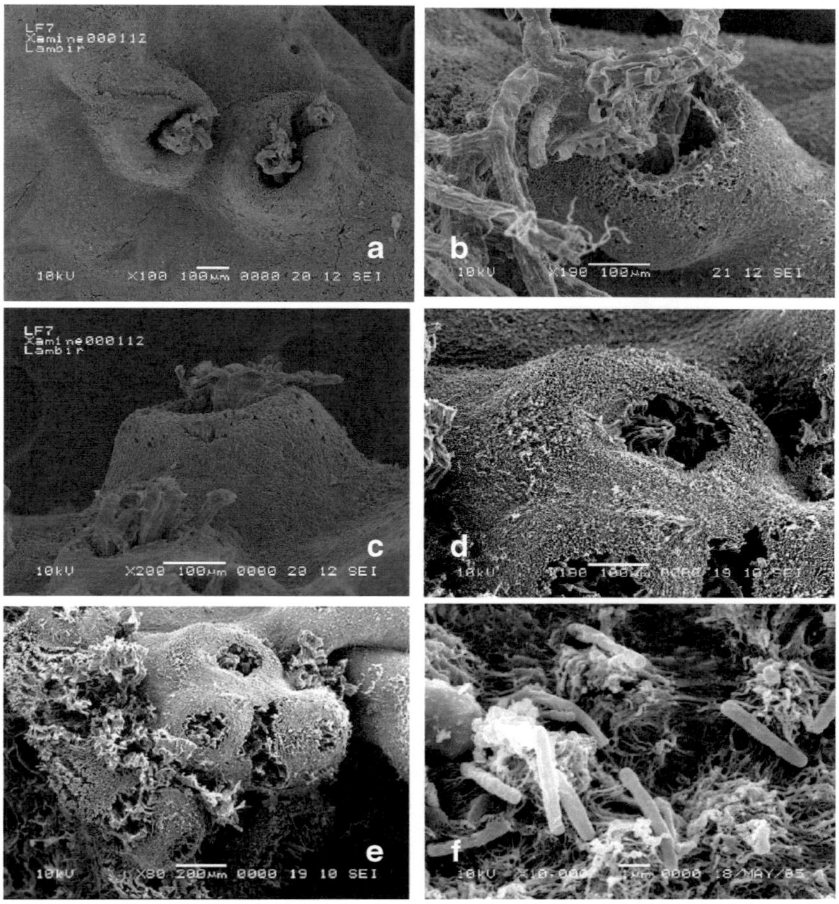

Fig. 3 Scanning electron microscopic micrographs showing phases of cellular decomposition of the epiphyte infected site. **a** "goose bump"-like epiphyte infected mounts are visible with epiphytes still attached to them. **b, c** Tissues in the mounts are showing signs of decomposition, **d** Epiphyte rhizoids drop off from the mount. **e** Mount starts to break up. **f** Bacterial infection sets in and the presence of bacteria becomes obvious

These were described by Kim and Lee as the characteristic aspects of genus *Neosiphonia* (Kim and Lee 1999; Choi et al. 2001; Masuda et al. 2001; Tani et al. 2003). The primary symptom of this infection was the presence of "goose bump"-like feature that was reported by Critchley et al. (2004) and Hurtado et al. (2006). The causative organism in the early report by Hurtado et al. (2006) was tentatively identified as *Polysiphonia*. In an effort to determine the identity of the epiphyte, multiple specimens

Table 1 Total culturable bacterial count (CFU cm^{-2}) and culturable bacterial species count (CFU cm^{-2}) on healthy and epiphyte infected seaweeds

Bacterial enumeration (CFU cm^{-2})	Healthy	Infected
Total bacterial count	2,420±98	8,620±106
Bacterial species count		
Alteromonas sp.	240±34	2,150 ±102
Escherichia coli	775±68	–
Flavobacterium sp.	330±12	2,930±184
Aeromonas sp.	678±45	–
Vibrio sp.	384±28	3,534±189

Values are means ± SD for five replicates of samples

from farms in the Philippines, Indonesia, Malaysia and Tanzania were studied in detail to determine the identity of the causative organism. Detailed descriptions and comparison of the epiphyte's morphological features with data reported by Tani et al. (2003) confirmed its identity as *N. apiculata*. Interestingly, symptoms observed on infected seaweeds in this study were different from those of specimens collected from Kudat, Sabah, in 2003 (Vairappan 2006). The causative epiphyte isolated from infected specimens in Kudat farms in 2003 was identified as *Neosiphonia savatieri*. Two prominent differences were the absence both of "goose bumps" and black spots at the end of the infection phase. Therefore, it could be suggested that *N. savatieri* and *N. apiculata* exhibits different symptoms upon infecting *K. alvarezii*. In 2005, the seaweed population in Kudat, Sabah (Malaysia) was badly infected by "ice-ice" disease and epiphyte, and in an effort to safe the industry seedlings were brought in from Southern Philippines. This desperate measure coupled with insufficient monitoring and quarantine lead to the introduction of *N. apiculata* to these farms. This new epiphyte species has caused severe outbreaks with dense epiphyte coverage, secondary infection and breaking-off of the seaweed thallus as compared to the earlier epiphyte. Presence of *N. apiculata* has also phased out the outbreak of *N. savatieri* in the culture farms of Kudat.

The early infection phases (Fig. 2a–d) of this species are similar to *N. savatieri*. However, when the epiphytes grew, the morphological features of the infected seaweed surface changed into "goose bumps" and the latter stage seems to lead to darkening of the "goose bumps", tissue disintegration and secondary bacterial infection (Fig. 3a–f). Infected specimens from all four countries showed similar morphological features. Hence, epiphyte infection with "goose bump" symptom and dark mounts during the latter phase could be associated with *N. apiculata* outbreaks.

Secondary bacterial infection further contributed to the disintegration of the infected tissue, leading to thallus rupture and breaking-off of seaweed from culture lines and contributing to about 20% of biomass loss in farms experiencing epiphyte outbreaks in Sabah, Malaysia. Generally, the total number of culturable bacteria isolated from the diseased thallus was about three times that of the number of culturable bacteria from the healthy thallus. Bacterial enumeration of the infected area only indicated the presence of *Alteromonas* sp., *Flavobacterium* sp. and *Vibrio* sp., as compared to five bacterial species in healthy specimens. Bacteria isolated from diseased seaweed are very closely related to or known as seaweed pathogens (Largo et al. 1995a, b; Sawabe et al. 1998, 2000; Vairappan et al. 2001).

A direct comparison of the relative bacterial abundance on SEM micrographs with the enumerated total number of culturable bacteria is not possible, since the specimens used for SEM observations were fixed in fixatives and dehydrated with graded acetone series. These steps would have washed away most of the surface bacterial flora. During this secondary infection, the microbial flora does not form microbial biofilm as is the case in primary bacterial infections (Vairappan and Chong 2006). The three dominant bacterial species in the infected thalli were swarmers, and this facilitated the spreading of the infection to other locations on the seaweed with relative ease.

Findings from this investigation revealed that the epiphyte outbreak associated with "goose bump"-like symptoms in all the four main carrageenophyte-producing countries are caused by *N. apiculata*. Infection of this species caused severe damage to the crops as compared to the milder outbreak of *N. savateri*. Final stages of infection lead to tissue darkening and secondary bacterial infection which caused further disintegrated of the infected site.

Epiphyte outbreak in major seaweed-producing farms has become a serious threat that warrants immediate attention from the seaweed farmers, seaweed purchasers and seaweed processing industries. Frequent epiphyte outbreaks have caused a severe drop in seaweed production, reduced carrageenan quality, decrease in the number of farmers, and outbreaks of "ice-ice" bacterial disease. Reduction in epiphyte outbreaks is possible if seaweed farmers practice better farm management with the use of healthy seedlings and better maintenance of their seaweed lines. Cultivation site selection criteria and cultivation procedure should be clearly transmitted to the farmers who are often illiterate and naive of good farming practices. Periodical visits and seminars by the local fisheries agencies should also be carried out to monitor farming progress and initial outbreaks of epiphyte, and to encourage the farmers. The role of the processing industry is equally important in educating the seaweed farmers and purchasers, in an effort to ensure constant supplies of quality raw material.

Acknowledgement We are grateful to Prof. Dr. Michio Masuda for his assistance in identifying the epiphytes. The main author is grateful to Sabah Parks Semporna for the permission to conduct surveys in seaweed culture farms in Semporna, Sabah. This study was supported by a Fundamental Research Grant (FRG0037-BD-1/2006) by the Malaysian Government, and Cargill Texturizing Solutions. C.S.C. is a post-graduate student registered at Borneo Marine Research Institute, Universiti Malaysia Sabah.

References

Baumann P, Gaunthier MJ, Baumann L (1972) Taxonomy of aerobic marine eubacteria. J Bact 110:402–429

Choi HG, Kim MS, Guiry MD, Saunders GW (2001) Phylogenetic relationship of *Polysiphonia* (Rhodomelaceae, Rhodophyta) and its relatives based on anatomical and nuclear small-subunit rDNA sequence data. Can J Bot 79:1465–1476

Critchley AT, Largo DB, Wee W, Bleicher LG, Hurtado AQ (2004) A preliminary summary on *Kappaphycus* farming and the impact of epiphytes. Jpn J Phycol 52:231–232

Doty MS, Alvarez VB (1975) Status, problems, advances and economics of *Eucheuma* farms. Mar Technol Soc J 9:30–35

Hurtado AQ, Critchley AT, Trespoey A, Lhonneur GB (2006) Occurrence of *Polysiphonia* epiphytes in *Kappaphycus* farms at Calaguas Is., Camarines Norte Phillippines. J Appl Phycol 18:301–306

Kim MS, Lee IK (1999) *Neosiphonia flavimarina* gen. et sp. nov. with a taxonomic reassessment of the genus *Polysiphonia* (Rhodomelaceae, Rhodophyta). Phycol Res 47:271–281

Kong MK, Chan KY (1979) A study on the bacterial flora isolated from marine algae. Bot Mar 22:83–97

Largo DB, Fukami F, Nishijima T, Olino M (1995a) Laboratory induced development of the ice-ice disease of the farmed red algae *Kappaphycus alvarezii* and *Eucheuma denticulatum* (Solieriaceae, Gigartinales, Rhodophyta). J Appl Phycol 7:539–543

Largo DB, Fukami F, Nishijima T (1995b) Occasional bacteria promoting ice-ice disease of the farmed red algae *Kappaphycus alvarezii* and *Eucheuma denticulatum* (Solieriaceae, Gigartinales, Rhodophyta). J Appl Phycol 7:545–554

Masuda M, Abe T, Kawaguchi S, Phang SM (2001) Taxonomic notes on marine algae from Malaysia VI. Five species of Ceramiales (Rhodophyceae). Bot Mar 44:467–477

Sawabe T, Ezura Y, Kimura T (1992) Characterization of an alginolytic marine bacterium from decaying Rishiri-konbu

Laminaria japonica var. ochotensis. Nippon Suisan Gakkaishi 58:141–145

Sawabe T, Makino H, Tatsumi M, Nakano K, Tajima K, Iqbal MM, Yumoto I, Ezura Y, Christen R (1998) *Pseudoalteromonas bacteriolytica* sp. nov., a marine bacterium that is the causative agent of red spot disease of *Laminaria japonica*. Int J Syst Bact 48:769–774

Sawabe T, Narita M, Tanaka R, Onji M, Tajima K, Ezura Y (2000) Isolation of *Pseudoalteromonas elyakovii* strains from spot-wounded fronds of *Laminaria japonica*. Nippon Suisan Gakkaishi 66:249–254

Tani M, Yamagishi Y, Masuda M, Kogame K, Kawaguchi S, Phang SM (2003) Taxonomic notes on marine algae from Malaysia. IX. Four species of Rhodophyceae, with the description of *Chondria decidua* sp. nov. Bot Mar 46:24–35

Trono GC Jr (1974) *Eucheuma* farming in the Philippines. U. P. Natural Science Research Center, Quezon City

Uyenco FR, Saniel LS, Jacinto GS (1981) The "ice-ice" problem in seaweed farming. Proc Int Seaweed Symp 10:625–630

Vairappan CS (2006) Seasonal occurrences of epiphytic algae on the commercially cultivated red alga *Kappaphycus alvarezii* (Solieriaceae, Gigartinales, Rhodophyta). J Appl Phycol 18:611–617

Vairappan CS, Chong SC (2006) Seaweed farming in Malaysia: challenges. In: Phang SM, Critchley AT, Ang PO Jr (eds) Seaweed utization in South East Asia. University of Malaya Maritime Research Centre (UMMReC), Kuala Lumpur, pp 161–170

Vairappan CS, Suzuki M, Motomura T, Ichimura T (2001) Pathogenic bacteria associated with lesions and thallus bleaching symptoms in the Japanese kelp *Laminaria religiosa* Miyabe (Laminariales, Phaeophyceae). Hydrobiologia 445:183–191

Vairappan CS, Tan KH, Rosnita R (2003) Seasonal fluctuation of carrageenan yield in *Eucheuma spinosum* (*Eucheuma denticulatum*) cultured in the waters of Pulau Balambangan, Kudat, Sabah. Borneo Sci 14:63–70

Vairappan CS, Chong SC, Lhonneur GB, Critchley AT (2007) Epiphyte outbreak in *Kappaphycus alvarezii* (Gigartinales, Rhodophyta) farms in South East Asia: Impact on the yield and Quality of carrageenan. XIXth International Seaweed Symposium, 26–31 March 2007, Kobe, Japan, pp 80

Werner A, Clarke D, Kraan S (2004) Strategic review of the feasibility of seaweed aquaculture in Ireland. Marine Institute Galway Technology Park, Productive Sector Operational Programme, National Development Plan 2000–2006, Ireland

The occurrence of pinhole disease in relation to developmental stage in cultivated *Undaria pinnatifida* (Harvey) Suringar (Phaeophyta) in Korea

Chan Sun Park · Kyung Yang Park · Jae Min Baek · Eun Kyoung Hwang

Originally published in the Journal of Applied Phycology, Vol 20, No 5, 35–40.
DOI: 10.1007/s10811-008-9329-1 © Springer Science + Business Media B.V. 2008

Abstract Infection rates of pinhole disease in farmed *Undaria pinnatifida* (Harvey) Suringar were examined between November 2002 and April 2003 at an *Undaria* farm at Wando on the southwestern coast of Korea. The growth stages of *U. pinnatifida* can be defined as: a growth phase from December to the end of February, and a decay phase from March to April. There was a significant increase in the level of infection with pinhole disease (measured as the number of pinholes per cm^2) between December and April ($P<0.01$). Infection progressed from the upper part of the frond in December to the basal part in April. The infection rate of pinhole disease in relation to growth phase on farms over time is limited by the harvest of the seaweed crop at the end of the cultivation period in April. The midrib, sporophyll and stipe showed no signs of infection at any time during the cultivation period. The reduced physiological activity and tissue aging that occurred over time in the sporophytic life phase is the probable cause of the increasing infection rate of the disease.

Keywords *Undaria pinnatifida* · Phaeophyta · Pinhole disease · Harpacticoid · Copepod · *Amenophia orientalis*

C. S. Park · K. Y. Park
Department of Marine Resources, Mokpo National University, Muan-gun, Chonnam 534-729, Korea

J. M. Baek · E. K. Hwang (✉)
Seaweed Research Center, National Fisheries Research & Development Institute,
1101 Okamdong,
Mokpo 530-831, Korea
e-mail: ekhwang@momaf.go.kr

Introduction

Seaweed cultivation is a major industry in many countries of the Asia-Pacific region, in particular Japan, Korea and China (Morita et al. 2003). In Korea, Miyok [*Undaria pinnatifida* (Harvey) Suringar] cultivation was largely developed, promoted and industrialized in the 1970s. Production continues to increase and approximately 305,672 tonnes wet weight was harvested in 2006 (MOMAF 2007). The need to increase output to meet demand for this seaweed has led to a substantial increase in the area being used for *U. pinnatifida* farming in Korea. *Undaria pinnatifida* production in Korea now accounts for 48% of total Korean seaweed production. In the last two decades, *U. pinnatifida* farms have suffered greatly from pinhole disease, which is characterized by the appearance of numerous small holes in the thallus of the alga. A thallus with pinhole disease is considered to be unsuitable for human consumption and therefore cannot command a good price in the market. *Undaria pinnatifida* affected with pinhole disease is, instead, used as feed for abalone.

Biotic pathogens known to infect macroalgae include other algae, bacteria, fungi, mycoplasma-like organisms, and viruses (Andrews 1976; 1979; Correa and Craigie 1991). Tseng (1981) recognized that algal diseases may be induced physiologically, by pathogenic organisms, and by pests such as grazing animals.

The causative agent of pinhole disease was first reported in Asia by Torii and Yammoto (1975) as thalestried copepods that were first identified from *U. pinnatifida* farms in Hakodate, Hokkaido, Japan. A similar infestation and occurrence of the disease was later noted by Kang (1981) from the *U. pinnatifida* beds along the southeast coast of Korea. Ho and Hong (1988) and Park et al. (1990)

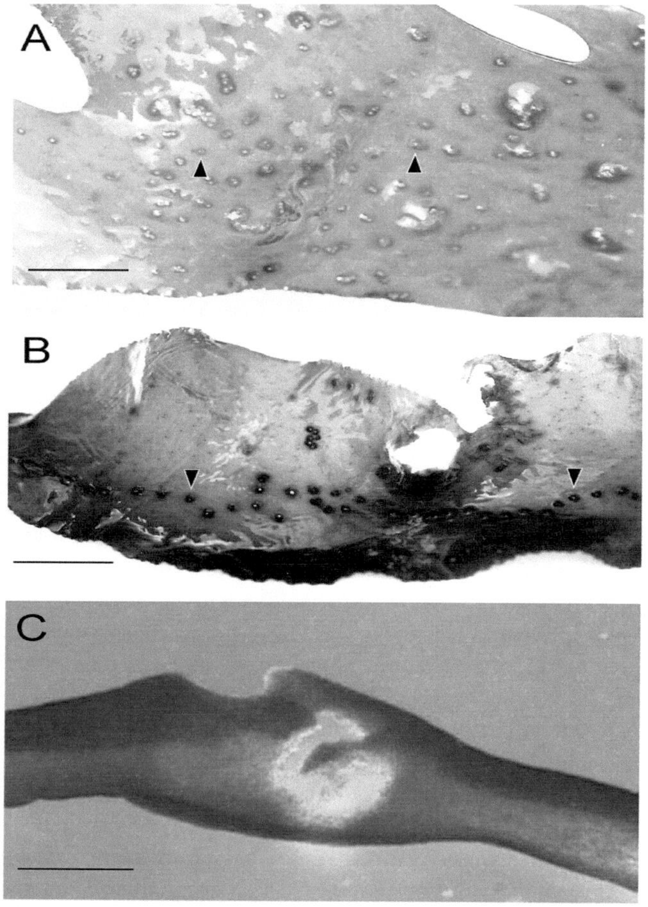

Fig. 1 Pinhole disease in sporophytes of *Undaria pinnatifida* (Harvey) Suringar. **a** Pinhole disease scattered irregularly (*arrowheads*). **b** Pinhole disease in a row (*arrowheads*). **c** Cross-section view of pinhole diseased frond. *Bars* **a, b** 1 cm; **c** 500 µm

both reported that the causative agent of pinhole disease was the harpacticoid copepod *Amenophia orientalis* and *Parathalestris infestus*. Torii and Yamamoto (1975) and Kang (1981) reported that infestation by *Amenophia* was greatest on farmed *U. pinnatifida* at the end of the cultivation period. Ho and Hong (1988) observed that the algal cells on the periphery of the excavated pinholes contained more brown pigments than normal cells in uninfested parts of the plants.

In order to reduce the economic impact of the disease, framers already harvest *U. pinnatifida* before it has become heavily infected with the disease. The objective of this study is to describe the occurrence of pinhole disease during the different developmental stages of *U. pinnatifida* on a seaweed farm. The results obtained from this study will enable farmers to better understand the optimum time for harvesting thalli prior to an outbreak of pinhole disease in *U. pinnatifida* cultivation operations.

Materials and methods

The research was conducted on an *Undaria pinnatifida* farm at Wando on the southwestern coast of Korea (34° 18.0′N, 126°45.6′E). This region of Korea is the most popular area for *U. pinnatifida* cultivation, but also includes farms for other commercially important algae, such as *Porphyra* spp., *Laminaria* spp., *Hizikia fusiformis*, *Sargassum fulvellum*, *Enteromorpha* spp. and *Capsosiphon fulvescens*.

Environmental factors such as water temperature and salinity were measured by a logging multi-parameter probe (YSI-85, YSI, Yellow Springs, USA), and current velocity at the study site was measured by a flow meter (BFM001, Valeport Lim, UK). Water temperature at the farm site decreased from 10°C in early December, reaching a minimum of 7°C during the middle of February, increasing again to 13°C in early April as solar radiation warmed the water. Salinity at 1 m depth at the research site ranged from 30 to 32 ppt. Variation in salinity was due to the input of fresh water from nearby rivers and land run-off. The mean current velocity at the site was 0.5 $m.s^{-1}$.

Cultivation of *Undaria*

In November 2002, young (5 mm in length, 15 days post nursery culture) *U. pinnatifida* sporophytes were trans-

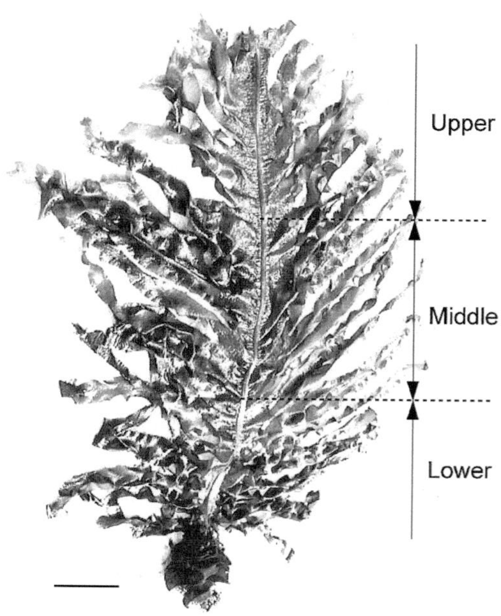

Fig. 2 Sporophyte of *U. pinnatifida* indicating the regions that the frond was divided into for analysis of pinhole disease occurrence (*upper*, *middle*, and *lower*). *Bar* 10 cm

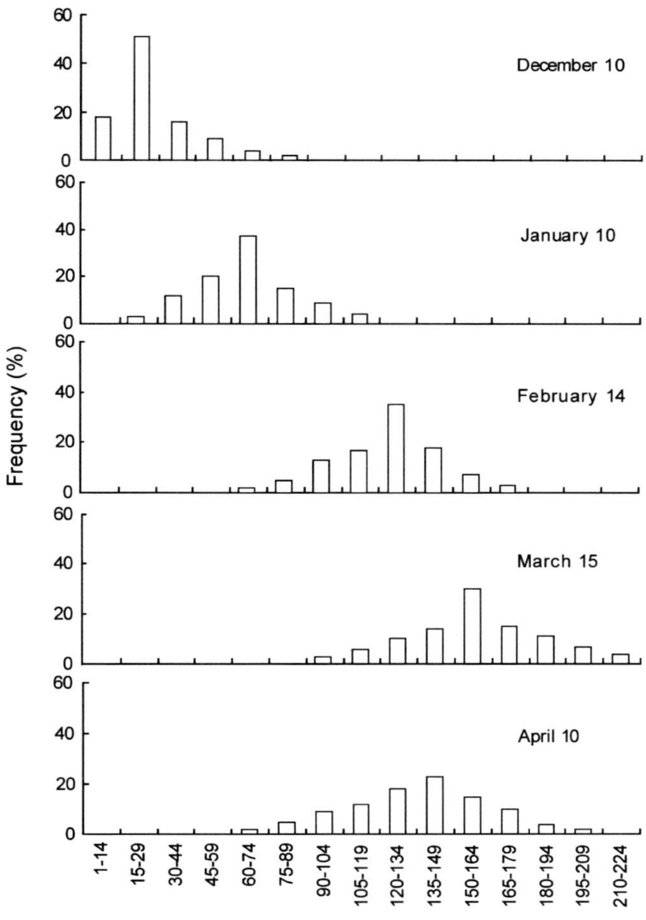

Fig. 3 Length frequency of *U. pinnatifida* sampled on 10 December 2002, and 10 January, 14 February, 15 March, and 10 April 2003

planted onto 100 m culture rope at an approximate density of 300–350 sporophytes per meter. On 15 November 2002 this rope was suspended at 1 m depth at the experimental site. On 5 December 2002 the rope was thinned to obtain a density of about 50 individuals per meter.

Algal samples

Each month between 10 December 2002 and 10 April 2003, samples of whole sporophytes of *U. pinnatifida* were collected from the experimental site. This encompassed both *Undaria*'s rapid growth phase and senescence. Each month 5 m culture rope containing about 250 individual plants was sampled. Total length (from apex to holdfast) was measured and the number of pinholes on each of the sporophytes was counted. Sporophytes were assigned to one of two categories (mature or young) depending on site and reproduction stage. Mature sporophytes were greater than 150 cm in length and possessed reproductive sporophylls. Young sporophytes were those plants less than 150 cm in length and having no reproductive sporophylls.

Pinhole disease

Examples of pinhole disease in *U. pinnatifida* from the experimental site are shown in Fig. 1a–c. The infection rate of pinhole disease on sporophytes of *U. pinnatifida* was measured as the percentage of infected thalli as follows: (number of thalli with pinhole × 100)/total inspected thalli ($n=250$). The difference in the infection rate on the upper, middle and lower parts of each thallus (Fig. 2) was quantified for each of the infected thalli. The number of pinholes per square centimeter of thallus was calculated based on the mean number of occurrence of pinhole disease

Table 1 Monthly infection rates and average number of pinholes per square centimeter on *Undaria pinnatifida* (Harvey) Suringar fronds sampled monthly between December 2002 to April 2003

Date	Infection rate of pinhole disease (%)	%	Infection in each section of the plant (%)	No. of pinhole per cm^2 of frond
10 December 2002	5±1.2	U	100	0.33±0.16 a
		M	0	
		L	0	
10 January 2003	10±1.4	U	100	1.66±0.66 a
		M	0	
		L	0	
14 February	30±3.2	U	95±3.6	3.66±1.33 a
		M	5±2.8	
		L	0	
15 March	80±3.8	U	86±1.4	14.0±2.33 b
		M	11±4.2	
		L	3±2.6	
10 April	95±4.5	U	79±1.2	18.33±3.33 c
		M	16±4.6	
		L	5±3.8	

Values are mean ± SE, $n=250$
Values in the same column with different lower case letters are significantly different ($P<0.01$)
Plant sections: *U* upper part of frond, *M* middle part of frond, *L* lower part of frond

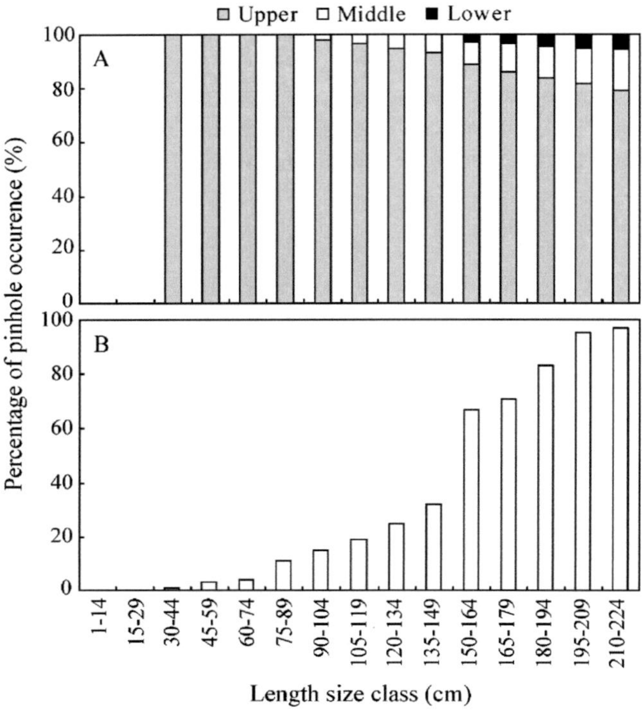

Fig. 4 Length frequency and partial difference of pinhole disease of *U. pinnatifida*. **a** Partial infection rate of pinhole disease from upper, middle and lower part of fronds. **b** Infection rate of pinhole disease from length size class of fronds

from 50 portions of 1 cm^2 randomly selected from each sporophyte (n=250).

Data analyses

The infection rate of pinhole disease was expressed as a percentage of plants exhibiting pinholes in each of five replicates of 50 plants. One-way ANOVA was used to test for significant differences in the recorded density of pinholes per square centimeter in each part on the kelp thallus among sampling date. When an ANOVA revealed a significant differences (P<0.05), an a posteriori Fisher's LSD test was applied to determine significant difference among pairwise months.

Results

The mean total length of *U. pinnatifida* at the experimental site increased linearly from 27 cm on 10 December to 160 cm on 15 March, and then remained relatively constant for the rest of April. Length frequency analysis of *U. pinnatifida* (Fig. 3) indicated the mean lengths of the population sample increased from 27±6 cm in December to 160±51 cm in March. In April, the mean length of the population was 145±47 cm, a decrease compared to the March sample. Sporophyll formation was first observed to occur in March (11% of plants) and gradually increased to a maximum (98% of plants) in April. A few young non-reproductive sporophytes were observed (2% of plants) in April and the mature reproductive plants exhibited increasing erosion in April.

Pinhole disease

Pinholes were first observed on the fronds in December 2002 and their occurrence persisted throughout the experimental period. In the early stage of pinhole disease, small brown spots appear on the thallus blade near the midrib, and these spots and holes subsequently spread to the other pinnate blades. Pinholes typically ranged from 0.5 to 1.5 mm in length and appeared to be of two types, present either as localized irregularities (Fig. 1a) or in a row (Fig. 1b). The algal cells on the periphery of the excavated pinholes presented a considerably darker brown color than the normal cells in the uninfested parts of the thallus. Tissue surrounding the pinholes were often swollen and raised (Fig. 1c). The number of pinholes observed on each frond, and the percentage of fronds with pinhole disease increased as *U. pinnatifida* thalli aged, with the infection rate of the fronds increasing from 5% of individuals in December to 95% at the end of April (Table 1). The average number of pinholes per square centimeter of thallus ranged from 0.33 in December 2002 to 18.33 in April 2003 (Table 1, one-way ANOVA, P<0.01). This result indicates that the incidence and severity of pinhole disease of *Undaria* were significantly higher in the late culture period than during the early culture period.

Pinholes occurred most frequently on the upper and middle parts of the fronds (Fig. 4a, b). In December, pinhole

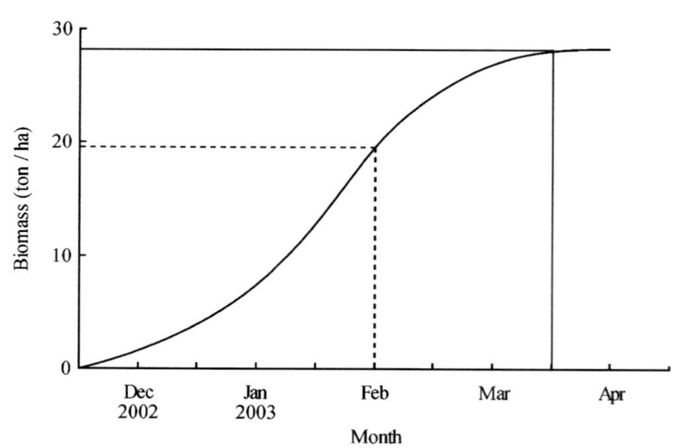

Fig. 5 Estimated biomass of *U. pinnatifida* during a culture period. *Solid line* Normal harvest amount, *dotted line* early harvest due to the pinhole disease occurrence

disease was found only on the upper part of the thallus, but by February the disease was present in the middle part of the thallus in 5% of diseased plants. By April the disease had also spread to the lower part of the frond in 5% of the diseased plants (Table 1). Pinholes were not observed on the midrib, stipe, sporophyll, or holdfast parts of fronds of any plants.

Discussion

Since the first report on harpacticoid copepods infesting *Undaria pinnatifida* cultivation areas in Korea by Kang (1981), studies have focused on the activity of the copepod rather than the direct impact on algal production (Ho and Hong 1988; Park et al. 1990),. Traditionally, the cultivation period for *U. pinnatifida* spanned the period from October to April, with harvest of thalli occurring between December and April. Recently, the harvest season has been shortened from December to February, so as to avoid the period when the most severe infestation of copepods and the resulting infection with pinhole disease occurs.

Our growth trials indicated that the maximum growth of *U. pinnatifida*, in terms of increase in biomass, occurs during the period between March and April and, therefore, reducing the harvest period leads to decreased production from farms. We estimate that the potential loss in harvest biomass resulting from harvesting plants early to avoid pinhole disease is one-third of total biomass (ca. 5–8 tons wet weight per hectare) in the region (Fig. 5).

The maximum infection of pinhole disease on a frond of *U. pinnatifida* occurs at the same time that the frond tissue is becoming reproductive, softening and starting to decay. Yoshikawa et al. (2001) and Kumura et al. (2006) reported that initiation of erosion coinciding with sporophyll formation indicates a close link between erosion and maturation of *U. pinnatifida* (Kurogi and Akiyama 1957; Nishikawa 1967). In addition, the increase in pinhole disease of *U. pinnatifida* thalli in April has also been attributed to the increasing water temperature (Rho et al. 1993).

The infection rate of pinhole disease was significantly lower at the middle and lower part of the thalli than in the upper part. Holes at the distal part of the fronds enlarge and coalesce as the disease progresses. This process accelerates decay at the distal ends of the fronds and, if the holes become sufficiently numerous, the weakened frond will fragment with concomitant loss of biomass. As the thalli age, the disease progresses from the tip of the plant toward the basal section. When the middle part of the thallus was further divided into three subparts; midrib, between basal top of the incision and midrib, and pinnate blade, the highest occurrence of pinholes was observed in the section between the basal top of the incision and the midrib. Also, fewer pinholes were observed at the pinnate blade and midrib. On the pinnate blade, more regular and larger pinholes were observed. The reason for these holes is unclear but they could be attributed to infection by another species of copepod.

The abnormalities observed in the algal cells are similar to those reported by Rho et al. (1993) and are attributable to the host plant's reaction to the mining actions of copepods. The pinholes on the thallus of the *U. pinnatifida* were observed to be used by *A. orientalis* as a place in which to copulate and hatch their young. Differences in occurrence of the disease on different parts of the frond probably reflect the plant's morphology. *U. pinnatifida* has a thin and soft pinnate blade, and thick and hard midrib. It is probable that the copepods prefer to bore into the softer parts of the plant. Although the initial damage to the fronds was caused by copepods, once the damage had been caused, several other biotic agents may combine to cause or accelerate decay of the frond. The fate of diseased *U. pinnatifida* fronds therefore probably depends on environmental (abiotic) as well as biotic factors.

We suggest that mechanical aging of *U. pinnatifida* frond tissue plays an important role in making the plant more susceptible to pinhole disease.

Acknowledgments This research was supported by a grant (E10600107A060000140) from the "Greenhouse gas removal by seaweed" Program funded by the Ministry of Maritime Affairs and Fisheries of Korean government and the National Fisheries Research and Development Institute (RP-2008-AQ-009). We thank Dr. Philip Heath (NIWA, New Zealand) for reviewing the English.

References

Andrews JH (1976) The pathology of marine algae. Bio Review 51:211–253

Andrews JH (1979) Pathology of seaweeds: current status and future prospects. Experientia 35:429–450

Correa JA, Craigie JS (1991) Algal pathology. In: Reina GG, Pedersen M (eds) Seaweed cellular biotechnology, physiology and intensive cultivation. Proceedings COST-48 (Subgroup 1) Workshop. Universidad De Las Palmas De Gran Canaria, pp 67–82

Ho JS, Hong JS (1988) Harpacticoid copepods (Thalestridae) infesting the cultivated Wakame (brown alga, *Undaria pinnatifida*) in Korea. J Nat His 22:1623–1637

Kang JW (1981) Some seaweed disease occurred at seaweed farms along the south-eastern coast of Korea. Bull Korean Fish Soc 14:165–170

Kumura T, Yasui H, Mizuta H (2006) Nutrient requirement for zoospore formation in two alariaceous plants *Undaria pinnatifida* (Harvey) Suringar and *Alaria crassifolia* Kjellman (Phaeophyceae: Laminariales). Fish Sci 72:860–869

Kurogi M, Akiyama K (1957) Studies of ecology and culture of *Undaria pinnatifida* (Sur.) Hariot. Bull Tohoju Reg Fish Res Lab 10:95–117

MOMAF (2007) Statistical year book of Ministry of Maritime Affairs and Fisheries. Ministry of Maritime Affairs and Fisheries, Korea. http://badarosi.momaf.go.kr. Cited 3 January 2007

Morita T, Jurashima A, Maegawa M (2003) Temperature requirements for the growth of young sporophytes of *Undaria pinnatifida* and *Undaria undarioides* (Laminariales, Phaeophyceae). Phycol Res 51:266–270

Nishikawa H (1967) Studies on the cultivation of *Undaria pinnatifida*: IV. Focused on growth and erosion (a prediction). Aquaculture 14:197–203

Park TS, Rho YG, Gong YG, Lee DY (1990) A harpacticoid copepod parasitic in the cultivated brown alga *Undaria pinnatifida* in Korea. Bull Korean Fish Soc 23:439–442

Rho YG, Gong YG, Lee DY, Cho YC, Jang JW (1993) On the parasitic copepod (Harpacticoida) in the cultivated brown alga, *Undaria pinnatifida* (Harvey) Suringar. Bull Nat Fish Res Dev Agency 47:197–210

Torii S, Yamamoto H (1975) *Thalestris* sp. a parasitic Copepoda on *Undaria pinnatifida* (a preliminary report). Hokusuishi Geppo 32: 23–34

Tseng CK (1981) Commercial cultivation. In: Lobban CS, Wynne MJ (eds) The biology of seaweeds. University of California Press, Berkley, pp 680–725

Yoshikawa T, Takeuchi I, Furuya K (2001) Active erosion of *Undaria pinnatifida* Suringar (Laminariales, Phaeophyceae) mass-cultured in Otsuchi Bay in northeastern Japan. J Exp Mar Biol Ecol 266: 51–65

The potential for seaweed resource development in subarctic Canada; Nunavik, Ungava Bay

G. Sharp · M. Allard · A. Lewis · R. Semple · G. Rochefort

Originally published in the Journal of Applied Phycology, Vol 20, No 5, 41–48.
DOI: 10.1007/s10811-008-9323-7

Abstract Ungava Bay is ice covered 6–7 months of the year and evidence of ice scouring of seaweeds is extensive in the intertidal and shallow subtidal. Maximum tidal amplitudes of 16 m, among the highest in Canadian waters, compound this impact. Despite this level of annual perturbation, very extensive and dense beds of fucoids in the intertidal and laminarians in the subtidal are common on the western shores of Ungava Bay. Ground surveys of 24 intertidal stations combined with satellite images delineated 82,000 tons standing crop of *Fucus vesiculosus* and *Fucus evanescens* in Payne Bay, of which 36,000 tons were considered harvestable. Subtidally, kelp cover reached peak biomass at 5–10 m consisting of three primary species, *Saccharina longicruris*, *Laminaria digitata*, and *Laminaria solidungula*. In the area of Payne Bay, kelp beds of 100 ha were common, averaging 9–12 kg m^{-2} wet weight. The productivity of brown algae at these latitudes has been assumed to be low relative to southern latitudes. Direct measurement of lineal growth indicates productivity is intermediate between arctic and temperate populations. The potential for medium level industrial harvest exists under conservative management strategies within the constraints of subarctic logistics.

G. Sharp (✉) · R. Semple
Fisheries and Oceans Canada,
1 Challenger Dr,
Dartmouth, NS, Canada B3Y 2A4
e-mail: sharpg@mar.dfo-mpo.gc.ca

M. Allard · A. Lewis · G. Rochefort
Nunavik BioScience,
1111 Dr. Frederik-Philips Blvd,
St. Laurent, QC, Canada H4M 2X6

Keywords Kelp · *Laminaria* · *Saccharina* · *Fucus* · Abundance · Harvest

Introduction

The Inuit communities of the Canadian north and more specifically Nunavik have made land claim agreements over the past 30 years that have provided more control over natural resources and a limited co-management role since the 1980s (JBNQA 1976). In 2006, an offshore resource claim was negotiated, and an agreement will be implemented for the coastal waters and islands of the near shore Nunavik region in 2007. As responsible stewards, the Inuit have initiated research to delineate, assess and evaluate these resources. Early marine research noted that seaweed was more common and abundant than first assumed in this region.

A number of macrophytes have been a part of the traditional food sources in the arctic and subarctic for the Inuit (Wein et al. 1996). These resources of the arctic and subarctic are now subject to the symptoms of global warming (Moritz et al. 2002). Although there is knowledge of algal diversity and distribution, we lack quantitative abundance and productivity information for this region (Wilce 1959; Lee 1980). The first priority has been to quantify those resources within a normal travel radius of isolated villages. These are also the resources that have the greatest potential for further management and possible development at a scale related to people and infrastructure available in a village. This study was designed to define the abundance, diversity, and distribution of dominant intertidal and subtidal brown macrophytes in the region of Payne Bay accessible to the Inuit village of Kangirsuk in Ungava Bay.

Materials and methods

Ungava Bay, located in the Nunavik region of northern Quebec, begins below the Arctic Circle at 58°20′N and extends to 61°00′N at the Hudson Straits. The bay and its inlets are covered by shore-fast ice from early December until open water leads develop in May, and by July open water predominates. The day length in mid-July is 18.4 h and in January it is 5.7 h. Water temperature ranges from −1.8°C in winter to a maximum of 5–8°C in summer in sheltered inlets. The bay is noted for its tidal amplitude ranging from an average of 10 m to extremes of 16 m creating currents of 5 to 16 km h^{-1}. During the period of ice break-up, the ice scouring of the intertidal extends down to −5 m (chart datum). Payne Bay is a 30-km inlet midway northward on the western side of Ungava Bay that receives the water of two rivers. The foreshore of the Payne Bay consists of bedrock ledges and mixed sand gravel flats with glacial till (Fulton 1995). Boulder foreshores are characterized by clusters of boulders of 0.3– 1.5 m diameter interdispersed with gravel or sand patches on gently sloping shores. Subtidally, the foreshore geology changes to sediment dominated bottoms at depths below 20 m.

Subtidal

Eleven diving stations were chosen to represent a range of environments from inner Payne Bay to the open waters of Ungava Bay. Diving safety issues dictated that each diving station had to be chosen in regard to currents and tide level. Transects started at 3 m below chart datum on a compass bearing perpendicular to the shoreline. Divers made qualitative observations of species, including an estimate of relative abundance of individuals (scale low <1 m^{-2}, moderate 1–5 m^{-2}, high >5 m^{-2}). The presence of other algae and bottom types were recorded at 3-m-depth intervals. Five 0.25 m^2 samples were taken at four stations at five depth intervals. All kelps were identified and total length, stipe length, blade width measure to 1 cm and wet weight to 1 g. GPS (Garmin™ position accuracy 15 m) was used to locate the transect beginning and end points.

At 11 stations a remote drop camera (Shark) was lowered in the shallow portion (<1–3 m) of kelp beds and the vessel motored slowly towards deeper water stopping at 25 m. The camera was held approximately 1 m off the bottom and the area observed was approximately 1 m^2. Low visibility forced us to move the camera lower, thus at times underestimating density. Observations of species and relative densities of individual kelp plants within the view of the camera (<1 m^{-2} low density, 1–5 m^{-2} moderate density, >5 m^{-2} high density) were made at three to five depth intervals ranging from −5 to −15 m. Since there was no recording of the image, an estimate of plant density and identity had to be made in real time while viewing a video display. At densities higher than 5 individuals m^{-2}, plant cover obscured any estimates of maximum density. The camera position was recorded with GPS and depth with a Raymarine™ digital sounder to 1 m.

Growth of kelps

At 8 m depth in a *Saccharina longicruris* (Bachelot de la Pylaie) Kuntze dominated ledge, a 10×10 m plot was marked with lead line. Within the plot, 40 *S. longicruris* plants were tagged from 30 cm to 2.0 m long and punched with a 2-cm hole 20 cm from the base of the blade on 10 August 2005. A 2 h search of the plot was conducted on 8 August 2006 for tagged plants. Recovered plants were removed, and stipe length, blade length to blade constriction, total blade length and distance from blade base to the punched hole was measured to the nearest cm.

Intertidal

The survey area was determined by the radius of daytime operation for standard freighter canoes, 10 km from the village of Kangirsuk (Fig. 1). These shorelines were divided into lines of longitude 30 s apart from 70°08′00″ N to 69°52′00″N on a hydrographic map. The north and south shores of Payne Bay each had a total of 66 points of intersection with the lines of latitude. The intersections with the shore were numbered and 30 sites were selected by random numbers for ground survey. These positions were entered as waypoints in a Garmin GPS. The survey began on 1 August 2005 and 24 of the 30 sites were completed by 15 August 2005 (Fig. 1).

The survey crews landed as close to the GPS coordinate as possible and began their survey either at the top of the intertidal or bottom depending on the state of the tide. Three temporary transects were established at each station; at the top of the intertidal zone (+8 to +9 m chart datum) in the highest area of continuous algal cover, midway (+5 to +6 m) in the zone and just above the low tide mark (0 to +2 m). Each transect was a 30 m line placed parallel to the shore. Along this line at 5 m intervals, a 0.25 m^2 quadrat was placed at the mid-point of the mark. All algal biomass was removed from the quadrat with a knife to a height of 2 cm. The biomass was separated into two fucoid species, *Fucus evanescens* C.Ag. and *F. vesiculosus* L. and the wet weight of each species was measured to 50 g. The type of substrate was classified into five sedimentry grades: sand (0.05–2 mm), gravel (2–64 mm), cobble (64–100 mm), rock (100–500 mm), boulder (>500 mm) or a mixture of these grades.

Payne Bay was subdivided into seven distinct areas defined by wave exposure and resource accessibility

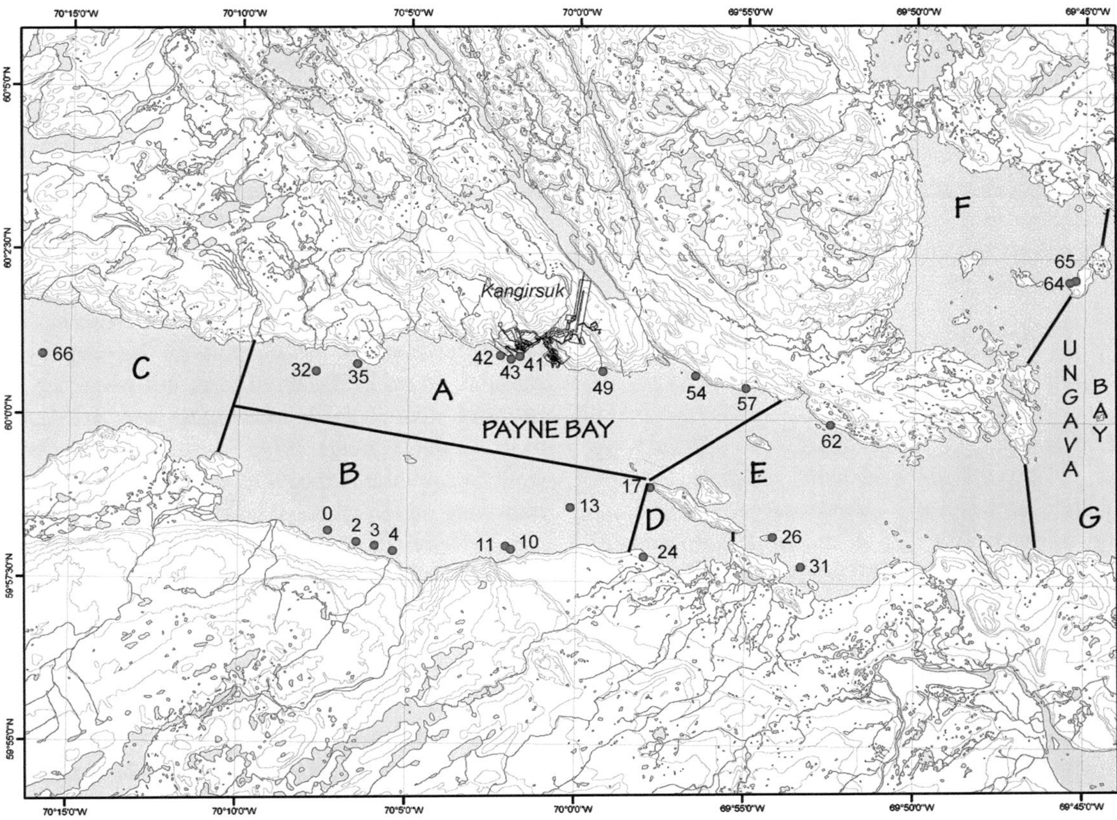

Fig. 1 Location of intertidal survey stations (•) in Payne Bay and Ungava Bay, Quebec, Canada. Subdivisions of the fucoid resources *A–G*

(Fig. 1). The index of wave exposure was a function of fetch (distance waves travel unimpeded) and open angle (the angle that waves travel over the fetch). Resource accessibility was a function of distance from the village of Kangirsuk, wave exposure, the necessity of an open water crossing by local harvest vessels and the existence of barrier reefs.

To calibrate the Landsat image, discrete areas of fucoid cover were outlined by a Trimble GeoXM (accuracy 1–2 m) with ArchPad software by walking their perimeter. These areas were converted from a line point shapefile to a line shapefile, and lines were formed from consecutive points and converted to polygons. The perimeters that represented fucoid cover were used in ERDAS Imagine remote sensing software as training areas for an unsupervised classification. The software identified areas in the Landsat image similar to those entered as training areas and calculated the total area of fucoid cover. The Landsat image resolution of 30 m provided an underestimate of true coverage since smaller fucoid stands would not have been identified.

Based on the analysis of the Landsat image the total standing and harvestable crop of fucoids was calculated for four of the seven harvest areas. These were the harvest areas closest to Kangirsuk with sheltered to semi-sheltered shores and where there was sufficient ground truthing data. To obtain the standing crop of fucoids for each harvest area, the fucoid cover was divided into three equal areas based on the three tidal strata: upper, middle and bottom. The average biomass for each species from all stations in the harvest area for each tide strata was multiplied by the total area of the strata. The total biomass of these three strata for each area was discounted according to accessibility. Discounts of the total wet biomass for each area were based on the proportion of harvest days during the summer when it would be safe to cross open water to access the harvest area. The low access areas distant from the village with high wave exposure had the harvestable biomass discounted by 75%, moderate access and low wave exposure areas were discounted by 50%, and areas of high accessibility and low wave exposure resources were not discounted. An exploitation rate of 10% was chosen as a precautionary level based on the annual rates for fucoid species from the Canadian Maritimes between 17 and 25% annually (Ugarte and Sharp 2001).

Growth of fucoids

In the mid-intertidal, six plots of 0.5×0.5 m were permanently marked with pins for tagging of individual *F.*

evanescens. Fully intact fronds were chosen that were in good condition. Fifty-six fronds were tagged from 3.0 cm to 30.0 cm long evenly distributed in the plots. A monofilament tag was used with a number on embossed Dymo™ labeling tape. The tag was held around the base of the frond with a slider made of neoprene rubber. The total length of each plant from base to the distal tip and the total number of dichotomous branches were recorded to 0.1 cm and 1 cm, respectively.

Results

On every dive and drop camera transect, kelp species were observed to −15 m if there was solid substrate. A high density of kelp plants (>5 m^{-2}) was observed on all but 2 of 11 diving transects (Fig. 2). A moderate kelp density (1–5 m^{-2}) was observed in 6 of 10 camera transects, while 2 had low kelp densities (Fig. 2).

In the shallow (<3 m) part of the subtidal, there were obvious signs of ice scouring at every station. This took the form of a cover of filamentous brown and green algal species and very young of the year kelp plants (Fig. 3). *Alaria esculenta* (L) Greville was common in this zone as small plants, but this species also extended down to 3–6 m in wave-exposed environments. Below this ice-impacted zone there were perennial stands of kelp. The most prolific species was *S. longicruris* particularly in the more sheltered and shallow portions of the beds, for example, Basking Island (Fig. 3). At locations like Pikyuluk Island, with very strong (5 km h^{-1}) currents, *Laminaria digitata* (Hudson) J.V. Lamouroux was the predominant species in the 3- to 6-m-depth range. *Laminaria solidungula* J. Ag was a common species in depths of 6–12 m. *Agarum clathratum* Dumortier was also associated with the deeper parts of the zone, at 10 m in sheltered sites but occurring at 3 m in exposed sites. Total kelp biomass was greatest in the 5- to 10-m-depth range (Fig. 3). Below 12 m, kelp species were in low abundance and a community of small tufted reds were mixed amongst sponges. *Saccharina groenlandica* (Rosenvinge) Lane, Mayes, Druehl & Saunders was only present in deeper water on the outer coast, at Tuvalik Point (Fig. 3).

The four kelp sampling stations represent the range of diversity and abundance of kelp species in the entire survey area (Table 1). The outer coast is represented by Tuvalik Point, where there was a mix of five kelp species. *S. longicruris* and *L. solidungula* together were 43% of the

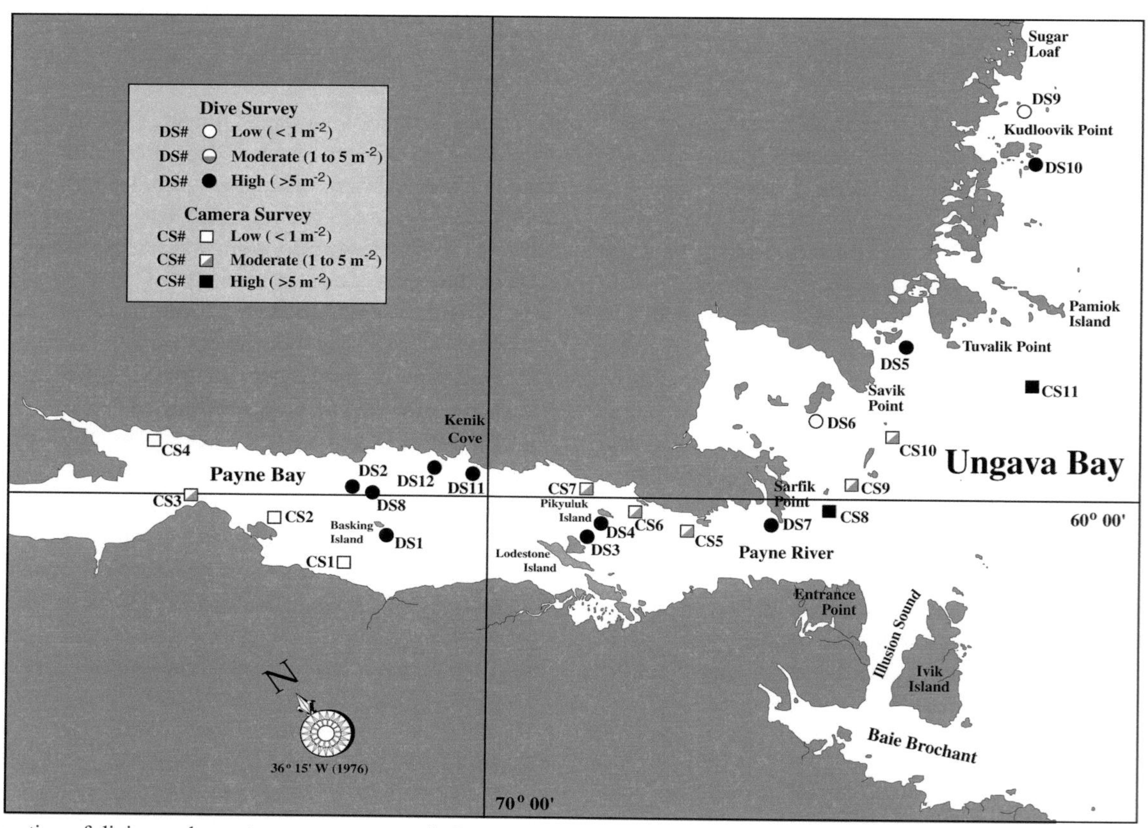

Fig. 2 Location of diving and remote camera transects in Payne Bay and Ungava Bay, Quebec, Canada, and relative densities of Laminariales recorded on each transect

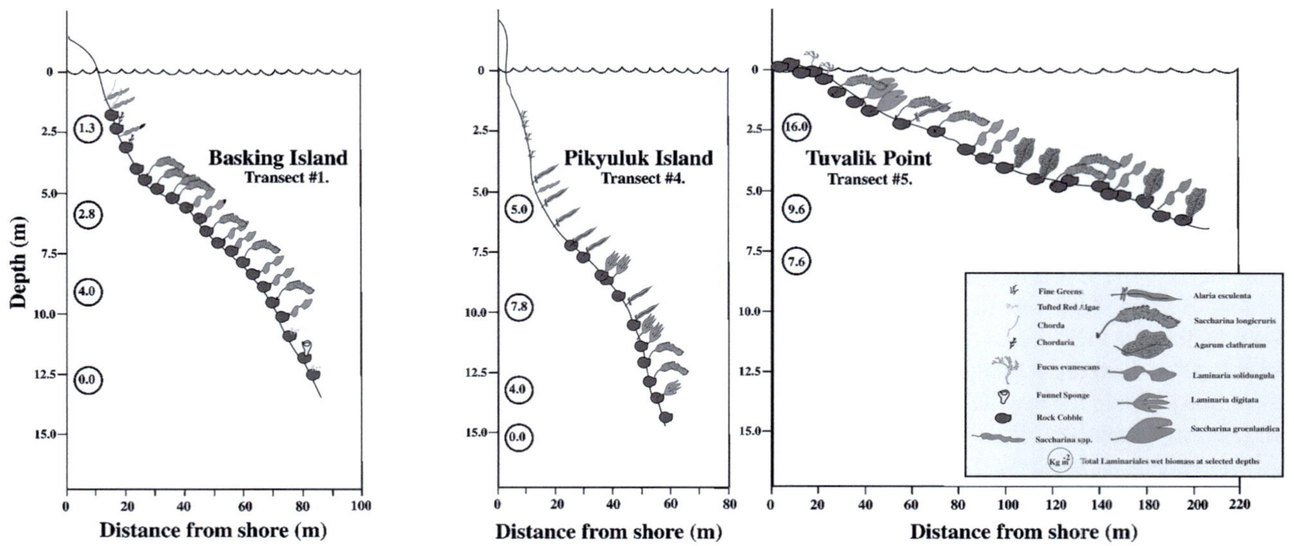

Fig. 3 Laminariales distribution in Payne Bay at three sites on depth profiles, and wet biomass kg m^{-2} (○) at selected depths

biomass (Table 1). The overall biomass was the second lowest of the sites but samples containing *S. longicruris* reached a maximum 16.7 kg m^{-2} and *L. solidungula* 10.4 kg m^{-2}. *Laminaria digitata* dominated the biomass at Pikyuluk Is., a site with very high current speeds (>10 km h^{-1}) (Canadian Hydrographic Service chart). The Kangirsuk shore site was a semi-sheltered ledge dominated by *S. longicruris* attaining the highest single sample weight of all stations, 19.0 kg m^{-2}, and highest mean biomass (Table 1). Basking Island was a very steeply sloping site and had the lowest biomass of all sites with a 50/50 mix of *S. longicruris* and *L. solidungula* (Table 1).

The total length of *S. longicruris* averaged 225±55 cm to a maximum of 900 cm. *L. digitata* was smaller at 88±29 cm as was *L. solidungula* at 121±42 cm. *Alaria esculenta* averaged 115±63 cm in the shallow ice scoured zone, it is usually less than 100 cm long, but below this depth it attained 300 cm. *Agarum clathratum* rarely reached 100 cm total length.

Growth of kelp

Thirteen of the original 40 tagged *S. longicruris* plants were recovered. The total length averaged 425±23 cm, the punched holes moved an average of 323±16.5 cm in one year (.9±.3 cm d^{-1}). The average distance to the blade constriction was 201±9 cm in the recovered plants.

Intertidal

In Payne Bay, *F. evanescens* averaged 1.47±.12 wet kg m^{-2} and *F. vesiculosus* averaged .28±.02 wet kg m^{-2} (n=460). The frequency distribution of biomass density of both species was skewed to the first quartile of distribution, the value 0. The greatest biomass of *F. evanescens* occurred on rock substrate reaching a maximum biomass of 13.1 kg m^{-2}. The second highest mean biomass occurred on boulders (Fig. 4). Boulders were ice scoured on the top and had 100% fucoid cover on their sides. Rocks were frequently protected

Table 1 Density and biomass (± SE) of kelp species at four sites in Western Ungava Bay

Station	Basking Is	Kangirsuk	Pikyuluk Is	Tuvalik Pt
Density m^{-2}	24±2	12±1	39±3	24±2
biomass g m^{-2}	2.9±.2	11.8±1.3	9.2±2.0	8.4±1.1
% species composition	*Sl* 53	*Sl* 87.5	*Ld* 75	*Ac* 40, *Ls* 35
	Ls 47	*Ls* 12.5	*Ae* 23	*Sl* 8, *Sg* 7
			Sl 2	*Ae* 5

Saccharina longicruris (Sl), Laminaria solidungula (Ls) Laminaria digitata (Ld) Saccharina groenlandica (Lg), Alaria esculenta (Ae) and *Agarum clathratum (Ac)*

from ice scour by the larger boulders in mixed substrates. While *F. evanescens* was abundant on boulder substrate it was absent on bedrock (Fig. 4). Bedrock was normally at the upper part of the intertidal where maximum *F. vesiculosus* biomass was 8.4 kg m^{-2}. Any area with a cobble substrate had very low fucoid biomass (Fig. 4). *F. vesiculosus* dominated the upper intertidal transects at each station while *F. evanescens* was the predominant component of the biomass in the lower station area (Fig. 5). Fucus biomass in the middle region of the zone was a mix of the two species (Fig. 5).

Most of Payne Bay was in the wave exposure catagory of semi-sheltered to sheltered (Table 2). However, based on distance from Kangirsuk, only Area A had high resource accessibility (Table 2). Area D was wave sheltered, but had two access problems: distance across open water and outlying barrier reefs exposed at low tide (Table 2). Areas E to G were over 10 km from Kangirsuk and required travel to the outer area of Ungava Bay.

Area A closest to the village of Kangirsuk was not discounted for access, but still had the lowest harvestable biomass of all areas due to lower fucoid cover (Table 3). Moderately accessible shoreline, Area B, discounted 50% for access, was still 3–4 times the annual harvestable biomass of the high access area. Area C and D, with low access due to travel distance and wave exposure had the standing crop discounted by 75% and also exceeded the high access area A in harvestable crop (Table 3). Total fucoid biomass in Payne Bay was 82,801 wet t 70% *F. evanescens* (Table 3). Following discounting for access, the total was 35,920 wet t harvestable biomass of fucoids.

Growth of fucoids

Thirty-one of the original 56 *F. evanescens* tags were recovered and measured. The mean length increment was

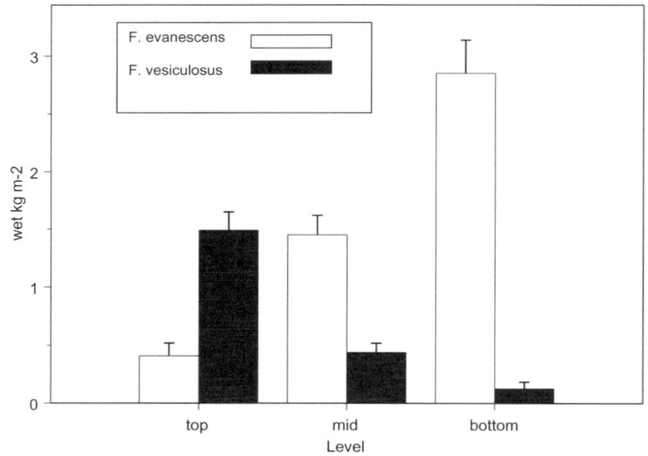

Fig. 5 Fucoid species biomass at three tide levels on transects in the intertidal of Payne Bay and environs, top (8–9 m), mid (5–6 m), bottom (1–2 m) (chart datum) (*bars* 1 SE)

3.6±0.7 cm and dichotomies increased by 3.0±1.0 over the year.

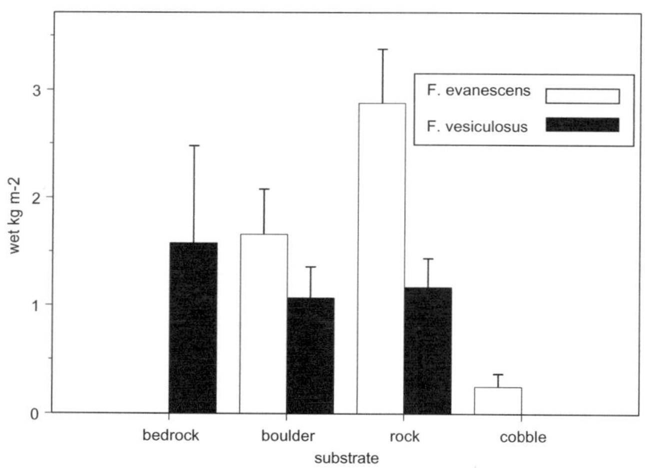

Fig. 4 Mean biomass density of fucoid species on five bottom types within 0.25 m^2 quadrats sampled from Payne Bay and environs in August 2005 (*bars* 1 SE)

Discussion

Although the fucoid resource base in Payne Bay is extensive and there is a moderately high standing crop and fucoid cover, it is very patchy. Patchiness is related directly to the presence or absence of optimal bottom type, rock and boulders. In the northern environment, the intertidal is more or less vulnerable to ice scour depending on the degree of substrate relief and the movement of ice by current and wind during ice break-up. The combination of boulders and rocks provide a high relief bottom protecting the fucoid biomass from the scouring effects of ice. Boulders scattered in piles on a gravel or sand matrix provide shelter within the piles from the pressure of moving ice. *Fucus vesiculosus* has a similar distribution in the St. Lawrence River estuary (Guichard et al. 2000). Small scale topographical heterogeneity ≤20 cm has a strong influence on relative abundance of seaweeds that are subjected to ice scour annually (Bourget et al. 1994). The mean biomass of *Fucus* spp. in the ice-scoured St. Lawrence River estuary, 1.5 kg m^{-2}, was within the range recorded in Ungava Bay (Archabault and Bourget 1996).

There may be other factors that allow fucoid species to develop significant biomass in the intertidal zone of ice scoured environments. *F. vesiculosus* in the Baltic has two strategies to survive ice scour. In sheltered waters, the plants lose buoyancy in the winter and sink below the ice level, while they also have a high vegetative regenerative capacity to allow recovery from annual ice scour even from small residual pieces of stipe and holdfast (Kiirikki and Ruuskanen 1996). The persistence of *F. vesiculosus* on

Table 2 Wave exposure (*Exp*), accessibility (*Acc*) and classification of seven potential harvest areas of the fucoid resource in Payne Bay and classification

Region	Fetch km	Open angle °	Distance to port	Open water	Barrier reefs	Classification Exp.	Acc.
A	≤10 >5	≤90	≤10	–	–	SS	HA
B	≤10 >5	≤90	≤10	+	–	SS	MA
C	≤10 >5	≤90	>10	+	–	SS	LA
D	<5	≤90	>10	+	+	S	LA
E	≤10	>90 ≤180	>10	+	–	SE	LA
F	<5	≤90	>10	+	+	S	LA
G	Unlimited	>180	>10	+	–	E	LA

bedrock substrate at the highest level of the intertidal in Payne Bay supports these observations. Severe ice scour events in temperate waters require less than 1 year for *F. vesiculosus* to recover density (McCook and Chapman 1997). Five years may be required for complete recovery of biomass after an intensive (>80%) harvest in the subarctic. Annual exploitation rates in temperate populations of fucoids are limited to 20–25% of harvestable biomass (Sharp and Semple 1997). Preliminary results of harvesting experiments and these growth measurements suggest 10% of the harvestable standing crop is a conservative limit for Payne Bay.

Robert Wilce (1959) found in a survey of seaweeds in the south and eastern parts of Ungava Bay in the mid-1950s found "immense kelp in large and more often discrete beds, the bulk of which consists of *S. longicruris*". Wilce could have been describing a general conclusion for our survey area on the western part of Ungava Bay. The list of kelp species found at five stations was very similar to our list with the exception of *L. nigripes*. *Laminaria digitata* has not been reported for this subarctic region, but has been reported for the Canadian arctic (Lee 1980).

The mean biomass of 1.2–7.5 kg m^{-2} in temperate waters is well within the range of values for Ungava Bay (Sharp and Carter 1986). At the southern limit of *S. longicruris* in Long Island Sound, USA, a maximum biomass was 24 kg m^{-2} compared to 18 kg m^{-2} in Payne Bay (Egan and Yarish 1990). *Laminaria saccharina* (=*Saccharina latissima*) populations in a high arctic Greenland fjord averaged 49 g m^{-2} (Borum et al. 2002). The maximum size of plants of *L. digitata* and *S. longicruris* was also very similar for these species from temperate waters (Sharp and Carter 1986).

The density and biomass of these populations may be similar in the subarctic and temperate waters but there is the question of annual productivity. On most dives, we observed an understory of small kelp plants that suggests recruitment is a regular if not annual event. Reduction in canopy plant density allows these shade inhibited plants to grow rapidly to replace the lost plants. There is also a high abundance of small (<1 m) Laminariales in the ice scour zone of the sub tidal suggesting that in this highly perturbed zone recruitment pulses are strong.

The limited data from Payne Bay tagging experiments indicate *S. longicruris* in Ungava Bay grows at yearly rates similar to rates recorded for this species or conspecifics in temperate waters 1.0–2.0 cm day^{-1} (Chapman and Craigie 1978; Egan and Yarish 1990; Parke 1948). In the high arctic, *L. saccharina (S. latissima)* retains old leaf blades, thus the specific growth rate in terms of growth per unit length of blade is lower than in southern populations (Borum et al. 2002). *S. longicruris* in Payne Bay only retained blades if over 200 cm and usually only a small portion of the blade, meaning that annual elongation did reflect specific growth rate in smaller plant sizes. The peak growth period for *L. solidungula* in arctic Alaska was from February to April and for *L. saccharina* (*S. latissima*) was during the ice-free period of late April to late July (Dunton 1985). At 74°N in Greenland, photosynthesis in *L. saccharina* (*S. latissima*) was able to balance and exceed respiration during the ice-covered period (Borum et al. 2002). The subarctic climate of Ungava Bay at 59°N

Table 3 Estimates of fucoid cover, total standing crop (wet weight) and harvestable standing crop discounted for access and annual harvestable tonnage of *Fucus evanescens* (*Fe*) and *Fucus vesiculosus* (*Fv*) in Payne Bay in four selected harvest areas, August 2005

Area	Discount %	Fucoid cover km^2	Total biomass wet t Fe	Total biomass wet t Fv	Harvestable wet t Fe	Fv	Annually harvestable 10% exploitation	
A	0	2.27	3,563	1,565	3,563	1,565	356	156
B	50	15.68	24,110	10,744	12,055	5,372	1,205	537
C	50	6.61	9,254	2,908	4,627	1,454	463	145
D	75	13.09	20,109	9,028	5,027	2,257	503	226
Σ		37.66	57,836	24,245	25,272	10,648	2,527	1,064

provides June to December as an ice-free growth period for *L. longicruris*. It appears the environment of Payne Bay yields standing crops and annual Laminariales growth rates intermediate between high arctic and temperate waters.

We did not present any estimates for standing crop for these species because the survey was not designed to estimate total area covered by kelp. However, we have confirmed by spot survey and aerial reconnaissance that 100s of km of the subtidal in Ungava Bay supports mature kelp beds. Recent aerial surveys of Leaf Bay, 200 km to the south of Payne Bay, indicate there is a similar distribution of kelp species.

It is unlikely that the Ungava Bay seaweed resources will follow the industrial development and exploitation path of the Canadian Maritime provinces (Ugarte and Sharp 2001). The remoteness of Ungava Bay from regular heavy transport prevents easy export of raw or processed material. The short open water period narrows the time for resource access. Similarly, the natural air drying season is short and affected by rapid changes in weather systems and mechanical methods incur very high fuel costs.

There are also important factors of logistics, culture and economics that will affect the successful utilization of these resources. A tidal range of over 10 m restricts the access time to shallow parts of the kelp resource from 1 to 2 h per day. Despite this limitation, the Inuit have continued to utilize traditional foods, and kelp species are among those consumed regularly (Wein et al. 1996). Gathering mussels and fresh seaweeds are as much a part of tradition as are gathering berries and hunting caribou. Fucoid species are accessible on all low tides by simply walking on the shore. Intertidal hand gathering of these species would be similar to shore based rockweed harvesting in Nova Scotia, Canada. The local vessel capacity and travel radius is limited by vessel size and power. It is more likely this resource will be exploited for a high unit value final product than any large industrial processing. A cottage type industry would have very little impact on the habitat and subarctic ecosystem as well as harmonize with local culture.

Acknowledgements We greatly appreciated the support and friendship of the people of Kangirsuk during this project. Thanks to Dr Wilce for his interest and good advice from his pioneering years of work in this region. Megan Wilson provided excellent assistance in the final preparation of the manuscript.

References

Archabault P, Bourget E (1996) Scales of coastal heterogeneity and benthic intertidal species richness, diversity and abundance. Mar Ecol Prog Ser 136:111–121

Borum J, Pedersen MF, Krause-Jensen D, Christensen PB, Nielsen K (2002) Biomass, photosynthesis and growth of *Laminaria saccharina* in a high-arctic fjord, NE Greenland. Mar Biol 141:11–19

Bourget E, DeGuise J, Daigle G (1994) Scales of substratum heterogeneity, structural complexity, and the early establishment of a marine epibenthic community. J Exp Mar Biol Ecol 181:31–51

Chapman ARO, Craigie JS (1978) Seasonal growth in *Laminaria longicruris*: relations with reserve carbohydrate production and storage. Mar Biol 46:209–213

Dunton KH (1985) Growth of dark exposed *Laminaria saccharina* (L.) Lamour and *Laminaria solidungula* J.Ag. (Laminariales: Phaeophyta) in the Alaskan Beaufort Sea. J Exp Mar Biol Ecol 94:181–189

Egan B, Yarish C (1990) Productivity and life history of *Laminaria longicruris* at its southern limit in the western Atlantic Ocean. Mar Ecol Prog Ser 67:263–273

Fulton RJ (1995) Surficial materials of Canada. Geological Survey of Canada, Ottawa, Ont. Map No. 1880A

Guilchard F, Bourget E, Agnard JP (2000) High-resolution remote sensing of intertidal ecosystems: A low-cost technique to link scale-dependent patterns and processes. Limnol Oceanogr 45:328–338

JBNQA (1976) James Bay and Northern Quebec Agreement http://www.gcc.ca/

Kiirikki M, Ruuskanen A (1996) How does *Fucus vesiculosus* survive ice scraping? Bot Mar 39:133–139

Lee RKS (1980) A catalogue of the marine algae of the Canadian Arctic. Nat Mus Can Publ Bot 9:1–82

McCook LJ, Chapman ARO (1997) Patterns and variations in natural succession following massive ice-scour of a rocky intertidal seashore. J Exp Mar Biol Ecol 214:121–147

Moritz REC, Bitz M, Steig EJ (2002) Dynamics of recent climate change in the Arctic. Science 297:1497–1502

Parke M (1948) Studies in British Laminariaceae. I. Growth in *Laminaria saccharina* (L) Lamour. J Mar Biol Assoc UK 27:651–709

Sharp GJ, Carter JA (1986) Biomass and population structure of kelp (Laminaria spp.) in southwestern Nova Scotia. Can Man Rep Fish Aquat Sci 1907:19

Sharp G, Semple R (1997) Rockweed *Ascophyllum nodosum*. DFO Can Stock Assess Sec Doc 97/31:12

Ugarte R, Sharp GJ (2001) A new approach to seaweed management in eastern Canada: The case of *Ascophyllum nodosum*. Can Biol Mar 42:63–70

Wein E, Freeman MR, Makus JC (1996) Preference for traditional foods among the Belcher Island Inuit. Arctic 49:256–264

Wilce RT (1959) The marine algae of the Labrador peninsula and northwest Newfoundland (ecology and distribution). Bull Nat Mus Can 158:1–103

The characterization of color mutations in Bangiaceae (Bangiales, Rhodophyta)

Jinfeng Wang · Pu Xu · Jianyi Zhu · Wei Zhou · Jianrong Xu · Xiugeng Fei · Xuecheng Zhang

Originally published in the Journal of Applied Phycology, Vol 20, No 5, 49–54.
DOI: 10.1007/s10811-007-9278-0 © Springer Science + Business Media B.V. 2007

Abstract The color mutations in Bangiaceae were investigated by treating the blades, conchocelis and conchospores phase of *Bangia* sp., *Porphyra yezoensis*, and *P. haitanensis* sampled in China with mutagen N-methyl-N′-nitro-N-nitrosoguanidine (MNNG). A high percentage of mutation in different expression characteristics in all three phases were shown within optimum mutagen concentrations. Among mutagenized blades, mutations occurred on single cells, which is a direct outcome of mutation of haploid cells. The mutation of mutagenized conchocelis resulted in a two-step process: low-level expression in conchocelis phase, and high-level expression in progeny, explaining that mutation took place in diploid cells. The mutations of conchospores were expressed immediately at germination of spores, indicating a change in ploidy. This paper reports the process of meiosis and its effect on frond development, and the relation between color mutations and morphological characteristics expressed by mutations in Bangiaceae.

Keywords Bangiaceae · Color mutations · Development · Meiosis · Mutagenesis

J. Wang · P. Xu (✉) · J. Zhu · W. Zhou · J. Xu
Department of Biology, Changshu Institute of Technology, Changshu 215500, China
e-mail: xupu66@sina.com

J. Wang
College of Life Science, Nanjing Normal University, Nanjing 210097, China

X. Fei
Institute of Oceanology, Chinese Academy of Sciences, Tsingdao 266071, China

X. Zhang
College of Marine Life Science, Chinese Ocean University, Tsingdao 266003, China

Introduction

The well-known genetic research on *Ulva mutabilis* development can be dated back to the 1960s (Fjeld and Løvlie 1976) for studying mutations of marine algae. A clear shift was made in the research field from green to red algae during the 1970s, during which genetic analysis was widely used as a tool for phycological studies, marked by a series of studies on *Gracilaria tikvahiae* by van der Meer (van der Meer and Bird 1977; van der Meer 1978, 1979; van der Meer et al. 1984; van der Meer and Zhang 1988).

In the early 1970s, spontaneous color mutations mutants of *Porphyra* were discovered. Researchers found red and green fronds in farmed populations of *P. yezoensis* and *P. tenera* successively (Miura et al. 1974; Aruga 1974; Miura 1975, 1976, 1977; Kobara et al. 1976), followed by the discoveries of mutants in the wild (Migita and Fujita 1983) and other mutants gained in laboratory (Miura 1985; Niwa et al. 1993). Some researchers wanted to increase the amount of survivable mutants by mutagenesis, but they did not succeed until 1994 when Mitman and van der Meer (1994) succeeded in *P. purpurea*. Afterwards, Yan and Aruga (1997) obtained some mutants by inducing the blades of *P. yezoensis*.

Among those studies, Miura and his coworkers used pigmentation markers as a tool for genetic analysis in *Porphyra*, and revealed genetic characteristics of the seaweed (Miura and Kunifuji 1980; Ohme et al. 1986; Ohme and Miura 1988; Miura and Ohme 1994), in which genetic principles were applied on biological issues (Miura 1990; Shin and Miura 1990; Miura et al. 1992). Xu et al. (2002a, b) mutagenized several species of *Porphyra* that are common in China, and attempted to study the mutation regularities of the algae.

In this research, color mutants in Bangiaceae and their characterization among generations were studied, especially at the stage of conchospore germination, by inducing the blades, conchocelis and conchospores of *Bangia* sp., *P. yezoensis* Ueda and *P. haitanensis* Chang et Zheng. Mutation regularities and biological characteristics of these species are introduced in this paper.

Materials and methods

Free-living conchocelis of *Bangia* sp., *Porphyra yezoensis* and *P. haitanensis* were sampled from a mature region or releasing carpospores of the foliose thalli, and deposited in the State Germplasm Bank for *Porphyra* in Marine Fisheries Institute of Jiangsu, China. The blades of *P. yezoensis* were collected from farming population in Nantong, Jiangsu of China, and farmed *Bangia* sp. and wild *P. haitanensis* were from Putian, Fujian, China.

A chemical mutagen, N-methyl-N′-nitro-N-nitrosoguanidine (MNNG), was weighted precisely and dissolved in formamide, and then the solution was diluted with sterilized seawater (the specific gravity is 1.020) to a concentration of 1 $g{\cdot}L^{-1}$ for later use.

Young blades of about 1 cm in length collected from wild or obtained via conchocelis cultivation in the laboratory were mutagenized with a 25 $\mu g{\cdot}mL^{-1}$– 30 min dose at 20°C per ten fronds each species. The blades were collected by bolting silk after induction, washed under flowing seawater 3 min, cultured individually in 500 mL Erlenmeyer flasks, and put in darkness for 1 day then under light. The seawater was daily refreshed, and PES (Provasoli 1966) was used as culture medium 3 days after mutagenesis. The blades were cultured at about 10-15°C and 40-50 μmol photons $m^{-2}{\cdot}s^{-1}$ according to different species under a 12–12 h light-dark photoperiod. The control was young blades without any mutagenic action.

Free living conchocelis of the three algae were kept in seawater (specific gravity 1.030) for 30 minutes and mutagenized with a dose of 25~50 $\mu g{\cdot}mL^{-1}$– 60 min MNNG at 20°C. The conchocelis were collected by bolting silk post-treatment, washed with flowing seawater for 3 min, cultured separately in 500 mL Erlenmeyer flasks, and then kept in darkness for the first 3 days. The culture medium was renewed twice a day. Afterwards, the filaments were chopped to 100~200 μm long and inoculated to shell, 1000 segments cm^{-2} in density. *P. haitanensis* was cultured at 18~29°C and 20~30 μmol photons $m^{-2}{\cdot}s^{-1}$, while the other species at 15~25°C and 20~40 μmol photons $m^{-2}{\cdot}s^{-1}$. The control group was free of mutagenic treatment one.

For the three species of algae, glass slides and bolting silk with conchospores in density of 30~50 individuals per vision scope (15×10) were observed, mutagenized with a dose of 25~50 $\mu g{\cdot}mL^{-1}$– 60 min at 20°C, washed in flowing seawater for 3 min, and then cultured separately in 500 mL Erlenmeyer flasks at 18~20°C and 50~60 μmol photons $m^{-2}{\cdot}s^{-1}$.

Three samples of mutagenized blades were randomly selected and observed under a light microscope for ten visions (15×40) a fortnight later. Percentage(s) of mutation was calculated by counting the number of mutated cells and surviving cells. Percentage(s) of mutation = Mutated cells / Survived cells. Conchocelis was cultured to release conchospores and the spores were collected on glass slides after periodically observing the surface of the shell in the next four weeks after mutagenesis, and then the germlings of the conchospores were examined and counted. Mutagenized conchospores were cultured for two weeks, and then 150 conchosporelings whose initial cells had divided more than twice were randomly examined.

Crossing tests and analyses were performed among the same species when the germlings grew up to 5 cm long. The blades were selected just before reproductive cells formed under a microscope, and then crossing materials were cultivated in the same containers under 15°C, 60~80 μmol photons $m^{-2}{\cdot}s^{-1}$ and 14L:10D. About two weeks after crossing with each other, the females were isolated and cultivated. Carpogonia and their offspring were observed to confirm whether the mutation is Mendelian or non-Mendelian transmission.

All the examinations and microscopic photography were carried out with a Nikon E800 microscope.

Results

Mutation of mutagenized blades Within one week, many light-colored cells appeared on blades after the conchosporelings were treated with MNNG. Most of the cells that distributed in patch with dispersive pigment died from suffering a fatal mutagenic effect. A few survived the treatment; they occurred as visible color mutations beside dead ones, and soon appeared in a dotted pattern of various colors. After division of these mutant cells, spotted and variegated blades with mutant sectors of various colors, shapes, and sizes formed. However, mutant cells in a sector had the same shape and color, and one sector corresponded to a single cell mutation (Figs. 1, a). In the control, none of the above-mentioned phenomena was observed (Table 1).

Mutation of mutagenized conchocelis Microscope observation showed that most conchocelis remained the wild type, and only a few changed color four weeks after being mutagenized and inoculated on shells. The difference between mutant spots and surrounding wild type spots of shell-boring conchocelis were clear as they grew (Fig. 3). Conchospores were almost all wild type when conchocelis matured. However, a few color mutants were also intermixed.

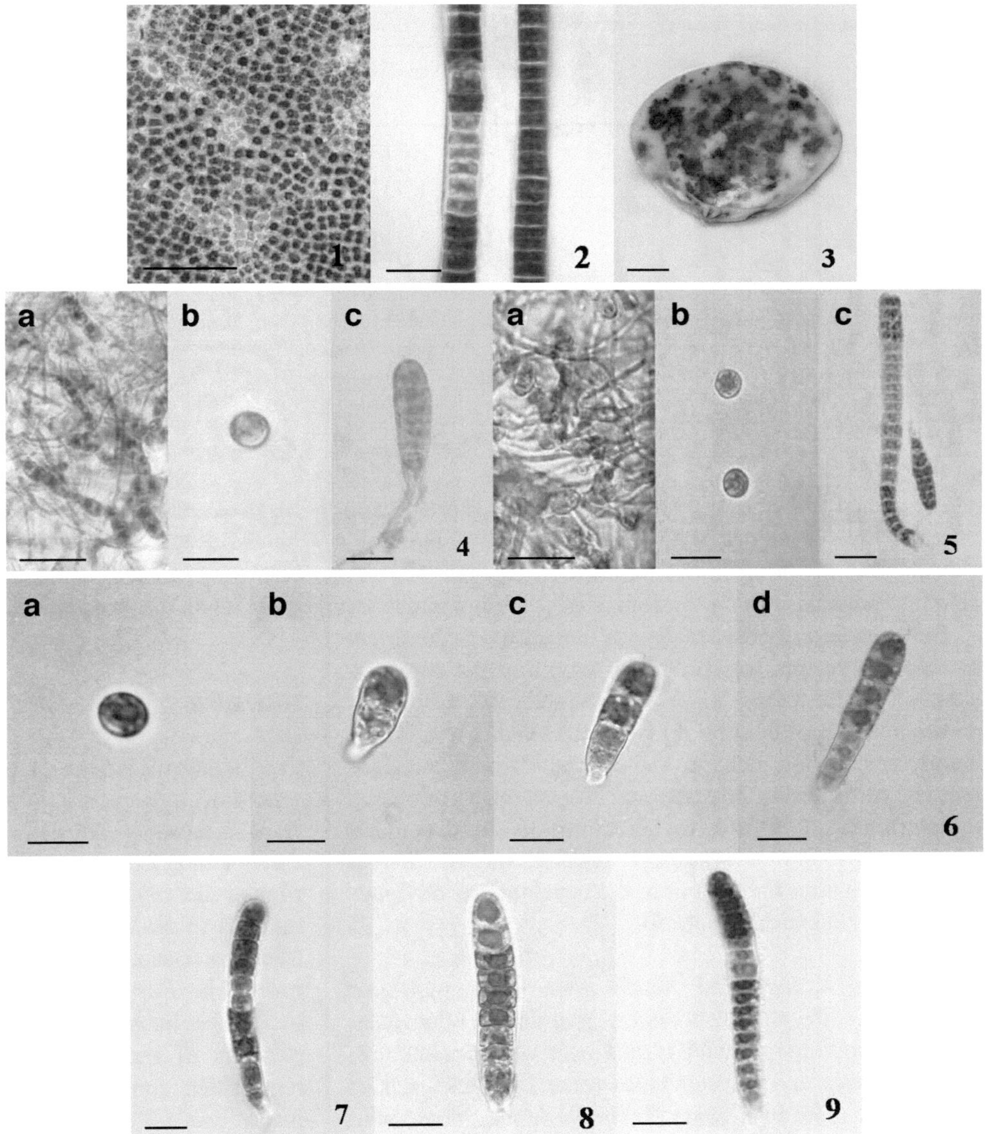

Figs. 1, 2 Mutagenized color mosaics of *Porphyra yezoensis* and *Bangia* sp. thallus
Fig. 3 Different colors in mutagenized shell-boring conchocelis of *Porphyra*
Figs. 4, 5 Cultivated results in single-spot of color mutation and wild-type conchocelis.
a. Conchocelis. **b**. Conchospore. **c**. Conchosporeling
Fig. 6 The course of conchospore germination in *Bangia* sp. **a**. Conchospore. **b**. Two-cell germling. **c**. Four-cell germling. **d**. Tetrad chimera of *Bangia* sp
Figs. 7, 8, 9 Tetrad chimeras in *Bangia* sp., *P. haitanensis*, and *P. yezoensis*.
Scale bar=20 μm in Fig. 1 and 5, and 1 cm in Fig. 2

The percentage of mutation in grown conchocelis was below 4.9%, which is much lower than that in blades. Surprisingly, mutations were highly expressed in the progenies of these mutagenized conchocelis, even though the mutagenic effects did not appear in the conchocelis phase (Table 2).

Seven, thinly growing spots of shell-boring conchocelis, were separated when they were tiny and cultured separately. The mutation in progeny is shown in Table 3.

Conchospores were released in the same mutation colors as their parents, and then germinated into a blade in own color (Fig. 4). Crossing experiments between mutant × wild type frond and mutant × mutant showed that there were no color differences between carpogonia and the maternal material. Phenotypes of F_1 conchocelis were similar to that of their maternal plant, as were the F_1 blades (Table 4).

Table 1 Effect of MNNG on young blades of *Porphyra*

*Frond	Treatment	Survive cells	Mutation cells	Percentages of mutation (%)
P. yezoensis	0	>10000	0	0
	25 μg·mL^{-1} –30 min	17890	2000	11.2
P.haitanensis	0	>10000	0	0
	25 μg·mL^{-1} –30 min	14820	1500	10.1
Bangia sp.	0	>10000	0	0
	25 μg·mL^{-1} –30 min	11137	653	5.9

*Blades were examined a fortnight after being mutagenized.

Table 2 Color mutation in mutagenized conchocelis

Frond	[1]Treatment	[2]Total Color sectors	[3]Sectors of Wt	Sectors of mutation	Percentages of mutation in M_1 blades (%)	Percentages of mutation in M_1 conchocelis (%)
P. yezoensis	0	150	150	0	0	0
	25 $\mu g \cdot mL^{-1}$–60 min	195	147	48	24.6	0
	50 $\mu g \cdot mL^{-1}$–60 min	264	136	128	48.5	4.9
P.haitanensis	0	150	150	0	0	0
	25 $\mu g \cdot mL^{-1}$–60 min	231	185	46	19.9	0
Bangia sp.	0	150	150	0	0	0
	50 $\mu g \cdot mL^{-1}$–60 min	254	164	90	35.5	0

[1] Conchospores were examined 10 days after releasing.
[2] Total color sectors of the 150 tested young blades
[3] Wt = Wild type

Additional to mutants, a large number of offspring were released from separated wild type or wild type-like conchocelis; the offspring showed a much higher rate of color chimeric mutants (Fig. 5), indicative of recessive mutation.

Development of chimeric fronds indicates that the origin of chimeric sectors was connected with the first two divisions of conchospores (Fig. 6). Each sector related only to an original cell of the first four, distinguishable in different colors among four-cell individuals. For the three algae species, cell division of early conchosporelings progressed in line (Figs. 7, 8, 9) in 2- to 4-color sectors. By examining conchospores of *P. yezoensis*, *P. haitanensis* and *Bangia* sp., ten days after the adhesion, the development of fronds were mainly identical (Table 5).

Mutation of mutagenized conchospores Color mutations occurred in some conchospores that germinated after mutagenesis. Germlings of wild type, single color mutants, 2-sector, 3-sector and 4-sector types came from such spores that divided more than twice (Table 6). Among them, wild type and single color mutants did not change phenotype in growth, but chimerical outcomes occurred following the first two divisions of conchospores. No individuals with more than four different sectors appeared, which indicates that the mutations had no relation with more than two divisions. The results are similar to the expression characteristics of color mutation in the progenies of mutagenized conchocelis.

Discussion

The mutagenic effect on blades demonstrated that dotted mutation took place in the unit of a single cell in Bangiaceae. A variegated sector originated from a mutant cell resulting from one mutation. Mutation of each cell only formed one-mutation tissue, and mutations of several cells resulted in many irregularly variegated tissues in blades. Mutation tissues distributed in segment in *Bangia*, while in pieces in *Porphyra*. Although the difference in the expression between *Bangia* and *Porphyra* was slight regardless of their different morphology, their mutation regularities appeared to be the same. The mutation in blades was a direct outcome of the haploid cells variation.

Mutation characteristics of mutagenized conchocelis and their offspring were expressed during the transformations of two generations in Bangiaceae. On one hand, color

Table 3 Color mutation in a spot of conchocelis and its progeny

Frond	[1]Number	M_1 conchocelis	M_1 blades
P. yezoensis	Yg-102	Green	All green
	Yr-201	Red	All red
	Yg-501	[2]Wt	Wt/green, Wt, green
P. haitanensis	Hg-107	Green	All green
	Hr-202	Red	All red
	Hg-501	Wt-Greenish	Wt/green, Wt, Green
Bangia sp.	Br-203	Red	All red

[1] Single spots which were pre-cultured
[2] Wt = Wild type

Table 4 Crosses of single-color mutants

[1]Frond	[2]Cross	F_1 conchocelis	F_1 blades
Yg-102	Green × Wt	Green	All green
	Wt × Green	Wt	All Wt
	Green × Green	Green	All green
	Green × Yr-201	Green	All green
Yr-201	Red × Wt	Red	All red
	Wt × Red	Wt	All Wt
	Red × Red	Red	All red
	Red × Yg-102	Red	All red

[1] Yg-102 and Yr-201 present a green mutant and a red mutant of *P. yezoensis*
[2] Female parents are shown at the left

Table 5 Color mutation in progenies of mutagenized conchocelis

Frond	[1]Treatment	Wt	%	[2]Single	%	2-Sector	%	3-Sector	%	4-Sector	%
P. yezoensis	0	150	100	0	0	0	0	0	0	0	0
	50 μg·mL^{-1}–60 min	106	70.7	11	7.3	22	14.7	10	6.7	1	0.7
P. haitanensis	0	150	100	0	0	0	0	0	0	0	0
	50 μg·mL^{-1}–60 min	117	78.0	2	1.3	26	17.3	5	3.3	0	0
Bangia sp.	0	150	100	0	0	0	0	0	0	0	0
	50 μg·mL^{-1}–60 min	103	68.7	4	2.7	19	12.7	17	11.3	7	4.7

[1] In each species, 150 germlings were tested.
[2] Single = Single color mutants

mutation occurred directly in the conchocelis phase, and would not change in the subsequent germination of conchospores or during their further development. Therefore, it is presumed that the mutation most likely belongs to cytoplasmic transmission, but a course of mitotic recombination was mentioned by van der Meer and Todd (1977) in *Gracilaria tikvahiae*. Nevertheless, mitotic recombination happens at a very low level. The crossing experiment proved that almost all of the mutants were inherited in a strictly maternal fashion. On the other hand, mutation in much more mutagenized conchocelis would not express in this phase. However, mutation took place after conchospore division. These mutations showed Mendelian single-gene transmission patterns, and correspond to the inheritance regularity of diploid cells. They are generally present in all three seaweeds.

During germination, the mutagenized conchospores are distinguished by color. This coincided with expression regularities of mutation when diploid cells converted into haploid ones. Mutation expressions of the spores were similar to the offspring's of mutagenized conchocelis.

The meiosis shown by color mutations As some surviving cells divided, variations took place in mutagenized blades of Bangiaceae, which manifests the characters of haploid gametophytes. Mutations detected in the conchocelis phase are generally relative to cytoplasmic transmission. When mutations are controlled by recessive nuclear genes, they would not be shown until the germination of conchospores released from mutagenized conchocelis. During generation alternation, colors would separate as the conchospore divided into the first four cells. The result is similar to that of hybridization among color mutants.

Crossing experiments were undertaken between different types of fronds by Miura et al. (1980). Their tests showed that most of the hybrid progeny appeared in chimeric types. The chimeras originating from conchosporelings resembled both colors of their parents in a 1:1 ratio, which is the characteristic of nuclear allele segregation. The production of chimera was considered similar to the undivided tetrasporophyte observed in *G. tikvahiae* by van der Meer (1977). However, more than 90% of germlings of *Porphyra* were chimeric fronds. It is supposed that conchospore germination was the site of meiosis. Moreover, cytological research by Ma and Miura (1984) show that meiosis took place during the germination of *P. yezoensis* conchospores, similarly to *P. torta*, *P. tenera* and *P. purpurea* (Burzycki and Waaland 1987; Tseng and Sun 1989; Mitman and van der Meer 1994).

In this experiment of mutagenesis in *Bangia* sp., *P. yezoensis*, and *P. haitanensis*, chimeras were present in all the above germlings. The origination of these chimeras demonstrated the occurrence of meiosis, and proved that conchospore germination of Bangiaceae is the site of meiosis. The results in this study echo the suggestion of van der Meer (1986) that meiosis takes place in germinating conchospore of at least some species of *Porphyra*. Furthermore, it could be common in Bangiaceae that meiosis occurred simultaneously with conchospore germination.

The development pattern displayed by the growth of germlings It is identical in all five algae of this experiment that

Table 6 Color mutation in germlings of mutagenized conchospores

Frond	*Treatment	Wt	%	Single	%	2-Sector	%	3-Sector	%	4-Sector	%
P. yezoensis	0	150	100	0	0	0	0	0	0	0	0
	25 μg·mL^{-1}–30 min	15	10.0	18	12.0	45	30.0	51	34.0	21	14.0
Bangia sp.	0	150	100	0	0	0	0	0	0	0	0
	25 μg·mL^{-1}–30 min	102	68.0	2	1.3	21	14.0	18	12.0	7	4.7

*In each species, 150 germlings were tested.

Mendelian mutants of conchocelis segregate their colors into different sectors as the germination of their offspring. These germlings were chimeras commonly in 2- to 4-color sectors, and there was no germling in over 4-color sectors, so were the mutagenized conchospores. The results indicate that the chimeric fronds resulted from the first two divisions of conchospores; and each sector of the chimera developed from one of the four cells during four-cell conchosporeling. Ohme and Miura (1988) described a tetrad configuration in conchosporelings that also came from a single-spot of recessive mutation. Similarly, in the seaweeds in the present study, linear tetrad configurations were shown in *Bangia* sp., *P. yezoensis*, and *P. haitanensis*.

Linear format is a ubiquitous type of development that exists in some species of Bangiaceae. The frond of Bangiaceae, which came from a zygote of different genetical backgrounds, is the productions of meiosis which developed sequentially in an individual.

Acknowledgements This research was supported by the National Sciences Foundation of China (NSFC, 40576074).

References

Aruga Y (1974) Color of the cultivated *Porphyra*. w *kushitachi* no Nori Kenkyu 23:1–14

Burzycki GM, Waaland JR (1987) On the position of meiosis in the life history of *Porphyra torta* (Rhodophyta). Bot Mar 30:5–10

Fjeld A, Lφvlie A (1976) Genetics of multicellular marine algae. In: Lewin RA (ed.). Botanical Monographs 12. The Genetics of Algae. Black Sci Pub, London UK, pp 219–235

Kobara T, Miura A, Aruga Y (1976) In vitro studies on the green type mutant of *Porphyra yezoensis* Ueda. La Mer 14:58–63

Ma JH, Miura A (1984) Observations of the nuclear division in the conchospores and their germlings in *Porphyra yezoensis* Ueda. Jpn J Phycol 32:373–378

Migita S, Fujita Y (1983) Studies on the color mutant types of *Porphyra yezoensis* Ueda, and thir experimental culture. Fac Fish Nagasaki Univ 54:55–60

Mitman GG, van der Meer JP (1994) Meiosis, blade development, and sex determination in *Porphyra purpurea* (Rhodophyta). J Phycol 30:147–159

Miura A (1975) Studies on the breeding of cultivated *Porphyra* (Rhodophyceae). Third International Ocean Development Conference (Aug.5–8, 1975) Tokyo. Marine Resources 3:81–93

Miura A (1976) Genetic studies of cultivated *Porphyra* (Nori) improvement. Mar Sci 8:15–21

Miura A (1977) Chimeral variegation found in *Porphyra*. J Phycol 13:45

Miura A (1985) Origin and Evolution of Diversity in Plants and Plant Communities. In: Hara H (ed). Genetic analysis of the variant color types of light red, light green and light yellow phenotypes of *Porphyra yezoensis* (Bangiales, Rhodophyta). Academic Scientific Book Inc, Tokyo, pp 270–284

Miura A (1990) Present trends and perspective in *Porphyra* (Nori) breeding-Genetics of pigmentation mutants in *Porphyra yezoensis*: Development origin of variegated gametophytic thalli. Suisan Ikushu 15:19–30

Miura A, Kunifuji Y (1980) Genetic analysis of the pigmentation types in the seaweed Susabi nori (*Porphyra yezoensis*). Iden 34(9): 14–20

Miura A, Ohme TM (1994) Mendelian inheritance of pigmentation mutant types in *Porphyra yezoensis* Ueda (Bangiales, Rhodophyta). Jap J Phycol 42:83–101

Miura A, Aruga Y, Fuseya M (1974) Thremmatological studies of the cultivated *Porphyra*. II. Effect of selection on the form and color of foliose thalli in *Porphyra yezoensis* f. *narawaensis*. Nippon Suisan Gakkai, Meeting abstract: 93

Miura A, Fu PF, Shin JA (1992) Interspecific cross experiments between *Porphyra yezoensis* Ueda and *P. tenera* Kjellman (Bangiales, Rhodophyta) by using pigmentation variants. J Tokyo Univ Fish 79(1):103–120

Niwa K, Miura A, Shin JA, Aruga Y (1993) Characterization and genetic analysis of the violet type pigmentation mutant of *Porphyra yezoensis* Ueda (Bangiales, Rhodophyta). Korean J Phycol 8:217–230

Ohme M, Miura A (1988) Tetrad analysis in conchospore germlings of *Porphyra yezoensis* (Rhodophyta, Bangiales). Plant Sci 57: 135–140

Ohme M, Kunifuji Y, Miura A (1986) Cross experiments of color mutants in *Porphyra yezoensis* Ueda. Jap J Phycol 34:101–106

Provasoli L (1966) Culture and Collections of Algae. In: A Watanabe, A Hattori (eds). Media and prospects for the cultivation. Proc U. S. pp 63–75

Shin JA, Miura A (1990) Estimation of the degree of self-fertilization in *Porphyra yezoensis* (Bangiales, Rhodophyta). Hydrobiologia 204/205:397–400

Tseng CK, Sun A (1989) Studies on the alternation of the nuclear phases and chromosome numbers in the life history some species of *Porphyra* from China. Bot Mar 32:1–8

van der Meer JP (1977) Genetics of *Gracilaria* sp. (Rhodophyceae, Gigartinales).II.The life history and genetic implications of cytokinetic failure during tetraspore formation. Phycologia 16:367–371

van der Meer JP (1978) Genetics of *Gracilaria* sp.(Rhodophyceae, Gigartinales). III. Non-Mendelian gene transmission. Phycologia 17:314–318

van der Meer JP (1979) Genetics of *Gracilaria* sp. (Rhodophyceae, Gigartinales). VI. Complementation and linkage analysis of pigmentation mutants. Can J Bot 57:64–68

van der Meer JP (1986) Genetic contributions to research on seaweeds. Porg Phycol Res 4:1–38

van der Meer JP, Bird NL (1977) Genetics of *Gracilaria* sp. (Rhodophyceae, Gigartinales).I. Mendelian inheritance of two spontaneous green variants. Phycologia 16:159–161

van der Meer JP, Todd ER (1977) Genetics of *Gracilaria* sp. (Rhodophyta, Gigartinales). IV. Mitotic recombination and its relationship to mixed phases in the life history. Can J Bot 55: 2810–2817

van der Meer JP, Zhang XC (1988) Similar unstable mutations in three species of Gracilaria. J Phycol 24:198–202

van der Meer JP, Patwary MU, Bird CJ (1984) Genetics of *Gracilaria tikvahiae* (Rhodophyceae).X. Studies on a bisexual clone. J Phycol 20:42–46

Xu P, Shen SD, Fei XG, Zhang XC, Zhu JY (2002a) Induction Effect and Genetic analysis of NG to Conchospores of *Porphyra*. Mar Sci Bull 4:68–75

Xu P, Shen SD, Fei XG, Zhang XC, Zhu JY (2002b) Induction Effect and Genetic analysis of NG to thallus of *Porphyra*. Mar Sci Bull 4:76–82

Yan XH, Aruga Y (1997) Induction of pigmentation mutants by treatment of monospore germlings with NNG in *Porphyra yezoensis* Ueda (Bangiales, Rhodophyta). Algae 12:39–52

Public health and carrageenan regulation: a review and analysis

Duika Burges Watson

Originally published in the Journal of Applied Phycology, Vol 20, No 5, 55–63.
DOI: 10.1007/s10811-007-9252-x

Abstract The status of carrageenan in the regulatory sphere influences how and where it may be used, with implications for seaweed farmers, carrageenan manufacturers and consumers. Over the period 1935 to the present the status of carrageenan has been effected by changes in the regulatory environment that reflect new understandings about carrageenan, health and health risks as well as broader trade, social and political changes. This paper reviews regulatory progress from the 1930s to the present. It reflects, in particular, the shifting priorities in public health and their effects on the regulatory status of carrageenan. Four case studies of public controversies about carrageenan safety are discussed in relation to regulatory responses and their public health significance. It is concluded that current assessments of risk associated with carrageenan have, in some contexts, failed to take into account the full spectrum of safety assessments that have been carried out and the maturing of food additive regulations thereby allowing a myth of risk to continue.

Keywords Carrageenan · Degraded carrageenan · Food regulation · Health risk · Poligeenan · Public health

Introduction

Carrageenan is approved and widely used as a food additive (Bixler 1996; Shah and Huffman 2003). Despite official sanction for use in food, over the last 50 years, carrageenan has been subjected to intense scrutiny for potential health risks associated with human consumption. Controversy and debate about carrageenan and human health have periodically flared in academic literature and the media (see for example Borthakur et al. 2007; Chapman 2001; Cohen and Ito 2002; Marcus and Watt 1969; Shah and Huffman 2003; Tobacman 1998, 2001; Tobacman et al. 2001a; Tomarelli et al. 1974).

Regulatory authorities, and independent scientific advisory committees, have maintained the position that carrageenan is safe for human consumption. Carrageenan is approved for use by the Food and Drug Administration (FDA) in the United States (Food and Drug Administration 2006) by the European Parliament and Council (Commission Directive 1995), and by the Joint World Health Organization and by the Food and Agriculture Organization Expert Committee on Food Additives (Cohen and Ito 2002). Yet in some jurisdictions precautionary measures have been instigated that place restrictions on where and how carrageenan may be used: Carrageenan is not permitted for use in infant formula in Europe (European Parliament and Council 1995) but is permitted for use in the US (Food and Drug Administration 2004).

The Scientific Committee on Food (SCF) of the European Commission recently endorsed a molecular weight distribution limit on carrageenan that is more restrictive than is the case in the US (Scientific Committee on Food 2003a). At the same time, the SCF acknowledges "there is no evidence ... that exposure to low molecular weight carrageenan from the use of food-grade carrageenan is occurring." This precautionary approach provides a rationale for adverse public and government response in Europe that is not necessarily matched elsewhere. This example illustrates the regulation of carrageenan is not uniform internationally; and controversy on the use of carrageenan has not been resolved to the satisfaction of all (Borthakur et al. 2007).

Legislative action in relation to food safety issues is progressive (Merrill 1997). This paper reviews the history of

D. Burges Watson (✉)
Institute of Health and Society, Newcastle University,
Newcastle upon Tyne NE2 4HH, UK
e-mail: duika.burges-watson@ncl.ac.uk

carrageenan from the time that it began to be industrially manufactured for food use in the US to the present. While differences still exist, there has been a growing move in recent years towards international harmonisation of food regulation through the Codex Alimentarius system and the EU Commission (Garrett et al. 1998; Livermore 2006; Millstone and Van Zwanenberg 2002; Veggeland and Borgen 2005).

Four examples are discussed in which carrageenan use in foods has been associated with risks to human health: carcinogenicity and ulceration in the 1960s; baby formula concerns in the 1980s, more recently controversy surrounding the publications of Dr. Joanne K. Tobacman and the appearance of weak mutagen in carrageenan.

Carrageenan as a regulated food additive in the US

Until the late 1930s, what is now recognised as carrageenan (extract) was not widely used as a substance added to food and the term carrageenan did not exist in regulations for foods. Decoctions of the seaweed *Chondrus crispus* had a long history of safe use in herbal and pharmaceutical preparations (See for example Council of the Pharmaceutical Society of Great Britain 1911; Felter and Lloyd 1898). Decoctions were made by macerating the seaweed with water or milk. The gelling properties of *C. crispus* had also been utilised for centuries in food products, such as the dessert blancmange (Stanley 1987). In 1862 Stanford coined the term "carrageenin" for the extract of *C. crispus*; the spelling was later changed to carrageenan by the American Chemical Society to reflect the use of -an as an affix denoting the presence of a polysaccharide (McHugh 2003).

Despite knowledge of the properties of carrageenan-bearing seaweed in the mid nineteenth century, it was almost a further century before a commercial industry based on the extracting carrageenan from certain seaweeds developed. In the US in 1937 ground *C. crispus* was found to stabilise dairy-produced chocolate milk. Three years later the Chicago based dairy company, Krim-ko, established a small plant near Boston to manufacture a water extract of *Chondrus* for chocolate milk and junket (Chapman 1950; Lewis et al. 1988). Algin Corp of America in Rockland, Maine (later to become Marine Colloids, Inc. and later still FMC Biopolymer) commercialised carrageenan extraction in the US in the 1950s. About the same time production of carrageenan began in Denmark and France.

Chondrus crispus (carrageenin) was regulated in food regulations in the US first as a GRAS substance, defined as a substance generally recognised as safe following the 1958 Miller amendment to the US Food and Drug Act of 1938 (See Table 1 for a chronology of regulation of carrageenan). Under the Miller amendment the FDA divided substances added to food into regulated food additives and substances that were GRAS either due to their history of use in food prior to the 1st January 1958 (the so-called grandfather clause) or on the basis of a consensus of expert opinion. The Miller amendment prohibited the use of any new food additives considered inadequately tested for safety.

Chondrus extract was grandfathered by the 1958 Amendment to the GRAS status through its previous uses. Under the GRAS listing Chondrus extract (carrageenin) was permitted for use in food under section 21CFR182.7255 of the US Code of Federal Regulations as harmless under prescribed conditions of use. However, for Marine Colloids Inc., the status of Chondrus extract limited production capability. The wording of the GRAS legislation allowed just one seaweed species from which carrageenan could be derived: *Chondrus crispus*.

During the early part of the 1950s a rapid increase in demand, and limited supply of *C. crispus*, led industry to explore other seaweed species as a source of the extract that was recognised by its chemical name carrageenan (Chopin 1998). From the 1950s into the 1960s the industry was experiencing other changes: increased knowledge of carrageenan chemistry and improved processes for optimising carrageenan performance in foods. Industry members became interested in how to develop the raw materials supply through the use of a broader range of seaweed species, and through increasing interest in cultivation. In 1960 Marine Colloids Inc. petitioned the FDA to alter the GRAS listing and to permit carrageenan to be defined by its chemistry, rather than the seaweed source. The FDA did not accept the chemistry definition but agreed to expand the seaweeds from which carrageenan could be extracted. To effect this change the FDA moved to list carrageenan as a regulated food additive under section 21CFR172.620, with the expanded, but limited, list of seaweed species and to establish a few purity criteria. Some confusion was created when Chondrus extract (carrageenin) was still listed as GRAS, and the extract of *C. crispus* - 'carrageenan' was listed as a regulated food-additive, a situation that still exists.

The 1938 Act (See Table 1) was essentially a 'policing statute' to ensure government had the authority to address dangers associated with foods or constituents of food that were considered 'injurious to health' (Merrill 1997). The origins of the Act can be traced to concerns about food adulteration in the late 19th and early 20th centuries (Coveney 2003). Following World War II significant advances in processing, preservation and packaging were developed that led to an increased interest and use of food processing aids (Atkins and Bowler 2000). The 1938 Act did not require advance approval for substances added to foods and processing aids could be widely adopted. In 1952 the US Congress established a special commission, chaired by

Table 1 Summary of US regulation of carrageenan

Year	Event	Status of carrageenan	Stated aims
1906	Federal Pure Food and Drugs Act and Federal Meat Inspection Act	Extract not yet developed. Carrageenan-bearing seaweeds little valued as food in US however long history of use in Ireland and elsewhere.	Regulates safety and quality of food. Defines official recipes for some products/prohibits food 'adulteration'.
1938	Federal Food, Drug and Cosmetics Act	1937, Krim-ko company, Chicago used ground Chondrus as ingredient for chocolate milk suspension. Not regulated.	Replaces Pure Food Act. Among provisions, labelling requirements were increased to include some 'truthful' listing of ingredients and additives on some products. Full ingredient listing not required on 'standardised products'.
1958	Food Additives Amendment Also Delaney clause	21CFR182.7255 Carrageenin (Chondrus extract) classified 'Generally Recognised as Safe' due to long history of use in foods.	Provides for a pre-market approval system for ingredients 'added to food'. Delaney clause gives FDA powers to ban food additives found to induce cancer in 'people or animals'. Classifies over 10,000 substances as food additives.
1960		MCInc. petition FDA to expand GRAS list. Carrageenan listed as a food additive under 21CFR172.620	Expanded list of carrageenan-bearing seaweeds. Carrageenan defined according to species list of eight seaweeds.
1969	White House Conference on Food, Nutrition and Health	1972 FDA review: molecular weight limitations proposed	Recommends a review of GRAS substances following FDA's ban of the artificial sweetener cyclamate as a potential carcinogen under the Delaney clause.
1973	FDA regulations for nutritional labelling of food introduced	1979 FDA move away from issuing molecular weight requirement and through Food Chemical Codex adopted water viscosity test to mimic molecular weight.	Voluntary for most foods, required for foods with added nutrients.
1990	Nutrition Labeling and Education Act	Carrageenan used for fat-replacement in meat (under auspices of USDA)	Requires mandatory labelling of nutrition of all processed foods, plus increased 'clarity' of ingredients labelling.

Congressman James Delaney, to consider the growing use of chemical processing aids in foods and the implications for consumer health. The resultant Food Additives Amendment (Food and Drug Administration 1958) was based on an assumption that anything added to food was adulterated unless it met with prior approval of the FDA, or unless there was already a long history of safe use (Merrill 1997). Proving safety of new additives became the responsibility of manufacturers (Food and Drug Administration 2004). The treatment of carrageenan was far from unique. Other GRAS substances came under specific regulations about this time (Food and Drug Administration 1958; Merrill 1997).

The passage of food regulations transferred more and more responsibility from the Federal Government as enforcer of honesty and fairness, to focus on pre-market controls. Regulation of new substances added to food provided a scientifically verified guarantee of safety that was not the case for GRAS category, however it also increased the surveillance of potential health risks by requiring manufacturers to prove, using scientific evidence, (usually from animal feeding studies) that any new additive was safe for human consumption.

The development of food regulations in the US was a prelude to similar actions in Europe and in the FAO/WHO Codex Alimentarius; although occurring somewhat later and usually by-passing the GRAS concept.

Four case studies of public health controversy

Case I: The origin of concerns and regulatory action: Ebimar

In the mid 1960s a UK company Glaxo (now GlaxoSmithKline Beecham) began to market a pharmaceutical product called *Ebimar* in France. It was based on 'carrageenin' which had been found to reduce the pain associated with peptic ulcers (Anderson and Hargreaves 1965; Bixler 1996; Piper and Fenton 1961). It was thought that stomach acids, including pepsin, contributed to the formation of ulcers (Anderson and Hargreaves 1965). Sulphated polysaccharides occurring naturally in the gastric mucous were found to inhibit the formation of pepsin. 'Carrageenin' was proposed as a cheap and natural alternative to other synthetic sulphates of polysaccharides that had been proposed as treatment agents.

At the doses required, the carrageenan was extremely viscous and difficult to consume in quantities thought necessary to produce a positive health benefit. It is well known that the viscosity of these solutions [carrageenin] could be reduced or destroyed by controlled heating with dilute mineral acid followed by purification. C16, the polygalactose sulfate component of Ebimar, was the product of this technology.

Given the limited knowledge of peptic ulcer causation, there was a correspondingly weak understanding of how or why the product worked. In the late 1960s, Marcus and Watt, two researchers under contract to the developers of Ebimar, undertook research to establish the mechanism through which low molecular weight carrageenan worked. During animal experimentation with both hydrolysed (C16) and unhydrolysed carrageenan, the latter being the type used in foods, Marcus and Watt (1969) discovered that, at very high doses, both could produce ulcerations in the cecum of the guinea pig, but ulcerations were more severe with the degraded form. Furthermore, the dose below which food-type carrageenan caused no further ulcers still showed considerable ulceration with C16. Marcus and Watt went on to speculate this ulceration as a precursor to the disease ulcerative colitis.

This is a good point to introduce some nomenclature that is common knowledge today among scientists and regulators, but was a source of confusion in the 1960s and continues to be one for the general public. C16 today would be called "poligeenan" a name provided by the US Adopted Name Council (USAN). While carrageenan is the raw material for producing poligeenan, the two different polymers have different properties and uses. Poligeenan with an average molecular weight of about 20,000 daltons has none of the food functions of carrageenan whose average molecular weight is never lower than 100,000 daltons and is usually much higher. The only application today for poligeenan is as a component of an X-ray imaging diagnostic product. Carrageenan for food use contains a very small fraction with a molecular weight in the range of that of poligeenan. Any suggestion that carrageenan with a small amount of low molecular weight matter could have the same ulcerating effect of poligeenan assumes equivalence between the substances.

Marcus and Watt concluded that the "...significance of our results in relation to human ulcerative colitis is at present only speculative and must await more comprehensive investigation" (Marcus and Watt 1969, p. 188S). In 1969, carrageenan use in food was already widespread and as Marcus and Watt reported, at the time of their publication there had been 'no reports of adverse effects' associated with carrageenan usage (Marcus and Watt 1969, p. 187S).

One effect of the Marcus and Watt publication was further surveillance and concern about carrageenan safety. Under FDA guidelines it was the responsibility of industry to prove carrageenan safe. Industry members conducted further studies involving other animal species. In 1972, the FDA reviewed carrageenan safety in light of Marcus and Watt's research and all the studies that had been generated as a result of their findings (Informatics Inc 1972).

On the basis of the Informatics report, the FDA arranged a meeting with health professionals and industry representatives to discuss a motion to modify the regulations for carrageenan in light of concerns about ulcerations found in the Marcus and Watt studies. It was agreed that the issue was of little significance to human health. Nevertheless, precautionary measures were proposed to limit the molecular weight to a minimum of 100,000 and to seek further animal studies to confirm this decision. Since 1969 scientific assessments of carrageenan have included short-term and long-term generational studies involving different dosages of degraded and non-degraded forms, and various animal studies including rats, mice, rabbits, rhesus monkeys, squirrel monkeys, pigs, gerbils, baboons, hamsters, ferrets, chick embryos and dogs (Cohen and Ito 2002; Greig 1999; JECFA 1974, 2001), While much of the above work was going on in the United States, various of the toxicology studies were also carried out in Europe.

All of the studies supported the safety of carrageenan for use in foods. Regulatory authorities saw no reason to question the safety of carrageenan as long as the average molecular weight was 100,000 daltons or higher.

Regulations were modified to insure that carrageenan used in foods would meet this limit, and a simple water viscosity measurement was adopted for this purpose.

As a further precautionary measure, Europeans limited the 'Acceptable Daily Intake' of 75 mg/kg body weight/day, an amount well above any average daily intake of even a diet high in carrageenan content (Bixler 1994). More recently JECFA increased the ADI to "not specified" meaning the use of carrageenan in foods allowed was for technical functionality reasons and the amount used did not have to be numerically specified (JECFA 2001).

While the above work should have settled the carrageenan safety for food processors, it did not. In 1984 labelling of food additives using the E number system came into effect among members of the European Economic Community. In the UK, additives had not previously been listed on food products. E numbers on food labels revealed additives to supermarket shoppers for the first time. The E numbers were widely interpreted by the public as new chemicals being added to food, rather than as existing additives being declared (London Food Commission 1988, p. 39). In May 1986, 89 MPs from across all parties signed a motion to seek a ban on all unnecessary additives in children's food. In January 1987 the Ministry of Agriculture, Fisheries and Food (MAFF) in the UK made public a survey of public attitudes to food additives. The survey revealed that one third of the British public could not see any justification for using additives at all (MAFF 1987).

The survey was significant in demonstrating the failures of policy communication about food additives, but also was linked to a more general decline of public trust in food policy and regulation (Lang 1999). With increasing public concern about additives, and a declining trust in public institutions, government regulators in the UK were under intense public scrutiny. Consumer groups were particularly active in voicing concern about baby foods and pre-schoolers diets (Bixler 1996).

Case II Infant formula and carrageenan

As already noted above, no foodstuff attracts more attention about safety than infant formula used in the first 4–6 months of a human life. It is considered particularly important because it may be the sole source of nutrition for infants over an important period of their physical development. Infant formula became a key policy concern early in the twentieth century along with population-based measures indicating declining rates of breastfeeding in industrialised countries, and increased scientific understanding of the importance of nutrition (Murphy 2004; Wolf 2003). Carrageenan had been used in the US since the late 1950s in liquid infant formulas to prevent fat separation and thereby assure more uniform nutrition. However, Marcus and Watt's research generated sufficient concern to warrant further investigation.

In the FDA's 1972 review of carrageenan, the use of carrageenan in liquid infant formula was considered, and it was decided the benefits of using carrageenan noted above outweighed any risk to the infant. Even though the terms of the 1958 Food Additive Amendment did not permit any public health benefits to be taken into account in determining the safe use of additives, practical considerations made it hard for regulatory authorities to exclude benefits in a risk assessment.

In Europe carrageenan is not permitted as a food additive for baby formula. The ban can also be traced to more general concerns about food additives that emerged in the 1980s. In 1992 the UK Food Advisory Committee in MAFF commissioned an extensive report on additives in baby milk and weaning formulas and concluded about carrageenan that:

> "Although there is no direct evidence of harm from carrageenan in infants and no toxicologically significant effects were seen in infant baboons fed carrageenan in commercial infant formulae for 16 weeks, high levels of reassurance are needed to permit additives in infant formulae. The Committee could not exclude the possibility of absorption of carrageenan by the immature gut or the possibility that absorbed material might affect the immune system in the infant. The Committee does not therefore consider carrageenan acceptable for use in infant formulae (European Commission 1992)."

The decision to ban the use of carrageenan in infant formulae, because of the 'high levels of assurance' needed, was subsequently reaffirmed by the EC-SCF (Scientific Committee on Food 2003b). The SCF suggested that because there was insufficient information on the effect of carrageenan on the immature gut of babies, its use was inadvisable but they had 'no objection to its use to the use of carrageenan, for technological reasons, in foods for older infants, such as follow-on formulae and weaning foods' (Scientific Committee on Food 2003b, p. 90). Industry has continued to support the benefits of using carrageenan in liquid infant formula and has sought to quantify risk and unequivocally assure safety to infants. New information has been provided to JECFA for their 2007 review of carrageenan.

Case III The 'Tobacman' controversy

Concerns about the use of carrageenan in food were not altogether assuaged despite the precautionary approaches adopted by various regulatory agencies and even though poligeenan, a suspected human carcinogen, is not and never has been used as a food additive. Poligeenan exists today specifically for diagnostic use only. The small amount of low molecular weight material present naturally in carrageenan is considered of no safety consequence by regulatory authorities. This is noteworthy because from time to time, academic researchers revisit the toxicity of poligeenan in relation to food grade carrageenan. The most recent and vocal researcher to enter this arena is Dr. Joanne K. Tobacman, currently at the University of Illinois in Chicago.

In *Environmental Health Perspectives*, a journal on environmental factors and human health, Dr. Tobacman reviewed all scientific literature relating to carrageenan safety (Tobacman 2001). The article reviewed 45 existing animal studies on poligeenan and carrageenan in relation to safety for food use. Critiques of the paper note how Dr. Tobacman ascribes results for poligeenan feeding studies to carrageenan and the disregard for how the method of administration to the animals can effect the results (Carthew 2002; Cohen and Ito 2002).

A consequence of this confusion of two different materials is exemplified in a quote from the author's interview of Dr. Tobacman in December 2003.

> 'I guess that underlying this issue is the consideration about how much data are sufficient to make a judgement about carcinogenicity? Many animal studies demonstrating ulcerations and neoplasms from carrageenan (sic) exposure were completed decades

ago. What evidence and how much evidence does it take to lead to changes in policy and behaviour?' (Tobacman 2003, pers. Comm., 3 December).

For Tobacman, rather than proving safety, the weight of evidence from past studies suggested that carrageenan in foods is a risk. As already discussed, shifts in regulatory policy in the 1950s required manufactures to produce 'data' as evidence of safety. Her concern was also fuelled by a clause in the 1958 Amendment that related specifically to carcinogenicity - the so-called 'Delaney clause'. The 'Delaney clause' was introduced as an additional clause to the Food Additives Amendment in 1958 and mandated that:

> '...no additive shall be deemed safe if it is found to induce cancer when ingested by man or animal, or if it is found, after tests which are appropriate for the evaluation of the safety of food additives, to induce cancer in man or animal...' (Quoted in Dean 1989, p. 6)

Tobacman's opponents have continued to stress that she is mixing up the suspected carcinogenicity of poligeenan with the lack of any such toxicity of carrageenan (Cohen and Ito 2002; Weiner et al. 2007).

The application of the Delaney clause may also be evaluated in a contemporary policy context. Merrill's (1997) review of regulatory policy demonstrates that the regulatory authorities did not anticipate, at the time the amendment was drafted, that the clause would be applied to a vast range of substances as subsequently occurred.

> 'In 1958, neither advocates nor opponents of the policy, including FDA officials, believed it would have broad application, for only a handful of chemicals had then been shown to be animal carcinogens' (Merrill 1997, p. 322)

Knowledge of cancer-producing substances in the 1950s was largely confined to polycyclic aromatic hydrocarbons; at that time not even cigarette smoke was accepted as carcinogenic (Weisburger 1994). Not surprisingly, the 'zero risk' tolerance applied to 'new' chemicals with potential carcinogenic effects (new including regulated food additives not considered GRAS) has been extremely controversial (Merrill 1997; Noah 1999; Noah and Merrill 1998; Vogel 2001; Weisburger 1994). Some argue that zero risk is necessary to protect the health of the public, others that carcinogens occur naturally in many foods and are of little risk (Ames and Gold 1997, 1998). Noah (1999, p. 34) suggests that the Delaney clause is used with less rigidity in food and drug law in the contemporary period because its wording preceded the development of more sensitive testing technologies and new medical knowledge about cancer. Where there is 'reasonable certainty of no harm', the Delaney clause is no longer used.

A second paper, Tobacman et al. (2001a) in the journal *Medical Hypotheses* used an epidemiological technique known as a 'time-trend' analysis to correlate the increased use of carrageenan in the twentieth century with the increased incidence of breast cancer. The authors wrote that 'although time-trend correlations represent a weak form of evidence, when significant positive correlations are found, they can support further evaluation' (Tobacman et al. 2001a, p. 596).

A further response to the Tobacman papers came from the European Commission Scientific Committee for Food. A Commission report, prepared to review and critique Tobacman's *EPH* and *Medical Hypotheses* papers, was critical of her findings in relation to the *Medical Hypotheses* article about which it concluded that it:

> ...did not support the hypothesis that breast cancer may be causally related to intakes of carrageenan and other water-soluble polymers used as food additives. The Committee noted that such correlations might be found for any dietary component or chemical to which there has been increasing exposure during the twentieth century (Scientific Committee on Food 2003a, p. 6).

Furthermore, the Commission found nothing new in Tobacman's *EPH* review that had not already been considered by the Scientific Committee for Food in determining the safety, purity criteria and ADI for carrageenan.

Although the report was critical of many of Tobacman's findings, there was one issue upon which the Committee felt further research should be undertaken - the possibility that 'native' carrageenan could create significant amounts of poligeenan either by processing techniques or by acids during digestion. Like earlier controversies over baby formula the perception of risk in Europe was deemed serious enough to warrant a response. Thus regardless of the deficiencies of Tobacman's research, or the diagnostic use of poligeenan for imaging, a precautionary approach was adopted. The Committee suggested 'if feasible, a molecular weight limit of not >5% below 50,000 should be introduced into the specification to ensure that the presence of any (low molecular weight) carrageenan [in food] is kept to a minimum' (Scientific Committee on Food 2003a, p. 6). This specification is enforced today although no appropriate, validated analytical method is available to quantify the percentage of low molecular weight material in carrageenan.

More recently Tobacman has published on bench top experiments on the interaction of carrageenan with various organ cells (Borthakur et al. 2007). The continued focus of her research has been to implicate carrageenan as a carcinogen by association. However, what happens *in vitro* does not provide a sufficient evidence base for what occurs *in vivo*. Even if a harmful relationship could be established in vitro, in vivo studies have demonstrated that ingested carrageenan does not pass the blood - gut barrier to interact with organ cells. Moreover, recent studies contradict cancer-producing effects of carrageenan and indicate that k-carrageenan may, though the enhancement of immune systems, actually inhibit tumours (Yuan et al. 2006). Work that seeks to establish a carrageenan/carcinogen link continues to be fuelled by suspicion.

Case IV Semicarbazide - A new problem handled quickly

Semicarbazide (SEM), a weak mutagen, was used for years in Europe as an indicator for the presence of the banned veterinary antibiotic, Nitrofurazone (de la Calle and Anklam 2005). The association between SEM and carrageenan came to attention through a circuitous route (summarised in Table 2).

The EC, EFSA and the World Health Organisation have declared that, based on levels reported in food, the health risk, if any, to consumers, including infants, appears to be very small (European Food Safety Authority 2003; World Health Organization 2007). Nevertheless, at the time the issue came to light it created concerns that carrageenan could pose a risk to human health. The carrageenan industry through its trade organisation, Marinalg International, was required to perform tests for SEM on representative commercial carrageenan products and issue a response (Marinalg International 2003).

At the time this issue arose there was a zero tolerance limits for SEM in foods, and this combined with a subsequent discovery of SEM in baby food prompted what Hoeneick and Gatermann (2006, p. 29) describes as 'violent discussions'. The discovery of SEM in carrageenan (Hoenicke et al. 2004) demonstrated to these researchers that it was not possible to differentiate between SEM resulting from Nitrofuran abuse; SEM occurring naturally or by bleaching processes involved in producing semi-refined carrageenan (PES). Subsequently tests demonstrated that SEM had weak mutagenic activity and posed no risk to humans in the amounts likely to be consumed (Abramsson-Zetterberg and Svensson 2005; AFC Panel 2005; Hoeneick and Gatermann 2006; Hoenicke et al. 2004).

The SEM carrageenan issue is a reminder of the constitutive power of public health regulation in initiating suspicions of risk. Combined with concerns about carrageenan associated with dated understandings of ulcerations and cancers and *potential* harm to infants in liquid infant formula, the SEM example demonstrates that understanding risk associated with carrageenan also requires an understanding of the different regulatory settings and contexts in which 'risks' emerge as public health issues. Despite all the concerns so far discussed, carrageenan is *still* regulated as safe.

Discussion and conclusion

After over fifty years of safe use of carrageenan in foods, some confusion and uncertainty in the public view still exists. One reason for this is that scientific accounts of carrageenan risk rarely demonstrate the historical, regulatory and public health contexts. Carrageenan risks are routinely taken out of context (See for example Tobacman 2001; Tobacman et al. 2001a; Tobacman and Walters 2001b).

Table 2 Public health concerns and semicarbazide

Year	Semicarbazide indicator
Pre 1950s	Azodicarbonamide (ADC) developed for use in plastics in Germany
1993	Nitrofuran drugs used in animal husbandry banned in the European Union following findings of mutagenic potential – zero-tolerance rules applied
2002	Semicarbazide (SEM), a known metabolite of nitrofurans used as marker for nitrofuran abuse – routine analysis begins with (more) sensitive methods
2003	SEM found in foods of non-animal origin including tomato sauce, egg, high levels found in baby food
2003	Denmark issues EC alert notification 2003/201 for 'carrageen derived from see weed' from Canada, Chile, Indonesia and Tanzania in relation to SEM and nitrofurans
2003	SEM linked to packaging and ADC used for PVC gaskets (extensive use in baby food)
2003	European Food Safety Authority finds SEM has weak mutagenic activity, low risk to human health. Nevertheless concerns about high levels in baby food
2004	Evidence of natural occurrence supports finding that SEM not a specific marker for nitrofuran abuse.
2004	Seaweed Industry Association Philippines announces budget to study SEM in carrageenan. Industry begins to assess alternate methods of bleaching/halt to bleaching.
2005	Scientific Panel on Food Additives, Flavourings, Processing Aids and Materials in Contact with Food concludes issue of carcinogenicity is not of concern for human health at the concentrations of SEM encountered in food.'
2005	ADC banned from use in plastics

A recent 90 day rat feeding study that was initiated to determine if a carrageenan near the lower molecular weight limit set by regulators (molecular weight of 257,000 daltons with <5% below 50,000 daltons with a range of 1.9% - 12%, determined by four different methods) would cause any toxicological responses (Weiner et al. 2007). The study concludes that carrageenan meeting regulatory purity criteria is safe for human consumption. On the basis of these findings the authors argue that the new molecular weight distribution specification for carrageenan in Europe is unnecessary. Studies of this type may help to clear up confusion from a scientific perspective, but they do not appear to resolve perceptions of safety in other contexts.

An alternative framework for assessing and communicating risk is to review issues of safety in their broader use and health contexts. For example, it is rarely mentioned that carrageenan was intentionally degraded to make C16 for use in a specific pharmaceutical product for peptic ulcer that is no longer on the market. The only use for this product today is industrial in nature (X-ray imaging diagnostics). However, renaming C16 poligeenan and the product for food use carrageenan may not be sufficient to allay public concerns. An extensive education program placing these different substances in a broader public health context may help. Without context a consumer could easily assume equivalence between carrageenan and poligeenan.

There is also the increased recognition of the potential value of carrageenan for antiviral, hypocholesterolaemic and hypoglycemic properties (Smit 2004). If current rounds of research prove successful, carrageenan will play an important part in sexual health applications, such as microbicides for HIV/AIDs prevention and use reduce the risk of cervical cancers by inhibiting the Human Papiloma Virus (Buck et al. 2006; Fernandez-Romero et al. 2007; Population Council 2006). This diversity of possible new uses for carrageenan across different regulatory settings provides a complex context for assessing health risks and benefits, but the positive nature of these applications will help the public and regulators to understand the potential health effects of different products.

This paper offers some insight into how various concerns within public health have influenced the risk regulatory process. At present regulation has provided support for the safe use of carrageenan in food. However, there is also the paradox that the more safety is proven, the more controversial that proof becomes.

Acknowledgements The author wishes to thank Dimitri Stancioff, retired senior scientist from FMC Biopolymer for review of the history of carrageenan production and use and certain regulatory matters prior to 1970.

References

Abramsson-Zetterberg L, Svensson K (2005) Semicarbazide is not genotoxic in the flow cytometry-based micronucleus assay in vivo. Toxicol Lett 155:211–217

AFC Panel (2005) Opinion of the scientific panel on food additives, flavourings, processing aids and materials in contact with foodon a request from the commission related to semicarbazide in food. EFSA J 219:1–36

Ames BN, Gold LS (1997) Environmental pollution, pesticides, and the prevention of cancer: misconceptions [published erratum appears in FASEB J 1997 Dec;11(14):1330]. FASEB J 11:1041–1052

Ames BN, Gold LS (1998) The causes and prevention of cancer: The role of environment. Biotherapy 11:205–220

Anderson W, Hargreaves GC (1965) Gastro-intestinal therapeutic. US Patent 3175942, March 13 1965

Atkins PJ, Bowler IR (2000) Food in society: economy, culture, geography. Arnold, London

Bixler HJ (1994) The carrageenan connection IV. Br Food J 96:12–17

Bixler HJ (1996) Recent developments in manufacturing and marketing carrageenan. Hydrobiologia 326/327:35–57

Borthakur A, Bhattacharyya S, Dudeja PK, Tobacman JK (2007) Carrageenan induces interleukin-8 production through distinct Bcl10 pathway in normal human colonic epithelial cells. Am J Physiol Gastrointest Liver Physiol 292:G829–G838

Buck CB, Thompson CD, Roberts JN, Müller M, Lowy DR, et al. (2006) Carrageenan is a potent inhibitor of Papillomavirus infection. PLoS Pathog 2(7): e69 doi:10.1371/journal.ppat.0020069

Carthew P (2002) Safety of carrageenan in foods. Environ Health Perspect 110:A176

Chapman VJ (1950) Seaweeds and their uses. 1st ed. Camelot Press, London

Chapman J.(2001) Cancer food scare. Daily Mail October 27;Sect. V2:21

Chopin T (1998) The seaweed resources of Eastern Canada. In: Critchley AT, Ohno M (eds.) Seaweed resources of the world. Japan International Cooperation Agency, Yokosuka, Japan, pp 273–302

Cohen S, Ito N (2002) A critical review of the toxicological effects of carrageenan and processed Eucheuma seaweed on the gastrointestinal tract. Crit Rev Toxicol 32:413–444

Commission Directive (1995) Directive No. 95/2/EC and 96/85EC. Office for Official Publications of the European Communities [serial online]. Available via: http://europa.eu.int/eur-lex/en/consleg/pdf/1995/en_1995L0002_do_001.pdf

Council of the Pharmaceutical Society of Great Britain (1911) The British Pharmaceutical Codex. The Pharmaceutical Press, London

Coveney J (2003) Why food policy is critical to public health. Crit Public Health 13:99–105

de la Calle MB, Anklam E (2005) Semicarbazide: occurrence in food products and state-of-the-art in analytical methods used for its determination. Analytical and Bioanalytical Chemistry 382: 968–977

Dean R (1989) Food additives: their regulatory status. In: Lewis RJ (ed.) Food additives handbook. Van Nostrand Reinhold, New York, pp 3–27

European Commission (1992) Opinion of the scientific committee on food on carrageenan. Brussels: Health and Consumer Protection Directorate General. (SCF/CS/ADD/EMU/199 Final)

European Food Safety Authority (2003) Press Release: EFSA gives update on semicarbazide: No reason to change current dietary habits including for babies. Precautionary action by industry recommended for baby foods. Available via: http://www.efsa.

europa.eu/etc/medialib/efsa/press_room/press_release/30.Par.0001.File.dat/15-10-2003_en1.pdf cited
European Parliament and Council (1995) Directive No. 95/2/EC and 96/85EC. Office for Official Publications of the European Communities
Felter HW, Lloyd JU (1898) King's American dispensatory. Ohio Valley Co., Cincinnatti
Fernandez-Romero JA, Thorn M, Turville SG, Titchen K, Sudol K, Li J et al. (2007) Carrageenan/MIV-150 (PC-815), a Combination Microbicide. Sex Transm Dis 34:9–14
Food and Drug Administration (1958) Federal food, drug, and cosmetic act: Chapter II - Definitions. Sec. 201. [321]
Food and Drug Administration.(2004) Code of Federal Regulations: CFR Title 21 Database. Available via: http://www.fda.gov/ohrms/dockets/dockets/03n0338/03n-0338-tr00003.htm cited September 8 2004
Food and Drug Administration (2006) Part 172 - food additives permitted for direct addition to food for human consumption. Code of Federal Regulations, USA
Garrett ES, Jahncke ML, Cole EA (1998) Effects of Codex and GATT. Food Control 9:177–182
Greig JB (1999) Carrageenan. Geneva. World Health Organisation: WHO safety evaluation of certain food additives. (Food Additive Series 42)
Hoenicke K, Gatermann R (2006) How can zero tolerances be controlled? The case study of Nitrofurans. Accreditation and quality assurance. J Qual Comparabili Reliabil Chem Meas 11:29–32
Hoenicke K, Gatermann R, Hartig L, Mandix M, Otte S (2004) Formation of semicarbazide (SEM) in food by hypochlorite treatment: is SEM a specific marker for nitrofurazone abuse? Food Additives & Contaminants 21:526–537
Informatics Inc (1972) Monograph on carrageenan. Maryland: Submitted under contract to the Food and Drug Administration
JECFA (1974) Seventeenth report of the joint FAO/WHO Expert committee on food additives, WHO Technical Series, No. 539
JECFA (2001) Evaluation of certain food additives specifications. Addendum 9. Joint FAO/WHO expert committee on food additives, WHO Technical Report series, No.891
Lang T (1999) The complexities of globalization: The UK as a case study of tensions within the food system and the challenge to food policy. Agric Human Values 16:169–185
Lewis JG, Stanley NF, Guist GG (1988) Commercial production and application of algal hydrocolloids. In: Lembi CA, Waaland JR (eds.) Algae and human affairs. Cambridge University Press, Cambridge, pp 205–236
Livermore MA (2006) Authority and legitimacy in global governance: Deliberation, institutional differentiation, and the codex alimentarius. New York University Law Review 81(2):766–801
London Food Commission (1988) Food adulteration and how to beat it. Unwin, London
Marcus AJ, Watt J (1969) Ulcerative colitis in the guinea-pig caused by seaweed extract. J Pharmaceut Pharmacol 21:187
Marinalg International (2003) Position statement on carrageenan and semicarbazide: EC alert notification 2003/201. Available via: http://www.marinalg.org/papers/c_papers.htm cited April 15 2007
McHugh DJ (2003) A guide to the seaweed industry. Rome: FAO Fisheries Technical Paper No. 441
Merrill RA (1997) Food safety regulation: reforming the Delaney Clause. Annu Rev Public Health 18:313–340
Millstone E, Van Zwanenberg P (2002) The evolution of food safety policy-making institutions in the UK, EU and Codex Alimentarius. Soc Policy Adm 36:593–609
Murphy E (2004) Risk, maternal ideologies and infant feeding. In: Williams L, Germov J (eds.) The social appetite: an introduction to the sociology of food and nutrition. Oxford University Press, Melbourne, pp 200–219
Noah L (1999) Legal aspects of the food additive approval process, in enhancing the regulatory decision-making approval process for direct food ingredient technologies: workshop summary. Institute of Medicine (IOM), Washington DC
Noah L, Merrill RA (1998) Starting from scratch? Reinventing the food additive approval process. Boston University Law Review 78:329–443
Piper D, Fenton B (1961) Effect of a sulfated polysaccharide on peptic digestion. A study in vitro of carrageenin (Ebimar). Gastroenterology 40:638–643
Population Council (2006) Microbicide Program. Available via: http://www.popcouncil.org/microbicides/index.html cited 15 September 2006
Scientific Committee on Food (2003a) Opinion of the Scientific Committee on Food on carrageenan. Brussels: European Commission; 5 March. (SCF/CS/ADD/EMU/199 Final)
Scientific Committee on Food (2003b) Report of the scientific committee on food on the revision of essential requirements of infant formulae and follow-on formulae. European Union.
Shah ZC, Huffman FG (2003) Current availability and consumption of carrageenan-containing foods. Ecol Food Nutr 42:357–371
Smit AJ (2004) Medicinal and pharmaceutical uses of seaweed natural products: A review. J Appl Phycol 16:245–262
Stanley N (1987) Production, properties and uses of carrageenan. In: McHugh DJ (ed.) Production and utilization of products from commercial seaweeds, vol 288. FAO, Rome
Tobacman JK (1998) (Effects of lambda-carrageenan on human mammary myoepithelial cells and relation to mammary carcinogenesis). In: International symposium on the impact of biotechnology on prediction, prevention and treatment of cancer. Nice, France
Tobacman JK (2001) Review of harmful gastrointestinal effects of carrageenan in animal experiments. Environ Health Perspect 109:983–994
Tobacman JK, Wallace RB, Zimmerman MB (2001a) Consumption of carrageenan and other water-soluble polymers used as food additives and incidence of mammary carcinoma. Med Hypotheses 56:589–598
Tobacman JK, Walters KS (2001b) Carrageenan-induced inclusions in mammary myoepithelial cells. Cancer Detec Prev 25:520–526
Tomarelli RM, Tucker Jr WD, Bauman LM, Savini S, Weaber JR (1974) Nutritional quality of processed milk containing carrageenan. J Agric Food Chem 22:819–824
Veggeland F, Borgen SB (2005) Negotiating international food standards: The World Trade Organization's impact on the Codex Alimentarius commission. Governance 18:675–708
Vogel D (2001) Ships passing in the night: The changing politics of risk regulation in Europe and the United States RSCAS working papers: EUI RSC 2001/16
Weiner ML, Nuber D, Blakemore WR, Harriman JF, Cohen SM (2007) A 90-day dietary study on kappa carrageenan with emphasis on the gastrointestinal tract. Food Chem Toxicol 45:98–106
Weisburger JH (1994) Does the delaney clause of the U.S. food and drug laws prevent human cancers? Toxicol Sci 22:483–493
Wolf JH (2003) Low breastfeeding rates and public health in the United States. Am J Public Health 93(12):2000–2010
World Health Organization (2007) Semicarbazide. Available via: http://www.who.int/foodsafety/chem/sem/en/index.html cited April 18 2007
Yuan H, Song J, Li X, Li N, Dai J (2006) Immunomodulation and antitumor activity of [kappa]-carrageenan oligosaccharides. Cancer Lett 243:228–234

Effect of alkali treatment time and extraction time on agar from *Gracilaria vermiculophylla*

Dora Luz Arvizu-Higuera ·
Y. Elizabeth Rodríguez-Montesinos ·
J. Iván Murillo-Álvarez · Mauricio Muñoz-Ochoa ·
Gustavo Hernández-Carmona

Originally published in the Journal of Applied Phycology, Vol 20, No 5, 65–69.
DOI: 10.1007/s10811-007-9258-4

Abstract The effects of alkali treatment time and extraction time of native agar and alkali treated agar obtained from *Gracilaria vermiculophylla* were studied. The response characteristics were mainly agar yield and gel strength. Alkali treatment was carried out at 0.5, 1.0, 1.5, 2.0, 2.5, and 3.0 h. Agar yield and gel strength decreased with the increase in the time of the alkali treatment. The highest yield (15.3%) and highest gel strength (1,064 g cm^{-2}) were obtained at 0.5 h, and therefore this time was used for the next experiment. The extraction of both native and alkali treated agars was carried out at 1.5, 2.0, 2.5, and 3.0 h. The best extraction time for alkali treated agar was 1.5 h, and for native agar 2.5 h. The alkali treated agar obtained with the different alkali treatment and extraction times showed higher melting (92.4–99.7°C) and gelling (35.7–39.6°C) temperatures. Native agar was lower in melting (60.2–64.1°C) and gelling (20.4–23.4°C) temperatures. The 3,6-anhydrogalactose content decreased with increasing alkali treatment time, with the opposite effect during the extraction of native and alkali treated agars.

Keywords Agar yield · 3,6-anhydrogalactose · Gel strength · Melting and gelling temperatures · Polysaccharide

D. L. Arvizu-Higuera (✉) · Y. E. Rodríguez-Montesinos ·
J. I. Murillo-Álvarez · M. Muñoz-Ochoa · G. Hernández-Carmona
Departamento de Desarrollo de Tecnologías,
Centro Interdisciplinario de Ciencias Marinas-IPN,
Apdo. Postal 592, La Paz, B.C.S., México 23000
e-mail: darvizuh@hotmail.com

Introduction

Agar is a polysaccharide comprised of two major components, agarose and agaropectin. Agarose is a neutral polysaccharide with a linear structure of repeating units of the disaccharide agarobiose, a dimmer of D-galactose and 3,6-anhydro-L-galactose. Agaropectin is an acid polysaccharide containing sulfate groups, pyruvic acid, and D-glucuronic acid conjugated to agarobiose (Araki 1966; Duckworth and Yaphe 1971; Yaphe 1984). Agar is obtained from some families of red algae (Rhodophyceae), mainly Gelidiaceae and Gracilariaceae. Agar is used as a food ingredient, accounting for 80% of its consumption. The remaining 20% is used for biotechnological applications. Agar was the first phycocolloid used in the human food industry (Armisen and Galatas 2000).

Agar is obtained by leaching of the alga in hot water, filtering off the extract, and separating the agar by freezing and thawing to eliminate the water (Armisen and Galatas 1987). The yield and quality of agar varies with the species and environmental factors related to the alga growth, light, nutrients, and temperature (Santelices 1988). Although the general aspects of the extraction process are known, studies are still needed to increase the yield and quality of the product. Agar of some algae, such as *Gelidium*, is easily extracted with boiling water. In other species, like *Gracilaria*, a pretreatment with alkali (NaOH) is needed for the desulphation of the native agar, causing the formation of a 3,6-anhydrogalactose bridge and an increase in the gel strength of the alkali treated agar (Freile-Pelegrín and Robledo 1997). The type of treatment before the extraction and the extraction time both affect the yield and quality of the agar.

The harvesting of red seaweeds started in Mexico in the 1940s. In 1960, AGARMEX set up an small agar extraction plant in Ensenada, Baja California. From 1970, the production was increased to process 1,200 t of dried seaweeds, and it remains the same at the present time. The raw material is *Gelidium robustum* (N.L. Gardner) Hollenberg & I.A. Abbott, which is harvested from Punta Descanso, Baja California, to Punta San Hipólito, Baja California Sur (Zertuche-González 1993). *Gelidium robustum* is the only species of red algae exploited commercially. However, Zertuche-González (1993) pointed out that extensive beds (60 ha) of another red seaweed, *Gracilaria vermiculophylla* (Ohmi) Pappenfus (former *pacifica*), were found in Laguna San Ignacio, Baja California Sur, with an average biomass ranging from 5 to 15 kg m^{-2} (wet weight). He estimated that 900 t per year (dry weight) could be harvested in that area. In 1993, 20 t of dried *G. vermiculophylla* were exported to China, but at present no one company is harvesting the beds. Recent studies confirmed the abundance of this species (Vergara-Rodarte 2006), but the characteristics of the agar remains unknown. The above data suggest that beds with economic potencial exist in Baja California, but the best extraction methods and chemical characteristic of its agar have not been studied in order to decide about industrial exploitation. Our objective was to study the effect of alkali treatment and extraction times on the yield and some physical and chemical properties of the resulting agar.

Material and methods

Gracilaria vermiculophylla was collected in Laguna San Ignacio, BCS, Mexico (26°45′10.6″N, 113°16′01.25″W and 26°35′50.7″N, 113°03′50.2″W) by diving at 3–5 m depth in April 2005. The plants were washed with 3% formaldehyde to prevent enzymatic hydrolysis, sun dried, and milled to 20-mesh size.

Effect of alkali treatment time

The alkaline treatment and the agar extraction were done according to the method described by Freile-Pelegrín and Robledo (1997), with some modifications. Twenty-five grams of *Gracilaria vermiculophylla* were soaked 12 h in 500 ml of 7% NaOH at room temperature (22°C) to rehydrate the seaweed. Then the solution was heated in a water bath at 85°C for 0.5, 1.0, 1.5, 2.0, 2.5, and 3.0 h, with constant agitation.

Agar extraction

After alkali treatment, each sample was washed three times with 500 mL of distilled water for 5 min, and then washed with 500 mL of 0.025% H_2SO_4 for 2 h with constant agitation. To remove excess acid, samples were washed with 500 mL of distilled water for 10 min. The algae were placed in 900 mL of hot distilled water with constant agitation. When the solution reached 80°C, the pH was adjusted to 6.2–6.5 with 10% phosphoric acid. The algae were boiled for 1.5 h. The extract was mixed with diatomaceous earth (Celite) and filtered using vacuum. The filtrate was allowed to gel at room temperature on plastic trays, frozen overnight, and thawed at room temperature. Finally, the agar was washed with ethanol, dried in an oven at 55°C for 24 h, cooled, and weighed to calculate the percentage yield of agar.

Effect of extraction time on alkali treated agar

Twenty-five grams of algae were soaked overnight in 500 mL of 7% NaOH at room temperature. The solution was heated in a water bath at 85°C for 0.5 h (this time was selected using the results obtained in the above experiment). Agar extraction was done as described above. The algae were boiled for four different times, 1.5, 2.0, 2.5, and 3.0 h.

Effect of extraction time on native agar

Twenty-five grams of algae were placed in 900 mL of hot distilled water. When the solution reached 80°C, the pH was adjusted to 6.2–6.5 with 10% phosphoric acid. The algae were boiled for four different times, 1.5, 2.0, 2.5, and 3.0 h. The extract was mixed with diatomaceous earth and filtered by using vacuum. The filtrate was allowed to gel at room temperature, frozen overnight, and then thawed. Finally, the agar was washed with 90% ethanol, dried in an oven at 55°C for 24 h, cooled, and weighed to calculate the percentage yield of agar.

Physical properties

A 1.5% agar solution (w/v) was prepared to measure the physical properties: gel strength, melting point, and gelling temperature. Agar solutions were allowed to gel at room temperature (22°C) overnight in plastic containers (3×9× 2.5 cm). Gel strength was measured using a modified Nikansui Shiki gelometer. The melting temperature was measured by preparing the agar gel in a test tube (1.7 cm diameter, 15 cm height) and placing an iron ball (7 mm diameter) on the surface of the gel. The test tube was clamped in a water-bath and the temperature increased gradually. The melting temperature was recorded with a precision thermocouple thermometer when the gel started to melt and the ball sank into the solution.

The gelling temperature was measured using the same hot solution in the tube above. The solution was allowed to cool down while moving the tube continuously, vertical to almost horizontal. The probe of the thermocouple was introduced in the solution when it ceased flowing, and temperature was recorded. Hysteresis was calculated as the difference between melting and gelling temperature.

Chemical properties

The percentage of 3,6-anhydrogalactose (3,6-AG) was determined by the resorcinol-acetal method (Yaphe and Arsenault 1965) using D-fructose as the standard. Absorbance was read at 555 nm.

Statistical analysis

All extractions and agar analysis were made in triplicate. Data were computed to obtain the average and standard error (±1 SE) with a significance level at 95%. An ANOVA was used to detect significant differences among treatments ($P<0.05$).

Results

Effect of alkali treatment time

Significantly higher yield (15.3%) and gel strength (1, 064 g cm^{-2}) were obtained at 0.5 h of alkali treatment, these values decreasing as the treatment time increased (Fig. 1). The melting temperature (94.2°C), gelling temperature (36.6°C) and hysteresis (57.6°C) were lower after 0.5 h of treatment. The maximum values were obtained at 3 h of alkali treatment for melting temperature (99.7°C) and hysteresis (61.4°C). The maximum gelling temperature (39.6°C) was at 1 h of treatment. No significant differences were found among the values obtained after 1 h. The highest value for 3,6-AG was at 1 h (48.9%), decreasing to a minimum (38.9%) at 3 h of alkali treatment (Table 1).

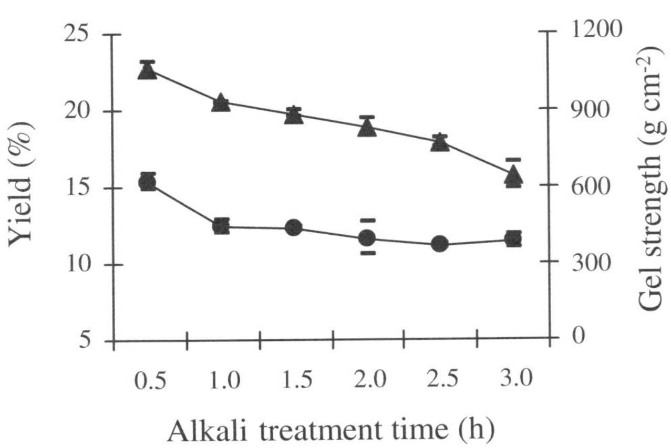

Fig. 1 Agar yield (●) and gel strength of 1.5% agar (▲) from *Gracilaria vermiculophylla* with different alkali treatment times. *Bars* represent standard error (±SE)

Effect of extraction time on alkali treated agar

Figure 2a shows an inverse relation between agar yield and gel strength, with the minimum yield (15.4%) and maximum gel strength (1,064 g cm^{-2}) obtained at 1.5 h of extraction. Significant differences were found between all treatment times for both characteristics. Higher values of melting temperature (98.1°C), gelling temperature (37.8°C), and hysteresis (60.3°C) were obtained after extraction for 2 h and the lower values were obtained at 3 h with 92.4°C, 35.7°C, and 56.7°C, respectively (Table 1). The 3,6-AG content showed a direct relationship with the extraction time, obtaining the minimum value of 42.7% at 1.5 h and the maximum of 45.7% at 3 h extraction time (Table 1).

Effect of extraction time on native agar

The yield of native agar was significantly higher than that of alkali treated agar. The agar yields for all treatments ranged from 29.7–34.6% and were significantly different at different extraction times. The highest yield was obtained at 1.5 h of extraction and the lowest at 2.5 h (Fig. 2b). Variation in gel strength was also significantly different among extraction times, and significantly lower than for alkali treated agar (Fig. 2b). The maximum value was 72 g cm^{-2} at 2.5 h of extraction and the minimum was 46 g cm^{-2} at 1.5 h. Melting temperature, gelling temperature, and hysteresis were significantly different among treatments. Higher values of melting (64.1°C) and gelling temperature (23.4°C) were obtained at 2.5 h, though maximum hysteresis (43.3°C) was obtained at 1.5 h (Table 1). These three characteristics had significantly lower values than those for alkali treated agar. The 3,6-AG content showed a similar trend to that for the extraction time in the alkali treated agar experiment, but significantly lower, from 30.5% at 1.5 h to 37.3% at 3 h (Table 1).

Discussion

From our results, the alkali treatment time had a significant effect on the agar yield. The yield was reduced with treatment time. However, we observed that during the washes of the product there was little agar loss by diffusion. The agar yield was lower than reported by Hurtado-Ponce and Umezaki (1988) for another species of *Gracilaria* from the Philippines, reported to be 23–48%.

Alkali treated agar from *G. vermiculophylla* had a gel strength from 644 to 1,064 g cm^{-2} similar to commercial agar (>800 g cm^{-2}). Mollet et al. (1998) reported lower values in the same species without gel strength improvement after alkali treatment. This suggests that yield and gel strength from some species are mainly dependent on the

Table 1 Physical properties and 3,6-anhydrogalactose content of agar from *Gracilaria vermiculophylla* with different treatments

Treatment	Time (h)	Melting temperature (°C)	Gelling temperature (°C)	Hysteresis temperature (°C)	3,6-AG (%)
Effect of alkali treatment time	0.5	94.2±0.7	36.6±1.1	57.6±1.3	42.7±0.6
	1.0	99.5±0.3	39.6±0.3	59.8±0.5	48.9±1.2
	1.5	98.3±0.6	39.1±0.7	59.1±0.6	46.1±2.6
	2.0	99.0±0.7	38.1±0.8	61.0±1.1	44.1±1.7
	2.5	99.4±0.3	39.6±0.3	59.8±0.5	45.5±0.5
	3.0	99.7±0.3	38.3±0.6	61.4±0.4	38.9±0.7
Effect of extraction time on alkali treated agar	1.5	94.5±0.4	36.5±1.3	58.0±1.2	42.7±0.6
	2.0	98.1±1.1	37.8±0.6	60.3±1.7	44.4±0.7
	2.5	95.6±1.2	35.7±0.4	59.9±1.2	44.8±1.6
	3.0	92.4±1.2	35.7±0.8	56.7±0.9	45.7±2.0
Effect of extraction time on native agar	1.5	63.7±0.5	20.4±0.3	43.3±0.6	30.5±0.3
	2.0	60.2±0.5	22.3±0.4	37.9±0.7	30.5±0.5
	2.5	64.0±1.4	23.3±1.6	40.7±0.5	34.4±0.5
	3.0	60.9±1.8	21.5±0.7	39.4±1.2	37.3±0.9

Data are mean ± SE

location, season, and processing methods (Hurtado-Ponce and Umezaki 1988; Ramalingam et al. 2002). For us, the shortest alkali treatment time significantly improved the gel strength, increasing it 14 times in comparison with native agar. Freile-Pelegrín and Murano (2004) found, for *G. cervicornis* (Turner) J. Agardh, a low gel strength (<50 g cm^{-2}) in native agar. They mentioned that the poor gelling ability of agarocolloids is because of the presence of D-galactose-6-sulphate residues and 4-0-methyl-α-L-galactose. The results of this study indicate that the gel strength of agar from *G. vermiculophylla* had high gel strength comparable to Superior Grade Japanese agar (Okazaki 1971).

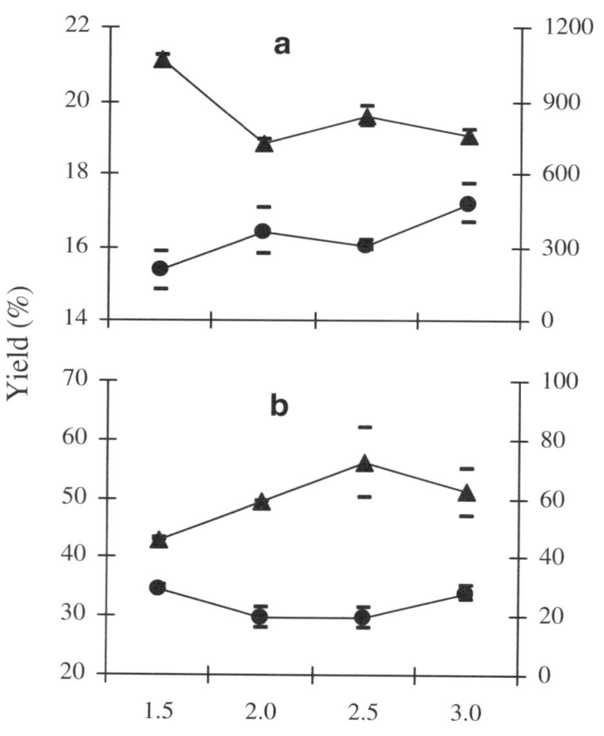

Fig. 2 Agar yield (●) and gel strength of 1.5% agar (▲) from *Gracilaria vermiculophylla* with different extraction times. **a** Alkali treated agar, **b** native agar. *Bars* represent standard error (±SE)

There was a direct relationship between the yield and gel strength with different times of alkali treatment. In contrast, during the extraction we found an inverse relation between the yield and gel strength in both native and alkali treated agar.

The alkali treated agar had a high melting temperature (92.4–99.7°C) that falls in the ranges of the United States Pharmacopeia (USP) standards (≥85°C). Melting temperatures of native agar for all extraction times (60.2–60.1°C) were lower than that measured by other authors (Freile-Pelegrín and Murano 2004; Rebello et al. 1997) in many species of *Gracilaria*.

The treatment of agar with alkali is usually done to eliminate the sulfate groups and increase the 3,6-AG content, consequently decreasing sulfation and increasing the gel strength (Chirapart et al. 1995). In our study, the native agar had lower 3,6-AG content than the alkali treated agar as a result of sulfate reduction (data not shown). Comparing the 3,6-AG content in native agar at 2 h extraction (30.5%) with the 3,6-AG of the agar with alkali treatment for the same time (44.4%), we obtained an increase of 45%. In contrast with the findings of Villanueva et al. (1997) for *G. eucheumoides* Harvey, we determined for *G. vermiculophylla* that longer times of alkali treatment reduced the 3,6-AG content.

Our results agree with Villanueva et al. (1997), confirming that optimum conditions to produce good quality agar from *Gracilaria* and other agarophytes are different from species to species, possibly caused by the variation of the specific seaweed to the treatment, hence our results may not be applicable to other *Gracilaria* species, but they are an useful tool for agar producers.

Our results showed a significantly higher native agar yield than alkali treated agar, although the gel strength of native agar was very low (maximum 72 g cm^{-2}) and not appropriate for the food industry. After alkali treatment, the yield was reduced by 50%, but the change on gel strength increased dramatically to 1,064 g cm^{-2}. Considering the gel strength as the main criteria to process *G. vermiculophylla*, the best conditions would be as follows: an alkali treatment with 7% NaOH at 85°C, for 0.5 h.; mild acid washing with 0.025% H_2SO_4 for 2 h, and extraction of the agar in boiling water for 1.5 h. However, it is possible to increase the yield to 16.5% by increasing the extraction time to 2 h. At this time, we can obtain an agar still with commercial gel strength (700 g cm^{-2}). Under this last condition, the physical properties of the agar were: melting temperature 98°C, gelling temperature 37°C, hysteresis 60°C and 3,6-AG 44%, which are in the range expected by the food market (Armisen 1995). Considering the abundance of raw material estimated by Zertuche-González (1993), the yield and the physical properties of the agar obtained, the commercial use of *G. vermiculophylla* from Laguna San Ignacio, Baja California Sur, is well justified, under an ecological management plan.

Acknowledgements We thank the Instituto Politécnico Nacional (CICIMAR-IPN) for the financial support, the "Comisión para el Fomento de Actividades Académicas (COFAA-IPN)" and the program "Estímulo al Desempeño de la Investigación (EDI-IPN)" for the incentives granted. Thanks to Dr. Ellis Glazier for editing the English language text.

References

Araki C (1966) Some recent studies on the polysaccharides of agarophytes. Proc Int Seaweed Symp 5:3–19

Armisen R (1995) World-wide use and importance of *Gracilaria*. J Appl Phycol 7:231–243

Armisen R, Galatas F (1987) Production, properties and uses of agar. In: McHugh DJ (ed) Production and utilization of products from commercial seaweeds. Fish Tech Paper 288. FAO

Armisen R, Galatas F (2000) Agar. In: Phillips GO, Williams PA (eds) Handbook of hydrocolloids. CRC, Boca Raton, Fla.

Chirapart A, Katou Y, Ukeda H et al (1995) Physical and chemical properties of agar from a new member of *Gracilaria, G. lemaneiformis* (Gracilariales, Rhodophyta) in Japan. Fish Sci 61: 450–454

Duckworth M, Yaphe W (1971) The structure of agar. Part 1. Fractionation of a complex mixture of polysaccharides. Carbohydr Res 16:359–366

Freile-Pelegrín Y, Robledo D (1997) Influence of alkali treatment on agar from *Gracilaria cornea* from Yucatan, Mexico. J Appl Phycol 9:533–539

Freile-Pelegrín Y, Murano E (2004) Agars from three species of *Gracilaria* (Rhodophyta) from Yucatán Peninsula. Bioresour Technol 96:295–302

Hurtado-Ponce AQ, Umezaki I (1988) Physical properties of agar gel from *Gracilaria* (Rhodophyta) of the Philippines. Bot Mar 31: 171–174

Mollet JC, Rahaoui A, Lemmoine Y (1998) Yield, chemical composition and gel strength of agarocolloids of *Gracilaria gracilis*, *Gracilariopsis longissima* and the newly reported *Gracilaria* c.f. *vermiculophylla* from Roscoff (Brittany, France). J Appl Phycol 10:59–66

Okazaki A (1971) Seaweeds and their uses in Japan. Tokai University Press

Ramalingam JR, Kaliaperumal N, Kalimuthu S (2002) Agar production from *Gracilaria* with improved qualities. Seaweed Res Utiln 24(1):25–34

Rebello J, Ohno M, Ukeda H, Sawamura M (1997) Agar quality of commercial agarophytes from different geographical origins: 1. Physical and rheological properties. J Appl Phycol 8:517–521

Santelices B (1988) Synopsis of biological data on the seaweed genera *Gelidium* and *Pterocladia* (Rhodophyta). FAO Fish Synop 145

Vergara-Rodarte MA (2006) Tendencias poblacionales de *Gracilaria vermiculophylla* (Ohmi) Papen. (Gracilariales, Rhodophyta) en la Laguna San Ignacio, B.C.S., durante un ciclo anual. Bachelor dissertation, Universidad Autónoma de Baja California Sur

Villanueva RD, Pagba CV, Montaño NE (1997) Optimized agar extraction from *Gracilaria eucheumoides* Harvey. Bot Mar 40: 369–372

Yaphe W (1984) Properties of *Gracilaria* agars. Hydrobiologia 116/117:171–186

Yaphe W, Arsenaul GP (1965) Improved resorcinol reagent for the determination of fructose and of 3,6 anhydrogalactose in polysaccharides. Anal Biochem 13:143–148

Zertuche-González JA (1993) Situación actual de la industria de las algas marinas productoras de ficocoloides en México. In: Zertuche-González JA (ed) Situación actual de la industria de macroalgas productoras de ficocoloides en América Latina y el Caribe. FAO Documento de campo No 13

Effects of extraction parameters on gel properties of carrageenan from *Kappaphycus alvarezii* (Rhodophyta)

Roike Iwan Montolalu · Yuri Tashiro · Shingo Matsukawa · Hiroo Ogawa

Originally published in the Journal of Applied Phycology, Vol 20, No 5, 71–76.
DOI: 10.1007/s10811-007-9284-2

Abstract Carrageenan was isolated under different extraction conditions from *Kappaphycus alvarezii* collected in North Sulawesi, Indonesia. Its gel properties include very strong elasticity even at low concentrations. Molecular weight and rheological properties were obtained by gel permeation chromatography and dynamic viscoelasticity measurements in order to clarify the average molecular weight at various extraction temperatures (50, 70, 90°C) and times (1, 3, 5 h), as well as gel formation ability. The results showed that both the weight-average and the number-average molecular weight decreased with increasing extraction temperature. However, the gelation rate of the carrageenan was found to be constant at around 40°C, whereas the storage modulus, G′, and loss modulus, G″, of the gels differed from each other.

Keywords Gel properties · Loss modulus · Number average molecular weight · Storage modulus · Weight average molecular weight

Abbreviations

M_w weight average molecular weight
M_n number average molecular weight

R. I. Montolalu · Y. Tashiro · S. Matsukawa · H. Ogawa (✉)
Department of Food Science and Technology,
Tokyo University of Marine Science and Technology,
Minato, Tokyo 108-8477, Japan
e-mail: ogawahi@kaiyodai.ac.jp

R. I. Montolalu
Faculty of Fisheries and Marine Science,
Sam Ratulangi University,
Jln. Kampus Unsrat Bahu,
Manado 95115, Indonesia

Symbols

G′ storage modulus
G″ loss modulus

Introduction

Carrageenans are natural polysaccharides obtained by extraction from certain species of red seaweed (Rhodophyta). They exist in three main forms: κ-carrageenan, ι-carrageenan and λ-carrageenan. Among these, κ-carrageenan is predominantly obtained by extraction from the tropical seaweed *Kappaphycus alvarezii*, known in the trade as *Eucheuma cottonii* (or simply cottonii) (Rudolph 2000). Studies interpreting the depolymerization kinetics of polysaccharides touch upon the importance of molecular characteristics and solvation quality. Depolymerization kinetics have been elucidated for several polysaccharides, including carrageenans, by measuring the molecular weight or the number of reducing ends (Lai et al. 2000).

Viscoelastic measurement is used extensively to study carrageenan gel viscoelasticity. The effects of salt, added milk system, temperature sweep, and concentration of samples on the rheological transition of carrageenan gel have been reported by several research groups (Zhang et al. 2004; Van de Velde et al. 2005; Michon et al. 2005). The rheological properties of κ-carrageenan have been extensively studied using dynamic measurement (MacArtain et al. 2003; Haug et al. 2004; Nunes et al. 2006). These investigations assessed the effect of κ-carrageenan concentration, added salt, added fish gelatin, and temperature on the mechanical properties of the gel formed. However, no data are available on the dynamic behavior of carrageenan extracts under different extraction conditions.

We are presently screening sulfated galactans from Indonesian red algae of the economically important Solieriaceae family, and carrageenophytes of the Gigartinales order, with a view towards identifying potential commercial sources of native polysaccharides. In this research, we extended our study to the carrageenan extracted from the commercially cultivated alga *K. alvarezii*. This seaweed is one of the main raw materials for the production of kappa-carrageenan, a widely used gelling sulfated galactan (Bixler 1996).

In this research, extraction parameters such as extraction temperature and extraction duration (time) were systematically varied and their effects on the resulting molecular characteristics [M_w (weight average molecular weight) and M_n (number average molecular weight)] and dynamic viscoelasticity were assessed. The purpose of this study was to isolate and characterize the polysaccharides present in the cell walls of the marine red alga *K. alvarezii*, as a function of different levels of extraction duration (hours) and temperature (°C). A set of parameters that have been optimized with respect to carrageenan yield and gelling properties is presented.

Materials and methods

Kappaphycus alvarezii (Doty) Doty ex P.C. Silva (Gigartinales, Solieriaceae) was collected by hand at the end of May 2005 on the coast of Nain Island located within the Bunaken National Marine Park, North Sulawesi Sea, Indonesia, which is in a tropical zone. Immediately after sampling, the seaweed was washed several times with clean seawater in order to remove nonalgal materials. No attempt was made to separate gametophyte thalli from carposporophyte thalli, as this process is difficult to implement in a plant-scale carrageenan production unit. The seaweed was then sun-dried, and stored under refrigeration.

Extraction

Dried algae (15 g) were soaked in 3 L water for about 24 h to remove adhering sand, salt and other impurities. The process was repeated twice. The washed algae were chopped into fragments approximately 1 cm in length. Extraction (15 g dried algae in 750 mL water) was performed at various temperatures (50, 70, 90°C). Each temperature was set for three different extraction times (1, 3, 5 h). The suspensions were centrifuged (12,000 rpm, 50°C, 30 min) and the supernatants designated as the supernatant fraction (C1). One volume of supernatant was poured into two volumes of 2-propanol whereby the polysaccharides precipitated as long fibers. The liquors were removed by centrifugation (12,000 rpm, 4°C, 30 min) and the samples were designated as the precipitated fraction (C2). The samples were dried under vacuum evaporator and recovered in a freeze-dried state. This product was designated as the yield fraction (C3).

Molecular weight distribution measurement

Sample preparation

Solution containing 0.3 % w/v of sample was made by dissolving in 0.2 M NaCl at room temperature with continuous stirring. The tubes were left overnight at 4°C to ensure complete hydration, and then heated at 80°C for 30 min and filtered through a 0.45-μm syringe filter into a scintillation vial before injection.

Determination of molecular weight

Number- and weight-average molecular weight (M_n and M_w, respectively) were obtained by gel permeation chromatography (GPC) on an HLC-8120GPC (Tosoh, Tokyo, Japan). Instrument control and data processing was performed using an SC-8020 system (Tosoh). The value of M_w and M_n was calculated using a calibration curve generated with eight monodisperse pullulan standards (Shodex, Showa Denko, Japan; in the range 0.59×10^4 to 78.8×10^4). The column set consisted of two TSK-GEL G6000PWXL columns of 7.8×300 mm with guard column PWXL 6.0×40 mm (Tosoh, Japan). Temperatures of the inlet, column oven and RI detector were set at 60, 65, and 45°C, respectively. The mobile phase used as eluent was 0.2 M NaCl at a flow rate of 1.0 mL min^{-1}. All measurements were performed at least twice.

Determination of dynamic rheology

Solutions of *K. alvarezii* polysaccharide were prepared in closed tubes at a total concentration of 1 % (w/v) by dissolving in distilled water at room temperature with stirring. The tubes were left overnight at 4°C, and were then heated at 90°C for 30 min while stirring.

Dynamical rheology measurement was carried out using a Rheo Stress RS-50-rheometer (Haake, Karlsruhe, Germany) with a PP35 Ti sensor (cone diameter 34.004 mm). Temperature dependence of storage (G′) and loss (G″) moduli was observed by using a cooling system to reduce the temperature from 85 to 10°C. The edge of the sample recipient was covered with a moistened sponge in order to minimize water evaporation. Cooling scans were performed at 1°C min^{-1}, with measurement of G′ and G″ at 1 Hz and the gap set to 1 mm. At least three replicates of each measurement were made. Prior to the temperature sweep, dynamic torque sweeps were conducted in order to select a common linear viscoelastic region for all system combinations.

Statistical analysis

One-way ANOVA was used to test the effects of extraction parameters on M_w and M_n of each fraction. Duncan's post hoc test was selected and the confidence interval was set to the 95% level. All statistical analyses were performed using SPSS (Version 12.0) software, (SPSS, Chicago, IL) for Windows.

Results

Average molecular weight—M_w and M_n

Each fraction (C1–C3) was submitted to GPC on HLC-8120GPC, using 0.2 M NaCl as eluent. The resulting molecular weight contents of all extracts and fractions analyzed are tabulated in Tables 1 (M_w) and 2 (M_n). The M_w of the extracts ranged from 0.22×10^6 (T=90°C, t=5 h, C3) to 8.54×10^6 (T=50°C, t=1 h, C2, Table 1), and the M_n ranged from 0.21×10^6 (T=90°C, t=3 h, C3) to 4.57×10^6 (T=50°C, t=5, C1, Table 2).

The results for M_w of supernatant fraction C1 were in the range 1.55 to 7.87×10^6 (Table 1), and those for M_n were in the range 0.63 to 4.57×10^6 (Table 2). The values of M_w and M_n for all fraction C1 samples decreased with increasing extraction temperature. No significant difference of the effect of treatment duration time was observed at any of the temperatures tested (Tables 1, 2).

Table 1 Values of weight average molecular weight (M_w) at varied temperatures and times in each fraction extraction

Extraction parameters		M_w ($\times10^6$)		
T (°C)	t (hours)	Fraction C1	Fraction C2	Fraction C3
50	1	6.22±0.19	8.54±1.38	2.28±0.08
	3	7.42±2.40	8.16±2.27	2.34±0.21
	5	7.87±0.93	5.31±0.41	2.28±0.04
		NS	NS	NS
70	1	5.26±2.19	8.14±2.26	3.48^a±0.76
	3	5.15±1.98	6.47±1.10	$1.7^{ab}0$±0
	5	3.58±1.41	6.29±1.57	2.22^b±0.14
		NS	NS	*
90	1	1.87±0.07	2.49^b±0.33	0.66^a±0.06
	3	1.84±0.18	0.52^a± 0.16	0.37^b±0.03
	5	1.55±0.02	0.38^a±0.01	0.22^c±0.01
		NS	**	**

Values are mean±SD; n=2.; means within the same column, for each temperature treatment, without a common superscript are significantly different ($P<0.05$). Each sampling temperature also shows the results of ANOVA in relation to treatment duration time.
* $P<0.1$, ** $P<0.05$, NS not significant ($P>0.1$)

Table 2 Values of number average molecular weight (M_n) at varied temperatures and times in each fraction extraction

Extraction Parameters		M_n ($\times10^6$)		
T (°C)	t (hours)	Fraction C1	Fraction C2	Fraction C3
50	1	2.61±0.42	4.40±2.74	0.92±0.04
	3	4.30±1.91	3.57±1.49	0.77±0.11
	5	4.57±1.50	2.53±0.11	0.90±0.04
		NS	NS	NS
70	1	2.52±0.20	4.20±0.50	1.47±0.42
	3	2.17±0.51	3.47±1.45	0.76±0.03
	5	3.89±0.61	2.63±0.39	0.73±0.20
		NS	NS	NS
90	1	0.87±0.01	0.75±0.36	0.32^a±0.03
	3	0.88±0.36	0.33±0.16	0.21^b±0.03
	5	0.63±0.07	0.23±0.01	0.09^c±0
		NS	NS	**

Mean±SD; n=2; means within the same column, for each temperature treatment, without a common superscript are significantly different ($P<0.05$). Each sampling temperature also shows the results of ANOVA in relation to treatment duration time
** $P<0.5$, NS not significant ($P>0.1$)

The results for M_w of supernatant fraction C2 were in the range 0.38 to 8.54×10^6 (Table 1), and those for M_n were in the range 0.23 to 4.4×10^6 (Table 2). The values of M_w and M_n at different extraction temperatures (50, 70, 90°C) decreased with increasing extraction time, except at the 5 h heating time. Also, the M_w values of supernatant fraction C2 indicate that the 1 h treatment duration time was significantly ($P<0.10$) different at the 90°C extraction temperature. For fraction C2, M_w values were not affected by treatment duration time at either 50 or 70°C. The M_n value of fraction C2 showed no significant differences between treatment duration times.

The results for M_w of yield extract C3 were in the range 0.22 to 3.48×10^6 (Table 1) and those for M_n were in the range 0.09 to 1.47×10^6 (Table 2). M_w and M_n values decreased with increasing extraction temperature, except for the 1-h heating time. At 90°C extraction temperature, the M_w and M_n values decreased with increasing heating time, with significantly different ($P<0.05$) for different treatment duration times. M_w and M_n of yield extract C3 were recorded for 50°C, with no significant difference among treatment duration times. The value of M_w of yield extract C3 showed a significant difference ($P<0.10$) between treatment duration times at 70°C, while M_n was not significantly ($P>0.10$) affected by treatment duration time.

Dynamic viscoelasticity

The viscoelastic properties of native carrageenan isolated from *K. alvarezii* at 50°C were studied using a constant polymer concentration of 1% w/v, at extraction times (t) of 1, 3, and 5 h. Figure 1 shows the temperature dependence of dynamic moduli (G′ and G″) during cooling from 85 to 10°C at different extraction duration times at a 50°C extraction temperature. The viscoelastic measurement on 1% w/v carrageenan at all extraction duration times showed a maximum in G′ and G″ as the temperature was reduced from 85 to 10°C, with G′ being larger than G″. At 1-h and 3-h extraction times (Fig. 1a,b), both G′ and G″ began to increase sharply at around 40°C and then G′ exceeded G″, indicating the rapid formation of a gel network. According to Lapasin et al. (1995), gels can be classified as strong or weak their G′/ G″ ratio. For the gel in Fig. 1c, as shown by the crossover of G′ and G″ at 43°C, the G′ and G″ started increasing sharply at around 40°C.

The viscoelastic measurement of G′ and G″ versus temperature sweep for the different extraction times at 70°C

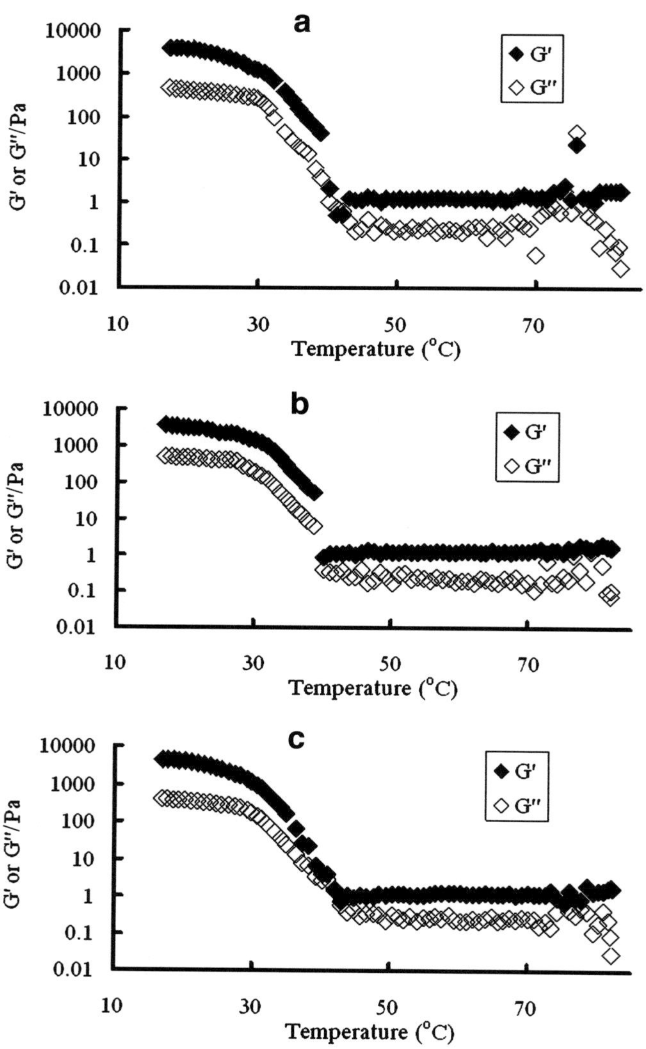

Fig. 1 a–c Temperature dependence of dynamic storage modulus G′ (♦) and dynamic loss modulus G″ (◇) of carrageenan at 50°C. **a** 1 h, **b** 3 h, **c** 5 h

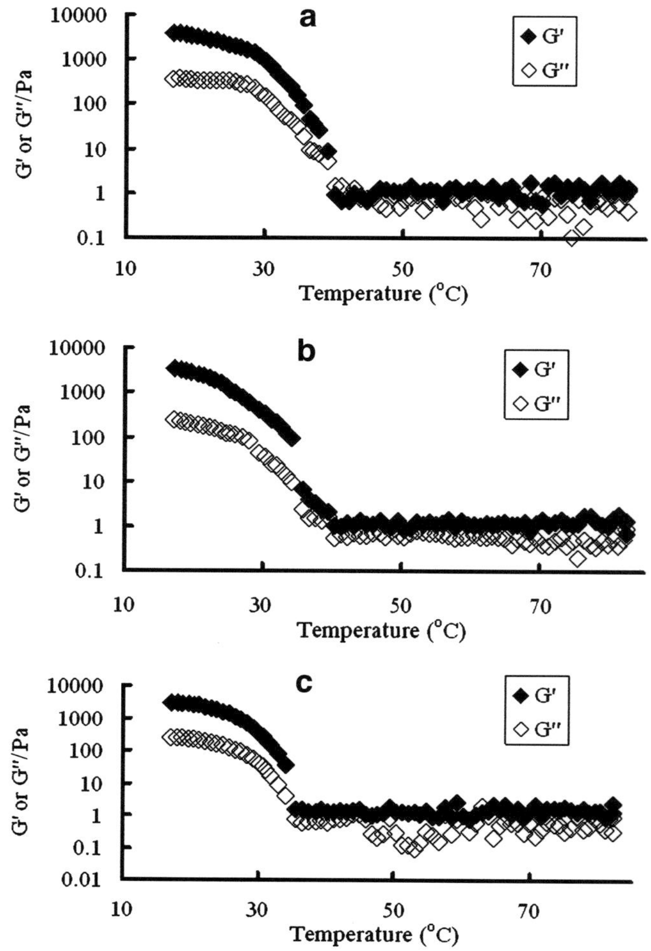

Fig. 2 a–c Temperature dependence of dynamic storage modulus G′ (♦) and dynamic loss modulus G″ (◇) of carrageenan at 70°C. **a** 1 h, **b** 3 h, **c** 5 h

extraction temperature are compared in Fig. 2a–c. It is readily seen that G′ remained higher than G″ for all extraction times. This indicates that the sample is indeed in a gel state. Similar behavior was also seen in Fig. 1a,b, where both G′ and G″ start increasing sharply at 40°C. Figure 2c shows the crossover of G′ and G″ at 35.6°C, and the measurements show the elastic behavior of gels with G′ values higher than G″ values, starting around 36°C. The effect on the moduli brought about by the order–disorder transition of extraction time is shown in Fig. 2 for three lengths of time. The rheological results are fully reversible, with no detectable thermal hysteresis. A sharp transition in the sample from a liquid-like solution to more-or-less elastic state occurred over a narrow temperature range.

Viscoelastic studies on native carrageenan isolated from *K. alvarezii* at 90°C were conducted using a constant polymer concentration of 1 % w/v, with extraction duration times (t) of 1, 3, and 5 h. Figure 3 shows the temperature dependence of dynamic moduli (G′ and G″) during cooling from 85 to 10°C, at different extraction duration times at

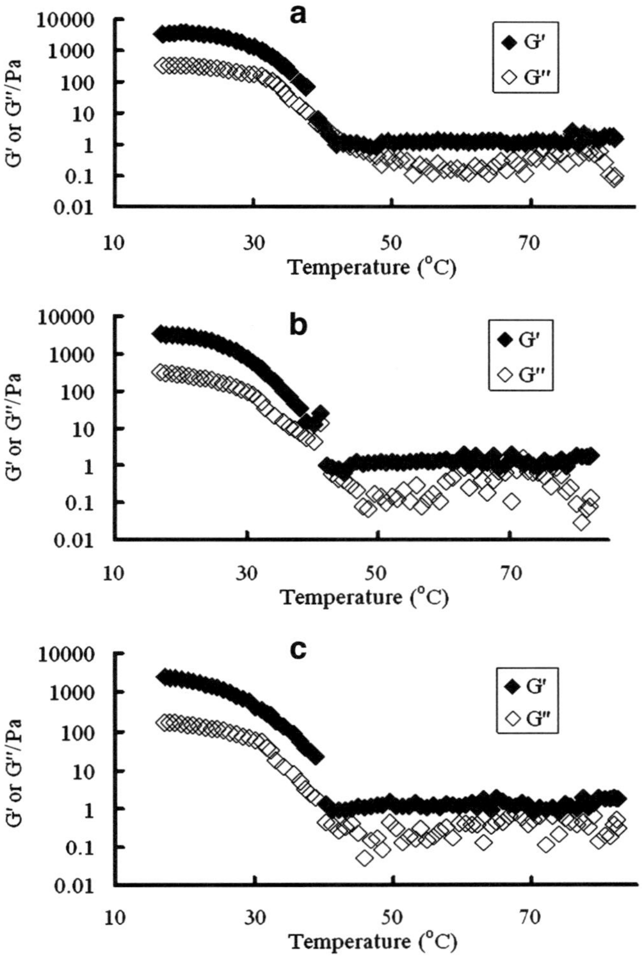

Fig. 3 a–c Temperature dependence of dynamic storage modulus G′ (◆) and dynamic loss modulus G″ (◇) of carrageenan at 90°C. **a** 1 h, **b** 3 h, **c** 5 h

90°C extraction temperature. The viscoelastic measurements of 1% w/v carrageenan at all extraction duration times (1, 3, 5 h) showed a maximum in G′ and G″ as the temperature was reduced from 85 to 10°C, with G′ being larger than G″. At the 1-h and 3-h extraction time (Fig. 3a,b), both G′ and G″ increased sharply around 42°C and then G′ exceeded G″, indicating the rapid formation of a gel network. Figure 3c shows the crossover of G′ and G″, which both increased sharply at around 40°C.

Discussion

Extraction parameters (temperature and duration time) were systematically varied to explore their effects on the gelling properties of the obtained carrageenan. The aim was to obtain gel properties (high molecular weight and high moduli) while maintaining an acceptable extraction yield. In the case of C3, the yield of carrageenan having the best gelling properties obtained in this study utilized the parameters 50°C extraction temperature and 5 h extraction duration, which led to the extraction of carrageenan having $M_w = 2.28 \times 10^6$ and $M_n = 0.9 \times 10^6$. This carrageenan presents optimized gelling properties when mixed at a concentration of 1% in distilled water, and does not exhibit any gel hysteresis. The gel showed a storage modulus G′=4,485 Pa. Hysteresis is defined as the temperature difference between the transition temperature measured upon heating and cooling (Van de Velde et al. 2005).

The maximum average molecular weight of carrageenan obtained from fractions C1, C2 and C3 was about 6×10^6, 8.5×10^6 and 3.5×10^6, respectively. The reason why C2 was larger than C1 is due to the removal of the lower molecular fraction by the addition of 2-propanol. The lower molecular weight was due to degradation from physical stress such as vacuum-drying, freezing, dehydration and so on.

Extraction conditions of fraction C2 and 1 h heating at 50 or 70°C produced more favorable carrageenan, i.e., higher molecular weight and higher yield. Extraction parameters were seen to affect the average molecular weight of carrageenan in this study; both M_w and M_n decreased with increasing extraction temperature because degradation of the molecular carrageenan can occur under the thermal movement of the molecule. Karlsson and Singh (1999) presented κ-carrageenan after acid hydrolysis at 35 and 55°C for 120 min, where the molecular weight decreased with increasing temperature. Reduction of molecular weight κ-carrageenan is due to the intense molecular movement itself because ester bonding between galactopyranose moieties is stable against heat stress.

The viscoelastic behavior of κ-carrageenan, which may be related to M_w, is generally distinguished by the same shape, high moduli likely resulting from its high molecular weight. For carrageenan yield from hybrid carrageenan from Indonesia, gel properties include very strong elasticity despite a low concentration. In all cases, the mechanical corresponding gel and thermal ramp of different heating times, which showed the same pattern for both G′ and G″, increased around 40°C with decreasing temperature. This suggests that carrageenan from hybrid carrageenan has a distinct sol–gel transition.

Our study on carrageenan extracted from Indonesia *K. alvarezii* revealed the following:

1. The biopolymers can be controlled by monitoring the extraction conditions.
2. Extraction parameters must be optimized in order to obtain carrageenan macromolecules with optimized gelling properties.
3. It is possible to obtain higher molecular weight carrageenan and higher yield under conditions of 50–70°C and 1 h heating without alkaline treatment.

References

Bixler HJ (1996) Recent developments in manufacturing and marketing carrageenan. Hydrobiologia 326/327:35–57

Haug IJ, Draget KI, Smidsrod O (2004) Physical behaviour of fish gelatin κ-carrageenan mixtures. Carbohydr Polym 56:11–19

Karlsson A, Singh SK (1999) Acid hydrolysis of sulphayted polysaccharides. Desulphation and the effect on molecular mass. Carbohydr Polym 38:7–15

Lai VMF, Lii Cheng L, Wei LH, Ting JL (2000) Kinetic compensation effect in depolymerisation of food polysaccharides. Food Chem 68:319–325

Lapasin R, De Lorenzi L, Pricl S, Torriano G (1995)Flow properties of hydroxypropylguar gum and its long chain hydrophobicderivatives.Carbohydr Polym 28:195

MacArtain P, Jacquier JC, Dawson KA (2003) Physical characteristics of calcium κ-carrageenan networks. Carbohydr Polym 53:395–400

Michon C, Chapuis C, Langendorff V, Boulenguer P, Cuvelier G (2005) Structure evolution of carrageenan/milk gels: effect of shearing, carrageenan concentration and nu fraction on rheological behavior. Food Hydrocolloids 19:541–547

Nunes MC, Raymundo A, Sousa I (2006) Rheological behavior and microstructure of pea protein/κ-carrageenan/starch gels with different setting conditions. Food Hydrocoll 20:106–113

Rudolph B (2000) Seaweed product: red algae of economic significance. In: Martin RE, Carter EP, Davis LM, Flich GJ (eds) Marine and freshwater products handbook. Technomic, Lancaster, pp 515–529

Van de Velde F, Antipova AS, Rollema HS, Burova TV, Grinberg NV, Pereira L, Gilsenan PM, Tromp RH, Rudolph B, Grinberg VY (2005) The structure of κ/ι-hybrid carrageenas II. Coil-helix transition as a function of chain composition. Carbohydr Res 340:1113–1129

Zhang Q, Matsukawa S, Watanabe T (2004) Theoretical analysis of water ^{1}H T_2 based on chemical exchange and polysaccharide mobility during gelation. Food Hydrocolloids 18:441–449

Gel-forming structures and stages of red algal galactans of different sulfation levels

Rando Tuvikene · Kalle Truus · Andres Kollist · Olga Volobujeva · Enn Mellikov · Tõnis Pehk

Originally published in the Journal of Applied Phycology, Vol 20, No 5, 77–85.
DOI: 10.1007/s10811-007-9229-9 © Springer Science + Business Media B.V. 2007

Abstract Sol-gel transition processes of algal galactans were studied using cryofixation method in combination with freeze-drying and scanning electron microscopy (SEM) techniques. The structures formed in successive stages of gelling process upon cooling were rapidly frozen at defined temperature points and viewed by SEM. It was established that in the case of both types of gelling galactans investigated, a fine honeycomb-like network exists for a wide range of solution temperatures. The formation and structure of this network depends on the structural type, gelling stage, and concentration of the galactan in solution. The honeycomb suprastructures exist also in carrageenan and agarose sols (at temperatures considerably exceeding the gelling temperatures). An additional helical network formed showed different behaviour in the case of carrageenan and agar-type polysaccharides. In the gel-formation process, tightening of the network takes place in both types of galactan gels; the honeycomb structures persist in carrageenan (furcellaran) but not in agarose gels.

Keywords Agarose · Furcellaran · Gel structure · Gelling process · Scanning electron microscopy

R. Tuvikene (✉) · K. Truus · A. Kollist
Department of Natural Sciences, Tallinn University,
Narva mnt 25,
10120 Tallinn, Estonia
e-mail: rtu@akvaarium.com

O. Volobujeva · E. Mellikov
Department of Materials Science,
Tallinn University of Technology,
Ehitajate tee 5,
19086 Tallinn, Estonia

T. Pehk
National Institute of Chemical Physics and Biophysics,
Akadeemia tee 23,
12618 Tallinn, Estonia

Introduction

Red algal galactans are sulfated polysaccharides that usually have a linear backbone built up of alternating 3-linked β-D-galactopyranose and 4-linked α-galactopyranose residues. The β-galactose residues always belong to D-series, whereas the α-galactose residues are D in carrageenans and L in agars. A substantial part of α-galactose may exist in the form of a 3,6-anhydro derivative (Fig. 1). Various hydroxy groups may be substituted with ester sulfate, methyl groups, pyruvic acid acetal and sometimes with additional monosaccharide residues (Painter 1983; Usov 1998).

The most important characteristic of the red algal galactans is their gel-forming ability, i.e. the ability to form well-ordered spatial structures during cooling of their hot polymeric solution. The gelling ability gives the basis for the vital functions of red algae as well as for their use in food industry, microbiology, chromatography, electrophoresis, etc. (Therkelsen 1993). Gel formation in an aqueous solution is a complex process that depends on polysaccharide structure, polymer concentration and temperature, but also on the presence of co- and counter-ions (Morris et al. 1980; Meunier et al. 2001). Certain cations (typically K^+ for κ-carrageenan and furcellaran, Ca^{2+} for ι-carrageenan) are found to induce conformational changes in the polymer with the initial coil-to-helix transition, which may be followed by subsequent aggregation of these helices to

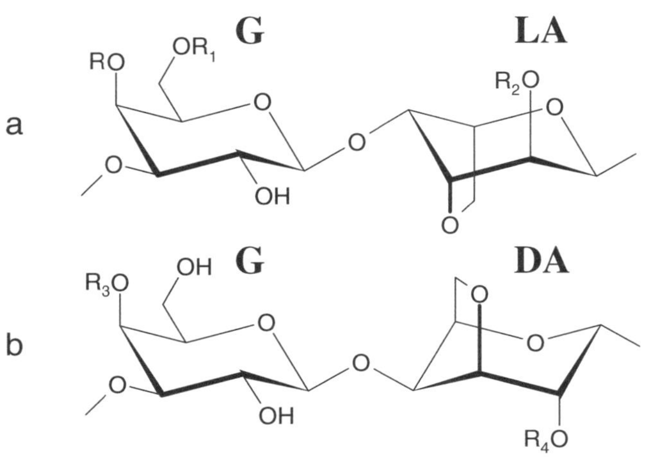

Fig. 1 Disaccharide repeating units of **a** agars and **b** carrageenans. **a** *R* is H or SO_3^-, R_1 is H or Me or SO_3^-, R_2 is H or Me. **b** R_3 is SO_3^-, R_4 is H or SO_3^- (κ- and ι-carrageenans, respectively). *G* is galactose and *A* is anhydrogalactose residue

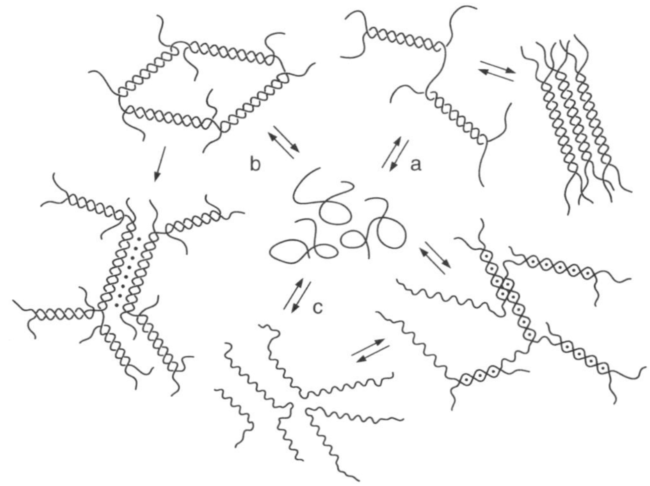

Fig. 2 Models of gel formation: *a* double-helical model described by Anderson et al. (1969), *b* domain model proposed by Morris et al. (1980) and Robinson et al. (1980), and *c* nested, single-helix model proposed by Smidsrød and Grasdalen (1982). *Dots* indicate gel-promoting cations

form a gel (Rochas and Rinaudo 1984; Paoletti et al. 1985). The final gel structure is often determined by mutual interactions between conformational transition, molecular cross-linking and phase-separation processes.

It is generally accepted that gelation mechanisms for agars and carrageenans follow different pathways. The association processes of polymer chains have been extensively investigated beginning with well-known research by Rees (1969, 1972). From this time on, the studies on gelation of carrageenans and agars have mainly been concerned with the structure details of the junction zones of the gel-forming elements (Morris 1986; Piculell 1995), and some different models of gel formation have been proposed (Fig. 2). Anderson and co-workers (1969) suggested that the junctions are formed by intertwined double helices (double-helical model), whereas Morris et al. (1980) and Robinson et al. (1980) proposed that they are caused by cation-mediated aggregates of double helices (domain model). According to the model of Smidsrød and Grasdalen (1982), the junctions in the carrageenan networks are formed by cation-specific salt bridges between ordered chain segments (nested, single-helix model). The sequence of steps leading to gelation and the type of conformational transition (coil-helix or coil-double helix) are still a matter of debate.

While elementary processes of gel formation have been intensively studied during the last decades, the macroscopic gel level that directly determines rheological properties and is responsible for gel strength has been investigated to a much lesser extent. This kind of information has mainly been obtained from light-scattering measurements (Mangione et al. 2003), from hysteresis in optical rotation between cooling and heating curves (Borgström et al. 1996), from calorimetric and conductivity measurements during melting of aggregated structure (Watase et al. 1992) and from studies of turbidity and birefringence (Aymard et al. 2001). Various microscopy techniques have gained importance in such kinds of investigations, including scanning electron microscopy (SEM) (MacArtain et al. 2003), transmission electron microscopy (Stokke et al. 1993; Hermansson 1990), atomic force microscopy (Morris et al. 1997), scanning tunnelling microscopy (Lee et al. 1992) and confocal laser scanning microscopy (Ould Eleya et al. 2006; Bourriot et al. 1999). These methods often involve dehydrating samples after freezing structures at preferred states. To avoid shear-induced damage of supramolecular structures by formation and growth of ice crystals, microscopy techniques often involve several cryofixation methods, such as rapid freezing of samples in liquid propane or nitrogen at normal pressure or under vacuum (Dunstan et al. 2001). Also cryo-SEM in combination with pressure-shift freezing is extensively used for direct imaging of the undisturbed real-space gel structures at the micron scale (Fuchigami et al. 2006).

Numerous investigations have focused on the macroscopic gel structure of carrageenan blends with sugars, proteins or other industrial gums, such as agarose (Norziah et al. 2006), locust bean gum (Dunstan et al. 2001) and konjac gum (Medina-Torres et al. 2006). It has been reported that at a high level of co-solutes, galactans are severely water deficient, resulting in the occurrence of polysaccharide chains as a sparsely cross-linked network (Evageliou et al. 1998; Whittaker et al. 1997). Other studies suggest that, under such conditions, the polymeric chains are present as small, ordered gel particulates embedded within viscous solution (Richardson and Norton 1998;

Nickerson et al. 2004). Similar characteristics have also been found in the case of "pure" solutions, where the gelation processes can be affected by spinodal demixing of sol. San Biagio and co-workers (1996) found that at agarose concentrations below 2 wt%, spinodal demixing into mesoscopic polymer-rich and solvent-rich regions takes place before coil-double helix transition, whereas at higher concentrations gelation appears to proceed directly from the homogeneous solution state. Recently, mixing and pressure influences of rheological properties and macroscopic structure of galactan gels have also been studied (Ross et al. 2006).

It should be mentioned that despite the numerous papers published about gelling processes so far, the crucial stages of macroscopic gel formation have been studied insufficiently. Also there has been no systematic investigation published on galactan gel formation by cooling. In this work, an undersulfated κ-carrageenan from *Furcellaria lumbricalis*, furcellaran, was investigated in comparison with other carrageenan and agarose preparations. Rheology of furcellaran gels is somewhat different from that of the other carrageenans, showing more elastic properties and higher gel strength.

The aim of this paper is to elucidate principal stages in the gel formation process of red algal galactans (carrageenans and agars) at the macroscopic level. Special attention is given to the structural changes accompanying gradual cooling of sol and gel formation steps. The solidification process is monitored using cryofixation method in combination with freeze-drying and SEM.

Experimental

Galactan samples

Commercial agarose sample (standard EEO) was obtained from *Serva* (Heidelberg, Germany); κ-carrageenan preparation was purchased from *Fluka*. Furcellaran was isolated (Tuvikene et al. 2006) from the red alga *Furcellaria lumbricalis* (Kassari Bay, the Baltic Sea, Estonia) by hot alkaline extraction in 0.02 M KOH solution (at 100°C, 4 h) and precipitated by isopropanol (four-fold volume per extract). Agarose from *Ahnfeltia tobuchiensis* (Truus et al. 2006) (various lots of commercial production) was obtained from the pilot production plant of the Institute of Chemistry, Estonian Academy of Sciences.

According to the labelling, the *Serva* agarose sample contains 0.06% sulfur and less than 0.01% salts. Furcellaran was found to contain (w/w) 5.8% K^+, 0.07% Na^+, 1.2% Ca^{2+} and 5.1% sulfur (bound to the matrix) as determined by ICP-OES method. This sample also contained a small amount of free salts.

SEM measurements

Scanning electron microscopy was carried out on a high-resolution LEO Supra 35 electron microscope equipped with Röntec EDX XFlash 3001 detector and Thermo Noran Maxray ER Parallel Beam spectrometer. Samples were prepared by inserting preheated (≈95°C) stainless steel capillary tubes (inner diameter 2 mm, length 60 mm) into hot (≈95°C) 2% galactan solution (w/w) and allowing them to fill up. After gelling for 4 h at 20°C, the filled tubes were tightly closed. The prepared tubes were then heated on a water bath at 98°C for 30 min, followed by slow (0.5°C/min) cooling to 20°C. The samples removed from the thermostat during the cooling stage (at predetermined temperature) were rapidly frozen in liquid nitrogen at different stages of gel formation (from 90 to 20°C), cryofractured twice to produce small (5 mm in length) open-ended tubes filled with frozen gel, and freeze-dried under vacuum at −60°C. The formed cryogenic gel surfaces were sputter-coated with platinum of about 1 nm thickness using a Polaron High Resolution Sputter Coater SC7640 and examined under an acceleration voltage up to 1.7 kV.

The highly concentrated (16%, w/w) agarose gel specimen was investigated without any precaution; the gel was freeze-dried after solidification, treated with liquid nitrogen and cryofractured.

Gel testing

For gel strength assessments, a simple gel tester equipped with a hemispherically tipped plunger (effective cross-section area 1 cm^2) was constructed. The gel strength measurements were done in triplicate for 1–2% gels (w/w) formed by dissolving dry galactan in hot water after gelling in an air thermostat at 20°C for 4 h. The cylindrical samples were 35 mm in diameter and 35 mm in height. The force needed to rupture the gel by the plunger was expressed in g/cm^2; a constant weight increase (350 g/min) was provided.

The melting temperature of the gel aged at 20°C for 4 hours was determined as the temperature at which a 4-mm glass bead fell down to the bottom of a screw-capped test tube (13 × 100 mm) during slow ($0.5°C.min^{-1}$) heating on a water bath. The tube was then cooled ($0.5°C.min^{-1}$) and turned horizontally every minute without removing it from the water bath. The gel-setting temperature was determined as the temperature at which the gel would no longer flow. All determinations of melting and gelling points were performed in duplicate.

Spectrophotometrical measurements

Optical density measurements were made on a temperature-controlled Shimadzu UV-1601 spectrophotometer at 260 or

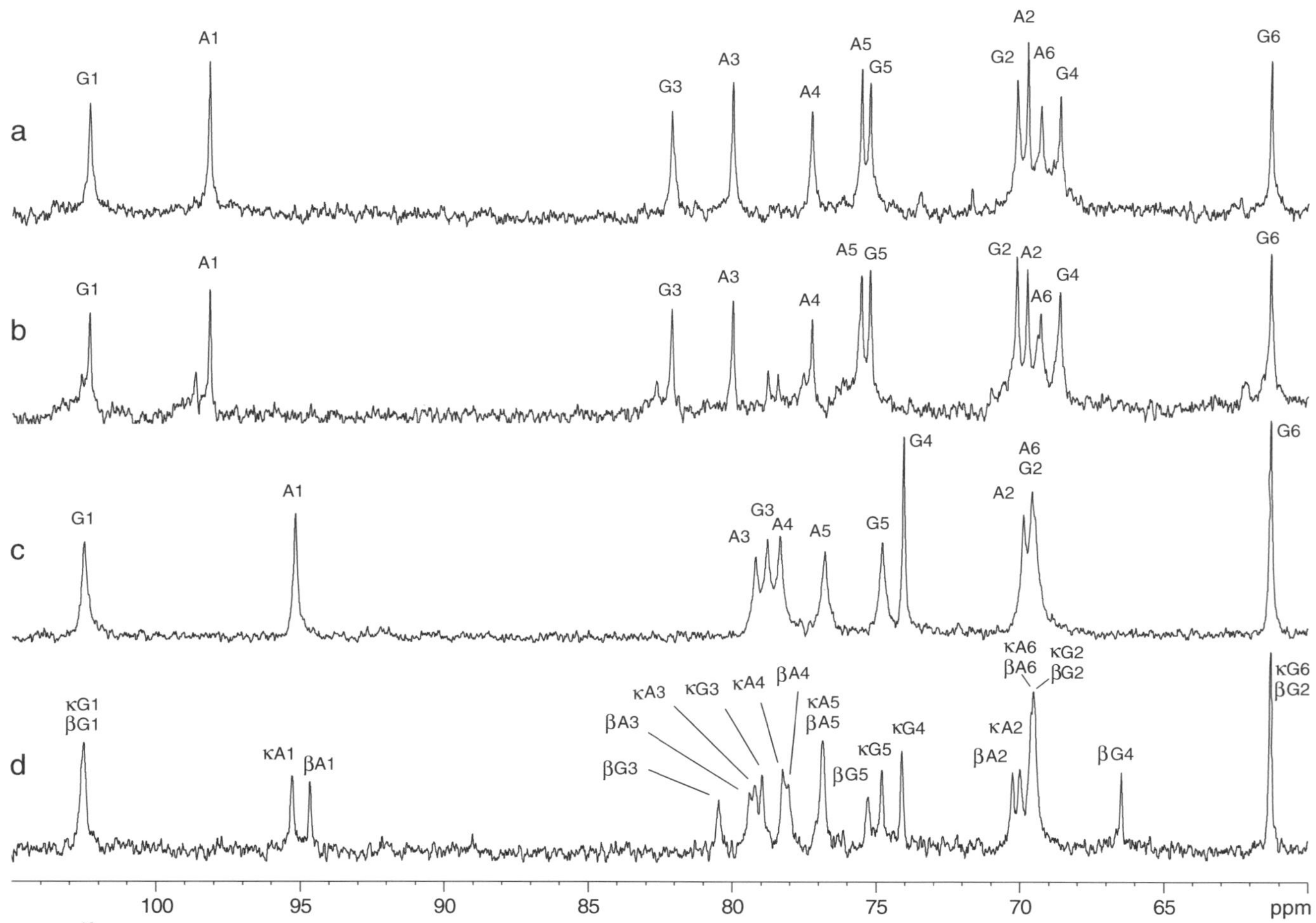

Fig. 3 ^{13}C-NMR spectra of galactan preparations. **a** Agarose *Serva*, recorded at 75°C, 6,700 transients. **b** Agarose from *A. tobuchiensis*, recorded at 75°C, 2,700 transients. **c** κ-Carrageenan *Fluka* purified by precipitation in isopropanol, recorded at 65°C, 5,700 transients. **d** Alkali-treated furcellaran, recorded at 85°C, 5,000 transients collected. β-Carrageenan signals correspond to the structure type in Fig. 1, part *b*, where R_3 and R_4 are H

400 nm against water as a blank. Hot (≈90°C) galactan solution was inserted into a 2-mm cuvette, sealed and thermostated in cuvette holder at 95°C for 5 min to allow sample equilibration. Thereafter, with continuous absorbance recording, the temperature was decreased slowly (0.5°C.min^{-1}) to 20°C, followed by isothermal hold for 50 min.

Table 1 ^{13}C-NMR chemical shifts (ppm) for basic signals of agar and carrageenan structures

Galactan	Unit	^{13}C chemical shift					
		C-1	C-2	C-3	C-4	C-5	C-6
Agarose	G	102.3	70.1	82.1	68.6	75.2	61.3
	LA	98.2	69.8	80.0	77.2	75.5	69.3
κ-Carrageenan	G	102.5	69.6	78.8	74.1	74.8	61.3
	DA	95.2	69.9	79.2	78.4	76.8	69.6
Furcellaran	κ G[a]	102.5	69.5	79.0	74.1	74.8	61.3
	κ DA[a]	95.3	70.0	79.2	78.2	76.9	69.5
	β G[b]	102.5	69.5	80.5	66.5	75.3	61.3
	β DA[b]	94.7	70.3	79.4	78.0	76.9	69.5

[a] Corresponding units from κ-carrageenan
[b] Corresponding units from β-carrageenan

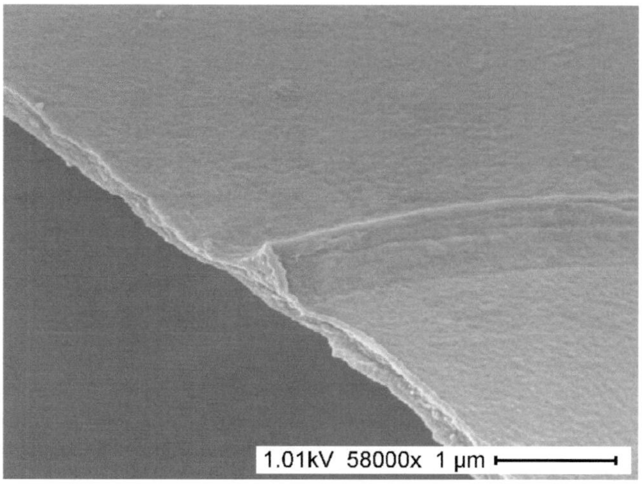

Fig. 4 SEM image of the wall of honeycomb structure (freeze-dried 2% κ-carrageenan gel)

NMR spectroscopy

Proton-decoupled ^{13}C nuclear magnetic resonance (NMR) spectroscopic analyses were carried out using a Bruker AMX-500 spectrometer. Spectra from 2% agarose or 4% carrageenan solution in D_2O (w/w) were obtained at 65–85°C, and about 2,700–6,700 transients were collected before the Fourier transform. Chemical shifts were converted to a tetramethyl silane scale on the basis of the C6 signal from the galactose subunit with a constant value (61.3 ppm) in these galactans (Usov and Shashkov 1985).

Results and discussion

Molecular structure

^{13}C-NMR spectra of four galactans investigated are presented in Fig. 3. The spectra allow (Usov et al. 1980) different galactan types (agaroses, carrageenans) to be distinguised on the basis of absolute configuration of the 1,4-linked anhydrogalactose ring. Undersulfation in furcellaran molecule causes additional resonance signals in ^{13}C-NMR spectrum (compared to commercial κ-carrageenan preparations) corresponding to carbons of non-sulfated β-carrageenan units (Knutsen and Grasdalen 1987). The values of chemical shifts for basic carbon signals of the galactans investigated are presented in Table 1. The values are in accordance with literature data (Van de Velde et al. 2002). The shift values of two investigated agaroses of different origin coincide, which is typical of commercial high-quality agaroses (Truus et al. 2006). Thus the structural type and characteristics of the galactans to be studied were established.

Scanning electron microscopy

To avoid structural damages to shock-sensitive SEM preparations during the rapid cooling process, galactan solutions were tightly enclosed in tubes (except in the case of very high concentrations, for which the mechanical strength of the sample is high). Material and dimensions of

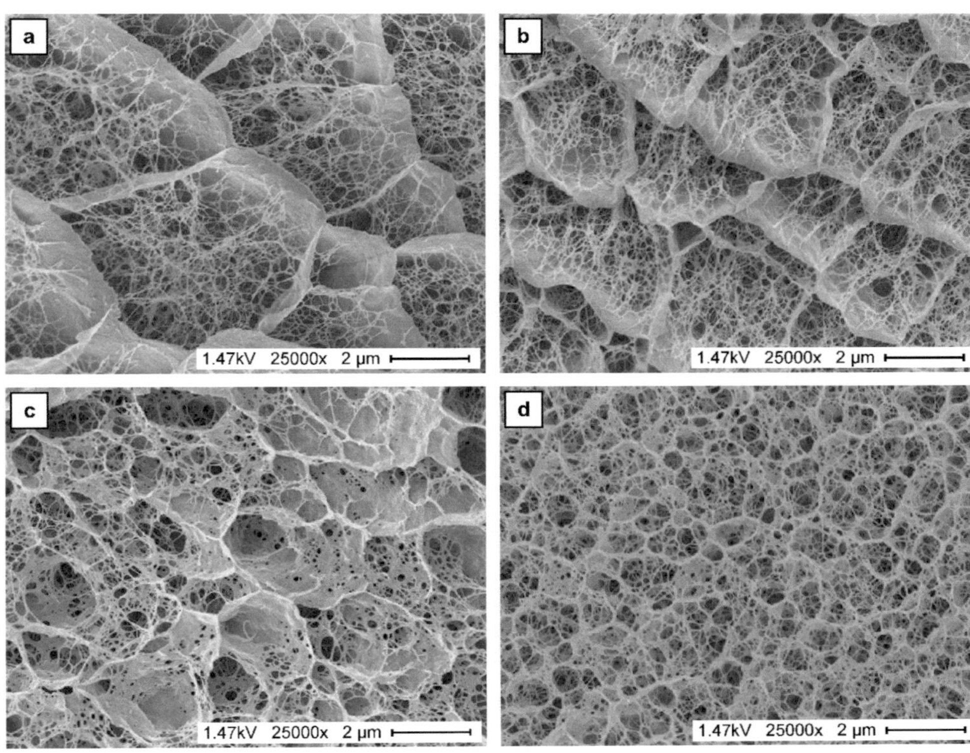

Fig. 5 SEM images of 2% agarose gel network structures at **a** 70°C, **b** 40°C, **c** 30°C and **d** 20°C

the tubes were chosen to guarantee the rapid freezing process and to handle samples without loss of the frozen state for any moment. Subsequent use of cryo-fracture and freeze-drying techniques were calculated to preserve the real structures corresponding to the desired gel-forming stage as well as to preserve the surface with an appropriate texture for SEM investigations. Generally, it has been accepted for some decades that ultrafast freezing techniques allow for the avoidance of distortions and defects in specimens even to the nanometer scale (Echlin 1978; Robards 1991). Also it was concluded from the gradual elaboration of the preparation technique and from the quality of the images achieved that any considerable damage to the specimens or any corresponding artefacts before direct SEM investigations had been avoided. From the investigation of the gelling processes based on SEM images, the following observations were made.

We observed novel precursor structures already at quite high (70°C) temperatures, significantly exceeding the macroscopic gelling temperature (which is the final result of numerous gel-forming microprocesses). Characteristic honeycomb structures appeared in agarose and furcellaran sols of low concentration already at high temperatures considerably exceeding the gelling point. The walls of the honeycomb structures were rather thin (about 100 nm in carrageenan gels, Fig. 4) for conventional SEM investigations. The quality of the images and high resolution are due to the new-generation microscope used. Similar supramolecular structures have been described in the case of mixed gels (Dunstan et al. 2001).

Compared to furcellaran, the structural units of the honeycomb network (cells) in agarose were considerably smaller, for example, at the gelling stage of 20°C, nearly a 10-fold difference existed. In the case of agarose gels, there

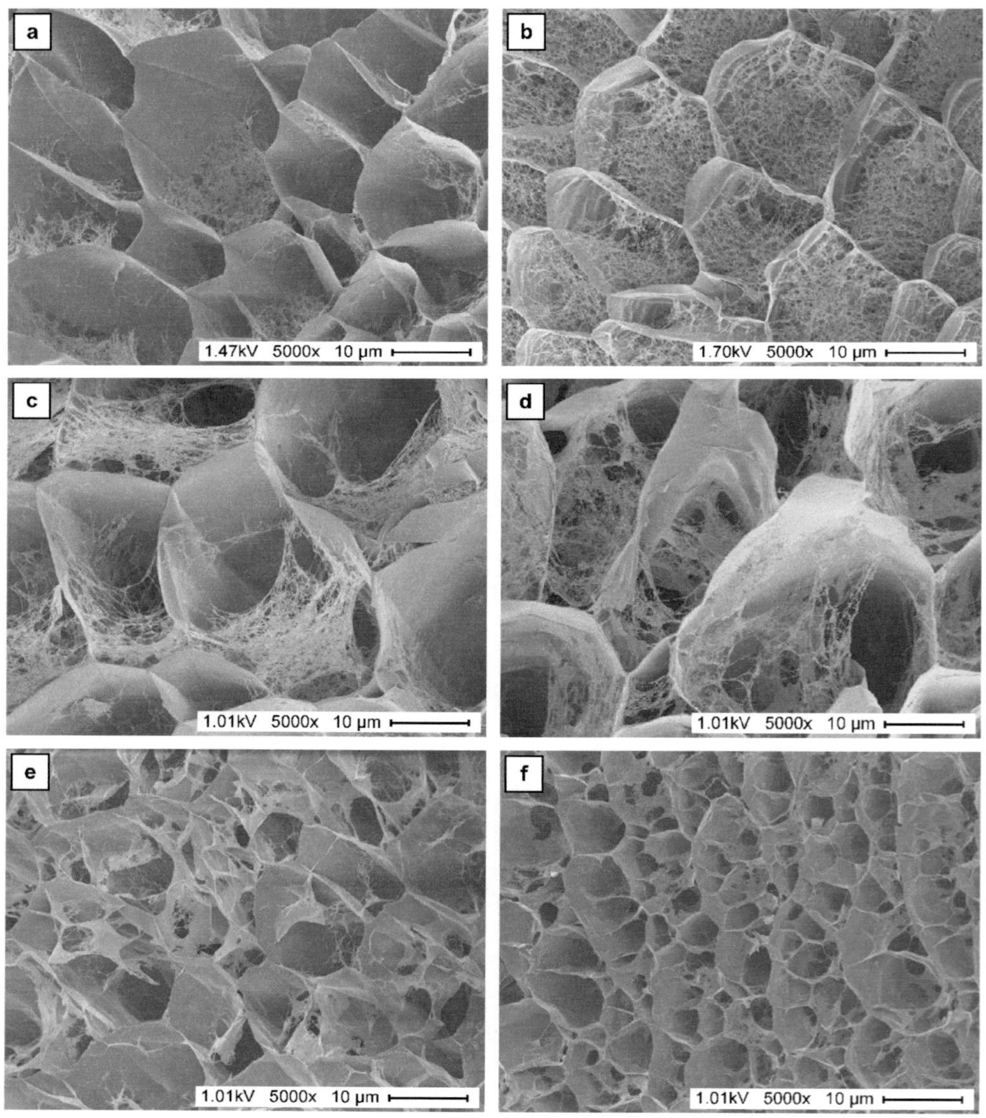

Fig. 6 SEM images of 2% furcellaran gel network structures at **a** 70°C, **b** 50°C, **c** 40°C, **d** 35°C, **e** 30°C and **f** 20°C

was a high dependence of cell diameter on the temperature of the gelation stage; furcellaran gels did not show such a strong correlation (not shown).

The honeycomb structures are filled with tight polysaccharide networks, detailed peculiarities of which depend on the gelling stage or temperature. In the case of agarose gels, the quota of reticulated (helical) structures increased through the cooling process during which honeycomb structures disappeared. An opposite tendency was observed in furcellaran gels, where the honeycomb pattern persisted also at low temperatures but the thin network disappeared (compare Figs. 5 and 6). This was the main qualitative difference in the gel-formation process between agar- and carrageenan-type polysaccharides. Also it was remarkable that the thickening of honeycomb structures in furcellaran gel occurred due to decreasing proportion of thin polysaccharide network; in this case, conversion of the network to honeycomb structure takes place (Fig. 6c,e).

The fact that honeycomb structures were able to change their shape and disappear (under the same sample preparation conditions) assures that there no substantial artefacts were being created. Also, the specimens did not show any cleavage or other damage. On the other hand, these honeycombs are not static walls (because of their liquid state) but rather potential and dynamic precursor structures reflecting the hidden collectivity tendencies of this type of macromolecules.

The greatest changes in gel structure during the cooling process took place considerably below the gelling temperature of the galactan sol. This is related to the thickening of structures, which caused a high increase in optical density (see Figs. 5, 6, 7). At that, some characteristic honeycomb structures existed in both types of galactan gels observed. The walls of honeycombs are quite specific structural units. These seem to be thin, even films without any noticeable structure. This seemingly plain structure is conditioned by the resolution limits of SEM techniques at the present time. As it appears from the images (Fig. 6c), wall formation is related to helical structure. However, it can be seen that the walls were double-layered (Fig. 4); compared to carrageenan gels, wall thickness was much less in agarose gels.

Depending on the structural characteristics of agarose, the macroscopic gel structure can vary considerably even in the case of preparations gelled under the same conditions; samples of lower gel strength (Table 2) showed sparser structural patterns (Fig. 8a,b). It seems to be evident that the course of gel formation as well as the final gel structure depends on numerous characteristics such as galactan structure, sol concentration, and gelling conditions. Also the cell size was highly dependent on galactan concentration (Fig. 8b,c). Gels of very high concentrations were of markedly different structure (Fig. 8d).

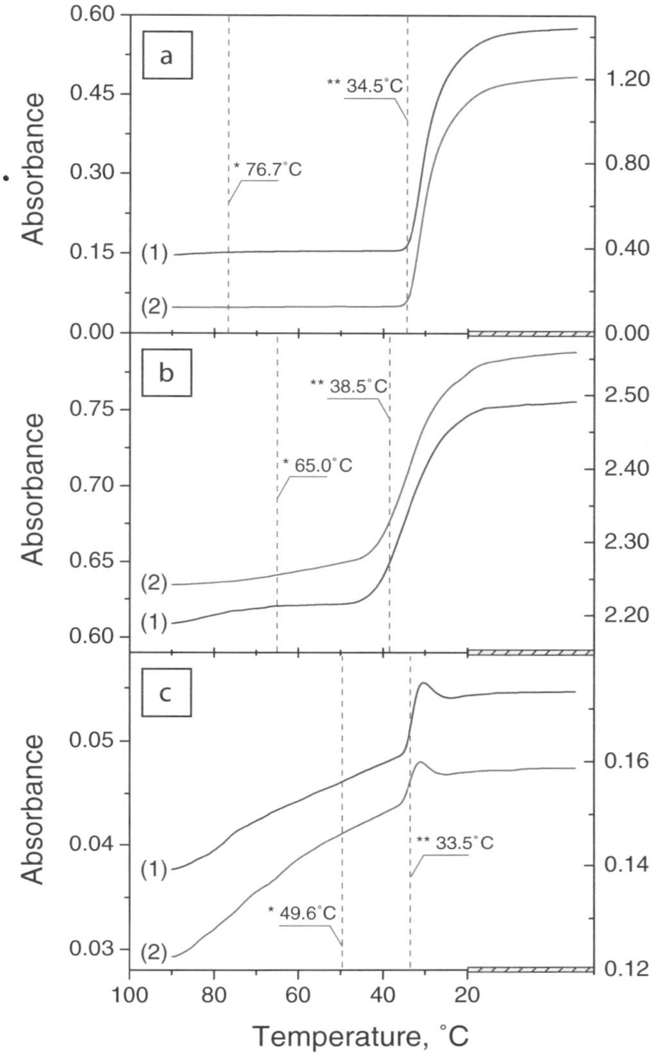

Fig. 7 Gelling profiles for galactan solutions (2%, w/w) at cooling rate of 0.5°C/min. Absorbance measured at 260 nm (*right scale*) and 400 nm (*left scale*) in 2-mm cuvette. *Two asterisks* indicate gelling temperature, *one asterisk* indicates gel melting temperature, (▨) indicates the isothermal region (20°C). **a** Agarose from *Ahnfeltia tobuchiensis*, **b** furcellaran, **c** κ-carrageenan (*Fluka*)

Spectrophotometrical and rheological investigations

The formation and stabilization of tightly packed structures is related to an increase in optical density (Fig. 7). For agarose gels this change is sharper and expresses well their gelling temperature. Although in the case of furcellaran, the tightening of large-cell honeycomb patterns occurs mainly in the gel state, the increase in optical density of their sols takes place well before reaching the gelling temperature.

The commercial κ-carrageenan preparation was evidently somewhat degraded (with a weak gelling ability at low concentration, Table 2) and contained impurities, probably

Table 2 Gel strength at various galactan concentrations

Galactan	Gel strength (g/cm^2)		
	1%	1.5%	2%
Agarose *Serva*	665	1,360	1,850
Agarose from *A. tobuchiensis*	260	500	725
κ-Carrageenan *Fluka*	15	380	1,350
Alkali-treated furcellaran[a]	445	800	1,110

[a] Extracted in 0.02 M KOH solution (at 100°C, 4 h)

some low-molecular carbohydrates, as confirmed by NMR spectra (not shown). The impurities affected the whole gelling process (Fig. 7c), and the preparations made were unsuitable for SEM investigations. In contrast, good gelling properties of furcellaran permitted preparation of high-quality SEM specimens.

Conclusions

In the gelling process of polysaccharides containing 3,6-anhydrogalactose, a fine honeycomb-like network participates as a "hidden" precursor structure. This network exists in galactan solutions at significantly higher temperatures than those of rheological changes related to gel formation.

Although the honeycomb structures exist in carrageenan as well as in agarose sols and gels, the behaviour of this fine network in gelling processes depends on the structural type of the galactan, i.e. on the configuration of the anhydrogalactose ring (AGR).

Carrageenan-type galactans (containing AGR of D-configuration) form gels mainly by tightening of the honeycomb network; other structures disappear in the course of the final gel formation. Correspondingly, carrageenan gels consist of a very fine network and are highly elastic. In contrast, agar-type galactans (containing L-AGR) gradually lose their honeycomb structure in the course of gel formation. Therefore the final structure of agar/agarose gels is built up of a more tight and homogeneous network, resulting in brittle, strong gels of high light absorbance.

The gel-forming process is highly sensitive to structural and compositional characteristics of galactans; gels of high galactan concentration may be of essentially different structure than those of low polysaccharide content.

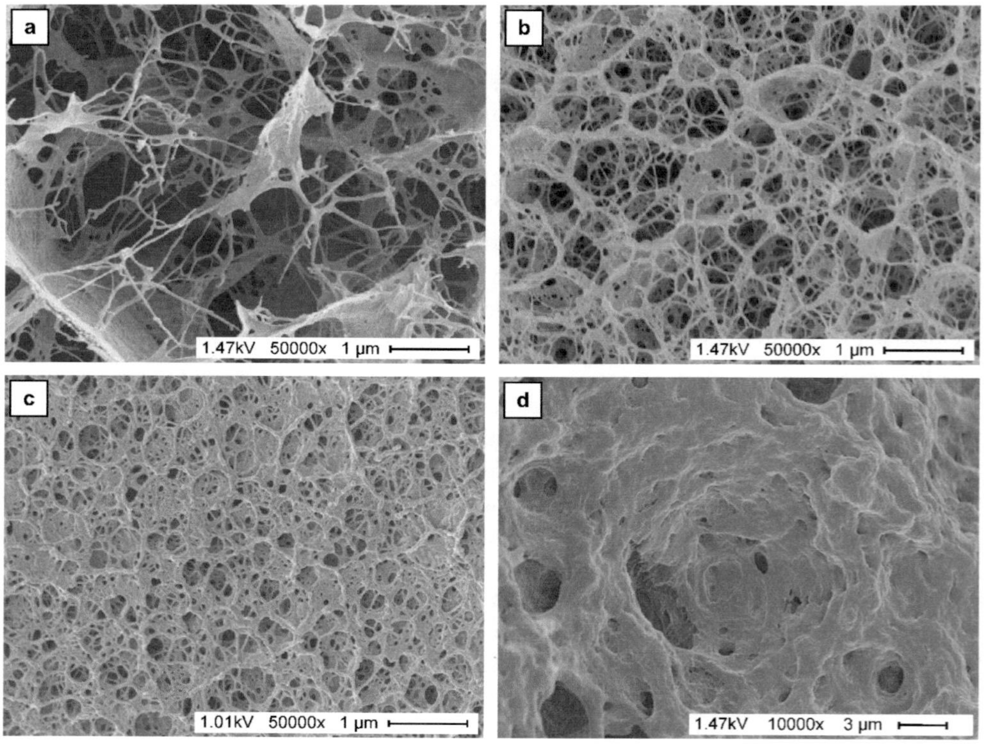

Fig. 8 SEM images of gel network structures at 20°C of **a** 2% agarose from *Ahnfeltia tobuchiensis*, and **b** 2%, **c** 6% and **d** 16% agarose (*Serva*)

References

Anderson NS, Campbell JW, Harding MM, Rees DA, Samuel JWB (1969) X-ray diffraction studies of polysaccharide sulfates: double helix model for κ- and ι-carrageenans. J Mol Biol 45:85–99

Aymard P, Martin DR, Plucknett K, Foster TJ, Clark AH, Norton IT (2001) Influence of thermal history on the structural and mechanical properties of agarose gels. Biopolymers 59:131–144

Borgström J, Piculell L, Viebke C, Talmon Y (1996) On the structure of aggregated kappa-carrageenan helices. A study by cryo-TEM, optical rotation and viscometry. Int J Biol Macromol 18:223–229

Bourriot S, Garnier C, Doublier J-L (1999) Micellar casein/kappa-carrageenan mixtures. 1. Phase separation and ultrastructure. Carbohydr Polym 40:145–157

Dunstan DE, Chen Y, Liao M-L, Salvatore R, Boger DV, Prica M (2001) Structure and rheology of the kappa-carrageenan/locust bean gum gels. Food Hydrocolloids 15:475–484

Echlin P (1978) Low-temperature biological scanning electron microscopy. In: Koehler KJ (ed) Advanced techniques in biological electron microscopy II. Springer, Berlin, pp 89–122

Evageliou V, Kasapis S, Hember MWN (1998) Vitrification of κ-carrageenan in the presence of high levels of glucose syrup. Polymer 39:3909–3917

Fuchigami M, Teramoto A, Jibu Y (2006) Texture and structure of pressure-shift-frozen agar gel with high visco-elasticity. Food Hydrocolloids 20:160–169

Hermansson A-M (1990) Structure and rheological properties of kappa-carrageenan gels. In: Burchard W, Ross-Murphy SB (eds) Physical networks: polymers and gels. Elsevier, London, pp 271–282

Knutsen SH, Grasdalen H (1987) Characterization of water-extractable polysaccharides from Norwegian *Furcellaria lumbricalis* (Huds.) Lamour. (Gigartinales, Rhodophyceae) by IR and NMR spectroscopy. Bot Mar 30:497–505

Lee I, Atkins EDT, Miles MJ (1992) Visualisation of the algal polysaccharide carrageenan by scanning tunnelling microscopy. Ultramicroscopy 42–44:1107–1112

MacArtain P, Jacquier JC, Dawson KA (2003) Physical characteristics of calcium induced κ-carrageenan networks. Carbohydr Polym 53:395–400

Mangione MR, Giacomazza D, Bulone D, Martorana V, San Biagio PL (2003) Thermoreversible gelation of kappa-carrageenan: relation between conformational transition and aggregation. Biophys Chem 104:95–105

Medina-Torres L, Brito-De La Fuente E, Gómez-Aldapa CA, Aragon-Piña A, Toro-Vazquez JF (2006) Structural characteristics of gels formed by mixtures of carrageenan and mucilage gum from *Opuntia ficus indica*. Carbohydr Polym 63:299–309

Meunier V, Nicolai T, Durand D (2001) Structure of aggregating κ-carrageenan fractions studied by light scattering. Int J Biol Macromol 28:157–165

Morris ER, Rees DA, Robinson G (1980) Cation-specific aggregation of carrageenan helices: domain model of polymer gel structure. J Mol Biol 138:349–362

Morris VJ (1986) Gelation of polysaccharides. In: Mitchell JR, Ledward DA (eds) Functional properties of food macromolecules. Elsevier, London, pp 121–170

Morris VJ, Gunning AP, Kirby AR, Round A, Waldron K, Ng A (1997) Atomic force microscopy of plant cell walls, plant cell wall polysaccharides and gels. Int J Biol Macromol 21:61–66

Nickerson MT, Paulson AT, Speers RA (2004) Time-temperature studies of gellan polysaccharide gelation in the presence of low, intermediate and high levels of co-solutes. Food Hydrocolloids 18:783–794

Norziah MH, Foo SL, Karim AA (2006) Rheological studies on mixtures of agar (*Gracilaria changii*) and κ-carrageenan. Food Hydrocolloids 20:204–217

Ould Eleya MM, Leng XJ, Turgeon SL (2006) Shear effects on the rheology of β-lactoglobulin/β-carrageenan mixed gels. Food Hydrocolloids 20:946–951

Painter TJ (1983) Algal polysaccharides. In: Aspinall GO (ed) The polysaccharides, vol. 2. Academic, New York, pp 195–285

Paoletti S, Delben F, Cesàro A, Grasdalen H (1985) Conformational transition of κ-carrageenan in aqueous solution. Macromolecules 18:1834–1841

Piculell L (1995) Gelling carrageenans. In: Stephen AM (ed) Food polysaccharides and their applications. Marcel Dekker, New York, pp 205–244

Rees DA (1969) Structure, conformation, and mechanism in the formation of polysaccharide gels and networks. Adv Carbohydr Chem Biochem 24:267–332

Rees DA (1972) Polysaccharide gels. A molecular view. Chem Ind (London) 19:630–636

Richardson RH, Norton IT (1998) Gelation behavior of concentrated locust bean gum solutions. Macromolecules 31:1575–1583

Robards AW (1991) Rapid-freezing methods and their applications. In: Hall JL, Hawks C (eds) Electron microscopy of plant cells. Academic, London, pp 257–312

Robinson G, Morris ER, Rees DA (1980) Role of double helices in carrageenan gelation: the domain model. J Chem Soc Chem Commun 1980:152–153

Rochas C, Rinaudo M (1984) Mechanism of gel formation in κ-carrageenan. Biopolymers 23:735–745

Ross KA, Pyrak-Nolte LJ, Campanella OH (2006) The effect of mixing conditions on the material properties of an agar gel. Food Hydrocolloids 20:79–87

San Biagio PL, Bulone D, Emanuele A, Palma-Vittorelli MB, Palma MU (1996) Spontaneous symmetry-breaking pathways: time-resolved study of agarose gelation. Food Hydrocolloids 10:91–97

Smidsrød O, Grasdalen H (1982) Some physical properties of carrageenan in solution and gel state. Carbohydr Polym 2:270–272

Stokke BT, Elgsaeter A, Kitamura S (1993) Macrocyclization of polysaccharides visualized by electron microscopy. Int J Biol Macromol 15:63–68

Therkelsen GH (1993) Carrageenan. In: Whistler RL, BeMiller JN (eds) Industrial gums: polysaccharides and their derivatives, 3rd ed. Academic, San Diego, pp 145–180

Truus K, Tuvikene R, Vaher M, Kailas T, Toomik P, Pehk T (2006) Structural and compositional characteristics of gelling galactan from the red alga *Ahnfeltia tobuchiensis* (Ahnfeltiales, the Sea of Japan). Carbohydr Polym 63:130–135

Tuvikene R, Truus K, Vaher M, Kailas T, Martin G, Kersen P (2006) Extraction and quantification of hybrid carrageenans from the biomass of the red algae *Furcellaria lumbricalis* and *Coccotylus truncatus*. Proc Estonian Acad Sci Chem 55:40–53

Usov AI (1998) Structural analysis of red seaweed galactans of agar and carrageenan groups. Food Hydrocolloids 12:301–308

Usov AI, Shashkov AS (1985) Polysaccharides of algae. XXIV. Detection of iota-carrageenan in *Phyllophora brodiaei* (Turn.) J. Ag. (Rhodophyta) using ^{13}C-NMR spectroscopy. Bot Mar 28:367–373

Usov AI, Yarotsky SV, Shashkov AS (1980) ^{13}C NMR spectroscopy of red algal galactans. Biopolymers 19:977–990

Van de Velde F, Knutsen SH, Usov AI, Rollema HS, Cerezo AS (2002) ^{1}H and ^{13}C high resolution NMR spectroscopy of carrageenans: applications in research and industry. Trends Food Sci Technol 13:73–92

Watase M, Kohyama K, Nishinari K (1992) Effects of sugars and polyols on the gel-sol transition of agarose by differential scanning calorimetry. Thermochim Acta 206:163–173

Whittaker LE, Al-Ruqaie IM, Kasapis S, Richardson RK (1997) Development of composite structures in the gellan polysaccharide/sugar system. Carbohydr Polym 33:39–46

Carrageenan of *Eucheuma isiforme* (Solieriaceae, Rhodophyta) from Nicaragua

Yolanda Freile-Pelegrín · Daniel Robledo

Originally published in the Journal of Applied Phycology, Vol 20, No 5, 87–91.
DOI: 10.1007/s10811-007-9270-8

Abstract The yield and physicochemical properties of native and alkali treated carrageenan from *Eucheuma isiforme* harvested from the Nicaraguan coast were investigated. The native carrageenan yield was 57.2% of dry weight and decreased to 43.5% when the alga was alkali treated. Native carrageenan viscosities showed significant differences between native (144.6±3.3 cPs) and treated carrageenan (113.9±2.6 cPs) (p<0.01). Alkali treatment reduced carrageenan sulphate content by 19.3% and increased 3,6 AG content by 13%. Alkali-treated carrageenan formed very weak gels in 1.5% solutions (<50 g cm^{-2}). Chemical analysis and FTIR spectra revealed that *Eucheuma isiforme* from Nicaragua is a good source of relatively pure iota-carrageenan with sufficient quality to serve as a substitute for traditional iota-carrageenan sources.

Keywords Carrageenan · *Eucheuma isiforme* · Extraction · Gel properties · Structure

Introduction

During the last 15 years, the carrageenan industry has been growing around 8% per annum producing 28 000 metric tonnes of carrageenan with a value of US$ 270 million (McHugh 2001). Increasing worldwide demand and development of new applications for carrageenan have added urgency to the search for new or additional raw material sources.

Y. Freile-Pelegrín (✉) · D. Robledo
Department of Marine Resources, Cinvestav,
Km 6 Carretera Antigua a Progreso Cordemex,
97310, A.P. 73 Mérida, Yucatán, México
e-mail: freile@mda.cinvestav.mx

The tropical seaweed *Eucheuma isiforme* (C. Agardh) J. Agardh has been described for the Gulf of Mexico and Caribbean coast (Cheney 1988). Previous studies on reproductive, biochemical aspects and carrageenan have been well documented for populations of *E. isiforme* from Florida (Dawes et al. 1974a, b; Dawes 1977) and recently from the Yucatan coast (Freile-Pelegrin et al. 2006; Freile-Pelegrin and Robledo 2006). The carrageenan content and properties, as well as the biochemical composition, of *E. isiforme* from Yucatan (Gulf of Mexico) differed from those of *E. isiforme* from Florida (US Gulf coast) mainly because of the higher seawater nutrient content on the Yucatan coast (Freile-Pelegrin and Robledo 2006). Environmental factors are known to influence phycocolloid yield and quality (Chopin et al. 1990, Brown 1995), and different physiological and environmental tolerances may also influence variation in carrageenan content. This has been reported for different algae species, between different life stages of the same species (Piriz and Cerezo 1991), and even between individuals of the same species growing under different environmental conditions.

In the Caribbean, *E. isiforme* has been reported as the most important carrageenophyte and was harvested commercially in Belize, Antigua and Barbuda for traditional food applications ('seamoss'). However, this was discontinued as a result of over-exploitation in the 1980s (Smith 1998). In spite of the richness of the Caribbean seaweed flora and the region's proximity to industrial processing facilities in North America and Europe, far less attention has been paid to this species as raw material for the phycocolloid industries, and its use at present has been limited to the preparation of traditional drinks and puddings (Espinosa-Avalos 1994). The scarcity of available data for *E. isiforme* from the Caribbean prevents any assessment of its potential use for carrageenan extraction. In particular, no previous reports on economic important seaweeds in Nicaragua have been produced. During a preliminary evaluation of seaweeds from the Caribbean coast of Nicaragua, *E. isiforme* was the most abundant species identified for potential use.

The present study was intended to improve understanding of carrageenan extracted from *E. isiforme* in preparation for its potential exploitation in Nicaragua. Determining biochemical composition and phycocolloid characteristics may help in understanding the physiological status of the algae, thus providing data for its use as raw material for industry. The carrageenan properties (i.e., yield, gel strength and viscosity) of native and alkali-treated carrageenans are described. Structural analyses were also done, including sulphate, 3,6-anhydro-D-galactose content and infrared spectroscopy.

Materials and methods

Twelve species of macroalgae, including *E. isiforme*, were collected from natural beds at Blue Fields (Nicaragua) in October 2004. The identification of algal material was done according to Wynne (2005). *Eucheuma isiforme* was washed thoroughly with tap water to remove excess salts and sand, oven-dried at 60°C and then milled prior to carrageenan extraction. *Eucheuma isiforme* biomass was free of epiphytes and thus considered 'pure seaweed'.

Ash content was determined according to Dawes (1977). Total protein was determined as proposed by Lowry et al. (1951), and total carbohydrate by the phenol sulphuric acid method (Dubois et al. 1956). All values are presented as percent dry weight.

The carrageenan from *E. isiforme* was obtained using the hot alkaline extraction method described by Freile-Pelegrin et al. (2006). Dry samples (5 g) were rehydrated at room temperature for 12 h in 500 mL of KOH solution (1% w/v), followed by the hot alkali extraction at 85°C during 3 h. The extract was mixed with diatomaceous earth (Celite), pressure filtered and the filtrate neutralized to pH 8.9 with 5 M HCl prior to the recovery of the carrageenan from the solution. Carrageenan was precipitated by slow addition of 250 mL of 2% CTAB (hexadecyl-trimethylamonium bromide) in 9:1 distilled water:acetone (Craigie and Leigh 1978; Chopin et al. 1990) and recovered over paper filter in vacuo. The fibrous carrageenan was carefully washed three times with 63 mL 95% ethanol nearly saturated with sodium acetate to remove CTAB residues. Sodium acetate was removed with three final washings with 95% ethanol and the carrageenan recovered in the same paper filter. The coagulum was dried for 24 h at 60°C, then weighed to calculate percent yield from dry and powdered seaweed. The same procedure without KOH was performed to obtain native carrageenan. All extractions were done in triplicate.

Rheological and chemical analysis

Rheological properties were measured for the native and alkali treated carrageenans. Water gel strength was determined according to Freile-Pelegrín and Robledo (1997) in a 1.5% w/v carrageenan solution using a Nikansui Shiki gelometer (1 cm^2 plunger). Viscosity was measured using a Cole Parmer Viscosimeter (Vernon Hills, Ill., USA) with a low centipoise adapter at 20 rpm (spindle number 8) on 18-mL samples of a 1.5% carrageenan solution, which were homogenized and allowed to stabilize in a recirculating bath at 75°C.

Sulfate content was measured turbidimetrically after hydrolyzing 25 mg carrageenan in sealed tubes for 12 h in 1 N HCl at 105°C (Jackson and McCandless 1978). The 3,6 anhydrogalactose content (3,6 AG) was determined following Matsuhiro and Zanlungo (1983). Molar ratios of galactose to 3,6 AG to ester sulfate were calculated based on the total carbohydrate content in the algae. Galactose content is expressed as total carbohydrate content in the algae minus the corresponding 3,6 AG.

Carrageenans extracted were analyzed by Fourier Transformed Infrared spectroscopy (FTIR). A commercial grade, Type II, predominantly iota carrageenan (SIGMA) was used as standard. About 4.0 mg of carrageenan were mixed thoroughly in a mortar with 200 mg of potassium bromide until homogenized. The infrared spectra of native and alkali-modified carrageenan were recorded on a Thermo-Nicolet Nexus 670 FT-IR spectrometer equipped with a DTGS KBr detector and purge gas generator at a spectral resolution of 0.09 cm^{-1} and a wave length precision of 0.01 cm^{-1}. Each spectrum (32 scans) was acquired at a resolution of 4 cm^{-1}

Statistical analysis

Data were tested for normality (Kolmogorov–Smirnov) and homogeneity of group variances (Bartlett's test) using statistical software (Statistica 6.0, Statsoft) and were treated statistically by one-way analysis of variance (ANOVA). Carrageenan characteristics from *E. isiforme* collected in Nicaragua and Yucatan were compared using a two-way ANOVA.

Results

The ash, protein and carbohydrate contents for *Eucheuma isiforme* from Nicaragua were 34.8±0.5%, 1.9±0.1% and 58.4±0.5%, respectively. The content and properties of native and alkali-treated carrageenan from *E. isiforme* from Nicaragua are summarized in Table 1. The native carrageenan yield was 57.2% of dry weight. A reduction in yield by 23.9% was observed after alkali treatment. Native and alkali-treated carrageenan formed very weak gels, <50 g cm^{-2} in 1.5% solutions. Carrageenan viscosity decreased after alkali treatment. Alkali treatment reduced carrageenan sulphate content by 19.3% and increased the 3,6 AG content by 13%. This is reflected in the molar ratios obtained for both carrageenans (Table 1). In order to compare carrageenan properties between Nicaraguan and Mexican material, data from Freile-Pelegrin et al. (2006) were also included in Table 1. In this regard, statistical analysis showed that all carrageenan properties had significant differences both between locations and between extraction conditions ($p<0.01$).

FTIR spectra are shown in Fig. 1. All spectra displayed an absorption band at 1,220-1,240 cm^{-1} related to sulphation level (Stancioff and Stanley 1969). The intense signal around 930 cm^{-1} was consistent with the presence of 3,6 AG (Stancioff and Stanley 1969). An increase at this peak was evident between the native and alkali-treated carrageenans implying the presence of the precursor, 1,4-linked galactose-6-sulfate. The spectrum of native carrageenan exhibited a shoulder at 867 cm^{-1}, indicating a sulphate group at C-6, and suggesting the presence of nu-carrageenan, considered the biological precursor to iota-carrageenan (Bodeau-Bellion 1983).

A particularly intense signal was recorded in all samples at 845–847 cm^{-1}, which is assigned to galactose-4-sulphate, and which corroborates the existence of a kappa- and iota-carrageenan mixture (Chopin et al. 1990). Another signal at 803–805 cm^{-1} was attributed to 3,6-anhydrogalactose-2-sulphate and specific to iota carrageenan. The peaks at 805 cm^{-1} and 845 cm^{-1} exhibited slight changes after alkaline transformation. The ratio between 805 and 845 cm^{-1} absorption bands in FTIR spectra was calculated and used as a qualitative parameter to determine the degree of iota/kappa hybridization (Rochas et al. 1986). The treated carrageenan and iota standard presented a similar ratio (0.86 and 0.82, respectively) greater than that obtained for native carrageenan (0.63). The increase in the ratio 805/845 in the alkali-treated carageenan corresponded to an increment of the iota faction relatively to kappa fraction.

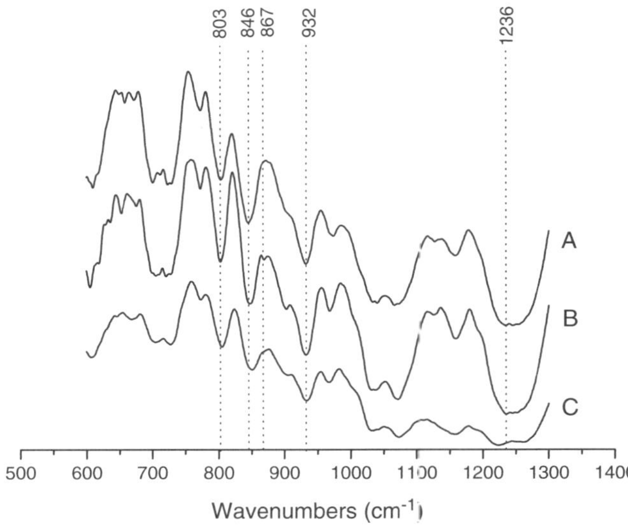

Fig. 1 FTIR spectra of native (**A**) and alkali treated carrageenan (**B**) from *Eucheuma isiforme* from Nicaragua, and commercial iota carrageenan (**C**)

Discussion

The carrageenan yields showed a comprehensible slight decrease since the hot alkaline extraction operations inevitably involve some degradation of the polysaccharide due to the rigors (heat, alkalinity) of processing (Stanley 1987). In this regard, yield reduction after alkali treatment for *E. isiforme* from Yucatan and Nicaragua were similar (ca. 24%), although the Nicaraguan material showed higher yields when compared to those reported by Freile-Pelegrin et al. (2006) (Table 1).

Table 1 Native and alkali-treated carrageenan properties of *Eucheuma isiforme* from Nicaragua compared with Yucatan

	E. isiforme from Nicaragua		*E. isiforme* from Yucatan	
	Native	Alkali-treated	Native	Alkali-treated
Carrageenan yield (%)	57.2±0.3[a]	43.5±0.7[b]	44.5±0.4[a]	33.8±0.5[b]
Sulfate (%)	32.6±0.13[a]	26.3±0.41[b]	31.0±1.65[a]	19.6±0.39[b]
3,6 AG (%)	19.4±0.13[a*]	22.0±0.04[b*]	27.5±0.30[a]	34.0±0.89[b]
Gal:3,AG:Sulfate	1:0.37:1.57	1:0.44:1.36	1:0.73:2.27	1:1.00:1.67
Viscosity (cps)	144.6±3.4[a]	113.9±2.6[b]	57.0±0.9[a]	160.0±0.0[b]

Carrageenan data from Yucatan (Freile-Pelegrin et al. 2006) included for comparison. Means ± SD indicated
Letters indicate statistical significance between native and alkali treated samples for each location. Different letters indicate significant differences at $p<0.01$; *$p<0.05$

Carrageenan viscosity values were within the ranges reported for other iota-producing species (Azanza-Corrales and Sa-a 1990; Brenden and Bird 1994; Freile-Pelegrin et al. 2006). The present FTIR spectra and chemical analysis results show that alkali treatment converted the precursors into 3,6 AG, diminishing viscosity. This is in agreement with Dawes (1977), who reported higher viscosity values for native carrageenan in *E. isiforme* and *E. nudum*. These authors argued that the high sulphate content in native carrageenan increased hydrophobicity and, therefore, viscosity. Native carrageenan sulphate contents are within the range for iota-producing *Eucheuma* species (Cheney et al. 1987; Santos 1989; Fostier et al. 1992). It is noticeable that reduction in sulphate after alkali treatment was higher for *E. isiforme* from Yucatan than for Nicaraguan specie (Table 1). The 3,6 AG contents in alkali-treated carrageenan from *E. isiforme* from Nicaragua was similar to that reported for Florida *E. isiforme* (19.4%), but lower than those previously reported for the same species (26%) by Fostier et al. (1992) and Freile-Pelegrin et al. (2006). It is also evident that the increase in 3,6 AG contents after alkali conversion was lower for the Nicaraguan *E. isiforme* (Table 1). It is well known that oceanographic and environmental conditions differ substantially between the Yucatan Peninsula and the Caribbean coast, with consequent differences in the physiological responses of seaweeds. A topographic upwelling on the Yucatan shelf has been described as one of the most important upwelling regions on the western oceanic margin (Merino 1997). These conditions may influence physiology and biochemical composition of *E. isiforme* found in Yucatan in a different way to those described for the Caribbean populations.

Very weak carrageenan gels were produced in *E. isiforme* from Nicaragua after alkali treatment (<50 g cm^{-2}). Similar values were reported by Santos (1989) for the same species (53 g cm^{-2}) and by Freile-Pelegrin et al. (2006). The FTIR spectra and molar ratios indicated that the phycocolloids extracted from *E. isiforme* from Nicaragua have a dominant iota-carrageenan which is similar to the same species from Yucatan (Freile-Pelegrín et al. 2006). The Galactose:3,6 AG ratio increased after alkali modification. This pattern has been described by Lawson et al. (1973), 1:1; Dawes (1977) 1:0.4 for the same species. The present FTIR spectra of *E. isiforme* native carrageenan exhibited a shoulder at 867 cm^{-1}, indicating the presence of the precursor nu-carrageenan (Chopin et al. 1990). This was corroborated through chemical analysis that showed an increase in 3, 6 AG content and a reduction in sulphate after alkali modification.

In conclusion, *E. isiforme* from Nicaragua is a good source of relatively pure iota-carrageenan with sufficient quality to serve as a substitute for traditional iota-carrageenan sources.

Acknowledgments We thank C. Chávez Quintal for her technical assistance and to Dr. JA Azamar for the FTIR spectra during laboratory analysis. We are grateful to Alvaro Mairena and Eduardo Siu from Blue Fields Indian and Caribbean University (BICU) for seaweed collection. This study was supported by the project CONACYT-SAGARPA 2002-C01-1057.

References

Azanza-Corrales R, Sa-a P (1990) The farmed *Eucheuma* species (Gigartinales, Rhodophyta) in Danajon Reef, Philippinines: carrageenan properties. Hydrobiologia 204/205:521–525

Bodeau-Bellion C (1983) Analysis of carrageenan structure. Physiol Vég 21:785–793

Brenden PCh, Bird KT (1994) Effects of environmental factors on carrageenan from *Gymnogongrus griffithsiae* (Gigartinales, Rhodophyta). J Appl Phycol 6:371–380

Brown MT (1995) Interactions between environmental variables on growth rate and carrageenan content of *Solieria chordalis* (Solieriaceae, Rhodophyceae) in culture. J Appl Phycol 7:427–432

Craigie JS, Leigh C (1978) Carrageenan and agar. In: Hellebust JA, Craigie JA (eds) Handbook of Phycological Methods: Physiological and Biochemical Methods. Cambridge University Press, Cambridge, pp 109–131

Cheney DP, Luistro AH, Bradley PM (1987) Carrageenan analysis of tissue cultures and whole plants of *Agardhiella subulata*. Hydrobiologia 151/152:161–166

Cheney DP (1988). The genus *Eucheuma* J. Agardh in Florida and the Caribbean. In: Abbott IA (ed) Taxonomy of economic seaweeds with reference to some Pacific and Caribbean species. La Jolla, Calif. 2:209–219

Chopin T, Hanisak MD, Koehn FE, Mollion J, Moreau S (1990) Studies on carrageenan and effects of seawater phosphorus concentration on carrageenan content and growth of *Agardhiella subulata* (C. Agardh) Kraft and Wynne (Rhodophyceae, Solieriaceae). J Appl Phycol 2:3–16

Dawes CJ (1977) Seasonal and reproductive aspects of plant chemistry and ι-carrageenan from Floridian *Eucheuma* (Rhodophyta, Gigartinales). Bot Mar 20:137–147

Dawes CJ, Mathieson AC, Cheney P (1974a) Ecological studies of Floridian *Eucheuma* (Rhodophyta, Gigartinales). I. Seasonal growth and reproduction. Bull Mar Sci 24:235–271

Dawes CJ, Lawrence JM, Cheney P, Mathieson AC (1974b) Ecological studies of Floridian *Eucheuma* (Rhodophyta, Gigartinales). III. Seasonal variation of carrageenan, total carbohydrate, protein and lipid. Bull Mar Sci 24:286–299

Dubois M, Gilles KA, Hamilton JK, Roberts PA, Smith P (1956) Colorimetric method for determination of sugar and related substances. Anal Chem 28:350–356

Espinosa-Avalos J (1994) Seaweed as food in the Caribbean. Appl Phycol Forum 11(3):13

Fostier AH, Kornprobst JM, Combaut G (1992) Chemical composition and rheological properties of carrageenans from two Senegalese Solieriaceae *Anatheca montagnei* Schmitz and *Meristotheca senegalensis* Feldmann. Bot Mar 35:351–355

Freile-Pelegrín Y, Robledo D (1997) Effect of season on the agar content and chemical characteristic of *Gracilaria cornea* from Yucatan, Mexico. Bot Mar 40:285–290

Freile-Pelegrin Y, Robledo D (2006). Carrageenan of *Eucheuma isiforme* (Solieriaceae, Rhodophyta) from Yucatán, Mexico. II. Seasonal variations in carrageenan and biochemical characteristics. Bot Mar 49:72–78

Freile-Pelegrin Y, Robledo D, Azamar JA (2006) Carrageenan of *Eucheuma isiforme* (Solieriaceae, Rhodophyta) from Yucatán, Mexico. I. Effect of extraction conditions. Bot Mar 49:65–71

Jackson SG, McCandless EL (1978) Simple, rapid, turbidimetric determination of inorganic sulfate and/or protein. Anal Biochem 90:802–808

Lawson CJ, Rees DA, Stancioff DJ, Stanley NF (1973) Carrageenans. Part VIII. Repeating structures of galactan sulphates from *Furcellaria fastigiata, Gigartina canaliculata, Gigartina chamissoi, Gigartina atropurpurea, Ahnfeltia durvillaei, Gymnogongrus furcellatus, Eucheuma cottonii, Eucheuma spinosum, Eucheuma isiforme, Eucheuma uncinatum, Aghardhiella tenera, Pachymenia hymantophora*, and *Gloiopeltis cervicornis*. J Chem Soc Perkin I: 2177–2182

Lowry OH, Rosebrough NJ, Farr AL, Randall RJ (1951) Protein measurement with the Folin phenol reagent. J Biol Chem 193:265–275

Matsuhiro B, Zanlungo A (1983) Colorimetric determination of 3,6-anhydrogalactose in polysaccharides from red seaweeds. Carbohydr Res 118:276–279

McHugh DJ (2001) Prospects for Seaweed Production in Developing Countries. FAO Fisheries Circular No 986. FAO (ed). Rome, Italy, p 28

Merino M (1997) Upwelling on the Yucatán shelf: hydrographic evidence. J Mar Syst 13:101–121

Piriz ML, Cerezo AS (1991) Seasonal variation of carrageenan in tetrasporic, cystocarpic and sterile stages of *Gigartina skottsbergii* S. et G. (Rhodophya, Gigartinales). Hydrobiologia 226:65–69

Rochas C, Lahaye M, Yaphe W (1986) Sulfate content of carrageenan and agar determined by infrared spectroscopy. Bot Mar 29:335–340

Santos GA (1989) Carrageenans of species of *Eucheuma* J. Agardh and *Kappaphycus* Doty (Solieriaceae, Rhodophyta). Aquat Bot 36:55–67

Smith A (1998) The seaweed resources of the Caribbean. In: Critchley A, Ohno M (eds) Seaweed Resources of the World. JICA, Japan, pp 324–330

Stancioff DJ, Stanley NF (1969) Infrared and chemical studies on algal polysaccharides. Proc Int Seaweed Symp 6:595–609

Stanley N (1987) Production, properties and uses of carrageenan. In McHugh DJ (ed) Production and Utilization of Products from Commercial Seaweeds. FAO Fisheries Technical Paper 288:116–146

Wynne MJ (2005) A checklist of benthic marine algae of the tropical and subtropical western Atlantic: second revision. Nova Hewigia Beih 129:1–152

Some properties of alginate gels derived from algal sodium alginate

H. Kakita · H. Kamishima

Originally published in the Journal of Applied Phycology, Vol 20, No 5, 93–99.
DOI: 10.1007/s10811-008-9317-5

Abstract Alginic acid in soluble sodium alginate turns to insoluble gel after contact with divalent metal ions, such as calcium ions. The sodium alginate character has an effect on the alginate gel properties. In order to prepare a suitable calcium alginate gel for use in seawater, the effects of sodium alginate viscosity and M/G ratio (the ratio of D-mannuronate to L-guluronate) on the gel strength were investigated. The wet tensile strengths of gel fibers derived from high viscosity sodium alginate were higher than those from low viscosity sodium alginate. The tensile strength increased with diminishing sodium alginate M/G ratio. Among the gel fibers tested, the gel fiber obtained from a sodium alginate I-5G (1% aqueous solution viscosity = 520 mPa·s, M/G ratio=0.6) had the highest wet tensile strength. After 13 days treatment in seawater, the wet tensile strength of the gel fiber retained 36% of the original untreated gel strength. For sodium alginates with similar viscosities, the seawater tolerance of low M/G ratio alginate was greater than that of the high M/G ratio one. This study enables us to determine a suitable calcium alginate gel for use in seawater.

Keywords Alginic acid · Brown algae · Gel property · Polysaccharide · Seawater · Seaweed

H. Kakita (✉) · H. Kamishima
AIST Shikoku, National Institute of Advanced Industrial Science and Technology,
Hayashi,
Takamatsu, Kagawa 761-0395, Japan
e-mail: h-kakita@aist.go.jp

Introduction

Seaweed can produce several polysaccharides (e.g., alginic acid, agar, and carrageenan) which differ from those of terrestrial plant origin. Most alginic acid is commercially prepared from brown algae, such as *Laminariales* and *Fucales* species, and used as a food viscosity stabilizer, textile printing glue, dental impression agent and sizing agents (Gacesa 1988). Alginic acid, an acidic polysaccharide, is a linear copolymer of D-mannuronate (M) and L-guluronate (G) and is composed of homo-polymer blocks of M and G, and a hetero-polymer block (Haug et al. 1967a). The main characteristics of alginic acid are as follows: (1) high viscosity solution and spin tolerant; (2) ability to form gels in the presence of certain divalent (or multivalent) cations, particularly calcium ions; (3) adhesive and biocompatible properties; and (4) ability to entrap other materials in the gel. These characteristics may help to develop new industrial materials.

Because the ratio of D-mannuronate to L-guluronate (M/G ratio) in the sodium alginate molecule affects the properties of calcium alginate gel (Haug et al. 1967b), the selection of suitable sodium alginates as raw materials fit for each purpose is important. For example, alginates rich in L-guluronate form strong but brittle gels, whereas those rich in D-mannuronate are weaker but more flexible (Penman and Sanderson 1972; Rees 1972).

We have already reported that a calcium alginate gel derived from sodium alginate I-7 (1% aqueous solution viscosity = 784 mPa·s, M/G ratio=1.3) was useful as a carrier of microorganisms in seawater (Li et al. 1994, 1995, 1996). For the efficient degradation of oil spilled on the sea surface, the alginate carrier was prepared incorporating both a marine oil-degrading yeast, *Candida* sp., and nutrients. In a biodegradation test in a flask with n-tetradecane as an oil component, this model system degraded up to 61% of the initial oil within 14 days at 20°C. But during the 14-day biodegradation test, the calcium alginate carriers decomposed. Improvement of alginate gel strength therefore still remains necessary.

To select a suitable calcium alginate gel for use in seawater, the wet tensile strengths of calcium alginate gel fibers derived from various sodium alginates were investigated with variations in the sodium alginate M/G ratio (0.4–1.3) and viscosity at 1% aqueous solution (16–994 mPa·s). We also evaluated the seawater tolerance of the gel fibers derived from various sodium alginates. In this paper, we describe a suitable calcium alginate gel fiber for use in seawater.

Materials and methods

The six sodium alginate samples [I-S (1% aqueous solution viscosity = 994 mPa·s, M/G ratio=1.3), I-7 (784 mPa·s, M/G ratio=1.3), I-5 (583 mPa·s, M/G ratio=1.3), I-3 (354 mPa·s, M/G ratio=1.3), I-1 (133 mPa·s, M/G ratio=1.3), and SI-L (16 mPa·s, M/G ratio=1.3)] were purchased from Kimica (Tokyo, Japan). The two sodium alginate samples [I-5M (565 mPa·s, M/G ratio=1.3) and I-5G (520 mPa·s, M/G ratio=0.6)] were kindly provided by Dr Hisashi Yamaguchi of Kimica. The four sodium alginate samples [350-M (370 mPa·s, M/G ratio=1.0), 150-M (171 mPa·s, M/G ratio=0.9), 350-G (350 mPa·s, M/G ratio=0.5), and 150-G (135 mPa·s, M/G ratio=0.4)] were purchased from Kibun Food Chemifa (Tokyo, Japan). Other reagents were of analytical grade.

Sodium alginate viscosity measurements

Viscosity measurements were made at 20°C on the sodium alginate solutions by means of a rotating cylinder viscometer, Model LVDV-I (Brookfield, Mass., U.S.A.). A range of alginate concentrations up to 2.5% w/v was used. Sodium chloride was used in all cases to adjust the ionic strength (Smidsrød et al. 1973). The final concentration of sodium chloride was adjusted to 0.1 M NaCl. In order to evaluate [η] values, (ln η)/c values against the alginate concentration (c) were plotted, and [η] values were determined from the intercept (Smidsrød 1970; Johnson et al. 1997). Molecular weights (Mw) were calculated using the formula: $[\eta]=2.0\times10^{-5}M_W$, where [η] is the intrinsic viscosity and M_W is the molecular weight (Smidsrød and Haug 1968).

Preparation of calcium alginate gel fibers

Five percent sodium alginate aqueous solutions (dope) were prepared by mixing sodium alginate powder in distilled water, followed by degassing with a centrifuge or vacuum pump. The dopes were extruded from a 0.84-mm (or 1.14-mm) diameter nozzle of an air-dispenser Model ML5000XII (Musashi Engineering, Tokyo, Japan) into a 0.5 M calcium chloride solution (as a coagulation reagent) at 30~200 kPa at room temperature. The extruded gel fibers were stretched up to 1.4-fold length of original gel length and then washed with water.

Gel fiber strength measurements

The wet tensile strength of the gel fiber was measured with a tensile tester Model Autograph 100A (Shimazdu, Kyoto, Japan), according to Japan Industrial Standard L1069. Measurement conditions were as follows: (1) fiber diameter, 20 mm; (2) speed, 20 mm/min; (3) temperature, 20± 2°C. The wet tensile strength (g d^{-1}) was defined as the minimum load (g) which caused a gel fiber to break, applied by a tensile tester / denier (d). One denier (d) is defined as dry weight of a fiber 9,000 m in length.

Effects of sodium alginate viscosity and M/G ratio on gel fiber strength

First, in the viscosity experiment, the wet tensile strengths of calcium alginate gel fibers were compared, using six sodium alginates (viscosity; 16, 133, 354, 583, 784, and 994 mPa·s, M/G ratio=1.3) as raw materials. Then the wet tensile strength of gel fiber prepared from a high M/G ratio alginate (viscosity; 565 mPa·s, M/G ratio=1.3) was compared with that with a low M/G ratio (viscosity; 520 mPa·s, M/G ratio=0.6). Finally, the wet tensile strengths of gel fiber derived from five sodium alginates [variations in viscosity (135, 171, 350, 370, and 520 mPa·s) and M/G ratio (0.4, 0.5, 0.6, 0.9, and 1.0)] were measured. For each experiment, six replicates of each gel fiber were measured.

Seawater tolerance of alginate gel fiber

The seawater tolerance of gel fiber prepared from a high M/G ratio alginate (I-5M; viscosity; 565 mPa·s, M/G ratio=1.3) was compared with that with a low M/G ratio (I-5G; viscosity; 520 mPa·s, M/G ratio=0.6). Five grams of gel fibers were immersed four times for 1 h each in 40 mL of soaking solution (either distilled water or seawater). The soaking solution was renewed each time. Finally the gel fibers were stored in the soaking solution. After soaking for 5 days, the wet tensile strength of each gel fiber (three replicates per fiber) derived from I-5M and I-5G alginates was measured.

In another experiment, the time-course (pre-soaking, and at 2 days, and 13 days) of the decrease in wet tensile strength of alginate gel fiber treated with seawater was measured. A low M/G ratio alginate I-5G (viscosity; 520 mPa·s, M/G ratio=0.6) was used as the raw material. In this experiment, five replicates of each gel fiber were measured.

To clarify the effect of additional calcium ions on the wet tensile strength of gel fiber soaked in seawater, calcium alginate gel fibers derived from I-5G (1% aqueous solution viscosity; 520 mPa·s, M/G ratio; 0.6) and I-S (994 mPa·s, M/G ratio; 1.3) were treated with various calcium concentrations in seawater for 5 days. After soaking, the wet tensile strengths of gel fibers (four replicates each) were measured.

Statistics

The data concerning gel fiber strength were analyzed by one-way ANOVA followed by Tukey's multiple comparison test (Winer et al. 1991) or *t* test.

Results

The viscosity of sodium alginate at a constant M/G ratio (M/G ratio=1.3) had a significant effect on the wet tensile strength of calcium alginate gel fiber (n=6, F=79.249, p< 0.001). The optimum sodium alginate viscosity (1% aqueous solution) for the wet tensile strength of calcium alginate gel fiber was 354–994 mPa·s (I-3, I-5, I-7, and I-S). The wet tensile strength ranged from 0.166 g d^{-1} at 16 mPa·s (SI-L) to 0.494 g d^{-1} at 994 mPa·s (I-S) (Fig. 1).

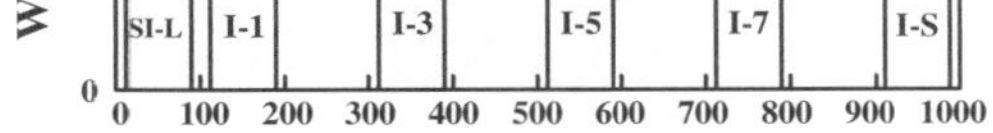

Fig. 1 Effect of sodium alginate viscosity on wet gel fiber strength. The wet tensile strengths of calcium alginate gel fibers prepared from alginates with different viscosity but constant M/G ratio (M/G ratio= 1.3) [SI-L (1% aqueous solution viscosity; 16 mPa·s), I-1(133 mPa·s), I-3 (354 mPa·s), I-5 (583 mPa·s), I-7 (784 mPa·s), I-S (994 mPa·s)]. Extrusion nozzle diameter: 0.84 mm. Wet tensile strengths of alginate gel fibers were measured according to Japan Industrial Standard L1069. Each data point is the mean of six replicates (mean ± SE). *Bars* marked with the same *letter* are not significantly different according to Tukey's multiple comparison test (p=0.05)

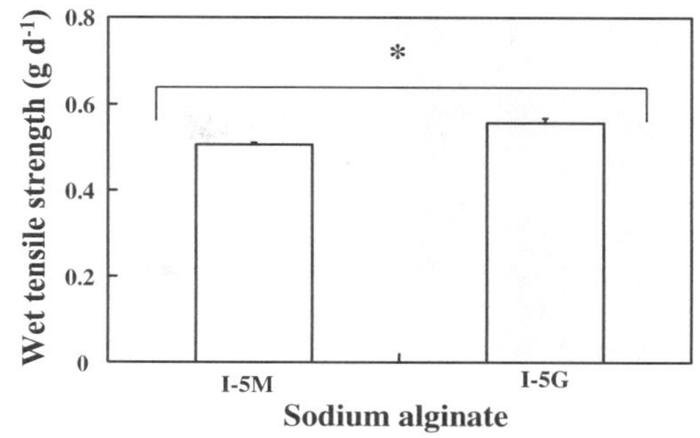

Fig. 2 Effect of M/G ratio on gel fiber strength. The wet tensile strengths of calcium alginate gel fibers derived from alginates with different M/G ratios but constant viscosity [I-5M (1% aqueous solution viscosity; 565 mPa·s, M/G ratio; 1.3) and I-5G (520 mPa·s, M/G ratio; 0.6)]. Extrusion nozzle diameter: 0.84 mm. Wet tensile strengths of alginate gel fibers were measured according to Japan Industrial Standard L1069. Each data point is the mean of six replicates (means ± SE). *The wet tensile strength of gel fiber derived from low M/G ratio alginate (I-5G) was statistically different to that of gel fiber from high M/G ratio alginate (I-5M)

At similar viscosities (520 and 565 mPa·s), the M/G ratio in the sodium alginate molecule also had a significant effect on the wet tensile strength of gel fibers (n=6, t=4.404, p< 0.005). Gel fiber strength derived from M/G ratio 0.6 (I-5G; 1% aqueous solution viscosity = 520 mPa·s, M/G ratio= 0.6) was stronger than that derived from M/G ratio 1.3 (I-5M; 1% aqueous solution viscosity = 565 mPa·s, M/G ratio= 1.3) (Fig. 2).

The wet tensile strength of each alginate gel fiber derived from five sodium alginates (150-G, 350-G, 150-M, 350-M, and I-5G) was measured (Fig. 3). Variations in sodium alginate had a significant effect on the wet tensile strength of calcium alginate gel fibers (n=6, F=14.671, p<0.001). Wet tensile strengths varied from 0.396 g d^{-1} (150-M; 1% aqueous solution viscosity = 171 mPa·s, M/G ratio=0.9) to 0.647 g d^{-1} (I-5G; 520 mPa·s, M/G ratio= 0.6). The strength of the alginate gel fibers, in order of derivation, were I-5G = 350-G (350 mPa·s, M/G ratio=0.5) > 150-G (135 mPa·s, M/G ratio=0.4) = 350-M (370 mPa·s, M/G ratio=1.0) > 150-M.

These results suggested that the wet tensile strengths of gel fibers derived from high viscosity sodium alginates were higher than those from low viscosity ones. The results also suggested that the tensile strength increased with decreasing sodium alginate M/G ratio. From these results, we selected I-5G sodium alginate as a raw material for alginate gel fibers in the subsequent experiment in this paper.

Soaking alginate gel fibers in seawater for 5 days had a significant effect on their wet tensile strengths [I-5M (n=3, t=25.829, p<0.001) and I-5G (n=3, t=11.193, p<0.001)] (Fig. 4). Although the wet tensile strength of the gel fiber derived from high M/G ratio alginate (I-5M) decreased with

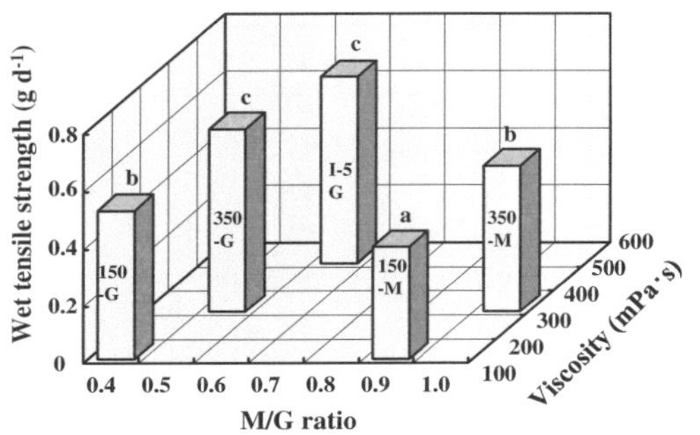

Fig. 3 Wet tensile strengths of alginate gel fibers derived from five sodium alginates, 150-G (1% aqueous solution viscosity; 135 mPa·s, M/G ratio; 0.4), 350-G (350 mPa·s, 0.5), 150-M (171 mPa·s, 0.9), 350-M (370 mPa·s, 1.0), and I-5G (520 mPa·s, 0.6). Extrusion nozzle diameter: 1.14 mm. Wet tensile strengths of alginate gel fibers were measured according to Japan Industrial Standard L1069. Each data point is the mean of six replicates (mean). *Bars* marked with the same *letter* are not significantly different according to Tukey's multiple comparison test ($p=0.05$)

5 days seawater treatment to 19% of the original untreated gel strength, the wet tensile strength of the gel fiber derived from low M/G ratio alginate (I-5G) retained 44% of the original untreated gel strength. For sodium alginates with similar viscosities, the seawater tolerance of low M/G ratio alginate was greater than that of the high M/G ratio one.

Alginate gel fibers subjected to seawater treatment for 2 days underwent significant change in their wet tensile strengths ($n=5$, $F=242.000$, $p<0.001$) (Fig. 5). There was no statistical difference in gel fiber strength between 2 days and 13 days of seawater soaking ($n=5$, $t=1.433$, $p>0.05$). After 2 days in seawater, the wet tensile strength of the gel fiber derived from low M/G ratio alginate (I-5G) decreased to 41% of the original untreated gel strength (Fig. 5). But

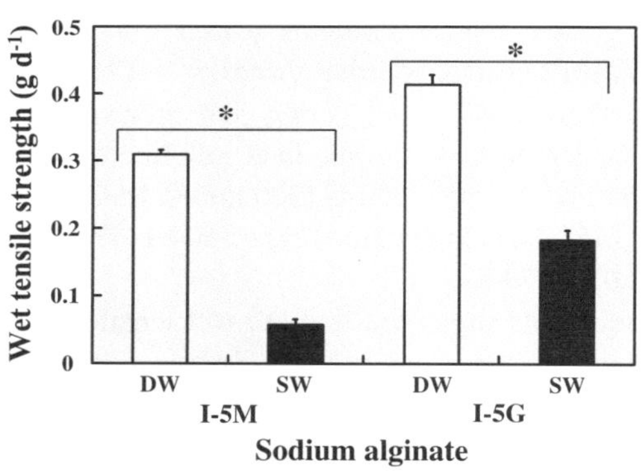

Fig. 4 Effect of M/G ratio on seawater tolerance of gel fiber. The wet tensile strengths of calcium alginate gel fibers derived from alginates with different M/G ratios but constant viscosity [I-5 M (1% aqueous solution viscosity; 565 mPa·s, M/G ratio; 1.3) and I-5G (520 mPa·s, M/G ratio; 0.6)]. Extrusion nozzle diameter: 1.14 mm. Gel fiber wet weight: 5 g. Soaking solution volume: 40 ml. Soaking solution renewal:, 1 h×4 on first day. Storage time: 5 days. Wet tensile strengths of alginate gel fiber were measured according to Japan Industrial Standard L1069. Each data point is the mean of three replicates (mean ± SE). *Significant statistical difference. *SW* Seawater, *DW* distilled water

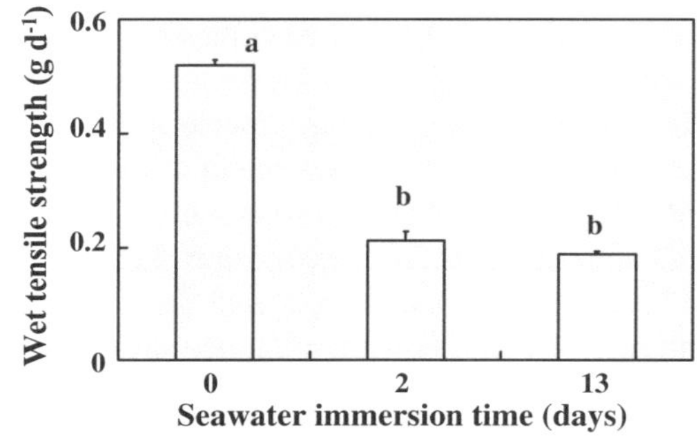

Fig. 5 The time-course of wet tensile strength of gel fiber immersed in seawater. The wet tensile strengths of calcium alginate gel fibers derived from I-5G (1% aqueous solution viscosity; 520 mPa·s, M/G ratio; 0.6)] with seawater immersion for 0, 2, and 13 days. Extrusion nozzle diameter: 0.84 mm. Gel fiber wet weight: 5 g. Soaking solution volume: 40 ml. Soaking solution renewal: 1 h×4 on first day. Wet tensile strengths of alginate gel fibers were measured according to Japan Industrial Standard L1069. Each data point is the mean of five replicates (mean ± SE). *Bars* marked with the same *letter* are not significantly different according to Tukey's multiple comparison test ($p=0.05$)

even after 13 days of seawater soaking, the gel fiber strength remained at 36% at the original untreated gel strength.

The concentration of calcium chloride in seawater had a significant effect on the wet tensile strength of alginate gel fiber ($n=4$, $F=46.591$, $p<0.001$).

After the gel fibers had been treated for 5 days with seawater, their wet tensile strengths decreased to 42% (gel fiber derived from sodium alginate I-5G) and 19% (gel fiber derived from sodium alginate I-S) of the original untreated gel strengths. There was no statistical difference in wet tensile strength between gel fibers (derived from sodium alginate I-5G) treated with distilled water and those treated with seawater containing 0.03 M calcium chloride ($n=4$, $t=2.198$, $p>0.05$). There was also no statistical difference in wet tensile strength between gel fibers (derived from sodium alginate I-S) treated with distilled water and those treated with seawater containing 0.05 M calcium chloride ($n=4$, $t=0.773$, $p>0.05$). Tukey's multiple comparison test indicated that there was no statistical difference in wet tensile strength between gel fibers (derived from sodium alginate I-S) treated with distilled water and those treated with seawater containing 0.04 M calcium chloride (Fig. 6), but the *t* test result indicated a significant statistical difference ($n=4$, $t=2.614$, $p<0.05$). Gel fiber (derived from I-5G) in seawater containing 0.005 M calcium chloride could retain 53% of the original untreated gel strength. From these results, adding 0.03 M calcium chloride seems to be enough to maximize the original gel fiber strength in seawater (for I-5G sodium alginate). For I-S sodium alginate, adding 0.05 M calcium chloride was necessary.

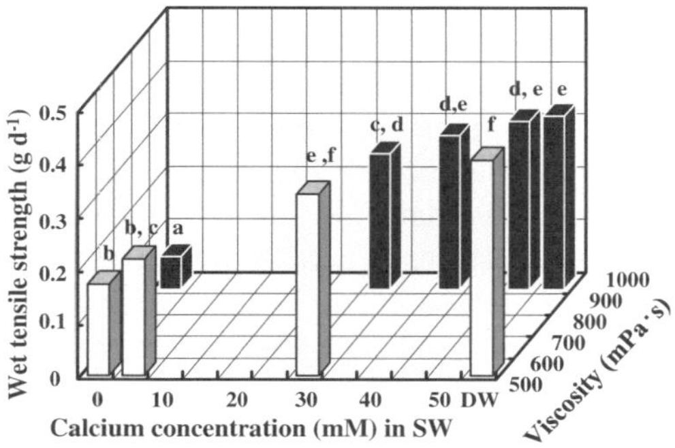

Fig. 6 Effect of calcium ions on the wet tensile strength of gel fiber soaked in seawater. Calcium alginate gel fibers derived from I-5G (1% aqueous solution viscosity; 520 mPa·s, M/G ratio; 0.6) and I-S (994 mPa·s, M/G ratio; 1.3) were treated in solutions with various calcium ion concentrations. Gel fiber wet weight: 5 g. Soaking solution volume: 40 ml. Soaking solution renewal: 1 h×4 on first day. Storage time;:5 days. *Unfilled bar* calcium alginate gel fibers derived from I-5G. *Filled bar* calcium alginate gel fibers derived from I-S. After soaking for 5 days, the wet tensile strengths of gel fibers were measured. Solutions included distilled water, seawater, seawater containing 0.005 M calcium chloride, seawater containing 0.03 M calcium chloride, seawater containing 0.04 M calcium chloride, and seawater containing 0.05 M calcium chloride. The wet tensile strengths of alginate gel fibers were measured according to Japan Industrial Standard L1069. Each data point is the mean of four replicates (mean). *Bars* marked with the same *letter* are not significantly different according to Tukey's multiple comparison test (p=0.05). *SW* Seawater, *DW* distilled water

Discussion

Under a constant sodium alginate M/G ratio of 1.3, the extent of sodium alginate viscosity in the range of about 16–994 mPa·s was proportional to the strength of calcium gel fiber. But in the high viscosity region above 354 mPa·s (I-3, I-5, I-7, and I-S), sodium alginate viscosity had no significant effect on the wet tensile strength of sodium calcium gel fiber (Fig. 1). Molecular weight (Mw) was calculated using the formula: $[\eta]=2.0\times10^{-5}M_W$. Molecular weights of sodium alginates tested were as follows: SI-L; 59×10^3, I-1; 136×10^3, I-3; 182×10^3, I-5; 206×10^3, I-7; 241×10^3, and I-S; 255×10^3. The correlation coefficient (r= 0.846) between the sodium alginate viscosities and the calcium gel fiber strengths was smaller than that (r=0.962) between the sodium alginate molecular weights and the fiber strengths. Thus, we infer that sodium alginate molecular weight was an influential factor in gel fiber strength. Because sodium alginate monomer (i.e. sodium mannuronate or guluronate) has a molecular weight of about 216, SI-L, I-1, I-3, I-5, I-7, and I-S corresponded to about 270, 630, 840, 950, 1,120, and 1,180 polymerization degrees, respectively. Therefore, we infer that an alginate with more than an 800 polymerization grade alginate (under a constant sodium alginate M/G ratio of 1.3) was needed to make a high strength gel fiber, although other parameters such as percentage and configuration of uronic acid blocks (GG-block, MM-block, and GM-block) also influence the polymer properties.

The influence of alginate molecular weight on the binding of calcium ions to alginate has been reported. Fang et al. (2007) revealed three distinct successive steps in the binding of calcium ions to alginate with increased concentration of calcium ions. The first step is the interaction of calcium ions with a single guluronate unit to form monocomplexes. The second step is the propagation and formation of egg-box dimers via pairing of these monocomplexes. The third step is the lateral association of the egg-box dimers, generating multimers. They also clarified that the third step has different association modes depending on the molecular weight of the alginate.

The influence of the percentage and configuration of uronic acid blocks (GG-block, MM-block, and GM-block) on the binding of calcium ions to alginate has also been reported. Donati et al. (2005) reported that the frictional force between the calcium–alginate gel network chains increases with increasing length of the alternating mannuronate–guluronate (MG) sequences. Li et al. (2007) reported that MG block or short G blocks in calcium–alginate junction zones contribute gel bond stability.

Gel fiber derived from low M/G ratio alginate was stronger than that from high M/G ratio alginate (Figs. 2, 3). The C-2 hydroxyl groups in the M (mannuronic acid) block are transformed from an equatorial conformation to an axial one, which reduces the flexibility of the polymer chain. But the whole structure is like cellulose in its conformation. In contrast, the G (guluronic acid) block has the axial–axial form of glycosides, which leads to a less flexible buckle-type ribbon structure. Because its molecular model illustrates the existence of large voids between the helices, the G-block has a higher affinity for cation ligands and causes a new configuration of the polymer chains which are brought together by the cations. As the arrangement of the model, in which cations are arrayed in the voids full of oxygen molecules between the chains, resembles an egg box holding eggs, it is named the 'egg box' structure (Rees et al. 1982). Braccini and Pérez (2001) developed a novel molecular modeling procedure and demonstrated that the popular "egg-box model" can still be referred to in the case of polyguluronate in alginates. The small-angel x-ray scattering profiles of the alginate gels depend on the alginate concentration, calcium ion concentration, and the alginate composition (Stokke et al. 2000). The small-angle x-ray scattering study also revealed that high-G alginates (proportion of guluronic acid residues fractions; 0.68) give alginic acid gels of a higher aggregate density compared to domains composed of loosely packed shorter junction zones in intermediate-G alginates (proportion of guluronic acid residues fractions; 0.50) or low-G alginates (proportion of guluronic acid residues fractions; 0.39) (Draget et al.

2003). We presume that low M/G ratio alginate could produce strong gel fibers because the guluronic acid polymer (poly-guluronic acid) adopts an egg-box structure, which is able to capture metal ions (calcium ions) and form a rigid gel.

The wet tensile strengths of both gel fibers derived from I-5M and I-5G were decreased by seawater treatment (Fig. 4). The percentage of the residual strength (44% of the original untreated gel strength) of the gel fiber derived from I-5G (G block rich alginate) was higher than that (19% of the original untreated gel strength) derived from I-5M (M block rich alginate). The concentrations of ions in typical seawater (salinity 34.5‰) are: sodium, 455 mM; magnesium, 52.5 mM; calcium, 10.2 mM; potassium, 9.2 mM; strontium, 0.15 mM (Lyman and Fleming 1940). We think that after calcium alginate gel fibers were immersed in seawater, sodium ions, which are the great majority of chemical components in seawater, replaced the calcium ions of alginate gel fibers, and gel fiber strength decreased. The high calcium ion capacity of G block rich alginate (i.e. I-5G) seems to be the reason for its high tolerance of seawater.

We used a microorganism carrier (a calcium alginate gel) prepared from an M block rich alginate I-7 alginate (1% aqueous solution viscosity = 784 mPa·s, M/G ratio=1.3) in our previous study. For preparation of gel fibers for use in seawater, G block rich alginates may be a more suitable raw material than M block rich ones. The change to G block rich alginate (i.e. I-5G) from M block rich alginate (i.e. I-7) causes the fiber strength improvement which was an unsolved problem in our previous study.

The results from the study of the time course of the fiber strength decrease on storage with seawater suggest that the fiber strength decrease was due to ion exchange from calcium ions to sodium ions, because the storage period (between 2 days and 13 days) in seawater had no significant effect on the wet tensile strength of alginate gel fiber ($n=5$, $t=1.433$, $p>0.05$) (Fig. 5).

To maximize the original gel fiber strength in seawater, an M block rich alginate (I-S) needed additional 0.05 M calcium ions (Fig. 6). In contrast, a G block rich alginate (I-5G) needed just 0.03 M calcium ions. We infer that the result arose from the difference in the calcium ion capacities of alginates.

Among the gel fibers tested those derived from G block rich alginate I-5G (1% aqueous solution viscosity = 520 mPa·s, M/G ratio=0.6) seem to be the best alginate gel fiber for use in seawater, because the gel fibers had the highest wet tensile strength and seawater tolerance.

This study provided some information on which to base the selection of a suitable alginate gel fiber for use in seawater. The development of alginate gel products for use in seawater would lead to extension of alginate application fields and contribute to the seaweed industry.

We conclude that this study enabled us to identify a suitable alginate gel fiber for use in seawater. A high viscosity sodium alginate with low M/G ratio seems to be the best raw material for deriving alginate gel fiber for use in seawater. Additional calcium ions were useful for preventing gel decomposition in seawater, a 0.03 M calcium chloride concentration being enough to maximize the original untreated gel strength in seawater.

Acknowledgements We gratefully acknowledge Dr Hisashi Yamaguchi of Kimica (Tokyo, Japan) for providing the sodium alginates (I-5G and I-5M). We are grateful to Mr Hideki Obika of AIST Shikoku for much helpful advice. We thank Ms. Satoko Tsukuda for her technical assistance.

References

Braccini I, Pérez S (2001) Molecular basis of Ca^{2+}-induced gelation in alginates and pectins: The egg-box model revisited. Biomacromolecules 2:1089–1096

Donati I, Holtan S, Mørch YA, Borgogna M, Dentini M, Skjåk-Bræk G (2005) New hypothesis on the role of alternating sequences in calcium-alginate gels. Biomacromolecules 6:1031–1040

Draget KI, Stokke BT, Yuguchi Y, Urakawa H, Kajiwara K (2003) Small-angle x-ray scattering and rheological characterization of alginate gels. 3. Alginic acid gels. Biomacromolecules 4:1661–1668

Fang YP, Al-Assaf S, Phillips GO, Nishinari K, Funami T, Williams PA, Li LB (2007) Multiple steps and critical behaviors of the binding of calcium to alginate. J Phys Chem B 111:2456–2462

Gacesa P (1988) Alginates. Carbohydr Polym 8:161–182

Haug A, Larsen B, Smidsrød O (1967a) Studies on sequence of uronic acid residues in alginic acid. Acta Chem Scand 21:691–704

Haug A, Myklestad S, Larsen B, Smidsrød O (1967b) Correlation between chemical structure and physical properties of alginates. Acta Chem Scand 21:768–778

Johnson FA, Craig DQM, Mercer AD (1997) Characterization of the block structure and molecular weight of sodium alginates. J Pharm Pharmacol 49:639–643

Li ZF, Obika H, Kamishima H, Fukuoka S, Kakita H, Kobayashi Y, Higashihara T (1994) Biodegradation of floating oil by a bio-system co-immobilizing a marine oil-degrading microorganism and nutrients. Seibutsu-kogaku 72:363–370 (in Japanese with English abstract)

Li ZF, Obika H, Kamishima H, Fukuoka S, Kakita H, Kobayashi Y, Higashihara T (1995) Improvement of immobilization conditions for biodegradation of floating oil by a bio-system co-immobilizing marine oil-degrading yeast *Candida* sp. and nutrients. Seibutsu-kogaku 73:295–299 (in Japanese with English abstract)

Li ZF, Obika H, Kamishima H, Fukuoka S, Kakita H, Kobayashi Y, Higashihara T (1996) Enhancement substances in chicken egg for the degradation of n-alkanes by *Candida* sp. Seibutsu-kogaku 74:163–166 (in Japanese with English abstract)

Li LB, Fang YP, Vreeker R, Appelqvist I (2007) Reexamining the egg-box model in calcium-alginate gels with X-ray diffraction. Biomacromolecules 8:464–468

Lyman J, Fleming RH (1940) Composition of sea water. J Mar Res 3:134–146

Penman A, Sanderson GR (1972) A method for the determination of uronic acid sequence in alginates. Carbohydr Res 25:273–282

Rees DA (1972) Shapely polysaccharides - Eighth Colworth Medal lecture. Biochem J 126:257–273

Rees DA, Morris ER, Thom D, Maddem JK (1982) Shapes and interactions of carbohydrate chains. In: Aspinall GO (ed) The polysaccharides, vol 1. Academic Press, New York, pp 224–235

Smidsrød O (1970) Solution properties of alginate. Carbohydr Res 13:359–372

Smidsrød O, Haug A (1968) A light-scattering study of alginate. Acta Chem Scand 22:797–810

Smidsrød O, Glover RM, Whittington SG (1973) The relative extension of alginates having different chemical composition. Carbohydr Res 27:107–118

Stokke BT, Draget KI, Smidsrød O, Yuguchi Y, Urakawa H, Kajiwara K (2000) Small-angle X-ray scattering and rheological characterization of alginate gels. 1. Ca-alginate gels. Macromolecules 33:1853–1863

Winer BJ, Brown DR, Michels KM (1991) Statistical principles in experimental design, third edition. McGraw-Hill, New York, pp 153–198

Growth and carrageenan quality of *Kappaphycus striatum* var. *sacol* grown at different stocking densities, duration of culture and depth

A. Q. Hurtado · A. T. Critchley · A. Trespoey · G. Bleicher-Lhonneur

Originally published in the Journal of Applied Phycology, Vol 20, No 5, 101–105.
DOI: 10.1007/s10811-008-9339-z

Abstract *Kappaphycus striatum* var. *sacol* was grown in two separate studies: (1) at two stocking densities, and (2) at four different depths, each for three different durations of culture (30, 45 and 60 days) in order to determine the growth rate of the seaweed and evaluate the carrageenan content and its molecular weight. The results demonstrated that stocking density, duration of culture and depth significantly ($P<0.01$) affected the growth rate, carrageenan content and molecular weight of *K. striatum* var. *sacol*. Decreasing growth rate was observed at both stocking densities and at four depths as duration of culture increased. A lower stocking density (500 g $m^{-1}line^{-1}$) showed a higher growth rate for the shortest durations, i.e. 30 days, as compared to those grown at a higher density. Likewise, decreasing growth rate was observed as depth increased, except at 50 cm after 60 days of culture. A 45-day culture period produced the highest molecular weight at both stocking densities (500 g $m^{-1}line^{-1}$ = 1,079.5±31.8 kDa, 1,000 g $m^{-1}line^{-1}$ = 1,167±270.6 kDa). 'Sacol' grown for 30 days at 50 cm (1,178 kDa) to 100 cm (1,200 kDa) depth showed the highest values of molecular weight of carrageenan extracted. The results suggested that *K. striatum* var. *sacol* is best grown at a stocking density of 500 g $m^{-1}line^{-1}$, at a depth of 50–100 cm, and for a duration of 30 days in order to provide the highest growth rate, carrageenan content and molecular weight.

Keywords *Kappaphycus striatum* var. *sacol* · Stocking density · Duration of culture and depths

A. Q. Hurtado (✉)
Aquaculture Department,
Southeast Asian Fisheries Development Center,
Tigbauan,
Iloilo 5021, Philippines
e-mail: aqhurtado@gmail.com

A. T. Critchley · A. Trespoey · G. Bleicher-Lhonneur
Raw Materials Procurement, Cargill Texturizing Solutions,
France SAS,
50500 Baupte, France

G. Bleicher-Lhonneur
e-mail: Bleicher_Lhonneur@cargill.com

Introduction

Kappaphycus, commercially known as 'cottonii', is composed of two major species, namely *K. alvarezii* (Doty) Doty and *K. striatum* (Schmitz) Doty. However, there are several different varieties within each species. The most popular variety is *K. alvarezii* var. *tambalang*, which is characterised by its long and cylindrical thallus and sparse branches with sharp pointed tips. "*Sacol"* is the most common variety of *K. striatum*, which is characterised by stubby and thick cylindrical branches with blunt and forked tips, which resemble a "cauliflower" shape. As a result, it is locally called 'flower-flower'. The *K. striatum* var. *sacol* variety took its name from an island near Zambonaga City, Philippines, called Sacol, where this seaweed is endemic and where the first commercial cultivation took place. There are three different colour morphotypes of sacol, viz. brown, green and red. Green is the most widely grown by farmers. *Sacol* is generally mixed with *tambalang* in the seaweed trade. It has the same chemical composition as *K. alvarezii* and *K. cottonii* (Aguillan et al. 2003), hence Sacol is fast gaining acceptance by the different stakeholders of the seaweed industry.

Since the onset of *Neosiphonia* epiphyte infestations in the late 1990s (Critchley et al. 2004; Hurtado et al. 2006), farmers have observed that *K. alvarezii* var. *tambalang* is more easily infected with 'ice-ice' and epiphytes than *K. striatum* var. *sacol*, especially during sudden changes in water conditions such as increases in temperature. As a result, farmers prefer to cultivate the latter because it is more resistant to epiphytes and 'ice-ice' infestation. This preference requires scientific study as this selection is mostly based on anecdotal recommendations.

This study was conducted to determine the effect of stocking densities, duration of culture and depth on growth, carrageenan content and molecular weight of *K. striatum* var. *sacol.*

Materials and methods

Thalli of *Kappaphycus striatum* var. *sacol* were grown at the Igang Marine Station of the Southeast Asian Fisheries Development Center Aquaculture Department (SEAFDEC), Igang Nueva Valencia, Guimaras (10°02′ N and 122°02′ E). In the first study, the seaweed was grown at two stocking densities (500 g and 1,000 g m^{-1}) for three different durations of culture (i.e. 30, 45 and 60 days) in order to determine daily growth rate, carrageenan content and molecular weight. A three-block PVC raft (1.5×3 m) was constructed to accommodate a total of 21-3 m long monofilament nylon cord (110 test lbs, 150 mm in diameter) at seven lines per block. The test plants were individually tied to one end of a 50 cm polyethylene rope (4 mm in diameter), the other end of which was finally tied to the monofilament nylon line of the PVC raft. To avoid grazing, the four corners of the PVC raft were tied to a floating net cage where the test plants were maintained 50 cm below the water surface. In the second study, seedlings were grown at 50, 100, 150 and 200 cm below the water surface. Test sample seedlings were tied individually to one end of a #7 polyethylene rope (4 mm diameter) and the other end was tied to a bamboo pole placed on top of a floating net cage. Three replicates for each stocking density, duration of culture and depth were performed.

Selected environmental parameters such as temperature and salinity were recorded daily using a thermometer and an Atago refractometer.

Native carrageenan was extracted from each treatment as described by Hayashi et al. (2007a, b) and the following parameters were determined: (1) carrageenan content was determined per dry weight (%) and (2) molecular weight (kDa) of the carrageenan was measured by size exclusion chromatography in a $LiNO_3$ solution (0.2 M, pH 7). The polymer concentration was measured by refractometry (Merck RI-71). All samples for carrageenan extraction and molecular weight measurement were replicated except for those of depth due to limited samples.

Results were analysed statistically (one-Way ANOVA, Tukey and Duncan Tests) at the 1% level of significance using SPSS ver. 10 software. A Pearson correlation was also made between growth rate, carrageenan content and molecular weight in order to determine any relationships between the variables.

Results and discussion

Stocking densities

Decreasing growth rate of *K. striatum* var. *sacol* was recorded as the duration of the culture period increased for both stocking densities (Fig. 1). A univariate ANOVA showed that differences between stocking densities and duration of culture resulted in significantly different growth rates ($P<0.01$). The highest growth rate (% increase in fresh weight day^{-1}) recorded was with the 500 g/30 day condition (4.0±0.75) and the lowest with 1,000 g/60 day (1.1±0.6). A significant difference in growth rate was observed between 30 and 60 days at a stocking density of 500 g $m^{-1}line^{-1}$. No significant differences in growth rate were observed between plants grown for 30 and 45 days, and between 45 and 60 days at 500 g $m^{-1}line^{-1}$. No significant differences in growth rate were observed among the three different durations of culture at a stocking density of 1,000 g m^{-1}. A significant difference in growth rate was observed between the two stocking densities at 30 days of culture. There were no significant differences recorded at either stocking density for the 45 and 60 day cultures. The results indicated that a lower stocking density (e.g. 500 g m^{-1}) and a shorter period of culture (e.g. 30 days) resulted in a higher growth rate than at 1,000 g m^{-1} stocking density with a longer duration of culture (45 and 60 days). These findings are similar to those of obtained with *K. alvarezii* in an earlier

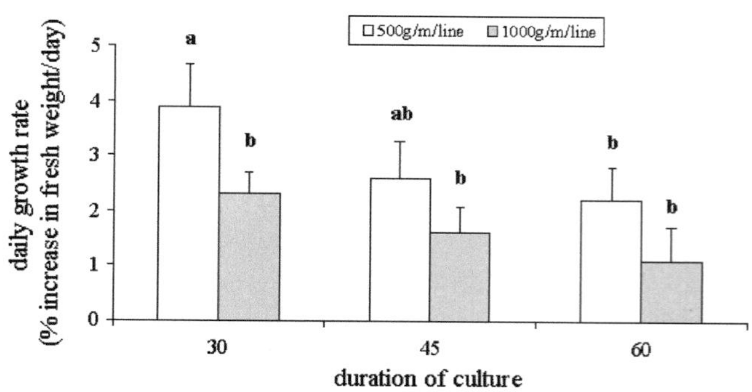

Fig. 1 Average (±SD) growth rate (% increase in fresh weigh day^{-1}) of *Kappaphycus striatum* var. *sacol* grown at two stocking densities and different durations of culture

report by Wu et al. (1989), who also reported a higher growth rate at a lower stocking density and shorter durations of culture (Hurtado et al. 2001).

Effect of depth

A decreasing growth rate of *K. striatum* var. *sacol* was recorded for 30 and 45 days of culture with increasing depth of culture. However, growth of seaweed after 60 days of culture at different depths seemed to have no discernable pattern (Fig. 2). Significant differences in growth rate between and among the depths and durations of culture were analysed (univariate ANOVA, $P<0.01$). The highest (4.7±0.7) and lowest (0.7±1.3) growth rate was recorded at 50 cm at 30 and 60 days of culture, respectively. Higher growth rates were observed near the water surface in the present study after 30 and 45 days at depths of 50–150 cm, similar to an earlier report by Gerung and Ohno (1997) in which they reported that *K. striatum* grew better at 1 m below the water surface. Similar reports on *K. alvarezii* were published by Hurtado-Ponce (1995) and Wu et al. (1989), but the opposite result was reported by Ohno et al. (1994), who found that *K. alvarezii* grew better at 3 m (4.2% day^{-1}) than at 1 m (3.7% day^{-1}) and 2 m (3.8% day^{-1}) below the water surface. At 60 days, some plants were missing and fragmented, probably due to the heavy biomass breaking up the plants. This is one reason why farmers prefer to cultivate the seaweed for shorter periods of time. Differences in growth rate may be attributed to individual conditions at locations and the performance of each species.

Water temperature and salinity readings ranged from 29.6 to 31.8°C and 29.5 to 32.5 ppt, respectively. Lower temperature and salinity readings were recorded from July to August, which coincided with the rainy months. Slightly higher temperatures and salinities were recorded in June, which coincided with higher growth of *K. striatum* var. *sacol* in the area.

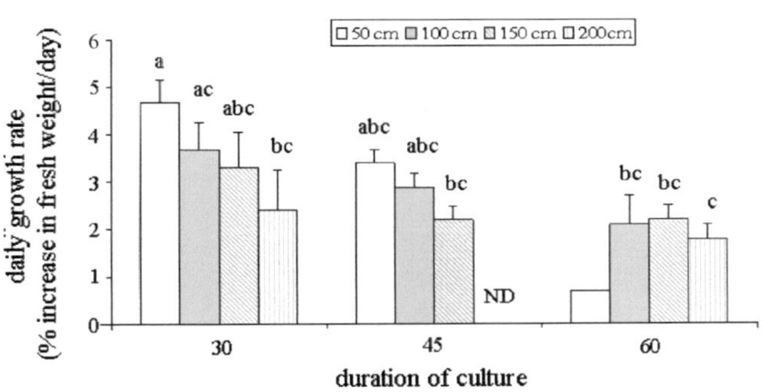

Fig. 2 Average (±SD) growth rate of *K. striatum* var. *sacol* grown at three depths and different durations of culture

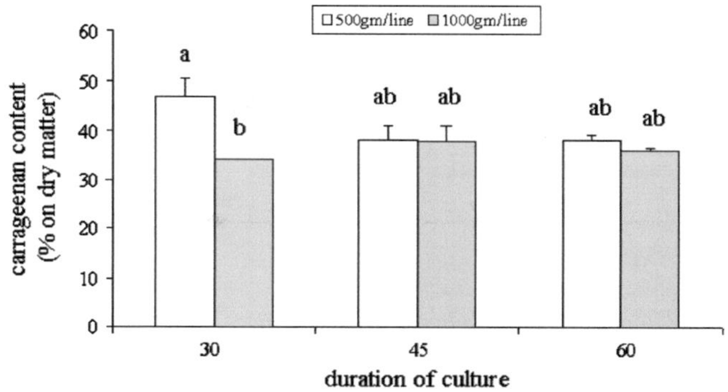

Fig. 3 Average (±SD) carrageenan content (% of dry weight) of *K. striatum* var. *sacol* grown at two stocking densities and different durations of culture

Carrageenan content (%) and molecular weight (kDa)

The highest carrageenan content (% of DW) was recorded when 'sacol' was grown at 500 g m^{-1} line^{-1} as compared to 1,000 g m^{-1}line^{-1}, at all three durations of culture except for 30 days. A significant difference in carrageenan content was observed only between 500 and 1,000 g m^{-1}line^{-1} after 30 days of culture. No significant difference in carrageenan content was observed at 45 and 60 days of culture at either stocking density (Fig. 3). The highest (1,167±270.6 kDa) and lowest (707±35.3 kDa) molecular weights were recorded after 45 and 60 days of culture, both at a stocking density of 1,000 g m^{-1} (Fig. 4). However, higher molecular weight values were obtained at both stocking densities at 45 days of culture. No definite pattern of carrageenan content was observed when '*Sacol'* was grown at different depths and days of culture (Table 1). However, the highest carrageenan content (38.1% of DW) was recorded after 60 days of culture at a depth of 100 cm; the lowest (31.2% of DW) was obtained after 45 days of culture at a depth of 50 cm. The highest molecular weight (viz. 1,200 kDa) was recorded after 30 days, the lowest (739 kDa) was recorded at 60 days of culture, both at a depth of 100 cm.

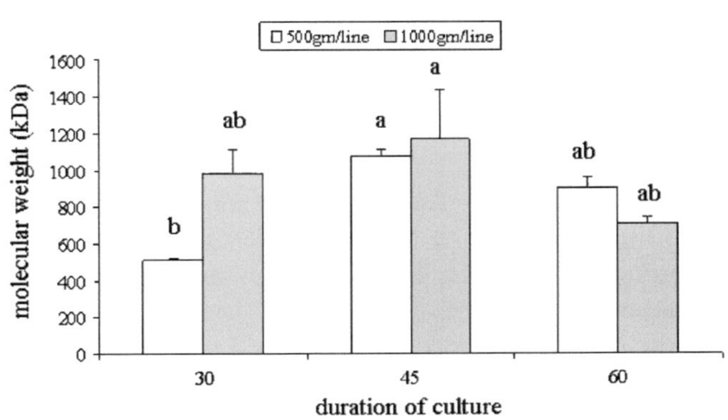

Fig. 4 Average (±SD) molecular weight (kDa) of *K. striatum* var. *sacol* grown at two stocking densities and different durations of culture

Table 1 Summary of averages (±SD) of daily growth rate, carrageenan content and molecular weight of *Kappaphycus striatum* var. *sacol* grown at two stocking densities, four depths, and three different durations of culture. Means in a row and column having the same superscripts are not significantly different at $P=0.01$

	Stocking density ($g\ m^{-1}line^{-1}$)	Duration of culture (days) 30	45	60
Daily growth rate (%/day)	500	4.0 ± 0.8^{a}	2.6 ± 0.7^{ab}	2.2 ± 0.6^{b}
	1,000	2.3 ± 0.4^{b}	1.6 ± 0.5^{b}	1.1 ± 0.6^{b}
Carrageenan content (% of DW)	500	46.6 ± 3.9^{a}	38.1 ± 2.6^{ab}	38.0 ± 1.1^{ab}
	1,000	33.9 ± 0.0^{b}	37.8 ± 3.0^{ab}	36.4 ± 0.6^{ab}
Molecular weight (kDa)	500	510 ± 9.19^{b}	$1,080\pm31.8^{a}$	908 ± 54.7^{ab}
	1,000	982 ± 130.6^{ab}	$1,167\pm270.6^{a}$	707 ± 35.3^{ab}
	Depth (cm)	Duration of culture (days) 30	45	60
Daily growth rate (%/day)	50	4.7 ± 0.5^{a}	3.4 ± 0.3^{abc}	0.7 ± 0^{d}
	100	3.7 ± 0.6^{ac}	3.0 ± 0.3^{abc}	2.1 ± 0.6^{bc}
	150	3.3 ± 0.8^{abc}	2.1 ± 0.3^{bc}	2.2 ± 0.3^{bc}
	200	2.4 ± 0.9^{bc}	ND	1.7 ± 0.3^{c}
Carrageenan content (% of DW)	50	35.3	31.2	34.8
	100	36.2	35.1	38.1
	150	37.4	34	35.2
	200	37.1	ND	36.5
Molecular weight (kDa)	50	1,178	1,119	661
	100	1,200	1,136	739
	150	861	974	918
	200	990	ND	815

The decreasing carrageenan content coincided with a decreasing growth rate as the duration of culture progressed when 'sacol' was grown at 500 g m^{-1} This indicated that, as the duration of cultivation was extended from 45 to 60 days, growth of 'Sacol' decreased, with an accompanying decreased carrageenan content. Younger thalli or tissues may indeed have a higher yield of the galactan, as reported by Mendoza et al. (2002) in 'sacol' and in other red seaweeds such as *Gelidium pristoides* (Carter and Anderson 1986) and *Gracilaria chilensis* (Hemmingson and Furneaux 1997).

Commercially, the average molecular weight of carrageenan ranges from 193 to 324 kDa and is presumed completely degraded if the average molecular weight is below 0.7 kDa (Uno et al. 2001). The present study obtained a range of carrageenan with a molecular weight from 707 to 1,200 kDa. This can be attributed to the absence of 'ice-ice' thalli and the harmful endophyte *Neosiphonia* as well as the favourable culture conditions (stocking density, duration of culture and depth). The occurrence of 'ice-ice' is linked to adverse ecological conditions (Uyengco et al. 1981) and the presence of pathogenic bacteria (Largo et al. 1995a, b, 1999). '*Sacol*' plants infected with 'ice-ice' have a distinctly lower carrageenan yield (Vairappan et al. 2007) and molecular weight, as seen in the reports of Mendoza et al. (2002).

Table 2 summarises the correlation between growth rate, carrageenan content and molecular weight in both stocking densities and duration of culture. Positive and negative correlations were measured between pairs of variables at *the* $P=0.01$ level of significance. The significantly high r value ($r=+1.00$, $r=-1.00$) indicates a strong correlation that growth rate:carrageenan, growth rate:molecular weight, and carrageenan:molecular weight are highly associated with one another in a linear way at 30 and 45 days of cultivation, both at the 500 and 1,000 g $m^{-1}line^{-1}$ stocking densities.

Table 2 Summary of correlation ($P=0.01$) between daily growth rate (DGR, %/day), carrageenan content (Cg, % of DW) and molecular weight (Mol.wt., kDa) of *K. striatum* var. *sacol* at two stocking densities and three durations of culture

Days of culture	Stocking density	DGR:Cg	DGR:Mol.wt.	Cg:Mol. wt.
30	500	−1.00*	−1.00*	−1.00*
	1000	−1.00*	1.00*	−1.00*
45	500	−1.00*	1.00*	−1.00*
	1000	1.00*	−1.00*	−1.00*
60	500	0.94	−0.48	−0.75
	1000	−0.99	−0.82	0.76

* Correlation is significant at the 0.01 level (2-tailed)

The results of the present study suggest that *K. striatum* var. *sacol* is best grown at a stocking density of 500 g $m^{-1}line^{-1}$, at a depth of 50–100 cm and for a duration of 30 days in order to yield the highest growth rate, carrageenan content and molecular weight. The results obtained in the present study provide information that will help understand and manage the growth performance of this seaweed and the carrageenan characteristics of '*Sacol*' as a significant source of kappa carrageenan in the global market. It is suggested, however, that a year-round study be made in order to fully assess the sustainability of '*Sacol*'.

Acknowledgements The senior author is sincerely grateful to Cargill Texturizing Systems, formerly Degussa Texturizing Systems, for their untiring financial support on seaweed R&D in the Philippines. Likewise, Keneth Tibubos is thanked for his unselfish technical assistance during the experimental stage.

References

Aguillan JT, Broom JE, Hemingson JA, Dayrit FM, Montaño MNE, Dancel MCA, Niñonuevo MR, Furneaux RH (2003) Structural analysis of carrageenan from farmed varieties of Philippine seaweed. Bot Mar 46:179–192

Carter AR, Anderson RJ (1986) Seasonal growth and agar contents in *Gelidium pristoides* (Gelidiales, Rhodophyta) from Port Alfred, South Africa. Bot Mar 29:117–123

Critchley AT, Largo D, Wee W, Bleicher-Lhonneur G, Hurtado AQ, Schubert J (2004) A preliminary summary on *Kappaphycus* farming and the impact of epiphytes. Jpn J Phycol 52:231–232

Gerung GS, Ohno M (1997) Growth rates of *Eucheuma denticulatum* (Burman) Collins et Harvey and *Kappaphycus striatum* (Schmitz) Doty under different conditions in warm waters of southern Japan. J Appl Phycol 9:413–415

Hayashi L, de Paula EJ, Fungyi C (2007a) Growth rate and carrageenan analyses in four strains of *Kappaphycus alvarezii* Rhodophyta, Gigartinales) farmed in the subtropical waters of São Paulo State, Brazil. J Appl Phycol 19:393–399

Hayashi L, Oliveira EC, Bleicher-Lhonneur G, Boulenguer P, Pereira RTL, von Seckendorff R, Shimoda VT, Leflamand A, Vallée P, Critchley AT (2007b) The effects of selected cultivation conditions on the carrageenan characteristics of *Kappaphycus alvarezii* (Rhodophyta, Solieriaceae) in Ubatuba Bay, São Paulo, Brazil. J Appl Phycol 19:505–511

Hemmingson JA, Furneaux RH (1997) Biosynthetic activity and galactan compostion in different regions of the thallus of *Gracilaria chilensis* Bird, McLachlan et Oliveira. Bot Mar 40:351–357

Hurtado-Ponce AQ (1995) Carrageenan properties and proximate composition of three morphotypes of *Kappaphycus alvarezii* Doty (Gigartinales, Rhodophyta) grown at two depths. Bot Mar 38:215–219

Hurtado AQ, Agbayani RF, Sanares R, de Castro-Mallare MTR (2001) The seasonality and economic feasibility of cultivating *Kappaphycus alvarezii* in Panagatan Cays, Caluya, Antique, Philippines. Aquaculture 199:295–310

Hurtado AQ, Critchley AT, Trespoey A, Bleicher-Lhonneur G (2006) Occurrence of *Polysiphonia* epiphytes in *Kappaphycus* farms at Calaguas Is., Camarines Norte, Philippines. J Appl Phycol 18:301–306

Largo DB, Fukami K, Nishijima T, Ohno M (1995a) Laboratory-induced development of the *ice-ice* disease of the farmed red algae *Kappaphycus alvarezii* and *Eucheuma denticulatum* (Solieriaceae, Gigartinales, Rhodophyta). J Appl Phycol 7:539–543

Largo DB, Fukami K, Nishijima T (1995b) Occasional pathogenic bacteria promoting *ice-ice* disease in the carrageenan-producing red algae *Kappaphycus alvarezii* and *Eucheuma denticulatum* (Solieriaceae, Gigartinales, Rhodophyta). J Appl Phycol 7:545–554

Largo DB, Fukami K, Nishijima T (1999) Time-dependant attachment mechanism of bacteral pathogen during ice-ice infection in *Kappaphycus alvarezii* (Gigartinales, Rhodophyta). Proc Int Seaweed Symp 16:643–650

Mendoza WG, Montaño NE, Ganzon-Fortes ET, Villanueva RD (2002) Chemical and gelling profile of *ice-ice* infected carrageenan from *Kappaphycus striatum* (Schmitz) Doty "sacol" strain (Solieriaceae, Gigartinales, Rhodophyta). J Appl Phycol 14:409–418

Ohno M, Largo DB, Ikumoto T (1994) Growth rate, carrageenan yield and gel properties of cultured kappa-carrageenan producing red alga *Kappaphycus alvarezii* (Doty) Doty in the subtropical waters of Shikoku, Japan. J Appl Phycol 6:1–5

Uno Y, Omoto T, Goto Y, Asai I, Nakamura M, Maitani T (2001) Molecular weight distribution of carrageenans studied by a combined gel permeation/inductively coupled plasma (GPC/ICP) method. Food Addit Contam 18:763–772

Uyengco FR, Saniel LS, Jacinto GS (1981) The 'ice-ice' problem in seaweed farming. Proc Int Seaweed Symp 10:625–630

Vairappan CS, Chung CS, Lhonneur GB, Critchley AT (2007) Epiphyte outbreak in Kappaphycus (Gigartinales, Rhodophyta) farms in Southeast Asia: Impact on the yield and quality of carrageenan. 19th International Seaweed Symposium. 26–31 March 2007, Kobe, Japan (Abstract only)

Wu C, Li J, Xia E, Peng Z, Tan S, Li J, Wen Z, Huang X, Cai Z, Chen G (1989) On the transplantation and cultivation of *Kappaphycus alvarezii* in China. Chin J Ocenaol Limnol 7:327–334

Purification and characterization of a *N*-acetylglucosaminidase produced by *Talaromyces emersonii* during growth on algal fucoidan

Elaine O'Connell · Patrick Murray · Charles Piggott · Franck Hennequart · Maria Tuohy

Originally published in the Journal of Applied Phycology, Vol 20, No 5, 107–115.
DOI: 10.1007/s10811-007-9291-3

Abstract A β-*N*-acetylglucosaminidase produced by a novel fungal source, the moderately thermophilic aerobic ascomycete *Talaromyces emersonii,* was purified to apparent homogeneity. Submerged fermentation of *T. emersonii,* in liquid medium containing algal fucoidan as the main carbon source, yielded significant amounts of extracellular *N*-acetylglucosaminidase activity. The *N*-acetylglucosaminidase present in the culture-supernatant was purified by hydrophobic interaction chromatography and preparative electrophoresis. The enzyme is a dimer with molecular weight and pI values of 140 and 3.85, respectively. Substrate specificity studies confirmed the glycan specificity of the enzyme for *N*-acetylglucosamine. Michaelis-Menten kinetics were observed during enzyme-catalyzed hydrolysis of the fluorescent substrate methylumbelliferyl-β-D-*N*-acetylglucosaminide at 50°C, pH 5.0 (K_m value of 0.5 mM). The purified *N*-acetylglucosaminidase displayed activity over broad ranges of pH and temperature, yielding respective optimum values of pH 5.0 and 75°C. The *T. emersonii* enzyme was less susceptible to inhibition by *N*-acetylglucosamine and other related sugars than orthologs from other sources. The enzyme was sensitive to Hg^{2+}, Co^{2+} and Fe^{3+}.

Keywords β-D-*N*-acetylglucosaminidase · Fucoidan · Algae · Fungi · *Aspergillus niger* · *Schizophyllum commune*

E. O'Connell · P. Murray · C. Piggott · F. Hennequart · M. Tuohy (✉)
Molecular Glycobiotechnology Group,
Department of Biochemistry, National University of Ireland, Galway, Ireland
e-mail: maria.tuohy@nuigalway.ie

Abbreviations
4-MU-GlcNAc Methylumbelliferyl-*N*-acetylglucosaminide

Introduction

Fucoidan is a complex sulphated fucan derived from brown algae and marine invertebrates that mediates a variety of significant biological effects (Berteau and Mulloy 2003). The cell walls of brown macroalgae are constituted of a matrix of cellulose, alginates and fucoidan. Although the individual roles of each of these polysaccharides, especially those of the fucoidans, are not well understood their function in maintaining the structural integrity of the cell is well recognized (Kloareg and Quatrano 1988; Nagumo and Nishino 1996). Studies have clearly shown that the composition and complexity of fucoidans from various brown macroalgae can vary considerably (Nishino et al. 1994; Berteau and Mulloy 2003). Various interesting bioactivities have been suggested for fucoidans, including effects on angiogenesis (Matsubara et al. 2005), coagulation (Senni et al. 2006), oxidative stress (Xue et al. 2001), inhibition of scavenger receptors (Carvalho et al. 2002), inhibition of viral infections (Ponce et al. 2003), parasite host infection (Maruyama et al. 2007) and enhanced immune control of cancer cells (Maruyama et al. 2006). As mentioned by Berteau and Mulloy (2003) in their excellent review of fucoidans, bioactivity can be correlated to the molecular structure of the fucoidan fraction. The structural basis for some of the biological and physiological properties of sulphated fucans depends on the presence of sulphate groups. Desulphation of fucoidan has lead to an understanding of the anti-coagulant property of the polysaccharide (Pomin et al. 2005). By isolating and identifying enzymes capable of cleaving this group of polysaccharides into smaller oligosaccharide and monosaccharide sugars, structural information can be obtained thereby providing a greater insight into their functions. In addition, it may be possible to alter and enhance the bioactivity of fucoidans selectively using purified enzymes.

Although almost nothing is known about the biosynthesis of algal and marine invertebrate fucoidans, enzymes capable of degrading these polysaccharides have been identified from several marine species (Berteau and Mulloy 2003). Two types of glycosidases act on fucose-rich regions of fucoidans, i.e. fucoidan hydrolase (EC 3.2.1.44) and α-L-fucosidase (EC 3.2.1.51). α-L-Fucosidase releases L-fucose from the non-reducing end of the fucoidan polysaccharide, while less is known about the mode of action of fucoidan hydrolase (also referred to as fucoidanase). One of the key difficulties in the identification (and purification) of fucoidan hydrolase and sources of this activity has been the lack of a simple assay method. While assays with aryl-glycosides (e.g. p-nitrophenyl-glycosides) are used routinely for the detection and quantification of exo- glycosidases, reducing sugar assay methods have been used to screen for and detect fucoidan hydrolase. Lack of sensitivity is the main disadvantage for this type of assay. To date only two fucoidan hydrolases, both from bacterial species, that share no common structural features, have been purified to apparent homogeneity (Furukawa et al. 1992; Kitamura et al. 1992). The mode of action of either type of enzyme is still largely unresolved; purification of the enzyme from other different sources is essential. Previous studies by Nishino and colleagues (1994) identified an amino sugar-containing fucan sulphate in a commercial preparation of fucoidan from the brown macroalga, *Fucus vesiculosus*. Previous work (unpublished) by this group has also identified an *N*-acetyl amino sugar following GCMS analysis of Sigma fucoidan, and in fucoidans isolated from another brown macroalga, *Ascophyllum nodosum*. In addition, acetyl groups have been found in almost all algal fucoidans studied in recent years (Berteau and Mulloy 2003) and may be present in other algal fucoidans, but are possibly removed during different chemical-based isolation procedures (Chizov et al. 1999). Consequently, hydrolysis of fucoidan to its monomeric sugars or selective modification may require a more complex repertoire of enzymes, including *N*-acetylglucosaminidase.

The present report describes the screening of three different fungi for enzymes capable of depolymerising fucoidan, i.e. the thermophile *Talaromyces emersonii* and two mesophiles *Schizophyllum commune* and *Aspergillus niger*, during growth on fucoidan as the main carbon sources. The specific aims were focused on identifying the types of enzymes produced in response to growth of these fungi on fucoidan, identifying novel sources of key enzymes and potentially novel enzymes which could ultimately be used to selectively modify the bioactivity of fucoidan preparations.

Materials and methods

Unless otherwise stated, all general research chemicals were purchased from Sigma-Aldrich (Dublin, Ireland). Crude fucoidan (*Fucus vesiculosus*), 4-nitrophenyl-glycosides, methylumbelliferyl-glycosides, disaccharides and cellooligosaccharides were from Sigma-Aldrich. Phenyl Sepharose 6 Fast Flow hydrophobic interaction matrix, standard molecular weight proteins and isoelectric point calibration kits were obtained from GE Healthcare (formerly Amersham Biosciences, Uppsala, Sweden). Sterile 0.22 μm filters and laboratory plastics were purchase from Sarstedt (Wexford, Ireland).

Microorganisms and culture conditions

Talaromyces emersonii IMI 392299 (derived from CBS 814.70), a moderately thermophilic, aerobic ascomycete was obtained from laboratory stocks and was routinely maintained on Sabouraud Dextrose agar (SDA). Submerged (or liquid) fermentation in 250 mL Erlenmeyer flasks was carried out at 45°C, pH 4.5 and 250 rpm (Tuohy and Coughlan 1992) in the mineral salts/inducing medium described by Moloney et al. (1983) using 0.25% (w/v) crude fucoidan as the carbon source. For enzyme screening studies, cultures were harvested periodically during a growth time period of 168 h and extracellular culture filtrates separated from the fungal mycelia by centrifugation at 5,761 *g* for 30 min in a Sorvall Superspeed centrifuge equipped with a GSA rotor. The supernatant was filtered through several layers of sterile fine-grade muslin to remove residual mycelia and particulate material. All time points were assayed for a variety of enzyme activities.

Schizophyllum commune (Basidiomycetes) and *Aspergillus niger* (Zygomycetes), both mesophilic fungi were originally obtained from CABI Biosciences (Surrey, UK; formerly the Imperial Mycological Institute). Both fungi were routinely sub-cultured in an identical manner to that described above for *T. emersonii* albeit at a different temperature, i.e. 32°C. The culture supernatant from time-course samples of both fungi was screened for various enzyme activities.

Measurement of enzyme activity and protein concentration

Unless stated otherwise, all enzyme assays were carried out at 50°C for 15 min in 100 mM sodium acetate buffer (NaOAc), pH 5.0, in final reaction volumes of 100–900 μL and final substrate concentrations of 0.66% (w/v) for endoglycanase (e.g. fucoidan hydrolase) and ~0.48–

0.95 mM for exoglycosidase activity. Fucoidanase activity was assayed by incubating 0.15 mL of a 1.0% (w/v) buffered fucoidan solution with 0.15 mL of enzyme with a final substrate concentration of 0.5% (w/v) at 50°C for 15 min. Reducing sugars released were quantified by dinitrosalicylic acid (DNS) method (Miller 1959) by reference to a fucose standard curve. Other endoglycanase activities were measured as described previously (Murray et al., 2001). Exoglycosidase activity was evaluated by monitoring the increase in absorbance at 405 nm (A_{405}) due to free 4-nitrophenolate anion following incubation of 5–10 μL of suitably diluted enzyme with 100 μL 1 mM 4-NP-glycosides for 15 min at 50°C in 100 mM NaOAc buffer, pH 5.0 (Tuohy et al., 2002). The reactions were terminated by the addition of 100 μL 1 M Na_2CO_3. Released *4*-nitrophenolate anion was measured using a Wallac Victor plate-reader equipped with a 405-nm filter. Appropriate blanks without substrate and/or enzyme were included in each assay and all assays were carried out in triplicate. A standard curve prepared with synthetic *4*-nitrophenol was used to quantify the colorimetric reaction product. In all cases, enzyme activity was expressed as $IU.mL^{-1}$ i.e. μmoles of reducing sugar equivalents or 4-nitrophenol released per min of reaction time per ml of enzyme solution.

Protein concentration in crude extracts was determined by the Bradford method (Bradford 1976) using Bovine Serum Albumin (BSA) fraction V as the protein standard. The protein concentration of purified (low abundance) protein samples was determined by measuring absorbance at 280 nm (A_{280}) using a nanodrop ND 1000 spectrophotometer (Labtech International, UK) since less sample was required and sample recovery was possible. Protein elution profiles were determined during chromatography by monitoring the A_{280} of the eluted fractions.

pH and temperature optima

The optimum pH for maximum enzyme activity of the purified *N*-acetylglucosaminidase was determined by monitoring enzyme activity, in the normal enzyme assay, over a pH range of 2.6–7.6, at 50°C, in the constant ionic strength citrate phosphate buffer (McIlvaine buffers) described by Elving et al. (1956). The optimum temperature for activity of the purified enzyme was carried out by conducting the normal assay at temperatures between 20 and 80°C (McIlvaine pH 5.0). Account was taken of the variation in pH as a function of temperature.

Electrophoresis and isoelectric focusing

The purity and composition of *N*-acetylglucosaminidase-rich protein samples were evaluated under denaturing conditions by SDS-PAGE, as described by Laemmli (1970). Protein bands were visualized using Coomassie Brilliant blue R250 (or Sypro Ruby). The molecular mass (m) of the purified protein was calculated by comparing the relative mobilities of standard and sample proteins (m range of standard proteins used was 53–212 kDa). Enzyme purity was also evaluated by isoelectric focusing and non-denaturing PAGE, followed by zymogram analysis. Isoelectric focusing (IEF) was used to determine the isoelectric point (pI) of the enzyme on ultrathin Ampholine PAG plates over the pH range 3.5–9.5, using ampholytes (pH 3.0–10.0) supplied by Amersham Biosciences according to procedures recommended in the manufacturer's information booklet 1804. Standard proteins of known pI (pH 3.5–9.3; Amersham Biosciences) were run in separate lanes in conjunction with unknown (*N*-acetylglucosaminidase-rich) samples. The location of the *N*-acetylglucosaminidase was determined by immersing the previously run IEF gel (or Native-PAGE) gel in 100 mM sodium acetate buffer, pH 5.0, containing 100 μM methylumbelliferyl-β-D-*N*-acetyl-glucosaminidine (4-MU-GlcNAc) for 5 min at 50°C and visualising release of methylumbelliferone (enzyme-active bands) under ultraviolet light. Images of PAGE and IEF gels were obtained using a BioRad Fluor-S MultiImager.

Enzyme purification

Crude extracellular filtrate was brought, by gradual addition of NaCl, to a final conductivity equivalent with a salt concentration of 1 M NaCl ('salted in') and pH 4.5 before application to a Phenyl Sepharose Fast Flow hydrophobic interaction chromatography column (3.0×8 cm) pre-equilibrated with 50 mM ammonium acetate buffer pH 4.5 and 1 M NaCl (buffer A). The column was washed after sample application with 2 column volumes of Buffer A. Adsorbed protein was eluted using decreasing salt (1 M–0 M Buffer A) and increasing ethylene glycol [0–15% ($^w/_v$)] gradients at a flow rate of 25 mL h^{-1}. A wash of double-deionised water (Millipore MillRX system) was used to elute the enzyme from the column. Fractions (3.13 mL) were collected throughout and analysed for *N*-acetylglucosaminidase activity and protein content. Due to the detection of high levels of β-glucosidase in the *T. emersonii* crude enzyme sample and the documented co-elution of both enzymes in previous reports (Jones and Kosman 1980), elution of this enzyme was also monitored from the Phenyl Sepharose chromatography step. Preparative electrophoresis was used to further purify the *N*-acetylglucosaminidase.

Fractions containing *N*-acetylglucosaminidase activity were pooled, concentrated by ultrafiltration (Amicon 5 kDa centricon), and applied to a Native PAGE gel (large Protean II electrophoresis system) using a preparative gel comb. One gel lane was stained for enzyme activity to locate the active enzyme using 100 μM 4-MU-GlcNAc for

5 min at 50°C; the resulting fluorescent band was used as a reference to excise the *N*-acetylglucosaminidase present in the remaining gel segment. The excised band was resuspended in 500 μL ammonium acetate buffer pH 4.5 and gently agitated overnight at 4°C to facilitate extraction of the *N*-acetylglucosaminidase protein from the gel. The enzyme sample obtained was passed through a sterile 0.22 μm PVDF filter to separate residual gel from the protein sample. Enzyme activity was confirmed by the normal assay method (4-MU-GlcNAc as substrate). Enzyme purity was determined by SDS-PAGE and IEF analysis.

Substrate specificity and estimation of kinetic constants

The specificity of the enzyme with respect to glycan and glycan moieties was investigated by measuring the release of methylumbelliferone (MU) by the purified enzyme from a range of MUα and β-D (or L) glycosides. Specifically, 5 μL of enzyme was combined with 50 μL of 0.5 mM MU-glycoside pH 5. Fluorescence was measured after a 30-min incubation at 50°C using a Wallac Victor multi-label counter, at excitation and emission wavelengths of 365 nm and 440 nm, respectively. Fluorescence units were converted to μmoles MU released during the assay using a standard curve prepared with free MU. In addition, the hydrolysis of a range of natural disaccharides was evaluated by monitoring the release of specific monosaccharides (e.g. glucose) as outlined elsewhere (Gilleran 2004). Kinetic studies were undertaken with a highly purified *N*-acetylglucosaminidase preparation using the assay procedure outlined above with increasing concentrations of 4-MU-GlcNAc. Experimental data generated from these analyses were fitted to a rectangular hyperbola. The computer program used was GraphPad Prism version 4.00 for Windows (GraphPad software, San Diego, Calif., USA). Standard formulae were used to determine K_m, V_{max}, k_{cat}, and k_{cat}/K_m.

Effect of metal ions and monosaccharides on enzyme activity

The effect of metal ions and monosaccharides on the activity of the purified *T. emersonii N*-acetylglucosaminidase was determined (Ulhoa and Peberdy 1991; Lisboa De Marco and Valadares-Inglis 2004) by incubating the enzyme with various metal salts, each at a final concentration of 100–1,000 μM, or monosaccharides at a final concentration of 0.5–10 mM, in the normal assay reaction cocktail. All assays were carried out in triplicate and appropriate controls in the absence of metal ions or monosaccharides were included. Residual activities (%) were calculated relative to a control reaction (no metal salt or monosaccharide added).

Results

Enzyme production during growth of thermophilic and mesophilic fungi on fucoidan

A 168-h time course study was undertaken to investigate enzyme production in submerged fermentation by *T. emersonii* (at 45°C), *S. commune* and *A. niger* (both at 32°C) during growth on 0.25% (w/v) fucoidan. Culture supernatants harvested at 12- to 24-h timed intervals were assayed for extracellular enzyme activities. As shown in Tables 1, 2 and 3, no α-fucosidase, fucoidan hydrolase or sulphatase activities were detected when the three fungi were grown on 0.25% (w/v) fucoidan. All three fungi produced *N*-acetylglucosaminidase and β-glucosidase ac-

Table 1 Enzyme production by *T. emersonii* grown on 0.25% (w/v) fucoidan for 168 h

Hour	Protein mg mL^{-1}	α-fuc	β-fuc	F-hydrolase	Sulphatase	β-glc	β-gal	*N*-acetyl	β-xyl
0	0.029	0	0	0	0	1.1	0	0	0
12	0.024	0	0	0	0	0	0	0	0
24	0.026	0	0	0	0	2.5	0	0	0
36	0.029	0	0	0	0	5.6	0	1.9	0
48	0.033	0	0	0	0	6.8	0	4.7	0
60	0.029	0	0	0	0	9.6	0	4.9	0
72	0.027	0	0	0	0	11.6	0	7.3	0
96	0.030	0	0	0	0	23.3	0	8.2	0
120	0.029	0	0	0	0	32.8	0	13.1	0
144	0.027	0	0	0	0	24.4	0	13.6	0
168	0.028	0	0	0	0	25.5	0	12.9	0

All enzyme activity results are specific activity $IU.mg^{-1}$

Enzyme abbreviations: *α-fuc* α-fucosidase; *β-fuc* β-fucosidase; *F-hydrolase* Fucoidan hydrolase; *β-glc* β-glucosidase; *β-gal* β-galactosidase; *N-acetyl N*-acetylglucosdaminidase; *β-xyl* β-xylosidase

Table 2 Enzyme production by *S. commune* grown on 0.25% (w/v) fucoidan for 168 h

Hour	Protein mg mL^{-1}	α-fuc	β-fuc	F- hydrolase	Sulphatase	β-glc	β-gal	*N*-acetyl	β-xyl
0	0.026	0	0	0	0	0	0	0	0
12	0.021	0	0	0	0	0	0	1.5	0
24	0.022	0	0	0	0	1.0	0	0	0
36	0.025	0	0	0	0	0	0	0	0
48	0.027	0	0	0	0	0	0	0	0
60	0.017	0	0	0	0	1.1	1.2	1.2	0
72	0.024	0	0	0	0	0	1.6	2.4	0
96	0.023	0	0	0	0	4.0	3.9	10.2	0
120	0.021	0	0	0	0	19.6	5.0	7.6	0
144	0.027	0	0	0	0	45.6	2.5	7.4	0
168	0.028	0	0	0	0	62.0	7.3	3.2	0

All enzyme activity results are specific activity IU.mg^{-1}
Enzyme abbreviations: see Table 1

tivities, while both mesophiles also produced β-galactosidase activity. In contrast, *T. emersonii* did not produce any β-galactosidase activity, even though the gene for β-galactosidase has been cloned from this source (Lydon 2005) and the corresponding native protein isolated (Thompson 2004). Secretion (and hence expression) of *N*-acetylglucosaminidase by *T. emersonii* occurred ~24 h earlier than for both mesophiles, suggesting that the activity may be induced rather than associated with fungal cell wall degradation (i.e. autolysis). In addition, appreciable levels of activity were obtained. Consequently, larger batches of culture were prepared and harvested at 120 h for enzyme purification studies.

Purification of β-D-*N*-acetylglucosaminidase

Crude extracellular filtrate was 'salted in' and purified as described in "Materials and Methods". Figure 1 illustrates the elution profile obtained showing that both the β-glucosidase and *N*-acetylglucosaminidase co-eluted. Fractions with the highest *N*-acetylglucosaminidase activity were pooled and purified further by preparative electrophoresis. (An exhaustive list of conventional chromatographic steps were tried, but the separation of the β-glucosidase and *N*-acetylglucosaminidase could not be achieved and significant losses of *N*-acetylglucosaminidase were noted; data not shown.)

Preparative electrophoresis indicated that the *N*-acetylglucosaminidase enzyme was purified to apparent homogeneity. The apparent homogeneity of the *T. emersonii* *N*-acetylglucosaminidase was verified by SDS-PAGE, Native-PAGE and IEF, using the appropriate protein detection and zymogram techniques. The purified enzyme preparation had a specific activity of 40.45 IU.mg^{-1}, and 63.9-fold purification factor.

Protein mass and pI values

SDS-PAGE analysis of the purified protein on a 10% polyacrylamide gel (Fig. 2a) indicated the presence of a

Table 3 Enzyme production by *A. niger* grown on 0.25% (w/v) fucoidan for 168 h

Hour	Protein mg mL^{-1}	α-fuc	β-fuc	F-hydrolase	Sulphatase	β-glc	β-gal	*N*-acetyl	β-xyl
0	0.026	0	0	0	0	0	0	0	0
12	0.024	0	0	0	0	0	0	0	0
24	0.026	0	0	0	0	1.3	0	0	0
36	0.027	0	0	0	0	2.0	1.8	0	0
48	0.031	0	0	0	0	2.2	3.0	0	0
60	0.028	0	0	0	0	2.9	5.0	1.3	0
72	0.028	0	0	0	0	3.5	5.3	3.0	0
96	0.034	0	0	0	0	18.7	13.6	9.5	0
120	0.032	0	0	0	0	26.2	26.0	23.2	0
144	0.030	0	0	0	0	36.0	28.7	26.8	0
168	0.032	0	0	0	0	32.1	26.4	27.9	0

All enzyme activity results are specific activity IU.mg^{-1}
Enzyme abbreviations: see Table 1

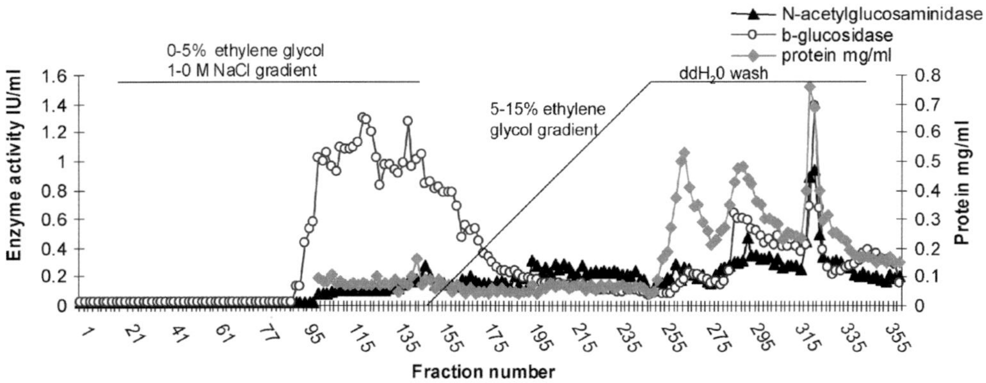

Fig. 1 The graph reflects the profile of two main exoglycosidase activities from the phenyl Sepharose fast flow chromatography step. A decreasing 1 M NaCl gradient and a simultaneous ethylene glycol [0–5% ($^{w}/_{v}$)] gradient were applied up to fraction 149. A further 5–15% ($^{w}/_{v}$) ethylene glycol gradient was applied from fractions 150–245 since the *N*-acetylglucosaminidase activity was unable to be detected up to this point. A further ddH$_2$O wash was needed to elute the tightly bound enzyme

single protein band with an estimated mass of 70 kDa. Activity staining of the purified enzyme following separation on a Native-PAGE gel using 4-MU-GlcNAc yielded a fluorescent band at a molecular mass of 140 kDa (Fig. 2b). Thus, the enzyme apparently is a dimer with subunit mass of 70 kDa and a native mass of 140 kDa.

Isoelectric focusing was used to establish the pI of the purified protein, which was determined to be at pH 3.85 (results not shown).

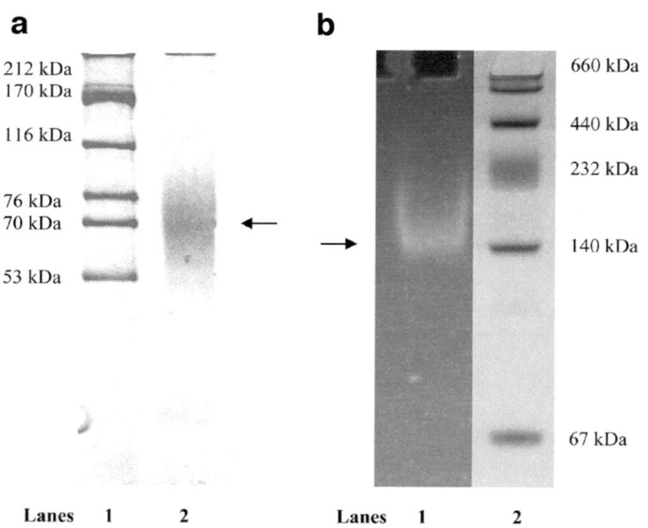

Fig. 2 Electrophoretic analysis of the purified *N*-acetylglucosaminidase from *T. emersonii.* **a** SDS-PAGE stained for proteins; lane 1 contains the high molecular mass standard proteins, lane 2 contains the purified enzyme with a single band at 70 kDa. **b** Native-PAGE stained for enzyme activity; lane 1 contains the single protein band stained using methylumbelliferyl-*N*-acetylglucosaminide with a single fluorescent band at 140 kDa, lane 2 contains the high molecular mass Native-PAGE protein standards stained with Coomassie brilliant blue

pH and temperature optima

Purified *N*-acetylglucosaminidase from *T. emersonii* exhibited activity over a broad pH range, i.e. activity was observed from pH 2.6–7.6, with optimum being detected at pH 5.0 (Fig. 3a). More than 50% of the activity measured at the optimum pH was observed at pH 3.8 and 7.6. At pH 5.0, the enzyme was active within the temperature range 35–75°C. Very low or no activity was found at 20, 25, 30 and 80°C (Fig. 3b), even though the *N*-acetylglucosaminidase was optimally active at 65–75°C.

Substrate specificity and kinetics

N-Acetylglucosaminidase purified from *T. emersonii* displayed strict specificity for 4-NP- and 4-MU-GlcNAc and exhibited <1% activity (relative to the GlcNAc derivative) against a range of 4-NP-glycosides and methylumbelliferyl-glycosides (Fig. 4); ~6% relative activity was observed against MU-β-Gal after prolonged incubation. The enzyme also showed no activity against a number of disaccharides including cellobiose, sucrose, lactose and maltose, thus confirming that the enzyme is not a cellobiase (or β-glucosidase). N-Acetylglucosaminidase purified from *T. emersonii* hydrolyzed the fluorescent substrate 4-MU-GlcNAc with Michaelis-Menten kinetics (data not shown). The K_m and V_{max} values for hydrolysis of this substrate at pH 5.0 and 50°C were 0.5 mM and 2.66 IU.mL^{-1} (44.33 IU. mg^{-1}), respectively. The turnover number for the enzyme (4-MU-GlcNAc as substrate) was 74,800 s^{-1}, while the catalytic efficiency (k_{cat}/K_m) was 149,600 s^{-1} mM^{-1}; both values indicate that the enzyme is catalytically very efficient.

Effect of metal ions and glycosides

The effect of metal ions on the action of the purified *N*-acetylglucosaminidase is shown in Table 4. The enzyme

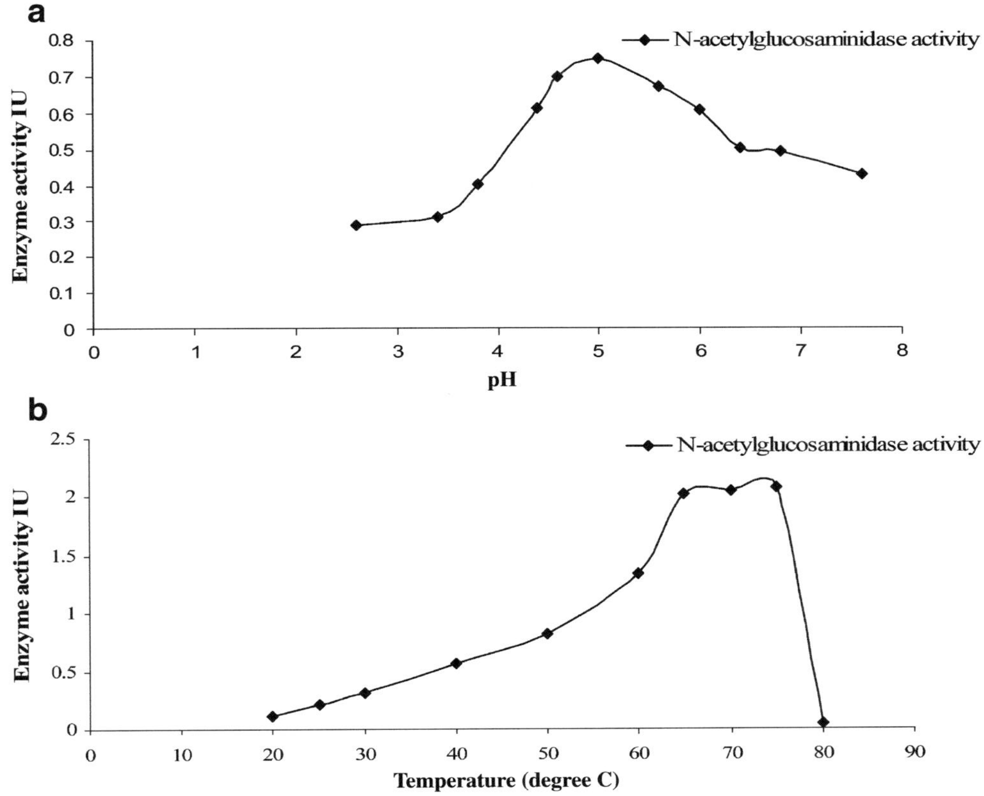

Fig. 3 Effect of pH (**a**) and temperature (**b**) on the activity of the purified *N*-acetylglucosaminidase

was sensitive to $Hg^{2+}$and Fe^{3+} at concentrations of 10 mM and slightly inhibited by Ca^{2+} and Co^{2+} ions. A range of different monosaccharides were included in the assay reaction cocktail to investigate potential effects (inhibitory or activatory) on the catalytic performance of the *T. emersonii N*-acetylglucosaminidase. In contrast to other *N*-acetylglucosaminidases, the *T. emersonii* enzyme displayed marked stability in the presence of the majority of sugars tested including the reaction product, GlcNAc, which was included at concentrations up to 10 mM and actually had a slight activating effect (~8.8–14.6% increase of activity). The other sugars included D-glucose, D-xylose, D-galacturonic acid and L-fucose, all of which have been identified to varying extents in different fucoidan fractions, also induced a slight activation of enzyme activity. Galactose was the only exception, as it decreased enzyme activity by ~15% of the control value at a concentration of 0.5 mM.

Discussion and conclusions

Our work reports the isolation of the first fungal enzyme produced during growth of the host organism on fucoidan as the main carbon source. In addition, to the authors' knowledge, it represents the first *N*-acetylglucosaminidase from a thermophilic fungus. Isolation of *T. emersonii N*-

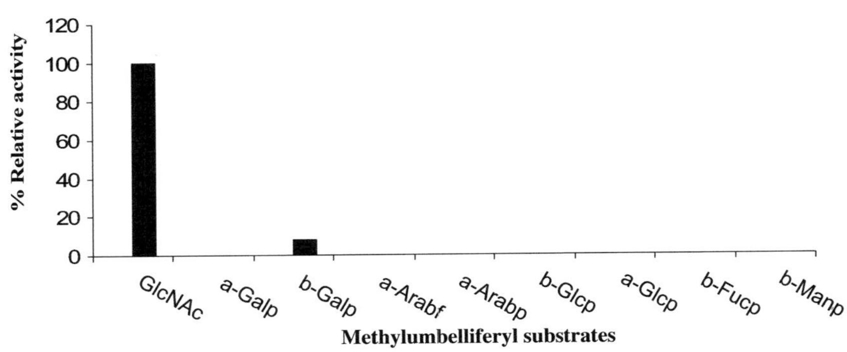

Fig. 4 Substrate screening using methylumbelliferyl substrates. The enzyme hydrolysed its fluorescent substrate methylumbelliferyl-*N*-acetylglucosaminidine (GlcNAc). Enzyme abbreviations: *GlcNAc N*-acetylglucosaminine; *α-Galp* α-galactopyranoside; *β-Galp* β-Galactopyranoside; *α-Arabf* α-Arabinofuranoside; *α-Arabp* α-Arabinopyranoside; *β-Glcp* β-Glucopyranoside; *α-Glcp* α-Glucopyranoside; *β-Fucp* β-Fucopyranoside; *β-Manp* β-Mannopyranoside

Table 4 Effect of metal ions and glucose on the activity of the purified *N*-acetylglucosaminidase isolated from *T. emersonii*

Reagents	Concentration (mM)	Relative activity %
Control	0	100
$CaCl_2$	10	96
$HgCl_2$	10	68
$FeCl_3$	0.1	111
$FeCl_3$	0.5	86
$FeCl_3$	10	24
$CoCl_3$	0.1	87
$CoCl_3$	0.5	95
$CoCl_3$	10	83
$ZnSO_4$	1.0	91
$ZnSO_4$	10	101
Glucose	0.5	101
Glucose	1.0	107
GlcNAc	0.5	109
GlcNAc	1.0	115
GlcNAc	10	109

acetylglucosaminidase yielded similar overall purification parameters (i.e. specific activity, fold purification) to data reported for other fungal hexosaminidases. The purified *T. emersonii* protein exhibited apparent homogeneity on denaturing and native polyacrylamide gel electrophoresis. The size and subunit structure of the *T. emersonii* enzyme shares many features in common with the *T. harazianum N*-acetylglucosaminidase. The *T. emersonii* enzyme is a dimer in solution with a native molecular mass of 140 kDa. The *T. harazianum* enzyme is also a dimer in solution but with a slightly lower molecular mass of 118 kDa (Ulhoa and Peberdy 1991). The purity of the *T. emersonii* enzyme was also confirmed by isoelectricfocusing, which revealed only one band specific for the enzyme. Its pI was found to be 3.85, similar to those for *N*-acetylglucosaminidases from the fungi *Sclerotinia fructigena* (pI 3.76) and *Aspergillus niger* (pI 4.4; Jones and Kosman 1980). The *T. emersonii* enzyme displays optimum activity at pH 5.0, while the *T. harazianum* counterpart is optimally active at pH 5.5 (Ulhoa and Peberdy 1991). However, the temperature optimum of the *T. emersonii* enzyme at 65–75°C is significantly higher than optima reported to date for other fungal *N*-acetylglucosaminidases. *N*-Acetylglucosaminidase purified from *T. harazianum* 39.1 had a temperature optimum of 50°C (Ulhoa and Peberdy 1991) and *N*-acetylglucosaminidase isolated from *T. harazianum* 1051 (Lisboa De Marco and Valadares-Inglis 2004) was found to be active within the temperature range 30–60°C and completely inactive at 70–80°C.

N-Acetylglucosaminidase from *Trichoderma harzianum* 39.1 displayed no activity against pNP-β-glucopyranoside, pNP-α- or -β-mannoside or cellobiose (Ulhoa and Peberdy 1991), similar to the *T. emersonii* enzyme. Thus, neither enzyme is a glucosidase. However, the *T. emersonii* enzyme displayed some activity, albeit small, towards 4-MU-β-galactopyranoside. Since glucose and galactose have similar structures and *N*-acetylglucosaminidases can often hydrolyse *N*-acetylgalactosaminide (Ulhoa and Peberdy 1991), the *T. emersonii* enzyme may also have the capacity to hydrolyse β-galactosamine slowly which may possibly be reflected in the observed hydrolysis of the β-galactopyranoside. However, by analogy, it should be noted that the enzyme did not hydrolyze MU or 4-NPGlc, even though 4-MU-GlcNAc was the preferred substrate of the enzyme, so the latter explanation may not be valid. Further studies are in progress to characterise this phenomenon. To date, mass spectrometry has not revealed the presence of contaminating β-galactosidase activity in the *T. emersonii N*-acetylglucosaminidase preparation, even at trace levels.

Inhibition by $Hg^{2+}$and Fe^{3+} suggests the presence of sulfhydryl or readily oxidizable group(s) that are essential for the stability or catalytically active structure and/or activity of the *T. emersonii N*-acetylglucosaminidase. Potent inhibitory effects have been reported for purified chitobiases from *Aeromonas hydrophilla* (Yabuki et al. 1986) and *Homarus americanus* (Lynn 1990), suggesting that the presence of sulfhydryl groups is critical for enzyme activity. Because more than 68% of the original activity of the *T. emersonii* enzyme remains in the presence of 10 mM $HgCl_2$, our results suggest that sulfhydryl groups may not be directly involved in catalysis, although some may be required to maintain the catalytically active conformation of the enzyme.

Glucose had no effect on the *T. emersonii* enzyme in concurrence with the report for *N*-acetylglucosaminidase from *Trichoderma harzianum* 39.1 (Ulhoa and Peberdy 1991) where glucose at a concentration of 10 to 20 mM had no effect on enzyme activity. These results contrast with the *N*-acetylglucosaminidase activity reported from *Trichoderma harzianum* 1051. In the latter case, glucose at a concentration of 1 mM inhibited 65% of the original activity of the purified enzyme (Lisboa De Marco and Valadares-Inglis 2004). In addition, the *T. emersonii* enzyme is not inhibited by GlcNAc, the reaction product, at concentrations up to 10 mM. This finding is most unusual for an exoglycosidase.

The catalytic properties of the *N*-acetylglucosaminidase from *T. emersonii* against native fucoidans, isolated from brown macroalgae using a virtually chemical-free procedure, are currently under investigation. The mode of action of this enzyme against fucoidan, chitin-derived poly- and oligosaccharides and fungal cell walls will be compared in order to identify its true potential function. The action of the enzyme in the presence of other fucoidan-modifying enzymes is being investigated. A *N*-acetylglucosaminidase induced on fucoidan is interesting and the results of the hydrolysis studies may provide a clue as to the abundance, location and function of hexosamines in these algal polysaccharides. Finally, as many of the potential applications of fucoidans are of a biopharmaceutical nature, the availability of an enzyme capable of acting at reaction temperatures that achieve pasteurization may also be important.

Acknowledgement This work was funded by a Department of Agriculture Food Institutional Research Measure award to M.T., under the National Development Plan (Ireland).

References

Berteau O, Mulloy B (2003) Sulfated fucans, fresh perspectives: structures, functions, and biological properties of sulfated fucans and an overview of enzymes active toward this class of polysaccharide. Glycobiology 13:29R–40R

Bradford M (1976) A rapid and sensitive method for the quantitation of microgram quantities of protein utilizing the principle of protein-dye binding. Anal Biochem 72:248–254

Carvalho M, Harada L, Gidlund M, Ketelhuth D, Boschcov P, Quintão E (2002) Macrophages take up triacylglycerol-rich emulsions at a faster rate upon co-incubation with native and modified LDL: An investigation on the role of natural chylomicrons in atherosclerosis. J Cell Biochem 84:309–323

Chizov A, Dell A, Morris H, Haslam S, McDowell R, Shashkov A, Nifant'ev N, Khatuntseva E, Usov A (1999) A study of fucoidan from the brown seaweed *Chorda filum*. Carbohydr Res 320: 108–119

Elving P, Markowitz J, Rosenthal I (1956) Preparation of buffer systems of constant ionic strength. Anal Chem 28:1179–1180

Furukawa S, Fujikawa T, Koga D, Ide A (1992) Purification and some properties of exo-type fucoidanases from *Vibrio* sp. N-5. Biosci Biotech Biochem 56:1829–1834

Gilleran CT (2004) An Environmentally-Benign Biotechnological Strategy to Utilize Carbohydrate Rich Waste. PhD Dissertation, National University of Ireland, Galway

Jones C, Kosman D (1980) Purification, properties, kinetics, and mechanism of β-*N*-acetylglucosamidase from *Aspergillus niger*. J Biol Chem 255:11861–11869

Kitamura K, Matsuo M, Yasui T (1992) Enzymatic degradation of fucoidan by fucoidanase from the hepatopancreas of *Patinopecten yessoensis*. Biosci Biotech Biochem 56:490–494

Kloareg B, Quatrano RS (1988) Structure of the cell-walls of marine-algae and ecophysiological functions of the matrix polysaccharides. Oceanogr Mar Biol 26:259–315

Laemmli UK (1970) Cleavage of structural proteins during the assembly of the head of bacteriophage T4. Nature 227:680–685

Lisboa De Marco J, Valadares-Inglis M (2004) Purification and characterization of an N-acetylglucosaminidase produced by a *Trichoderma harzianum* strain which controls *Crinipellis perniscosa*. Appl Microbiol Biotechnol 64:70–75

Lydon A (2005) Molecular characterization of the gene encoding a thermostable β-galactosidase from *Talaromyces emersonii*. PhD Dissertation, National University of Ireland, Galway

Lynn K (1990) Chitinase and chitobiases from the American lobster (*Homarus americanus*). Compar Biochem Physiol 96B:761–766

Maruyama H, Tamauchi H, Iizuka M, Nakano T (2006) The role of NK cells in antitumor activity of dietary fucoidan from *Undaria pinnatifida* sporophylls (Mekabu). Planta Med 15:1415–1417

Maruyama H, Tanaka M, Hashimoto M, Inoue M, Sasahara T (2007) The suppressive effect of Mekabu fucoidan on an attachment of *Cryptosporidium parvum* oocysts to the intestinal epithelial cells in neonatal mice. Life Sci 80:775–781

Matsubara K, Xue C, Zhao X, Mori M, Sugawara T, Hirata T (2005) Effects of middle molecular weight fucoidans on in vitro and ex vivo angiogenesis of endothelial cells. Int J Mol Med 15:695–699

Miller GL (1959) Use of dinitrosalicylate acid reagent for the determination of reducing sugar. Anal Chem 31:426–428

Moloney A, Hackett T, Considine P, Coughlan M (1983) Isolation of mutants of *Talaromyces emersonii* CBS 814.70 with enhanced cellulose activity. Enzyme Microbial Technol 5:260–264

Murray P, Grassick A, Laffey C, Cuffe M, Higgins T, Savage A, Planas A, Tuohy M (2001) Isolation and characterization of a thermostable endo-β-glucanase activity on 1,3–1,4– β-D-glucans from the aerobic fungus *Talaromyces emersonii* CBS 814.70. Enzyme Microb Tech 29:90–98

Nagumo T, Nishino T (1996) Fucan sulfates and their anticoagulant activities. In: Dumitriu S (ed) Polysaccharides in Medicinal Applications. Maecel Dekker, New York

Nishino T, Nishioka C, Ura H, Nagumo T (1994) Isolation and partial characterization of a novel amino sugar-containing fucan sulfate from commercial *Fucus vesiculosus* fucoidan. Carbohydr Res 225:213–224

Pomin V, Pereira M, Valente A, Tollefsen M, Pavao M, Mourao P (2005) Selective cleavage and anticoagulant activity of a sulfated fucan: stereospecific removal of a 2-sulfate ester from the polysaccharide by mild acid hydrolysis, preparation of oligosaccharides, and heparin cofactor II-dependent anticoagulant activity. Glycobiology 15:369–381

Ponce NM, Pujol CA, Damonte EB, Flores ML, Stortz CA (2003) Fucoidans from the brown seaweed *Adenocystis utricularis*: extraction methods, antiviral activity and structural studies. Carbohydr Res 338:153–165

Senni K, Gueniche F, Foucault-Bertaud A, Igondjo-Tchen S, Fioretti F, Colliec-Jouault S, Durand P, Guezennec J, Godeau G, Letourneur D (2006) Fucoidan a sulfated polysaccharide from brown algae is a potent modulator of connective tissue proteolysis. Arch Biochem Biophys 445:56–64

Thompson R (2004) Biochemical and structural studies on glycosyl hydrolases from a thermophilic eukaryote. PhD Dissertation, National University of Ireland, Galway

Tuohy M, Coughlan M (1992) Production of thermostable xylan-degrading enzymes by *Talaromyces emersonii*. Biores Technol 39:131–137

Tuohy M, Walsh D, Murray P, Claeyssens M, Cuffe M, Savage A, Coughlan M (2002) Kinetic parameters and mode of action of the cellobiohydrolases produced by *Talaromyces emersonii*. Biochim Biophys Acta 1596:336–380

Ulhoa C, Peberdy J (1991) Purification and characterization of an extracellular chitobiase from *Trichoderma harzianum*. Cur Microbiol 23:285–289

Xue C-H, Fang Y, Lin H, Chen L, Li Z-J, Deng D, Lu C-X (2001) Chemical characters and antioxidative properties of sulphated polysaccharides from *Laminaria japonica*. J Appl Phycol 13:67–70

Yabuki M, Mizushina K, Amatatsu T, Ando A, Fujii T, Shimada M, Yamashida M (1986) Purification and characterization of chitinase and chitobiase produced by *Aeromonas hydyrophila* subsp. *Anaerogenes* A52. J Gen Appl Microbiol 32:25–32

Essential role of seaweed cultivation in integrated multi-trophic aquaculture farms for global expansion of mariculture: an analysis

Amir Neori

Originally published in the Journal of Applied Phycology, Vol 20, No 5, 117–120.
DOI: 10.1007/s10811-007-9206-3

Abstract Ecologically friendly aquaculture crops, such as seaweeds, herbivores, omnivores, and detritivores can be cultured using relatively less of our limited natural resources and produce relatively less pollution. They also top FAO's estimates of aquaculture crops for the 21st century. These crops already comprise nearly 90% of global aquaculture tonnage, >90% of all aquaculture production in China and >60% of production even in North America. Consumers prefer them, most likely due to their low prices. Production costs of organisms low on the food chain are low due to the ability of these organisms to efficiently utilize low-cost, mostly plant-based diets and to recycle their own waste. Thus, ecologically friendly aquaculture is not a dream but a dominant global reality. The less ecologically-friendly aquaculture of salmon, sea bream, fed shrimp, among others, has attracted public opposition to aquaculture, but these crops totaled approximately only 10% of global production in 2004. The profitability of industrialized monocultures of these crops is threatened further by rising costs of energy and feed, environmental regulation compliance, disease, and public opposition. Current monoculture practices and perceptions intrinsic to the aquaculture industry can be turned around into a vision of sustained profitable expansion of carnivores production with trophically lower organisms in ecologically-balanced aquaculture farms. This category of aquaculture, which is the modern intensive form of polyculture practiced in Asia, feeds the waste of carnivore culture to lower trophic level organisms, primarily algae and mollusks. Species are selected based on their ecological functions in addition to their economic potential. Ecologically-balanced farms turn the costly treatment of carnivore waste outside the farm to a revenue-generating process of biofiltration, conversion, and resource recovery into plant and mollusk crops inside the farm. In doing so, they solve several of the major problems faced by modern aquaculture. The aquaculture industry can protect its own interests – and reap major benefits – by understanding the importance of ecological balance, the potential of seaweeds as components in feeds, and the importance of the culture and R&D of low trophic level organisms. The industry should also accept the relevance of environmental, social, and image aspects of aquaculture to its success. Governments have the tools to reward multi-trophic farms with seaweeds by means of tax credits and nutrient credits and to penalize unbalanced monoculture approaches by means of 'polluter pays' fines, thereby providing the multi-trophic farms with a significant economic advantage. Such measures have been discussed, but their implementation has been slow.

Keywords Aquaculture crops · Fish · Integrated multi-trophic aquaculture · Profitability · Seafood · Seaweeds · Sustainable

Introduction

Despite mounting criticism of the ecological and social un-sustainability of aquaculture, in the very near future it will surpass capture fisheries as the main source of seafood on the global market (Delgado et al. 2002; FAO 2006). Criticism of modern aquaculture practices can be expected

A. Neori (✉)
Israel Oceanographic and Limnological Research,
The National Centre for Mariculture,
P.O. Box 1212, Eilat 88112, Israel
e-mail: aneori@gmail.com

to persist; however, as long as the industry delivers good products at reasonable prices, the market for cultivated marine organisms will continue to expand. The challenge facing the industry is how to keep expanding sustainably while meeting the concerns expressed by a wide spectrum of interested parties on a wide variety of issues (Neori et al. 2007). Even if most of today's customers as yet do not care whether the fish they buy come from sustainable culture, the industry should. It should learn how to expand in the face of a deteriorating environment, rising feed and energy prices, and the increasing sustainability-awareness of the consumer.

Analysis

The two main issues that modern aquaculture has not yet come to grips with are (1) the ability of present technological developments in aquaculture to significantly expand fish supplies with large quantities of carnivorous fish at prices that compete with the prices of capture fisheries and (2) the capacity of these technologies to be sustainable and to meet the environmental, ecological and social concerns of the public. These two questions are intertwined and can be resolved by ecologically balanced aquaculture.

The basic problem to be considered by the industry is the ecological imbalance in its development of industrial-scale aquaculture, the so-called "Blue Revolution" (White et al. 2004). On a purely ecological basis, animal culture requires plant culture, a factor apparently side-stepped by the proponents of the "Blue Revolution". It is a solidly based fact that, given current aquaculture practices, carnivorous fish culture still consumes – in the form of fish meal and fish oil – more fish biomass than it produces (Tacon 2004). To meet this challenge, the industry has been developing "fishmeal replacement" – feed which replaces the proteins and oil ingredients from fish sources in commercial fish feeds by land plant commodities such as soy meal and canola oil. However, this course of action is in a socio-economical sense not promising for several reasons, one of which is that it competes with more traditional uses of land plant resources for direct human consumption. The limited economic logic of the fishmeal and fish oil replacement approach is exacerbated by a complete ignorance of the environmental, social, economical, and health-related benefits to the cultured animals and the consumers associated with the cultivation in aquaculture farms of seaweeds, herbivores, and detritivores. Therefore, replacements of fish meal and fish oil by land crops are unlikely to resolve the main obstacles facing expansion of mono-aquaculture, including the presently debated "off-shore mariculture" (which merely moves the monoculture's negative environmental implications further off-shore (Naylor 2006)). One remedy is straightforward: if the aquaculture industry wants to double or even triple carnivorous fish production in a couple of decades, as the FAO predicts, it should shift to an ecologically balanced expansion, i.e. cultivate plants, herbivores, and detritivores at quantities several folds the desired quantities of carnivorous fish (Neori et al. 2007). Herbivorous and omnivorous fish are ecologically superior to carnivorous fish since they waste less nitrogen than species higher up the food chain, and some of them can recycle their own waste by consuming *in situ* populations of algae, bacteria, and other organic particles that develop in the culture water (see, for example, Avnimelech 2005). Aquatic plants can provide fish culture with three necessary sustainability features – financial diversity, environmental compatibility, and social relevance. Seaweeds increase and diversify the income of aquaculture farms, treat their fish waste, supply essential protein and oil for the feed industry (see, for example, Valente et al. 2006), provide nutritious human food, and provide jobs in aquatic plant and mollusk culture for the poor and the displaced (Neori and Troell in press; Troell et al. 2006). The combination of low production costs and high health value is likely to allow the development and expansion of the market for seaweeds for human consumption in the Western world. Of course, seaweeds already constitute popular foods in large regions of the world, with the main consumers being Chinese, Japanese and Koreans in their homelands and wherever they have migrated to (McHugh 2003). The demand for seaweeds by the phycocolloid, food additive, cosmetics and medicine markets has been growing steadily too.

Aquaculture itself has had for generations the technological capability to take care of both its environmental impact and its ecological imbalance. The largest aquaculture sector, inland fish farming, is one example. Carps have for centuries been cultivated in polyculture ponds that do not pollute but instead recycle farm and household waste by means of bacteria, algae, and aquatic plants to feed the fish (Edwards 2004). Modern aquaponics recycles the wastewater from intensive fish culture, such as that of the omnivorous tilapia, for vegetable hydroponics (http://www.aquaponics.com/InfoAquaponics.htm; Rakocy et al. 1992). In marine integrated multi-trophic aquaculture, farming of fish, seaweeds, and mollusks in cages and ponds takes place side-by side, mostly in South-East Asia but also in Chile and Canada (Neori et al. 2004; Neori and Troell in press). The seaweeds or 'sea vegetables', being primary producers, add oxygen to the seawater while removing the nutrient wastes from seawater of the nearby fish farms. Some cultivated seaweeds are also net bio-energy producers (Troell et al. 2004), a characteristic of increasing relevance with the rising fossil energy prices. Mollusk bivalves, being omnivorous filter-feeders, eat microalgae, whose ecological function in cleaning the seawater is similar to that of the seaweeds (Shumway et al. 2003).

Fish and shrimp are the most noticeable seafood crops in Western countries and are the main aquaculture crops in several countries, including Norway, Chile, Scotland, Canada and Thailand. However, these crops have consistently constituted only a small fraction of the global production in marine aquaculture (Fig. 1), and according to the Food and Agricultural organization (FAO), they will remain doing so (see forecast in Fig. 1).

Recommendations for action

As the FAO data trends clearly suggest, without major technological breakthroughs, crops belonging to the lowest trophic levels – seaweeds, mollusks, and freshwater herbivore-omnivore fish – will remain the main constituents of aquaculture production well into the 21st century. This fact should act as an incentive to the modern aquaculture establishment – the industry, its regulators, and research and development (R&D) funding agencies – to devote much of their attention and funds toward lower trophic level ecologically friendly crops and sustainable culture technologies. In order to evolve without crises and without an excessive waste of R&D resources on unsustainable 'super carnivore' species and monoculture technologies, and to alleviate the concerns of the different parties involved, including those of environmentalists, modern aquaculture should follow several key principles.

Firstly, it has to comprehend the 'essentiality' of linking its expansion ecologically by balancing the cultured species on the regional and local levels. The economic reasons that make sustainable species – plants, herbivores and omnivores – leaders in global aquaculture should be researched

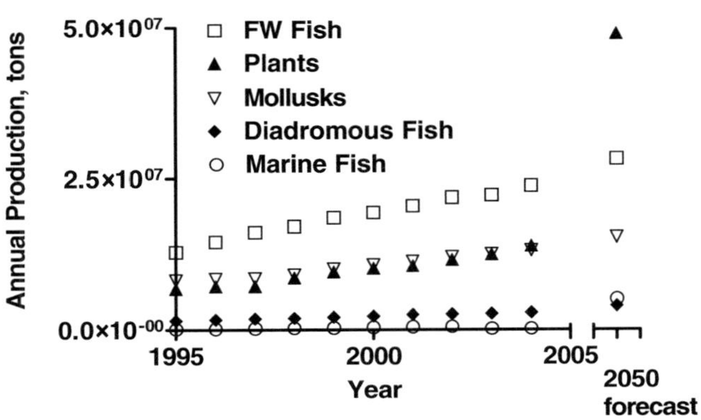

Fig. 1 FAO statistics for global aquaculture production by main species groups. *Right* an aquaculture production forecast for the same species groups for the year 2050 (by J. Muir, in Delgado et al. 2002)

and understood. The industry would then be able to recognize how integrated multi-trophic aquaculture farms with seaweeds and mollusks can solve many of its sustainability problems and help maintain competitive prices of cultivated carnivorous fish and shrimp.

Secondly, the single grower has to become a full-range producer of all trophic levels and, consequently, to become ecologically balanced on the local scale. The growers should familiarize themselves with the approach of fish polyculture with plants, including seaweeds, that is available 'off the shelf' for all water types under names such as 'integrated aquaculture', 'aquaponics', 'partitioned aquaculture' and, most recently, 'integrated multi-trophic aquaculture' or 'IMTA' (Chopin et al. 2001; Neori et al. 2004).

Thirdly, the industry can begin to seriously consider seaweeds as major ingredients in its fish and shrimp feed (Powell 2003); it can also expand its R&D on herbivorous and omnivorous fish (for example carps, tilapias, milkfish, gray mullet, catfish) and enlarge further the production share of bivalve mollusks, not only because these crops are ecologically-friendly, but also because at the present time these are the only crops that are net producers of biomass, relative to the 'high-value' carnivorous marine fish (Tacon 2004). Low trophic level organisms can be cultured by applying present day low-cost technologies, close to shore and, wherever low-cost land is available, in simple ponds, rather than in costly off-shore cages and water-recirculating technologies.

Fourthly, the environmental, social, and image aspects of any aquaculture technology should be considered as an integral part of the deal. Consumers will expect more of seafood than price and taste. The global market is becoming more interested in where the product is coming from, how it is produced, its availability to the poor, and even the living conditions of the people who work on the farms or who are displaced by them (Martinez 1998; White et al. 2004). Aquaculture will have to provide more than profitability, with the environmental and social considerations becoming more important. While the Blue Revolution aquaculture industry has become excellent at producing high-quality and high-value fish and shrimp, it has to become equally good at accepting the responsibility of the environmental and social implications of its activities. Cooperation on these aspects with governments and the FAO will provide aquaculture with the impetus to maintain itself as a growth industry.

Acknowledgements The contributions in the form of discussions and publications to the development of the ideas in this article by colleagues and friends around the world, in particular M. Troell, are gratefully acknowledged. Two anonymous reviewers and an editor have helped to significantly improve the manuscript.

References

Avnimelech Y (2005) Using the pond as a biofilter: review of theory and practice. Int J Recirculat Aquacult 6:1–17

Chopin T, Buschmann AH, Halling C, Troell M, Kautsky N, Neori A, Kraemer G, Zertuche-Gonzalez J, Yarish C, Neefus C (2001) Integrating seaweeds into aquaculture systems: a key towards sustainability. J Phycol 37:975–986

Delgado CL, Rosegrant MW, Meijer S, Wada N, Ahmed M (2002) "Fish as Food: Projections to 2020," paper presented in the 2002 Biennial Meeting of International Institute for Fisheries Economics and Trade (IIFET). Available at: http://www.fishforall.org/media/release/fish2020.pdf

Edwards P (2004) Traditional Chinese aquaculture and its impact outside China. World Aquacult 35:24–27

FAO Fisheries Department (2006) State of world aquaculture 2006. FAO Fisheries Technical Paper No. 500, FAO, Rome, Italy. Available at: http://www.fao.org/docrep/009/a0874e/a0874e00.htm

Martinez AR (1998) Engineering the Blue Revolution. Samudra Triannual Report of ICSF 20:45–55

McHugh DJ (2003) A guide to the seaweed industry. FAO Fisheries Technical Paper No. 441. FAO, Rome, Italy

Naylor R (2006) Editorial: offshore aquaculture legislation. Science 313:1363

Neori A, Troell M (in press) Ecologically balanced aquaculture for the rehabilitation of Asian fisheries. World Aquacult

Neori A, Chopin T, Troell M, Buschmann AH, Kraemer GP, Halling C, Shpigel M, Yarish C (2004) Integrated aquaculture: rationale, evolution and state of the art emphasizing seaweed biofiltration in modern mariculture. Aquaculture 231:361–391

Neori A, Troell M, Chopin T, Yarish C, Critchley A, Buschmann AH (2007) The need for a balanced ecosystem approach to "Blue Revolution" aquaculture. Environment 49:36–43

Powell K (2003) Eat your veg. Nature 426:378–379

Rakocy JE, Losordo TM, Masser MP (1992) Recirculating aquaculture tank production systems: integrating fish and plant culture. SRAC Publication No. 454. Available at: http://www.lsuagcenter.com/NR/rdonlyres/CAED7908-B3EC-4CCD-A85E-347A54F51D1B/306/fish_plant.pdf

Shumway SE, Davis C, Downey R, Karney R, Kraeuter J, Parsons J, Rheault R, Wikfors G (2003) Shellfish aquaculture – in praise of sustainable economies and environments. World Aquacult 34:15–17

Tacon AGJ (2004) Use of fish meal and fish oil in aquaculture: a global perspective. Aquat Resource Cult Dev 1:3–14.

Troell M, Tyedmers P, Rönnbäck P, Kautsky N (2004) Aquaculture – energy use. In: Cleveland C (ed) Encyclopedia of energy. Elsevier, New York, pp 97–108

Troell M, Robertson-Andersson D, Anderson R, Bolton J, Maneveldt G, Halling C, Probyn T (2006) Abalone farming in South Africa: an overview with perspectives on kelp resources, abalone feed, potential for on-farm seaweed production and socio-economic importance. Aquaculture 257:266–281

Valente LMP, Gouveia A, Rema P, Matos J, Gomes EF, Sousa-Pinto I (2006) Evaluation of three seaweeds *Gracilaria bursa-pastoris*, *Ulva rigida* and *Gracilaria cornea* as dietary ingredients in European sea bass (Dicentrarchus labrax) juveniles. Aquaculture 252: 85–91.

White K, O'Neill B, Tzankova Z (2004) At a crossroads: will aquaculture fulfill the promise of the blue revolution? Silver Spring Meriland SeaWeb: http://www.seaweb.org/resources/documents/reports_crossroads.pdf

Opportunities and challenges for the development of an integrated seaweed-based aquaculture activity in Chile: determining the physiological capabilities of *Macrocystis* and *Gracilaria* as biofilters

Alejandro H. Buschmann · Daniel A. Varela · María Carmen Hernández-González · Pirjo Huovinen

Originally published in the Journal of Applied Phycology, Vol 20, No 5, 121–127.
DOI: 10.1007/s10811-007-9297-x © Springer Science + Business Media B.V. 2007

Abstract Seaweed production is a reality in Chile. More than ten species are commercially used to produce phycocolloids, fertilizers, plant growth control products, human food or animal fodder and feed additives. These multiple uses of algae offer a number of possibilities for coupling this activity to salmon, abalone and filter-feeder farming. In this context, different experiments carried out in Chile have demonstrated that *Gracilaria chilensis* and *Macrocystis pyrifera* have great potential in the development of an integrated aquaculture strategy. The present Integrated Multi-Trophic Aquaculture (IMTA) approach study showed that *Gracilaria* can be cultured best at 1 m depth whereas *Macrocystis* has an especially good growth response at 3 m depth. Both species use available nitrogen efficiently. On the other hand, high intensities of solar radiation (UV and PAR) can be critical at low depths of cultivation, and our results indicate that both species show photosynthetic susceptibility mainly at noon during the summer. The demand of *Macrocystis* for abalone feeding is increasing, thus improving the opportunity for developing an integrated nutrient waste recycling activity in Chile. Although *Gracilaria* shows a higher nitrogen uptake capacity than *Macrocystis*, its market value does not yet allow a massive commercial scaling.

Keywords Chile · *Gracilaria* · Integrated multi-trophic aquaculture · *Macrocystis* · Physiology · Seaweeds as biofilters

A. H. Buschmann (✉) · D. A. Varela · M. C. Hernández-González · P. Huovinen
Centro I-Mar, Universidad de Los Lagos,
Camino Chinquihue km 6,
Puerto Montt, Chile
e-mail: abuschma@ulagos.cl

Introduction

Aquaculture development has strongly increased in Chile over the last 20 years. Aquaculture production increased from 7,499 tonnes of fish, mussel and seaweeds in 1985 to 739,368 tonnes in 2005 (Sernapesca 1985, 2005), making Chile the leading marine aquaculture producer country in the western world. Nevertheless, the increase of production, especially salmon farming, is causing increasing concern related to environmental impacts, such as benthic effects (Soto and Norambuena 2004), fish escapees (Soto et al. 2001), introduction of toxic chemicals, such as copper, from cages (Buschmann and Fortt 2005), and the effects of antibiotics (Miranda and Zemelman 2002; Fortt et al. 2007). An additional problem mentioned in a recent review (Sellner et al. 2003) indicates that massive introduction of dissolved nutrients, principally nitrogen, into the water column may result in a potential enhancement of risks of harmful algal blooms. Recently, Buschmann et al. (2006) indicated an increase in dinoflagellate abundance associated with salmon farms. Furthermore, Arzul et al. (2001) suggested that marine animal excretion in the surrounding waters could increase phytoplankton biomass, modify the structure of phytoplanktonic communities and promote the risk of harmful algal blooms. In consequence, the literature is emphasizing the importance of developing a nutrient-balanced aquaculture concept, by co-culturing in the same embayment fed organisms with extractive ones (Chopin et al. 2001; Troell et al. 2003; Neori et al. 2004), thus allowing the recycling of wastes produced by fed aquaculture. This strategy should move aquaculture towards a more sustainable activity. Because of the great importance of fed aquaculture in Chile, the introduction of seaweed cultivation seems an important development to sustain salmon production in the future. Seaweed aquaculture is of great importance in oriental countries (Sahoo and Yarish 2005), but in western countries its activity is almost exclusively carried out in Chile on commercial scales (Chopin et al. 2007).

The selection of seaweed species for their commercial use as biofilters depends principally on two aspects. The interest of investors for a new species to be incorporated depends firstly on the commercial value and, secondly, on the physiological capabilities for growth in culture conditions together with its capacity to remove dissolved nutrients (Buschmann et al. 2001; Chopin et al. 2001). At present, the commercial interest and the farming capabilities of *Macrocystis pyrifera* and *Gracilaria chilensis* have been studied, and they are considered the best candidates for commercial seaweed farming in Chile (Buschmann et al. 2007). The efficiency of seaweeds as biofilters depends on several culture conditions, such as the water flow and nutrient concentration (Buschmann et al. 2001). Therefore, for this reason, it is of great importance to establish the physiological capabilities of the selected seaweeds in order to design an efficient integrated aquaculture system. In this study, we describe the physiological capabilities of the two algae cultivated in an integrated farm near an Atlantic salmon farm in southern Chile. Data from this study are discussed with previously published results on the physiological status of these species in natural conditions in southern Chile, and the feasibility of an efficient Integrated Multi-Trophic Aquaculture (IMTA) model for Chilean aquaculture is critically analyzed.

Materials and methods

An IMTA culture approach was carried out in Calbuco (41° 48′S, 73°11′W) in the X region in southern Chile (Fig. 1). A 100-m long-line was installed about 100 m from a salmon farm that produces a gross biomass of 2500–3000 t of fish per year (Fig. 2). The seaweed long-line was installed in the main water flow (average currents: 7.6 and 2.6 cm s^{-1} during the flood and ebb period, respectively) in the bay receiving most of the nutrient discharges during the flood current period (Fig. 1). The environmental conditions over 2 years of research are shown in Table 1. The temperature and the photosynthetically active radiation

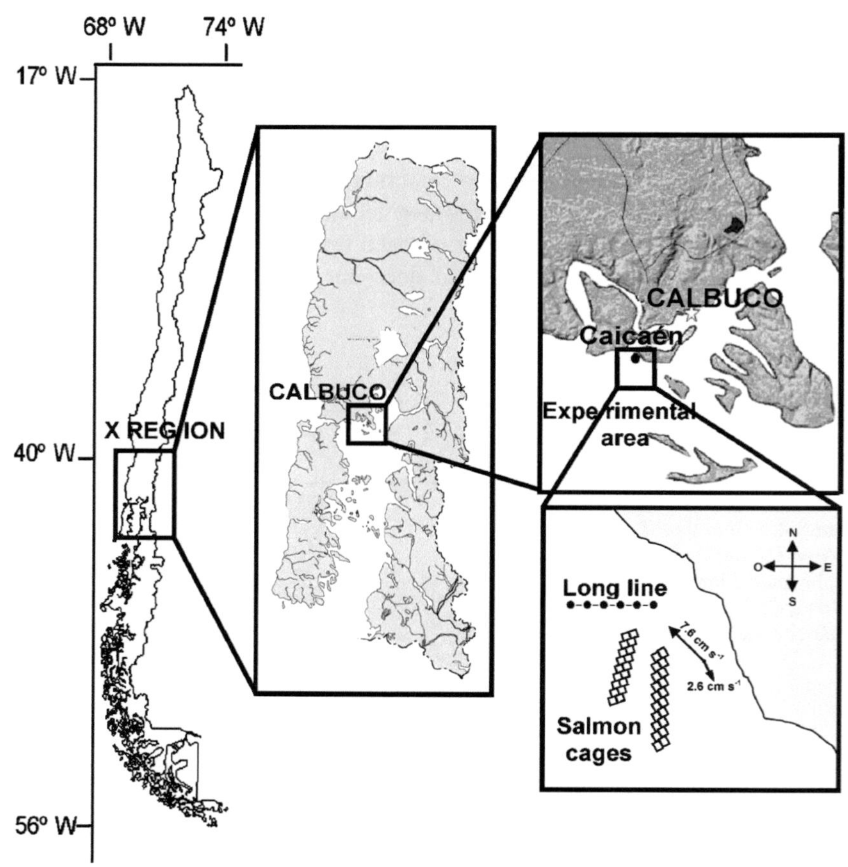

Fig. 1 Map showing the experimental location in Calbuco in the X Region in southern Chile and the long-line position near the salmon cages. The *arrows* show the main directions of ebb and flood currents and the *numbers* show the average current velocities

Fig. 2 Integrated suspended farming of *Gracilaria* and *Macrocystis* in Calbuco. **a** Salmon farming operation; **b** experimental long-line for seaweed production; **c** cultivated *Macrocystis* plant; **d** *Gracilaria* bundles during the seeding process

(PAR) were measured with a diving PAM (Diving-PAM, Walz, Effeltrich, Germany), pH with a Hanna 211 pH-meter and the salinity with a CTD (Seacat Profiler 19 Plus, Seabird). Macronutrients (nitrate, ammonium and phosphate) were measured following the methods of Strickland and Parsons (1972).

The experimental 100-m long-line was installed during the winter and fall of 2004. The long-line design permitted the cultivation of *Macrocystis pyrifera* (L.) C. Agardh and *Gracilaria chilensis* Bird, McLachlan & Oliveira at three depths, while the culture lines were divided into six sections that allowed three randomly distributed culture sections for *Macrocystis* and another three for *Gracilaria* at each culture depth. The extension of one replicate culture line was 10 m for *Macrocystis* and 3 m for *Gracilaria*. This design permitted the calculation of the specific growth rate (SGR) of 10–18 plants of *Macrocystis* by section, measuring the length of individuals for each treatment monthly as follows: SGR=Ln (final length/initial length)/growth period. *Macrocystis* production was calculated after determining the wet biomass obtained at the final harvest, after 9 months of culture, which is the growth period to reach a harvestable size. In the case of *Gracilaria*, a monthly harvesting strategy was followed (Halling et al. 2005) and, therefore, 10 bundles of thalli by section with an initial weight of 100 g were weighed every month to obtain the monthly biomass increment and to calculate the SGR and the

Table 1 Range of annual values of environmental parameters at the three culture depths in Calbuco, southern Chile

Environmental variables	Depth		
	1 m	3 m	6 m
Temperature (°C)	10.0–15.0	10.1–15.0	10.1–14.9
PAR (maximum range) (μmol e^{-1} m^{-2} s^{-1})	80–1,160	27–404	13–142
pH	7.9–8.1	7.9–8.1	7.9–8.1
Salinity (‰)	32.4–32.7	32.4–32.7	32.4–32.7
NO_3 (μM)	0.049–23.9	0.21–23.7	0.9–24.0
NH_4 (μM)	0.028–6.11	0.028–6.11	0.028–6.11
PO_4 (μM)	0.51–1.03	0.51–1.03	0.51–1.03

biomass production (g produced per month in 1-m culture line).

To determine the physiological performance of cultivated *Macrocystis* and *Gracilaria*, photosynthetic parameters, nitrogen uptake rate and nitrate reductase enzyme activity, as well as total nitrogen and protein content were determined. The quantum efficiency (α), the electron transport rate (ETR_{max}) and the saturating irradiance for photosynthesis (E_k) were measured with a pulse amplitude modulation fluorometer (Diving-PAM) under laboratory conditions within 24 h of the sampling.

Nitrogen uptake measurements were carried out in the laboratory using the collected seaweed tissues of both species according to a perturbation method (sensu Lobban and Harrison 1994) over a 10-h period in μM NO_3 g $(DW)^{-1}$ h^{-1}. Nitrate was measured using cadmium columns following the method of Strickland and Parsons (1972). Furthermore, algal tissues were analyzed bimonthly for nitrate reductase activity (μmol NO_2 g^{-1}(FW) min^{-1}) following the methodology described by Hurd et al. (1995) for brown algae and Chow et al. (2004) for red algae. The total nitrogen and protein content in the tissues were measured bimonthly from frozen samples using the Kjeldahl method (Bradstreet 1965). All the experiments and analyses were carried out in triplicate.

Results and discussion

Growth rates reached on average 6% day^{-1} in *Macrocystis pyrifera* and 4% day^{-1} in *Gracilaria chilensis* in a suspended integrated aquaculture system in southern Chile. Both species show the strongest growth during the spring season (Table 2). The growth capacity of the algal species in this study in an integrated suspended salmon–seaweed culture system allows an annual production of over 25 kg m^{-1} of *Macrocystis* during a 9-month production period, and an average of 2.8 kg m^{-1} $month^{-1}$of *Gracilaria* in spring (Table 2). The results correspond to the optimal growing conditions during springtime in the suspended culture system, which for *Macrocystis* was at 3 m and for *Gracilaria* at 1 m water depth (Table 2). In general, the production values of *Macrocystis* in our integrated culture system were higher than in previous experiments in a monoculture condition (see Gutierrez et al. 2006), but significantly lower than the experimental scale monoculture results obtained by Westermeier et al. (2006). The growth values for *Gracilaria* are similar to previous results obtained for integrated cultures (Troell et al. 1997; Halling et al. 2005) indicating that in integrated culture conditions this red alga can increase over 30% of its production capacity in the presence of salmon culture installations. In the case of *Macrocystis*, epiphytism is not a major issue and only slight cover of bryozoans on the thalli could be found. On the other hand, we have been able to control epiphytism on *Gracilaria*. In the case of this red alga, repeated cuts of the algal bundles can enhance epiphytism (Halling et al. 2005), and for this reason we have developed a strategy of harvesting the complete bundle in a similar way to that used in suspended cultivation of *Kappaphyccus* (Trono 1993).

Macrocystis pyrifera and *G. chilensis* had mean values of quantum efficiency (α) over 25% and 14%, ETR_{max} values of 36 and 13 μmol e^- m^{-2} s^{-1}, and the E_k values of 134 and 84 μmol photons m^{-2} s^{-1}, respectively (Table 2), values that are similar to previously published data (Table 3). These data show that *Macrocystis* (optimal growing depth at 3 m) requires less light energy than *Gracilaria* (optimal growing depth at 1 m) to reach the biomass production reported in this study. Despite *Gracilaria* grows at higher rates at 1 m depth than deeper in the water column, it has a lower light saturation value than *Macrocystis* at a depth of 3 m (Table 2). Nevertheless, the optimal growing depth or season

Table 2 Summary of the growth and physiological responses of *Macrocystis pyrifera* and *Gracilaria chilensis* in a suspended culture system associated with a commercial salmon farm in Calbuco, southern Chile

	Macrocystis	*Gracilaria*
Optimal culture depth (m)	3	1
Highest productivity period	Spring	Spring
Growth rate (% day^{-1})	0.06±0.01	0.04±0.01
Maximum biomass production (kg (wet) m^{-1})	25.0 (in 9 months)	2.8 (by month)
Quantum efficiency (α) (μmol e^{-1} m^{-2} s^{-1}) (μmol m^{-2} s^{-1})$^{-1}$	0.27±0.03 (0.072–0.27)	0.14±0.03 (0.013–0.23)
Electron transport rate (ETR_{max}) (μmol e^- m^{-2} s^{-1})	36±9 (4.2–36)	13±3 (2.2–33)
Saturating irradiance for photosynthesis (E_k) (μmol photons m^{-2} s^{-1})	134±23.7 (58–218)	84±11 (12–250)
Nitrate uptake rate (μM NO_3 $g(DW)^{-1}$ h^{-1})	11.8±4.5 (3.7 –16.9)	4.9±2.3 (1.2–35.6)
NR activity (μmol NO_2 g^{-1} (FW) min^{-1})	1.8±0.08 (0.61–1.8)	2.13±0.16 (0.66–2.13)
Nitrogen content (% DW)	2.31–2.42	4.52–4.60
Protein content (% DW)	13.03–15.08	28.24–28.75

Mean and standard deviation values represent the spring average values, and the data in parenthesis the annual range.

Table 3 Physiological characteristics of intertidal *Macrocystis pyrifera* and *Gracilaria chilensis* in summer conditions in southern Chile

Physiological parameters	*Macrocystis*	*Gracilaria*	Reference
Photosynthetic characteristics			
Quantum efficiency (α) [μmol e^{-1} m^{-2} s^{-1}]	0.22±0.01	0.15±0.004	Gómez et al. 2004
[μmol m^{-2} s^{-1}]$^{-1}$		0.05–0.08	Gómez et al. 2005
Electron Transport Rate (ETR_{max}) μmol e^{-} m^{-2} s^{-1}	72.3±8.1	28.2±4.6	Gómez et al. 2004
Saturating irradiance for photosynthesis (E_k) μmol photons m^{-2} s^{-1}	327±45	183±34	Gómez et al. 2004

is not necessarily optimal for photosynthesis. Highest quantum efficiency and the maximum photosynthesis, as estimated from ETR, in the case of *Macrocystis* coincide with the period (spring) of maximal growth rate, in contrast to *Gracilaria*. Solar UV radiation can be high in this geographical location (Huovinen et al. 2006a) and the importance of photoprotective compounds, such as mycosporine-like amino acids (MAAs) found in red algae, including *Gracilaria* from southern Chile (Huovinen et al. 2004; Gómez et al. 2005), may play an important role in increasing the physiological capabilities of *Gracilaria* to be farmed at a depth of only 1 m. Based on their physiological characteristics, *Macrocystis* and *Gracilaria* in the intertidal zone in southern Chile exhibit sun-adapted strategies (Gómez et al. 2004). On the other hand, *Gracilaria* in the Quempillen River estuary showed more shade-adapted characteristics (Gómez et al. 2005). The data of the present study indicate that *Macrocystis* is a sun-adapted species that can be cultivated near the surface (3 m optimal growth depth), but care must be taken at the end of the spring and early summer when light intensity increases (Gutierrez et al. 2006).

Nitrate uptake rate, nitrate reductase activity, nitrogen and protein tissue content of *Gracilaria* and *Macrocystis* are shown in Table 2. In spring, *Macrocystis* has a higher nitrate uptake capacity than *Gracilaria*, but considering the seasonal data, *Gracilaria* showed the highest maximal nitrate uptake rates. The nitrate reductase activity does not show significant differences, but the nitrogen and protein content are significantly higher in *Gracilaria* than in *Macrocystis* (Table 2). Higher nitrogen content in red algae may be explained by the presence and accumulation of nitrogen containing compounds such as phycobiliproteins and MAAs (Korbee et al. 2005; Huovinen et al. 2006b), and can indicate that *Gracilaria* may be the best choice to be used as a biofilter. Recently, it has been demonstrated that different algal species have a differential capacity to absorb nitrate and ammonium from seawater (Bracken and Stachowicz 2006), indicating that the use of several species can be complementary from a biofiltering point of view. *Gracilaria* has been shown to use ammonium efficiently, reducing 90% of the total dissolved nitrogen discharges in tank cultures (Buschmann et al. 1994; Buschmann et al. 1996a, b). Therefore, the relative ability of *Macrocystis* integrated with *Gracilaria* to remove ammonium and nitrate from the seawater must be established in the near future considering seasonal nitrogen outputs, targeting its more effective removal.

It is relevant and important to reduce the seaweed farming area required to increase economic profitability and reduce interactions with other human activities in the coastal areas (Buschmann et al. 2001). In this study, we found that the optimal growing depths for *Macrocystis* and *Gracilaria* are different. In this context we propose, as a hypothesis, that both species should be introduced simultaneously at two different depths to increase the nutrient removal effectiveness from the environment. This idea requires the engineering development of new culture systems.

Light and nutrients as such affect the physiological performance of algae but may also interact. Therefore, these environmental factors and their consequences must be included in the development of a culture model. Photosynthetic activity may be affected not only by high solar radiation (UV and PAR), but also by nutrient availability. For example, in some red algae like *Porphyra* and *Grateloupia*, elevated nitrogen (ammonium) concentrations have been shown to mitigate the photosynthetic activity decrease under high UV radiation conditions and increase the content of photosynthetic pigments, soluble proteins and internal nitrogen, as well as photoprotective compounds (such as MAAs, in red algae: Huovinen et al. 2006b; Korbee et al. 2005). Inorganic nitrogen may thus modify the responses of *Gracilaria* under high irradiance conditions, with beneficial effects such as improved recovery and photoprotective processes under enriched nitrogen conditions. How this works in high nitrogen environments like those found near salmon cages (Buschmann et al. 2006) remains to be elucidated, but it is clear that *Macrocystis* in summer conditions with higher temperatures, higher solar irradiances and lower salinities (Buschmann et al. 2004) can reduce the productivity of juvenile sporophytes.

Conclusion

As finfish aquaculture uses the water column, a high fish biomass can be produced in a small sea surface area. On the other hand, seaweed farming depends on sunlight and therefore the primary productivity is related to the light reaching the sea surface and not to a culture volume. Therefore, a major goal in the development of IMTA is the reduction of the seaweed farming size while maintaining their effectiveness. Incorporating two species with different light requirements to an IMTA system opens a possibility that seaweeds can use the water column thus increasing their efficiency and effectiveness as biofilters.

Acknowledgments The authors acknowledge the financial support of FONDECYT 1050550 and Programa Bicentenario de Inserción Académica from CONICYT-Chile. We recognize the great support of the following people: Verónica Riquelme, Adrián Villarroel, Luis Henríquez, Pablo Leal, Pamela Fernández, Mónica Maldonado and Helena Abreu. The English review by Sandra Pereda is also acknowledged.

References

Arzul G, Seguel M, Clement A (2001) Effect of marine animal excretions on differential growth of phytoplankton species. ICES J Mar Sci 58:386–390

Bracken MES, Stachowicz JJ (2006) Seaweed diversity enhances nitrogen uptake via complementary use of nitrate and ammonium. Ecology 87:2397–2403

Bradstreet RB (1965) The Kjeldahl Method for Organic Nitrogen. Academic Press, New York

Buschmann A, Fortt A (2005) Efectos ambientales de la acuicultura intensiva y alternativas para un desarrollo sustentable. Ambiente y Desarrollo 20:58–64

Buschmann AH, Mora O, Gómez P, Böttger M, Buitano S, Retamales CA, Vergara PA, Gutierrez A (1994) *Gracilaria chilensis* outdoor tank cultivation in Chile: use of land-based salmon culture effluents. Aquat Eng 13:283–300

Buschmann AH, Troell M, Kautsky N, Kautsky L (1996a) Integrated tank cultivation of salmonids and *Gracilaria chilensis* (Rhodophyta). Hydrobiologia 326/327:75–82

Buschmann AH, López DA, Medina A (1996b) A review of the environmental effects and alternative production strategies of marine aquaculture in Chile. Aquat Eng 15:397–421

Buschmann AH, Troell M, Kautsky N (2001) Integrated algal farming: a review. Cah Biol Mar 42:83–90

Buschmann AH, Vásquez JA, Osorio P, Reyes E, Filún L, Hernández-González MC, Vega A (2004) The effect of water movement, temperature and salinity on abundance and reproductive patterns of *Macrocystis* spp (Phaeophyta) at different latitudes in Chile. Mar Biol 145:849-862

Buschmann AH, Riquelme VA, Hernández-González MC, Varela DA, Jiménez JE, Henríquez L, Vergara PA, Guíñez R, Filún L (2006) A review of the impacts of salmon farming on marine coastal ecosystems in the southeast Pacific. ICES J Mar Sci 63:1338–1345

Buschmann AH, Hernández-González MC, Varela DA (2007) Seaweed future cultivation in Chile: perspectives and challenges. International Journal Environment and Pollution (in press)

Chopin T, Buschmann AH, Halling C, Troell M, Kautsky N, Neori A, Kraemer GP, Zertuche-González JA, Yarish C, Neefus C (2001) Integrating seaweeds into aquaculture systems: a key towards sustainability. J Phycol 37:975–986

Chopin T, Robinson SMC, Troell M, Neori A, Buschmann AH, Fang J (2007) Ecological engineering: Multi-trophic integration for sustainable marine aquaculture. In: Encyclopedia of Ecology, Elsevier (in press)

Chow F, De Oliveira MC, Pedersén M (2004) In vitro assay and light regulation of nitrate reductase in red alga *Gracilaria chilensis*. J Plant Physiol 161:769–776

Fortt A, Cabello F, Buschmann AH (2007) Residuos de tetraciclina y quinolonas en peces silvestres en una zona costera donde donde se desarrolla la acuicultura del salmón en Chile. Rev Chil Infect 24:8–12

Gómez I, Figueroa FL, Ulloa N, Morales V, Lovengreen C, Huovinen P, Hess S (2004) Patterns of photosynthesis in 18 species of intertidal macroalgae from southern Chile. Mar Ecol Progr Ser 270:103–116

Gómez I, Figueroa FL, Huovinen P, Ulloa N, Morales V (2005) Photosynthesis of the red alga *Gracilaria chilensis* under natural solar radiation in an estuary in southern Chile. Aquaculture 244:369–382

Halling C, Aroca G, Cifuentes M, Buschmann AH, Troell M (2005) Comparison of suspended cultivation methods of *Gracilaria chilensis* in an integrated seaweed and fish cage culture. Aquacult Int 13:409–422

Huovinen P, Gómez I, Figueroa FL, Ulloa N, Morales V, Lovengreen C (2004) Ultraviolet-absorbing mycosporine-like amino acids in red macroalgae from Chile. Bot Mar 47:21–29

Huovinen P, Gómez I, Lovengreen C (2006a) A five-year study of solar ultraviolet radiation in southern Chile (39°S): potential impact on coastal marine algae? Photochem Photobiol 82: 515–522

Huovinen P, Matos J, Sousa-Pinto I, Figueroa FL (2006b) The role of ammonium in photoprotection against high irradiance in the Mediterranean red alga *Grateloupia lanceola*. Aquat Bot 84: 308–316

Hurd CL, Berges JA, Osborne JO, Harrison PJ (1995) An in vitro assay for marine macroalgae: optimization and characterization of the enzyme for *Fucus gardneri* (Phaeophyta). J Phycol 31:835–843

Gutierrez A, Correa T, Muñoz V, Santibáñez A, Marcos R, Cáceres C, Buschmann AH (2006) Farming of the giant kelp *Macrocystis pyrifera* in southern Chile for development of novel food products. J Appl Phycol 18:259–267

Korbee N, Huovinen P, Figueroa FL, Aguilera J, Karsten U (2005) Availability of ammonium influences photosynthesis and the accumulation of mycosporine-like amino acids in two *Porphyra* species (Bangiales, Rhodophyta). Mar Biol 146:645–654

Lobban CS, Harrison P (1994) Seaweed Ecology and Physiology. Cambridge University Press, Cambridge

Miranda CD, Zemelman R (2002) Bacterial resistance to oxytetracycline in Chilean salmon farming. Aquaculture 212:31–47

Neori A, Chopin T, Troell M, Buschmann AH, Kraemer GP, Halling C, Shpigel M, Yarish C (2004) Integrated aquaculture: rationale, evolution and state of the art emphasizing seaweed biofiltration in modern aquaculture. Aquaculture 231:361–391

Sahoo D, Yarish C (2005) Mariculture of seaweeds. In: Andersen R (ed) Phycological Methods: Algal Culturing Techniques. Academic Press, New York, pp 219–237

Sellner KG, Doucette GJ, Kirkpatrick GJ (2003) Harmful algal blooms: causes, impacts and detection. J Ind Microbiol Biotechnol 30:383–409

Sernapesca (1985) Anuario Estadístico de Pesca 1985. Servicio Nacional de Pesca, Ministerio de Economía Fomento y Reconstrucción, Valparaíso

Sernapesca (2005) Publicaciones, Anuarios Estadísticos de Pesca; Desembarques y Acuicultura. http://www.sernapesca.cl.

Soto D, Norambuena F (2004) Evaluation of salmon farming effects on marine systems in the inner seas of southern Chile: a large-scale mensurative experiment. J Appl Ichthyol 20:493–501

Soto D, Jara F, Moreno CA (2001) Escaped salmon in the inner seas, Southern Chile: Facing ecological and social conflicts. Ecol Appl 11:1750–1762

Strickland JDH, Parsons TR (1972) A practical handbook of sea-water analysis (2nd edn). Fish Res Bd Canada Bull 167:311

Troell M, Halling C, Nilsson A, Buschmann AH, Kautsky N, Kautsky L (1997) Integrated open sea cultivation of *Gracilaria chilensis* (Gracilariales, Rhodophyta) and salmons for reduced environmental impact and increased economic output. Aquaculture 156:45–62

Troell M, Halling C, Neori A, Buschmann AH, Chopin T, Yarish C, Kautsky N (2003) Integrated mariculture: asking the right questions. Aquaculture 226:69–80

Trono Jr, GC (1993) *Eucheuma* and *Kappaphycus*: taxonomy and cultivation. In: Ohno M, Critchley AT (eds) Seaweed Cultivation and Marine Ranching. JICA, Yokosuka, pp 75–88

Westermeier R, Patiño D, Piel MI, Maier I, Müller DG (2006) A new approach to kelp mariculture in Chile: production of free-floating sporophyte seedlings from gametophyte cultures of *Lessonia trabeculata* and *Macrocystis pyrifera*. Aquacult Res 37:164–171

Integrated seaweed cultivation on an abalone farm in South Africa

Deborah V. Robertson-Andersson · Michelle Potgieter · Joakim Hansen · John J. Bolton · Max Troell · Robert J. Anderson · Christina Halling · Trevor Probyn

Originally published in the Journal of Applied Phycology, Vol 20, No 5, 129–145.
DOI: 10.1007/s10811-007-9239-7

Abstract Land-based abalone aquaculture in South Africa, based on the local species *Haliotis midae*, started in the early 1990s and has grown rapidly in the last decade, with 13 commercial farms now producing over 850 t per annum. Over 6,000 t per annum of kelp *Ecklonia maxima* are now harvested for this purpose, and some kelp beds are reaching maximum sustainable limits. Research into seaweed aquaculture as feed (*Ulva* and some *Gracilaria*) for abalone started in the late 1990s on the southeast coast (where there are no kelp beds) using abalone waste water. A growing body of evidence suggests that a mixed diet of kelp plus other seaweeds can give growth rates at least as good as compound feed, and can improve abalone quality and reduce parasite loads. A pilot scale *Ulva lactuca* and abalone integrated recirculation unit using 25% recirculation was designed and built on the south west coast of South Africa using one 12,000-L abalone tank containing 13,200 15± 2.5 g abalone, connected to two 3,000-L seaweed tanks containing an initial starting biomass of 10 kg of seaweed, replicated 3 times. In an 18-month period, there were no significant differences in abalone health or growth rates, sediment build up and composition, mobile macro fauna densities and species between the recirculation or the flow-through units. Transfer of oxygen generated by the seaweeds to the abalone tanks was poor, resulting in the recirculated abalone tanks having lower (33%) dissolved oxygen concentrations than a comparable flow-through abalone unit. Seaweed nutrient content and specific growth rates in the units were comparable to seaweeds cultivated in fertilized effluent (SGR=3.2±3.4%.day^{-1}; Yield=0.2±0.19 kg.m^{2}.day^{-1}). Indications were that at this low recirculation ratio the seaweeds in the units were nutrient limited and that there were no negative effects to the abalone being cultivated in such a recirculation unit at this recirculation ratio.

D. V. Robertson-Andersson · M. Potgieter · J. J. Bolton
Botany Department, University of Cape Town,
Rondebosch 7701, South Africa

J. Hansen · C. Halling
Department of Systems Ecology, Stockholm University,
SE 106 91 Stockholm, Sweden

M. Troell
Department of Systems Ecology, Stockholm Resilience Centre,
Stockholm University,
SE-106 91 Stockholm, Sweden

M. Troell
The Beijer International Institute of Ecological Economics,
The Royal Swedish Academy of Sciences,
Box 50005, S-104 05 Stockholm, Sweden

R. J. Anderson
Seaweed Unit, Marine and Coastal Management,
Pvt Bag X2 Roggebaai, South Africa

T. Probyn
Aquaculture Unit, Marine and Coastal Management,
Pvt Bag X2 Roggebaai, South Africa

D. V. Robertson-Andersson (✉)
Biodiversity and Conservation Biology,
University of the Western Cape,
Bellville 7535, South Africa
e-mail: drobertson-andersson@uwc.ac.za

Keywords Integrated · Abalone · Seaweed · Aquaculture · *Ulva* · Recirculation

Abbreviations

25% Recirc.	25% Recirculation abalone units
BOD	Biological oxygen demand

DIP	Dissolved inorganic P
FAN	Free (unionised) ammonia nitrogen
FTU	Flow-through abalone units
SGR	Specific growth rate
SST	Sea surface temperature
TAN	Total ammonia nitrogen

Introduction

The South African abalone (*Haliotis midae* L) cultivation industry started developing in the 1990s and developed rapidly and is now the largest producer outside Asia (FAO 2004). Wild harvested kelp (*Ecklonia maxima* (Osbeck) Papenfuss) constitutes the major feed for farmed abalone in South Africa, and 5,447 t was harvested during 2003 (Anderson 2003). This resource is now approaching limits of sustainable harvesting and the annual maximum sustainable yield (6–10% of total standing crop) has been reached in kelp concession areas with high abalone farm concentrations (Troell et al. 2006). The use of kelp for abalone feed has decreased from 5,800 t in 2005 to 3,800 t in 2006 as farms are utilizing more compound feeds and some farms now cultivate *Ulva spp.* as feed. Abalone cultivators saw the need for alternative feed sources for the abalone and this has seen research driven mainly in two main areas: compound feeds (see review by Sales and Britz 2001), and seaweed feeds. Initial research had shown good potential for on-farm integrated seaweed cultivation in South Africa (Fourie 1994; Smit 1997; Hampson 1998; Morgan 2000; Steyn 2000; Miller 2001; Robertson-Anderson 2003; Njobeni 2006). All of the research into land-based seaweed cultivation prior to this was by placing the seaweeds in the effluent flow with no recirculation of the seaweed outflow water. A large body of literature has shown that integration of seaweeds with the primary culture organism (see Brzeski and Newkirk 1997; Troell et al. 1999a, b; Buschmann et al. 2001) and in particular abalone, can bring many benefits to overall farm performance (Neori et al. 1991, 1996, 1998, 2004).

The main aim of this study was to test a fully integrated recirculating seaweed/abalone unit. Such a unit would have hypothesized advantages and disadvantages for the primary cultured organism. Recirculation could increase the temperature of the water and positively affect the primary organism's growth rates. Recirculation may increase sediments, parasites and pest species in the system, adversely affecting the primary organism's health and growth rates. Water quality in an integrated system could also be poorer than a flow though system.

Flow rates, stocking densities and other parameters relevant to *Ulva lactuca* L cultivation in abalone effluent in the available seaweed tanks had been determined by Robertson-Andersson (2003). The data from this study were used in the design and set-up of the integrated system in conjunction with abalone cultivation parameters as determined by the existing abalone cultivation conditions. The integrated units were set up to a ratio 20 kg of abalone to 1 kg of seaweed based on the limitations of the abalone cultivation and optimization of seaweed growth and biofiltering efficiency as determined by Robertson-Andersson (2003).

Materials and methods

Experimental set up

Research took place at an abalone farm (34°37′S, 19°28′E), at Danger Point, Gansbaai, approximately 140 km east of Cape Town. The farm I & J Cape Abalone Mariculture Pty Ltd is a land-based intensive mariculture operation and cultivates primarily abalone (*H. midae*) (~ 240 t per annum), which range from 1- to 6-year-old animals. The abalone were cultivated in concrete flow-through tanks (6.6 × 2.08 m × 0.88 m deep;±12,000-L). The experimental units were designed and built in August 2003, and consisted of three separate integrated units and three separate flow-through units. In September 2003, the units were stocked with abalone 15±2.5 g in weight, from the same brood stock. The abalone culture tanks contained 24 Ivey Blue up™ baskets with 550 abalone per basket. In September 2004, the entire experiment was graded and the stocking density reduced to 450 animals per basket.

The 90 kg of fresh kelp per week, was fed to the animals, for the duration of the experiment (60 kg on Mondays, 30 kg on Fridays). The average protein content of the kelp was 7.8 % (Simmons 1990; Simpson 1994). The abalone tanks were drained and cleaned once a week as per farm cleaning procedures.

The six experimental seaweed tanks were 5 × 1 m surface area and 0.6 m deep with an outlet 17 cm from the top. They were made of white PVC lining supported on a frame. The PVC lining was rounded on the bottom. The tanks were aerated by a 30-mm PVC pipe that ran along the bottom centre of the tank. Holes (3 mm) were spaced evenly every 250 mm along the pipe and the air was supplied by a blower. The seaweed tanks were stocked with a starting biomass of 10 kg (2 kg m^{-2} stocking density) of *U. lactuca* per tank and were harvested every 14 days to measure growth rates. The tanks were shaded with 50% shade cloth from September to February. Four additional seaweed tanks that were being fed fertilized abalone waste water at 12 volume exchanges per day without recirculation (approximately double the water volume that each integrated seaweed tanks received), were used to compare seaweed growth rates and seaweed tissue properties. The fertilizer consisted of a 100 g mixture of Maxiphos® and ammonium sulphate in a ratio of 1:6.

Three abalone tanks were on a flow-through system receiving 6,000 L h^{-1} (Fig. 1). Three additional abalone tanks had their waste water gravity-fed to two seaweed tanks each. After the water passed through the two seaweed tanks, 25% (1,500 L h^{-1}) was pumped by a 2-kW pump back to the abalone tank from which it originated. Then, 4,500 L h^{-1} (75 %) of seawater was supplied from the same source as the flow-through units (Fig. 1). The overflow water exited the units from the seaweed tanks. The recirculation ratio of 25% was chosen for the long term monitoring as a small scale study, carried out over 72 h investigating physico-chemical variables at 50% recirculation, showed that the dissolved oxygen concentration was very low and this may have detrimental effects on the abalone in a long term study (see Harris et al. 1999).

Abalone sub-samples

Samples of abalone were taken from tanks from January 2004 to May 2005. At each sampling, 33 animals were randomly removed from each tank, blotted dry to remove excess water and weighed to the nearest 0.01 g. Shell length was measured along the longest axis to the nearest 0.01 mm.

Grading was done in September 2004 as per normal farm grading procedures. Animals were harvested from a tank and the basket number and age group recorded. Animals were then mechanically graded into different size classes and the weights of these size classes were recorded for a total biomass value.

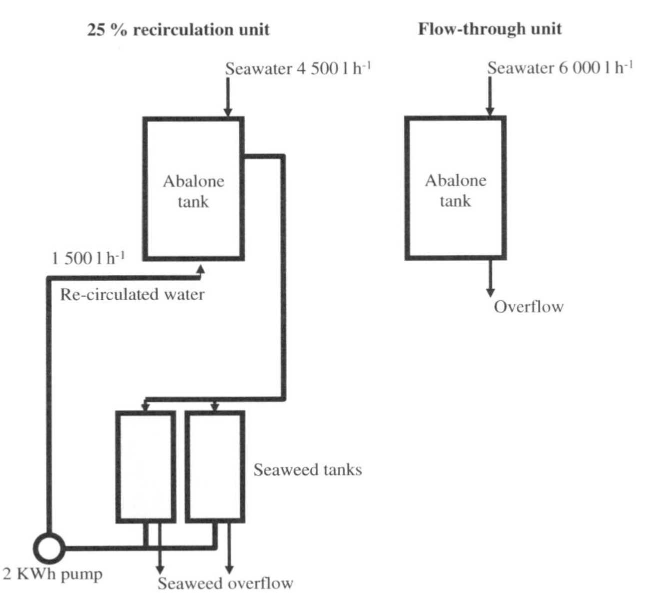

Fig. 1 Schematic diagram of 25% recirculation unit and flow-through unit

Physiochemical variables and water quality

Temperature, pH and dissolved oxygen were recorded daily at 0800, 1200 and 1600 hours for long term monitoring (16 months) in the flow-through abalone tanks, the seaweed tanks, incoming seawater and the integrated seaweed/abalone tanks. Additional intensive short term monitoring was done during three 36-h periods in September 2003 - spring, January 2004 - summer and June 2004 - winter. During this period, water samples for ammonium, phosphate, nitrate and nitrite were taken every 4th hour and analyzed according to the methods below.

Ammonium concentration was determined according to Grasshoff et al. (1976), scaled down to a sample volume of 5 mL and reagent additions of 0.2 mL. Dissolved Inorganic Phosphate (DIP - PO_4^{3-}) concentration was determined using the method described by Grasshoff et al. (1976), with a slight modification in that samples and reagent amounts were reduced by a factor of 10. Nitrate (NO_3^-) and Nitrite (NO_2^-) concentrations were determined using the method described by Nydahl (1976). Ammonium and phosphate samples were taken in triplicate. Nitrate and nitrite were analyzed from a single sample. Testing was done for Total Ammonia Nitrogen (TAN) and then free (unionised) ammonia nitrogen (FAN) was calculated using the TAN concentrations, pH, temperature and salinity values following Bower and Bidwell (1978).

Abalone health

Five sub-samples of abalone were removed from every experimental tank every 3 months by Dr A. Mouton (Abalone Farmers Association of Southern African veterinarian). The health of these animals were analyzed using the standard South African veterinarian procedures and data reported in commercial health reports (Mouton 2004). Veterinary aspects investigated included counting the number of sabellid *Terebrasabella heterouncinata* Fitzhugh, Rouse tunnels on the growing edge of the shell, investigating the gonad histology to determine gonad maturity, a histological examination of all organs to determine parasite status (expressed as the percentage of sample infected) and whether or not the animals were experiencing some form of environmental stress (expressed as either being present or absent), a histological examination of the digestive gland to determine the general condition of the abalone. In addition, animals from both the flow-through units and the 25% recirculation units were compared with animals from the farm's own health reports.

Seaweed growth

Seaweed Specific Growth Rate (in % wet wt day^{-1}) and yield (Y = g wet wt m^{-2} day^{-1}) was determined according to Evans (1972) and calculated as:

$$SGR\% = 100x[ln(W_t/W_0)]/(t_t - t_0)$$
$$Y = [(W_t - W_0)/t]/SA$$

where W_0 and W_t are initial and final wet weights (wt) in grams and t_0, t_t are initial and final times in days, respectively, and SA is the surface area.

Because a brown algal epiphyte *Myrionema strangulans* Greville was found detrimental to seaweed SGR if present in high numbers by Robertson-Andersson (2003), a modification of the Braun-Blanquet (B-B) percentage cover-abundance scale was developed to measure percentage cover of *M. strangulans* on individual thalli. An *Ulva* thallus was placed on a white board and then the percentage coverage of *M. strangulans* was estimated. Five thalli were taken from each seaweed tank. The range of the B-B values given as percentage cover is as follows: 0% cover=1; 1–10% cover=2; 11–25% cover=3; 26–50% cover=4; > 50% cover=5.

Seaweed tissue nitrogen and phosphorus

Seaweed samples were taken from the 25% seaweed recirculation tanks and from fertilized tanks to record dry to wet weight ratios and for biochemical analysis. After each weighing, the seaweed samples collected were washed in distilled water, and visible epiphytes and epifauna were removed. After washing, the samples were spun in a salad spinner for 1 min, weighed on an electronic balance to 2 decimal places, oven dried (70°C, 72 h) and then re-weighed. The dried seaweed was ground using a mechanical grinder with a maximum mesh size of 1 mm. The powder was stored in sealed glass jars in a dessicator at room temperature.

Total nitrogen was determined using the micro-Kjeldahl technique (Solorzano 1969). The protein content was determined by multiplying the N concentration obtained from the micro-Kjeldahl technique by a factor of 6.25, based on the protein N content of 0.16 g.g^{-1} from methods described by Fleurence et al. (1995). This conversion factor, although commonly accepted, includes N not only in the form of protein but includes intracellular reserve pools of N (Fleurence et al. 1995) and therefore the micro-Kjeldahl method would tend to overestimate the actual protein content.

Phosphate (PO_4^{3-}) concentration was determined using the tri acid digestion method described by Murphy and Riley (1962).

Environmental events

Long term temperature monitoring showed that the units were vulnerable to external environmental events that were explained using a series of satellite images, data and tools obtained from http://www.rsmarinesa.org.za of Sea Surface Temperatures (SST).

Sediments

Size distribution of suspended particulate matter as well as the total sediment concentration (wet weight) were measured over a 4 day period. Samples were taken from three locations in the tank; incoming water, water column (30 cm from the surface) and bottom sediment samples. Samples for suspended particulate fractions and total suspended particulate matter were sub-sampled from a 10 L volume siphoned from the tanks. Whatman G/F Filter papers (25 mm) of 1.0 μm were used for syringe filtering and the larger (110 mm) filter papers were used when filtering the particulate fraction samples. Filters were pre-ashed for 2 h at 400°C and weighed. A volume of 5 L was filtered through a series of mesh filters, 50, 40, 30, 20 μm. The filtrate was then transferred to pre-weighed filters paper using de-ionized water. It was assumed that the use of distilled water did not decrease the filtrate and that is was necessary for removing salt. Three sub-samples of 100 mL were taken from the 10 L sample and then syringe-filtered through 25 mm Whatman Glass Fiber filter for analysis of total particle concentration. Sediment accumulation was measured during the cleaning of the abalone tanks. Water from the tank was drained leaving an accumulated layer of sediment at the bottom of the tank. Sediment deposits from baskets and feeder trays were washed down with water, side walls were scrubbed. The sediment rich bottom water was then transferred to 500 L tanks and from this three 20 mL sub-samples and a 5 L sample were collected after vigorous mixing and handled as described for the water column samples. In addition, sediment build-up in the water column of the abalone tanks over the 7-day cleaning cycle was also measured by siphoning 10 l of water and measuring total particle concentration.

Mobile macro-fauna

Samples of mobile macro-invertebrate fauna were taken on one occasion in September 2004 during tank cleaning. Water was drained through a draining hole at the bottom of the tanks and a plankton net (<100 μm mesh) was held in the water flow to collect fauna. After being almost emptied, the drain hole was blocked and the tank walls were brushed and washed. The remaining water and sediment was then sucked out via a pump filling up a 400 L container. From this container a 50 L sample was taken after rigorous mixing. In addition to the tank sampling intake, seawater to the farm was sampled at three occasions from September to October 2004 during low, mid and high tide (n=9). The pumped seawater was flushed through a plankton net for about 1 h, which correlates to about 2.9–4.1 m^3 depending on flow rate (measured at each occasion). The sediment samples and the material collected in the plankton-net were sieved through a 1 mm mesh to separate the macrofauna (>1 mm). A sub-sample was taken when the density of animals and detritus load was very high. The animals were then sorted to higher taxonomic groups alive and then preserved in 95% ethanol for more detailed identification. The sorted and identified animals were dried in an oven at 60°C to constant weight and weighted. Taxa with extremely low biomass were assigned a minimum weight of 0.1 mg. Air-breathing isopods (e.g., *Ligia dilatata* Brandt), which easily escaped during sampling, had to be excluded. Sessile macrofauna that had to be scraped off the tank walls to be sampled were also excluded. These tank-fouling species included hydrozoans, bryozoans, actiniarias, porifera, bivalves as well as gastropods. Because the herbivorous gastropod the Cape keyhole limpet *Fissurella mutabilis* (Sow.) was found to be so devastating to the seaweed in previous studies (see Robertson-Andersson 2003; Hansen et al. 2006; Njobeni 2006), a modification of the Braun-Blanquet percentage cover-abundance scale was developed to compare numbers of key hole limpets, rather than measuring weights and lengths. A 25 cm quadrat was placed on the side walls of the recirculation tanks and kelp fed tanks and numbers of limpets in the quadrats were noted. The range of the BB values given as percentage cover is as follows: 0 limpets present=1; 1–3 limpets=2; 4–7 limpets=3; 8–10 limpets=4; > 10 limpets=5.

Statistical analysis

All data are expressed as means±standard errors. Analyses were done using STATISTICA V6.1. For abalone growth data, an initial analysis of co-variance was first tested with the baseline value of the outcome, i.e. either length or weight used as a covariate. This was done to account for any differences in starting values. To test for actual differences ANOVAs were performed on the data. Other data was analysed ANOVAs after verifying normal distribution and homogeneity of variances. All data was regarded significant at $p<0.05$.

Results

Abalone growth rates

From September to November 2003 there were no differences in growth or length between abalone in the two systems, indicating that the experimental animals had a 2 month acclimatization phase (Figs. 2, 3). Average length increases in the 25% recirculation units (25% recirc.) were 1.42 $mm.month^{-1}$ vs 1.38 $mm.month^{-1}$ for the flow-through units (FTU) for the duration of the experiment. The 25% recirc. units performed best in autumn and winter, with growth rates of over 4 mm a month. Weight increases in the 25% recirc. units were 3.1 $g.month^{-1}$ vs 2.8 g. $month^{-1}$ in the flow-through units. There were no significant differences between the systems in either length or weight from September 2003 to September 2004. In October and November 2004, abalone in the 25% recirc. units had significantly higher growth rates than those in the flow-through units (ANOVA; $p<0.05$; df=197). This was 2 months after the September grading where the stocking density was reduced in the baskets. At the end of the experiment there was no significant difference in weight or length between abalone in the two systems (Figs. 2, 3).

The farm-graded biomass increased from September 2003 to September 2004 by 275% (144 kg) in the 25% recirc. and 256% (134 kg) in the flow-through units, respectively. A frequency distribution of weights from the grading done in September 2004 and the final weight

Fig. 2 Length (mm) of abalone in 25% recirc. units vs flow-through units on a kelp-only diet from September 2003 to May 2005 (n=4,000)

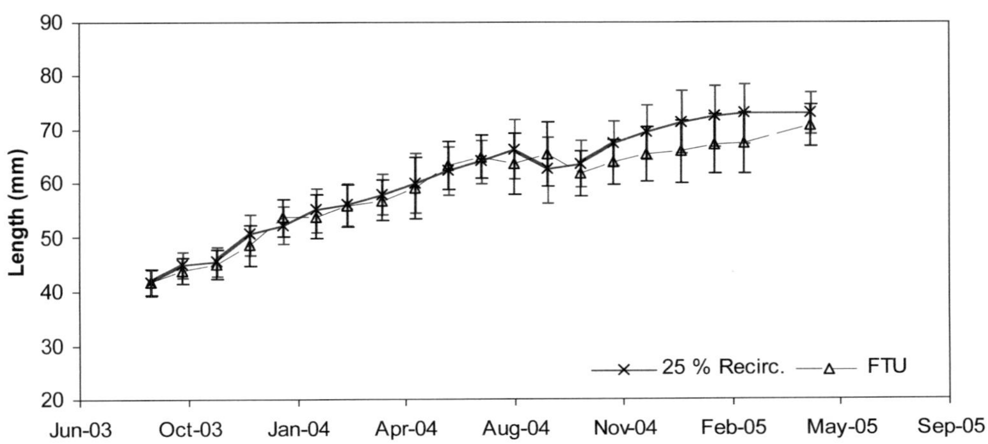

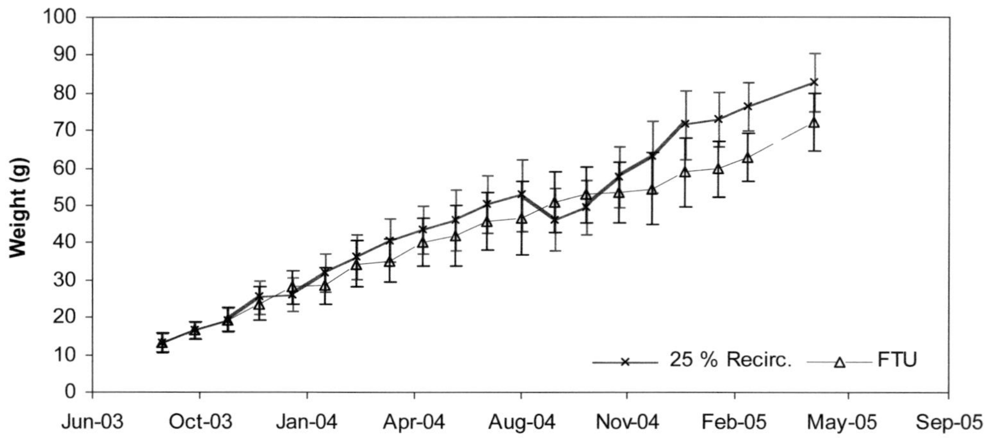

Fig. 3 Weight (g) of abalone in 25% recirc. units vs flow-through units on a kelp-only diet from September 2003 to May 2005 (n=4,000)

measurements showed that there was no significant difference in the mean greatest size of the abalone.

Abalone health

There were no significant differences in the counts of the sabellids between the two treatments over the experimental period. There was also no evidence of shell-boring polychaetes. The animals in both systems were in a good condition for the duration of the experiment. In December 2003, animals in the 25% recirc. units showed signs of environmental stress in their digestive gland histology this corresponded to high temperatures (20.5°C) and high pH (8.00) in the 25% recirc. system.

Gonad condition is a measure of the sexual maturity of the animals and is of critical importance for management. Compared with animals on the rest of the farm whose gonads showed a mixture of developmental stages of sex cells, the gonad development was lower for recirculation units with only immature sex cells present, while the animals in the flow though units remained immature. Histological examinations of the animals showed no differences between units, with both having gut protozoa present (10–50% infected—due to the kelp diet) and the presence of a *Rickettsia*-like intracellular prokaryote (10–30% infected). Histologically, the animals in both treatments were in a better condition than the rest of the farm animals (50–100% infected with *Coccidia, Rickettsia*, gut protozoa and gill inflammation).

Water quality and physiochemical variables

Temperature

In January 2003 (summer), the seaweeds and 25% recirc. units showed a clear diurnal temperature rhythm with the highest temperatures in the units being recorded at 1600 hours and lowest temperatures being recorded at 0000 hours (Table 1). There was a very close correlation between temperatures in the 25% recirc. abalone units and the seaweed tanks, with the 25% recirc. abalone tanks being affected by the temperature in the seaweed tanks (df=42; r=0.85; p<0.0001). The range of temperature was 5°C for both units.

In June 2004 (winter), during the day, the seaweed tanks helped to increase the temperature of the 25% recirc. abalone units above that of the incoming seawater and the flow-through units. At night, the seaweed tanks lost a lot of heat, and this helped to significantly decrease the temperature in the 25% recirc abalone units (ANOVA; df=42; p<0.05) compared to the flow-through units. The maximum daily temperature range was greater in the 25% recirc. units (3°C) compared to the flow-through units (1.9°C) (Table 1).

September 2003 (spring) temperatures were similar to winter and showed that the recirculation lost heat through the seaweed tanks at night, while during the day they imparted heat to the recirculation units (Table 1).

Dissolved oxygen

In January, the dissolved oxygen in the 25% recirc. abalone units (6.7 mg L^{-1} average) were significantly lower than the flow-through units (7.7 mg L^{-1} average) or the seaweed units (8.9 mg L^{-1} average) (ANOVA, df=42; p<0.05). Although the dissolved oxygen in the seaweed tanks was high (8.9 mg L^{-1}) this oxygen was not being transferred to the 25% recirc. abalone units. In addition, during periods of darkness and particularly around 0400 hours in the 25% recirc. abalone units the dissolved oxygen concentration decreased to below 5.4 mg L^{-1}.

In June the dissolved oxygen in the 25% recirc. abalone units was significantly lower compared to the flow-through units (ANOVA; df 28; p<0.05). The oxygen in the seaweed tanks had a significantly higher concentration than both abalone containing units and the ambient seawater (ANOVA;

Table 1 Temperature, pH and dissolved oxygen from three 36-h intensive sampling periods in September 2003, January 2004 and June 2004, in the 25% recirc. abalone units, flow-through units, seaweed tanks and ambient seawater

	September 2003	January 2004	June 2004
	(min) Mean (max)	(min) Mean (max)	(min) Mean (max)
Temperature (°C)			
Flow-through units	(13.6) 15.9 (18.4)	(14.5) 16.6 (19.5)	(14.8) 15.4 (16.7)
25% recirc. units	(13.3) 16.4[a] (19.1)	(15.6) 18.1[a] (20.6)	(13.4) 14.6[a] (16.4)
Seaweed tanks	(12.7) 16.9[a] (19.3)	(16.0) 18.8[a] (24.0)	(13.3) 14.5[a] (16.8)
Ambient seawater	(14.2) 15.6 (16.7)	(14.6) 16.3 (19.5)	(15.0) 15.4 (16.3)
pH			
Flow-through units	(7.4) 8.1 (8.4)	(7.6) 7.9 (8.1)	(7.7) 7.9 (8.2)
25% Recirc. units	(7.8) 8.0 (8.3)	(7.4) 7.6 (7.9)	(7.8) 8.0 (8.2)
Seaweed tanks	(7.8) 8.1 (8.5)	(7.6) 8.0 (9.3)	(7.8) 8.0 (8.4)
Ambient seawater	(8.1) 8.2 (8.3)	(7.9) 8.1 (8.3)	(8.1) 8.2 (8.4)
Dissolved oxygen (mg L^{-1})			
Flow-through units	(6.8) 7.7[b] (8.5)	(6.7) 7.7[b] (8.6)	(6.8) 7.6[b] (8.3)
25% recirc. units	(5.1) 6.9 (8.5)	(5.3) 6.7 (7.9)	(6.3) 6.9 (7.5)
Seaweed tanks	(7.2) 8.4[a] (10.4)	(7.0) 9.0[a] (13.0)	(7.0) 8.3[a] (10.1)
Ambient seawater	(8.0) 8.4[a] (8.9)	(8.0) 9.1[a] (10.3)	(8.0) 8.5[a] (9.0)

Letters indicated values significantly different between treatments (ANOVA, $p<0.05$)

$df=28$; $p<0.05$). The dissolved oxygen values were higher than those found in summer and the range for all treatments was smaller when compared to the summer values. Spring dissolved oxygen concentrations were similar to winter and the seaweeds showed a diurnal rhythm in oxygen production. Between 2000 and 0400 hours the 25% recirc. units had significantly lower dissolved oxygen values compared to the flow-through units (ANOVA; $df=28$; $p<0.05$) (Table 1).

pH

The pH in the seaweeds tanks in January and June showed a clear diurnal rhythm due to photosynthesis. In both abalone treatments pH showed a diurnal rhythm but this was exacerbated in the 25% recirc. abalone units. There were no significant differences in pH between the flow-through units and the 25% recirc. abalone units in any of the seasons (Table 1).

Long term physiochemical variables

Long term measurements of dissolved oxygen at 25% recirculation was on average 4±0.5 % lower than the flow-through abalone units. Temperature was 1±0.2 % higher and pH was 1±0.1 % lower in the 25% recirc. abalone units compared to the flow-through units over the whole experimental period. There was an inverse linear relationship between temperature and oxygen in the 25% recirc. and flow-through units with less oxygen being available in the water at higher temperatures ($y=-0.3985\times+1.3383$; $df=24834$; $r=0.2$; $p<0.05$). The dissolved oxygen in the 25% recirc abalone units was 33±6 % lower than in the seaweed tanks with the pH being 2±0.6 % lower. There was a positive linear correlation between temperature and dissolved oxygen ($y=0.1244\times+0.6401$; $df=24834$; $r=0.25$; $p<0.05$), due to the production of oxygen by seaweed photosynthesis.

Environmental events

Six high temperature environmental events occurred (Fig. 4). These events can be divided into two types: advection events and warm water intrusion events. Both are caused by a decrease in wind-forcing. The more common event (4 out of 6) is a decrease in upwelling and increase in solar advection resulting in increased temperatures and high nutrient availability of coastal waters. The other type of event is a warm water intrusion over the Western Agulhas Bank and Agulhas eddies. A warm water intrusion over the Western Agulhas Bank, such as Event C, is seen as a rise in temperature of 3.7°C (Fig. 4), decrease in dissolved oxygen by 1 mg L^{-1}, a 0.3 increase in pH, and a 7-kg decrease in seaweed production per seaweed tank (Fig. 8). In Event D, the farm experienced ambient seawater temperature increasing 4.8°C above the average to 23.4°C (the highest recorded temperature for 2005) (Fig. 4). Coupled with this, the pH and the dissolved oxygen in the water decreased by 1.06 units and 1.8 mg l^{-1} respectively at the height of the event. Seaweed production decreased by 4.5 kg per tank (Fig. 8). In Event F, the local SST increased to a high of 19°C (Fig. 4). This was more due to advection of surface waters and northwards displacement of surface waters from the front back to the inshore region. There was no intrusion of water over the Western Agulhas Bank and the warm water still remained nutrient rich. Seaweed production during this event increased by 2.5 kg per tank.

Upwelling events can be seen by sharp decreases in temperatures over a short time scale (Fig. 4). The lowest temperature recorded at Danger Point (during this study)

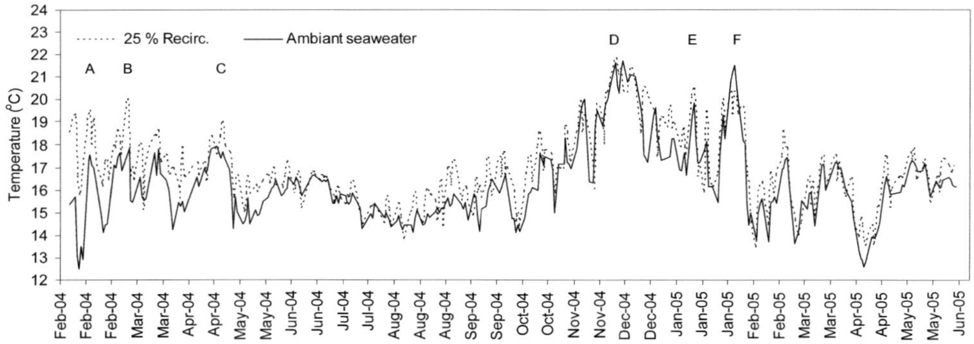

Fig. 4 Daily average long term temperature measurements from February 2004 to May 2005 of the incoming seawater and 25% recirc. abalone units for comparison. *Letters* denote large scale environmental events: Event *A*: from 26 February until 10 March 2004 - advection; Event *B*: 18 March to 26 April 2004 - advection; Event *C*: 21 April to 4 May 2004 - warm water intrusion; Event *D*: 24 January to 7 February 2005 - warm water intrusion; Event *E*: 22 February to 3 March 2005 - advection; Event *F*: 17 March until 1 April 2005 - advection. Sharp decreases in ambient seawater temperatures are indicative of upwelling events

occurred in an upwelling event on 1 August 2003 with an ambient sea temperature of 11.5°C.

Water nutrients

Mean total ammonia nitrogen (TAN) did not accumulate in the recirculation units. The flow-through units had significantly higher TAN (3.3±1.3 μmol) than the 25% recirc. units (2.0±1.1 μmol) (ANOVA; *df*=12; $p<0.05$). The seaweed tanks had significantly lower TAN (4.7±1.3 μmol) than the 25% recirc. (ANOVA; *df*=12; $p<0.01$). Two peaks in TAN production occurred in both the 25% recirc. and the control abalone treatments between 2000 and 0400 hours. Ambient seawater TAN averaged 3.9±0.8 μmol. There was no seasonal pattern in any of the data.

Table 2 lists ammonia concentrations for summer and winter. In summer the pH and temperatures were high (Table 1) resulting in greater ammonia conversion, compared to winter which had lower temperatures and lower ammonia concentrations.

There was no accumulation of DIP under recirculation. Mean DIP for the flow-through units were 0.88±0.39 μmol, for the 25% recirc units 1.17±0.24 μmol, for the seaweed tanks 1.28±0.36 μmol and for the incoming seawater 0.64±0.41 μmol. The seaweed tanks had significantly lower DIP compared to the flow-through units (ANOVA; *df*=12; $p<0.05$). Results were similar in the three seasons tested.

Table 2 Ammonia concentrations in summer and winter of in a 25% recirc. units and flow-through units fed a kelp only diet

	Mean (μmol)	Max (μmol)	Min (μmol)
January 2004 – Summer			
Flow-through units	0.10±0.07	0.29	0.03
25% recirc.	0.08±0.05	0.18	0.02
June 2004 – Winter			
Flow-through units	0.16±0.009	0.34	0.08
25% recirc.	0.06±0.03	0.12	0.04

There was no accumulation of nitrate under recirculation. Mean nitrate concentrations for the flow-through units were 3.47±0.89 μmol, for the 25% recirc. units 1.17±0.22 μmol, for the seaweed tanks 0.25±0.27 μmol and for the incoming seawater 3.07±0.92 μmol. Mean nitrate in the flow-through units was significantly higher (ANOVA; *df*=12; $p<0.05$) than 25% recirc. units, which in turn was significantly higher than the seaweed tanks (ANOVA; *df*=12; $p<0.05$).

Mean nitrite concentrations for the flow-through units were 0.63±0.46 μmol, for the 25% recirc. units 0.22±0.05 μmol, for the seaweeds 0.08±0.08 μmol and for the incoming seawater 0.84±0.50 μmol. There was more fluctuation in the flow-through units with small peaks in concentrations at 0000 and 0400 hours. The flow-through units had significantly (ANOVA; *df*=6; $p<0.05$) higher average concentrations than the seaweeds and the 25% recirc. abalone units. The seaweeds tanks had significantly lower average concentration than the 25% recirc. abalone units (ANOVA; *df* = 6; $p<0.05$). There was no accumulation of nitrite under recirculation.

The nutrient uptake efficiency was measured as the average percentage decrease in nutrients (Table 3) (defined by Buschmann et al. 2001; Chopin et al. 2001) it can be seen that phosphate was not removed by the seaweeds. Nitrogen in the seaweed tanks running at 25% recirculation was almost completely removed in nitrate and nitrite forms. There was also a large difference in the uptake efficiency between the abalone units on flow-through vs recirculation, showing that the amount of nutrients removed by the seaweeds decreased nutrients in the whole system compared to a flow-through system.

Table 3 Percentage of dissolved nutrients removed between the flow-through units compared to the 25% recirc. units and between the seaweeds tanks compared to the waste water existing the 25% recirc. units

	NH_4	NO_3	NO_2
25% recirc. abalone	38.06 ±18	64.8±11.1	48.8±25.2
vs flow-through abalone units	32.5–85.4	33.1–77.9	5.1–89.4
Seaweed tanks vs 25% recirc.	80.8±22.6	77.7±24.5	75.7±13.7
abalone units	24.3–99.5	30–100	53.4–100

Seaweed biomass in this study was 2 kg m^{-2}. Mean±SD and range values are shown

Suspended particle loads

There was no significant difference in the suspended solids in the water column between the two systems either in the total load (105.3 mg L^{-1} flow-through units vs 90.6 mg L^{-1} 25% recirc. units) or in the particle size fractionation (Fig. 5). There was no significant difference in the particle fractions between the two systems with the larger size class (> 50 μm) having the highest percentage of the total load. Over a 7-day period there was no significant difference in the suspended solid loads between the two systems (0.085–0.11 g L^{-1} flow-through units and 0.082–0.11 g L^{-1} 25% recirc. units). Regression analysis showed that the concentration of particles in the flow-through units and the 25% recirc. units remained constant over the 7 day period (p= 0.75, r=0.149). Particle fractionation over the 7-day period in the two systems showed no significant differences. The greatest load in both systems was in the largest particle size class (> 50 μm). In a 7 day period between 55–72 kg of sediments would be released through the overflow water form both systems.

Sediment accumulation

The amounts of accumulated bottom sediments were significantly more than the suspended particle load (ANOVA, df=12; p<0.05). There was a significant difference between the bottom sediment loads of the two systems (ANOVA; df=72; p<0.001). The particle fractionation showed that the flow-through units had significantly more of the larger particles (> 50 μm and 40–50 μm and 20–30 μm) (p<0.05), while the 25% recirc. abalone units had significantly more particles in the 20–30 μm range) (Fig. 6). The flow-through units would produce 45.45 kg of sediment a year vs the 41.64 kg produced by the 25% recirc. units in bottom sediments alone.

Mobile macro-fauna

The mean dry weights and numbers of mobile macro fauna were respectively, 2.19±0.67 g.DW or 8,232 individuals per abalone tank in the 25% recirc. units and 2.17±1.75 g. DW or 11,648±10,960 individuals per abalone tank in the flow-through units, with no significant difference in either dry weight or numbers. In total, 28 faunal taxa were identified in all the samples. The diversity (including unidentified taxa) was no significantly different between the 25% recirc. and the flow-through units. Thirty-four taxa of mobile and sessile fauna were found in the seaweed tanks alone. The combined system supports approximately 48 taxa.

The total faunal density and diversity was significantly higher in the abalone tanks compared to an equal volume of intake seawater (ANOVA; df=9; p<0.001). The most abundant taxa in the abalone tanks were the amphipod *Paramoera capensis* (Dana) and tanaids (Tanaidacea spp.) together with polychaetes, especially Nereid species. Other abundant taxa were the amphipod *Hyale* sp. and the isopod *Janiopsis palpalis* Barnard, as well as other (unidentified) amphipods. These taxa contributed 81–99 % of the total mobile macrofauna biomass in the abalone tanks.

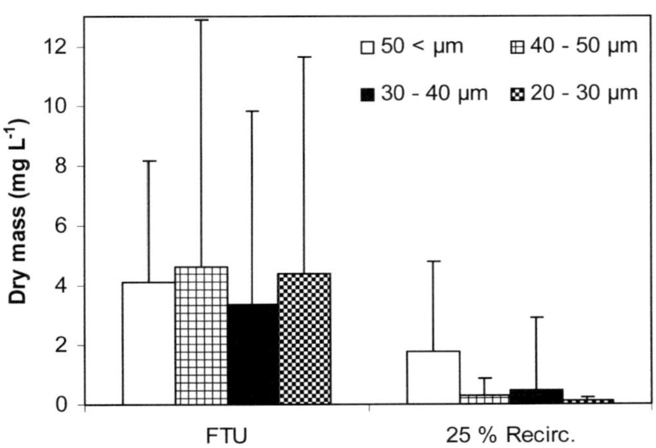

Fig. 5 Particle fraction of water column suspended particles between the flow-through units and the 25% recirc. units

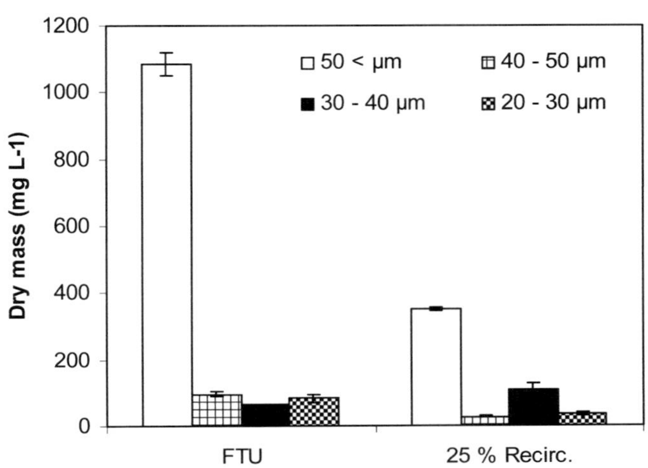

Fig. 6 Particle fraction of bottom sediments between the flow-through units and the 25% recirc. units

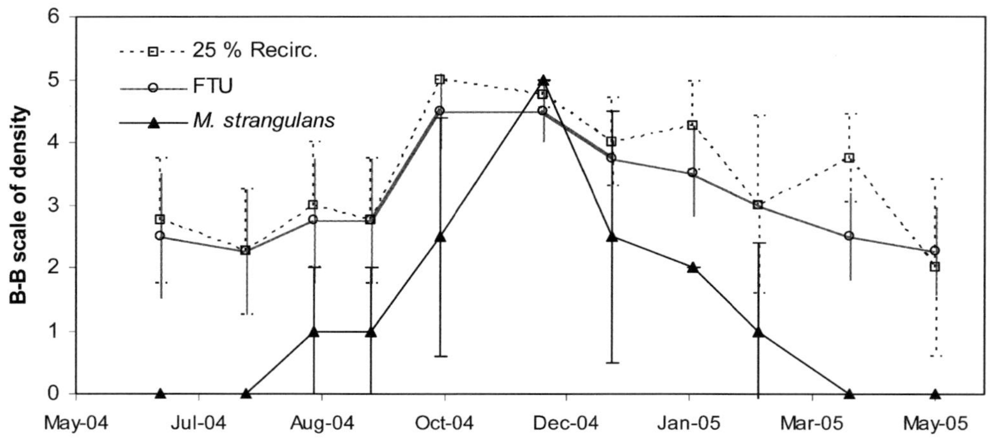

Fig. 7 Density of keyhole limpets in the 25% recirc. seaweed tanks (n=6) and the flow-through seaweed tanks (n=4) and percentage cover of *Ulva* thalli by *M. strangulans* from the seaweed tanks

Keyhole limpet numbers showed a seasonal trend in number with an increase around September (Fig. 7). Although there were slightly higher numbers in the 25% recirc. units in the winter months, this was not significantly different from the flow-through units.

Seaweed production, nutrient and water content

There was a seasonal trend in SGR of the seaweeds with highest SGR (6.25±3.4 day^{-1}) and yields (0.12–2 kg m^{-2} day^{-1}) occurring in September through to February in the 25% recirc. seaweed tanks. The high standard error was due to low growth rates in some tanks which experienced a *M. strangulans* infestation (Fig. 7). SGR in June and July was lower (1.59±1.1% day^{-1}) with an average of 3.2% day^{-1} with yields of 0.0–0.03 kg m^{-2} day^{-1} over the experimental period. *Ulva* production per tank (Fig. 8) shows that during the winter months the algae did not increase in biomass but rather declined as the amount harvested per tank was lower than the initial 10 kg stocking density. There was no significant difference in algal production between fertilized flow-though seaweed tanks and the 25% recirc. seaweed tanks.

Table 4 lists the proportions of N and P in the seaweeds at different seasons in the spring and summer months. The N content in the seaweeds was low and this corresponded with periods of high SGR. In addition, the water content in the seaweeds was higher from October 2001 to February 2002 (late spring to late summer) and decreased in March 2002 to August 2003 (autumn to winter). Phosphate content in the seaweeds increased in the autumn and winter months. Average tissue protein values obtained for this study were 36.6%, from the 25% recirc. seaweed tanks and 33.4% from the fertilized seaweed tanks.

Discussion

Abalone growth rates

This study did not find a significant difference in abalone growth rates at the conclusion of the experiment. Thus, abalone growth rates were not affected by being cultivated in a recirculation system and were similar to those obtained on the farm under normal culture conditions. Fleming et al.

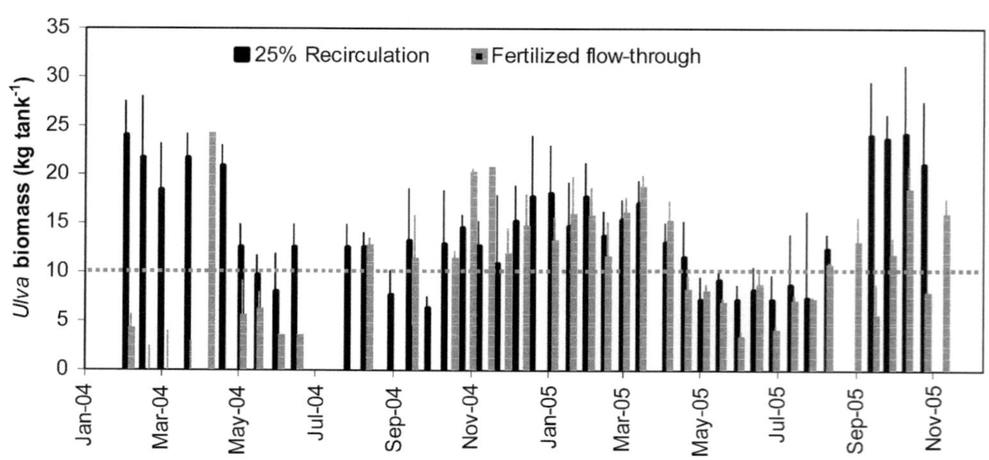

Fig. 8 *Ulva* biomass harvested per tank (n = 6) from January 2004 to May 2005 in the 25% recirc. seaweed tanks (n=6) and the flow-through seaweed tanks (n=4). The fertilized flow-through tanks are for comparison and received 12 volume exchanges per day while the 25% recirc. seaweed tanks received 7.2 volume exchanges per day

Table 4 Tissue N and P concentrations in seaweed samples from 25% recirc. abalone units and fertilized flow-through seaweed tanks

		September 2003	December 2003	April 2004	July 2004
P ($mg.g^{-1}$)	25% recirc.	0.65± 0.40	0.80*±0.01	2.27^{b}±0.20	3.78^{a}±0.66
	Fertilized flow-through units	1.05^{c}±0.13	0.66±0.03	2.33^{b}±0.08	3.15^{a}±0.49
N ($mg.g^{-1}$)	25% recirc.	11.51±1.91	13.98±2.44	45.02^{a}*±6.06	43.80^{a}±5.13
	Fertilized flow-through units	22.54^{c}*±2.48	14.96±6.6	36.04^{b}±0.43	51.64^{a}*±2.27
N:P ratio	25% recirc.	17.71±2.66	17.53±2.75	20.05*±4.42	11.67±0.64
	Fertilized flow-through units	21.46±1.4	22.48±8.84	15.48±0.36	11.59±7.76
SGR ($\%.d^{-1}$)	25% recirc.	2.33±3.79	3.53±0.73	1.62±0.88	− 0.85±1.71
	Fertilized flow-through units	− 0.89[1]±2.96	3.28±0	2.14±3.69	− 5.73±2.15
% water	25% recirc.	88.06±0.45	88.31±0.12	85.93±0.38	85.89±1.15
	Fertilized flow-through units	87.99±0.65	89.25±0.92	85.62±0.17	85.79±0.85

* Values significantly higher between treatments (ANOVA, $p<0.05$)
Letters indicated values significantly different within treatments
(n=6 for 25% recirc. seaweed tanks and n=4 for fertilized flow-through seaweed tanks. (Mean±s std SD)
[1] Low growth rates reported due to a loss of seaweed biomass due to a *M. strangulans* infestation

(1996) and Mai et al. (2001) state that most abalone farmers would be satisfied with a growth of 2–3 mm per month. This however relates to tropical abalone which grow much faster than the temperate species *H. midae*. There was some indication that growth rates were significantly different at certain times of the year in the different units, but this did not affect the outcome of the experiment. It is possible that the 1 and 2% increase in temperature in the recirculation units was too low to positively affect growth rates. The low recirculation ratio could also account for the lack of negative effects on abalone growth rates.

Gonad index

This is a measure of the sexual maturity of the animals and is of critical importance for management, as a mature gonad may make up 15–20% of an abalone's body weight in a fully mature individual (Hahn 1989). One of the hypotheses was that a recirculation system would cause an increase in the gonad index due to increased temperatures, meaning that animal energy was going to reproductive growth and not somatic growth. The increase in gonad index in the recirculation units also corresponded with a natural increase due to wild abalone spawning seasons (Hahn 1989) and thus was not an effect of the animals being under recirculation.

Seaweed nutrient content

The protein content of the *Ulva* grown in the recirculation units is less than the 44% reported by Goldburg and Triplett (1997) in their recirculation system. Our results fall within the protein range found to be most beneficial to *H. midae* (36–44% protein) (Shipton 1999; Britz 1996; Sales and Britz 2001). Average tissue protein values obtained for this study were much higher than wild harvested *Ulva* (3.7–24%) (Smith and Young 1954; Nisizawa et al. 1987; Simpson 1994; Castro-Gonzales et al. 1996; Simpson and Cook 1998; Wilkinson 2001; Wong and Cheung 2001) and has positive effects on abalone growth when used as a food source (Naidoo et al. 2006). During summer, the seaweeds experienced higher SGR but lower tissue N. This finding agrees with previous reports (Duke et al. 1986, 1987, 1989a, b), of an inverse relationship between SGR and tissue nitrogen, but is in contrast with studies by Neori et al. (1991). This result may have been confounded as a result of the *M. strangulans* infestation.

Work done by Björnsäter and Wheeler (1990) showed that there is a relationship which exists between tissue N and P for *Ulva* in the North-east Pacific with a ratio of 16:24 indicating that the seaweeds were nutrient sufficient and our cultivated seaweeds ratios fall within the low end of this range. This relationship however, needs to be tested to see if it holds true for the *Ulva* cultivated in this experiment.

Seasonal changes in tissue N and P can be explained by coastal oceanographic process. Boyd et al. (1985) showed that there is an increase in nitrogen values in the seawater in the Walker Bay area (close to I & J) in autumn and winter. Phosphate content in the seaweeds increased in the autumn and winter months, as found by Robertson-Andersson (2003) and corresponds with a seasonal increase in phosphate concentrations in the Walker Bay area (Hutchings and Andrews 1980; Chapman and Shannon 1985; Mitchell-Innes and Walker 1991).

Seaweed SGR

This research has shown how seaweed SGR is either positively or negatively affected by large scale (nutrient upwelling and advection or intrusion) environmental events in the adjacent environment. In addition, their SGR was also affected by events on a tank scale. One of the most notable negative effects on seaweed growth rates was the presence of a

M. strangulans infestation which occurred in spring and persisted over the summer months. This algae forms small round epiphytic crusts ('brown spots') on the surface of the *Ulva* and, as the *Myrionema* ages, the centre of the crusts disintegrates, apparently encouraging break-up of the host thallus. The decrease in *Ulva* SGR was accompanied by increasing occurrence of *M. strangulans* on the algae (a relationship previously described in Robertson-Andersson 2002, 2003; Robertson-Andersson and Wilson 2004). The increase in *M. strangulans* presence is known to correlate with a linear decrease in SGR (Robertson-Andersson 2003). Another negative effect on seaweed growth rates in the same period was the increased presence of a grazer, the Cape key-hole limpet (Hansen et al. 2006).

The research in this paper is complicated due to the effects of shading the seaweed treatments with a 50% shadecloth. Robertson-Andersson (2003) studied effects of seasonal shading in the same seaweed tanks over the same seasonal period and compared this to unshaded tanks and found that the tanks that were shaded had a significantly higher SGR than the unshaded tanks, mainly due to an infestation of *M. strangulans* causing the complete loss of the seaweed culture in the unshaded tanks. There were no significant differences in any of the physiochemical variables tested or differences in the water quality, when seaweed cultures were shaded or unshaded. The maximum benefit in terms of SGR and decreased *M. strangulans* presence occurred when the seaweed tanks were shaded from September to February (Robertson-Andersson 2003). Additional experiments looked at the type of shade cloth used. No change in *M. strangulans* percentage cover was obtained using a 20% shadecloth while an 80% shadecloth decreased SGR significantly (Robertson-Andersson 2003). As the primary goal of this research was to maintain the seaweed as biofilters to enable the recirculation system to continue functioning efficiently over the period tested, it was deemed necessary to shade the seaweed tanks.

SGR showed a seasonal trend and was highest in summer. SGRs observed in this study were similar to those found by Robertson-Andersson (2003) in the abalone treatment but lower than those obtained in the literature, although in most cases, smaller tank sizes were used in the latter studies (Lapointe and Tenore 1981; Duke et al. 1989a, 1989b; Björnsäter and Wheeler 1990; Neori et al. 1991). Robertson-Andersson (2003), showed that there is a decrease in SGR when scaling up tank sizes. Growth rates for *U. fenestrata* Postels et Ruprecht under experimental conditions were 16% wet $wt.day^{-1}$ (Björnsäter and Wheeler 1990). This is lower than those obtained for *U. curvata* (Kützing) De Toni, 52% wet $wt.day^{-1}$ in short-term laboratory studies (Duke et al. 1989a, b); *U. lactuca*, 18.6% wet $wt.day^{-1}$ (Neori et al. 1991) and *U. fasciata* Delile, 36% wet $wt.day^{-1}$ (Lapointe and Tenore 1981). The low growth rate for *U. fenestrata* may be attributed to the lower water temperature (13°C) compared to other experiments (Björnsäter and Wheeler 1990). The SGR obtained by Neori et al. (1991) were low and could be due to the scaling up of the tank sizes as the other studies were in experimental, small-scale setups.

Seaweed water content

The tissue water content for both systems falls within the range of 80–90% which is the case for many macro-algae (DeBoer 1981). Our data seem to suggest that at high SGR with correspondingly low tissue protein there is more cellular water in the seaweed, although this relationship has not been tested.

Physiochemical variables

Over the long term, the integrated system raised the water temperature by 1% at 25% recirculation. During the short term studies in September, and in the study by Lindström (2006) using the same experimental set-up during spring 2005, the flow-through units had higher night-time temperatures while the 25% recirc. abalone tanks were losing heat at night, by being connected to the seaweed tanks which were cooler (minimum of 13.3°C). This means that the seaweed tanks were losing more heat or were less able to retain the heat at higher recirculation in the winter periods, due to the seaweed tanks having a lower latent heat capacity and no insulation. They were thus more vulnerable to cooler night time temperatures. During the day, the seaweed tanks heated up very quickly and were able to raise the temperature of the recirculating abalone tanks, while the flow-through units were dependant on incoming seawater temperatures and would remain low in an upwelling event. Schuenhoff et al. (2003) showed that an integrated fish/abalone/seaweed culture system with 50% re-circulation in Israel using *U. lactuca* as the biofilter also experienced heat loss during winter re-circulation. The construction of a greenhouse cover over the biofilter tanks prevented this heat loss. They found maximum heat loss in early mornings and maximum heat-gain around 1600 hours, the same pattern experienced in this study and by Lindström (2006). The temperature ranges experienced in the seaweed and abalone tanks were within the optimum physical range for *H. midae* (Britz et al. 1997).

It is generally thought that connected seaweed tanks should be able to provide higher dissolved oxygen concentrations to the re-circulating culture system (Troell et al. 2003; Neori et al. 2004). Schuenhoff et al. (2003) found that water re-circulated through seaweed tanks raised dissolved oxygen levels and that this could support fish-tanks without additional oxygen "at times of highest oxygen demand". Neori et al. (1996) studied a land-based fish/seaweed re-circulation system and found that connected seaweed tanks contributed to the oxygen balance in the fish compartment by significantly slowing the rate of oxygen depletion. However, in the current study at 25% recirculation dissolved oxygen was 4% lower compared to flow-through units. This was despite the fact that the seaweed tanks in the re-circulation system had produced oxygen even at high temperatures and had 33% more oxygen than the 25% recirc. abalone units. The data showed that the dissolved oxygen in the seaweed tanks was not being transferred to the abalone tanks. Various theories were proposed and changes were made to the system from August 2003 to December 2003 to try to improved oxygen concentrations. The loss of oxygen was not due to bacterial depletion (unpublished data). Two mixing systems were designed and built (1) to increase the mixing time to allow the oxygen to become more saturated and (2) to increase oxygen through tumbling. Both designs failed in their objectives and it is thought that the seaweeds were supersaturating the oxygen in the tanks and that during the pumping this oxygen was being blown off and did not remain in the recirculated water. A larger system with a longer water retention time might improve these values. At higher recirculation ratios in spring and summer, when the temperature in incoming seawater is higher (intrusion event) and air temperature is also high (the abalone system picks up heat from warm air being blown into the tank for aeration) and at night when there is a large amount of seaweed in system with the oxygen holding capacity being low, this is likely to be a critical failing of the system. Robertson-Andersson (2003) showed that a high stocking density (4.5 kg m^{-2} in the same system), compared to lower densities (0.5–2 kg m^{-2}), exhibited the highest concentrations of dissolved oxygen at day time and the lowest at night time, due to photosynthesis and respiration.

Water nutrients

The low ammonium, nitrate and nitrite concentrations in the seaweed tanks under recirculation indicates that biofiltration by the seaweeds is occurring. The low nitrate and nitrite concentration in the seaweed tanks supports the notion of N-starvation, since seaweeds are known to prefer ammonium before nitrate (Lobban et al. 1985; Loban and Harrison 1994), and a good supply of ammonium would result in less reduction of nitrate and nitrite. Phosphate was the only nutrient analyzed that showed significantly higher concentrations in re-circulating tanks compared to controls, which is an indication of N-limitation in the system (Troell et al. 2003). Schuenhoff et al. (2003) show reduced N/P ratios in an integrated fish/abalone/seaweed culture and suspected N-limitation in the system.

If water quality biofiltration is the main goal of integrated aquaculture using seaweeds, it is often necessary to starve the seaweeds to obtain maximum reduction of the water nutrient concentrations (Troell et al. 2003; Neori et al. 1998, 2003, 2004). In a 1-stage biofilter such as the one in our study, a large biofilter area is required to strip nutrients efficiently, hence the high seaweed to abalone ratio. Our system could have been made more efficient if we had used a multi-stage biofilter with the two seaweed tanks in series rather than in parallel as in the study by Neori et al. (2003). Reasons for having the seaweed tanks in parallel and not in series were to prevent transfer of potential diseases, pathogens and epiphytic algae harming the biofilter. *Myrionema strangulans* can completely decimate the seaweed biofilter and, if the seaweed tanks were in series, transfer of this alga would result in both seaweed tanks having infestations. By having the two tanks in parallel one tank could be infected while the other would remain unaffected, thus maintaining the biofiltration capacity of the system. As Lindström (2006) found and with the results presented by Robertson-Andersson (2003)—in the same system—it is important to optimize seaweed densities to obtain maximum nutrient reduction. Nitrogen-starved seaweeds do not provide a good quality feed (Neori et al. 1998, 2004; Troell et al. 2003). By altering the seaweed stocking density one can change the products of a recirculation system to optimise either seaweed biofiltration efficiency or seaweed quality and growth. Biofiltration efficiency can change over the cultivation period of the seaweeds, this was one of the reasons for the short harvesting period of 14 days, Robertson-Andersson (2003) found optimum nutrient efficiency occurred when the seaweeds were stocked at 3.5 kg m^{-2} and that nutrient efficiency decreased at 4.5 kg m^{-2}. By stocking the tanks with an initial biomass of 2 kg m^{-2} and harvesting them more frequently, we were able to maintain a high nutrient efficiency over the 14-day period regardless of season. Another means of changing biofiltering efficiency is to alter the water flow rates, with low flow rates having high biofiltering efficiencies and low biomass production (Buschmann et al. 2001). The flow rates in the seaweed tanks in our recirculation system were low due to the low recirculation ratio. When flow rates were increased by increasing the recirculation ratio, biofiltration efficiency decreased (Lindström 2006). For normal operating conditions a balance that provides for both a high biofiltation capacity and a high biomass needs to be found (Neori et al. 1998; Troell et al. 2003).

Sediments

Suspended solids management is a key factor in determining the success of re-circulation systems and suspended solids concentration should not exceed 15 mg L^{-1} for re-circulating systems (Chen et al. 1993). In this study, the surface particle concentration was greater than 90 mg L^{-1}, which is six times this limit. The Chen et al. (1993) study was based on a gravel bed drawdown biofilter which is highly sensitive to clogging by particulates. In our study, the total tank accumulations of sediments in the flow-through units and re-circulation systems were not significantly different from each other. This is an important result because it provides evidence to dismiss the notion that the re-circulation systems accumulate more particles than a flow-through unit.

Our results showed that there no significant differences in particle size fractionation in the water column, total particle load, or the loading of 35 µm or smaller fractions between a flow-through system and a recirculation system. Brandt (2006) showed that this trend remained even when the recirculation ratio was changed to 50 and 75%. Since sabellids prefer organic particles smaller than 35 µm as feed (Chalmers 2002), these results indicate that, at least from a feed perspective, kelp fed re-circulating systems do not favour sabellids by increasing their available feed. This is supported in the sabellid count data, as higher counts of sabellids would indicate a higher food availability in the recirculation systems.

Mobile and sessile macro-fauna

A mobile macro-fauna population can easily become established in an abalone culture system via association with the cultured species, feed input (fauna coming in with the kelp as it is not cleaned before being placed in the tanks), or introduction from the surrounding ecosystems through the seawater intakes as few filters are used. Within re-circulation systems there is, theoretically, a higher potential for build up of mobile macro-fauna densities, since the fauna can circulate with the water and may stay long enough to reproduce. This could have positive and negative impacts on the target aquaculture species. In integrated animal-seaweed cultivation, positive effects may be anticipated from the conversion of particles to dissolved nutrients that can be taken up as a nutrient source for the cultivated algae, or through grazing on epiphytes and microfilms (Shacklock and Doyle 1983; Brawley and Fei 1987; Anderson et al. 1998; Klamermans et al. 2002), increasing seaweed growth and reducing the need for cleaning. No significant differences in densities or taxa of mobile macrofauna between the 25% recirc. and the flow-through units were found. In addition, numbers of *F. mutabilis* were not significantly different between treatments, contrary to what was expected. Thus, recirculation at this low water exchange rate does not influence mobile fauna densities. The cleaning of the abalone tanks in the system every 7th day and the harvesting of the seaweed tanks every 14th day is also likely to prevent the build up of a mobile fauna population in these tanks. None of the taxa found in the tanks are pest species and most are detritivores. However, the amphipod *Hyale* sp. is largely a herbivore and several of the other amphipod species found may graze on both micro and macro algae. The keyhole limpet *F. mutabilis* is known to graze by trapping floating thalli in the cultivated seaweed tanks (Robertson-Andersson 2003). It becomes more problematic in the system in September when large numbers are found (Robertson-Andersson 2003, 2007; Hansen et al. 2006). Fortunately, it can be controlled through the use of fresh water washing of the tanks and seaweeds for longer than 20 min, which is not harmful to the seaweeds (Robertson-Andersson 2003; Smit et al. 2003; Hansen et al. 2006). Other problematic species, such as *Paridotea reticula* Barnard, have been found in *G. gracilis* seaweed tanks on the farm (Smit et al. 2003; Njobeni 2006; Hansen et al. 2006) A 3 h freshwater treatment has been successfully used to control this pest isopod (Smit et al. 2003). This exposure had a minor effect on seaweed growth.

Environmental events

The I & J farm, located on the Western Agulhas Bank, is subject to intermittent wind-driven coastal upwelling, particularly northward of prominent capes (including Danger Point) (Boyd et al. 1985; Probyn et al. 1994). On the Western Agulhas Bank, the upwelling season starts later (i.e., summer) (Jury 1988) and is episodic in nature (Largier et al. 1992). There is an 18°C thermal front moving westwards from Cape Agulhas, which contains colder water and nutrients inshore of it (Largier et al. 1992). This front clearly marks longshore variations in upwelling associated with Cape Agulhas. There is a measured jet in roughly the same position (Bang and Andrews 1974). Upwelling and subsequent containment of water inshore results in higher nutrient levels inshore of the front with lower to negligible levels above the front and offshore of the front (Largier et al. 1992). Although the upwelling is episodic on the Western Agulhas Bank the subsurface thermoclines remain tilted upwards towards the coast throughout the upwelling season (Boyd et al. 1985). When the wind-forcing weakens in this upwelled system, warm surface water moves shorewards at the surface, but the isotherm still remains tilted. The nutrient rich waters are still contained inshore of this front. This results in warm nutrient rich water in the Walker Bay area. If the thermocline deepens, then a warm water intrusion of the Agulhas current water occurs particularly over the outer shelf (Boyd et al. 1985; Probyn et al. 1994). This results in very high temperatures with low nutrient concentrations (Largier et al. 1992). The SGR of seaweed in the recirculation system was either positively or negatively affected by these environmental events. This was surprising as the system was designed on data from a previous study (Robertson-Andersson 2003) and these effects had not been investigated in that study.

Conclusions

All studies performed on 25% recirc. units showed that the dissolved nutrient concentration, particle concentration, abalone SGR, abalone health, bacterial and mesoherbivore abundance are similar in the 25% re-circulating units and the flow-through units (Flodin 2005; Hansen 2005; Potgieter 2005; Brandt 2006; Lindström 2006; Robertson-Andersson 2007). The lack of transfer of oxygen between the seaweed tanks and the abalone tanks in the recirculation system is problematic, especially at night. There were no adverse effects on the abalone with respect to health or growth rate or the seaweeds by running the system at 25% recirculation, with both cultured organisms behaving similarly to ones cultivated in a flow-through system. Possible environmental impacts from accumulated and suspended sediment build up over the 7 day cleaning cycle which would decrease water quality, increase bacteria loading and meso-herbivore feed availability particularly sabellids were all non existent. The relatively low recirculation rate is probably the reason for this. And is an indication that aside from problems with dissolved oxygen this system can be run at a higher recirculation ratio.

Acknowledgements The authors would like to thank SIDA (Swedish International Development Agency), the South African National Research Foundation, the SANCOR student travel grant and the DEAT Frontier programme for providing funding for this research. MC&M and DEAT for providing research facilities. A special thank you to the I & J farm and staff who without their time, effort and facilities, this research would have been impossible.

References

Anderson RJ (2003) Seaweed. In: Research highlights 2001–2002. Department of Environmental Affairs and Tourism and Marine and Coastal Management Cape Town, pp 31–32

Anderson RJ, Smit AJ, Bolton JJ (1998) Differential grazing effects by isopods on *Gracilaria gracilis* and epiphytic *Ceramium diaphanum* in suspended raft culture. Aquaculture 169:99–109

Bang ND, Andrews WRH (1974) Direct current measurements of a shelf edge frontal jet in the southern Benguela system. J Mar Res 32:405–417

Björnsäter BO, Wheeler PA (1990) Effect of nitrogen and phosphorus supply on growth and tissue composition of *Ulva fenestrata* and *Enteromorpha intestinalis* (Ulvales, Chlorophyta). J Phycol 26:603–611

Bower CE, Bidwell JP (1978) Ionization of ammonia in seawater: effects of temperature, pH and salinity. J Fish Res Board Can 35:1012–1016

Boyd AJ; Tromp BBS, Horstman DA (1985) The hydrology off the South African south western coast between Cape Point and Danger Point in 1975. S Afr J Mar Sci 3:145–168

Brandt M (2006) A re-circulating integrated abalone/seaweed culture system in South Africa: effects on particulate and sediment load. MFS Dissertation, Stockholm University, Stockholm

Brawley SH, Fei XG (1987) Studies of meisoherbivory in aquaria and an un-barricaded mariculture farm on the Chinese coast. J Phycol 23:614–623

Britz PJ (1996) Effect of dietary protein level on growth performance of South African abalone, *Haliotis midae*, fed fishmeal-based semi-purified diets. Aquaculture 140:55–61

Britz PJ, Hecht T, Mangold S (1997) Effect of temperature on growth, feed consumption and nutritional indices of *Haliotis midae* fed a formulated diet. Aquaculture 152:191–203

Brzeski V, Newkirk G (1997) Integrated coastal food production systems — a review of current literature. Ocean Coast Manage 34:66–71

Buschmann A, Troell M, Kautsky N (2001) Integrated algal farming: a review. Cah Biol Mar 43:615–655

Castro-Gonzales MI, Perez-Gil Romo F, Perez-Estrella S, Carillo-Dominguez SD (1996) Chemical composition of the green alga *Ulva lactuca*. Ciencas Marinas 22:205–213

Chalmers R (2002) An investigation into the feeding biology and factors influencing the population dynamics of *Terebrasabella heterouncinata* (Polychaeta: Sabellidae), a problematic tube welling polychaete in farmed abalone in South Africa. M.Sc. Dissertation, Rhodes University, Grahamstown

Chapman P, Shannon LV (1985) The Benguela ecosystem part II: Chemistry and related processes. Oceanogr Mar Biol Ann Rev 23:183–251

Chen S, Timmons MB, Aneshansley DJ, Bisogni JJ (1993) Suspended solids characteristics from re-circulating aquacultural systems and design implications. Aquaculture 112:143–155

Chopin T, Buschmann AH, Halling C, Troell M, Kautsky N, Neori A, Kraemer GP, Zertuche-González JA, Yarish C, Neefus C (2001) Integrating seaweeds into marine aquaculture systems: a key towards sustainability. J Phycol 37:975–986

DeBoer JA (1981) Nutrients. In: Lobban CS, Wynne MJ (eds) The biology of seaweeds. Botanical monographs Vol. 17. Blackwell, Oxford, pp 356–392

Duke CS, Lapointe BE, Ramus J (1986) Effects of irradiance on growth, RuBPCase activity and chemical composition of *Ulva* species (Chlorophyta). J Phycol 22:362–370

Duke CS, Litaker W, Ramus J (1987) Seasonal variation in RuBPCase activity and N allocation in the Chlorophyte seaweeds *Ulva curvata* (Kutz.) and *Codium decorticatum* (Woodw) Howe. J Exp Mar Biol Ecol 112:145–164

Duke CS, Litaker W, Ramus J (1989a) Effects of the temperature, nitrogen supply and tissue nitrogen on ammonium uptake rates of the Chlorophyte seaweeds *Ulva curvata* and *Codium decorticatum*. J Phycol 25:113–120

Duke CS, Litaker W, Ramus J (1989b) Effect of temperature on nitrogen limited growth rate and chemical composition of *Ulva curvata* (Ulvales, Chlorophyta). Mar Biol 100:143–150

Evans GC (1972) The quantitative analysis of plant growth. Studies in Ecology. Blackwell, Oxford

FAO (2004) The state of the world fisheries and aquaculture 2004. Rome, FAO.

Fleming AE, Van Barneveld RJ, Hone PW (1996) The development of artificial diets for abalone: a review and future directions. Aquaculture 140:5–53

Fleurence J, Le Coeur C, Mabeau S, Maurice M, Landrein A (1995) Comparison of different extractive procedures for proteins from the edible seaweeds *Ulva rigida* and *Ulva rotundata*. J Appl Phycol 7:577–582

Flodin JMT (2005) Bacterial water quality in South African abalone (*Haliotis midae*) tank culture: effects on bacterial levels from seaweed integration and feed choice. MFS Dissertation. Stockholm University, Stockholm

Fourie AM (1994) Tank culturing of *Gracilaria verrucosa* for use as a food source for abalone. MSc Dissertation. University of Port Elizabeth, Port Elizabeth

Goldburg R, Triplett T (1997) Murky Waters: Environmental Effects of Aquaculture in the United States. Environmental Defense Fund, Washington, DC

Grasshoff K, Ehrhardt M, Kremling K (1976) Methods of seawater analysis. Weinheim, Germany

Hahn KO (1989) CRC Handbook of culture of abalone and other marine gastropods. CRC Press, Boca Raton

Hansen J (2005) Densities of mobile macrofauna and control of the herbivorous Keyhole limpet *Fissurella mutabilis* in an integrated abalone-seaweed tank-culture in South Africa. MFS Dissertation, Stockholm University, Stockholm

Hansen J, Robertson-Andersson DV, Troell M (2006) Control of the herbivorous gastropod *Fissurella mutabilis* (Sow.) in a land-based integrated abalone-seaweed culture. Aquaculture 255:384–388

Hampson S (1998) Evaluation of effluent water from an abalone mariculture system as a culture medium for two strains of *Gracilaria gracilis*. M.Sc. Dissertation. University of Port Elizabeth, Port Elizabeth

Harris JO, Maguire GB, Edwards SJ, Johns DR (1999) Low dissolved oxygen reduces growth rate and oxygen consumption rate of juvenile greenlip abalone, *Haliotis lae igata* Donovan. Aquaculture 174:265–278

Hutchings L, Andrews WRH (1980) Upwelling in the southern Benguela Current, South Africa. Prog Oceanog 9:l–81

Jury MR (1988) A climatological mechanism for wind-driven upwelling near Walker Bay and Danger Point, South Africa. S Afr J Mar Sci 6:175–181

Klamermans P, Malta J - E, Verschuure JM, Schrijvers L, Lentz LF, Lien AT (2002) Effects of grazing by isopods & amphipods on growth of *Ulva* spp. (Chlorophyta). Aquat Ecol 36:425–433

Lapointe BE, Tenore KR (1981) Experimental outdoor studies with *U. fasciata* Delile, I: interaction of light and nitrogen on nitrogen uptake, growth and biochemical composition. J Exp Mar Biol Ecol 53:135–152

Largier JL, Chapman P, Peterson WT, Swarts VP (1992) The western Agulhas bank: circulation, stratification and ecology. S Afr J Mar Sci 12:319–339

Lindström A (2006) A re-circulating integrated abalone/seaweed culture system in South Africa: effects on water quality. MFS Dissertation. Stockholm University, Stockholm

Lobban CS, Harrrison PJ (Eds) (1994) Seaweed ecology and physiology. Cambridge University Press, Cambridge

Lobban CS, Harrrison PJ, Duncan MJ (1985) The Physiological Ecology of Seaweeds. Cambridge University Press, Cambridge

Mai K, Wu G, Zhu W (2001) Abalone, *Haliotis discus hannai* Ino, can synthesize Myo-Insitol de novo to meet physiological needs. J Nutrition 131:2898–2903

Miller D (2001) Outdoor cultivation of *Gracilaria* in pond and raceway systems at Jacobsbaai Sea Products (Pty) Ltd., South Africa. Honours Dissertation, University of Cape Town, Cape Town

Mitchell-Innes BA, Walker DR (1991) Short term variability during an anchor station study in the Southern Benguela upwelling system: phytoplankton production and biomass in relation to species changes. Prog Oceanog 28:65–89

Morgan D (2000) The potential for *Gracilaria* polyculture at Jacobs Bay Sea Products Pty. Honours Dissertation. University of Cape Town, Cape Town

Mouton A (2004) Veterinary health reports for Abalone farms details of methods used and results determination. AFASA report afa03074

Murphy J, Riley JP (1962) A modified single-solution method for the determination of phosphate in natural waters. Analytical Chim Acta 27:31–36

Naidoo K, Maneveldt G, Ruck K, Bolton JJ (2006) A comparison of various seaweed-based diets and formulated feed on growth rate of abalone in a land-based aquaculture system. J Appl Phycol 18:437–443

Neori,A, Cohen I, Gordin H (1991) *Ulva lactuca* biofilters for marine fish pond effluents. II. Growth rate, yield and C:N ratio. Bot Mar 34:483–489

Neori A, Krom M, Ellner S, Boyd C, Popper D, Rabinovitch R, Davison P, Dvir O, Zuber D, Ucko M, Angel D, Gordin H (1996) Seaweed biofilters as regulators of water quality in integrated fish-seaweed culture units. Aquaculture 141:183–199

Neori A, Ragg NLC, Shpigel M (1998) The integrated culture of seaweed, abalone, fish and clams in modular intensive land-based systems: II. Performance and nitrogen partitioning within an abalone (*Haliotis tuberculata*) and macroalgae culture system. Aquac Eng 17:215–239

Neori A, Msuya FE, Shauli L, Schuenhoff A, Kopel F, Shpigel M (2003) A novel three stage seaweed (*Ulva lactuca*) biofilter design for integrated mariculature. J Appl Phycol 15:543–553

Neori A, Chopin T, Troell M, Buschmann AH, Kraemer GP, Halling C, Shpigel M, Yarish C (2004) Integrated aquaculture: rationale, evolution and state of the art emphasizing seaweed biofiltration in modern aquaculture. Aquaculture 231:361–391

Nisizawa K, Noda H, Kikuchi R, Watanabe T (1987) The main seaweed foods in Japan. Hydrobiologica 151/152:5–29

Njobeni A (2006) The cultivation of *Gracilaria gracilis* (Rhodophyta) in an integrated aquaculture system, for the production of abalone feed and the bioremediation of aquaculture effluent. MSc Dissertation, University of Cape Town, Cape Town

Nydahl F (1976) On the optimum concentrations for reduction of nitrate to nitrite by cadmium. Talanta 23:349–357

Potgieter M (2005) A comparison of suspended particle size and sediment loading produced by artificial and seaweed diets in integrated flow-trough and re-circulating aquaculture systems on a commercial South African abalone farm. MSc Dissertation, University of Cape Town, Cape Town

Probyn TA, Mitchell-Innes BA, Brown PC, Hutchings L, Carter RA (1994) A review of primary production and related processes on the Agulhas Bank. S A J Sci 90:166–173

Robertson-Andersson DV (2002) The Brown Strangler. 50th Newsletter of the Phycological Society of Southern Africa, Cape Town. 11–12.

Robertson-Andersson DV (2003) The cultivation of *Ulva lactuca* (Chlorophyta) in an integrated aquaculture system, for the production of abalone feed and the bioremediation of aquaculture effluent. MSc Dissertation, University of Cape Town, South Africa

Robertson-Andersson DV (2007) Biological and economical feasibility studies of using seaweeds *Ulva lactuca* (chlorophyta) in recirculation systems in abalone farming. PhD Dissertation, University of Cape Town, South Africa

Robertson-Andersson DV, Wilson D (2004) Nitrogen and thallus colour. 58th Newsletter of the Phycological Society of Southern Africa, Cape Town. 15–17.

Sales J, Britz PJ (2001) Review; Research on abalone (*Haliotis midae* L.) cultivation in South Africa. Aquaculture Res 32:863–874

Schuenhoff A, Shpigel M, Lupatsch I, Ashkenazi A, Msuya FE, Neori A (2003) A semi-recirculating, integrated system for the culture of fish and seaweed. Aquaculture 221:167–181

Shacklock PF, Doyle RW (1983) Control of epiphytes in seaweed cultures using grazers. Aquaculture 31:141–151

Shipton TA (1999) The protein requirements of the South African abalone, *Haliotis midae*. PhD Dissertation, Rhodes University, Grahamstown

Simmons R (1990) Abalone farming: how should the stock be fed? In: Cook P (ed) Proceedings of a workshop convened by The Mariculture Association of Southern Africa. The Mariculture Association of Southern Africa, Cape Town, pp 33–42

Simpson BJA (1994) An investigation of diet management strategies for the culture of the South African abalone, *Haliotis midae*. MSc Dissertation, University of Cape Town, Cape Town

Simpson BJA, Cook PA (1998) Rotation diets: a method of improving growth of cultured abalone using natural algal diets. J Shellfish Res 17:635–640

Smit AJ (1997) Optimising *Gracilaria verrucosa* production in a laboratory based culture system. MSc Dissertation, University of Port Elizabeth

Smit AJ, Fourie AM, Robertson BL, Du Preez DR (2003) Control of the herbivorous isopod *Paridotea reticulata* in *Gracilaria gracilis* tank cultures. Aquaculture 217:385–393

Smith DG, Young EG (1954) Amino acids of marine algae. J Biochem 217:845–853

Solorzano L (1969) Determination of ammonium in natural waters by the phenol-hypochlorite method. Limnol Oceanogr 14: 799–801

Steyn PP (2000) A comparative study of the production and suitability of two *Ulva* species as abalone fodder in a commercial mariculture system. M. Sc. Dissertation. University of Port Elizabeth, South Africa

Troell M, Kautsky N, Folke C (1999a) Applicability of integrated coastal aquaculture systems. Ocean Coast Manag 42:63–69

Troell M, Rönnback P, Halling C, Kautsky N, Buschmann A (1999b) Ecological engineering in aquaculture: use of seaweeds for removing nutrients from intensive mariculture. J Appl Phycol 11:89–97

Troell M, Halling C, Neori A, Chopin T, Buschmann AH, Kautsky N, Yarish C (2003) Integrated mariculture: asking the right questions. Aquaculture 226:69–90

Troell M, Robertson-Andersson DV, Anderson RJ, Bolton JJ, Maneveldt G, Halling C, Probyn T et al (2006). Abalone farming in South Africa: perspectives on kelp resources, abalone feed, potential for on-farm seaweed production and socio-economic benefits. Aquaculture 257:266–281

Wilkinson S (2001) An investigation of protein variation in *Ulva lactuca* and *Ulva rigida* under high nitrogen culture conditions. Honours Dissertation, University of Cape Town, Cape Town

Wong KH, Cheung PCK (2001) Nutritional evaluation of some subtropical red and green seaweeds. Part II. In vitro protein digestibility & amino acid profiles of protein concentrates. Food Chem 72:11–17

Determining the most appropriate feeding regime for the South African abalone Haliotis midae Linnaeus grown on kelp

Tamson L. Francis · Gavin W. Maneveldt · Jonathan Venter

Originally published in the Journal of Applied Phycology, Vol 20, No 5, 147–152.
DOI: 10.1007/s10811-007-9266-4

Abstract Beach-cast kelp (the most widely used feed for commercially grown South African abalone) is plentiful during winter months when periodic storms cause kelp to wash ashore. During summer, however, this resource is not always readily available and farmed abalone are often starved for short periods. The aim of this research was to assess how periodic kelp starvation influences growth of the commercially grown abalone, *Haliotis midae* Linnaeus. Growth of grow-out abalone was monitored on a commercial abalone farm over a period of six months and consisted of 3 treatments with 2 replicates (n = ±250 abalone per replicate). The treatments were: Control (abalone always given more kelp than what they typically needed); Treatment 1 (abalone fed their weekly ration only once a week); Treatment 2 (abalone fed half their weekly ration every 3 and then 4 days respectively). While the data at first suggest that the control animals outperform the treatment animals, after undergoing an initial adjustment period to the new feeding regime, the treatment animals perform better. Weight gain and feed conversion efficiencies show that the treatment animals perform better overall. The control animals generally required much more feed to produce comparable increases in both length and weight compared to the treatment animals. This study has shown that periodic bouts of starvation is beneficial to *Haliotis midae*, allowing variable growth spurts when returned to full feed rations.

Keywords Abalone · Compensatory growth · Feed conversion efficiency · Growth · *Haliotis midae* · Kelp · Starvation

This paper has not been submitted elsewhere in identical or similar form.

T. L. Francis · G. W. Maneveldt (✉)
Department of Biodiversity and Conservation Biology,
University of the Western Cape,
Private Bag X17,
Bellville 7535, South Africa
e-mail: gmaneveldt@uwc.ac.za

J. Venter
Jacobsbaai Sea Products Farm,
Private Bag X2, Rhine Road,
Jacobsbaai 8050, South Africa

Introduction

Abalone (*Haliotis* spp.) are in considerable demand in the Far East where they are treated as part of traditional cuisine and ceremony (Chen 1989; Sales 1999; Sales and Britz 2001). Consequently, they have been exploited for centuries because of their food value (Barkai and Griffiths 1986). Presently there is an increasing demand for small, cocktail size abalone of 40–70 mm shell length in the international market (Jarayabhand and Paphavasit 1996; Najmudeen and Victor 2004).

Worldwide there are approximately 90 species of abalone, of which 15 are harvested commercially (Sales and Janssens 2004). In South Africa only six Haliotid species occur, and of these, *Haliotis midae* Linnaeus is the only one that is commercially exploited (Cook 1998; Sales and Britz 2001; Evans et al. 2004). Although the South African abalone fishery has existed since 1949, the first attempt to cultivate *H. midae* was in 1981 (Genade et al. 1988; Sales and Britz 2001; Steinberg 2005). Since then, South Africa has become the largest abalone producer outside Asia (FAO 2004). This emerging market has largely been driven by over exploitation of the wild abalone stocks by poaching and by high market prices (Cook 1998; Troell et al. 2006).

The growing South African abalone industry depends largely on a steady supply of feed resources, particularly fresh kelp (Troell et al. 2006). Currently more than 7000 tons of fresh kelp fronds are harvested annually in South Africa to feed cultured abalone and this figure is expected to increase as the growing demand for kelp by local abalone farmers also increases (Anderson et al. 2006). During the winter months, beach-cast kelp is plentiful as storms cause kelp to wash ashore. In summer, however, kelp has to be harvested at sea and delivered to the abalone farms (Anderson et al. 2006; Rothman et al. 2006). The amount of kelp delivered to an abalone farm varies from day to day and often there are days where no kelp may be available. This then means that abalone are often starved of food for short periods. Thus far, no research has focused on the effects of periodic kelp starvation on the growth of farmed *H. midae*. The aim of this research was therefore to determine the most appropriate feeding regime for the South African abalone, *H. midae* and to assess how periodic starvation influences its growth.

Materials and methods

The research was conducted at the Jacobsbaai Sea Products (17°53′ 12.5″ E, 32° 58′ 2.5″ S, Western Cape, South Africa) abalone farm. Flow through seawater with a flow rate of 850–1300 L.h^{-1} and moderately aerated, was supplied at a temperature of 13.8±0.76°C in the holding tanks. The abalone were grown in culture baskets (80×57×25 cm; length, width and depth respectively) subdivided with feeding plates to increase the surface area.

Grow-out abalone (abalone with a shell length >20 mm) were supplied by the Jacobsbaai Sea Products abalone farm. Since growth of abalone is variable, individuals of similar size and of the same gene pool (spawned May 2002) were used. These animals were subdivided into two replicate baskets of approximately 12.5 kg (± 250 individuals). The initial weight and shell length of the abalone were measured at 46.47 g ± 0.25 and 64.31 mm ± 0.75 respectively. Both body weight and shell length were measured monthly.

Daily feed consumption

A pre-investigation feeding trial was initiated two weeks prior to commencement of the study to ascertain the daily feed consumption of grow-out abalone of the size used in the experiment. It was found that ±12.5 kg of abalone of the weight and length above consumed roughly 500 g of kelp per day. This data was used as the basis for the various treatments that followed and a base weight of 550 g per day (realized feed consumption) was established to compensate for variable daily growth between baskets and between individuals within baskets. The pre-investigation feeding trial ran concurrently with the experiment in order to monitor the on-going consumption as abalone grew throughout the experiment; the feed values were then adjusted accordingly.

Treatments

Three treatments (each with two replicates) were tested.

Control Considered the ideal conditions in which kelp was always available, i.e. always more than the daily "realised feed consumption" (initially 550 g + 50 g) of kelp in the basket. Monitored daily, deteriorating kelp was always removed and fresh kelp topped up with as much as needed to always have 600 g in the baskets.

Treatment 1 Initially fed 3.85 kg (550 g × 7) of kelp on day 1 and then again 3.85 kg of kelp every 7 days later; i.e. bulk feeding at 7 day intervals. This assumed that if the abalone consumed more than the daily "realised feed consumption" of 550 g, they would at one stage or another have to starve until day eight when the next bulk feeding occurred.

Treatment 2 This treatment was included to compensate for any feeding routine that might arise from treatment 1. Abalone were initially fed 1.925 kg [(550 g × 7)/2] on day 1 and then again 1.925 kg, 3 and then 4 days later for the duration of the experiment. The 3–, 4-day cycle was maintained throughout this treatment. As half (1.925 kg) of the weekly "realised feed consumption" amount of kelp was given over 2 different cycles, it was assumed that there would be both periods of sufficient kelp available (i.e. 3-day cycle), as well as periods where the abalone would be slightly starved of kelp (i.e. 4-day cycle).

By month 2 the "realised feed consumption" was increased by 50 g to 600 g of kelp as dictated by the separate, on-going pre-investigation experiment. This new value substituted the previous 550 g values in all the treatments above and quantities were adjusted accordingly. Hereafter, the "realized feed consumption" was kept constant at 600 g for both treatments 1 and 2 to build in greater starvation periods into the experiment for the latter four months, while that of the control was increased by 50 g every second month.

Sampling and data collection

The experiment was conducted over six months. Representative animals were randomly selected from each treatment (N=30 per replicate at 0–3 months; N=40 per replicate at

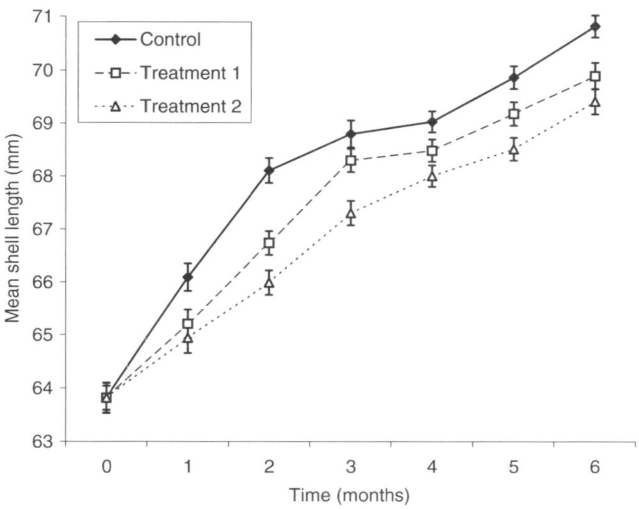

Fig. 1 Abalone growth in shell length

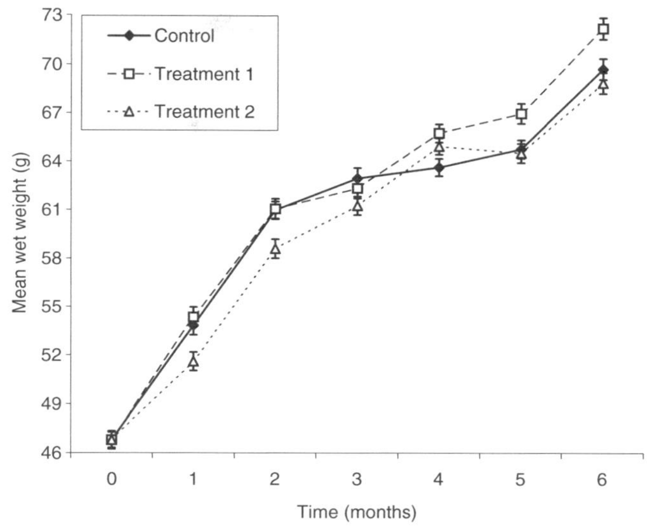

Fig. 2 Abalone growth in body weight

4–6 months to compensate for later differential growth). Before all measurements, abalone were blotted dry to remove excess water. Body weight was recorded to the nearest 0.01 g using an electronic balance. Shell length was measured along the longest axis of the abalone shell to the nearest 0.1 mm with a vernier callipers.

Daily increment in shell length (DISL) was calculated using the formula of Zhu et al. (2002): $DISL(\mu m/day) = [(SL_t - SL_i)/t] \times 1000$, where SL_t = the final mean shell length (mm), SL_i = the initial mean shell length (mm), and t = the feeding trial period in days.

Specific growth rate (SGR in % body weight.day^{-1}) was calculated using the formula of Britz (1996): $SGR = [(\ln(Wf) - \ln(Wi))/t] \times 100$, where ln(Wf) = the natural log of the final mean weight of abalone, ln(Wi) = the natural log of the initial mean weight of abalone, and t = the feeding trial period in days.

The feed conversion efficiency (FCE) was calculated for each treatment, using the formula of Simpson and Cook (1998): FCE=(growth/ration)×100, where growth = the blotted wet weight (g) gained per day and ration = the blotted wet feed (g) intake.

Table 1 Growth parameters of abalone from the three treatments

Treatment	Control	Treatment 1	Treatment 2
DISL $^{(0–6)}$	38.123 ^{a}	33.102 ^{b}	30.478 ^{b}
DISL $^{(2–6)}$	22.124 ^{b}	25.725 ^{ab}	27.896 ^{a}
SGR $^{(0–6)}$	0.222 ^{b}	0.235 ^{a}	0.207 ^{b}
SGR $^{(2–6)}$	0.112 ^{b}	0.136 ^{a}	0.130 ^{ab}
FCE $^{(0–6)}$	0.020 ^{b}	0.025 ^{a}	0.022 ^{ab}
FCE $^{(2–6)}$	0.010 ^{b}	0.015 ^{a}	0.014 ^{a}

Daily increment increase in shell length (DISL - μm.d^{-1}), specific growth rate (SGR - % body weight.day^{-1}) and feed conversion efficiency (FCE) are provided for the periods 0–6 months (0–6) and 2–6 months (2–6). Means with the same letter are not statistically different.

Statistical analyses

Unless otherwise stated, all graphed data are expressed as means ± se. Data for all experimental replicates were pooled as no significant differences were found between them. As no adjustment period was allowed before the start of the experiment, the first two months were considered to be the time necessary for the treatment animals to adjust to

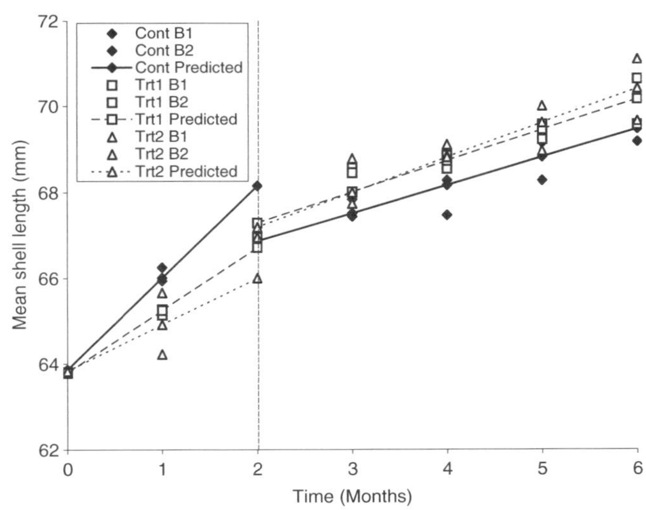

Fig. 3 Abalone growth in shell length with linear regression functions fitted for the periods 0–2 and 2–6 months. Cont = Control; Trt 1 = Treatment 1; Trt 2 = Treatment 2; B1 = mean for basket 1; B2 = mean for basket 2; Predicted = the statistically determined treatment mean determined by the split plot function

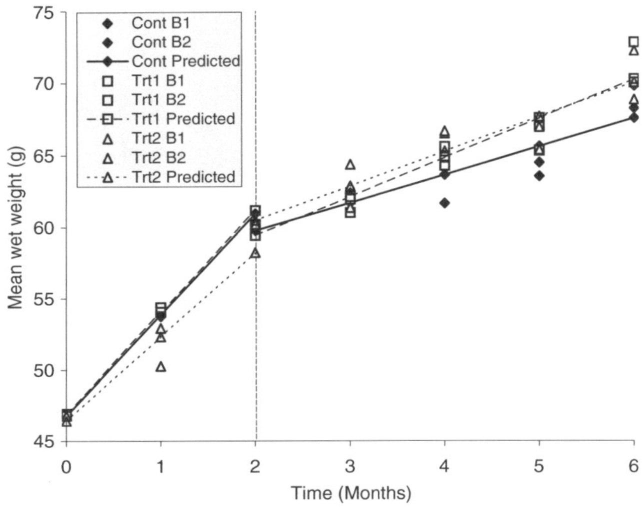

Fig. 4 Abalone growth in body weight with linear regression functions fitted for the periods 0–2 and 2–6 months. Cont = Control; Trt 1 = Treatment 1; Trt 2 = Treatment 2; B1 = mean for basket 1; B2 = mean for basket 2; Predicted = the statistically determined treatment mean determined by the split plot function

the new feeding regimes. For this reason, and to monitor the ongoing change over time, the observations were combined as a split plot analysis of variance with treatments (control, treatments 1 and 2) as main plot factors and months as split plot factors (see Little and Hills 1972). For each treatment and basket combination, linear regression functions were fitted on mean abalone weight and length change over the periods 0–2 and 2–6 months. To do this, mean abalone weight and length were adjusted to be equal at the start of each analysis. Results from the linear regression analyses were then subjected to a one-way analysis of variance to compare treatment regression parameters. Differences among treatments were considered statistically significant at $P<0.05$.

Results

Abalone fed a diet of kelp will typically invest more energy into length, quite often at the expense of weight (Troell et al. 2006). Growth trends in length are therefore often considered more reliable when determining farm management procedures for abalone fed kelp. This considered, the data suggest that the control animals outperform the treatment animals (Fig. 1) with a DISL of 38.123 $\mu m.d^{-1}$ (Table 1). No significant differences ($P=0.0975$) were found between treatments 1 and 2 with regards to mean shell length gain for the entire experimental period (Fig. 1, Table 1). The performance in length in the control animals, however, comes at the expense of weight (Fig. 2) and the control is not the better performing treatment in terms of specific weight gain (SGR values in Table 1).

Upon closer examination of the actual rates of growth (visible by the slopes of the graphs) during the latter months (2–6) of the experiment, it is evident that something quite different is happening (see Figs. 3 & 4, Table 1). The data for the latter 2–6 months show that there are still no significant differences between treatments 1 and 2 for both mean shell length ($P=0.6046$) and weight ($P=0.2063$); both treatments, however, perform better than the control (Figs. 3 & 4). This is clearly evident from the DISL and SGR values obtained for the period 2–6 months (Table 1). Also, the treatment animals have higher feed conversion efficiencies (FCE) despite the control animals receiving more feed overall.

DISL, SGR and FCE values often mean little to the abalone farmer. When the data are re-interpreted into values

Table 2 Length and weight gained by the animals from the three treatments and the amount of kelp required to produce 1 mm of shell length and 10 g of body weight

Treatment	Control	Treatment 1	Treatment 2
Mean length (mm) gain $^{(0–6)}$	7.015 ^{a}	6.091 ^{b}	5.608 ^{b}
Mean length (mm) gain $^{(2–6)}$	2.721 ^{b}	3.164 ^{a}	3.431 ^{a}
Mean weight (g) gain $^{(0–6)}$	22.968 ^{b}	25.441 ^{a}	22.101 ^{b}
Mean weight (g) gain $^{(2–6)}$	8.775 ^{b}	11.179 ^{a}	10.293 ^{a}
Mean kelp consumed (g) by each individual abalone $^{(0–6)}$	499.214 ^{a}	443.8 ^{b}	443.8 ^{b}
Mean kelp consumed (g) by each individual abalone $^{(2–6)}$	376.614 ^{a}	331.8 ^{b}	331.8 ^{b}
Amount of kelp (g) required to produce 1 mm shell length $^{(0–6)}$	71.168 ^{b}	72.864 ^{b}	79.138 ^{a}
Amount of kelp (g) required to produce 1 mm shell length $^{(2–6)}$	138.397 ^{a}	104.862 ^{b}	96.699 ^{b}
Amount of kelp (g) required to produce 10 g of abalone $^{(0–6)}$	217.348 ^{a}	174.442 ^{b}	200.806 ^{a}
Amount of kelp (g) required to produce 10 g of abalone $^{(2–6)}$	429.186 ^{a}	296.802 ^{b}	322.351 ^{b}

All data are provided for the periods 0–6 months (0–6) and 2–6 months (2–6). Means with the same letter are not statistically different.

that the abalone farmer can better understand, we see clearly than the treatment animals perform better that the control animals (Table 2). Despite the control animals gaining more length over the entire experimental period, much more kelp was required to produce an equivalent amount of length in the control animals compared to the treatment animals. In terms of weight gain and the amount of kelp required to produce a net gain in weight, both sets of treatment animals perform better than the control animals. Overall, the treatment animals have higher feed conversion efficiencies that are even more striking for the latter four months (2–6) of the experiment.

Discussion

While abalone characteristically invest more into shell length than into body weight when fed kelp (because of the high ash and low protein contents - Troell et al. 2006), to the abalone farmer weight gain is more important. This is because revenue is generated on a per-weight-basis. Also, it is generally assumed that providing abalone with more kelp feed than they typically require, will produce optimum growth. This study has, however, shown that periodic bouts of starvation actually benefit the abalone and consequently the abalone farmer as well.

Starvation or restricted feeding is not unusual to marine invertebrates as food periodically becomes scarce or unavailable to them (see e.g. Durazo-Beltrán et al. 2004). It has been documented (e.g. Carefoot et al. 1993; Takami et al. 1995) that abalone can withstand long periods of starvation before body reserves are depleted. This is usually achieved by first metabolising carbohydrates and lipid stores, and then later body proteins (Roberts et al. 2001). Segawa (1991) suggested that carbohydrates and lipids were utilized during normal feeding, and that after about two weeks' starvation, proteins were metabolised as the main energy source. This, however, did not apply to the present study as abalone were never starved for periods longer than 1 to 2 days.

In starved molluscs, metabolic rates tend to decrease as starvation progresses (Gaty and Wilson 1986; Carefoot 1987). Many studies (e.g. Quinton and Blake 1990; Jobling and Koskela 1996) have shown that like dietary composition, reproductive state, and unfavourable environments, food restriction or starvation often causes an animal to display compensatory growth. Compensatory growth may be defined as the ability to display a rapid growth spurt when returned to full rations following brief periods of food restriction or starvation (Weatherly and Gill 1981; Miglavs and Jobling 1989; Rueda et al. 1998). Compensatory growth is usually accompanied by hyperphagia (i.e. an increase in food intake - Gurney 2004) and displays itself in improved feed conversion efficiency values (Greef et al. 1986; Miglavs and Jobling 1989). This is clearly evident in the high FCE values obtained by the treatment animals.

In conclusion, once the treatment animals overcame the adjustment period, their presumed slowed metabolism resulted in rapid weight gain through compensatory growth. Periodic kelp starvation is thus not necessarily detrimental to the South African abalone *H. midae* and may indeed be beneficial as this study has shown, no doubt because of the positive effects of compensatory growth. It should, however, be stressed that were the growth rates not examined so closely, one could so easily have missed the growth spurts evident in the latter months of the experiment. If the experiment were, however, run for a longer period, this trend would have been evident for the full data set. In addition to the main outcome of this research, this study has highlighted the importance of relatively long-term experimentation on relatively slow-growing organisms.

Acknowledgements We thank the Department of Biodiversity and Conservation Biology at the University of the Western Cape and the Jacobsbaai Sea Products farm for providing funding, research facilities and technical support. We also thank the South African National Research Foundation (NRF) and the South African Department of Environmental Affairs and Tourism (DEAT) for additional research funding. Deborah Robertson-Andersson and Cliff Jones provided valuable discussion and comments. We thank Max Troell (University of Stockholm, Sweden) and John Bolton (University of Cape Town, South Africa) for the collaboration that made this project a reality.

References

Anderson RJ, Rothman MD, Share A, Drummond H (2006) Harvesting of the Kelp *Ecklonia maxima* in South Africa affects its three obligate, red algal epiphytes. J Appl Phycol 18:343–349

Barkai R, Griffths CL (1986) Diet of the South African abalone *Haliotis midae*. S Afr J Mar Sci 4:37–44

Britz PJ (1996) The suitability of selected protein sources for inclusion in formulated diets for South African abalone, *Haliotis midae*. Aquaculture 140:63–73

Carefoot TH (1987) Diet and its effect on oxygen uptake in the sea hare *Aplysia*. J Exp Mar Biol Ecol 114:275–287

Carefoot TH, Qian P-Y, Taylor BE, West T, Osborne J (1993) Effect of starvation on the energy reserves and metabolism in the Northern abalone, *Haliotis kamtschatkana*. Aquaculture 118: 315–325

Chen HC (1989) Farming the small abalone, *Haliotis diversicolor supertexta*, in Taiwan. In: Hahn KO (ed) Handbook of culture of abalone and other marine gastropod. CRC Press, Florida, pp 265–283

Cook P (1998) The current status of abalone farming in South Africa. J Shellfish Res 3:601–602

Durazo-Beltrán E, Viana MT, D'Abramo LR, Toro-Vazquez JF (2004) Effects of starvation and dietary lipid on the lipid and fatty acid composition of muscle tissue of juvenile green abalone (*Haliotis fulgens*). Aquaculture 238:329–341

Evans BS, Sweijd NA, Bowie RCK, Cook PA, Elliott NG (2004) Population genetic structure of the perlemoen *Haliotis midae* in

South Africa: evidence of range expansion and founder events. Mar Ecol Prog Ser 270:163–172

FAO (2004) The State of the World Fisheries and Aquaculture 2004. Food and Agriculture Organization, Rome, pp 153

Gaty G, Wilson JH (1986) Effect of body size, starvation, temperature and oxygen tension on the oxygen consumption of hatchery-reared ormers *Haliotis tuberculata* L. Aquaculture 56:229–237

Genade AB, Hirst AL, Smit CJ (1988) Observations on the spawning, development and rearing of the South African abalone, *Haliotis midae* Linn. S Afri J Mar Sci 6:3–12

Greef JC, Messner HH, Roux CZ, Janse Van Rensburg RJ (1986) The effect of compensatory growth on feed intake, growth rate and efficiency of feed utilization in sheep. S Afr J Anim Sci 16:155–161

Gurney WS (2004) Resource Allocation, Hyperphagia and Compensatory Growth. Bull Math Biol 66:1731–1753

Jarayabhand P, Paphavasit N (1996) A review of the culture of tropical abalone with special reference to Thailand. Aquaculture 140:159–168

Jobling M, Koskela J (1996) Interindividual variations in feeding and growth in rainbow trout during restricted feeding and in a subsequent period of compensatory growth. J Fish Biol 49:658–667

Little TM, Hills FJ (1972) Statistical methods in agricultural research. University of California Davis, Berkeley, California, pp 100

Miglavs I, Jobling M (1989) Effects of feeding regime on food consumption, growth rates and tissue nucleic acids in juvenile Arctic charr, *Salvelinus alpinus*, with particular respect to compensatory growth. J Fish Biol 34:947–957

Najmudeen TM, Victor AC (2004) Seed production and juvenile rearing of the tropical abalone *Haliotis varia* Linnaeus 1758. Aquaculture 234:277–292

Quinton JC, Blake W (1990) The effects of feeding cycling and ration level on the compensatory growth response in rainbow trout, *Oncorhynchus mykiss*. J Fish Biol 37:34–41

Roberts RD, Lapworth C, Barker RJ (2001) Effect of starvation on the growth and survival of post-larval abalone (*Haliotis iris*). Aquaculture 200:323–338

Rothman MD, Anderson RJ, Smit AJ (2006) The effects of harvesting of the South African kelp (*Ecklonia maxima*) on kelp population structure, growth rate and recruitment. J Appl Phycol 18:335–341

Rueda FM, Martinez FJ, Zamora S, Kentouri M, Divanach P (1998) Effect of fasting and refeeding on growth and body composition of red porgy, *Pagrus pagrus* L. Aquac Res 29:447–452

Sales J (1999) Digestibility of feedstuffs in abalone (*Haliotis* species). AFMA MATRIX (Journal of the South African Feed Manufacturers Association) 8:17–21

Sales J, Britz PJ (2001) Research on abalone (*Haliotis midae* L.) cultivation in South Africa. Aquac Res 32:863–874

Sales J, Janssens GPJ (2004) Use of feed ingredients in artificial diets for abalone: a brief update. Nutr Abstr Rev: Series B74: 13N–21N

Segawa S (1991) Oxygen consumption and ammonia excretion by the abalone *Sulculus diversicolor aquatilis* in starved condition. Nippon Suisan Gakkaishi 57:2001–2006

Simpson BJA, Cook PA (1998) Rotation diets: A method of improving growth of cultured abalone using natural diets. J Shellfish Res 17:635–640

Steinberg J (2005) The illicit abalone trade in South Africa. ISS Occasional Paper No. 105, pp 11

Takami H, Yamakawa H, Nakano H (1995) Survival and physiological stress of juvenile Disk Abalone *Haliotis discus discus* during long-term starvation. Fish Sci 61:111–115

Troell M, Robertson-Anderson D, Anderson R, Bolton J, Maneveldt G, Halling C, Probyn T (2006) Abalone farming in South Africa: An overview with perspectives on kelp resources, abalone feed, potential for on-farm seaweed production and socio-economic importance. Aquaculture 257:266–281

Weatherly AH, Gill HS (1981) Recovery growth following periods of restricted ration and starvation in rainbow trout, *Salmo gairdneri* Richardson. J Fish Biol 18:195–208

Zhu W, Mai K, Wu G (2002) Thiamin requirement of juvenile abalone, *Haliotis discus hannai* Ino. Aquaculture 207:331–343

Palm oil mill effluent (POME) cultured marine microalgae as supplementary diet for rotifer culture

Charles Santhanaraju Vairappan · Ang May Yen

Originally published in the Journal of Applied Phycology, Vol 20, No 5, 153–158.
DOI: 10.1007/s10811-007-9305-1

Abstract Malaysia is the world's leading producer of palm oil products that contribute US$ 7.5 billion in export revenues. Like any other agro-based industries, it generates waste that could be utilized as a source of organic nutrients for microalgae culture. Present investigation delves upon *Isochrysis* sp. culture in POME modified medium and its utilization as a supplement to *Nanochloropsis* sp. in rotifer cultures. The culture conditions were optimized using a 1 L photobioreactor (Temp: 23°C, illumination: 180~200 μmol photons $m^{-2}s^{-1}$, n=6) and scaled up to 10 L outdoor system (Temp: 26–29°C, illumination: 50~180 μmol photons $m^{-2}s^{-1}$, n=3). Algal growth rate in photobioreactor (μ= 0.0363 h^{-1}) was 55% higher compared to outdoor culture (μ=0.0163 h^{-1}), but biomass production was 1.3 times higher in outdoor culture (Outdoor=91.7 mg $m^{-2}d^{-1}$; Photobioreactor=69 mg $m^{-2}d^{-1}$). Outdoor culture produced 18% higher lipid; while total fatty acids (FA) was not significantly affected by the change in culture systems as both cultures yield almost similar concentrations of fatty acids per gram of sample (photobioreactor=119.17 mg g^{-1}; outdoor culture=104.50 mg g^{-1}); however, outdoor cultured *Isochrysis* sp. had 26% more polyunsaturated fatty acids (PUFAs). Rotifers cultured in *Isochrysis* sp./ *Nanochloropsis* sp. (1:1, v/v) mixture gave similar growth rate as 100% *Nanochoropsis* sp. culture (μ=0.40 d^{-1}), but had 45% higher counts of rotifers with eggs (t=7, maximum). The *Isochrysis* sp. culture successfully lowered the nitrate (46%) and orthophosphate (83%) during outdoor culture.

C. S. Vairappan (✉) · A. M. Yen
Laboratory of Natural Products Chemistry,
Institute for Tropical Biology and Conservation,
University Malaysia Sabah,
88999 Kota Kinabalu,
Sabah, Malaysia
e-mail: csvairappan@yahoo.com

Keywords Live feed · *Isochrysis* · Microalgae · *Nanochloropsis* · POME · Rotifer

Introduction

For the last ten years, Malaysia has been acknowledged as the world's leading producer and exporter of palm oil products. Based on the statistic obtained from the Malaysian Palm Oil Board, Malaysia controls about 45% of total palm oil production in the world. It has been reported that in 2005, there was a total of 423 mills with a production capacity of approximately 89 million tonnes of fresh fruit bunch (FFB) per year. The total export revenue in the year 2005 was about US$ 7.5 billion. Although the expansion of this industry contributes to the growth and development of the national agri-industry, but the by-product generated has the potential to deteriorate the aquatic environment.

The main by-product in crude palm oil production is liquid sludge waste known as palm oil mill effluent (POME). This waste is rich in organic matter and highly polluting; for every FFB processed, 0.75 tonnes of POME are generated (Department of Environmental (DEO) 1999). Hence, in 2005, 66.8 million tones of POME were generated. Current biological treatment technologies consist of conventional oxidation ponds (anaerobic and aerobic), open and closed tank digesters with biogas recovery and land application (Ma 1999; Kennedy and Hishamuddin 2001).

When managed at optimal levels, these treatments are able to meet the basic requirement of the Environmental Quality Act 1974 and the specific regulations of palm oil mill effluent 1977. However more often than not, seepage and overloading occurs causing accidental release into the aquatic environment as most of the palm oil mills in Malaysia are situated near water sources such as rivers and coastal areas (Kennedy and Hishamuddin 2001). Therefore, a tertiary treatment that is practical and beneficial should be devised to help solve this problem.

It is generally known that microalgae plays an important role in self-purification of natural waters, and therefore offer an alternative means as a tertiary treatment of organic wastewater. The initial idea to incorporate microalgae as an agent of bioremediation of wastewater was first proposed by Oswald and Gotaas in 1957 (Vílchez et al. 1997). In recent years, both freshwater and marine microalgae have been studied for their application in bioremediation. The ability of microalgae to assimilate organic waste to produce biomass helps reduce the biomass production cost. This is vital in industries such as aquaculture that use large amounts of microalgae biomass daily as live feeds (Paniagua-Michel et al. 1987; Phang and Ong 1988; Villasclaras et al. 1996; Jones 1998; Valenzuela-Espinoza et al. 1999; Phang et al. 2000; Kennedy and Hishamuddin 2001).

The aim of this investigation was to study the bioremediation potential and nutritional properties of the POME-cultured marine microalgae *Isochrysis* sp., and its utilization as a diet supplement for rotifer cultures. This study was conducted to support and promote the current practice of a few traditional coastal aquaculturists who are using POME fortified seawater to culture the microalgae used in live feed culture.

Materials and methods

Palm oil mill effluent (POME) was collected from a private mill in Kota Marudu, Sabah, Malaysia. This aerobic lagoon liquor was divided into three equal 25 L containers and digested aerobically for 4 weeks at the Laboratory of Natural Product Chemistry, University Malaysia Sabah (UMS) at 28°C under shaded conditions. Heavy aeration by an air stone was provided throughout the digestion period. The changes in chemical characteristics of the POME were determined in accordance with established analytical methods. Biochemical oxygen demand (BOD_3) (DOE 1999) was analyzed using the Winkler titration method (APHA Standard Methods 1995). Nitrate and orthophosphate were analyzed by using a Spectroquant NOVA 60 photometer (UK). Sampling was carried out in triplicate for each container of liquor (n=9).

This study was performed using *Isochrysis* sp. (UMS0501) obtained from the Malaysian Fisheries Department Kota Kinabalu, Sabah, Malaysia. These particular species are widely used as live feeds in shrimp and marine fish cultures. Main stocks were maintained in f/2 medium (Borowitzka 1988) while working stocks were cultured using POME modified media. Based on our ongoing research, it was noted that 5% aerobically digested POME with 0.075% of NPK inorganic fertilizer (QUICK, Associated Manufactures Traders (M) Sdn. Bhd; N-20%, P-10%, K-10%) in seawater with the salinity of 30 ppt and pH of 8.3–8.5 was suitable for the growth of *Isochrysis* sp. at 23°C±1°C and an irradiance of 180~200 μmol photons $m^{-2}s^{-1}$ (surface). This formulation was used in both the photobioreactor and outdoor systems. Optimum conditions were obtained and maintained in the photobioreactor throughout the semi-continuous culture while the outdoor culture was subjected to changes in temperature (26–29°C) and light intensity (50~180 μmol photons $m^{-2}s^{-1}$) due to the changes in weather. The photobioreactor cultures were repeated six times to observe repeatability of growth rates and fatty acid profile (n=6).

Initial cell density at start of the semi-continuous photobioreactor (1 L) culture was 1.0×10^8 cells L^{-1}. Biomass was harvested at the early stationary phase for every replicate. As for the outdoor culture, the inoculum was first cultured in the photobioreactor until it reaches late exponential phase and transferred into 9 L of POME modified media in a 30×30×30 cm glass aquarium. Initial cell densities inoculated were 4.0×10^9 cells L^{-1}. The outdoor cultures were harvested at the early stationary phase. The outdoor culture was carried in triplicate simultaneously to ensure an identical culture environment throughout the investigation. The mean growth rate average was calculated according to Renaud et al. (2002) while gross biomass production was based on the following equation (Phang and Ong 1988):

$$P_{GB} = (C/1000) \times (V/T) \times (1/A)$$

where, P_{GB} = gross biomass productivity, C = concentration of algae dry weight (chlorophyll $a\times67$) in mgL^{-1}, V = total volume of culture (L), T = retention time (d) and A = surface area (m^2).

Biomass from each replicate (from both culture systems) was harvested by centrifugation at 8000 rpm for 20 min. Samples for biochemical analysis were first washed twice with 0.5 M ammonium formate to remove salts, and freeze dried and stored at −80°C until analysis. The samples used for the rotifer bioassay were only resuspended in seawater after centrifuging and used fresh.

Bioremediation study Following biomass collection, the supernatant of each replicate was analyzed to determine the biochemical oxygen demand (BOD_3), and the nitrate and orthophosphate concentrations in the POME-modified media after culturing. Reduction rates were calculated based on the concentrations of the pollutants before and after culturing.

Biochemical analyses Chlorophyll *a* was determined by the method of Hansmann (1973) while total lipid was quantified gravimetrically following extraction with chloroform: methanol (2:1; v/v). The obtained lipid faction was then methylated using sodium methylate to transform the fatty acids to fatty acids methyl esters (FAME). The FAME was analyzed using a Shimadzu 2010 Gas Chromatograph (FID detector, auto injector, AOC-20i) with a capillary column BPX 70 (0.25 mm i.d×60 m). The injection and detector temperature was 260°C. Injection volume was 1 μL with the split ratio of 1:100 and the carrier gas used was Helium. The column temperature regime was 140°C for 4 min, followed by 4°C min^{-1} until 240°C (hold 30 min). The fatty acid profile was identified by comparing the FAME peaks to the retention time of the samples with that of a standard mixture (Spelco 37 FAME).

Bioassay: Rotifer trial feeding The feeding assay involved the feeding of the zooplankton *Brachionus plicatilis* Müller (collected from UMS hatchery) with the harvested biomass grown on the POME alternative media. Both the photobioreactor and outdoor culture systems produced equal quality *Isochrysis* (Table 3 & Fig. 2); however the outdoor culture system requires less management and produces more biomass. Therefore, the feeding trails were conducted using the biomass harvested from the outdoor system.

The rotifers were maintained using *Nanochloropsis* sp. before the feeding trials. Two sets of feeding trial were conducted. In the control experiment, the rotifers were fed with 100% *Nanochloropsis* sp. (collected from the UMS hatchery) (maintaining a final density of 1×10^6 cells mL^{-1} as advised by the hatchery staff) while the other set involves the feeding of rotifer with a ratio of 1:1 of *Isochrysis* sp. and *Nanochloropsis* sp. (1:1 based on biomass of the microalgae). All sets were fed three times a day and carried out in triplicate, rotifer density counting was done three times per replicate (n=9).

Initial rotifer density was 50 rotifers mL^{-1}, with the total solution volume of 200 mL (carried out in 400 mL beakers with added aeration). Daily, 1 mL of the culture was pipette from the centre of these beakers into a concave glass dish. Few drops of formalin (10%) were added to inhibit rotifer movements. The determination of rotifer density and eggs production was conducted under a stereo microscope at 40× magnification (Stemi 2000, Carl Zeis). Rotifer density counting was carried out three times per replicate and counted daily as rotifer mL^{-1}, while growth rate average was calculated as mentioned in Douillet (2000). Rotifer feeding experiments were carried out in a wet laboratory equipped with temperature (26°C) and light (L:D; 12:12) control facilities.

Statistical analysis

The differences between the properties of biomass cultured in both photobioreactor and outdoor culture system were tested using the independent Student t-test (P<0.05) using SPSS (version 11.0).

Results

Isochrysis sp. was successfully cultured in both the photobioreactor and the outdoor system using this modified medium. The indoor culture was conducted under optimum culture conditions in a 1 L photobioreactor and was repeated to ensure reproducibility (n=6). The growth rate of this species in the controlled environment was better (about 55% increase; $p<0.05$) than that of the upscaled 10 L outdoor system (Fig. 1). The outdoor gross biomass production was 91.7 mg m^{-2} day^{-1} while the photobioreactor culture only produced about 69 mg m^{-2} day^{-1}. Both systems reduced the pollutant levels of the secondary treated POME. Table 1 shows pollutant levels of the secondary treated POME before and after 4 weeks laboratory digestion, while Table 2 gives the detail reductions of each major pollutant following algae culture. The highest reductions were noted for nitrate and orthophosphate for both systems (nitrate: photobioreactor= 38.1%, outdoor=46.4%; orthophosphate: photobioreactor= 86.6%, outdoor=83.3%).

The outdoor system also produced significantly ($p<0.05$) higher percentage of lipid compared to the photobioreactor system (photobioreactor=44.5±3.4 %; outdoor= 52.8±2.4%). However, there were no significant differences ($p>0.05$) in total fatty acid concentrations in both systems (Photobioreactor=119.2±7.3 mg g^{-1}; Outdoor= 104.5±16.6 mg g^{-1}). By comparing the major fatty acids groups, it was observed that only the polyunsaturated fatty acids had significant differences ($p<0.05$) (Fig. 2). A detailed list of fatty acids profiles are shown in Table 3.

As the outdoor systems are more cost effective and produce a higher percentage of polyunsaturated fatty acids, the outdoor produced biomass was used for the bioassay. The rotifer feeding trials were successfully accomplished with a mixture of *Isochrysis* sp. and *Nanochloropsis* sp. (1:1, v/v) and 100% *Nanochloropsis* sp. (control). Both of these feeding regimes had similar specific growth rates

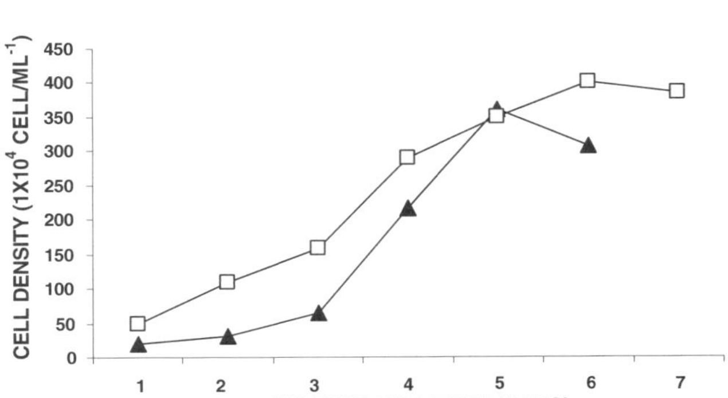

Fig. 1 Growth curves of *Isochrysis* sp. cultured in photobioreactor (▲, μ=0360 h^{-1}) and outdoors (□, μ=0.0163 h^{-1}) in POME modified media. Standard deviations were less than 5%, and are not shown

Table 1 Pollutant levels before and after aerobic digestion for 28 days at 28°C (shaded condition)

Pollutant (mg L^{-1})	Secondary treated POME	
	Before digestion	After digestion
Nitrate	16.167±0.288	90.333±0.153
Orthophosphate	15.8±0.361	11.267±0.288
BOD_3	135.567±8.422	77.3±4.309

Values are means±S.D. for nine replicates of samples

(mix=0.401 d^{-1}; *Nanochloropsis* sp.=0.400 d^{-1}). Detail daily growth rates are given in Table 4.

The total egg count for both feeding trials changes from day to day for the entire duration of the bioassay (Fig. 3). Overall, the feeding trial using the mixed algae had a higher number of eggs mL^{-1} daily. The highest difference was on day 7 where the mixed algae trial had 45% higher eggs ratio compared to the control.

Discussion

The differences in growth rates in the culture systems can be attributed to the differences in temperature and irradiance. Based on our research and others (Zhu et al. 1997; Renaud et al. 2002; Tzovenis et al. 2003a), growth of *Isochrysis*, like other algae, is dependent on temperature and irradiance. In this case, *Isochrysis* sp. exhibited optimum growth at 23°C with an irradiance of 180~200 μmol photons $m^{-2}sec^{-1}$, which was successfully maintained in the photobioreactor. However, in the outdoor system both temperature and light fluctuated due to weather and diurnal changes. The adaptation of the microalgae cells to these changes could delay growth leading to lower growth rates.

Even though the photobioreactor culture achieved a higher growth rate, the outdoor system reached a higher cell density and gross biomass. During outdoor culture, organic constituents from POME could have undergone further degradation, which could have contributed a slight increase in inorganic nutrients, resulting in an increase in microalgae cell density.

Table 2 Pollutant level before and after culturing of *Isochrysis* sp. in the photobioreactor and in outdoor culture

Pollutant (mg L^{-1})	Before microalgae culture	After microalgae culture	
		Photobioreactor	Outdoor
Nitrate	6.155±0.193	3.811±0.299	3.300±0.200
Ortho-phosphate	0.207±0.023	0.0278±0.005	0.0367±0.011
BOD_3	3.867±0.217	3.042±0.154	2.996±0.204

Values are means±S.D. for six replicates

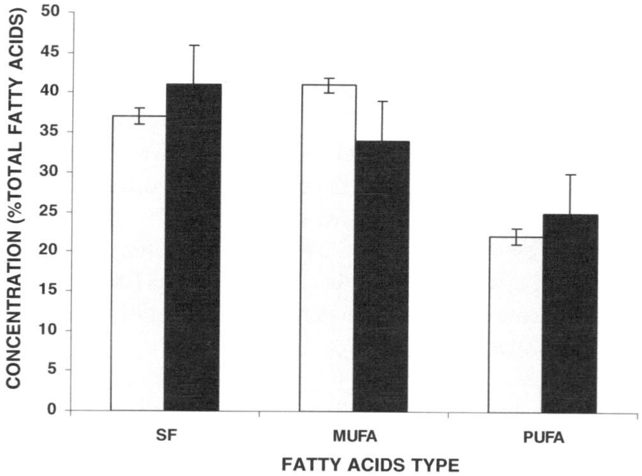

Fig. 2 Percentages of fatty acids groups for *Isochrysis* sp. culture in photobioreactor (□) and outdoor culture (■). (SF: Saturated fatty acids; MUFA: Monounsaturated fatty acids; PUFA: Polyunsaturated fatty acids)

Raw POME had high concentrations of nitrate, orthophosphate and BOD_3. These inorganic components are suitable to be used as nutrients for the culturing of microalgae. Due to the nature of the POME, being a freshwater waste and its dark colour, the suitable percentage to be used in formulation for the optimum growth of the marine *Isochrysis* was only 5%. The reduction in nitrate and orthophosphate are due to the utilization by the microalgae. At the end of the culture, the pollutants in both systems were successfully reduced and there were no significant differences between the amounts of pollutants reduced in these two types of culture system.

The differences in lipid percentage and fatty acid composition between the two systems could also be due to the different temperature and irradiance conditions (Renaud et al. 2002; Tzovenis et al. 2003b). Previous studies of *Isochrysis* (Volkman et al. 1989; Renaud et al. 1999, 2002; Tzovenis et al. 2003b) found that regardless of culture conditions, there were several major fatty acids that were always present in this genus (C14:0, C16:0, C18:1n9, C18:4n-3, C18:5n-3 and C22:6n-3). All of these fatty acids were detected in both culture systems except for C18:4n-3 and C18:5n-3 which could not be detected as they were not part of the 37 component FAME standard that was used as standard in this analysis. Therefore the ratio of n-3/n-6 cannot be calculated accurately.

One of the major factors in determination of suitable culture systems and conditions for the purpose of live feeds is the fatty acid profile (Fidalgo et al. 1998; Renaud et al. 1999). The important polyunsaturated fatty acids for marine organism are generally C22:6n-3, C20:5n-3 and C20:4n-6, which are important for the development and survival of

Table 3 Fatty acid profile of *Isochrysis* sp. biomass cultured in photobioreactor and outdoor culture

Fatty acid	Culture system	
	Photobioreactor	Outdoor
Saturated		
C10:0*	2.31±0.35	0.20±0.18
C14:0	12.80±0.82	12.86±0.48
C15:0*	0.22±0.04	0.68±0.12
C16:0	18.68±0.31	18.94±1.18
C17:0*	0.50±0.06	1.41±0.016
C18:0	1.83±0.43	1.03±0.33
C20:0	0.19±0.08	0.14±0.12
C21:0	0.35±0.14	0.47±0.41
C22:0	0.00	1.29±1.12
C24:0	2.46±0.26	2.46±0.44
C23:0	0.49±0.04	0.63±0.13
Monounsaturated		
C14:1	1.51±0.45	1.42±0.38
C16:1*	3.47±0.31	5.51±0.47
C17:1	0.26±0.22	0.49±0.00
C18:1n9t*	0.00	0.66±0.24
C18:1n9c*	31.59±0.46	23.47±0.66
C20:1	1.18±0.24	0.62±0.54
C24:1	1.94±0.03	1.68±0.21
Polyunsaturated		
C18:2n6t*	3.38±0.85	5.33±0.72
C18:2n6c*	4.55±0.38	3.25±0.39
C18:3n6*	1.96±0.26	3.07±0.23
C18:3n3*	4.57±0.70	7.23±0.59
C20:2*	0.15±0.01	0.00
C20:3n6*	0.48±0.02	0.65±0.00
C20:3n3	0.24±0.02	0.25±0.02
C20:4n6	0.22±0.03	0.25±0.02
C20:5n3*	0.31±0.05	0.00
C22:2	0.24±0.02	1.10±0.72
C22:6n3	5.94±1.34	5.62±0.18

Values are means±S.D. for three replicates. Data presented as percentage of total fatty acids

* ~ significant differences ($p < 0.05$)

larvae, while short chain saturated fatty acids (C14:0 and C16:0) are reported to supply high energy for rapid growing bivalves (Brown et al. 1997; Renaud et al. 1999). *Isochrysis* is also known to have a wide range in the percentage of PUFA as part of the total fatty acids such as C20:5n-3 (traces-27%), C22:6n-3 (traces-19%) and C20:4n-6 (traces-1%). This mainly depends on the strain used and culture conditions (Volkman et al. 1989; Fidalgo et al. 1998; Renaud et al. 1999, 2002; Tzovenis et al. 2003b). Based on the fatty acid profile from both culture systems in the present study, the *Isochrysis* biomass produced using POME-modified medium is suitable as a quality live feed for aquaculture.

Other aspects that influence the suitability of a culture system for mass production are the cost of management and

Table 4 Daily growth rates of the rotifers fed with *Nanochloropsis* sp. (Control) and a 1:1 mixture of *Nanochloropsis* sp. and *Isochrysis* sp

Duration of culture (day)	Growth rate (d^{-1})	
	Control	Mixture
1	−0.020±0.029	0.192±0.163
2	−0.002±0.116	−0.006±0.109
3	0.315±0.108	0.195±0.276
4	−0.660±0.182	0.017±0.024
5	1.131±0.046	0.609±0.095
6	1.831±0.196	1.644±0.026
7	0.413±0.021	0.557±0.117
8	0.194±0.012	0.186±0.084
9	−0.317±0.153	−0.225±0.005

Values are means±S.D. for nine replicates

quality of biomass required. While biomass harvested from photobioreactor systems is of high quality, the production costs are often high due to the required sterilization and artificial light source. These systems are also sensitive, require high maintenance and are difficult to up-scale. Due to the high cost, most operations that require high quality axenic cells for the purpose of secondary metabolites and fatty acids isolation are cultured using closed photobioreactors (Olaizola 2003; Borowitzka 1999) while outdoor systems are more feasible and cost effective for the production of live feeds where large quantities of non-axenic biomass are needed daily. The 10 L culture tank used for this research can be considered as an open culture system as it was conducted outdoors in an uncontrolled environment (Kennedy and Hishamuddin 2001).

The ability of *Isochrysis* to grow well in tropical outdoor cultures provides a cheaper alternative to produce live feeds. While both systems produce *Isochrysis* biomass that is suitable as a live feed, the outdoor system is more manageable and cost effective. Moreover, the outdoor

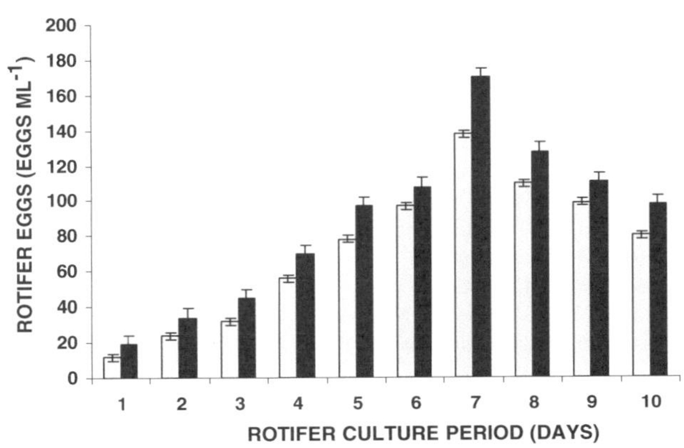

Fig. 3 Total number of eggs mL^{-1} produced daily during feeding trials; (1) *Nanochloropsis* sp. (□), and (2) 1:1 mixture of *Nanochloropsis* sp. and *Isochrysis* sp. (■)

system also produces a higher cell content of polyunsaturated fatty acids that are important for the development of marine organisms.

The rotifers that were fed with 100% *Nanochloropsis* sp., grew more irregularly and produced less eggs per mL compared to the cultures that were fed with the mixed *Isochrysis* and *Nanochloropsis* diet. This showed that *Isochrysis* grown using the alternative POME-modified medium is suitable to be used as a supplement to further enrich and improve the rotifer cultures.

In summary, data obtained from this study describes the practical application of aerobically digested palm oil mill effluent in improving the quality and quantity of live feed in aquaculture efforts in East Malaysia. Local small traditional aquaculturists have started using POME-fortified medium since only relatively little POME (5%) is required to improve their microalgae production significantly. The present investigation is the beginning of our attempts to understand and increase the amount of aerobically digested POME in microalgae culture media as we strive to educate the traditional aquaculturists to use a new nutrient source to produce better quantity and quality of live feed.

Acknowledgement The authors would like to thank Kota Marudu Palm Oil Mill and Ms Chin Ne Ne for the effluents. AMY is a post graduate registered at Borneo Marine Research Institute, Universiti Malaysia Sabah, and would like to acknowledge Prof. Dr. Maryati Mohamed and Nagao for scholarship during her post graduate studentship. The authors would also like to thank the Director of Institute for Tropical Biology and Conservation for support and facilities made available during the course of this investigation.

References

American Public Health Association (APHA) (1995) Standard methods for the examination of water and wastewater, 19th edn. American Public Health Association, Washington D.C.

Borowitzka MA (1988) Algal growth media and sources of algal culture. In: Borowitzka MA, Borowitzka LJ (ed) Micro-algal Biotechnology. Cambridge University Press, Cambridge, pp 456–465

Borowitzka MA (1999) Commercial production of microalgae: ponds, tanks, tubes and fermenters. J Biotechnol 70:313–321

Brown MR, Jeffrey SW, Volkman JK, Dunstan GA (1997) Nutritional properties of microalgae for mariculture. Aquaculture 151:315–331

Department of Environment (DOE) (1999) Industrial processes & the environment: Crude palm oil industry. Handbook No. 3. Aslita Sdn. Bhd., Kuala Lumpur, pp 5–22

Douillet PA (2000) Bacterial additives that consistently enhance rotifer growth under synxenic culture conditions. 1. Evaluation of commercial products and pure isolates. Aquaculture 182:249–260

Fidalgo JP, Cid A, Torees E, Sukenik A, Herrero C (1998) Effects of nitrogen source and growth phase on proximate biochemical composition, lipid classes and fatty acid profile of the marine microalga *Isochrysis galbana*. Aquaculture 166:105–116

Hansmann E (1973) Pigment analysis. In: Stein JR (ed) Handbook of Phycological Methods. Cambridge University Press, Cambridge, pp 359–368

Jones WR (1998) Practical applications of marine bioremediation. Cur Opin Biotech 9:300–304

Kennedy AG, Hishamuddin O (2001) Bioremediation of treated and raw POME (Palm Oil Mill Effluent) using *Spirulina plantensis*. In: Japar Sidik B, Arshad A, Tan SG, Dacud SK, Jambari HA, Sugiyama S (ed) Aquatic resources and environment studies of the straits of Malacca: current research and reviews. Kuala Lumpur, pp 203–210

Ma AN (1999) Treatment of palm oil mill effluent. In: Singh G, Huan LK, Teo L, Lee DK (ed) Oil palm and the environment. a Malaysian perspective. Malaysian Oil Palm Grower' Council, Malaysia, pp 113–123

Olaizola M (2003) Commercial development of microalgal biotechnology: from the test tube to the marketplace. Biomol Eng 20:459–466

Paniagua-Michel J, Farfan BC, Bückle-Ramirez F (1987) Culture of marine microalgae with natural biodigested resources. Aquaculture 64:249–256

Phang SM, Ong KC (1988) Algal biomass production in digested palm oil mill effluent. Biol Waste 25:177–191

Phang SM, Miah MS, Yeoh BG, Hisham MA (2000) *Spirulina* cultivation in digested sago starch factory wastewater. J Appl Phycol 12:395–400

Renaud SM, Thinh L, Parry DL (1999) The gross chemical composition and fatty acid composition of 18 species of tropical Australian microalgae for possible use in mariculture. Aquaculture 170:147–159

Renaud SM, Thinh L, Lambrinidis G, Parry DL (2002) Effect of temperature on growth, chemical composition and fatty acid composition of tropical Australian microalgae grown in batch cultures. Aquaculture 211:195–214

Tzovenis I, Pauw ND, Sorgeloss P (2003a) Optimisation of T-ISO biomass production rich in essential fatty acids, I. Effect of different light regimes on growth and biomass production. Aquaculture 216:203–222

Tzovenis I, Pauw ND, Sorgeloss P (2003b) Optimisation of T-ISO biomass production rich in essential fatty acids, II. Effect of different light regimes on production of fatty acids. Aquaculture 216:223–242

Valenzuela-Espinoza E, Millán-Núñez R, Núñez-Cebrero F (1999) Biomass production and nutrient uptake by *Isochrysis* aff. *galbana* (Clone T-ISO) cultured with a low cost alternative to the f/2 medium. Aquacult Eng 20:135–147

Vílchez C, Garbayo I, Lobato MV, Vega JM (1997) Microalgae-mediated chemicals production and waste removal. Enzyme Microb Technol 20:562–572

Villasclaras SS, Sancho MEM, Caballero MTE, Pérez AD (1996) Production of microalgae from olive mill wastewater. Int Biodet Biodeg 245–247

Volkman JK, Jeffrey SW, Nichols PD, Rogers GI, Garland CD (1989) Fatty acids and lipid composition of 10 species of microalgae used in mariculture. J Exp Mar Biol Ecol 128:219–240

Zhu CJ, Lee YK, Chao TM (1997) Effects of temperature and growth phase on lipid and biochemical composition of *Isochrysis galbana* TK1. J Appl Phycol 9:451–457

Seaweed micropropagation techniques and their potentials: an overview

C. R. K. Reddy · Bhavanath Jha · Yuji Fujita · Masao Ohno

Originally published in the Journal of Applied Phycology, Vol 20, No 5, 159–167.
DOI: 10.1007/s10811-007-9205-4

Abstract The seaweed industry worldwide uses 7.5–8.0 million tonnes of wet seaweeds annually with a majority of it derived from cultivated farms, as the demand for seaweed based-products exceeds the supply of seaweed raw material from natural stocks. The main advantage of cultivation is that it not only obviates overexploitation of natural populations but also facilitates the selection of germplasm with desired traits. To enhance the economic prospects of seaweed cultivation, varied practices, such as simple and cost effective cultivation methods, use of select germplasm as seed stock coupled with good farm management practices, etc., are adopted. Nevertheless, in vitro cell culture techniques have also been employed as they facilitate development and propagation of genotypes of commercial importance. There are more than 85 species of seaweeds for which tissue culture aspects have been reported. Although the initial aim of these techniques focuses mostly on genetic improvement and clonal propagation of seaweeds for mariculture, recently the scope of these techniques has been extended for use in bioprocess technology for production of high value chemicals of immense importance in the pharmaceutical and nutraceutical sectors. Recently, there has been a phenomenal interest in intensifying seaweed tissue and cell culture research to maximize the add-on value of seaweed resources. This paper deals with the status of seaweed micropropagation techniques and their applications in the context of the marine biotech industry. Further, it also provides an analysis of the problems to be resolved for removing the barriers that are impeding the true realization of potentials offered by these techniques for sustainable development and utilization of seaweed resources.

Keywords Callus · Clonal propagation · Plant growth regulators · Polyamines · Somaclones · Tissue culture

C. R. K. Reddy (✉) · B. Jha
Discipline of Marine Biotechnology and Ecology,
Central Salt and Marine Chemicals Research Institute,
Bhavnagar 364 002, India
e-mail: crk@csmcri.org

Y. Fujita
Faculty of Fisheries, Nagasaki University,
1-14, Bunkyo-machi,
Nagasaki 852-8521, Japan

M. Ohno
Usa Marine Biological Institute, Kochi University,
Usa-Cho, Tosa,
Kochi 781-1164, Japan

Introduction

The development of techniques for the culture of isolated plant organs, tissues and cells have led to several exciting opportunities in the area of plant biotechnology, and allowed widespread use of cell cultures for *in vitro* genetic manipulation, plant propagation and production of commercially useful products (Cocking 1990). Following the success achieved by application of these techniques in higher plants, tissue culture of seaweeds was initiated as early as 1978 (Chen and Taylor 1978) with considerable interest and optimism to further enhance the economic prospects of seaweed resources as a whole. Since then, concerted efforts were made to develop basic background technologies for consistent production and regeneration of calluses from diverse group of seaweeds (see reviews by Polne-Fuller 1988; Butler and Evans 1990; Garcia-Reina et al. 1991; Aguirre-Lipperheide et al. 1995; Rajakrishna

Kumar 2002). Several protocols for routine callus induction and regeneration are now available in the literature for a wide range of seaweeds (Polne-Fuller and Gibor 1987; Huang and Fujita 1997; Reddy et al. 2003; Rajakrishna Kumar 2002; Rajakrishna Kumar et al. 2004, 2007). Recently, the scope of these techniques has been extended for use in bioprocess technology for production of high value chemicals of immense commercial value in the pharmaceutical and nutraceutical sectors (Rorrer 2000; Rorrer and Cheney 2004; Munoz et al. 2006). Therefore, *in vitro* cell culture technology facilitates development of new generation technologies in the area of seaweed cultivation and utilization, that collectively help intensive cultivation of higher yielding strains, coupled with green processing technologies with controlled production of products of potential commercial applications at competitive rates.

The earlier articles have dealt at great length about the status, applications, potentials and needs in tissue culture of seaweeds (see reviews of Butler and Evans 1990; Garcia-Reina et al. 1991; Aguirre-Lipperheide et al. 1995). Therefore, a review of basic aspects of seaweed tissue culture is beyond the scope of this article. Instead, this article provides a brief review of the current status of the micropropagation techniques and their potential applications in overall seaweed biotechnology, including the cultivation, secondary metabolite production and genetic improvement. Further, it also provides an analysis of the problems to be resolved for removing the barriers that are impeding the true realization of potentials offered by these techniques for sustainable development and utilization of seaweed resources.

Clonal propagation and selection of strains with superior traits

The earlier studies have used clonal propagation as the commonest and simplest approach to select superior strains from wild populations to improve the performance of cultivated crops (Santelices 1992). There are several studies which have exploited the organogenetic potential of seaweeds for isolating superior clones from several economically important seaweeds like *Chondrus* (Cheney et al. 1981), *Gigartina* (Sylvester and Waaland 1983), *Gracilaria* (Patwary and van der Meer 1982, 1983) and *Kappaphycus* (Doty and Alvarez 1973). Most of the seaweed selection studies have been approached from an empirical perspective for a single trait and a few have been concerned with superior growth potential combined with either sterility or with high polysaccharide yields and gel strength. Although this approach is generally regarded as successful (van der Meer and Patwary 1983: van der Meer 1986), the findings suggest the need for continuous monitoring and isolation of clones with superior quality for a given trait, and finally the maintenance of the selected strain, eliminating the variants arising spontaneously from the selected clones (Santelices 1992).

Most recently, Titlyanov et al. (2006a, b) while working on *Gelidium* described methods for mass generation of planting material for tank-bubbling and field cultivation using fragments and cell aggregates of apical meristem. Further, freeze-thawing of apical meristem tissues enabled the production of plantlets producing rhizoids that could be used for cultivation in the sea. Both these methods could effectively maximize the number of propagules per donor plant and further facilitate mass production of seed stock required for mariculture round the year. A similar approach has also been successfully demonstrated for producing mass plantlets and tetraspores from fragments and cell aggregates of meristematic and submeristematic tissue of *Palmaria palmata* (Titlyanov et al. 2006a, b).

Simple vegetative propagation of thallus segments (2–3 cm long) in laboratory culture is also utilized to ascertain various culture conditions and media required for successful tissue and protoplast culture in seaweeds. This practice is very useful, especially when one deals with species which are non-responsive to tissue and protoplast culture. The culture media and conditions generally employed for tissue and cell cultures are similar to those optimized for growing intact plants.

Nevertheless, the selection of elite germplasm through clonal propagation is a continuous process and a substantial amount of harvest is utilized as seed material for subsequent cultivation. Further, isolation of useful mutants for cultivation through this process is very cumbersome and less likely to be successful as genetic variant cells in thallus are small in number and are veiled by more common non-variant cells (Garcia-Reina et al. 1991). This has eventually led to the exploration of the possibilities of developing in vitro cell culture technology for this group of plants.

Seaweed tissue and callus culture

The seaweed tissue culture is, relatively, a developing area of research with just three decades of development and lags far behind that of higher plants, which got started more than a century ago. The revolution made in plant cell and tissue culture studies has formed the foundation for the genetic engineering of crop plants (Cocking 1990). Micropropagation using tissue culture methods are currently used for the large scale propagation of clones with superior traits in a number of crop species. Furthermore, cell culture offers several advantages for the isolation of mutants over conventional selection procedures using whole plant material. Cell suspensions allow a very large number of cells to be screened simultaneously for a desired trait in a reasonable time frame and reproducible manner. In addition, regeneration of plants from callus can result in the recovery of new genetic variants as a consequence of the well known phenomenon of somaclonal variation (Evans and Sharp 1983). Unlike higher plants, seaweeds show distinct alternation of generations (haploid and diploid) that can be effectively utilized for genetic improvement. Haploid tissues enable easy detection of mutants, while subsequent chromosome doubling produces fertile individuals homozygous at all loci, and these provide pure breeding lines for selection and hybridization.

The methods employed in seaweed tissue culture are essentially the same as described for higher plant tissue culture. It generally involves preparation of axenic explants and their culture on solid agar medium enriched with a range of macro- and micro-nutrients, vitamins, sugars (as carbon source), and plant growth regulators (auxins and cytokinins). The initial studies carried out from 1980 to 1990 have mostly dealt with fundamental aspects of seaweed tissue culture (Aguirre-Lipperheide et al. 1995). These studies have largely concentrated on the development of basic techniques for routine tissue culture of different seaweeds, with a few attempting to enhance the callus induction rate and growth by using different plant growth regulators and carbon sources in explant culture medium (Lawlor et al. 1989; Kaczyna and Megnet 1993; Yokoya and Handro 1996, 2002; Yokoya et al. 2004; Rajakrishna Kumar et al. 2004). A list of seaweed species in which plants are reported to have been regenerated from tissue or callus culture is provided in Table 1. The benefits that were expected as a result of application of these techniques to

Table 1 Seaweed species from which tissue culture have been accomplished

Species	Status of success	Reference
Chlorophyta		
Enteromorpha sp.	–	Polne-Fuller and Gibor 1986
Enteromorpha intestinalis	CI	Polne-Fuller and Gibor 1987
Ulva angusta	CI	Polne-Fuller and Gibor 1987
Ulva sp.	–	Polne-Fuller and Gibor 1986
Rhodophyta		
Agardhiella subulata	PR	Cheney et al. 1987; Bradley and Cheney 1990; Huang et al. 1998
Ahnfeltiopsis flabelliformis	PR	Huang and Fujita 1997
Ceramium kondoi	CI	Gusev et al. 1987
Chondrus crispus	CI	Chen and Taylor 1978; Chen 1982
Carpopeltis affinis	CI & PR	Huang and Fujita 1997
C. prolifera	CI & PR	Huang and Fujita 1997
Chondrcanthus tenellus	CI & PR	Huang and Fujita 1997
Eucheuma alvarezii	PR	Polne-Fuller and Gibor 1987; Dawes and Koch 1991; Dawes et al. 1993
	SE	Reddy et al. 2003
	CI	Munoz et al. 2006
E. denticulatum	PR	Dawes and Koch 1991; Dawes et al. 1993
E. uncinatum	PR	Polne-Fuller and Gibor 1987
Furcellaria fastigata	CI	Gusev et al. 1987
Gelidiella acerosa	PR	Rajakrishna Kumar et al. 2004
G. robustum	PR	Polne-Fuller and Gibor 1987
Gelidium sp.	PR	Titlyanov and Titlyanova 2006
G. vagum	CI	Gusev et al. 1987
Gigartina exasperata	PR	Polne-Fuller and Gibor 1987
Gloiopeltis tenax	CI	Huang and Fujita 1997
Gracillaria acuminata	CI	Huang and Fujita 1997
G. chilensis	CI	Collantes et al. 2004
G. corticata	CI & PR	Rajakrishna Kumar et al. 2007
G. papenfussii	PR	Polne-Fuller and Gibor 1987
G. perplexa	CI	Yokoya et al. 2004
G. tenuistipitata	CI	Yokoya et al. 2004
G. tenuifrons	PR	Yokoya 2000
G. textori	CI	Huang and Fujita 1997
G. verrucosa	CI & PR	Gusev et al. 1987; Kaczyna and Megnet 1993
Grateloupia doryphora	BF	Robaina et al. 1990
G. dichotoma	CI	Yokoya and Handro 1996
G. filiformis	CI & PR	Yokoya et al. 1993
G. filicina	CI	Huang and Fujita 1997

Table 1 (continued)

Species	Status of success	Reference
G. turutura	CI	Huang and Fujita 1997
G. imbricata	CI	Huang and Fujita 1997
Hypnea musciformis	CI	Rajakrishna Kumar et al. 2007
Laurencia sp.	PR	Garcia-Reina et al. 1988
	SD	Robaina et al. 1992
	CI	Huang and Fujita 1997
Meritotheca papulosa	CI	Huang and Fujita 1997
Ochtodes secundriramea	PR	Malaikal et al. 2001; Rorrer and Cheney 2004
Phyllophora nervosa	CI & PR	Gusev et al. 1987
Pionitis crispata	CI	Huang and Fujita 1997
Porphyra perforata	PR	Polne-Fuller and Gibor 1987
P. lenceolata	PR	Polne-Fuller and Gibor 1987
P. nereocystis	PR	Polne-Fuller and Gibor 1987
P. umbilicalis	CI	Liu and Kloareg 1992
P. yezoensis	PR	Yamazaki et al. 1998; Hafting 1999
Pterocladia capillacea	CI	Liu and Gordon 1987
	CI & BS	Liu et al. 1990
Ptilophora subcostata	CI	Huang and Fujita 1997
Smithora naiadum	PR	Polne-Fuller and Gibor 1987
Soleria filiformis	CI	Robeldo and Garcia-Reina 1993
Phaeophyta		
Cystoseira expensa	SD	Lawlor et al. 1990
C. osmundacea	CI	Polne-Fuller and Gibor 1987
C. retorta	SD	Lawlor et al. 1990
C. retroflexa	SD	Lawlor et al. 1990
C. siliquosa	SD	Lawlor et al. 1990
Dictyosiphon foeniculaceus	CI	Saga et al. 1982
Ecklonia cava		Notoya and Aruga 1989, 1992a; Kawashima and Tokuda 1990
E. radiata	PR	Notoya 1988; Lawlor et al. 1989
E. stolonifera	PR	Notoya 1988
Eisenia bicyclis	PR	Notoya and Aruga 1990
Egregia menziesis	CI	Polne-Fuller and Gibor 1987
Fucus sp.	PR	Fries 1985
Laminaria angustata	PR	Saga et al. 1978; Saga and Sakai 1983
L. digitata	PR	Fries 1980; Asensi et al. 2001
L. hyperborea	PR	Fries 1980
L. japonica	PR	Yan 1984
L. saccharina	PR	Lee 1985
L. setchellii	CI	Qi et al. 1995
L. sinclairii	CI	Polne-Fuller and Gibor 1987
Macrocystis pyrifera	CI	Polne-Fuller and Gibor 1987
Prionitis lanceolata	CI	Polne-Fuller and Gibor 1987
Pelvetia fastigiat	CI	Polne-Fuller and Gibor 1987
Sargassum confusm	PR	Kirihara et al. 1997
S. flutians	PR	Polne-Fuller and Gibor 1987
S. heterophyllum	PR	Mooney and Staden 1984
S. muticum	PR	Polne-Fuller and Gibor 1987
S. polycystum	PR	Rao and Rao 1996
Sargassum sp.	AE	Yoshida et al. 1999
S. tenerrimum	CI	Rajakrishna Kumar et al. 2007
Stylopodium sp.	CI	Polne-Fuller and Gibor 1987
Turbinaria conoides	CI	Rajakrishna Kumar et al. 2007
Undaria pinnatifida	PR	Zhang 1982; Yan 1984; Kawashima and Tokuda 1993; Notoya and Aruga 1992b
U. undarioides	PR	Kimura and Notoya 1997

BF, bud formation; CI, callus induction; PR, plant regeneration; SE, somatic embryogenesis; AE, adventitve embryogenesis; BS, biochemical study; SD, shoot development seaweeds are minuscule and incomparable with that of the higher plants. This is presumably due to many factors. Seaweed calli, compared to those of higher plants, are generally slow growing and small in size. Furthermore, occurrence of calli in some seaweeds is sporadic, and the percentages obtained are often very low (e.g. 4% in *Gracilaria verrucosa*: Gusev et al. 1987; 1% in *Sargassum muticum*: Polne-Fuller and Gibor 1986; 15% in *Gelidium vagum*: Gusev et al. 1987). In addition, the role of plant growth regulators and carbon sources on callus formation in seaweed explants is variable, and hence the occurrence of callus is suggested to be due to internal factors inherent to the explant rather than to the culture conditions employed (Aguirre-Lipperheide et al. 1995).

Plant growth regulators (PGRs) and callus induction

The inconsistency in responses to PGRs by seaweeds could be mostly due to lack of understanding of the physiological role of these substances in their growth and differentiation. Tarakhovskaya et al. (2007) reviewed phytohormones in algae in general and reported the occurrence of essentially all known phytohormones in concentrations comparable with that of higher plants, indicating their function in various complex metabolisms. Recently, Stirk et al. (2005) identified 19 endogenous cytokinins, including 1 new aromatic cytokinin (benzyladenine and toplin derivatives), in 31 seaweed species. This work reveals some interesting differences in the cytokinin composition in seaweeds compared to higher plants. These differences included an abundance of iP (isopentenyladenine) and *cZ* (*cis*-zeatin) conjugates in the seaweed samples and no DHZ (dihydrozeatin). The latter group of cytokinins is more commonly occurring in higher plants than in the former groups (Stirk et al. 2005).

The role of plant growth regulators on callus formation in seaweed explants is debatable and showed no definite trend. Nevertheless, indole-3-acetic acid (IAA), 2,4-dichlorophenoxyacetic acid (2,4-D) and kinetin had stimulatory role in callus formation, growth and regeneration both in intercalary and apical explants of *Gracilaria tenuistipitata*, *G. perplexa* and *Grateloupia dichotoma* (Yokoya and Handro 1996; Yokoya et al. 2004). Kaczyna and Megnet (1993) also succeeded in induction of callus in *Gracilaria verrucosa* using a combination of auxins and cytokinins. The effect of PGRs on callus formation also varied with the seaweed and photon irradiance used during explant culture. The brown color morph of *Hypnea musciformis* had the highest rate of callus formation in high photon irradiance with low concentration of IAA, while green color morph produced calli with 2,4-D irrespective of concentration studied and photon flux densities used. A water soluble extract from *Laurencia* sp. increased callus formation on explants of the same species (Robaina et al. 1992). However, the type of substances involved in these responses and their mode of action are unknown.

In the light of these conflicting findings, it would be an interesting aspect to investigate the use of some PGR conjugates, such as iP and cZ, which have recently been detected in several seaweeds for their abilities to induce callus formation and multiplication, rather than extensively examining the higher plant phytohormones.

Production of micropropagules from callus culture as a means of seed stock

A number of studies have shown regeneration of micropropagules directly from callus and sometimes from explants of some red seaweeds (Dawes and Koch 1991; Reddy et al. 2003; Rorrer and Cheney 2004; Rajakrishna Kumar et al. 2004, 2007). Dawes and Koch (1991), Reddy et al. (2003) and Rajakrishna Kumar et al. (2004) have used this phenomenon as a means of maintenance and clonal propagation of seed stock for mariculture of economically important seaweeds. Rorrer and Cheney (2004) employed micropropagules in bioreactor cultivation for production of low volume high value products from a variety of seaweeds. Considering their utility in wide range of applications, some studies attempted to enhance the quantity of micropropagules production by developing innovative in vitro culture methods. Subculture of thin slices of pigmented callus as embedded culture in 0.4% solidified Provasoli enriched seawater (PES) medium enabled mass production of micropropagules from *Kappaphycus alvarezii* which is commercially cultivated world over for κ-carrageenan.

Development of somaclones from callus culture

Successful regeneration of fresh plants from several seaweed species has been reported with a few dealing with inadvertent selection resulting from somaclonal variation. Yan (1984) obtained tissue culture progeny from calli of *Laminaria* and *Undaria* with rapid growth and high temperature tolerance for longer periods than the normal sporophytes. The *in vitro* culture of *Laminaria digitata* stipe medullary explants produced cell-filament suspension cultures which gave rise to normal sporophytic plants along with other forms. The genetic identity of these plants was confirmed with cell-filament suspension using nine polymorphic micro-satellite markers, although flow cytometric analysis of nuclei of cell-filament and regenerated plants displayed 2C and 4C level (Asensi et al. 2001). Garcia-Reina et al. (1988) described two types of callus of *Laurencia* obtained from the same plant with identical appearance in terms of pigmentation and texture. These differed markedly in regeneration potential and growth rate of the regenerated plants. The tissue culture progeny obtained from pigmented callus of *Kappaphycus* also showed enhanced growth rate as high as 1.5–1.8 times the rate of farmed plants propagated through vegetative means (Reddy et al. 2003). The ploidy and RAPD analysis of tissue culture progeny did not show any evidence for genetic variability (Reddy, unpublished data). The tissue culture of *Gelidiella acerosa*, *Gracilaria corticata* and *Turbinaria conoides* also produced calli which showed regeneration patterns quite similar to that of *Kappaphycus*, but the amount of propagules produced is comparatively less (Rajakrishna Kumar et al. 2004, 2007).

Polyamines as growth promoters in seaweeds

Like hormones, another group of PGRs, namely polyamines (PAs), are implicated in many biological processes in higher plants, including cell division, root and floral initiation, fruit development, senescence and abiotic stress responses. PAs, spermidine, spermins and their diamine obligate precursor putrescine belong to a class of aliphatic amines that are commonly present in all living organisms and are labeled as a new class of growth substances or hormonal messengers. PAs are present in plants in amounts varying from micromolar to millimolar and thus their levels are significantly higher than those of plant hormones (Kakkar and Sawhney 2002). Recently, occurrence of PAs has been reported in several intertidal marine macroalgae subjected to lethal hypo-saline stress (Lee 1998). Generally, putrescine and spermidine are the most abundant while spermine only occurs in trace levels. Changes in environmental conditions influence the biosynthesis and accumulation of PAs in algae. This is an important consideration as many marine macroalgae grow in intertidal habitats where there are many environmental fluctuations that they need to tolerate in order to grow. For example, at a low salinity, six green macroalgae accumulated higher levels of putrescine (Lee 1998). As in higher plants, PAs play an important role in cell growth and development of seaweeds, with an increase in endogenous concentrations just prior to cell division (Lee 1998). These few examples suggest that PGRs (both hormones and polyamines) can potentially play a similar role in seaweed growth and development as in higher plants.

Cell suspension cultures from seaweed callus

Callus development in multicellular marine macroalgae has been related to the anatomical structure and cellular organization of the thallus (Aguirre-Lipperheide et al. 1995). The earlier studies have often used the term "callus-like formations" to distinguish the callus of macroalgae (Garcia-Reina et al. 1991; Yokoya et al. 1993) from that of higher plants where the callus (disorganized growth of cells) develops from differentiated tissues as a result of wounding (Yeoman 1987). Despite the differences in cellular organization among the red (pseudo-parenchymatous type tissue) and brown (parenchymatous type tissue) seaweeds, the callus obtained from both groups showed more or less similar morphology, with uniseriate, pigmented and branched filamentous outgrowths from both cortical and medullary tissues of explants. The callus in higher plants is friable type and forms numerous cell aggregates when transferred to agitated liquid cultures. In contrast to this, the filamentous callus observed in seaweeds is rigid (non-friable) and regenerates rapidly into full plants when transferred to agitated liquid cultures and thereby limiting the scope of applications offered by cell suspension cultures. However, the subcultured callus tufts excised from *Turbinaria conoides* explants produced tiny somatic embryo-like colonies on pigmented filamentous cells with occasional friable callus clumps when cultured in soft agar (0.4% agar). This finding perhaps highlights the need of innovative culture techniques for obtaining all categories of in vitro cell technology for seaweeds (Rajakrishna Kumar et al. 2007)

Bioprocess engineering of cell and tissue culture of seaweeds

Bioprocess engineering of seaweeds is one of the most recent and exciting developments of seaweed biotechnology. This development has opened up new opportunities to produce and recover seaweed products directly from cells and cell aggregates in photo-bioreactors (Huang and Rorrer 2002; Rorrer and Cheney 2004; Munoz et al. 2006) which dispense with the use of whole plants. The main advantage of photo-bioreactor cultivation of cell and tissue cultures of macroalgae is the enablement of continuous, steady and defined production of high yields of quality product, thereby circumventing the barriers of seasonality. The optimization of culture conditions for each cell system can also be undertaken with ease and precision for deriving maximum benefits from photo-bioreactor grown cell cultures (Rorrer and Cheney 2004). Further, the downstream process used for recovery of products from cell culture is more environmentally friendly as compared to the conventional process that utilizes the whole plants as a source of raw material for extraction. The bioprocess technology consists of three main components: cell and tissue culture development, photo-bioreactor design, and identification of strategies for eliciting secondary metabolite biosynthesis. The former two components have been investigated using selective species known for biosynthesis of novel oxylipins and halogenated terpenoids (Rorrer and Cheney 2004). Rorrer and Cheney (2004) described a variety of methods to develop phototropic suspension culture suitable for *in vitro* cultivation in photo-bioreactor systems and assessed the factors limiting their process cultivation performance using different photo-bioreactor configurations. The photo-bioreactor cultivation of cell and tissue culture will no doubt form a potential technology platform for the controlled production of low volume high value products from macroalgal resources, including pharmaceuticals and nutraceuticals. Consequently, the macroalgal culture development providing a controlled growth environment suitable for secondary metabolite biosynthesis may outweigh the need for optimization of growth rate or minimize the inputs required for biomass production, thereby enhancing the efficiency of performance of photo-bioreactor cultivation.

Conclusion

Development of in vitro cell culture technology is of fundamental importance if seaweed biotechnology is to play a central role in the growth of global seaweed industry in future. Considering the present status of progress made in tissue culture of seaweeds, it would be realistic to believe that these techniques can help to the extent of generating genetic mutants of commercial importance. The techniques which have been so far described for propagation of seaweeds through tissue culture have been tested at laboratory scale and have not been validated for their suitability in commercial scale production. The following aspects have to be critically studied if the economic prospects associated with in vitro cell culture technology is to be realized:

- Methods for producing viable axenic material with greater consistency are required.
- New tissue and callus culture methods need to be developed for obtaining callus from both pigmented and non-pigmented parts of thallus.
- Physiology of plant growth regulators in plant cell division and development needs to be studied if high callus induction rate and growth are required.
- Genotyping and selection of tissue culture progeny with desired stable traits must be studied.
- Reports describing truly cell suspension cultures from seaweeds are rare and limited to the genus *Porphyra*. Techniques for developing cell suspension culture have to be carried out if high value secondary metabolites to be produced in vitro.
- It is imperative to establish homozygous lines of economically important seaweeds for growth and phycocolloid yields. The benefits of in vitro cell manipulations techniques can be more effective and realized if such select genetic lines are used for improving the traits.

Acknowledgments The first two authors (C.R.K. and B.J.) are grateful to Dr P.K. Ghosh, Director, CSMCRI for facilities and encouragement. We would also like to thank Mr. Manoj K. Gupta for assisting us in preparing this manuscript and DST, New Delhi. and INSA, New Delhi. for providing partial international travel support for presenting this work in 19th International Seaweed Symposium held during 26–31 March 2007 at Kobe, Japan.

References

Aguirre-Lipperheide M, Estrada-Rodriguez FJ, Evans LV (1995) Facts, problems and needs in seaweed tissue culture: an appraisal. J Phycol 31:677–688

Asensi A, Gall EA, Marie D, Billot C, Dion P, Kloareg B (2001) Clonal propagation of *Laminaria digitata* (Phaeophyceae) sporophytes through a diploid cell-filament suspension. J Phycol 37:411–417

Bradley PM, Cheney DP (1990) Some effects of plant growth regulators on tissue cultures of the red alga *Agardhiella subulata* (Gigartinales, Rhodophyta). Hydrobiologia 204/205:353–360

Butler DM, Evans LV (1990) Cell and tissue culture of macroalgae. In: Akatsuka I (ed) Introduction to applied phycology. SPB Academic Publishing, The Hague, The Netherlands, pp 629–645

Chen LCM (1982) Callus-like formations from Irish moss. Biol Bull Nat Taiwan Nor Uni 17:63–67

Chen LCM, Taylor ARA (1978) Medullary tissue culture of the red alga *Chondrus crispus*. Can J Bot 56:883–886

Cheney DP, Mathiesen A, Schubert D (1981) The application of genetic improvement techniques to seaweed cultivation: I. strain selection in the carrageenophyte *Chondrus crispus*. Proc Int Seaweed Symp 10:559–567

Cheney DP, Luistro AH, Bradley PM (1987) Carrageenan analysis of tissue culture and whole plants of *Agardhiella subulata*. Hydrobiologia 151/152:161–166

Cocking EC (1990) All sorts of plant genetic manipulation. In: Lycett GW, Grierson D (eds) Genetic engineering of crop plants. Butterworths, London, pp 1–11

Collantes G, Melo C, Candia A (2004) Micropropagation by explants of *Gracilaria chilensis* Bird, McLachlan and Oliveira Gloria. J App Phycol 16:203–213

Dawes CJ, Koch EW (1991) Branch, micropropagules and tissue culture of the red algae *Eucheuma denticulatum* and *Kappaphycus alvarezii* farmed in the Philippines. J Appl Phycol 6:21–24

Dawes CJ, Trono GC, Lluisma AO (1993) Clonal propagation of *Eucheuma denticulatum* and *Kappaphycus alvarezii* for Philippine seaweed farms. Hydrobiologia 260/261:379–383

Doty MS, Alvarez V (1973) Seaweed farms: a new approach for U.S. industry. Marine Technology Society Annual Conference Paper 9:701–708

Evans DA, Sharp WR (1983) Single gene mutations in tomato plants regenerated from tissue culture. Science 221:949–951

Fries L (1980) Axenic tissue cultures from the sporophytes of *Laminaria digitata* and *Laminaria hyperborea*. J Phycol 16:475–477

Fries L (1985) Propagation of *Fucus* (Fucales, Phaeophyta) by hairs with trichothallic growth. Phycologia 24:481–484

Garcia-Reina GL, Romero RR, Luque A (1988) Regeneration of thalliclones from *Laurencia* sp, (Rhodophyta). In: Paris MSS, Mavituna F, Novais JM (eds) Plant cell biotechnology. Springer, Berlin Heidelberg New York, pp 81–86

Garcia-Reina GL, Gomez-Pinchetti JL, Robeldo DR, Sosa P (1991) Actual potential and speculative applications of seaweed cellular biotechnology: some specific comments on *Gelidium*. Hydrobiologia 221:181–194

Gusev MV, Tambiev AH, Kirikora NN, Shelyastina NN, Aslanyan RR (1987) Callus formation in seven species of agarophyte marine algae. Mar Biol 95:593–597

Hafting JT (1999) A novel technique for propagation of *Porphyra yezoensis* Ueda blades in suspension cultures via monospores. J Appl Phycol 11:361–367

Huang W, Fujita Y (1997) Callus induction and thallus regeneration in some species of red algae. Phycol Res 45:105–111

Huang YM, Rorrer GL (2002) Optimal temperature and photoperiod for cultivation of *Agardhiella subulata* microplantlets in a bubble-column photobioreactor. Biotechnol Bioeng 79:135–144

Huang YM, Maliakal S, Cheney D, Rorrer G (1998) Comparison of development, photosynthesis and growth of filament clump and regenerated microplantlet cultures of *Agardhiella subulata* (Rhodophyta, Gigartinales). J Phycol 34:893–901

Kaczyna F, Megnet R (1993) The effects of glycerol and plant growth regulators on *Gracilaria verrucosa* (Gigartinales, Rhodophyceae). Hydrobiologia 268:57–64

Kakkar RK, Sawhney VK (2002) Polyamine research in plants-a changing perspective. Physiol Plant 116:281–292

Kawashima Y, Tokuda H (1990) Callus formation in *Ecklonia cava* Kjellman (Laminariales, Phaeophyta). Hydrobiologia 204/205:151–159

Kawashima Y, Tokuda H (1993) Regeneration from the callus of *Undaria pinnatifida* (Harvey) Suringar (Laminariales, Phaeophyta). Hydrobiologia 260/261:385–389

Kimura H, Notoya M (1997) Tissue culture of *Undaria undariodes* (Yendo) Okamura (Phaeophyta, Laminariales). J Mar Biotechnol 5:100–102

Kirihara S, Fujikawa Y, Notoya M (1997) Axenic tissue culture of *Sargassum confusum* C.Agardh (Phaeophyta) as a source of seeds for artificial marine forests. J Mar Biotecnol 5:142–146

Lawlor HJ, McComb JA, Borowitzka MA (1989) Tissue culture of *Ecklonia radiata* (Phaeophyceae, Laminariales): effects on growth of light, organic carbon source and vitamins. J Appl Phycol 1:105–112

Lawlor HJ, Borowitzka MA, McComb JA (1990) Tissue culture of brown seaweeds. Aust J Biotechnol 4:260–264

Lee TF (1985) Aposporous gametophyte formation in stipe explants from *Laminaria saccharina* (Phaeophyta). Bot Mar 28:179–185

Lee TM (1998) Investigations of some intertidal green macroalgae to hyposaline stress: Detrimental role of putrescine under extreme hyposaline conditions. Plant Sci 138:1–8

Liu XW, Gordon ME (1987) Tissue and cell culture of New Zealand *Pterocladia* and *Porphyra* species. Twelfth international seaweed symposium. Hydrobiologia 151/152:147–154

Liu XW, Kloareg B (1992) Explant axenisation for tissue culture in marine macro algae. Chin J Oceanol Limnol 10:268–277

Liu XW, Rochas C, Kloareg B (1990) Callus culture of *Pterocaldia capillacea* (Rhodophyta) and analysis of cell wall polysaccharides. J Appl Phycol 2:297–303

Malaikal S, Cheney D, Rorrer GL (2001) Halogenated monoterpenes production in regenerated plantlet cultures of *Ochtodes secundiramea* (Rhodophyta, Cryptonemiales). J Phycol 37:1010–1019

Mooney PM, Staden JV (1984) *In vitro* plantlet formation and multiple shoot induction in *Sargassum heterophyllum*. S Afr J Bot 51:41–44

Munoz J, Armando C, Cahue-Lopez, Patino R, Robledo D (2006) Use of plant growth regulators in micropropagation of *Kappaphycus alvarezii* (Doty) in airlift bioreactors. J Appl Phycol 18:209–218

Notoya M (1988) Tissue culture from the explant of *Ecklonia stolonifera* Okamura (Laminariales, Phaeophyta). Jpn J Phycol 36:175–177

Notoya M, Aruga Y (1989) Tissue culture from the explant of *Ecklonia cava* Kjellman (Laminariales, Phaeophyta). Jpn J Phycol 37:302–304

Notoya M, Aruga Y (1990) Tissue culture from the explant of stipe of *Eisenia bicyclis* (Kjellman) Setchell (Laminariales, Phaeophyta). Jpn J Phycol 38:387–390

Notoya M, Aruga Y (1992a) Tissue culture of parthenogenetic and normal young sporophytes of *Ecklonia cava* Kjellman (Laminariales, Phaeophyta). Jpn J Phycol 40:53–55

Notoya M, Aruga Y (1992b) Tissue culture of *Undaria pinnatifida* (Harvey) Suringar (Laminariales, Phaeophyta). Jpn J Phycol 40:393–395

Patwary MU, van der Meer JP (1982) Genetics of *Gracilaria tikvahiae* (Rhodophyceae). VIII. Phenotypic and genetic characterization of some selected morphological mutants. Can J Bot 60:2556–2564

Patwary MU, van der Meer JP (1983) Genetics of *Gracilaria tikvahiae* (Rhodophyceae) IX: Some properties of agars extracted from morphological mutants. Bot Mar 26:295–299

Polne-Fuller M (1988) The past, present and future of tissue culture and biotechnology of seaweeds. In: Stadler T, Mollion J, Verdus MC, Karamanos Y, Morvan, Christiaen D (eds) Algal biotechnology. Elsevier, London, pp 7–31

Polne-Fuller M, Gibor A (1986) Algal cell, callus and tissue cultures and selection of algal strains. Beih Nova Hedwig 83:30–36

Polne-Fuller M, Gibor A (1987) Calluses and callus -like growth in seaweeds: induction and culture. Hydrobiologia 151/152:131–138

Qi H, Rorrer KA, Jiang ZD, Rorrer GL (1995) Kinetics of non-photosynthetic callus production from the brown algae *Laminaria setchellii* (Laminariales). Phycol Res 43:179–182

Rajakrishna Kumar (2002) Studies on the tissue culture of economic seaweeds. PhD thesis, Bhavnagar University, Bhavnagar, India

Rajakrishna Kumar G, Reddy CRK, Ganesan M, Tiruppathi S, Dipakkore S, Eswaran K, Subba Rao PV, Bhavanath Jha (2004) Tissue culture and regeneration of thallus from callus of *Gelidiella acerosa* (Gelidiales, Rhodophyta). Phycologia 43:596–602

Rajakrishna Kumar G, Reddy CRK, Bhavanath Jha (2007) Callus induction and thallus regeneration from the callus of phycocolloid yielding seaweeds from the Indian coast. J Appl Phycol 19:15–25

Rao UM, Rao AS (1996) Callus formation in *Sargassum polycystum*. Phykos 35:147–150

Reddy CRK, Raja Krishna Kumar G, Eswaran K, Siddhanta AK, Tewari A (2003) *In vitro* somatic embryogenesis and regeneration of somatic embryos from pigmented callus of *Kappaphycus alvarezii* (Gigartinales, Rhodophyta). J Phycol 39:610–616

Robaina RR, Garcia JP, Garcia-Reina G, Luque A (1990) Morphogenetic effect of glycerol on tissue cultures of the red seaweed *Grateloupia doryphora*. J Appl Phycol 2:137–143

Robaina RR, Garcia JP, Luque A (1992) The growth pattern and structure of callus from the red alga *Laurencia* sp. (Rhodophyta,

Ceramiales) Compared to shoot regeneration. Bot Mar 35:267–272

Robeldo DR, Garcia-Reina G (1993) Apical callus formation in *Solieria filiformis* (Gigartinales, Rhodophyta) cultured in tanks. Hydrobiologia 260/261:401–406

Rorrer GL (2000) Cell and tissue cultures of marine seaweeds. In: Spier RE (ed) Encyclopedia of cell technology. Wiley, New York, pp 1105–1116

Rorrer GL, Cheney DP (2004) Bioprocess engineering of cell and tissue cultures for marine seaweeds. Aquacult Eng 32:11–41

Saga N, Sakai Y (1983) Axenic tissue culture and callus formation of the marine brown alga *Laminaria angustata*. Bull Jpn Soc Sci Fish 49:1561–1563

Saga N, Uchida T, Sakai Y (1978) Clone *Laminaria* from single isolated cell. Bull Jpn Soc Sci Fish 44:87

Saga N, Motomura T, Sakai Y (1982) Induction of callus from the marine brown alga *Dictyosiphon foeniculaceus*. Plant Cell Physiol 23:727–730

Santelices B (1992) Strain selection of clonal seaweeds. In: Round FE, Chapman DJ (eds) Progress in phycological research, vol. 8. Biopress, Bristol, pp 86–115

Stirk WA, Gold JD, Novak O, Strnad M, Van Staden J (2005) Changes in endogenous cytokinins during germination and seedling establishment of *Tagetes minuta* L. Plant Growth Regul 47:1–7

Sylvester AW, Waaland R (1983) Cloning the red alga *Gigartina exasperata* for culture on artificial substrates. Aquaculture 31:305–318

Tarakhovskaya ER, Maslov YI, Shishova MF (2007) Phytohormones in Algae. Russ J Plant Physiol 54:163–170

Titlyanov E, Titlyanova T (2006) Production of plantlets of the red alga of *Gelidium* genus (Rhodophyta) from thallus fragments. Russ J Mar Biol 32:284–288

Titlyanov EA, Titlyanova TV, Kadel P, Luning K (2006a) Obtaining of Plantlets from Apical Meristem of the Red Alga *Gelidium sp.* J App Phycol 18:167–174

Titlyanov EA, Titlyanova TV, Kadel P, Luning K (2006b) New methods of obtaining plantlets and tetraspores from fragments and cell aggregates of meristematic and submeristematic tissue of the red alga *Palmaria palmata*. J Exp Mar Biol Ecol 339:55–64

van der Meer JP (1986) Genetics of *Gracilaria tikvahiae* (Rhodophyceae). XI. Further characterization of a bisexual mutant. J Phycol 22:151–158

van der Meer JP, Patwary MU (1983) Genetic modification of *Gracilaria tikvahiae* (Rhodophyceae). The production and evaluation of polyploids. Aquaculture 33:311–316

Yamazaki A, Nakanishi K, Saga N (1998) Axenic tissue culture and morphogenesis in *Porphyra yezoensis* (Bangiales, Rhodophyta). J Phycol 34:1082–1087

Yan ZM (1984) Studies on tissue culture of *Laminaria japonica* and *Undaria pinnatifida*. Hydrobiologia 116/117:314–316

Yeoman MM (1987) Bypassing the plant. Ann Bot 60:157–174

Yokoya NS (2000) Apical callus formation and plant regeneration controlled by plant growth regulators on axenic culture of the red alga *Gracilariopsis tenuifrons* (Gracilariales, Rhodophyta). Phycol Res 48:133–142

Yokoya NS, Handro W (1996) Effects of auxins and cytokinins on tissue culture of *Grateloupia dichotoma* (Gigartinales, Rhodophyta). Hydrobiologia 326/327:393–400

Yokoya NS, Handro W (2002) Effects of plant growth regulators and culture medium on morphogenesis of *Solieria filiformis* (Rhodophyta) cultured in vitro. J Appl Phycol 14:97–102

Yokoya NS, Guimaraes MPBS, Handro W (1993) Development of callus like structures and plant regeneration in thallus segments of *Grateloupia filiformis* Kutzing (Rhodophyta). Hydrobiologia 260/261:407–413

Yokoya NS, West JA, Luchi AE (2004) Effects of plant growth regulators on callus formation, growth and regeneration in axenic tissue culture of *Gracilaria tenuistipitata* and *Gracilaria perplexa* (Gracilariales, Rhodophyta). Phycol Res 52:244–254

Yoshida G, Uchida T, Arima S, Terawa T (1999) Development of adventitive embryos in cauline leaves of *Sargassam macrocarpum* (Fucales,Phaeophyta). Phycol Res 47:61–64

Zhang J (1982) Some experiments and observations on the tissue and cell cultured of *Undaria pinnatifida*. J Shandong Coll Oceanol 12:29–38

Seaweed protoplasts: status, biotechnological perspectives and needs

C. R. K. Reddy · Manoj K. Gupta · Vaibhav A. Mantri · Bhavanath Jha

Originally published in the Journal of Applied Phycology, Vol 20, No 5, 169–182.
DOI: 10.1007/s10811-007-9237-9 © Springer Science + Business Media B.V. 2007

Abstract Protoplasts are living plant cells without cell walls which offer a unique uniform single cell system that facilitates several aspects of modern biotechnology, including genetic transformation and metabolic engineering. Extraction of cell wall lytic enzymes from different phycophages and microbial sources has greatly improved protoplast isolation and their yield from a number of anatomically more complex species of brown and red seaweeds which earlier remained recalcitrant. Recently, recombinant cell wall lytic enzymes were also produced and evaluated with native ones for their potential abilities in producing viable protoplasts from *Laminaria*. Reliable procedures are now available to isolate and culture protoplasts from diverse groups of seaweeds. To date, there are 89 species belonging to 36 genera of green, red and brown seaweeds from which successful protoplast isolation and regeneration has been reported. Of the total species studied for protoplasts, most belonged to Rhodophyta with 41 species (13 genera) followed by Chlorophyta and Phaeophyta with 24 species each belonging to 5 and 18 genera, respectively. Regeneration of protoplast-to-plant system is available for a large number of species, with extensive literature relating to their culture methods and morphogenesis. In the context of plant genetic manipulation, somatic hybridization by protoplast fusion has been accomplished in a number of economically important species with various levels of success. Protoplasts have also been used for studying foreign gene expression in *Porphyra* and *Ulva*. Isolated protoplasts are also exploited in numerous miscellaneous studies involving membrane function, cell structure, bio-chemical synthesis of cell walls etc. This article briefly reviews the status of various developments in seaweed protoplasts research and their potentials in genetic improvement of seaweeds, along with needs that must to be fulfilled for effective realization of the objectives envisaged for protoplast research.

Keywords Cell wall lytic enzymes · Marine macroalgae · Protoplast isolation · Protoplast fusion · Somatic hybridization · Selection of transformants

C. R. K. Reddy (✉) · M. K. Gupta · V. A. Mantri · B. Jha
Discipline of Marine Biotechnology and Ecology,
Central Salt and Marine Chemicals Research Institute,
Bhavnagar 364 002, India
e-mail: crk@csmcri.org

Introduction

Protoplasts are naked living plant cells devoid of cell walls which provide a unique uniform cell system underpinning several aspects of modern biotechnology. Recently, due to public antagonism, especially in Europe, to recombinant DNA technologies (GM crops) there has been renewed interest in exploiting protoplasts in somatic hybridization, cybridization, protoclonal variation studies, proteomics and metabolomics in higher plants (Davey et al. 2005). Interestingly, it was the observation of natural enzymatic degradation of cell walls during fruit ripening that stimulated interest in investigations on preparation of protoplasts using cell wall lytic enzymes in higher plants. Treatment of either cells or tissues with specific cell wall lytic enzymes results in total removal of their rigid and complex polysaccharide cell walls. Although protoplasts isolation from macrophytic benthic marine algae was reported as early as 1970 using mechanical methods (Tatewaki and Nagata 1970; Enomoto and Hirose 1972; Kobayashi 1975), the success in producing a large number of viable protoplasts became possible only after the development of an enzymatic method by Millner et al.

(1979) for *Enteromorpha intestinalis* (Linnaeus) Nees. Since then, considerable efforts have been made to isolate and culture protoplasts from a wide variety of multicellular marine macrophytic algae (see reviews by Polne-Fuller and Gibor 1987; Evans and Butler 1988; Butler et al. 1990; Reddy 1991; Reddy et al. 1994; Aguirre-Lipperheide et al. 1995). Nevertheless, regeneration of protoplasts to complete thalli, especially from anatomically complex seaweeds as well as giant kelps, is a relatively recent development (Sawabe et al. 1997; Matsumura et al. 2000). Thus, protoplasts can now be prepared and regenerated routinely from a diverse group of seaweed species with ease and success. This article briefly reviews the status of various developments in seaweed protoplasts research and their potential in genetic improvement of seaweeds, along with needs that must to be fulfilled for effective realization of the objectives envisaged for protoplast research.

Enzymes used for protoplast isolation

Seaweed protoplasts are prepared mainly by employing enzymatic methods. The enzymatic methods require a fair understanding of chemical composition of cell walls. The seaweed cell walls are structurally complex, diverse in chemical composition and differ from land plant cell walls by the abundance of amorphous matrix components as compared to the crystalline skeletal components (Kloareg and Quatrano 1988). As in higher plants, the most common skeletal polysaccharide is cellulose, a linear polymer of D-glucose with a β-1,4 linkages (Kreger 1962). The cellulose content of cell walls in Phaeophyta and Rhodophyta ranges from 1 to 8% of the dry weight of thallus (Black 1950; Ross 1953) and in some Chlorophyta members contain up to 70% (Frei and Preston 1961). In contrast, the members of Caulerpales and Dichotomosiphonales contain skeletal polysaccharides based on β-1,3 xylans, while for members of Codiales and Dasycladales they are based on β-1,4 mannans in place of the most common cellulose framework (Frei and Preston 1961; Mackie and Preston 1968).

Similarly, the matrix polysaccharides of seaweeds show great structural variability with significant eco-physiological and commercial interest. In Chlorophyta, the matrix polysaccharides are either xylogalactoarabinans or glucuronoxylorhamnans with varying sulphate contents (Percival and Young 1971; Medcalf et al. 1972). In Rhodophyta, mainly linear sulphated galactans are found composed of two regularly repeated galactose units alternatively linked by β(1–4) and α(1–3) linkages or D-galactose alternatively linked by β(1–4) and α(1–3) linkages. The major matrix component in brown seaweeds (10–45% of the thallus dry wt) is a polyuronide made up of two β-1,4-D-mannuronic acid and α-1,4-L-guluronic acid.

Thus, seaweeds, unlike higher plants, exhibit a variety and complexity in their cell wall composition and, accordingly, a combination of specific enzymes are needed to digest the cell walls completely to isolate protoplasts from different groups. Protoplasts from green seaweeds can be prepared using commercially available cellulases alone or in combination with macerozyme, whereas brown and red seaweeds require alginase and agarase/carragenase, respectively, in addition to commercial cellulases. The latter groups of enzymes are not commercially available or are cost prohibitive. The most common sources of noncellulolytic seaweed cell wall degrading enzymes are either herbivorous marine invertebrates which feed on them or culture filtrates of marine bacteria (Table 1).

Most recently, Inoue et al. (2007) cloned for the first time cDNA encoding abalone alginate lyase (HdAly) from hepatopancreas of Abalone in to baculovirus, and confirmed production of recombinant enzyme (rHdAly) following infection with Sf9 insect cell system. The purified enzyme thus produced was as effective in digesting the cell wall matrix as native HdAly. Further, the protoplast yield obtained from *Laminaria japonica* using rHdAly was also comparable with those yields obtained using native HdAly. The advantage of using recombinant enzymes for protoplast preparation is manifold. First recombinant enzymes are pure and free from several contaminant enzymes, like lipases, proteases and nucleases that are reported to have deleterious effect on protoplast viability. Secondly, the enzyme activity is consistent, free from variations and can be produced at will throughout the year.

Protoplast isolation, culture and regeneration

The yield, viability and regeneration rate of isolated protoplasts is dependent on several factors, including the enzyme constituents and their concentrations, pH, osmotic conditions and ionic strength of protoplast isolation medium, incubation temperature, physiological state and age of donor plant, and protoplast culture medium and its culture conditions (Polne-Fuller and Gibor 1984; Cheney et al. 1986; Bjork et al. 1990; Butler et al. 1990; Chen 1998; Krishna Kumar et al. 1999; Chen and Shih 2000; Matsumura et al. 2000; Shikh et al. 2005; Reddy et al. 2006). Pretreatment of thallus with proteases (protease or papain) or plasmolytic solutions was also carried out prior to enzymatic treatment to obtain enhanced yields of viable protoplasts from some members of Chlorophyta and Rhodophyta (Kawashima et al. 1989; Yamaguchi et al. 1989; Fujita and Saito 1990; Bjork et al. 1992; Amano and Noda 1992; Chen and Chiang 1994a, b; Wakabayashi et al. 1999). The protoplast yields have also been reported to vary with pH, osmoticum type and its concentration in protoplast

Table 1 Enzymes isolated from Phycophages and micro-organism for seaweeds protoplast isolation

Enzymes	Source	Reference
Agarase	*Alteromonas* sp.	Quatrano and Caldwell 1978
	Aplysia dactylomela	Gomez-Pinchetti and Garcia Reina 1993
	Arthrobacter	Quatrano and Caldwell 1978
	Aspergillus niger; A.tamari	Nikolaeva et al. 1999; Patel 1979
	Chaetomium globosum	Nikolaeva et al. 1999
	Cytophaga diffluens	Nikolaeva et al. 1999
	Delesseria sanguinea	Le Gall et al. 1990
	Littorina littorea; L. striata	Nikolaeva et al. 1999; Gomez-Pinchetti and Garcia Reina 1993
	Pseudomonas atlantica	Chen et al. 1994
	Pleurotus ostreatus	Nikolaeva et al. 1999
	Rhizopus arrhizus	Payton and Roberts 1979
	Stachybotrys atra	Nikolaeva et al. 1999
	Trichoderma viride	Nikolaeva et al. 1999
	Uronema marinum	Nikolaeva et al. 1999
Alginate lyases	*Alginovibrio aquatilis*	Stevens and Levin 1979
	Alteromonas sp. H-4	Sawabe et al. 1997
	Aplysia dactylomela; A. depilans; A. juliana	Gomez-Pinchetti et al. 1989; Boyen et al. 1990a; Wakabayashi et al. 1999
	Diadema antillarum	Gomez-Pinchetti et al. 1989
	Dolabella auricula	Nisizawa et al. 1968
	Haliotis tuberculata	Boyen et al. 1990a
	Littorina striata	Gomez-Pinchetti et al. 1989
	Patella vulgata	Ducreux et al. 1987
	Pseudomonas alginovora	Boyen et al. 1990b
	Pseudoalteromonas sp.	Cited from Takamistu et al. 2006
	Sphingomonas sp.	Wong et al. 2000
	Strongylocentrous intermedius	Saga and Sakai 1984
	Turbo coronatus	Tang 1982
	Vibrio sp. Strain-510	Hu et al. 2006
Carrageenase I	*Pseudomonas carrageenovra*	Le Gall et al. 1990; Smith and Bidwell 1989
Carrageenase II	*Delesseria sanguinea*	Le Gall et al. 1990
Cellulase	*Alteromonas* sp. ND137	Yamasaki et al. 1998
	Aplysia juliana	Wakabayashi et al. 1999
	Chaetomium globosum	Nikolaeva et al. 1999
	Pseudomonas sp. P-1; P.sp. ND 137	Fujita and Migita 1987; Yukihisa and Yuto 2006
	Trichoderma viride	Nikolaeva et al. 1999
Laminarinase	*Aplysia juliana*	Wakabayashi et al. 1999
Mannase	*Aeromonas* sp. F-25	Araki et al. 1987
Porphyranase	*Acinetobacter* sp.	Yamasaki et al. 1998
	Alteromonas sp. ND137	Yamasaki et al. 1998
	Aspergillus nidulans-051 biA1	Packer 1994
	Flavobacterium sp.	Yamasaki et al. 1998
	Haliotis iris	Packer 1994
	Pseudomonas sp. ND 137	Yukihisa and Yuto 2006
	Vibrio sp. AP-2	Araki et al. 1987
Protease	*Aspergillus niger*	Nikolaeva et al. 1999
	Chaetomium globosum	Nikolaeva et al. 1999
	Pleurotus ostreatus	Nikolaeva et al. 1999
	Trichoderma viride	Nikolaeva et al. 1999
Xylanase	*Acinetobacter* sp.	Yamasaki et al. 1998
	Alcaligenes sp. XY-234	Araki et al. 1998
	Alteromonas sp. ND137	Yamasaki et al. 1998
	Aspergillus niger	Nikolaeva et al. 1999
	Chaetomium globosum	Nikolaeva et al. 1999
	Flavobacterium sp.	Yamasaki et al. 1998
	Pleurotus ostreatus	Nikolaeva et al. 1999
	Pseudomonas sp. ND 137;	Yukihisa and Yuto 2006;
	Pseudomonas sp. PT-5	Yamaura et al. 1990
	Trichoderma viride	Nikolaeva et al. 1999
	Vibrio sp. AX-4	Araki et al. 1987

isolation solution. At higher pH (>7), the initial protoplast release in both *Ulva* and *Enteromorpha* was delayed and subsequent yield was less than that obtained for pH 5 and 6 (Reddy, unpublished data). Butler et al. (1990) observed decreased protoplast yields with increased pH values for *Laminaria* and attributed the low yields to the decreased activity of cellulase. Although mannitol or sorbitol is commonly used as osmotic stabilizer in protoplast isolation and culture solutions, the inorganic osmoticum, NaCl, gave improved yield of protoplasts in *Laminaria* (Butler et al. 1989).

Recently, Uppalapati and Fujita (2002) and Reddy et al. (2006) further simplified the protoplast isolation method by significantly reducing the ionic strength of protoplast isolation medium. These studies revealed that the simple dissolution of cell wall lytic enzymes (2% Cellulase Onozuka R-10) in low ionic medium (1% NaCl in deionised water) provided substantially high yields of viable protoplasts in shorter durations in different green seaweeds than the one prepared with normal seawater. Such enzyme preparations in low ionic strength medium together with macerozyme (2%), abalone acetone powder (1%) and agarase (50 units), also provided high yields of protoplasts when tested for *Porphyra* pretreated with 1% protease (Shikh et al. 2005).

To date, there are 89 species belonging to 36 genera of green, red and brown seaweeds from which successful protoplast isolation and regeneration has been reported (Table 2). Of the total species studied for protoplasts, most belonged to Rhodophyta with 41 species (13 genera) followed by Chlorophyta and Phaeophyta with 24 species each belonging to 5 and 18 genera, respectively. Moreover, the species for which protoplast technology has been developed are primarily either edible or of phycocolloid importance. Many of these studies are developmental and mainly aimed at development of fundamental techniques for successful protoplast isolation and regeneration (rarely for foreign gene expression).

Seaweed protoplasts are potentially totipotent. When given the correct chemical and physical stimuli, each protoplast is capable of regenerating a new wall and undergoing repeated mitotic cell division to produce a complete plant. Algal cells and protoplasts are most frequently cultured in a medium which is used to culture the same whole intact thallus. Unlike higher plants and their protoplasts, all cells and protoplasts of seaweeds, with their simple morphological and anatomical thallus structure, regenerate and differentiate directly into a complete thallus, in many cases even without the addition of any growth regulators to the culture medium. Maximum regeneration rate (>90%) of protoplasts in *Ulva* and *Enteromorpha* was observed in cultures grown at 20°C and 25°C, while at 30°C there was a lack of differentiation of fronds from the rhizoidal system which grew prominently (Reddy and Fujita 1991). The continued presence of osmoticum >0.4 M in culture medium hampered cell division and further growth in both *Ulva* and *Porphyra* (Reddy et al. 1989; Polne-Fuller and Gibor 1990).

The list of species capable of regeneration into complete plant from protoplast is steadily increasing. The plant regeneration from protoplasts of most anatomically complex seaweeds, like *Laminaria*, *Undaria* and *Gracilaria*, has been achieved recently. It is anticipated that the cellular biotechnology, which has been progressing tardily due to failure of regeneration of protoplasts in many seaweed species, may gain momentum and can now be effectively applied for economic gains of seaweed resources.

Protoplasts as seed stock for cultivation

Several studies have attempted to diversify the application of protoplasts for as many applications as possible. In order to provide continuous seed stock for cultivation of green seaweeds, Chen (1998) and Chen and Shih (2000) developed protoplast-based methods for producing stocks of seedlings (micro thalli) with survival abilities of many years in an incubator while maintaining their potential to develop into leafy thalli of *Ulva* and *Monostroma*. Protoplasts from *Monostroma* and *Porphyra* have been successfully tested for their seeding and regeneration in laboratory conditions (Reddy et al. 2006; Shikh et al. 2005). The protoplast-seeded nylon threads contained homogenous establishment of seedlings throughout its surface, due to secondary seeding of spores released from protoplast-derived sporangia developed during regeneration stages of protoplasts. As a result, threads seeded with protoplasts of *M. oxyspermum* always had a dense growth of germlings derived from both primary and secondary seeding (Reddy et al. 2006).

Biochemical studies involving protoplasts

Protoplasts have been used for physiological studies in higher plants and are potentially invaluable tools for similar studies in seaweeds. Protoplasts may have advantages over intact tissue due to absence of anionic cell wall which in turn promotes the easier transport of inorganic ions across the membrane. Protoplasts have been used to investigate the mechanism of inorganic carbon uptake in *Chondrus crispus* (Smith and Bidwell 1989), *Ulva rigida* (Bjork et al. 1992; Haglund et al. 1992), *Gracilaria tenuestipitata* (Haglund et al. 1992), and it has been reported that the involvement of an extra-cellular carbonic anhydrase localized either in the cell wall or outside of plasma membrane

Table 2 Seaweed species from which protoplast isolation and regeneration have been accomplished

Species	Status of success	Reference
Chlorophyta		
Enteromorpha bulbosa	PI	Yamaguchi et al. 1989
E. compressa	PR	Chou and Lu 1989; Reddy and Fujita 1991; Uppalapati and Fujita 2002; Reddy et al. 2006
E. flexuosa	PR	Reddy et al. 2006
E. intestinalis	PI	Millner et al. 1979; Saga et al. 1986
	PR	Rusig and Cosson 2001
E. linza	PI	Saga 1984
	PR	Fujita and Migita 1985; Chou and Lu 1989; Reddy and Fujita 1991;Uppalapati and Fujita 2002
	BS	Amano and Noda 1992
E. prolifera	PR	Kawashima et al. 1989; Reddy and Fujita 1991; Uppalapati and Fujita 2002
E. tubulosa	PR	Reddy et al. 2006
Ulva angusta	PR	Polne-Fuller and Gibor 1987
U. conglobata	PR	Reddy and Fujita 1989; Uppalapati and Fujita 2002; Reddy et al. 2006
U. fasciata	PR	Reddy et al. 1989; Chen and Shih 2000; Uppalapati and Fujita 2002; Reddy et al. 2006
	BS	Beer and Bjork 1994
U. lactuca (wild)	BS	Amano and Noda 1992
	PI	Reddy et al. 2006; Yamaguchi et al. 1989
	PR	Chou and Lu 1989; Reddy et al. 2006
U. lactuca (mutant)	PR	Zhang 1983; Fujimura et al. 1989; Reddy et al. 1989; Uchida et al. 1992; Uppalapati and Fujita 2002
U. pertusa	PI	Saga 1984; Yamaguchi et al. 1989
	BS	Fujimura and Kajiwara 1990; Amano and Noda 1992
U. pertusa (mutant)	PR	Fujita and Migita 1985; Reddy et al. 1989; Uppalapati and Fujita 2002
U. reticulata	PR	Reddy et al. 2006
	PI	Bjork et al. 1992
U. rigida	PR	Reddy et al. 2006
	PI&GS	Corzo et al. 1995
Monostroma angicava	PR	Zhang 1983; Saga and Kudo 1989
M. latissimum	PR	Chen and Chiang 1994a, b
	PI	Chen 1998
M. nitidum	PR	Fujita and Migita 1985; Uppalapati and Fujita 2002
	PI	Yamaguchi et al. 1989
	BS	Amano and Noda 1992
M. oxyspermum	PR	Krishna Kumar et al. 1999; Uppalapati and Fujita 2002; Reddy et al. 2006
M. zostericola	PI	Saga 1984; Yamaguchi et al. 1989
Chaetomorpha aerea	BS	Klotchkova et al. 2003
Bryopsis plumosa	BS	Kim et al. 2005
Phaeophyta		
Cladosiphon okamuranus	PR	Uchida and Arima 1992
Calpomenia bullosa	PI	Yamaguchi et al. 1989
Dictyopteris undulata	PI	Kajiwara et al. 1988
D. prolifera	PI	Kajiwara et al. 1988
Dictyota dichotoma	PI	Kajiwara et al. 1988
Durvillaea potatorum	PI	Kevekordes et al. 1993
Ecklonia radiata	PI	Kevekordes et al. 1993
Ectocarpus siliculosus	PR	Kuhlenkamp and Muller 1994
Eisenia bycyclis	PI	Yamaguchi et al. 1989; Wakabayashi et al. 1999
Endarachne binghamiae	PI	Wakabayashi et al. 1999
Fucus distichus	PR	Kloareg and Quatrano 1987
F. serratus	PI	Mussio and Rusig 2006
F. vesiculosus	PI	Mussio and Rusig 2006
Laminaria japonica	PI	Saga and Sakai 1984; Sawabe et al. 1997
L. digitata	CW	Butler et al. 1990
	BS	Rodde and Larsen 1997
L. saccharina	CW	Butler et al. 1990
	PI	Benet et al. 1994
	PR	Benet et al. 1997
	BS	Rodde and Larsen 1997
Macrocystis pyrifera	CW	Saga et al. 1986; Kloareg et al. 1989; Polne-Fuller et al. 1990
M. angustifolia	PI	Kevekordes et al. 1993
Padina arborescens	PI	Yamaguchi et al. 1989

Table 2 (continued)

Species	Status of success	Reference
Pilayella littoralis	PR	Mejjad et al. 1992
Sargassum muticum	PI	Saga et al. 1986; Polne-Fuller et al. 1990
Scytosiphon lomentaria	PI	Yamaguchi et al. 1989
Sphacelaria sp.	PR	Ducreux and Kloareg 1988; Ducreux et al. 1987; Rusig et al. 1994
Undaria pinnatifida	PI	Fujita and Migita 1985; Tokuda and Kawashima 1988; Wakabayashi et al. 1999
Rhodophyta		
Acrosorium polyneurum	PI	Yamaguchi et al. 1989
Bangia atropurpurea	PI	Araki et al. 1994
Chondrus crispus	BS	Smith and Bidwell 1989
	PI	Le Gall et al. 1990
Gelidium robustum	PI	Coury et al. 1993
Gracialria asiatica	PI	Yan and Wang 1993
G. chilensis	PR	Cheney 1990
G. chorda	PI	Chou and Lu 1989
G. filicina	PI	Yamaguchi et al. 1989
G. gigas	PI	Chou and Lu 1989
G. lemaneiformis	CW	Cheney et al. 1986
	PI	Bjork et al. 1990
G. salicornia	PI	Chou and Lu 1989
G. sordida	PI	Bjork et al. 1990
G. tenuestipitata	PI	Chou and Lu 1989; Bjork et al. 1990
G. tikvahiae	PR	Cheney et al. 1986; Cheney 1990
G. verrucosa	PI	Bjork et al. 1990; Araki et al. 1998
	PI&CW	Mollet et al. 1995
Grateloupia sparsa	PI	Chen and Chiang 1994a, b
	CW	Chen and Chiang 1995
G. filicina	PI	Yamaguchi et al. 1989; Chen and Chiang 1994a, b
G. turuturu	PI	Yamaguchi et al. 1989
Halymenia formosa	PI	Chou and Lu 1989
Kappaphycus alvarezii	PI	Zablackis et al. 1993
	BS	Zablackis et al. 1993
	PR	Salvador and Serrano 2005
Laurencia obtusa	PI	Balestri et al. 1989
Palmaria palmata	PI	Liu et al. 1992; Nikolaeva et al. 1999
	PS	Gall et al. 2004
Plocamium cartilagineum	PI	Balestri et al. 1989
Porphyra.sp (wild)	PI	Packer 1994
P. angusta	PI	Chou and Lu 1989
P. crispata	PI	Chou and Lu 1989; Gall et al. 1993
P. dentata	PI	Gall et al. 1993
P. lanceolata	PR	Polne-Fuller et al. 1984
P. leucosticta	PR	Chen 1987
P. linearis	PR	Chen et al. 1988
P. nereocystis	PR	Waaland et al. 1990
P. okamurae	PI	Fujita and Saito 1990
P. okhaensis	PI	Shikh et al. 2005
P. perforata	PR	Polne-Fuller et al. 1984, 1990; Saga et al. 1986
P. pseudolinearis	PR	Fujita and Saito 1990
P. seriata	PI	Fujita and Saito 1990
P. suborbiculata	CW	Tang 1982
	PI	Fujita and Saito 1990
	PR	Araki et al. 1987
P. tenera	PI	Song and Chung 1988
	PI	Fujita and Saito 1990
P. tenuipedalis	PR	Saga and Sakai 1984; Fujita and Migita 1985; Araki et al. 1987; Yamaguchi et al. 1989
P. yezoensis	PI	Fujita and Saito 1990
Solieria filiformis	PI	Gomez-Pinchetti and Garcia Reina 1993

BS, Biochemical study; CW, cell wall formation; PI, protoplast isolation; PR, plant regeneration; GS, genetic study is responsible for uptake of HCO_3^- by activating CO_2 transporter at low CO_2 concentration which in turn increases the efficiency of transporter by providing a faster conversion of HCO_3^- to CO_2.

Reciprocally, several studies have attempted to measure the oxygen evolution rate to evaluate whether the isolated protoplasts are in the same physiological state as cells in an intact plant. Protoplasts of *Enteromorpha intestinalis* had a slightly higher rate of oxygen evolution than thallus fragments (Millner et al. 1979), while protoplasts of *Ulva rigida* (Bjork et al. 1992), *Macrocystis pyrifera* (Davison and Polne-Fuller 1990) and *Laminaria saccharina* (Benet et al. 1994) retained about 65%, 40%, 100% oxygen evolution, respectively, for at least one day. The proteins pattern in protoplast and thalli of some species of green, red and brown seaweed showed variation, although amino acid composition remained the same (Amano and Noda 1992).

Fujimura and Kajiwara (1990) were the first to report the production of biflavour compounds from regenerating protoplast of *Ulva pertusa*. The bioflavours were found to be derivatives of long chain aldehydes that were responsible for typical seaweed odor. Seceration of é-carrageenan was detected in culture media from the protoplasts of *Kappaphycus alvarezii* var. *tambalang* after 24 h of culture in the absence of K^+ ions (Zablackis et al. 1993).

Most recently, Kim et al. (2005) isolated and characterized a novel lectin (Bryohealin) exerting a characteristic activity in the cellular wound healing process during protoplast formation from *Bryopsis plumosa*, that appeared to have quite different amino acid composition (about 6%) to that of higher plants (less than 1%) and have a molecular mass of 53.8 kDa. Thus, protoplasts provide unique opportunities to study various fundamental aspects of cell wall biochemical studies which are otherwise impossible with normal intact cells of thallus.

Protoplast fusion and somatic hybridization

Fusing protoplasts of different origins together to produce novel somatic hybrids provides new opportunities to study somatic cell genetics as well as new approaches to genetic manipulation of plants. Table 3 shows the status of protoplast fusion accomplished for seaweeds. The first report of protoplast fusion and fusion product regeneration between two color morphs of *Porphyra* was that of Fujita and Migita (1987) using PEG method. Although they were able to produce several green and red chimeric thalli, the genetic nature of somatic hybrids has not been reported. Since then, several attempts have been made to accomplish interspecific and intergeneric protoplast fusions in several algae employing both PEG and electrofusion methods (Reddy and Fujita 1989; Cheney 1990; Fujita and Saito 1990; Reddy et al. 1992; Cheney and Kurtzman 1992; Dai et al. 1993; Mizukami et al. 1995). In most cases, the frequency of heterokaryon formation and subsequent recovery of somatic hybrids or cybrids following the protoplast fusions was very low. In order to circumvent this problem, some researchers used freshly released spores as protoplasts to develop hybrid plants. Kapraun (1989, 1990) demonstrated the regeneration of intraspecific-parasexual hybrids from fusion of zoospores of *Enteromorpha* and *Ulvaria*. The cytological and karyological studies of germlings regenerated from presumptive fusion products of *Enteromorpha* and some *Ulvaria* indicated nuclear fusion following plasmogamy of fused zoospores. The chromosome counts in field collected plants and also 10 week-old control plants showed haploid nuclei with 10 chromosomes, while diploid nuclei with 20 chromosomes were found in germlings of fusion products. Nevertheless, some germlings and fusion products of *Ulvaria* contained multinucleate cells, presumably resulting from zoospore fusion without nuclear fusion.

Recently, Kito et al. (1998) for the first time demonstrated successful trans-divisional protoplast fusion and regeneration of somatic hybrids between *Monostroma* and *Porphyra*. The resultant hybrid progeny have also been found to have characteristic fatty acid components as well as DNA of either parent indicating possible genetic recombination in hybrid progeny.

In most of the studies, the heterokaryons (heteroplasmic fusion products) have been identified from homokaryons and un-fused parental protoplasts based on visual observation, since the seaweeds and their cells have fascinating colors. Nevertheless, this method of screening is less reliable particularly when identifying the heterokaryons of similar color cells. Therefore, it is essential to develop reliable screening methods, such as differential fluorescence labeling, genetic complementation, physiological complementation or restoration of growth capacity in hybrids upon fusion of inactivated cells with chemical poisons (ethidium bromide, sodium azide or irradiation with X- and gamma rays), for selection of heterokaryons as in higher plants.

Genetic transformation

The development of genetic transformation procedures for macroalgae is in its inception and in formative stages as compared to crop plants. As expected, the red and brown seaweeds with commercial importance have become the first preferred choice of target species for carrying out genetic transformation studies. The seaweed species for

Table 3 Status of protoplasts fusion and regeneration of somatic hybrids in seaweeds

Fusion species	Status	Reference
Ulva linza×*Monostroma angicava*	Callus formation	Zhang 1983
U. pertusa×*U. conglobata*	Plant development	Reddy and Fujita 1989
U. pertusa×*Enteromorpha prolifera*	Plant development	Reddy et al. 1992
Enteromorpha×*Porphra yezoensis*	Protoplast fusion	Saga et al. 1986
E. linza×*E. linza*	Plant development	Kapraun 1989[a]
Gracilaria chilensis×*G. tikvahiae*	Plant development	Cheney 1990
Ulvaria oxysperma×*U. oxysperma*	Plant development	Kapraun 1990[a]
Porphyra yezoensis (red)×*P. yezoensis* (green)	Plant development	Fujita and Migita 1987
P. yezoensis×*P. pseudolinearis*	Plant development	Fujita and Saito 1990
P. yezoensis×*P. haitanensis*	Callus formation	Dai et al. 1993
P. yezoensis×*P. tenera* (green)	Callus formation	Araki and Morishita 1990
P. yezoensis (green)×*P. suborbiculata*	Callus formation	Mizukami et al. 1995
P. yezoensis×*P. vietnamensis*	Callus formation	Matsumoto et al. 1995
P. tenera×*P. suborbiculata*	Callus formation	Matsumoto et al. 1995
P. linearis×*P. miniata*	Callus formation	Chen et al. 1995
P. suborbiculata×*P. tenuipedalis*	Callus formation	Achiha and Nakajima, 1995
P. yezoensis×*Bangia atropurpurea*	Callus formation	Fujita 1993
P. pseudolinearis×*B. atropurpurea*	Callus formation	Fujita 1993
P. yezoensis×*Monostroma nitidum*	Plant development	Kito et al. 1998

[a] Zoospore fusion

which genetic transformation has been undertaken are shown in Table 4. The first report of accomplishment of genetic engineering in a macro alga was that of Kurtzman and Cheney (1991), who successfully demonstrated the transient expression of the bacterial β-glucoronidase gene (*gus*) in the carrageenan-producing red alga *Kappaphycus alvarezii*. The *gus* gene was fused to the CaMV 35S promoter and introduced into the algal tissues by micro-particle bombardment. Similar results of transient expression of 35S-*gus* gene construct in the electroporated protoplasts of *Porphyra* (Kubler et al. 1994) and *Ulva* (Huang et al. 1996) have also been reported. Subsequently, a number of such studies have been accomplished in different species of economically important seaweeds, such as *Porphyra*, *Gracilaria*, *Grateloupia*, *Laminaria* and *Undaria*, by utilizing various gene transfer methods reported in the literature (Kuang et al. 1998; Qin et al. 1994, 1998a, 2003). Liu et al. (2003) for the first time attempted to use homologous promoters for foreign gene expression in *Porphyra* protoplasts using a portion of 18SrDNA in a vector pQD-GUS, and compared the expression of GUS protein with that of the parent pBS-GUS vector.

Table 4 Status of genetic engineering in marine macroalgae

Species	Method of gene transfer	Status of expression	Reference
Acetabularia mediterranea	Micro-injection	Stable	Neuhaus et al. 1986
Chondrus crispus	Vector mediated	stable	Collen et al. 2006
Gracilaria changii	Biolistic particle	Transient	Gan et al. 2003
Kappaphycus alvarezii	Biolistic particle	Transient	Kurtzman and Cheney 1991
Laminaria japonica	Biolistic particle	Transient	Qin et al. 1998a
L. japonica	Biolistic particle	Transient	Qin et al. 1998b
L. japonica	Biolistic particle	Stable	Jiang et al. 2003
L. digitata	Vector mediated	Stable	Roeder et al. 2005
Porphyra leucosticta	Electroporation	Transient	Lin et al. 2001
P. miniata	Electroporation	Transient	Kubler et al. 1994
P. tenuipedalis	Electroporation	Transient	Achiha 1995
P. yezoensis	Electroporation	Transient	Kuang et al. 1998
P. yezoensis	Electroporation	Transient	He et al. 2001
P. yezoensis	Vector mediated	Stable	Cheney et al. 2001
P. yezoensis	Vector mediated	Transient	Liu et al. 2003
P. yezoensis	Electroporation	Transient	Mizukami et al. 2004
Ulva lactuca	Electroporation	Transient	Huang et al. 1996
Undaria pinnatifida	Vector mediated	Transient	Yu et al. 2002
U. pinnatifida	Biolistic particle	Stable	Qin et al. 2003

The resultant transformants showed increased GUS activity with pQD-GUS compared with those of pBS-GUS. Bernasconi et al. (2004) described *Agrobacterium*-mediated transformation of *Porphyra yezoensis* expressing the bacterial nitroreductase gene (nfsI) and capable of detoxifying trinitrotoluene (TNT). Of the methods used for gene transfer to macroalgae, biolistic bombardment method has been reported to be more effective when thallus explants were used. The Chinese scientists have achieved success in stable transformation of *Laminaria japonica* using biolistic method. The transformants showed stable expression of LacZ gene (*lacZ*) and CAT gene (*cat*) in parthenogenetically regenerated sporophytes (Qin et al. 2005). However, the SV40 promoter reported to be more efficient than CaMV35S in expression of above mentioned genes in *Laminaria*. Endogenous tubulin promoter has also been used to determine its transcription promoting abilities for foreign gene *gusA* in protoplasts of *Porphyra yezoensis* using vector pATubGUS (Gong et al. 2007).

For the first time, the Expressed sequence tags (ESTs) from protoplasts and thalli have been used to identify the genes involved in cell wall regeneration and stress responses in *Chondrus crispus* and *Laminaria digitata* (Collen et al. 2006; Roeder et al. 2005). The fraction of stress-related ESTs were five times higher in the protoplast than in the thallus library of *C. crispus* (Collen et al. 2006). The ESTs that were statistically over-represented in protoplasts included: glutathione S-transferases, heat shock proteins, and vanadium bromoperoxidase along with several genes of unknown function, whereas in thallus NADH dehydrogenase, peroxidase ESTs were over-represented together with several genes of unknown function (Collen et al. 2006; Roeder et al. 2005).

Since, the protoplast regeneration systems are constantly being developed for all the important seaweeds, it would become much easier and more reliable to select transformants from untransformed ones if protoplasts are used rather than seaweed explants whose regeneration protocols are under-developed.

Promoter selection

Efficient expression of foreign genes in transformants depends on the promoter used and is crucial for genetic engineering. Promoters from seaweeds or seaweed associated infective virus or bacteria have rarely been isolated and studied. The major drawback in developing reliable methods for transformation of macroalgae is the lack of availability of native algal promoters. The two promoters that have been used most often for constitutive expression are the NOS (nopaline synthase) promoter from *A. tumefaciencs* T-DNA and 35S promoter from cauliflower mosaic virus (CaMV). The studies carried out to find better promoters for kelp transformation revealed that CaMV35S and fcp (from diatom fucoxanthin, chlorophyll a/c binding protein gene) promoter were found to be more efficient in stable expression of foreign genes than Ubiquitin (from maize) and amt (adenine methyltransferase) promoter (Table 5). Another study using SV40 promoter (from simian virus) resulted in uniform expression of *lacZ* reporter gene in regenerated *Laminaria* sporophytes indicating its high transcription recognition efficiency irrespective of tissue specificity (Jiang et al. 2003). The same promoter has also been confirmed to work well with *Undaria* for both transient and stable expression. There is considerable work going on to understand different gene regulatory domains in algal system for using such regulatory domains as possible algal promoters for efficient stable expression of foreign genes.

Table 5 Promoters and reporters genes used for genetic transformation of macroalgae

Species	Promoter	Reporter gene	Marker gene	Reference
Acetabularia mediterranea	SV40	Nil	G418 (pSVneo)	Neuhaus et al. 1986
Kappaphycus alvarezii	CaMV35S	*gus*	—	Kurtzman and Cheney 1991
Porphyra miniata	CaMV35S	*gus*	*npt-II (pB1121)*	Kubler et al. 1994
Ulva lactuca	CaMV35S	*gus*	*npt-II (pB1121)*	Huang et al. 1996
Laminaria japonica	CaMV35S/SV40	*lacZ/cat*	—	Qin et al. 1998a
L. japonica	CaMV35S/SV40	*lacZ/cat*	—	Qin et al. 1998b
P. yezoensis	CaMV35S	*gus*	*npt-II*	Kuang et al. 1998
P. tenera	CaMV35S	*gus*	—	Okauchi and Mizukami, 1999
P. leucosticta	CaMV35S	*lac Z*	—	Lin et al. 2001
P. yezoensis	CaMV35S	*gus/gfp*	—	Cheney et al. 2001
Gracilaria changii	SV40	*lacZ*	—	Gan et al. 2003
P. yezoensis	Rubisco	*gus*	—	Hado et al. 2003
L. japonica	SV40	*lacZ/gal*	—	Jiang et al. 2003
Undaria pinnatifida	SV40	*lacZ/gus*	—	Qin et al. 2003
P. yezoensis	18S rDNA	*gus*	—	Liu et al. 2003
P. yezoensis	pYez-Rub	*gus/gfp/luc*	—	Mizukami et al. 2004
P. yezoensis	Tubulin	*gus*	amp-R	Gong et al. 2007

Screening for transformants

Another important factor in genetic transformation studies is the selection system for recovery of transgenic cells which depends on the gene present in the expression cassette and selective agent employed. Kanamycin and the hygromycin are two of the commonly used selective agents for transgenic plant cells. The selection is based upon the expression of bacterial neomycin phosphotransferase gene (*nptII*) and the hygromycin phosphotransferase gene (*hptII*). Herbicides and other phytotoxic chemicals (bialaphos, chlorsulfuron, sulfonamide, etc.) have also been employed in a few cases. For example, bialaphos resistance gene (*bar*) from fungus encodes a phosphinothricin acetyltransferase that is able to detoxify the herbicide.

Susceptibility of *Laminaria* and *Undaria* to nine antibiotics (lincomycin, ampicillin, streptomycin, kanamycin, neomycin, chloramphenical, hygromycin, zeocin, G-418) and one herbicide (basta) has been studied (Qin et al. 1998b). The results indicated that both the species were sensitive only to chloramphenical, hygromycin and basta, and subsequently either one has been used as selectable markers for screening kelp transformants. The recent work in our laboratory with protoplasts of green seaweeds such as *Ulva* and *Monostroma* have also shown considerable resistance to antibiotics (hygromycin and kanamycin). This finding has further expanded the scope of the studies to screen for more broad-spectrum antibiotics and herbicides for developing reliable selection systems for recovery of transformants following transformation.

Conclusion

Reliable procedures are now available to isolate and culture protoplasts-to-plant regeneration from a wide range of seaweeds including anatomically most complex taxa like *Laminaria*, *Undaria* and *Gracilaria* that are among the commercially important genera. Further, the success in protoplast isolation and regeneration of protoplasts provided a tool and impetus to study of somatic cell hybridisation. The studies on interspecific and intergeneric protoplast fusion clearly demonstrated the possibilities of developing new and in certain cases improved strains in seaweeds. The natural colors of seaweeds, to some extent, facilitate direct visual identification of heterokaryons, leading to misidentification of protoplast products of similar color. Hence, it is essential to develop reliable methods, such as differential fluorescence labeling, genetic complementation or physiological complementation, for isolating bi-parental fusion products or heterokaryons. Although protoplasts have been successfully used for transient expression of foreign genes in some seaweeds, the evidence for efficient stable expression is lacking due to poor understanding of different gene regulatory domains in algal systems. Nevertheless, the following aspects have to be critically studied for effective realization of benefits associated with exploitation of protoplasts for genetic improvement and metabolic engineering of seaweeds for sustainable yields and products:

- Methods for producing viable axenic material with greater consistency are required.
- Most of the seaweed tissue and protoplast culture studies have so far been developmental studies and have not been applied to select genotypes. It is imperative that homozygous lines for fast growth and high yielding strains of quality gel forming polysaccharides are established.
- Protoplast fusion studies require development of new methods for recovery of hetero-karyons from those of homo-karyons.
- It is also necessary to generate data on susceptibility of seaweeds to different broad-spectrum antibiotics and herbicides. This information is useful for selection and recovery of transformants following the genetic engineering techniques.
- The progress achieved so far with genetic transformation studies provides a useful groundwork for continued research into development of transgenic macroalgae with useful traits. The precise genetic improvement of selected alga will require long term study, including selection of algal promoter elements as well as development and evaluation of appropriate genetic manipulation techniques specifically suitable for the seaweeds.

Acknowledgments The authors are grateful to Dr P.K. Ghosh, Director, CSMCRI for support and encouragement. The first author (C.R.K.) would also like to thank DST, New Delhi and INSA, New Delhi for providing partial international travel support for presenting this work in 19th International Seaweed Symposium held during March 26-31, 2007 at Kobe, Japan.

References

Achiha H (1995) Crossbreeding of *Porphyra tenuipedalis* by electroporation. Kaiyo Monthly 27:671–677

Achiha H, Nakajima Y (1995) interspecific hybrid of isolated protoplasts between *Porphyra suborbiculata* and *P. tenuipedalis* by electrofusion method. Jpn J Phycol (Sôrui) 43:223–226 (in Japanese)

Aguirre-Lipperheide M, Estrada-Rodriguez FJ, Evans LV (1995) Facts, problems and needs in seaweed tissue culture: an appraisal. J Phycol 31:677–688

Amano H, Noda H (1992) Proteins of protoplasts from several seaweeds. Nippon Suisan Gakkaishi 58(2):291–299

Araki T, Morishita T (1990) Fusion of protoplasts from wild type *Porphyra yezoensis* and green type *P. tenera* thalli (Rhodophyta). Nippon Suisan Gakkaishi 56:1161

Araki T, Aoki T, Kitamikado M (1987) Preparation and regeneration of protoplasts from wild type of *Porphyra yezoensis* and green variant of *P. tenera*. Nippon Suisan Gakkaishi 53:1623–1627

Araki T, Hayakawa M, Tamaru Y, Yoshimatsu K, Morishita T (1994) Isolation and regeneration of haploid protoplasts from *Bangia atropurpurea* (Rhodophyta) with marine bacterial enzymes. J Phycol 30:1040–1046

Araki T, Inoue N, Morishita T (1998) Purification and characterization of β-1,3-xylanase from a marine bacterium, *Alcaligenes* sp. XY-234. J Gen Appl Microbiol 44:269–274

Balestri E, Della Pieta F, Cinelli F (1989) Production of protoplasts from two red marine algae: *Laurencia obtusa* (Hudson) Lamourax and *Plocamium cartilagineum* (L.) Dixon, from Mediterranean Sea. In: Miyachi S. Karube I, Ishida T (eds) Current topics in marine biotechnology. The Jpn Soc Mar Biotechnol, Tokyo, pp 375–378

Beer S, Bjork M (1994) Photosynthetic properties of protoplast as compared with thalli of *Ulva fasciata* (Chlorophyta). J Phycol 30:633–637

Benet H, Bruss U, Duval JC, Kloareg B (1994) Photosynthesis and photoinhibition in protoplasts of the marine brown alga *Laminaria saccharina*. J Exp Bot 45:211–220

Benet H, Gall AE, Asensi A, Kloareg B (1997) Protoplast regeneration from gametophytes and sporophytes of some species in the order Laminariales (Phaeophyceae). Protoplasma 199:39–48

Bernasconi P, Cruz-Uribe T, Rorrer G, Bruce N, Cheney D (2004) Development of a TNT-detoxifying strain of the seaweed *Porphyra yezoensis* through genetic engineering. J Phycol 40(suppl 1):31

Bjork M, Ekman P, Wallin A, Pedersen M (1990) Effect of growth rate and other factors on protoplast yield from four species of *Gracilaria* (Rhodophyta). Bot Mar 33:433–439

Bjork M, Gomez-Pinchetti JL, Garcia-Reina G, Pedersen M (1992) Protoplasts isolation from *Ulva rigida* (Chlorophyta). Br Phycol J 27:401–407

Black WAP (1950) The effect of the seasonal variation in the cellulose content of the common Scottish Laminariaceae and Fucaceae. J Mar Bio Assoc UK 29:379–387

Boyen C, Kloareg B, Polne-Fuller M, Gibor A (1990a) Preparation of alginate lyases from marine molluscs for protoplast isolation in brown algae. Phycologia 29:173–181

Boyen C, Bertheau Y, Barbeyron T, Kloareg B (1990b) Preparation of guluronate lyase from *Pseudomonas alginovora* for protoplast isolation in *Laminaria*. Enzyme Microb Technol 12:885–890

Butler DM, Ostgaard K, Boyen C, Evans LV, Jensen A, Kloareg B (1989) Isolation condition for high yields of protoplasts from *Laminaria saccharina* and *L. digitata* (Phaeophyceae). J Exp Bot 40:1237–1246

Butler DM, Evans LV, Kloareg B (1990) Isolation of protoplasts from marine macroalgae. In: Akatsuka I (ed) Introduction to applied phycology. SPB Academic Publication, The Hague, The Netherlands, pp 647–666

Chen LCM (1987) Protoplasts morphogenesis of *Porphyra leucosticta* in culture. Bot Mar 30:399–403

Chen YC (1998) Development of protoplasts from holdfasts and vegetative thalli of *Monostroma latissimum* (Chlorophyta, Monostromatacae) for algal seed stock. J Phycol 34:1075–1081

Chen YC, Chiang YM (1994a) Development of protoplasts from *Grateloupia sparsa* and *G. filicina* (Halymeniaceae, Rhodophyta). Bot Mar 37:361–366

Chen YC, Chiang YM (1994b) Isolation and regeneration of protoplasts of *Monostroma latissimum* Wittrock (Monostromataceae, Chlorophyta). Bot Bull Acad Sin 35:45–51

Chen YC, Chiang YM (1995) Ultra structure of cell wall regeneration from isolated protoplasts of *Grateolopia sparsa* (Halymeniaceae, Rhodophyta). Bot Mar 38:393–399

Chen YC, Shih HC (2000) Development of protoplasts of *Ulva fasciata* (Chlorophyta) for algal seed stock. J Phycol 36:608–615

Chen LCM, Hong MF, Craigie JS (1988) Protoplasts development from *Porphyra linearis* — an edible marine red alga. In: Puite KJ, Dons JJM, Huizing HJ, Kool AJ, Koornneef M, Krens FA (eds) Progress in plant protoplasts research, vol 7. Kluwer Academic Publishers, Dordrecht, pp 123–124

Chen LCM, Craigie JS, Xie ZK (1994) Protoplast production from *Porphyra linearis* using a simplified agarase procedure capable of commercial application. J Appl Phycol 6:35–39

Chen LCM, McCracken IR, Xie ZK (1995) Electrofusion of protoplasts of two species of *Porphyra* (Rhodophyta). Bot Mar 38:335–338

Cheney D (1990) Genetic improvement of seaweeds through protoplasts fusion. In: Yarish C, Penniman C, Van Patten P (eds) Economically important marine plants of the Atlantic: their biology and cultivation. Univ. Conn Sea Grant Program, Storrs, pp 15–25

Cheney DP, Kurtzman (1992) Progress in protoplast fusion and gene transfer in red algae. In: Absrtracts of the XIV International Seaweed Symposium, Brest, France, 16–21 August, 1992

Cheney DP, Mar E, Saga N, Van der Meer J (1986) Protoplasts isolation and cell division in the agar producing seaweed *Gracialria* (Rhodophyta). J Phycol 22:238–243

Cheney D, Metz B, Stiller J (2001) *Agrobacterium*-mediated genetic transformation in the macroscopic marine red alga *Porphyra yezoensis*. J Phycol 37(Suppl):11

Chou HN, Lu HK (1989) Protoplasts from seaweeds: isolation, culture and fusion. In: Miyachi S, Karube I, Ishida Y (eds.) Current topics in marine biotechnology. Jap Soc Mar Biotechnol, Tokyo, pp 227–230

Collen J, Roeder V, Rousvoal S, Collin O, Kloareg B, Boyen C (2006) An Expressed Sequence Tag analysis of thallus and regenerating protoplasts of *Chondrus crispus* (Gigartinales, Rhodophyceae). J Phycol 42:104–112

Corzo A, Vergara JJ, Garcia-Jimenez M (1995) Isolation and flow cytometric characterization of protoplasts from marine macroalgae. J Phycol 31:1018–1026

Coury DA, Naganuma T, Polne-Fuller M, Gibor A (1993) Protoplasts of *Gelidium robustum* (Rhodophyta). Hydrobiologia 260/261:421–427

Dai J, Zhang Q, Bao Z (1993) Genetic, breeding and seedling raising experiments with *Porphyra* protoplasts. Aquaculture 111:139–145

Davey MR, Anthony P, Power JB, Lowe KC (2005) Plant protoplasts: status and biotechnological perspectives. Biotechnol Adv 23:131–171

Davison IR, Polne-Fuller M (1990) Photosynthesis in protoplasts of *Macrocystis pyrifera* (Phaeophytae). J Phycol 26:384–387

Ducreux G, Kloareg B (1988) Plant regeneration from protoplasts of *Sphacelaria* (Phaeophyceae). Planta 174:25–29

Ducreux G, Maillard P, Sihachakr D (1987) Patterns of differentiation and regeneration in single cells and protoplasts of *Sphacelaria* (Phaeophyceae). In: Stadler T, Mollion J, Verdus MC, Karamanos Y, Morran H, Christiaen D (eds) Algal biotechnology. Elsevier Applied Science Press, London, pp 129–139

Enomoto K, Hirose H (1972) Culture studies on artificially induced aplanospores in the marine alga *Boergesenia forbesii* (Harvey) Feldman (Chlorophyceae, Siphonocladales). Phycologia 11:119–122

Evans LV, Butler DM (1988) Seaweed biotechnology-current status and future prospects. In: Roger LJ, Gallon JR (eds) Biochemistry of the algae and cyanobacteria. Clarendon Press, Oxford, pp 335–350

Frei E, Preston RD (1961) Variants in the structural polysaccharides of algal cell walls. Nature 192:939–943

Fujimura T, Kajiwara T (1990) Production of bioflavor by regeneration from protoplasts of *Ulva pertusa* (Ulvales, Chlorophyta). Hydrobiologia 204–205(1):143–149

Fujimura T, Kawai T, Shiga M, Kajiwara T, Hatanaka A (1989) Regeneration of protoplasts into complete thalli in the marine green alga *Ulva pertusa*. Nippon Suisan Gakkaishi 55:1353–1359

Fujita Y (1993) Crossbreeding of green and red algae by cell fusion. Kaiyo Monthly 25:690–695

Fujita Y, Migita S (1985) Isolation and culture of protoplasts from seaweeds. Bull Fac Fish Nagasaki Univ 57:39–45

Fujita Y, Migita S (1987) Fusion of protoplasts from thalli of two different color types in *Porphyra yezoensis* Ueda and development of fusion products. Jpn J Phycol 35:201–208

Fujita Y, Saito M (1990) Protoplasts isolation and fusion in *Porphyra* (Bangiales, Rhodophyta). Hydrobiologia 204/205:161–166

Gall E, Chiang YM, Kloareg B (1993) Isolation and regeneration of protoplasts from *Porphyra crispate* and *Porphyra dentate*. Eur J Phycol 28:277–283

Gall Le L, Rusig AM, Cosson J (2004) Organisation of the microtubular cytoskeleton in protoplasts from *Palmaria palmata* (Palmariales, Rhodophyta). Bot Mar 47:231–237

Gan YS Qin S, Othman RY, Yu DZ, Phang SM (2003) Transient gene expression of LacZ in partical bombared Gracilaria Changii (Gracilariales, Rhodophyta). J Appl Phycol 15:345–349

Gomez-Pinchetti JL, Garcia Reina G (1993) Enzymes from marine phycophages that degrade cell walls of seaweeds. Mar Biol 116:553–558

Gomez-Pinchetti JL, Robaina R, Garcia Reina G (1989) Quantification of the agarolytic activity from marine herbivores: its potential use for protoplast isolation. In Abstracts of the XIII International Seaweed Symposium, Vancouver, Canada, 13–18 August 1989

Gong QH, Yu WG, Dai J, Liu HQ, Rifu XU, Guan H, Pan K (2007) Efficient *gusA* Transient Expression in *Porphyra yezoensis* Protoplasts Mediated by Endogenous Beta-tubulin Flanking Sequences. J Ocean Univ China 6:21–25

Hado M, Okauchi M, Murase N, Mizukami Y (2003) Transient expression of GUS gene using Rubisco gene promoter in the protoplasts of *Porphyra yezoensis*. Nippon Suisan Zoshoku 51:355–360 (in Japanese with English abstract)

Haglund K, Bjork M, Ramazanov Z, Garcia-Reina G, Pederson M (1992) Role of carbonic anhydrase in photosynthesis and inorganic carbon assimilation in the red alga *Gracilaria tenuistipitata*. Planta 187:275–281

He P, Yao Q, Chen Q, Guo M, Xiong A, Wu W, Ma J (2001) Transfering and expression of glucose oxidase gene-Gluc in *Porphyra yezoensis*. J Phycol 37:23

Hu X, Jiang X, Hwang HM (2006) Purification and characterization of an alginate lyase from marine *Bacterium Vibrio* sp. mutant strain 510-64. Curr Microbiol 53:135–140

Huang X, Weber JC, Hinson TK, Mathieson AC, Minocha SC (1996) Transient expression of the GUS reporter gene in the protoplasts and partially digested cells of *Ulva lactuca* L. (Chlorophyta). Bot Mar 39:467–474

Inoue A, Kagaya M, Ojima T (2007) Preparation of protoplasts from Laminaria japonica using native and recombinant abalone alginate lyases. J Appl Phycol DOI 10.1007/s10811-007-9199-y

Jiang P, Qin S, Tseng CK (2003) Expression of the *lacZ* reporter gene in the sporophytes of the seaweed *Laminaria japonica* (Phaeophyceae) by gametophyte-targeted transformation. Plant Cell Rep 21:1211–1216

Kajiwara T, Hatanaka A, Fujimura T, Kawai T, Irie M (1988) Isolation of protoplasts from marine brown algae *Dictyotaceae* plants. Nippon Suisan Gakkaishi 54:1255

Kapraun DF (1989) Recovery and development of parasexual fusion products in *Enteromorpha linza* (L.) J Ag (Ulvales, Chlorophyta). J App Phycol 1:239–246

Kapraun DF (1990) Parasexual fusion products in green algae: *Enteromorpha* and *Ulvaria* (Ulvales, Chlorophyta). Hydrobiologia 204/205:151–159

Kawashima Y, Matsumoto M, Tokuda H (1989) Protoplasts isolation and regenration of a green alga, *Enteromorpha prolifera*. In: Miyachi S, Karube I, Ishida Y (eds) Current topics in marine biotechnology. Jap Soc Mar Biotechnol, Tokyo, pp 231–234

Kevekordes K, Beardall J, Clayton MN (1993) A novel method for extracting protoplasts from large brown algae. J Exp Bot 44:1587–1593

Kim GH, Tatiana AK, Yoon KS, Song YS, Lee KP (2005) Purification and characterization of a lectin, bryohlealin, involved in the protoplast formation of a marine green alga *Bryaopsis plumosa* (Chlorophyta). J Phycol 42:86–95

Kito H, Kunimoto M, Kamanishi Y, Mizukami Y (1998) Protoplasts fusion between *Monostroma nitidum* and *Porphyra yezoensis* and subsequent growth of hybrid plants. J Appl Phycol 10:15–21

Kloareg B, Quatrano RS (1987) Isolation of protoplasts from zygotes of *Fucus distichus* (L.) Powell (Phaeophyta). Plant Sci 50:189–194

Kloareg B, Quatrano RS (1988) Structure of the cell walls of marine alage and ecophysiological functions of the matrix polysaccharides. Oceanogr Mar Biol Annu Rev 26:259–315

Kloareg B, Polne-Fuller M, Gibor A (1989) Mass production of viable protoplasts from *Macrocystis pyrifera* (L.) C. Ag. (Pheophyta). Plant Sci 62:105–112

Klotchkova TA, Ok KC, John AW, Kim GH (2003) Cytochemical and ultrastructural studies on protoplast formation from disintegrated cells of the marine alga *Chaetomorpha aerea* (Chlorophyta). Eur J Phycol 38:205–216

Kobayashi K (1975) Growth of extra cellular protoplast of *Bryopsis maxima* in an agar medium. Bull Tokyo Gakugei Univ Ser 27:1–5

Kreger DR (1962) Cell walls. In: Lewin RD (ed) Physiology and biochemistry of algae. Academic Press, New York, pp 315–335

Krishna Kumar GR, Addepalli MK, Reddy CRK (1999) Regeneration of the thallus of *Monostroma oxyspermum* (Chlorophyta) from protoplasts in axenic culture. Phycologia 38:503–507

Kuang M, Wang SJ, Li Y, Shen DL, Tseng CK (1998) Transient expression of exogenous GUS gene in *Porphyra yezoensis* (Rhodophyta). Chin J Oceanol Limnol 16:56–61

Kubler JE, Minocha SC, Mathieson AC (1994) Transient expression of the GUS reporter gene in protoplasts of *Porphyra miniata* (Rhodophyta). J Mar Biotechnol 1:165–169

Kuhlenkamp R, Muller D G (1994) Isolation and regeneration of protoplasts from healthy and virus infected gametophytes of *Ectocarpus siliculosus* (Phaeophyceae). Bot Mar 37:525–530

Kurtzman AL, Cheney DP (1991) Direct gene transfer and transient gene expression in a marine red alga using the biolistic method. J Phycol 27(Suppl.):42 pp

Le Gall Y, Braud JP, Kloareg B (1990) Protoplast production in *Chondrus crispus* gametophytes (Gigartinales, Rhodophyta). P1 Cell Rep 8:582–585

Lin CM, Larsen J, Yarish C, Chen T (2001) A novel gene transfer in *Porphyra*. J Phycol 37(Suppl):29–30

Liu QY, Chen LCM, Taylor ARA (1992) Ultrastructure of cell wall regeneration by isolated protoplasts of *Palmaria palmata* (Rhodophyta). Bot Mar 35:21–33

Liu HQ, Yu WG, Dai JX, Gong QH, Yang KF, Zang Y (2003) Increasing the transient expression of GUS gene in *Porphyra yezoensis* by 18S rDNA targeted homologous recombination. J Appl Phycol 15:371–377

Mackie W, Preston RD (1968) The occurrence of mannan microfibrils in the green algae *Codium fragile* and *Acetabularia crenulata*. Planta 79:249–253

Matsumoto M, Kawashima Y, Fukui E (1995) Transduction of high temperature-tolerant to *Porphyra* by cell fusion. Kaiyo Monthly 27:677–682

Matsumura W, Yasui H, Yamamoto H (2000) Mariculture of *Laminaria japonica* (Laminariales, Phaeophyceae) using protoplast regeneration. Phycol Res 48:169–176

Medcalf DG, Root CF, Gravey CL, Mukhopadhyay D, Miller CK, Hopewell WD (1972) Chemical characterization of mucilaginous polysaccharides from Ulvaceae species native to the Puget Sound. In: Nisizawa K, Arasaki S, Chihara M, Hirose H, Nakamura Y, Tsuchiya Y (eds) Proceedings of the 7th Int Seaweed Symp., Univ. Tokyo, pp 541–547

Mejjad M, Loiseaux-de-Goer S, Ducreux G (1992) Protoplast isolation, development and regeneration in different strains of *Pilayella littoralis* (L.) Kjellm. (Phaeophyceae). Protoplasma 169:42–48

Millner PA, Callow ME, Evans LV (1979) Preparation of protoplasts from the green alga *Enteromorpha intestinalis* (L.). Planta (Berlin) 147:174–177

Mizukami Y, Okauchi M, Kito H, Ishimoto S, Ishida T, Fuseya M (1995) Culture and development of electrically fused protoplasts from red marine algae, *Porphyra yezoensis* and *P. suborbiculata*. Aquaculture 132:361–367

Mizukami Y, Hado M, Kito H, Kunimoto M, Murase N (2004) Reporter gene introduction and transient expression in protoplasts of *Porphyra yezoensis*. J Appl Phycol 16:23–29

Mollet JC, Verdus MC, Kling R, Morvan H (1995) Improved protoplast yield and cell wall regeneration in *Gracilaria verrucosa* (Huds.) Papenfuss (Gracilariales, Rhodophyta). J Exp Bot 46:239–247

Mussio I, Rusig AM (2006) Isolation of protoplasts from *Fucus serratus* and *F. vesiculosus* (Fucales, Phaeophyceae): factors affecting protoplast yield. J Appl Phycol 18:733–740

Neuhaus G, Neuhaus UG, Groot EJ, Schweiger HG (1986) High yield and stable transformation of the unicellular green alga *Acetabularia* by microinjection of SV40 DNA and pSV2neo. EMBO J 5:1437–1444

Nisizawa K, Fujibayashi S, Kashiwabara Y (1968) Alginate lyases in the hepatopancreas of a marine mollusc, *Dolabella auricula* Solander. J Biochem 64:25–37

Nikolaeva EV, Usov AI, Sinitsyn AP, Tambiev AH (1999) Degradation of agarrophytic red algal cell wall components by new crude enzyme preparations. J Appl Phycol 11:385–389

Okauchi M, Mizukami Y (1999) Transient β-Glucuronidase (GUS) gene expression under control of CaMV 35S promoter in *Porphyra tenera* (Rhodophyta). Bull Natl Inst Aquacult 1:13–18

Packer MA (1994) Protoplast isolation from single cells and small tissue fragments of wild *Porphyra* fronds (Rhodophyta). Bot Mar 37:101–108

Patel KS (1979) A new record of agar digesting microflora. Geobios 6:222

Payton MA, Roberts CF (1979) Agar as carbon source in relation to the isolation of lactose non-utilizing mutants of *Aspergillus nidulans*. J Gen Microbiol 110:475–478

Percival E, Young M (1971) Low molecular weight carbohydrates and water-soluble polysaccharide metabolism by the Cladophorales. Phytochemistry 10:807–810

Polne-Fuller M, Gibor A (1984) Developmental studies in *Porphyra*. I. blade differentiation in *Porphyra perforata* as expressed by morphology, enzymatic digestion and protoplast regeneration. J Phycol 20:609–616

Polne-Fuller M, Gibor A (1987) Calluses and callus like growth in seaweeds: induction and culture. Hydrobiologia 151/152:131–138

Polne-Fuller M, Gibor A (1990) Developmental studies in *Porphyra* (Rhodophyceae) III. Effect of culture conditions on wall regeneration and differentiation of protoplasts. J Phycol 26:674–682

Polne-Fuller M, Biniaminov M, Gibor A (1984) Vegitative propagation of *Porphyra perforata*. Proc Int Seaweed Symp 11:308–313

Polne-Fuller M, Rogerson A, Amano H, Gibor A (1990) Digestion of seaweeds by the marine amoeba *Trichosphaerium*. Hydrobiologia 204/205:409–413

Qin S, Zhang J, Li WB, Wang XH, Tong S, Sun YR, Tseng CK (1994) Transient expression of GUS gene in phaeophytes using biolistics particle delivery system. Oceanol Limnol Sin 25:353–356

Qin S, Jiang P, Li XP, Tseng CK (1998a) The expression of *lacZ* in regenerated sporophytes of parthenogenetic *Laminaria japonica*. In: Proc. of the 2nd Asia-Pacific Marine Biotechnology Conference and 3rd Asia-Pacific Conference on Algal Biotechnology, Pahokee (Thailand), pp 205–208

Qin S, Wu JQ, Wang XH, Li XP, Jiang P, Tseng CK (1998b) The expression of foreign gene in *Laminaria japonica* (Laminariales, Phaeophyta). In: Mortan B (ed) The marine biotechnology of the South China Sea. Hong Kong University Press, Hong Kong, pp 209–217

Qin S, Yu DZ, Jiang P, Teng CY, Zeng CK (2003) Stable expression of *lacZ* reporter gene in seaweed *Undaria pinnatifida*. High Technol Lett 13:87–89 (in Chinese with English abstract)

Qin S, Jiang P, Tseng CK (2005) Transforming kelp into a marine bioreactor. Trends Biotechnol 23:264–268

Quatrano RS, Caldwell BA (1978) Isolation of a unique marine bacterium capable of growth on a wide variety of polysaccharides from macroalgae. Appl Environ Microbiol 36:979–981

Reddy CRK (1991) Studies on the protoplast preparation, fusion and regeneration of fusion products from members of the Ulvaceae (Chlorophyta). PhD thesis, Nagasaki University, Nagasaki, Japan, 129 pp

Reddy CRK, Fujita Y (1989) Protoplast isolation and fusion of *Ulva pertusa* and *U. conglobata*. In: Miyachi S, Karube I, Ishida Y (eds) Current topics in marine biotechnolgy. Jpn Soc Mar Biotechnol, Tokyo, pp 235–238

Reddy CRK, Fujita Y (1991) Regeneration of plantlets from *Enteromorpha* (Ulvales, chlorophyta) protoplasts in axenic culture. J Appl Phycol 3:265–275

Reddy CRK, Migita S, Fujita Y (1989) Protoplasts isolation and regeneration of three species of *Ulva* in axenic culture. Bot Mar 32:483–490

Reddy CRK, Iima M, Fujita Y (1992) Induction of fast growing and morphologically different strains through intergeneric protoplasts fusion of *Ulva* and *Enteromorpha* (Ulvales, chlorophyta). J Appl Phycol 4:57–65

Reddy CRK, Fujita Y, Bajaj YPS (1994) Somatic hybridisation in algae. In: Bajaj YPS (ed) Biotechnology in agriculture and forestry 27. Somatic hybridisation in crop improvement I. Springer-Verlag, Berlin Heidelberg, pp 483–502

Reddy CRK, Dipakkore S, Rajakrishan Kumar G, Jha B, Cheney DP, Fujita Y (2006) An improved enzyme preparation for rapid mass production of protoplast as seed stock for aquaculture of macrophytic marine green algae. Aquaculture 260:290–297

Rodde HRS, Larsen B (1997) Protoplasts of *Laminaria digitata* and *Laminaria saccharina* (Phaeophyta)-Cultivation and biosynthesis of alginate. Bot Mar 40:391–395

Roeder V, Collen J, Rousvoal S, Corre E, Leblanc BC (2005) Identification of stress gene transcripts in *Laminaria digitata* (Phaeophyceae) protoplast cultures by expressed sequence tag analysis. J Phycol 41:1227–1235

Ross AG (1953) Some typical analyses of red seaweeds. J Sci Food Agric 4:333–335

Rusig A, Cosson J (2001) Plant regeneration from protoplasts of *Enteromorpha intestinalis* (Chlorophyta, Ulvophyceae) as seed stock for macroalgal culture. J Appl Phycol 13:103–108

Rusig AM, Le Guyader H, Ducreux G (1994) Differentiation and microtubule reorganization in the apical cell protoplast of *Sphacelaria* (Phaeophyceae). Protoplasma 179:83–94

Saga N (1984) Isolation of protoplasts from edible seaweeds. Bot Mag Tokyo 97:423–427

Saga N, Sakai Y (1984) Isolation of protoplast from *Laminaria* and *Porphyra*. Nippon Suisan Gakkaishi 50:1085

Saga N, Kudo T (1989) Isolation and culture of protoplasts from the marine green alga *Monostroma angicava*. J Appl Phycol 1:25–30

Saga N, Polne-Fuller M, Gibor A (1986) Protoplasts from seaweeds: production and fusion. Beih Nova Hedweg 83:37–43

Salvador RC, Serrano AE (2005) Isolation of protoplasts from tissue fragments of Philippine cultivars of *Kappaphycus alvarezii* (Solieriaceae, Rhodophyta). J Appl Phycol 17:15–22

Sawabe T, Ezura Y, Yamamoto H (1997) Plant regeneration from protoplasts of *Laminaria japonica* Areschoug (Laminariales, Phaeophyceae) in a continuous-flow culture system. Plant Cell Rep 17:109–112

Shikh D, Reddy CRK, Jha Bhavanath (2005) Production and seeding of protoplasts of *Porphyra okhaensis* (Bangiales, Rhodophyta) in laboratory culture. J Appl Phycol 17:331–337

Smith RG, Bidwell RGS (1989) Inorganic carbon uptake by photosynthetically active protoplasts of the red macroalga *Chondrus crispus*. Mar Biol 102:1–4

Song Ho S, Chung GH (1988) Isolation and purification of protoplasts from *Porphyra tenera* thalli. Aquaculture 1:103–108

Stevens PT, Levin RE (1979) Purification and characteristics of an alginase from *Alginovibrio auatilis*. Appl Environ Microbiol 33:1156–1161

Takamistu C, Hideto K, Akihiro S, Takashi B (2006) Detection of alginate oligosaccharides from mollusk. Biosci Biotechnol Biochem 70:2793–2796

Tang Y (1982) Isolation and cultivation of the vegetative cells and protoplasts of *Porphyra suborbiculata* Kjellm. J Shandong Coll Oceanol 12:37–50

Tatewaki M, Nagata K (1970) Surviving protoplasts in vitro and their development in *Bryopsis*. J Phycol 6:401–403

Tokuda H, Kawashima Y (1988) Protoplast isolation and culture from a brown alga, *Undaria pinnatifida*. In: Stadler T, Mollion J, Verdus MC, Karamanos Y, Morvan H, Christaen D (eds) Algal biotechnology. Elsevier Applied Science, London, pp 151–157

Uchida T, Arima S (1992) Regeneration of protoplasts isolated from the sporophyte of *Cladosiphon okamuranus* Tokida (Chordariaceae, Phaeophyta). Jpn J Phycol 40:261–266

Uchida A, Yoshikawa T, Ishida Y, Saga N (1992) Stable protoplasts isolation and its regeneration into thallus of the marine green alga *Ulva pertusa*. Nippon Suisan Gakkaishi 58:153–157

Uppalapati SR, Fujita Y (2002) A simple method for mass isolation of protoplasts from species of *Monostroma*, *Enteromorpha* and *Ulva* (chlorophyta, Ulvales). J Appl Phycol 14:165–168

Waaland JR, Dickson LG, Watson B A (1990) Protoplasts isolation and regeneration in the marine red alga *Porphyra nereocystis*. Planta 181:522–528

Wakabayashi T, Kuboi T, Tuboi MK, Hara M (1999) Preparation of high yields of algal protoplasts using buccal juice of sea hare and commercial cellulase. Mar Biotechnol 1:407–410

Wong TY, Preston LA, Schiller NL (2000) Alginate Lyase: Review of major sources and enzyme characteristics, structure-function analysis, biological roles, and applications. Annu Rev Microbiol 54:289–340

Yamaguchi K, Araki T, Aoki T, Tseng C, Kitamikado M (1989) Algal cell wall degrading enzymes from viscera of marine animals. Nippon Suisan Gakkaishi 55:105–110

Yamasaki T, Miyazaki Y, Kamei Y (1998) Isolation of bacteria that decompose major polysaccharides in the cell wall of the marine red alga *Porphyra yezoensis* and their application for protoplast production. Can J Microbiol 44:789–794

Yamaura I, Matumoto T, Funatsu M, Murai E (1990) Purification and some properties of endo-b-1,3-xylanase from *Pseudomonas* sp. PT-5. Agric Biol Chem 54:921–926

Yan XH, Wang SJ (1993) Regeneration of whole plants from *Gracilaria asiatica* Chang et Xia protoplasts (Gracilariaceae, Rhodophyta). Hydrobiologia 260/261:429–436

Yu DZ, Yang LL, Jiang P, Qin S, Zeng CK (2002) Transient expression of LacZ reporter gene in the economic seaweed *Undaria pinnatifida*. High Technol Lett 12:93–95

Yukihisa A, Yuto K (2006) Preparation of recombinant polysaccharide-degrading enzymes from the marine bacterium, Pseudomonas sp. ND137 for the production of protoplasts of *Porphyra yezoensis*. Eur J Phycol 41:321–328

Zablackis E, Vreeland V, Kloareg B (1993) Isolation of protoplasts from *Kappaphycus alvarezii* var. tambalang (Rhodophyta) and secretion of carrageenan fragments by cultured cells. J Exp Bot 44:1515–1522

Zhang D (1983) Study on the protoplasts preparation, culture and fusion of somatic cells from two species of green algae *Ulva linza* and *Monostroma angicava* Kjellm. J Shandong Coll Oceanol 13:57–65

Preparation of protoplasts from *Laminaria japonica* using native and recombinant abalone alginate lyases

Akira Inoue · Mayu Kagaya · Takao Ojima

Originally published in the Journal of Applied Phycology, Vol 20, No 5, 183–190.
DOI: 10.1007/s10811-007-9199-y

Abstract *Laminaria japonica* protoplasts were released with high yields using the abalone alginate lyase HdAly in combination with a cellulase and chelating agents. Addition of EDTA at concentrations higher than 10 mM to *Laminaria* thalli which had been preincubated with HdAly and Cellulase Onozuka, dramatically improved the yield of protoplasts. EDTA was far more effective than EGTA, indicating that chelating divalent metal ions such as Mg^{2+} and Sr^{2+} in addition to Ca^{2+} is a key factor for high-yield production of *Laminaria* protoplasts. Protoplasts had a mean diameter of 27 μm, suggesting that most protoplasts were derived from cortical cells rather than epidermal layer cells. Recombinant HdAly (rHdAly) was produced from a cDNA clone in the Sf9 insect cell expression system. rHdAly had substantially the same enzymatic properties and protoplast-producing ability as did native HdAly. The optimal conditions for high yield production of protoplasts from *Laminaria* using native and recombinant HdAlys were investigated.

Keywords Abalone · Alginate lyase · Brown algae · *Laminaria japonica* · Protoplast

Introduction

Brown algae are one of the most valuable seaweeds, and have been used as not only food but also as source of industrial and pharmaceutical commodities. *Laminaria* species are distributed worldwide, and its 45 species are among the most prolific of the macrophytic algae (Kain 1971). *Laminaria japonica* Areschoug is one of at least 17 *Laminaria* species that thrive in East Asia, and is commonly found in the waters of Japan. *L. japonica* is harvested both from the wild and under domestic cultivation. According to recent taxonomic research based on multi-gene sequencing analysis (Yoon et al. 2001; Lane et al. 2006), the former genus *Laminaria* has been divided into two major clades, with *L. japonica* reassigned to genus *Saccharina*.

The techniques necessary for cellular engineering of brown algae, such as protoplast fusion and genetic manipulation are fundamental to the improvement of frond quality and thus the commercial value of algae. However, these methods have yet to be firmly established. Some studies have shown that a mixture of the polysaccharide-degrading enzymes, cellulase and alginate lyase, are useful for isolating protoplasts from brown algae (Saga and Sakai 1984; Saga 1984; Saga et al. 1986; Kloareg and Quatrano 1987a,b; Polne-Fuller and Gibor 1987; Fisher and Gibor 1987; Ducreux and Kloareg 1988; Kajiwara et al. 1988; Tokuda and Kawashima 1988; Kloareg et al. 1989; Butler et al. 1989; Sawabe et al. 1993; Wakabayashi et al. 1999; Matsumura et al. 2000). Cellulose and alginate are cell wall components of *Laminaria*. Thus, a mixture of cellulase and alginate lyase could be expected to degrade brown algal cell-wall structure, allowing the release of protoplasts. Crude or purified preparations of alginate lyase from marine invertebrates or bacteria, along with fungal cellulase have typically been used for the preparation of protoplasts from brown algae.

The first report of the production of protoplasts from *Laminaria* species using crude extracts from sea urchin (*Strongylocentrotus intermedius*) was published in 1984 (Saga & Sakai). Later, Butler et al. (1989) produced

A. Inoue · M. Kagaya · T. Ojima (✉)
Laboratory of Marine Biotechnology and Microbiology,
Graduate School of Fisheries Sciences,
Hokkaido University, 3-1-1 Minato-cho,
Hakodate, Hokkaido 041-8611, Japan
e-mail: ojima@fish.hokudai.ac.jp

protoplasts from *L. saccharina and L. digitata* using mannuronan lyases from abalone (*Haliotis tuberculata*) and guluronan lyase from *Pseudomonas alginovora*, along with a partially purified cellulase (CELF) from *Trichoderma reesei*, and demonstrated that pretreatment of thallus with EGTA increased protoplast yield. Sawabe et al. (1993) showed that purified alginate lyase from a culture supernatant of *Alteromonas* sp. strain H-4, along with a commercially available cellulase could yield protoplasts from *L. japonica*. However, acetone powder extracts from abalone pancreas did not yield any protoplasts under otherwise identical conditions (Sawabe and Ezura 1996), but Matsumura et al. (2000) were able to isolate protoplasts using abalone acetone powder extracts. Although the latter experiment used EGTA thallus pretreatment, this discrepancy may be due to differences in alginate lyase activity in the extracts. It is well known that abalone acetone powder contains not only alginate lyase but also polysaccharide-degrading enzymes such as cellulase and fucosidase, but their composition and activities strongly depend on the season when the abalone are harvested, and on the production of enzyme in the individual abalone. An alginate lyase, HdAly, from the pacific abalone (*H. discus hannai*) has been purified and characterized (Shimizu et al. 2003). Mature HdAly is an endolytic mannuronan lyase that consists of 256 amino acids with a molecular weight of 28,896.

Although abalone acetone extract powder has been used for the production of protoplasts from *L. japonica* as described above, it is still unknown if HdAly efficiently degrades *Laminaria* cell wall and cell wall matrix components. It is also not clear how important HdAly is in the production of protoplasts from *L. japonica*. In the present paper the effectiveness of purified HdAly and recombinant HdAly (rHdAly) expressed in insect cell line Sf9 is evaluated in the preparation of protoplasts from *L. japonica*.

Materials and methods

Laminaria japonica Areschoug was harvested at Nanaehama, Hakodate or Usujiri in Hokkaido, Japan. Restriction and modifying enzymes were purchased from Takara (Otsu, Japan). *Pfu* Ultra High-Fidelity DNA polymerase (Stratagene, La Jolla, CA, USA) was used for PCR-based mutagenesis. *Escherichia coli* TOP10 and insect cell line Sf9 were purchased from Invitrogen (Carlsbad, USA). Native HdAly and its substrate sodium alginate were prepared by the method of Shimizu et al. (2003).

Isolation of protoplasts Thallus of *L. japonica* was gently cleaned with 70% ethanol, cut into pieces (0.01 g), and incubated in artificial sea water (SEA LIFE, Marine Tech, Tokyo, Japan) containing 0.6 M mannitol, Cellulase Onozuka R-10 (Yakult Honsha Co. Ltd., Tokyo, Japan), and HdAly on an orbital shaker (60 rpm). Residual thallus debris was removed by filtration through an 80 μm nylon net. Protoplasts were collected by centrifugation at 350 × g for 10 min and observed microscopically. Protoplasts were counted using a hemacytometer (Erma Inc., Tokyo, Japan). Data are shown as means ± SD from three experiments.

Alginate lyase activity assay The alginate lyase activity of HdAly was measured as described previously (Shimizu et al. 2003). Lyase activity was assayed in a 1 mL reaction mixture containing 0.1% sodium alginate and 10 mM sodium phosphate (pH 7.0) at 30°C unless mentioned otherwise. One unit of lyase activity is defined as the amount of enzyme that increases absorbance at 235 nm by 0.01 in 1 min.

Cellulase activity assay Cellulase activity was assayed in artificial sea water containing 0.5% carboxy methyl cellulose (CMC) and Cellulase Onozuka R-10 at 30°C. The reducing sugar liberated by hydrolysis of CMC was determined by the method of Park and Johnson (1949). One unit of cellulase was defined as the amount of enzyme that liberates reducing sugars equivalent to 1.0 μmol of glucose per min under the conditions described above.

SDS-polyacrylamide gel electrophoresis and western blotting SDS-polyacrylamide gel electrophoresis (SDS-PAGE) was done by the method of Porzio and Pearson (1977). For western blotting, protein was electroblotted to a nitrocellulose membrane after electrophoresis. Anti-c-*myc*- or Anti-6×His-tag antibodies (Santa Cruz Biotechnology, Santa. Cruz, CA) and alkaline phosphatase-conjugated anti-rabbit IgG goat antibody (Sigma, St. Louis, MO) were used as primary and second antibodies, respectively. Signal detection was performed with Super signal west pico chemiluminescent substrate (Pierce, Rockford, IL).

Recombinant HdAly expression and purification A cDNA encoding the mature form of HdAly, which corresponds to residues 18–273, except for the signal peptide, was subcloned into the *Bam*H I and *Xho* I sites of pFastBac1 (Invitrogen). To facilitate detection and purification, a c-*myc* epitope tag, -EQKLISEEDL-, and octa-histidine tag, -HHHHHHHH, were introduced at the C-terminus by site-directed mutagenesis. The nucleotide sequence of the clone was confirmed using an automated DNA sequencer (type 3130 xl, Applied Biosystems, Foster City, CA). Bacmid DNA and recombinant baculovirus were prepared according to the manufacturer's protocol. Sf9 cells were infected with recombinant virus at an m.o.i. of 5. Cells were cultured at 27°C for 3 days. Harvested cells were sonicated in a solution containing 50 mM sodium phosphate (pH 8.0), 0.5 M NaCl, and 0.1% Triton-X100. After centrifugation at 12,000 × g for 15 min, supernatant was applied to a Ni-NTA agarose column (0.5×2 cm; Qiagen, Hilden, Germany). Resin was washed with 30 mM imidazole (pH 8.0), 0.5 M NaCl buffer. Protein was eluted with a buffer containing 100 mM imidazole (pH 8.0), 0.5 M NaCl and separated by SDS-PAGE. Fractions containing recombinant HdAly were pooled, stored on ice, and used within 3 days. Protein concentrations were determined by the Bradford (1976) method using bovine serum albumin as a standard protein.

Results

Isolation of protoplasts with purified HdAly. Laminaria japonica thalli incubated in artificial seawater containing 0.6 M mannitol, 1% Cellulase Onozuka R-10, and 100 U mL^{-1} native HdAly at 17°C yielded 0.7–1.0×10^3 protoplasts g^{-1} fresh weight and largely retained their original structure even after 24 hours. Addition of chelating agents to the enzyme solution markedly increased the yield of protoplasts (Fig. 1). The number of protoplasts was greatest (2×10^7 protoplasts g^{-1} fresh weight) at a final EDTA concentration within the range of 50–100 mM. Concentrations above 125 mM reduce protoplast yield. Protoplast yields with EGTA as the chelating agent were only 2–4×10^5 protoplasts g^{-1} fresh weight between 10–150 mM. These results suggest that chelating divalent cations significantly increases the effectiveness of enzymatic protoplast release

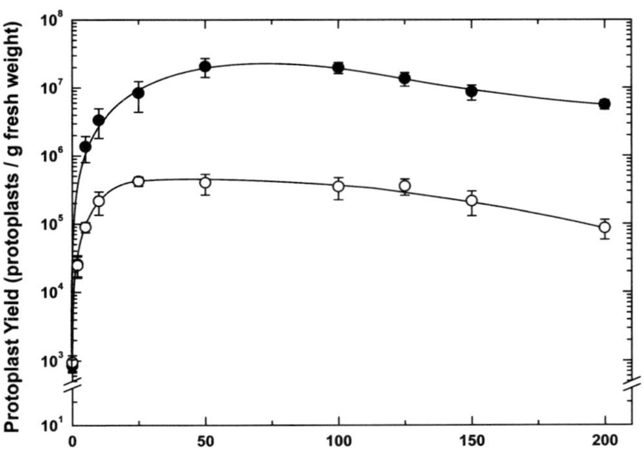

Fig. 1 Effect of chelating agents on the yield of protoplasts from *L. japonica.* Thallus pieces were incubated in artificial seawater containing 1% Cellulase Onozuka R-10, 0.6 M mannitol, and 100 U mL^{-1} HdAly at 17°C for 4 hours. EDTA or EGTA were added, and incubation continued for 4 hours. EDTA and EGTA are represented by closed and open circles, respectively

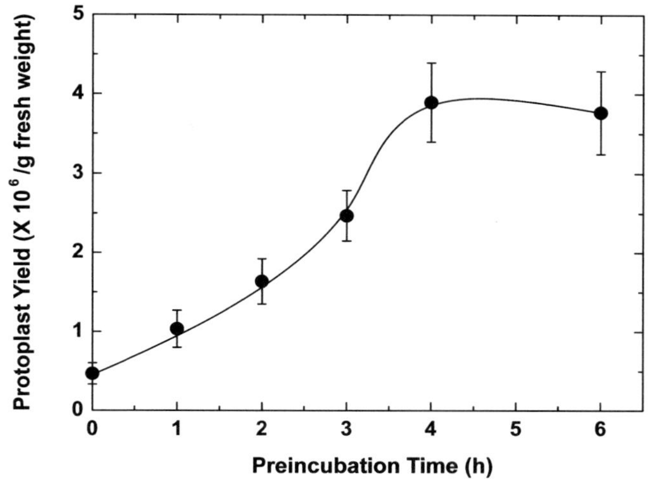

Fig. 2 Effect of HdAly and Cellulase Onozuka R-10 preincubation times on protoplast yield. Thallus pieces were preincubated with digestive enzymes before the addition of 10 mM EDTA. Other conditions were the same as described for Fig. 1

from *L. japonica* thallus. The higher release of protoplasts with the relatively non-specific divalent binding of EDTA than with EGTA, which tends to chelate Ca^{2+} rather more specifically, suggests that chelating cations other than Ca^{2+} is crucial for high-efficiency isolation of protoplasts from *L. japonica*.

Protoplast optimization with polysaccharide-degrading enzymes and EDTA To determine the most efficient incubation time for protoplasting before the addition of EDTA, thalli were incubated with HdAly and Cellulase Onozuka R-10. The number of protoplasts increased linearly up to 4 hours (Fig. 2). No alteration of frond structure was observed even after 6 hours, but the addition of EDTA caused the rapid decomposition of thallus, mostly in the cortical and medullary layers. Protoplast release after addition of 10 mM EDTA reached a maximum at 3 hours (Fig. 3). Although protoplasts were released at temperatures ranging from 5°C to 22°C, maximum protoplast yield was at 17°C, with a notable decline in the number of viable protoplasts at 27°C (Fig. 4). To test the effect of cation chelation on HdAly, alginate lyase activity was assayed in concentrations of EDTA from 0–100 mM. EDTA had no effect on HdAly activity at any concentration (data not shown).

Optimum enzyme concentration Protoplast yields increased hyperbolically with an increase of HdAly concentration (Fig. 5). Because of the diminishing returns, concentrations of 80 to 150 U mL^{-1} of HdAly are sufficient to yield nearly maximum amounts of protoplasts. Yields with Cellulase Onozuka R-10 were maximized at between 2.5 and 10 U mL^{-1} (Fig. 6). Concentrations greater than 12.5 U mL^{-1} decreased yield.

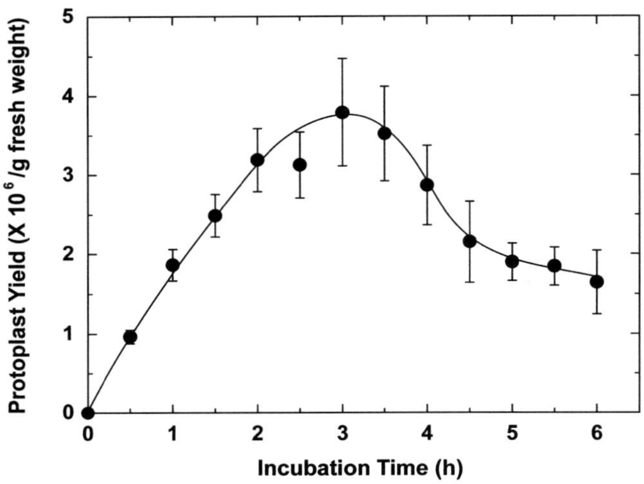

Fig. 3 Time-dependent protoplast yield after addition of EDTA. *L. japonica* protoplasts were released with cellulase and HdAly, followed by 10 mM EDTA and collected at the indicated times. Other conditions were the same as described for Fig. 1

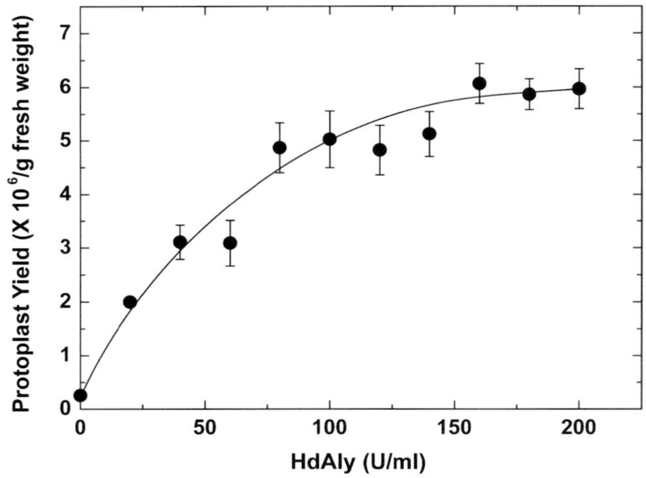

Fig. 5 Effect of alginate lyase activity of HdAly on the yield of protoplast. Thallus pieces were incubated in the same condition with Fig. 1 except for various concentrations of HdAly. After EDTA was added to be 10 mM, incubation continued for more 4 hours

Protoplast size To determine the tissue of origin of protoplasts released with HdAly, the distribution of protoplast diameters was determined (Fig. 7). More than 90% of the protoplasts ranged from 20 to 40 μm diameter, with a mean of 27 μm. This result indicates that almost all of the protoplasts originated from cortical layer rather than from the epidermal layer.

Expression and characterization of recombinant HdAly, and its utilization for protoplast preparation Recombinant HdAly fused with an octahistidine tag at the C-terminus was expressed in Sf9 cells. After Ni-NTA affinity chromatography, rHdAly was detected on SDS-PAGE as single band which was recognized with anti-c-*myc* and anti-6×His-tag antibodies (Fig. 8). Lyase activities of native and recombinant enzymes were compared with temperature and pH as variables, and alginate as substrate (Fig. 9). The alginate lyase activity of rHdAly was 1260 U mg^{-1} and activity of native enzyme was 1320 U mg^{-1} in 10 mM phosphate buffer (pH 7.0). Optimal temperature and pH of rHdAly were 30°C and 7.6, respectively, which were 5°C and 0.4 lower than native HdAly. Half-maximal activity of rHdAly was 20 min at 35°C but native enzyme still had 70% of its activity after the same incubation. These results indicate that rHdAly is not as heat stabile as native HdAly, and has slightly different optima. These differences may be caused by the absence of some post-translational modification (*e.g.*, carbohydrate chain anchoring) in the recombinant form.

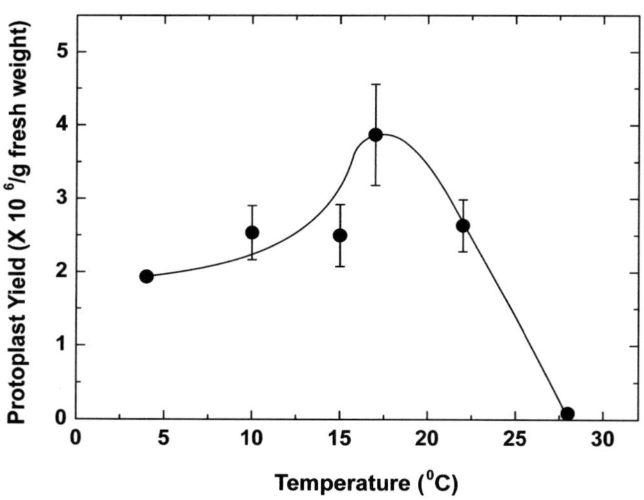

Fig. 4 Temperature-dependent protoplast yield. Thallus pieces were preincubated at temperatures ranging from 4 to 27°C before addition of 10 mM EDTA. Incubation continued for 4 hours at the preincubation temperature. Other conditions were the same as described for Fig. 1

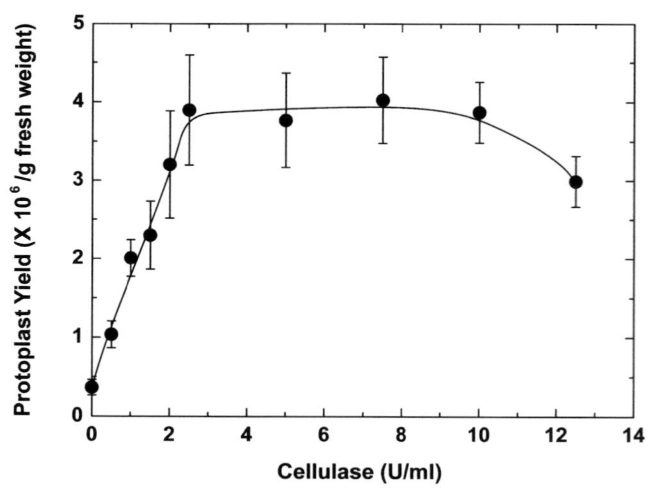

Fig. 6 Effect of cellulase activity on the yield of protoplast. Protoplasts were isolated in the same condition with Fig. 5 except for 150 U/mL HdAly and different concentrations of Cellulase Onozuka R-10. Cellulase activity of 1% Cellulase Onozuka R-10 was 4.8 U mL^{-1}

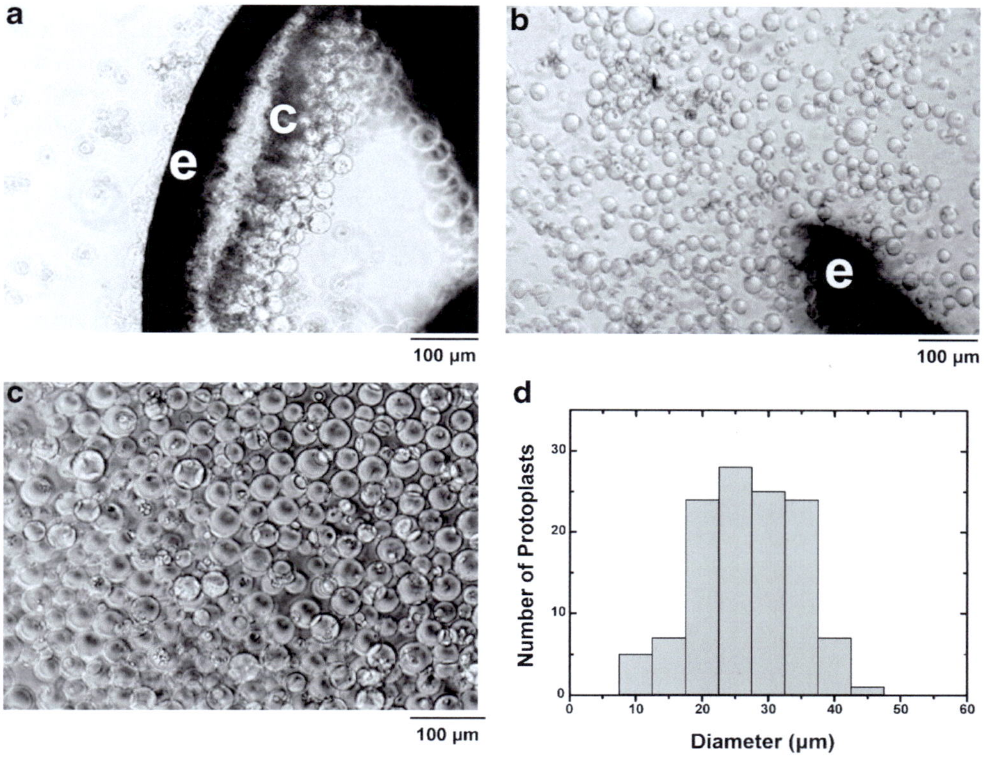

Fig. 7 Protoplasts size distribution. **a** *and* **b**, thallus and isolated protoplasts after incubation for 4 hours with 10 mM EDTA. Epidermal and cortical layers are shown in "e" and "c", respectively. **c**, isolated protoplasts after centrifugation. **d**, distribution of protoplast diameters

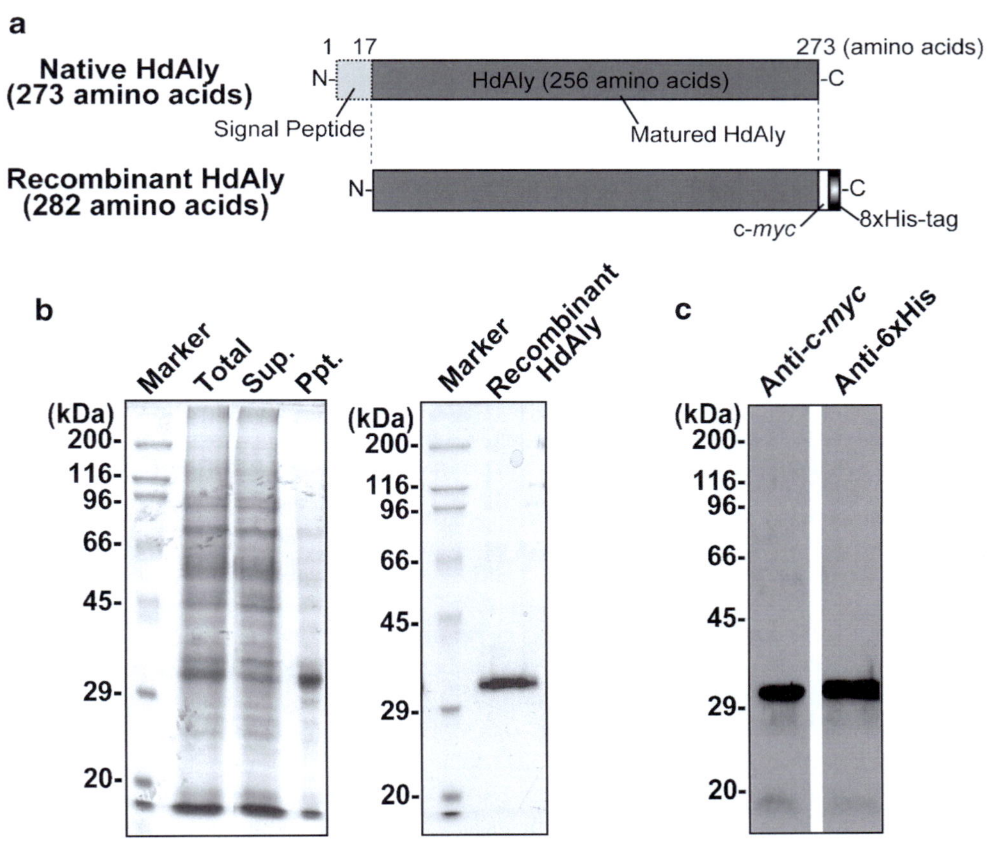

Fig. 8 Baculoviral expression of rHdAly. **a**, schematic figure of expressed rHdAly. **b**, purification of rHdAly. *Total*, total cell extracts; *Sup.*, supernatant after centrifugation; *Ppt.*, precipitate after centrifugation; *Recombinant HdAly,* purified rHdAly eluted from the Ni-NTA resin. **c**, western blotting analysis of purified rHdAly. Anti-c-*myc* and anti-6×His-tag antibodies were used as primary antibodies

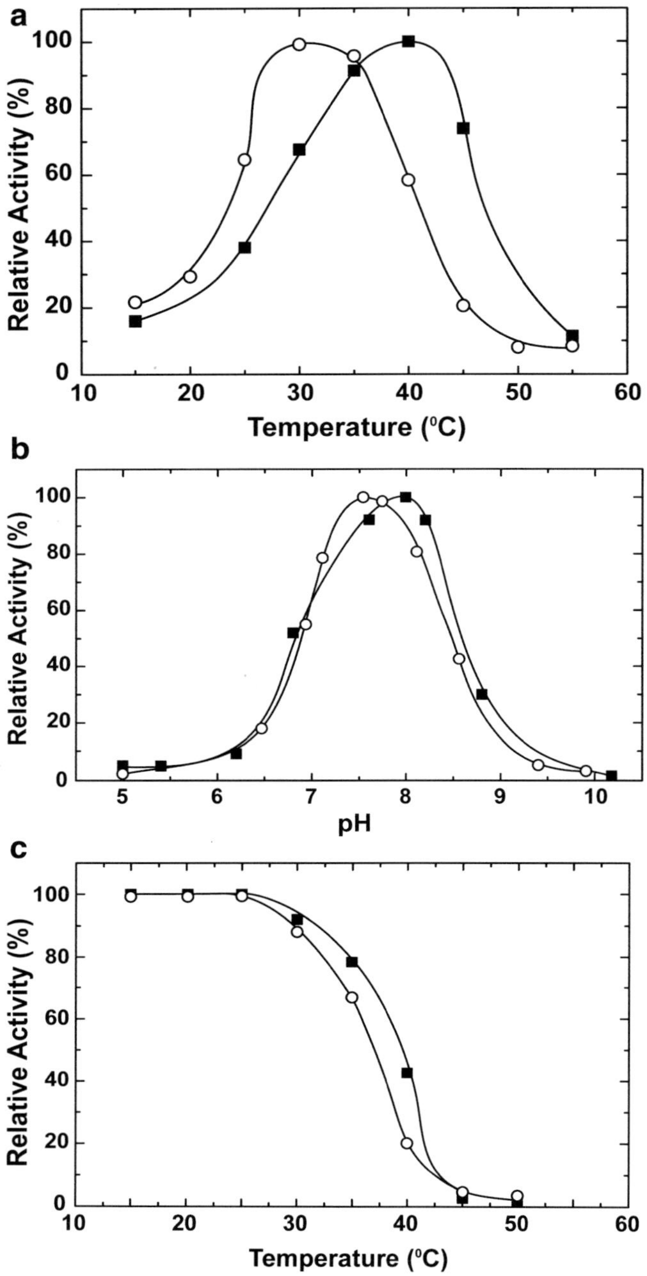

Fig. 9 Enzymatic properties of rHdAly. Closed square and open circle represent native HdAly and rHdAly, respectively. **a**, temperature dependence of native and rHdAly. Lyase activity was measured at the indicated temperature. **b**, pH dependence of native and rHdAlys. **c**, heat stability of native and rHdAlys. Samples were incubated at the indicated temperature for 20 min, and then activity was measured

rHdAly was also used for protoplasting at the optimum conditions described above, *i.e.*, 3 h preincubation of thallus pieces in artificial seawater containing 150 U mL^{-1} rHdAly, 1% Cellulase Onozuka R-10, and 0.6 M mannitol at 17°C, followed by addition of 10 mM EDTA for 3 more hours. The number of isolated protoplasts (6×10^6 cells/g protoplasts g^{-1} fresh weight) and mean diameter (28 μm) were indistinguishable from of protoplasts isolated with the native form. These results demonstrate that rHdAly can substitute for the native enzyme in the preparation of protoplasts from *L. japonica.*

Discussion

Protoplasts can be isolated from *Laminaria japonica* using native HdAly or rHdAly. The greatest advantage of using purified alginate lyase is that protoplasts are not damaged by contaminating enzymes. Reduced protoplast yields (Saga et al. 1986) and viability (Butler et al. 1989) were caused by contaminating proteases, lipases and ribonucleases in crude extracts from abalone acetone powder. The decrease in protoplast yield at high concentrations of Cellulase Onozuka R-10, but not HdAly in the present study may be due to contaminants in the crude cellulase preparation.

Early reports of successful protoplast isolation by pretreating thallus with the calcium chelator EGTA were confirmed in this work, but we found that EDTA was clearly superior to EGTA, indicating that Ca^{2+} is not the only cation involved in enzymatic release of *L. japonica* cells. Divalent metal ions, such as Mg^{2+} and Sr^{2+}, which could be taken up by the alginate of *L. japonica* in the wild may influence cell-wall stability. It is well known that metal accumulation in algae varies with season and location. In some brown algae, the content and proportion of mannuronic and guluronic acids vary with the season and frond portion (Saraswathi et al. 2003; Jothisaraswathi et al. 2006). In this study, EDTA was added after preincubation with polysaccharide-degrading enzymes. Thallus shape remained relatively intact during preincubation with enzymes without EDTA, though the combination of HdAly and cellulase digested alginate and cellulose enough to release some protoplasts. Considering that alginate can form a gel by the adsorption of divalent metal ions, partially digested alginate by HdAly still may form gel through metal ions in the absence of EDTA, allowing thalli to maintain their original structure. The optimum incubation temperature to produce protoplasts from *L. japonica* was 17°C, which is inconsistent with the optimum temperature of native and recombinant HdAly. Under these conditions, HdAly may nick rather than digest or decompose alginate. Protoplast release was maximized at 50 mM EDTA, a concentration at which divalent metal ions in artificial sea water should be almost entirely chelated by EDTA. Accordingly, metal ion chelation from not only thallus but also sea water could be a significant factor for isolating protoplasts from *L. japonica*.

The duration of incubation after addition of EDTA also affected protoplast yields, indicating that HdAly continues to digest alginate even after addition of the chelating agent. Sequestration of metal ions from alginate may cause not only decomposition of the gel matrix but may also expose other alginate molecules, which can then be digested by HdAly.

Cultured young and small thalli (0.5–2 cm in length) of *L. japonica* have been successfully used for protoplast preparation in some cases (Sawabe and Ezura 1996; Matsumura et al. 2000). These thalli have undeveloped cell walls. In our experiments, thalli of mature sporophytes (50–200 cm), which were harvested from February through August at Nanaehama and Usujiri in Hokkaido, Japan, were used. Protoplasts could be isolated not only from fronds, but also from stipes and holdfasts with the method described above (data not shown). More than 8×10^7 protoplasts g^{-1} fresh weight were isolated from stipe tissue and young thalli with the method described above. Thus, under optimized enzymatic conditions, this method gives efficient protoplast yields independent of the tissue of origin and size of thallus.

Almost all of the released protoplasts are apparently derived from cortical cells, according to the mean diameters of observed protoplasts. The epidermal layer hardly decomposed in our experiments. Isolated protoplasts from the epidermal cells of *L. japonica* divide in culture, but those from cortical and medullary cells do not (Matsumura et al. 2000). Although we have not examined regeneration of protoplasts, it is possible that these cells have a low ability to regenerate. Abnormal developmental patterns have also been reported in neophytes regenerated from *L. japonica* protoplasts (Matsumura et al. 2000). If these abnormalities are shown to be caused by cells that are damaged by contaminating enzymes in abalone acetone powder extracts, the use of purified alginate lyase would be especially useful for protoplast preparation.

rHdAly was expressed in Sf9 cells and purified, and had lyase activity identical to native enzyme except for subtle differences in optimum temperature, pH, and heat stability. Purified rHdAly can thus be used for protoplast preparation, indicating the possibility of production of plentiful and pure recombinant enzymes for protoplast preparation without depending on wild invertebrates as a source. The expression of an active recombinant cellulase may provide a superior substitute for Cellulase Onozuka R-10, which is currently the most popular enzyme in protoplast isolation. One candidate may be an endo-β-1,4-D-glucanase, HdEG66, which was isolated from abalone (*H. discus hannai*) and its cDNA cloned (Suzuki et al. 2003). We are now in the process of constructing a baculoviral expression system for HdEG66. Construction of an expression system for recombinant enzymes and minimizing the number and quantity of enzymes used to obtain protoplasts should improve protoplast yield and viability.

Total RNA has been isolated from *L. japonica* protoplasts by conventional methods, allowing the construction of a cDNA library. One of our goals is to identify proteins involved in cellular processes and to clarify their structure and function in *L. japonica*. Because of its importance to a wide variety of domestic industries, and its role in coastal ecosystems, understanding the biochemistry and molecular biology of *L. japonica* could have significant benefits. Many of the fundamental techniques necessary for proteomic research, such as cDNA cloning, protein expression studies, and functional protein analysis are currently underway.

Acknowledgments This work was supported by a Grant-in-Aid for Young Scientists (B), No. 17780164, and a 21st century COE program grant of the international COE of "Marine Bio-Manipulation Frontier for Food Production" both from the Ministry of Education, Culture, Sports, Science and Technology (MEXT), Japan, and the Akiyama Foundation.

References

Bradford MM (1976) A rapid and sensitive method for the quantitation of microgram quantities of protein utilizing the principle of protein-dye binding. Anal Biochem 72:248–254

Butler DM, Østgaard K, Boyen C, Evans LV, Jensen A, Kloareg B (1989) Isolation conditions for high yields of protoplasts from *Laminaria saccharina* and *L. digitata* (Phaeophyceae). J Exp Bot 40:1237–1246

Ducreux G, Kloareg B (1988) Plant regeneration from protoplasts of *Sphacelaria* (Phaeophyceae). Planta 174:25–29

Fisher DD, Gibor A (1987) Production of protoplasts from the brown alga, *Sargassum muticum* (Yendo) Fensholt (Phaeophyta). Phycologia 26:488–495

Jothisaraswathi S, Babu B,Rengasamy R (2006) Seasonal studies on alginate and its composition II: *Turbinaria conoides* (J.Ag.) Kütz. (Fucales, Phaeophyceae). J Appl Phycol 18:161–166

Kain JM (1971) Synopsis of biological data on *Laminaria hyperborea*. FAO Fish Synop 84:1–74

Kajiwara T, Hatanaka A, Fujimura T, Kawai T, Irie M (1988) Isolation of protoplasts from marine brown algae Dictyotaceae plants. Nippon Suisan Gakkaishi 547:1255

Kloareg B, Quatrano RS (1987a) Enzymatic removal of the cell walls from zygotes of *Fucus distichus* (L.) Powell (Phaeophyta). Hydrobiologia 151/152:123–129

Kloareg B, Quatrano RS (1987b) Isolation of protoplasts from zygotes of *Fucus distichus* (L.) Powell (Phaeophyta). Plant Sci 50:189–194

Kloareg B, Polne-Fuller M, Gibor, A (1989) Mass production and regeneration of protoplasts from *Macrocystis pyrifera*. Plant Sci 62:105–112

Lane CE, Mayes C, Druehl LD, Saunders GW (2006) A multi-gene molecular investigation of the kelp (Laminariales, Phaeophyceae) supports substantial taxonomic re-organization. J Phycol 42:493–512

Matsumura W, Yasui H, Yamamoto H (2000) Mariculture of *Laminaria japonica* (Laminariales, Phaeophyceae) using protoplast regeneration. Phycol Res 48:169–176

Park JT, Johnson MJ (1949) A submicrodetermination of glucose. J Biol Chem 181:149–151

Polne-Fuller M, Gibor A (1987) Calluses and callus-like growth in seaweeds, induction and culture. Hydrobiologia 151/152:131–138

Porzio MA, Pearson AM (1977) Improved resolution of myofibrillar proteins with sodium dodecyl sulfate-polyacrylamide gel electrophoresis. Biochim Biophys Acta 490:27–34

Saga N (1984) Isolation of protoplasts from edible seaweeds. Bot Mag Tokyo 97:423–427

Saga N, Sakai Y (1984) Isolation of protoplasts from *Laminaria* and *Porphyra*. Bull Jap Soc Sci Fish 50:1085

Saga N, Polne-Fuller M, Gibor A (1986) Protoplasts from seaweed: production and fusion. Beih Nova Hedwigia 83:37–43

Saraswathi SJ, Babu B, Rengasamy R (2003) Seasonal studies on the alginate and its biochemical composition I: Sargassum polycystum (Fucales), Phaeophyceae. Phycol Res 51:240–243

Sawabe T, Ezura Y, Kimura T (1993) Application of an alginate lyase from *Alteromonas* sp. for isolation of protoplasts from a brown algae *Laminaria japonica*. Bull Jap Soc Sci Fish 59:705–709

Sawabe T, Ezura Y (1996) Regeneration from *Laminaria japonica* Areschoug (Laminariales, Phaeophyceae) protoplasts isolated with bacterial alginase. Plant Cell Rep 15:892–895

Shimizu E, Ojima T, Nishita K (2003) cDNA cloning of an alginate lyase from abalone, *Haliotis discus hannai*. Carbohydr Res 338:2841–2852

Suzuki K, Ojima T, Nishita K (2003) Purification and cDNA cloning of a cellulase from abalone *Haliotis discus hannai*. Eur J Biochem 270:771–778

Tokuda H, Kawashima Y (1988) Protoplast isolation and culture of a brown alga, *Undaria pinnatifida*. In: Stadler T, Mollion J, Verdus MC, Karamanos Y, Morvan H, Christiaen D (eds) Algal biotechnology. Elsevier Applied Science, London, pp 151–159

Wakabayashi T, Kuboi T, Tuboi T, Kaji M, Hara M (1999) Preparation of high yields of algal protoplasts using buccal juice of sea hare and commercial cellulase. Mar Biotechnol 1:407–410

Yoon HS, Lee JY, Boo SM, Bhattacharya D (2001) Phylogeny of Alariaceae, Laminariaceae, and Lessoniaceae (Phaeophyta) based on plastid-encoded RuBisCo spacer and nuclear-encoded ITS sequence comparisons. Mol Phylogenet Evol 21:231–243

Protoplast isolation and regeneration from *Gracilaria changii* (Gracilariales, Rhodophyta)

Hui-Yin Yeong · Norzulaani Khalid · Siew-Moi Phang

Originally published in the Journal of Applied Phycology, Vol 20, No 5, 191–201.
DOI: 10.1007/s10811-007-9249-5

Abstract The tropical agarophyte *Gracilaria changii* has been much researched and documented by the Algae Research Laboratory, University of Malaya, especially with regards to its potential as a seaweed bioreactor for valuable compounds. Protoplast regeneration of this seaweed was developed following the optimization of protoplast isolation protocol. Effect of the concentration and combination of isolating enzymes, incubation period, temperature, enzyme solution pH, tissue source on the protoplast yields were used to optimize the isolation protocol. The enzyme mixture with 4% w/v cellulase Onozuka R-10, 2% w/v macerozyme R-10 and 1 unit mL^{-1} agarase was found to produce the highest yield of protoplast at 28°C and 3 h incubation period. Thallus tips gave higher yields of protoplasts than middle segments. Freshly isolated *G. changii* protoplasts were cultured in MES medium. Regeneration of protoplast cell walls after 24 h was confirmed by calcofluor white M2R staining under UV fluorescence microscopy. The protoplasts with regenerated cell walls then underwent a series of cell division to produce callus-like cell masses in MES medium. Following this, juvenile plants of *G. changii* were obtained.

Keywords Callus · Cell wall degrading enzymes · *Gracilaria changii* · Protoplast isolation · Protoplast regeneration

H.-Y. Yeong
Institute of Postgraduate Studies, University of Malaya,
50603 Kuala Lumpur, Malaysia

N. Khalid · S.-M. Phang (✉)
Institute of Biological Sciences, University of Malaya,
50603 Kuala Lumpur, Malaysia
e-mail: phang@um.edu.my

Introduction

Protoplast technology has many potential applications especially in seaweeds. Protoplasts serve as excellent experimental materials for research ranging from basic biochemical and morphogenetic studies (Cocking 2000) to expressed sequence tag (EST) analysis (Collén et al. 2006) and genetic transformation (Dipakkore et al. 2005). Over the last two decades, protocols for successful isolation and regeneration of commercial seaweeds have been published (Cheney et al. 1986; Kumar et al. 1999; Salvador and Serrano 2005; Dipakkore et al. 2005). However, parameters for both isolation and regeneration are very much species-specific and depend on various factors ranging from complexity of thallus anatomy, cell wall composition, age and environmental history of the plant.

Gracilaria species contribute 60% of the world's agar with about US$ 82.2 million in the world market value (McHugh 2003). Of the 22 species of *Gracilaria* in Malaysia, *Gracilaria changii* Abbott, Zhang & Xia has the highest potential for commercial development (Phang et al. 1996; Lim and Phang 2004; Phang 2006). *Gracilaria changii* contains high yields of agar (12 to 25% dry weight) and agarose (13 to 16% dry weight) with high gel strength (up to 563 $g.cm^{-2}$ for agar & 950 $g.cm^{-2}$ for agarose) (Phang et al. 1996). *Gracilaria changii* contains high amounts of β-carotene, essential fatty acids (20:5ω3), protein (Norziah and Chio 2000; Marinho-Soriano et al. 2001; Chu et al. 2003) and bioactive compounds (Wong et al. 1994). The Algae Research Laboratory of the University of Malaya has been researching on *G. changii* for the last 20 years and has accumulated extensive information on the species including proteomics (Wong et al. 2006), functional genomics (Chan et al. 2004), expressed sequence tags (ESTs) (Teo et al. 2007) and genetic transformation (Gan et al. 2003, 2006). Therefore, *G. changii* has the potential to be commercialised especially through bioprocess development for the pharmaceutical, nutraceutical and phycocolloid industries. With our success in transformation of *G. changii* (Gan et al. 2003, 2006) we now have to optimise tissue and protoplast culture as a step towards providing a platform for metabolic engineering of this seaweed. Clones with desirable qualities obtained through this technology will provide high quality planting stocks for mariculture. Protoplast culture in enclosed bioreactors will avoid the concerns of GMO release into the marine environment.

To date, protoplasts have been successfully isolated from nine genera of red seaweeds, *Porphyra* (Kito et al. 1998; Dipakkore et al. 2005), *Kappaphycus* (Salvador and Serrano 2005), *Grateloupia* (Chen and Chian 1994), *Bangia* (Araki et al. 1994), *Gelidium* (Coury et al. 1993), *Solieria* (Pinchetti et al. 1993), *Palmaria* (Liu et al. 1992), *Chondrus* (Zhang 1991) and *Gracilaria* (Cheney et al. 1986; Björk et al. 1990; Chu et al. 1998; Araki et al. 1998). Few reported the success of protoplast regeneration into whole plants especially for the anatomically complex red seaweeds such as *Gracilaria* species. Yan and Wang (1993) published the only report of protoplast regeneration into whole plant for *G. asiatica*.

Protoplasts are generally obtained through enzymatic isolation which is dependent on many factors such as source of donor tissue (Pinchetti et al. 1993) and its physiological status (Björk et al. 1990), enzymes and isolation conditions. In this paper, we report a procedure for protoplast isolation and successful regeneration into whole plants for *G. changii*.

Materials and methods

Whole healthy plants (tetrasporophyte, gametophyte and carposporophyte) of *G. changii* plants were collected from the mangroves at Morib, Selangor, Malaysia, and kept cool in an ice chest with seawater to be transported back to the laboratory.

In the laboratory, the seaweed samples were cleaned of mud and epiphytes by using a soft brush and filtered seawater. The samples were maintained as unicultures in tanks of filtered and aerated seawater in both the hatchery and laboratory (25±2°C, 17.30 μmol photons m^{-2} s^{-1}, 12 h light: 12 h dark, 28±2 ppt).

Thalli of *G. changii* were cut into 5 cm pieces and sterilised using the following protocol. The thalli were thoroughly cleared of epiphytes using forceps and autoclaved seawater under the dissecting microscope. The thalli were then rinsed several times with autoclaved sterile seawater followed by sonication in autoclaved seawater for 3 min. The thalli were then surface sterilised with 1% w/v povidone-iodine for 30 s before incubation in seawater containing antibiotics for 12 h. The antibiotic mixture comprised of 0.1 gL^{-1} kanamycin, 0.3 gL^{-1} penicillin G, 0.02 gL^{-1} polymyxin B sulphate, 0.001 gL^{-1} nalidixic acid, 1 gL^{-1} streptomycin sulphate and 0.02 gL^{-1} cefotaxine (Tan 1999). Sterility was tested using LB broth (Merck, Germany) and Marine Agar 2216E (Difco) medium.

Optimal osmoticum for *G. changii* cells was determined by soaking five thin cross-sections of thalli in 60% autoclaved seawater (30 ppt) and 40% ultra-high quality water (UHQ) containing different concentrations of mannitol (0.1–1.3 M) for 30 min. Number of plasmolysed cells per 10 cells of both cortical and epidermal cells in each cross section were counted. The percentage of plasmolysed cells were used to determine the suitable mannitol concentration for protoplast isolation of *G. changii*.

Protoplast isolation and purification

Axenic thalli of *G. changii* were chopped into smaller explants (1–2 mm^2) and rinsed twice with autoclaved sterile seawater (30 ppt) followed by rinsing in high salinity (45 ppt) MES medium to prepare the cells for plasmolysis. The explants were immersed in a medium containing 60% autoclaved seawater (30 ppt) and 40% ultra-high quality water (UHQ) containing 1.0 M mannitol for 15 min in the dark (Yan and Wang 1993); 0.5 g fresh wt of the treated explants were removed and incubated in 5 mL of protoplast isolating solution in petri dishes placed in the dark and under shaking conditions (orbital shaker; 100 rpm) for 3 h. The isolating medium contained various concentrations and combinations of enzymes, 0.85 M mannitol and 5 mM $CaCl_2$, in 60% sterile seawater (30 ppt) and 40% UHQ water. The solutions containing the protoplasts were then filtered through 70 μm nylon mesh to remove the undigested tissue and debris. The remaining filtrate containing isolated protoplasts was diluted one-fold with high salinity (45 ppt) MES medium and centrifuged at 600 g for 5 min. After that, 75% of the supernatant was discarded and the protoplast pellets were resuspended with fresh culture medium and this procedure was repeated twice. The final protoplast pellets were resuspended in 1 mL culture medium (MES, PES) and the protoplast yield was determined by Neubauer improved bright-line Haemocytometer cell count (Guillard 1973).

Optimisation of protoplast isolation conditions

Enzyme solution The commercial enzymes agarase, abalone acetone powder (Sigma, USA), cellulase Onozuka R-10 and macerozyme R-10 (PhytoTechnology, USA) were dissolved in 60% sterile seawater (30 ppt) and 40% UHQ water with 0.85 M mannitol, 5 mM $CaCl_2$ as the standard protoplast isolation solution. To select the suitable enzymes for protoplast isolation of *G. changii*, various concentrations of enzymes, singly or in combination, were used. Single enzyme solutions were: 1, 2, 3% w/v abalone acetone powder; 1, 2, 4% w/v cellulase Onozuka R-10 and macerozyme R-10; and 1, 10, 20 units mL^{-1} agarase. Combination enzyme solutions were: Combination A (2% w/v cellulase Onozuka R-10 and 5 units mL^{-1} agarase), B (2% w/v cellulase Onozuka R-10, 1% w/v Macerozyme R-10 and 5 units mL^{-1} agarase), C (2% w/v cellulase Onozuka R-10, 1% w/v Macerozyme R-10 and 10 units mL^{-1} agarase) and D (4% w/v cellulase Onozuka R-10, 2% w/v Macerozyme R-10 and 1 units mL^{-1} agarase).

Enzyme incubation period, temperature, pH and tissue source All experiments were conducted using Combination D isolation medium. The explants were incubated for various periods (2, 3, 4, 6 and 24 h), temperature (22, 25 and 28°C), pH (5.9, 6.1, 6.3, 6.5 and 6.7) in the dark with shaking. Two different sources of *G. changii* explant were used, namely the tips (within 2 cm from apex) and middle segments (6 to 10 cm from apex). The protoplasts obtained were harvested and counted as described above.

Protoplast cell wall formation

The viability of freshly isolated protoplasts was determined by using 0.01% w/v fluorescein diacetate staining (FDA) (Sigma, USA) and observed under a UV fluorescence microscope (DMLB, Leica). To confirm that the protoplasts were true protoplasts without cell walls, they were stained with 0.01% Calcofluor White M2R (Fluorescent brightener 28, Sigma, USA) and examined under UV fluorescence microscope (DMLB, Leica).

Protoplast culture and regeneration

Freshly isolated protoplasts were harvested and dispensed into 2 mL culture medium in 24-well tissue culture test plates (TPP, Switzerland) based on protoplast density ranging from 7.0×10^4 to 7.5×10^4 cells mL^{-1}. In an earlier experiment, a range of protoplast density (1.45×10^5, 7.28×10^4, 3.64×10^4, 1.82×10^4, 9.1×10^3, 4.55×10^3 protoplasts mL^{-1}) was investigated. After 60 days, only the set containing 7.28×10^4 protoplasts mL^{-1} produced dividing cells. Two enrichment culture media, MES and PES with different pH (pH 6.5, 7.0, 7.5 and 8.0) were tested for protoplast regeneration. The 24-well tissue culture test plates containing the protoplast cultures were placed in the dark (wrapped loosely in aluminium foil) for two days before gradual exposure to a final irradiance of 17.30 μmol photons m^{-2} s^{-1} at 12 h light : 12 h dark photoperiod at 25 to 27°C. Half the culture medium was replaced with fresh medium every five days.

To differentiate protoplast regeneration from spore germination, mature cystocarps of *G. changii* were collected from the field and carpospore release induced (Santelices et al. 2003). The carpospores were incubated in MES medium in the same way as the protoplasts. The carpospores were monitored until germination into whole plants.

Results

The best parameters for protoplast isolation were determined for producing the highest yield of protoplast. Parameters include osmoticum, concentrations and combinations of enzymes, enzyme incubation period, temperature, pH and tissue source.

Osmoticum

No plasmolysed cell was observed from the thallus cross-sections soaked in solutions containing 0.7 M mannitol and below. The number of plasmolysed cells increased with increasing concentrations of mannitol (Fig. 1) and at 1.2 M mannitol, cells were overplasmolysed (Fig. 2). Thus solution with 1.0 M mannitol was considered suitable osmoticum where 50% of the cells were plasmolysed and none were overplasmolysed.

Protoplast isolation using enzymes

Single enzymes Of the four enzymes tested, 4% cellulase Onozuka R-10 produced the highest number of protoplasts ($93.07\pm14.54\times10^3$ protoplasts g^{-1} fresh wt), followed by 2% cellulase Onozuka R-10 ($48.93\pm1.29\times10^3$ protoplasts g^{-1} fresh wt) and 1% cellulase Onozuka R-10 ($28.80\pm2.80\times10^3$ protoplasts g^{-1} fresh wt) (Fig. 3). Protoplast

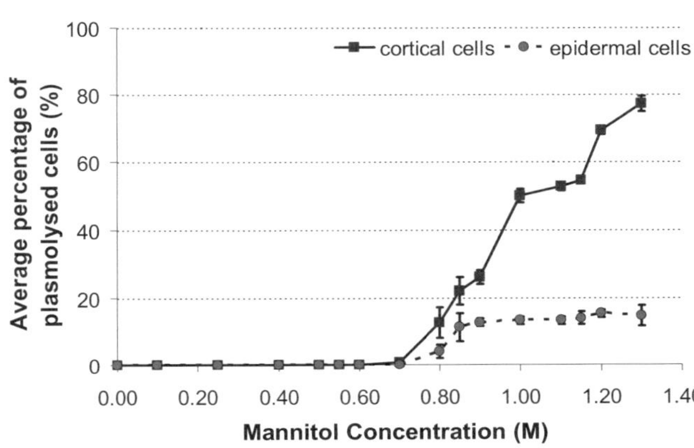

Fig. 1 Average percentage of plasmolysed cortical and epidermal cells of the thallus cross section in different mannitol concentrations

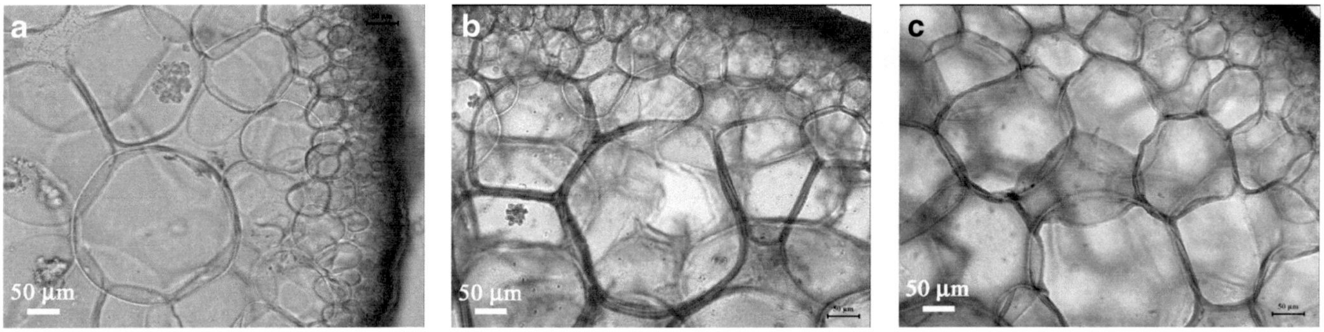

Fig. 2 Thin sections of *Gracilaria changii* thallus in different mannitol concentrations. (**a**) Cells remained normal in mannitol concentration at 0.6 M. (**b**) Sufficiently plasmolysed cells obtained in 1 M mannitol. (**c**) Over-plasmolysed cells in mannitol concentration above 1.2 M

yields from the other enzymes ranged from as low as $0.27 \pm 0.23 \times 10^3$ protoplasts g^{-1} fresh wt (2% macerozyme R-10) to $7.33 \pm 5.77 \times 10^3$ protoplasts g^{-1} fresh wt (3% w/v abalone acetone powder).

Combination of enzymes Use of various combinations and concentrations of three enzymes (cellulase, macerozyme and agarase), showed that Combination C (2% cellulase, 1% macerozyme & 10 units mL^{-1} agarase) produced the highest protoplast yield ($81.48 \pm 0.85 \times 10^4$ protoplasts g^{-1} fresh wt) ($p<0.05$) (Fig. 4). Combination B ($72.22 \pm 0.56 \times 10^4$ protoplasts g^{-1} fresh wt in 2% cellulase, 5 units mL^{-1}agarase & 1% macerozyme) produced 60% more protoplasts than Combination A ($45.00 \pm 2.00 \times 10^4$ protoplasts g^{-1} fresh wt in 2% cellulase & 5 units per mL^{-1}agarase) in the presence of macerozyme (Fig. 4).

Enzyme incubation period, temperature, pH and tissue source 3 to 4 h enzyme incubation period produced the best protoplast yield, $9.73 \pm 0.83 \times 10^5$ protoplasts g^{-1} fresh wt ($p<0.05$) (Fig. 5). Longer incubation period reduced yield to 3.47 ± 0.61 (6 h) and $2.27 \pm 0.46 \times 10^5$ protoplasts g^{-1} fresh wt (24 h). Among the temperatures tested, no significant difference was observed (Fig. 6). The most suitable pH was found to be 6.3 giving a yield of $18.17 \pm 1.65 \times 10^5$ protoplasts g^{-1} fresh wt ($p<0.05$) (Fig. 7). Protoplast yields decreased as the pH increased or decreased from the optimum of pH 6.3.

Tips of the thallus were found to produce more protoplasts ($14.46 \pm 3.90 \times 10^5$ protoplasts g^{-1} fresh wt) than middle segments ($3.53 \pm 0.70 \times 10^5$ protoplasts g^{-1} fresh wt) ($p<0.05$).

Protoplast cell wall formation

Freshly isolated protoplasts most probably from cortical cells, were spherical in shape, 8–35 µm in diameter with a central vacuole and surrounded by cytoplasm bounded by

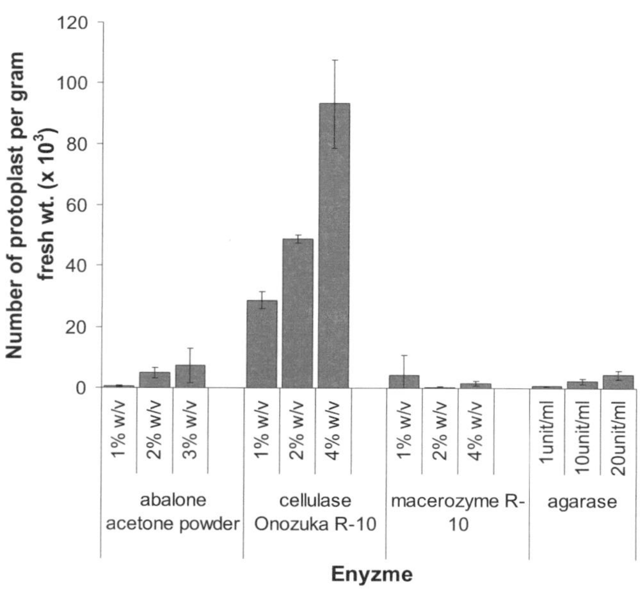

Fig. 3 Effect of the different concentrations of single enzyme on the protoplast yield of *Gracilaria changii* ($p<0.05$)

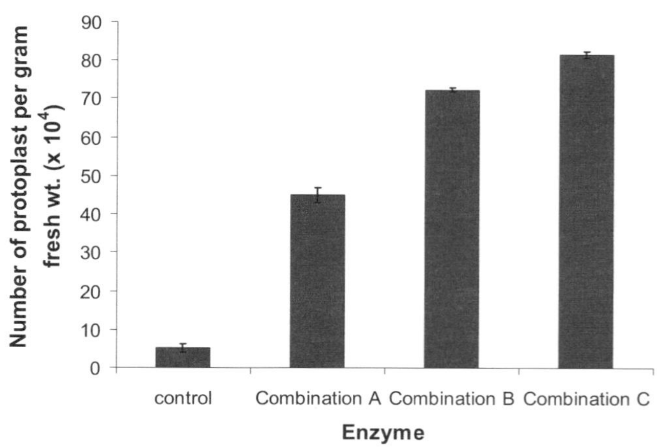

Fig. 4 Effect of combinations and concentrations of enzymes on the protoplast yield of *Gracilaria changii*. The combinations and concentration of enzyme mixtures are: Combination A: 2% w/v cellulase Onozuka R-10 & 5 units mL^{-1} agarase; Combination B: 2% w/v cellulase Onozuka R-10, 1% w/v macerozyme R-10 & 5 units mL^{-1} agarase; Combination C: 2% w/v cellulase Onozuka R-10, 1% w/v macerozyme R-10 & 10 units mL^{-1} agarase ($p<0.05$)

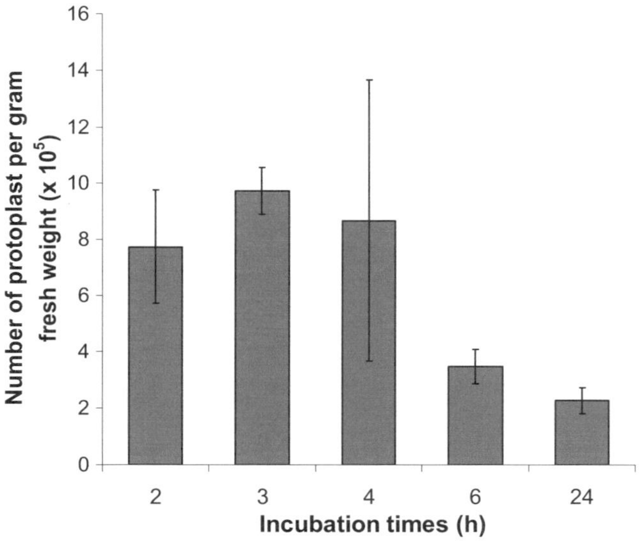

Fig. 5 Effect of the various enzyme incubation times on the protoplast yield of *Gracilaria changii* ($p<0.05$)

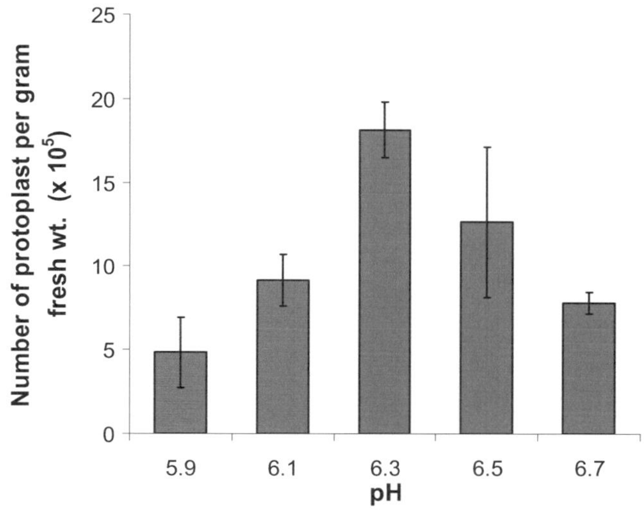

Fig. 7 Effect of different enzyme incubation pH on the protoplast yield of *Gracilaria changii* ($p<0.05$)

the plasma membrane (Fig. 8a). Larger protoplasts with a translucent appearance, probably from medullary cells, 45–55 μm in diameter were rarely obtained. Viable protoplasts were stained florescent yellow-orange with FDA stain (Fig. 8b). The freshly isolated protoplasts when stained with Calcofluor White M2R showed no trace of blue fluorescence when examined under UV florescence microscope. This signified that true protoplasts without cell walls were successfully obtained. After 14 h, some blue fluorescent 'protoplasts' were observed indicating the regeneration of cell walls. However, it was only after 24 h that the highest percentage of stained cells was obtained (Fig. 8c).

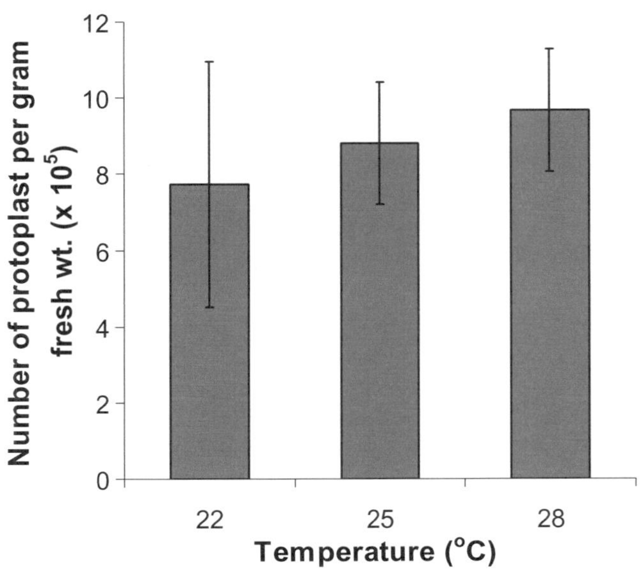

Fig. 6 Effect of different enzyme incubation temperature on the protoplast yield of *Gracilaria changii*

Protoplast regeneration

In both MES and PES medium, protoplasts after cell wall regeneration underwent cell division. Preliminary observations showed cell division until the 16-cell stage occurred only in MES medium (Fig. 9a). Cell division continued to develop macrocolonies and eventually callus formation. Regenerated juvenile plants were obtained in MES medium at pH 7.5 after 60 days culture (Fig. 9b). Further development produced up to 12 plants from the same callus-like cell masses after 135 days (Fig. 9c) and elongated further.

To ensure that the plants regenerated were from protoplast and not carpospore, the regeneration process from both protoplast and spores were compared. Carpospores from *G. changii* were observed to germinate by a process of cell division forming first the holdfast (after 35 days). The shoot then developed from this holdfast (50 days) (Fig. 10 a–h). The process was different from the protoplast regeneration process.

Discussion

Prior to enzyme incubation, explants were placed in a solution which could induce plasmolysis. Plasmolysed cells will ease protoplast release in enzyme isolation solution since the plasma membranes were already detached from the cell walls (Collin and Edwards 1998). Normally, cell plasmoylsis is obtained by manipulating the osmotic concentration which could be achieved by adding sugars (glucose and sucrose) or sugar alcohols (mannitol and sorbitol) to the isolating solution (Evans and Bravo 1983).

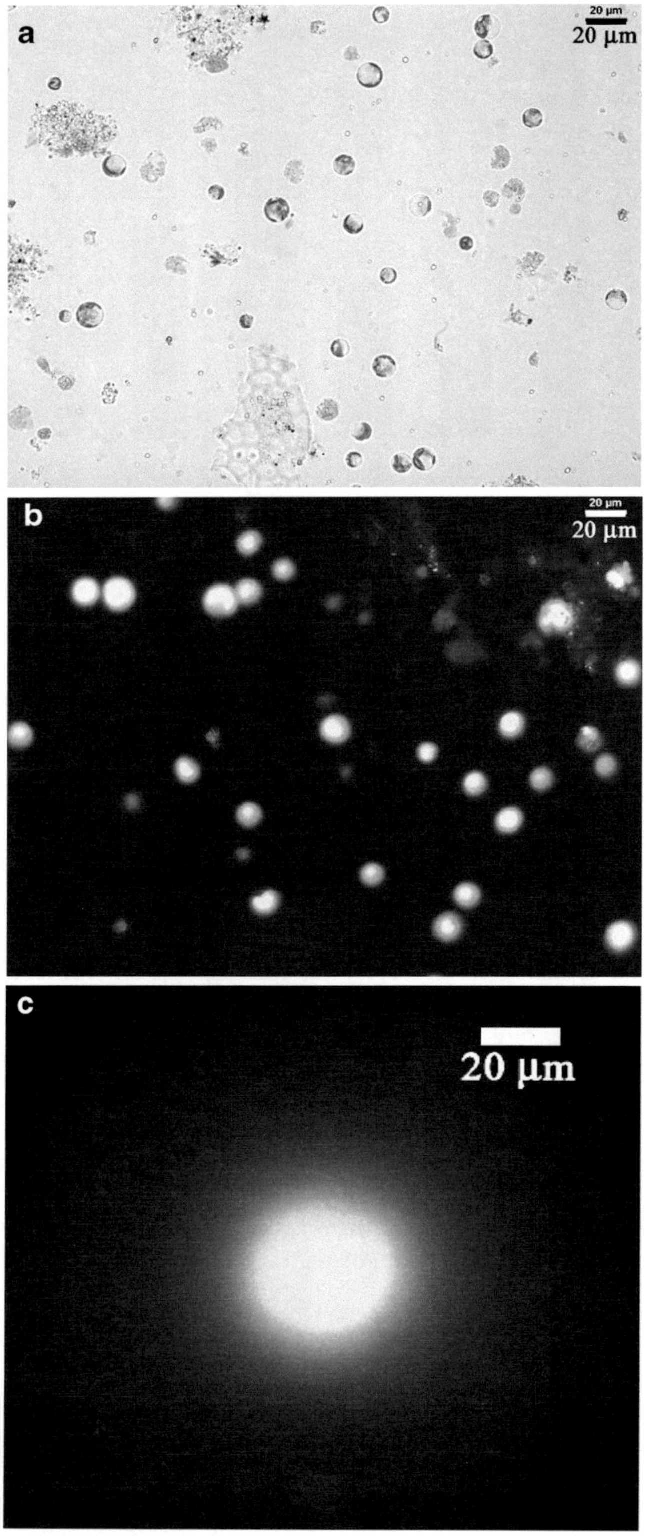

Fig. 8 Protoplasts of *Gracilaria changii*. (**a**) Freshly isolated protoplasts. (**b**) Viable protoplasts stained with FDA under UV fluorescence microscope. (**c**) One-day-old protoplasts stained with calcofluor white under UV fluorescence showing presence of regenerated cell wall

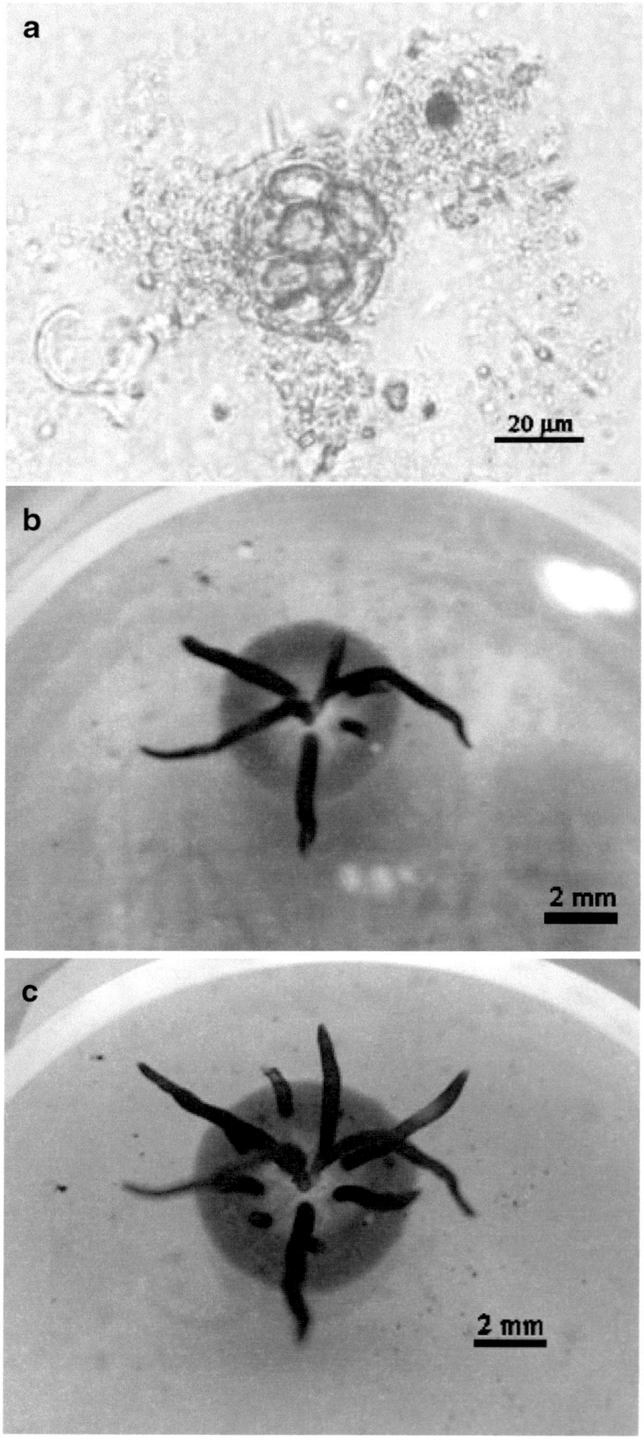

Fig. 9 Protoplast regeneration of *Gracilaria changii*. (**a**) Dividing cell of protoplast after 5 days in culture. (**b**) Plants regenerated from protoplast callus-like-cell-masses after 60 days. (**c**) Elongation and increased plant formation from protoplast callus-like-cell-masses after 135 days

In this study, mannitol was used as the standard osmotic stabiliser and the osmotic concentration of the solution was standardised as 1 M mannitol based on the optimal concentration of mannitol in the plasmolysis study. Similarly, 1 M mannitol was used as the pre-plasmolysis concentration by Yan and Wang (1993) in the protoplast isolation of *G. asiatica*. Butler et al. (1990) reported that direct plasmolysis of cells in the enzyme isolation solution will cause accumulation of contaminants into the cells, which may produce a toxic effect on the protoplast. This may be avoided by pre-plasmolysing the cells before enzyme digestion. In seaweeds, similar observations were obtained by Pinchetti et al. (1993) who reported that tissue source which is pre-plasmolysed before treating by enzyme mixture produced improved protoplast yields and enhanced the viability of the isolated protoplasts. Furthermore, plasmolysis will help protoplasts to maintain their integrity after the cell walls were removed (Compton et al. 2000).

Commercial cell wall degrading enzymes were assessed for protoplast isolation from *G. changii*. When the enzymes cellulase, macerozyme, agarase and abalone acetone powder were used individually, protoplast yields were low, ranging from $0.27\pm0.23\times10^3$ to $93.07\pm14.54\times10^3$ protoplasts g^{-1} fresh wt. Similar results were obtained from *G. tikavahiae* (cellulase Onozuka R-10, macerozyme, abalone gut enzyme or agarase), *Solieria filiformis* (cellulase, cellulysin or abalone acetone powder) and *Kappaphycus alvarezii* (cellulase or abalone acetone powder) (Cheney et al. 1986; Pinchetti et al. 1993; Salvador and Serrano 2005). These results showed that the use of individual enzymes were not effective in producing good protoplast yields. This is because enzymes are substrate specific and the enzyme activities are dependent on the type of substrate present in the tissue (Chen et al. 1994). The understanding of cell wall structure and biochemical composition is important when applying enzymes for protoplast release. The anatomically complex seaweeds like *Gracilaria* have cell walls composed of cellulosic microfibrillar network (the skeleton) associated with a more amorphous matrix material (the matrix) such as sulphated polygalactans (agarcolloids or agars), mucilages and cellulose (Graham and Wilcox 2000). Bellanger et al. (1990) reported that the cell wall of *G. verrucosa* contains an agar matrix and a fibrillar part which has a cellulose type polysachaaride comprising 53% of glucose and therefore an enzyme mixture containing agarase and cellulase is best for isolating protoplasts from it. Therefore, an enzyme mixture which digests both the skeleton and the matrix components is necessary for complete removal of the cell walls (Butler et al. 1990; Chen et al. 1994).

As a result, combinations of cellulase (for algal cellulose digestion), macerozyme (for pectin digestion which assists in cell separation) and agarase (for agar digestion) were used in the present study to improve yields. Results showed that combinations of enzymes produce higher protoplasts yields ($81.48\pm0.85\times10^4$ protoplasts g^{-1} fresh wt) then use of single enzymes ($93.07\pm14.54\times10^3$ protoplasts g^{-1} fresh wt).

Combination B produced 60% more protoplasts than Combination A in the presence of macerozyme. Similar results were obtained for *Solieria filiformis*, *Porphyra linearis* and *P. okhaensis* (Pinchetti et al. 1993; Chen et al. 1994; Dipakkore et al. 2005). This suggests that the addition of macerozyme into the enzyme mixture enhanced the protoplast yields because macerozyme digests pectin and polygalactans (Evans and Bravo 1983). Combinations of cellulases or hemicellulases, pectinases and polysaccharide-degrading enzymes have been successfully used in protoplast isolation of other *Gracilaria* species such as *G. asiatica, G. lemaneiformis, G. sordida, G. tenuistipitata, G. tikvahiae, G. verrucosa* (Cheney et al. 1986; Björk et al. 1990; Yan and Wang 1993; Chu et al. 1998). In contrast, use of single enzymes produced good protoplast yields with anatomically simple green seaweeds such as *Enteromorpha compressa, E. flexuosa, E. tubulosa, Monostroma angicava, M. oxyspermum, Ulva reticulata, U. conglobata and U. fasciata* (Saga et al. 1986; Saga and Kudo 1989; Kumar et al. 1999; Reddy et al. 2006). All these reports supported the suggestion that the simpler the cell wall composition the less the requirement of degrading enzymes needed during protoplast isolation.

Besides cellular structure and biochemical composition of the seaweed, other characteristics, such as physiological status of seaweed tissue source are important for improving protoplast isolation and regeneration. In this study, effect of the donor tissue source (and therefore thallus age) on the protoplast yields was evaluated. Results showed that protoplast yield obtained from the younger tissue at the tips of the thallus was significantly ($p<0.05$) higher than from the older middle segment tissue with $14.46\pm3.90\times10^5$ and $3.53\pm0.70\times10^5$ protoplasts g^{-1} fresh wt, respectively. Sporelings of *G. tikvahiae* were also reported to be a better source of protoplasts than the mature plant (Cheney et al. 1986), while apices cells gave higher yields then mature plants of *Solieria filiformis* (Pinchetti et al. 1993) and *G. verrucosa* (Araki et al. 1998). Younger tissue have less secondary thickening of the cell wall, contributing to easier cell separation and therefore higher protoplast yield (Collin and Edwards 1998). Björk et al. (1990) reported that higher protoplast yields were obtained from *G. sordida* plants with higher growth rates compared to the one with low growth rate. Similarly in this study, protoplast yields were found higher in apical tissue which are actively growing, as reported by Björk et al. (1990).

Enzyme activity is influenced by many physical-chemical factors. The pH optima for the enzymes used, range from pH 5.0 (cellulase) to 6.0 (agarase). When used in combination, the enzyme mixtures had pH ranging from pH 6.2 to 6.5. *Gracilaria* tissue extract has a pH ranging from pH 6.5 to 6.9 while the seawater used has pH ranging from 7.5–7.8. Results showed that the mixed enzyme

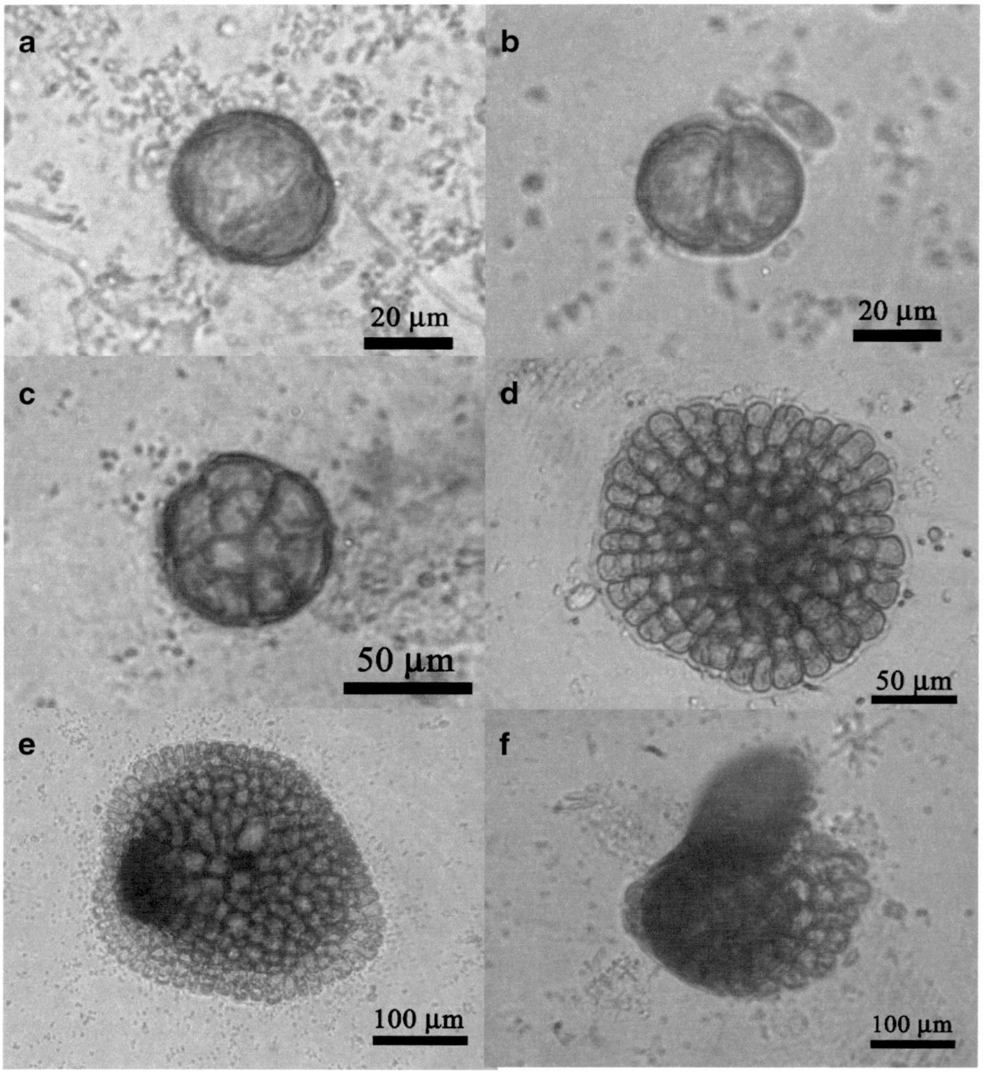

Fig. 10 Carpospore germination and growth from *Gracilaria changii*. (**a**) Carpospore of *Gracilaria changii*. (**b**) Dividing carpospore of *Gracilaria changii* after 5 days in culture. (**c**) Spore division after 21 days in culture. (**d**) Holdfast formation after 35 days in culture. (**e**) Holdfast formation after 45 days in culture. (**f**) Shoot development from the holdfast after 50 days in culture. (**g**) Shoot development from the holdfast after 52 days in culture. (**h**) Elongation of newly developed shoot from holdfast. (**i**) Development of juvenile plants from carpospore. (**j**) Branching of juvenile plants from carpospore

isolation solution at pH 6.3 and 28°C incubation temperature provided the most suitable conditions for protoplast isolation. In general, pH used in published reports for seaweed protoplast isolation ranged between pH 6.0 and 7.0 (Cheney et al. 1986; Björk et al. 1990; Yan and Wang 1993; Chu et al. 1998). Dipakkore et al. (2005) reported pH 6 produced the highest protoplast yield from *Porphyra okhaensis* using similar enzyme combination (Combination D). A 3 h incubation period was found to produce the highest protoplast yield. Incubation times longer than 6 h caused the cells to be over digested and therefore reduced the protoplast yield significantly ($p<0.05$).

Regeneration of cell wall from *G. changii* was observed starting from 14 h till 24 h. Protoplasts left in 24-well tissue culture plates containing MES medium at pH 7.5 were observed to have divided into 16 cells after 5 days and 32 celled callus-like cell mass after 10 days. Germlings were produced from the callus-like cell mass after 60 d. Up to 12 plants were produced from each callus-like cell mass. In this study, protoplast regeneration was only obtained in MES medium and not in PES medium. This may due to the sensitivity of the *G. changii* protoplasts to the presence of ammonium and nitrate in the PES medium. Collin and Edwards (1998) and Compton et al. (2000) reported that the presence of ammonium nitrate in culture medium is essential to enhance cell division but is toxic to protoplasts at 20 mM in higher plants. Dipakkore et al. (2005) reported that three types of protoplast regeneration have been obtained for *Porphyra* by various workers; callus, filamentous conchocelis and leafy blade types. Yan and Wang (1993) reported on protoplast regeneration of *Gracilaria* sp. into a whole plant. Here, the protoplast divided and formed a callus-like cell mass which then proceeded to produce whole plants. This is the only published report of regeneration of *Gracilaria* protoplasts into whole plants. To date, our report is the first on protoplast regeneration

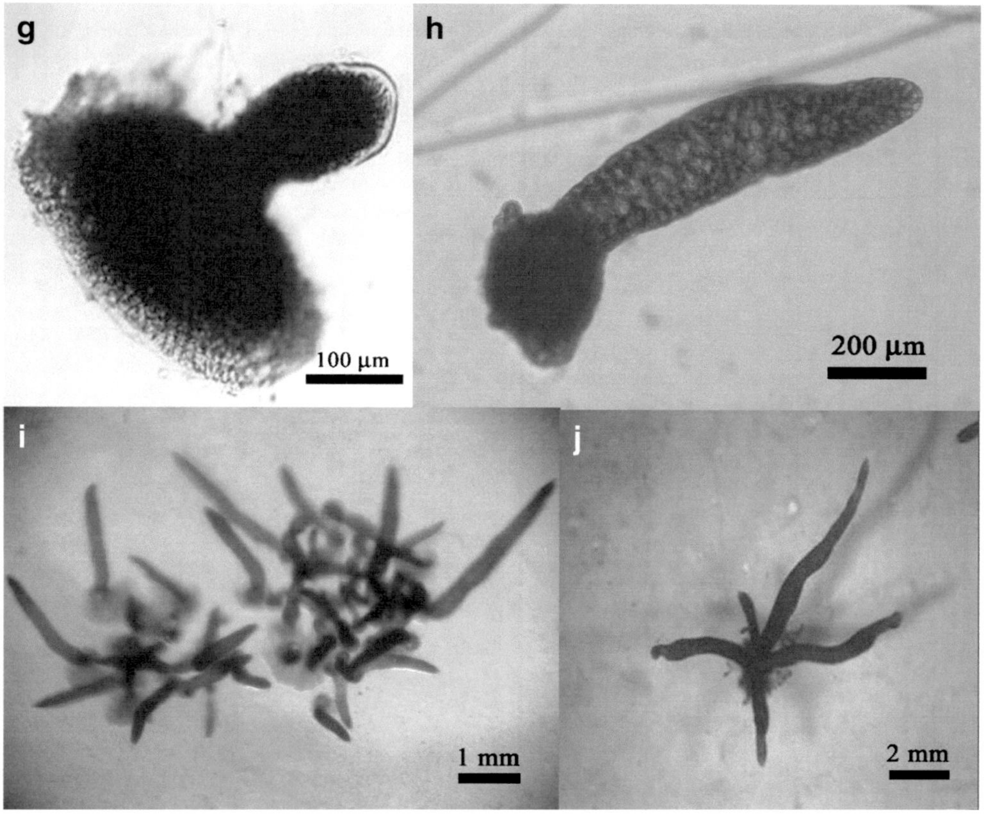

Fig. 10 (continued)

into whole plants for the tropical *G. changii*. Whole plants have only been reported for *Gracilaria* spp. through tissue culture (Kumar et al. 2007) while previous reports have only achieved protoplast regeneration into the callus stage (Cheney et al. 1986; Björk et al. 1990; Chu et al. 1998).

In our experiment, the protoplasts were cultured in liquid culture and at a density of approximately 7.5×10^4 cells mL^{-1}. Protoplasts of higher plants may be cultured using various techniques including plating in agar medium, hanging/sitting drop culture, liquid on agar culture and agarose bead culture (Ochatt and Power 1992). However, while agar and agarose may be suitable for callus culture, it may be toxic to protoplasts especially those found difficult to regenerate (Hunter 1994). Protoplast density is important for recalcitrant species where 'protoplast to protoplast' relationship is important (Ochatt and Power 1992). Adequate 'neighbour' cells provide undefined diffusible biochemicals which stimulate cell wall regeneration and cell division. In higher plant culture, co-culturing with fast growing protoplasts act as 'nurse-cells' and can enhance protoplast division (Ochatt and Power 1992). Evans and Bravo (1983) reported that a protoplast density of 5×10^3 to 1×10^5 protoplasts mL^{-1} was required for most of the protoplast culture system for higher plants in order to enhance the cell to cell communication. With the *G. changii*, a density of 7.28×10^4 protoplasts mL^{-1} produced dividing cells. A higher protoplast density (1.45×10^5 protoplasts mL^{-1}) did not produce cell division, probably because of competition for resources and toxins excreted by cell undergoing necrosis.

The difference between protoplast regeneration and carpospore germination in *G. changii* is that protoplast regenerates into a callus-like cell mass which produces several shoots but each carpospore only produces one holdfast prior to a single shoot formation which may form branches. This proved that the prior cultures were not contaminated with spores and the plants regenerated were from protoplasts and not carpospores.

Conclusion

We have succeeded in the isolation and regeneration of protoplasts from *G. changii* in liquid culture. Callus-like cell masses were obtained, which later developed further into whole plants. This provides not only a gene expression system for confirming stable expression of the genetic transformation protocols developed by us, but also provides a process to propagate cells in enclosed bioreactors for metabolic engineering. This research has added to our collection of scientific information on an important tropical agarophyte which also has potential as a source of bioactive metabolites.

Acknowledgments This research was funded by the Ministry of Science, Technology and Innovation, Malaysia (Grant No: 39-02-03-9002/Oracle 8301902; 12-02-03-2028/Oracle 8422028), University of Malaya Short-term Research Grant No. F0140/2002D, F0133/2004A, F0233/2004D and P0114/2006A. Ms Yeong conducted this research under a PASCA scholarship from the University of Malaya. The authors gratefully acknowledge the following for their advice and kind assistance: Professor He Peimin, Dr. Gan Sook Yee, Dr. Tan Eng Lai, Professor Qin Song and Dr. Jiang Peng.

References

Araki T, Hayakawa M, Tamaru Y, Yoshimatsu K, Morishita T (1994) Isolation and regeneration of haploid protoplasts from *Bangia atropurpurea* (Rhodophyta) with marine bacterial enzymes. J Phycol 30:1040–1046

Araki T, Lu Z, Morishita T (1998) Optimization of parameters for isolation of protplasts from *Gracilaria verrucosa* (Rhodophyta). J Mar Biotechnol 6:193–197

Bellanger F, Verdus MC, Henocq V, Christiaen (1990) Determination of the composition of the fibrillar part of *Gracilaria verrucosa* (Gracilariales, Rhodopyta) cell wall in order to prepare protoplasts. Hydrobiologia 204/205:527–531

Björk M, Ekman P, Wallin A, Pedersén M (1990) Effects of growth rate and other factors on protoplast yield from four species of *Gracilaria* (Rhodophyta). Bot Mar 33:433–439

Butler DM, Evans LV, Kloareg B (1990) Isolation of protoplasts from marine macroalgae. In: Akatsuka I (ed) Introduction to applied phycology. SPB Academic, Netherlands, pp 647–668

Chan CX, Teo SS, Ho CL, Rofima Yasmin O, Phang SM (2004) Optimisation of RNA extraction for *Gracilaria changii* (Gracilariales, Rhodophyta). J Appl Phycol 16:297–301

Chen YC, Chiang YM (1994) Development of protoplast from *Grateloupia sparsa* and *Grateloupia filicina* (Halymeniaceae, Rhodophyta). Bot Mar 37:361–366

Chen LC-M, Craigie JS, Xie ZK (1994) Protoplast production from *Porphyra linearis* a simplified agarase procedure capable of commercial application. J Appl Phycol 6:35–39

Cheney DP, Mar E, Saga N, Meer J (1986) Protoplast isolation and cell division in the agar-producing seaweed *Gracilaria* (Rhodophyta). J Phycol 22:238–243

Chu J, Liu W, Zhang Z (1998) A preliminary study on the isolation and cultivation of protoplast from *G. verrucosa*. Mar Sci Bull 17 (6):17–20 (In Chinese with English abstract)

Chu WL, Norazmi M, Phang SM (2003) Fatty acid composition of some Malaysian seaweeds. Malays J Sci 22(2):21–27

Cocking EC (2000) Turning point article plant protoplasts. In Vitro Cell Dev Biol Plant 36:77–82

Collén J, Roeder V, Rousvoal S, Collin O, Kloareg B, Boyen C (2006) An expressed sequence tag analysis of thallus and regenerating protoplasts of *Condrus crispus* (Gigartinales, Rhodophyceae). J Phycol 42:104–112

Collin HA, Edwards S (1998) Protoplast culture. In: Rickwood D, Howe C (eds) Plant cell culture. The introduction to biotechniques series. Bios Scientific, Oxford UK, pp 71–81

Compton ME, Saunders JA, Veilleux RE (2000) Use of protoplast for plant improvement. In: Trigiano RN, Gray DJ (eds) Plant tissue culture concepts and laboratory exercise. CRC, USA, pp 249–261

Coury DA, Naganuma T, Polne-Fuller M, Gibor A (1993) Protoplasts of *Gelidium robustum* (Rhodophyta). Hydrobiologia 260/261:421–427

Dipakkore S, Reddy CRK, Jha B (2005) Production and seeding of protoplast of *Porphyra okhaensis* (Bangials, Rhodophyta) in laboratory culture. J Appl Phycol 17:331–337

Evans DA, Bravo JE (1983) Protoplast isolation and culture. In: Evans DA, Sharp WR, Ammirato P, Yamada Y (eds) Techniques gor propagation and breeding. Handbook of plant cell culture, vol 1. Technique for propagation and breeding. Macmillan, New York, USA, pp 124–176

Gan SY, Qin S, Rofina YO, Yu D, Phang SM (2003) Transient expression of *lacZ* in particle bombarded *Gracilaria changii* (Gracilariales, Rhodophyta). J Appl Phycol 15:315–353

Gan SY, Rofina YO, Qin S, Phang SM (2006) Crop improvement in seaweed cultivation. In: Phang SM, Crichery A, Ang P (eds) Advances in seaweed cultivation and utilization in Asia. University of Malaya Maritime Research Centre, Kuala Lumpur, Malaysia, pp 81–103

Graham LE, Wilcox LW (2000) Red algae. In: Algae. Prentice-Hall, Upper Saddle River, NJ, pp 343–396

Guillard RRL (1973) Division rates. In: Stein JR (ed) Culture methods and growth measurements. Handbook of Phycological Methods. Cambridge University Press, London, pp 289–311

Hunter CF (1994) Protoplast and haploid cultures. In: Hunter CF (ed) In vitro cultivation of plant cells. Butterworth-Heinemann, Oxford, pp 87–112

Kito H, Kunimoto M, Kamanishi Y, Mizukami Y (1998) Protoplast fusion between *Monostroma nitidum* and *Porphyra yezoensis* and subsequent growth of hybrid plants. J Appl Phycol 10:15–21

Kumar GRK, Addepalli MK, Reddy CRK (1999) Regeneration of the thallus of *Monostroma oxyspermum* (Chlorophyta) from protoplasts in axenic culture. Phycologia 38:503–507

Kumar GR, Reddy CKR, Jha B (2007) Callus induction and thallus regeneration from callus of phycocolloid yielding seaweeds from the Indian coast. J Appl Phycol 19:15–25

Lim PE, Phang SM (2004) Gracilaria species (Gracilariales, Rhodophyta) of Malaysia including two new recods. Malays J Sci 23(2):71–80

Liu QY, Chen LC-M, Taylor ARA (1992) Ultrastructure of cell wall regeneration by isolated protoplasts of *Palmaria palmata* (Rhodophyta). Bot Mar 35:21–33

Marinho-Soriano, Silva ETSF, Moreira WSC (2001) Seasonal variation in the biomass and agar yield from *Gracilaria cervicornis* and *Hydropuntia cornea* from Brazil. Bioresour Technol 77 (2):115–120

McHugh DJ (2003) A guide to the seaweed industry. FAO Fisheries Technical Paper No 441. FAO, Rome, p 105

Norziah MH, Chio YC (2000) Nutritional composition of edible seaweed *Gracilaria changii*. Food Chem 68:69–76

Ochatt SJ, Power JB (1992) Plant regeneration from cultured protoplasts of higher plants. In: Fowler MW, Warren GS (eds) Plant biotechnology. Comprehensive Biotechnology, Second supplement. Pergamon, Oxford, pp 99–127

Phang SM (2006) Seaweed resources in Malaysia: current status and future prospects. Aquat Ecosyst Health Manag 9:185–202

Phang SM, Shaharuddin S, Noraishah H, Sasekumar A (1996) Studies on *Gracilaria changii* (Gracilariales, Rhodophyta) from Malaysian mangroves. Hydrobiologia 326/327:347–352

Pinchetti JLG, Björk M, Pedersén M, Reina GG (1993) Factors affecting protoplast yield of the carrageenophyte *Solieria filiformis* (Gigartinales, Rhodophyta). Plant Cell Rep 12:541–545

Reddy CRK, Dipakkore S, Kumar R, Jha B, Cheney DP, Fujita Y (2006) An improved enzyme preparation for rapid mass production of protoplast as seed stock for aquaculture of macrophytic marine green algae. Aquaculture 260:290–297

Saga N, Kudo T (1989) Isolation and culture of protoplast from the marine green alga *Monostroma angicava*. J Appl Phycol 1:25–30

Saga N, Polne-Fuller M, Gibor A (1986) Protoplast from seaweed: production and fusion. Beih Nova Hedwig 83:37–43

Salvador RC, Serrano AE (2005) Isolation of protoplasts from tissue fragments of Philippine cultivars of *Kappaphycus alvarezii* (Solieriaceae, Rhodophyta). J Appl Phycol 17:15–22

Santelices B, Correa JA, Hormazábal M, Flores V (2003) Contact responses between spores and sporelings of different species, karyological phases and cystocarps of coalescing Rhodophyta. Mar Biol 143:381–392

Tan EL (1999) Tissue culture and protoplast isolation of *Gracilaria changii* (Rhodophyta). Dissertation, University of Malaya

Teo SS, Ho CL, Teoh S, Lee WW, Tee JM, Raha AR, Phang SM (2007) Analyses of expressed sequence tags from an agarophyte, *Gracilaria changii* (Gracilariales, Rhodophyta). Eur J Phycol 42:41–46

Wong WH, Goh SH, Phang SM (1994) Antibacterial properties of Malaysian seaweeds. In: Phang SM, Lee YK, Borowitzka MA, Whitton BA (eds) Algal biotechnology in the Asia-Pacific region. University of Malaya, Kuala Lumpur, pp 75–81

Wong PF, Tan LJ, Nawi H, AbuBakar S (2006) Proteomics of the red alga, *Gracilaria changii* (Gracilariales, Rhodophyta). J Phycol 42:113–120

Yan XH, Wang SJ (1993) Regeneration of whole plants from *Gracilaria asiatica* Chang et Xia protoplasts (Gracilariaceae, Rhodophyta). Hydrobiologia 260/261:429–436

Zhang QQ (1991) Studies on the isolation, culture and regeneration of protoplasts in *Chondrus ocellatus* Holm. J Qingdao Oceanol Univ 21:52–62

Callus induction and micropropagation improved by colchicine and phytoregulators in *Kappaphycus alvarezii* (Rhodophyta, Solieriaceae)

Leila Hayashi · Nair S. Yokoya · Daniela M. Kikuchi · Eurico C. Oliveira

Originally published in the Journal of Applied Phycology, Vol 20, No 5, 203–209.
DOI: 10.1007/s10811-007-9234-z

Abstract Tissue culture techniques were applied for micropropagation of the red alga *Kappaphycus alvarezii* in order to select the best strain and experimental system for *in vitro* culture. Five strains were tested: brown (BR), green (GR) and red (RD) tetrasporophytes, brown female gametophyte (BFG), and a strain originating from tetraspore germination ("Edison de Paula", EP). The effects of three culture media were tested on callus formation, regeneration from explants and from callus in the three tetrasporophytic and EP strains: seawater enriched with half-strength of von Stosch's (VS 50) and Guillard & Ryther's (F/2 50) solutions, plus synthetic ASP 12-NTA medium, with or without gelling agent. Explants of the EP strain were treated with glycerol and the phytoregulators indole-3-acetic acid (IAA); 2,4-diclorophenoxyacetic acid (2,4-D); and benzylaminopurine (BA), alone or in combination. The effects of colchicine (0.01%) during 24, 48, 72 hours and 14 days were analyzed in the BFG and EP strains. The EP strain showed the highest percentage of explants forming callus and regeneration from explants in VS 50, indicating its high potential for micropropagation in comparison to the other strains. Regeneration from callus was very rare. Treatments with glycerol and IAA:BA (5:1 mg L^{-1}) stimulated the regeneration from explants. Significant differences were observed in the percentages of regeneration of EP strain explants treated with colchicine for 14 days. Our results indicate that IAA and BA stimulated the regeneration process, and that colchicine produced explants with high potential for regeneration, being useful for improving the micropropagation of *K. alvarezii*.

Keywords Colchicine · *Kappaphycus alvarezii* · Micropropagation · Strain selection · Tissue culture

L. Hayashi (✉) · D. M. Kikuchi · E. C. Oliveira
Departamento de Botânica, Instituto de Biociências,
Universidade de São Paulo,
Caixa Postal 11.461,
05422-970 São Paulo, Brazil
e-mail: leilahayashi@hotmail.com

N. S. Yokoya
Instituto de Botânica, Secretaria de Meio Ambiente do Estado de São Paulo, Seção de Ficologia,
Caixa Postal 3005,
01061-970 São Paulo, Brazil

Introduction

Since the beginning of its commercial cultivation in the 1970s, *Kappaphycus alvarezii* (Doty) Doty ex P. C. Silva has been propagated vegetatively. Repeated clonal propagation reduces genetic variability which may be the reason for the observed decrease in growth rates, carrageenan yield and gel strength, and the increase of susceptibility to diseases (Trono and Lluisma 1992; Hurtado and Cheney 2003). According to Ask and Azanza (2002), those problems could be mitigated by strain selection, genetic manipulation and transgenic production. The cultivation could also be improved by careful selection of specimens from previous cultivated generations or in seed stocks from wild strains (van der Meer 1987).

The application of micropropagation in commercially important seaweeds may help to multiply selected strains, to increase seed stock production for cultivation and to produce a high number of uniform specimens with desired characteristics in a short period of time. Previous studies showed that micropropagation of *K. alvarezii* can be successful, with production of propagules within 4 to 8 weeks (Dawes and Koch 1991). Reddy et al. (2003) observed that plants originating from tissue culture had growth rates 1.5 to 1.8 times higher than field plants when cultivated in sea farms in India. According to these authors, the high potential of development of adventitious plantlets is a promising tool to the fast clonal production of stocks for commercial cultivation.

Despite the advantages, seaweed tissue culture has faced several problems including the lack of good protocols to obtain axenic cultures, callus induction and differentiation of whole plants with the application of plant growth regulators (Collantes and Melo 1995).

Organic carbon sources and plant growth regulators play an important role on *in vitro* propagation of macroalgae, although the optimum concentrations of these compounds vary from species to species. The carbon sources support growth and development, while the plant growth regulators control the growth pattern and development (Garcia-Jimenez et al. 1998). Several studies reported the occurrence of hormones in seaweeds, *e.g.* auxins (Jacobs et al. 1985; Bradley 1991; Jacobs 1993; Stirk et al. 2004), and cytokinins (Bradley 1991; Zhang et al. 1991; Stirk and van Staden 1997; Stirk et al. 2003). The effects of exogenous applications of auxins and cytokinins in seaweeds are documented in literature (Dawes and Koch 1991; Huang and Fujita 1997; Yokoya et al. 1999; Yokoya 2000; Yokoya et al. 2004).

Based on several techniques, including intensive cultivation and selection, mutation induced by X-rays and treatments with colchicine, new and improved strains of *Laminaria* J.V. Lamouroux were produced, resulting in higher production and geographic expansion of cultivation industries in China (Cheney 1984). The effect of colchicine is well known as an inhibitor of spindle fibers formation during cell division, giving rise to cells with double the number of chromosome, stimulating polyploidy (De Robertis and De Robertis 1980).

In this study we tested the technique of tissue culture in the micropropagation of *K. alvarezii* aiming to select the best strain and experimental system for *in vitro* culture, to assess the effects of phytoregulators on micropropagation, and to produce new strains by application of colchicine.

Materials and methods

Strains studied

Five strains of *Kappaphycus alvarezii* were tested: three tetrasporophytic strains (brown, BR, green, GR, and red, RD), a brown female gametophyte (BFG) and a brown strain derived from tetraspore germination isolated by Paula et al. (1999), named "Edison de Paula" strain (EP). All were cultivated in sterilized seawater (32 psu) enriched with 25% Guillard & Ryther's solution (F/2), according to McLachlan (1973), photoperiod of 14 h, temperature of 25°C (±1°C) and irradiance of 200 (±10) µmol photons m^{-2} s^{-1}. Cultures were aerated in alternating periods of 30 minutes.

Axenic cultures

Axenic cultures of the five strains of *Kappaphycus alvarezii* were obtained according to the methodology described by Yokoya (1996). Branches of *K. alvarezii* (5 cm) were incubated in an antifungal and antibiotic solution (including 10,000 units penicillin G, 10 mg streptomycin sulphate and 25 µg amphotericin B *per* liter, Sigma) and 0.1 µg mL^{-1} nystatin (Sigma) added to sterilized seawater enriched with 50% von Stosch's solution (VS 50, according to Edwards 1970) for 48 h. After this period, branches were transferred to an aseptic chamber and washed in sterilized seawater with 0.05% sodium hypochlorite and 2% organic detergent (Amway), followed by successive washes with autoclaved seawater. Thallus segments 3 mm in length were cut and inoculated into the culture medium described in each experiment.

Axenic cultures were grown under the following culture conditions: salinity of 30 psu, photoperiod of 14 h, temperature of 23°C (±2°C) and irradiance of 40 (±10) µmol photons m^{-2} s^{-1}.

Effects of culture medium

Three media were tested: autoclaved seawater (30 psu) enriched with 50% of von Stosch's solution (VS 50) or 50% Guillard & Ryther solution (F/2 50), and synthetic ASP 12-NTA medium (ASP 12, according to Iwasaki 1961, Sigma), with addition of 0.5% agar (Merck). For each treatment, 3 replicates with 8 to 10 explants from the BR, GR, RD and EP strains were inoculated into 30 mL of culture medium. Calluses were induced in 45 days, and the culture medium was renewed at intervals of 15 days.

After the period of callus induction, explants were transferred to liquid culture medium (30 mL) under the same culture conditions. Each treatment had three replicates with the surviving explants from the previous experiment (from 1 to 9 explants *per* replicate). Explants were cultured during 35 days and culture medium was renewed weekly.

Effects of phytoregulators on direct regeneration

In our study, direct regeneration is defined as the regeneration originating from explant cells. The experimental design for testing the phytoregulators followed the methodology described by Yokoya (1996) and the phytoregulator concentrations were chosen based on results of Yokoya et al. (1999) and Yokoya (2000). Explants from the EP strain were cultured in VS 50 medium with glycerol

(90 mM, Synth) previously autoclaved, with the addition of 2,4-dichlorophenoxyacetic acid (2,4-D, Sigma) and 6-benzylaminopurine (BA, Sigma) in concentrations of 0.5 and 5.0 mg L^{-1}. Combination of IAA:BA (indole-3-acetic acid, IAA - Sigma) was tested in concentration of 5:1 (mg L^{-1}). Culture medium without phytoregulators was used as the control. For each treatment, three replicates with 5 explants in each were inoculated into 30 mL of culture medium. Explants were cultured for 61 days and the renewal of culture medium was made fortnightly.

Effects of colchicine on direct regeneration

Explants from the BFG and EP strains were cultured for 24, 48, 72 h and 14 days in VS 50 medium with glycerol (90 mM), previously autoclaved, and 0.01% colchicine (Sigma). VS 50 culture medium without colchicine was used as the control.

Three replicates with 5 explants in each were used in the control, while 10 replicates with 6 to 8 explants in each were used in the treatments with colchicine. Explants were inoculated into 30 mL of culture medium and cultured for 73 days. The culture medium was renewed weekly.

Data and statistical analysis

The percentage of explants forming callus and the number of explants with direct or indirect regeneration (i.e. developing upright axes from the explant or callus) was recorded.

Data were submitted to one-way, many effects or repeated measurements analysis of variance followed by *a posteriori* tests of Tukey (one-way), Duncan (many effects), and Unequal N (repeated measurements with different n) considering $p<0.05$. Analysis was using *Statistica* Software (release 6.0).

Results

Effects of culture medium

Two types of callus were observed in the four strains of *Kappaphycus alvarezii* studied: a) filamentous callus, observed in the explant surface exposed to the air, composed of filaments derived from divisions of cortical and medullary cells (Fig. 1a); and b) compact callus, growing on the explant surface in contact with culture medium, formed by cortical and medullary cells and having organized medullary and cortical regions (Fig. 1a and b). The EP strain also presented filament proliferation growing from the explant region in contact with culture medium (Fig. 1c). In all strains, callus formation was observed after 15 days of cultivation, and callus growth was observed during the entire experimental period.

Explants of the BR strain had the highest percentage of explants forming callus when cultured in F/2 50 in comparison to others culture media. The EP strain showed higher levels of explants forming callus in VS 50 and F/2 50 media

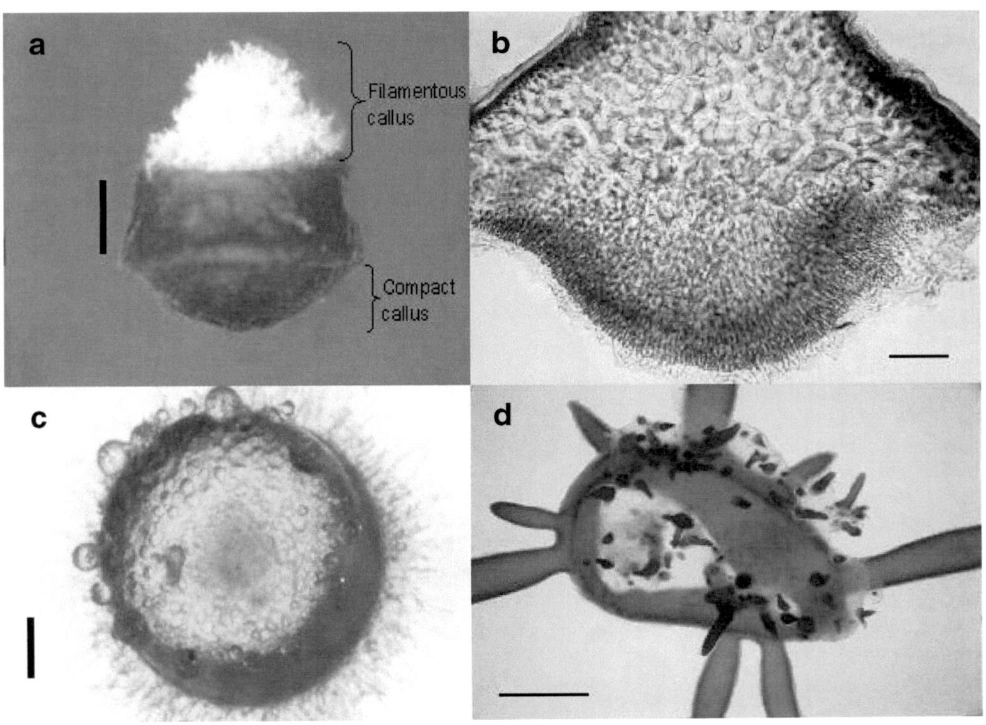

Fig. 1 Callus formation and indirect regeneration in *Kappaphycus alvarezii*. **a)** Filamentous and compact calluses from green tetrasporophytic strain induced by VS 50 medium with 0.5% agar after 45 days (scale bar= 1 mm); **b)** Longitudinal section of compact callus from green tetrasporophytic strain (scale bar =200 μm); **c)** Callus from "Edison de Paula" strain growing on the surface of the explant in contact with VS 50 medium with 0.5% agar after 45 days (scale bar=1 mm); **d)** Indirect regeneration from filamentous callus of the brown tetrasporophytic strain cultured at VS 50 liquid medium and agitation during 30 days (scale bar=500 μm)

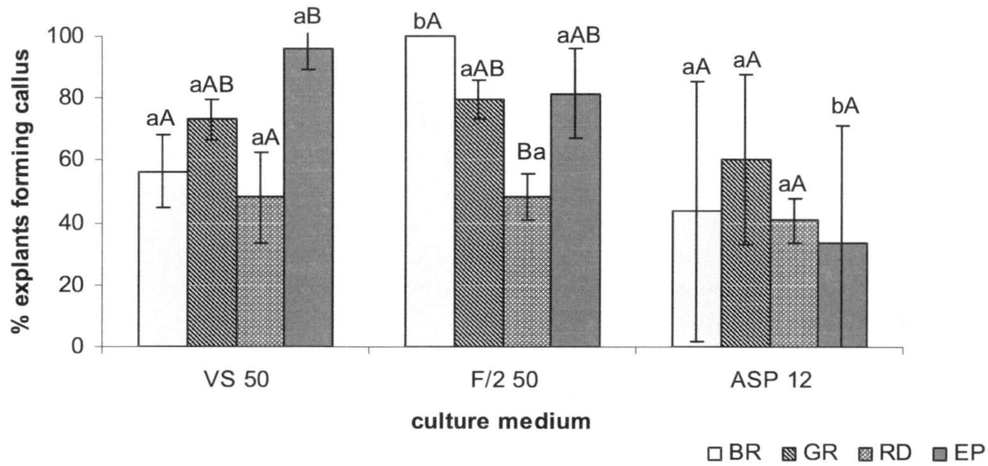

Fig. 2 Effects of culture medium on callus formation in brown (BR), green (GR) and red (RD) tetrasporophytic strains, and "Edison de Paula" (EP) strain of *Kappaphycus alvarezii* cultured in VS 50, F/2 50 or ASP 12 with 0.5% agar for 45 days. Values presented as average (n=3) and vertical bars represent confidence intervals. The uppercase letters indicate significant differences among culture medium for each strain and the lowercase letters, the significant differences among strains, according to Duncan's *a posteriori* test ($p<0.05$)

than in ASP 12 medium. For the other strains, significant differences were not observed among the treatments (Fig. 2).

Callus production from the EP strain was also significantly different than those presented by the BR and RD strains when cultured in VS 50 medium, and significantly different than the RD strain when cultured in F/2 50 medium, while the BR strain showed significant differences only in relation to the RD strain when cultured in F/2 50 medium (Fig. 2).

During the period of callus induction, direct regeneration occurred in some explants independently of strain or type of culture medium. In these explants, calluses ceased from growing.

Despite the calluses transferring to liquid medium, indirect regeneration was very rare, in all culture medium tested, with the exception of one explant among approximately 100 from all strains tested in our study. The explant of BR strain showed several micropropagules (approximately 90) regenerating from filamentous callus, when cultured in VS 50 medium with glycerol and mechanical agitation for 30 days (Fig. 1d).

Effects of phytoregulators on direct regeneration

Explants cultured in VS 50 medium with glycerol and IAA:BA 5:1 (mg L^{-1}) showed significant differences in the number of upright axes *per* explant than the control or treatment with 2,4-D 0.5 (mg L^{-1}) (Fig. 3). In the others treatments, no significant differences were observed. The same explants, when transferred to VS 50 without glycerol and cultured for 23 days, presented significant differences only in treatment with IAA:BA (5:1 mg L^{-1}) in comparison to the control (Fig. 3).

Effects of colchicine on direct regeneration

Explants of EP strain of *Kappaphycus alvarezii* cultured in VS 50 medium with glycerol, in different incubation periods with colchicine, showed direct regeneration in all treatments. No significant differences in the number of upright axes *per* explant among treatments with colchicine were observed. However, the level of direct regeneration

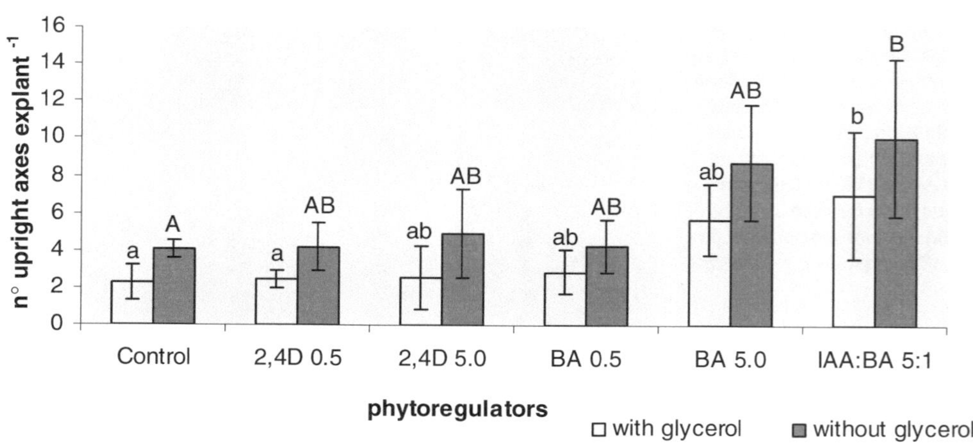

Fig. 3 Effects of glycerol and phytoregulators in direct regeneration of "Edison de Paula" (EP) strain of *Kappaphycus alvarezii* cultured for 58 days in liquid VS 50 medium. Values presented as average (n=3) and vertical bars represents confidence intervals. The uppercase letters indicate significant differences among treatments with glycerol and the lowercase letters, the significant differences among treatments without glycerol, according to Tukey's test ($p<0.05$)

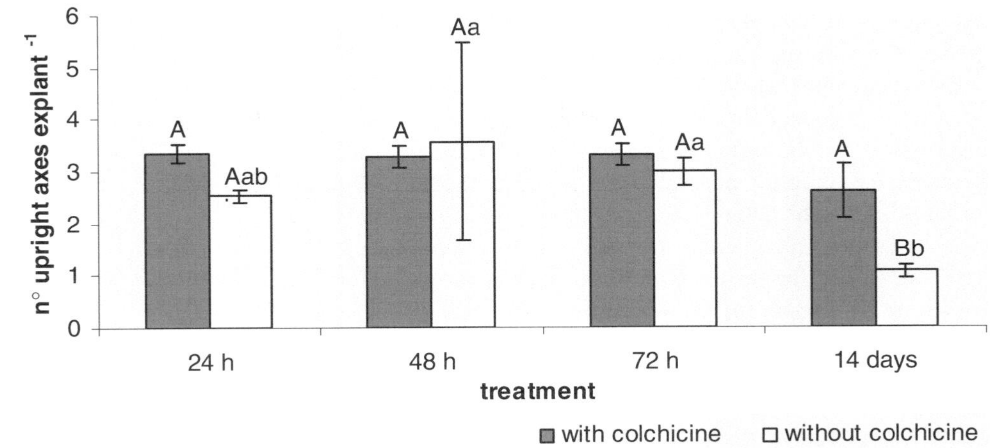

Fig. 4 Effects of colchicine on direct regeneration of "Edison de Paula" (EP) strain of *Kappaphycus alvarezii* cultured in liquid VS 50 medium with glycerol for 67 days. Values presented as average (n=3 for control; n=10 for treatments) and vertical bars represent confidence intervals. The uppercase letters indicate significant differences in the treatments and the lowercase letters, the significant differences among treatments, according to Unequal N's *a posteriori* test ($p<0.05$)

was significantly different in explants incubated for 14 days with colchicine than the control (Fig. 4).

Explants of BFG strain incubated with colchicine for 24, 48, 72 h and 14 days died after 15 days.

Discussion

Callus formation in *Kappaphycus alvarezii* strains was observed after 15 days in all culture media tested, corroborating data obtained by Reddy et al. (2003) for the same species. These authors observed percentages of callus induction higher than 80% in explants cultured in Provasoli culture medium with 1.5% of bacto-agar, without addition of other compounds, as phytoregulators or carbon organic compounds. In our experiments, the suitable medium for callus induction was F/2 50 in the *K. alvarezzi* BR strain, while in the EP strain higher levels of callus induction were presented in VS 50 and F/2 50 media. Although no significant differences were observed among the culture media, explants cultured in ASP 12 medium had lower percentage of explants forming callus and presented thallus bleaching. Based on these results, VS 50 medium was selected for subsequent experiments.

Some explants cultured in 0.5% agar showed direct regeneration from the explant surface in contact with the medium. When this occurred, the growth of filamentous callus ceased. The EP strain showed the formation of filamentous callus in the same region, corroborating the observation made for the same species by Polne-Fuller and Gibor (1987).

In general, the regeneration occurred directly from explant cells or in the compact callus region, as observed by Azanza-Corrales and Dawes (1989), who studied the wounding process in thallus fragments of *K. alvarezii* var. *tambalang*. Reddy et al. (2003) observed percentages of indirect regeneration lower than 10%, while Polne-Fuller and Gibor (1987) observed regeneration from calluses and explants in *Eucheuma* and *Kappaphycus* cultured in medium with agar as well as in liquid medium, although they did not quantify the number of regenerations.

According to Yokoya et al. (2004), the physical state of culture medium may affect the indirect regeneration process depending on the species. These authors observed that calluses of *Gracilaria tenuistipitata* (Chang & Xia) regenerated when cultured in solid medium, while the regeneration of *G. perplexa* (Byrne, Zuccarello, West, Liao & Kraft) occurred only after transferring from medium with agar to liquid medium, indicating a morphogenetical function of physical state of the medium. In *Solieria filiformis*, the solid medium induced filament growth, while adventitious plantlets were regenerated in liquid medium (Yokoya and Handro 2002). In the present work, the physical state of the medium did not induce the regeneration from calluses, and indirect regeneration was very rare. The exception was the BR strain, which showed indirect regeneration from only one callus. Despite Reddy et al.'s (2003) observation that mechanical agitation could stimulate the regeneration of adventitious plantlets, which gave rise to micropropagules after 40 days, this response was not observed in the present study. Only one callus among 15 (6.7%) of the BR strain regenerated when cultured under mechanical agitation (data not shown).

In relation to the effects of phytoregulators on direct regeneration, explants treated with IAA:BA (5:1 mg L^{-1}) regenerated up to 7 upright axes per explant when cultured with glycerol for 35 days, and up to 10 upright axes per explant when cultured without glycerol for 23 days. These results are not in agreement to the ones obtained by Yokoya (2000), where the direct regeneration of *Gracilariopsis tenuifrons* (Bird & Oliveira; Fredericq & Hommersand) was promoted by IAA:BA (1:5 mg L^{-1}). Probably there are some interactions of exogenous phytoregulators and endogenous substances that promote similar effects in some species and different effects in other species.

Explants incubated with colchicine for 14 days showed more direct regeneration than the control. However, in our experiment, the chromosome numbers were not checked, so this response could not be related to a possible doubling of chromosomes. The chromosome number of *K. alvarezii* requires further investigation. Kapraun and Lopez-Bautista (1997) described the presence of 10 bivalents in prophase I (n=10) of *Kappaphycus alvarezii* from the Philippines, while Contador (2001) observed up to 6 bivalents in some cells and more than 10 in others from the same material studied in the present work. Further studies should be conducted in order to verify the effects of colchicine on the ploidy of *K. alvarezii*.

Based on results obtained in the present study, the EP strain is more tolerant and showed higher potential for micropropagation than the other strains tested in our experiments. Previous studies with the same strains of *K. alvarezii* demonstrated that EP strain had a lower growth rate than the tetrasporophytic strain when cultivated in the sea, but produced higher carrageenan content (Hayashi 2001; Hayashi et al. 2007).

In conclusion, the micropropagation of EP strain of *Kappaphycus alvarezii* was stimulated by IAA:BA (5:1 mg L^{-1}), and treatments with colchicine also stimulated the regeneration of upright axes. Both factors could be useful to enhance the micropropagation of *K. alvarezii*.

Acknowledgements This study was supported by the Conselho Nacional de Desenvolvimento Científico e Tecnológico (CNPq) and the Fundação de Amparo à Pesquisa do Estado de São Paulo (FAPESP). We are thankful to José Domingos and Rosário Petti for helping in laboratory.

References

Ask EI, Azanza RV (2002) Advances in cultivation technology of commercial eucheumatoid species: a review with suggestions for future research. Aquaculture 206:257–277

Azanza-Corrales R, Dawes CJ (1989) Wound healing in cultured *Eucheuma alvarezii* var. *tambalang* (Doty). Bot Mar 32:229–234

Bradley P (1991) Plant Hormones do have a role in controlling growth and development of algae. J Phycol 27:317–321

Cheney DP (1984) Genetic modification in seaweeds: applications to commercial utilization and cultivation. In: RR Colwell, AJ Sinskey, ER Parisier (eds) Biotechnology in the marine sciences. Wiley, New York, pp 161–175

Collantes G, Melo C (1995) Cultivo de tejidos y celulas de algas marinas. In: K Alveal, ME Ferrario, EC Oliveira, E Sar (eds) Manual de Métodos Ficológicos. Universidad de Concepción, Concepción, Chile, pp 457–477

Contador CRB (2001) Reproductive and biological aspects of *Kappaphycus alvarezii* (Doty) Doty ex Silva and *Kappaphycus striatum* (Schmitz) Doty (Gigartinales, Rhodophyta). Bases for introduction and cultivation of exotics species in Brazilian littoral (in Portuguese). Dissertation, University of São Paulo

Dawes CJ, Koch EW (1991) Branch, micropropagule and tissue culture of the red algae *Eucheuma denticulatum* e *Kappaphycus alvarezii* farmed in Philippines. J Appl Phycol 3:247–257

De Robertis E, De Robertis EM Jr (1980) Cell and molecular biology, 7th edn. Saunders College, Philadelphia

Edwards P (1970) Illustrated guide to the seaweeds and seagress in the vicinity of Porto Aransas, Texas. Contribution of Marine Sciences Austin 15:1–228

Garcia-Jimenez P, Rodrigo M, Robaina RR (1998) Influence of plant regulators, polyamines and glycerol interaction on growth and morphogenesis of carposporelings of *Grateloupia* cultured *in vitro*. J Appl Phycol 20:95–100

Hayashi L (2001) Extraction, yield and properties of *Kappaphycus alvarezii* (Doty) Doty ex P. Silva carrageenan, in experimental cultivation in Ubatuba, São Paulo (in Portuguese). Master Degree Dissertation, University of São Paulo

Hayashi L, Paula EJ, Chow F (2007) Growth rates and carrageenan analysis of four strains of *Kappaphycus alvarezii* (Rhodophyta, Gigartinales) farmed in the subtropical waters of São Paulo State, Brazil. J Appl Phycol (in press). DOI 10.1007/s10811-006-9135-6

Huang W, Fujita Y (1997) Callus induction and thallus regeneration of the red alga *Meristotheca papulosa* (Rhodophyta, Gigartinales). Bot Mar 40:55–61

Hurtado AQ, Cheney DP (2003) Propagule production of *Eucheuma denticulatum* (Burman) Collins *et* Harvey by tissue culture. Bot Mar 46:338–341

Iwasaki H (1961) The life cycle of *Porphyra tenera in vitro*. Biol Bull 121:173–187

Jacobs WMP (1993) A search for some angiosperm hormones and their metabolites in *Caulerpa paspaloides* (Chlorophyta). J Phycol 29:595–600

Jacobs WMP, Falkenstein K, Hamilton RH (1985) Nature and amount of auxin in algae: IAA from extracts of *Caulerpa paspaloides* (Siphonales). Pl Physiol 78:844–848

Kapraun DF, Lopez-Bautista J (1997) Karyology, nuclear genome quantification and characterization of carragenophytes *Eucheuma* and *Kappaphycus* (Gigartinales). J Appl Phycol 8:465–471

McLachlan J (1973) Growth media-marine. In: JR Stein (ed) Handbook of phycological methods. Culture methods and growth measurements. Cambridge University Press, Cambridge, pp 25–51

Paula EJ, Pereira RTL, Ohno M (1999) Strain selection in *Kappaphycus alvarezii* var. *alvarezii* (Solieriaceae, Rhodophyta) using tetraspore progeny. J Appl Phycol 11:111–121

Polne-Fuller M, Gibor A (1987) Calluses and callus-like growth in seaweeds: induction and culture. Hydrobiologia 151/152:131–138

Reddy CRK, Kumar GRKK, Siddhanta AK, Tewari A (2003). *In vitro* somatic embryogenesis and regeneration of somatic embryos from pigmented callus of *Kappaphycus alvarezii* (Doty) Doty (Rhodophyta, Gigartinales). J Phycol 39:610–616

Stirk WA, van Staden J (1997) Isolation and identification of cytokinins in a new commercial seaweed product made from *Fucus serratus* L. J Appl Phycol 9:327–330

Stirk WA, Novák O, Strnad M, Van Staden J (2003) Cytokinins in microalgas. Plant Growth Regul 41:13–24

Stirk WA, Arthur GD, Lourens AF, Novák O, Strnad M, Van Staden J (2004) Changes in cytokinin and auxin concentrations in seaweed concentrates when stored at an elevated temperature. J Appl Phycol 16:31–39

Trono GC, Lluisma AO (1992) Differences in biomass production and carrageenan yields among four strains of farmed carrageenophytes in Northern Bohol, Phillipines. Hydrobiologia 247: 223–227

van der Meer JP (1987) The development and genetic improvement of marine crops. In: T Stadler, J Mollion, M-C Verdus, Y Karamanos, H Morvan, D Christiaen (eds) Algal biotechnology. Elselvier Applied Science Publishers, New York, pp 1–15

Yokoya NS (1996) Grow and morphogenesis control by auxins and citocinin in three species of rodophyceae: *Gracilariopsis tenuifrons, Grateloupia dichotoma* e *Solieria filiformis*. Dissertation (in Portuguese). University of São Paulo

Yokoya NS (2000) Apical callus formation and plant regeneration controlled by plant growth regulators on axenic culture of the red alga *Gracilaria tenuifrons* (Gracilariales, Rhodophyta). Phycol Res 48:133–142

Yokoya NS, Handro W (2002) Effects of plant growth regulators and culture medium on morphogenesis of *Solieria filiformis* (Rhodophyta) cultured *in vitro*. J Appl Phycol 14:97–102

Yokoya NS, Kakita H, Obika H, Kitamura T (1999) Effects of environmental factors and plant growth regulators on growth of the red alga *Gracilaria vermiculophylla* from Shikoku Island, Japan. Hydrobiologia 398/399:339–347

Yokoya NS, West JA, Luchi AE (2004) Effects of plant growth regulators on callus formation, growth and regeneration in axenic tissue cultures of *Gracilaria tenuistipitata* and *Gracilaria perplexa* (Gracilariales, Rhodophyta). Phycol Res 52:244–254

Zhang W, Chapman DJ, Phinney BO, Spray CR (1991) Identification of cytokinins in *Sargassum muticum* (Phaeophyta) and *Porphyra perforata* (Rhodophyta). J Phycol 27:87–91

Bioactive volatile compounds from marine algae: feeding attractants

Yoshihiko Akakabe · Tadahiko Kajiwara

Originally published in the Journal of Applied Phycology, Vol 20, No 5, 211–214.
DOI: 10.1007/s10811-007-9309-x

Abstract Chemical communications play an important role in plants, fungi, and algae. Volatile organic compounds in marine algae are released into the seawater. These compounds play a role as either pheromones or allelochemicals. We observed that the turbinid gastropod *Lunella coronata coreensis* inhabits the intertidal zone and often grazes the green alga *Ulva pertusa*. Feeding tests and feeding preference studies were performed with green, brown and red algae or by using the powdered freeze-dried seaweed in agar. The snails fed on *U. pertusa* preferentially compared to the other marine algae, and recognized chemoreception compounds from the alga but not their structural or morphological differences. From feeding tests using artificial foods, it is suggested that the feeding attractants are in the essential oil of the alga *U. pertusa*.

Keywords Herbivores · Chemical communications · Pheromones · Allelochemicals · Chemoreception compounds · Feeding attractant

Introduction

Volatile organic compounds in marine algae are released into the seawater. We have demonstrated that long-chain aldehydes (C_{15}, C_{17}) are formed from fatty acids through the formation of the corresponding 2-hydroperoxy acids in the green alga *Ulva pertusa*, and that these hydropeoxides exist in a variety of marine algae (Akakabe et al. 1999, 2000, 2001). We have also found that short-chain aldehydes (C_6, C_9) and middle-chain aldehydes (C_{10}) are formed from fatty acids (C_{20}) in marine algae, whereas they are formed from C_{18} in higher plants (Boonprab et al. 2003, 2006; Akakabe 2003). The volatile compounds play important roles as chemical communications in marine ecosystems. The compounds act as pheromones (Pohnert and Boland 2002) or allelochemicals, chemical defenses such as deterrence against herbivores (Hay et al. 1998; Schnitzler et al. 2001; Pelletreau and Muller-Parker 2002; Pohnert 2002), inhibition of bacterial and fungal fouling (Bakus et al. 1986; Davis et al. 1989; Melton and Bodnar 1988), and suppression of competing neighbors (Suzuki et al. 1998). We have already investigated the pheromone-mediated mating process in the brown algae *Dictyopteris* spp during sexual reproduction and elucidated the mechanistic pathway for the biosynthesis (Yamamoto et al. 2003; Kajiwara et al. 2006). In the green alga *Bryopsis maxima*, one major volatile compound was released by the mechanical wounding and was identified as (*Z*)-8-heptadecene by GC-MS analysis and synthesis (submitted). It was suggested that these compounds might act as allelochemicals.

On the other hand, small marine herbivores (molluscs, crabs and shrimp) with limited mobility are primary consumers in the intertidal zone, and act as a link between producers (phytoplankton and marine algae) and secondary consumers (fish, birds, etc.). Among these, the turbinid gastropod *Lunella coronata coreensis* inhabits shallow areas and tide pools in the intertidal zone of Japan, and often grazes marine algae such as the green alga *U. pertusa*. In general, feeding behaviors of lower herbivores such as *L. coronata coreensis* are divided into three steps: (1) searching marine algae, (2) taking into a mouth, and (3) consuming the algae. These behaviors are called chemore-

Y. Akakabe (✉) · T. Kajiwara
Department of Biological Chemistry, Faculty of Agriculture, Yamaguchi University,
1677-1 Yoshida,
Yamaguchi 753-8515, Japan
e-mail: akakabe@yamaguchi-u.ac.jp

ception processes. These processes are classified as attractant, arrestant, incitant, and stimulant. Attractant is an effect of inducing herbivores to foods. Arrestant is an effect of resting near the foods. Incitant is reception of starting nutrition. Stimulant is reception advancing to graze continuously. From these results, it was suggested that these behaviors are at least directed with these compounds. In the course of investigation of the allelochemicals in marine ecosystems, an interaction between seaweeds and the snail *L. coronata coreensis* was examined.

Materials and methods

The herbivore gastropod *Lunella. coronata coreensis*, was collected in the intertidal zone in Yamaguchi Prefecture, Japan, in 2003 to 2004. The snails were kept at 20°C under a 12/12 h light/dark cycle in filtered seawater and fed the fresh green alga *Ulva pertusa*. Several marine algae, the green *U. pertusa*, *Caulerpa okamurae* and *Codium fragile*, the brown *Dictyopteris prolifera* and *Undaria pinnatifida*, and the red *Gelidium amansi* and *Carpopeltis affinis*, were collected in the same place and were cultured at 13°C until being used in the bioassays. The marine algae for preparation of essential oils were immediately frozen at −20°C until use.

Preparation of essential oils of *U. pertusa*

In order to examine whether the feeding behaviors are controlled either by volatile organic compounds or other compounds, an essential oil from marine alga was prepared by a simultaneous distillation extraction apparatus. CH_2Cl_2 / pentane (1 : 2, 50 mL) was used as an extraction solvent. The alga (100 g) was homogenized with phosphate buffer (pH 7.0, 250 ml), and the homogenate was placed in a flask (1,000 ml). Steam distillation was continued for 2 h, while the solvent extraction was continued for further 10 min. The extract was dried over anhydrous Na_2SO_4 and concentrated under reduced pressure to give an essential oil (29.7 mg, 2.97×10^{-2}% yield).

Preparation of artificial foods

In order to examine whether the feeding behaviors are controlled either by morphological properties or by allelochemicals, artificial foods using an algal powder were prepared. the algae were freeze-dried for 24 h and ground finely with a blender. Agar (0.39 g) was mixed with distilled water (7 ml) and the solution was heated at 120°C for 20 min. The hot solution was stirred and then poured into cold water (7 ml) containing freeze-dried marine algae (1 g fresh weight) or essential oil (0.3 mg) in ether. The solution was stirred homogeneously and spread on a culture dish.

Feeding test

Each alga (1 g fresh weight) was placed at the center of a culture dish (9 cm diameter) with seawater (50 mL) and 3 snails (n=3). The snails were starved for 24 h before the feeding tests. In order to estimate changes in the alga due to factors other than feeding, the same amount of alga was placed in another dish without the snails as a control. The snails were allowed to feed on the algae for 48 h before measuring the remaining algae. The amount of consumption was defined by the following the equation (Hay et al. 1998): $consumption(\%) = \left[\left(H_0 \times C_f/C_0\right) - H_f\right]$, where H_0 and H_f are mass of the initial alga and mass of the remaining alga after 48 h, respectively, for feeding tests, and where C_0 and C_f are mass of the initial alga and mass of the remaining alga after 48 h, respectively, for controls.

In order to examine the palatability of the snails, the pairwise feeding trials were carried out between all 7 species of algae. Two algae species were placed symmetrically on each side of a 21-cm diamter tray with 200 mL of seawater. Ten snails starved for 24 h were centered in the tray (n=3). The snails were allowed to feed for 48 h before measuring the mass of remaining algae.

To investigate the existence of chemoreception compounds (attractants, arrestants, insitants, and stimulants) in the algae, consumption of an artificial food containinig essential oil of *U. pertusa* was also examined. These foods were supplied to 3 snails (n=3). Each culture dish (9 cm in diameter) was placed on a paper with 678 grids. Feeding rates were measured by counting the number of mesh squares from which the food was consumed.

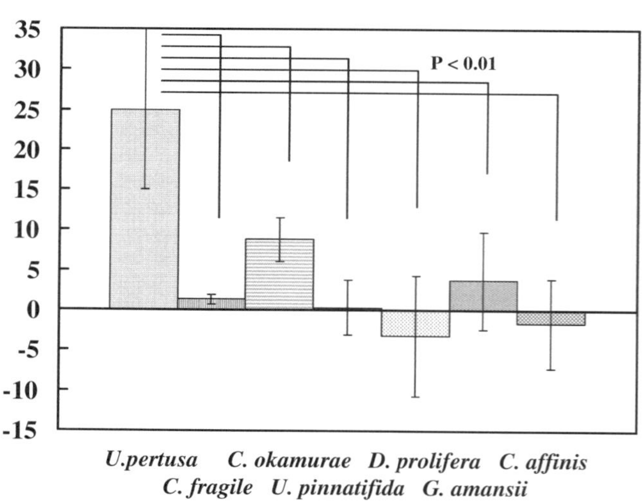

Fig. 1 Feeding preference of the herbivorous gastropod *L. coronata coreensis* for fresh marine algae

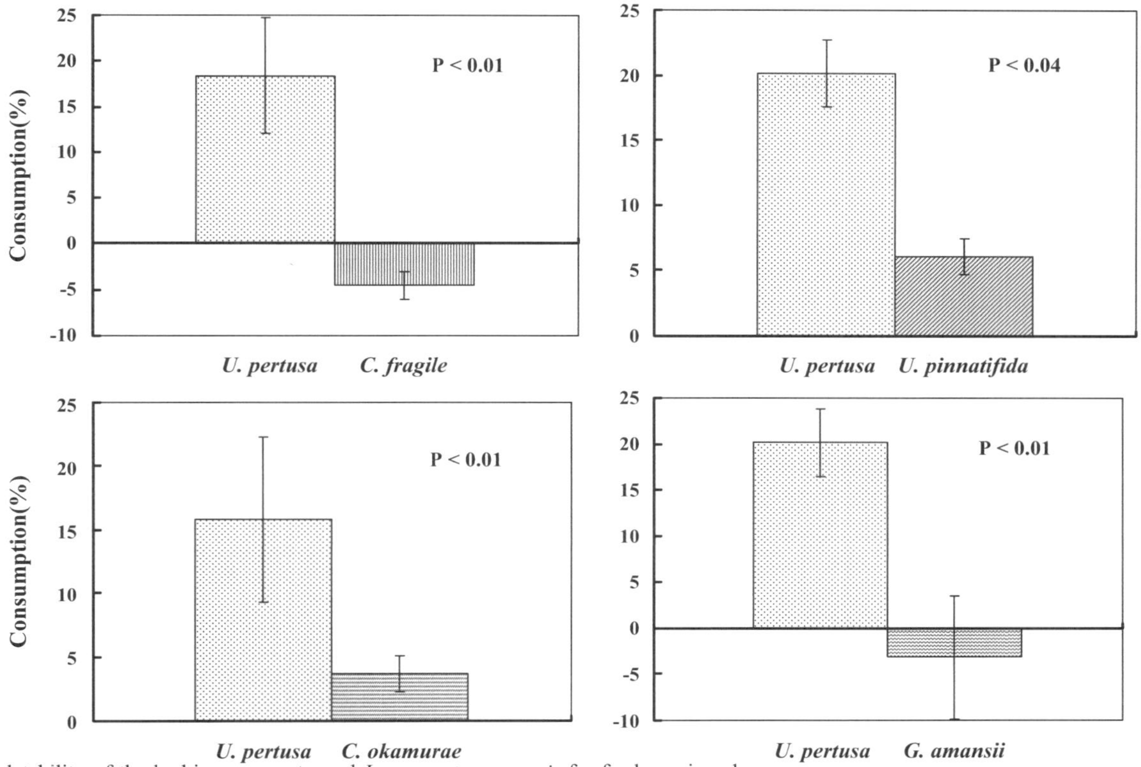

Fig. 2 Palatability of the herbivorous gastropod *L. coronata coreensis* for fresh marine algae

Results

In order to investigate feeding behaviors of the herbivorous gastropod *Lunella coronata coreensis* for fresh marine algae, the feeding preference test was performed with 7 species of algae which inhabited the same intertidal zone where the snails coexist (Fig. 1). Among of these, the green alga *Ulva pertusa* was consumed (25.0±10.1%), preferentially, followed by *Caulerpa okamurae* (8.8±2.7%) and *Gelidium amansii* (3.73±6.2%). Furthermore, statistical analyses that the snails fed significantly more on *U. pertusa* than on the other algae ($p<0.01$).

When two species of algae were supplied simultaneously to the snails (pairwise feeding trial), the patterns of palatability remained the same. Thus, the snails were more attracted toward *U. pertusa* and fed significantly more on this alga than the other alga in all cases (Fig. 2). One suggestion for these differences in feeding preference is that it is a result of

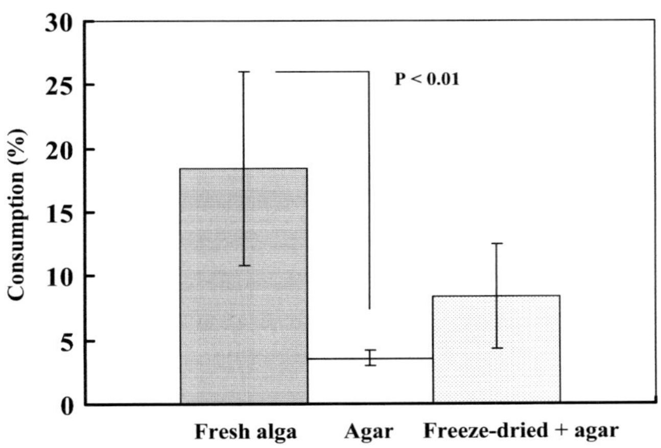

Fig. 3 Feeding preference of the herbivorous gastropod *L. coronata coreensis* for artificial foods (freeze-dried)

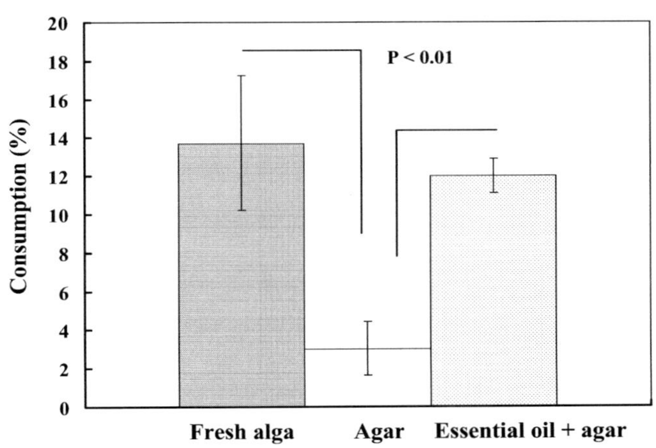

Fig. 4 Feeding preference of the herbivorous gastropod *L. coronata coreensis* for artificial foods (essential oil)

differences in the morphology of the algae. To exclude this possibility, feeding preference was examined using freeze-dried powdered *U. pertusa* incorporated in agar (Fig. 3). As shown in Fig. 3, the artificial food evoked a positive response in the snails, there was no significant difference between the food and the fresh alga. This result suggests that the feeding behaviors were dependent on not only on any morphological characteristics but also on other factors such as allelochemicals. However, the consumption of artificial foods was relatively low (8.4±4.1%). From these results, it was anticipated that volatile compounds of the alga might be lost in the freeze-drying process and these compounds might be of importantce to feeding behaviors.

To investigate the role that chemoreception compounds might play in the feeding behaviors of the snails, further feeding tests using the essential oil as an artificial food were carried out (Fig. 4). When the essential oil was assayed in the feeding test, a strong positive response was found (12.0±0.9 %) equivalen to that of the fresh alga (13.7±3.5%). It is therefore suggested that the essential oil contains the chemoreception compounds that enhance the feeding behaviors.

Discussion

In general, it is considered that lower herbivores such as the gastropod *Lunella coronata coreensis* are under the control of allelochemicals in their behaviors. Sakata et al. (1988) reported that methanol extracts of *Ulva pertusa*, which contained four types of lipids, were active as feeding stimulants for marine gastropods such as *Haliotis discus*, *Turbo cornutus*, and *Omphalius pfeifferi*. On the other hand, dimethylsulfide and acrylic acid in the alga can deter feeding by some herbivores (Van Alstyne and Houser 2003).

We have shown here that the herbivorous gastropod *L. coronata coreensis* fed preferentially on *U. pertusa* compared with other algae inhabiting the same intertidal zone by feeding tests and pairwise feeding trials. The feeding tests indicated that the snails recognized chemoreception compounds from the algae rather than their structural or morphological differences, because artificial foods containing freeze-dried alga evoked strongly positive responses to the snails compared with only agar. It is suggested that volatile compounds in the essential oil of the alga play important roles in feeding behaviors of the snails. We have already found that long-chain aldehydes that have marine odors were major components in the essential oil and were increased by a mechanical wounding (unpublished results). Consequently, it is indicated that the feeding behaviors of the snails are controlled by the chemoreception compounds as feeding attractants, and that the volatile compounds in the oils of the alga would play important roles in these behaviors.

References

Akakabe Y, Matsui K, Kajiwara T (1999) Enantioselective α-hydroperoxylation of long-chain fatty acids with crude enzyme of marine green alga *Ulva pertusa*. Tet Lett 40:1137–1140

Akakabe Y, Matsui K, Kajiwara T (2000) α-Oxidation of long-chain unsaturated fatty acids in the marine green alga *Ulva pertusa*. Biosci Biotechnol Biochem 64:2680–2681

Akakabe Y, Matsui K, Kajiwara T (2001) Enantioselective 2-hydroperoxylation of long-chain fatty acids in marine green algae. Fisheries Sci 67:328–332

Akakabe Y, Matsui K, Kajiwara T (2003) 2,4-Decadienals are produced *via* (*R*)-11-HPITE from arachidonic acid in marine green alga *Ulva conglobata*. Bioorg Med Chem 11:3607–3609

Bakus GJ, Targett NM, Schulte B (1986) Chemical ecology of marine organisms: an overview. J Chem Ecol 12:951–987

Boonprab K, Matsui K, Akakabe Y, Yotsukura N, Kajiwara T (2003) Hydroperoxy-arachidonic acid mediated *n*-hexanal and (*Z*)-3- and (*E*)-2-nonenals formation in *Laminaria angustata*. Phytochem 63:669–678

Boonprab K, Matsui K, Akakabe Y, Yoshida M, Yotsukura N, Chirapart A, Kajiwara T (2006) Formation of aldehyde flavor (*n*-hexanal, 3*Z*-nonenal and 2*E*-nonenal) in the brown alga, *Laminaria angustata*. J Appl Phycol 18:1573–1576

Davis AR, Targett NM, McConnell OJ, Young CM (1989) Epibiosis of marine algae and benthic invertebrates: natural products chemistry and other mechanisms inhibiting settlement and overgrowth. In: Scheuer PJ (ed), Bioorganic Marine Chemistry, vol 3. Springer, Berlin, pp 85–114

Hay ME, Piel J, Boland W, Schnitzler I (1998) Seaweed sex pheromones and their degradation products frequently suppress amphipod feeding but rarely suppress sea urchin feeding. Chemoecology 8:91–98

Kajiwara T, Yamamoto Y, Akakabe Y, Matsui K, Shimizu H, Kawai T (2003) Biomimetic conversion of (*3S*)-(−)-neodictyoprolenol to optically pure (*1S,2R*)-(−)-dictyopterene B, marine algal sex pheromone. Z Naturf 58c:109–112

Melton T, Bodnar JW (1988) Molecular biology of marine micro-organisms: biotechnological approaches to naval problems. Nat Res Rev 40:24–39

Pelletreau K, Muller-Parker G (2002) Sulfuric acid in the phaeophyte alga *Desmarestia munda* deters feeding by the sea urchin *Strongylocentrotus droebachiensis*. Mar Biol 141:1–9

Pohnert G (2002) Phospholipase A2 activity triggers the wound-activated chemical defense in the diatom *Thalassiosira rotula*. Plant Physiol 129:103–111

Pohnert G, Boland W (2002) The oxylipin chemistry of attraction and defense in brown algae and diatoms. Nat Prod Rep 19:108–122

Sakata KN, Sakura T, Ina K (1988) Algal phagostimulants for marine herbivorous gastropods. J Chem Ecol 14:1405–1416

Schnitzler I, Pohnert G, Hay ME, Boland W (2001) Chemical defense of brown algae (*Dictyopteris spp.*) against the herbivorous amphipod *Ampithoe longimana*. Oecologia 126:515–521

Suzuki Y, Takabayashi T, Kawaguchi T, Matsunaga K (1998) Isolation of an allelopathic substance from the crustose coralline algae, *Lithophyllum spp.*, and its effect on the brown alga, *Laminaria religiosa* Miyabe (*Phaeophyta*). J Exp Mar Biol Ecol 225:69–77

Van Alstyne KL, Houser LT (2003) Dimethylsulfide release during macroinvertebrate grazing and its role as an activated chemical defense. Mar Ecol Prog Ser 250:175–181

Yamamoto Y, Akakabe Y, Matsui K, Shimizu H, Kajiwara T (2000) Neodictyoprolenol and dictyoprolenol, the possible biosynthetic intermediates of dictyopterenes, in the Japanese brown algae *Dictyopteris*. Z Naturf 56c:6–12

Liquid seaweed extracts identified using ^{1}H NMR profiles

James S. Craigie · Shawna L. MacKinnon · John A. Walter

Originally published in the Journal of Applied Phycology, Vol 20, No 5, 215–221.
DOI: 10.1007/s10811-007-9232-1

Abstract Aqueous extracts of *Ascophyllum nodosum* and several other brown seaweeds are manufactured commercially and widely distributed for use on agricultural crops. The increasingly regulated international trade in such products requires that they be standardized and defined to a degree not previously required. We examined commercially available extracts using quantitative ^{1}H NMR and principal components analysis (PCA) techniques. Extracts manufactured over a 4-year period using the same process exhibited characteristic profiles that, on PCA, clustered as a discrete group distinct from the other commercial products examined. In addition to recognizing extracts made from different seaweeds, analysis of the ^{1}H spectra in the 0.35–4.70 ppm region allowed us to distinguish amongst extracts produced from the same algal species by different manufacturers. This result established that the process used to make an extract is an important variable in defining its composition. A comparison of the ^{1}H NMR integrals for the regions 1.0–3.0 ppm and 3.0–4.38 ppm revealed small but significant changes in the *A. nodosum* spectra that we attribute to seasonal variation in gross composition of the harvested seaweed. Such changes are reflected in the PCA scores plots and contribute to the scatter observed within the data point cluster observed for Acadian soluble extracts when all data are pooled. Quantitative analysis using ^{1}H NMR (qNMR) with a certified external standard (caffeine) showed a linear relationship with extract concentration over at least an order of magnitude (2.5–33 mg/mL; R^2>0.97) for both spectral regions integrated. We conclude that qNMR can be used to profile (or "fingerprint") commercial seaweed extracts and to quantify the amount of extract present relative to a suitably chosen standard.

Keywords *Ascophyllum nodosum* · Brown algae · Nuclear magnetic resonance spectra · Quantitative NMR · Standards

Abbreviations

PCA	Principal components analysis
qNMR	Quantitative nuclear magnetic resonance spectroscopy
TSP	Sodium 3-trimethylsilylpropionate-2,2,3,3-d_4

Introduction

Seaweeds have been used to improve soils and foster plant growth since ancient times (Newton 1951; Chapman and Chapman 1980). This practice, however, has been limited to agricultural regions contiguous to marine coastlines due to the cost of transporting raw seaweed to distant sites. The concept of converting seaweed into a readily transportable form of fertilizer was developed almost a century ago (Penkala 1912), but the advent of a practical process for liquefying seaweed is more recent (Milton 1952). The latter process involves a mild alkaline hydrolysis at elevated temperatures under pressure and was applied to *Ascophyllum nodosum* (L) Le Jolis. The product, "Maxicrop", became the first widely available seaweed extract that could be exported

Issued as NRCC no. 42,652.

J. S. Craigie
Acadian Seaplants Limited, 30 Brown Ave.,
Dartmouth, NS B3B 1X8, Canada

J. S. Craigie (✉) · S. L. MacKinnon · J. A. Walter
Institute for Marine Biosciences,
National Research Council of Canada,
1411 Oxford Street, Halifax, NS B3H 3Z1, Canada
e-mail: james.craigie@nrc-cnrc.gc.ca

as a dry powder (Stephenson 1974). Interest in the application of seaweed extracts to amend irrigation water for crops and in foliar sprays quickly followed and continues to the present (Senn and Kingman 1978; Verkleij 1992; Ayad et al. 1997; Norrie and Keathley 2006; Stirk and van Staden 2006).

Extracts are made commercially from a number of seaweeds, usually brown algae, and may be manufactured by various processes, either with alkali or acid at high or low temperatures. Other aqueous extracts are prepared by "cell-burst" techniques at ultra-low temperatures, while enzymatic and fermentative methods also have been used. A feature common to the various preparations is the extreme complexity of the organic matrix of the extracts. Typically, the organic composition of seaweed extracts remains poorly defined and standards for commercial products are minimal or lacking entirely. A parallel exists with the complex of natural metabolites occurring in biofluids, cosmetics, functional foods and nutraceuticals, where the number of liquid-soluble components can be large, variable, and frequently ill-defined or unknown. The increasing worldwide trade in extracts and biologicals is driving a demand from consumers and national regulatory agencies for improved standardization and better characterization of the products.

Similarly, seaweed extracts are traded globally; for example, Acadian Seaplants' soluble *Ascophyllum* extract is marketed in 70 countries. The increasingly stringent regulations being developed by individual countries and the European Union require that such biological products be defined as unambiguously as possible. Fortunately, quantitative NMR techniques have been developed to the point where they now can be used to obtain "biological signatures" or "fingerprints" of complex mixtures of natural products, the composition of which may not be known (Pauli et al. 2005, 2007). Examples for marine matrices include the analysis of phycotoxins in shellfish and dinoflagellate preparations (Burton et al. 2005), and carrageenan mixtures in red algal extracts (van de Velde et al. 2002; Tojo and Prado 2003). It appeared reasonable, therefore, to expect that these new analytical tools might be used to recognize and characterize commercial seaweed extracts in a reasonably definitive manner. The objectives of the present study were to demonstrate that (1) useable ^{1}H NMR spectra could be obtained routinely from a variety of commercial seaweed extracts, (2) such spectra would be sufficiently distinctive to be used as product fingerprints, and (3) the extracts could be quantified relative to certified standards.

Materials and methods

The 38 Acadian *Ascophyllum* soluble seaweed extracts examined were kindly provided by Acadian Seaplants Limited (ASL). The extracts were production lots manufactured for commercial purposes using standardized company protocols. A pilot-plant trial conducted by ASL with *Sargassum* sp. was not included in the PCA analysis. Nineteen other seaweed extracts were competitor samples obtained in the marketplace from various locations. In only four of these was the seaweed species defined. The processes by which the extracts were manufactured obviously vary, but technical details remain undisclosed.

Liquid extracts were centrifuged at 15,000 × *g* for 30 min to ensure that only water-soluble constituents would be analyzed. In cases where samples were presented in powder or flake form, they were dissolved in distilled water at 9.0% w/v and clarified by centrifugation as above. The pH of all solutions was measured, and triplicate subsamples of the extracts (ca. 0.5 g) were weighed accurately into tared 10-mL beakers and dried (forced draft oven, 120°C, 22 h) to determine the dry mass present. The remaining clarified solutions were freeze-dried. Triplicate subsamples of the freeze-dried products were used to determine the moisture content and were then discarded.

Solutions were prepared for NMR spectroscopy by transferring accurately weighed (4.50–99.00 mg), freeze-dried extracts to 3 mL volumetric flasks. Final concentrations of extract typically were 10 mg mL^{-1}, but useable spectra were obtained from 1.5 to 33 mg mL^{-1}. Care must be taken to ensure that solvation of the dry extract is complete as undissolved matter cannot be seen due to the intense color of the solution. A deuterium oxide (99.9%) solution containing reference TSP (1.00 mg mL^{-1}) was added (1.50 mL), and the flasks were stoppered and vigorously swirled by hand for 2–3 min. The solutions were made to volume (3.00 mL) with neat D_2O, a magnetic flea was added, and the contents vigorously stirred for 5 min. Finally, each flask was mixed by inverting the contents 10 times. No pH adjustment was required for diluted solutions in the pH range of 8.3–8.7. Concentrated Acadian samples (33 mg/mL) had a pH of ca. 10. Seven of the competitor samples were acidic products, and their spectra were obtained without further pH adjustment (designated by unfilled symbols in Fig. 1). A duplicate of one of these was adjusted to ca. pH 8.5 with 1 M NaOH in D_2O before the solution was made to volume. Its spectrum was included in the PCA analysis (Fig. 1: filled inverted triangle).

The NMR experiments were performed with a Bruker DRX-500 spectrometer at 500.13 MHz (^{1}H) at either 20 or 25°C. Samples of seaweed extract in D_2O (0.6 or 0.7 mL, sufficient to fill the sensitive volume of the tube) were placed in Wilmad 535-ppm 5 mm glass NMR tubes. Spectra were obtained with a triple-band inverse triple-axis gradient probe, tuned and matched, using a sequence of a 90° pulse followed by a 2.18 s acquisition time, a 2 s relaxation delay, and accumulating 64 or 128 scans after 4

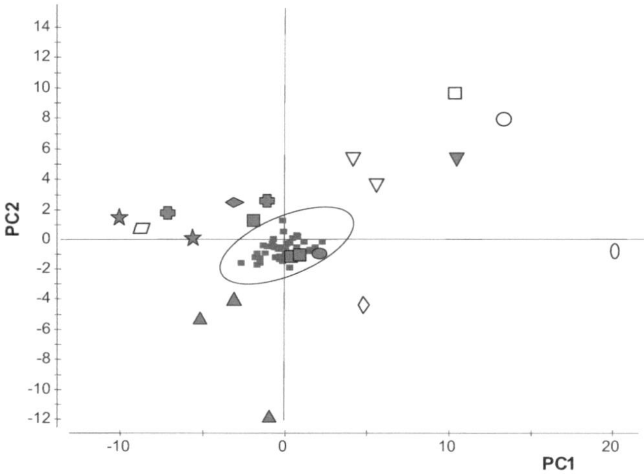

Fig. 1 A PCA scores plot showing the close grouping of 38 Acadian *Ascophyllum* extracts (*small squares*) and the relationship of a number of other commercial seaweed extracts. Each *different symbol* represents products from a single manufacturer. *Unfilled symbols* represent samples recorded under acidic conditions (pH 4.3–6.7). The *ellipse* encloses all Acadian *Ascophyllum* extracts plus two products that are known to incorporate Acadian extract. *Large squares* within the ellipse are replicate spectra of the same extract recorded under the same conditions apart from magnet shimming (one broad and one sharp spectrum), showing the relatively small changes in the PCA scores plot that result from instrument variations

dummy (unrecorded) scans into 32K data points. Spectral width was 7,508 Hz (15.011 ppm). The receiver gain was set to the largest power of two that would not lead to receiver saturation and kept constant for quantitative comparisons. Spectra were referred to TSP at 0 ppm in the D_2O solvent. Spectra were processed with Bruker XWinNMR 2.6 using an exponential window function, line broadening 0.3 Hz, processed with 64K data points. Baseline correction of phased spectra was performed by subtracting a straight line passing through the centre of the noise band at 10.0 and −1.5 ppm. Spectra for quantitation were integrated with no further corrections over the ranges 3.0 to 4.30 ppm and −1.0 to 3.0 ppm using a constant integral scale. Intensities were calibrated against spectra of a separate certified standard (41.3 mM caffeine solution in D_2O) recorded under identical conditions. To compensate for probe damping effects (Burton et al. 2005), the quantity compared was the product of the integral × p360, where p360 is the 360° pulse length measured for each sample.

Principal components analysis over the chemical shift range 4.675–0.325 ppm was performed using Bruker Amix, Windows Excel, and Umetrics SIMCA 10.0 software. For each spectrum, the summed signals within chemical shift regions of uniform width (bins of 0.05 ppm), centered at 0.05 ppm intervals, comprised the variables used in the PCA. This information was imported into Excel, and each spectrum was normalized such that the total area in the bins was equal to 1.0. The normalized binned spectra were imported into SIMCA for analysis with autoscaling and mean-centering. Two-dimensional "scores" plots of the first and second principal components were developed to show similarities or differences amongst the extracts.

Results

The ^1H NMR data obtained for the 38 Acadian *Ascophyllum* soluble seaweed extracts and the 19 competitor products were analyzed together in order to obtain the broadest possible overview of these types of extracts. The ASL spectra were processed without regard for the site of harvest in Nova Scotia or the season or year in which the seaweed was collected. The resolution obtained using two principal components is displayed as a PCA scores plot (Fig. 1). Small changes in common variables such as temperature and pH were found not to affect the analysis in any important way. Minor temperature change (293 to 298 K) did not affect the chemical shifts of the organic constituents relative to TSP. The well-known effect of temperature on the chemical shift for HDO was eliminated as the HDO integrals were excluded from the analysis. A pH change from ca. 6.2 to ca. 10 affected the chemical shifts of only a

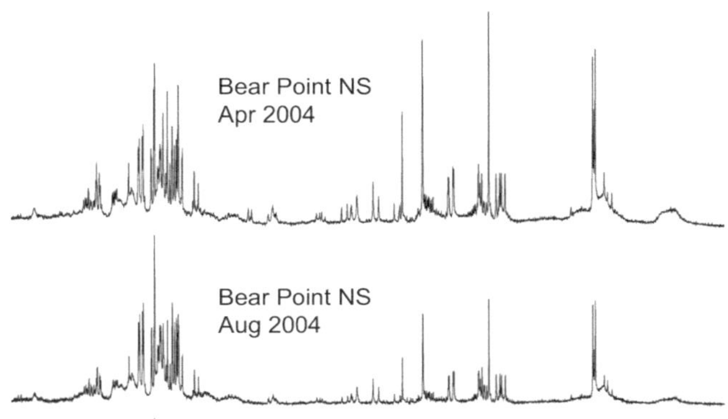

Fig. 2 ^1H NMR spectra showing the extremes in variation encountered for the Acadian *Ascophyllum* extracts as revealed in the PCA scores plot. The low field portions of the spectra (not shown) contain a singlet at 8.45 ppm and a few very minor resonances (5.4–5.9 ppm) as in Fig. 4 below. The algae were harvested from sites in Nova Scotia as follows: 09-Apr-04, Bear Pt.; 31-Aug-04, Bear Pt.; 16-Sep-04, Ingomar; and 04-Nov-04, Barrington

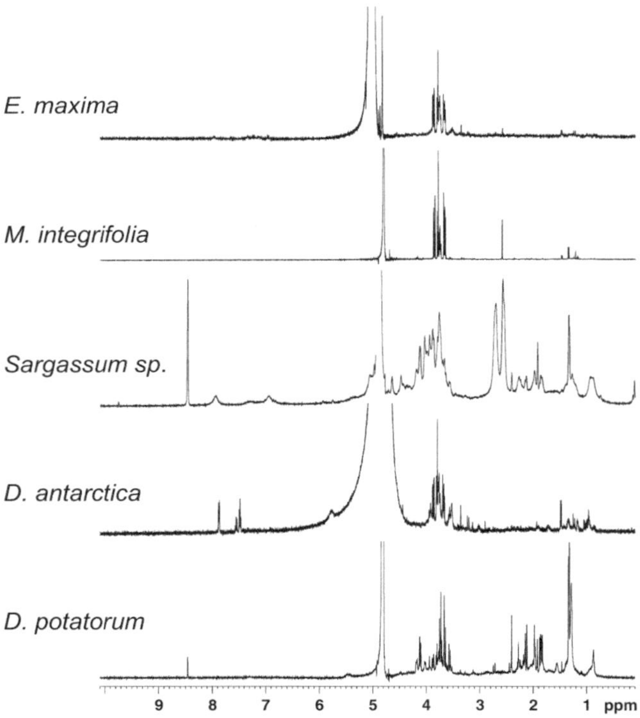

Fig. 3 Spectra for commercial products manufactured from specified seaweeds, *Durvillaea antarctica* (Cham.) Hariot, *D. potatorum* (Labill.) Aresch., *Ecklonia maxima* (Osbeck) Papenfuss and *Macrocystis integrifolia* Bory, and for an experimental extract made from *Sargassum* sp.

few resonances and did not materially alter the PCA analysis. The relative integrals over the two ranges were also unaffected by pH as resonances did not cross the 3.00-ppm dividing point.

The PCA scores plot (Fig. 1) demonstrates the remarkably tight clustering of the Acadian *Ascophyllum* soluble extracts (covering a range of concentrations from 33 to 1.7 mg mL^{-1}, and pH from ca. 8 to 10) compared to the relatively broadly scattered data points obtained for various other commercial products shown. In one case, an acidic competitor product also was evaluated at ca. pH 8.5 (filled inverted triangle in Fig. 1). This pH change shifted the PC1 value to a higher positive number, but had no effect on the PC2 value (unfilled inverted triangle). The change in pH did not move the value closer to the ASL data cluster. Repeated use of the same geometric symbol in Fig. 1 indicates that a separate sample from the same manufacturer was analyzed. We have not, however, analyzed a sufficient number of competitor samples to address the question of scatter within product lines produced by other manufacturers. Spectra having the most extreme variation within the Acadian extracts as shown by PCA analysis are presented in Fig. 2. The seaweed from which these extracts were manufactured was harvested from three sites in Nova Scotia in different months of the year as shown in the figure. The close similarity of their spectra establishes that the major components of the organic matrix of the extracts vary only within quite narrow limits.

The ^{1}H NMR spectra shown in Fig. 3 are, with the exception of the *Sargassum* sp., of aqueous extracts from specified brown seaweeds that are marketed by other manufacturers. Although the four extracts were included in the PCA scores plot (Fig. 1), they clearly lay outside of the *A. nodosum* cluster and did not form a discernable grouping. The *Sargassum* sp. aqueous extract was from a pilot-scale experiment conducted by Acadian Seaplants Limited. Its spectrum differed markedly from those of the four other species shown and was not included in the PCA analysis. Inspection of these spectra reveals that each has characteristic features that distinguish it from the others and from all the *Ascophyllum nodosum* spectra.

Figure 4 shows a typical ^{1}H NMR spectrum for an Acadian *Ascophyllum* extract (ASL 4009) and four other commercial extracts for which the seaweed source was unspecified. Again, a unique spectrum was obtained for each product, and all differed from those generated from the Acadian *Ascophyllum* extracts.

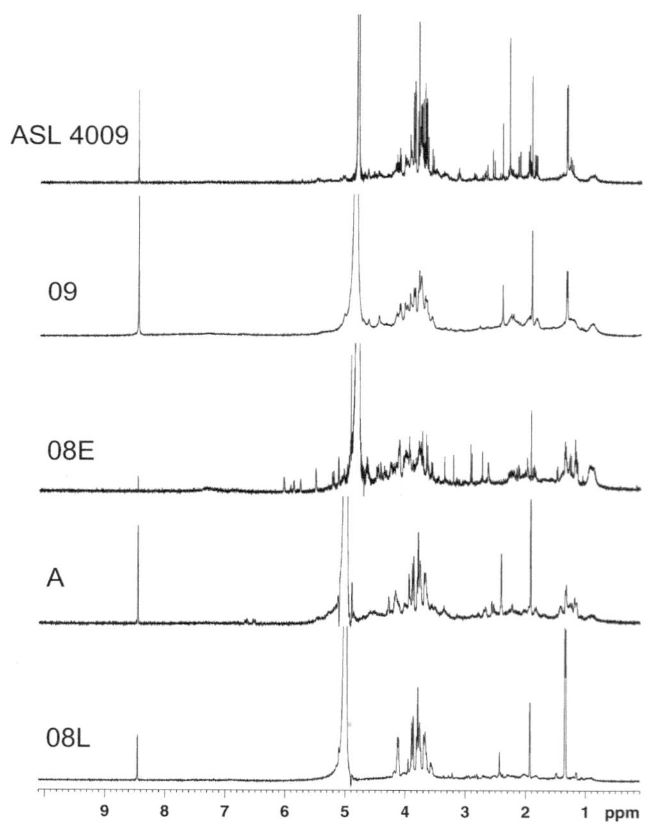

Fig. 4 Spectra for four commercial seaweed extracts from different manufacturers (09, 08E, A and 08L) in which the algal source was unspecified, compared to a typical Acadian *Ascophyllum* extract (ASL 4009)

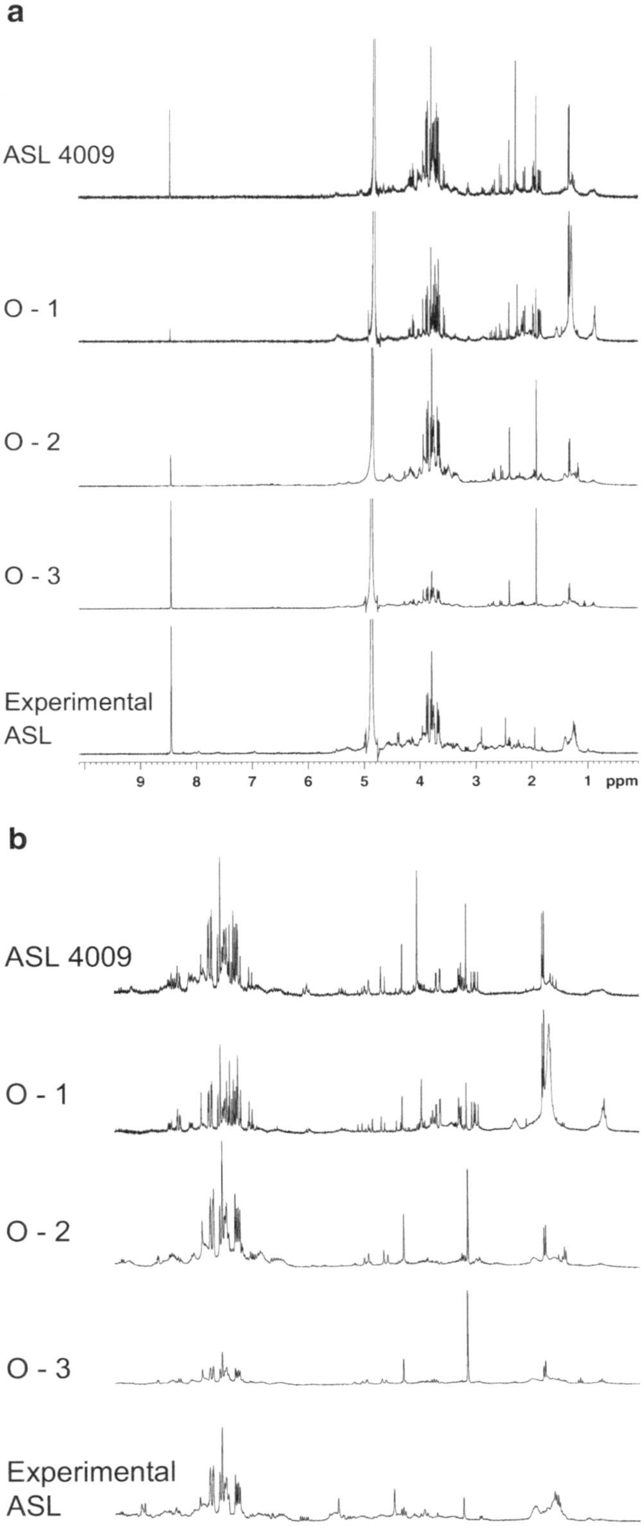

Fig. 5a, b Examples of spectra for three commercial *Ascophyllum nodosum* products O-1 to O-3 compared with Acadian *Ascophyllum* extract (*top*) and a pilot-scale preparation made with *A. nodosum* using a non-alkaline procedure (*bottom*)

The effect of varying the extraction conditions was investigated with five aqueous extracts made from a single species, *A. nodosum* (Fig. 5). The first four were alkaline extracts: ASL 4009 and extracts O-1 to O-3 manufactured by other commercial competitors. The last was an experimental ASL extract from a pilot-scale trial conducted by Acadian Seaplants Limited without the use of alkali. The ^{1}H NMR spectra of the commercial extracts show a general resemblance to each other, but they are not identical, especially in the low frequency region. The experimental ASL sample differed markedly from the others as might be expected as it was produced using a non-alkaline process.

Figure 6 establishes that plots of the ^{1}H NMR integrals as a function of extract concentration were linear (R^2>0.97) over the concentration range tested (2.5–33 mg mL^{-1}) for the two spectral regions compared. When data for the integral ratios for these regions (3.00–0.20 ppm and 4.38–3.00 ppm) were arranged into early-season and late-season

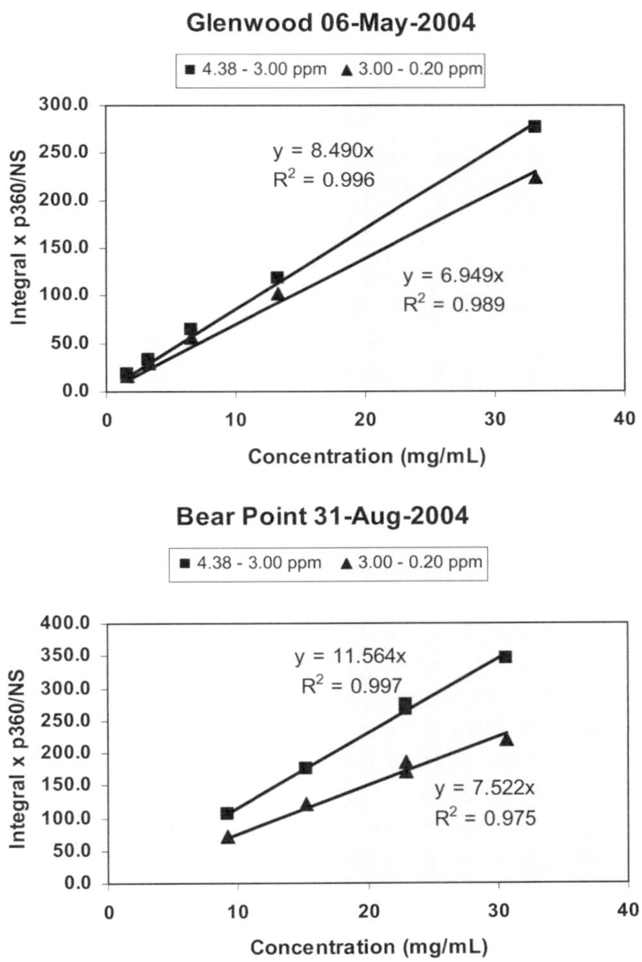

Fig. 6 The relationship of ^{1}H NMR integration with concentration of seaweed extracts for two regions of the qNMR spectra. Data shown are for Acadian *Ascophyllum* extracts manufactured in two different seasons

Table 1 Ratios of integrals for ^{1}H NMR spectral regions 3.0–0.2 ppm to 4.38–3.0 ppm for extracts made from early- and late-season harvests of *Ascophyllum nodosum*

Early season		Late season	
Date and site in NS	Ratio	Date and site in NS	Ratio
06 May 04, Glenwood	0.818	24 Dec 03, Sluice Point	0.762
19 May 04, Barrington	0.848	31 Aug 04, Bear Point	0.650
18 Apr 05, Bear Point	0.871	16 Sep 04, Ingomar	0.702
10 May 05, Sluice Point	0.917	04 Nov 04, Barrington	0.738
Average ±standard deviation	0.864 ±0.0417		0.713±0.0487

sets (Table 1), a change in the proportion of major constituents was revealed.

Ratios for the four commercial Acadian extracts manufactured from seaweed harvested in the spring months were significantly higher than those from extracts made from late-season seaweed. The minimum ratio within the latter group occurred in late summer (31-Aug) and was highest for the early winter product (24-Dec).

Discussion

Analysis of the qNMR spectra of a variety of commercial seaweed extracts and their comparisons using principal components analysis revealed that the techniques we have applied are relatively robust. Indeed, we found that the PCA results are not materially affected by sample pH over the range of ca. 6.3 to ca. 10. Neither do small changes in the recording temperature for the sample (20 to 25°C) compromise the results. Even a slight line broadening of the NMR spectrum due to inaccurate shimming of the magnet, as can occur during automatic sample analysis, did not interfere with accurate integration of the spectra.

It is remarkable, considering their long history of use, that commercial seaweed extracts remain so ill-characterized. Chemical analyses of extracts generally have focused on those components known to act as plant growth regulators such as auxins, betaines, cytokinins and gibberellins, either in the commercial products or in the native seaweeds (Senn and Kingman 1978; Jameson 1993; Craft et al. 2007; Hiltz et al. 2007). Recent interest in bioactive β-1,3-oligoglucans (Patier et al. 1993) and a report that humic acids may occur in brown seaweeds (Radwan et al. 1997) has broadened the focus of work. None of these studies permit the identification of a particular seaweed extract or allow for the quantitative determination of extract in a commercial product. Knowledge of the composition of the parent seaweed is useful, but insufficient, as the use of alkali is common in the liquifaction process and can greatly increase the complexity of the resulting chemical matrix (Niemela and Sjostrom 1985).

Seasonal variation in the composition of plants is a normal phenomenon governed by environmental variables such as changes in temperature, length of photoperiods, irradiance levels, and the availability of fertilizer nutrients. Changes in the bulk constituents of seaweeds such as *Ascophyllum nodosum* have long been documented (Black 1948), so it was not surprising to see this reflected in the qNMR spectra of commercial products manufactured from seaweed harvested at various times of the year. Fructification of *A. nodosum* intensifies during the winter in the Canadian Maritimes, reaching a climax in May following which the receptacles are shed, often within a 10- to 14-day period. Mature receptacles can account for up to approximately 40% of the total fresh plant biomass when the alga is harvested during its peak of reproduction (R. Ugarte and A. Sams, pers. comm., 2007). The data shown in Table 1 are from products manufactured in different years and harvested from several sites. However, grouping the data in this simplified manner ignores the fact that microenvironments can modify the timing and extent of physiological and biochemical changes in the alga. In the three cases where the harvesting sites were common to both seasonal groupings, differences remained clear. The minimum ratio was recorded for a product from plants harvested in late August after they had experienced an extended period of high irradiance and elevated seawater temperature combined with a seasonally low available nutrient supply. While a more extensive investigation is required for seaweed extracts, it appears that qNMR will prove a useful tool for recognizing changes in metabolite composition as has been demonstrated for developing tomatoes (Le Gall et al. 2003).

It is notable, however, that the 38 samples of Acadian extract made from *Ascophyllum nodosum* harvested in all seasons over a 4-year period exhibited such a relatively dense cluster when examined by PCA. This result indicates that variations in gross composition are small. We can conclude that, as remarked by Black (1948), this species can be harvested for any particular (major) constituent at any time of the year.

Acknowledgements We thank Ian Burton, who was most helpful in conducting the qNMR experiments. We appreciate the valuable discussion and comments of Alan Critchley, Tobias Karakach, Jeff Norrie and Kirty Solanky.

References

Ayad JY, Mahan JE, Allen VG, Brown CP (1997) Effect of seaweed extract and the endophyte in tall fescue on superoxide dismutase, glutathione reductase, and ascorbate peroxidase under varying levels of moisture stress. Proc Am Forage Grassland Council 6:173–177

Black WAP (1948) The seasonal variation in chemical composition of some of the littoral seaweeds common to Scotland. Part I. *Ascophyllum nodosum*. J Soc Chem Ind 67:355–357

Burton IW, Quilliam MA, Walter JA (2005) Quantitative ^{1}H NMR with external standards: use in preparation of calibration solutions for algal toxins and other natural products. Anal Chem 77:3123–3131

Chapman VJ, Chapman DJ (1980) Seaweeds and their uses, 3rd edn. Chapman & Hall, London, pp 30–61

Craft CA, Hiltz D, Hankins S, MacKinnon SL (2007) Detection of plant growth hormones in *Ascophyllum nodosum* and seaweed products. MaNaPro XII Int Symp Mar Nat Products 12:PO74

Hiltz D, MacKinnon SL, Hankins S, Stefanova R, Craft CA (2007) Improved method of analysis of betaines in *Ascophyllum nodosum* and commercial seaweed. MaNaPro XII Int Symp Mar Nat Products 12:PO76

Jameson PE (1993) Plant hormones in the algae. Prog Phycol Res 9:239–279

Le Gall G, Colquhoun IJ, Davis AL, Collins GJ, Verhoeyen ME (2003) Metabolic profiling of tomato (*Lycopersicon esculentum*) using ^{1}H NMR spectroscopy as a tool to detect potential unintended effects following a genetic modification. J Agric Food Chem 51:2447–2456

Milton RF (1952) Improvements in or relating to horticultural and agricultural fertilizers. British Patent 664,989

Newton L (1951) Seaweed utilization. Samson Low, London, pp 31–57

Niemela K, Sjostrom E (1985) Alkaline degradation of alginates to carboxylic acids. Carbohydr Res 144:241–249

Norrie J, Keathley JP (2006) Benefits of *Ascophyllum nodosum* marine-plant extract applications to 'Thompson seedless' grape production. Acta Hortic 727:243–245

Patier P, Yvin J-C, Kloareg B, Lienart Y, Rochas C (1993) Seaweed liquid fertilizer from *Ascophyllum nodosum* contains elicitors of plant D-glycanases. J Appl Phycol 5:343–349

Pauli GF, Jaki B, Lankin D (2005) Quantitative ^{1}H NMR: development and potential of a method for natural products analysis. J Nat Prod 68:133–149

Pauli GF, Jaki BU, Lankin DC, Walter JA, Burton IW (2007) Quantitative NMR of bioactive natural products. In: Colgate SM, Molyneaux RJ (eds) Bioactive natural products: detection, isolation and structural determination. CRC, Boca Raton, in press

Penkala L (1912) Method of treating seaweed. British Patent 27,257

Radwan A, Davies G, Fataftah A, Ghabbour EA, Jansen SA, Willey RJ (1997) Isolation of humic acid from the brown algae *Ascophyllum nodosum, Fucus vesiculosus, Laminaria saccharina* and the angiosperm *Zostera marina*. J Appl Phycol 8:553–562

Senn TL, Kingman AR (1978) Seaweed research in crop production 1958-1978. Project no. 99-6-09329-2. Economic Development Administration, US Department of Commerce, Washington DC

Stephenson WA (1974) Seaweed in agriculture & horticulture, 3rd edn. Barglya & Gylver Rateaver, Pauma Valley, CA

Stirk WA, van Staden J (2006) Seaweed products as biostimulants in agriculture. In: Critchley AT, Ohno M, Largo DB (eds) World seaweed resources: an authoritative reference system. A multimedia, interactive DVD ROM. ETI Bioinformatics, Amsterdam

Tojo E, Prado J (2003) A simple ^{1}H NMR method for the quantification of carrageenans in blends. Carbohydr Polym 53: 325–329

Van de Velde F, Knutsen SH, Usov AI, Rollema HS, Cerezo AS (2002) ^{1}H and ^{13}C high resolution NMR spectroscopy of carrageenans: application in research and industry. Trends Food Sci Technol 13:73–92

Verkleij FN (1992) Seaweed extracts in agriculture and horticulture: a review. Biol Agric Hort 8:309–324

The use of HRMAS NMR spectroscopy to study the *in vivo* intra–cellular carbon/nitrogen ratio of *Solieria chordalis* (Rhodophyta)

S. Bondu · N. Kervarec · E. Deslandes · R. Pichon

Originally published in the Journal of Applied Phycology, Vol 20, No 5, 223–229.
DOI: 10.1007/s10811-007-9231-2 © Springer Science + Business Media B.V. 2007

Abstract The red alga *Solieria chordalis* (J. Agardh) C. Agardh (Rhodophyta) was used as a model to investigate the effects of changes in seawater salinity on the carbon/nitrogen ratio. Carbohydrates and nitrogenous metabolites are major components of this alga and their metabolisms are intimately linked. Previous publications have provided a solid database for these two primary metabolic pathways from experiments and observations *in situ*. Storage products (e.g. floridean starch), cell wall polysaccharides (carrageenan) and low molecular weight carbohydrates such as floridoside and digeneaside are major compounds constituting the pool of available carbon. Compounds such as amino acids and peptides, constitute the pool of nitrogen. This study focuses on the intracellular C/N ratio inside the pool of low molecular weight compounds. This C/N ratio can be defined as the balance between carbohydrates and amino acids. High–resolution magic angle spinning nuclear magnetic resonance spectroscopy (HRMAS NMR) provides a powerful approach for *in vivo* analysis of the pool of intracellular organic compounds. These *in vivo* results were complimented with quantitative data obtained from high performance liquid chromatography (HPLC). *In vivo* and *in vitro* experimental analyses provided a intracellular molecular balance and defined the C/N ratio. In order to study the effect of salt stress on the carbon/nitrogen ratio, *S. chordalis* was cultured under controlled conditions. Effects of hyposalinity and hypersalinity stresses (low 22 ‰ and high 50 ‰ salinity) were tested. Both HPLC and NMR data, obtained on stressed and unstressed algae, generated insights into variations of carbonated and nitrogenous metabolites, involving changes of the C/N ratios, and demonstrated the adaptive responses of the seaweed.

Keywords Carbon/nitrogen ratio · High resolution magic angle spinning · Salt stress · *Solieriaceae*

Introduction

Biological materials are heterogeneous, complex matrices containing different kinds of compounds with different molecular weights. Acquiring reliable spectra on these semi–solid matrices by solution state NMR spectroscopy is severely limited due to differences in magnetic susceptibilities. These factors are drastically reduced when samples are placed in a rotor spinning around an axis which is oriented at the socalled "magic" angle of 54.7° with the respect to the magnetic field B_0. This NMR technique is called high resolution magic angle spinning (HRMAS) nuclear magnetic resonance (NMR) spectroscopy and can be use to quantify some intracellular metabolites (Weybright et al. 1998, Bollard et al. 2000). This is also a nondestructive method, which can be utilised in intact biological material without extraction and purification steps and provides high resolution spectra with narrow peaks approaching that of liquid samples.

Whereas a number of HRMAS NMR studies have been reported on animal tissue (Moka et al. 1997, Bollard et al.

19th International Seaweed Symposium KOBE JAPAN.

S. Bondu (✉) · E. Deslandes
Laboratoire d'Ecophysiologie et de Biotechnologie des Halophytes et des Algues Marines (LEBHAM), Université de Bretagne Occidentale–IUEM, Technôpole de Brest Iroise, 29280 Plouzané, France
e-mail: stephanie.bondu@univ-brest.fr

N. Kervarec · R. Pichon
Laboratoire de Résonance Magnétique Nucleaire, Université de Bretagne Occidentale, UFR Sciences et Techniques, Avenue Le Gorgeu, 29200 Brest, France

2000, Wang et al. 2003), higher plant material (Ni and Eads 1993) and bacteria cells (Gudlavalleti et al. 2006, Li 2006), few reports have dealt with the use of the technique on algae. By using this *in vivo* NMR technique Chauton et al. (2003) classified several microalgal species and Broberg et al. (2000), quantified some metabolites in the red macroalga *Gracilariopsis lemaneiformis*. Furthermore, only a few minutes are required in order to produce well resolved ^1H NMR spectra of intracellular compounds, and this allows for an accurate analysis from inside selected cells. With this reduction of analysis time and high quality spectra, it is also possible to follow the physiological responses of an organism to changing environmental parameters. The only report of HRMAS NMR spectroscopy using macroalgae was used to investigate metabolism (Broberg et al. 1998). In their report, HRMAS NMR analyses provided insights into the increase or decrease of certain compounds inside the cells due to light and salt stress.

The aim of this study was the novel application HRMAS NMR spectroscopy to easily determine the relative carbon/nitrogen ratio in the red alga *Solieria chordalis*. These results were complemented with quantitative data obtained from HPLC and colorimetric experiments. *S. chordalis*, cultured under controlled conditions, demonstrated metabolic changes in intracellular low molecular weight compounds as the result of salt stresses. Adaptive responses of this seaweed were defined.

Materials and methods

The red alga *Solieria chordalis* (J. Agardh) C. Agardh (Rhodophyta) used for this study was collected at Trez Hir (Brittany, France) in February 2005. Thalli were kept in aquaria in the laboratory, under continuous–flow system, at 12 °C, in a photon flux density of 50 µmol photons.$m^{-2}.s^{-1}$ with a 16:8 (light:dark) photoperiod. Thalli of *Solieria chordalis* were kept under these conditions until experiments.

Some thalli were randomly removed from the aquaria and were put into three culture tanks (500 g fresh weight, 10 L) at different salinities : Hyposalinity (22 ‰), Normal salinity (34 ‰), Hypersalinity (50 ‰). The three salinities were prepared with natural seawater at 34 ‰ that was either diluted with distilled water or supplied with NaCl to yield 22 ‰ and 50 ‰, respectively. These three tanks were placed in darkness for three days after which they were transferred to continuous light (under artificial cool–white fluorescent lights at a photon flux density of 50 µmol photons.$m^{-2}.s^{-1}$). All cultures were incubated at 12 °C, tanks were constantly aerated. Fresh material (20 g) was removed from each tank after 8, 24 and 48 h in dark and 76, 81, 96, 104 and 127 h during the light periods. HRMAS NMR analyses were performed immediately on these excised tissues. The materials were then freeze dried and ground to powder with a ball mill. These samples were used for the floridoside, digeneaside and amino acids quantification.

In vivo analyses by HRMAS ^1H NMR Spectroscopy

All NMR experiments were recorded on a BRUKER DRX 500 equipped with an indirect HRMAS $^1H/^{13}C/^{15}N$ probehead. About 4 or 5 mg of living samples were loaded in 4 mm ZrO2 cylindrical rotors with 10 µL of D_2O, at a spinning rate of 5.000 HZ. The spectra were performed with a 30° pulse, with a presaturation of the HDO signal. The temperature was controlled at 30°C. On all HRMAS ^1H spectra, the signal of taurine (at 3.3 ppm) has been used to standardise spectra. In order to interpret the HRMAS NMR spectra, comparisons were made with a pure sample of floridoside and digeneaside.

Determination of floridoside, digeneaside and amino acid contents in *S. chordalis*

Preparation of pure floridoside: Pure floridoside was extracted and isolated from the red alga *Palmaria palmata* (Lamouroux) Silva collected at Rosporden, Brittany, France. It was purified as described elsewhere (Simon–Colin et al. 2002).

Preparation of pure digeneaside: Digeneaside was extracted and isolated from the red alga *Ceramium botryocarpum* Griffiths ex Harvey. The aqueous methanolic phase resulting from the extraction of MeOH–$CHCl_3$–water 12:5:3 (v/v/v), was concentrated on a rotary evaporator and freeze dried. The crude extract, dissolved in water, was applied to an ion exchange column (Cl–form, 1 X8, 100–200 mesh, Dowex–Fluka Chemika), converted to OH–form. Following washing with distilled water, the column was eluted with a solution of NaOH 1 M. After neutralisation, the eluent was desalted by gel–filtration through a Sephadex G10 column (100*1.5 cm i.d.). Elution profiles were determined by the phenol–sulphuric method for sugar detection and salt detection by conductivity monitor. The digeneaside fraction was purified by high performance liquid chromatography on a organic acids columns (ICSep ICE–COREGEL–87H3 column, Transgenomic) as shown in Ascencio and Gudlavalleti et *al.* (2006).

Sample preparation: Triplicate were realized. Dry weight of powder of *Solieria chordalis* (80 mg) extracted twice with 12:5:3 MeOH–$CHCl_3$–water (5 mL) under magnetic stirring for 2 hours. The aqueous methanolic phase was freeze dried and store at 4 °C before fractionation on an ion exchange chromatography. Freeze dried crude extract were dissolved in 1 mL of water and applied to an ion exchange column (Cl^-form, 1 X8, 100–200 mesh,

Dowex–Fluka Chemika, 3 mL), converted to OH–form. This system was rinsed with 12 mL distilled water. The neutral effluent was freeze dried. The anion exchanger column was eluted by 10 mL of 1 M NaOH. The eluent was freeze dried. The neutral and alkaline fractions were analysed by NMR spectroscopy (one–dimensional ^{1}H, recorded on a Brucker DRX–400 spectrometer) after desalting on a Sephadex G10 column. Floridoside and digeneaside were identified in neutral and eluent fraction, respectively. The recovery yield of floridoside and digeneaside was estimated to be 85.3±3.4% and 64.32±6.6%, respectively by using the standards, treated in the same conditions as the samples.

Floridoside was quantified on neutral fraction by high performance liquid chromatography through a Suger–Pack1 (Waters) column (6.5 mm ID * 300 mm L), placed in a oven at 95°C and eluted by solution of Ca^{2+}EDTA (50 mg.L^{-1}) in isocratic mode at a flow rate of 0.5 mL.min^{-1}. Floridoside was eluted at T_R 10.3 min. Injection volumes were 50 μL. Quantification of floridoside was achieved by comparing peak integration with those of pure floridoside solutions of variable concentrations in the range 0–2 mg mL^{-1} prepared in the same solvent.

For all these studies, the HPLC system (Waters) constituted a pump 600E and a 717plus autosampler injector. The detection system was a refractive index detector (Jasco–930).

Due to the basic pH of eluent sample, digeneaside content could not be quantify correctly by HPLC. Indeed, the neutralization of the samples generate a high salt content affecting the chromatographic separation. The total sugar content in eluent fraction, corresponding to the amount of digeneaside, was also determined by the phenol–sulphuric acid method using pure digeneaside as standard (coefficient of determination R_2=0.9985).

Amino acid contents were quantified by colourimetric method with ninhydrin reagent on crude extract. (coefficient of determination R_2=0.9606).

Results and discussion

Content of low molecular weight compounds in *S. chordalis* and determination of C/N ratio

Figure 1 presents the HRMAS ^{1}H NMR spectrum for *S. chordalis*. To interpret the spectra, comparisons were made with floridoside and digeneaside purified from *Palmaria palmata* and *Ceramium botryocarpum* respectively; isethionic acid and taurine standards were purchased from Sigma. 2D NMR spectroscopy (^{1}H–^{1}H COSY, ^{1}H–^{13}C HMBC, ^{1}H–^{13}C HMQC) on a Bruker Advance 500 spectrometer, was also performed on a crude sample of

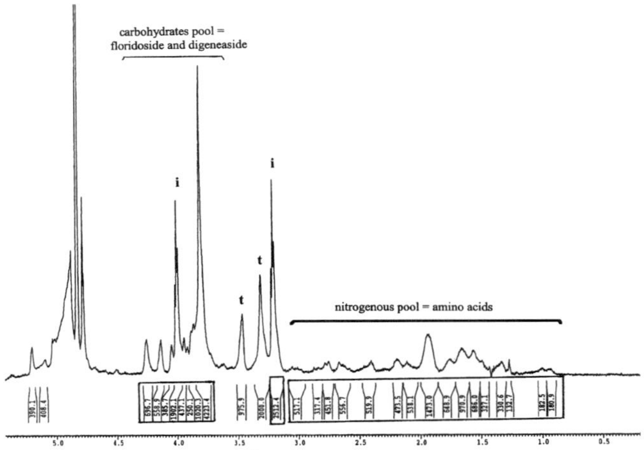

Fig. 1 *In vivo* HRMAS ^{1}H NMR spectra obtained from *Solieria chordalis* in reference salinity (34‰). "i" and "t" refer to the resonance of isethionic acid and taurine, respectively

S. chordalis in order to verify assignments made by HRMAS NMR spectra. Signals between 3.8 ppm and 4.3 ppm were assigned to floridoside (2–*O*–α–D–galactopyranosylglycerol) and digeneaside (2–O–α–D–mannopyranosyl–D–glyceric acid) indicated the carbonated pool. Anomeric protons of these carbohydrates to 5.2 ppm and 4.85 ppm respectively, could not be integrated because of the residual HDO signal at 4.8 ppm. The presence of digeneaside may be considered surprising in the Solieriaceae. However, its use as a chemosystematic marker in Ceramialean taxa (Kirst 1980) has been criticised by several authors since this compound has been found in members of the orders Gracilariales and Bangiales (Karsten et al. 1999, Wilcox et al. 2001).

On the HRMAS spectrum, signals between 0 ppm to 3.0 ppm represented free amino acids or constitute small peptides. These signals represented the nitrogenous pool available within the seaweed. The other major signals were assigned to taurine (3.3 ppm and 3.5 ppm) and isethionic acid (3.2 ppm and 4.0 ppm). The taurine is a sulphur containing amino acid and is linked with the metabolism of isethionate. So it can't be considered inside the amino acids pool. We have chosen it like a internal reference to integrate signals on spectra.

Similar NMR assignments for all these compounds have already been reported by Broberg and al. (1998) using HRMAS ^{1}H NMR spectra from *Gracilariopsis lemaneiformis*.

It was possible to estimate the relative C/N ratio by integrating signals from the HRMAS ^{1}H NMR spectra since signals reflect the relative concentration of the integrated compound. So, C/N ratio can be determined.

$$\text{C/N ratio} = \frac{\sum \text{signal integrals(carbonated pool)} - \text{signal integral (isethionic acid at 3, 2 ppm)}}{\sum \text{signal integrals (nitrogenous integrals)}}$$

Care must be taken when calculating this ratio due to the uncertainty of signal integrations and also possible overlapping signals. Well–resolved spectra are necessary to decrease the uncertainty on integration. It is not possible to remove totally the effects of overlapping signals. However, this problem is negligible in this study because it concerns only the signals for the α–protons of amino acids (overlapping with signals from carbohydrates).

This calculated C/N ratio gives a global idea of the molecular balance within the algal and reflects the metabolic *in vivo* situation. We considered that if the calculated C/N ratio is less than 1, the metabolic balance favours the nitrogen pool. On the other hand, if this ratio is gretare than 1, the molecular balance is dominated by carbonated compounds.

Effects of changing culture conditions

The HRMAS spectra, performed on salt–stressed and no salt–stressed algae after 8, 24 and 48 h in the dark and 76, 81, 96, 104 and 127 h during light period, generate insights on the internal metabolism of the seaweed. By integrating signals on the whole these spectra, it was possible to determinate the C/N ratio inside cells of *S. chordalis* subjected to light and salt stresses. These results are presented in Fig. 2 and must be supported by accurate quantitative data on the levels of floridoside, amino acids and digeneaside (Fig. 3 (A,B,C respectively)).

Culture of S. chordalis for three days in the dark at different salinities

When *S. chordalis* was kept in the dark at 34 ‰ salinity for three days, the C/N ratio remained near to 1 (0.89<R<1.16), thus the carbon–nitrogen balance was equilibrated. Digeneaside, floridoside and amino acids levels varied between 6.3 and 10.49 $mg.g^{-1}$ DW, between 4 and 7 $mg.g^{-1}$ DW and between 31.3 and 41.2 $mmol.g^{-1}$DW, respectively.

When *S. chordalis* was kept in the dark at low salinity (22 ‰) and high salinity (50 ‰) for three days, the C/N ratio of these salt–stressed algae remained stable (R=1.2) until 24 h after which it decreased drastically to reach a value of 0.66 in under 48 h. However, the dark treatment and salt shock constantly reduced the level of floridoside and amino acid by 53 % under hypersalinity and by 42 % under hyposalinity compared to the level of digeneaside

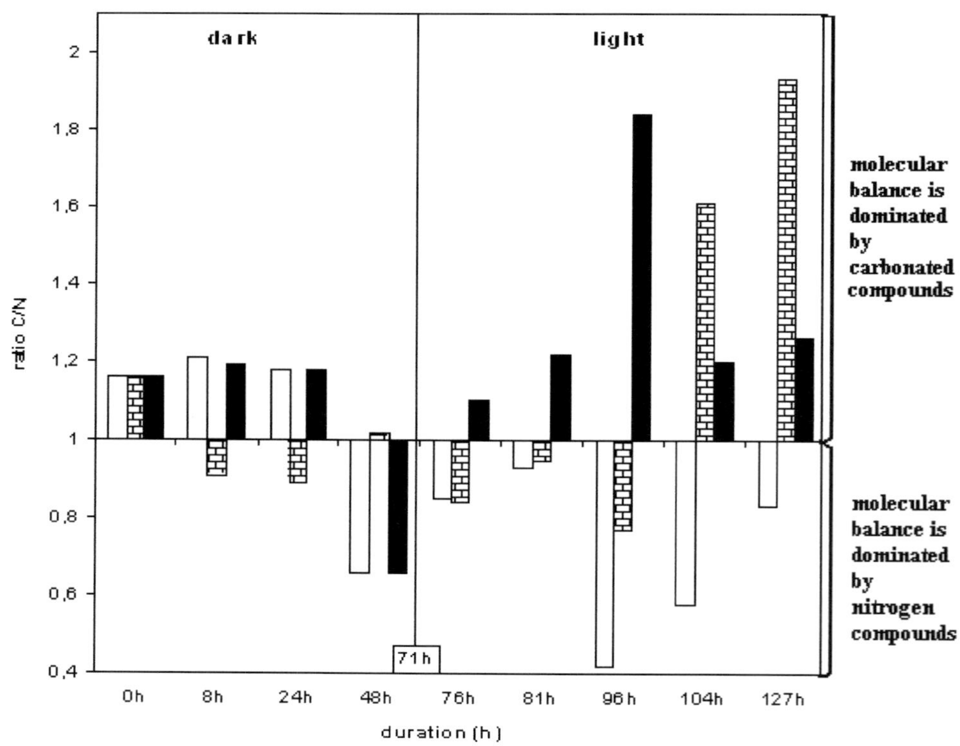

Fig. 2 Relative C/N ratio in *S. chordalis* subjected to light and salt stresses (white bar : Hyposalinity 22‰, squared bar : Normal salinity 34‰ , dark bar : Hypersalinity 50‰) calculated from the signal integrals on HRMAS proton NMR spectra

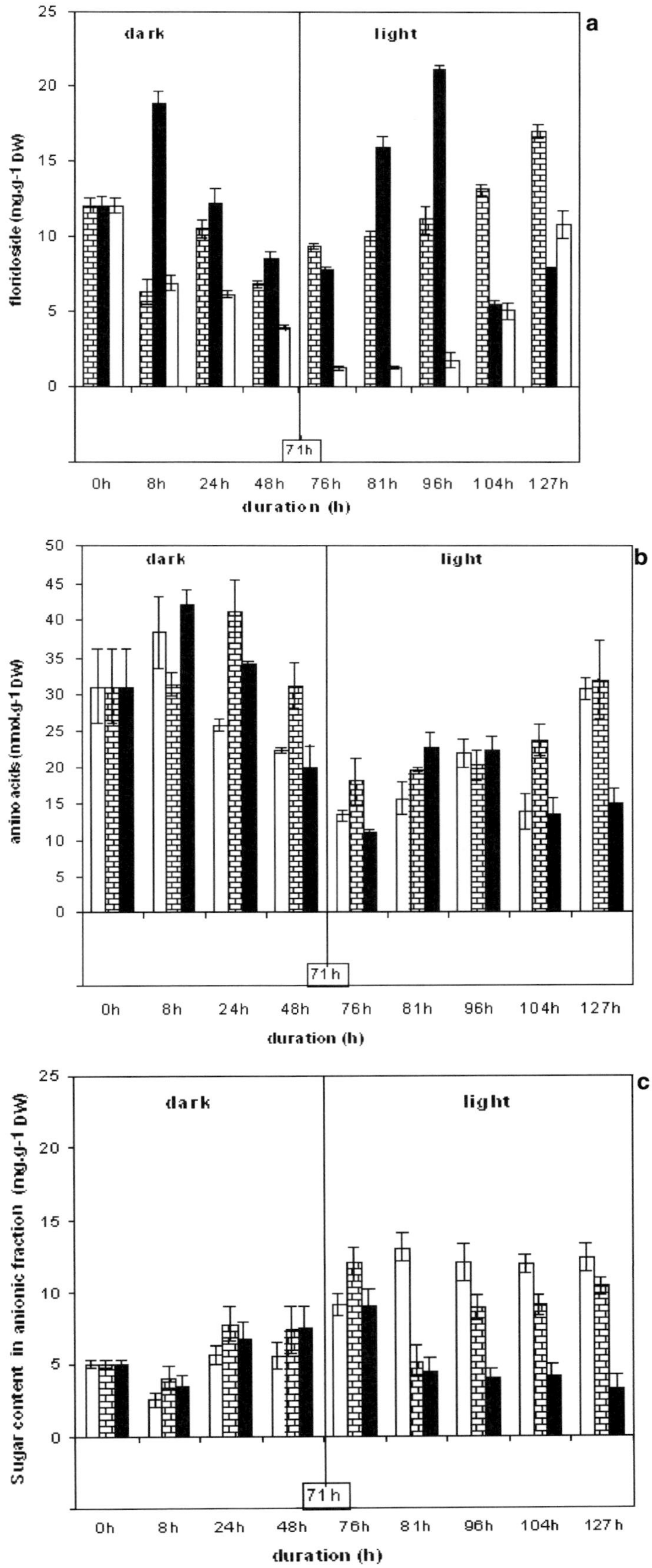

Fig. 3 Changes in floridoside (**a**) amino acids (**b**) and digeneaside levels (**c**), in *S. chordalis* according to culture conditions : From 0 to 71 h, algae were cultivated in the dark at different salinities. From 71 to 127 h the algae were cultivated the light at the same salinities (white bar : Hyposalinity 22‰, squared bar : Normal salinity 34‰ , dark bar : Hypersalinity 50‰). Data shown are mean values of three replicates (±S.D.)

which is multiplied by 2. Thus, only a synthesis of small peptides could explain the high decrease C/N ratio and amino acid level after 48 h.

After dark treatment, when *S. chordalis* was kept three days in continuous light under hyposalinity, the C/N ratio remained always less than 1 (0.42 <R<0.93). The minimum value of C/N ratio was reached (R=0.42) after 24 h of light. This decrease of C/N ratio was mainly due to an increase of nitrogen level as the carbon levels remained stable (average 1.5 $mg.g^{-1}$ DW for floridoside and 12 $mg.g^{-1}$ DW for digeneaside). Hyposalinity orientates the molecular balance towards synthesis of nitrogen compounds. However, no increase of the amount of amino acids was noticed under these culture conditions compared to normal salinity. We hypothesized a possible small peptide synthesis, explaining the ratio less than 1.

After dark treatment, when *S. chordalis* was kept three days in light under hypersalinity, the C/N ratio remained always greater than 1 (1.1<R<1.84). The maximum was reached under 24 h under of light (R=1.84). This ratio followed the variation of level of floridoside which rose between 7.75 $mg.g^{-1}$ DW and 21.12 $mg.g^{-1}$ DW during the first 24 h of light exposure. The molecular balance is oriented towards a synthesis of carbonated compound, the floridoside. This is consistent with the theory that floridoside is the main photosynthetic and osmoregulator compound in the red seaweeds (Reed 1990). In these culture conditions, the synthesis of floridoside was stimulated comzpared to digeneaside. This result supported the hypothesis that digeneaside do not play more than a minor role in osmotic balance (Kirst and Bisson 1979).

After dark treatment, when *S. chordalis* was kept for three days in light without salt–stress, the C/N ratio remained relatively unchanged and near to 1 (0.77 <R< 0.95) during the first 24 h of light exposure. Thus, the photosynthetic carbon was not oriented towards the synthesis of low–molecular–weight compounds but certainly aimed towards storage product like floridean starch or carrageenan. This hypothesis is supported by the study of Fournet et al. (1999) which showed that *S. chordalis* synthesized floridean starch under the same conditions of irradiance when the alga was post–conditioned one week in the dark. After this first 24 h, the C/N ratio increased strongly to reach 1.93 (after 2.5 days) resulting in the orientation of photosynthetic carbon towards a carbonated compound, the floridoside for which the level was multiplied by 53 %.

Conclusion

This study showed that HRMASS NMR experiments can serve as a powerful, analytical tool to define the carbon/nitrogen ratio and also provide information on the intracellular metabolism of *S. chordalis*. Our results showed that two metabolic *in vivo* pathways were distinguishable due to salt stress after post–conditioning in the dark. Hypersalinity (or high seawater salinity stress) tended to increase the carbonated pool (particular floridoside) in order to equilibrate the osmotic pressure inside cells whereas hyposalinity (or low seawater salinity stress) stimulated the nitrogenous metabolism with deprivation of carbonated metabolic pathways. These results supported labelling experiments performed with *Grateloupia doryphora* which demonstrated an increase of floridoside level under hyper–osmotic conditions and a high incorporation of photosynthetic carbon into synthesized amino acids under hypo–osmotic conditions (Simon Colin et al. 2004). In *S. chordalis*, we suggest that there is a synthetic pathway for small peptides under hypo osmotic conditions. Future work in this field must concentrate on the pool of oligo– and dipeptides in order to strengthen these preliminary results.

In vivo and *in vitro* experimental analyses are both highly desirable to provide a molecular balance of the internal operations of the inside seaweed cells and demonstrate the adaptive response of *S.chordalis* when exposed to salt–stress.

Acknowledgements This work was supported by a fellowship from the French Ministry of Education and Research.
These authors thank Dr. Critchley.A for assistance with the English text contribution n°1042 of the IUEM European Institute for Marine Studies (Plouzane, France).

References

Bollard ME, Garrods S, Holmes E, Lindon JC, Humpfer E, Spraul M, Nicholson JK (2000) High–resolution ^{1}H and ^{13}C magic angle spinning NMR spectroscopy of rate liver. Magn Reson Med 44:201–207

Broberg A, Kenne L (2000) Use of high–resolution magic angle spinning nuclear magnetic resonance spectroscopy for *in situ* studies of low–molecular–mass compounds in red algae. Analyt Biochem 284:367–374

Broberg A, Kenne L, Pederson M (1998) In situ identification of major metabolites in the red alga Gracilariopsis lemaneiformis using high–resolution magic angle spinning nuclear magnetic resonance spectroscopy. Planta 206: 300–307

Chauton MS, Optun OI, Bathen TF, Volent Z, Gribbestad IS, Johnsen G (2003) HR MAS ^{1}H NMR Spectroscopy analysis of marine microalgal whole cells. Marine Ecol Prog Series 256:57–62

Fournet I, Zinoun M, Deslandes E, Diouris M, Floc'h JY (1999) Floridean starch and carrageenan contents as responses of the red alga *Solieria chordalis* to culture conditions. Euro J Phycol 34:125–130

Gudlavalleti SK, Szymanski CM, Jarell HC, Stephens DS (2006) In vivo determination of *Neisseria meningitidis* serogroup a capsular polysaccharide by whole cell high–resolution magic angle spinning NMR Spectroscopy. Carbohydrate Res 341:557–562

Karsten U, West JA, Zuccarello GC, Nixdorf O, Barrow KD, King RJ (1999) Low molecular weight carbohydrate patterns in the Bangiophyceae (Rhodophyta). J Phycol 35:967–976

Kirst GO (1980) Low mw carbohydrates and ions in Rhodophyceae: quantitative measurement of floridoside and digeneaside. Phytochemistry 19:1107–1110

Kirst GO, Bisson MA (1979) Regulation of turgor pressure in marine algae : ions and low–molecular–weight organic compounds. Australian J Plant Physiol 6:539–556

Moka D, Vorreuther R, Schicha H, Spraul M, Humpfer E, Lipinski M, Foxall PJD, Nicholson JK, Lindon JC (1997) Magic angle spinning proton nuclear magnetic resonance spectroscopic analysis of intact kidney tissue samples. Analyt Commun 34:107–109

Ni QX, Eads TM (1993) Analysis by Proton NMR of changes in Liquid–Phase and Solid–Phase components during ripening of banana. J Agricol Food Chem 41:1035–1040

Reed RH (1990) Solute accumulation and osmotic adjustment. In: Cole KM, Sheath RG (eds) *Biology of the red algae*. Cambridge University Press, Cambridge, pp 147–170

Simon–Colin C, Kervarec N, Pichon R, Deslandes E (2002) Complete ^{1}H and ^{13}C spectral assignment of floridoside. Carbohydrate Res 337:279–280

Simon–Colin C, Kervarec N, Pichon R, Deslandes E (2004) NMR ^{13}C–isotopic enrichment experiments to study carbon–partitioning into organic solutes in the red alga *Grateloupia doryphora*. Plant Physiol Biochem 42:21–26

Wang Y, Bollard ME, Keun H, Antti H, Beckonert O, Ebbels TM, Lindon JC, Holmes E, Tang H, Nicholson JK (2003) Spectral editing and pattern recognition methods applied to high–resolution magic angle spinning ^{1}H nuclear magnetic resonance spectroscopy of liver tissues. Analyt Biochem 323:26–32

Li Wei (2006) Multidimensional HRMAS NMR : a platform for *in vivo* studies using intact bacterial cells. Analyst 131:777–781

Weybright P, Millis K, Campbell N, Cory DG, Singer S (1998) Gradient, high–resolution, magic angle spinning ^{1}H nuclear magnetic resonance spectroscopy of intact cells. Magnet Res Med 39(3):337–344

Wilcox SJ, Bloor SJ, Hemmingson JA, Furneaux RH, Nelson WA (2001) The presence of gigartinine in a New Zealand *Gracilaria* species. J Appl Phycol 13:409–413

Growth, pigments, UV-absorbing compounds and agar yield of the economic red seaweed *Gracilaria lemaneiformis* (Rhodophyta) grown at different depths in the coastal waters of the South China Sea

Juntian Xu · Kunshan Gao

Originally published in the Journal of Applied Phycology, Vol 20, No 5, 231–236.
DOI: 10.1007/s10811-007-9247-7

Abstract The economic red alga, *Gracilaria lemaneiformis* Bory, was grown at different depths in the coastal waters of the South China Sea, and its growth, pigments, ultra-violet (UV)-absorbing compounds and agar yield were investigated in order to see the impacts of depth change. *Gracilaria lemaneiformis* grew slower at greater depths in March, while the highest relative growth rate (RGR) was found at about 1.0 m depth in April, about 9% higher than that at surface water (0.5 m below the surface). The RGR increased with the increasing daily photosynthetically active radiation (PAR) dose received by the thalli at different depths. The contents of phycoerythrin and chlorophyll *a* increased, while that of UV-absorbing compounds (UVAC, absorption peak at 325 nm) decreased with increased depth. The highest levels of the UVAC in the thalli grown in surface seawater played a protective role against solar UV radiation (280–400 nm). The content of UVAC declined at deeper depths and under indoor low PAR. The agar yield of the thalli increased with the increasing depths, with the highest content found at 3.5 m depth.

Keywords *Gracilaria lemaneiformis* · Growth · Depth · Solar radiation · UV-absorbing substances · Photosynthetic pigments

J. Xu · K. Gao (✉)
Marine Biology Institute, Shantou University,
Shantou, Guangdong 515063, China
e-mail: ksgao@stu.edu.cn

K. Gao
State Key Laboratory of Marine Environmental Science,
Xiamen University,
Xiamen 361005, China

Introduction

Macroalgae play an important role in marine primary production in coastal waters. Biomass production of economically important species has potential for remediation of CO_2 and nutrients (Gao and McKinley 1994), and *Gracilaria* is commercially important, accounting for more than 50% of the seaweed used for agar production (Armisen 1995). *Gracilaria lemaneiformis* has been cultivated on large scales in both the southern and the northern parts of China for food and agar production, playing an effective role against coastal eutrophication.

When *G. lemaneiformis* is farmed in the sea, vegetative branches are sandwiched and fixed in ropes which are then hung and floated in surface seawater. As the plants grow with increasing numbers of branches and mass, they are exposed to different quantity and quality of sunlight due to their sinking and expansion to deeper depths. From a photobiological point of view, photosynthetically active radiation (PAR, 400–700 nm) and ultra-violet radiation (UVR) (280–400 nm) attenuate disproportionally with depth, thus different ratios of PAR to UV-A (315–400 nm) or UV-B (280–315 nm) would exert different impacts on *G. lemaneiformis*.

Macroalgae growing at greater depths usually increase their photosynthetic pigments to compensate for the low light availability (Ramus et al. 1976), while those inhabiting shallower water or upper intertidal zone often accumulate UV-absorbing compounds to screen off some extent of UVR (Franklin and Foster 1997). High levels of solar PAR and UVR at noon inhibit photosynthetic performance of macroalgae (Häder et al. 2001). UVR is known to damage DNA (Buma et al. 2001; Zacher et al. 2007) and enzymes (Bischof et al. 2002) of macroalgae. However, positive effects of UV-A, such as aiding in the recovery process of the inhibited photosynthesis (Flores-Moya et al. 1999), DNA repair (Pakker et al. 2000a,b) and the enhancement of photosynthesis (Viñegla et al. 2006) and growth (Henry and Van Alstyne 2004) have also been suggested.

Previous studies indicate that macroalgae show different ecophysiological characteristics in response to change in growth depth (Ramus et al. 1976; Molloy and Bolton 1996). However, little is known about the relationship between growth and levels of PAR and UVR at different depths, where solar radiation is attenuated to different extents for different wavelengths. With the intention of exploring such a relationship to optimize the cultivation conditions for *G. lemaneiformis*, we investigated its growth rate and contents of photosynthetic pigments, UV-absorbing compounds and agar while growing the thalli at different depths.

Materials and methods

Gracilaria lemaneiformis Bory was collected from its cultivation area at Nanao (116.6°E, 23.3°N), Shantou, China, in March and April 2004. Young plants, about 30 cm long, were selected and maintained in sand-filtered seawater for a night before being returned to the sea and fixed at different depths. The growth experiments were carried out for 12 days in the end of March and 10 days in the middle of April, while the experiments for pigment analyses lasted 45 days from 20 March to 3 May. The daily mean temperature of the surface seawater was 18–19°C in March and 20–22°C in April; salinity of the seawater was constant with a mean value of about 32 psu. During the experiments, the daily dose of solar PAR reaching the sea surface changed from 1.0 to 12.1 MJ m^{-2}. At the site where the plants were hung at different depths, no biomass loss due to grazing was observed.

Measurement of growth

The study area is in a bay with low wave agitation. Thalli of about 10 g (fresh wt) were fixed in nylon ropes and placed at 0.5, 1.0, 1.5, 2.5 and 3.5 m depth by tying the ropes to a floating system supported by float-balls. Triplicate cultures were carried out at each depth, and each culture had 3–4 individuals. The thalli were wrapped in a net (mesh 0.4 cm) and examined by comparing their morphological features at regular intervals; there were no visible losses of the biomass of each individual. Biomass changes were monitored every 3 days for 12 days from 20 March to 2 April and every 2 days for 10 days from 11 to 21 April 2004, respectively. Fresh weight of the thalli was measured after lightly blotting with tissue paper. The relative growth rate (RGR, % day^{-1}) was estimated as follows: $RGR = 100 \times (lnN_t - lnN_0)/t$, where N_0 is the initial fresh weight and N_t is the fresh weight after t days.

Determination of UV-absorbing compounds and photosynthetic pigments

Concentrations of UV-absorbing compounds (UVAC, absorption peaked at 325 nm) were estimated for the thalli grown at different depths in the sea as well as for those grown under indoor conditions. About 200 mg (fresh wt) thallus was extracted in 20 mL absolute methanol for 24 h at 4°C in darkness. This extract was centrifuged at 5,000 *g* for 10 min and then used to determine the amounts of UV-absorbing compounds and the photosynthetic pigments using a scanning spectrophotometer (UV 530; Beckman Coulter, USA). The contents of UVAC per fresh mass were estimated by determining the ratio of UVAC to chlorophyll *a* (Chl *a*) according to Dunlap et al. (1995) and multiplying the ratio of Chl *a* content to fresh weight of the sample used. Chl *a* concentration was estimated according to Wellburn (1994). For the determination of phycoerythrin, about 200 mg of thallus was ground thoroughly in 2 ml 0.1 M phosphate buffer (pH 6.8) with quartz sand, and then further extracted in 10 ml 0.1 M phosphate buffer for 0.5 h (pH 6.8). The extracts were centrifuged for 15 min at 5,000 *g* and the phycoerythrin concentration determined according to Beer and Eshel (1985).

Agar extraction

Thalli of *G. lemaneiformis* grown at different depths for about 45 days (from 20 March to 3 May 2004) were collected for the analysis of their agar content. Agar extraction was carried out according to Marinho-Soriano et al. (1999). Dried samples of about 5 g were immersed in 500 ml of distilled water (pH 6.0) in a flask and heated for 1h at 110°C. The extract was filtered at 70°C. The residue was re-extracted under the same conditions. The filtrates were gelled at room temperature and then frozen at −20°C overnight. The frozen gel was thawed, washed with distilled water and dried for 24 h at 60°C. The agar yield was calculated as the percentage of dry mass.

Evaluation of UVAC concentrations under indoor conditions

To examine the effect of low light in the absence of UVR on the contents of UVAC, an indoor experiment was carried out under very low PAR (~ 20 μmol photons m^{-2} s^{-1}) for 3 months (from 19 February to 16 May 2004). Thalli of about 10 g were maintained in a glass tank (~ 5 L) with filtered natural seawater. The contents of UVAC were measured every 3 days for the first 2 weeks and at the end of the culture.

Incident solar irradiance was continuously monitored using a broad band filter radiometer (Eldonet; Real Time Computer, Möhrendorf, Germany), which has 3 channels respectively for photosynthetically active radiation (PAR, 400–700 nm), ultraviolet-A radiation (UV-A, 315–400 nm) and ultraviolet-B radiation (UV-B, 280–315 nm) (Häder et al. 1999). For underwater irradiances, a diving ELDONET was used. The daily dose of solar radiation was estimated at different depths in the testing periods. The solar PAR doses that the thalli received at different depths were estimated according to the measured diurnal solar irradiances and the PAR attenuation coefficients.

Statistical analysis

Differences among the treatments were tested using one-way analysis of variance (ANOVA) or *t*-test with SPSS (version 11.5). A confidence level of 95% was used in all analyses.

Results

From 20 March to 2 April 2004, the daily doses of solar PAR reaching the sea surface ranged 1 to 8.6 MJ m^{-2}, with the daily average of 4.04 MJ m^{-2}; while that of UVR ranged from 0.15 to 1.18 MJ m^{-2}, with the daily average of 0.57 MJ m^{-2} (Table 1). During this period, the thalli of *Gracilaria lemaneiformis* in the surface water (about 50 cm below the surface) grew fastest compared with greater depths to 3.5 m ($P<0.01$). The relative growth rate (RGR) in the surface water was 18 times higher than that at 2.5 m. At the depth of 3.5 m, negative RGR was observed, indicating that respiratory losses exceeded the photosynthetic gain (Fig. 1a, c).

From 11 to 21 April 2004, the daily solar PAR dose ranged from 2.9 to 11.8 MJ m^{-2} with the average value of 8.2 MJ m^{-2} per day, while that of UVR ranged from 0.48 to 1.76 MJ m^{-2}, with the average value of 1.27 MJ m^{-2} (Table 1). The thalli of *G. lemaneiformis* showed the highest growth rate at 1 m depth. The RGR at 0.5 m was 8% lower ($P<0.05$) than that at 1 m, reflecting a photoinhibition in the surface seawater caused by the high levels of solar radiation. Compared with the period at the end of March, *G. lemaneiformis* grew much faster in the middle of April at all depths ($P<0.01$) (Fig. 1b, c).

Despite the surface seawater temperature change, the RGR increased with increased daily solar PAR doses that the thalli received at different depths. Daily PAR dose of 4.3 MJ m^{-2} resulted the highest growth rate (Fig. 1d). At the highest levels of daily PAR of 5.8 MJ m^{-2} and UVR of 0.46 MJ m^{-2} the RGR decreased by 9.7%.

The content of both Chl *a* and phycoerythrin (PE) increased with increasing depth (Fig. 2). The highest contents of Chl *a* and PE were found at day 3, and the lowest at day 45 for depths of 1.5 m or less. At the deepest depth of 3.5 m, PE content was the highest at day 45, while Chl *a* reached similar level with days 3 and 21.

The content of UVAC at 0.5 m depth showed a significant ($P<0.05$) increase after 3 days in relation to the initial value, and reached the highest value at the end of the experiment. At other depths, the UVAC content decreased with increasing depth and with time and showed the lowest values at the end of the experiment. After 45 days, the UVAC concentrations of the thalli cultured at 0.5 m were higher ($P<0.05$) than that at days 3 and 21 (Fig. 3a, b). To test the change in UVAC content under indoor PAR without UVR, the thalli were cultured under low PAR (~20 μmol photons m^{-2} s^{-1}). The content of UVAC significantly deceased from 134.1±7.4 to 98.2±6.2 μg g^{-1}(f.w.) after 3 days ($P<0.01$), and then gradually decreased to a steady value of 51.2±4.3 μg g^{-1}(f.w.) after 2 weeks until the end of the culture period (90 days) (Fig. 3c).

Agar yield of *G. lemaneiformis* increased with the increasing depths (Fig. 4). The greatest agar yield was found in thalli cultured at 3.5 m (19.9±0.5%) and the least in the thalli cultured at 0.5 m (14.2±0.8%). There are

Table 1 Averaged daily doses of PAR, UV-A and UV-B during the growth periods of *Gracilaria lemaneiformis*

Daily dose (MJ m^{-2})	Depth (m) 0	0.5	1.0	1.5	2.5	3.5
March (20 March to 2 April 2004)						
PAR	4.04	1.34	0.45	0.15	0.016	0.002
UV-A	0.55	0.67	0.008	0.001	1.51×10^{-5}	2.27×10^{-7}
UV-B	0.015	6.51×10^{-4}	2.80×10^{-5}	1.20×10^{-6}	2.18×10^{-9}	4.00×10^{-12}
April (11 to 21 April 2004)						
PAR	8.24	5.78	4.05	2.84	1.40	0.69
UV-A	1.24	0.45	0.16	0.06	0.007	0.001
UV-B	0.036	0.014	0.005	0.002	3.13×10^{-4}	4.69×10^{-5}

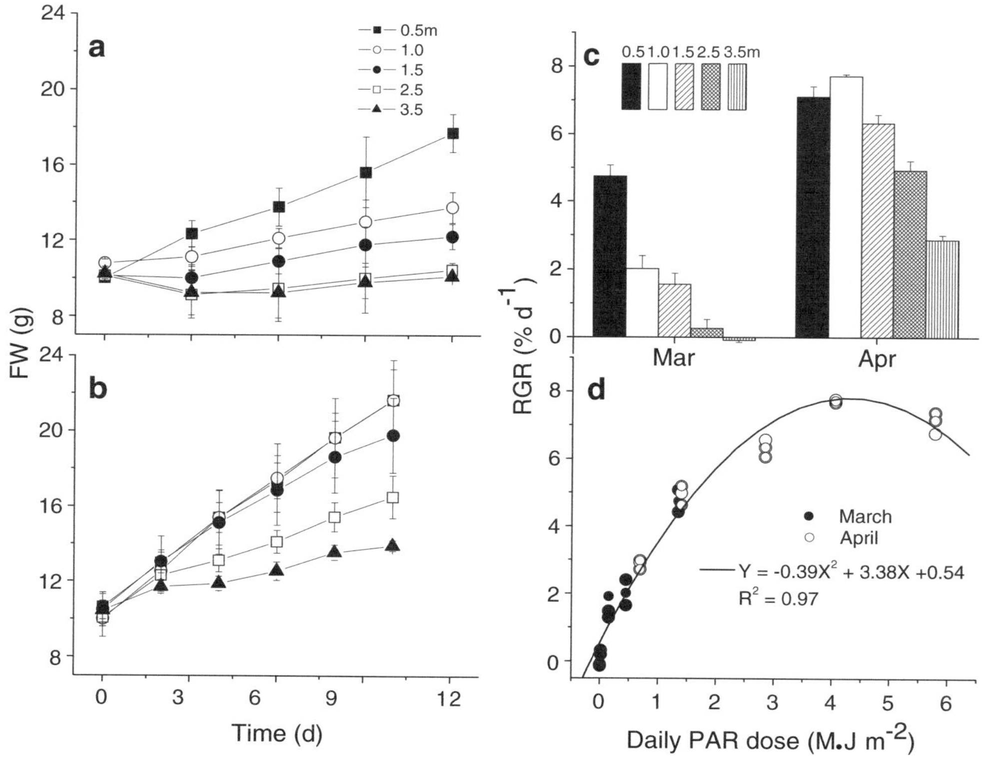

Fig. 1 Change in fresh biomass with time in *Gracilaria lemaneiformis* grown at different depths from 20 March to 2 April (**a**) and from 11 to 21 April 2004 (**b**), the relative growth rate (RGR) at different depths (**c**) and the relationship of the RGR with daily PAR dose (**d**). Daily PAR doses at deferent depths were calculated from the measured diurnal irradiances and the mean attenuation coefficients (*K*) for PAR wavebands. *Vertical bars* represent ± SD for the means (n=3)

significant differences (P<0.05) among depths except for the comparison between 1.0 and 1.5 m depths.

Discussion

Solar PAR, UV-A and UV-B attenuated disproportionatey with increasing depth. The averaged attenuation coefficients were 1.5 m^{-1} for PAR, 3.1 m^{-1} for UV-A and 4.1 m^{-1} for UV-B during the experimental periods, respectively. Solar irradiances were reduced to 1% of the sea surface at 2.3 m for PAR, 1.1 m for UV-A and 0.3 m for UV-B during the March experiment and at 6.5 m for PAR, 2.3 m for UV-A and 2.4 m for UV-B during the April culture. Our results showed that the growth rate of *G. lemaneiformis* was closely related to the daily PAR dose, while UVR showed insignificant effects on the growth except for the thalli grown at surface water during the April experiment. High levels of UVR at surface water during this period could be responsible for the observed decrease in the RGR, since UV-B is known to damage DNA (Zacher et al. 2007) and reduce photosynthetic capacity (Hanelt et al. 1997; Han et al. 2003) in macroalgae. On the other hand, *G. lemaneiformis* had higher Chl *a* and PE contents at deeper depth. Cells grown at deeper water layers can make more efficient use of low irradiances by increasing the contents of photosynthetic pigments (Malta et al. 2003), but

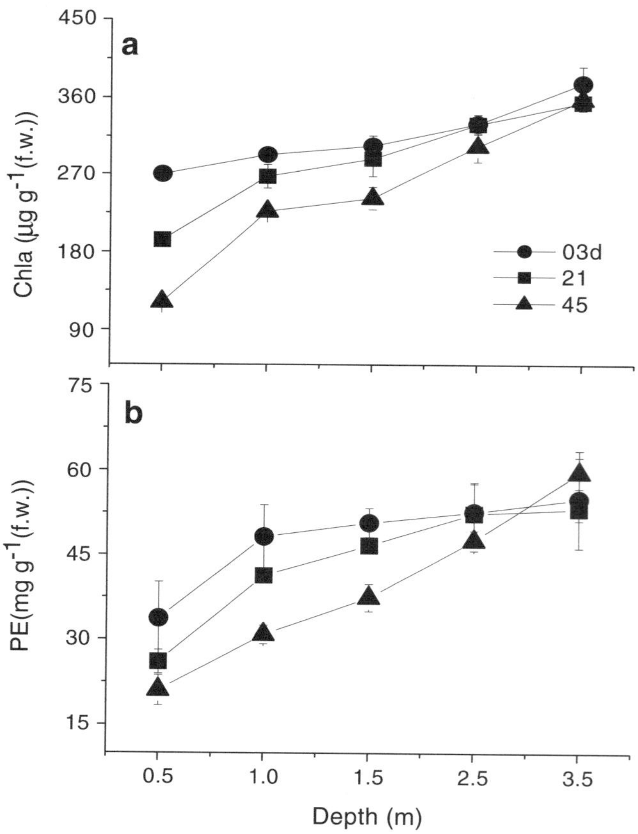

Fig. 2 The contents of Chl *a* (**a**) and PE (**b**) in *G. lemaneiformis* thalli grown at different depths during the period of 19 March to 3 May 2004. *Vertical bars* represent ± SD for the means (n=3)

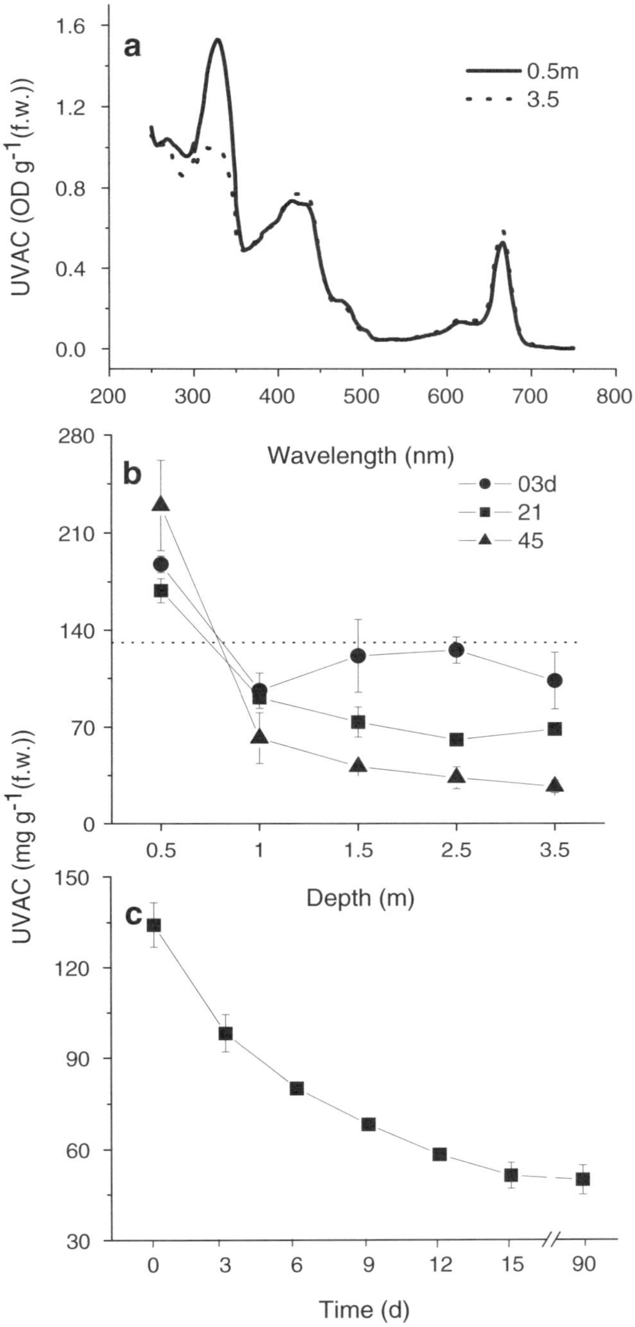

Fig. 3 Absorption spectra of *G. lemaneiformis* thalli grown at 0.5 and 3.5 m depths after 21 days (**a**), the contents of UVAC in the thalli at different depths (**b**) and the changes in the UVAC content of the thalli grown under low PAR from 19 February to 16 May 2004. *Dotted line* indicates the initial level of UVAC concentration. *Vertical bars* represent ± SD for the means (n=3)

severe shortage of light at deeper depths can still result in reduction of growth rate as shown in our study.

At the surface water or shallower layers where UVR may result in harmful effects, *G. lemaneiformis* thalli accumulated higher concentrations of UVAC. Algae inhabiting shallow water are more likely to contain higher levels of UVAC which plays a protective role against UVR (Maegawa et al. 1993; Karsten et al. 1998). In the present study, the thalli grown under indoor low PAR without UVR had decreased contents of UVAC in about 2 weeks, indicating that decreased UVAC contents at deeper depths were related to the lower solar irradiance and the negligible amount of UVR.

Gracilaria lemaneiformis responded to reduced solar radiation at increased depths by increasing cellular contents of the photosynthetic pigments, Chl *a* and PE and decreasing the content of UVAC. This strategy can enable the alga to capture more light for photosynthesis and to reduce the metabolic cost of the production of the secondary products, UVACs. Because formation of secondary products and cell growth are antagonistic processes and substantial metabolic investment might be required for the production of UV-absorbing compounds (Raven 1991), the decrease of UVAC concentrations can reduce the metabolic investment and ensure the limited photosynthetic products are preferentially provided for growth.

High levels of UV-B can reduce production of agar (Eswaran et al. 2002). In the present study, we observed the lowest agar yield in thalli grown in surface water, and the agar content increased with depth. However, at depths of 2.5 or 3.5 m, where the UVR was negligible (less than 1% of the surface UVR), agar yield was significantly higher than at other depths. While UVR may negatively affect agar yield in shallower layers, decreased growth rate at greater depths might contribute to the accumulation of agar due to reduced cell division.

Gracilaria lemaneiformis grew fastest in surface seawater in March and at 1 m depth in April. This implied that, in the coastal waters of the South China Sea, *G. lemaneiformis* would be better grown in surface seawater in February and

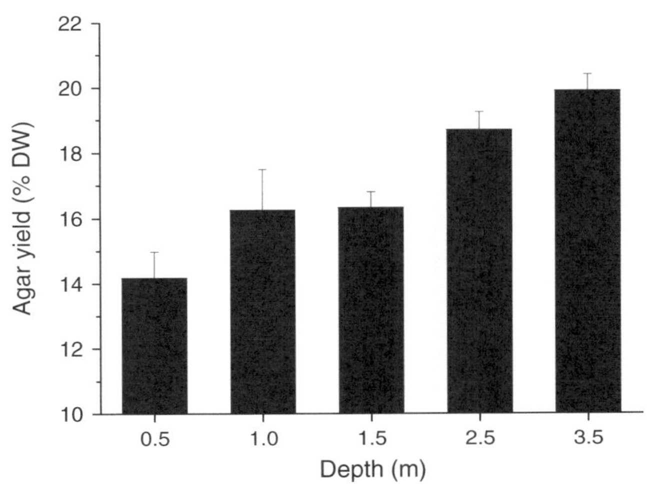

Fig. 4 The agar yield, as percentage of the dry mass of *G. lemaneiformis* grown at different depths. *Vertical bars* represent ± SD of the means (n=3)

March and at a depth of about 1 m in April and May, when solar radiation is high. The results of this study can help us understand to what extent changes in solar PAR or UVR can affect the growth, pigments and agar yield of *G. lemaneiformis*. Stocking density may also have and important effect on growth rate since it affects the intensity of either PAR or UVR. In view of the RGR–PAR dose relationship (Fig. 1d), increased stocking density will certainly reduce the optimum growth depth

Acknowledgement This study was funded by "863" project (2006AA10A413) from Ministry of Science and Technology and National Natural Science Foundation of China (Key Project No. 90411018).

References

Armisen R (1995) World-wide use and importance of *Gracilaria*. J Appl Phycol 7:231–243

Beer S, Eshel A (1985) Determining phycoerythrin and phycocyanin concentrations in aqueous crude extracts of red algae. Aust J Mar Freshw Res 36:785–792

Bischof K, Kräbs G, Wiencke C, Hanelt D (2002) Solar ultraviolet radiation affects the activity of ribulose−1,5-bisphosphate carboxylase-oxygenase and the composition of photosynthetic and xanthophylls cycle pigments in the intertidal green alga *Ulva lactuca* L. Planta 215:502–509

Buma AGJ, De Boer MK, Boelen P (2001) Depth distributions of DNA damage in Antarctic marine phyto- and bacterioplankton exposed to summertime UV radiation. J Phycol 37:200–208

Dunlap WC, Rae GA, Helbling EW, Villafañe VE, Holm-Hansen O (1995) Ultraviolet-absorbing compounds in natural assemblages of Antarctic phytoplankton. Antarct J 30:323–326

Eswaran K, Mairh OP, Subbarao PV (2002) Inhibition of pigments and phycocolloid in a marine red alga *Gracilaria edulis* by Ultraviolet-B radiation. Biol Plantarum 45:157–159

Gao K, McKinley KR (1994) Use of macroalgae for marine biomass production and CO_2 remediation: a review. J Appl Phycol 6:45–60

Franklin LA, Forster RM (1997) The changing irradiance environment: consequences for marine macrophyte physiology, productivity and ecology. Eur J Phycol 32:207–232

Flores-Moya A, Hanelt D, Figueroa F, Altamirano M, Vinegla B, Salles S (1999) Involvement of solar UV-B radiation in recovery of inhibited photosynthesis in the brown alga *Dictyota dichotoma* (Hudson) Lamouroux. J Photochem Photobiol B Biol 49:129–135

Häder DP, Lebert M, Marangoni R, Colombetti G (1999) ELDONET-European light dosimeter network hardware and software. J Photochem Photobiol B Biol 52:51–58

Häder DP, Porst M, Lebert M (2001) Photoinhibition in common Atlantic macroalgae measured on site in Gran Canaria. Helgol Mar Res 55:67–76

Han T, Han YS, Kain JM, Häder DP (2003) Thallus differentiation of photosynthesis, growth, reproduction, and UV-B sensitivity in the green alga *Ulva pertusa* (Chlorophyceae). J Phycol 39:712–721

Hanelt D, Wiencke C, Nultsch W (1997) Influence of UV radiation on the photosynthesis of arctic macroalgae in the field. J Photochem Photobiol B Biol 38:40–47

Henry BE, Van Alstyne KL (2004) Effects of UV radiation on growth and phlorotannins in *Fucus gardneri* (Phaeophyceae) juveniles and embryos. J Phycol 40:527–533

Karsten U, Sawall T, Wiencke C (1998) A survey of the distribution of UV-absorbing substances in tropical macroalgae. Phycol Res 46:271–279

Maegawa M, Kunieda M, Kida W (1993) Difference of the amount of UV-absorbing substance between shallow and deep water red algae. Jpn J Phycol 41:351–354

Malta M, Rijstenbil JW, Brouwer PEM, Kromkamp JC (2003) Vertical heterogeneity in physiological characteristics of *Ulva* spp. mats. Mar Biol 143:1029–1038

Marinho-Soriano E, Bourret E, De Casabianca ML, Maury L (1999) Agar form the reproductive and vegetative stages of *Gracilaria bursa-pastoris*. Bioresour Technol 67:1–5

Molloy FJ, Bolton JJ (1996) The effect of season and depth on the growth of *Gracilaria gracilis* at Luderitz, Namibia. Bot Mar 39:407–413

Pakker H, Beekman CAC, Breeman AM (2000a) Efficient photoreactivation of UVB-induced DNA damage in the sublittoral macroalga *Rhodymenia pseudpalmata* (Rhodophyta) in relation to ultraviolet-B-exposure. Eur J Phycol 35:109–114

Pakker H, Martins R, Boelen P, Buma AGJ, Nikaido O, Breeman AM (2000b) Effects of temperature in the photoreactivation of ultraviolet-B induced DNA damage in *Palmaria palmate* (Rhodophyta). J Phycol 36:334–341

Ramus J, Beale SI, Mauzerall D, Haward KL (1976) Changes in photosynthetic pigment concentration in seaweeds as a function of water depth. Mar Biol 37:223–229

Raven JA (1991) Responses of aquatic photosynthetic organisms to increased solar UV-B. J Photochem Photobiol B Biol 9:239–244

Viñegla B, Segovia M, Figueroa FL (2006) Effect of artificial UV radiation on carbon and nitrogen metabolism in the macroalgae *Fucus spiralis* L. and *Ulva olivascens* Dangeard. Hydrobiologia 560:31–42

Wellburn AR (1994) The spectral determination of chlorophylls a and b, as well as total cartenoids, using various solvents with spectrophotometers of different resolution. J Plant Physiol 144:307–313

Zacher K, Roleda MY, Hanelt D, Wiencke C (2007) UV effects on photosynthesis and DNA in propagules of three Antarctic seaweeds (*Adenocystis utricularis*, *Monostroma hariotii* and *Porphyra endiviifolium*). Planta (In press)

Changes of growth characteristics and free amino acid content of cultivated *Porphyra yezoensis* Ueda (Bangiales Rhodophyta) blades with the progression of the number of harvests in a nori farm

Kyosuke Niwa · Hirofumi Furuita · Takeshi Yamamoto

Originally published in the Journal of Applied Phycology, Vol 20, No 5, 237–243.
DOI: 10.1007/s10811-007-9273-5

Abstract It is essential for successful *Porphyra* breeding to understand the growth characteristics and the quantitative aspect of the taste of gametophytic blades. However, there have been only a few studies on such characteristics with the progression of the number of harvests in nori farms. In this study we examined the changes of blade shape, blade thickness and free amino acid (FAA) content of two *P. yezoensis* f. *narawaensis* pure lines (HG-4 and HG-5) with the progression of the number of harvests in a nori farm. With the progression of the number of harvests, the blade length-to-width ratio of both pure lines decreased, while the blade thickness increased. From the results, it is suggested that the increase of blade length is significantly higher in HG-4 than in HG-5, and that the blade is thicker in HG-5 than in HG-4. Total FAA contents between the two lines were not significantly different throughout the investigation period. Among the major four FAA (aspartic acid, glutamic acid, alanine and taurine), only glutamic acid content decreased with the progression of the number of harvests in both pure lines. These results suggest that the superior taste of dried nori sheets made of the first harvest of blades is due to thinner blades and higher content of glutamic acid.

Keywords Blade shape · Breeding · Free amino acid · Gametophytic blade · Seaweed cultivation

K. Niwa (✉)
Fisheries Technology Institute, Hyogo Prefectural Technology Center for Agriculture, Forestry and Fisheries,
Akashi, Hyogo 674-0093, Japan
e-mail: kyousuke_niwa@pref.hyogo.jp
e-mail: kyousuke_niwa@pref.hyogo.lg.jp

H. Furuita · T. Yamamoto
Inland Station, National Research Institute of Aquaculture,
Fisheries Research Agency,
Tamaki, Mie 519-0423, Japan

Introduction

The red algal genus *Porphyra* (nori) has a dimorphic life cycle with gametophytic haploid blades and sporophytic diploid filaments referred to as the conchocelis phase. In this genus, *P. yezoensis* is the world's most lucrative marine algal crop, and is extensively cultivated in Japan, Korea and China. Since the 1960's the rapid increase of nori production has been based on several technical innovations such as: (i) development of artificial seeding by using oyster shells with conchocelis fragments; (ii) extension of nori farms by the floating culture method; (iii) frozen storage of nori nets with conchospores attached or young blades grown to 1–3 cm long; (iv) invention of rapid-harvesting boat for cultivated *Porphyra*; and (v) introduction of automatic processing machines for dried nori sheets (Miura 1992). Furthermore, selective breeding of cultivated *Porphyra* has been intensively attempted by nori cultivators, and the two vigorously growing cultivars, *P. yezoensis* f. *narawaensis* and *P. tenera* var. *tamatsuensis* are bred in Japan (Miura 1984). Through these nori culture techniques and genetic improvement, the total annual production of nori in Japan amounts to approximately US $1 billion. Although there are many strains of *P. yezoensis* f. *narawaensis* produced by selective breeding (Niwa and Aruga 2003, 2006; Niwa et al. 2004), nori cultivators have always requested superior strains with higher growth rate, better flavor, darker color, and higher tolerance to diseases. Hence, it is essential to understand the characteristics with regard to growth and taste of the gametophytic blades in nori farms for the establishment of such superior strains.

In nori farming, nori cultivators usually harvest several times from the same nori net, because the remaining basal portions of the blades on the nori nets grow to harvestable size again in about a week. However, only a few studies have been conducted on the changes with regard to growth characteristics and taste of cultivated *Porphyra* blades with the progression of the number of harvests (Miura 1979, Sakai and Kasai 2000). Among the substances giving taste in the gametophytic blades, free amino acids (FAA) are known to be the most important substances (Noda et al. 1975; Harada et al. 1990; Yoshie et al. 1993). In the present study, we investigated the changes of blade shape, blade thickness and FAA contents of two *P. yezoensis* f. *narawaensis* pure lines with the progression of the number of harvests in a nori farm. In addition, we also carried out a laboratory culture experiment to examine whether cut harvesting influences the increase of the blade width.

Materials and methods

Experiment in a nori farm

The two pure lines HG-4 and HG-5 of *Porphyra yezoensis* f. *narawaensis* were used in the present study. The origin and isolation procedure of HG-4 were as described in Niwa et al. (2004). HG-5 was established by a slightly modified procedure of Niwa et al. (2004) by archeospore propagation from the wild-type portion of an F_1 blade (consisting of the wild and green type sectors) which was derived from the heterozygous conchocelis obtained by the cross between a green mutant (G-1) and a wild type (0110) of *P. yezoensis* f. *narawaensis* (Niwa et al. 2002; Niwa and Aruga 2003).

Free-living conchocelis fragments of the two pure lines were propagated into oyster shells (Fig. 1a). With these conchocelis-penetrated shells, the seeding of nori nets with conchospores of the two pure lines was separately carried out on land (Fig. 1b). After seeding, the nori nets with attached conchospores were frozen at -25°C in a large freezer until the germling culture in a nori farm. The germlings (gametophytic blades) on the nori nets were cultivated by the floating culture facility in a nori farm at Akashi, Hyogo Prefecture, Japan. Experimental culture steps are shown in Fig. 2. For germling culture, the nori nets were raised on the floating culture structures with nursery "U"-shaped frames. The nets were emersed on these floating structures for about 2–4 hours every morning (except for rainy days) to minimize diatom attachment and disease (Fig. 1c). After the germling culture, the gametophytic blades reached approximately 1–2 cm long. The nori nets were dried and stored at -25°C again. The nori nets were then placed again on the floating culture structures without nursery frames in the nori farm, and the gametophytic blades on the nori nets were grown without emersion. Harvesting was carried out five times by using a rapid-harvesting boat through the investigated period (Fig. 1d).

The samples of gametophytic blades of the two pure lines were collected immediately before the first, third, and fifth harvests (Fig. 2). The length and width of the sample blades were measured using vernier calipers. The blade thickness was measured under a compound light microscope after being sectioned by hand using a razor blade. The FAA contents of gametophytic blades were determined as described from Niwa et al. (2003). Briefly, freeze-dried sample blades were cut into small pieces and extracted three times with 75 % (v/v) hot ethanol. The extracts were concentrated using a rotary evaporator and defatted with diethyl ether. The aqueous phase was dried in a vacuum and redissolved in citrate buffer (pH=3). The samples were analyzed using an amino acid analyzer (Hitachi L-8500).

Experiment in the laboratory

The pure line HG-4 was cultured in the laboratory to examine the influence of harvesting on the blade width. A mature conchocelis colony was transferred into a 300 mL flask with a culture medium agitated by aeration at 15°C under 80 μmol photons · m^{-2} · s^{-1} in 10:14 h (light:dark). Vinylon monofilaments (about 6 cm long) were put into the flask for conchospore attachment. After attachment of conchospores, the conchocelis colony was removed from the flask. The attached conchospores were cultured to obtain experimental gametophytic blades. The blades were detached from the monofilaments when they grew to 2–3 cm long. The detached blades were transferred to a 1 L flask and further cultured. The culture medium was a modified Provasoli Enriched Seawater (Provasoli 1966) without soil extract and vitamins, and was renewed twice a week.

Forty days after conchospore attachment, the developed gametophytic blades were divided into two groups (control and experimental groups) consisting of ten individuals each. After the blade length and width were measured, each blade of the experimental group was cut in two, the tip and basal portions. The basal portions were further cultured in a 1 L flask. The blades in the control group were not cut, and were also cultured under the same culture conditions. After two weeks (54 days after conchospore attachment) the blade width of both groups was measured.

Statistical analysis

Data of blade length, blade length-to-width ratio and blade thickness between HG-4 and HG-5 were statistically tested using Student's *t*-test. When standard deviation (SD) was significantly different, Welch-*t* test was applied. Data of the length-to-width ratio, blade thickness and glutamic acid content among the three different harvest times were statistically treated using one-way ANOVA. If the variances

Fig. 1 (**a**) Oyster shells with propagated conchocelis for the artificial seeding of *Porphyra*. (**b**) Land seeding facility with a seawater tank and a large reel for nori nets to attach conchospores of *Porphyra*. Oyster shells with propagated conchocelis are put into the seawater tank, and nori nets wound on the reel are rotated to soak the nets in seawater containing conchospores. (**c**) Germling culture in the sea by the floating culture method. Conchospore germlings on nets are raised and emersed for approximately 2–4 hours every day (except for rainy days) on these floating structures with nursery frames. (**d**) Harvesting by a rapid-harvesting boat of cultivated *Porphyra* blades on floating culture structures. The boat lifts the nori nets up from the sea, and the harvester mounted on the roof rapidly cuts the gametophytic blades grown on the nori nets

were normally distributed, Tukey-Kramer multiple comparisons test was applied. When variances were not normally distributed, the Kruskal-Wallis non-parametric test was applied. Differences of blade width between the experimental and control groups in the laboratory culture were determined by Student's *t*-test.

Results

Growth characteristics of gametophytic blades in the nori farm

Figure 3 shows the gametophytic blades of the two pure lines (HG-4 and HG-5) attached to the nori net at the three different harvest times in the nori farm. At the first harvest, blades of both pure lines were mostly long and linear; the mean blade length was 29.6±4.8 cm in HG-4 whereas in HG-5 22.7±3.5 cm (average ± SD, n=30). Hence, the blade length was significantly longer in HG-4 than in HG-5 at the first harvest (Student's *t*-test, $p<0.01$), and the blade length-to-width ratio was significantly higher in HG-4 than in HG-5 (Welch-*t* test, $p<0.01$). With the progression of the number of harvests, the width of the gametophytic blades of the two pure lines became broader relative to the length. The blade length-to-width ratios of both HG-4 and HG-5 were highest at the first harvest and lowest at the fifth harvest (Kruskal-Wallis test, $p<0.05$, Fig. 4).

The blade thickness of HG-4 and HG-5 at the three different harvest times is shown in Fig. 5. With the progression of the

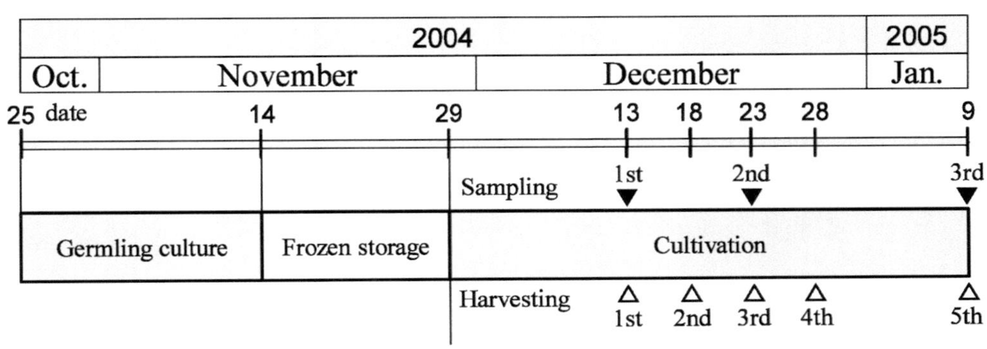

Fig. 2 Process of *Porphyra* cultivation in the present study

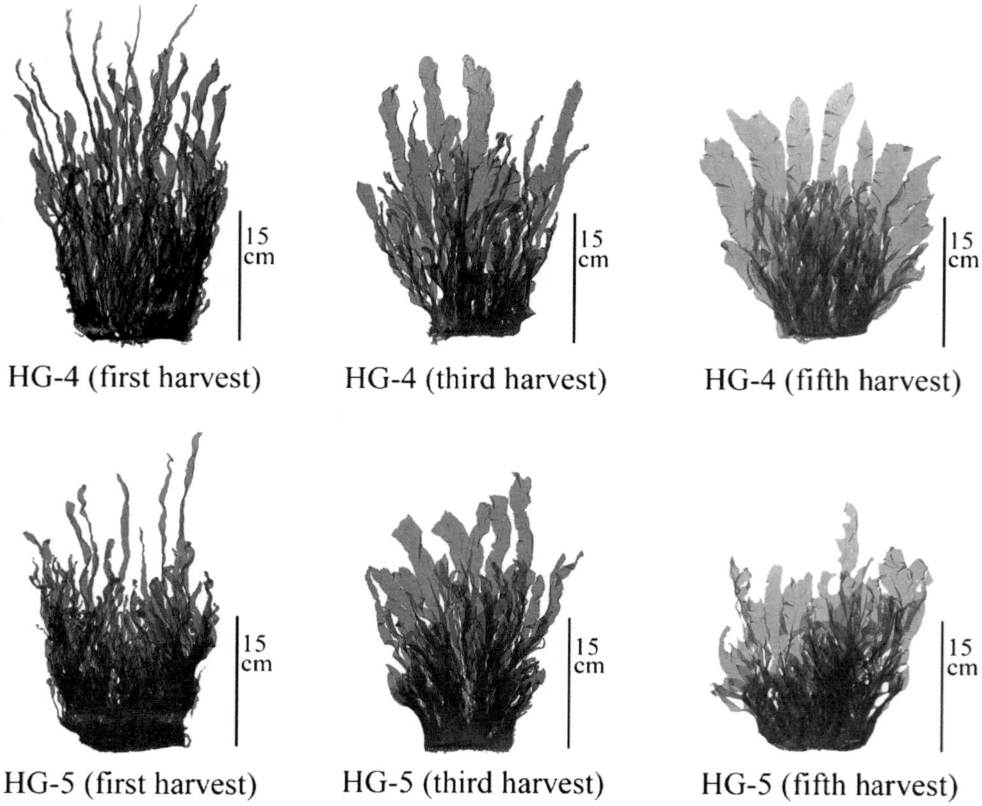

Fig. 3 Gametophytic blades grown on nori nets of the two pure lines HG-4 and HG-5 of *P. yezoensis* f. *narawaensis* sampled at the first, third, and fifth harvest

number of harvests, the blade thickness of both pure lines increased (Tukey-Kramer test, $p<0.01$). The blades of HG-5 were significantly thicker than those of HG-4 at all the harvest times examined (Student's *t*-test, $p<0.05$).

Free amino acid contents of gametophytic blades in the nori farm

Figure 6 shows total free amino acid (FAA) contents of HG-4 and HG-5 at the three different harvest times in the nori farm. Although the total FAA content of HG-5 tended to increase with the progression of the number of harvests, no clear relationship was observed between the total FAA content and the harvest time in HG-4, and the total FAA contents between the two lines were not significantly different throughout the investigation period. The changes of the four major FAA, aspartic acid (Asp), glutamic acid (Glu), alanine (Ala) and taurine (Tau), in the gametophytic blades of HG-4 and HG-5 at the three different harvest times are shown in Figs. 7 and 8, respectively. The Asp,

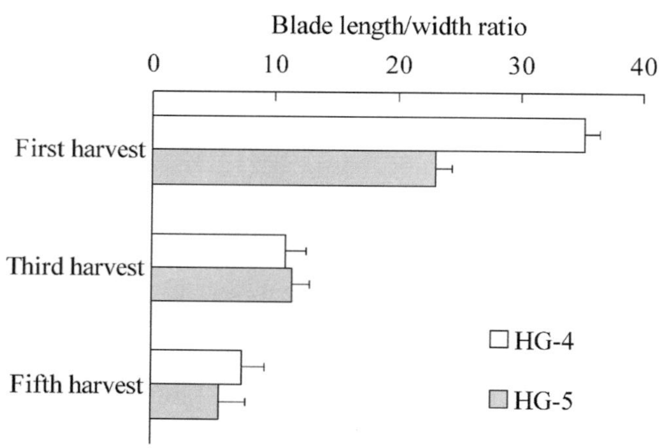

Fig. 4 Blade length-to-width ratios of the two pure lines HG-4 and HG-5 of *P. yezoensis* f. *narawaensis* at the first, third, and fifth harvest. Average ± SD (n=30)

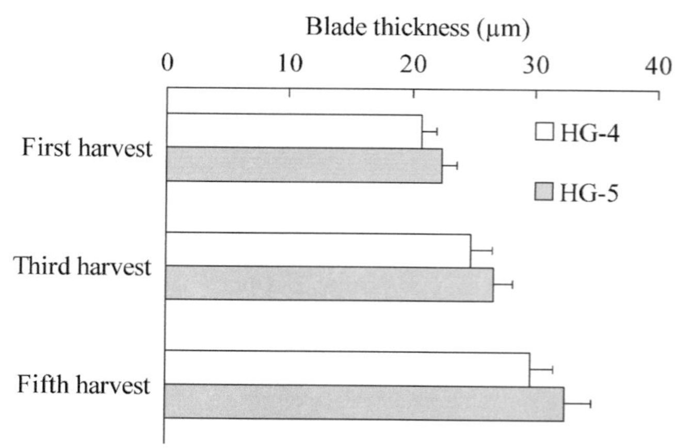

Fig. 5 Blade thickness of the two pure lines HG-4 and HG-5 of *P. yezoensis* f. *narawaensis* at the first, third, and fifth harvest. Average ± SD (n=10)

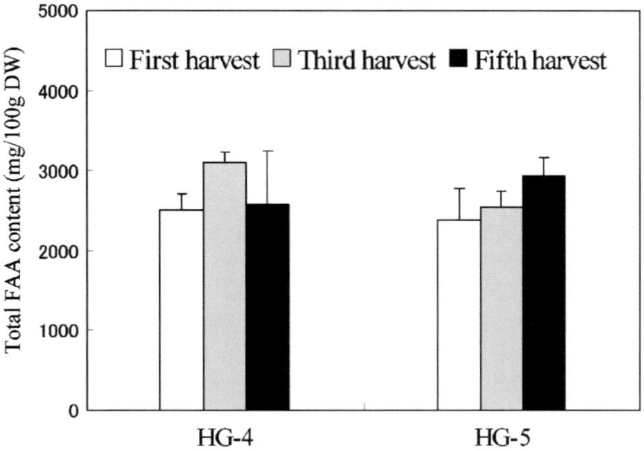

Fig. 6 Total free amino acid contents of the two pure lines HG-4 and HG-5 of *P. yezoensis* f. *narawaensis* at the first, third and fifth harvest. Average ± SD (n=3 except for n=2 in HG-4 at the fifth harvest and HG-5 at the third harvest)

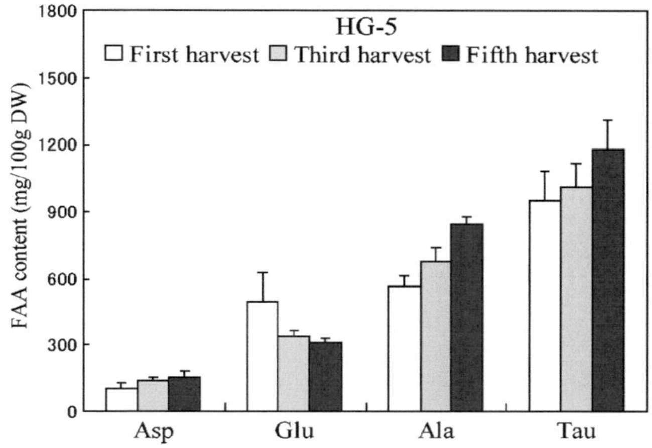

Fig. 8 Free amino acid contents of the pure line HG-5 of *P. yezoensis* f. *narawaensis* at the first, third, and fifth harvest. Average ± SD (n=3 for the first and fifth harvest, and n=2 for third harvest)

Ala, and Tau contents of the two pure lines were not highest at the first harvest, and the changes of the three FAA contents in both of the lines correlated with those of the total FAA contents. However, the Glu content alone decreased with the progression of the number of harvests in both pure lines, even though a significant difference was not found in HG-5 (Tukey-Kramer test: HG-4, $p<0.05$; HG-5, $p>0.05$). Therefore, the ratio of Glu-to-total FAA contents also decreased with the progression of the number of harvests (Table 1).

Changes of blade width in the laboratory culture

Gametophytic blades of HG-4 were cultured for 54 days after conchospore attachment in the laboratory. After 40 days in culture immediately prior to being cut, the blade length was 52.8±6.4 mm in the experimental group and 51.0±5.6 mm in the control group, whereas the blade width was 4.6±0.7 mm in the former and 4.7±0.6 mm in the latter (Fig. 9; average ± SD, n=10). After two weeks (after 54 days in culture), the blade width of the experimental group was higher than that of the control group (Student's *t*-test, $p<0.05$): the width was 15.0±2.1 mm in the former and 13.0±2.1 mm in the later (Fig. 9; average ± SD, n=10). Representative samples of the gametophytic blades of the experimental and the control groups after 54 days in culture are shown in Fig. 10.

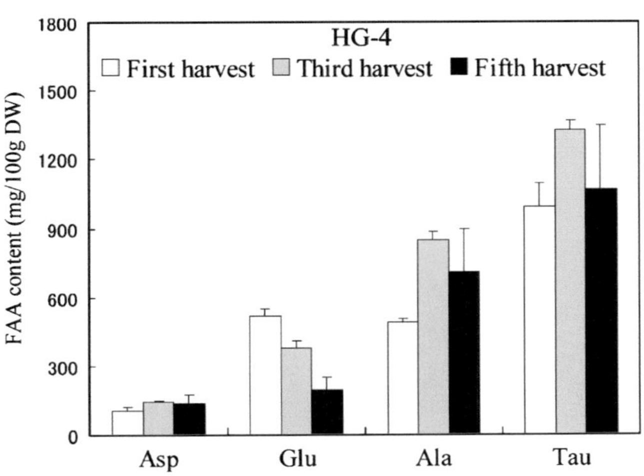

Fig. 7 Free amino acid contents of the pure line HG-4 of *P. yezoensis* f. *narawaensis* at the first, third, and fifth harvest. Average ± SD (n=3 for the first and third harvest, and n=2 for the fifth harvest)

Discussion

In the present study, both pure lines HG-4 and HG-5 showed long linear blades in the nori farm (Fig. 3), indicating that their blades represent the same characteristic shape of *P. yezoensis* f. *narawaensis* (Miura 1984; Niwa et al. 2004). In nori farms, gametophytic blades are harvested usually by cutting, and the remaining sections of blades on nori nets are shortened to various different lengths by the cut harvesting. Therefore, if the blade length and shape among different strains are compared, it is necessary to cultivate their blades under the same conditions, and to

Table 1 The ratios of glutamic acid-to-total free amino acid contents of the two pure lines HG-4 and HG-5 in *P. yezoensis* f. *narawaensis* at the first, third, and fifth harvest in the nori farm

	First harvest	Third harvest	Fifth harvest
HG-4	20.6%	12.2%	7.6%
HG-5	20.7%	13.3%	10.7%

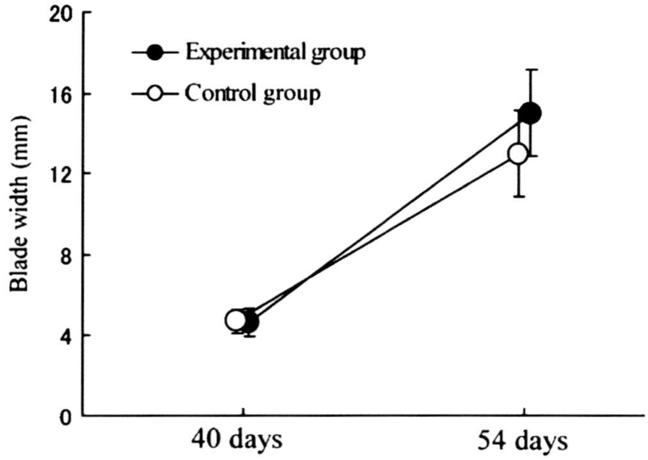

Fig. 9 Growth of the blade width of the pure line HG-4 of *P. yezoensis* f. *narawaensis* in the laboratory culture. Blades in the experimental group were cut in two, the tip and basal portions, 40 days after the conchospore attachment. The basal portions were further cultured until 54 days after the conchospore attachment. The blades in the control group were not cut and cultured until 54 days after the conchospore attachment under the same culture conditions. Average ± SD (n=10)

examine those before cutting at the first harvest. In the present study, it is inferred that the increase of blade length is higher in HG-4 than in HG-5 based on the blade length and the length-to-width ratio at the first harvest.

Hannach and Waaland (1989) reported that environmental conditions, such as photon flux density and water motion, affected the blade shape of *P. abbottae* in laboratory culture. Niwa et al. (2004) confirmed that different blade shapes also occurred in genetically different strains of *P. yezoensis* f. *narawaensis* in laboratory culture. In the present study, the blade length-to-width ratios of both HG-4 and HG-5 decreased with the progression of the number of harvests in the nori farm. This trend is in agreement with a previous study (Miura 1979). The main reason is that the decrease of the blade length-to-width ratio is due to the increase of the blade width with aging as shown in Fig. 9, although the blade length was shortened by harvesting. On the other hand, it has been unclear whether the cut harvesting also increases the blade width. Our result in the laboratory culture experiment indicates that the cut harvesting increases the blade width.

From the results of the blade thickness in HG-4 and HG-5, it was clarified that the blade thickness increased with the progression of the number of harvests (Fig. 5), and it was inferred that the thickness increases with aging. Therefore, if the blade thickness of different strains is compared, it is necessary to cultivate them under the same conditions and to collect samples at the same harvest time. In the present study, the blade of HG-5 was always thicker than that of HG-4 at all harvest times examined. Hence, this difference is probably due to genetic variation. The blade thickness is also regarded as one of the taxonomic features especially for identification of cultivated *Porphyra* species (e.g., Ueda 1932; Kurogi 1961; Miura 1984; Niwa et al. 2005). However, because the thickness also changes with aging, it is necessary to compare blade thickness data with consideration of the blade age as shown in Niwa et al. (2005).

The dried nori sheets made of the first harvested blades are well known to have superior taste in the nori industry. It is considered that the better flavor is attributable to the blade softness at the first harvest (Harada et al. 1990). Because the softness of dried nori is important to allow flavors to be readily released while in the mouth when eating nori and is closely related to the thickness (Niwa et al. 2005), our results indicate that the superior taste of nori at the first harvest is due to the fact that the blades are thinnest at the first harvest among those of all the harvest time as mentioned above. In determining the taste of dried nori sheets, free amino acid (FAA) contents have also been considered to be an important factor. Among FAA, aspartic acid (Asp), glutamic acid (Glu), alanine (Ala), and taurine (Tau) are predominant in the total FAA (Noda et al. 1975; Harada et al. 1990; Yoshie et al. 1993; Niwa et al. 2003). Noda et al. (1975) considered that Glu and Ala, two of the

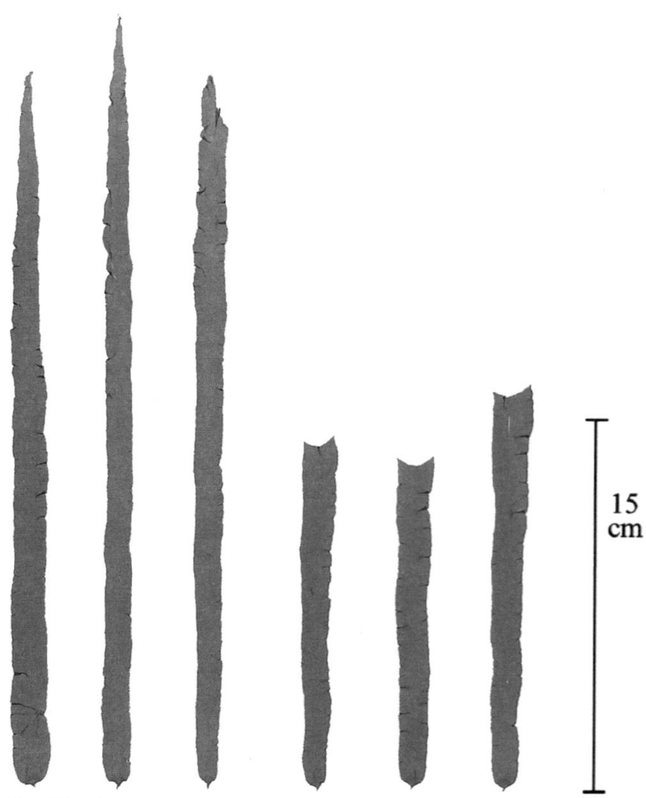

Fig. 10 Gametophytic blades of the pure line HG-4 of *P. yezoensis* f. *narawaensis* 54 days after conchospore attachment in the laboratory culture. Three blades on the left side, control group; three blades on the right side, experimental group

four major FAA, were the most important FAA in determining the taste of dried nori sheets. However, there have been only a few studies on changes of FAA contents in the gametophytic blades of *Porphyra* with the progression of the harvest month in nori farms (Harada et al. 1990, Sakai and Kasai 2000). Sakai and Kasai (2000) reported that FAA content including Glu increased with the progression of the harvest in Hokkaido. However, in the present study, Glu content alone decreased with the progression of the number of harvests among the four major FAA contents in both HG-4 and HG-5 (Figs. 7 and 8), although a significant difference was not found in HG-5. In a preliminary experiment carried out in the previous year to the current experiment, the Glu content of blades was also highest at the first harvest (data not shown). In addition, the data of Harada et al. (1990) also indicated that Glu content tended to decrease with the progression of the harvest month in Hyogo Prefecture. Therefore, although it is necessary to accumulate the data in different nori farms and it is possible that the change of the FAA contents depends on environmental conditions in nori farms, our results of the FAA contents suggest that the highest Glu content and Glu-to-total FAA contents ratio are one of the important factors in determining the superior taste of the first harvest. Therefore, in order to isolate superior strains with regard to the taste of nori, it is also necessary to direct our attention to gametophytic blades with higher Glu content and higher ratio of Glu-to-total FAA contents as one of the indicators for superior taste of nori, in addition to the blade thinness.

Conclusion

With the progression of the number of harvests in a nori farm, the blade length-to-width ratio of two pure lines (HG-4 and HG-5) of *P. yezoensis* f. *narawaensis* decreased, while the blade thickness increased. The comparisons of these growth characteristics between HG-4 and HG-5 suggest that the increase of blade length is significantly higher in HG-4 than in HG-5, and that the blade is thicker in HG-5 than in HG-4. Total FAA contents between the two lines were not significantly different throughout the investigation period. Among the four major FAA, the glutamic acid content alone decreased with the progression of the number of harvests in both pure lines. These results suggest that the superior taste of dried nori sheets made of the first harvest blades is due to the thinner blades and higher content of glutamic acid.

Acknowledgement We thank the Hyogo Nori Institute for seeding nori nets with conchospores of the two pure lines, and 'Shinmei-Suisan' of Akashiura Fishermen's Cooperative for cultivating the gametophytic blades in their nori farm. We also thank Dr. Yusho Aruga for valuable comments and critical reading of the manuscript. Partial financial support by the Fisheries Agency of Japan is gratefully acknowledged.

References

Hannach G, Waaland JP (1989) Growth and morphology of young gametophytes of *Porphyra abbottae* (Rhodophyta): effects of environmental factors in culture. J Phycol 25:247–254

Harada K, Osumi Y, Fukuda N, Amano H, Noda H (1990) Change of amino acid compositions of 'Nori', *Porphyra* spp. during storage. Nippon Susisan Gakkaishi 56:607–612

Kurogi M (1961) Species of cultivated *Porphyras* and their life histories. Bull Tohoku Reg Fish Res Lab 18:1–115

Miura A (1979) Studies on genetic improvement of cultivated *Porphyra* (laver). Proceedings of the 7th Japan-Soviet Joint Symposium on Aquaculture. Tokai Univ Press, Tokyo, pp 161–168

Miura A (1984) A new variety and a new form of *Porphyra* (Bangiales, Rhodophyta) from Japan: *Porphyra tenera* Kjellman var. *tamatsuensis* Miura, var. nov. and *P. yezoensis* Ueda form. *narawaensis* Miura, form. nov. J Tokyo Univ Fish 71:1–14, pl. I-XI

Miura A (1992) Nori. In Miura A (ed.), Cultivation of edible algae in Japan. Koseisha Koseikaku, Tokyo, pp 11–24 (in Japanese).

Niwa K, Aruga Y (2003) Rapid DNA extraction from conchocelis and ITS-1 rDNA sequences of seven strains of cultivated *Porphyra yezoensis* (Bangiales, Rhodophyta). J Appl Phycol 15:29–35

Niwa K, Aruga Y (2006) Identification of currently cultivated *Porphyra* species by PCR-RFLP analysis. Fish Sci 72:143–148

Niwa K, Mizuta A, Aruga Y (2002) Genetic characterization of a spontaneous green-type pigmentation mutant of *Porphyra yezoensis* and the significance of using heterozygous conchocelis in nori farming. Fish Sci 68:729–735

Niwa K, Furuita H, Aruga Y (2003) Free amino acid contents of the gametophytic blades from the green mutant conchocelis and the heterozygous conchocelis in *Porphyra yezoensis* Ueda (Bangiales, Rhodophyta). J Appl Phycol 15:407–413

Niwa K, Kikuchi N, Iwabuchi M, Aruga Y (2004) Morphological and AFLP variation of *Porphyra yezoensis* Ueda form. *narawaensis* Miura (Bangiales, Rhodophyta). Phycol Res 52:180–190

Niwa K, Kikuchi N, Aruga Y (2005) Morphological and molecular analysis of the endangered species *Porphyra tenera* (Bangiales, Rhodophyta). J Phycol 41:294–304

Noda H, Horiguchi Y, Araki S (1975) Studies on the flavor substances of 'Nori', the dried laver *Porphyra* spp. -II. Free amino acids and 5'-nucleotides. Nippon Suisan Gakkaishi 41:1299–1303

Provasoli L (1966) Media and prospects for the cultivation of marine algae. In: Watanabe A, Hattori A (eds) Cultures and collections of algae. Proc. U.S.-Japan Conf. Hakone, Sept 1966. Jap Soc Plant Physiol, pp 63–75

Sakai H, Kasai T (2000) Fatty acids, free amino acids and 5'-nucleotides of Dried Laver, Hoshi-nori, harvested in different months in Hokkaido and produced under different drying conditions. Nippon Shokuhin Kagaku Kaishi 47:327–332 (in Japanese with English abstract)

Ueda S (1932) A systematic study of genus *Porphyra* found on the Japanese coast. J Imp Fish Inst 28:1–45, pl. I–XXIV (in Japanese)

Yoshie Y, Suzuki T, Shirai T, Hirano T (1993) Free amino acids and fatty acid composition in dried nori of various culture locations and prices. Nippon Suisan Gakkaishi 59:1769–1775 (in Japanese with English abstract)

Environment determines nitrogen content and stable isotope composition in the sporophyte of *Undaria pinnatifida* (Harvey) Suringar

M. C. Carvalho · K. Hayashizaki · H. Ogawa

Originally published in the Journal of Applied Phycology, Vol 20, No 5, 245–253.
DOI: 10.1007/s10811-007-9271-7

Abstract Nitrogen content and $\delta^{15}N$ (nitrogen stable isotope ratio) were measured in different parts of the sporophyte of *Undaria pinnatifida* (Harvey) Suringar from two bays in the northeastern Japan. There were clear differences between the thalli collected in winter and in summer: high nitrogen content and low average $\delta^{15}N$ were observed for the thalli collected in winter, while the opposite pattern was found for the thalli from summer. In addition, the pattern of internal $\delta^{15}N$ distribution in the thallus changed with season. It is possible that the cause for these seasonal differences in the algae was the seasonal change in environmental conditions, because in winter water is normally rich in nitrogen, while in summer it is poor. *U. pinnatifida* sporophyte may be useful as an indicator of nitrogen sources in coastal waters, but consideration must be given to the effect of isotope fractionation on $\delta^{15}N$ of the plant, especially in winter. Potential may exist for the use of different parts of the thallus to indicate nitrogen sources at different periods, but more investigation is necessary to accomplish this.

Keywords $\delta^{15}N$ · Bio-indicator · Macroalgae · Seasonal variation · Translocation

Introduction

Anthropogenic nitrogen inputs have occurred in coastal waters around the world due to urbanization, industrialization, agriculture and aquaculture. These inputs have often been considered the probable causes of environmental problems such as eutrophication, and their influence on the coastal environment has often been examined in near shore waters (Vitousek 1994; Howarth et al. 1996). To evaluate the influence of anthropogenic nitrogen inputs in coastal areas, it is essential to differentiate the natural nitrogen from the anthropogenic one, and stable isotopes have been successfully employed for this purpose (Lindau et al. 1989). Particularly, the nitrogen stable isotope composition of macrophytes has widely been used to observe the dispersal pattern of anthropogenic nitrogen inputs in coastal areas (McClelland and Valiela 1998; Costanzo et al. 2001; Umezawa et al. 2005).

When employing nitrogen stable isotopes to trace nitrogen sources, it is necessary to know how the stable isotope compositions of the plant and sources are related. In the case of marine macrophytes, it has been assumed that their nitrogen stable isotope composition and nitrogen sources are similar, when nitrogen is a limiting nutrient in marine environments (Peterson and Fry 1987). However, the stable isotope composition of the plant may differ from that of the source if the nitrogen is abundant for the plant, because plants fractionate the nitrogen stable isotopes in this case (Handley and Raven 1992). Moreover, it has been observed that different parts of the thallus in large bodied seaweeds can show different nitrogen stable isotope composition (Fredriksen 2003). This variation may complicate their use as a bio-indicator of nitrogen sources, but it has not been evaluated.

Undaria pinnatifida is a brown alga heavily consumed as food in Japan (Ogawa 2004). The occurrence of this alga as an invasive species around the world has been reported (Perez et al. 1981; Hay and Luckens 1987; Casas et al. 2004). The thallus is morphologically and functionally

M. C. Carvalho (✉) · K. Hayashizaki · H. Ogawa
Kitasato University, School of Fisheries Sciences,
160–4 Utou, Okkirai, Sanriku-cho,
Ofunato, Iwate 022-0101, Japan
e-mail: matheus@samerica.com

differentiated in pinnate blade, midrib and sporophyll, and each part can easily be identified. These characteristics allow the determination of the internal distribution of the nitrogen stable isotope composition in a thallus. In addition, it has been demonstrated that nitrogen fertilizer addition to culture water improves the production of this species without environmental pollution (Ogawa and Fujita 1997), which suggests that this cultured alga probably absorbed most of the added nitrogen very quickly. Then, the thallus could also remain a temporal record of isotopic signatures from nitrogen sources during its growth, which could possibly supply information of nitrogen source changes. In order to evaluate these hypotheses, in this paper we investigate the temporal and spatial variation of the internal distribution of nitrogen content and nitrogen stable isotopes in a sporophytic thallus of *U. pinnatifida*.

Materials and methods

Sporophytic thalli of *U. pinnatifida* were collected from two bays, Okkirai and Otsuchi Bay, Sanriku Ria Coast, in northeastern Japan (Fig. 1). Sample collections were conducted in July 2005 and January, March and June 2006. The collected materials were brought back to the laboratory, where their length were measured, and pieces with about 3 cm in length were removed at the positions shown in Fig. 2.

Sampling was conditioned to the life history of *U. pinnatifida.* Wild *U. pinnatifida* sporophytes develop from January to July, and grow in the low intertidal zone (Tokuda et al. 1986). On the other hand, cultured specimens are placed in the culture field in December, on ropes that are kept submersed all the time, and are harvested in April.

At each sampling occasion, a different purpose was researched:

1) In July 2005, adult thalli of the wild *U. pinnatifida* population were collected from different sampling stations (T1, T2 and T3) in Otsuchi Bay. The several individuals collected at a single station (T1: 6; T2: 8; T3: 9) were pooled in a single sample. Duplications were made for each station. This was a preliminary sampling to observe the general trend of internal nitrogen stable isotope distribution in the thallus.
2) In January 2006, we collected young cultured thalli from station K1 in Okkirai Bay to be compared to those formerly collected. This was a preliminary sampling with only 2 specimens. No sample was available for the position S and samples from the upper parts of the thallus were not collected (Fig. 2).
3) Since the internal distribution of nitrogen stable isotopes was very different in thalli from January 2006 compared to thalli from July 2005 (details in the Results), we decided to do another sampling in Okkirai Bay in March 2006. This time, 19 specimens including adult and young thalli were collected from station K2. All the individuals were sampled for the positions MM and MB, but only 5 were for positions LM and S (abbreviations in Fig. 2). We chose March because at this time adult cultured thalli is present and they could be directly compared with those collected from Otsuchi Bay in July 2005.
4) The stable isotope composition of thalli collected in July 2005 from Otsuchi bay were very different from that of thalli collected in January and March 2006 from Okkirai Bay (details in the Results). It was not clear whether this was due to the sampling bay or time of the year. In order to clarify that, thalli were collected from

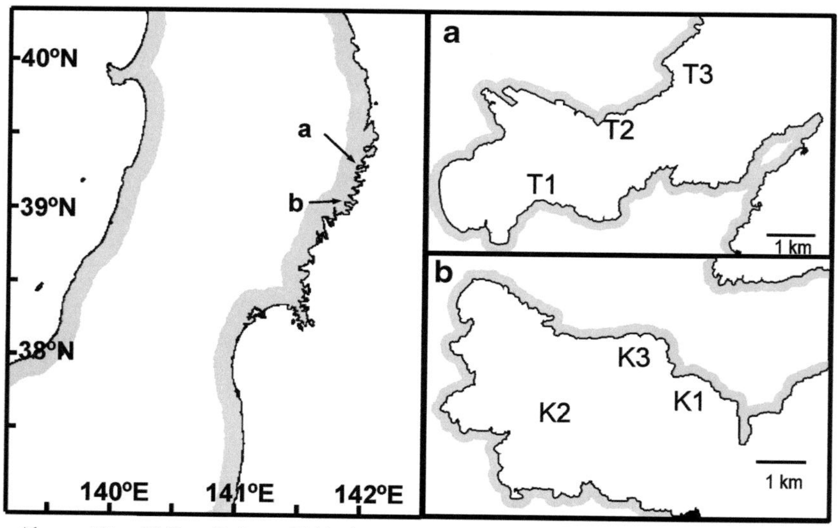

Fig. 1 Sampling locations. Above: Otsuchi Bay. Below: Okkirai Bay

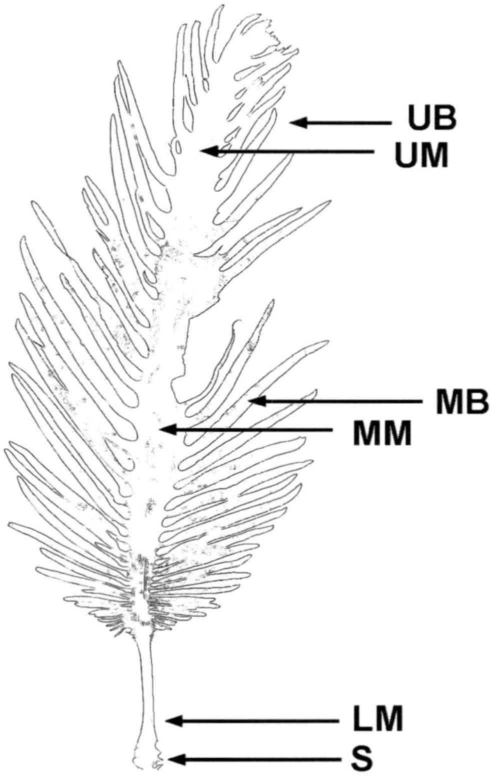

Fig. 2 Samples for nitrogen content and $\delta^{15}N$ measurements collected in a single thallus. UM: upper midrib; UB: upper pinnate blade; MM: middle midrib; MB: middle pinnate blade; LM: low midrib; S: sporophyll

both bays, 15 from Okkirai, station K3, and 18 from Otsuchi, station T2, in June 2006. All specimens were sampled for positions MM and MB, but only 5 from Okkirai Bay and 6 from Otsuchi Bay were sampled for LM and S (abbreviations in Fig. 2). At this time, it was not possible to collect samples from station K2, as it would be better to be consistent with the sampling in March 2006. This was because adult wild *U. pinnatifida* (which occurs in summer) grows in the intertidal zone, and station K2 was submerse, with cultured specimens during winter.

Measurements of nitrogen content and nitrogen stable isotope ratio were conducted in a stable isotope mass spectrometer ($Delta^{plus}$ Xp, Thermo Finnigan) equipped with an elemental analyzer (Flash EA, Thermo Finnigan) connected by a flow controller (Conflo III, Thermo Finnigan). Aliquots (about 1.5 mg) of the powdered samples folded with tin capsules were put in the elemental analyzer where they were burned. The N_2 from reduced combustion gases flowed into the mass spectrometer, and the stable isotope composition of N_2 was measured. Nitrogen content was calculated based on the integrated amplitude of the chromatogram generated by the mass spectrometer at the time of isotopic measurement. A working standard (L-alanine) was used as a reference for sample measurements for both nitrogen content and isotopic value. Three to five replications of the standard measurement for each series of measurements showed a precision around 0.2‰ for the stable isotope ratio and around 0.2% for nitrogen content. Stable isotope ratios are presented following the usual delta notation:

$$\delta^{15}N = \left[\left(R_{sample} - R_{reference}\right) - 1\right] \times 10^3\ ‰ \qquad (1)$$

Where: $\delta^{15}N$ is the value for nitrogen; ^{15} is the mass of the heavy stable isotope of nitrogen; R is the $^{15}N/^{14}N$ ratio; sample refers the analyzed sample; reference for nitrogen is atmospheric N_2 (Peterson and Fry 1987).

Parametric statistical tests were performed. The strategy was to do the tests for normality (goodness of fit test) and homoscedasticity (Barttlet test), followed by transformation of the data (square root or logarithm) and adjustment of sample size by random elimination when necessary, and then to conduct two-way ANOVA followed by Tukey's post hoc tests (Sokal and Rohlf 1994). Details of these preliminary statistical procedures for each data set are shown in Table 1. Regression lines had their correlation coefficients tested for significance (different from 0 or not) using a Student *t* test. Most of the statistical procedures were done using KyPlot software (Kyens Lab, Japan). The tests for normality and for regression line significance were done using the XLStats add-in for Microsoft Excel by Rodney Carr. The power achieved with ANOVA was calculated using the software G*Power 3 (Faul et al. 2007).

Results

Overview of the statistical results

Before analyzing the results of the measurements, it is important to evaluate the validity of the statistical tests. In Table 1, the results for the assumptions of population normality and homoscedasticity are shown for each two-way ANOVA test. *P* values for normality varied from 0.17 to 1.00 (Table 1). Higher values indicate higher chance of the population being normal. Only a few data sets showed $P<0.50$. Homoscedadsticity was the general rule (most of the $P<0.15$), but there were exceptions (Table 1). Although some violations occurred, the general condition was of data sets in accordance with the specifications for the test. ANOVA is regarded as a robust test, which means that small deviations of the assumptions do not influence the overall results (Sokal and Rohlf 1994). Thus, it was employed to analyze the data.

Table 1 Details of 2 way ANOVA

Common factors	Tested factors	Data number	Df	Transformation	NP (*P* range) HV (*P*)	*F* crit	*F*	*P*	Power (1 - β)
Month: June 2006	*1) Position*: LM, S, MM, MB *2) Bay*: Okkirai and Otsuchi	Ok Ot LM 5 6(1) S 5 6(1) MM 16 16(2) PB 16 16(1)	1) 3 2) 1 int.) 3 err.) 76	%N: Sqrt δ^{15}N: ln n adjustment (Otsuchi)	%N: 1.00~0.35 0.07 δ^{15}N: 1.00~0.56 0.00	1) 2.73 2) 3.97 int.) 2.73	1) 225.58; 53.77 2) 0.00; 8.01 int.) 0.55; 1.11	1) 0.00; 0.00 2) 0.97; 0.01 int.) 0.65; 0.35	1) 1.00; 1.00 2) 0.05; 0.84 int.) 0.17; 0.32
Bay: Otsuchi Bay	*1) Position*: LM, S, MM, MB *2) Year*: 2005 and 2006	05 06 LM 6 6 S 6 6 MM 6 6(12) PB 6 6(12)	1) 3 2) 1 int.) 3 err.) 40	%N: Sqrt δ^{15}N: none n adjustment (2006)	%N: 1.00~0.46 0.09 δ^{15}N: 1.00~0.46 0.00	1) 2.84 2) 4.09 int.) 2.84	1) 149.89; 20.07 2) 0.02; 3.73 int.) 0.77; 1.15	1) 0.00; 0.00 2) 0.90; 0.06 int.) 0.52; 0.34	1) 1.00; 1.00 2) 0.05; 0.54 int.) 0.24; 0.33
Bay: Okkirai Bay	*1) Position*: LM, S, MM, MB *2) Month*: March and June (Okkirai Bay)	03 06 LM 5 5 S 5 5 MM 16(3) 16 PB 16(3) 16	1) 3 2) 1 int.) 3 err.) 76	%N: Sqrt δ^{15}N: none n adjustment (March)	%N: 1.00~0.17 0.01 δ^{15}N: 1.00~0.71 0.00	1) 2.73 2) 3.97 int.) 2.73	1) 191.68; 2.35 2) 112.46; 147.53 int.) 10.30; 7.06	1) 0.00; 0.08 2) 0.00; 0.00 int.) 0.00; 0.00	1) 1.00; 0.62 2) 1.00; 1.00 int.) 1.00; 0.98
Bay: Otsuchi Bay *Month*: July 2005	*1) Position*: LM, S, MM, MB, UM, UB *2) Station*: T1, T2, T3.	2 for each group, no sample reduction, 18 groups.	1) 5 2) 2 int.) 10 err.) 18	None	%N: 1.00~1.00 0.37 δ^{15}N:1.00~1.00 0.03	1) 2.77 2) 3.55 int.) 2.41	1) 205.33; 6.70 2) 43.15; 18.66 int.) 2.73; 0.56	1) 0.00; 0.00 2) 0.00; 0.00 int.) 0.03; 0.82	1) 1.00; 1.00 2) 1.00; 1.00 int.) 0.99; 0.40

Abbreviations: LM, S, MM, PB, UM and UB: refer to Fig. 2; Ok: Okkirai; Ot: Otsuchi; T1, T2, T3: see Fig. 1; Df: degrees of freedom; int: interaction term; err: error; %N: nitrogen content; n: number of samples; NP: normality of population; HV: homogeneity of variances; sqrt: square root transformation; ln: natural logarithmic transformation; power: power of the test for α=0.05. The columns for *F*, *P* and Power are arranged with %N results to the left and δ^{15}N to the right.

The significance level of the ANOVA tests was set at 5% (α=0.05). The fixing of the significance level may bring concern on the power (1 - β) of the test. Then, a post-hoc calculation of test power was done for ANOVA. Results are in Table 1. Most of the tests showed high power (>0.95), and most of the tests with lower power were those with high *P* values for the treatment effects (Table 1). This is in fact the natural trend of the hypothesis test, and common sense supports it: if the sampled means are in fact very similar to the true mean, it is very difficult to reject H_0 when it is wrong (Sokal and Rohlf 1994). In other words, it is very difficult in these cases to have both small α and high power. Therefore, in this report we assume all the results of ANOVA as valid, regardless of the low power that they may have shown at times.

Spatial and temporal variations of nitrogen content

Nitrogen content averages and standard deviations of each part of all thalli are shown in Fig. 3a. Samples collected in January and March were very distinct from those collected in June and July. The high nitrogen content in the pinnate blade was remarkable, with values near 4% in the cold months, and about 2% in the warm months. A two-way ANOVA with season and part of the thallus as factors followed by Tukey *post hoc* tests confirmed the seasonal differences (Table 1, line 3). This procedure was conducted for the samples collected in March and June 2006 from Okkirai Bay. The difference between two seasons was highly significant, as well as the difference among the parts of the thallus. The interaction term was also significant because the internal distribution of nitrogen changes according to the season: in summer, S and MB were relatively homogeneous, as well as LM and MM; in winter, MB stood out and LM, S and MM were not so different. In other words, the sporophyll in March contained similar level of nitrogen content as in summer season, while all the other parts contained more nitrogen in winter than in summer.

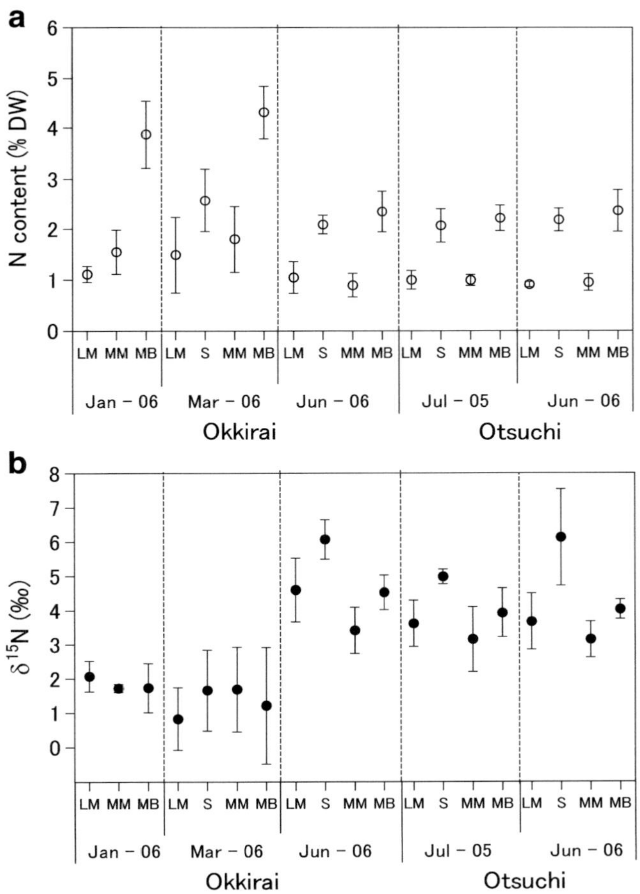

Fig. 3 Nitrogen content and $\delta^{15}N$ averages of the different parts of the thalli collected from Otsuchi Bay and Okkirai Bay in winter and summer. Vertical bars show standard deviation

Figure 3a also shows that nitrogen content was similar for samples collected from both bays. Two-way ANOVA confirmed this, as the influence of Bay was found to be not significant (Table 1, line 1). Looking only at Otsuchi Bay results, it is clear that data from July 2005 and June 2006 are also very similar and two-way ANOVA showed that year was not a significant factor influencing the results (Table 1, line 2).

Spatial and temporal variations of $\delta^{15}N$

Figure 3b shows the temporal and spatial variation of average $\delta^{15}N$ of each part of the thallus. As nitrogen content, $\delta^{15}N$ showed remarkable seasonal difference between samples collected in January and March and those collected in June and July. While in the former months the average values for the thalli were between approximately 1 and 2‰, in the latter the averages distributed between 3 and 6‰. A two-way ANOVA with season and part of the thallus as factors followed by Tukey *post hoc* tests was conducted for the samples collected in March and June 2006 from Okkirai Bay. Significant difference was observed between seasons and for interaction term, but not for parts of thallus (Table 1, line 3). The significance of the interaction term reflected the isotopic rearrangement in the thallus at the different seasons. From pairwise post hoc tests no significant differences were observed among parts of the thallus in March samples. When both July 2005 and June 2006 data from Otsuchi Bay were compared by a two-way ANOVA, significant difference was only observed among parts of thallus, and not for temporal or interaction term (Table 1, line 2). When both Okkirai and Otsuchi Bay data taken in June 2006 were compared, significant difference was observed among parts of thallus and also between bays, but not for interaction term (Table 1, line 1). In summer sporophyll showed higher $\delta^{15}N$ than the other parts

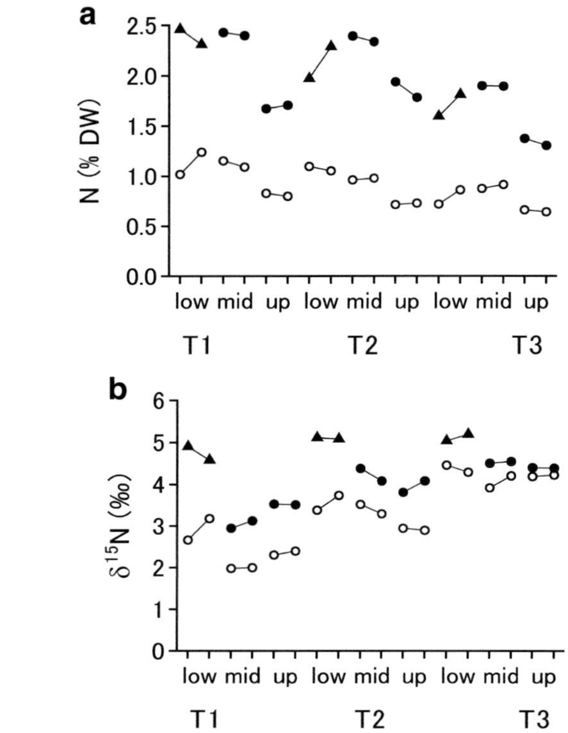

Fig. 4 Nitrogen content and δ^{15}N in different parts of the thallus collected from Otsuchi Bay in July 2005 at different sampling stations (T1, T2 and T3). Filled triangles: sporophyll. Filled circles: pinnate blades. Empty circles: midrib. Linked symbols are replicated samples

of thallus. Pinnate blade and lower midrib showed lower and similar values, although midrib tended to show the lowest values.

Spatial variation in a single bay

Thalli were collected at three stations (T1, T2 and T3) from inner to outer portion of Otsuchi Bay in July 2005 (Fig. 4a). The upper part of the thallus was comparatively poor in nitrogen than other parts, and the mid-rib contained less nitrogen at the same level through the stations. From a two-way ANOVA for nitrogen content with factors of stations and parts of the thallus, all the factors including interaction term were significant (Table 1, line 4).

In Fig. 4b, analogous analysis is presented for δ^{15}N. The δ^{15}N of sporophyll was very similar among the sampling stations, and higher than the other parts of the thallus. Midrib was the part of the thallus with lowest values, while pinnate blade showed intermediate values. All the parts except for sporophyll tended to show higher value from inner to outer part of the bay, and approached to the value of sporophyll at T3. When a two-way ANOVA for δ^{15}N with factors of stations and parts of the thallus was done, the two main factors were significant, while the interaction term was not (Table 1, line 4).

It may not be immediately clear the reason why the interaction term in two-way ANOVA was significant for nitrogen content and not for δ^{15}N, due to the large number of groups (18). However, a detailed observation of the results (Fig. 4) shows that the relative nitrogen content distribution in the thallus changes according to the station: at station T1, the sporophyll showed practically the same nitrogen content of middle pinnate blades, while at the other stations the value of the sporophyll was somewhat lower. Similar case was that of low midrib: at station T1, its value was practically the same of middle midrib, but at station T2 it was higher and at station T3 it was lower. On the other hand, the relative δ^{15}N distribution in the thallus was very uniform among the stations, with the exceptions of middle pinnate blade values at station T1, which were lower than those of upper pinnate blades, while at the other stations values were similar, and of middle midrib at station T2, which was higher than upper midrib, while at the other stations it was lower.

Size influence

In Fig. 3, the results of thalli with a wide size span are pooled together, and then it is necessary to verify whether or not this affected the results. Thus, nitrogen content and δ^{15}N were plotted against thallus size in Fig. 5. Thalli

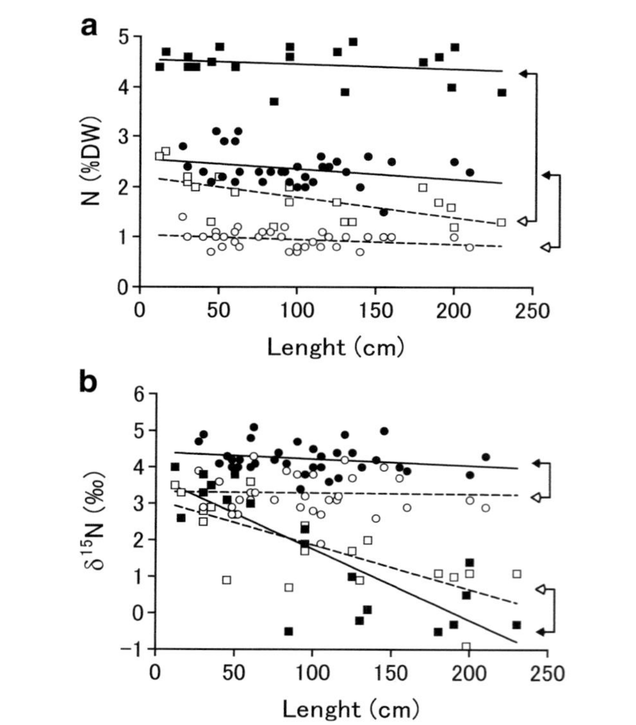

Fig. 5 Nitrogen content and δ^{15}N of MB and MM (abbreviations in Fig. 2) plotted against the thallus length. Squares: thalli collected in January and March 2006. Circles: thalli collected in June 2006. Open symbols: MM. Filled symbols: MB. Full lines are linear regression of the full symbols. Dashed lines are linear regressions of the open symbols

collected in March 2006 from Okkirai Bay were used to represent winter samples, and thalli collected in June 2006 from both Okkirai and Otsuchi Bays were used to represent summer. Data from both bays were pooled together due to their verified similarity (Fig. 3, Table 1). In Fig. 5, it can be readily seen that $\delta^{15}N$ of thalli sampled in winter was the variable with the clearest influence from the size. Accordingly, only the two regression lines of $\delta^{15}N$ of MM and MB against the length of the thallus were statistically significant at the level of $\alpha=0.01$. These two correlations showed respectively $r=-0.74$ and -0.82. Considering that the total length measurement is subject to a considerable error due to the loss of the older and upper parts, these correlation coefficients can be considered excellent. The correlation between total length and nitrogen content of MM in winter was also significant, but for $\alpha=0.05$, with $r=-0.63$. It should be noted that the relative distribution of nitrogen content and $\delta^{15}N$ at the different parts of the thallus were not affected by thallus length (Fig. 5).

Discussion

The role of fractionation

In this study the difference in nitrogen contents of the thalli from warm and cold months was very clear (Fig. 3a). A correspondence seems to exist between dissolved inorganic nitrogen concentration [DIN] and nitrogen content of *U. pinnatifida*: when [DIN] is high, nitrogen content in thallus is high, and when [DIN] is low, nitrogen content in thallus is also low (Kumura et al. 2006), which seems to be the normal characteristic for seaweeds in general (Lobban and Harrison 1994). Thus, the results found here suggest that [DIN] was high in winter and low in summer, independent of the bay (Fig. 3a). Previous investigations confirm this probable pattern for Otsuchi Bay (Iizumi et al. 1990) and Okkirai Bay (Sawada and Hayakawa 1997). Phytoplankton dynamics, wind and the mixing of warm Kuroshio and cold Oyashio currents shape nutrient concentration patterns in this region (Furuya et al. 1993; Sawada and Hayakawa 1997).

[DIN] is related to the $\delta^{15}N$ of the marine plant body during nitrogen uptake through isotope fractionation. Under nutrient rich condition, plants can fractionate the uptaken nitrogen and absorb the lighter isotope more intensively, which makes plant $\delta^{15}N$ lower than $\delta^{15}N$ of the source (Handley and Raven 1992). On the contrary, when the inorganic nitrogen supply is scarce, isotopic fractionation is small or does not exist, and the $\delta^{15}N$ in the primary producer becomes similar to the value of the source. This phenomenon has been verified for several groups of primary producers (Handley and Raven 1992). As previously discussed, it is very probable that [DIN] was high in winter (Fig. 3a). In this condition it is possible that *U. pinnatifida* fractionated the nitrogen stable isotopes during uptake, which may have led to its relatively low $\delta^{15}N$. On the other hand, [DIN] was probably low in summer (Fig. 3a), which suggests that fractionation was probably small or null during nitrogen assimilation. More intensive uptake leads to lower fractionation (Peterson and Fry 1987), and *U. pinnatifida* uptakes nitrogen more intensely when [DIN] is low (Campbell 1999). This reinforces the previous hypothesis of fractionation in winter but not in summer.

However, even in summer there was a difference in $\delta^{15}N$ among several parts of the thallus (Fig. 3b). It has been proposed in similar cases that the cause for the internal differences in stable isotope composition would be fractionation during the allocation of substances in the thallus (Stephenson et al. 1984; Fenton and Ritz 1989; Fredriksen 2003). Translocation of carbon from the pinnate blades to the meristem has been shown to exist for *U. pinnatifida* thallus using ^{14}C tracer (Wu and Meng 1997), and it is known that aminoacid transport exists in laminarian kelps (Lobban and Harrison 1994). This suggests that nitrogenated compounds may be transported from the pinnate blades to the meristem, and possibly to the sporophyll, when the thallus is matured. It could be suggested that pinnate blades absorb most of nitrogen and then they distribute it to the other parts of the thallus. Since the other parts of the thallus show different $\delta^{15}N$ value from pinnate blades, nitrogen could possibly be fractionated during the translocation process. If this was the case here, then the direction of the fractionation was the opposite between midrib and sporophyll (Fig. 3b), which may imply that different substances are transported for these two different parts of the thallus.

Nitrogen source influence

There was a clear correlation between $\delta^{15}N$ and the size of the thalli collected in March in Okkirai Bay (Fig. 5b). Size of the thallus in this sample probably indicated its age. Accordingly, the algae larger than 150 cm in length were all matured and smaller algae less than 70 cm in length were immature. Older algae had been absorbing nitrogen for longer time than the younger, and the nitrogen available for them was not the same. The increase of $\delta^{15}N$ from larger algae to smaller algae may indicate that the nitrogen source changed with the time. Data of the small thalli in winter tended to be similar to the data of thalli caught in summer, which may indicate that the main nitrogen source in winter shifted to another in summer.

The influence of time on $\delta^{15}N$ of DIN might have affected the internal $\delta^{15}N$ distribution in the thallus. Each part of the thallus is formed at a different time, and then they are likely to have access to different nitrogen sources. Particularly, the sporophyll should be formed after the other parts are already formed. In Fig. 4b, it is seen that the $\delta^{15}N$ composition of the sporophyll was the same for all the sampled thalli, but the values of the other parts of the thallus were not the same among the various samples. It is possible that these differences reflect a change in nitrogen sources showing different $\delta^{15}N$. If so, it seems that the sporophyll of all thalli appeared at approximately the same time, and were formed from the same nitrogen source.

Thalli from station T3 were smaller in length than those from other stations, which suggest that they could have been younger, even though they had sporophyll (Kurogi and Akiyama 1957).The relatively homogeneous $\delta^{15}N$ composition in the different parts of the thalli from station T3 may reflect the almost exclusive use of the newer nitrogen source for thallus formation. Thalli from station T3 were poorer in nitrogen (Fig. 4a), which can be expected for younger thalli in the environmental conditions of the study. In the same sense, the irregular $\delta^{15}N$ distribution in thalli in winter (Fig. 3b) may have been a reflex of nitrogen source change through the growth of the plant. This may have masked the "normal" internal $\delta^{15}N$ distribution pattern (the one observed in summer, Fig. 3b). The possible fractionation during uptake may also have contributed for the irregular $\delta^{15}N$ distribution in the thallus.

Implications for the use of U. pinnatifida as bio-indicator

In this study, explanations have been proposed for the internal $\delta^{15}N$ distribution changes according to season so that it could be possible to employ this information for a potential search for nitrogen sources using *U. pinnatifida*. Unfortunately, the possible causes of variation confound each other, especially in winter. It may be prudent to use the entire-thallus average $\delta^{15}N$ to obtain the average DIN $\delta^{15}N$ used up to that point in the life of the thallus. There may also be the potential to use $\delta^{15}N$ of the sporophyll to determine the sources of newer nitrogen in summer, because the sporophyll is younger than the other parts of the thallus and the internal $\delta^{15}N$ distribution was clear at this season. However, these are preliminary conclusions and laboratory studies should be conducted for a better understanding of the possible factors behind the internal $\delta^{15}N$ distribution in *U. pinnatifida*, and the consequences for the use of this species as an indicator of nitrogen sources.

Acknowledgements We are grateful to Dr. H. Otobe, from the Oceanic Research Institute of the University of Tokyo for the use of the installations at the center, to Mr. K. Morita, Mr. M. Kurosawa and Mr. J. Takada for sample collection in Otsuchi Bay, and to Mr. K. Niita and Mr. K. Niinuma for sample collection in Okkirai Bay.

References

Campbell SJ (1999) Uptake of ammonium by four species of macroalgae in Port Phillip Bay, Victoria, Australia. Mar Freshwat Res 50:515–520

Casas G, Scrosati R, Piriz ML (2004) The invasive elp *Undaria Pinnatifida* (Phaeophyceae, Laminariales) reduces native seaweed diversity in Nuevo Gulf (Patagonia, Argentina). Biol Invas 6:411–416

Costanzo SD, O'Donohue MJ, Dennison WC, Loneragan NR, Thomas M (2001) A New Approach for Detecting and Mapping Sewage Impacts. Mar Poll Bull 42:149–156

Faul F, Erdfelder E, Lang A-G, Buchner A (2007) G*Power 3: A flexible statistical power analysis program for the social, behavioral, and biomedical sciences. Behavior Research Methods 39:175–191

Fenton GE, Ritz DA (1989) Spatial variability of 13C:12C and D: H in *Ecklonia radiata* (C. Ag.) J. Agardh (Laminariales). Est Coast Shelf Sci 28:95–101

Fredriksen S (2003) Food web studies in a Norwegian kelp forest based on stable isotope (δ ^{13}C and δ ^{15}N) analysis. Mar Ecol Prog Ser 260:71–81

Furuya K, Takahashi K, Iizumi H (1993) Wind-dependent formation of phytoplankton spring bloom in Otsuchi Bay, a ria in Sanriku, Japan. J Oceanog 49:459–475

Handley LL, Raven JA (1992) The use of natural abundance of nitrogen isotopes in plant physiology and ecology. Pl, Cell Environ 15:965–985

Hay CH, Luckens PA (1987) The Asian kelp *Undaria pinnatifida* (Phaeophyta: Laminariales) found in a New Zealand harbour. New Zealand J Bot 25:329–332

Howarth RW, Billen G, Swaney D, Townsend A, Jaworski N, Lajtha K, Downing JA, Elmgren R, Caraco N, Jordan T, Berendse E, Freney J, Kudeyarov V, Murdoch P, Zhao-Liang Z (1996) Regional nitrogen budgets and riverine N & P fluxes for the drainages to the North Atlantic Ocean: Natural and human influences. Biogeochem 35:75–139

Iizumi H, Furuya K, Takeuchi I, Kustsuwada K, Tada A, Kawaguchi K (1990) Dynamics of nutrients and chlorophyll a at Otsuchi Bay: Analyses of the data of monthly observation. Marine Research Center Report 16:63–65

Kumura T, Yasui H, Mizuta H (2006) Nutrient requirement for zoospore formation in two alariaceous plants *Undaria pinnatifida* (Harvey) Suringar and *Alaria crassifolia* Kjellman (Phaeophyceae: Laminariales). Fisheries Sci 72:860–869

Kurogi M, Akiyama K (1957) Studies of ecology and culture of *Undaria pinnatifida* (Sur.) Hariot. Bull Tohoku Regional Fish Res Lab 10:95–117

Lindau CW, Delaune RD, Patrick WH Jr, Lambremont EN (1989) Assessment of stable nitrogen isotopes in fingerprinting surface water inorganic nitrogen sources. Water Air Soil Pollut 48:489–496

Lobban CS, Harrison PJ (1994) Seaweed ecology and physiology. Cambridge Univ Press, New York

McClelland JW, Valiela I (1998) Linking nitrogen in estuarine producers to land-derived sources. Limnology and Oceanography 43:577–585

Ogawa H (2004) Wakame. In: Masao O (ed) Biology and Technology of Economic Seaweeds. Uchida Rokakuho, Tokyo, pp 42–58

Ogawa H, Fujita M (1997) The effect of fertilizer application on farming of the seaweed *Undaria pinnatifida* (Laminariales, Phaeophyta). Phycol Res 45:113–116

Perez R, Lee JY, Juge C (1981) Observations sur la biologie de l'algue japonaise *Undaria pinnatifida* (Harvey) Suringar introduite accidentellement dans l'étang de Thau. Science et Peche 325:12

Peterson BJ, Fry B (1987) Stable isotopes in ecosystem studies. Ann Rev Ecol Syst 18:293–320

Sawada A, Hayakawa Y (1997) Mean seasonal changes and recent trends of dissolved inorganic nutrients in Okkirai Bay, Japan. Nippon Suisan Gakkaishi 63:152–159

Sokal RR, Rohlf FJ (1994) Biometry: the principles and practice of statistics in biological research. Freeman, New York

Stephenson RL, Tan FC, Mann KH (1984) Stable carbon isotope variability in marine macrophytes and its implications for food web studies. Mar Biol 81:223–230

Tokuda H, Ono M, Ogawa H (1986) The resources and cultivation of seaweeds. Midorishobou, Tokyo

Umezawa Y, Miyajima T, Yamamuro M, Kayanne H, Koike I (2005) Fine-scale mapping of land-derived nitrogen in coral reefs by $\delta^{15}N$ in macroalgae. Limnol Oceanog 47:1405–1416

Vitousek PM (1994) Beyond global warming: Ecology and global change. Ecology 75:1861–1876

Wu C, Meng J (1997) Translocation of assimilates in *Undaria* and its cultivation in China. Hydrobiologia 352:287–293

Antioxidant activities of phlorotannins isolated from Japanese Laminariaceae

Toshiyuki Shibata · Kanji Ishimaru · Shigeo Kawaguchi · Hiromichi Yoshikawa · Yoichiro Hama

Originally published in the Journal of Applied Phycology, Vol 20, No 5, 255–261.
DOI: 10.1007/s10811-007-9254-8

Abstract The antioxidant activities of brown algal phlorotannins were evaluated using the inhibition of phospholipid peroxidation in the liposome system, and by determining radical scavenging activities against the superoxide anion and 1,1-diphenyl-2-picrylhydrazyl (DPPH). Oligomers of phloroglucinol (1,3,5-trihydroxybenzene), eckol (a trimer), phlorofucofuroeckol A (a pentamer), dieckol and 8,8′-bieckol (hexamers), isolated from the Laminarian brown algae *Eisenia bicyclis*, *Ecklonia cava* and *Ecklonia kurome*, showed potent inhibition of phospholipid peroxidation at 1 μM in the liposome system. The phlorotannins had significant radical scavenging activities against the superoxide anion (50% effective concentration values: 6.5–8.4 μM) and DPPH (50% effective concentration values: 12–26 μM), and were more effective compared to ascorbic acid and α-tocopherol. For the purpose of using phlorotannins as functional food ingredients, the antioxidant activity of a complex of crude phlorotannins and soybean protein was examined. The complex had a pronounced DPPH-radical scavenging activity. These results suggest that phlorotannins are potent anti-inflammatory substances, and that the Laminariaceous brown algae, which are abundant in phlorotannins, may be useful as a new functional foodstuff or supplement with anti-inflammatory activity.

T. Shibata (✉)
Laboratory of Quality in Marine Products,
Department of Life Sciences,
Graduate School of Bioresources, Mie University,
1577 Kurimamachiya,
Tsu 514-8507, Japan
e-mail: shibata@bio.mie-u.ac.jp

K. Ishimaru
Department of Applied Biological Sciences,
Faculty of Agriculture, Saga University,
1Honjo,
Saga 840-8502, Japan

S. Kawaguchi
Department of Animal and Marine Bioresource Sciences,
Faculty of Agriculture, Kyushu University,
6-10-1 Hakozaki, Higashi-ku,
Fukuoka 812-8581, Japan

H. Yoshikawa
Department of Life, Environment and Materials Sciences,
Faculty of Engineering, Fukuoka Institute of Technology,
3-30-1 Wajiro-Higashi, Higashi-ku,
Fukuoka 811-0295, Japan

Y. Hama
Department of Applied Biochemistry and Food Sciences,
Faculty of Agriculture, Saga University,
1 Honjo,
Saga 840-8502, Japan

Keywords Brown algae · DPPH-radical · Liposome · Phlorotannin–soy bean protein complex · Phaeophyta · Polyphenols · Superoxide anion radical

Introduction

It is well-known that the oxidative damage to cell membranes, DNA and/or proteins induced by free radicals and reactive oxygen species may play a causative role in ageing and lifestyle diseases. Therefore, it is believed that inhibiting such damage by taking antioxidant supplements could be an effective therapy for such diseases. Recent epidemiological and clinical studies have indicated that consumption of plant foods and drinks (e.g., tea, red wine and soybean products) moderate the risk of lifestyle diseases

(Messina et al. 1994; Sasazuki et al. 2000; Wu et al. 2001). The beneficial effects of such foods and drinks are thought to be due to the presence of polyphenols with antioxidant activity, and many active substances (e.g., catechins, resveratrol and isoflavones) have been isolated (Jang et al. 1997; Esaki et al. 1999a, b; Zhao et al 2001; Havsteen 2002; Aggarwal and Shishodia 2006). Polyphenols—common secondary metabolites—are found universally in plants (Haslam 1989; Herbert 1989). However, in stark contrast to polyphenols in terrestrial plants, the literature on the properties of marine algal polyphenols from a human physiological viewpoint is very sparse. Marine algal polyphenols—phlorotannins—which have been found to exist only within brown algae, are formed by the polymerization of phloroglucinol (Ragan and Glombitza 1986; Nakamura et al. 1996). Previously, we isolated the phlorotannins eckol (a phloroglucinol trimer), phlorofucofuroeckol A (a pentamer), dieckol and 8,8′-bieckol (hexamers) from the Japanese Laminariaceous brown algae *Eisenia bicyclis* (Kjellman) Setchell, *Ecklonia cava* Kjellman and *Ecklonia kurome* Okamura, and reported their distribution (Shibata et al. 2004) and physiological properties (Shibata et al. 2002a, b, 2003). However, the effectiveness of phlorotannins against free radicals is still obscure.

The aim of this study was to evaluate the antioxidant and free radical scavenging activities of phlorotannins isolated from Laminariaceous brown algae. As a first step, the activities of the phlorotannins were determined using the inhibition of phospholipid peroxidation in the liposome system, and their scavenging activities against the superoxide anion and 1,1-diphenyl-2-picrylhydrazyl (DPPH) radicals. Thereafter, a phlorotannin–soybean protein complex was prepared and its function as an antioxidant foodstuff assessed.

Materials and methods

The brown algae *Eisenia bicyclis*, *Ecklonia cava* and *Ecklonia kurome* were collected from along the coasts of the Itoshima Peninsula (33°37′N, 130°10′E) in Fukuoka Prefecture and the Miura peninsula (35°08′N, 130°36′E) in Kanagawa prefecture, Japan. The algae were washed with filtrated seawater, air-dried and pulverized, and the algal powders were then stored at −40°C until use.

Catechin and epigallocatechin gallate (EGCG) were kindly donated by Kurita Water Industries (Tokyo, Japan).

Extraction of phlorotannins from algal powders was carried out according to the modified method of Folch et al. (1957) described in previous reports (Nagayama et al. 2002, 2003; Shibata et al. 2006b). Each of the phlorotannins in the crude extracts was purified on a column (1.5 cm i.d. × 150 cm) of Wakogel C-300HG (Wako, Tokyo, Japan) with $CHCl_3$:methanol (MeOH):water (80:20:2, v/v) as the eluent (Shibata et al. 2002b, 2003). The purity of the phlorotannins was checked by thin-layer chromatography (TLC) and reversed-phase three-dimensional high-performance liquid chromatography (RP-3D-HPLC). TLC plates (Silica Gel 60 F_{254}, 0.25 mm; Merck, Darmstadt, Germany), which had been activated at 120°C for 5 min before use, were developed with $CHCl_3$:MeOH:water:acetic acid (65:25:4:3, v/v) (Shibata et al. 2002a, 2004, 2006b). Fifty percent H_2SO_4 (Krebs et al. 1969), vanillin-H_2SO_4 (Krebs et al. 1969), and paprika pigment (Nakamura et al. 1991), were used as the detecting agents for organic compounds, for phenolic compounds, and for antioxidant substances, respectively. The HPLC system consisted of a Waters 600 pump (Waters, Milford, MA), a MCPD-3600 UV detector (Otsuka Electronics, Osaka, Japan) and an Inertsil ODS-3 column (4.6 mm i.d. ×

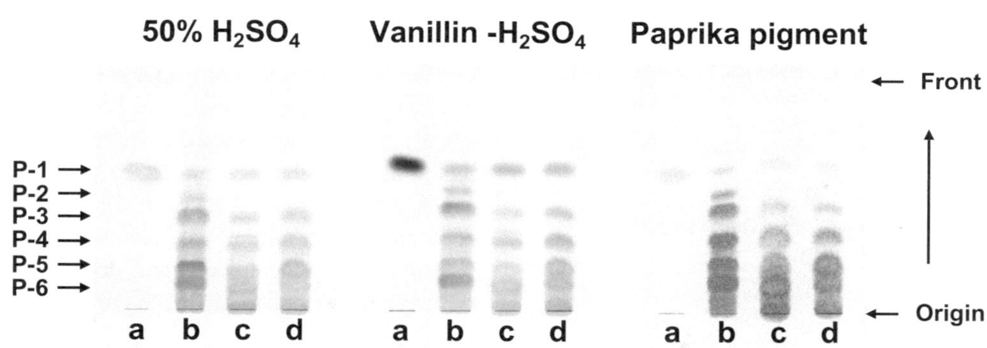

Fig. 1 Thin-layer chromatography (TLC) of phlorotannins from the Laminariaceous brown algae *Eisenia bicyclis*, *Ecklonia cava* and *Ecklonia kurome*. *P-1* Phloroglucinol, *P-2* phloroglucinol tetramer, *P-3* eckol, *P-4* phlorofucofuroeckol A, *P-5* dieckol, *P-6* 8,8′-bieckol. *a* Phloroglucinol, *b* crude phlorotannins from *Eisenia bicyclis*, *c* crude phlorotannins from *Ecklonia cava*, *d* crude phlorotannins from *Ecklonia kurome*

250 mm, GL Science, Tokyo, Japan). Elution was performed at a flow-rate of 1.0 mL min^{-1} with a linear gradient from 30% MeOH/water (eluent A) to 100% MeOH (eluent B) for 20 min, and was maintained for 25 min (Shibata et al. 2002a, b, 2003, 2004, 2006b). The photodiode array detector was set to a wavelength range of 190–400 nm.

Antioxidant activity in the liposome system

A large unilamellar vesicle was prepared by the extrusion method described by MacDonald et al. (1991) with some modifications. Briefly, egg yolk lecithin (Wako, Tokyo, Japan) in chloroform (10 mg mL^{-1}) was dried under nitrogen gas and kept under high vacuum for 1 h. The lipid was suspended in 10 mM phosphate buffer (pH 7.4) and freeze-thawed 5 times in a MeOH bath. The multilamellar vesicle suspension was transferred into a Lipofast Apparatus (Avestin, Ottawa, Canada) and passed thorough a polycarbonate membrane (pore size 100 nm) 19 times to obtain unilamellar liposomes (10 mg mL^{-1}). The antioxidant activity of the sample to be tested in the liposome system was measured by the method of Esaki et al. (1997). The test samples were dissolved in 100 μL ethanol (EtOH). In the control test, 100 μL EtOH was used instead of the sample solution. The amount of thiobarbituric acid-reacting substance was determined by measuring the absorbance at 535 nm.

Radical scavenging activities against DPPH and superoxide anion

The DPPH-radical scavenging activity was measured using the modified method of Yamaguchi et al. (1998) described in a previous report (Shibata et al. 2006a). The samples were dissolved in 100 μL EtOH, and the solution was mixed with 100 μL 300 mM DPPH (Wako) in EtOH. In controls, 100 μL EtOH was used instead of the sample solution. After the reaction mixture had been stored at room temperature for 30 min, its absorbance was measured at 520 nm. The superoxide anion radical scavenging activity was determined with the SOD assay kit-WST (Dojindo Laboratories, Kumamoto, Japan) (Ukeda et al. 1999). Samples were dissolved in 20 μL EtOH. In controls, 20 μL EtOH was used instead of the sample solution. The production of water-soluble tetrazolium salt was compared with that of controls.

The 50% effective concentration (EC_{50}) value was defined as the amount of sample required to scavenge 50% DPPH radical or superoxide anion radical. Terrestrial polyphenols (resveratrol, catechin and EGCG) and antioxidant vitamins (ascorbic acid and α-tocopherol) were used as positive controls.

Preparation of phlorotannin–soybean protein complex

A phlorotannin–soybean protein complex was prepared by the method of Ishimaru and Nonaka (2001). Briefly, defatted soybean (200 g) was steeped in water (1 L) and

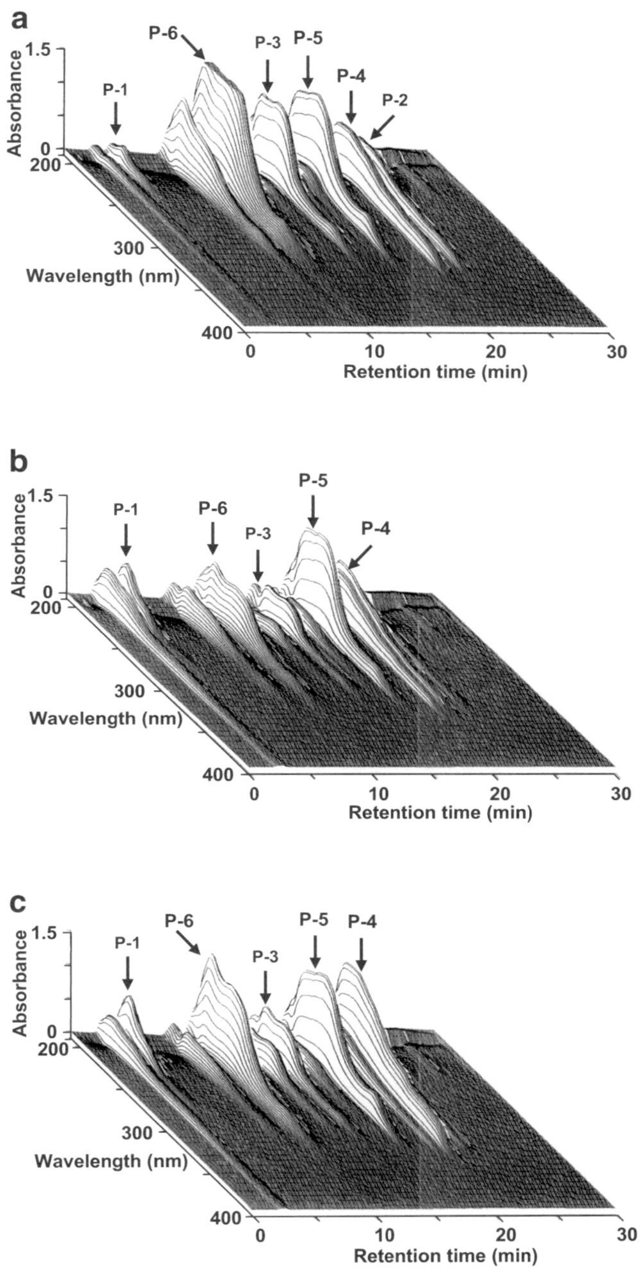

Fig. 2 Reversed-phase three-dimensional high-performance liquid chromatography (RP-3D-HPLC) of crude phlorotannins from the Laminariaceous brown algae *Eisenia bicyclis* (**a**), *Ecklonia cava* (**b**) and *Ecklonia kurome* (**c**). *P-1* Phloroglucinol, *P-2* phloroglucinol tetramer, *P-3* eckol, *P-4* phlorofucofuroeckol A, *P-5* dieckol, *P-6* 8,8′-bieckol

autoclaved at 121°C for 20 min. After filtration, the soybean protein extract (500 mL) was mixed with 100 mL of the crude phlorotannins from *Eisenia bicyclis* in water (10 mg/mL), and the pH of the mixture was adjusted to 4.5 with 2 *N* HCl. The phlorotannin–soybean protein complex that precipitated in the acidic solution was lyophilized. The amount of phlorotannins conjugated to soybean protein was determined by RP-3D-HPLC. The conjugated amount (%) was calculated using the following formula:

Conjugation ratio (%) = (1− the amount of phlorotannins detected in the supernatant of the phlorotannin–soybean protein complex / the initial content of each compounds in the crude phlorotannins) × 100.

The DPPH-radical scavenging activity of the complex was measured using the method of Huang et al. (2004). In the control test, lyophilized soybean protein extract was used instead of the sample.

Results

The yield of crude phlorotannins from *Eisenia bicyclis*, *Ecklonia cava* and *Ecklonia kurome* was about 3% of the algal powder (dry weight) for all three algae. The crude phlorotannins from the brown algae were analyzed by TLC and RP-3D-HPLC. As shown in Fig. 1, five to six spots detected in all three algal fractions were stained by paprika pigment, indicating the presence of antioxidant substances. Judging from previous reports (Nakamura et al. 1996; Shibata et al. 2004, 2006b), the spots can be identified as follows: phloroglucionl (P-1), unknown phloroglucinol tetramer (P-2), eckol (P-3), phlorofucofuroeckol A (P-4), dieckol (P-5) and 8,8′-bieckol (P-6). In RP-3D-HPLC analysis, peaks in the fractions of crude phlorotannins were identified by comparing their retention times with those of purified phlorotannins and their on-line ultraviolet spectra. Within 30 min, a distinct separation of each of the components in the brown algal extracts was obtained by a linear gradient system of 30–100% MeOH (Fig. 2). Each individual phlorotannin was purified by silicic acid column chromatography. The purity of each oligomer in this experiment exceeded 90%.

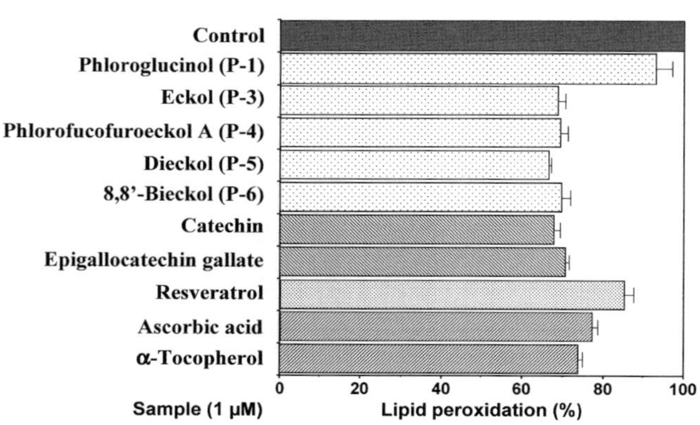

Fig. 3 Antioxidant activities of phlorotannins, terrestrial polyphenols and antioxidant vitamins in the liposome system. Lipid peroxidation was induced by 4 mM 2, 2′-azobis (2-amidinopropane) dihydrochloride. The analytical data are presented as mean ± SD (*n*=5)

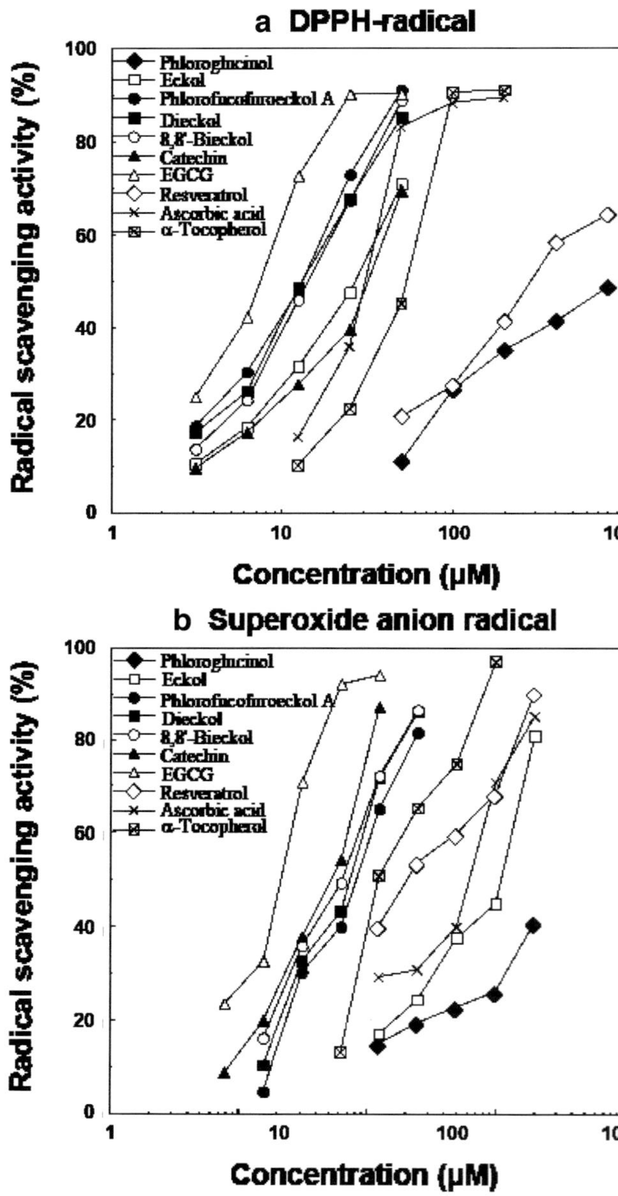

Fig. 4a,b Radical scavenging activities of phlorotannins, terrestrial polyphenols and antioxidant vitamins. **a** 1-Diphenyl-2-picrylhydrazyl (DPPH)-radical, **b** superoxide anion radical. All analytical data are the means of three determinations

Table 1 Conjugation of phlorotannins to soybean protein. Reversed-phase three-dimensional high-performance liquid chromatography (RP-3D-HPLC) conditions are described in Materials and methods. All analytical data in the table are the mean of three determinations

Compound	Molecular weight	Conjugation (%)
Eckol (P-3)	372	27.5
Unknown phloroglucinol tetramer (P-2)	478	98.2
Phlorofucofuroeckol A (P-4)	602	100
Dieckol (P-5)	742	100
8,8′-Bieckol (P-6)	742	100

Antioxidant activity of phlorotannins in the liposome system

Figure 3 shows the inhibitory effects of phlorotannins, terrestrial polyphenols and antioxidant vitamins on phospholipid peroxidation in a liposome system initiated by 2,2′-azobis (2-amidinopropane) dihydrochloride. At a final concentration of 1 μM, eckol, phlorofucofuroeckol A, dieckol and 8,8′-bieckol each showed about 30% inhibition of phospholipid peroxidation. Thus, the oligomeric phloroglucinols were found to possess potent antioxidant activity, comparable in magnitude to that of well-known antioxidants such as catechins and vitamins.

Radical scavenging activities of phlorotannins

A further experiment was carried out using the DPPH-radical and the superoxide anion radical scavenging assays to determine the antioxidant activity of the phlorotannins. With the exception of phloroglucinol and resveratrol, the phlorotannins, terrestrial polyphenols and antioxidant vitamins scavenged the DPPH-radical in a dose-dependent manner (Fig. 4a). The EC_{50} values of the phlorotannins eckol, phlorofucofuroeckol A, dieckol and 8,8′-bieckol were 26, 12, 13 and 15 μM, respectively. In contrast, those of catechin, EGCG, ascorbic acid and α-tocopherol were 32, 7.4, 30, and 52 μM, respectively. The phlorotannins were potent free radical scavengers and about twice as effective as catechin, ascorbic acid and α-tocopherol. Figure 4b shows the superoxide anion scavenging activities of the phlorotannins, terrestrial polyphenols and antioxidant vitamins. The EC_{50} values of eckol, phlorofucofuroeckol A, dieckol, 8,8′-bieckol, catechin, EGCG, resveratrol, ascorbic acid and α-tocopherol were 107, 8.4, 7.6, 6.5, 5.2, 2.1, 21, 62, and 12 μM, respectively. With the exception of eckol, the phlorotannins were found to have very potent superoxide anion radical scavenging activity, being about 2–10 times more active than ascorbic acid and α-tocopherol.

Phlorotannin–soybean protein complex

In order to use phlorotannins as functional food ingredient, a complex of crude phlorotannins and soybean protein was prepared. As shown in Table 1, phlorotannins of high molecular weight—the pentamer and hexamers of phloroglucinol—showed pronounced affinity for the soybean protein. About 5 g phlorotannin–soybean protein complex was obtained from 1 g crude phlorotannins and 200 g defatted soybean. The DPPH-radical scavenging activity of the complex was about four times stronger than that of the lyophilized soybean protein extract alone (Fig. 5).

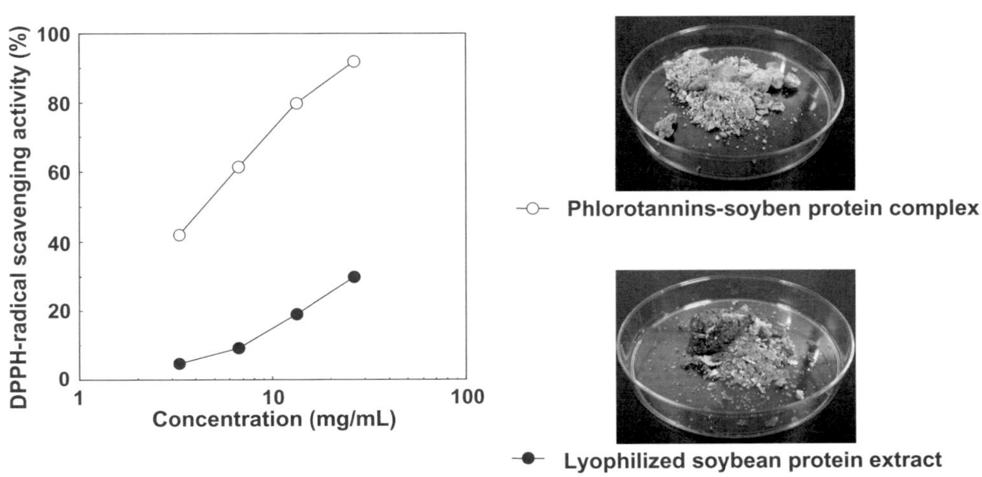

Fig. 5 DPPH-radical scavenging activity of phlorotannin–soybean protein complex. The complex was steeped in acetate buffer (pH 4.5) overnight at room temperature before being homogenized and centrifuged. The supernatant was mixed with the DPPH in EtOH. All analytical data are the means of three determinations

Discussion

A main target of free radical damage in the body is the cell membrane, which contains abundant unsaturated lipids. The liposome has been used extensively as a model of the cell membrane for in vitro biochemical research. Therefore, the unilamellar liposome lipid/aqueous system was used to evaluate the antioxidative properties of phlorotannins. In the present study, the oligomeric phloroglucinols were found to possess potent antioxidant activity in the liposome system. Catechins have been reported to show strong interactions with lipid bilayers and to exert various physiological activities (Kondo et al. 1999). Like catechins, it may be that phlorotannins have high affinity for lipid bilayers and inhibit phospholipid peroxidation. Most of the phlorotannins tested had more potent reducing activities toward the DPPH and superoxide anion radicals than antioxidant vitamins. Generally, it is believed that the radical scavenging activities of polyphenols such as flavonoids are ascribed to the phenolic hydroxy groups attached to the ring structure (Kondo et al. 1999). Although relationships between the structure of phlorotannins and their radical scavenging activities are unclear, it may be that phenolic hydroxy groups attached to the eckol skeleton play an important role in the radical scavenging activities. The phlorotannin–soybean protein complex had pronounced DPPH-radical scavenging activity; the activity was higher than that of soybean protein, which is rich in isoflavones. This complex may have a potential use as a new functional material in the food industry.

Previously, we described the inhibitory effect of phlorotannins on hyaluronidase (Shibata et al. 2002b) and lipoxygenases (Shibata et al. 2003). We also showed that phlorotannins had no toxicity following oral administration to mice (Nagayama et al. 2002). In addition, we reported that the yield of phlorotannins among *Eisenia bicyclis*, *Ecklonia cava* and *Ecklonia kurome* was about 3.0–4.0% of the algal powder (dry weight), with no clear seasonal variation in phlorotannin content being observed (Shibata et al. 2004). Recently, Sugiura et al. (2006) and Tsukui et al. (2006) isolated phlorotannins from the Laminarian brown alga *Eisenia arborea*. Laminariaceous brown algae are one of the most common algae in Japan. Thus, the evidence obtained in the previous and present studies suggests that phlorotannins are potent anti-inflammatory substances and that the Laminarian brown algae, which are rich in phlorotannins, may be useful as a novel functional foodstuff or supplement with anti-inflammatory activity.

Acknowledgments We thank grateful to Dr. Takashi Nakamura of the Graduate school of Kyushu University for his cooperation throughout this study. This work was supported in part by the Comprehensive Research Laboratory of Fukuoka Institute of Technology.

References

Aggarwal BB, Shishodia S (2006) Molecular targets of dietary agents for prevention and therapy of cancer. Biochem Pharmacol 71:1397–1421

Esaki H, Onozaki H, Kawakishi S, Osawa T (1997) Antioxidant activity and isolation from soybeans fermented with *Aspergillus* spp. J Agric Food Chem 45:2020–2024

Esaki H, Watanabe R, Onozaki H, Kawakishi S, Osawa T (1999a) Formation mechanism for potent antioxidative *o*-dihydroxyisoflavones in soybeans fermented with *Aspergillus saitoi*. Biosci Biotechnol Biochem 63:851–858

Esaki H, Kawakishi S, Morimitsu Y, Osawa T (1999b) New potent antioxidative *o*-dihydroxyisoflavones in fermented Japanese soybean products. Biosci Biotechnol Biochem 63:1637–1639

Folch J, Less M, Sloan Stanley GH (1957) A simple method for the isolation and purification of total lipids from animal tissues. J Biol Chem 226:497–509

Haslam E (1989) Plant polyphenols, Cambridge University Press, Cambridge

Havsteen BH (2002) The biochemistry and medical significance of the flavonoids. Pharmacol Ther 96:67–202

Herbert RB (1989) The biosynthesis of secondary metabolites, 2nd edn. Chapman and Hall, New York

Huang S-M, Shigetomi N, Tanaka N, Terahara N, Fujioka T, Yoshida M, Ishimaru K (2004) Preparation and functional evaluation of plant polyphenol-bean protein complex. Nippon Shokuhin Kagaku Kogaku Kaishi 51:626–632

Ishimaru K, Nonaka G (2001) Rapid purification of catechins using soybean protein. Nippon Shokuhin Kagaku Kogaku Kaihi 48:664–670

Jang M, Cai L, Udeani GO, Slowing KV, Thomas CF, Beecher CWW, Fong HHS, Farnsworth NR, Kinghorn AD, Mehta RG, Moon RC, Pezzuto JM (1997) Cancer chemopreventive activity of resveratrol, a natural product derived from grapes. Science 275:218–220

Kondo K, Kurihara M, Miyata N, Suzuki T, Toyoda M (1999) Mechanistic studies of catechins as antioxidants against radical oxidation. Arch Biochem Biophys 362:79–86

Krebs KG, Heusser D, Wimmer H (1969) Spray reagents. In: Stahl E (ed) Thin-layer chromatography. Springer, Heidelberg, pp 854–909

MacDonald RC, MacDonald RI, Menco BPhM, Takeshita K, Subbarao NK, Hu L-R (1991) Small-volume extrusion apparatus for preparation of large, unilamellar vesicles. Biochim Biophys Acta 1061:297–303

Messina MJ, Persky V, Setchell KDR, Barnes S (1994) Soy intake and cancer risk: a review of the in vitro and in vivo data. Nutr Cancer 21:113–131

Nagayama K, Iwamura Y, Shibata T, Hirayama I, Nakamura T (2002) Bactericidal activity of phlorotannins from the brown alga *Ecklonia kurome*. J Antimicrob Chemother 50:889–893

Nagayama K, Shibata T, Fujimoto K, Honjo T, Nakamura T (2003) Algicidal effect of phlorotannins from the brown alga *Ecklonia kurome* on red tide microalgae. Aquaculture 218:601–611

Nakamura T, Mukaiyama T, Nagayama K (1991) A rapid and simple method to detect antioxidant substances on a thin-layer chromatography plate. J Fac Agric Kyushu Univ 36:93–98

Nakamura T, Nagayama K, Uchida K, Tanaka R (1996) Antioxidant activity of phlorotannins isolated from the brown alga *Eisenia bicyclis*. Fisheries Sci 62:923–926

Ragan MA, Glombitza K-W (1986) Phlorotannins, brown algal polyphenols. Progr Phycol Res 4:129–241

Sasazuki S, Kodama H, Yoshimasu K, Liu Y, Washio M, Tanaka K, Tokunaga S, Kono S, Arai H, Doi Y, Kawano T, Nakagaki O, Takada K, Koyanagi S, Hiyamuta K, Nii T, Shirai K, Ideishi M, Arakawa K, Mohri M, Takeshita A (2000) Relation between green tea consumption and the severity of coronary atherosclerosis among Japanese men and women. Ann Epidemiol 10:401–408

Shibata T, Yamaguchi K, Nagayama K, Kawaguchi S, Nakamura T (2002a) Inhibitory activity of brown algal phlorotannins against glycosidases from the viscera of the turban shell *Turbo cornutus*. Eur J Phycol 37:493–500

Shibata T, Fujimoto K, Nagayama K Yamaguchi K, Nakamura T (2002b) Inhibitory activity of brown algal phlorotannins against hyaluronidase. Int J Food Sci Technol 37:703–709

Shibata T, Nagayama K, Tanaka R, Yamaguchi K, Nakamura T (2003) Inhibitory effects of brown algal phlorotannins on secretory phospholipase A_2s, lipoxygenases and cyclooxygenases. J Appl Phycol 15:61–66

Shibata T, Kawaguchi S, Hama Y, Inagaki M, Yamaguchi K, Nakamura T (2004) Local and chemical distribution of phlorotannins in brown algae. J Appl Phycol 16:291–296

Shibata T, Nagayama K, Kawaguchi S, Hama Y (2006a) Antioxidant and radical scavenging activities of brown algal phlorotannins. ITE Lett Batter New Tecnol Med 7:69–72

Shibata T, Hama Y, Miyasaki T, Ito M, Nakamura T (2006b) Extracellular secretion of phenolic substances from living brown algae. J Appl Phycol 18:787–794

Sugiura Y, Matsuda K, Yamada Y, Nishikawa M, Shioya K, Katsuzaki H, Imai K, Amano H (2006) Isolation of a new anti-allergic phlorotannin, phlorofucofuroeckol-B, from an edible brown alga, *Eisenia arborea*. Biosci Biotechnol Biochem 70:2807–2811

Tsukui K, Kakinuma M, Amano H, Uchiyama Y, Shioya K, Katsuzaki H, Imai K (2006) Identification of elastase inhibitors from brown alga, *Eisenia arborea*. ITE Lett Batter New Tecnol Med 7:616–619

Ukeda H, Kawana D, Maeda S, Sawamura M (1999) Spectrophotometric assay for superoxide dismutase based on the reduction of highly water-soluble tetrazolium salts by xanthine–xanthine oxidase. Biosci Biotechnol Biochem 63:485–488

Wu JM, Wang ZR, Hsieh TC, Bruder JL, Zou JG, Huang YZ (2001) Mechanism of cardioprotection by resveratrol, a phenolic antioxidant present in red wine. Int J Mol Med 8:3–17

Yamaguchi T, Takamura H, Matoba T, Terao J (1998) HPLC method for evaluation of the free radical-scavenging activity of foods by using 1, 1-diphenyl-2-picrylhydrazyl. Biosci Biotechnol Biochem 62:1201–1204

Zhao B-L, Guo Q, Xin W-J (2001) Free radical scavenging by green tea polyphenols In: Packer L (ed) Flavonoids and other polyphenols. Methods in enzymology, vol 335. Academic, New York, pp 217–231

Algicidal activity of polyunsaturated fatty acids derived from *Ulva fasciata* and *U. pertusa* (Ulvaceae, Chlorophyta) on phytoplankton

Mochammad Amin Alamsjah · Shotaro Hirao · Fumito Ishibashi · Tatsuya Oda · Yuji Fujita

Originally published in the Journal of Applied Phycology, Vol 20, No 5, 263–270.
DOI: 10.1007/s10811-007-9257-5 © Springer Science + Business Media B.V. 2007

Abstract Isolation of algicidal compounds from *Ulva fasciata* revealed that the algicidal substances were the polyunsaturated fatty acids (PUFAs) as hexadeca-4,7,10,13-tetraenoic acid (HDTA) C16:4 n-3, octadeca-6,9,12,15-tetraenoic acid (ODTA) C18:4 n-3, α-linolenic acid (ALA) C18:3 n-3 and linoleic acid (LA) C18:2 n-6. The fatty acid composition of four species of Ulvaceae (*U. fasciata*, *U. pertusa*, *U. arasakii* and *U. conglobota*) was analyzed by capillary gas chromatography to investigate the relationship with the algicidal activity. The results indicate that highly algicidal species, *U. fasciata* and *U. pertusa*, showed higher contents of C16:4 n-3, C18:3 n-3, and C18:4 n-3. Concentrations of these PUFAs released from the seaweed in the culture medium were also analyzed. These PUFAs were found to be significantly active against *Chattonella antiqua, C. marina, Fibrocapsa japonica*, *Heterosigma akashiwo*, *Karenia mikimotoi*, moderately effective against *Heterocapsa circularisquama*, *Prorocentrum minimum*, *P. sigmoides*, *Scrippsiella trochoidea*, whereas low effective against *Alexandrium catenella* and *Cochlodinium polykrikoides*. It is suggested that the PUFAs are useful mitigation agents to remove several harmful effects without causing detrimental effects on surrounding marine living organisms.

Keywords Algicidal compound · Red tide phytoplankton · *Ulva fasciata* · *Ulva pertusa*

M. A. Alamsjah · S. Hirao · F. Ishibashi (✉) · Y. Fujita
Graduate School of Science and Technology, Nagasaki University,
1-14 Bunkyo-machi,
Nagasaki 852-8521, Japan
e-mail: fumito@net.nagasaki-u.ac.jp

T. Oda
Faculty of Fisheries, Nagasaki University,
1-14 Bunkyo-machi,
Nagasaki 852-8521, Japan

Introduction

The red tide phytoplankton known as harmful algal blooms (HAB) is reported to occur in world-wide, with warnings for public health and fisheries industries. Chemical agents such as copper sulfate (Steidinger 1983) and hydrogen peroxide (Ryu et al. 1998) have been proposed as algicidal substances but are known to have a wide-spectrum in sterilizing effects. Viruses (Garry et al. 1998), bacteria (Imai et al. 1995), planktonic ciliates (Kamiyama 2002) and heterotrophic dinoflagellates (Nakamura 2002) also have been proposed as biological control agents. However, these methods also have the potential of disastrous environmental consequences (Jeong et al. 2000).

Several studies have been done aimed at development of a novel, low cost, environmentally benign method to control HAB by utilizing macroalgae. Studies on the allelochemicals of macroalgae to microalgae were initially done by Kakisawa et al. (1988), in which the algicidal substance of octadeca-6,9,12,15-tetraenoic acid (ODTA) C18:4 n-3 from the brown alga *Cladosiphon okamuranus* was demonstrated to be toxic against 21 phytoplankton species. Nagayama et al. (2003) have investigated toxic effect of the brown alga *Ecklonia kurome* on red tide phytoplankton and several phlorotannins were isolated as the active principles. Jeong et al. (2000) also screened extracts of macroalgae from the coast of Korea for growth inhibition of *Cochlodinium polykrikoides* and found that *Corallina pilulifera*, *Ulva pertusa*, *Ishige foliacea* and *Endarachne binghamiae* were significantly active. Jin and Dong (2003) carried out comparative studies on the algicidal effects of two different strains of *U. pertusa* on *Heterosigma akashiwo* and *Alexandrium tamarense* and Nan et al. (2004) reported on the algicidal interaction between green alga *U. pertusa* and eight phytoplankton species. Recently, Wang et al. (2006) showed and effect of the green alga *U. pertusa* and the red alga *Gracilaria lemaneiformis* on the growth of *H. akashiwo* in co-culture.

We screened the methanol extracts of thirty-seven species of macroalgae (10 Chlorophyta, 13 Phaeophyta and 14 Rhodophyta) collected from the coast of Nagasaki Prefecture, Japan, for algicidal activity against red tide phytoplankton, and found that the green algae *U. fasciata* and *U. pertusa* (Ulvaceae, Chlorophyta) had a higher algicidal activity than the other macroalgae tested (Alamsjah et al. 2005). We also isolated and identified hexadeca-4,7,10,13-tetraenoic acid (HDTA) C16:4 n-3, C18:4 n-3 (ODTA), α-linolenic acid (ALA) C18:3 n-3 and linoleic acid (LA) C18:2 n-6 as the active principles. These polyunsaturated fatty acids (PUFAs) are found to be significantly active against several red tide phytoplankters at low concentrations and are promising chemical agents for HAB control. In this paper we summarize our study on the algicidal effects of *Ulva* species involving isolation and structure determination of active principles, structure-activity relationship (SAR) study for fatty acids, quantitative analysis of the algicidal compounds released from the plant body, and comparison of fatty acid composition among green seaweeds of the family of Ulvaceae.

Materials and methods

Green algae of the Ulvaceae (*Ulva fasciata*, *U. pertusa*, *U. arasakii*, and *U. conglobota*) were collected in May 2005—April 2006. Ecological damage during harvesting was prevented by not removing the algae stems. All samples were brought to the laboratory in plastic bags containing seawater to prevent evaporation, and then washed with distilled water to remove potential contaminants. For convenience, macroalgae tissues were dried completely for 1 d at room temperature and then ground to powder using a blender.

Axenic phytoplankton cultures of *Alexandrium catenella* NIES-677, *Chattonella antiqua* NIES-1, *C. marina* NIES-3, *Eutreptiella gymnastica* NIES-381, *Fibrocapsa japonica*, NIES-462, *Heterocapsa triquetra* NIES-7, *Heterosigma akashiwo* NIES-4, *Karenia mikimotoi* NIES-680, *Nephroselmis astigmatica* NIES-252, *N. viridis* NIES-486, *Oltmannsiellopsis unicellularis* NIES-359, *Prorocentrum sigmoides* NIES-683, *Pterosperma cristatum* NIES-221, *Pyramimonas* aff. *amylifera* NIES-251, *P. parkeae* NIES-254, *Scrippsiella sweeneyae* NIES-684 and *S. trochoidea* NIES-369 were obtained from the National Institute for Environmental Studies, Japan. The other species, *Nannochloropsis oculata* ST-6, *Cochlodinium polykrikoides* ND-14, *Heterocapsa circularisquama* ND-12, *Prorocentrum minium* ND-34, *Pinnularia* sp. NB-1, *Skeletonema costatum* NB-2, *Stephanopyxis palmeriana* NB-3 and *Rhizosolenia hebetata f. semispina* NB-4 had been maintained at Marine Plant Science Laboratory, Faculty of Fisheries, Nagasaki University of Japan. The phytoplankton species were cultured aseptically in Guillard's f/2 medium. These cultures were maintained at 20°C, 40 μmol photons m^{-2} s^{-1} using 40 W fluorescent lamps (FL40SD, Toshiba) with a 12L:12D cycle.

Extraction and isolation of algicidal compounds from U. fasciata

Ulva fasciata (4.36 kg, dry wt) was crushed into small pieces using a blender and extracted twice with MeOH (10.5 L, 4 d and 12.5 L, 2 d). After removing the MeOH under reduced pressure, the residue, containing about 300 mL of water, was partitioned between water (700 mL) and hexane (1,200 mL). The hexane extract was concentrated, taken up in hexane (350 mL), and extracted with 4 M aqueous HCl (100 mL×3) followed by 50% aqueous MeOH (150 mL×2 and 300 mL). The hexane layer was dried (Na_2SO_4) and concentrated to leave an deep green oily residue (12.95 g), which was chromatographed on silica gel (63–230 μm, 240 g) eluted with a gradient of hexane-EtOAc (9:1, 5:1, 1:1, then 0:1) to give four fractions (FL1-FL4). The active fraction, FL3 (2.35 g), eluted with hexane:EtOAc=1:1, was then chromatographed on Octadecyl-S (50 μm, 38 g) eluted with 90% MeOH to give four fractions (FL3-1 to FL3-4). FL3-2 (319.4 mg) and FL3-3 (548.8 mg) were active. HPLC purification of FL3-2 (319.4 mg) (Cosmosil 5C18-MS-II, 20×250 mm, 80% CH_3CN, flow rate 5 mL min^{-1}) yielded compound 1 (73.5 mg). Part of FL3-3 (82 mg) was also separated by HPLC in the same manner to yield compound **1** (13.1 mg), compound **2** (21.4 mg), and compound **3** (22.1 mg). FL3-1 (158 mg) was purified with TLC RP-18 (Merck, 20×20 cm, layer thickness 0.25 mm) using 85% CH_3CN as the solvent to give compound **4** (16.9 mg) along with compound **3**.

Spectral data (IR, EIMS, HRMS, ^{1}H and ^{13}C NMR) for compounds **1**–**3**: see Alamsjah et al. (2005).

Compound **4** (linoleic acid). EIMS 281 (M^+). ^{1}H NMR (400 MHz, $CDCl_3$) δ 0.89 (3H,t, J=6.6 Hz), 1.23–1.41 (14H, m), 1.63 (2H, m), 2.01–2.10 (4H, m), 2.34 (2H, t, J= 7.6 Hz), 2.77 (2H, m), 5.28–5.43 (4H, m), 7.26 (1H, br.). ^{13}C NMR (100 MHz, $CDCl_3$) δ 14.1 (CH_3), 22.7 (CH_2), 24.8 (CH_2), 25.7 (CH_2), 27.2 (CH_2×2), 29.1 (CH_2), 29.2 (CH_2), 29.3 (CH_2), 29.4 (CH_2), 29.7 (CH_2), 31.5 (CH_2), 33.9 (CH_2), 127.9 (CH=C), 128.2 (CH=C), 130.0 (CH=C), 130.2 (CH=C), 178.7(C=O).

Measurements of fatty acid composition

The dried sample of each green macroalga (10 g) was extracted using 150 mL of methanol (Wako Pure Chemical) for 3 d. The extraction procedure was repeated twice and the extracts were combined. The 50 mg of the crude extract was esterified by exposure to 2.5 mL of 3% HCl/MeOH overnight. After concentrating under reduced pressure, the residue was taken up in CH_2Cl_2 and washed with 5% $NaHCO_3$ solution. The CH_2Cl_2 layer was separated and dried over Na_2SO_4. After concentration, the residue was passed through a short column on silica gel (63–230 μm, 500 mg) using CH_2Cl_2 as the eluent. Finally, the methyl ester fraction was diluted in 1 mL of hexane and analyzed using a GC-2014 gas chromatograph (Shimadzu) equipped with a capillary column (CP-Sil 88 for FAME fused silica WCOT, 50 m×0.25 mm i.d., 0.2 μm film thickness, Chrompack, Hewlett-Packard) and a flame ionization detector (FID). The detector and the injector temperature were maintained at 300°C. The initial oven temperature was programmed at 170°C for 15 min, and then, was allowed to rise to 230°C at a rate of 5°C min^{-1}, and was kept at the final temperature for 5 min. Nitrogen was used as the carrier gas at a flow rate of 24 mL min^{-1}. A standard fatty acid methyl ester mixture C8–C22 and C14–C22 (Supelco) was used for identification of the peaks.

Structure activity relationship study for fatty acids (C16, C18, C20 and C22)

The fatty acids, palmitic acid C16:0, palmitoleic acid *cis*-C16:1 n-7, stearic acid C18:0, oleic acid C18:1 n-9, α-linolenic acid C18:3 n-3, arachidic acid C20:0, arachidonic acid C20:4 n-6, eicosapentaenoic acid C20:5 n-3, behenic acid C22:0, and docosahexaenoic acid C22:6 n-3 were purchased from Tokyo Kasei Kogyo. The other fatty acids, palmitelaidic acid *trans*-C16:1 n-7, linolic acid C18:2 n-6, eicosenoic acid C20:1 n-9 and erucic acid C22:1 n-9 were from MP Biomedicals, Wako Pure Chemical, Sigma and Alfa Aesar, respectively. C16:4 n-3 (HDTA) and C18:4 n-3 (ODTA) used for the algicidal assay were the natural compounds isolated from *U. fasciata*.

Algicidal assay of C16-C22 fatty acids against H. akashiwo and C. marina

Heterosigma akashiwo and *C. marina* were cultured aseptically in sterilized Guillard's f/2 medium at 20°C under illumination of a fluorescent lamp (40 μmol photons m^{-2} s^{-1}) under a 12 h light: 12 h dark cycle. Cells in the exponential growing phase were used throughout the experiments. To each of the phytoplankton cell suspension (cell density ca. 3×10^4 cells mL^{-1}) in a 24-well micro plate (Iwaki), varying amounts of the fatty acid solution in MeOH (<2 μL) were added to make the final concentration of 100, 75, 50, 25, 15, 5, 2, 1 and 0.5 μg mL^{-1}. After 4 h cultivation, survivability and mortality of cells were calculated from microscope observations (×20). Algicidal activity was calculated using the formula: Algicidalactivity = (dead cells/living and dead cells) × 100% . All data in this study were tested by means of ANOVA ($p<0.05$) and by multiple range test for comparing means in an analysis of variance. Essentially the same procedure was used for the algicidal assay of C16:4 n-3, C18:3 n-3 and C18:4 n-3 on various phytoplankton (Table 4).

Algicidal compounds released from dry powder and fresh tissue of U. fasciata in seawater

The dry powder of *U. fasciata* (1.5 g L^{-1}, size 150 μm, autoclaved and non-autoclaved) were mixed in autoclaved of seawater (ASW) using a stirrer for 6, 12, 24, 36 and 48 h on 20°C in the dark and filtered through a No. 2 filter paper. The filtrate (8 mL) and 2 mL of *H. akashiwo* (cell density 3×10^5 cells mL^{-1}) were mixed in a 5 cm diameter Petri dish to check algicidal activity at 4 h. The water solution was then adjusted to pH 2 using 1 M HCl, extracted 3 times with CH_2Cl_2 (100 mL each). The extracts were combined, dried over Na_2SO_4, and concentrated. The extract was esterified as described above.

For analysis of algicidal compounds released from fresh tissue of *U. fasciata*, 50 g of whole material or cut tissue (0.5×0.5 cm^2 size segments) were used. The fresh tissue sample was rinsed with ASW and were further maintained in 1 L flat bottom aeration flasks with ASW containing 1% of antibiotic mixture (penicillin G 1 g and streptomycin sulfate 2 g in distilled water 100 mL) for 2 d. The water rinsed from the cut tissue was collected and then analyzed for fatty acid composition as described for the dry powder.

Statistics

All the experiments in this study were done separately in at least triplicate and tested by ANOVA ($p<0.05$). Simple regression and correlation analysis was used to examine data for possible relationships and to obtain LC_{50} (median lethal concentration) values (Nakai et al. 1999). For sigmoid curves, the y values (percentage) were converted by probit transformation using Microsoft Excel®.

Results

Structural determination of algicidal compounds from U. fasciata

Four algicidal compounds (compounds **1**–**4**) were isolated from *U. fasciata* and their structures were elucidated as C16:4 n-3 (HDTA), C18:4 n-3 (ODTA), C18:3 n-3 (ALA) and C18:2 n-6 (LA) on the basis of spectroscopic information (Alamsjah et al. 2005). Structures of all compounds isolated were verified by comparison of authentic samples obtained from commercial sources or by synthesis (unpublished data).

Fatty acid composition of macroalgae

Fatty acid compositions of methanol extracts of four macroalgae of the Ulvaceae were analyzed using GLC and the results are shown in Table 1. All the macroalgae showed similar fatty acid composition, palmitic acid C16:0, C16:4 n-3 (HDTA), C18:3 n-3 (ALA) and C18:4 n-3 (ODTA) being predominant, except for *U. arasakii,* which had higher concentration of C18:2 n-6 (LA) and lower ratio of C18:4 n-3.

Correlation between total active PUFAs amount and algicidal activity

Table 2 shows quantitative analysis of the three highly active compounds, C16:4 n-3 (HDTA), C18:3 n-3 (ODTA), and C18:4 n-3 (ALA) in the dry tissue of four macroalgae and the algicidal activity of their methanol extracts. Total average of active PUFAs of C16:4 n-3, C18:3 n-3 and C18:4 n-3 from *U. fasciata* and *U. pertusa* were 1338.21 and 1094.44 mg 100 g^{-1}, respectively, which were 2 to 10-fold of those of the other two species.The table also shows that the algicidal activities correlate well with the amount of total active PUFAs in dry tissues. Dry powder of *U. fasciata* and *U. pertusa* also showed potent activity on *H. akashiwo*, with 4 h LC_{50} values of 0.4 and 0.7 g L^{-1}, respectively. Furthermore, algicidal activity of *U. fasciata* and *U. pertusa* based on seasonal variation were checked (Fig. 1). Figure 1a and c show the monthly change in the total amount of the active PUFAs and algicidal activity to *H. akashiwo* expressed as the LC_{50} value of *U. fasciata* (1A) and *U. pertusa* (1C). There seems to be a trend that the amount of active PUFAs produced by the green algae declines in June and July, and consequently the algicidal activities decreases. However, the correlation analyses (Fig. 1b & d) indicated that the algicidal activity of *U. fasciata* and *U. pertusa* were not significantly different ($p > 0.05$) based on seasonal variation.

Structure activity relationship study for fatty acids

Sixteen fatty acids of the numbers of carbon from C16 to C22 were evaluated for algicidal activity against *H. akashiwo* and *C. marina*. Highly unsaturated fatty acids, C16:4 n-3 (HDTA), C18:3 n-3 (ALA), C18:4 n-3 (ODTA), arachidonic acid C20:4, eicosapentaenoic acid C20:5 and docosahexaenoic acid C22:6 showed strong activity with 4 h LC_{50} values less than 5 and 8 μg mL^{-1} on *H. akashiwo* and *C. marina*, respectively (Table 3). In this study, C18:2 n-6 (LA) also showed strong activity against *H. akashiwo*, however it showed lower activity against *C. marina*. Among the fatty acids of the same carbon number, the activity increased as the number of unsaturation increased, while the number of carbons was irrelevant. Interestingly, palmitoleic acid C16:1 n-7 had moderate activity while longer fatty acids having the same unsaturation (oleic acid C18:1, eicosenoic acid C20:1 and eruric acid C22:1) were almost or completely inactive.

Table 1 Fatty acid composition (mean ± SD, % of the total composition) of four *Ulva* species from Nagasaki beach (collected May 2005–April 2006)

Fatty acids		*Ulva fasciata* n=13	*Ulva pertusa* n=14	*Ulva arasakii* n=1	*Ulva conglobota* n=2
decanoic	C10:0	0.78±0.68	0.96±0.94	0.29	0.55±0.39
myristic	C14:0	0.70±0.36	0.68±0.46	0.44	1.03±1.02
myristoleic	*cis*-C14:1 n-9	1.81±1.44	2.07±1.19	0.87	3.35±0.53
palmitic	C16:0	29.32±5.41	27.36±4.10	25.43	34.16±0.33
hexadecatetraenoic	C16:4 n-3	10.57±2.53	12.73±2.66	11.53	9.28±1.87
stearic	C18:0	0.91±0.42	1.03±0.59	nd	2.39±0.43
elaidic	*trans*-C18:1 n-9	1.66±1.30	1.64±1.5	nd	3.35±0.53
oleic	*cis*-C18:1 n-9	5.12±3.52	5.15±3.53	7.40	6.31±2.52
linolic	C18:2 n-6	7.87±1.72	8.09±2.46	21.00	6.81±2.09
α-linolenic	C18:3 n-3	17.25±1.76	17.96±2.79	22.98	14.37±0.09
octadecatetraenoic	C18:4 n-3	20.49±7.39	20.56±6.81	5.36	11.84±7.14
eicosenoic	C20:1 n-9	1.98±1.93	1.35±0.89	0.42	1.77±0.21
behenic	C22:0	0.80±0.80	0.65±0.55	2.87	2.50±0.62
erucic	C22:1 n-9	2.46±1.49	2.88±1.59	1.42	2.57±0.37
PUFA:FA		0.56±0.09	0.59±0.10	0.61	0.42±0.07

nd: not detected; PUFA: FA: ratio of total polyunsaturated fatty acids to total fatty acids.

Table 2 The active compounds of HDTA, ALA, and ODTA/dry tissue (mean ± SD, mg 100 g^{-1}), 4 h LC_{50} of algicidal activity of methanol extracts (mean ± SD, mg L^{-1}) and dry powder (mean ± SD, g L^{-1}) from *Ulva fasciata, Ulva pertusa, Ulva arasakii and Ulva conglobota* on lysis of *Heterosigma akashiwo*

	Ulva fasciata	*Ulva pertusa*	*Ulva arasakii*	*Ulva conglobota*
HDTA, ALA, ODTA/dry tissue	1338.21± 358.02	1094.44± 525.98	267.15	119.00± 84.90
4 h LC_{50} of algicidal activity of methanol extracts	64.20± 17.32	81.91± 24.49	161.07	526.28± 89.78
4 h LC_{50} of algicidal activity of dry powder	0.40±0.01	0.70± 0.01	1.50	2.30±0.13

Algicidal activity on phytoplankton

Algicidal activities of C16:4 n-3, C18:3 n-3, and C18:4 n-3 were evaluated against 25 species of phytoplankton including red tide phytoplankton species (Table 4). Algicidal profiles of these fatty acids were almost similar. C16:4 n-3, C18:3 n-3, and C18:4 n-3 were found to be significantly active against *C. antiqua*, *C. marina*, *Fibrocapsa japonica*, *H. akashiwo*, *Karenia mikimotoi*, moderately effective against *Heterocapsa circularisquama*, *Prorocentrum minimum*, *P. sigmoides*, *Scrippsiella trochoidea*, whereas low effective against *A. catenella* and *Coch. polykrikoides*.

Algicidal compounds (C16:4 n-3, C18:3 n-3 and C18:4 n-3) released from U. fasciata

C16:4 n-3, C18:3 n-3, and C18:4 n-3 released from autoclaved and non-autoclaved dry powder of *U. fasciata* in seawater were different significantly ($p<0.05$) (Table 5). Released amount of the total PUFAs reached maximum at 6 h, and afterward, gradually decomposed with time. At 6 h, 46.72% of the total active PUFAs contained in the dry powder were released into the seawater. After 48 h, remaining active PUFAs decreased to 2% of the maximum amount. No significant difference was observed in the released amount between whole and cut tissues (Table 6).

Discussion

The algicidal principles of *Ulva fasciata* and *U. pertusa* were found to be PUFAs such as C16:4 n-3, C18:3 n-3, and C18:4 n-3 (Alamsjah et al. 2005), one of which, C18:4 n-3, was the same compound that had been isolated as the allelopathic substance of *C. okamuranus* (Kakisawa et al. 1988). Other researchers have also reported that green algae such as *U. pertusa*, *Monostroma lattisium* and *Enteromorpha*

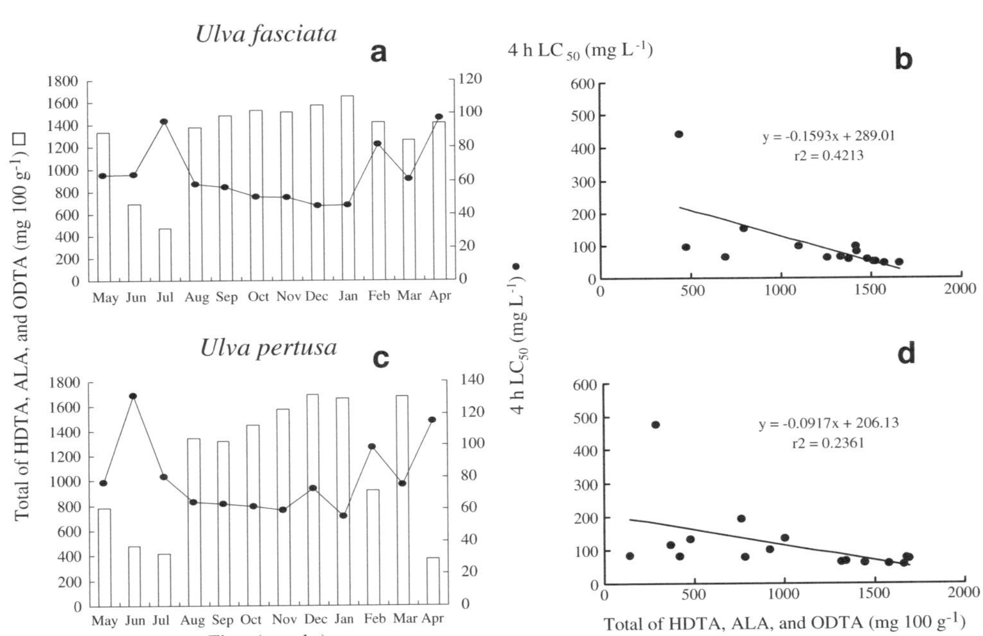

Fig. 1 **a** and **c**: seasonal variation of total C16:4 n-3 (HDTA), C18:3 n-3 (ALA) and C18:4 n-3 (ODTA) concentration (mg 100 g^{-1}, white bars) in dry tissue of *Ulva fasciata* (**a**) and *Ulva pertusa* (**c**) and algicidal activity [4 h LC50 (mg L^{-1}), •] on lyses of *Heterosigma akashiwo*. **b** and **d**: correlation analysis between total of C16:4 n-3 (HDTA), C18:3 n-3 (ALA) and C18:4 n-3 (ODTA) (mg 100 g^{-1}) and algicidal activity [4 h LC50 (mg L^{-1})] for *Ulva fasciata* (**b**) and *Ulva pertusa* (**d**)

Table 3 Algicidal activity (4 h LC_{50} in μg mL^{-1}) of C16–C22 fatty acids against *Heterosigma akashiwo* and *Chattonella marina*

Fatty acids		*Heterosigma akashiwo*	*Chattonella marina*
palmitic	C16:0	79.28	29.50
palmitelaidic	*trans*-C16:1 n-7	>100	43.75
palmitoleic	*cis*-C16:1 n-7	7.28	20.31
octadecatetraenoic	C16:4 n-3	1.93	5.34
stearic	C18:0	77.63	29.24
oleic	C18:1 n-9	34.01	24.85
linoleic	C18:2 n-6	3.46	23.90
α-linolenic	C18:3 n-3	2.59	7.11
hexadecatetraenoic	C18:4 n-3	0.83	4.02
arachidic	C20:0	94.38	34.07
eicosenoic	C20:1 n-9	>100	43.59
arachidonic	C20:4 n-6	2.34	6.85
eicosapentaenoic	C20:5 n-3	2.11	6.79
behenic	C22:0	>100	52.97
eruric	C22:1 n-9	>100	45.56
docosahexaenoic	C22:6 n-3	4.37	7.19

sp. obtained in Japanese water contain large amount of PUFA composed mainly of C16:4 n-3, C18:3 n-3, and C18:4 n-3 (Takagi et al. 1985).

The higher toxicity of C16:4 n-3, C18:3 n-3, and C18:4 n-3 may be also reflected by the amphiphatic property of PUFAs, which probably disrupts the membrane integrity of phytoplankton cells. Chemical structural feature such as the number of unsaturated double bonds may also be involved in the biological activity of these unsaturated fatty acids. These results are supported by earlier experiments by Kakisawa et al. (1988), Suzuki et al. (1996) and Sellem et al. (2000). Chiang et al. (2004) showed that the highest toxicity from the green colonial alga *Botryococcus braunii* was observed with C18:3 n-3 (ALA), followed by C18:2 n-6 (LA) and C18:1 n-9. In this study, PUFAs were also more toxic than saturated fatty acids against *H. akashiwo* and *C. marina*. Morohashi et al. (1991) determined that unsaturated fatty acids of the *cis*-configuration were more effective than saturated acids. Interestingly, C18:3 n-3 (ALA) and C16:4 n-3 (HDTA) also showed lethal effect on phytoplankton, to the same extent as C18:4 n-3 (ODTA)

Table 4 Algicidal activity of HDTA, ALA, and ODTA (μg mL^{-1}) against phytoplankton

Plant family	Species	HDTA		ALA		ODTA	
		25	5	25	5	25	5
Raphidophyceae	*Chattonella antiqua* NIES-1	+	±	+	±	+	±
	Chattonella marina NIES-3	+	+	+	+	+	+
	Fibrocapsa japonica NIES-462	+	+	+	+	+	+
	Heterosigma akashiwo NIES-4	+	+	+	+	+	+
Dinophyceae	*Alexandrium catenella* NIES-677	–	–	–	–	±	–
	Cochlodinium polykrikoides ND-14	–	–	–	–	–	–
	Karenia mikimotoi NIES-680	+	+	+	+	+	+
	Heterocapsa circularisquama ND-12	±	±	±	±	±	±
	Heterocapsa triquetra NIES-7	±	–	±	–	+	±
	Prorocentrum minimum ND-34	±	–	±	–	±	–
	Prorocentrum sigmoides NIES-683	±	–	±	–	±	–
	Scrippsiella sweeneyae NIES-684	±	–	±	–	+	±
	Scrippsiella trochoidea NIES-369	+	±	+	±	+	±
Prasinophyceae	*Nephroselmis astigmatica* NIES-252	+	±	+	±	+	±
	Nephroselmis viridis NIES-486	+	±	+	±	+	±
	Pterosperma cristatum NIES-221	±	±	±	±	+	±
	Pyramimonas aff. *amylifera* NIES-251	+	+	+	+	+	+
	Pyramimonas parkeae NIES-254	+	±	+	±	+	±
Ulvophyceae	*Oltmannsiellopsis unicellularis* NIES-359	+	±	+	±	+	±
Euglenophycea	*Eutreptiella gymnastica* NIES-381	±	–	±	–	±	–
Bacillariophyceae	*Pinnularia* sp. NB-1	–	–	–	–	–	–
	Skeletonema costatum NB-2	–	–	–	–	–	–
	Stephanopyxis palmeriana NB-3	–	–	–	–	–	–
	Rhizosolenia hebetata f. semispina NB-4	–	–	–	–	–	–
Eustigmatophyceae	*Nannochloropsis oculata* ST-6	–	–	–	–	–	–

Each concentration of fatty acid was added to f/2 medium inoculated with approximately 3×10^4 cells mL^{-1} of phytoplankton for 4 h. A "+" symbol indicates more than 70% of mortality cells; "±" symbol indicates moderate of mortality cells, i.e. between 30 and 69%; "–" symbol indicates less than 30% of mortality cells. "NIES" strains were obtained from the National Institute for Environmental Studies, Japan. "ND", "NB", and "ST" strains were maintained at Marine Plant Science Laboratory, Faculty of Fisheries, Nagasaki University, Japan

Table 5 Algicidal compounds [C16:4 n-3 (HDTA), C18:3 n-3 (ALA), and C18:4 n-3 (ODTA)] released from dry powder of *Ulva fasciata* in seawater (mg L^{-1}) and algicidal activity (percentage of mean ± SD) on *Heterosigma akashiwo*

Time	Autoclaved		Non-autoclaved	
	HDTA, ALA, ODTA	Activity	HDTA, ALA, ODTA	Activity
1 h	–	–	4.82 (20.43)	100.00±0.00
2 h	–	–	5.18 (21.96)	100.00±0.00
3 h	–	–	5.68 (24.08)	100.00±0.00
4 h	–	–	7.38 (31.28)	100.00±0.00
5 h	–	–	9.20 (39.00)	100.00±0.00
6 h	11.90 (50.45)	100.00± 0.00	11.02 (46.72)	100.00±0.00
12 h	7.84 (33.24)	100.00± 0.00	1.32 (5.60)	63.51±4.75
24 h	5.26 (22.30)	100.00± 0.00	0.64 (2.71)	10.93±1.33
36 h	3.62 (15.35)	100.00± 0.00	0.56 (2.37)	7.08±3.35
48 h	2.84 (12.04)	100.00± 0.00	0.26 (1.10)	2.77±0.87

– : not tested; ( ): percentage of HDTA, ALA, and ODTA released in sea water; 1.5 g L^{-1} of dry powder of *Ulva fasciata* in sea water contained 23.59 mg L^{-1} of HDTA, ALA, and ODTA.

(Kakisawa et al. 1988), which is active against phytoplankton without cell coverings and inactive to ones with rigid cell walls. Kakisawa et al. (1988) also showed that C18:4 n-3, the allelochemical of the brown alga *C. okamuranus*, had different responses against phytoplankton.The fatty acid composition profiles of each macroalga of the family Ulvaceae were almost similar, having a high concentration of C16 and C18 PUFAs. These data are in agreement with other researchers which found that a dominance of C16 and C18 PUFAs is characteristic of green macroalgae (Jamieson and Reid 1972; Ackman and McLachlan, 1977). In this study, the ratio of total PUFAs to total fatty acids (PUFA:FA ratio) of the family Ulvaceae showed high growth rates, as Ahlgren et al. (1992) suggested that PUFA:FA ratio could be useful indicator of the physiological status of the algae. This is because PUFAs are usually stable major components of cell membranes while the saturated fatty acids are environmentally-sensitive storage products (Sargent et al. 1987). Thus, the biosynthetic substances of *Ulva* species are really promising for practical harmful algal bloom control. Ikawa (2004) tried to explain the inhibitory effects of PUFAs on phytoplankton, which it may have a disrupting influence on the lipid bilayer of natural biological membranes of eukaryotic cells. Murata et al. (1989) and Oda et al. (1992) also reported that the toxic activity of PUFAs may be due to oxidation products derived through photo oxidation or metabolic processes, whereas Kogteva and Bezuglov (1998) determined that the potent activity of PUFAs may act as second messengers to modulate functionally active proteins.

Table 6 Algicidal compounds [C16:4 n-3 (HDTA), C18:3 n-3 (ALA), and C18:4 n-3 (ODTA)] released from fresh tissue of *Ulva fasciata* in seawater (mg L^{-1}) and algicidal activity (percentage of mean ± SD) on *Heterosigma akashiwo*

Antibiotic	Whole tissue		Cutting tissue	
	HDTA, ALA, ODTA	Activity	HDTA, ALA, ODTA	Activity
Untreated	0.22 (0.28)	1.63±0.74	0.32 (0.41)	5.00±2.65
Treated	0.26 (0.34)	4.67±2.89	0.48 (0.62)	6.63±2.10

( ): percentage of HDTA, ALA, and ODTA released in sea water; 50 g L^{-1} of fresh tissue of *Ulva fasciata* in sea water contained 77.45 mg L^{-1} of HDTA, ALA, and ODTA.

Although fatty acids are hydrophobic substances with a very limited solubility in water, Kattner et al. (1983), studying the lipid concentration during a spring phytoplankton bloom in the northern North Sea in Europe, estimated that fatty acids constituted about 3% of total dissolved organic matter. The total fatty acid concentration amounted to about 1.15 μmol L^{-1} before the bloom and increased to a maximum of 5 μmol L^{-1} during a bloom. In the present study, the C16:4 n-3, C18:3 n-3, and C18:4 n-3 from autoclaved of dry powder of *U. fasciata* were werer release in seawater more than non-autoclaved dry powder of *U. fasciata* in time-dependent manner. This means decomposition of algicidal activity of *U. fasciata* was caused by biological agent (e.g. bacteria) and chemical reaction such as photooxidation. Similarly, algicidal activity of autoclaved dry powder of *U. fasciata* also was higher than non-autoclaved dry powder of *U. fasciata* in time-dependent manner under controlled conditions in the laboratory, while Ikawa (2004) have mentioned that effective levels of PUFA soluble in long term exposure to low levels under natural conditions can have effects which are only observed with higher level acute doses. Scutt (1964) also reported that fresh filtrates of *Chlorella vulgaris* did not contain autoinhibitors and that, only after storage for several days, auto inhibitory activity appeared. THis was probably because fatty acids from fresh tissues of macroalgae in seawater were released gradually.

The main factors generally considered when selecting harmful algal bloom control methods are effectiveness, toxicity, cost and practicability. Taking into account the characteristics described so far, algicidal compounds of *Ulva* species are promising for controlling HAB species because (1) *U. fasciata* and *U. pertusa* have high inhibition effect on several red tide phytoplankton without recovery at a relatively low concentration, (2) they have high biodegradability, ecological acceptability, and a relatively environmentally benign nature. In conclusion, the PUFAs may useful mitigation agents to remove harmful red tide phytoplankton such as *C. marina* and *H. akashiwo* without causing detrimental effects on surrounding marine living organisms.

The world-wide distribution of *Ulva* species will also give an opportunity to evaluate the genetic variability for their toxic effects on HAB. Further research on the precise mechanism and mode of action of the algicidal substances from *Ulva* species to cause death of HAB species may shed new light on the algicidal activity and application of *Ulva* species as effective control agents.

Acknowledgements We thank to the Ministry of Education, Culture, Sports, Science and Technology of Japan having given a part of these research funds to study. A part of this study was also supported financially by the Nagasaki Prefecture Collaboration of Regional Entities for the Advancement of Technological Excellence, Japan Science and Technology Agency (JST). We also thank to member of research group of Kashima-shi Fishery Cooperation, Saga Prefecture which had cooperate to study it.

References

Ackman RG, McLachlan J (1977) Fatty acids in some Nova Scotian marine seaweeds: a survey for octadecapentaenoic and other biochemically novel fatty acids. Proc NS Inst Sci 28:47–64

Ahlgren G, Gustaffson IB, Boberg M (1992) Fatty acid content and chemical composition of freshwater microalgae. J Phycol 28:37–50

Alamsjah MA, Hirao S, Ishibashi F, Fujita Y (2005) Isolation and structure determination of algicidal compounds from *Ulva fasciata*. Biosci Biotechnol Biochem 69:2186–2192

Chiang IZ, Huang WY, Wu JT (2004) Allelochemicals of *Botryococcus braunii* (Chlorophyceae). J Phycol 40:474–480

Garry RT, Hearing P, Cosper EM (1998) Characterization of a lytic virus infectious to the bloom-forming microalga *Aureococcus anophagefferens* (Pelagophyceae). J Phycol 34:616–621

Ikawa M (2004) Algal polyunsaturated fatty acids and effects on plankton ecology and other organisms. UNH Center Freshwat. Biol Res 6:17–44

Imai I, Ishida Y, Sakaguchi K, Hata Y (1995) Algicidal marine bacteria isolated from northern Hiroshima Bay. Fish Sci 61:624–632

Jamieson GR, Reid EH (1972) The component fatty acids of some marine algal lipids. Phytochem 11:1423–1432

Jeong JH, Jin HJ, Sohn CH, Suh KH, Hong YK (2000) Algicidal activity of the seaweed *Corallina pilulifera* against red tide microalgae. J Appl Phycol 12:37–43

Jin Q, Dong S (2003) Comparative studies on the allelopathic effects of two different strains of *Ulva pertusa* on *Heterosigma akashiwo* and *Alexandrium tamarense*. J Exp Mar Biol Ecol 293:41–55

Kakisawa H, Asari F, Kusumi T, Toma T, Sakurai T, Oohusa T, Hara Y, Chiharai M (1988) An allelopathic fatty acid from the brown alga *Cladosiphon okamuranus*. Phytochem 27:731–735

Kamiyama T (2002) Regulation of harmful algal blooms by planktonic ciliates. In: Hiroishi S, Imai I, Ishimaru T (eds) Prevention and extermination strategies of harmful algal blooms. Nihon Suisan Gakkai Kansyu 134, Kouseisya Kouseikaku, Tokyo, Japan, pp 89–101 (in Japanese)

Kattner G, Gercken G, Hammer KD (1983) Development of lipids during a spring plankton bloom in the northern North Sea. II. Dissolved lipids and fatty acids. Mar Chem 14:163–173

Kogteva GS, Bezuglov VV (1998) Unsatutared fatty acids as endogenous bioregulators. Biochem (Moscow) 63:6–15

Morohashi M, Tsuchiya K, Mita T, Kawamura M (1991) Identification of (Na, K) ATPase inhibitor in brine shrimp, *Artemia salina*, as long chain fatty acids. J Comp Physiol B 161:69–72

Murata H, Sakai T, Endo M, Kuroki A, Kimura M, Kumanda K (1989) Screening of removal agents of a red tide plankton *Chattonella marina* with special reference to the ability of the free radicals derived from the hydrogen peroxide and polyunsaturated fatty acids. Nippon Suisan Gakkaishi 55:1075–1082

Nagayama K, Shibata T, Fujimoto K, Honjo T, Nakamura T (2003) Algicidal effect of phlorotannins from the brown alga *Ecklonia kurome* on red tide microalgae. Aquacult 218:601–611

Nakai S, Inoue Y, Hosomi M, Murakami A (1999) Growth inhibition of blue-green algae by allelopathic effects of macrophytes. Water Sci Technol 39:47–53

Nakamura Y (2002) Control of red tides by heterotrophic dinoflagellates. In: Hiroishi S, Imai I, Ishimaru T (eds) Prevention and extermination strategies of harmful algal blooms. Nihon Suisan Gakkai Kansyu 134, Kouseisya Kouseikaku, Tokyo, Japan, pp 81–88 (in Japanese)

Nan C, Zhang H, Zhao G (2004) Allelopathic interactions between the macroalga *Ulva pertusa* and eight microalgal species. J Seares 52:259–268

Oda T, Ishimatsu A, Shimada M, Takeshita S, Muramatsu T (1992) Oxygen-radical-mediated toxic effects of the red tide flagellate *Chattonella marina* on *Vibrio alginolyticus*. Mar Biol 112:505–509

Ryu HY, Shim JM, Bang JD, Lee C (1998) Experimental chemical treatments for the control of dinoflagellate, *Cochlodinium polykrikoides* in the land-based culture of olive flounder *Paralichthys olivaceus*. Kor J Aquacult 11:285–294

Sargent JR, Parkes RJ, Mueller-Harvey I, Henderson J (1987) Lipid biomarkers in marine ecology. In: Sleigh MA (ed) Microbes in the sea. Ellis Horwood, Chichester, United Kingdom, pp 119–138

Scutt JE (1964) Autoinhibitor production by *Chlorella vulgaris*. Am J Bot 51:581–584

Sellem F, Pesando D, Bodennec G, Abed AE, Girard JP (2000) Toxic effects of *Gymnodinium cf. mikimotoi* unsaturated fatty acids to gametes and embryos of the sea urchin *Paracentrotus lividus*. Water Res 34:550–556

Steidinger KA (1983) A re-evaluation of toxic dinoflagellate biology and ecology. Prog Phycol Res 2:147–188

Suzuki M, Wakana I, Denboh T, Tatewaki M (1996) An allelopathic polyunsaturated fatty acid from red algae. Phytochem 43:63–65

Takagi T, Asahi M, Itabashi Y (1985) Fatty acid composition of twelve algae from Japanese waters. Yukagaku 34:1008–1012

Wang Y, Yu ZM, Song XX, Zhang SD (2006) Effects of *Ulva pertusa* and *Gracilaria lemaneiformis* on growth of *Heterosigma akashiwo* (Raphidophyceae) in co-culture. Huan Jing Ke Xue 27:246–252 (in Chinese)

Potential use of seaweeds in the laying hen ration to improve the quality of n-3 fatty acid enriched eggs

S. Carrillo · E. López · M. M. Casas · E. Avila · R. M. Castillo · M. E. Carranco · C. Calvo · F. Pérez-Gil

Originally published in the Journal of Applied Phycology, Vol 20, No 5, 271–278.
DOI: 10.1007/s10811-008-9334-4

Abstract An association between dietary fish oil and decreased yolk weight and reduced sensory quality of eggs has been reported when eggs are enriched with n-3 FA from fish oil. Seaweeds are an important source of compounds that seem to increase egg weight when included in the laying hen diet. The objectives of this study were to determine the influence of the dietary seaweeds *Macrocystis pyrifera, Sargassum sinicola* and *Enteromorpha* sp. on the physical quality, lipid composition and consumer acceptability of n-3 FA enriched eggs. One-hundred and forty-four 35-week-old Leghorn hens were randomly distributed in four treatments that consisted of the inclusion of 2% of sardine oil (SO) and 10% of each marine alga (MA) in laying hens' diets; a control diet (C) was also prepared. The study lasted 8 weeks and egg physical quality, egg lipids and sensory attributes were evaluated. The results showed that incorporation of 10% *M. pyrifera* in the diets is an effective way of increasing the n-3 FA content, the albumen height and yolk color, but not the egg weight, when these are enriched with n-3 FA from fish oil. The egg flavor was also not affected.

Keywords Marine algae · Fish oil · n-3 fatty acids enriched eggs · Egg quality · Laying hens

S. Carrillo (✉) · E. López · R. M. Castillo · M. E. Carranco · C. Calvo · F. Pérez-Gil
Depto. Nutrición Animal,
Instituto Nacional de Nutrición Salvador Zubirán,
Vasco de Quiroga No. 15,
14000 Distrito Federal, Mexico
e-mail: silvicarrillo3@hotmail.com

M. M. Casas
Centro Interdisciplinario de Ciencias Marinas,
Baja California Sur,
La Paz, Mexico

E. Avila
Facultad de Medicina Veterinaria y Zootecnia,
Universidad Nacional Autónoma de México,
D.F., Mexico

Introduction

It is known that the n-3 fatty acids—alpha-linolenic acid (ALA, 18:3n-3), eicosapentaenoic acid (EPA, 20:5n-3) and docosahexaenoic acid (DHA, 22:6n-3)—are essential for normal growth and development and may play an important role in the prevention and treatment of coronary heart disease, hypertension, diabetes, arthritis, other inflammatory and autoimmune disorders and some types of cancer, as well as having an important role in neuronal development (Simopoulos 2000; Kris-Etherton et al. 2002).

These findings have stimulated interest by consumers to look for food of superior quality. Dietary fatty acid (FA) modification has proved to be a viable method for adding value to foods for the health conscious consumer. Recent dietary fat studies have centered on the manipulation of specific fatty acids. In an attempt to increase n-3 FA content in eggs, the utilization of fish oil and flaxseed as feed ingredients has been a common practice (Baucells et al. 2000; González-Esquerra and Leeson 2001; Castillo et al. 2001; Castillo-Badillo et al. 2005)

When eggs are to be enriched with ALA, flaxseed, flaxseed oil and to a lesser extent canola can be considered. When the aim is to enrich eggs with the n-3 FAs EPA and DHA, fish oil, fish meal and marine microalgae are ingredients frequently used (González-Esquerra and Leeson 2000). It is very important to point out that ALA is not equivalent in its biological effects to the long chain n-3 fatty acids found in marine oils, EPA and DHA. This is because ALA is not efficiently converted to EPA and DHA in either chickens or humans. These are more rapidly incorporated into plasma and membrane lipids and produce more rapid effects than does ALA (Simopoulos 2000). However, an association among dietary fish oil and decreased yolk weight and reduced sensory quality of eggs has been reported (Van Elswyk et al. 1995; Marshall and Van Elswyk 1994; Van Elswyk 1997; González-Esquerra and Leeson 2000; Castillo-Badillo et al. 2005).

The reduced sensory quality of eggs from hens fed fish oil caused by the presence of off-flavors and their lipid stability has long been recognized. Lipids are not the only substances responsible for fishy flavors; it might also be a non-lipid fraction that, interacting with some lipid compounds, produces fishy flavors. For example, amines can impart a fishy taint (González-Esquerra and Leeson 2001; Honkatukia et al. 2005). Also, oxidation products could be responsible in part for fishy odors found in eggs, but the use of stabilized n-3 FA sources in poultry diets could in theory ameliorate this problem (Van Elswyk et al. 1995; Cherian et al. 1996; Leskanich and Noble 1997).

On the other hand, it has been observed that egg weight increases when seaweeds have been included in the laying hens' rations (Meza et al. 1996; Ramos et al. 1997; Rendon et al. 2003).

The objectives of the current study were to determine the influence of dietary seaweeds *Macrocystis pyrifera, Sargassum sinicola and Enteromorpha* spp. on physical quality, lipid composition and consumer acceptability of n-3 FA enriched eggs.

Materials and methods

The seaweeds were collected in Baja California Sur, Mexico: *Macrocystis pyrifera* (Mp) in Bahía Tortugas, *Sargassum sinicola* (Ss) and *Enteromorpha* spp (Espp) in the beaches along La Paz Bay. Samples were collected manually from the intertidal zone at depths between 60 cm and 1.50 m. Afterwards, they were spread over a cement slab and sun-dried for 2 days. Whole plants were ground in a hammer mill and then in a knife mill using a 1 mm mesh. A sample of 1 kg was taking by quartering, to grind it finely and carry out chemical analysis by triplicate (Table 1).

The following analysis were carried out on the algae meals using standard methods (AOAC, 2000): moisture content (over-drying at 60°C to constant weight, AOAC method 976.05), ash (ignition at 550°C in an electric furnace, AOAC method 923.03), crude fiber (Soxhlet apparatus, AOAC method 962.09) and ether extract (Soxhlet apparatus, AOAC method 920.39), nitrogen content was determined using the micro-Kjeldahl method (AOAC method 976.05), and a conversion factor of 6.25 was used to calculate protein content. To determine Ca, Mg, Na, K, and Fe samples of algae meals were subjected to acid digestion and analyzed by atomic absorption spectrophotometry following the procedures described by AOAC method 968.08. P was obtained by colorimetric method (AOAC 2000); Fe, Zn and Cu by AOAC method 999.11. Gross energy was determined using a calorimetric Parr Bomb.

Total lipids were determined following the method described by AOAC (AOAC method 923.07). Cholesterol was determined using KOH, ethanol and 5-α-cholestane (as

Table 1 Chemical composition of seaweed meal

Nutrient (g 100 g^{-1})	*S. sinicola*	*M. pyrifera*	*Enteromorpha* spp.
Moisture	7.40	9.27	6.70
Ash	38.35	33.47	32.64
Crude protein	6.57	10.50	14.10
Crude fiber	6.55	4.32	5.17
Ether extract	1.05	0.61	2.20
Metabolizable energy (Kcal g^{-1})	2.52	2.22	2.39
Calcium	3.21	1.24	2.49
Phosphate	0.01	2.57	3.50
Sodium	20.07	31.11	9.20
Potassium	5.77	5.55	1.80
Magnesium	0.90	0.49	0.71
Total lipids (g 100 g^{-1})	1.93	2.88	2.27
Cholesterol (mg 100 g^{-1})	4.00	5.60	8.00
Fatty acids (% total fatty acids)			
Linoleic acid (C18:2 LA)	6.99	8.40	6.93
α-linolenic acid(C18:3ALA)	2.65	1.67	6.39
Arachidonic acid (C20:4 AA)	9.83	13.55	1.32
Eicosapentaenoic acid (C20:5 EPA)	3.53	4.87	2.88
Docosahexaenoic acid (C22:6 DHA)	0.60	1.51	0.92

internal standard) (Fenton and Sim 1991). The analysis was carried out by gas chromatography in a Varian 3400 CX chromatograph equipped with a DB-5 capillary column (3 m×0.25 mm i.d.), autosampler and a flame-ionization detector. Nitrogen was used as carrier gas at a flow rate of 30 mL min^{-1}. Temperatures were: column 280°C, injector 260°C and detector 280°C. The fatty acid profile of the seaweeds and fish oil (Table 2) were analyzed by gas chromatography using a DB23 column (30 m×0.25 mm i.d.) on a Varian 3400 CX gas-liquid chromatograph equipped with an auto-sampler and flame-ionization detector. Nitrogen was used as carrier gas at a flow rate of 30 mL min^{-1}. Operation temperatures were: column 230°C, injector 150°C, detector 300°C. Retention times were compared with methyl-fatty acids standards (AOAC method 969.33).

Experimental design One-hundred and forty-four 35-week-old Leghorn hens were randomly distributed in four treatments with three replicates of 12 hens each. The treatments consisted in the inclusion of 2% of sardine oil (SO) and 10% of each alga (Mp, Ss and Espp) into the diet of the hens; a control diet (C) was also prepared. Algae partially substituted sorghum, soybean and salt in the diets (Table 3). Birds were lodged in cages. The study lasted 8 weeks, with water and feed ad libitum.

From each treatment, feed intake and egg production were recorded daily. At the end of the 8th week, the average of these parameters was calculated and feed conversion ratio estimated. Egg physical quality was measured at the end of the 8th week; 15 eggs were randomly selected from each treatment, albumen height was measured with a tripod micrometer. Haugh Units were estimated taking into consideration the height of albumen corrected by egg weight. Shell weight was determined and shell thickness was measured near the equator of the egg with an electronic micrometer. Yolk color was measured with a Roche Color Fan. Readings were performed over a white background with artificial light of the same color (semi-automated egg quality system of the Technical Service and Supplies was used).

A total of 15 eggs per treatment were randomly collected from each treatment at the end of the 8th week. The 15 eggs were divided into five samples (3 eggs per sample). They were broken, and the yolk and albumen were mixed homogeneously. Five 1-g aliquots of whole egg were extracted using chloroform-ethanol (1:1) for lipid analysis (method 923.07 of AOAC 2000). The percentage of fat was determined gravimetrically. Cholesterol was determined by adding KOH, ethanol and 5-α-cholestane (as internal standard) to five 1-g aliquots of whole egg (Fenton and Sim 1991). Cholesterol was determined by gas chromatography in a Varian 3400 CX chromatograph equipped with a DB-5 capillary column (3 m×0.25 mm i.d.), autosampler and a flame-ionization detector. Nitrogen was used as carrier gas at a flow rate of 30 mL min^{-1}. Temperatures were: column 280°C; injector 260°C and detector 280°C.

Esters of fatty acids from all lipid extracts were prepared using 0.2-g aliquot of each extract. Methanolic boron-trifloride was employed as the transesterifying agent. The fatty acid methyl esters were quantified using a DB23 column (30 m×0.25 mm i.d.) on a Varian 3400 CX gas-liquid chromatograph. Nitrogen was used as carrier gas at a flow rate of 30 mL min^{-1}. Temperatures were: column 230°C, injector 150°C, detector 300°C. Retention times were compared with methyl-fatty acids standards (Sigma). Fatty acid peaks were identified by comparison with retention times of fatty acid methyl ester standards, and quantified by comparison with an internal standard area (AOAC method 969.33)

Sensory Evaluation At the end of the 8th week, 32 untrained panelists, from both sexes, usual consumers of eggs were set up in a sensory laboratory, equipped with single cabins under red light to rule out any visual influence. A hedonic acceptance test was used (Watts et al. 1992). Four samples of fried eggs, one for each treatment, labeled with random numbers, were provided to the panelists. Salt was not added. They received a questionnaire to indicate the degree of acceptance of each sample choosing from the following options: "like it a lot", "like it somewhat", "neither like it nor dislike it", "dislike it somewhat" and "dislike it a lot". They also received bread and water to be consumed before testing each sample in order to eliminate any residual flavor.

Table 2 Fatty acid composition of fish oil (% total of fatty acids)

Fatty acid	%	Fatty acid	%
C4:0	0.06	C18:2 (LA)	1.25
C6:0	0.01	C18:3	0.29
C8:0	0.03	C18:3 (ALA)	1.02
C10:0	0.01	C24:4	2.37
C11:0	0.01	C20:1	2.14
C12:0	0.09	C20:2	0.21
C13:0	0.05	C20:3	0.05
C14:0	7.43	C20:3	0.20
C15:0	0.63	C20:0 (AA)	1.26
C15:1	0.001	C21:0	0.13
C16:0	13.83	C22:0	0.79
C16:1	10.32	C20:5 (EPA)	15.54
C17:0	1.30	C22:1	2.36
C17:1	1.33	C22:2	0.02
C18:0	2.03	C23:0	0.77
C18:1	5.48	C24:0	0.07
C18:1 n-9	15.58	C24:1	2.43
C18:2	0.08	C22:6 (DHA)	10.70

Table 3 Experimental diets supplemented with sardine oil and seaweeds (%)

Ingredients	C	SO+Mp	SO+Ss	SO+Espp
Sorghum (9%)[a]	62.69	50.14	50.14	52.16
Soybean (48%)[a]	23.83	24.07	24.84	23.35
Seaweed	0	10	10	10
Calcium carbonate	8.82	8.66	7.93	8.42
Soybean oil	2.35	3.05	2.99	2.35
Sardine oil	0	2.00	2.00	2.00
Calcium orthophosphate 1820 [b]	1.45	1.42	1.49	1.45
Salt (NaCl)	0.36	0.15	0.15	0.15
Vitamin and mineral premix[c]	0.25	0.25	0.25	0.25
NR micoad (klin-sil)[d]	0.10	0.10	0.10	0.10
Choline chloride 60	0.05	0.05	0.05	0.05
Methionine 98	0.05	0.05	0.05	0.05
Furacyl[e]	0.001	0.001	0.001	0.001
Avelut powder[f]	0.005	0.005	0.005	0.005
Avired[g]	0.003	0.003	0.003	0.003
Composition (g 100 g^{-1})				
Crude protein	16.46	16.46	17.57	19.00
Ether extract	4.42	6.81	7.12	6.54
Crude fiber	2.10	2.91	2.45	3.97
Crude ash	7.97	14.04	13.21	13.72
Metabolizable energy (Mj g^{-1})	11.92	11.50	11.50	11.56

[a] Crude Protein Content
[b] Calcium orthophosphate P 21% min, Ca 13% min, F0.21% max
[c] Vitamins and mineral mix, per kg, for laying hens: 12,000 UI vit. A; 2500 UIP vit. D_3; 30 UI vit. E; 2 mg vit. K_3; 2.25 mg vit B_1; 7.5 mg B_2; 3.5 mg B_6; 0.02 mg B_{12}; 12.5 mg D-pantothenic acid; 0.125 mg biotin; 1.5 mg folic acid. Mineral mix (mg.kg^{-1} of complete diet): Zinc, 50; copper, 12; iodine, 0.3; cobalt, 0.2; iron, 110; selenium, 0.1; manganese, 110.
[d] Mycotoxin sequestrant
[e] Furazolidon-bacitracin-zinc
[f] Saponified xanthophylls of Aztec marigold (yellow, 15 ppm)
[f] Red xanthophylls (canthaxanthin, 10 ppm)
C Control, *SO* sardine oil, *Mp M. pyrifera*, *Ss S. sinicola*, *Espp Enteromorpha* spp.)

Yolk Color This test was performed in single cabins and under white light, using a preference test to evaluate this attribute (Watts et al. 1992). Thirty-two untrained panelists, from both sexes, and usual consumers of eggs participated. They received 4 yolks from fresh egg, one for each treatment, labeled with random numbers and the same questionnaire as used for the sensory evaluation.

Statistical analysis All data were by ANOVA for a completely random design. The differences among means were compared by Tukey's test ($P \leq 0.05$). The GLM procedure of SAS v.6.12 (SAS Institute) was used.

Results

Egg production, feed intake, feed conversion as well as egg weight and shell thickness were not affected ($P>0.05$) by the inclusion of sardine oil (SO) and marine algae (MA) in the laying hen diet (Table 4). The treatment SO+Mp showed higher ($P<0.05$), albumen height and Haugh Units than SO+Espp. The egg yolk colour was lower ($P<0.05$) in the treatment SO+Espp than in all the other treatments, including the control group. The shell weight was lower in SO+Mp than the control group ($P<0.05$).

The egg total lipid content was only affected in the brown alga *M. pyrifera* treatment ($P<0.05$) showing a value slightly higher than the other groups (Table 5). Egg cholesterol content decreased only in the SO+Espp treatment ($P>0.05$). The eggs of the treatments that included SO+MA, mainly SO+Mp, had a higher saturated fatty acid (SFA) content (Table 5) than the control group ($P<0.05$). The monounsaturated fatty acids (MUFA), palmitoleic, cis-10-heptadecenoic, erucic and nervonic acid showed an increase in all treatments that included SO+MA, mainly SO+Mp; only the oleic and the cis-11-eicosenoic acids decreased. The highest content of the polyunsaturated acids (PUFA), linolelaidic, cis-11,14-eicosadienoic, tricosanoic, eicosapentaenoic (EPA) and docosahexaenoic (DHA)

Table 4 Production parameters and egg physical quality from hens fed with sardine oil and seaweeds

	Control	SO + Mp	SO + Ss	SO + Espp
Egg production (%)	87.3±5.6 [a]	81.4±1.8[a]	77.4±6.2[a]	82.9±2.85[a]
Feed intake ($g.hen^{-1}.d^{-1}$)	107.4±3.9 [a]	107.6±4.1[a]	109.6±4.1[a]	112.8±4.6[a]
Feed conversion	2.0±0.2 [a]	2.3±0.1[a]	2.35±0.1[a]	2.3±0.2[a]
Egg weight (g)	61.9±3.6 [a]	59.8±2.8 [a]	61.6±4.6 [a]	61.0±3.0 [a]
Yolk Colour	10.2±0.6 [a]	10.9±0.6 [a]	9.5±0.9 [b]	8.3±0.9 [c]
Albumen height (mm)	6.6±1.0 [ab]	7.5±0.8b [a]	6.7±0.9 [ab]	5.8±1.5 [b]
Haugh Units (HU)	79.9±6.9 [ab]	86.6±4.5 [a]	80.3±6.0 [ab]	72.5±16.5 [b]
Shell thickness (μm)	323±28 [a]	299±33 [a]	314±25 [a]	318±27 [a]
Shell weight (g)	5.5±0.5 [a]	5.0±0.5 [b]	5.2±0.5 [ab]	5.3±0.4 [ab]

[a,b,c] Different letters in each row indicate significant differences ($P<0.05$)
SO Sardine oil, *Mp M. pyrifera*, *Ss S. sinicola*, *Espp Enteromorpha* spp.

Table 5 Egg yolk lipids from hens fed with sardine oil and seaweeds

	Control	SO+Mp	SO+Ss	SO+Espp
Total lipids ($g.100g^{-1}$)	10.6±0.3 [b]	11.5±0.2 [a]	10.3±0.2 [b]	10.3±0.4 [b]
Cholesterol ($g.100g^{-1}$)	381±15 [a]	399±15[a]	385±8[a]	362±10[b]
Fatty acids (% total FA)				
Saturated total	32.62	35.25	34.57	34.56
C14:0	0.26	0.76	0.66	0.62
C15:0	0.04	0.14	0.12	0.12
C16:0	23.15	25.46	25.11	25.10
C17:0	0.19	0.45	0.43	0.38
C18:0	8.73	8.34	8.13	8.25
C20:0	0.03	0.03	0.06	0.07
C22:0	0.0	0.03	0.03	0.01
C24:0	0.22	0.04	0.03	0.03
Monounsaturated	41.47	41.01	41.91	41.96
C15:1	0.07	0.08	0.07	0.07
C16:1	1.89	3.76	3.35	3.46
C17:1	0.11	0.26	0.25	0.23
C18:1	39.10	35.96	37.14	37.43
C20:1	0.21	0.24	0.23	0.21
C22:1	0.00	0.03	0.62	0.02
C24:1	0.09	0.64	0.25	0.54
Polyunsaturated	25.98	23.74	23.62	23.33
C18:2	0.05	0.11	0.09	0.08
C18:2 n-6	18.80	10.49	11.37	11.48
C20:2	0.01	0.05	0.04	0.04
C18:3	0.16	0.07	0.07	0.09
C18:3 n-3	1.21	0.78	0.76	0.73
C20:3	0.20	0.08	0.09	0.10
C20:3	0.25	0.12	0.13	0.14
C24:3	0.08	0.04	0.03	0.02
C20:4 n-6	3.01	0.88	0.88	0.99
C20:5 n-3.	0.03	1.52	1.40	1.11
C22:6 n-3	2.18	9.60	8.76	8.55
Total n-6 FA	21.81	11.37	12.25	12.47
Total n-3 FA	3.42	11.9	10.92	10.39
n-6:n3 ratio	7:1	1:1	1:1	1:1
I:S ratio	1:2	1:2	1:2	1:2

[a,b] Different letters in each row indicate significant differences ($P<0.05$)
C Control, *SO* sardine oil, *Mp M. pyrifera*, *Ss S. sinicola*, *Espp Enteromorpha* spp.

acids, was in the treatments with SO+AM, mainly SO+Mp ($P<0.05$)

Sensory evaluation

When the marine algae *M. pyrifera, S. sinicola and Enteromorpha* spp. were included in the laying ration, the flavor of n-3 FA enriched eggs was not affected ($P>0.05$). In regard to yolk colou, only the treatment that included the green alga (SO+Espp) showed a lower score ($P<0.05$) than the other treatments.

Discussion

In general, under our conditions, the seaweeds *M. pyrifera, S. sinicola* and *Enteromorpha* spp. can be incorporated into the laying hens' diets, but is important to avoid the use of higher levels than 10% as this could affect the acceptance of the feed, lower the egg production or cause aqueous faeces in the birds, because of the high mineral content in the seaweed (Carrillo et al. 2002).

Several researchers have reported impaired production parameters in hens fed fish products, particularly reduced egg weight (Marshall and Van Elswyk 1994; González-Esquerra and Leeson 2000; Castillo et al. 2001; Castillo-Badillo et al. 2005). This observation seems to be a consequence of (1) the effect of n-3 FA on lipid metabolism and circulating oestradiol; (2) the lower serum lipids apparent in birds fed high amounts of n-3 FA, thus limiting the availability of lipids for yolk formation; and (3) a greater concentration of AGw3 which can reduce the concentration of vitamin E in plasma and tissue, which affects the formation of the egg as to produce an egg enriched with n-3 FA the hen requires twice the vitamin E that the liver normally has as a reserve (Van Elswyk 1997; González-Esquerra and Leeson 2000; Surai and Sparks 2000). In our study, we observed no change in egg weight, when the seaweeds were add to laying hens rations.

The reduction in the yolk color in the treatment that included *Enteromorpha* spp. (SO+Espp) could be a consequence of the amount and type of carotenoids in this alga. It is possible that the pigments of *Enteromorpha* spp. (lutein, neoxanthin, violaxanthin, zeaxanthin, antheraxanthin and cryptoxanthin) are more sensitive to sunlight as well as to the drying and storage, compared to those of the brown seaweeds (violaxanthin, flavoxanthin, fucoxanthin, zeaxanthin, neoxanthin, fucoxantinol and occasionally lutein) (Ragan 1981).

The differences observed in the shell weight could be a consequence of the form in which this measurement was made; it is possible that the shell still had albumen remains at the time of weighing. Other authors (Rojkind 1977; Rodriguez et al. 1995; Ramos et al. 1997; Rendón et al. 2003) that have used up to 10% of seaweed in the ration have not detected any effect on shell weight.

The observed increase in the total lipid content present in the SO+Mp treatment could be related to the highest egg cholesterol and fatty acid content. Nevertheless, there is no clear explanation for this. We had hoped to decrease the cholesterol concentration with the addition of seaweed and fish oil as a result of the hypocholesterolemic properties attributed to them; the seaweed, by action of its sterols (ergosterol, chondrosterol, fucosterol, sitosterol, campesterol) and polysaccharides (alginic acid, fucoidan cellulose, rhamnose, xylose and glucuronic acid) with hypocholesterolemic activity (Jiménez-Escrig and Goñi 1999; Freile 2001); and the fish oil by action of the n-3 FA. Nevertheless, in this current study, the only effect was observed in the SO+Espp treatment, with a 5% reduction compared to the control group. It is possible that the mineral salts and dietary fiber from seaweeds caused aqueous faeces, which did not allow an efficient use of seaweed compounds and of n-3 FA. It could also be that, as Noble (1999) indicates, it is not easy to reduce the cholesterol, since unlike fatty acids, the cholesterol has proven to be extremely resistant to any change. For that reason, the results obtained until now have not been very successful.

As far as fatty acids go, the results in this study confirm that this fraction is one of the easiest to be modified through the amount and type of fat that is incorporated in the diet of the birds. Some authors (Summers et al. 1966; Noble 1999) maintain that the changes that can be obtained in the composition of the saturated fatty acids (SFA) are minimal compared to the changes that can be observed in monounsaturated fatty acids (MUFA) and the polyunsaturated fatty acids (PUFA). Other authors such as Simopoulos (2000), Pappas et al. (2005) and Grobas and Mateos (1996), note an increase in all saturated fatty acids, when fish oil or fish meal have been incorporated into the poultry rations. The results obtained in this study agree with those reported by Simopoulos (2000) and Grobas and Mateos (1996). Grobas et al. (2001) observed a reduction in the concentration of oleic acid and cis-11-eicosenoic acid in diets that included flaxseed oil as a source of n3 FA.

Is interesting to indicate that, although in the fish oil used the amounts of EPA and DHA are similar (15 and 11%, respectively), the proportion of these in the egg was not the same. The egg DHA concentration was 6 times greater than EPA; these findings were also observed by Huang and Liu (1999). The explanation for this could be related to the metabolism of the n-3 FA in the birds, although there can be a conversion of DHA to EPA and vice versa, there is a preference for the deposition of DHA in tissues (González-Esquerra and Leeson 2001).

The total concentration of n-3 FA obtained in this study was superior in all the treatments that included SO+MA. An average ratio n-6 FA:n-3 FA from 1:1 was kept in treatments with MA. Simopoulos (2000) suggests ideal proportions of 2:1 or 1:1, since the n-6 FA and the n-3 FA are equally implied in the production of eicosanoids, prostaglandins, thromboxans and leucotrienes. Therefore, to keep a balance is essential for a good health state.

One of the aspects why the use of fish oil or fish meal in the diets of poultry is often avoided, is that it eventually transmits an off flavor to the egg, which reduces the sensorial quality. In the current study, the flavor of the egg was not affected. On many occasions, the oxidation products can be responsible for this unpleasant flavor, a reason why the use of antioxidants has been suggested to reduce this effect. Future work on oxidation and antioxidants in trials comparing fish oil and fish oil with addition of seaweed is suggested because the marine algae have compounds with antioxidant activity (Herber-McNeill and Van Elswyk 1998; Freile 2001; Allen et al. 2001).

The lower preference by egg yolk color of the SO+Espp treatment was possibly related to the lighter coloration of the yolks in comparison with the other treatments.

It is concluded that the incorporation of 10% *M. pyrifera* in the laying hens' diets is an effective way to increase the n-3 FA content, the albumen height and yolk color, but not the egg weight, when these are enriched with n-3 FA from fish oil.

References

Allen VG, Pond KR, Sakers KE, Fontenot JP, Bagley CP Ivy RL, Evans RR, Schmidt RE, Fike JH, Zhang X, Ayad JY, Brown CP, Millar MF, Montgomery J, Mahan J, Wester DB, Melton C (2001) Tasco: Influence of a brown seaweed on antioxidant in forages and livestock- A review. J Anim Sci 79(E Suppl):E21–E31

AOAC (2000) Official methods of analysis of AOAC internacional. AOAC Internacional, USA, pp 19–24 Chapter 34 (3), 41

Baucells MD, Crespo N, Barroeta AC, López-Ferrer S, Grashorn MA (2000) Incorporation of different polyunsaturated fatty acids into eggs. Poult Sci 79:51–59

Carrillo DS, Casas VMM, Ramos RF, Pérez-Gil RF, Sánchez RI (2002) Algas marinas de Baja California Sur, México: Valor nutrimental. Arch Lat Nutr 52:400–405

Castillo DRM, Carrillo DS, Pérez-Gil RF, Avila GE, Cassis NL (2001) Aceite de sardina como fuente de ácidos grasos omega-3 y su efecto sobre la calidad y sabor del huevo para plato. XXVI Convención Anual ANECA, 25–28 abril 2001, Acapulco, Guerrero, pp 67–69

Castillo-Badillo C, Vazquez-Valladolid JL, González-Alcorta M, Morales-Barrera E, Castillo-Domínguez RM, Carrillo-Domínguez S (2005) El aceite de atún como fuente de ácidos grasos w-3 en el huevo para consumo. Rev Grasas Aceites 56:154–162

Cherian G, Wolfe FW, Sim JS (1996) Dietary oils added tocopherols: effects on egg or tissue tocopherols, fatty acids, and oxidative stability. Poult Sci 75:423–431

Fenton M, Sim JS (1991) Determination of egg yolk cholesterol content by on-column capillary gas chromatography. J Chromatogr 540:323–329

Freile PY (2001) Las algas en la Botica. Avance Perspect 20:283–292

González-Esquerra R, Leeson S (2000) Effect of feeding hens regular or deodorized menhaden oil on production parameters, yolk fatty acid profile, and sensory quality of eggs. Poult Sci 79: 1597–1602

González-Esquerra R, Leeson S (2001) Alternatives for enrichment of eggs and chicken meat with omega-3 fatty acids. Can J Anim Sci 81:295–305

Grobas S, Mateos GG (1996) Influencia de la Nutrición sobre la composición Nutricional del huevo. Curso de Especialización FEDNA Madrid España

Grobas S, Méndez J, Lázaro R, De Blas C, Mateos GC (2001) Influence of source and percentage of fat added to diet on performance and fatty acid composition of egg yolks of two strains of laying hens. Poultry Sci 80:1171–1179

Herber-McNeill SM, Van Elswyk ME (1998) Dietary marine algae maintains egg consumer acceptability while enhancing yolk color. Poult Sci 77:493–496

Honkatukia M, Reese K, Preisinger R, Tuiskula-Haavisto M, Weigend S, Roito J, Mäki-Tanila A, Vilkki J (2005) Fishy taint in chicken eggs is associated with a substitution within a conserved motif of the *FMO3* gene. Genomics 86:225–232

Huang YS, Liu JW (1999) Fatty acids. Metabolism. In: Sadler MJ, Strain JJ, Caballero B (eds) Encyclopedia of Human Nutrition. Academic Press, London, pp 730–737

Jiménez-Escrig A, Goñi CI (1999) Evaluación Nutricional y efectos fisiológicos de microalgas marinas comestibles. Arch Lationoam Nutrición 49:114–119

Kris-Etherton PM, Williams RD, Harris S, Lawrence JA (2002) Fish consumption, fish oil, omega 3 fatty acids, and cardiovascular disease. Circulation 106:2747–2757

Leskanich CO, Noble RC (1997) Manipulation of the n-3 polyunsaturated fatty acid composition of avian eggs and meat. J Food Sci 54:1457–1460

Marshall AC, Van Elswyk ME (1994) Oxidative stability and sensory quality of stored eggs from hens fed 1.5% menhaden oil. J Food Sci 59:261–263

Meza AML, Carrillo DS, Pérez-Gil RF, Avila GE, Casas VM (1996) Albumen quality in eggs from hens fed rations containing marine algae. 45th Western Poultry Disease Conference. May 1–5, 1996, Cancun, México

Noble RC (1999) Manipulation of the nutritional value of eggs. In: Wiseman J, Garnsworthy PC (eds) Recent developments in Poultry Nutrition 2. Nottingham University Press, UK, pp 251–268

Pappas AC, Acamovic-Sparks NCH, McDvitt RM (2005) Effects of supplementing broiler breader diets with organic selenium and polyunsaturated fatty acids on egg quality during storage. Poult Sci 84:865–874

Ragan MA (1981) Chemical constituents of seaweeds. In: Lobban CS, Wynne MJ (eds) The biology of seaweeds. University of California Press, USA, pp 589–626

Ramos F,Carrillo S, Pérez-Gil F, Avila E, Casas M (1997) Egg quality when including *Ulva lactuca* and *Macrocystis pyrifera* in laying hens rations. 16th International Congress of Nutrition, Montreal, Canada. pp 358

Rendón LU, Carrillo DS, Arellano MLG, Casas VMM, Pérez-Gil RF, Avila GE (2003) Chemical composition of the residue of alginates (*Macrocystis pyrifera*) extraction. Its utilization in laying hens feeding. Cuban J Agric Sci 37(3):287–293

Rodríguez BM, Carrillo DS, Pérez-Gil RF, Avila GE, Casas VM (1995) Efecto sobre la calidad del huevo y cascarón al incluir algas marinas *Sargassum sinicola* y *Ulva lactuca* en raciones para ponedoras. Memorias XX Convención de la Asociación

Nacional de Especialistas en Ciencias Avícolas, ANECA. Acapulco, Gro. México, pp 291–297
Rojkind AR (1977) Algas marinas bentónicas como suplemento en la alimentación animal 1: Ensayos con pollos y gallinas ponedoras. Contribución técnica No. 19. Centro de Investigación de Biología Marina, Estación Puerto Deseado y Estación Austral. Buenos Aires, Argentina, p 24
Simopoulos AP (2000) Human requirement for n-3 polyunsaturated fatty acids. Poult Sci 79:961–970
Summers JD, Slinger SJ, Anderson WJ (1966) The effect of feeding various fats and fat by-products on the fatty acids and cholesterol composition of eggs. Br Poult Sci 7:127–134
Surai PF, Sparks NHC (2000) Tissue specific fatty acids and alfa-tocopherol profiles in male chickens depending on dietary tuna oil and vitamin E provision. Poult Sci 79:1132–1142
Van Elswyk ME (1997) Comparision of n-3 fatty acids sources in laying hen rations for improvement of whole egg nutritional quality: a review. Br J Nutr 78(Suppl 1):S61–S69
Van Elswyk ME, Dawson PL, Sams AR (1995) Dietary menhaden oil influences sensory characteristics and headspaces volatiles of shell eggs. J Food Sci 60:85–89
Watts BM, Ylimaki GL, Jeffrey LE, Elías LM (1992) Métodos sensoriales básicos para la evaluación de alimentos. Centro Internacional de Investigación para el Desarrollo. Ottawa, Canadá, pp 110

Physiological differences in the growth of *Sargassum horneri* between the germling and adult stages

Han Gil Choi · Ki Hoon Lee · Hyun Il Yoo · Pil Jun Kang · Young Sik Kim · Ki Wan Nam

Originally published in the Journal of Applied Phycology, Vol 20, No 5, 279–285.
DOI: 10.1007/s10811-007-9281-5

Abstract The effects of temperature, irradiance, and daylength on *Sargassum horneri* growth were examined at the germling and adult stages to discern their physiological differences. Temperature–irradiance (10, 15, 20, 25, 30°C × 20, 40, 80 µmol photons $m^{-2}s^{-1}$) and daylength (8, 12, 16, 24 h) experiments were carried out. The germlings and blades of *S. horneri* grew over a wide range of temperatures (10–25°C), irradiances (20–80 µmol photons $m^{-2}s^{-1}$), and daylengths (8–24 h). At the optimal growth conditions, the relative growth rates (RGR) of the germlings were 21% day^{-1} (25°C, 20 µmol photons $m^{-2}s^{-1}$) and 13% day^{-1} (8 h daylength). In contrast, the RGRs of the blade weights were 4% day^{-1} (15°C, 20 µmol photons $m^{-2}s^{-1}$) and 5% day^{-1} (12 h daylength). Negative growth rates were found at 20 µmol photons $m^{-2}s^{-1}$ of 20°C and 25°C treatments after 12 days. This phenomenon coincides with the necrosis of *S. horneri* blades in field populations. In conclusion, we found physiological differences between *S. horneri* germlings and adults with respect to daylength and temperature optima. The growth of *S. horneri* germlings could be enhanced at 25°C, 20 µmol photons $m^{-2}s^{-1}$, and 8 h daylength for construction of *Sargassum* beds and restoration of barren areas.

H. G. Choi (✉) · K. H. Lee · H. I. Yoo
Faculty of Biological Science and Research Institute for Basic Science, Wonkwang University, Iksan, Jeonbuk 570-749, South Korea
e-mail: hgchoi@wonkwang.ac.kr

P. J. Kang · K. W. Nam
Department of Marine Biology, Pukyong National University, Busan 608-737, South Korea

Y. S. Kim
School of Marine Life Science, Kunsan National University, Kunsan, Jeonbuk 573-701, South Korea

Keywords Daylength · Growth · Irradiance · Korea · *Sargassum horneri* · Temperature

Introduction

The genus *Sargassum* (Fucales, Phaeophyceae) is distributed worldwide and forms dense monspecific or mixed stands in sublittoral zones (Yoshida 1983). Several species of *Sargassum* form underwater forests that provide habitat and spawning grounds to marine invertebrates and fish, thus playing an important role as a primary producer (Tsukidate 1992; Martin-Smith 1993; Choi et al. 2003).

Recently, *Sargassum* and *Ecklonia* beds have declined markedly, as a result of being covered by crustose red algae, and they have not recovered naturally in coastal areas of Korea. Such barren areas have been slowly expanding, and are known as "Isoyake" areas in Japan. Barren areas are created mainly by the intensive grazing pressures of herbivorous marine animals such as sea urchins and abalones (Harrold and Pears 1987; Largo and Ohno 1993). This increase in barren grounds in coastal areas has resulted in the decrease of marine resources such as fish, shellfish, and abalone. Thus, the construction of artificial underwater forests and the restoration of seaweed beds with *Eisenia*, *Ecklonia*, and *Sargassum* has been received particular attention from ecologists and phycologists (Yamauchi 1984; Largo and Ohno 1993; Terawaki et al. 2001; Choi et al. 2003).

Sargassum horneri (Turner) C. Agardh is a large dioecious alga growing up to 3–5 m long that forms underwater forest along the coasts of Korea (Kang 1968; Choi et al. 2003). Like other *Sargassum* spp., *S. horneri* is very important for marine ecosystems and is also used for fertilizer and food such as salads (Kang 1968). In coastal areas of Korea, *S. horneri* grows rapidly from November to March, produces receptacles from late March to July, and necroses from late July to September. Thus, due to its fast growth rate, *S. horneri* could be used to construct underwater forests in seaweed barrens. For example, *S. horneri* seedlings grow up to 2–3 m long within 10 months after transplantation (Yamauchi 1984; Choi et al. 2003). To date, the life cycle and reproduction period of *S. horneri* have been examined by Japanese phycologists (Uchida 1993; Uchida and Arima 1993). In Korea, however, the effects of environmental factors on the growth of *S. horneri* germlings have not been examined, even though it is essential to know the optimal growth conditions for producing seedlings in order to construct *Sargassum* beds. Uchida and Arima (1993) observed two types of *S. horneri*: one produced embryos in the spring and the other in the autumn, and their growth responses to environmental factors were different. Thus, *S. horneri* growing on the rocky shores of Korea might have different physiological characteristics compared to Japanese *S. horneri*, and these algae may also differ in their germling and adult stages, since they grow during different seasons in the field. Therefore, the aim of this study was to examine the effects of environmental factors on the growth of *S. horneri* at the germling and adult stages, in order to determine the physiological differences between these stages.

Materials and methods

Germling experiments

Fertile female and male plants of *S. horneri* were taken from the intertidal zones of Huppo, Wuljin, in June 2005 for preliminary culture experiments. The present data were obtained from fertile plants of *S. horneri* collected at Jeonchonri, Wolsung, Kyungbuk, Korea (35° 47′ N, 129° 29′ E) in May 2006. To eliminate the effect of genetic difference on the growth of *S. horneri* germlings, one female and one male plant were used to release embryos, and their lengths were 175 cm and 135 cm, respectively. Fifty female receptacles with embryos, as well as male receptacles, were excised from parent plants and their lengths and widths were measured. They were then dried in an oven at 80°C for 5 days and weighed. The lengths, widths and dry weights of 50 air bladders of *S. horneri* were also measured by the same method.

For the germling culture, the female and male receptacles were cleaned several times with filtered seawater to remove diatoms and detritus attached to the receptacles. They were then put in a plastic tank to induce fertilization and release of embryos. After 24 h the receptacles were removed and a clean suspension of eggs was prepared as described by Creed et al. (1996).

A zygote suspension was inoculated into 20 Petri dishes, each containing eight glass slides cut to 2.5×2.0 cm and 30 mL autoclaved seawater. After 24 h, we selected the 114 slides on which the germlings were most evenly distributed to use in the daylength and temperature–irradiance experiments. The settlement density of the germlings was 16.6± 5.74 germlings cm^{-2} (mean±SD, n=30).

Two experiments were carried out in the laboratory to investigate the influence of environmental factors on the growth of *S. horneri*. In the temperature–irradiance experiment, a total of 45 beakers were used and each beaker contained 150 mL PESI medium (Tatewaki 1966) to which GeO_2 (2 mL) per 1 L PESI had been added to inhibit the growth of diatoms and protozoa. Two slides with germlings were put into each beaker, and the beakers were then placed within five incubators that had been set up at 10, 15, 20, 25, and 30°C and 20, 40, and 80 μmol photons m^{-2}s^{-1} with a 16:8 h light:dark cycle, and cultured for 12 days. Light was provided by 40 W cool-white fluorescent tubes and the level of irradiance was measured using a digital illumination meter (DX-200). Different irradiance levels (20, 40, and 80 μmol photons m^{-2}s^{-1}) were created by covering the beakers with black plastic mesh. This experiment was repeated three times and the culture medium was changed every 4 days during the 12-day culture period.

After 12 days, the lengths of 30 germlings, excluding the rhizoids, were measured in each replicate beaker. The relative growth rate (RGR % day^{-1}) was calculated with the mean length for each replicate using the following equation (Rueness and Tananger 1984):

$$RGR(\% day^{-1}) = 100 \ln (L_t/L_o)/t$$

where L_o is the initial plant length, L_t is the final length after t days, and t is the number of days.

For the daylength experiment, 12 beakers, each containing two cut slides with attached germlings and 150 mL PESI medium with GeO_2 (2 mL), were prepared as described above. The beakers were placed in four incubators that had been set up at 8, 12, 16, and 24 h light per 24 h. The plants were cultured for 12 days at 20±1°C and 40 μmol photons m^{-2}s^{-1}. The experiments were replicated three times and the culture medium changed every 4 days throughout the experimental period.

Adult stage experiments

Twenty vegetative plants of *S. horneri* were collected from the intertidal zone of Jeonchonri, Wolsung, Kyungbuk, Korea in October 2006. The plants were transported to the laboratory and their lengths and wet weights were measured.

The vegetative plants ranged between 23 and 59 cm in length and 2.84–23.34 g in wet weight. The average length and wet weight of the vegetative plants were 35.90±9.95 cm and 9.96±6.17 g (mean±SD, *n*=20 individuals), respectively. The apical blades from one vegetative plant were cut and rinsed several times with filtered seawater to remove diatoms and detritus attached to the blades. The healthy apical blades, 10 mm in length and 5.66±1.51 mg (mean±SD, *n*=30 individuals) in wet weight, were excised and kept for 2 days to reduce any negative effects of the cutting.

We used a total of 36 beakers, each of which contained 150 mL PESI medium to which GeO_2 (2 mL per 1 L PESI) had been added to inhibit the growth of diatoms and protozoa. Three apical blades were put into each beaker and placed in four incubators that had been set up at 10, 15, 20, and 25°C and 20, 40, and 80 µmol photons $m^{-2}s^{-1}$ and 16:8 h LD (light:dark), and cultured for 12 days. The experiments were replicated three times and the culture medium was changed every 4 days throughout the experimental period.

To examine the effects of daylength on blade growth we prepared 12 beakers that contained three apical blades (10 mm long) and 150 mL PESI medium with GeO_2. Three apical blades were put into each beaker and placed in four incubators that had been set up at 8, 12, 16, and 24 h light/24 h. The plants were cultured for 12 days at 40 µmol photons $m^{-2}s^{-1}$ and 20±1°C. The RGR (% day^{-1}) was calculated using the mean blade length and weight of each replicate with the equation described above.

Statistical analyses

Statistical analyses were carried out using STATISTICA version 5.0 software. A two-way ANOVA was used to test the effects of temperature and irradiance on the RGR of germling lengths, blade lengths, and blade weights. A one-way ANOVA was used to determine the differences in RGR for germling lengths, blade lengths, and blade weights of specimens cultured at four different daylengths. A Tukey test was applied when significant differences were detected between the means (Sokal and Rohlf 1995). Homogeneity of the variance was tested using Cochrane's test (Underwood 1997).

Results

Characteristics of parent plants

The female receptacles of *S. horneri* were cylindrical but much thicker and shorter than the male receptacles. The lengths of the female receptacles ranged between 13 and 24 mm and the average length was 19.46±1.80 mm (mean±SD, *n*=50 receptacles). The widths of the female receptacles with attached germlings were 3–5 mm and the average width was 3.70±0.51 mm (*n*=50 receptacles). After germling release, the receptacle width was about 2 mm. The dry weight of each receptacle was 10.47±7.82 mg (mean±SD, *n*=50). The male receptacle was 31.16±13.40 mm (*n*=50) in length and was 2.77±0.72 mm (*n*=50) in width.

The air bladder of *S. horneri* was also cylindrical with a length of 10.48±1.63 mm (mean±SD, *n*=50 individuals), and a width of 1.26±0.31 mm (*n*=50). The mean dry weight of the air bladders with a serrated leaf attached to the top was 1.90±0.86 mg (mean±SD, *n*=50). The mean length of the serrated leaves of the air bladders was 12.94±3.11 mm (mean±SD, *n*=50).

Effects of temperature and irradiance on growth

Germling stages

The germlings of *S. horneri* grew at temperatures ranging from 10 to 25°C and irradiances between 20 and 80 µmol photons $m^{-2}s^{-1}$ after 12 days (Fig. 1). Germling growth was inhibited between 25–30°C and, at 30°C, approximately 85% of the germlings died when no rhizoids were produced within 12 days, irrespective of the irradiance levels. Thus, all data obtained from 30°C were excluded from the statistical analyses. The germling lengths ranged between 0.5 and 2.93 mm with a maximal length of 2.93±0.14 mm (mean±SE, *n*=3 replicates) at 25°C and 20 µmol photons $m^{-2}s^{-1}$. The relative growth rates of *S. horneri* were between 6.22 and 21.17 % day^{-1}, with a maximal rate of 21.17±0.38% day^{-1} at 25°C and 20 µmol photons $m^{-2}s^{-1}$.

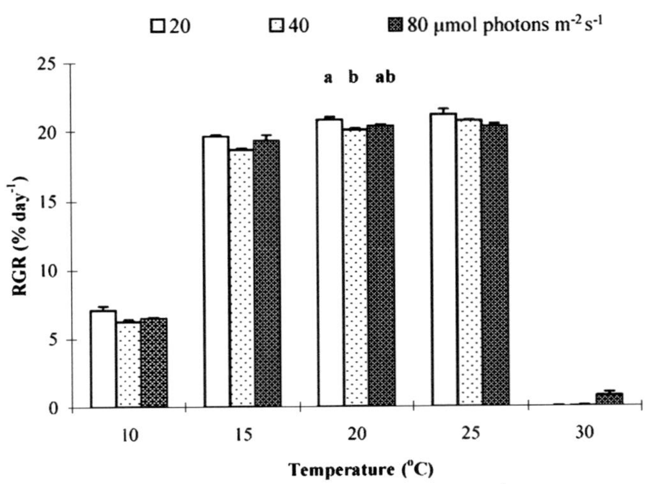

Fig. 1 Mean(±SE) relative growth rate (RGR; % day^{-1}) of *Sargassum horneri* germlings. Seaweeds were cultured at various temperatures and irradiances for 12 days. Each mean value is based on three replicates. Lower case letters indicate significant differences based on a Tukey test

The RGRs of the *Sargassum* germlings were greater at higher temperatures (15–25°C) than at 10°C, and were found to be significantly different among the various temperatures ($P<0.001$). A Tukey test revealed that the RGRs were significantly different between temperatures, except between 20 and 25°C (Fig. 1). The germlings of *S. horneri* grew faster at 20 μmol photons $m^{-2}s^{-1}$ than at 40 and 80 μmol photons $m^{-2}s^{-1}$ ($P<0.001$), but no differences were found at the higher irradiances ($P>0.05$). There was no significant interaction between the effects of temperature and irradiance levels on the RGR of the *Sargassum* germlings ($P=0.34$).

Adult stages

After 12 days, the RGRs of *S. horneri* were very low, ranging from 0.08 to 1.88 % day^{-1} for blade length and between 0.11 and 4.46% day^{-1} for blade weight (Fig. 2a,b). For blade length, we found significant differences in the RGRs of *S. horneri* between irradiance levels, but not for temperature (Table 1). The RGR of *S. horneri* for blade length was greater at low irradiances (20, 40 μmol photons $m^{-2}s^{-1}$) than at 80 μmol photons $m^{-2}s^{-1}$ (Table 1). For *S. horneri* blade length, a negative growth rate was found at 25°C and 80 μmol photons $m^{-2}s^{-1}$. This indicates that the growth of *S. horneri* blades is inhibited at high irradiance levels and high seawater temperatures. The RGR for blade length was significantly greater at 40 than at 80 μmol photons $m^{-2}s^{-1}$. For blade weight, the RGRs were greater than those of blade length. This result indicates that *Sargassum* blades grew mainly in width and thickness at this life stage. Also, there were significant differences in the relative growth rates of blade weight between temperatures and irradiance levels (Table 1). The growth in blade weight was superior at lower temperatures (10, 15°C) than at higher temperatures (20, 25°C). The RGR for blade weight was significantly greater at 15°C, the optimal temperature, than at 20 and 25°C (Tukey test). With respect to the adults, negative RGR values were observed at an irradiance of 80 μmol photons $m^{-2}s^{-1}$ and temperatures of 20 and 25°C after 12 days. In the temperature-irradiance experiment, maximal growth occurred at 15°C and 20 μmol photons $m^{-2}s^{-1}$ for both blade length and weight.

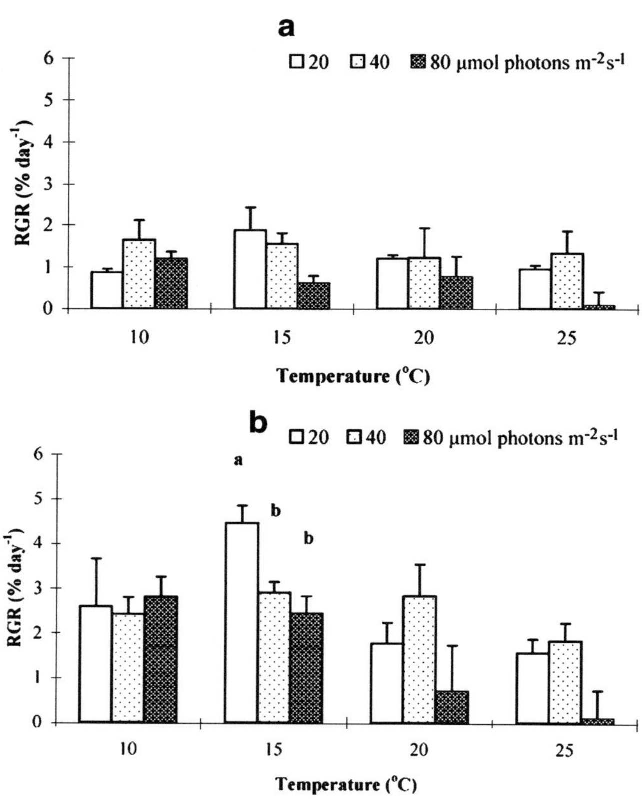

Fig. 2 Mean (±SE) RGR (% day^{-1}) of *S. horneri* for blade length (**a**), and blade weight (**b**). Seaweeds were cultured at various temperatures and irradiances for 12 days. Each mean value is based on three replicates. Lower case letters indicate significant differences based on a Tukey test

Effects of daylength on growth

Germling stages

The initial length of the *S. horneri* germlings was 230.55± 41.88 μm (mean±SD, $n=80$ spores). The germlings grew well in a wide range of daylengths ranging between 8 and 24 h and grew to lengths of 1.02–1.28 mm after 12 days. The RGRs for germling length were from 11.53 to 13.42% day^{-1} (Fig. 3), and the RGRs of germling length were significantly different among daylengths ($F_{3,8}=29.51$, $P<0.001$). The optimal daylength for germling growth was 8 h, and the growth of the germlings was slowed at 24 h of daylength. A Tukey test showed that the RGRs of the germlings were significantly greater at 8 h than at 24 h of daylength (Fig. 3).

Adult stages

After 12 days the RGRs of the *S. horneri* blades were between 0.35 and 2.72 % day^{-1} for length, and 2.25 and 4.93% day^{-1} for weight in the range of daylengths between 8 h and 24 h (Fig. 4a,b). The maximal RGR was observed at 12 h of daylength for both blade length and weight. A Tukey test revealed that the blade length RGR decreased significantly only at 24 h of daylength ($F_{3,8}=7.18$, $P<0.05$), but no differences were found for blade weight among the daylengths ($F_{3,8}=6.48$, $P<0.05$). After 16 days the blades of *S. horneri* began to necrose from the apical area, most severely at 24 h of daylength as compared to the blades of other daylengths.

Table 1 Results of two-way ANOVA and Tukey tests for the effects of temperature and irradiance on relative growth rate (RGR, % day^{-1}) of *Sargassum horneri* blades grown at various temperature and irradiance levels

Factor	df	MS	F	*P*	MS	F	*P*
		Length			Weight		
Temperature	3	0.53	1.20	0.33	7.67	7.26	<0.01
Irradiance	2	1.91	4.25	<0.05	4.33	4.09	<0.05
Interaction	6	0.42	0.93	0.49	1.71	1.62	0.18
Residuals	24	0.45			1.06		
Tukey test (*P*=0.05)							
Temperature		10=15=20=25			10=15=20>25, 15>20>25, 20=25		
Irradiance		20=40=80, 40>80			20=40>80, 40=80		

Discussion

The growth of *Sargassum horneri* was affected by daylength at both the germling and adult stages. The germlings of *S. horneri* grew better under 8 h of daylength than under constant light. This growth trend of the *S. horneri* germlings was similar to shoot formation of *S. horneri* (Uchida 1993), and to the main branch elongation and receptacle growth of *S. muticum* (Hales and Fletcher 1990; Uchida et al. 1991). Also, the length and weight growths of *S. horneri* blades were maximal at 12 h of daylength (neutral day). These results indicate that the effects of daylength on the growth of *S. horneri* are slightly different between the germling and adult stages. Also, these growth responses to daylength are very important for understanding the growth patterns of natural *S. horneri* populations. In Korea, daylength ranges between 9.5 and 14.5, being maximal in June and minimal in December (Choi 1996). In field populations of *S. horneri*, the main growth period for the juveniles and adults is from October to March, with 9.5–11.5 h daylength. Thus, the results found with culture experiments are very similar to the daylength growth responses of germlings and adults in natural populations.

Temperature and irradiance are very important environmental factors influencing the growth of germlings and adults in *Sargassum* spp. (Hales and Fletcher 1989, 1990). In the present study, the maximum growth of *S. horneri* germlings was 20.74% day^{-1}(RGR) at 25°C, just as in *S. muticum* (Hales and Fletcher 1989). On the other hand, the average RGRs of *S. horneri* for blade length and blade

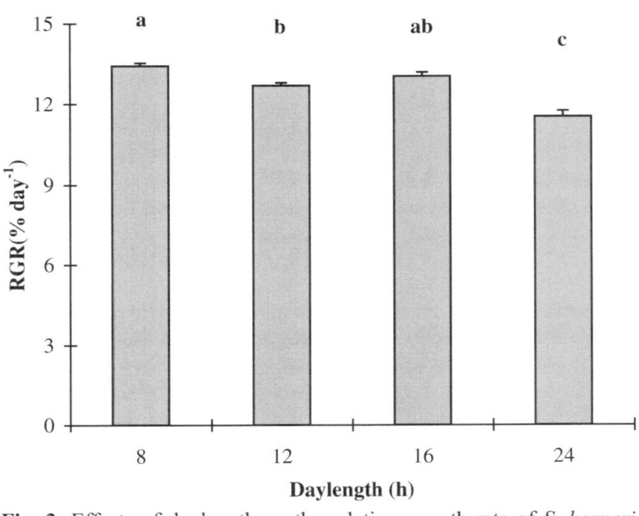

Fig. 3 Effects of daylength on the relative growth rate of *S. horneri* germlings. Seaweeds were cultured at various irradiance levels for 12 days. *Bars* show standard errors (*n*=3 replicates). Lower case letters indicate significant differences based on a Tukey test

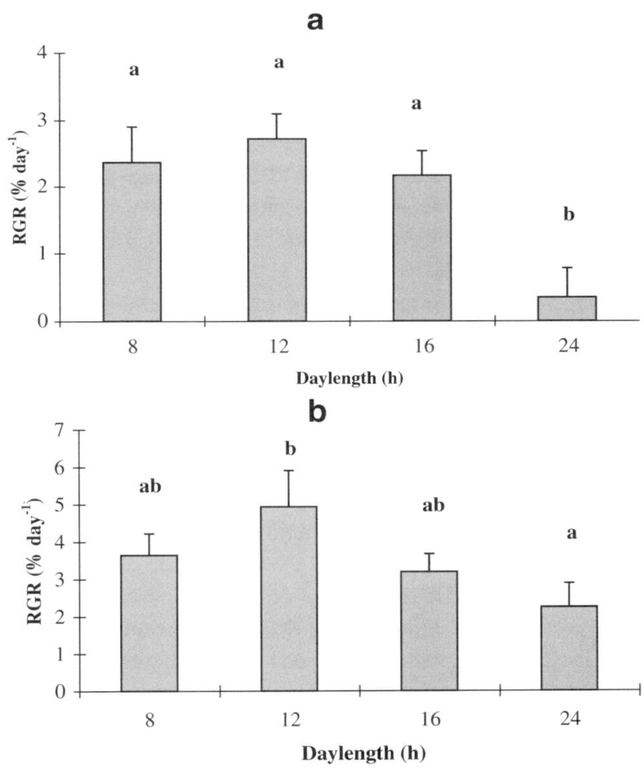

Fig. 4 Effects of daylength on the relative growth rate of *S. horneri* for blade length (**a**), and blade weight (**b**). Seaweeds were cultured at various irradiance levels for 12 days. *Bars* show standard errors (*n*=3 replicates). Lower case letters indicate significant differences based on a Tukey test

weight were very low, and the maximal value for blade weight was 4.46% day^{-1}(RGR) at 15°C. Hales and Fletcher (1989, 1990) found that maximal growth of *S. muticum* germlings and adults occurred at different salinity and irradiance conditions, but their optimal temperature for growth was exactly the same. Such different temperature requirements for optimal growth at germling and adult stages of *S. horneri* may closely relate to seawater temperature. The average seawater temperature of the sampling area ranged from 7.62°C in February to 23.94°C in September 2006. Thus, the fronds of *S. horneri* could grow well at 15°C, which corresponds to the seawater temperature of the sampling area in November. Germling growth was significantly greater at 20 μmol photons m^{-2}s^{-1} than at 40 and 80 μmol photons m^{-2}s^{-1}, indicating that *S. horneri* germlings require a very low irradiance for growth. This coincides with the light conditions of sublittoral zones, where *S. horneri* germlings grow in the field.

In the present study we found remarkable physiological differences in the growth of *S. horneri* between the germling and adult stages. At optimal growth conditions the RGRs of the germlings were 21% day^{-1} (25°C, 20 μmol photons m^{-2}s^{-1}) and 13% day^{-1} (8 h daylength), whereas those of the blade weights were 4% day^{-1} (15°C, 20 μmol photons m^{-2}s^{-1}) and 5% day^{-1} (12 h daylength), indicating that the RGRs of the germlings were 3–5 times greater than those of the blades. Similar growth rates were also found in transplanted *S. horneri* individuals as 3.3% day^{-1} for length and 4.7% day^{-1} for frond weight (Yamauchi 1984). *Sargassum* germlings grew faster than adult plants. For instance, the RGRs of transplanted *S. muticum* juveniles were 2.4% in length and 0.3% in weight, but its germlings grew 28% day^{-1}(Yamauchi 1984; Hales and Fletcher 1989).

Sargassum germlings grew faster than adults but this does not mean that germlings will be more successful for construction of *Sargassum* beds because germlings are vulnerable to invertebrate grazers (Choi et al. 2003). Transplantation of live adults to barren areas might be a better method to establish seaweed beds despite their lower growth rates, because they are more resistant to grazing, provide immediate habitat to marine animals, and act as a seed bank producing embryos. However, transplanting live adults is very difficult because of cost and labor requirements. Largo and Ohno (1993) described various construction methods of artificial seaweed beds. They are divided into spore techniques (spore dispersal, spore-bag, and rope-seeding method) and vegetative plant transplanting techniques (concrete block, threading, and gravel-bag methods). To date, the spore-bag and rope-seeding methods have been used in order to create *Sargassum* beds (Largo and Ohno 1993; Choi et al. 2003). Choi et al. (2003) found that *S. horneri* juveniles transplanted on tiles were completely consumed by herbivores (sea urchins and snails), but juveniles transplanted on iron pipes grew up to 313 cm in length and formed underwater forests. Thus, the construction of *S. horneri* beds by transplantation of juveniles is possible under lower grazer density as on iron pipes (Choi et al. 2003). In conclusion, we found physiological differences between *S. horneri* germlings and adults with respect to daylength and temperature optima. The growth of *S. horneri* germlings for construction of *Sargassum* beds and restoration of barren areas could be enhanced by the application of the following conditions: 25°C, 20 μmol photons m^{-2}s^{-1} and 8 h daylength.

Acknowledgments This paper was partially supported by a grant from KORDI (BSPM33000-1710-3) and from Wonkwang University in 2007.

References

Choi HG (1996) Life history and temperature and light responses in growth and reproduction of *Caulacanthus okamurai* (Gigartinales, Rhodophyta) in Korea. Dissertation, University of Pukyong

Choi CG, Kim HG, Sohn CH (2003) Transplantation of young fronds of *Sargassum horneri* for construction of seaweed beds. J Korean Fish Soc 36:469–473

Creed JC, Norton TA, Kain JM (1996) Are neighbours harmful or helpful in *Fucus vesiculosus* populations? Mar Ecol Prog Ser 133:191–201

Hales JM, Fletcher RL (1989) Studies on the recently introduced brown alga *Sargassum muticum* (Yendo) Fensholt. IV. The effect of temperature, irradiance and salinity on germling growth. Bot Mar 32:167–176

Hales JM, Fletcher RL (1990) Studies on the recently introduced brown alga *Sargassum muticum* (Yendo) Fensholt. V. Receptacle initiation and growth, and gamete release in laboratory culture. Bot Mar 33:241–249

Harrold C, Pears JP (1987) The ecological role of echinoderms in kelp forest. In: Jangoux M, Lawrence JM (eds) Echinoderms study, vol 2. Balkema, Rotterdam, pp 137–233

Kang JW (1968) Illustrated encyclopedia of fauna and flora of Korea, vol 8. Marine algae. Samhwa, Seoul

Largo DB, Ohno M (1993) Constructing an artificial seaweed bed. In: Ohno M, Critchley AT (eds) Seaweed cultivation and marine ranching. Japan International Cooperation Agency pp 113–130

Martin-Smith KM (1993) The phenology of four species of *Sargassum* at Magnetic Island, Australia. Bot Mar 36:327–334

Rueness J, Tananger T (1984) Growth in culture of four red algae from Norway with potential for mariculture. Hydrobiologia 116/117:303–307

Sokal RR, Rohlf FJ (1995) Biometry, 3rd edn. Freeman, New York

Tatewaki M (1966) Formation of a crustaceous sporophyte with unilocular sporangia in *Scytosiphon lomentaria*. Phycologia 6:62–66

Terawaki T, Hasegawa H, Arai S, Ohno M (2001) Management-free techniques for restoration of *Eisenia* and *Ecklonia* beds along the central Pacific coast of Japan. J Appl Phycol 13:13–17

Tsukidate J (1992) Ecology of *Sargassum* spp. and *Sargassum* forest formation. NOAA Technical Report NMFS 106:63–72
Uchida T (1993) The life cycle of *Sargassum horneri* (Phaeophyta) in laboratory culture. J Phycol 29:231–235
Uchida T, Arima S (1993) Crossing experiments between autumn- and spring-fruiting types of *Sargassum horneri* (Phaeophyta). Nippon Suisan Gakkaishi 59:1685–1688
Uchida T, Yoshikawa K, Arai A, Arai S (1991) Life cycle and its control of *Sargassum muticum* (Phaeophyta) in batch cultures. Nippon Susan Gakkaishi 57:2249–2253
Underwood AJ (1997) Experiments in ecology: their logical design and interpretation using analysis of variance. Cambridge University Press, Cambridge
Yamauchi K (1984) The formation of *Sargassum* beds on artificial substrata by transplanting seedlings of *S. horneri* (Turner) C. Agardh and *S. muticum* (Yendo) Fensholt. Bull Jpn Soc Sci Fish 50:1115–1123
Yoshida T (1983) Japanese species of *Sargassum* subgenus Bactrophycus (Phaeophyta, Fucales). J Fac Sci Hokkaido Univ Ser V (Botany) 13:99–246

Effects of temperature, photosynthetic photon flux density, photoperiod and O_2 and CO_2 concentrations on growth rates of the symbiotic dinoflagellate, *Amphidinium* sp.

Y. Kitaya · L. Xiao · A. Masuda · T. Ozawa · M. Tsuda · K. Omasa

Originally published in the Journal of Applied Phycology, Vol 20, No 5, 287–292.
DOI: 10.1007/s10811-008-9331-7

Abstract Symbiotic dinoflagellates of the species *Amphidinium* are expected to be pharmaceutically useful microalgae because they produce antitumor macrolides. A microalgae production system with a large number of cells at a high density has been developed for the efficient production of macrolide compounds. In the present study, the effects of culture conditions on the cellular growth rate of dinoflagellates were investigated to determine the optimum culture conditions for obtaining high yields of microalgae. *Amphidinium* species was cultured under conditions with six temperature levels (21–35°C), six levels of photosynthetic photon flux density (15–70 μmol photons m^{-2} s^{-1}), three levels of CO_2 concentration (0.02–0.1%), and three levels of O_2 concentration (0.2–21%). The number of cells cultured in a certain volume of solution was monitored microscopically and the cellular growth rate was expressed as the specific growth rate. The maximum specific growth rate was 0.022 h^{-1} at a temperature of 26°C and O_2 concentration of 5%, and the specific growth rate was saturated at a CO_2 concentration of 0.05%, a photosynthetic photon flux density of 35 μmol photons m^{-2} s^{-1} and a photoperiod of 12 h day^{-1} upon increasing each environmental parameter. The results demonstrate that *Amphidinium* species can multiply efficiently under conditions of relatively low light intensity and low O_2 concentration.

Keywords Algae · *Amphidinium* sp. · Photosynthetic photon flux density · Specific growth rate · Symbiotic dinoflagellate · Temperature

Y. Kitaya (✉) · L. Xiao
Osaka Prefecture University,
1-1, Gakuen-cho,
Sakai, Osaka 599-8531, Japan
e-mail: kitaya@envi.osakafu-u.ac.jp

A. Masuda · T. Ozawa
Yanmar Co.,
3286, Itohara, Musashi-machi,
Kunisaki, Oita 873-0421, Japan

M. Tsuda
Kochi University,
Monobe B200, Nankoku,
Kochi 783-8502, Japan

K. Omasa
The University of Tokyo,
Yayoi 1-1-1, Bunkyo,
Tokyo 113-8657, Japan

Introduction

The amphidinolides are a series of unique cytotoxic macrolides isolated from the dinoflagellate *Amphidinium* sp. that has been separated from marine acoel flatworms of the genus *Amphiscolops* (Kobayashi and Tsuda 2004; Kobayashi and Kubota 2007). Up to now, 39 amphidinolides have been reported (Oguchi et al. 2007). Amphidinolides H (Kobayashi et al. 1991, 2000) and B (Ishibashi et al. 1987; Kobayashi et al. 1989; Bauer et al. 1994), which were initially isolated from species of the marine dinoflagellate *Amphidinium* (strains Y-25 and Y-5, respectively), are 26-membered macrolides possessing unique structures such as an allyl epoxide and an *S*-cis-diene, and exhibit potent cytotoxicity (IC_{50} 0.0045–0.00014 mg mL^{-1}) against cultured tumor cells in vitro and antitumor activities in vivo. More recently, a new series of potently cytotoxic *Amphidinium* macrolides, designated iriomoteolides, have been isolated from the benthic *Amphidinium* strain HYA024 (Tsuda et al. 2007a, b). These macrolides are, therefore,

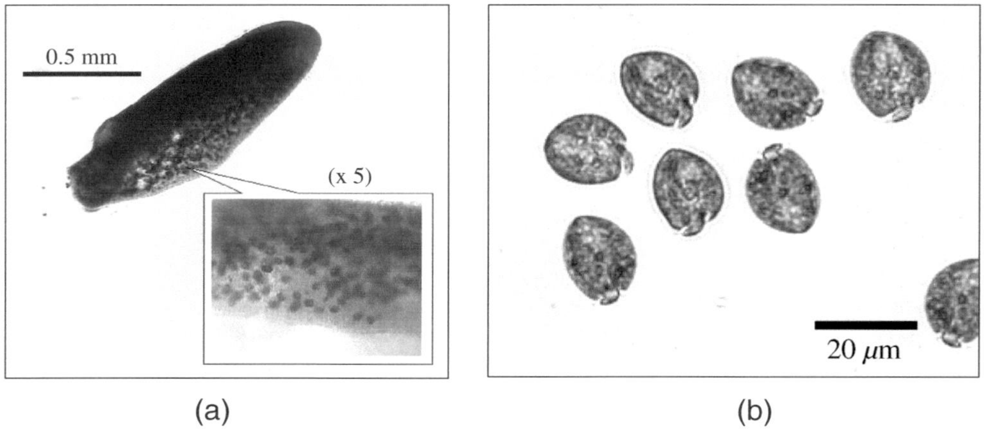

Fig. 1 **a** *Amphiscolops* species containing dinoflagellate cells. **b** Dinoflagellates of *Amphidinium* sp. separated from *Amphiscolops*

expected to be good candidates for antitumor drugs with a novel mechanism of action because their chemical structure is unlike that of conventional anticancer materials. Nevertheless, development of these compounds as anticancer drugs has not been smooth due to limited sample amounts.

Development of a production system for *Amphidinium* cells separated from *Amphiscolops* is desirable in order to produce macrolide compounds efficiently. However, the construction of a mass culture system for *Amphidinium* production has not yet been established because environmental control of the mass culture of *Amphidinium* cells at a high density is difficult using conventional culture methods. Microalgal cell growth rates are affected by environmental parameters such as temperature, light intensity and period, and gas composition in the culture system, as well as the combination of all these factors. Few papers have focused on environmental control of cell cultures of symbiotic *Amphidinium*. In order to establish a symbiotic *Amphidinium* production system with a large number of cells and a rapid turnover rate, the appropriate combination of environmental parameters needs to be determined.

This research was initiated to optimize culture condition for symbiotic *Amphidinium* production. The present study is a starting point for the optimization of the culture. Effects of variable culture conditions on the cell growth rate of dinoflagellates were investigated to determine the potentially optimum culture conditions for obtaining high yields of this microalga. Effects of temperature, photosynthetic photon flux density, photoperiod and CO_2 and O_2 concentrations on growth rates of dinoflagellate cells were assessed using a hanging-drop culture system.

Materials and methods

The dinoflagellate *Amphidinium* sp. (strain HYA002) was separated from the internal cells of the marine acoel

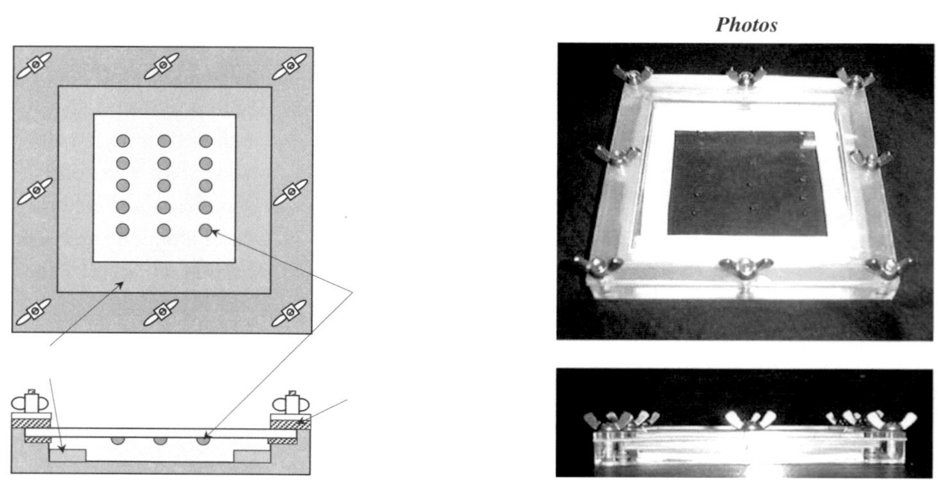

Fig. 2 Apparatus of the culture vessel with water droplets

flatworm *Amphiscolops* sp. (Fig. 1), which was collected off Sunabe beach, Okinawa, Japan (Oguchi et al. 2007). The voucher specimen and the SSU rDNA gene are deposited at the Center for Advanced Marine Core Research, Kochi University.

The experimental system for evaluating the effects of environmental variables on the growth rate of microalgal cells is shown in Fig. 2. The culture system was set up with a hanging-drop culture device according to the procedure reported by Kitaya et al. (1998, 2005). *Amphidinium* cells were cultured in water droplets (3 μL in volume each) in a vessel (100 mL in volume). Each water droplet is held on the surface of a translucent plastic plate and contains initially about ten *Amphidinium* cells. The culture vessel is completely sealed after placing water droplets with *Amphidinium* cells and controlling the internal CO_2 and O_2 concentrations at set levels. The vessel was sealed in a box in which the gas composition had been controlled beforehand. Changes in the gas composition in the vessel were almost negligible during the experiment. The number of water droplets in each treatment was at least five. The vessel was submerged in water at constant temperature. The water layer above the vessel cut out long-wave radiation from the light source, which would otherwise cause a heat load inside the vessel.

Wet paper inside the vessel kept the relative humidity at almost 100% and thus inhibited evaporation from the water droplets. The basal culture solution was sea water. Provasoli's Marine Water Enrichment solution (Provasoli 1963) as a supplemental nutrition, and $NaHCO_3$ as a carbon source for photosynthesis, were mixed with the basal culture solution. The initial pH in the culture solution was adjusted to 7.5. The salinity of the culture solution was adjusted to 3.3%.

The number of dinoflagellate cells in each water droplet was counted with a microscope at 70x magnification. The specific growth rate was used as a standardized index for the growth rate of cells. The specific growth rate (μ) was determined in the logarithmic multiplication stage by the following equation; $\mu = (\ln N_2 - \ln N_1) / (T_2 - T_1)$, where N_1 and N_2 are the cell numbers at times T_1 and T_2, respectively. In the present experiment, T_1 and T_2 were 48 and 144 h, respectively, after the start of the culture. The doubling time or the mean generation time (t_d), is given by the equation $t_d = \ln 2 / \mu$.

Cell growth rates determined in the hanging-drop culture method were compared with those in a conventional culture method to certify the hanging-drop culture method. In the conventional culture method, *Amphidinium* was cultured in aerated containers (210×140×70 mm depth and 800 mL medium volume) with the same culture condition as that in the water droplet method except culture vessel size and medium volume.

In a preliminary experiment, *Amphidinium* was cultured in solutions with four levels of Provasoli's Marine Water Enrichment concentrations (0–2%) and four levels of $NaHCO_3$ concentrations (0–10 mmol) to determine the chemical composition of the basal culture solution.

In the main experiment, *Amphidinium* was cultured at six different temperatures (21–35°C), four photosynthetic photon flux densities (PPFD, 15–70 μmol photons m^{-2} s^{-1}; measured on the inner surface of plastic cover of the hanging-drop culture device), three photoperiods (6–24 h), three CO_2 concentrations (0.02–0.1%), and three O_2 concentrations (0.2–21%). The light source was fluorescent lamps (FPL55EX-N, Matsushita Electric, Osaka, Japan). PPFD was controlled by covering each vessel with neutral shading films.

Results

Cells multiplied almost exponentially by cell division in both the conventional culture and the hanging-drop culture (Fig. 3). The cell density reached about 6.6×10^4 cells mL^{-1} in the conventional culture, while the cell density and total number of cells per hanging droplet were about 3.3×10^4 cells mL^{-1} and 98 cells per hanging droplet, respectively, 144 h after starting the culture. Specific growth rates determined with the hanging-drop culture and the conventional culture methods showed good correlation (r=0.97) as shown in Fig. 4. It was confirmed that the specific growth rate can be estimated accurately with the water droplet method. Specific growth rates increased significantly with increasing concentration of Provasoli's Marine Water Enrichment solution from 0 to 0.5% and then gradually

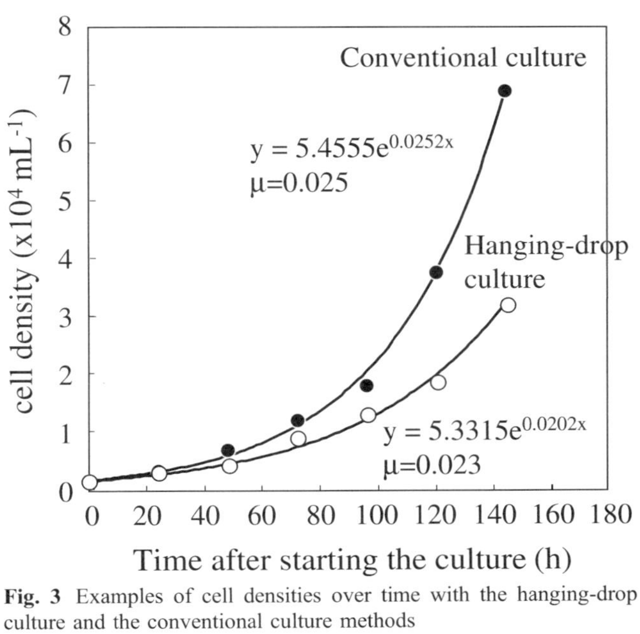

Fig. 3 Examples of cell densities over time with the hanging-drop culture and the conventional culture methods

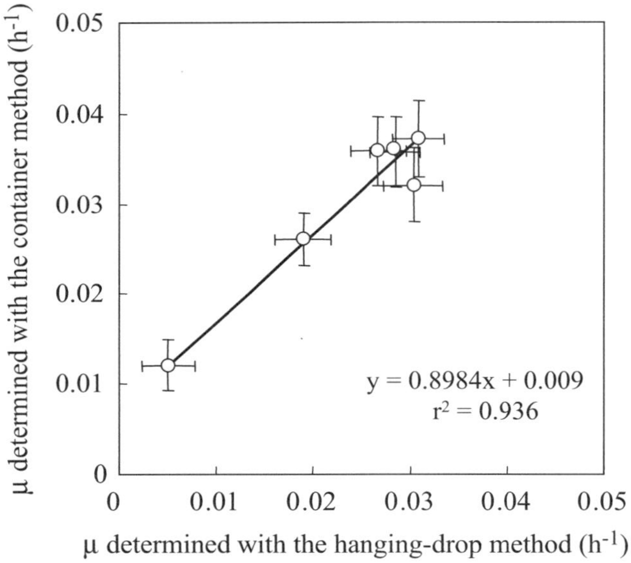

Fig. 4 Specific growth rates (µ) of *Amphidinium* cells determined with the hanging-drop culture and the conventional container culture methods. Volumes of culture media were 3 µL and 800 mL in the hanging-drop culture method and the conventional culture method, respectively. Means with standard deviations are shown (n=3–10)

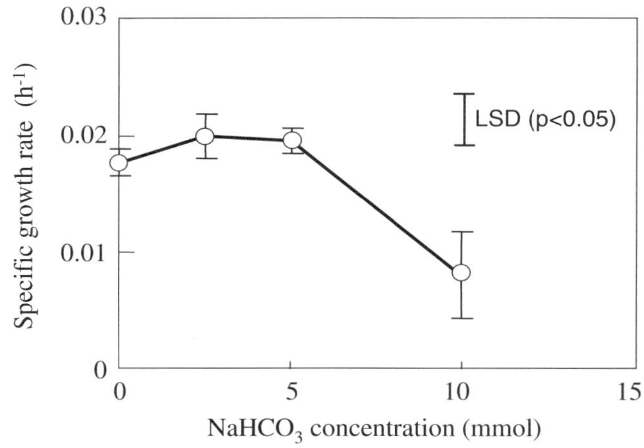

Fig. 6 Effects of $NaHCO_3$ concentrations on specific growth rates of *Amphidinium* cells. Vertical bars on plots indicate standard deviations (n=10). Temperature: 28°C, O_2 concentration: 5%, CO_2 concentration: 0.05%, PPFD: 50 µmol photons m^{-2} s^{-1}, photoperiod: 24 h day^{-1}

increased at concentrations higher than 0.5%, although there were no significant differences among the treatments (Fig. 5). Specific growth rates slightly increased with increasing $NaHCO_3$ concentration from 0 to 2.5 mmol and then decreased significantly at $NaHCO_3$ concentrations higher than 5 mmol (Fig. 6). The optimum $NaHCO_3$ concentration for cell division was thus in the range of 2.5–5 mmol. Therefore, 1% Provasoli's Marine Water Enrichment solution and 5 mmol $NaHCO_3$ were used as basic supplements in the experiment for assessing appropriate culture conditions.

Specific growth rates doubled with increasing temperature from 21 to 26°C, and decreased with increasing temperature from 29 to 35°C (Fig. 7). The optimum temperature was 24–29°C .

Specific growth rates increased significantly by 50% with increasing PPFD from 15 to 35 µmol photons m^{-2} s^{-1} and reached a constant value at PPFD values greater than 35 µmol photons m^{-2} s^{-1} under continuous light (Fig. 8). Specific growth rates increased significantly by 80% with increasing photoperiod from 6 to 12 h day^{-1} and reached an almost constant value at photoperiods of 12 to 24 h day^{-1} (Fig. 9).

Specific growth rates tended to increase slightly with increasing CO_2 concentration from 0.002 to 0.05%, and

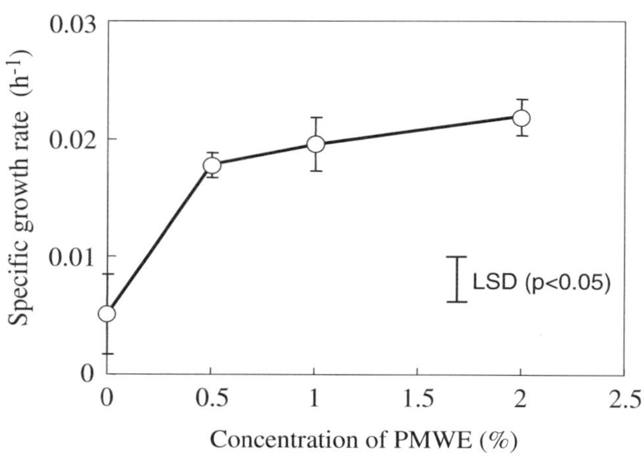

Fig. 5 Effects of Provasoli's Marine Water Enrichment concentration on specific growth rates of *Amphidinium* cells. Vertical bars on plots indicate standard deviations (n=10). Temperature: 28°C, O_2 concentration: 5%, CO_2 concentration: 0.05%, photosynthetic photon flux density (PPFD): 50 µmol photons m^{-2} s^{-1}, photoperiod: 24 h day^{-1}

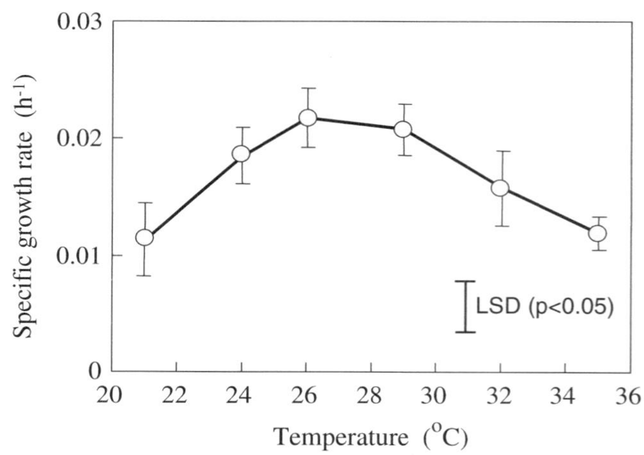

Fig. 7 Effects of medium solution temperatures on specific growth rates of *Amphidinium* cells. Vertical bars on plots indicate standard deviations (n=10). O_2 concentration: 5%, CO_2 concentration: 0.05%, PPFD: 50 µmol photons m^{-2} s^{-1}, photoperiod: 24 h day^{-1}

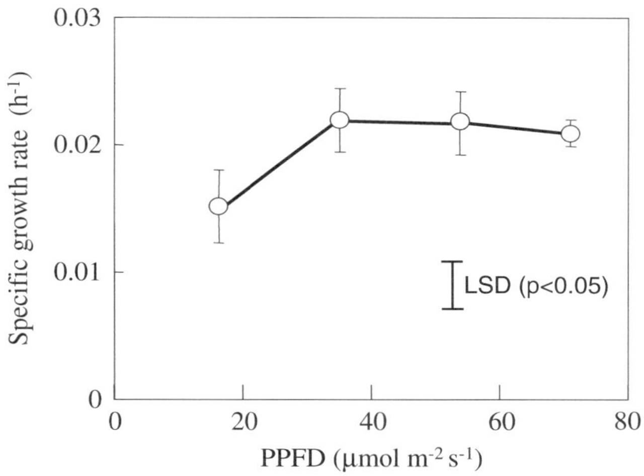

Fig. 8 Effects of PPFD on specific growth rates of *Amphidinium* cells. Vertical bars on plots indicate standard deviations (n=10). Temperature: 26°C, O_2 concentration: 5%, CO_2 concentration: 0.05%, photoperiod: 24 h day^{-1}

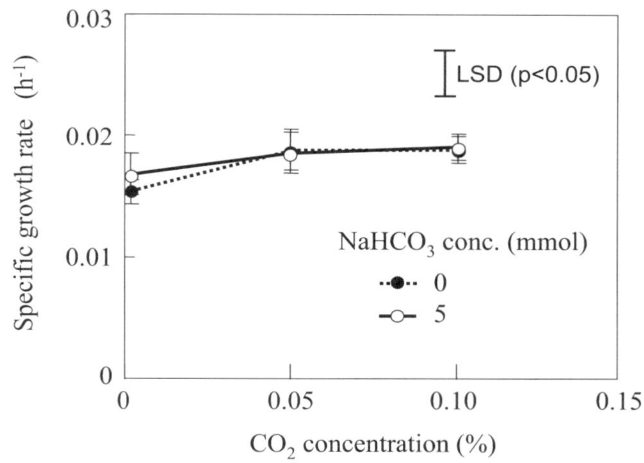

Fig. 10 Specific growth rates of *Amphidinium* cells as affected by CO_2 concentrations with or without $NaHCO_3$ in the medium solution. Vertical bars on plots indicate standard deviations (n=10). Temperature: 26°C, O_2 concentration: 5%, PPFD: 40 µmol photons m^{-2} s^{-1}, photoperiod: 24 h day^{-1}

reached a constant value at CO_2 concentrations of 0.05 to 0.1% regardless of $NaHCO_3$ concentration, although there were no significant differences among the 0.002 to 0.1% treatments (Fig. 10). Specific growth rates tended to increase slightly with increasing O_2 concentration from 0.3 to 5%, and then decreased with increasing O_2 concentration from 5 to 21% regardless of CO_2 concentration, although there were no significant differences among the treatments of 0.3–21% (Fig. 11).

The maximum specific growth rate was 0.022 h^{-1} at a temperature of 26°C and O_2 concentration of 5%, and the specific growth rate was saturated at a CO_2 concentration of 0.05%, a photosynthetic photon flux density of 35 µmol photons m^{-2} s^{-1} and a photoperiod of 12 h day^{-1}.

Discussion

The symbiotic dinoflagellates used in this study exhibited a maximum specific growth rate equivalent to that of asymbiotic dinoflagellates, e.g., *Gyrodinium aureolum* (μ=0.015) (Dixon and Syrett 1988), *Ceratium furca* (μ=0.03) and *Ceratium fusus* (μ=0.023) (Baek et al. 2008). The conditions that provided the highest growth rate of the symbiotic dinoflagellates in this study would be similar to those inside the body of the host animal, *Amphiscolops*. The doubling time (t_d) was 31 h at a specific growth rate of 0.022 h^{-1}. The results demonstrate that symbiotic *Amphidinium* species can multiply efficiently under conditions of relatively low light intensity and low O_2 concentration.

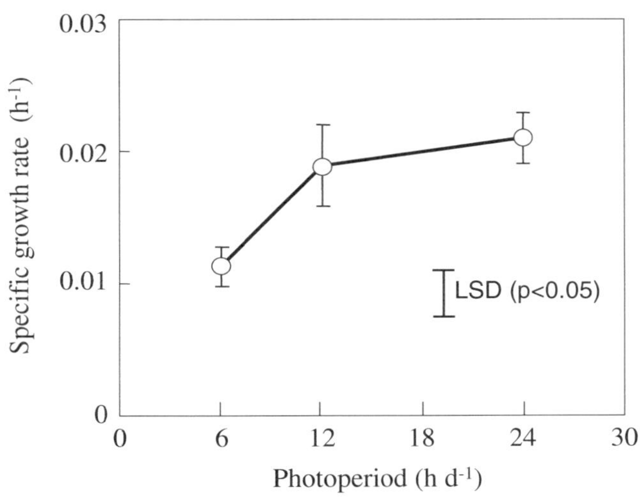

Fig. 9 Effects of photoperiods on specific growth rates of *Amphidinium* cells. Vertical bars on plots indicate standard deviations (n=10). Temperature: 26°C, O_2 concentration: 5%, CO_2 concentration: 0.05%, PPFD: 50 µmol photons m^{-2} s^{-1}

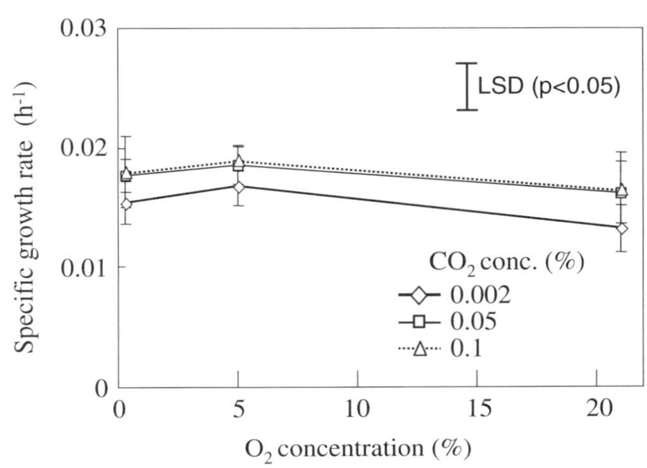

Fig. 11 Specific growth rates of *Amphidinium* cells as affected by O_2 concentrations at different CO_2 concentrations. Vertical bars on plots indicate standard deviations (n=10). Temperature: 28°C, PPFD: 50 µmol photons m^{-2} s^{-1}, photoperiod: 24 h day^{-1}

This experiment was conducted with algal cultures at low cell densities. Cell productivity is dependent not only on the growth rate but also on the population density. Moreover, the fragility of cells represents one of the biggest obstacles to the cultivation of dinoflagellates (e.g., Dixon and Syrett 1988). In order to establish a system to produce secondary metabolites such as amphidinolides, it is necessary to determine appropriate conditions for cellular multiplication at high cell densities and also for the efficient production of secondary metabolites in cells.

Acknowledgment This work was supported by a Grant-in-Aid for Scientific Research from the Ministry of Education, Culture, Sports, Science, and Technology of Japan.

References

Baek SH, Shimode S, Kikuchi T (2008) Growth of dinoflagellates, *Ceratium furca* and *Ceratium fusus* in Sagami Bay. Harmful Algae 7:163–173

Bauer I, Maranda L, Shimizu Y, Peterson RW, Cornell L, Steiner JR, Clardy J (1994) The structures of amphidinolide B isomers: strongly cytotoxic macrolides produced by a free-swimming dinoflagellate *Amphidinium* sp. J Am Chem Soc 116:2657–2658

Dixon GK, Syrett PJ (1988) The growth of dinoflagellates in laboratory cultures. New Phytol 109:297–302

Ishibashi M, Ohizumi Y, Hamashima M, Nakamura H, Hirata Y, Sasaki T, Kobayashi J (1987) Amphidinolide B, a novel macrolide with potent antineoplastic activity from the marine dinoflagellate *Amphidinium* sp. J Chem Soc Chem Commun 1987:1127–1129

Kitaya Y, Kibe S, Oguchi M, Tanaka H, Miyatake H, Nakano Y (1998) Effects of CO_2 and O_2 concentration and light intensity on growth of microalgae (*Euglena gracilis*) in CELSS. Life Support Biosphere Sci 5:243–247

Kitaya Y, Azuma H, Kiyota M (2005) Effects of temperature, CO_2/O_2 concentrations and light intensity on cellular multiplication of microalgae. *Euglena gracilis*. Adv Space Res 35:1584–1588

Kobayashi J, Tsuda M (2004) Amphininolides, bioactive macrolides from symbiotic marine dinoflagellate. Nat Prod Rep 21:77–93

Kobayashi J, Kubota T (2007) Bioactive macrolides and polyketides from marine dinoflagellates of the genus *Amphidinium*. J Nat Prod 70:451–460

Kobayashi J, Ishibashi M, Nakamura H, Ohizumi Y, Yamasu T, Hirata Y, Sasaki T, Ohta T, Nozoe S (1989) Cytotoxic macrolides from a cultured marine dinoflagellate of the genus *Amphidinium*. J Nat Prod 52:1036–1041

Kobayashi J, Shigemori H, Ishibashi M, Yamasu T, Hirota H, Sasaki T (1991) Amphidinolies H and G, new potent cytotoxic macrolides from the cultured symbiotic dinoflagellate *Amphidinium* sp. J Org Chem 56:5221–5224

Kobayashi J, Shimbo K, Sato M, Shiro M, Tsuda M (2000) Absolute stereochemistry of amphidinolides G and H. Org Lett 2:2805–2807

Oguchi K, Tsuda M, Iwamoto R, Okamoto Y, Endo T, Kobayashi J, Ozawa T, Masuda A (2007) Amphidinolides B6 and B7, cytotoxic macrolides from a symbiotic dinoflagellate *Amphidinium* species. J Nat Prod 70:1676–1679

Provasoli L (1963) Growing marine seaweeds. Proceedings of the 4th International Seaweed Symposium. Pergamon, Oxford, pp 9–17

Tsuda M, Oguchi K, Iwamoto R, Okamoto Y, Kobayashi J, Fukushi E, Kawabata J, Ozawa T, Masuda A, Kitaya Y, Omasa K (2007a) Iriomoteolide-1a, a potent cytotoxic 20-membered macrolide from a benthic dinoflagellate *Amphidinium* species. J Org Chem 72:4469–4474

Tsuda M, Oguchi K, Iwamoto R, Okamoto Y, Fukushi E, Kawabata J, Ozawa T, Masuda A (2007b) Iriomoteolides-1b and 1c, 20-membered macrolides from a marine dinoflagellate *Amphidinium* species. J Nat Prod 70:1661–1663

Photosynthetic and respiratory responses of *Gracilaria vermiculophylla* (Ohmi) Papenfuss collected from Kumamoto, Shizuoka and Iwate, Japan

S.lPhooprong · H. Ogawa · K. Hayashizaki

Originally published in the Journal of Applied Phycology, Vol 20, No 5, 293–300.
DOI: 10.1007/s10811-007-9253-9 © Springer Science + Business Media B.V. 2007

Abstract The photosynthetic and respiratory responses to irradiance, salinity and temperature of the red alga, *Gracilaria vermiculophylla*, collected from Kumamoto, Shizuoka and Iwate in Japan were studied using an electronic Dissolved Oxygen sensor. The parameters derived from the photosynthesis versus irradiance relationship indicated the potential to acclimate to broad irradiance variations in all of the populations of *G. vermiculophylla* collected from these three sites. In addition, the light-saturated photosynthesis rate (P_{max}) and the dark respiration rate of all populations increased with increasing temperature up to 20–30°C, while the P_{max} decreased at 35°C. All populations also showed a broad variation of photosynthetic responses to salinity changes in the range from 10 to 30 psu. On the other hand, the population from Iwate showed high photosynthetic efficiency, especially in the temperature range of 5–10°C, and showed low values of saturation irradiance compared to the populations from Shizuoka and Kumamoto. These results suggest that there is greater potential to acclimate to low irradiance and low temperature in the population from Iwate compared to those from the Shizuoka and Kumamoto populations. However, the P_{max} of the populations from Iwate and Shizuoka was reached at 20°C and 25°C, respectively, while the Kumamoto population reached P_{max} at 30°C. This implies that the latter population has greater potential to tolerate higher temperatures than the former. Such characteristics in photosynthesis and respiration of *G. vermiculophylla* collected from the three locations probably indicate an acclimation to prevailing environmental conditions in their respective habitats.

S. Phooprong (✉) · H. Ogawa · K. Hayashizaki
School of Fisheries Science, Kitasato University,
Iwate 022-0101, Japan
e-mail: spooprong@yahoo.com

Keywords *Gracilaria vermiculophylla* · Photosynthesis · Respiration · Salinity · Temperature · Irradiance

Introduction

The economic importance of gracilarioid algae (species of *Gracilaria*) as the main global sources of agar and food additives has been well documented (Bird 1995; Yokoya et al. 1999). As the demand for the algae increases with the growing world population, large-scale cultivation of *Gracilaria*, using advanced agricultural technologies, is necessary to ensure sufficient supply. Understanding the nature of the algae and how they interact with various environmental factors is therefore essential for future development of such technologies. Environmental factors, namely temperature, salinity and light, are the main contributors to growth and distribution of marine macroalgae, since these factors directly influence their photosynthetic ability as well as respiration (Gessner and Schramm 1971; Lobban and Harrison 1994). Laboratory study of algae behavior under imitated natural environment, where these environmental factors can be well controlled, is certainly one of the most practical approaches that will lead to such future development.

The species *Gracilaria vermiculophylla* was originally described from Akkeshi Lagoon, eastern Hokkaido, Japan (Ohmi 1956), which is considered one of the countries with high diversity of natural environment (Yoshida 1998). The species was subsequently reported in tropical and subtropical regions in the northwestern Pacific (Tseng and Xia 1999) and the northeastern Pacific (Bellorin et al. 2004). The broad distribution of this alga over various ecological conditions of the region has led to the hypothesis for this study that *G. vermiculophylla* from the coasts of Japan has a high potential to tolerate a wide range of environmental changes. In order to verify the validity of the hypothesis proposed, a laboratory research of the algae's responses to environmental changes were conducted.

Although a number of photosynthetic studies on *Gracilaria* have been done in the past (Rosenberg and Ramus 1982; Lapointe et al. 1984; Orduna-Rojas et al. 2002), only a few studies have focused on the combined effects of irradiance, salinity and temperature upon the photosynthetic response (Penniman and Mathieson 1985; Dawes et al. 1999; Phooprong et al. 2007). The present study, therefore, is intended to conduct comparative analyses of photosynthetic performance in response to irradiance, salinity and temperature variations of *G. vermiculophylla* found in different parts of Japan under laboratory conditions. The samples were collected from Kumamoto, Shizuoka and Iwate in southwestern Japan, central Japan and northeastern Japan, respectively. The natural surroundings of the sample site locations are estuarine habitats where the salinity ranges are 3–33 psu, 20–33 psu and 2–33 psu, and the temperature ranges are 9–30°C, 6–30°C and 4–22°C for Kumamoto, Shizuoka and Iwate, respectively. However, light intensity is probably limited by extreme fluctuation in water turbidity. Although it would be worthwhile to include *Gracilaria* from the western Pacific and from southern and northern parts of Japan in this study, the lack of information about the algae in these habitats from the previous studies and the timeframe for this study were the major restraints. Nevertheless, the natural conditions of *G. vermiculophylla* sample sites selected for this study should be sufficiently diverse to represent the environmental conditions for most algae sources in Japan.

Based on the surrounding conditions of their habitats, various photosynthetic responses of the algae samples were expected from this experimental study. It was also anticipated that some photosynthetic performance of Kumamoto population might be similar to that of Shizuoka population and some could be similar to that of Iwate population, considering the temperature and salinity ranges of their habitats.

Materials and methods

Samples of *G. vermiculophylla* were collected from intertidal habitats at three different sites in Japan: Yatsushiro in Kumamoto Prefecture (southwestern Japan), Hamamustu in Shizuoka Prefecture (central Japan), and Yamada in Iwate Prefecture (northeastern Japan) (Fig. 1). *G. vermiculophylla* samples were repeatedly washed using filtered seawater and then cut into 2.5 to 3.0 cm segments. In order to ensure the uniformity of thallus age, only the apical and intercalary parts of the algae were selected for the experiments. The samples were kept for 3 days in an acclimation bottle wherein running seawater provided sufficient nutrients to keep them fresh and

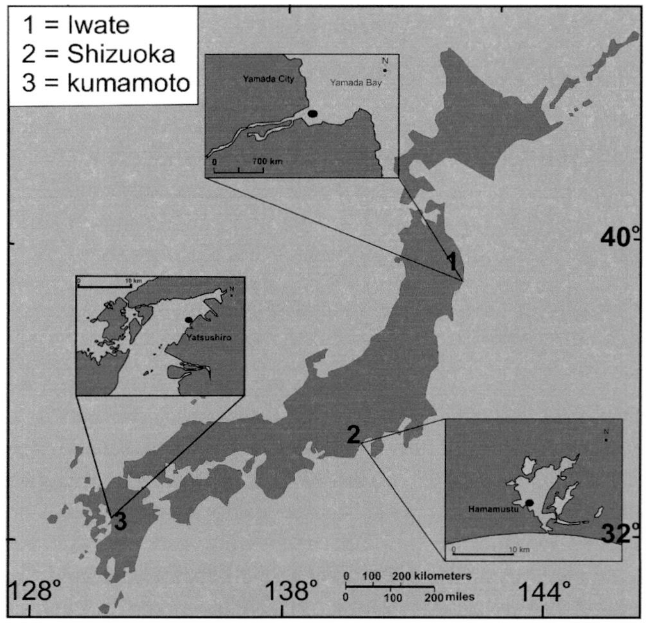

Fig. 1 Map of Japan showing the locations of the three collection sites for *Gracilaria vermiculophylla*

healthy under artificial white light with 100 μmol photon $m^{-2}s^{-1}$ PFD (photons flux density), and a 12:12 h (L:D) photoperiod.

On the day of the experimental run, each dish of the test samples was transferred into the reaction chamber previously filled with 150 mL of autoclaved seawater. The reaction chamber was hermetically closed and was constantly stirred by a magnetic stirrer, so that oxygen evolved (photosynthesis) or consumed (respiration) during incubation could be measured using an oxygen analyzer (Obisphere Laboratories series 3600, Switzerland).

To investigate the sole effect of temperature variation on photosynthesis of the algae, the P-I experiments were conducted for 7 temperature levels, 5, 10, 15, 20, 25, 30 and 35°C, under ambient salinity of 33 psu. These temperatures represent temperature ranges at the algal samples collection sites. The light intensity was set to vary from 0 to 710 μmol photons $m^{-2}s^{-1}$. For each temperature level, 4 replicates of experimental runs were conducted. Thus, 28 replicates, for which approximately 400 segments of algal samples were used (10–15 segments of alga samples were required for one replicate).

The combined effects of temperature and salinity variation were investigated for 3 temperatures, 5, 15 and 25°C, at 3 different salinities, 10, 20 and 33 psu, which resulted in 9 experimental combinations. Similar to that of the investigation on sole temperature effect, the light intensity variation used was from 0 to 710 μmol photons $m^{-2}s^{-1}$, and 4 replicates were run for each experimental combination and, therefore, 36 replicates, which required more than 500 segments of algal samples in total.

During the experiments, salinity was adjusted by diluting the ambient seawater with distilled water, while variation of PFD light intensity was achieved by placing filters between the light source (SERIC XC-100 xenon-discharge lamp) and the reaction chamber. Each irradiation took 30 min, with photosynthetic measurement done in the last 15 min. Respiration was measured in complete darkness. After the experiment, the samples were oven dried at 60°C for 24 h to constant weight. Finally, the biomass dry weight and the amount of oxygen production per dry weight biomass per unit time were then calculated.

Calculations for photosynthetic parameters were achieved by fitting the *P-I* curve to the modified non linear regression equation of Honda (1999):

$$\upsilon \approx \Phi_T I^2 / \left(\Psi_T + \Omega_T I + I^2\right)$$

where υ is the gross photosynthesis rate (P_g), I is the irradiance (E), Φ is the light-saturated, photosynthesis rate (P_{max}), Ω is irradiance at initial photosynthetic saturation (E_k), and constant value, Ψ_T=1/1,000,000. The compensation point (E_c) was calculated using equation, $\Omega \times R_d/(\Phi - R_d)$, (where R_d was oxygen consumed during dark respiration). Finally, photosynthetic efficiency was obtained by using the equation, $\alpha = \Phi/\Omega$.

A two-way ANOVA followed by Bonferroni's post hoc test was performed in order to verify the differences in photosynthetic and respiration responses to salinity and temperature for algae from each site. In addition, a one-way ANOVA followed by Tukey's Multiple Comparison Test was performed for comparing saturated photosynthesis rates from all three sites at different temperatures for each light treatment. A separate set of one-way ANOVA was also performed in order to compare photosynthetic parameters (P_{max}, E_k, E_c and α) and dark respiration (R_d) of the algae from all three sites at different temperatures. Before performing ANOVAs, the data sets for multiple measurement comparisons were tested to ensure their equal variances using Bartlett's method. All tests were performed using Graphpad Prism ver 4.

Results

Photosynthesis-irradiance curves measured at 20 and 30°C showed a wide range of responses of *G. vermiculophylla* from the three different habitats. In Fig. 2, a rapid increase of photosynthesis rates following the increase of irradiance up to 200 μmol photons $m^{-2}s^{-1}$ was observed in all populations, but the rates of increase dropped significantly beyond this point up to 710 μmol photons $m^{-2}s^{-1}$. Under the highest irradiance of 710 μmol photons $m^{-2}s^{-1}$, photoinhibition was not absent. Initial saturation irradiance (E_k) and compensation point (E_c) increased when the temperature was increased from 5 to 30°C in all populations (Table 2). The ranges of E_k and E_c were 73–498 and 1–30 μmol photons m^{-2} s^{-1}, respectively. P_{max} of the Kumamoto population significantly ($p<0.05$) increased with a temperature rise from 5 to 30°C (Table 2). However, for the Shizuoka and Iwate populations, the significant increase of P_{max} ($p<0.05$) was found only from 5 to 20°C (Table 2).

Optimum temperature at which the highest light-saturated photosynthesis rate was observed was 30°C in the Kumamoto population, while it was in the range 20–30°C in the Shizuoka and Iwate populations. Additionally, the light-saturated photosynthesis rates for the all three populations decreased at 35°C (Table 2). Figure 3a shows the combined effects of temperature and salinity on P_{max} for all three populations. The interaction between salinity and temperature was found to be significantly different ($p<$ 0.05) in all populations (Fig. 3a; Table 1). The lowest observed P_{max} values for the three populations were in low salinity (10 psu) and low temperature (5°C), and they were significantly different from other treatment ($p<0.05$). These values increased proportional with temperature and salinity. For all populations, except the Kumamoto population at 5°C, the highest value of P_{max} was reached at the ambient salinity of 33 psu (Fig. 3a). For other combinations of increasing temperature and salinities, the algae from all three populations were still able to produce oxygen through photosynthesis. The combined effect of temperature and salinity on R_d was found to be insignificant ($p>0.05$) in all

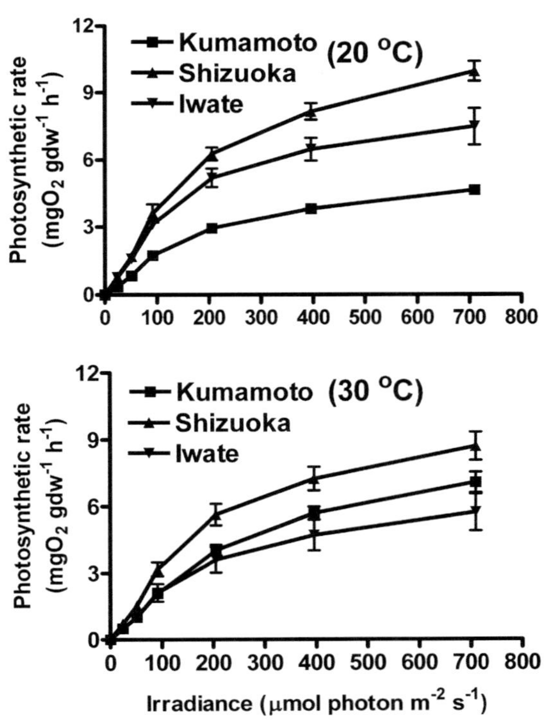

Fig. 2 *P-I* curves of *Gracilaria vermiculophylla* collected from the three sites in response to temperature at 20 and 30°C (± SD, n=4)

Fig. 3 Effect of temperature and salinity variation on (**a**) P_{max} and (**b**) R_d of *Gracilaria vermiculophylla* from the three sites (± SD, n=4)

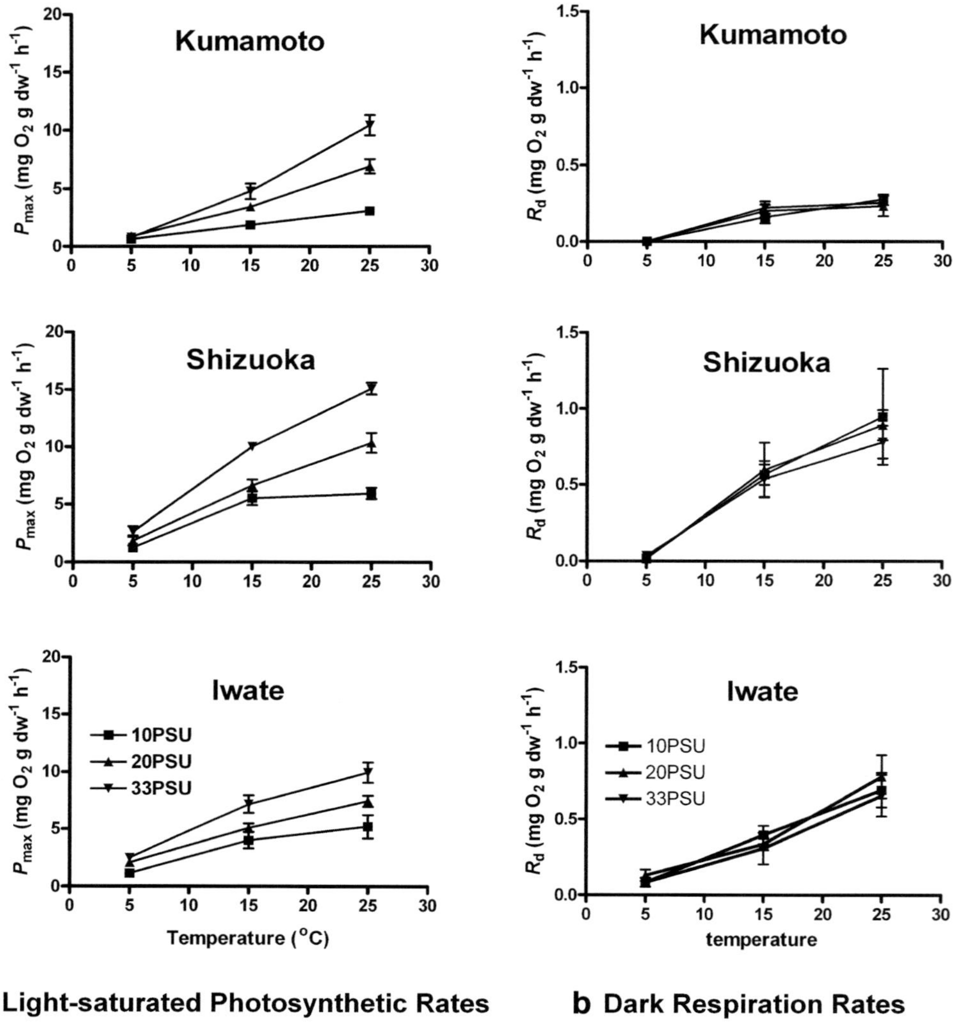

three populations (Fig. 3b; Table 1). In general, their R_d significantly ($p<0.05$) increased with the increase of temperature from 5 to 25°C at all salinity levels (Fig. 3b). The R_d values for the Kumamoto and Shizuoka populations could not be recorded at 5°C because they were too close to zero (Fig. 3b; Table 2). However, R_d values for the Iwate population could be recorded because they were 2.5 times that of the Kumamoto population and 30 times that of the Shizuoka population, respectively (Fig. 3b; Table 2).

Comparison of plants from the three sites

Photosynthesis–irradiance curves of the three populations measured at 20 and 30°C showed significant differences in saturated photosynthesis rates (Fig. 2). The saturated photosynthesis rates for the Kumamoto population at 20°C were in the irradiance range of 395 to 710 μmol photons $m^{-2}s^{-1}$ and were significantly ($p<0.05$) lower than those of the Shizuoka and Iwate populations. However, the rate for the Kumamoto population was significantly higher at 30°C than that of the Iwate population (Fig. 2). Nevertheless, the rates for the Shizuoka population at both temperatures were significantly higher ($p<0.05$) among the three populations. Photosynthetic efficiency (α) of the Iwate and Shizuoka populations were significantly higher than that of the Kumamoto population from 10 to 25°C (Table 2). E_k values of the Iwate population were significantly lower than those of the Kumamoto and Shizuoka populations from 10 to 25°C ($p<0.05$), while the Kumamoto population showed the significantly ($p<0.05$) highest E_k values at 25 to 30°C among the three populations (Table 2). E_c values of three populations were not significantly ($p>0.05$) different from 5 to 35°C (Table 2).

Table 1 The level of significance of photosynthetic responses to the combined effect of salinity and temperature of *Gracilaria vermiculophylla* from the three sites following two-way ANOVA

	P_{max}				R_d			
	DF	*MS*	*F*	*p-values*	*DF*	*MS*	*F*	*p-values*
Kumamoto								
a) Salinity	2	36	159	<0.0001	2	0.0008	0.6	0.54
b) Temperature	2	110	485	<0.0001	2	0.21	166	<0.0001
Interaction of a × b	4	13	58	<0.0001	4	0.002	1.9	0.14
Shizuoka								
a) Salinity	2	76	279	<0.0001	2	0.01	0.64	0.54
b) Temperature	2	225	828	<0.0001	2	2.2	115	<0.0001
Interaction of a × b	4	16	58	<0.0001	4	0.01	0.53	0.71
Iwate								
a) Salinity	2	29	72	0.0003	2	0.01	1.71	0.20
b) Temperature	2	97	243	<0.0001	2	1.14	143	<0.0001
Interaction of a × b	4	3	8	<0.0001	4	0.007	0.89	0.48

Discussion

The photosynthetic and respiratory responses of *G. vermiculophylla* from Japan indicate that all populations have good tolerance to irradiance without photoinhibition even at 710 μmol photons $m^{-2}s^{-1}$. This is in contrast to shade-adapted *G. foliifera* from Beaufort, North Carolina, in which photosynthesis declined at irradiances above 600 μmol photons $m^{-2}s^{-1}$ (Rosenberg and Ramus 1982). However, Dawes et al. (1999) have observed that the photosynthesis of *G. cornea* from Florida Keys, Florida, declined at irradiances above 1,000 μmol photons $m^{-2}s^{-1}$, while Penniman and Mathieson (1985) found that no photoinhibition occurred of *G. tikvahiae* from Great Bay

Table 2 Effect of varying temperature on photosynthesis and respiration of *Gracilaria vermiculophylla* from the three sites

	Population	P_{max}	R_d	E_k	E_c	α
5°C	Kumamoto	0.8 ± 0.3^{a}	0.005 ± 0.004^{a}	73 ± 27^{a}	1 ± 1^{a}	0.012 ± 0.005^{a}
	Shizuoka	2.6 ± 0.5^{ab}	0.04 ± 0.03^{a}	143 ± 37^{a}	2 ± 2^{a}	0.019 ± 0.003^{a}
	Iwate	2.5 ± 0.3^{b}	0.1 ± 0.03^{a}	87 ± 18^{a}	3 ± 1^{a}	0.029 ± 0.003^{b}
10°C	Kumamoto	3.1 ± 0.3^{a}	0.1 ± 0.04^{a}	239 ± 35^{a}	4 ± 2^{a}	0.013 ± 0.003^{a}
	Shizuoka	6.1 ± 0.9^{b}	0.2 ± 0.05^{b}	208 ± 34^{a}	6 ± 1^{a}	0.030 ± 0.009^{b}
	Iwate	5.5 ± 0.9^{b}	0.3 ± 0.03^{c}	125 ± 30^{b}	6 ± 1^{a}	0.044 ± 0.006^{c}
15°C	Kumamoto	4.8 ± 0.7^{a}	0.2 ± 0.04^{a}	268 ± 66^{a}	9 ± 2^{a}	0.018 ± 0.003^{a}
	Shizuoka	10.0 ± 0.4^{b}	0.5 ± 0.12^{b}	220 ± 41^{ab}	12 ± 3^{a}	0.047 ± 0.01^{bc}
	Iwate	7.2 ± 0.8^{c}	0.3 ± 0.10^{a}	184 ± 8^{b}	8 ± 3^{a}	0.039 ± 0.004^{c}
20°C	Kumamoto	6.9 ± 0.4^{a}	0.2 ± 0.02^{a}	338 ± 10^{a}	13 ± 2^{a}	0.020 ± 0.001^{a}
	Shizuoka	13.2 ± 1.6^{b}	0.7 ± 0.04^{b}	330 ± 16^{a}	18 ± 3^{a}	0.040 ± 0.007^{bc}
	Iwate	9.9 ± 1.3^{c}	0.6 ± 0.08^{b}	208 ± 27^{b}	19 ± 7^{a}	0.047 ± 0.003^{c}
25°C	Kumamoto	10.4 ± 0.9^{a}	0.3 ± 0.03^{a}	449 ± 32^{a}	17 ± 3^{a}	0.023 ± 0.001^{a}
	Shizuoka	15.1 ± 0.5^{b}	0.8 ± 0.11^{b}	357 ± 29^{b}	19 ± 3^{a}	0.042 ± 0.004^{bc}
	Iwate	10.0 ± 0.9^{a}	0.7 ± 0.14^{b}	258 ± 11^{c}	17 ± 4^{a}	0.039 ± 0.003^{c}
30°C	Kumamoto	12.2 ± 1.6^{a}	0.6 ± 0.08^{a}	498 ± 101^{a}	24 ± 2^{a}	0.025 ± 0.003^{a}
	Shizuoka	13.0 ± 0.8^{a}	0.8 ± 0.07^{b}	337 ± 21^{b}	23 ± 5^{a}	0.039 ± 0.004^{b}
	Iwate	8.6 ± 1.2^{b}	0.7 ± 0.06^{ab}	348 ± 62^{b}	30 ± 8^{a}	0.025 ± 0.005^{a}
35°C	Kumamoto	8.1 ± 0.5^{a}	0.7 ± 0.13^{a}	392 ± 54^{a}	37 ± 8^{a}	0.021 ± 0.004^{a}
	Shizuoka	8.8 ± 1.3^{a}	1.0 ± 0.07^{a}	476 ± 60^{a}	59 ± 6^{a}	0.018 ± 0.001^{a}
	Iwate	7.1 ± 0.8^{a}	0.8 ± 0.21^{a}	395 ± 68^{a}	45 ± 19^{a}	0.020 ± 0.003^{a}

P_{max} = maximum photosynthesis; R_d = dark respiration (mgO_2 $gdw^{-1}h^{-1}$; E_k = initial saturation irradiance; E_c = compensation point (μmol photons $m^{-2}s^{-1}$); α = photosynthetic efficiency (mgO_2 gdw^{-1} h^{-1}/μmol photons $m^{-2}s^{-1}$)

Significant differences ($p<0.05$) among populations in each temperature test are indicated by different superscript letters

Mean ± SD ($n=4$)

Estuary, New Hampshire, at a high irradiance of 1,400 μmol photons $m^{-2}s^{-1}$. All three populations were light-saturated up to the irradiance of 200 μmol photons $m^{-2}s^{-1}$, according to the results from *G. tikvahiae* (Penniman and Mathieson 1985; Ramus and van der Meer 1983), *G. foliifera* (Rosenberg and Ramus 1982) and *G. salicornia* (Phooprong et al. 2007). The E_k values of the all three populations studied here were similar to that of intertidal red algae studied by Dawes et al. (1999) and Mathieson and Norall (1975). On the other hand, the E_c values of the three populations ranged from 4 to 30 μmol photons $m^{-2}s^{-1}$. Dawes et al. (1999) also found similar values with the E_c values of *G. cornea* ranging between 11 and 38 μmol photons $m^{-2}s^{-1}$. However, Penniman and Mathieson (1985) reported a lower value of 8–10 μmol photons $m^{-2}s^{-1}$.

The maximum photosynthetic rate (P_{max}) and initial saturation irradiance (E_k) of all three populations were considerably lower at 5°C than those at 15°C and 25°C. Davison (1991) has established that lower temperature limits electron transfer and carbon fixation and this, therefore reduces the ability of the algae to use light. The optimum temperatures of photosynthesis in the three populations were between 20 to 30°C, depending upon their habitats, but their photosynthesis rates remarkably decreased at 35°C. This is probably because the water temperature in Japan usually does not reach 35°C, and the enzymes involved in electron transfer may have been denatured at higher temperatures. Their response to temperature is similar to that of *Laurencia brongniartii* from Japan growing in subtidal zones and having a photosynthetic optimum temperature at 26–28°C (Nishihara et al. 2004). However, *G. tikvahiae* from Great Bay Estuary, New Hampshire, showed that the optimum temperatures range was between 25 and 35°C and that photosynthesis rates decreased above 37.5°C (Penniman and Mathieson 1985). This was confirmed by Dawes et al. (1999), who observed high photosynthesis rates for *G. cornea* from Florida between 25 and 35°C.

Photosynthetic response to salinity of the three populations showed that hyposaline treatment reduced the light saturated photosynthesis rate at all temperature levels. According to the studies of Penniman and Mathieson (1985) and Koch and Lawrence (1987), *Gracilaria* species have a photosynthesis salinity tolerance after 48 h of acclimation. In this study, the *G. vermiculophylla* plants were acclimated in low salinity of 10 to 20 psu for 48–72 h before undergoing the measurement of photosynthesis. However, the P_{max} values for the three populations were higher in the range of 10 to 20 psu (1.9–6.6 mgO_2 gdw^{-1} h^{-1} at 15°C and 3.1–10.3 mgO_2 gdw^{-1} h^{-1} at 25°C) than those of *G. cornea* growing in oceanic salinity conditions, where P_{max} values were found to be 3.6–4.0 mgO_2 gdw^{-1} h^{-1} at 23°C (Dawes et al. 1999). Furthermore, the P_{max} values of *G. vermiculophylla* observed here were significantly higher at 10 psu than that of *G. salicornia* growing in subtropical to tropical waters, which usually grows in the salinity range of 30 to 35 psu (Phooprong et al. 2007). These results suggest that algae growing in estuarine conditions have a broad potential to salinity change, especially in hyposaline conditions, because their photosynthetic activity is still higher than that of the algae from ocean salinity habitats. Therefore, this photosynthetic characteristic verifies the hypothesis of this study that plants of *G. vermiculophylla* can respond to broad salinity changes. Similar results were observed in *G. verrucosa* from mangrove and salt marsh estuaries by Dawes et al. (1978) where the photosynthesis rates of both algae were higher at 30 psu than the at 10 psu. Gessner and Schramm (1971) reported that the photosynthesis rate decreased in seaweeds as a direct reflection of the decrease in salinity because of ionic loss, Cl^- in particular, rapidly occurring when marine algae are exposed to a hypo-osmotic condition. Lapointe et al. (1984) supported the hypothesis of Gessner and Schramm (1971) proposing that this condition probably relates to ex-osmosis of ions required as co-factors in photosynthesis. This hypothesis was confirmed by the study of Imai et al. (1997), in which they observed a reduction of photosynthesis rate of the green alga *Rhizoclonium riparium* in diluted seawater, corresponding to a decrease in the concentration of the bicarbonate ion (HCO_3^-).

Although temperature is one of the major environmental factors controlling growth and distribution of the plants (Lobban and Harrison 1994), the importance of the interaction between two or more of the major factors should also be considered. Species found in subtidal and intertidal zones are usually influenced by a combination of two or more ambient factors (Dawes and Kovach 1992). In this study, P_{max} of *G. vermiculophylla* was affected by the interaction of salinity and temperature, and the photosynthesis rates of the alga exposed to 33 psu and 25°C was higher than when exposed to 10 psu and 5°C. These interactive effects on the photosynthesis of the algae may be more significant than the effects of each single factor.

High photosynthetic and respiration rates in *G. vermiculophylla* from the Iwate population exposed to 5–25°C in situ water temperature were measured at 5–10°C. This result indicates that the Iwate population is probably capable of carrying out the photosynthesis in low temperatures and, in fact, this population is abundant from early winter to late spring when temperature is 4–10°C and 15–20°C, respectively, but declines in abundance in summer (Phooprong 2005). On the other hand, the Kumamoto population showed higher photosynthesis and respiration rates at higher temperature (30°C), but lower photosynthesis and respiration rates from 5 to 20°C among the three populations. However, the photosynthesis rates of the Kumamoto population increased 3 times, while the increases in the photosynthesis rates of the Iwate and Shizuoka populations were about 2.5

and 2 times from 10 to 25°C, respectively. In addition, the P_{max} of the Kumamoto population was similar to that of the Iwate and Shizuoka populations at the temperatures of 25 and 30°C, respectively (Table 2). These results imply that the Kumamoto population is more sensitive to temperature changes than the Iwate and Shizuoka populations and is probably capable of carrying out photosynthesis at higher temperatures because the temperature there is normally around 30°C in summer.

At the temperature range of 15–30°C, the Iwate and Shizuoka populations showed the α value ranges of 0.025–0.047 mgO_2 gdw^{-1} h^{-1}/μmol photon m^{-2} s^{-1} and 0.039–0.047 mgO_2 gdw^{-1} h^{-1}/μmol photon m^{-2} s^{-1}, and the E_k values ranges of 184–348 μmol photon $m^{-2}s^{-1}$ and 220–357 μmol photon $m^{-2}s^{-1}$, respectively. These results are similar to that obtained in the study on *G. salicornia* from Okinawa, Japan and Rayong, Thailand (Phooprong et al. 2007), wherein the values of α and E_k for Okinawa population are 0.036–0.041 mgO_2 gdw^{-1} h^{-1}/μmol photon m^{-2} s^{-1} and 144–345 μmol photon $m^{-2}s^{-1}$, respectively, while that for Rayong population are 0.044–0.057 mgO_2 gdw^{-1} h^{-1}/μmol photon m^{-2} s^{-1} and 184–422 μmol photon $m^{-2}s^{-1}$, respectively. These test results indicate that Rayong population is subjected to greater fluctuations in irradiance than the Okinawa population. Furthermore, the Kumamoto population in this study showed lower α and higher E_k values in the temperature range of 20 and 30°C than those of the Iwate and Shizuoka populations, which is an indication that Kumamoto population can acclimate better to higher light intensity than the other two populations. Thus, for the Iwate and Kumamoto populations, the influences of environmental condition of their habitats were clearly reflected through their photosynthetic performances.

Prior to the experiment, it was anticipated that some photosynthetic behavior of Kumamoto population should be similar to that of Shizuoka population and some should be similar to that of Iwate population, as the temperature range of its habitat was slightly different from that of Shizuoka population, while the salinity range was almost similar to that of Iwate population. Surprisingly, from the experimental results, similar photosynthetic responses were obtained for populations from Shizuoka and Iwate while the Kumamoto population exhibited the most unique photosynthetic behavior among the three algae populations. This unexpected outcome has raised another question, that there might be some other factors that significantly influence photosynthetic behaviors of the algae populations, which were not in the original scope of this research study. Two factors that have come to the authors' attention are the difference in their genetics and the difference in their acclimation condition. Considering the geography of Japan and the flow directions of the ocean currents, there might be a good possibility that algae populations from Shizuoka and Iwate are genetically very close. The algae probably originated in Iwate and then spread south to Shizuoka by the movement of ocean currents. The Kumamoto population, whose photosynthetic responses were different from the others, may have come from a different origin. On the other hand, if the algae from Kumamoto have the same origin as the populations from Iwate and Shizuoka, it may have settled in this present habitat for so long that complete phenotypic acclimation and genetic adaptation could have taken place, which finally led to the change in physiological characteristics and hence photosynthetic response of the alga. This hypothesis, however, requires a significant further research.

In conclusion, the wide range of photosynthetic responses to temperature variation and the ability to perform photosynthesis even in very low salinity of all three algae populations have justified that *G. vermiculophylla* from Japan's coasts have high tolerance to changes in environmental conditions. These characteristics help explain why this type of algae is found in distinctly differing physical environments. Thus, the main hypothesis of this study has been verified. The results from this present study could be helpful for future researches in determining the optimum conditions for propagation of this type of algae.

Acknowledgement The first author is very grateful to Monbukagakusho scholarship for funding this research study in Japan. Special thanks go to Mr. Rob Fitch for his valuable time and effort in reviewing the manuscript.

References

Bellorin AM, Oliveira MC, Oliveira EC (2004) *Gracilaria vermiculophylla*: A western Pacific species of Gracilariaceae (Rhodophyta) first recorded from the eastern Pacific. Phycol Res 52:69–79

Bird CJ (1995) A review of recent taxonomic concepts and developments in the Gracilariaceae (Rhodophyta). J Appl Phycol 7:255–267

Davison IR (1991) Environmental effects on algal photosynthesis: temperature. J Phycol 27:2–8

Dawes CJ, Kovach CW (1992) Ecology of the algae of a Florida Key II. Effects of irradiance, salinity and desiccation on intertidal and subtidal populations of seven macroalgae. Bull Mar Sci 50:165–170

Dawes CJ, Moon RE, Davis MA (1978) Photosynthetic and respiratory rates and tolerances of benthic algae from a mangrove and salt marsh estuary: A comparative study. Estuar Coast Mar Sci 6:175–185

Dawes CJ, Orduna-Rojas J, Robledo D (1999) Response of the tropical red seaweed *Gracilaria cornea* to temperature, salinity and irradiance. J Appl Phycol 10:419–425

Gessner F, Schramm W (1971) Salinity In: Kinne O (Ed), Marine Ecology, vol I (2). Wiley Interscience, New York, pp 705–820

Honda M (1999) A theoretical analysis and field evaluation of light and temperature model of production by *Ecklonia cava*. Hydrobiologia 398/399:361–374

Imai M, Katayama N, Yamaguchi Y (1997) Effect of salinity on growth, photosynthesis and respiration in freshwater alga *Rhizoclonium riparium* (Chlorophyceae, Cladophorales). Phycol Res 45:233–237

Koch EW, Lawrence J (1987) Photosynthetic and respiratory responses to salinity changes in the red alga *Gracilaria verrucosa*. Bot Mar 30:327–329

Lapointe BE, Rice DL, Lawrence JM (1984) Responses of photosynthesis, respiration, growth and cellular constituents to hypo-osmotic shock in the red alga *Gracilaria tikavahiae*. Comp Biochem Physiol 77:127–132

Lobban CS, Harrison PJ (1994) Seaweed ecology and physiology. Cambridge University Press, New York

Mathieson AC, Norall TL (1975) Photosynthetic studies of *Chondrus crispus*. Mar Biol 33:207–213

Nishihara GN, Terada R, Noro T (2004) Photosynthesis and growth rates of *Laurencia brongniartii* J. Agardh (Rhodophyta, Ceramiales) in preparation for cultivation. J Appl Phycol 16:303–308

Ohmi H (1956) Contributions to the knowledge of Gracilariaceae from Japan, II. On a new species of the genus *Gracilariopsis* with some considerations on its ecology. Bull Fac Fish Hokkaido Univ 6:271–279

Orduna-Rojas J, Robledo D, Dawes CJ (2002) Studies on the tropical agarophyte *Gracilaria cornea* J. Agardh (Rhodophyta, Gracilariales) from Yucatan, Mexico. I. Seasonal physiological and biochemical responses. Bot Mar 45:453–458

Penniman CA, Mathieson AC (1985) Photosynthesis of *Gracilaria tikvahiae* MacLachlan (Gigartinales, Rhodophyta) from the Great Bay Estuary, New Hampshire. Bot Mar 28:427–435

Phooprong S (2005) Seasonal variation of reproductive stage of *Gracilaria vermiculophylla* (Gracilariales) from Yamada Bay, Japan. Master Dissertation, Kitasato University

Phooprong S, Ogawa H, Hayashizaki K (2007) Photosynthetic and respiratory responses of *Gracilaria salicornia* (C. Agardh) Dawson (Gracilariales, Rhodophyta) from Thailand and Japan. J Appl Phycol (in press)

Ramus J, van der Meer JP (1983) A physiological test of the theory of complementary chromatic adaptation. I. Color mutants of a red seaweed. J Phycol 19:86–91

Rosenberg G, Ramus J (1982) Ecological growth strategies in the seaweeds *Gracilaria foliifera* (Rhodophyceae) and *Ulva* sp. (Chlorophyceae): Photosynthesis and antenna composition. Mar Ecol Prog Ser 8:133–241

Tseng CK, Xia B (1999) On the *Gracilaria* in the western Pacific and the southeastern Asia region. Bot Mar 42:209–217

Yokoya NS, Kakita H, Obika H, Kitamura T (1999) Effects of environmental factors and plant growth regulators on growth of the red alga *Gracilaria vermiculophylla* from Shikoku Island, Japan. Hydrobiologia 398/399:339–347

Yoshida T (1998) Marine algae of Japan. Uchida Roukakuho, Tokyo. (in Japanese)

In situ study of relative electron transport rates in the marine macroalga *Fucus vesiculosus* in the Baltic Sea at different depths and times of year

N. G. A. Ekelund · C. A. Nygård · R. Nordström · A. M. Gylle

Originally published in the Journal of Applied Phycology, Vol 20, No 5, 301–306.
DOI: 10.1007/s10811-007-9288-y

Abstract The brown alga *Fucus vesiculous* is one of the few marine species in the Baltic Sea. *Fucus vesiculosus* shows high morphological and physiological variability as a response to its environmental conditions. The salinity in the Baltic Sea is 4–5 psu, compared to 35 psu in the Atlantic. Photosynthesis of algae is usually measured after collection and transportation to constant culture conditions. However, in this study, relative photosynthetic electron transport rates, calculated from chlorophyll *a* fluorescence parameters were compared in algae collected from 1 and 4 m depths by SCUBA divers. Measurements of light response curves from the same individuals of *F. vesiculosus* at different depths and times of the year have, to our knowledge, not been made previously. Measurements were performed on four different occasions during the spring of 2005 (25 February, 3 and 29 April, and 26 May) in the Baltic Sea, using rapid light curves generated with a Diving PAM. In addition, samples were collected for photoinhibition studies in the laboratory. The light response curves obtained in situ at 1 and 4 m depths for *F. vesiculosus* showed lower values of light saturation with depth. When algae from 1 and 4 m depths were exposed to high irradiances of photosynthetically active radiation (1,400 μmol photons m^{-2} s^{-1}), algae from 1 m depth showed a higher degree of photoinhibition in comparison to algae from 4 m depth.

Keywords Diving PAM · Fluorescence · Photoinhibition · Photosynthesis

N. G. A. Ekelund (✉) · C. A. Nygård · R. Nordström · A. M. Gylle
Department of Natural Sciences, Mid Sweden University, 851 70 Sundsvall, Sweden
e-mail: nils.ekelund@miun.se

Introduction

Photosynthesis is the process by which photoautotrophs convert light energy into chemical energy. The process consists of many different chemical and physical reactions, which can be quantified using various methods (Franklin and Badger 2001). Typically, methods involve the collection of marine algae from their natural environments, after which they are transported to the laboratory where the measurements are carried out. In earlier studies, specimens were often incubated for a certain time before photosynthetic measurements were conducted in the laboratory (Brinkhuis 1977). However, with techniques such as pulse amplitude modulated (PAM) chlorophyll (Chl) fluorescence it is now possible to make qualitative photosynthetic measurements as well as quantifying aquatic productivity (Franklin and Badger 2001; Raniello et al. 2006). Electron transport rates (ETR) can be estimated using Chl *a* fluorescence, either from a spot on a surface of the thallus or as an integrated measure of evolved oxygen from a sample (Carr and Björk 2003; Longstaff et al. 2002; White and Critchley 1999). Light response curves obtained from *Gelidium sesquipedale* showed a similar pattern independent of which method was used (Silva et al. 1998). A study by Longstaff et al. (2002) showed a significant correlation between O_2 evolution and ETR in *Ulva lactuca* at low to medium levels of irradiance; at higher irradiances ETR continued to increase steadily, while O_2evolution tended towards an asymptote.

Macroalgae play an important role in primary biomass production in aquatic ecosystems, especially in coastal areas (Häder et al. 2003). However, macroalgae are also important for other organisms living in the water column as they provide shelter and food to fish and many invertebrates. Macroalgae behave differently in comparison to phytoplankton as they are sessile and cannot migrate in order to find less stressful conditions. The Baltic Sea is a unique ecosystem and differs from the Atlantic ecosystem with constant low salinity, a lack of tides, occurrences of ice-cap and a long cold winter (Nygård 2005). In the most northern part, the salinity is in the range 1–3 practical salinity units (psu). The most important species of macroalgae in the Baltic Sea is the bladderwrack *F. vesiculosus*, which is smaller and has no bladders in comparison to the marine *F. vesiculosus* found in the North Atlantic (Kautsky 1992; Wallentinus 1991). Due to the fluctuation of environmental physical conditions such as temperature and irradiance, many macroalgae have to adjust both morphologically and physiologically in order to sustain growth. Similar adaptations are found in seaweeds from polar, i.e. Antarctic and Arctic, regions (Wiencke et al. 2007).

One of the most important factors for the establishment of macroalgae in the water column is the amount of incident solar radiation. The vertical distribution of macroalgae in the water column is controlled by light penetration. Nygård (2005) showed that the photosynthetic properties of *F. vesiculosus* correlated well with the attenuation coefficients at different sites in the Bothnian Sea. The results of Nygård (2005) demonstrated that *F. vesiculosus* showed acclimatisation to low light by an increase in Chl *a* and a lower maximum photosynthetic capacity. This means that species adapted to lower irradiances thrive best at deeper depths. The possibility of plants to adjust to different light conditions is often used to categorise species as either shade- or sun-adapted species. A study by Silva et al. (1998) showed that the photosynthetic capacity of *Gelidium sesquipedale* decreased in deeper waters due to low irradiance and a 'shade' acclimation pattern. Obligate shade species have a rather limited capacity for electron transport but have a large antenna for photon capture; hence they may be limited by electron transport under conditions of high irradiance (Demmig-Adams and Adams 1996).

The objectives of this study were (1) to assess in situ differences in light responses of *F. vesiculosus* at different depths on four occasions during the spring of 2005, and (2) to compare differences in photoinhibition and recovery between algae collected at 1 and 4 m depth after exposure to high irradiance in the laboratory.

Material and methods

Plant material and collection

This study was carried out on 25 February, 3 and 29 April, and 26 May 2005. *F. vesiculosus* was collected from 1 and 4 m depths in the Bothnian Sea (the Bothnian Sea is in the northern arm of the Baltic proper) to the east of Åstön (Sweden; 62°24′N; 17°45′E; 4.5–5 psu). In situ temperature and quantum irradiance were measured by means of specific sensors housed on the DIVING-PAM fluorometer used for fluorescence measurements (Table 1). *F. vesiculosus* was collected and transported (1 h) from the Bothnian Sea in plastic bags to the laboratory in Sundsvall (Sweden).

In situ studies

Relative electron transport rates (rETR) were measured by SCUBA divers with a PAM fluorometer (Diving PAM, Walz, Effeltrich, Germany) to generate rapid light curves (RLCs), which are based on the relationship between ETR and irradiance. When fluorescence is measured under given actinic irradiance under a flash of saturating light, F′ is the value that corresponds to fluorescence under actinic light, and Fm′ is the maximum fluorescence when all photosystem II (PSII) reaction centres are closed

Table 1 Temperatures and irradiances from 1 and 4 m depths in the Bothnian Sea to the east of Åstön (Sweden; 62°24′N; 17°45′E; 4,5-5 psu) on 25 February, 3 and 29 April, and 26 May 2005

	25 February	3 April	29 April	26 May
Temperature at 1 m depth, (°C)	-	2	3.5	5
Temperature at 4 m depth, (°C)	-	2	4	3
Air temperature (°C)	−1.6	-	8	12
Irradiance at 1 m depth (μ mol photons m^{-2} s^{-1})	26	55	35	25
Irradiance at 4 m depth (μ mol photons m^{-2} s^{-1})	-	43	10	12
Sunrise	0700 hours	0602 hours	0435 hours	0330 hours
Sunset	1700 hours	1948 hours	2103 hours	2208 hours

(Suggett et al. 2003). The ETRs were calculated using the equation:

$$ETR\left(\mu mol\, e^{-} m^{-2} s^{-1}\right) = \Delta F / Fm' \times Irradiance \times 0,5 \times Absorbance\ factor$$

where, ΔF=Fm′−F′, ΔF/Fm′ is the maximum quantum yield of PSII photochemistry measured in light, 0.5 indicates that 50% of the absorbed quanta are distributed to photosystem II, and the absorbance factor used was 0.74 indicating the absorbance of the thallus. The absorbance factor was estimated in an earlier investigation by Nygård (2005) by measuring the mean absorbance of the thallus tips.

At each measuring period, three consecutive RLCs were measured from different places on intact thalli of three individuals of *F. vesiculosus* at depths of 1 and 4 m. Each individual of *F. vesiculosus* at each depth was marked with clips of different colours in order that the same individual could be recognised when coming back at the different measuring times (25 February, 3 and 29 April, and 26 May) in 2005. Each RLC started with an initial fluorescence measurement under dim light (Fo) followed by a saturating light pulse (>8,000 μmol photons m^{-2} s^{-1}). The saturating pulse was followed by a 10 s exposure to actinic light followed by another saturating light pulse. This sequence was repeated a total of nine times with increasing irradiance, 0, 2.5, 22, 40, 110, 146, 258, 372 and 606 μmol photons m^{-2} s^{-1}.

Laboratory studies

At every occasion when in situ measurements were performed (25 February, 3 and 29 April, and 26 May) algae were collected by SCUBA diving at 1 and 4 m depth. At each depth, three individuals of *F. vesiculosus* were collected and placed in dark plastic bags to prevent exposure to direct solar radiation. The plants were transported to the laboratory (1 h), where the plants were maintained in aerated aquariums with ambient seawater in a cold room at a temperature of 6–8 °C for 4 days to acclimatise to the same irradiance and light rhythm as for the corresponding day of collection (Table 1).

The plants were divided into two parts: one part served as a control and one part was photoinhibited at 1,400 μmol photons m^{-2} s^{-1} (Philips 400W MASTER SON-T) for 4 h. The same RLCs as for the in situ studies were performed before and after photoinhibition and after 24 h of recovery. The algae were dark-adapted for 5 min before the fluorescence measurements and three measurements were performed with each individual thallus.

The response of photosynthesis was modelled by a non-rectangular hyperbola, where the initial slope represents the apparent quantum efficiency of photosynthesis (Henley 1993). The value of the light saturation point was calculated by extrapolating the linear function described by the slope to its intersection with the light-saturated maximum. T-tests between treatments were used in a statistical analysis with a 95% significance level. The results represent mean values from nine replicates of each thallus at each depth.

Results

The in situ temperatures, irradiances and photoperiods at the collection sites are listed in Table 1. The temperatures were similar and consistent during the investigation for each date and at each depth (Table 1). The light was attenuated by approximately 15–50% at a depth of 4 m relative to 1m (Table 1). Of all days tested during the investigation period, the irradiance was highest on 3 April at both depths (Table 1).

The light response curves obtained in situ at 1 and 4 m depths for *F. vesiculosus* showed significantly lower levels of light saturation with depth (Table 2). The initial slope showed no significant variation between the different dates and depths (Table 2). Data is missing from the 4 m depth on 25 February due to experimental failure.

Samples of *F. vesiculosus* collected on all of the dates from 1 m depth and photoinhibited at 1,400 μmol photons m^{-2} s^{-1} for 4 h showed lower saturation values (Fig. 1a–c). A recovery of the saturation level toward the control level following photoinhibitory treatment was seen only in *F. vesiculosus* from 29 April and 26 May (Fig. 1c,d). The initial slope decreased significantly following photoinhibitory treatment of *F. vesiculosus* from 1 m depth in all experiments, but recovered well within 24 h (Table 3). *F. vesiculosus* from 4 m depth followed a similar trend to that observed in algae collected from a depth of 1 m (Table 3).

Table 2 Calculated parameters of the light response curves obtained with the DIVING-PAM (pulse amplitude modulated) fluorometer measured in situ in the Bothnian Sea to the east of Åstön (Sweden; 62°24′N; 17°45′E; 4,5-5 psu) at 1 and 4 m depth on 25 February, 3 and 29 April, and 26 May 2005. Significant difference between values for surface and depth is indicated by a lower case a

Day	Surface		Depth	
	Slope	Saturation[b]	Slope	Saturation
25 February	0.16	127	-	-
3 April	0.21	110	0.19	66 a
29 April	0.08	183	0.20	73
26 May	0.13	160	0.19	85 a

[b] The unit for saturation is μmol photons m^{-2} s^{-1}

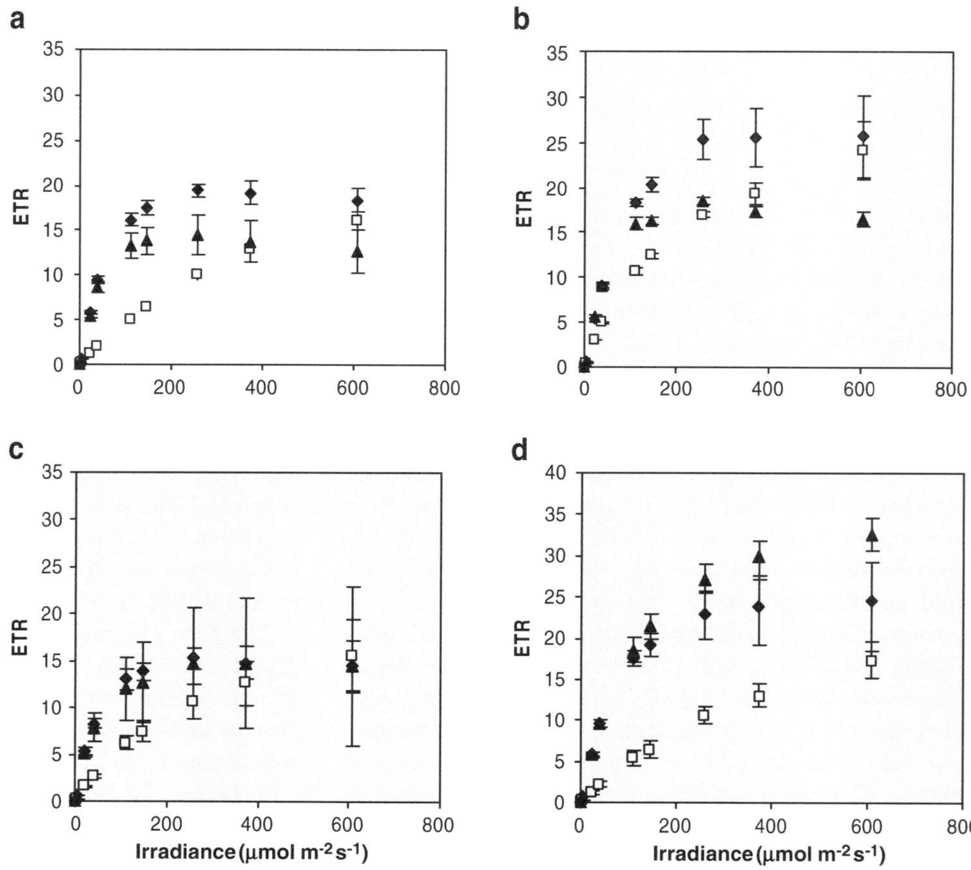

Fig. 1 Light response curves for *Fucus vesiculosus* collected at a 1 m depth in the Bothnian Sea to the east of Åstön (Sweden; 62°24′N; 17°45′E; 4,5-5 psu) on 25 February 25 (**a**), 3 April **b**), 29 April (**c**) and 26 May **d**) in 2005. The calculated ETR values are obtained from control (♦) and stressed (□; 1,400 μmol photons m^{-2} s^{-1}) algae and after 24 h of recovery (▲) in the laboratory. *Error bars* Standard deviation from three replicates

However, relative to the 1 m results, the slope after high irradiance treatment did not decrease as drastically as with algae from 4 m (Table 3). Only the initial slope on 29 April failed to show significantly lower values in comparison to control values (Table 3). The saturation values of ETR for algae collected at 1 and 4 m depths from control and recovered samples varied between 61 and 139 μmol photons m^{-2} s^{-1} (Table 3). However, the saturation values after high irradiance reached values between 300 and 500 μmol photons m^{-2} s^{-1} (Table 3).

Table 3 Calculated parameters of the light response curves for *Fucus vesiculosus* collected at 1 and 4 m depths in the Bothnian Sea to the east of Åstön (Sweden; 62°24′N; 17°45′E; 4,5-5 psu) on 25 February, 3 and 29 April, and 26 May 2005. The data are obtained from control and stressed (1,400 μmol photons m^{-2} s^{-1}) algae and after 24 h of recovery. Significant differences between values of control and stressed samples at the different depths are indicated by lower case a, and between stressed and recovered by lower case b

Day		Surface slope			Surface saturation[c]	
	Control	Stressed	Recover	Control	Stressed	Recover
25 February	0.25	0.03 a	0.21 b	78	416 a	63 b
3 April	0.24	0.08 a	0.25 b	112	289 a	71 b
29 April	0.25	0.07 a	0.29 b	61	331 a	54 b
26 May	0.26	0.04 a	0.27 b	96	454 a	139 b
Day		Depth Slope			Depth saturation	
	Control	Stressed	Recover	Control	Stressed	Recover
3 April	0.28	0.12 a	0.25 b	85	270 a	87 b
29 April	0.16	0.10	0.21	70	167 a	61 b
26 May	0.26	0.09 a	0.25 b	88	348 a	106 b

[c] The unit for saturation is μmol photons m^{-2} s^{-1}

Discussion

To our knowledge, measurements of light response curves from the same individuals of *F. vesiculosus* at different depths and times of the year have never been made before. Most in situ studies are performed over 1 day, and even if several days are used for the investigations, different individuals are often used. The fluorescence method using a Diving-PAM has the advantage that the in situ photosynthetic capacity of macroalgae can be assessed with a high degree of accuracy. The most important factor for in situ ETR measurements of several individuals during the same dive is that it is done quickly, as a diver normally has air for approximately 1 h. In addition, the weather conditions are of vital importance to successful measurements. Unfortunately, poor weather conditions affected the first measurement in February, when conditions were too harsh to collect data from 4 m depth. The general trend in the present in situ study was that the saturation levels of ETR were lower at 4 m depth in comparison to 1 m depth (Table 2). Similar results were observed for *Gelidium sesquipedale* at different depths, i.e. ETR values were lower at 22 m depth in comparison to 10 m (Silva et al. 1998). In the study by Silva et al. (1998) the algae were collected at different depths and brought ashore in dim light before measurements. Nygård (2005) used a method similar to that of Silva et al. (1998), but measured oxygen evolution instead of ETRs. Nygård (2005) showed that *F. vesiculosus* from the site used in the present study responded with a higher photosynthetic maximum in algae at lower depths and Kd (attenuation coefficient). This indicates that *F. vesiculosus* in the northern Baltic is acclimatised to its light environment, and shows a similar response independent of the measuring technique used (fluorescence or oxygen evolution) or whether in situ or laboratory measurements are taken. The reason for the nonsignificance between the saturation levels at 1 m in comparison to 4 m depth on 29 April is presumably due to the spring bloom of phytoplankton, which affected levels of irradiance (Table 1). The difference in light regimes at different depths will probably increase as the days lengthen and solar irradiance increases. A study by Longstaff et al. (2002) measured the ETR of *Ulva lactuca* at a depth of 1.8 m with an irradiance of 400–600 μmol photons m^{-2} s^{-1} at midday, which cannot be compared with the irradiance of 10–55 μmol photons m^{-2} s^{-1} in the present study. Another important factor for photosynthesis is temperature, which was only 2–5 °C in the present study, i.e. differing from other studies where the temperature was normally much higher. The optimum temperature for ETR has been shown to be 4–10°C for *F. vesiculosus* from the Baltic, in comparison to *F. vesiculosus* from the Irish Sea, where the optimum was 15–20°C (Nygård 2005).

Earlier studies have shown that surface-adapted algae recover faster from exposure to high irradiance than algae living in deeper water (Häder and Figueroa 1997). The present study showed that algae from 1m recovered in a similar way as algae from 4m despite higher values of light saturation and lower slopes after high irradiance (Table 3). One difference between the deeper and the surface algae is the amount of Chl *a*, which increases with increasing water depth (Nygård 2005). According to Dring (1982) a higher level of Chl *a* will result in similar slopes but different maxima of photosynthesis when it is related to chlorophyll. Nygård (2005) also proposed that this variation is probably a result of larger photosynthetic units (PSU) rather than a higher density of PSU. When both Atlantic and Baltic *F. vesiculosus* were treated with UV-B radiation for 5 h, the results showed that Baltic algae were more sensitive, indicating a faster recovery in Atlantic algae (Nygård and Ekelund 2006). Many different mechanisms operating in the down regulation of PSII have been proposed in order to explain the difference between plants or algae growing under different light conditions. Critchley and Russell (1994) suggested that reactions centres of PSII have two forms, with an active form carrying out linear electron transport and a second form acting as a downregulating dissipative form. Another important factor in the regulation of photosynthesis is the structural organisation of thylakoid membrane components. In comparison to surface algae, deep water algae resemble shade plants, with larger grana stacks than high-light-adapted algae. In *Macrocystis pyrifera* it was shown that the size of PSU units were much larger at a depth of 20 m in comparison to those of surface algae (Smith and Melis 1987). The size of PSU units and structure of thylakoid membranes suggests an increase in the number of dissipative centres that could be mobilised during high irradiance to regulate the rate of recovery after photoinhibition.

Runcie and Riddle (2006) showed that *Iridaea mawsonii* growing in low light was well suited to a rapid increase in high light exposure. This tolerance was due to the rapid development of high levels of non-photochemical quenching (NPQ), which is in contrast to the suggestion by Hanelt (1998) that algae from deeper water at low light intensities have less possibility to induce NPQ. A lower level of NPQ will make the recovery from high irradiance slower in deep water algae in comparison to shallow algae. Brown algae from surface waters have been shown to react faster to high irradiance by a decrease in photosynthetic capacity and a faster recovery than algae from deep waters (Hanelt 1998). In the present study, a similar response was shown in *F. vesiculosus* at 1 m depth, with a fast recovery even if the stress after high irradiance showed higher saturation values in comparison to algae at 4 m depth (Table 3). A factor that is related to the NPQ value is the minimal fluorescence value F_o, which is very low in *F. vesiculosus* living in the Baltic compared to *F. vesiculosus* living in the Atlantic at high salinities (A.M. Gylle et al., manuscript in preparation). A low F_o value indicates a greater peripheral antenna compared to core antenna and this might induce a higher efficiency of excitation energy utilisation (Zhu et al. 2005), although was not shown that F_o values differ between different depths in the present study. Slow reaction kinetics from algae living at deep waters has also been shown in *Sphacelaria plumosa* and *Desmarestia viridis* (Hanelt 1998). In *Ecklonia cava*, the possibility of recovery from high irradiance is related to life history (Altamirano et al. 2004). Young sporophytes are more sensitive to high irradiance in comparison to older gametophytes with higher levels of photo-damage. This difference is due to the more complex structure and higher pigment content in gametophytes compared to the younger sporophytes (Altamirano et al. 2004). In summary, this study indicates higher saturation levels from in situ measurements of *F. vesiculosus* at 1 m depth in comparison to algae living at 4 m depth. However, the photoinhibitory effects from high irradiance are greater in algae from 1 m depth in comparison to those from 4 m depths. Overall, the first part of this study differs from many other studies in that individuals of *F. vesiculosus* have not previously been transferred from different depths to the surface before taking measurements.

Acknowledgements The authors would like to thank Dr. Stefan Falk and Dr. Tessa Pocock for help, critical reading of the manuscript and comments on the language.

References

Altamirano M, Murakami A, Kawai H (2004) High light stress in the kelp *Ecklonia cava*. Aquat Bot 79:125–135

Brinkhuis BH (1977) Seasonal variations in salt-marsh macroalgae photosynthesis. II. *Fucus vesiculosus* and *Ulva lactuca*. Mar Biol 44:177–186

Carr H, Björk M (2003) A methodological comparison of photosynthetic oxygen evolution and estimated electron transport rate in tropical *Ulva* (Chlorophyceae) species under different light and inorganic carbon conditions. J Phycol 39:1125–1131

Critchley C, Russell AW (1994) Photoinhibition of photosynthesis in vivo: the role of protein turnover in photosystem II. Physiol Plant 92:188–196

Demmig-Adams B, Adams WW III (1996) The role of xantophyll cycle carotenoids in the protection of photosynthesis. Trends Plant Sci 1:21–26

Dring MJ (1982) The biology of marine plants. Cambridge University Press, Cambridge

Franklin LA, Badger MR (2001) A comparison of photosynthetic electron transport rates in macroalgae measured by pulse amplitude modulated chlorophyll fluorometry and mass spectrometry. J Phycol 37:756–767

Häder D-P, Figueroa FL (1997) Photoecophysiology of marine macroalgae. Photochem Photobiol 66:1–14

Häder D-P, Kumar HD, Smith RC, Worrest RC (2003) Aquatic ecosystems: effects of solar ultraviolet radiation and interactions with other climate change factors. Photochem Photobiol Sci 2:39–50

Hanelt D (1998) Capability of dynamic photoinhibition in Arctic macroalgae is related to their depth distribution. Mar Biol 131:361–369

Henley WJ (1993) Measurement and interpretation of photosynthetic light-response curves in algae in the context of photoinhibition and diel changes. J Phycol 29:729–739

Kautsky H (1992) The impact of pulp-mill effluents on phytobenthic communities in the Baltic Sea. Ambio 21:308–313

Longstaff BJ, Kildea T, Runcie JW, Cheshire A, Dennison WC, Hurd C, Kana T, Raven JA, Larkum WD (2002) An in situ study of photosynthetic oxygen exchange and electron transport rate in the marine macroalga *Ulva lactuca* (Chlorophyta). Photosynth Res 74:281–293

Nygård CA (2005) Ecophysiological responses of Baltic and Atlantic Fucus vesiculosus to environmental factors. Thesis, Mid Sweden University. ISBN 91-85317-15-2

Nygård CA, Ekelund NGA (2006) Photosynthesis and UV-B tolerance of the marine alga *Fucus vesiculosus* at different sea water salinities. J Appl Phycol 18:461–467

Raniello R, Lorenti M, Brunet C, Buia MC (2006) Photoacclimation of the invasive alga *Caulerpa racemosa* var. *cylindica* to depth and daylight patterns and a putative new role for siphonaxanthin. Mar Ecol 27:20–30

Runcie JW, Riddle MJ (2006) Photosynthesis of marine macroalgae ice-covered and ice-free environments in East Antarctica. Eur J Phycol 41:223–233

Silva J, Santos J, Serodio J, Melo RA (1998) Light response curves for *Gelidium sesquipedale* from different depths, determined by two methods: O_2 evolution and chlorophyll fluorescence. J Appl Phycol 10:295–301

Smith BM, Melis A (1987) Photosystem stoichiometry and excitation distribution in chloroplasts from surface and minus 20 meter blades of *Macrocystis pyrifera*, the giant kelp. Plant Physiol 84:1325–1330

Suggett DJ, Oxborough K, Baker NR, McIntyre HL, Kana TM, Geider RJ (2003) Fast repetition rate and pulse amplitude modulation chlorophyll a fluorescence measurements for assessment of photosynthetic electron transport in marine phytoplankton. Eur J Phycol 38:371–384

Wallentinus I (1991) The Baltic Sea gradient In: Mathieson AC, Nienhuis PH (eds) Ecosystems of the world. Elsevier, Amsterdam, pp 83–108

White AJ, Critchley C (1999) Rapid light curves: A new fluorescence method to assess the state of the photosynthetic apparatus. Photosynth Res 59:63–72

Wiencke C, Clayton MN, Gómez I, Iken K, Lüder UH, Amsler CD, Karsten U, Hanelt D, Bischof K, Dunton K (2007) Life strategy, ecophysiology and ecology of seaweeds in polar waters. Rev Environ Sci Biotechnol 6:95–126

Zhu XG, Govindjee, Baker NR, deSturler E, Ort DR, Long SP (2005) Chlorophyll a fluorescence induction kinetics in leaves predicted from a model describing each discrete step of excitation energy and electron transfer associated with photosystem II. Planta 223:114–133

Effects of photon flux density and agricultural fertilizers on the development of *Sarcothalia crispata* tetraspores (Rhodophyta, Gigartinales) from the Strait of Magellan, Chile

C. Werlinger · A. Mansilla · A. Villarroel · M. Palacios

Originally published in the Journal of Applied Phycology, Vol 20, No 5, 307–315.
DOI: 10.1007/s10811-007-9301-5

Abstract Tetraspores of *Sarcothalia crispata* from San Juan Bay, Strait of Magellan, Chile, were cultivated under different combinations of photon flux densities and agricultural fertilizers in the laboratory. In the experiment, the *S. crispata* specimens were cultured in combinations of different photon flux densities (50, 100, 150 μmol photons m^{-2} s^{-1}) and enriched seawater solutions (sodium nitrate + monocalcium phosphate, urea + monocalcium phosphate, ammonium nitrate + monocalcium phosphate), always adjusting the N and P concentrations to 10 and 3 mg L^{-1}, and in sea water as control. After 45 days, the tetrasporeling plants were found to be larger at photon flux densities of 50 and 100 μmol photons m^{-2} s^{-1} in the nutrient enrichment experiments; growth was greatest in the sea water enriched with ammonium nitrate and urea. An analysis of the combined effect of the photon flux density and nutrients revealed that the best combination for sporeling growth was the ammonium nitrate and urea solution at 50–100 μmol photons m^{-2} s^{-1}.

Keywords Early development · Sporeling culture · Mass cultivation · Nutrients · Growth

C. Werlinger (✉) · A. Villarroel
Departamento de Oceanografía, Universidad de Concepción,
Casilla 160-C Concepción, Chile
e-mail: cwerling@udec.cl

A. Mansilla · M. Palacios
Departamento de Ciencias y Recursos Naturales,
Universidad de Magallanes,
Punta Arenas, Chile

Introduction

Phycocolloid extraction and macroalgae farming are currently important productive activities in Chilean fisheries, demonstrating sustained growth and development in recent years. Phycocolloid production using carrageenan-producing algae from the genera *Iridaea*, *Gigartina*, *Mazzaella*, and *Sarcothalia* is second only to that using *Gracilaria chilensis* Bird, McLachlan & Oliveira, the most economically important macroalgae group in Chile.

Sarcothalia crispata (Bory) Leister, commonly known as "luga negra" and previously called *Iridaea ciliata*, is endemic to Chile. This red alga is distributed along the cold temperate coasts of central to southern Chile, from its northern limit in Valparaíso (36°S) to Tierra del Fuego (54°S; Ramírez and Santelices 1991; Hommersand et al. 1993; Hoffmann and Santelices 1997). Commercial extraction of this alga began in Chile in the 1960s (Romo et al. 2001).

Uninterrupted fishery statistics from 1986 are available for *S. crispata*. The first landing record reports 23,449 wet tons of this alga, and landings in 2005 reached 24,942 wet tons. *S. crispata* harvests are largest in central-southern Chile, between approximately 36° and 44°S (Sernapesca 2003). However, in spite of continuous landing records for over 20 years, landings that included *S. crispata* through the year 2000 were listed under the common name "luga-luga", which included at least two other carrageenophytes (*Gigartina skottsbergii* or "luga roja" and *Mazzaella sp.* or "luga cuchara"). Therefore, prior to 2000, the exact volume of *S. crispata* landings is not known, making it difficult to evaluate the resource's current state of sustainability based on landing records.

The opinions of experts and subsistence fishers, as well as information found in the fishery statistics, allow an estimate of steady production levels that have remained at around 20,000–30,000 wet tons per year during the 30 years of "luga negra" extraction. However, with its increased national importance, populations of this species have begun to present symptoms of overexploitation. Two pieces of evidence indicate a decline in this resource: (1) although the extraction effort has increased during recent years, the volume of landings has remained approximately the same for several years, and (2) currently, populations available for extraction are found in harder to reach areas (Avila et al. 1994, 1996). Consequently, if the current extraction pressure continues, it is highly likely that the natural *S. crispata* populations will collapse.

Considering the increasing vulnerability of natural populations of this and other species of high commercial interest, together with the growing demand for the product on national and international markets, mass farming of *S. crispata* is a highly interesting alternative for maintaining—or even increasing—the productivity of this raw material (Mansilla et al. 2002). Experiments in the laboratory and in the field have shown that culturing algae from spores can be an effective method of mass production (Alveal et al. 1991, 1994, 1995, 1997; Candia et al. 1993) as long as aquaculturists are able to produce large enough quantities of initial stages under adequate conditions to assure good initial growth and the subsequent development of adult algae.

For this, certain aspects of the culturing methodology are particularly relevant for the successful implementation of new farming techniques; these aspects require research and innovation. As stated by many authors, the sporeling supply creates a bottleneck in successful commercial cultivation methodologies (Buschmann et al. 1999; Liu et al. 2004). Therefore, much research has been done in this area. Zhao et al. (2006) reported on early sporeling development in *Gracilaria asiatica*, and Mansilla et al. (2006), Avila et al. (2003), and Romo et al. (2001) on the effects of environmental factors on germination and growth in initial stages of commercial species from southern Chile. Other topics of study in the area include the effect of light period on *Undaria pinnatifida* gametophyte egg-discharge (Liu et al. 2004), the use of *Undaria* gametophyte clones in sporeling cultures (Wu et al. 2004), new methods of rearing sporelings using gametophyte clones and strain selection in *Laminaria japonica* (Li et al. 1999; Zhang et al. 2007), vegetative propagation through tissue fragmentation in *Gigartina skottsbergii* (Buschmann et al. 2001; Hernandez-Gonzalez et al. 2007), and spore viability and survivorship (Buschmann et al. 1999; Romo et al. 2001).

When cultivating algae from spores, low-cost agricultural fertilizers can be used to stimulate sporeling and germling growth during the industrial seed production stage (Werlinger 1998), replacing the expensive nutrients that are used in the experimental phase and design of the culture methodology. The most appropriate environmental conditions can also be estimated for this stage. According to Avila et al. (1999b), innovations in this phase would unquestionably have an important impact on the amount of money required to produce these seaweeds, as well as affecting the technological aspects of their mass production under farmed conditions.

The objective of this study is to improve the method for culturing *S. crispata* from spores, focusing on sporeling production using spores of individuals collected from natural populations in the Strait of Magellan that have not yet been subjected to exploitation. We used studies on the improvement of methodologies for the commercial cultivation of *G. chilensis* as references (Alveal et al. 1994, 1997).

Materials and methods

Mature tetrasporic fronds of *S. crispata* were collected at the end of September 2004 (austral spring) from natural intertidal algae beds in San Juan Bay. These beds are located in the western section of the Strait of Magellan (53° 37′S, 70°59′W), Chile, approximately 70 km southwest of the city of Punta Arenas. Following collection, the specimens were transported in glass containers with seawater to the marine biology laboratory of the Departamento de Ciencias Naturales y Recursos Marinos, Facultad de Ciencias, Universidad de Magallanes, Chile.

We performed laboratory experiments using a combination of four different culture medium under three different photon flux densities. Three of the media used were different enriched seawater solutions and one was non-enriched seawater (control); the combinations and concentrations of agricultural fertilizers used are indicated in Table 1. Each of the media was made using filtered and sterilized natural seawater (provided by the "Centro Cultivos Marinos"). The nitrogen and phosphorous concentrations were adjusted to 10 and 3 mg L^{-1}, respectively, based on the optimal values detected in similar experiments with *Gracilaria chilensis*; sodium nitrate values were 0.06 L^{-1} and sodium monocalcium phosphate values were 0.01 L^{-1} (Alveal et al. 1991).

In addition, we used three photon flux densities (50, 100, 150 μmol photons m^{-2} s^{-1}) on the experiments. The different levels were obtained by using different combinations of PAR Phillips TLT 20W/54 daylight fluorescent tubes. The culture chamber was maintained at 8°C±1°C, with a 12 h:12 h light:dark photoperiod.

Experimental treatments

Fertile frond fragments collected from natural populations were carefully washed under tap and distilled water to

Table 1 Types of agricultural fertilizers used in laboratory experiments, with corresponding concentrations of nitrogen and phosphorous

Fertilizer	Control	Solution 1	Solution 2	Solution 3
Sodium nitrate (Chilean nitrate)	-	0.061 g.L^{-1} (10 mg.L^{-1} nitrogen)	-	-
Urea	-	-	0.043 g.L^{-1} (10 mg.L^{-1} nitrogen)	-
Ammonium nitrate	-	-	-	0.057 g.L^{-1} (10 mg.L^{-1} nitrogen)
Monocalcium phosphate	-	0.007 g.L^{-1} (3 mg.L^{-1} phosphorous)	0.007 g.L^{-1} (3 mg.L^{-1} phosphorous)	0.007 g.L^{-1} (3 mg.L^{-1} phosphorous)

remove epiphytes and organic matter remains. To stimulate the opening of the sporangia and induce the spore release according to the methodology of Romo et al. (2001) and Avila et al. (2003), we desiccated mature fronds fragments (1–2 cm^2), placing them in trays with 200 mL of 0.45 μm filtered and sterilized seawater with continuous shaking for 24 h. The resulting spore suspension (average density: 33,000 spores mL^{-1}) was then used to inoculate 36 glass slides.

After the spores had settled onto the glass slides (average density: 133 spores cm^{-2}), they were separated into 12 containers, each having three slides and one of the four culture media. In the incubation chamber, the containers were separated into three compartments corresponding to the three photon flux densities. In summary, each compartment held four containers, corresponding to the four culture media treatments. Seawater and culture media were replaced once a week for a month and a half, according to Romo and Paula (1995).

Growth measurements

We quantified tetrasporeling growth using digital images of the plants on the glass slides, which were obtained on six occasions over 45 days (days 10, 17, 24, 31, 38, 45) using a digital camera attached to a stereoscopic microscope. Images were processed using Image-Pro Plus, version 4.1. We took five photos in the center of each of the 36 slides, maintaining a uniform distance for each of the photographed areas. Using these photographs, we measured the diameter of all sporelings present in the image, analyzing a total of 3,600–4,600 sporelings per sample to obtain the average sporeling size.

Statistical analysis

A three-way Analysis of Variance (ANOVA) was used to evaluate differences in the sporeling growth under different nutrient and photon flux density treatments and given different time periods, as well as the interactions between these factors (Sokal and Rohlf 1979). When significant differences were found, we applied the Tukey test (Zar 1999) to evaluate which experimental treatment accounted for the observed differences (Sokal and Rohlf 1979; Zar 1999). Prior to the ANOVA, assumptions of a normal distribution and homogeneity of variance were evaluated using the Shapiro-Wilks test and Levene's tests, respectively (Sokal and Rohlf 1979; Zar 1999).

Results

Sporeling growth

During the 45 days of culturing, *S. crispata* tetrasporelings tended to increase steadily in size. This size increase was greater towards the end of the experiment, when we also observed differences in growth with the different nutrient media. However, these differences were not as evident under the different photon flux densities (Fig. 1).

The analysis of variance (Table 2) indicated significant differences in the *S. crispata* sporeling growth between the culture media ($p<0.0001$), under different photon flux densities ($p<0.0001$), and for different times ($p<0.0001$). We observed a significant interaction between the treatment factors ($p<0.0001$), indicating a synergistic effect of nutrients and light on growth over time (Table 2).

With respect to the different culture media, differences in the sporeling development were apparent in the first measurements, and the sporelings generally presented sustained, clear differences up to the penultimate measurement (day 38) with respect to the different treatments (Fig. 1). The cultivated sporelings in the control seawater treatment (i.e. no added nutrients) were smallest, reaching an average maximum size of 30.05±3.78 μm (mean ± SD) on day 45. The largest sporelings were those cultivated in media enriched with urea and ammonium nitrate, reaching average maximum sizes of 35.99±4.55 μm and 35.97± 4.44 μm, respectively, after 38 days of culture. Sporelings cultivated in sodium nitrate reached intermediate sizes, averaging maximum sizes of 33.28±4.17 μm after 38 days.

A post-hoc Tukey test (Table 3, Fig. 1) indicated that individuals in the seawater and sodium nitrate + mono-

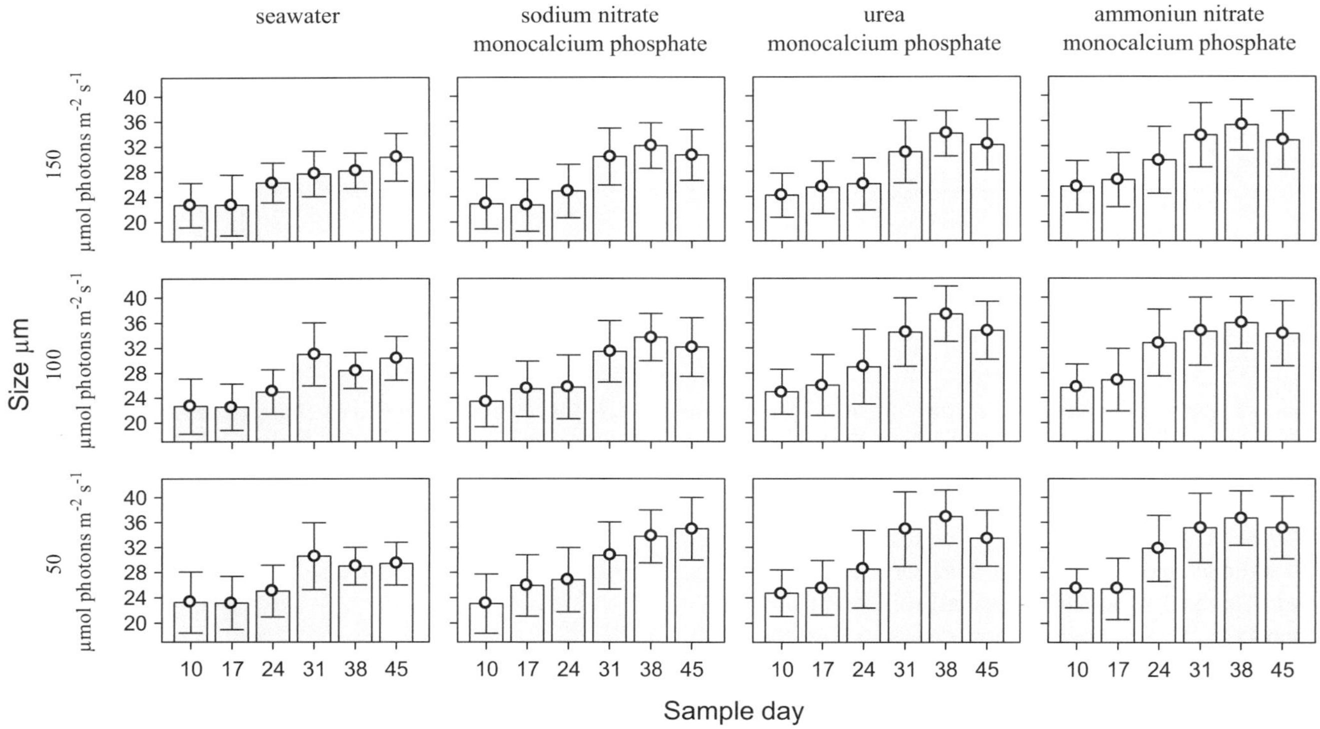

Fig. 1 Average size of *Sarcothalia crispata* sporelings grown under different photon flux density and in different cultura media; data are average values±S.D

calcium phosphate treatments were the smallest on average during almost the entire experimental period; they were significantly smaller than the sporelings cultivated in media enriched with urea or ammonium nitrate. There were no significant differences in the sizes of sporelings cultivated in the two media that most stimulated growth (urea + monocalcium phosphate and ammonium nitrate + monocalcium phosphate) for the first measurements (day 10) or the last two measurements (days 38 and 45). However, on three other occasions, the sporelings in ammonium nitrate enriched seawater (ammonium nitrate + monocalcium phosphate) were significantly larger than the urea enriched sporelings (urea + monocalcium phosphate), presenting average sizes of 25.56 μm±3.89, 31.17 μm±5.72 and 34.45 μm±5.63, respectively.

Table 2 Three-way analysis of variance for average size of *Sarcothalia crispata* sporelings with culture medium, photon flux density and time

Source of variation	SS	*df*	MS	*F*	*p*
Culture medium	91,556	3	30519	1,383	0.00001
Photon flux density	11,016	2	5508	250	0.00001
Time	402,396	5	80479	3647	0.00001
Interactions					
Medium × photon flux density	3,221	6	537	24	0.00001
Medium × time	22,687	15	1960	69	0.00001
Photon flux density × time	2,371	10	237	11	0.00001
Medium × photon flux density × time	9,564	30	319	14	0.00001

A visual inspection of germling growth under the different photon flux densities (Fig. 1) showed that all treatment groups increased in size over time, with differences between treatments beginning on day 17. Growth slowed on day 31 and tended to stabilize on the final two measurement dates (days 38 and 45; Fig. 1). Results from the Tukey test (Table 4) indicated that the greatest increase in growth occurred for sporelings submitted to the low (50 μmol photons m^{-2} s^{-1}) and medium (100 μmol photons m^{-2} s^{-1}) photon flux densities, as compared to the high photon flux density treatment (150 μmol photons m^{-2} s^{-1}); no significant differences were found between the low and medium photon flux density treatments. This pattern was observed throughout the entire experiment and became more evident over time. The largest sporelings were observed around 38 days of culture in the low and medium photon flux density treatments, averaging 34.10 μm±3.9 and 33.88 μm±4.0, respectively, whereas sporelings under the high photon flux density reached average values of 32.89 μm±3.9.

Combined effect of photon flux density and enriched seawater

Considering the effect of the different culture media with respect to the photon flux density (Table 5), the Tukey test

Table 3 Tukey test for differences in the growth of *S. crispata* sporelings grown under different combinations of enriched seawater on different sample days

Day	Culture medium: Seawater	Compared	Sodium nitrate + monocalcium phosphate	Compared	Urea + monocalcium phosphate	Compared	Ammonium nitrate + monocalcium phosphate
10	22.87	=	23.10	≠	24.63	≠	25.56
17	22.72	≠	24.61	≠	25.73	=	26.29
24	25.47	=	25.87	≠	27.93	≠	31.17
31	29.77	≠	30.88	≠	33.50	≠	34.45
38	28.46	≠	33.28	≠	35.98	=	35.97
45	30.05	≠	32.58	≠	33.59	=	34.16

revealed significant differences in growth between tetraspores deposited in seawater without added nutrients and those in enriched seawater (ammonium nitrate + monocalcium phosphate). The sporelings maintained under low and medium photon flux densities did not present significant differences in size, but were significantly larger than the sporelings maintained under the high photon flux density. The sporelings cultivated in the other two culture media (sodium nitrate + monocalcium phosphate, urea + monocalcium phosphate) also presented significant differences between the three photon flux density treatments (Table 5).

Considering the effect of the photon flux density relative to the different culture media (Table 6), we observed differences in the average sporeling size between all the enriched seawater treatments except for the sporelings maintained under medium light intensities and cultivated in seawater enriched with urea + monocalcium phosphate or ammonium nitrate + monocalcium phosphate; the average sizes were not significantly different for these treatment groups.

We observed a significant interaction of nutrient and photon flux density treatments on sporeling growth. The largest sporelings were observed in treatments with low and medium light intensities in seawater enriched with urea + monocalcium phosphate and ammonium nitrate + monocalcium phosphate (Tables 5 and 6).

Table 4 Tukey test for differences in the size of *S. crispata* sporelings grown under different treatments of photon flux density on different sample days

Day	Photon flux density (PAR) μmol photons m^{-2} s^{-1}: 50	Compared	100	Compared	150
10	24.17	=	24.18	=	23.96
17	25.20	=	25.19	≠	24.53
24	28.22	=	27.74	≠	27.08
31	32.79	=	32.78	≠	30.86
38	34.43	=	33.70	=	32.56
45	33.43	=	32.74	=	31.44

Discussion

The most successful culture media for the development of *Iridaea*, under laboratory conditions, are enriched seawater, following the formula of Provasoli (McLachlan 1973) and employed by Infante and Candia (1988), and the Swm-3 medium modified by McLachlan (1973) and employed by Hannach and Santelices (1985) and Luxoro and Santelices (1989). For both *Iridea laminarioides* and *I. cilata*, good development of juvenile forms is favored by very complete nutrient media. A comparison done with *I. ciliata* revealed that the Provasoli medium promotes greater growth than that obtained with seawater enriched only with nitrate and orthophosphate, with respect to concentrations similar to the Erdschreiber medium (McLachlan 1973) and the addition of EDTA-Fe.

Nevertheless, today, the search for ways to reduce the culturing costs and times makes animal-based organic wastes and agricultural fertilizers a much more convenient option for obtaining nutritive solutions. Ambler et al. (1988) conducted experiments utilizing water with *Argopecten purpuratus* excrement as an additive for culturing *Gracilaria chilensis* in aquaria; they observed positive effects on growth. In contrast, Buschmann et al. (1994) and Martínez and Buschmann (1996) studied the effect of effluents from salmon tanks on biomass production and agar quality in *G. chilensis* cultures, but did not observe significant stimuli for either parameter of the alga. Yoneshigue-Braga and Neves (1981) studied the use of agricultural fertilizers for the culture of *Gracilaria* and other algae in tanks, finding that commercial preparations can be adequately used to fertilize algae, as previously proposed by authors such as Alveal et al. (1991, 1999a). The most-used fertilizers are ammonium sulfate (Tseng 1981; Friedlander 2001), urea, and pig manure or fermented bird guano (Lipkin 1985), but sodium nitrate and ammonium nitrate are also used (Buschmann et al. 2004). Furthermore, Alveal et al. (1991) indicated that concentrations of 0.06 g L^{-1} of sodium nitrate and 0.01 g L^{-1} of sodium phosphate are the most appropriate concentrations for the treatment of *G. chilensis* germlings.

Table 5 Tukey test for differences in the growth of *S. crispata* sporelings under different densities of photon flux for each type of culture medium

Culture medium	Photon flux density (PAR) μmol photons m^{-2} s^{-1}				
	50	Compared	100	Compared	150
Seawater	26.91	=	26.99	≠	26.43
Sodium nitrate + monocalcium phosphate	29.80	≠	29.01	≠	27.39
Urea + monocalcium phosphate	30.80	≠	31.56	≠	29.16
Ammonium nitrate + monocalcium phosphate	31.56	=	31.77	≠	30.55

The results obtained from this study indicate that all three nutrient solutions had positive effects on the tetraspore growth, although the magnitude of the effect varied between solutions. The most effective solutions were those containing a combination of urea-monocalcium phosphate and ammonium-monocalcium phosphate, differing partially from the results of Alveal et al. (1995), who evaluated the effect of the same nutrients on the microthalli of *Gracilaria* obtained from carpospores. These authors observed a greater increase in germling size when maintained in a solution of "Chilean nitrate" (sodium nitrate) + monocalcium phosphate, followed by those fertilized with a solution of ammonium nitrate + monocalcium phosphate; the smallest size increase occurred in germlings maintained in a solution of urea + monocalcium phosphate. Nevertheless, our results agree with the proposal by Santelices (1996) that, although algae can use organic nitrogen, particularly urea, they generally prefer ammonium to nitrate. Other authors, such as Haglund and Pedersén (1993), Lobban and Harrison (1994), and Chopin et al. (1990) have also indicated that algae probably conserve energy by using NH_4^+ since their cells must reduce NO_3^- to ammonium before it can be incorporated into amino acids.

With respect to the requirements of radiant energy, in general, the early phases of macroalgae development are well adapted to low light intensities since the reproductive structures liberated in winter must grow under the canopy of the adult algae (Gómez and Wiencke 1996; Hanelt and Nultsch 1997). This could explain the significantly lower growth of *S. crispata* tetraspores at 150 μmol photons m^{-2} s^{-1} as compared to the lower photon flux densities of 100 μmol photons m^{-2} s^{-1} and 50 μmol photons m^{-2} s^{-1} (Romo et al. 2001; Avila et al. 2003). These findings agree with those of Ugarte (1982), who found significant differences in the development of *Mazzaella laminarioides* tetraspores and carpospores under different irradiances. In the same study, after 5 days of culturing, the individuals developed from tetraspores grew faster at light intensities of 77 μmol photons m^{-2} s^{-1}, whereas individuals developed from carpospores grew better when cultivated at 55 μmol photons m^{-2} s^{-1}. In contrast, Avila et al. (1999a) did not observe differences between tetraspores and carpospores of *G. skottsbergii* grown at 5 and 10°C, using three irradiances (6, 18, and 25 μmol photons m^{-2} s^{-1}); the best results for both tetraspores and carpospores occurred at 10°C and a photon flux density of 6 μmol photons m^{-2} s^{-1}. Different studies conducted on *I. ciliata* have indicated different optimum ranges of photon flux density for the development of initial stages. Hannach and Santelices (1985) found, for example, that the optimal development of this species occurred at a photon flux density of 35–60 μmol photons m^{-2} s^{-1} (12 h:12 h light:dark photoperiod, temperature 15°C) for both gametophytes and tetrasporophytes. According to Romo and Alveal (1995), the photon flux density for the incubation of *Iridaea* sp. spores is restricted to moderate and low light intensities (12–60 μmol photons m^{-2} s^{-1}) in combination with temperatures of 10–20°C; at higher illumination values, the growth rate decreases and a photon flux density above 120 μmol photons m^{-2} s^{-1} is inhibitory. Avila et al. (2003) stated that the basic requirement for *S. crispata* cultivation at 15°C is an irradiance range of 30–50 μmol photons m^{-2} s^{-1}. On the other hand, for the same species under different laboratory conditions (16 h:8 h light:dark photoperiod, 15°C), the

Table 6 Tukey test for differences in the size of *S. crispata* sporelings grown in different culture media for each density of photon flux density

Photon flux density (μmol photon m^{-2} s^{-1})	Culture medium						
	Seawater	Compared	Sodium nitrate + monocalcium phosphate	Compared	Urea + monocalcium phosphate	Compared	Ammonium nitrate + phosphate monocalcium
50	26.91	≠	29.80	≠	30.80	≠	31.56
100	26.99	≠	29.01	≠	31.56	=	31.77
150	26.43	≠	27.39	≠	29.16	≠	30.55

greatest growth was observed at a photon flux density of 60 μmol photons m^{-2} s^{-1} (Romo et al. 2001).

A comparison of the results from different authors indicates a photon flux density range of 12–60 μmol photons m^{-2} s^{-1} for optimum growth in several red macro-algae. Nevertheless, all of the previously cited studies indicate that the different developmental stages require different temperatures, illumination, and photoperiods (Henkel 1952; Boalch 1961; Fries 1963). Differences may also be found between individuals from different populations of the same species, or between species that inhabit the same region but in areas with different light conditions (Gantt 1990).

In our study, the photon flux density that produced the greatest effect on sporeling development (50–100 μmol photons m^{-2} s^{-1}, 12 h:12 h light:dark photoperiod) was in the upper part of the range reported by earlier studies. Nevertheless, the temperature used in our study was approximately 8°C, which is several degrees cooler than the temperatures used by other researchers. This is an important point to consider, given that, in the Strait of Magellan, seawater temperatures are much lower than in the zones inhabited by the species used in previous studies.

In summary, the combined effect of enriched seawater solutions and photon flux density on the growth of *S. crispata* tetraspores was positive and synergistic. A significant interaction between different parameters in different species has been mentioned by authors such as Buschmann et al. (2004), in studies regarding the effects produced by nutrients (phosphorous, nitrogen, ammonia), photon flux density, and temperature in a culture of *Gigartina skottsbergii*. These researchers observed an interaction among these factors, which, jointly, produced up to 2% greater growth in this carrageenophytic algae. Furthermore, Navarro-Angulo and Robledo (1999) observed that a combination of two nutrients (nitrogen, phosphorous) produced a synergistic increase in the development of both carpospores and tetraspores of *Gracilaria cornea* (27±1.3°C, 12 h:12 h light:dark photoperiod, 100 μmol photons m^{-2} s^{-1}) with respect to the separate effects of each fertilizer. In addition, studies by Mansilla et al. (2002) clearly indicate an interaction between photoperiod and photon flux density in the development of tetraspores and carpospores of *Mazzaella laminarioides*.

The results from our study support the effect of differential stimulation on the growth of *S. crispata* sporelings as a consequence of different culture media prepared from agricultural fertilizers, combined with the use of different photon flux densities. Our results agree in general with those obtained by Romo et al. (2001) and Avila et al. (2003) for the same species. The combined effect of these variables produced a synergistic effect on sporeling development. We identified two culture media that were clearly responsible for stimulating germling growth (with no significant differences between media), as well as two photon flux density treatments that significantly increased the sporeling size. The best treatment combinations for germling growth in *S. crispata* correspond to a combination of seawater enriched with ammonium nitrate + monocalcium phosphate and urea + monocalcium phosphate, cultivated under low (50 μmol photons m^{-2} s^{-1}) or intermediate (100 μmol photons m^{-2} s^{-1}) photon flux density.

Acknowledgements This research was partially funded by the Dirección de Investigación of the Universidad de Concepción through Project 201.112.073-1.1, the Universidad de Magallanes, and Project MECESUP USC0303 (UCO) "Programa para la movilidad de estudiantes y coordinación académica en las carreras de Biología Marina". We also thank Paula Neill and Danielle Barriga for critically reviewing the manuscript. Criticisms and comments by the anonymous reviewers are acknowledged with gratitude.

References

Alveal K, Romo H, Werlinger C, Nuñez M (1991) Utilización de esporas para el cultivo de *Gracilaria*. Informe final Proyecto PNUD CHI/87/011 (Programa de las Naciones Unidas para el Desarrollo) Universidad de Concepción

Alveal K, Romo H, Werlinger C, Nuñez M (1994) Uso de esporas como alternativa de propagación masiva de macroalgas. Revista Invest Cient Tecnoló. Ser Cienc Mar 3:77–87

Alveal K, Romo H, Werlinger C (1995) Cultivo de *Gracilaria* a partir de esporas. In: Alveal K, Ferrario M, Oliveira EC, Sar E (eds) Manual de Métodos Ficológicos, Universidad de Concepción, Editorial Aníbal Pinto, Concepción, Chile, 599–609

Alveal K, Romo H, Werlinger C, Oliveira EC (1997) Mass cultivation of the agar-producing alga *Gracilaria chilensis* (Rhodophyta) from spores. Aquaculture 148:77–83

Ambler R, Edding M, Rodríguez A, Escobar S (1988) Excreción de amonio por *Argopecten purpuratus* (Mollusca, Pectinidae) y su utilización por *Gracilaria* sp. (Rhodophyta, Gracilariaceae) : Una posibilidad de policultivo. Gayana Bot 45:309–315

Avila M, Otaíza R, Norambuena M, Nuñez M, Candia A, Poblete A (1994) Desarrollo de tecnología de cultivo y repoblación de luga negra en la X Región. Instituto de Fomento Pesquero-Corporación de Fomento a la Producción. SGI. 94/9:97 pp

Avila M, Otaíza R, Nuñez M (1996) Biological basic for the management of "luga negra" *Sarcothalia crispata*, (Gigartinales, Rhodophyta) in southern Chile. Hydrobiología 326/327:235–244

Avila M, Candia A, Nuñez M, Romo H (1999a) Reproductive biology of *Gigartina skottsbergii* (Gigartinaceae, Rhodophyta) from Chile. Hydrobiologia 398/399:149–157

Avila M, Ask E, Rudolph B, Nuñez M, Norambuena R (1999b) Economic feasibility of *Sarcothalia* (Gigartinales, Rhodophyta) cultivation. Hydrobiologia 398/399:435–442

Avila M, Pavez H, Candia A, San Martin R, Caceres J (2003) Effects of temperature and irradiance on the germination of *Sarcothalia crispata* in southern Chile. In: Chapman ARO, Anderson RJ, Vreeland VJ, Davison IR (eds) Proceedings of the 17th International Seaweed Symposium. Oxford University Press, New York, pp 27–34

Boalch GT (1961) Studies on *Ectocarpus* in culture. II. Growth and nutrients of a bacteria free culture. J Mar Biol Assoc UK 41:287–304

Buschmann A, Mora O, Gomez P, Böttger M, Buitano S, Retamales C, Vergara P, Gutierrez A (1994) *Gracilaria chilensis* outdoor tank cultivation in Chile: use of land based salmon culture effluents. Aquacul Engin 13:283–300

Buschmann AH, Correa JA, Westermeier R (1999) Recent advances in the understanding of the biological basis for *Gigartina skottsbergii* (Rhodophyta) cultivation in Chile. Hydrobiologia 398/399:427–434

Buschmann AH, Correa JA, Westermeier R, Paredes MA, Aedo D, Potin P, Aroca G, Beltran J, Hernández-Gonzalez MC (2001) Cultivation of *Gigartina skottsbergii* (Gigartinales, Rhodophyta): Recents advances and challenges for the future. J Appl Phycol 13:255–266

Buschmann AH, Varela D, Cifuentes M, Hernández M, Henríquez L, Westermeier R, Correa JA (2004) Experimental indoor cultivation of the carrageenophytic red alga *Gigartina sottsbergii.* Aquaculture 241:357–370

Candia A, Gónzalez A, Poblete A, Otaíza R, Avila M (1993) Efecto de factores ambientales sobre la esporulación y viabilidad de esporas en *Iridaea ciliata* Kutzing (Rhodophyta, Gigartinales). Libro de Resúmenes, Symposium de algas marinas chilenas y III Encuentro de Macroalgólogos Iquique, Chile, p 28

Chopin T, Hanisak MD, Koehn FE, Mollion J, Moreau S (1990) Studies of carrageenans and effects of seawater phosphorus concentration on carrageenan content and growth of *Agardhiella subulata* (C. Agardh) Kraft and Wynne (Rhodophyceae, Solieriaceae). J Appl Phycol 2:3–16

Friedlander M (2001). Inorganic nutrition in pond cultivated *Gracilaria conferta*: nitrogen, phosphate and sulfate. J Appl Phycol 13:279-286

Fries L (1963) On the cultivation of axenic red algae. Physiol Plant 16:695–705

Gantt E (1990) Pigmentation and photoacclimatation. In: Cole KM, Sneath RG (eds) Biology of the red algae. Cambridge University Press, Cambridge 203–219

Gómez I, Wiencke C (1996) Photosynthesis, dark respiration and pigment contents of gametophytes of the Antartic brown algae *Desmarestia menziesii*. Bot Mar 39:149–157

Haglund K, Pedersén M (1993) Outdoor pond cultivation of the subtropical marine red alga *Gracilaria tenuistipitata* in brackish water in Sweden: growth, nutrient uptake, co-cultivation with rainbow trout and epiphyte control. J Appl Phycol 5:271–284

Hanelt DC, Nultsch W (1997) Influence of UV radiation on the photosynthesis of Arctic macroalgae in the field. J Photobiol 38:40–47

Hannach G, Santelices B (1985) Ecological differences between the isomorphic reproductive phases of two species of *Iridaea* (Rhodophyta. Gigartinales). Mar Ecol Progr Ser 22:291–303

Henkel R (1952) Ernahrungs physiologish untersuchungen an meeresalgen ins besondere an *Bangia pumila*. Kiel Meeresforsch 8:192–211

Hernández-Gonzalez MC, Buschmann AH, Cifuente M, Correa JA, Westermeier R (2007) Vegetative propagation of the carrageno-phytic red algae *Gigartina skottsbergii* Setchel et Gardner: Indoor and filed experiments. Aquaculture 262:120–128

Hoffmann AJ, Santelices B (1997) Flora marina de Chile central. Ediciones Universidad Cátolica de Chile, Santiago

Hommersand MH, Guiry M, Frederiq S, Leister G (1993) New perspectives in the taxonomy of the Gigartinaceae (Gigartinales, Rhodophyta). In: Chapman ARO, Brown MT, Lahaye M (eds) Proceedings of the 14th International Seaweed Symposium. Kluwer, Belgium, 260/261:105–120

Infante R, Candia A (1988) Cultivo de *Gracilaria verrucosa* (Hudson) *Pappenfuss* e *Iridaea ciliata* Kutzing, (Rhodophyta, Gigartinaceae), en laboratorio: Esporulación inducida y colonización de esporas en diferentes sustratos. Gayana Bot 45(1-4):297–304

Li D, Zou ZG, Liu H, Wu CY (1999) A new method of *Laminaria japonica* strain selection and sporeling raising by the use of gametophyte clone. Hydrobiologia 398/399:473–476

Lipkin Y (1985) Outdoor cultivation of sea vegetables. Plant Soil 89:159–183

Liu J, Dong L, Shen Y, Wu C (2004) Effect of the light period on egg-discharge of gametophyte clones of *Undaria pinnatifida* (Phaeophyta). J Appl Phycol 16:519–521

Lobban CS, Harrison PJ (1994) Seaweed ecology and physiology. Cambridge University Press, Cambridge, pp 366

Luxoro C, Santelices B (1989) Additional evidence for ecological differences among isomorphic reproductive phases of *Iridaea laminarioides*. (Rhodophyta, Gigartinales). J Phycol 25:206–212

Mansilla A, Werlinger C, Navarro N (2002) Efectos del fotoperíodo e intensidad luminosa sobre el desarrollo de tetrásporas y carpósporas de *Mazzaella laminarioides* (Bory) Fredericq (Gigartinales-Rhodophyta). Anales Instituto Patagonia, Serie Cs Nat (Chile), 30:49–60

Mansilla A, Werlinger C, Palacios, M, Navarro NP, Cuadra P (2006) Effects of UVB radiation on the inicial stages of growth of *Gigartina skottsbergii*, *Sarcothalia crispata* and *Mazaella laminariodes* (Gigartinales, Rhodophyta). J Appl Phycol 18:451–459

Martínez LA, Buschmann AH (1996) Agar yield and quality of *Gracilaria chilensis* (Gigartinales, Rhodophyta) in tank culture using fish effluents. Hydrobiologia 326/327:341–345

McLachlan J (1973) Growth media-marine. In: Stein JR (ed) Handbook of Phycological Methods. Cambridge University Press, Cambridge, pp 25–52

Navarro-Angulo L, Robledo D (1999) Effects of nitrogen source, N:P and N-pulse concentration and frequency on the growth of *Gracilaria cornea* (Gracilariales, Rhodophyta) in culture. Hydrobiologia 398/399:315–320

Ramírez M, Santelices B (1991) Catálogo de algas marinas bentónicas de las costas templadas del Pacífico de Sudamérica. Monografías Biológicas 5. Pontificia Universidad Católica de Chile, Santiago, Chile

Romo H, Alveal K (1995) Técnicas para el cultivo experimental, medición de crecimiento, manejo de poblaciones de *Iridaea (Mazzaella)*. In: Alveal K, Ferrario M, Oliveira EC, Sar E (eds) Manual de Métodos Ficológicos. Universidad de Concepción, Concepción, pp 563–576

Romo H, Paula EJ (1995) Métodos experimentales para el cultivo de *Porphyra*. In: Alveal K, Ferrario M, Oliveira EC, Sar E (eds) Manual de Métodos Ficológicos, Universidad de Concepción, Editorial Anibal Pinto S. A., Chile, pp 555–562

Romo H, Alveal K, Werlinger C (2001) Growth of the commercial carrageenophyte *Sarcothalia crispata* (Rhodophyta, Gigartinales) on suspended culture in central Chile. J Appl Phycol 13:227–232

Santelices B (1996) Seaweed research and utilization in Chile: moving into a new phase. Hydrobiología 326/327:1–14

Sernapesca (2003) Anuario estadístico de pesca 2003. Servicio Nacional de Pesca. Ministerio de Economía, Fomento y Reconstrucción-Chile

Sokal PR, Rohlf FJ (1979) Biometría: Principios y Métodos Estadísticos en la Investigación Biológica. Blume, Madrid

Tseng CK (1981) Commercial Cultivation. In: Lobban CS, Wynne MJ (eds) The biology of seaweeds. Blackwell, Oxford, pp 680–725

Ugarte R (1982) Germinación de tetrásporas y carpósporas en 3 especies del género *Iridaea* Bory (Rhodophyta, Gigartinales). Thesis, Universidad de Concepción

Werlinger C (1998) Studies on the agarophyte *Gracilaria chilensis* Bird, McLachlan and Oliveira (Rhodophyta, Gracilariales):

growth, photosynthesis and mass cultivation from spores. PhD Thesis, Tokyo University of Fisheries
Wu C, Li D, Liu H, Peng G, Liu J (2004) Mass culture of *Undaria* gametophyte clones and their use in sporeling culture. Hydrobiologia 512:153–156
Yoneshigue-Braga Y, Neves MH (1981) Preliminary studies on mass culture of *Gracilaria* sp. using different nutrient media. In: Levring T (ed) Proceedings of Xth International Seaweed Symposium. Pergamon, Paris, pp 295–299
Zar JH (1999) Biostatistical Analysis, 4th edn. Prentice Hall, New Jersey
Zhang QS, Qu SC, Cong YZ, Luo SJ, Tang XX (2007) High throughput and gametogenesis induction of *Laminaria japonica* gametophyte clones. J Appl Phycol DOI 10.1007/s1081100792205
Zhao F, Wang A, Liu J (2006) New phenomenon in early development of sporelings in *Gracilaria asiatica* Chang et Xia (Gracilariaceae, Rhodophyta). Chinese J Oceanol Limnol 24(4):364–369

Effects of kinetin and nitrogen on growth rates, pigment and protein contents in wild and phycoerythrin-deficient strains of *Hypnea musciformis* (Rhodophyta)

Aline P. Martins · Nair S. Yokoya · Maria Angela M. Carvalho · Estela M. Plastino

Originally published in the Journal of Applied Phycology, Vol 20, No 5, 317–323.
DOI: 10.1007/s10811-007-9289-x

Abstract *Hypnea musciformis* (Wulfen in Jacqu.) J.V. Lamour. is the main source for carrageenan production in Brazil and strains with selected characteristics could improve the production of raw material. The effects of kinetin on growth rates, morphology, protein content, and concentrations of pigments (chlorophyll *a*, phycoerythrin, phycocyanin, and allophycocyanin) were assessed in the wild strain (brown phenotype) and in the phycoerythrin-deficient strain (green phenotype) of *H. musciformis*. Concentrations of kinetin ranging from 0 to 50 μM were tested in ASP 12-NTA synthetic medium with 10 μM nitrate (N-limited) and 100 μM nitrate (N-saturated). In N-limited condition, kinetin stimulated growth rates of the phycoerythrin-deficient strain and formation of lateral branches in both colour strains. Kinetin stimulated protein biosynthesis in both strains. However, differences between both nitrogen conditions were significant only in the phycoerythrin-deficient strain. In the wild strain, effects of kinetin on concentrations of phycobiliproteins were not significant in both nitrogen conditions, except for chlorophyll content. However, the phycoerythrin-deficient strain showed an opposite response, and kinetin stimulated the phycobiliprotein biosynthesis, with the highest concentrations of phycoerythrin in N-saturated medium, while the highest concentrations of allophycocyanin and phycocyanin were observed in N-limited medium. These results indicate that the effects of kinetin on growth, morphology, protein and phycobiliprotein contents are influenced by nitrogen availability, and the main nitrogen storage pools in phycoerythrin-deficient strain of *H. musciformis* submitted to N-limited conditions were phycocyanin and allophycocianin, the biosynthesis of which was enhanced by kinetin.

Keywords Colour strains · *Hypnea musciformis* · Kinetin · Nitrogen · Pigments · Proteins

Introduction

The genus *Hypnea* (Gigartinales, Rhodophyta) includes 50 species distributed in warm waters (Masuda et al. 1997), many of them with economic importance in several countries (Critchley and Ohno 1998). In Brazil, *Hypnea musciformis* (Wulfen in Jacqu.) J.V. Lamour is the main raw material for carrageenan production (Oliveira 1998).

Along the Brazilian coast, populations of *Hypnea musciformis* usually have a brown phenotype. However, a brown plant with only one green branch was collected in Espirito Santo State, Southeastern Brazil, and a phycoerythrin-deficient strain (green phenotype) was originated from this green branch (Yokoya et al. 2003). In the laboratory, brown and green specimens were isolated and cultured in different environmental conditions, such as irradiance levels, photoperiods, temperature and salinity, in order to check if the colour variation could be a response to environmental factors. The colour of these *H. musciformis* isolates is a stable characteristic, and is not a result of photoacclimation processes (Yokoya et al. 2003, 2006a).

A. P. Martins · N. S. Yokoya (✉)
Seção de Ficologia, Instituto de Botânica,
P.O. Box 3005, 01061-970 São Paulo, SP, Brazil
e-mail: nyokoya@pesquisador.cnpq.br

M. A. M. Carvalho
Seção de Fisiologia e Bioquímica de Plantas,
Instituto de Botânica,
P.O. Box 3005, 01061-970 São Paulo, SP, Brazil

E. M. Plastino
Instituto de Biociências, Universidade de São Paulo,
P.O. Box 11461, 05421-970 São Paulo, SP, Brazil

The occurrence of colour variants in Brazilian red algae was reported for *Gracilaria* species: *G. birdiae* Plastino and E. C. Oliveira, *G. cornea* J. Agardh, and *G. domingensis* (Kützing) Sonder ex Dickie (Plastino 2003). The first report on colour variants of *Hypnea musciformis* was described by Yokoya et al. (2003), and the green strain had lower phycoerythrin concentration than the brown strain, characterising a phycoerythrin-deficient strain.

The major nitrogen storage pools in seaweeds are amino acids and proteins, and in nitrogen-limited condition, phycoerythrin is very important as a source of nitrogen (Bird et al. 1982). Therefore, the phycoerythrin-deficient strain of *H. musciformis* could provide an interesting biological system to study nitrogen metabolism as well as other physiological processes. Moreover, there has been an increasing interest in the selection of seaweed strains not only for the improvement of cultivation for commercial purposes but also for their potential to produce different compounds with biological activities.

Cytokinins play an important role in a number of physiological processes, such as cell division and differentiation, morphogenesis, senescence and photosynthesis in different species belonging to seven algal divisions (Mooney and van Staden 1986). There are relatively few reports on the presence of hormones in seaweeds, but recently, these compounds have been studied more intensively. Stirk et al. (2003) identified 19 endogenous cytokinins, including the first report of aromatic cytokinins (benzyladenine and topolin derivatives) in 31 macroalgal species. Moreover, this study highlighted some differences on cytokinin composition in seaweeds in comparison with the higher plants. These differences include an abundance of isopentenyladenine (iP) and *c*-zeatin (*c*Z) conjugates in the seaweed samples and no dihydrozeatin (DHZ), while in higher plants, the most common cytokinin types are DHZ derivatives, and iP and *c*Z conjugates are not often detected (Stirk et al. 2003).

The wild and phycoerythrin-deficient strains of *H. musciformis* showed differences on cytokinin metabolism (Yokoya et al. 2006b). The phycoerythrin-deficient strain showed more cytokinin-like activities (measured by soybean callus biossay with kinetin standards) than the wild strain, with the highest activity co-chromatographing with zeatin (Z) and a low activity co-chromatographing with monophosphates. The zeatin-like activity was equivalent to 5 μg l^{-1} kinetin in the phycoerythrin-deficient strain, while the wild strain presented a lower activity (equivalent to 1 μg l^{-1} kinetin), in addition to a low iP-like activity.

Based on these reports, the aims of our study were to determine the effects of kinetin in both N-limited and N-saturated culture media on growth rates, morphogenesis, protein content and concentrations of pigments in the wild and in the phycoerythrin-deficient strains of *H. musciformis*.

Materials and methods

Phycoerythrin-deficient strain (green phenotype) of *Hypnea musciformis* was originated from only one green branch, which itself originated as a spontaneous mutation from a wild plant (brown phenotype) collected from Ponta da Baleia, Espirito Santo State, Brazil. In the laboratory, unialgal cultures of the brown and green strains were initiated by clonal propagation using segments (5-10 mm) of brown and green branches isolated from the same mother-plant. Culture medium was composed of sterilised seawater (salinity of 30±1 psu) enriched with von Stosch's solution at half strength (VSES/2) following Edwards (1970), with vitamin concentrations reduced to 50%. Medium renewal was carried out every 2 weeks. Cultures were incubated under 24±2 °C, irradiances of 40.0-

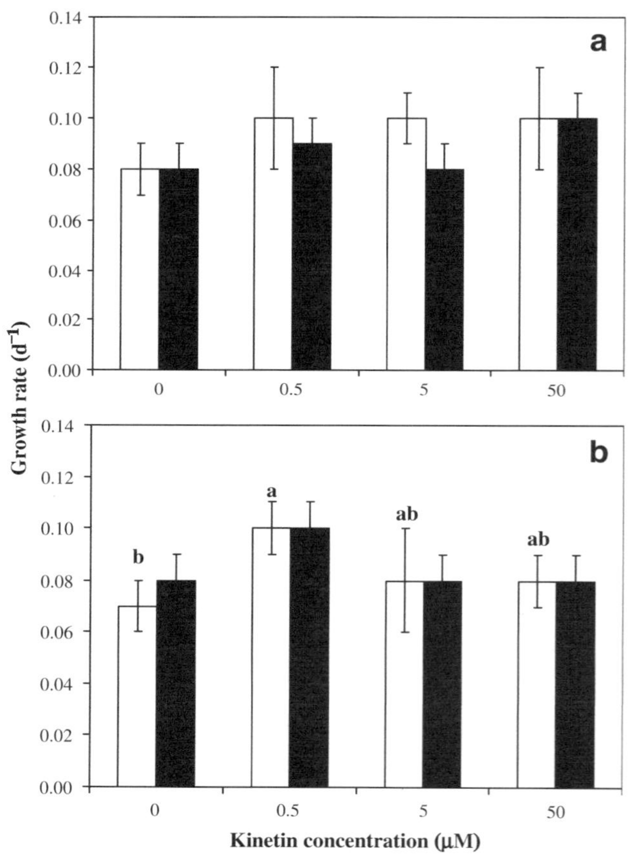

Fig. 1 Growth rates of wild strain (**a**) and phycoerythrin-deficient strain (**b**) of *Hypnea musciformis* cultured in ASP 12-NTA medium in N-limited (10 μM nitrate) and N-saturated (100 μM nitrate) conditions with different kinetin concentrations. Each data point is the mean of six replicates with four apical segments (10 mm). Different *letters* indicate significant differences according to Student-Newman-Keuls multiple comparison test (p=0.05). □ N-limited, ■ N-saturated

50.0 μmol photons m^{-2} s^{-1}, provided by cool-white fluorescent lamps with 14:10 h light:dark cycle, without aeration. Irradiance was measured with a quantum photometer (LI-250; Li-Cor, USA) equipped with underwater quantum sensor (LI-192 SA; Li-Cor).

Kinetin and nitrogen experiments

Concentrations of kinetin ranging from 0 to 50 μM were added to ASP12-NTA synthetic medium (Iwasaki 1961) prepared with 10 μM nitrate (N-limited condition) and 100 μM nitrate (N-saturated condition), with pH of 8.0 and salinity of 30 psu. Both nitrogen concentrations were determined by a previous study on the effects of nitrate availability on the growth of both strains of *H. musciformis*. Each treatment was tested with six replicates of four apical segments (10 mm). These explants were cultured in Erlenmeyer flasks with 150 ml of medium. Other experimental conditions were the same as described for unialgal cultures. Experiments with brown and green strains were conducted simultaneously. Fresh biomass was registered weekly at the same intervals of medium renewal. Growth rates were calculated as $\left[\ln\left(B_f.B_o^{-1}\right).t^{-1}\right]$, where B_o is the initial fresh biomass, B_f is the fresh biomass after t days, and t corresponds to the experimental period (Yokoya et al. 2003).

Morphological analysis included the records of the total number of lateral branches (first order) and of branchlets of the last order. Photographs of specimens from all treatments were taken with a digital camera connected to a stereomicroscope, and morphological parameters were recorded using Axion Vision 3.1 software (Carl Zeiss Vision GmbH).

Pigment and total soluble protein contents

Pigment extractions were carried out at 4°C and in darkness, according to Kursar et al. (1983) with modifications (Plastino and Guimarães 2001). The algal mass

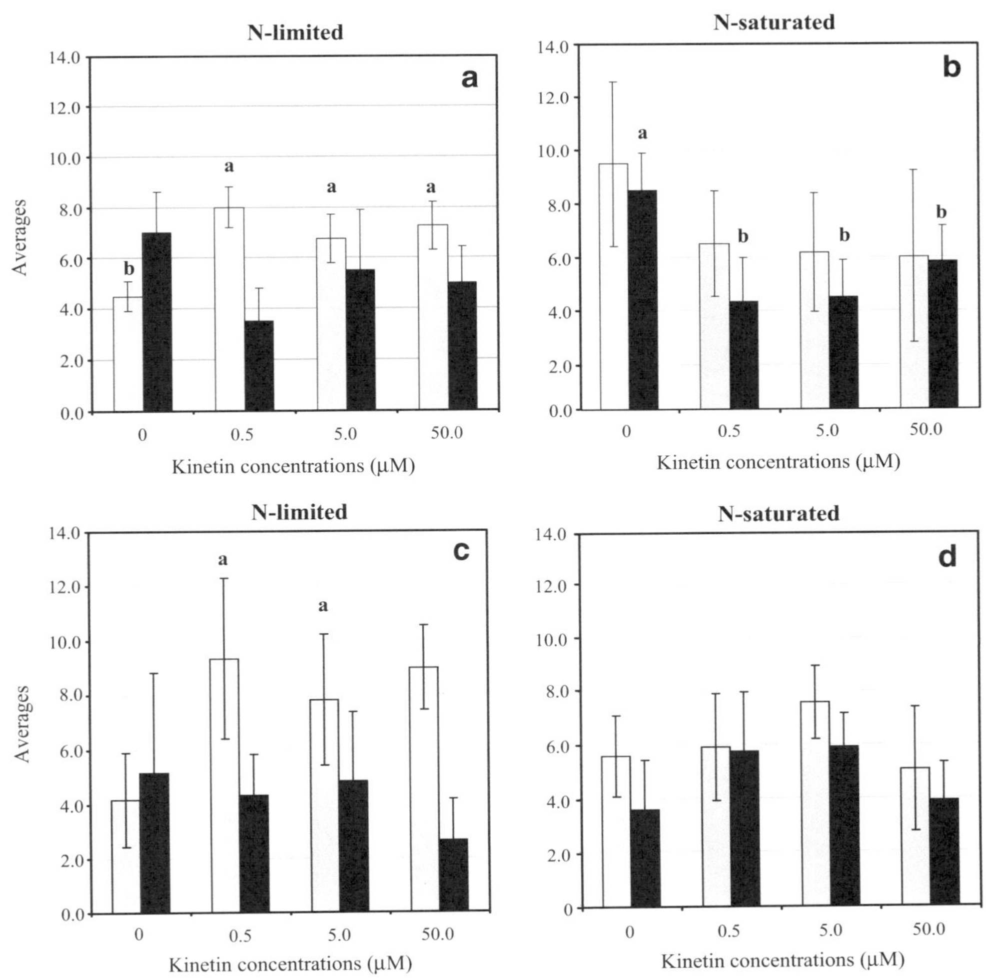

Fig. 2 Branching development of wild strain (**a**–**b**) and phycoerythrin-deficient strain (**c**–**d**) of *Hypnea musciformis* cultured in ASP 12-NTA medium with N-limited (10 μM nitrate) and N-saturated (100 μM nitrate) conditions with different kinetin concentrations. Each data point is the mean of six replicates with four apical segments (10 mm). Different *letters* indicate significant differences according to Student-Newman-Keuls' multiple comparison test (p= 0.05). □ Number of lateral branches, ■ number of branchlets

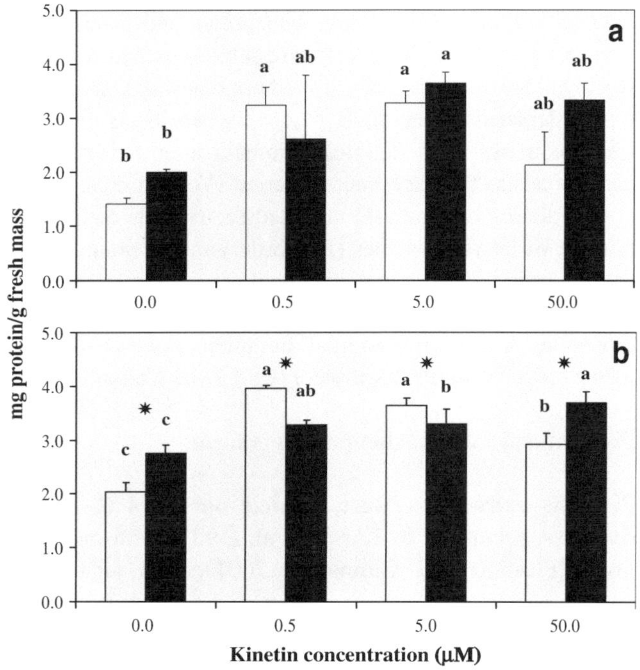

Fig. 3 Total protein content in wild strain (**a**) and phycoerythrin-deficient strain (**b**) of *Hypnea musciformis* cultured in ASP 12-NTA medium in N-limited (10 μM nitrate) and N-saturated (100 μM nitrate) conditions with different kinetin concentrations. Each data point is the mean of three replicates with four apical segments (10 mm). Different *letters* and * indicate significant differences among kinetin and nitrate concentrations, respectively, according to Student-Newman-Keuls' multiple comparison test (p=0.05). □ N-limited, ■ N-saturated

(300 mg fresh mass for each triplicate) was ground to a powder with liquid nitrogen and mixed with phosphate buffer 50 mM (pH 5.5). The homogenates were centrifuged at 36,000 *g* for 25 min for separation of the phycobiliproteins present in the supernatants. Chlorophyll *a* was extracted after dissolving the pellet in acetone 90%, and centrifuging at 12,000 *g* for 15 min. Pigments were quantified by spectrophotometry (HP 8452A), and concentrations were calculated according to Kursar et al. (1983) for phycobiliproteins, and Jeffrey and Humphrey (1975) for chlorophyll *a*.

For total soluble protein analyses, the algal biomass was ground with liquid nitrogen, and extractions were carried out at 4°C, using 0.2 M phosphate buffer (pH 7.95) containing 5 mM EDTA and 2 mM β-mercaptoetanol. Buffer was added in a proportion of 10 ml g^{-1} fresh biomass, and the homogenates were centrifuged at 12,000 *g* for 20 min. Total soluble protein contents were determined according to Bradford (1976), using a Bio-Rad protein assay kit and BSA as standard.

Data analysis

Data were analysed by unifatorial and bifatorial analysis of variance (ANOVA), and Student-Newman-Keuls' multiple comparison test was conducted to distinguish significantly different results (Zar 1999).

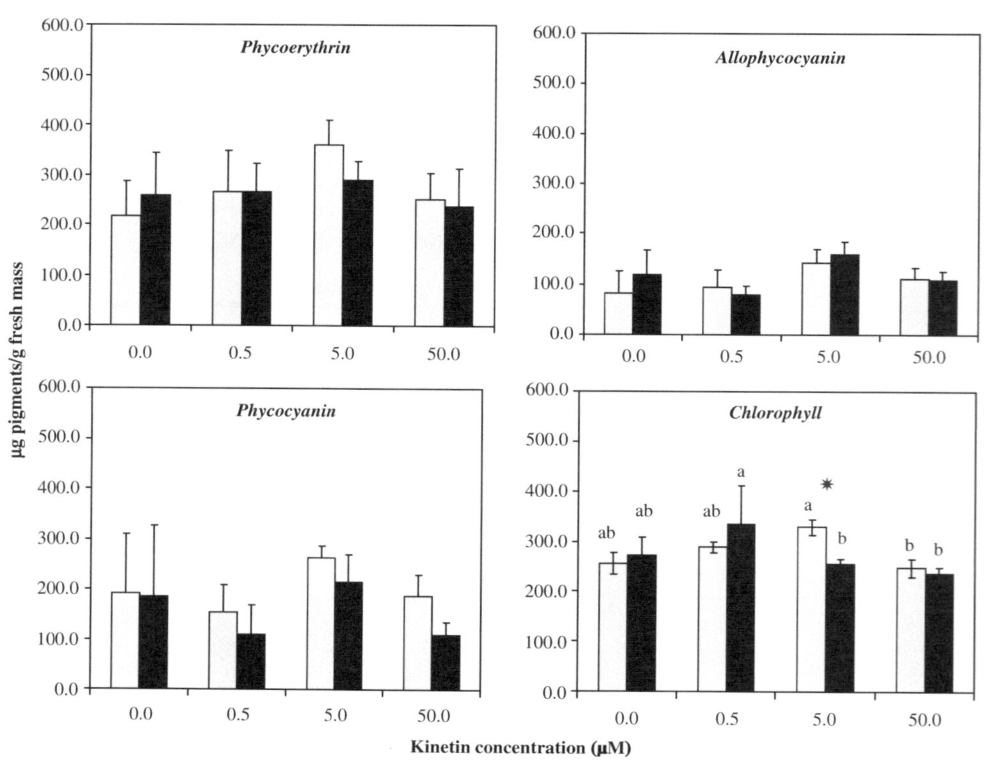

Fig. 4 Pigment content in wild strain of *Hypnea musciformis* cultured in ASP 12-NTA medium in N-limited (10 μM nitrate) and N-saturated (100 μM nitrate) conditions with different kinetin concentrations. Each data point is the mean of three replicates with four apical segments (10 mm). Different *letters* and * indicate significant differences among kinetin and nitrate concentrations, respectively, according to Student-Newman-Keuls' multiple comparison test (p=0.05). □ N-limited, ■ N-saturated

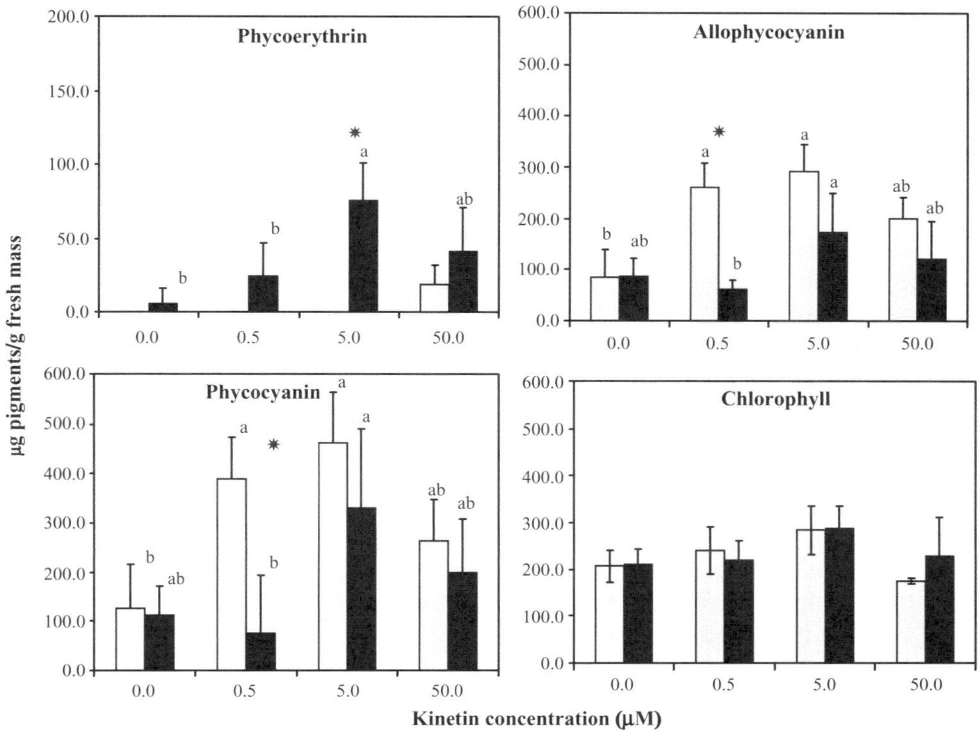

Fig. 5 Pigment content in phycoerythrin-deficient strain of *Hypnea musciformis* cultured in ASP 12-NTA medium in N-limited (10 μM nitrate) and N-saturated (100 μM nitrate) conditions with different kinetin concentrations. Each data point is the mean of three replicates with four apical segments (10 mm). Different *letters* and * indicate significant differences among kinetin and nitrate concentrations, respectively, according to Student-Newman-Keuls' multiple comparison test (p=0.05). □ N-limited, ■ N-saturated

Results

The effect of kinetin on growth rates of the wild strain of *Hypnea musciformis* in N-limited and N-saturated conditions were not significant (Fig. 1a) while kinetin stimulated growth rates of the phycoerythrin-deficient strain only in N-limited condition (Fig. 1b).

Morphological analysis showed that under N-limited condition, kinetin stimulated the development of lateral branches in both strains of *H. musciformis*, doubling the number when low concentrations were used (Fig. 2a,c). However, the development of branchlets of the last order in both strains was not affected by kinetin under N-limited condition (Fig. 2a,c). The number of branchlets was reduced to half with 0.5 and 5.0 μM kinetin in the wild strain cultured under N-saturated (Fig. 2b), while kinetin had no significant effect in the phycoerythrin-deficient strain in the same condition (Fig. 2d).

Addition of kinetin stimulated significantly protein biosynthesis in both wild and phycoerythrin-deficient strains, and the highest protein concentrations were detected in both strains cultured in N-limited and N-saturated media with the addition of 0.5 and 5.0 μM of kinetin (Fig. 3a,b). The phycoerythrin-deficient strain of *H. musciformis* showed significant differences between N-limited and N-saturated conditions in all kinetin concentrations (Fig. 3b), and presented higher protein concentration than the wild strain (Fig. 3a,b). In the wild strain of *H. musciformis*, addition of kinetin had no significant effect on the phycobiliprotein content neither in N-limited nor in N-saturated conditions, except for chlorophyll (Fig. 4), with the highest values in treatments with 0.5 and 5.0 μM of kinetin in N-saturated and N-limited conditions, respectively. However, phycoerythrin-deficient strain showed opposite response in comparison to the wild strain, and kinetin had a strong effect on phycobiliprotein biosynthesis, with the highest concentrations of phycoerythrin in treatment of 5.0 μM of kinetin in N-saturated medium. Allophycocyanin and phycocyanin contents were higher in treatments with 0.5 and 5.0 μM of kinetin in N-limited conditions (Fig. 5).

Discussion

Although the role of hormones in vascular plants has been extensively studied, there are relatively few reports on their occurrence and physiological functions in algae. The role of hormones, including cytokinins, in algal development is far from being understood (Chapman 1999). Several factors could interfere with the hormonal effects, such as the plasticity on the hormonal control (Leopold and Noodén 1984), the complex interactions of factors, including the endogenous characteristics of explant (Endreß 1994), as well as the culture conditions (temperature, light, nutrients, salinity, pH, physical state of the medium), and differential sensitivity to the hormones (Trewavas 1982; Davies 1995).

In the present study, kinetin stimulated growth rates of the phycoerythrin-deficient strain of *Hypnea musciformis* cultured in N-limited condition, while no significant effects were observed in the wild strain. These differences between both strains might be related to different levels of endogenous substances, which could influence the sensitivity to the plant growth regulators (Davies 1995). Yokoya et al. (2006b) reported that the phycoerythrin-deficient strain of *H. musciformis* showed more cytokinin-like activity than the wild strain with most of the activity (equivalent to 5 $\mu g\ l^{-1}$ kinetin) co-chromatographing with Z and just a small fraction of the activity with the monophosphates. The wild strain presented less activity (equivalent to 1 $\mu g\ l^{-1}$ kinetin) which co-chromatographed with Z, and a little with iP. Cytokinin also stimulated growth of excised branches in *H. musciformis* (Jennings 1969), similar to the response observed in the phycoerythrin-deficient strain of the same species.

Kinetin stimulated the development of lateral branches in both strains of *H. musciformis* under N-limited condition. Similar response was also described in studies on apical dominance effect which was partially controlled by auxin and cytokinins in *Pterocladiella capillacea* and *Sphacelaria furcigera* (Dworetzky et al. 1980; Scrosati 2002).

The most significant effect of kinetin was to stimulate protein biosynthesis in the wild and phycoerythrin-deficient strains of *H. musciformis*. This observation is in accordance with the main role played by cytokinins in plant development (Benková et al. 1999). The stimulatory effects of cytokinin on cell division are also related to the protein biosynthesis (Evans 1984).

A positive correlation between kinetin and phycoerythrin concentrations was observed in the phycoerythrin-deficient strain of *H. musciformis* cultured in N-saturated condition, indicating that kinetin could play an important role in stimulating not only phycoerythrin biosynthesis but also allophycocianin and phycocyanin biosynthesis. These results could explain the higher protein content observed in phycoerythrin-deficient strain than in the wild strain of *H. musciformis*.

In conclusion, our results indicated that the effects of kinetin on growth, morphology, protein and phycobiliprotein contents could be influenced by nitrogen availability, and that the main nitrogen storage pools in phycoerythrin-deficient strain of *H. musciformis* submitted to N-limited conditions are phycocyanin and allophycocyanin, which biosynthesis was enhanced by kinetin.

Acknowledgements The authors thank Noemy Yamaguishi-Tomita for finding the mother-plant, which originated the brown and green strains of *Hypnea musciformis*, and José Domingos for technical support. This research was supported by fellowships and grants from Conselho Nacional de Desenvolvimento Científico e Tecnológico (PIBIC, CNPq, Brazil), and from Fundação de Amparo à Pesquisa de São Paulo (Fapesp, Brazil).

References

Benková E, Witters E, Dongen WV et al (1999) Cytokinins in tobacco and wheat chloroplasts. Occurrence and changes due to light/dark treatment. Plant Physiol 121:245–252

Bird KT, Habig C, Debusk T (1982) Nitrogen allocation and storage patterns in *Gracilaria tikvahiae* (Rhodophyta). J Phycol 18:344–348

Bradford MM (1976) A rapid and sensitive method for the quantitation of microgram quantities of protein utilizing the principle of protein-dye binding. Anal Biochem 72:248–254

Chapman DJ (1999) Auxins and algal galls: a new perspective on phytohormones in algae. J Phycol 35:445–446

Critchley AT, Ohno M (1998) Seaweed resources of the world. Japan International Cooperation Agency, Yokosuka

Davies PJ (1995) The plant hormone concept: concentration, sensitivity and transport. In: Davies PJ (ed) Plant hormones-physiology, biochemistry and molecular biology. Kluwer, Dordrecht, pp 13–38

Dworetzky B, Klein RM, Cool PW (1980) Effect of growth substances on "apical dominance" in *Sphacelaria furcigera* (Phaeophyta). J Phycol 16:239–242

Edwards P (1970) Illustrated guide to the seaweeds and seagrasses in the vicinity of Porto Aransas, Texas. Contrib Mar Sci 15:1–228

Endreß R (1994) Plant cell biotechnology. Springer, Berlin Heidelberg New York

Evans ML (1984) Functions of hormones at the cellular level of organization. In: Scott TK (ed) Encyclopedia of plant physiology. New series, vol 10. Springer, Berlin Heidelberg New York, pp 23–79

Iwasaki H (1961) The life cycle of *Porphyra tenera* in vitro. Biol Bull 121:173–187

Jeffrey SW, Humphrey GF (1975) New spectrophotometric equations for determining chlorophylls a, b, c_1 e c_2 in higher plants, algae and natural phytoplankton. Biochem Physiol Pflanzen 167:191–194

Jennings RC (1969) Cytokinins as endogenous growth regulators in the algae *Ecklonia* (Phaeophyta) and *Hypnea* (Rhodophyta). Aust J Biol Sci 22:621–627

Kursar TA, van Der Meer J, Alberte RS (1983) Light-harvesting system of red alga *Gracilaria tikvahiae*. I. Biochemical analyses of pigment mutations. Plant Physiol 73:353–360

Leopold AC, Noodén LD (1984) Hormonal regulatory systems in plants. In: Scott TK (ed) Encyclopedia of plant physiology. New series, vol 10. Springer, Berlin Heidelberg New York, pp 4–22

Masuda J, Yamagishi Y, Chiang Y-M et al (1997) Overview of *Hypnea* (Rhodophyta, Hypneaceae). In: Abbott IA (ed) Taxonomy of economic seaweeds, vol 6. California Sea Grant College System, University of California, La Jolla, pp 127–133

Mooney PA, van Staden J (1986) Algae and cytokinins. J Plant Physiol 123:1–21

Oliveira EC (1998) The seaweed resources of Brazil. In: Critchley AT, Ohno M (eds) Seaweed resources of the world. Japan International Cooperation Agency, Yokosuka, pp 366–371

Plastino EM (2003) Cultivo en laboratorio y selección de cepas en algas rojas (Rhodophyta). In: Mansilla A, Werllinger C, Navarro N (eds) Memorias de Curso Internacional de Posgrado y Especialización de Macroalgas em Ambientes Subantárticos. Ediciones Universidad de Magallanes, Universidad de Magallanes, Punta Arenas, pp 53–58

Plastino EM, Guimarães M (2001) Diversidad intraespecífica. In: Alveal K, Antezana T (eds) Sustentabilidad de la biodiversidad, un problema actual. Bases científico-técnicas, teorizaciones y proyecciones. Universidad de Concepción, Concepción, pp 19–27

Scrosati R (2002) Morphological plasticity and apparent loss of apical dominance following natural loss of main apex in *Pterocladiella capillacea* (Rhodophyta, Gelidiales) fronds. Phycologia 41:96–98

Stirk WA, Novák O, Strnad M et al (2003) Cytokinins in macroalgae. Plant Growth Regul 41:13–24

Trewavas AJ (1982) Growth substance sensitivity: The limiting factor in plant development. Physiol Plant 55:60–72

Yokoya NS, Plastino EM, Artel R (2003) Physiological responses and pigment characterization of two colour strains of the carrageenophyte *Hypnea musciformis* (Rhodophyta). In: Chapman ARO, Anderson RJ, Vreeland VJ, Davison R (eds) Proceedings of the 17th International Seaweed Symposium. Oxford University Press, New York, pp 425–433

Yokoya NS, Necchi Jr O, Martins AP et al (2006a) Growth responses and photosynthetic characteristics of wild and phycoerythrin-deficient strains of *Hypnea musciformis* (Rhodophyta). J Appl Phycol 19:197–205

Yokoya NS, Stirk WA, van Staden J (2006b) Cytokinin-like and auxin-like activities and the effects of plant growth regulators in wild and phycoerythrin-deficient strains of *Hypnea musciformis* (Rhodophyta). In: Abstracts of the XI Brazilian Phycological Congress and Latin-American Symposium of Harmful Algae. University of Vale do Itajaí, Itajaí, 27 March-1 April 2006

Zar JH (1999) Biostatistical analysis. Prentice Hall, N.J.

Rapid and slow modulation of nitrate reductase activity in the red macroalga *Gracilaria chilensis* (Gracilariales, Rhodophyta): influence of different nitrogen sources

Fungyi Chow · Mariana Cabral de Oliveira

Originally published in the Journal of Applied Phycology, Vol 20, No 5, 325–332.
DOI: 10.1007/s10811-008-9310-z

Abstract Nitrate is one of the most important stimuli in nitrate reductase (NR) induction, while ammonium is usually an inhibitor. We evaluated the influence of nitrate, ammonium or urea as nitrogen sources on NR activity of the agarophyte *Gracilaria chilensis*. The addition of nitrate rapidly (2 min) induced NR activity, suggesting a fast post-translational regulation. In contrast, nitrate addition to starved algae stimulated rapid nitrate uptake without a concomitant induction of NR activity. These results show that in the absence of nitrate, NR activity is negatively affected, while the nitrate uptake system is active and ready to operate as soon as nitrate is available in the external medium, indicating that nitrate uptake and assimilation are differentially regulated. The addition of ammonium or urea as nitrogen sources stimulated NR activity after 24 h, different from that observed for other algae. However, a decrease in NR activity was observed after the third day under ammonium or urea. During the dark phase, *G. chilensis* NR activity was low when compared to the light phase. A light pulse of 15 min during the dark phase induced NR activity 1.5-fold suggesting also fast post-translational regulation. Nitrate reductase regulation by phosphorylation and dephosphorylation, and by protein synthesis and degradation, were evaluated using inhibitors. The results obtained for *G. chilensis* show a post-translational regulation as a rapid response mechanism by phosphorylation and dephosphorylation, and a slower mechanism by regulation of RNA synthesis coupled to *de novo* NR protein synthesis.

F. Chow (✉) · M. C. de Oliveira
Instituto de Biociências, Universidade de São Paulo,
CP 11461, CEP 05422-970 São Paulo, SP, Brazil
e-mail: fchow@usp.br

Keywords Enzyme regulation · *Gracilaria chilensis* · Nitrate reductase · Nitrogen sources · Rhodophyta

Introduction

Macroalgae growth is often limited by the availability of nitrogen during at least part of the year. Nitrogen assimilation begins with nitrate uptake, the main nitrogen source in seawater, from the external media to the cytoplasm, followed by its reduction to nitrite via the enzyme nitrate reductase (NR). Subsequently nitrite is transported into the chloroplast and reduced to ammonium by the enzyme nitrite reductase (NiR). The first reduction step by NR activity is thought to limit the overall rate of nitrate assimilation (Solomonson and Barber 1990; Berges 1997), and to serve as a central control point of nitrogen metabolism (Crawford and Arst 1993; Campbell 1999).

Nitrate reductase activity can be regulated at different levels, including gene transcription, protein synthesis and degradation, redox and allosteric modulation, besides phosphorylation and dephosphorylation of the NR protein. The synthesis and activation of NR is regulated primarily by nitrate and light. However, ammonium, circadian rhythms, iron, carbon and other nitrogen metabolites can also modulate NR activity (Solomonson and Barber 1990; Crawford 1995; Ramalho et al. 1995; Milligan and Harrison 2000). Post-translational mechanisms, such as phosphorylation and dephosphorylation, can rapidly modulate NR activity in response to alterations in the availability of nitrate and light (Solomonson and Barber 1990; Corzo and Niell 1992; Huber et al. 1992a, b).

Previous studies have focused on nitrate and ammonium effects on NR activity. In general, nitrate stimulates NR activity (Crawford and Arst 1993), while ammonium is known to decrease NR activity in certain algal groups (Solomonson and Barber 1990; Campbell 1999; Gao et al. 2000). However, the action of ammonium on NR activity is still unclear. Ammonium can be indirectly associated to NR activity levels by preferential ammonium uptake, repression of nitrate transport, ammonium toxicity by producing pH changes, or negative feedback regulation of NR expression; or directly by degradation of the enzyme or post-translational modifications (Flynn 1991).

The rapid regulation of NR activity is well known for light:dark transition (Huber et al. 1992a, b). Nitrate reductase activity is higher during the light period and decreases to lower levels during the dark period of the day. Different mechanisms of regulation have been postulated for this light:dark transition in NR activity, including phosphorylation and dephosphorylation (Solomonson and Barber 1990; Huber et al. 1992a, b; Kaiser and Huber 1994), besides protein synthesis and degradation (Deng et al. 1990; Solomonson and Barber 1990; Vergara et al. 1998; Lopes et al. 2002).

The use of specific inhibitors is a common strategy to study the regulatory mechanisms of enzymes. The phosphorylation and dephosphorylation of proteins are major intracellular control mechanisms in eukaryotes and are associated to rapid regulation of NR activity levels. Okadaic acid and calyculin A are inhibitors of protein phosphatase 1 and 2A, and have been shown to frequently prevent the activation of NR (Huber et al. 1992a, b; MacKintosh 1992; Kaiser and Huber 1994). To study the slower regulation mechanism through *de novo* NR protein synthesis, actinomycin D, a RNA synthesis inhibitor, is commonly used (Vergara et al. 1998).

Gracilaria chilensis Bird, McLachlan *et* Oliveira has been widely cultivated in Chile for agar production, and successfully tested as a biofilter in polyculture systems with fish and mollusks to reduce the excess of dissolved nitrogen (Buschmann et al. 1994; Chow et al. 2001). The purpose of this study was to investigate the response of NR activity from *G. chilensis* to different nitrogen sources, specifically to assess possible regulation at short- and long-term by nitrogen and to verify the role of phosphorylation/dephosphorylation and synthesis/degradation mechanisms in the modulation of NR activity.

Materials and methods

Tetrasporophytic plants of *Gracilaria chilensis*, collected in La Herradura Bay, Chile (29°58′30″S; 71°22′30″W), were transported to the Laboratory of Marine Algae Édison José de Paula of the University of São Paulo, and grown under the following culture conditions: 5 g of fresh weight (FW) per 1 L of sterilized seawater enriched with 100% von Stosch (VS) (containing 500 μM nitrate; Edwards 1970), salinity of 32 psu, 15±2°C, 12:12 h (light:dark; LD), 65± 6 μmol photons m^{-2} s^{-1} and intermittent air-bubbling (30 min on, 30 min off). The culture medium was replaced weekly. The biological material for the following experiments came from this culture stock.

For each experiment, apical portions of 1.5 cm were cut from stocks seven days after the addition of 100% VS, and cultured under specific treatment conditions in a fresh culture media at 1 g FW L^{-1}. From each experiment, algal samples collected at midday were immediately frozen in liquid nitrogen and stored at –80°C for later *in vitro* NR activity assay, as described by Chow et al. (2004). All reagents were analytical grade, being provided by SIGMA (MA, USA) or Life & Technology-Gibco BRL (MD, USA).

Effect of different nitrogen sources on NR

Apical portions as described above were exposed for 24 h to different concentrations of nitrate (0.1, 0.2, 0.4, 0.6, 0.8, 1, 2, 4, 10, 15, 20 and 40 mM $NaNO_3$), ammonium (0, 2.5, 5, 15, 45 and 100 μM $NH_4(SO_4)_2$), or urea (0, 2.5, 5, 15 and 45 μM), and assayed for *in vitro* NR activity.

Based on the results from concentration curves above, nitrogen concentrations that stimulate NR activity were chosen for long-term exposure: 500 μM nitrate, 5 μM ammonium and 15 μM urea. Apical portions of 1.5 cm were cut from stocks which were in culture for seven days after the addition of 100% VS, and mixed together. From this mix, three samples were frozen in liquid nitrogen to estimate initial NR activity (Fig. 2, time 0). The remaining apical fragments were divided into nine culture flasks (three for each nitrogen source), and placed in fresh seawater enriched with each one of the different nitrogen source (changed every seven days). Algal samples (collected at midday) for the NR assay were taken from the culture flasks every 2 days for 11 days. The ratio of biomass/ culture media (1 g FW L^{-1}) was kept by removing the appropriate volume from the experimental flask each time a sample was taken.

To verify the short-term (24 h treatment) effect on NR activity of nitrate addition as the sole nitrogen source versus the VS enrichment medium, apical portions were disposed as described above and exposed to 500 μM of nitrate (equal amount of nitrate as found in a 100% VS medium) added as $NaNO_3$ or 100% VS. Based on the results of this experiment, a VS enrichment medium was selected for short-term (2–120 min) and long-term (1–21 days) experiments for measuring NR activity. For the long-term experiment, the VS was added twice, once at the beginning of the experiment and again at day 14. To determine the concentration of dissolved nitrate in the seawater and estimate the nitrate uptake by *G. chilensis*, 50 mL of culture medium was collected immediately after each algal sampling, and frozen at –20°C until measurement. Nitrate uptake was measured as reminiscent nitrate concentration in the external medium using an AutoAnalyzer II Technicon following the method described in Grasshoff et al. (1983).

The use of inhibitors to assess NR regulation

The regulatory mechanism of phosphorylation and dephosphorylation in NR activity was evaluated using 4 nM of calyculin A and 5 nM of okadaic acid which are inhibitors of protein phosphatase 1 and 2A. The regulation by protein synthesis and degradation was evaluated using 2.5 μg mL^{-1} of actinomycin D, a RNA synthesis inhibitor. These concentrations were selected after previous experiments. Inactivation of NR activity by phosphorylation occurs during the light:dark transition, and the NR activity is kept low until the beginning of the light phase. Consequently, the effect of calyculin A or okadaic acid was evaluated for 2 h during the dark phase. A light-pulse of 80 μmol photons m^{-2} s^{-1} (Dark + light pulse treatment) was applied 15 min before the alga sample was taken, once the addition of inhibitors should affect the light activation of NR activity by dephosphorylation. Actinomycin D effect was evaluated at midday and after 16.5 h of exposure to the inhibitor which was added 30 min before the beginning of dark phase. To verify the effect of the different inhibitors under light, dark or dark + light pulse, three controls were performed: one at midday (light control), at dark phase (dark control), and at dark phase + light pulse (dark + light pulse control).

NR in vitro assay

The optimized extraction and *in vitro* NR assay for *G. chilensis* were performed according to Chow et al. (2004). The samples were ground in liquid nitrogen, suspended in a 0.2 M phosphate buffer (pH 8.0), 5 mM EDTA, 1 mM DTT, and 0.3% w/v BSA, and centrifuged at 12,000 *g* for 15 min at 4°C. The supernatant was recovered and assayed for NR activity at 20°C in a 0.2 M phosphate buffer (pH 8.0), 6 mM KNO_3, and 0.5 mM $MgSO_4$, after the addition of 0.04 mM NADH. Assay controls without NADH were prepared for each experiment. The enzymatic reaction was stopped after 5 min by adding 1.4 mM $ZnSO_4$ and 43% v/v ethanol, nitrite concentration being determined at 543 nm after the addition of 9.6 mM sulphanilamide and 0.7 mM *n*-(1-naphtyl)ethylenediamine dihydrochloride. One unit of NR activity (U) is defined as the amount of enzyme required to produce 1 μmol of nitrite per minute at 20°C (Chapman and Harrison 1988).

Statistical analyses

All experiments were performed in triplicate with independent replication of data (n=3; each replicate being in a different flask). The results were analyzed using the single factor ANOVA test using STATISTICA 5.0 software (StatSoft, Inc, USA). Prior to ANOVA analysis, normality and variance homogeneity were tested. If the ANOVA test suggested differences between means, the Newman-Keuls test was performed to verify pair wise significance with a probability of 95% (Zar 1999).

Results

To study the short-term effect (24 h exposure) of nitrate, ammonium or urea on NR activity, different concentrations of each nitrogen source were evaluated. For algae cultivated under nitrate, NR activity significantly increased from 0.2 to 4 mM nitrate ($P<0.001$; Fig. 1a), 5 to 45 μM ammonium ($P<0.05$; Fig. 1b) and 5 to 45 μM urea ($P<0.05$; Fig. 1c) when compared to the enzymatic activity without nitrogen supplement (concentration zero).

In the long-term experiment, the addition of 5 μM ammonium or 15 μM urea as the sole source of nitrogen, increased NR activity after 24 h. Nevertheless, in the subsequent days the activity decreased to lower levels (Fig. 2).

The effect of nitrate alone or 100% VS enrichment medium (the same used for stock culture, see Material and methods) on NR activity was tested 24 h after addition, both at the final concentration of 500 μM nitrate, and a control without nitrogen enrichment (only seawater) was used. A significant ($P<0.0001$) increase of NR activity was observed with nitrogen enrichment (nitrate treatment = $120.46\pm2.77\times10^{-3}$ U g^{-1}; VS treatment = $136.47\pm14.18\times10^{-3}$ U g^{-1}), compared to the control without nitrogen enrichment (= $48.99\pm6.74\times10^{-3}$ U g^{-1}). The treatment using VS medium showed higher NR activity than nitrate treatment, but no significant difference was observed between the treatments.

Both in short (Fig. 3) and long-term (Fig. 4a) experiments, the addition of 100% VS rapidly (*ca.* 2 min) increased NR activity around 2.5 fold. In the short-term experiment 120 min after the addition of VS, NR activity was maintained higher than initial activity (before the addition of VS; time 0, Fig. 3).

In the long-term experiment (21 days), NR activity gradually decreased, finally reaching the observed initial values before the addition of VS after five days (Fig. 4a). After seven days NR activity showed fluctuations at lower levels. The external concentration of nitrate also presented a gradual decrease, reaching initial values before the addition

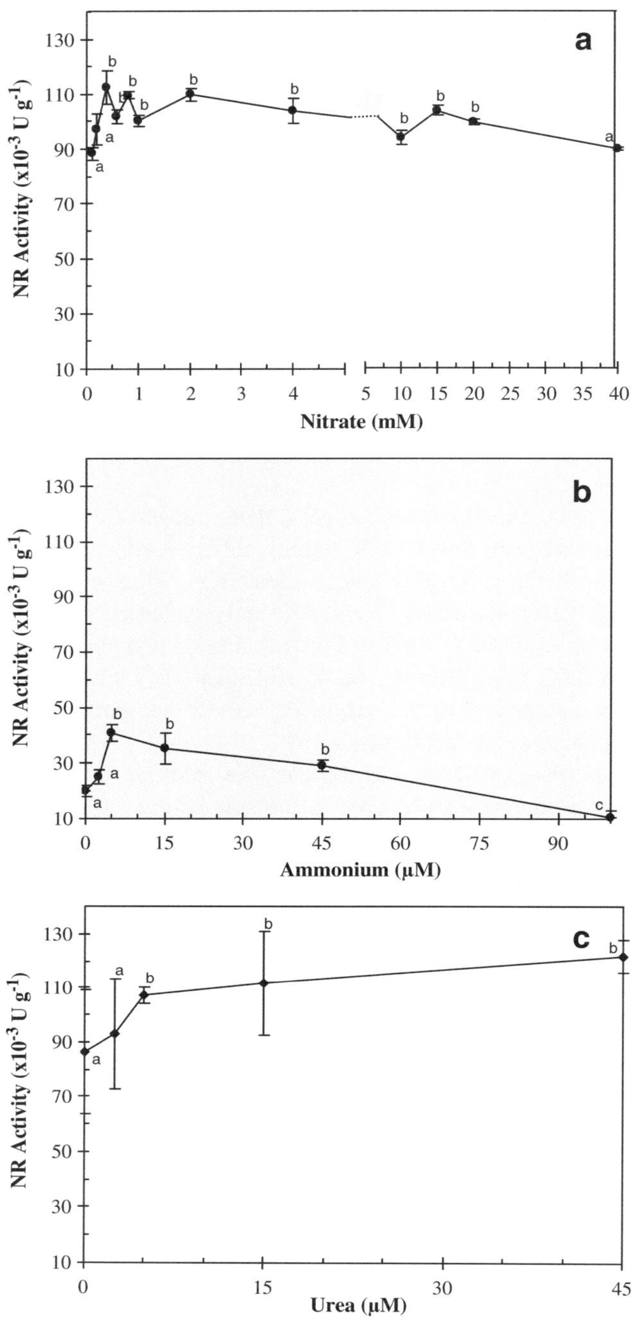

Fig. 1 Nitrate reductase (NR) activity in *Gracilaria chilensis* exposed for 24 h under different concentrations of (**a**) nitrate, (**b**) ammonium, or (**c**) urea. Each data point represents the mean ± SD (n=3). Concentrations of particular nitrogen sources (X axis) are on different scales. Different letters represent statistically different values between the treatments

of VS after ten days. The addition of VS medium to the samples after 14 days in the same culture medium did not increase NR activity (Fig. 4a). On the other hand, the uptake of nitrate by *G. chilensis* was rapid and nitrate concentration was similar to initial values before the addition of VS after seven days (Fig. 4b). After the initial ten days of treatment discoloration of the thalli was observed, this discoloration was only reversed five days after VS addition on the 14th day.

Higher values of NR activity were observed during the light phase (L), while lower activities were observed during the dark phase (Dark; Fig. 5a and b). A pulse of 80 μmol photons m^{-2} s^{-1} for 15 min during the dark phase (Dark + light pulse) was capable of increasing 1.5 fold NR activity in relation to dark phase activity (Fig. 5a and b). When calyculin A, an inhibitor of phosphatase, was added, a significant decline in NR activity was observed when compared to dark + light pulse treatment without the inhibitor (Fig. 5a). The addition of okadaic acid, also an inhibitor of phosphatase, presented no statistically significant effect (Fig. 5a). A significant decrease in NR activity in the light phase was observed with the addition of 2.5 μg mL^{-1} of the RNA synthesis inhibitor actinomycin D (Fig. 5b).

Discussion

To study the effect of different nitrogen sources on NR activity, nitrate, ammonium or urea were added to the seawater as the sole nitrogen source. At the location where *G. chilensis* thalli were collected, ammonium concentration in seawater is approximately 0.4 μM (Chow et al. 2001). Therefore, the concentration used in our experiment (5 μM) is 12 fold higher than in seawater. Even so, the addition of ammonium or urea in the laboratory experiment did not initially inhibit NR activity, on the contrary; NR activity was stimulated after 24 h of addition, similar to that observed when nitrate was added. However, after three days, NR activity decreased drastically in samples containing either ammonium or urea, remaining low until the end

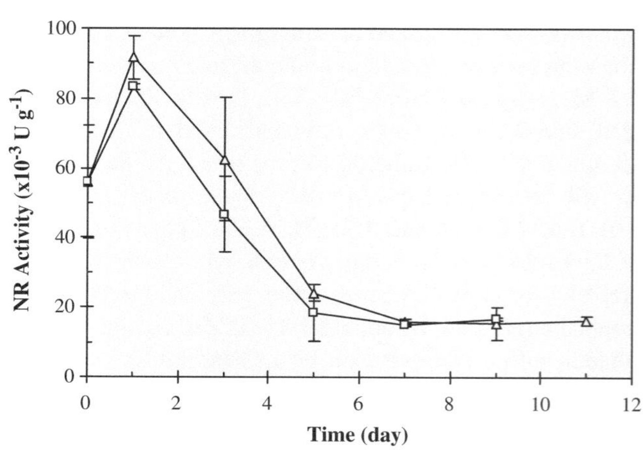

Fig. 2 Nitrate reductase (NR) activity in *Gracilaria chilensis* under 5 μM ammonium (Δ) or 15 μM urea (□). Each data point represents the mean ± SD (n=3)

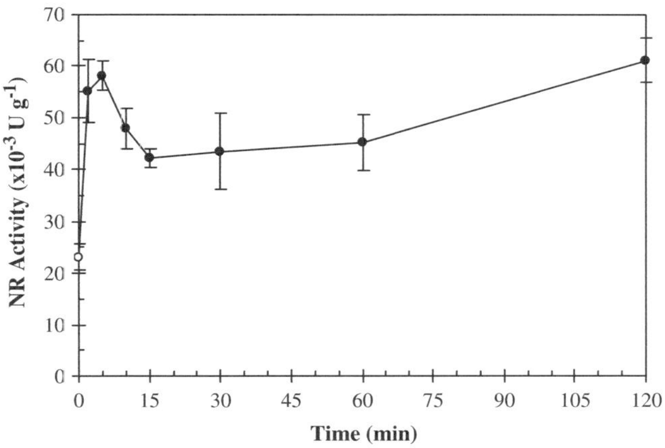

Fig. 3 Short-term (2 – 120 min) effect of the addition of a 100% von Stosch (VS) enrichment medium (equivalent to 500 μM nitrate) on nitrate reductase (NR) activity in *Gracilaria chilensis*. The white circle at time zero represents NR activity immediately before VS addition. Each data point represents the mean ± SD (n=3)

of the experiment (11 days). The observed results are different from the literature, for other algae the addition of ammonium to culture media rapidly inhibits NR activity (Solomonson and Barber 1990; Crawford and Arst 1993; Campbell 1999; Gao et al. 2000). Many of these studies use extremely high concentrations of ammonium (0.1–10 mM); therefore, the reduction of NR activity could be due to toxic levels of this compound. One of the central questions on NR inhibition by ammonium is whether it is caused by a toxic effect on the alga or whether ammonium specifically affects the NR protein. However, there are cases where NR activity was not inhibited at high ammonium concentrations, particularly in dinoflagellates (Collos and Slawyk 1980; Hochman 1982).

One hypothesis to explain the higher NR activity observed in *G. chilensis* during the first day with the addition of ammonium or urea is that the addition of these nutrients stimulated the metabolism of the algae, indirectly increasing NR activity. The biomass increase under ammonium or urea observed along the experimental period for *Gracilaria chilensis* support this assumption.

Nitrate is considered one of the main factors regulating NR activity (Crawford 1995). The addition of nitrate alone, or together with other nutritional compounds like vitamins, phosphate, and metals in the VS enrichment medium induced NR activity in a similar way in *G. chilensis*, indicating that the nitrate was sufficient to induce high NR activity. Nitrate reductase of *G. chilensis* is substrate-inducible, in agreement with what was observed for NR from other species (Crawford 1995; Lopes et al. 1997).

In the long-term (21 days) experiment, the low levels of NR activity in *G. chilensis*, after seven days in the same culture medium, indicate the need to add nutrients weekly for this species under the culture conditions as described in Material and methods. The decolouration of the thalli observed after the initial ten days could be related to limitation of nitrate in the culture medium and the possible utilization of nitrogen assimilated reserves such as phycobiliproteins (Hafting 1999). Nitrate can be accumulated and stored inside vacuoles in the cells, thus maintaining NR activity until the internal nitrate is depleted (Demanche et al. 1979). Therefore, the low NR activity in nitrate-deficient algae could be related to the absence or limitation of this compound, both in the external medium and in the internal pool, indicating that nitrate is necessary for the biosynthesis of NR protein and to protect the NR enzyme against degradation processes.

The addition of fresh VS medium to *G. chilensis* after 14 days, rapidly stimulated nitrate uptake without a concomitant induction of NR activity, suggesting that nitrate uptake system can operate independent of nitrate

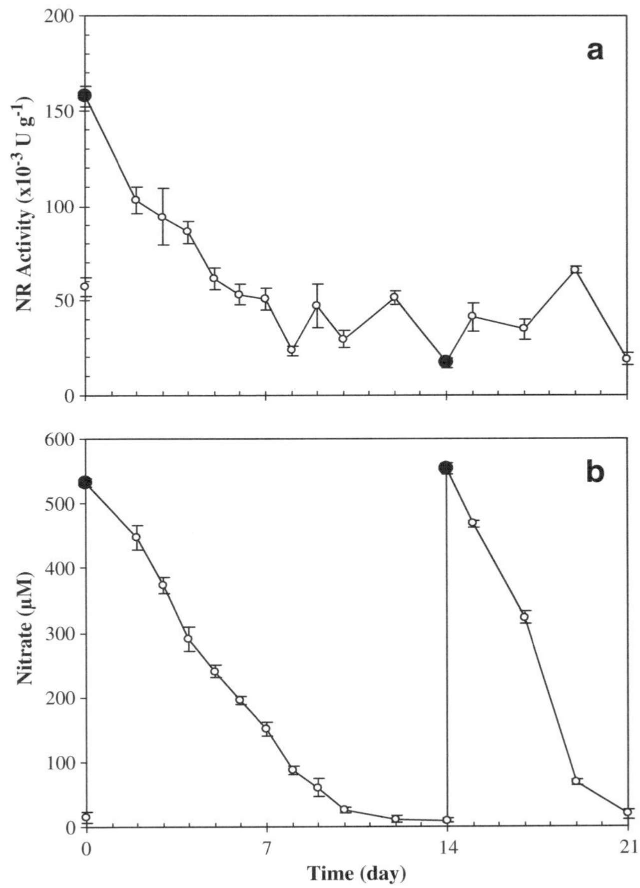

Fig. 4 Long-term (1 – 21 days) effect of the addition of a 100% von Stosch (VS) enrichment medium (equivalent to 500 μM nitrate) in *Gracilaria chilensis*: (**a**) Nitrate reductase (NR) activity. (**b**) Nitrate uptake shown as external nitrate concentration. White circles at time zero represent NR activity or external nitrate concentration immediately before VS addition. Black circles at time zero and time 14 represent NR activity or external nitrate concentration immediately after VS addition. Each data point represents the mean ± SD (n=3)

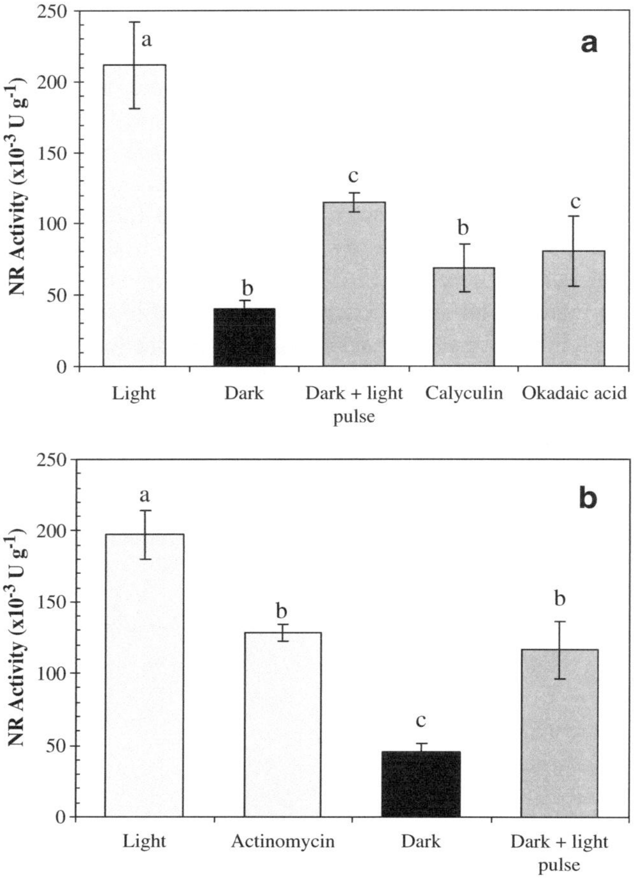

Fig. 5 Effect of (**a**) 4 nM calyculin A or 5 nM okadaic acid and (**b**) 2.5 μg mL^{-1} actinomycin D on nitrate reductase (NR) activity in *Gracilaria chilensis* during different phases of the light:dark cycle. White bars represent treatments during light phase, dark bars represent treatments during dark phase, and grey bars represent treatments during dark phase plus light pulse. Light correspond to control treatment during the light phase, Dark correspond to control treatment during the dark phase, and Dark + light pulse correspond to control treatment during the dark phase plus a light-pulse of 15 min at 80 μmol photons m^{-2} s^{-1}. Different letters on the top of columns represent statistically different values (or treatments). Each data point represents the mean ± SD (n=3)

reduction. In the absence of nitrate, the nitrate uptake system is active and ready to operate as soon as nitrate is available in the external medium, whereas NR activity is low and does not promptly recover. NR activity is affected by the lower concentration of nitrate, but the nitrate uptake system does not appear to be negatively affected suggesting that these systems are differentially regulated. This assumption is supported by results from Lartigue and Sherman (2005) for *Enteromorpha* sp., where nitrate uptake peak was observed under initial exposure to external nitrate, and then uptake declines as internal nitrate pool is filled. As an intracellular nitrate pool is available, NR continues to reduce nitrate to nitrite, therefore, NR is the rate-limiting step in nitrate assimilation. Lopes et al. (1997) in turn showed that for *G. tenuistipitata* nitrate uptake occurred mainly during the night, whereas nitrate reduction occurred mainly during the day, suggesting that both processes were independent. The absence of NR activity induction on the long-term experiment after the addition of VS on the 14th day indicates that most of the NR protein is degraded.

Inducible and constitutive NR isoforms have been found in higher plants (Ferrario et al. 1995). The inducible NR protein is regulated by environmental factors (*e.g.* nitrate and light), whereas the constitutive enzyme maintains a low constant rate of NR activity (Lillo 1994). In *G. tenuistipitata* grown under continuous darkness, a low constant activity independent of light and lacking endogenous rhythm was observed, in nitrate starved algae a similar low constant activity also being noted (Lopes et al. 1997). The authors suggested that this low constant activity could reflect a putative constitutive NR enzyme isoform. Similarly, for *G. chilensis* low constant NR activity was observed after five days when ammonium or urea was used as a nitrogen source.

Nitrate reductase activity can be regulated by phosphorylation and dephosphorylation of the NR protein adjusting rapidly to specific environmental conditions (Solomonson and Barber 1990; Kaiser et al. 1992; Huber et al. 1992a, b). The rapid induction (2 min) of NR activity in *G. chilensis*, observed after the addition of nitrate, suggests fast post-translational regulation. Protein synthesis and degradation, and transcriptional regulation mechanisms, require some hours (*ca.* 4–6 h) to be functional (Deng et al. 1990; Lillo 1994).

The rapid regulation of NR activity has been better studied during the light:dark transition (Huber et al. 1992a, b; Kaiser et al. 1992; MacKintosh 1992). There are several hypotheses about the NR activity modulation during this period: (a) a control mediated by NR protein biosynthesis and degradation cycles; (b) inhibition by Mg^{+2} during the dark phase; (c) inactivation by the binding of an inhibitory protein that involves ATP and phosphorylation; and (d) regulation by structural modification of NR mediated by the ATP/AMP proportion and enzyme phosphorylation, using MgATP as a substrate (Kaiser and Spill 1991).

During the dark phase, *G. chilensis* NR activity is low compared to that of the light phase. A light pulse of 15 min during the dark phase induced NR activity in *G. chilensis*. To investigate the possible roles of phosphorylation and dephosphorylation, and protein synthesis and degradation on the light:dark transition of NR activity in *G. chilensis*, three different inhibitors were tested. Since the fluctuations of *G. chilensis* NR are too rapid to be explained only by enzyme synthesis and degradation, other regulation processes must be involved. Our hypothesis was that dephosphorylated active NR was present during the light phase, whereas during the dark phase, the protein was phosphorylated and inactive; phosphatase activity reversed NR to the active dephosphorylated form, after being exposed to light for 15 min. Therefore, calyculin A, a phosphatase inhibitor, was added 2 h before the light pulse to verify whether this reversing of NR to the active form was blocked.

The transient induction of NR activity by the light pulse, during the dark phase was inhibited by the addition of calyculin A. These results suggest a rapid regulation of NR by phosphorylation and dephosphorylation mechanism, as was reported for higher plants (Kaiser and Spill 1991; Huber et al. 1992a, b).

Nitrate reductase protein synthesis and degradation has been reported in *G. tenuistipitata* during the light:dark transition (Lopes et al. 2002), and a similar regulatory mechanism involving enzyme degradation was observed for algae under extreme conditions of stress or substrate deficiency (Berges 1997). When *G. chilensis* was kept for 14 days in the same culture medium, most of the NR protein was probably degraded, once the addition of nitrate did not rapidly stimulate NR activity.

In higher plants, maximal accumulation of NR mRNA during the dark phase is followed by NR protein synthesis during the light phase (Deng et al. 1990). Actinomycin D, a RNA synthesis inhibitor, was added 30 min before the beginning of the dark phase, and after 16.5 h of exposure, samples were collected at midday to allow enough time for the inhibitor to block mRNA synthesis and, consequently, NR protein synthesis. The use of actinomycin D had a significant effect on NR activity, indicating that *de novo* protein synthesis and degradation also plays a role in the regulation of the daily fluctuations observed in the activity of NR in *G. chilensis* (Chow et al. 2004).

Therefore, from the results obtained for *G. chilensis* post-translational regulation was evidenced as a rapid phosphorylation and dephosphorylation mechanism modulated by nitrate and light responses, and a slower mechanism by regulation of RNA synthesis coupled to *de novo* NR protein synthesis are also relevant for NR regulation.

Acknowledgments Funding for investigation and scholarships was obtained from Conselho Nacional para o Desenvolvimento Científico e Tecnológico (CNPq), Fundação de Amparo à Pesquisa do Estado de São Paulo (FAPESP) and International Foundation for Science (IFS). We are grateful to Dr. Elisabete Braga for the seawater analyses in the uptake experiments and to Rosário Petti for technical assistance.

References

Berges JA (1997) Algal nitrate reductases. Eur J Phycol 32:3–8

Buschmann AH, Mora OA, Gómez P, Böttger M, Buitano S, Retamales C, Vergara PA, Gutierres A (1994) *Gracilaria chilensis* outdoor tank cultivation in Chile: Use of land-based salmon culture effluents. Aquac Eng 13:283–300

Campbell WH (1999) Nitrate reductase structure, function and regulation: bridging the gap between biochemistry and physiology. Annu Rev Plant Physiol Plant Mol Biol 50:277–303

Chapman DJ, Harrison PJ (1988) Nitrogen metabolism and measurement of nitrate reductase activity. In: Lobban CS, Chapman DJ, Kremer B (eds) Experimental phycology. A laboratory manual. Cambridge University Press, Cambridge

Chow F, Macchiavello J, Santa-Cruz S, Fonck E, Olivares J (2001) Utilization of *Gracilaria chilensis* (Rhodophyta; Gracilariaceae) as a biofilter in the depuration of effluents from tank cultures of fish, oysters, and sea urchins. J World Aquacul Soc 32:215–220

Chow F, Oliveira MC, Pedersén M (2004) *In vitro* assay and light regulation of nitrate reductase in red alga *Gracilaria chilensis*. J Plant Physiol 161:769–776

Collos Y, Slawyk G (1980) Nitrogen uptake and assimilation by marine phytoplankton. In: Falkowski PG (ed) Primary productivity in the sea. Plenum, New York

Corzo A, Niell FX (1992) Inorganic nitrogen metabolism in *Ulva rigida* illuminated with blue light. Mar Biol 112:223–228

Crawford NM (1995) Nitrate: nutrient and signal for plant growth. Plant Cell 7:859–868

Crawford NM, Arst HN Jr (1993) The molecular genetics of nitrate assimilation in fungi and plants. Annu Rev Genet 27:115–146

Demanche JM, Curl HC Jr, Lundy DW, Donaghay PL (1979) The rapid response of the marine diatom *Skeletonema costatum* to changes in external and internal nutrient concentration. Mar Biol 53:323–333

Deng MD, Moureaux T, Leydecker M, Caboche M (1990) Nitrate-reductase expression is under the control of a circadian rhythm and is light inducible in *Nicotiana tabacum* leaves. Planta 180:257–261

Edwards P (1970) Illustrated guide to the seaweeds and sea grasses in the vicinity of Porto Aransas, Texas. Contr Mar Sc Austin 15:1–228

Ferrario S, Valadier M-H, Morot-Gaudry J-F, Foyer CH (1995) Effects of constitutive expression of nitrate reductase in transgenic *Nicotiana plumbaginifolia* L. in response to varying nitrogen supply. Planta 196:288–294

Flynn KJ (1991) Algal carbon-nitrogen metabolism: a biochemical basis for modelling the interactions between nitrate and ammonium uptake. J Plankton Res 13:373–387

Gao Y, Smith GJ, Alberte RS (2000) Temperature dependence of nitrate reductase activity in marine phytoplankton: biochemical analysis and ecological implications. J Phycol 36:304–313

Grasshoff K, Enhardt M, Kremling K (1983) Methods of sea water analysis. Verlag Chemie, Weinheim

Hafting JT (1999) Effect of tissue nitrogen and phosphorus quota on growth of *Porphyra yezoensis* blades in suspension cultures. Hydrobiologia 398/399:305–314

Hochman A (1982) Studies of nitrate reductase in the fresh water dinoflagellate *Peridinium cinctum*. Arch Microbiol 133:62–65

Huber JL, Huber SC, Campbell WH, Redinbaugh MG (1992a) Reversible light/dark modulation of spinach leaf nitrate reductase activity involves protein phosphorylation. Arch Biochem Biophys 296:58–65

Huber SC, Huber JL, Campbell WH, Redinbaugh MG (1992b) Apparent dependence of the light activation of nitrate reductase and sucrose-phosphate synthase activities in spinach leaves on protein synthesis. Plant Cell Physiol 33:639–646

Kaiser WM, Huber SC (1994) Posttranslational regulation of nitrate reductase in higher plants. Plant Physiol 106:817–821

Kaiser WM, Spill D (1991) Rapid modulation of spinach leaf nitrate reductase by photosynthesis. II. *In vitro* modulation by ATP and AMP. Plant Physiol 96:368–375

Kaiser WM, Spill D, Brendle-Behnisch E (1992) Rapid light-dark modulation in leaves and roots: indirect evidence for the involvement of protein phosphorylation/dephosphorylation. Physiol Plant 89:557–562

Lartigue J, Sherman TD (2005) Response of *Enteromorpha* sp. (Chlorophyceae) to a nitrate pulse: nitrate uptake, inorganic nitrogen storage and nitrate reductase activity. Mar Ecol Prog Ser 292:147–157

Lillo C (1994) Light regulation of nitrate reductase in green leaves of higher plants. Physiol Plant 90:616–620

Lopes PF, Oliveira MC, Colepicolo P (1997) Diurnal fluctuation of nitrate reductase activity in the marine red alga *Gracilaria tenuistipitata* (Rhodophyta). J Phycol 33:225–231

Lopes PF, Oliveira MC, Colepicolo P (2002) Characterization and daily variation of nitrate reductase in *Gracilaria tenuistipitata* (Rhodophyta). BBRC 295:50–54

MacKintosh C (1992) Regulation of spinach-leaf nitrate reductase by reversible phosphorylation. Biochim Biophys Acta 1137:121–126

Milligan AJ, Harrison PJ (2000) Effects of non-steady-state iron limitation on nitrogen assimilatory enzymes in the marine diatom *Thalassiosira weissflogii* (Bacillariophyceae). J Phycol 36:78–86

Ramalho CB, Hastings JW, Colepicolo P (1995) Circadian oscillation of nitrate reductase activity in *Gonyaulax polyedra* is due to changes in cellular protein levels. Plant Physiol 107:225–231

Solomonson LP, Barber MJ (1990) Assimilatory nitrate reductase: functional properties and regulation. Annu Rev Plant Physiol Plant Mol Biol 41:225–253

Vergara JJ, Berges JA, Falkowski PG (1998) Diel periodicity of nitrate reductase activity and protein levels in the marine diatom *Thalassiosira weissflogii* (Bacillariophyceae). J Phycol 34:952–961

Zar JH (1999) Biostatistical analysis. 4th edn. Prentice-Hall, Englewood Cliffs, NJ

Molecular characterization of four genes involved in sulfur metabolism in *Porphyra purpurea* (Roth) C. Agardh

Subhash C. Minocha · Matt Hunt · Dennis E. Mathews

Originally published in the Journal of Applied Phycology, Vol 20, No 5, 333–345.
DOI: 10.1007/s10811-008-9319-3

Abstract Sulfated polysaccharides (carrageenans and agars) are among the most important products of red algae that are used as food additives as well as in molecular biology research. The quality and value of the product is greatly dependent on the levels and sites of sulfation of the polysaccharides. Little information is currently available on the molecular details of sulfur metabolism in red algae. Considering the economic importance of sulfated polysaccharide, elucidating the molecular details of sulfur metabolism in these organisms could help in future endeavors to improve algal commercial value, e.g., through genetic engineering. A cDNA library from the red alga *Porphyra purpurea* (Roth) C. Agardh was used to isolate four cDNAs with homology to genes encoding known sulfur assimilation enzymes: sulfate adenyltransferase (ATP sulfurylase), adenosine 5′-phosphosulfate kinase (APSK), sulfite reductase, and cysteine synthase. These cDNAs were characterized with respect to their molecular properties and a cDNA with homology to APSK was used to functionally complement an *Escherichia coli* auxotroph APSK^{-} mutant. The other cDNAs are being similarly characterized with respect to their ability to produce functional enzymes. Elucidation of the regulation of expression of these genes will aid in future research to determine the biochemical and genetic details of the sulfate assimilation pathway as well as its genetic manipulation in red algae.

Keywords Carrageenan · cDNA library · Porphyra · Red algae · Sulfate assimilation

S. C. Minocha (✉) · M. Hunt · D. E. Mathews
Department of Plant Biology, University of New Hampshire, Durham, NH 03824, USA
e-mail: sminocha@unh.edu

Present address:
M. Hunt
University of Illinois at Urbana-Champaign, 384 ERML, 1201 W. Gregory Drive, Urbana, IL 61801, USA

Abbreviations

APS	Adenosine 5′-phosphosulfate
APSK	APS kinase
APSR	APS reductase
ATPS	ATP sulfurylase
CS	Cysteine synthase
Cys	Cysteine
GSH	Glutathione
Met	Methionine
OAS	O-Acetylserine
PAPS	3′-Phosphoadenosine-5′-phosphosulfate
SAT	Sulfate adenyltransferase
SeT	Serine acetyltransferase
SR	Sulfite reductase
ST	Sulfotransferase

Introduction

Red algae (Rhodophyta) are a diverse group of eukaryotic organisms that range in size and complexity from a single cell to multicellular branched filaments. They are found in almost all aquatic ecosystems, from Antarctic waters under sheets of ice to hot springs (Craigie 1990). One of the most economically important features of red algae is the presence of sulfated polysaccharides in their cell walls. Sulfated polysaccharides play important physiological roles in the life of algae, including energy storage, water binding to protect against desiccation, protection against damage by wave force, acting as antioxidants, and in ion exchange and binding (McCandless 1981; Cosson et al. 1995; Rocha de Souza et al. 2007). Their commercial usefulness lies in their ability to form semisolid gels of varying strength depending on the specific quantity and the site of sulfation.

The sulfated polysaccharides of red algae are classified into two broad categories: carrageenans and agars. Carrageenans have a repeating α-(1→3) and β-(1→4) linkages of D-galactose and 3,6-anhydrogalactose, and are used as thickening agents in foods, e.g., ice creams, jams, and chocolate milk. This backbone can be substituted by various groups, e.g., methyl, pyruvate, and sulfate, at various positions along the chain. The location and type of substitution determines the family and specific type of a carrageenan chain. For example, the κ-carrageenan family, which includes κ, ι, μ, and ν carrageenans, is always sulfated at the C-4 position of galactose (Cosson et al. 1995; http://www.lsbu.ac.uk/water/hycar.html). Mu carrageenan is considered to be the direct biological precursor of κ carrageenans. The λ family of carrageenans is defined by being sulfated on the C-2 position of both the galactose and the 3,6-anhydrogalatose monomers of the chain (Cosson et al. 1995). Included in the family of λ carrageenans are λ, ξ, and π carrageenans. The β-carrageenan family, which is unusual in that its members are not sulfated, contains β-carrageenan and its precursors α, δ, and γ carrageenans.

Different types of carrageenans vary in their ability to form gels. The general property of carrageenans is that the more residues sulfated along the polysaccharide chain, the less gel-forming strength that polysaccharide will have. Kappa and iota carrageenans are highly desired types because they form relatively high strength gels. Lambda carrageenan, unlike κ and ι carrageenans, does not form a gel. Several genera of red algae are used for the commercial production of carrageenans.

Agars can be divided into three categories: non-ionized agarose, ionized agarose, and galactans (Cosson et al. 1995). Non-ionized agarose is completely non-sulfated and has very high gelling strength. In contrast, ionized agarose has a higher degree of sulfation; as with carrageenans, the more sulfation along the polysaccharide chain, the lower is the gelling strength of the polysaccharide. The main difference between agars and carrageenans is their type and degree of substitution; while agars are low in sulfate group substitutions (less than 5%), carrageenans have a higher degree of methylation (Craigie 1990; Cosson et al. 1995).

Uptake and assimilation of sulfur

Sulfur is one of the macronutrients necessary for life in all organisms. Although it is abundant in nature, human contributions of sulfur into the environment have increased tremendously over the past century, largely due to consumption of fossil fuels (Brimblecombe et al. 1989; Driscoll et al. 2003). The utilization of sulfur in cellular metabolism is dependent on activities of the transporters and the enzymes involved in its assimilation (Giordano et al. 2005; Saito 2000, 2004). The amount of information available on the genetics and biochemistry of sulfur assimilation and metabolism in algae, particularly the red algae, is rather limited in comparison to that in higher plants and bacteria (Leustek et al. 2000; Giordano et al. 2005; Kopriva 2006). With a few exceptions (Ikeda and Moritani 1994; Toda et al. 2001), it appears that algal sulfur assimilation is more like that in higher plants than that in prokaryotes.

Sulfate is imported into higher plant cells by plasma membrane sulfate transporters (Yildiz et al. 1994). Once in the cell, sulfate has several possible fates: it can remain in the cytosol, be transported into the plastid, or be stored in the vacuole. For metabolic assimilation, sulfate is first activated by the enzyme sulfate adenyltransferase (SAT), a.k.a. ATP sulfurylase (ATPS: EC 2.7.7.4) by addition of an AMP moiety from ATP (Fig. 1; Saito 2000; Leustek et al. 2000). The product, adenosine 5′-phosphosulfate (APS), is either phosphorylated by APS kinase (APSK: EC 2.7.1.25) or reduced by glutathione (GSH)-dependent APS reductase (APSR: EC 1.8.4.9, Kopriva and Koprivova 2004; Kopriva 2006). In algae, both pathways are apparently active because the end products are needed in large quantities for the production of sulfated polysaccharides and reduced sulfur compounds (Gao et al. 2000). The phosphorylated product of APS kinase, 3′-phosphoadenosine-5′-phosphosulfate (PAPS), is used as a substrate for cytosolic and membrane bound sulfotransferases. Sulfotransferases in plants sulfonate flavonols, desulfoglucosinolates, choline, and gallic acid glucosides; they are also thought to be involved in the sulfonation of polysaccharides (Varin et al. 1997; Bowman and Bertozzi 1999; Kopriva 2006). Since sulfite (the product of APSR) is toxic, its production is highly regulated in the cell (Saito 2000, 2004); it is reduced to sulfide by a ferredoxin-dependent sulfite reductase (SR: EC 1.8.7.1), an enzyme that is quite similar to nitrite reductase. Sulfide is used in the production of cysteine (Cys) by addition to O-acetylserine (OAS) by the enzyme OAS (thiol)lyase (OASTL: 2.5.1.47). This enzyme plays a key role in overall regulation of the sulfate assimilation pathway (Saito 2000; Kopriva 2006). OAS is synthesized by serine acetyltransferase (SeT: EC 2.3.1.30) from serine and acetyl-Co-enzymeA. These two enzymes (SeT and OASTL) constitute the cysteine synthase (CS) complex (Hell et al. 2002).

Depending on the species, prokaryotes can utilize sulfur in many forms, including sulfate, sulfite, thiosulfate, sulfide, L-Cys, and glutathione (Kredich 1996). Sulfate is

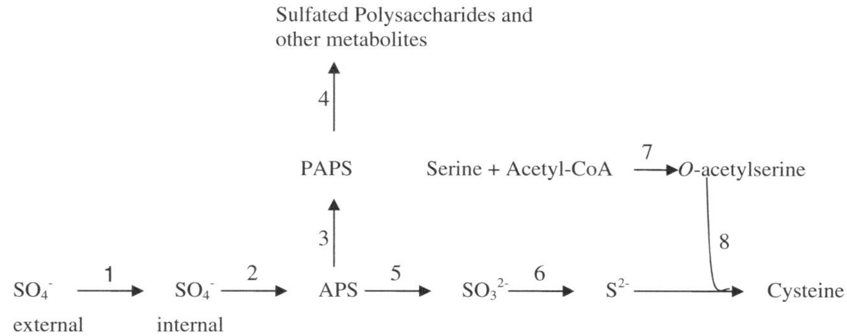

Fig. 1 The sulfate assimilation pathway in plants including the enzymes: *1* sulfate transporter (ST), *2* sulfate adenyltransferase (SAT), *3* adenosine 5′-phosphosulfate kinase (APSK), *4* sulfotransferases (STs), *5* adenosine 5′-phosphosulfate reductase (APSR), *6* sulfite reductase (SR), *7* serine acetyltransferase (SeT), *8* cysteine synthase [CS—*O*-acetylserine (thiol) lyase]. *APS* adenosine 5′-phosphosulfate, *PAPS* 3′-phosphoadenosine-5′-phosphosulfate

imported into prokaryotic cells by a sulfate-permease multiprotein ABC transporter complex (Higgins 1992). It is then activated by a heterodimeric SAT (Fig. 2). Prokaryotic SATs, unlike their eukaryotic counterparts, are GTP-dependent. The product, APS, is further activated by APSK, which is a highly efficient enzyme in *Escherichia coli*, and perhaps compensates for a relatively inefficient SAT (Satishchandran and Markham 1989). The product of the APSK reaction, PAPS, is not used for sulfate ester formation as in plants, fungi, and animals (Kredich 1996); rather it is used in the reductive sulfur pathway. To achieve this, prokaryotes have an enzyme called PAPS reductase that reduces PAPS to sulfite directly in a thioredoxin/glutaredoxin-dependent mechanism (Krone et al. 1991; Lillig et al. 1999). Sulfite is further reduced to sulfide by a NADPH-dependent SR (Eschenbrenner et al. 1995). The mechanism of Cys formation from OAS and sulfide is similar to that in plants (Kredich 1996); OAS is formed from Ser and acetyl-CoA by SeT. Cys synthase, a.k.a. OAS (thiol)-lyase, catalyzes the reaction between OAS and sulfide to form Cys.

Characterization of red algal sulfur assimilation

As mentioned above, several genes and proteins that participate in the sulfate assimilation pathway have been isolated from algae. However, a comprehensive understanding of the pathway and its regulation in any single species is lacking. Details of the identity and function of sulfotransferases have not been deduced although these enzymes have been postulated to function in the Golgi apparatus (Craigie 1990). The molecular biology of sulfur assimilation in red algae is in its infancy, and data for any red algal species are not extensive enough to allow detailed examination of the metabolism of sulfur in that species (Kopriva 2006). The cloning and characterization of all genes encoding enzymes involved in the reactions of this pathway from a single macroalgal species will allow us to understand the regulation of this pathway and its potential genetic manipulation to alter the quantity or the site of sulfation in the cell wall polysaccharides. The present study represents a first step in that effort for *Porphyra purpurea*. Although *P. purpurea* (Roth) C. Agardh is not itself a commercial source of cell wall polysaccharides, the life cycle of the genus *Porphyra* is very well characterized (Mitman and van der Meer 1994; Nelson et al. 1999) and members of the genus contain several sulfated polysaccharides (Craigie 1990). The specific objectives of our research were: (1) to isolate *P. purpurea* cDNAs with homology to known sulfate assimilation pathway genes; (2) to characterize the cDNAs with respect to sequence analysis; and (3) to determine if these cDNAs encode functional proteins.

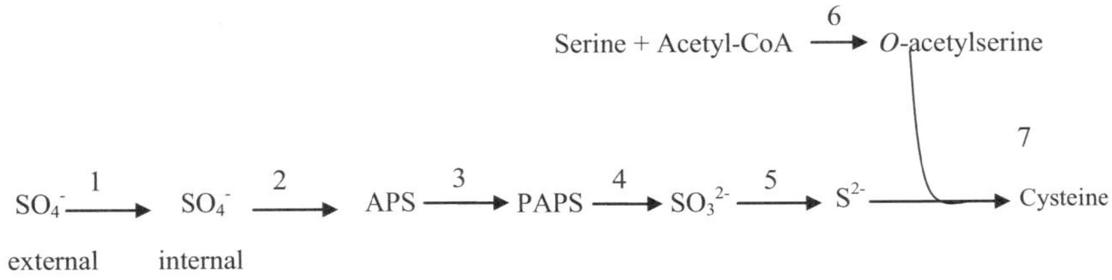

Fig. 2 The sulfate assimilation pathway in prokaryotes. The enzymes are: *1* sulfate permease, *2* SAT, *3* APSK, *4* 3′-phosphoadenosine-5′-phosphosulfate reductase (PAPSR), *5* SR, *6* SeT, *7* = CS

Materials and methods

Gametophytic thalli of *P. purpurea* were collected from Dover Point, (Dover, NH). Species identification was determined by assessing morphological and molecular characteristics. Identifying morphological characteristics included the single cell layer thick thallus and the segregated carposporangial and spermatangial halves of the thallus, separated by a distinct longitudinal margin. Molecular characterization involved PCR amplification and DNA sequence determination of a portion of the nuclear 18S ribosomal RNA gene. Using the BLAST Search program to compare sequences in the GenBank DNA database showed a very close match to an 'Avonport strain' of *P. purpurea* (Accession # L26201).

A cDNA library from *P. purpurea* gametophytic tissue was constructed using the ZAP Express cDNA Synthesis Kit (Stratagene, La Jolla, CA). Total RNA was used for isolation of poly-A mRNA (PolyATract, Promega, Madison, WI); the latter was subjected to reverse transcription to produce first strand cDNA using an oligo-dT primer with a 5′*Xho*I restriction site (5′GAGAGAGAGAGAGAGAGA GAGAACTAGTCTCGAGTTTTTTTTTTTTTTTTTT-3′). The second cDNA strand was synthesized with DNA polymerase I. Next, *Pfu* DNA polymerase was used to create blunt ends, and *Eco*RI adapters were ligated onto both sides of the cDNA. The cDNA was digested with *Xho*I to remove the unwanted *Eco*RI adapter from the 3′ end of the cDNA. The final product was ligated into the pBK-CMV vector and packaged into bacteriophage λ using the manufacturer's protocol.

Genomic DNA was isolated from gametophytic tissue (0.2 g) of *Porphyra yezoensis* using the hexadecyltrimethylammonium bromide (CTAB) method (Porebski et al. 1997). PCR was performed using the various primers listed in Table 1. Fresh PCR products were cloned into the pCR2.1-TOPO vector following the manufacturer's protocol (Invitrogen, Carlsbad, California). The plasmid was electroporated into electrocompetent *E. coli* TOP10 cells. The open reading frame (ORF) of the cDNA of interest was excised from the TOPO vector and cloned into the expression vector pFLAG-Shift$_{12c}$ (Sigma, St Louis, MO). The ligation mix was used to transform *E. coli*. The *E. coli* strain JM81A(Yale *E. coli* Stock Center, New Haven, Connecticut) was made competent for Z-competent cell transformation (Zymogen, Orange, California) using the manufacturer's specifications.

Plasmid DNA was digested with various restriction enzymes (New England Biolabs), and the DNA digests or PCR products were separated by agarose gel electrophoresis. The gels were stained in 0.5 $\mu g \cdot mL^{-1}$ ethidium bromide and visualized with a NucleoTech Gel-Doc system. DNA sequencing was performed at the UNH Hubbard Center for Genome Studies using the DYEnamic (Amersham-Pharmacia) ET terminator Cycle Sequencing premix and an ABI/PE 377 sequencer. Randomly selected clones was sequenced using either of the vector (M13 and T7) sequencing primers; other primers used for sequencing various DNAs were insert specific and are listed in Table 1.

Table 1 A list of the specific oligonucleotide primers used for PCR amplification or sequencing analysis: *SAT* sulfate adenyltransferase, *APSK* adenosine 5′-phosphosulfate kinase, *ST* sulfate transporter, *APSR* adenosine 5′-phosphosulfate reductase, *SR* sulfite reductase, *CS* cysteine synthase

Primer name	Primer sequence[a]
SAT.5′ 162	5′-CCCCTCACCGGCTCCCTCATTC-3′
SAT/Kpn	5′-CCTACCATGGCCAAGGTCTGCCCCGTC-3′
SAT/Eco	5′-GAATTCTTAGGCGCTGGCCTCCTGTTC-3′
SAT3′ (5′UTR)	5′-CAGGCAGGGAGAGCGGAATGGT-3′
SAT5′ (3′UTR)	5′-CGGAAAACAAGGGCAAGACGAC-3′
PAPS.5′ 28	5′-CCCACCCGACCGTCACACGCCGTTAC-3′
PAPS.3′ 550	5′-GCCACGTCCAGCGGGCAGTCCACAAA-3′
PAPS.5′ 125	5′-GGCCCTCACCTTCGCATTCCCTACAA-3′
PAPS.3′ 541	5′-AGCGGGCAGTCCACAACCACCTCAAT-3′
APSK.5′ B	5′-GATCCACAAGGCCGCCAGCAT-3′
APSK.5′ 55	5′-TGTCAGTCGCCGTCGTCAAGTG-3′
APSK.3′ B	5′-GCGTCAATGCCCGTAAAGTTCT-3′
APSK/Kpn	5′-GGTACCATATGTCGACCAACGTCGTCTGGCA-3′
APSK/Hind	5′-AAGCTTTAGACGGCGGCGGGGACCTTC-3′
FSR	5′-GTATACATGGCGTTCGTG-3′
RSR	5′-TAGCTATGAAACGTGTTG-3′
FSR9	5′-CGCCCTTCCCTTCTCCCACCATC-3′
RSR355	5′-GCGGTCGTCCTGCTGGTAAGAAC-3′
SRR175	5′-AAGAAAAACGGATGAAGC-3′
SRF239	5′-CTCCCGTTTCCGCTCCTC-3′
SRR2119	5′-ACTCACATCACACCAACA-3′
SRR2458	5′-CAACGCAACGACCAGGAG-3′
SRR2648	5′-CCAGCGAGGAGCGGAAAC-3′
SRFN1331	5′-AACGACACCCACCTGACC-3′
SRFN388	5′-TGTGGTTGGGCGGATGTC-3′
FOASL	5′-CGTTCATCTTCTCCACACAT-3′
ROASL	5′-CCACACTGTTTCCCCGTTCC-3′
FOAL 185	5′-TGCGGTGCTCGTTCCTCCTT-3′
ROAL 892	5′-TCTCGGGGTTCTTCTCCTTG-3′
CSR45	5′-TTTCTCCTTCTCCTCTGTTC-3′
CSF921	5′-ATGGGTTAGAGGGGAAAAGT-3′
CSF964	5′-AAGGAGAAGAACCCAGAG-3′
CSF1309	5′-GACGAACAGCGGGAAGAG-3′
CSF1316	5′-AGCGGGAAGAGGCGTATG-3′
CSR212	5′-ACACGAAAGCCATCAAAG-3′
CSR964	5′-CTCTGGGTTCTTCTCCTT-3′
CSR1309	5′-CTCTTCCCGCTCTTCGTC-3′
CSR1335	5′-CTCTACTCCGCCTTCATT-3′

[a] Primers were prepared by Integrated DNA technologies Inc. (Coralville, IA)

PCR products with homology to target gene sequences were labeled with DIG-dUTP according to the manufacturer's protocol (Roche, Indianapolis, IN). The *P. purpurea* phage cDNA library was titrated according to the manufacturer's protocol (Stratagene). The cDNA library was diluted to a titer of 50,000 pfu μL^{-1} and plated on NZY top agar in large 150 × 15 mm NZY Petri dishes.

Functional complementation and protein characterization

Functional complementation of *E. coli* auxotroph JM81A (*cysC92, tfr-8*) was carried out in Z-competent transformants grown on M9 minimal medium with 1 μM IPTG (isopropyl-thio-β-D-galactoside) to induce the pFLAG-$Shift_{12c}$ *tac* promoter in the absence of sulfur amino acids. Growth was visualized after 24–48 h of incubation at 37°C. Crude protein lysates from selected *E. coli* transformants were separated by sodium dodecyl sulfate polyacrylamide gel electrophoresis (SDS-PAGE) and subjected to western blot analysis to determine if a differentially expressed protein was present in transformed auxotrophs. Following electrophoresis, the proteins were transferred to a nylon membrane. The membrane was incubated in primary antibody solution (FLAG antibody; Sigma). Further processing with the secondary antibody (1:1,000 diluted anti-murine HRP-bound antibody) followed the supplier's instructions (Amersham Biosciences, Piscataway). The membrane was washed and exposed to 2 mL of 1:1 ECL detection reagents (Amersham Biosciences) for 1 min, and exposed to photographic film.

Results

In order to carry out the objectives of this study, the general approach involved construction of a cDNA library from *P. purpurea*, production of DIG-labeled probes from amplified *P. purpurea* or *P. yezoensis* DNA with homology to known sulfate assimilation genes, probing the *P. purpurea* cDNA library with the labeled probes to select cDNA clones of interest, sequencing of the cDNAs, and complementation of *E. coli* auxotroph strains. Data are presented here in the order of the most to the least information available for the various genes in the pathway (Fig. 1).

Adenosine 5′-phosphosulfate kinase

The red alga *P. yezoensis* is closely related to *P. purpurea*. A searchable EST database was available for *P. yezoensis* (www.kazusa.or.jp/en/plant/porphyra/EST/) which at the time of starting this project had about 10,000 ESTs (Nikaido et al. 2000; now there are over 20,000 ESTs—Asamizu et al. 2003). Multiple overlapping ESTs were identified for APSK, which, upon alignment (Seqman; DNASTAR, Madison, WI) produced a single contig. Primers (PAPS.5′ 28 and PAPS.3′ 550; Table 1) based on this contig were used for PCR amplification of an APSK gene fragment from *P. yezoensis* genomic DNA. The 900 bp product was excised from the gel, purified, and used as template in a second PCR reaction with nested primers (PAPS.5′125 and PAPS.3′541 – Table 1). Sequencing of the resulting 850 bp PCR product cloned in the TOPO vector, showed that the amplified *P. yezoensis* genomic DNA (py*APSK*) contained an apparent intron of 520 bp in the 5′ coding region. A DIG-labeled probe based on the py*APSK* genomic fragment was prepared and used to probe the *P. purpurea* cDNA library. Initially, five selected phagemid clones were sequenced, and the complete sequence of one (pp*APSK*-6) cDNA was confirmed by sequencing it three times with the primers listed in Table 1. The 1,045 bp cDNA of *P. purpurea* thus identified (pp*APSK*; Accession # AY325892) is made up of a 239 bp 5′UTR, a 702 bp ORF, and a 104 bp 3′UTR. While it is not known if the 5′UTR sequence represents the complete 5′ end of the cDNA, the 3′ end is presumed to be complete due to the presence of Poly(A). The deduced protein from the ORF of *P. purpurea* APSK shows homology to several known APSK proteins found in the NCBI database.

In order to confirm that the cloned pp*APSK* cDNA produces a functional enzyme, the ORF was cloned into an expression vector. The primers *APSK*/Hind and *APSK*/Kpn (Table 1) were used to amplify the ORF from pAPSK-BK-CMV. Sequences for *Hin*dIII and *Kpn*I restriction sites were added to the primer sequence to aid in cloning. The resulting PCR product was initially cloned into the pCR2.1-TOPO vector and then moved into the *Eco*RI site of the pFLAG-$Shift_{12c}$ expression vector. Plasmid DNA was prepared from several transformed colonies and used to confirm that it contained a pp*APSK* cDNA of the proper size and inserted in the proper orientation. Confirmed p*APSK*-FLAG plasmid DNA was used to transform Z-competent JM81A *E. coli* (*cysC92, tfr-8*), an *APSK* knockout that is unable to grow without supplementation of the medium with Cys and Met. Four transformed lines of JM81A containing pp*APSK*-FLAG in the correct orientation grew well on minimal medium without sulfur amino acids, while the non-transformed JM81A auxotroph did not grow (Hunt 2003).

To determine whether the transformed JM81A bacteria produced a protein of the expected size, western blot analysis using a FLAG epitope-specific antibody was performed on lysates from both non-transformed and transformed bacteria. A differentially expressed protein of about 35 kDa was seen in IPTG-induced transformed JM81A cells (data not presented, see Hunt 2003). Thus, it appears that the pp*APSK* cDNA encodes a functional enzyme.

Sulfate adenyltransferase

Random sequencing of several cDNAs from the *P. purpurea* cDNA library using M13 and T7 primers led to identification of a clone called M8 as having homology to known SAT genes using BLASTx (Hunt 2003). However, this cDNA contained only a partial ORF (missing ~250 bp from the 5′ end). A DIG-labeled probe made from this cDNA was used to identify several phagemid plaques in the cDNA library. Three selected clones were sequenced with the primers listed in Table 1; the sequence of the clone M8–3, which contained a complete pp*SAT* ORF, was submitted to NCBI (Accession # AY325891). The pp*SAT* cDNA ORF-encoded polypeptide shows homology to known SAT proteins. The *P. purpurea* SAT cDNA is made up of a 197 bp 5′UTR, a 1,263 bp ORF, and a 170 bp 3′UTR.

Cysteine synthase

Several ESTs with homology to known *CS* genes were found in the *P. yezoensis* EST database. The aligned sequences resulted in a single contig that was used to design PCR primers (Table 1). A PCR product of ~750 bp amplified from *P. yezoensis* genomic DNA using the FOAL185 and the ROAL892 primers was sequenced and found to have a sequence identical to that of the *P. yezoensis* CS contig. The PCR product (py*CS*) was cloned into the TOPO vector. A DIG-labeled probe made from the cloned py*CS* gene fragment was used to isolate ten phagemid clones from the *P. purpurea* cDNA library. The DNA sequence of two of the *P. purpurea* cDNA clones showed homology with other *CS* cDNAs in the NCBI database. The cDNA sequence (pp*CS*; Accession # AY325893) is made up of a 148 bp 5′UTR, a 1,134 bp ORF, and 183 bp 3′UTR.

Sulfite reductase

The PCR primers FSR9 and RSR 355 (Table 1) based on the ESTs of *P. yezoensis* py*SR* were successful in amplifying a predicted 600 bp product from *P. yezoensis* genomic DNA. The PCR product was cloned into the TOPO vector. Following confirmation of the insert by M13 and T7 sequencing, a DIG-labeled probe made from this fragment was used to identify phagemid clones in the *P. purpurea* cDNA library. Several selected clones were sequenced and found to be homologous to the py*SR* of *P. yezoensis* as well as to other known *SR* genes in the NCBI database. The final sequence of clone SR2–1 (pp*SR*; Accession # AY325894) contains an ORF that encodes a polypeptide of 546 amino acids and shows a high degree of homology (38–53% identity and 55–70% similarity) to various SR proteins. The transcript appears to have a rather long 5′UTR of 990 bp and a 3′UTR of 105 bp. Within the 5′UTR there are six ORFs of 222, 189, 114, 132, 78, and 213 bp, which do not have high homology to any known gene sequence in the database. It is possible that one or more of these ORFs may function in post-transcriptional regulation of the SR gene in a manner analogous to plant S-adenosylmethionine decarboxylase (*SAMDC*) genes (Hanfrey et al. 2002).

Sulfate transporter

A cDNA with homology to the ABC transporter protein in the sulfate transporter permease complex in *E. coli* had been reported from *Acetabularia acetabulum* by Ikeda and Moritani (1994). The Kazusa *P. yezoensis* EST database did not have any ESTs homologous to known plant *ST* genes, but several ESTs had homology to the *Acetabularia* cDNA sequence. Based on the alignment of these ESTs, a single contig was constructed and two PCR primers FST and RST (Table 1) were used for PCR with *P. yezoensis* genomic DNA. Gradient PCR with these primers yielded a distinct product of around 450 bp (py*ST*; Hunt 2003). Following its cloning and sequencing, the putative py*ST* product seemed to have high homology with bacterial ABC transporters (48–57% identity; 62–65% homology) and also with the *Acetabularia cys*A cDNA sequence. However, this homology was in a region common to all ABC transporter genes and not unique to the sulfate permease ABC transporters. Because using a probe based upon a sequence of such low specificity on the *P. purpurea* cDNA library might result in many false positives, this clone was not analysed further.

Adenosine 5′-phosphosulfate reductase

No ESTs with homology to *APSR* were found in the *P. yezoensis* database; however, the APSR proteins have multiple cross-kingdom consensus regions. Two of these regions were chosen to design degenerate primers APSRF and APSRR (Table 1). Gradient PCR with *P. yezoensis* genomic DNA using these primers yielded several products. The two most prominent products (of 0.5 and 1.0 kb in size) were gel purified and used for cloning and sequencing. The smaller of these fragments did not have significant homology with any identified gene in the NCBI database, while the larger fragment had high homology with a non-sulfur pathway 14–3–3 protein (data not shown). Isolation of a *P. purpurea* APSR cDNA has not yet been achieved.

Discussion

Giordano et al. (2005) and Kopriva (2006) have recently reviewed the current state of knowledge on assimilation of sulfur in plants and bacteria. While the review illustrates great progress in the biochemistry and molecular biology of sulfur metabolism and its role in the cellular physiology of higher plants, particularly Arabidopsis, the molecular biology of the sulfur assimilation pathway in algae remains largely unknown. While information from Arabidopsis and other plants will be useful in helping to understand sulfur metabolism in unrelated species, delineating the sulfur metabolic pathway and its control in red algae will undoubtedly strengthen our ability to target genetic manipulation of the products of this metabolism, i.e., cell wall polysaccharides, in commercially important marine algae.

The overall goal of this study was to clone and characterize cDNAs representing sulfate assimilation genes from the red alga *P. purpurea*. The general approach was to identify EST sequences for sulfate assimilation genes in the closely related species *P. yezoensis* using an available database. PCR primers designed based on these ESTs could then be used to generate probes to screen the *P. purpurea* cDNA library. At the time of starting this project, the *P. yezoensis* database (www.kazusa.or.jp/en/plant/porphyra/EST/) had about 10,000 ESTs. The database has since been upgraded to 20,000+ ESTs (Asamizu et al. 2003). The information obtained for the different target genes varied as some attempts at isolating cDNAs with homology to the genes of interest were unsuccessful, while others have been fully sequenced and one cDNA has been functionally characterized. The sequences of four cDNAs with homology to known sulfate assimilation genes have been determined: pp*SAT*, pp*APSK*, pp*SR*, and pp*CS*. The ORFs for pp*SAT*, pp*APSK*, and pp*CS* were cloned in an expression vector but, to date, only the protein product resulting from the pp*APSK* cDNA has been shown to functionally complement an *E. coli* auxotroph deficient in this step. The following discussion summarizes the information gained from each gene in this study and places that information in the context of genes in the sulfate assimilation pathway for other plants and algae.

Adenosine phosphosulfate kinase

Two separate enzymatic pathways metabolize the substrate APS (Fig. 1) in higher plants; APS is either activated by APSK to form PAPS, which is used as a substrate by sulfotransferases, or it is reduced by APSR to sulfite, for use in Cys biosynthesis. The reaction catalyzed by APSK is:

$$\text{APS} + \text{MgATP} \rightleftharpoons \text{PAPS} + \text{MgADP}$$

Plant APSKs are very efficient enzymes (Lillig et al. 2001) that are not affected by the relative concentrations of Cys, GSH or OAS. The enzyme is inhibited by high concentrations of its substrate APS, but is activated by thioredoxins (Arz et al. 1994; Lillig et al. 2001). Four potential *APSK*s have been identified in *A. thaliana* (Kopriva 2006), two of which have been shown to encode functional enzymes (Arz et al. 1994; Lillig et al. 2001). The three sulfate activation genes (*cysD*, *cysN*, and *cysC*) involved in the formation of PAPS in *E. coli* are encoded by a single operon (Leyh et al. 1992).

Arz et al. (1994) showed that an Arabidopsis *APSK* cDNA was able to produce a protein that was enzymatically active; this cDNA was then used by Schiffmann and Schwenn (1994) to functionally complement the APSK$^-$ *E. coli* strain JM81A, the same strain used in our study. Lillig et al. (2001) isolated a second APSK cDNA from Arabidopsis that was able to functionally complement the APSK$^-$ strain *MET14* of *Saccharomyces cerevisiae*. Both cDNAs encoded proteins with putative plastid transit peptides. The two other sequences in Arabidopsis that may encode cytosolic APSKs, because they lack a transit peptide sequence (Lillig et al. 2001), have not been functionally characterized.

The 1,045 bp *APSK* cDNA cloned from *P. purpurea* contains a putative 5′UTR of 239 bp, a 702 bp ORF, and a 104 bp 3′UTR. There are several criteria that support the argument that the 702 bp ORF identified here is the complete ORF for a cytoplasmic APSK. The alignment of amino acid sequence of ASPKs from various sources suggests that the ppAPSK encoded by the 702 bp ORF would have an amino-terminus similar to APSKs from fungi, animals and bacteria. Several plant APSKs have a longer amino-terminus because of the presence of a chloroplast transit peptide. However, ASPK activity in plants is localized in the cytoplasm as well as the chloroplast. The py*APSK* and pp*APSK* cDNA sequences are 90% identical following the putative start codon but diverge significantly in the presumptive 5′UTR.

The ability of the *APSK* cDNA to encode a functional protein was demonstrated by complementation of the APSK$^-$ *E. coli* strain JM81A, in a method similar to that described by Schiffmann and Schwenn (1994). The results of western blot analysis of pp*APSK*-FLAG-transformed and non-transformed JM81A cell extracts further showed a differentially expressed polypeptide of the predicted size in the pp*APSK*-FLAG transformed cells that reacted with the FLAG epitope antibody (Hunt 2003).

The *P. purpurea* APSK polypeptide contains sequences that are important in APSK enzyme catalysis. The two motifs that have been identified as important in substrate binding, the ATP binding P-loop GLSGSGKST and the ATP/(P)APS binding motif DPKGLYKKARSGEIKNFT (Satishchandran et al. 1992; MacRae et al. 1998, 2000), are both present in the cDNA-encoded polypeptide.

Sulfate adenyltransferase

Sulfate adenyltransferase (SAT) occupies a key position in the sulfate assimilation pathway as it is the sole enzyme responsible for activation of inorganic sulfur in eukaryotes (Gao et al. 2000; Crawford et al. 2000; Kopriva 2006). The reaction catalyzed by SAT is:

$$SO_4^{2-} + MgATP \rightleftharpoons MgPP_i + 5' - \text{adenylylsulfate(APS)}$$

Since the forward reaction is energetically unfavorable, the cellular concentrations of APS are rather low (≥0.1 μM; Crawford et al. 2000). The reaction is "pushed" forward in plants by the breakdown of PP_i via pyrophosphatase as well as rapid utilization of APS (Saito 2004). Sulfur availability has been shown to affect *SAT* transcripts in *Zea mays* and *Chlamydomonas reinhardtii* (Bolchi et al. 1999; Yildiz et al. 1996). In both cases, *SAT* transcripts increased in response to reduced sulfate concentrations in the growth medium. The enzyme is upregulated by high levels of OAS and downregulated by high levels of GSH, which point to an overall sulfate assimilation regulatory circuit (Saito 2000; Clarkson et al. 1999; Lappartient and Touraine 1996; Hirai et al. 2003). Because of the location of SAT at a branch point in the sulfate assimilation pathway (Fig. 1), the regulation of the genes and their encoded proteins allows for control of sulfur flow through the two branches.

To compensate for the high energy costs of this reaction, animals and prokaryotes have different mechanisms; while the former encode SAT and APSK on a single polypeptide that carries out both reactions to make PAPS in a coordinate manner (Kurima et al. 1999), the latter combine GTPase activity with SAT activity in a heterodimeric protein complex to reduce the overall ΔG_{rxn} (Leyh et al. 1988). Different SAT isoforms localize either to the cytosol or plastids in plants (Leustek et al. 1994; Murillo and Leustek 1995; Crawford et al. 2000; Saito 2000; Rotte and Leustek 2000). The *SAT* cDNAs from Arabidopsis and *C. reinhardtii* (Leustek et al. 1994; Murillo and Leustek 1995; Yildiz et al. 1996; Hatzfeld et al. 2000; Kopriva 2006) were characterized by complementation of the SAT-deficient yeast auxotroph S5 or the *E. coli* auxotroph JM221. A striking feature of these experiments was that, despite the plant proteins being only 25–40% identical to the yeast or bacterial enzymes, they were functionally active in these organisms.

The pp*SAT* cDNA (accession no. AY325891) is made up of a 197 bp 5′UTR, a 1,263 bp ORF, and a 170 bp 3′UTR. When the general structural features of other plant SAT cDNAs in the NCBI database are compared with that of pp*SAT* cDNA, it is apparent that the 5′UTR of the latter is the largest of them all, and the ORF is smaller than most others. One likely explanation for the shorter ORF in pp*SAT* (and also in py*SAT*) is the apparent lack of a chloroplast transit peptide coding sequence (Sato et al. 1987). A cDNA for a cytosolic SAT that complemented the SAT knockout yeast W303Met3 has been isolated from *Solanum tuberosum*. The pp*SAT* cDNA-derived polypeptide shows homology to known plant, bacterial and fungal SAT proteins. The *SAT* sequences of the two *Porphyra* species are 84% identical over the entire length of the ORF.

The crystal structure of SAT from yeast (Ullrich et al. 2001) shows that two of the most important sites necessary for substrate binding and catalysis, which include the ATP binding motif VxAFQxRNP and the GRD-loop APS binding motif VGRDHAG, appear to be present in the ppSAT. The structure of the ATP binding site of the SAT is different from that of APSK because the two use ATP for different functions. Whereas APSK removes a single Pi from ATP, SAT uses the AMP moiety from ATP to form APS and releases PPi.

Cysteine synthase

Cysteine synthase (CS) and serine acetyltransferase (SeT) form a bi-enzyme complex in plants and bacteria (Bogdanova and Hell 1997; Droux et al. 1998; Mino et al. 2000). While CS controls the final reaction in the biosynthesis of Cys from assimilated sulfate, SeT catalyzes the production of OAS, which is the substrate for CS. The reaction catalyzed by the CS/SeT complex is:

$$OAS + H_2S \rightleftharpoons \text{cysteine} + \text{acetic acid}$$
$$\text{Serine} + \text{acetyl} - CoA \rightleftharpoons OAS + CoA - SH$$

Plant, algal and bacterial CSs can be functionally grouped together by virtue of utilizing the same reaction mechanism and containing the same prosthetic group, i.e., pyridoxal 5′-phosphate (Fujimori et al. 1991; Rao et al. 1993; Burkhard et al. 1998; Nakamura et al. 1999). Cysteine synthases in plants as well as algae have isoforms localizing to the cytosol (Hell et al. 1994; Takahashi and Saito 1996; Nakamura et al. 1999), mitochondria (Nakamura et al. 1999; Jost et al. 2000), and plastids (Hell et al. 1994; Ravina et al. 1999). In both plants and algae, different isoforms of SeT also localize to the same three sites (Ruffet et al. 1995; Roberts and Wray 1996; Saito et al. 1997; Howarth et al. 1997; Toda et al. 1998, 2001; Urano et al. 2000). The cytosolic isoform is inhibited by Cys, while the organelle-localized isoforms are not subject to this inhibition (Takagi et al. 1999; Inoue et al. 1999). cDNAs encoding functional SeTs have been isolated from various species including the red alga *Cyanidioschyzon merolae*, and from Arabidopsis (Toda et al. 1998; Howarth et al. 1997). No sequence homology was found in the *P. yezoensis* EST database for this enzyme.

Cysteine synthase is the final enzyme in the sulfate assimilation/reduction pathway in both prokaryotes and eukaryotes. In plants, isoforms of this enzyme are present in the cytosol, plastids as well as in mitochondria. Multiple cDNA have been cloned from *Oryza sativa* and Arabidopsis that encode CS isoforms localized to all three sites in the cell (Hell et al. 1994; Nakamura et al. 1999; Jost et al. 2000; Kopriva 2006). The pp*CS* cDNA (accession no. AY325893) contains a 148 bp 5′UTR, a 1,134 bp ORF, and a 183 bp 3′UTR. Both the size of the 5′UTR and the length of the ORF of pp*CS* cDNA are within the size range of CSs that are organelle targeted. The iPSORT protein analysis program predicts a mitochondrial transit peptide to be present in the protein encoded by the pp*CS* ORF. Again, the *P. yezoensis* EST database allowed for the assembly of a py*CS* contig containing a complete ORF. The pp*CS* cDNA has 85% sequence identity with the py*CS* contig over the length of the ORF (1,153 bp). The translated pp*CS* cDNA ORF contains the highly conserved pyridoxal 5′-phosphate binding residues SVKDR (Ravina et al. 1999; Nakamura et al. 1999; Hunt 2003).

Sulfite reductase

In plants, SR is a plastid-localized enzyme that reduces sulfite to sulfide by a six electron reduction reaction [similar to that of nitrate reductase (NR)] utilizing reduced ferredoxin (Brühl et al. 1996). The reaction is:

$$SO_3^{2-} + 6\ \text{Ferredoxin}_{red} \rightleftharpoons S^{2-} + 6\ \text{Ferredoxin}_{ox}$$

Because SR has homology to NR, in both peptide sequence and protein functionality (Crane et al. 1995), it also has the ability to reduce nitrite to ammonium, but with a significantly higher K_m (McRee et al. 1986). The protein has been purified from a number of plants, algae, and cyanobacteria, including *P. yezoensis* (Aketagawa and Tamura 1980; Koguchi and Tamura 1988, 1989; Gisselmann et al. 1993; Takahashi et al. 1996a, 1996b; Brühl et al. 1996). Perhaps due to the toxicity of sulfite (Crawford et al. 2000), SR is encoded by a constitutively expressed gene. cDNAs for *SR* have been cloned from Arabidopsis and *Nicotiana tabacum*. Complementation analysis to confirm the function of the cDNAs is unavailable due to the lack of SR knockouts in bacteria (Brühl et al. 1996; Yonekura-Sakakibara et al. 1998; Kopriva 2006).

The cloned pp*SR* cDNA (accession no.AY325894) has a long 990 bp 5′UTR, larger than that in most other plant SR cDNAs. The sequence of the 5′UTR does not show a high degree of homology with any other sequence in the NCBI database, although the ~170 bp at the 3′ end do show homology to other SR sequences in the database. The long 5′UTR contains six ORFs of 222, 189, 114, 132, 78, and 213 bp. The large size of the pp*SR* 5′UTR and the presence of ORFs within it, show that it may be involved in post-transcriptional regulation of the *SR* transcript, as seen for other enzymes (Pesole et al. 2001; Hanfrey et al. 2002). The pp*SR* cDNA shares 87% sequence identity with a py*SR* EST contig over its limited length (Hunt 2003); the encoded protein has high homology to other known SR proteins (Hunt 2003). However, the ppSR protein sequence derived from the ORF lacks a 39–100 amino acid N-terminal portion that is present in most other proteins. It is possible that this region is simply not present in the pp*SR*; however, this is inconsistent with the fact that all higher plant SRs are plastid localized (Crawford et al. 2000). The pp*SR* cDNA ORF itself is, on average, 300 bp shorter than the other higher plant SR ORFs. The ppSR ORF-encoded polypeptide contains several highly conserved residues that are necessary for enzymatic function, including three Arg residues (R28, R62, and R98) and two lysines (K189 and K191) important in substrate binding, and four Cys residues (C413, C418, C458, and C462) involved in 4Fe-4S binding to the protein (Yonekura-Sakakibara et al. 1998; Nakayama et al. 2000). There are no bacterial or yeast strains available for complementation with the pp*SR* clone.

Sulfate transporters

In higher plants, sulfate transporter (ST) proteins typically have 12 membrane domains; they are found in plasma membranes, chloroplast outer membranes, and perhaps also in tonoplast membranes (Takahashi et al. 1999; Hawkesford 2003). The structure and function of STs are similar to those of phosphate transporters (Smith et al. 2000; Saito 2004). Both high affinity and low affinity STs have been identified (Vidmar et al. 2000; Khurana et al. 2000; Kopriva, 2006).

Several cDNAs encoding STs have been identified from plants (Smith et al. 1995, 1997, 2000; Bolchi et al. 1999, Vidmar et al. 2000; Kopriva 2006). The expression of *ST* genes is regulated by the overall sulfur status of the cell as well as the availability of sulfate in the environment; e.g., ST expression is up-regulated by low sulfur and high amounts of OAS and down-regulated by high sulfate in the environment and high levels of cellular GSH (Vidmar et al. 2000; Hirai et al. 2003). To determine the ability of plant *ST* cDNA ORFs to encode functional protein, a complementation approach has been used with the ST-deficient yeast auxotroph *ysd*1 (Smith et al. 1995, 1997; Vidmar et al. 2000). The present study did not succeed in cloning any *ST* cDNA from *P. purpurea*. There are no ESTs with homology to known eukaryotic or bacterial *ST*s in the *P. yezoensis* database.

Adenosine 5′-phosphosulfate reductase

The APSR reaction (shown below) represents the committing step in sulfate reduction in higher plants and algae.

$$\mathrm{APS} + 2\ \mathrm{glutathione_{red}} \rightleftharpoons \mathrm{SO_3^{2-}} + \mathrm{glutathione_{ox}} + \mathrm{AMP} + 2\mathrm{H^+}$$

The enzyme is localized in the plastids, where readily available reducing power can be exploited for enzyme function (Gutierrez-Marcos et al. 1996; Gao et al. 2000). The activity of APSR in plants and algae increases in response to sulfur starvation and decreases with relatively high concentrations of reduced sulfur (Gutierrez-Marcos et al. 1996; Ravina et al. 2002). The APSR mRNA transcript was found to increase upon increased supply of OAS, which itself increases in response to sulfur stress (Saito 2000). It is postulated that APSR is highly regulated in order to avoid cytotoxicity due to excess sulfite and sulfide (Saito 2000, 2004).

The *APSR* gene has been isolated from Arabidopsis and the green alga *Enteromorpha intestinalis* (Gutierrez-Marcos et al. 1996; Gao et al. 2000). The ability of the cDNA ORFs to encode a functional APSR protein was demonstrated by complementation of the PAPS reductase mutant *E. coli* strain JM96 (Gutierrez-Marcos et al. 1996; Gao et al. 2000). Since there was no EST with homology to *APSR* in the initial *P. yezoensis* gametophytic database, we used degenerate primers based on conserved amino acid regions of this protein in several plants. These primers, however, did not yield a homologous cDNA from our library. In the new sporophytic *P. yezoensis* EST database, however, it appears that there is at least one clone that has high homology to known plant *APSR* sequences.

Recently, it has also been shown that the sulfate reduction pathway in some plants has an alternative to the reduction of APS via APSR. *Physcomitrella patens* and *Selaginella lepidophylla* apparently contain a PAPS reductase (Kopriva and Koprivova 2004; Kopriva 2006) that reduces PAPS to sulfite. A pathway in which APS is phosphorylated to PAPS, which is then reduced to sulfite is the exclusive sulfate reduction pathway in fungi and some bacteria (Fig. 2). A gene for PAPS reductase has not been identified in any higher plant (Kopriva 2006). It would be of interest to determine whether the PAPS reduction pathway is common to the algae as well as the lower plants.

Overall, this project was successful in isolating four cDNAs from a *P. purpurea* cDNA library that showed high homology to four different genes encoding proteins of the sulfate assimilation pathway. All four cDNAs have been fully sequenced, and one (pp*APSK*) has been used to successfully complement an *E. coli* strain deficient in the production of APSK. This represents a first step in the process of characterizing the sulfur assimilation pathway in red algae. However, much information still has to be gathered regarding the other genes in this pathway. Sulfotransferases are probably a family of proteins that are vital to the production of sulfated polysaccharides in red algae, and not much is known about them. Until *Porphyra* sulfotransferase EST sequences can be identified, isolating members of this gene family will have to rely on degenerate primers designed from the gene sequences of other organisms. The substrate specificity of the sulfotransferases that have been studied so far in higher plants and other organisms suggests that generating reliable assays for molecular targets of these enzymes will be an involved process. However, isolation and characterization of sulfotransferases that specifically sulfonate carrageenans or agars will greatly aid in the overall understanding of sulfur flow in red algae and the sulfonation of their polysaccharide targets.

Acknowledgments We thank Dr. Estelle Hrabak for her expert advice in several protocols, Troy Bray for providing *Porphyra yezoensis* tissue, and the Yale *E. coli* stock center for providing the JM81A *E. coli* auxotroph strains. The help of Dr. Andrew Laudano and his students was invaluable for the western blot analysis, and the staff at the UNH Hubbard Center for DNA sequencing. This project was partially funded by the UNH Marine Program and the Center for Marine Biology.

References

Aketagawa J, Tamura G (1980) Ferredoxin-sulfite reductase from spinach. Agric Biol Chem 44:2371–2378

Arz H, Gisselmann G, Schiffmann S, Schwenn J (1994) A cDNA for adenylyl sulphate (APS)-kinase from *Arabidopsis thaliana*. Biochim Biophys Acta 1218:447–452

Asamizu E, Nakajima M, Kitade Y, Saga N, Nakamura Y, Tabata S (2003) Comparison of RNA Expression profiles between the two generations of *Porphyra yezoensis* (Rhodophyta), based on Expressed Sequence Tag Frequency Analysis. J Phycol 39:923–930

Bogdanova N, Hell R (1997) Cysteine synthesis in plants: protein-protein interactions of serine acetyltransferase from *Arabidopsis thaliana*. Plant J 11:251–262

Bolchi A, Petrucoo S, Tenca P, Forani C, Ottonello S (1999) Coordinate modulation of maize sulfate permease and ATP sulfurylase mRNAs in response to variations in sulfur nutritional status: stereospecific down-regulation by L-cysteine. Plant Mol Biol 39:527–537

Bowman K, Bertozzi C (1999) Carbohydrate sulfotransferases: mediators of extracellular communication. Chem Biol 6:R9–R22

Brimblecombe P, Hammer C, Rohde H, Ryaboshapko A, Boulron CF (1989) Human influence on the sulphur cycle. In: Brimblecombe P, Lein AY (eds) Evolution of the global biogeochemical sulphur cycle. Wiley, Chichester

Brühl A, Haverkamp T, Gisselmann G, Schwenn JD (1996) A cDNA clone from *Arabidopsis thaliana* encoding plastidic ferredoxin: sulfite reductase. Biochem Biophys Acta 1295:119–124

Burkhard P, Rao G, Hohenester E, Schnackerz K, Cook P, Jansonius J (1998) Three-dimensional structure of *O*-acetylserine sulfhydrylase from *Salmonella typhimurium*. J Mol Biol 283:121–133

Clarkson D, Diogo E, Amâncio S (1999) Uptake and assimilation of sulphate by sulphur deficient *Zea mays* cells: the role of *O*-acetyl-L-serine in the interaction between nitrogen and sulphur assimilatory pathways. Plant Physiol Biochem 37:283–290

Cosson J, Deslandes E, Zinoun M, Mouradi-Givernaud A (1995) Carrageenans and agars, red algal polysaccharides. Prog Phycol Res 11:269–324

Craigie J (1990) Cell walls. In: Cole M, Sheath R (eds) Biology of the red algae. Cambridge University Press, New York, pp 241–242

Crane B, Siegel L, Getzoff E (1995) Sulfite reductase structure at 1.6 Å: evolution and catalysis for reduction of inorganic anions. Science 270:59–67

Crawford N, Kahn M, Leustek T, Long S (2000) Nitrogen and sulfur. In: Buchanan B, Gruissem W, Jones R (eds) Biochemistry and molecular biology of plants. American Society of Plant Physiologists, Rockville, pp 786–849

Driscoll CT, Driscoll KM, Mitchell MJ, Raynal DJ (2003) Effects of acidic deposition on forest and aquatic ecosystems in New York State. Environ Pollution 123:327–336

Droux M, Ruffet M, Douce R, Job D (1998) Interactions between serine acetyltransferase and *O*-acetylserine (thiol) lyase in higher plants: structural and kinetic properties of the free and bound enzymes. Eur J Biochem 255:235–245

Eschenbrenner M, Covès J, Fontecave M (1995) NADPH-sulfite reductase flavoprotein from *Echerichia coli*: contribution to the flavin content and subunit interaction. FEBS Lett 374:82–84

Fujimori Y, Nakamura K, Tamura G (1991) Cysteine synthase from a red alga *Porphyra yezoensis*: purification and properties. Jpn J Phycol 39:7–14

Gao Y, Schofield M, Leustek T (2000) Characterization of sulfate assimilation in marine algae focusing on the enzyme 5′-adenylylsulfate reductase. Plant Physiol 123:1087–1096

Giordano M, Norici A, Hell R (2005) Sulfur and phytoplankton: acquisition, metabolism and impact on the environment. New Phytol 166:371–382

Gisselmann G, Klausmeier P, Schwenn J (1993) The ferredoxin: sulfite reductase gene from *Synechococcus* PCC7942. Biochim Biophys Acta 1144:102–106

Gutierrez-Marcos J, Roberts M, Campell E, Wray J (1996) Three members of a novel small gene-family from *Arabidopsis thaliana* able to complement functionally an *Escherichia coli* mutant defective in PAPS-reductase activity encode proteins with a thioredoxin-like domain and "APS reductase" activity. Proc Natl Acad Sci USA 93:13377–13382

Hanfrey C, Franceschetti M, Mayer MJ, Illingworth C, Michael AJ (2002) Abrogation of upstream open reading frame-mediated translational control of a plant S-adenosylmethionine decarboxylase results in polyamine disruption and growth perturbations. J Biol Chem 277:44131–44139

Hatzfeld Y, Lee S, Lee M, Leustek T, Soito K (2000) Functional characterization of a gene encoding a fourth ATP sulfurylase isoform from *Arabidopsis thaliana*. Gene 248:51–58

Hawkesford M (2003) Transporter gene families in plants: the sulphate transporter gene family—redundancy or specialization? Physiol Plant 117:155–163

Hell R, Bork C, Bogdanova N, Frolov I, Hauschild R (1994) Isolation and characterization of two cDNAs encoding for compartment specific isoforms of *O*-acetylyserine (thiol) lyase from *Arabidopsis thaliana*. FEBS Lett 351:257–262

Hell R, Jost R, Berkowitz O, Wirtz M (2002) Molecular and biochemical analysis of the enzymes of cysteine biosynthesis in the plant *Arabidopsis thaliana*. Amino Acids 22:245–257

Higgins C (1992) ABC transporters: from microorganisms to man. Annu Rev Cell Biol 8:67–113

Hirai M, Fujiwara T, Awazuhara M, Kimura T, Noji M, Soito K (2003) Global expression profiling of sulfur-starved *Arabidopsis* by DNA macroarray reveals the role of *O*-acetyl-L-serine as a general regulator of gene expression in response to sulfur nutriton. Plant J 33:651–663

Howarth J, Roberts M, Wray J (1997) Cysteine biosynthesis in higher plants: a new member of the *Arabidopsis thaliana* serine acetyltransferase small gene-family obtained by functional complementation of an *Escherichia coli* cysteine auxotroph. Biochim Biophys Acta 1350:123–127

Hunt M (2003) Genes of sulfur assimilation in *Porphyra purpurea* (MSc thesis). University of New Hampshire, Durham, NH

Ikeda M, Moritani C (1994) Roles of anions in cells: studies on a Cl (−)-translocating ATPase and sulfate uptake system in *Acetabularia acetabulum*. Yakugaku Zasshi 114:539–549

Inoue K, Noji M, Saito K (1999) Determination of the sites required for the allosteric inhibition of serine acetyltranferase by L-cysteine in plants. Eur J Biochem 266:220–227

Jost R, Berkowitz O, Wirtz M, Hopkins L, Hawkesford M, Hell R (2000) Genomic and functional characterization of the *oas* gene family encoding *O*-acetylserine (thiol) lyases, enzymes catalyzing the final step in cysteine biosynthesis in *Arabidopsis thaliana*. Gene 253:237–247

Khurana O, Coupland L, Shelden M, Howitt S (2000) Homologous mutations in two diverse sulphate transporters have similar effects. FEBS Lett 477:118–122

Koguchi O, Tamura G (1988) Ferredoxin-sulfite reductase from a cyanobacterium, *Spirulina platensis*. Agric Biol Chem 52:373–380

Koguchi O, Tamura G (1989) Isolation and partial characterization of homogeneous ferredoxin-sulfite reductase from a red alga, *Porphyra yezoesnsis*. Agric Biol Chem 53:1653–1662

Kopriva S (2006) Regulation of sulfate assimilation in *Arabidopsis* and beyond. Ann Bot 97:479–495

Kopriva S, Koprivova A (2004) Plant adenosine 5′-phosphosulphate reductase: the past, the present, and the future. J Exp Bot 55:1775–1783

Kredich N (1996) Biosynthesis of cysteine. In: Neidhardt F, Curtiss R, Ingraham J, Lin E, Low K, Magasanik B, Reznikoff W, Riley M, Schaechter M, Umbarger H (eds) *Echerichia coli* and *Salmonella*: cellular and molecular biology. AMS, Washington, DC, pp 514–527

Krone F, Westphal G, Schwenn J (1991) Characterisation of the gene *cys*H and its product phospho-adenylylsulphate reductase from *Echerichia coli*. Mol Gen Genet 225:314–319

Kurima K, Singh B, Schwartz N (1999) Genomic organization of the mouse and human genes encoding the ATP sulfurylase/adenosine 5′-phosphosulfate kinase isoform SK2. J Biol Chem 274:33306–33312

Lappartient A, Touraine B (1996) Demand-driven control of root ATP sulfurylase activity and SO_4^{2-} uptake in intact canola: the role of phloem-translocated glutathione. Plant Physiol 111:147–157

Leustek T, Martin MN, Bick JA, Davies JP (2000) Pathways and regulation of sulfur metabolism revealed through molecular and genetic studies. Annu Rev Plant Physiol Plant Mol Biol 51:141–165

Leustek T, Murrillo M, Cervantes M (1994) Cloning of a cDNA encoding ATP sulfurylase from *Arabidopsis thaliana* by functional expression in *Saccharomyces cerevisiae*. Plant Physiol 105:897–902

Leyh T, Taylor J, Markham G (1988) The sulfate activation locus of *Escherichia coli* K12: cloning, genetic, and enzymatic characterization. J Biol Chem 263:2409–2416

Leyh T, Vogt T, Suo Y (1992) The DNA sequence of the sulfate activation locus from *Escherichia coli* K12. J Biol Chem 267:10405–10410

Lillig C, Prior A, Schwenn J, Åslund F, Ritz D, Vlamis-Gardikas A, Holmgren A (1999) New thioredoxins and glutaredoxins as electron donars of 3′-phosphoadenylylsulfate reductase. J Biol Chem 274:7695–7698

Lillig C, Schiffmann S, Berndt C, Berken A, Tischka R, Schwenn J (2001) Molecular and catalytic properties of *Arabidopsis thaliana* adenylyl sulfate (APS)-kinase. Arch Biochem Biophys 392:303–310

MacRae I, Rose A, Segel I (1998) Adenosine 5′-phosphosulfate kinase from *Penicillium chrysogenum*: site-directed mutagenesis at putative phosphoryl-accepting and ATP P-loop residues. J Biol Chem 273:28583–28589

MacRae I, Segel I, Fisher A (2000) Crystal structure of adenosine 5′-phosphosulfate kinase from *Penicillium chrysogenum*. Biochemistry 39:1613–1621

McCandless E (1981) Polysaccharides of the seaweeds. In: Lobban C, Wynne M (eds) The biology of seaweeds. University of California Press, Los Angeles, pp 559–588

McRee D, Richardson D, Richardson J, Siegel L (1986) The heme and Fe_4S_4 cluster in the crystallographic structure of *Escherichia coli* sulfite reductase. J Biol Chem 261:10277–10281

Mino K, Hiraoka K, Imamura K, Sakiyama T, Eisaki N, Matsuyama A (2000) Characteristics of serine acetyltransferase from *Escherichia coli* deleting different lengths of amino acid residues from the C-terminus. Biosci Biotechnol Biochem 64:1874–1880

Mitman G, van der Meer J (1994) Meiosis, blade development, and sex determination in *Porphyra purpurea* (rhodophyta). J Phycol 30:147–159

Murillo M, Leustek T (1995) Adenosineo-5′-triphosphate-sulfurylase from *Arabidopsis thaliana* and *Echerichia coli* are functionally equivalent but structurally and kinetically divergent: nucleotide sequence of two adenosine-5′-triphosphate-sulfurylase cDNAs from *Arabidopsis thaliana* and analysis of a recombinant enzyme. Arch Biochem Biophys 323:195–204

Nakamura T, Yamaguchi Y, Sano H (1999) Four rice genes encoding cysteine synthase: isolation and differential responses to sulfur, nitrogen and light. Gene 229:155–161

Nakayama M, Akashi T, Hase T (2000) Plant sulfite reductase: molecular sturcture, catalytic function and interaction with ferredoxin. J Inorg Biochem 82:27–32

Nelson W, Brodie J, Guiry M (1999) Terminology used to describe reproduction and life history stages in the genus *Porphyra* (bangiales, rhodophyta). J Appl Phycol 11:407–410

Nikaido I, Asamizu E, Nakajima M, Nakamura Y, Saga N, Tabata S (2000) Generation of 10,154 Expressed Sequence Tags from a leafy gametophyte of a marine red alga, *Porphyra yezoensis*. DNA Res 7:223–227

Pesole G, Mignone F, Gissi C, Grillo G, Licciulli F, Liuni S (2001) Structural and functional features of eukaryotic mRNA untranslated regions. Gene 276:73–81

Porebski S, Bailey LG, Baum BR (1997) Modification of a CTAB DNA extraction protocol for plants containing high polysaccharide and polyphenol components. Plant Mol Biol Rep 15:8–15

Rao G, Mottonen J, Goldsmith E, Cook P (1993) Crystallization and preliminary X-ray data for the A-isozyme of *O*-acetylserine sulfhydrylase from *Salmonella typhimurium*. J Mol Biol 231:1130–1132

Ravina C, Barrosa C, Vega J, Gotor C (1999) Cysteine biosynthesis in *Chlamydomonas reinhardtii*. Eur J Biochem 264:848–853

Ravina C, Chang C-I, Tsakraklides GP, McDermott JP, Vega JM, Leustek T, Gotor C, Davies JP (2002) The sac mutants of *Chlamydomonas reinhardtii* reveal transcriptional and posttranscriptional control of cysteine biosynthesis. Plant Physiol 130:2076–2084

Roberts M, Wray J (1996) Cloning and characterization of an *Arabidopsis thaliana* cDNA clone encoding an organellar isoform of serine acetyltransferase. Plant Mol Biol 30:1041–1049

Rocha de Souza MC, Marques CT, Dore CMG, Ferreira da Silva FR, Rocha FAO, Leite EL (2007) Antioxidant activities of sulfated polysaccharides from brown and red seaweeds. J Appl Phycol 19:153–160

Rotte C, Leustek T (2000) Differential subcellular localization and expression of ATP sulfurylase and 5′-adenylylsulfate reductase during ontogenesis of arabidopsis leaves indicates that cytosolic and plastid forms of ATP sulfurylase may have specialized functions. Plant Physiol 124:715–724

Ruffet M, Lebrun M, Droux M, Douce R (1995) Subcellular distribution of serine acetyltransferase from *Pisum sativum* and characterization of an *Arabidopsis thaliana* putative cytosolic isoform. Eur J Biochem 227:500–509

Saito K (2000) Regulation of sulfate transport and synthesis of sulfur-containing amino acids. Curr Opin Plant Biol 3:188–195

Saito K (2004) Sulfur assimilatory metabolism. The long and smelling road. Plant Physiol 136:2443–2450

Saito K, Inoue K, Fukushima R, Noji M (1997) Genomic structure and expression analysis of serine acetyltransferase gene in *Citrullus vulgaris* (watermelon). Gene 189:57–63

Satishchandran C, Markham G (1989) Adenosine-5′-phosphosulfate kinase from *Escherichia coli* K12: purification, characterization, and identification of a phosphorylated enzyme intermediate. J Biol Chem 264:15012–15021

Satishchandran C, Hickman Y, Markham G (1992) Characterization of the phosphorylated enzyme intermediate formed in the adenosine 5′-phosphosulfate kinase reaction. Biochemistry 31:11684–11688

Sato Y, Kanno N, Sato M (1987) Activation of sulphate in red alga *Porphyra yezoensis*. Kitasato Arch Exp Med 60:25–32

Schiffmann S, Schwenn J (1994) APS-sulfotransferase activity is identical to higher plant APS-kinase (EC 2.7.1.25). FEBS Lett 355:229–232

Smith F, Ealing P, Hawkesford M, Clarkson D (1995) Plant members of a family of sulfate transporters reveal functional subtypes. Proc Natl Acad Sci USA 92:9373–9377

Smith F, Hawkesford M, Ealing P, Clarkson D, Vanden-Berg P, Belcher A, Warrilow A (1997) Regulation of expression of a cDNA from barley roots encoding a high affinity sulphate transporter. Plant J 12:875–884

Smith F, Rae A, Hawkesford M (2000) Molecular mechanisms of phosphate and sulfphate transport in plants. Biochim Biophys Acta 1465:236–245

Takagi H, Awano N, Kobayashi S, Noji M, Asoito K, Nakamori S (1999) Overproduction of L-cysteine and L-cystine by expression of genes for feedback inhibition-insensitive serine acetyltransferase from *Arabidopsis thaliana* and *Escherichia coli*. FEBS Lett 179:453–459

Takahashi H, Saito K (1996) Subcellular localization of spinach cysteine synthase isoforms and regulation of their gene expression by nitrogen and sulfur. Plant Physiol 112:273–280

Takahashi H, Asanuma W, Saito K (1999) Cloning of an *Arabidopsis* cDNA encoding a chloroplast localizing sulphate transporter isoform. J Exp Bot 50:1713–1714

Takahashi S, Sakata T, Tamura G (1996a) Ferredoxin-linked sulfite reductase from turnip roots. Biosci Biotechnol Biochem 60:142–144

Takahashi H, Sasakura N, Masaaki N, Saito K (1996b) Isolation and characterization of a cDNA encoding a sulfate transporter from *Arabidopsis thaliana*. FEBS Lett 392:95–99

Toda K, Takano H, Miyagishima S, Kuroiwa H, Kuroiwa T (1998) Characterization of a chloroplast isoform of serine acetyltransferase from the thermo-acidiphilic red alga *Cyanidioschyzon merolae*. Biochim Biophys Acta 1403:72–84

Toda K, Takano H, Nazaki H, Kuroiwa T (2001) The second acetyltransferase, bacterial-type *O*-acetylserine (thiol) lyase and eukaryotic-type *O*-acetylserine (thiol) lyase from the primative red alga *Cyanidioschyzon merolae*. J Plant Res 114:291–300

Ullrich T, Blaesse M, Huber R (2001) Crystal structure of ATP sulfurylase from *Saccharomyces cerevisiae*, a key enzyme in sulfate assimilation. EMBO J 20:316–329

Urano Y, Manabe T, Joji M, Saito K (2000) Molecular cloning and functional characterization of cDNAs encoding cysteine synthase and serine acetyltransferase that may be responsible for high cellular cysteine content in *Allium tuberosum*. Gene 257:269–277

Varin L, Marsolais F, Richard M, Rouleau M (1997) Biochemistry and molecular biology of plant sulfotranferases. FASEB J 11:517–525

Vidmar J, Tagmount A, Cathala N, Touraine B, Davidian J (2000) Cloning and characterization of a root specific high-affinity sulfate transporter from *Arabidopsis thaliana*. FEBS Lett 475:65–69

Yildiz F, Davies J, Grossman A (1994) Characterization of sulfate transport in *Chlamydomonas reinhardtii* during sulfur-limited and sulfur-sufficient growth. Plant Physiol 104:981–987

Yildiz F, Davies J, Grossman A (1996) Sulfur availability and the *SAC1* gene control adenosine triphosphate sulfurylase gene expression in *Chlamydomonas reinhardtii*. Plant Physiol 112:669–675

Yonekura-Sakakibara K, Ashikari T, Tanaka Y, Kusumi T, Hase T (1998) Molecular characterization of tobacco sulfite reductase: enzyme purification, gene cloning, and expression analysis. J Biochem (Tokyo) 124:615–621

Floating period of Sargassacean thalli estimated by the change in density

Kousuke Yatsuya

Originally published in the Journal of Applied Phycology, Vol 20, No 5, 347–350.
DOI: 10.1007/s10811-007-9293-1

Abstract The density (fresh weight/volume) of the detached thalli of four sargassacean species (*Myagropsis myagroides*, *Sargassum horneri*, *S. patens*, *S. siliquastrum*) was measured weekly to determine their floating period. Thalli of the four species were collected with respect to the difference in their reproductive stage, and their floating abilities at different stages were determined. Floating thallus was ballasted with pre-weighed leads until the final addition of a weight caused sinking. The density and floating period of sargassacean thalli were dependent on the species and the specimens' reproductive stage. The floating period of thalli detached before maturation was estimated to be between 4 and 14 weeks. However, thalli detached during maturation had a floating period of between 1 and 8 weeks, and those detached after maturation sank within 2 weeks. *Sargassum horneri* had the lowest density and the longest floating period among the four species investigated and this may explain the high abundance of *S. horneri* in floating seaweeds observed in the Sea of Japan.

Keywords Density of seaweed · Floating period · Floating seaweed · Sargassaceae

K. Yatsuya
Kyoto Institute of Oceanic and Fishery Science,
Miyazu, Kyoto 626-0052, Japan

Present address:
K. Yatsuya (✉)
Seikai National Fisheries Research Institute,
1551-8 Taira-machi,
Nagasaki 851-2213, Japan
e-mail: yatsuya@affrc.go.jp

Introduction

Forests of sargassacean seaweeds are among the most productive communities along temperate coasts (Taniguchi and Yamada 1978; Mann 2000; Yatsuya et al. 2005a). Most of the thalli produced are detached from the forests as floating seaweed (Yatsuya 2005), which is known to be an important food and habitat for fish larva and invertebrates (e.g., Senta 1965; Kingsford 1995).

Floating seaweeds usually originate from coastal seaweed beds and eventually sink to the bottom or are stranded on shore (Yoshida 1963); however, pelagic *Sargassum* on the surface of the central North Atlantic Ocean is adapted to survive over long periods of time (Parr 1939). In Japan, seasonal changes in the distribution and species composition of floating seaweeds have previously been investigated (Yoshida 1963; Ohno 1984; Yatsuya et al. 2005b). However, their floating period has rarely been examined, even though it should affect the ranges of dispersal of these seaweeds.

There have been rough estimates of the floating period of sargassacean thalli. Yoshida (1963) inferred from the recovery rate of labeled floating seaweeds that they kept afloat for 2 months. Ohno (1984) investigated the abundance of floating seaweeds and the deterioration of seaweed beds, suggesting a floating period of approximately 1 month. The floating period of seaweed is affected by its density; however, the change in the density during the floating period has never been investigated. In the present study, the density of four dominant sargassacean species was measured weekly to determine their floating ability (i.e., density and floating period) with respect to the differences in species and reproductive stage.

Materials and methods

Myagropsis myagroides (Mertens ex Turner) Fensholt, *Sargassum horneri* (Turner) C. Agardh, *Sargassum patens* C. Agardh, *Sargassum siliquastrum* (Turner) C. Agardh, used in this experiment, are dominant species in seaweed beds and floating seaweeds off the central part of Honshu Island facing the Sea of Japan (Ikehara and Sano 1986; Yatsuya 2005).

Attached thalli of the four sargassacean species were collected by a scuba diver from January to June 2004, according to their reproductive stages (Table 1), e.g., immature (sterile specimens without receptacles), during maturation (from receptacle formation to egg liberation), and post-mature (no egg and embryo on the receptacles). No collection was carried out in the other seasons, when floating seaweeds in this region are rarely observed (Yoshida 1963; Yatsuya et al. 2005a). *Myagropsis myagroides*, *Sargassum patens*, and *S. siliquastrum* were collected at a depth of 2 m on Yoro coast and *Sargassum horneri* was collected at a depth of 2 m on Odasyukuno coast. At each sampling, three individuals of similar size were collected for each species. The thalli were cut at the base of the main branches, since most of the thalli usually detach at that portion in the natural detaching process (Yatsuya 2005). The total length of the thalli collected for the experiments ranged from 1.7 to 3.7 m.

The density of thalli was determined as follows. The mesh bag with the thalli inside was placed in seawater. Then the bag was ballasted with lead weights until a final addition of the weight caused sinking. Since buoyancy of the mash bag was preliminarily corrected, the combined weight of the immersed thalli plus the ballasts was equal to the weight of displaced seawater. Therefore, the density of ambient seawater (d_t) was derived in the following manner:

$$dt = \frac{Sw + bw}{Sv + bv} \qquad (1)$$

where S_w=thallus weight (g), b_w=ballast weight (g), S_v= thallus volume (cm^3), and b_v=ballast volume (cm^3). The density of ambient seawater was measured by ACL215-DK (Alec Electronics, Kobe, Japan). The weight and volume of the ballast were known before measurement, therefore, the volume (S_v) was determined by the equation (1) and the density of the thallus (g cm^{-3}) was determined as (S_w/S_v).

After the measurement, each thallus was stored in a mesh plastic container (52 cm in length×77 cm in width×41 cm in depth, mesh size approximately 1 cm), floated and moored at a pier. The density and weight of stored thalli were measured every week until they sank, and these measurements were expressed by the mean of three individuals. Water temperature in the container was measured at 60 minute intervals by RMT thermometer (Rigo, Tokyo, Japan).

Table 1 Density and maturation stages of attached sargassacean thalli used in the experiments (n=3)

Species	Sampling date				
	January 28	March 10	April 20	May 19	June 18
Myagropsis myagroides	0.929^a	0.941^b	0.952^c	N.D.	N.D.
Sargassum siliquastrum	0.929^a	N.D.	0.973^c	0.987^c	N.D.
S. patens	0.906^a	N.D.	0.993^b	0.916^c	0.992^c
S. horneri	0.761^a	0.874^b	N.D.	0.987^c	N.D.

See the text for the definition of the maturation stages
N.D.: no data
^{a} Immature, ^{b} mature, ^{c} post-mature

Results

The temperature of surface water was lowest in February and early March, varying around 11°C, and had increased gradually to 23–25°C in late June. Although the density of water deeper than 2 m was almost constant at 1.025, the density of surface water occasionally decreased to 1.016.

The density of attached sargassacean thalli was different according to species and their reproductive stage (Table 1). On 28 January, the density of immature *Sargassum horneri* was 0.761, the lowest density measured in this study. The density of attached thalli increased as they matured, ranging from 0.761 to 0.929 for immature thalli, 0.874 to 0.941 for mature thalli, and 0.916 to 0.992 for post-mature thalli (Table 1).

Figure 1 shows the changes in the density and fresh weight of detached thalli. The density of all sargassacean species continuously increased through the time. Increase in the density of thalli detached before maturation was slower than those detached during or after maturation. In all cases, except for Experiment 5, which was started on 18 June, the fresh weight of thalli gradually increased from the start of the experiments, and it decreased for the latter part of the floating period (Fig. 1). In Experiment 5, the weight of thalli rapidly decreased from the start of the experiment, due to deterioration of the thalli in the water of higher temperature.

Discussion

Floating ability is a well-known characteristic of sargassacean species, which aids the dispersal of those species (Paula and Eston 1987). This study reports quantitative information on this characteristic with respect to the

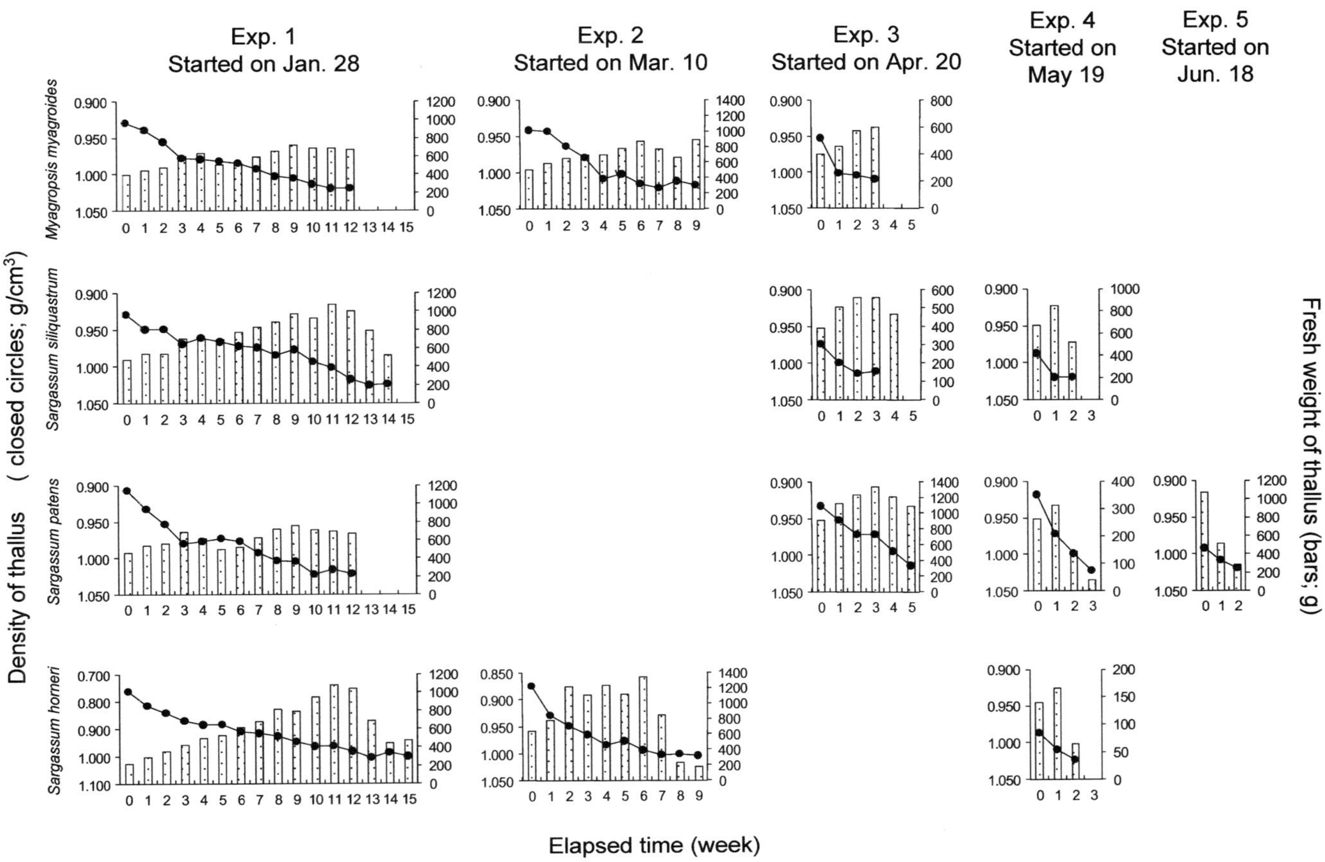

Fig. 1 Changes in density (*closed circles*) and fresh weight (*bars*) of sargassacean thalli, measured weekly in the experiments (n=3)

differences in species and reproductive stage of floating seaweeds.

At first, the floating period of sargassacean thalli was estimated as follows. Woodcock (1950) showed that floating seaweeds were concentrated at convergence lines in which water was descending during higher wind speed. He inferred that a descending current of 3 cm/s would carry downward any plants with a density higher than 1.0. Thalli

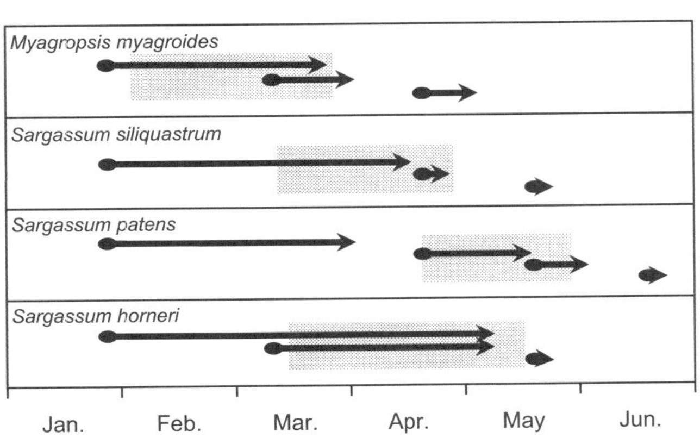

Fig. 2 Estimated floating periods (*arrows*) of four sargassacean species that were detached at different maturation stages. Their maturation period is expressed by the shading

with a density higher than 1.005 in the experiments of the present study were so fragmented and damaged that most of the leaves and vesicles on the thalli were lost. Such fragmented and deteriorated thalli have never been observed in floating seaweed in this region (K. Yatsuya, unpublished data 2004). Therefore, the density of thalli when they start sinking was tentatively supposed to be 1.005 in this study.

Estimated floating periods ranged from less than 1 week to 14 weeks, varying among species and reproductive stages (Fig. 2, Table 2). Thalli of *S. horneri*, detached on 10 March, had an 8-week floating period, while those of *M. myagroides* floated for 3 weeks (Table 2). For thalli

Table 2 Estimated floating period (weeks) of sargassacean thalli

Species	Date of detachment				
	28 January	10 March	20 April	19 May	18 June
Myagropsis myagroides	8	3	2	N.D.	N.D.
Sargassum siliquastrum	11	N.D.	1	<1	N.D.
S. patens	9	N.D.	4	2	<1
S. horneri	14	8	N.D.	<1	N.D.

N.D.: no data, <1: less than 1 week

detached on 20 April, the estimated floating period of *S. patens* was longer than that of the other two species (Table 2). The longest floating period (14 weeks) was estimated for *S. horneri*, which detached on 28 January. The estimated floating periods decreased as the maturation stage progressed. The thalli detached before maturation were estimated to float for 8–14 weeks, while those detached during and after maturation floated for 3–8 weeks and 2 weeks or less respectively (Fig. 2, Table 2). Immature thalli, except for *S. patens,* detached on 28 January, formed receptacles during their floating period.

The weight and density of floating thalli varied with the time. A gain in the weight may be due to an increase in water content, because few epizoa and epiphytes were attached, and little growth was observed. A decrease in the weight was attributed to fragmentation and deterioration of the thalli. An increase in the density may be the result of degradation and detachment of vesicles and of an increase in the water content of thalli.

Our results show that the floating period of sargassacean thalli is species-specific. *Sargassum horneri* has the lowest density and the longest floating period among the four sargassacean species investigated. The high floating ability of *S. horneri* may support the frequent observations and high abundance in floating seaweeds in the Sea of Japan (Yoshida 1963; Ikehara and Sano 1986; Yatsuya et al. 2005b). *Sargassum horneri* is an annual seaweed, while the other three species are perennials. Annual seaweeds have to settle on a new substratum every year; thus *S. horneri* has the advantage of having propagules with a high dispersal ability.

The floating ability (i.e., thallus density and floating period) of sargassacean thalli seems to decrease as the reproductive stage progresses. The relative abundance of vesicles and receptacles may be related to the change in the floating ability. Yatsuya et al. (2005a) showed that the weight ratio of vesicles to whole thalli was constant before and during maturation, while those of receptacles increased for the same period. This result suggests that buoyancy provided by vesicles was constant, while the formation of receptacles would lead to an increase in the density of the thalli. After maturation, the deterioration of the thalli, especially vesicles, should also increase density.

According to the estimated floating period of the thalli detached at various reproductive stages, quantitative aspects of floating seaweeds could be inferred. In *Sargassum* beds in the Sea of Japan, most thalli detach after maturation, while few thalli detach before maturation (Taniguchi and Yamada 1978; Murase et al. 2000; Yatsuya et al. 2005a). This implies that most of the detached sargassacean thalli in nature will float for 2 weeks or less, and that few of them have more than 8 weeks longevity in floating conditions. On the basis of the fact that floating seaweeds in the Sea of Japan are transported a maximum of 17 km/day (Yoshida 1963), it may be possible for floating seaweeds to be transported such long distances as from Kyusyu Island to Tohoku district (approximately 1,000 km); however, this may be on rare occasions since those thalli must be detached before maturation and be floating for more than 8 weeks.

Acknowledgements I thank Dr. H. Nakahara and Dr. T. Komatsu for the critical comments on the manuscript, and T. Nishigaki, A. Douke, M. Itani, and Y. Wada who helped with the experiment. This is contribution no. 157 from the Kyoto Institute of Oceanic and Fishery Science.

References

Ikehara K, Sano O (1986) Distribution and species composition of floating seaweeds collected in the Sado Straits of the Japan Sea. Bull Jap Sea Reg Fish Res Lab 36:59–75

Kingsford MJ (1995) Drift algae: a contribution to near-shore habitat complexity in the pelagic environment and an attractant for fish. Mar Ecol Prog Ser 116:297–301

Mann KH (2000) Ecology of coastal waters, 2nd edn. Blackwell Science, Massachusetts

Murase N, Kito H, Mizukami Y, Maegawa M (2000) Productivity of a *Sargassum macrocarpum* (Fucales, Phaeophyta) population in Fukawa Bay, Sea of Japan. Fish Sci 66:270–277

Ohno M (1984) Algological observation on the floating seaweeds of offshore water of Shikoku Island in Japan. Bull Jap Soc Sci Fish 50:1653–1656

Parr AE (1939) Quantitative observations on the pelagic *Sargassum* vegetation of the western north Atlantic. Bull Bingham Oceanogr Coll 6(7):1–90

Paula EJ, Eston VR (1987) Are there other *Sargassum* species potentially as invasive as *S. muticum*? Bot Mar 30:405–410

Senta T (1965) Importance of drifting seaweeds in the ecology of fishes. Japan Fisheries Resource Conservation Association, Tokyo

Taniguchi K, Yamada Y (1978) Ecological study on *Sargassum patens* C. Agardh, and *S. serratifolium* C. Agardh in the sublittoral zone at Iida Bay of Noto Peninsula in the Japan Sea. Bull Jap Sea Reg Fish Res Lab 29:239–253

Woodcock AH (1950) Subsurface pelagic *Sargassum*. J Mar Res 9 (2):77–92

Yatsuya K (2005) Studies on production and outflow process of *Sargassum* forest. Spec Rep Kyoto Inst Ocean Fish Sci 7:1–41

Yatsuya K, Nishigaki T, Douke A, Wada Y (2005a) Annual net production of the five Sargassaceae species in Yoro, western Wakasa Bay, Sea of Japan. Fish Sci 71:1098–1106

Yatsuya K, Nishigaki T, Douke A, Wada Y (2005b) Species composition of the drifting seaweed at the western part of Wakasa Bay. Bull Kyoto Inst Ocean Fish Sci 27:13–18

Yoshida T (1963) Studies on the distribution and drift of the floating seaweeds. Bull Jap Sea Reg Fish Res Lab 23:141–186

Abundance of drifting seaweeds in eastern East China Sea

Teruhisa Komatsu · Daisuke Matsunaga · Atsuko Mikami · Tatsuyuki Sagawa · Etienne Boisnier · Kenichi Tatsukawa · Masakazu Aoki · Tetsuro Ajisaka · Shinya Uwai · Katsuhiko Tanaka · Kenichi Ishida · Hideaki Tanoue · Takashige Sugimoto

Originally published in the Journal of Applied Phycology, Vol 20, No 5, 351–359.
DOI: 10.1007/s10811-007-9302-4

Abstract Drifting seaweeds play important ecological roles in offshore waters. Recently, large amounts of drifting seaweed rafts were found in the eastern East China Sea between the continental shelf and the oceanic front of the Kuroshio Current. However, so far there have been no quantitative reports about this particular area. Two research cruises were organized to survey abundance and standing crop of drifting seaweeds in eastern East China Sea in May 2002 and March 2004, using visual census and net sampling of drifting seaweeds. Visual census data were composed of drifting seaweed raft diameter, perpendicular distance from the transect (navigation course of the research vessel) to the raft, and positions. Using these data, we calculated the "effective stripe width" using the DISTANCE software. Drifting seaweed abundance (composed exclusively of *Sargassum horneri*) in waters located between the continental shelf peripheral area and the Kuroshio oceanic front was estimated to be higher than in any other area within eastern East China Sea in March and May. Abundance means in May 2002 and March 2004 were 6.14 and 29.05 rafts km^{-2}, respectively, while standing crop reached 126.81 and 20.35 kg km^{-2} (wet weight). Mean diameter and drifting seaweed rafts in May 2002 were significantly greater than in March 2004, reflecting seasonal growth of *Sargassum horneri*.

Keywords Drifting seaweed · *Sargassum horneri* · East China Sea · Biomass · Abundance · Kuroshio

Abbreviations

AIC Akaike's Information Criterion
ESW Effective Strip half-Width

Introduction

On rocky coasts along the northwestern Pacific, *Sargassum* species form a luxuriant forest in spring and a scanty one in summer (Komatsu et al. 1982). In spring, the *Sargassum* forest has a great influence on marine environments such as

T. Komatsu (✉) · D. Matsunaga · A. Mikami · T. Sagawa · E. Boisnier · K. Tatsukawa · K. Ishida · H. Tanoue
Ocean Research Institute, The University of Tokyo,
1-15-1, Minamidai,
Nakanoku, Tokyo 164-8639, Japan
e-mail: komatsu@ori.u-tokyo.ac.jp

M. Aoki
Shimoda Marine Research Center, University of Tsukuba,
5-10-1, Shimoda,
Shizuoka 415-0025, Japan

T. Ajisaka
Division of Applied Biosciences, Graduate School of Agriculture,
Kyoto University,
Kitashirakawa Oiwakecho, Sakyoku,
Kyoto 606-8502, Japan

S. Uwai
Department of Natural Science, Faculty of Science,
Niigata University,
8050 Ikarashi 2-no-cho,
Niigata City 950-2181, Japan

K. Tanaka
Environmental Science Research Laboratory, Central Research
Institute of Electric Power Industry,
1646, Abiko, Abikoshi,
Chiba 270-1194, Japan

T. Sugimoto
Institute of Oceanic Research and Development, Tokai University,
3-20-1, Orido, Shimizu-ku,
Shizuoka 424-8610, Japan

water temperature (Komatsu et al. 1982, 1990, 1994; Komatsu 1985), pH (Komatsu and Kawai 1986), dissolved oxygen concentration (Komatsu 1989), downward illumination (Komatsu 1990) and water flow (Komatsu and Murakami 1994). Commercially important fish spawn in the *Sargassum* forest in spring, while abalone and turban shells are traditionally associated with this particular habitat. Therefore, *Sargassum* forests play very important ecological roles in nearshore coastal waters. Their stems or lateral branches generally attain several meters in length in spring when most of these species mature. From winter to spring, when they become longer, waves and currents detach the seaweeds from the benthic substratum (Yoshida 1963). *Sargassum* species have many vesicles filled with gas to provide buoyancy and, hence, *Sargassum* species can float after being detached. Once detached, they become drifting seaweeds and are common in Japanese waters (Yoshida 1963), as in the Sargasso Sea. In this particular area, drifting seaweeds spend their entire floating-life stage in a vegetative reproductive state (Parr 1939).

Drifting seaweeds play key ecological roles in offshore waters as well, as they play host to attaching or accompanying flora and fauna. Hence, drifting seaweeds constitute moving ecosystems (e.g., Cho et al. 2001). Thus, they serve as means of dispersal for littoral animals such as intertidal animals (e.g., Ingólfsson 1995). Moreover, seaweeds themselves also enlarge their habitats by means of drifting rafts. For example, Hernández-Carmona et al. (2006) reported that sporophytes of driftng *Macrocystis pyrifera* (Linnaeus) C. Agardh remained fertile with high germination success as long as sori were present (125 days). Since drifting fragments of *Sargassum muticum* (Yendo) Fensholt which invaded European waters in the 1980s could continue growth and become fertile, they have enlarged their geographical distribution (Rueness 1989).

With regard to fisheries, they are spawning habitats for flying fish (e.g., *Hirundichthys oxycephalus* Bleeker) (Ichimaru et al. 2006), Japanese halfbeak (*Hyporhamphus sajori* Temminck & Schlegel) and Pacific saury (*Cololabis saira* Brevoort) (Ikehara, 1986). They also serve as nursery habitats for larvae and juveniles of some commercially important pelagic fish species, such as yellowtail (*Seriola quinqueradiata* Temminck & Schlegel,) or jack mackerel (*Trachurus japonicus* Temminck & Schlegel) (Senta 1965), which spawn in the East China Sea (Fig. 1). Consequently, investigating the distribution and abundance of drifting seaweeds in this specific area is of primary interest for fishery purposes.

Yoshida (1963) reported that most of Japanese drifting seaweeds are found in nearshore coastal waters (within 20 km from the shore) in the vicinity of oceanic fronts west or north of Kyushu Island in the East China Sea (Fig. 1). Yoshida (1963) also underlined that the abundant season of

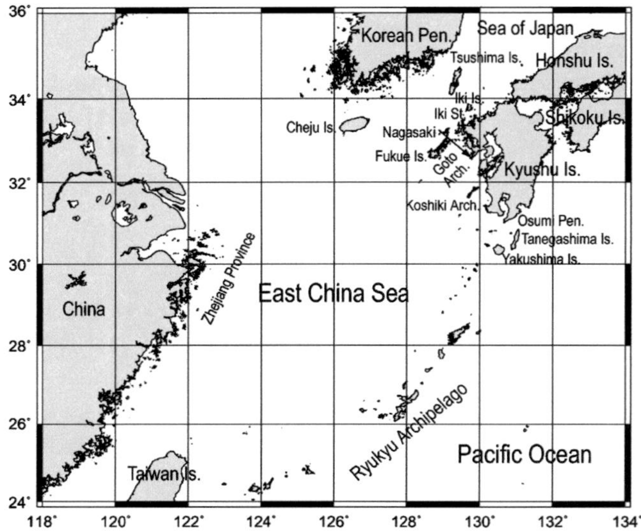

Fig. 1 Map showing the geographical positions of the East China Sea

drifting seaweeds stretches from March to May and that abundance becomes generally significantly lower in waters located south of the Osumi Peninsula, especially south of Tanegashima Island and Yakushima Island. Senta (1965) reported that drifting seaweeds were not present in the oceanic front, between continental shelf waters and the Kuroshio Current. Mitani (1965a) investigated distribution of drifting seaweeds off the west coast of Kyushu Island, based on results of Segawa et al. (1959a, b, 1960, 1961b, 1964) and Yoshida (1963). He recorded drifting seaweeds mainly in the Iki Strait and off Fukue Island, as predicted by Segawa et al. (1961a). As a result, it has been reported since the 1960s that drifting seaweeds are hardly present in offshore East China Sea waters. However, data about geographical distributions of saury eggs and vesicles of *Sargassum* species reported by Shojima (1981) studying the distribution of Pacific saury eggs, which are usually associated with drifting seaweeds, questioned this.

Recently, Konishi (2000) and Komatsu et al. (2007) reported that drifting seaweeds are found on the continental shelf along the East China Sea oceanic front (between Kuroshio and continental shelf waters) between March and May. These drifting seaweeds were composed of only one species, *Sargassum horneri* (Turner) C. Agardh. Even though it is now admitted that floating seaweeds occur in this specific area, little is known about their abundance. The Kagoshima Prefectoral Fisheries Technology and Development Center investigated this question, mainly by visual survey, in order to fix an open season on yellowtail juveniles (Kubo 2004, 2006). However, efforts were concentrated on coastal waters and the abundance of drifting seaweeds in deeper waters remained unstudied.

Quantitative studies dealing with drifting seaweeds are generally based on either neuston net sampling or visual survey. The former method was, for instance, applied to drifting seaweeds of *Sargassum fluitans* (Børgesen) Børgesen and *S. natans* (Linnaeus) Gaillon in the northwestern Atlantic (Stoner 1983). This method is quantitative because the net mouth frame is constant. It enables the sampling of small floating objects, but surveys have to be limited to very narrow areas. Conversely, visual surveys have been used to estimate the biomass of the giant kelp *Macrocystis pyrifera* in the northeast Pacific (e.g., Kingsford 1995; Hobday 2000) and *Carpophyllum* species in waters around New Zealand (Kingsford 1992). Thiel and Gutow 2005 suggest selecting a sampling method with reference to both the size of the target and the area under scrutiny. As a result, visual survey was considered a suitable method in our case.

This study aims at acquiring details about drifting seaweeds abundance in March and May along the continental shelf and the Kuroshio Current in eastern East China Sea using the visual survey method.

Materials and methods

Visual censuses were conducted at sea, in good conditions (i.e. wave height less than 1 m) and from sunrise to sunset. Two sessions were organized: from 11 to 19 May 2002 (R/V Hakuho Maru, KH02-1 leg 1) and from 9 to 16 March 2004 (R/V Hakuho Maru, KH04-1). Observations were taken from the vessel deck at a height of 11.5 m above the sea surface. The vessel navigated along designed transects. The following elements were recorded at each floating seaweed occurrence: time, position, the perpendicular distance from the boat, seaweed diameter. Distance and diameter were estimated by four trained researchers (two teams of two people), operating alternately. Some rafts were sampled randomly every day using an ORI ring net with a diameter of 1.6 m. Samples were identified and weighted (wet weight).

Drifting seaweed abundance was estimated with reference to the Line Transect Method (Buckland et al. 2001) and calculated with the software DISTANCE 4.1. Release 2 (Thomas et al. 2002, 2004). Following Thomas et al. (2002), we estimated the parameter $\widehat{\mu}$, called the effective strip half-width (ESW), and defined as the maximum limit for distinction.

$$\widehat{D} = \frac{n}{2\widehat{\mu}L}, \qquad (1)$$

where $\widehat{D}$, n and L are respectively the abundance of drifting seaweed rafts, the total number of rafts of drifting seaweeds along a transect and the length of the transect. $\widehat{\mu}$ was estimated via three different models fitted to the frequency distribution of drifting seaweeds per distance classes. In order to take into account the lack of homogeneity in sea and meteorological conditions between transects, the model was fitted to data separately for each transect. Akaike's Information Criterion (AIC) provides an objective, quantitative method for model selection. The model which displayed the lowest AIC [defined by eq. (2)] was adopted as the reference model (Buckland et al. 2001).

$$AIC = -2 \cdot \log_e(\Lambda) + 2(q) \qquad (2)$$

where Λ is the maximized likelihood and q the number of estimated parameters. This number of estimated parameters was the number of parameters based on combination key-function and adjustment-term provided by the program DISTANCE4.1. The maximized likelihood was obtained from maximum likelihood estimation (Buckland et al. 2001).

Rafts of drifting seaweeds were classified into classes (0–10 m, 10–20 m, 20–30 m and 30–40 m), this distance representing the perpendicular distance between the boat and the raft. We ignored those situated further than 40 m as it became impossible to estimate the distance and difficult to observe the seaweed. Calculation results for each model are indicated in Tables 5 and 6. The number of rafts within EWS was obtained.

The ratio of wet weight/diameter (g m^{-1}) of drifting seaweed rafts in May 2002 was estimated by dividing the mean wet weight of rafts by the mean diameter of rafts sampled in May 2002. Then the mean wet weight of rafts along each transect was calculated by multiplying the mean diameter of rafts along this transect by the ratio. The upper and lower values of wet weight of drifting seaweeds were calculated using the standard deviation of the diameter of the rafts. The population mean for the wet weight along each transect was estimated by using the relation between diameter and wet weight of drifting seaweed rafts collected by the ORI ring net during the observation in March 2004. Mean diameter of drifting seaweed rafts along each transect was converted to wet weight using the equation representing this relation. The upper and lower values of wet weight of drifting seaweeds were calculated using the standard deviation of the diameter of the rafts.

Standing crop of drifting seaweeds along a transect was obtained by multiplying the density of rafts obtained by the program DISTANCE4.1 by the weighted mean of the rafts collected along this particular transect. Then, the total biomass along each transect was calculated by multiplying the standing crop by the survey area obtained from $\widehat{\mu}$ and L.

Results

Distribution of drifting seaweeds

May 2002

In May 2002, lots of rafts were distributed in the area located between the continental shelf waters and the

Table 1 Results for distribution of drifting seaweed rafts observed along the transects 2002. Results for distributions of drifting seaweed rafts observed along the transects from 11 to 19 May 2002

Date	Length of transect (km)	Number of rafts	Mean diam. (m)	SD of diam.	Mean perpen. dist. (m)	SD of perpen. dist.
11 May	179.64	14	2.00	0.55	11.43	9.49
12 May	212.31	40	1.03	0.16	13.93	9.87
14 May	216.83	9	1.56	0.53	4.44	5.27
15 May	151.14	2	1.00	0.00	10.00	0.00
16 May	173.61	0	–	–	–	–
18 May	202.82	12	1.33	0.49	10.83	11.645
19 May	201.32	35	2.71	1.20	11.61	10.08

diam Diameter of raft, *perpen dist* perpendicular distance

oceanic front, especially in the East China Sea (Table 1, Fig. 2). Surface water temperatures in this area ranged from 20 to 24°C while those in the Kuroshio Current went beyond 24°C (11th Regional Coast Guard Headquarters 2002). We collected 20 rafts during the research cruise using the ORI ring net. Although one individual of *Sargassum macrocarpum* C. Agardh was observed near Kyushu Island on 13 May 2002, the other samples of drifting seaweeds were all *S. horneri*. Diameter mean and wet weight mean for drifting seaweed rafts along the different transects (Tables 1 and 2) were higher than those in March 2004 (see Tables 3 and 4). A one-way analysis of variance (ANOVA) indicated that this difference was statistically significant ($F_{0.05(1),1,318}=3.87$; $p<0.001$). Amongst the 2002 transects, wet weight mean was the highest on 19 May (Table 2).

Table 2 Estimated wet weight mean for drifting seaweed rafts observed along the transects from 11 to 19 May 2002

Date	Mean wet weight (g) of rafts	Lower value of wet weight of raft (g)	Higher value of wet weight of raft (g)
11 May	1,524 (264)	1,105 (191)	1,943 (336)
12 May	7,849 (1,359)	6,630 (1,148)	9,069 (1,570)
14 May	1,189 (206)	785 (136)	1,593 (276)
15 May	7,621 (1,320)	7,621 (1,320)	7,621 (1,320)
16 May	0 (0)	0 (0)	0 (0)
18 May	10,136 (1,755)	6,402 (1,109)	13,870 (2,402)
19 May	20,653 (3,576)	11,508 (1,993)	29,798 (5,160)

Higher and lower values of wet weight of drifting seaweeds were calculated using the standard deviation of the diameter of the rafts. Numbers inside parentheses are dry weights calculated from wet weights based on a relation between wet weight and dry weight of *S. horneri* (Mikami 2007)

March 2004

Figure 3 shows observation transects (solid line), observed rafts of drifting seaweeds (closed marks) and the Kuroshio Current position (dark stripe) during the March 2004 cruise. Drifting seaweeds were clearly present in waters located between the continental shelf and the oceanic front of the Kuroshio Current. Most of them were found in an area with sea surface temperatures ranging from 14 to 17°C (11th Regional Coast Guard Headquarters 2004). We collected 14 rafts during the cruise using the ORI ring net. Analysis of species composition of the rafts showed that *S. horneri* was the only species recorded. Mean diameter of drifting seaweed rafts was less than 1 m in March (Table 3). Many

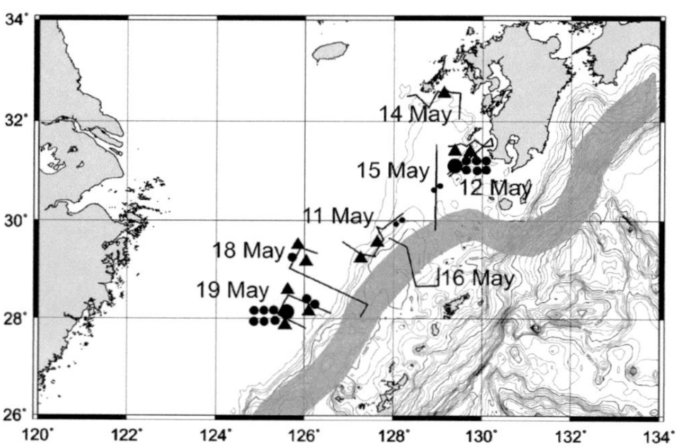

Fig. 2 Visual survey transects for drifting seaweed rafts (*solid lines*) in the East China Sea during R/V Hakuho Maru cruise, KH-02-1 leg1, and the distribution of the latter (*closed circles and triangles*) in May 2002. *Large, medium and small closed circles* are drifting seaweed rafts with diameters over 3 m, of 2–3 m, and 1–2 m, respectively. Triangles are groups of drifting seaweed rafts with diameters below 2 m. *Dark stripe* represents the path of Kuroshio Current (source: Hydrographic and Oceanographic Department of Japan Coast Guard, 2002)

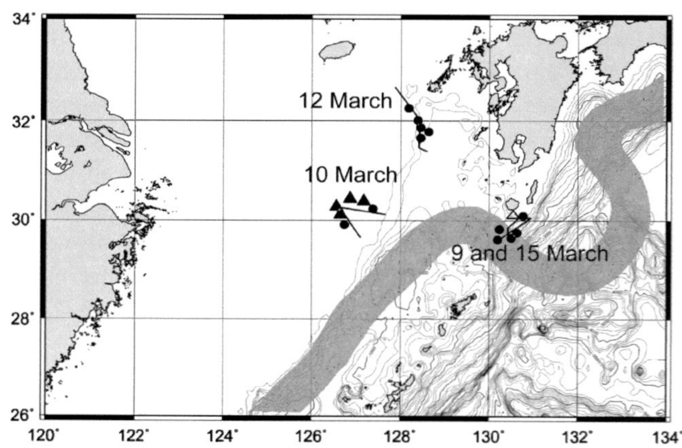

Fig. 3 Visual survey transects for drifting seaweed rafts (*solid lines*) in the East China Sea during R/V Hakuho Maru cruise, KH-04-1, and the distribution of the latter (*closed circles and triangles*) in March 2004. *Large, medium and small closed circles* are drifting seaweed rafts with diameters over 3 m, of 2–3 m, and 1–2 m, respectively. *Triangles* are groups of drifting seaweed rafts with diameters below 2 m. *Dark stripe* represents the path of Kuroshio Current (Source: Hydrographic and Oceanographic Department of Japan Coast Guard, 2004)

Table 3 Results for distributions of drifting seaweed rafts observed along the transects on 9, 10, 12, and 15 March 2004

Date	Length of transect (km)	Number of rafts	Mean diam. (m)	SD of diam.	Mean perpen. dist. (m)	SD of perpen. dist. (m)
9 and 15 Mar	143.14	27	0.69	0.44	7.19	4.40
10 Mar	187.85	1887	0.79	0.58	11.46	8.51
12 Mar	183.10	50	0.66	0.44	8.48	10.46

diam Diameter of raft, *perpen dist* perpendicular distance

drifting seaweeds were found along the 10 March transect. Since the 9 March and 15 March transects overlapped (Fig. 3), we considered the results as one unique transect (Table 3). The exponential model appears as a suitable model to explain the relation between diameter (D_i) and wet weight (W_t) of drifting seaweed rafts: W_t=1.2222 D_i^2−0.0787D_i (Fig. 4; R^2=0.962). Combined with mean diameter data for each transect, this relation enables to estimate the mean wet weight for the drifting seaweed rafts observed. Amongst transects, the mean wet weight was the lowest along the 12 March transect (Table 4).

Generally receptacles are formed apically on a branchlet in *S. horneri*. However, we found a branchlet growing from the apex of mature receptacle of *S. horneri* during both research cruises of May 2002 and March 2004 in the East China Sea. *Sargassum horneri* found in the East China Sea has a unique development feature different from those distributed in near shore coastal waters.

Abundance and biomass of drifting seaweeds

May 2002

In May 2002, ESW values for all the transects ranged from 8.4 to 13.8 m, but on 11 May reached 18.0 m (Table 5), due

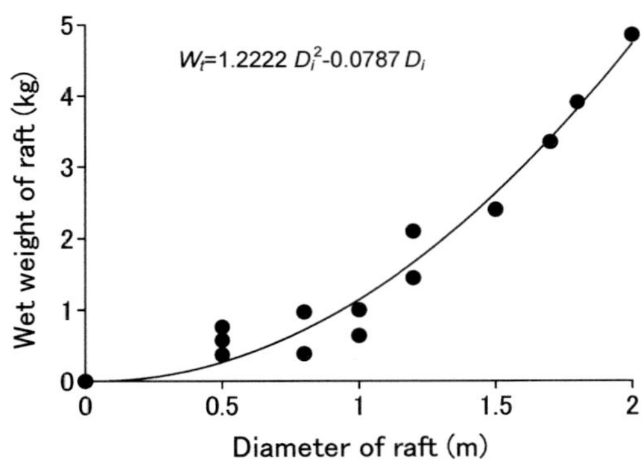

Fig. 4 Regression curve (R^2=0.962) showing the relation between diameter (D_i) and wet weight (W_t) of drifting seaweeds rafts collected during the R/V Hakuho Maru cruise, KH-04-1. This relation is expressed by the equation: W_t=1.2222 D_i^2−0.0787D_i

Table 4 Estimated wet weight mean for drifting seaweed rafts observed along the transects on 9, 10, 12 and 15 March 2004

Date	Mean wet weight (g) of rafts	Lower limit of wet weight of raft (g)	Higher limit of wet weight of raft (g)
9&15 Mar	528 (91)	326 (56)	730 (126)
10 Mar	701 (121)	335 (58)	1,066 (185)
12 Mar	480 (83)	278 (48)	682 (118)

Higher and lower values of wet weight of drifting seaweeds were calculated using the standard deviation of the diameter of the rafts. Numbers inside parentheses are dry weights calculated from wet weights based on a relation between wet and dry weight of *S. horneri* (Mikami 2007)

probably to extremely calm sea conditions. Raft densities of drifting seaweed along the 12 and 19 May transects were extremely high (Table 2, Fig. 5). No drifting seaweed was observed in Kuroshio Current waters (Figs. 2 and 5). Mean diameter and mean wet weight for drifting seaweed rafts were the highest on 19 May (Tables 1 and 2). Similarly, this transect displayed the highest standing crop (126.81 kg km^{-2} wet weight (WW)) (Fig. 6). Using the relation between wet weight and dry weight of *S. horneri* (Mikami 2007), wet weight was converted into dry weight. Hence, the highest standing crop equals 21.99 kg km^{-2} dry weight (DW). Total biomass along this particular transect was estimated at 660.87 kg WW (observed area = 5.21 km^{-2}).

March 2004

In March 2004, ESW values for the three transects ranged from 9.6 m to 12.1 m (Table 6). Raft density was the

Table 5 Results for models applied to the distribution of drifting seaweed rafts along each transect during May 2002 by the program DISTANCE 4.1

Date	Model (key + adjustment)	AIC	ESW	$\hat{n}$	$\hat{D}$	$\hat{D}_{LCL}$	$\hat{D}_{UCL}$	$\hat{D}_{CV}$
11 May	Half +Hermite	56.9	18.04	14	2.16	1.05	4.43	0.342
12 May	Half +Hermite	127.8	13.81	37	6.31	4.14	9.62	0.210
14 May	Unifo+Poly	24.7	8.44	9	2.46	0.99	6.13	0.412
15 May	Unifo+Poly	8.6	12.47	2	0.53	0.00	0.00	0.873
18 May	Half +Hermite	34.4	12.47	10	1.98	0.83	4.71	0.398
19 May	Half +Hermite	107.6	12.94	32	6.14	3.92	9.64	0.224

AIC Akaike's Information Criteria, *ESW* effective strip width, $\hat{n}$ total number of rafts within *ESW*, $\hat{D}$ density of rafts, $\hat{D}_{LCL}$ lower confidence limit of $\hat{D}$, $\hat{D}_{UCL}$ upper confidence limit of $\hat{D}$, $\hat{D}_{CV}$ coefficient of variation of $\hat{D}$ Half, Unifo, Hermite and Poly are Half-normal, Uniform, Hermite and Polynomial distributions, respectively

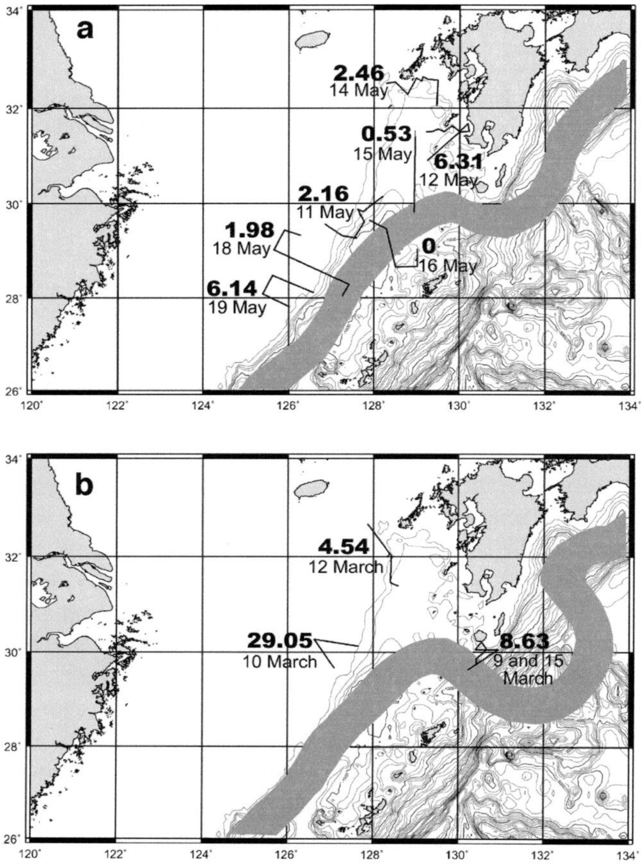

Fig. 5 Estimated raft densities of drifting seaweeds (in *bold*; rafts km^{-2}) along the observation transects (*solid lines*) in the East China Sea during R/V Hakuho Maru cruises, KH-02-1 leg1, in May 2002 (**a**) and KH-04-1, in March 2004 (**b**). *Dark stripe* represents the path of Kuroshio Current

highest along the 10 March transect (Table 3, Fig. 5). Mean wet weight for this transect, *M*, equals 701 g. This was the highest mean (Table 4). This transect displays the highest standing crop (20.35 kg WW km^{-2}) (Fig. 6). This is equivalent to 3.56 kg DW km^{-2}. Total biomass along this transect was estimated at 92.39 kg WW (observed area = 4.54 km^{-2}).

Discussion

Some common characteristics between drifting seaweeds collected in May 2002 and March 2004 occurred: most of them were distributed in the East China Sea, between the continental shelf and the oceanic front of the Kuroshio Current (125–128 E, 28–31 N), and not in coastal waters. This contradicts previous conclusions with regard to the East China Sea (Senta 1965).

The number of drifting seaweed rafts in March 2004 was greater than in May 2002, while the mean for drifting seaweed diameter in May 2002 was significantly higher than in March 2004. In the East China Sea, drifting seaweeds consisted almost exclusively of *S. horneri*, which are generally small in winter before growing up in spring, when they mature along the Japanese coast (e.g., Umezaki 1984). Drifters, whose position was tracked by satellite, attached to drifting seaweed rafts deployed near Chinese coasts in March 2004, traveled up to waters located in the east coast of Kyushu Island, under the influence of Kuroshio Current, and Cheju Island, Korea, under the influence of Tsushima Warm Current, where we lost contact in June (Komatsu et al. 2007). Mikami (2007) reported that drifting seaweeds keep on growing once detached from the bottom, despite not being mature. Smaller sizes reported in March 2004 are consistent with *S. horneri* living on the benthic substratum around Japanese (e.g., Umezaki 1984) and Chinese coastlines without the influence of warm water brought from the Kuroshio Current. On the other hand, *S. horneri* distributed along the coastline of Honshu Island facing the Kuroshio Current matures in March when it attains its maximum branch length (e.g., Mikami 2007).

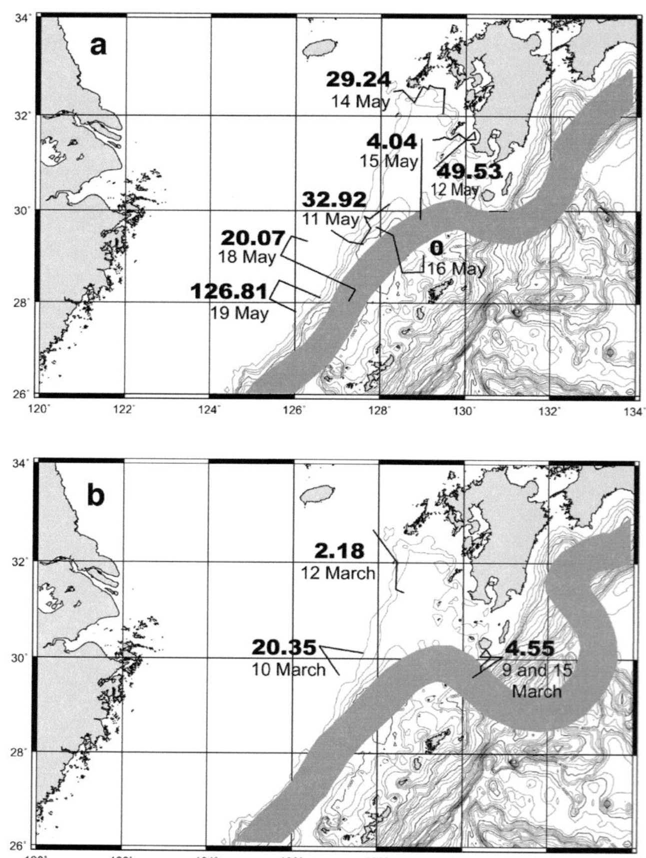

Fig. 6 Estimated standing crop of drifting seaweeds (kg WW km^{-2}) along the observation transects (*solid lines*) during R/V Hakuho Maru cruises, KH-02-1, in May 2002 (**a**) and KH04-1 in March 2004 (**b**). *Dark stripe* represents the path of Kuroshio Current

Table 6 Results for models applied to the distribution of drifting seaweed rafts along each transect during March 2004 by the program DISTANCE 4.1 Abbreviations are identical to the ones in Table 5

Date	Model (key + adjustment)	*AIC*	*ESW*	$\hat{n}$	$\hat{D}$	$\hat{D}_{LCL}$	$\hat{D}_{UCL}$	$\hat{D}_{CV}$
10 Mar	Half+Hermite	423.2	12.10	132	29.05	5.23	161.24	0.210
9&15 Mar	Half+Hermite	88.2	11.34	28	8.63	5.35	13.91	0.236
12 Mar	Half+Hermite	46.4	9.62	16	4.54	2.11	9.80	0.302

In 1959, Segawa et al. (1961b) investigated abundance and distribution of drifting seaweed between north Kyushu Island and Tsushima Island (Fig. 1). They used classes for wet weight (below 1 kg, from 1 to 10 kg, from 10 to 50 kg, from 50 to 100 kg) and estimated it through visual observation. They also estimated standing crop of drifting seaweeds at about 1,670 kg WW km^{-2} in June, when the highest abundance of drifting seaweeds was observed. This result is almost identical to Parr's study (1939) (5,367 kg WW nautical mile^{-2}: 1,565 kg WW km^{-2}) in Sargasso Sea, using a neuston net. Yoshida (1963), who investigated biomass of drifting seaweed around Japan from May to June using the Segawa et al. (1961b) method, reported that standing crop of drifting seaweed in south of Kyushu (Fig. 1) was about 900 kg WW km^{-2}, about 100 kg WW km^{-2} off Tohoku and about 500 kg WW km^{-2} around the Noto Peninsula in the Sea of Japan. The differences in standing crop estimation between our study and the previous ones may be due to sampling method, as well as the weight estimation method.

Mitani (1965b) estimated biomass of drifting seaweed based on the relationship between the diameter of the drifting seaweeds observed from an airplane and the diameter/wet weight of drifting seaweeds sampled from the research vessel. He estimated the standing crop at about 114.1 kg WW km^{-2} in the area located southeast off Goto Archipelago (northern limit of East China Sea), about 101.1 kg WW km^{-2} in the area located between south of Goto Archipelago and south of the Nagasaki Peninsula, about 10.9 kg WW km^{-2} located in the area located north of Koshiki Archipelago, and about 0.3 kg WW km^{-2} in the area located south of Koshiki Archipelago (Fig. 1). Since the estimation method used by Mitani (1965b), where the airplane was maintained at constant altitude and speed, is much more accurate than the one employed by Segawa et al. (1961b) and Yoshida (1963), his estimation of standing crop, though much smaller, looks more realistic.

In May 2002, south of Goto Archipelago, the standing crop of drifting seaweeds was about 29.24 kg WW km^{-2}, and about 49.53 kg WW km^{-2} in the area located between Koshiki Archipelago and south Kyushu Island (Fig. 1). Our estimation is almost equal to Mitani's (1965b). This may indicate that the method used by Buckland et al. (2001) and Thomas et al. (2002) is efficient.

The highest standing crop of drifting seaweeds, in March as in May, was located between the continental shelf and the oceanic front of the Kuroshio Current (20.35 kg WW km^{-2} in March and 126.81 kg WW km^{-2} in May). With reference to current maps and buoys tracking data, these drifting seaweeds are likely to come from Chinese coasts, especially Zhejiang Province (Fig. 1), where *Sargassum* beds were mainly composed of *S. horneri* (Komatsu et al. 2005). In May, the standing crop in this area is higher than in March simply because *S. horneri* generally reaches its maximum size (and biomass) in late spring (Umezaki 1984; Komatsu et al. 2005) in areas without influence of warm water brought by the Kuroshio Current in winter, such as the coast of the Sea of Japan and west coast of the East China Sea.

There was little difference in ESW between the two sampling sessions. This suggests that visual surveys were correctly conducted during both cruises. During both cruises, a peak of seaweed biomass was recorded in the center of East China Sea, where yellowtail and jack mackerel larvae usually accompany drifting seaweeds in spring and early summer (Konishi 2000).

Interestingly, as pointed out by Yoshida (1963), drifting seaweeds did not appear in Kuroshio Current waters but between the coast and the oceanic fronts in the East China Sea. It is estimated that most of drifting seaweeds are slowly transported to the east along the Japanese coasts by weak coastal currents. Therefore, yellowtail juveniles that accompany drifting seaweeds have a high possibility of leaving them in coastal waters off Honshu Island (Yamamoto et al. 2007) rather than in offshore waters of the Pacific Ocean.

Drifting seaweeds were then found in the area, including the continental coastal waters and the continental shelf part of East China Sea, in February (Shojima 1981), March (this study), April (Konishi 2000) and May (this study). The origin of drifting seaweeds is partially clarified by Komatsu et al. (2007). However, it is unknown where drifting seaweeds found in February originate.

Our results suggest large amounts of drifting seaweeds are distributed in the peripheral area of continental shelf waters in the East China Sea. They are probably transported from Chinese coasts. During the economic development of Japan, seagrass and seaweed beds have been lost in nearshore shallow waters due to pollution and reclamation (Komatsu 1997). In China, developing industries and increasing population produce water pollution and need land reclamation to build factories, ports and home sites, especially in coastal waters as in those of Japan from the 1960s to 1980s. It is necessary to investigate and monitor distribution, biomass and origin coasts of drifting seaweed in the East China Sea in order to preserve and manage them in a sustainable way.

Acknowledgements The authors thank to the captain and crew of R/V Hakuho Maru, Ocean Research Institute, the University of Tokyo. They express sincere appreciation to the researchers who helped them for observation and sampling of drifting seaweeds. This work was supported partially by the grant in aid for scientific research of Japan Science Promotion Society, No. 15405003 and 19405033.

References

11th Regional Coast Guard Headquarters (2002) Oceanographic Rapid Report, No. 10, 1–2

11th Regional Coast Guard Headquarters (2004) Oceanographic Rapid Report, No. 5, 1–2

Buckland ST, Anderson DR, Burnham KP, Laake JL, Borchers DL, Thomas L (2001) Introduction to distance sampling - estimating abundance of biological populations. Oxford University Press, New York, 432pp

Cho SH, Myoung JG, Kim JM (2001) Fish fauna associated with drifting seaweed in the coastal area of Tongyeong, Korea. Trans Am Fish Soc 130:1190–1202

Hernández-Carmona G, Hughes B, Graham MH (2006) Reproductive longevity of drifting kelp *Macrocystis pyrifera* (Phaeophyceae) in Monterey Bay, USA. J Phycol 42:1199–1207

Hobday AJ (2000) Abundance and dispersal of drifting kelp *Macrocystis pyrifera* rafts in the Southern California Bight. Mar Ecol Prog Ser 195:101–116

Ichimaru T, Mizuta K, Nakazono A (2006) Studies on the egg morpgology and spawning season in the mirror-finned flying fish *Hirundichthys oxycephalus* in the waters near Kyushu, Japan. Nippon Suisan Gakkaishi 72:21–26. (in Japanese with English abstract and figure legends)

Ingólfsson A (1995) Floating clumps of seaweed around Iceland: natural microcosms and a means of dispersal for shore fauna. Mar Biol 122:13–21

Kingsford MJ (1992) Drift algae and small fish in coastal waters of northeastern New Zealand. Mar Ecol Prog Ser 80:41–55

Kingsford MJ (1995) Drift algae: a contribution to near-shore habitat complexity in the pelagic environment and an attractant for fish. Mar Ecol Prog Ser 116:297–301

Komatsu T (1985) Temporal fluctuations of water temperature in a *Sargassum* forest. J Oceanogr Soc Jpn 41:235–243

Komatsu T (1989) Day-night reversion in the horizontal distributions of dissolved oxygen content and pH in a *Sargassum* forest. J Oceanogr Soc Jpn 45:106–115

Komatsu T (1997) Long-term changes in the *Zostera* bed area in the Seto Inland Sea (Japan), especially along the coast of the Okayama Prefecture. Oceanol Acta 20:209–216

Komatsu T, Kawai H (1986) Diurnal changes of pH distributions and the cascading of shore water in a *Sargassum* forest. J Oceanogr Soc Jpn 42:447–458

Komatsu T, Murakami S (1994) Influence of a *Sargassum* forest on the spatial distribution of water flow Fish Oceanogr 3:256–266

Komatsu T, Ariyama H, Nakahara H, Sakamoto W (1982) Spatial and temporal distributions of water temperature in a *Sargassum* forest. J Oceanogr Soc Jpn 38:63–72

Komatsu T, Kawai H, Sakamoto W (1990) Influence of *Sargassum* forests on marine environment. Bull Coast Oceanogr 27:115–126, 1990. (in Japanese with English abstract)

Komatsu T, Murakami S, Kawai H (1994) Some features of jump of water temperature in a *Sargassum* forest. J Oceanogr 52:109–124

Komatsu T, Tatsukawa K, Wang W, Liu H, Ajisaka T, Zhang S, Tanaka K, Zhou M, Uwai S, Sugimoto T (2005) Species composition and biomass of *Sargassum* beds in Gouqi Island, Zhejiang Province, China. Jpn J Phycol 53(1):103 (in Japanese)

Komatsu T, Tatsukawa K, Filippi JB, Sagawa T, Matsunaga D, Mikami A, Ishida K, Ajisaka T, Tanaka K, Aoki M, Wang WD, Liu HF, Zhang SD, Zhou MD, Sugimoto T (2007) Distribution of drifting seaweeds in eastern East China Sea. J Mar Syst 67:245–252

Konishi Y (2000) Drifting seaweeds coming from China too. Seikai Fisheries Research Institute News 11–15 (in Japanese)

Kubo M (2004) The present situation report about the floating seaweeds investigation for the yellowtail larva fishery in Kagoshima Prefecture. Kaiyo Monthly 36(6):458–463 (in Japanese)

Kubo M (2006) Passing year change of flow algae appearing in Kagoshima sea area. Kaiyo Monthly 38(8):575–582 (in Japanese)

Mikami A (2007) Estimating the net primary production of *Sargassum* species (Phaeophyceae) throughout different life stages: from fixation to drifting. Doctor Thesis, University of Tokyo, Tokyo (in Japanese)

Mitani F (1965a) At attempt to estimate the population size of the "Mojako", the juvenile of the yellowtail, *Seriola quinqueradiata* Temminck et Schlegel, from the amount of the floating seaweeds based on the observations made by means of airplane and vessels -I. State of distribution of the floating seaweeds. Bull Jpn Soc Sci Fish 31(6):423–428 (in Japanese with English abstract)

Mitani F (1965b) At attempt to estimate the population size of the "Mojako", the juvenile of the yellowtail, *Seriola quinqueradiata* Temminck et Schlegel, from the amount of the floating seaweeds based on the observations made by means of airplane and vessels -II. Estimate of the amount of the floating seaweeds. Bull Jpn Soc Sci Fish 31(6):429–434 (in Japanese with English abstract)

Parr AE (1939) Quantitative observations on the pelagic *Sargassum* vegetation of the western North Atlantic. Bull Bingham Oceanogr Coll 6:1–94

Rueness J (1989) *Sargassum muticum* and other introduced Japanese macroalgae: biological pollution of European coasts. Mar Poll Bull 20:173–176

Segawa S, Sawada T, Higaki M, Yoshida T (1959a) Studies on the floating seaweeds-I. Annual vicissitude of floating seaweeds in the Tsuyazaki region. Sci Bull Fac Agri Kyushu Univ 17:83–89 (in Japanese with English abstract)

Segawa S, Sawada T, Higaki M, Yoshida T (1959b) Studies on the floating seaweeds-II. Floating seaweeds found in Iki-Tusima region. Sci Bull Fac Agri Kyushu Univ 17:291–297 (in Japanese with English abstract)

Segawa S, Sawada T, Yoshida T (1960) Studies on the floating seaweeds-V. Seasonal change in amount of the floating seaweeds off the coast of Tsuyazaki. Sci Bull Fac Agri Kyushu Univ 17:437–441 (in Japanese with English abstract)

Segawa S, Sawada T, Higaki M, Yoshida T, Kamura S (1961a) Studies on the floating seaweeds-VI. The floating seaweeds of the West Kyushu region. Sci Bull Fac Agri Kyushu Univ 18:411–417 (in Japanese with English abstract)

Segawa S, Sawada T, Higaki M, Yoshida T (1961b) Studies on the floating seaweeds-VII. Seasonal changes in the amount of the floating seaweeds found on Iki- and Tsusima Passages. Sci Bull Fac Agri Kyushu Univ 19:125–133 (in Japanese with English abstract)

Segawa S, Sawada T, Yoshida T (1964) Studies on the floating seaweeds-IX. The floating seaweeds found on the sea around Japan. Sci Bull Fac Agri Kyushu Univ 21:111–115 (in Japanese with English abstract)

Senta T (1965) Importance of drifting seaweeds in the ecology of fisheries. A series of studies on fisheries 13. Japan Fisheries Resource Conservation Association, Tokyo, pp 1–54 (in Japanese)

Shojima Y (1981) Studies about collection spawn of Pacific saury in Kyushu waters and East China Sea. In Fishing resource research conference 13th Float Fish sectional meeting minutes. Fishing Resource Workshop Executive Office, Tokyo, pp 88–99 (in Japanese)

Stoner AW (1983) Pelagic *Sargassum*: evidence for a major decrease in biomass. Deep Sea Res 30:1259–1264

Thiel M, Gutow L (2005) The ecology of rafting in the marine environment. 1. The floating substrata. Oceanogr Mar Biol 42:181–263

Thomas L, Buckland ST, Burnham KP, Anderson DR, Laake JL, Borchers DL, Strindberg S (2002) Distance sampling. In: El-Shaarawi AH, Piegorsch WW (eds) Encyclopedia of Environmetrics Volume 1. Wiley, Chichester, pp 544–552

Thomas L, Laake JL, Strindberg S, Marques FFC, Buckland ST, Borchers DL, Anderson DR, Burnham KP, Hedley SL, Pollard JH, Bishop JRB (2004) Distance 4.1. Release 2. Research Unit for Wildlife Population Assessment, University of St. Andrews, UK. Release 2. http://www.ruwpa.st-and.ac.uk/distance/)

Umezaki I (1984) Ecological studies of *Sargassum horneri* (Turner) C. Agardh in Obama Bay, Japan Sea. Bull Jpn Soc Sci Fish 50:1193–1200

Yamamoto T, Ino S, Kuno M, Sakaji H, Hiyama Y, Kishida T, Ishida Y (2007) On the spawning and migration of yellowtail *Seriola quinqueradiata* and research required to allow catch forecasting. Bull Fish Res Agen 21:1–29. (in Japanese)

Yoshida T (1963) Studies on the distribution and drift of the floating seaweed. Bull Tohoku Reg Fish Res Lab 23:141–186 (in Japanese with English abstract)

Diversity, biomass and distribution pattern of *Sargassum* beds in the South West lagoon of New Caledonia (South Pacific)

L. Mattio · G. Dirberg · C. E. Payri · S. Andréfouët

Originally published in the Journal of Applied Phycology, Vol 20, No 5, 361–373.
DOI: 10.1007/s10811-008-9318-4

Abstract *Sargassum* (Phaeophyceae, Fucales) is a genus of worldwide distribution recently recognised to proliferate in several regions of the South Pacific. In New Caledonia, species of this genus naturally structure one of the major lagoon habitats but their extent, composition and biomass remain largely unknown. To fill these gaps in our knowledge over large areas, we applied a combination of remote sensing and in situ methods applied to the Neo Caledonian South West Lagoon. Space borne high resolution Landsat (30-m resolution) and Quickbird (2.4-m resolution) images of the Neo Caledonian South West lagoon were used to estimate the spatial extent of the different beds of interest. Species composition was determined for 11 *Sargassum* beds and seasonal variations were investigated for four representative beds over an 18 month period using quadrats and transects. Relationships between surface cover and biomass were estimated from seasonal field data sets. Finally, four methods requiring variable levels of sampling effort were designed to estimate the total biomass at the scale of each bed, considering (or not) the specific composition, and spatial and temporal variations. Seven Sargassaceae species were identified. Mean surface cover (24.4–51.6%) and total biomass [3.4–1,461.9 t dry matter (DM)] varied widely between beds. Overall, biomass temporal variations were not significantly different, but species-level variations seemed to be bed-specific. The extent of the 11 beds was 9 km^2; their total biomass was estimated and compared using each of the four methods, and the most precise method provided an estimate of 2,900 t DM. This study demonstrates a way of characterising *Sargassum* beds, efficiently and on a large scale, using a combination of remotely sensed and in situ data. These methods should be useful for possible future biomonitoring of *Sargassum* beds in New Caledonia, and in other areas worldwide.

Keywords Large brown algae · Sargassaceae · Remote sensing · Habitat management · Coral reef

Abbreviations

DM	Dry matter
QB	Quickbird
L7	Landsat7

L. Mattio (✉) · C. E. Payri
UMR7158 "SAE",
Institut de Recherche pour le Développement,
BPA5, 98848 Nouméa, New Caledonia
e-mail: lydiane.mattio@noumea.ird.nc

L. Mattio
UMR 6540 DIMAR COM, Université de la Méditerranée,
13288 Marseille cedex 9, France

G. Dirberg
UMR 6535 LOB COM, Université de la Méditerranée,
13288 Marseille cedex 9, France

G. Dirberg · S. Andréfouët
UR 128, Institut de Recherche pour le Développement,
BPA5, 98848 Nouméa, New Caledonia

C. E. Payri
Unversité de la Polynésie Française,
BP6570 Faa'a,
Tahiti, French Polynesia

Introduction

Sargassum C. Agardh (1820) is the richest genus of the Sargassaceae, accounting for almost 80% of the species recorded for this family by Guiry and Nic Dhonncha (2007). This brown algal genus is widespread all over the globe (Okamura 1932; Yoshida 1989), especially in tropical and sub-tropical regions (Nizamuddin 1962; Womersley 1987; Phillips 1995), where they form large beds that play an important role in coral reef ecosystem functioning (Rossier and Kulbicki 2000; Godoy and Coutinho 2002; Tanaka and Leite 2003).

In New Caledonia, about 46 taxa have been recorded (Grunow 1915, 1916a, b; May 1953; Garrigue and Tsuda 1988; Noro and Abbott 1994); however, since the pioneering work of Grunow (1915, 1916a, b), no taxonomic revision has been done. To date, *Sargassum* is one of the most conspicuous alga forming large beds, especially in the South West Lagoon. Together with seagrass and *Halimeda* (calcareous green algae) beds, they determine one of the three major lagoon habitats structured by macrophytes. *Sargassum* bed canopies can reach 1.5 m high, and host a wide diversity of invertebrates (Mukai 1971; Tanaka and Leite 2003) or act as sheltering nursery grounds for several fish species (Ornellas and Coutinho 1998), thus playing an important role in their life history (McClanahan et al. 1994). *Sargassum* species are also of interest for pharmaceutical and cosmetic applications due to their natural substance contents (Zubia et al. 2003; Stiger et al. 2004). In New Caledonia, they could be used for their properties of toxic heavy metal biosorption (Davis et al. 2000; Sheng et al. 2004), especially in local mining industries as part of their waste management strategies.

On the negative side, recent biological invasion of *S. muticum* on the Pacific coast of northern America and on the Atlantic coast of Europe have been reported (Critchley et al. 1990). In various tropical Pacific localities, coral reefs have been faced with a proliferation of *Sargassum* populations following reef degradation. These expansions have been reported in French Polynesia (Stiger and Payri 1999) and in the Fijian archipelago (Moseley and Aalbersberg 2003). *Sargassum* beds can be affected by changes in environmental factors, e.g. *Sargassum* populations in the Great Barrier Reef experience fluctuations in biomass in relation to enhanced siltation and nutrient loads (McCook 1999; Schaffelke 1999). The apparent links between proliferation or reduction and environmental changes make *Sargassum* populations potentially good indicators to monitor ecosystem changes, including increases of anthropogenic pressure. This has interesting potential for the New Caledonian South West Lagoon, which currently faces increases in nutrient and sediment inputs (Tenorio et al. 2005; Jacquet et al. 2006) as well as the development of huge mining projects in upstream watersheds (Labrosse et al. 2000; Bellwood et al. 2006).

Despite their ecological role, their biotechnological potential, and the warning signals they could send, *Sargassum* remain poorly studied in New Caledonia. Biodiversity conservation, integrated coastal management and bioprospecting request accurate, spatially explicit data on diversity, distribution extent and biomass. This is true for most taxa, but particularly for *Sargassum* beds given their importance in New Caledonia (Payri et al. 2007). In the South West lagoon, the accurate characterisation of the spatial and temporal variations in diversity, extent and biomass of *Sargassum* beds is a new priority, but poses a significant challenge given the extent of the area of interest (the New Caledonian lagoon covers 19,000 km^2) and the expected seasonal variations (Hurtado and Ragaza 1999; Stiger and Payri 1999).

Studies using satellite and digital airborne remote sensing have proven efficient for mapping marine habitats, including coral reef ecosystems (Mumby et al. 1998; Andréfouët et al. 2003; White et al. 2003). Remote sensing has been used to estimate biomass over large areas, thus avoiding time-consuming, expensive and sometimes destructive field sampling (Belsher and Mouchot 1992; Andréfouët et al. 2004). Andréfouët et al. (2004) found that Sargassaceae surface cover for all reef habitats and algal biomass samples collected in the field were necessary to convert biomass to the scale of the reef. Conversely, the measurement of seagrass standing crop could be estimated quasi directly from the linear relationship between biomass and the signal measured by the sensor after depth correction (Mumby et al. 1997; Riegl et al. 2005), thus avoiding the need for systematic field surveys. While the latter direct techniques have been performed using only optical or acoustic data in a variety of studies on seagrass beds, we are not aware of any similar *Sargassum*-specific studies in coral reef environments

The objectives of this work were to assess the feasibility of characterising *Sargassum* beds in space and time across the New Caledonian South West Lagoon using a combination of remote sensing data and field and laboratory work. The variables of interest were species diversity, extent and biomass. For the latter variable, sampling all beds over an extended period of time to assess spatial and temporal variability was not an option. Thus, in order to define a rapid, inexpensive and simple way of estimating Sargassaceae biomass over large areas, we compared a variety of methods requiring different levels of sampling effort. These methods included mapping using remote sensing techniques, and biomass estimates based on both destructive quadrats samples and non-destructive transect data.

Materials and methods

New Caledonia is located in the south western Pacific Ocean between 18°–23° S and 162°–168° E, 2,500 km east of Australia (Fig. 1). The main island, "Grande Terre", is surrounded by a 1,600 km-long barrier reef enclosing a wide lagoon. The south west region includes a vast lagoon area, which harbours a large diversity of marine habitats within a mean depth of 20 m. We investigated a total of 11 different *Sargassum* beds distributed in the south west lagoon area, between the capital city of Nouméa and the barrier reef located 17 km offshore. *Sargassum* beds were defined as any lagoon area displaying notable perennial Sargassaceae populations.

Selection and extent of *Sargassum* beds

Our archives included several Landsat 7 (L7) images at 30 m spatial resolution, acquired at different seasons and in different years. To guide site selection, we selected an image acquired on the 3 August 2002, which was notable for its very high water clarity conditions. A total of 11 areas visually interpreted as *Sargassum* beds were selected using this image, and quick preliminary surveys were performed to confirm the dominance of *Sargassum* species in those algal beds.

Simple Landsat image contrast enhancement defined the boundaries of the algal beds on a suite of different Landsat images. Specifically, we used three Landsat images from 8 November 1999, 30 September 2000, and 26 April 2001. In addition, three images acquired on 28 March 2002, 4 September 2002 and 22 October 2002 were also used to assess annual seasonal variations. Outer bed boundary locations were confirmed in situ in late 2006 by field checks (SCUBA, spot dives and/or manta tows) and using acoustic data. Results were visually compared with boundary limits defined from Landsat images. A full presentation of the acoustic data processing is beyond the scope of this study but readers can refer to Chevillon (2005) and Payri et al. (2007) for further information.

The surface area of each targeted *Sargassum* bed (*S*, in km^2) was estimated by counting the number of Landsat pixels (900 m^2 each) located within the dark patches (ENVI toolbox, Research Systems, Boulder CO).

Typology

For each of the 11 *Sargassum* beds, field stations were selected depending on the Landsat image contrast variability and total surface area to account for the heterogeneity of the bed (Table 1). At all stations, geomorphology (reef flat or lagoon floor), substratum (sandy rocky floor, rubble and/or maerl), and depth were described. From these observations, an algal bed typology was obtained. Species diversity, relative species surface cover and overall Sargassaceae species surface cover were considered afterwards to finalise the typology.

Species identification

For species identification, raw material was taken from saxicolous populations, air-dried and pressed as herbarium voucher specimens prior to taxonomical work. Receptacles (fertile organs), if any, or apical growth parts were preserved in silica gel (Chase and Hills 1991) for later molecular analyses. *Sargassum* morphotype classifications were based on morphological analysis of holdfast, axis, leaves (laterals), vesicles and receptacles (Yoshida 1989).

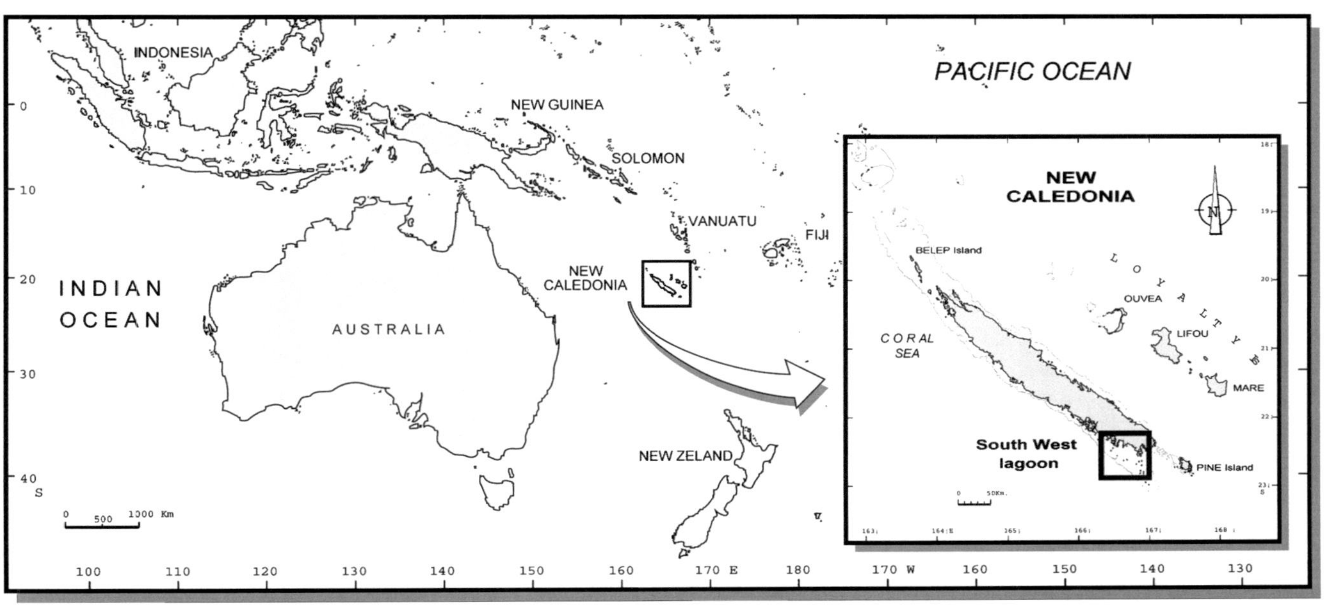

Fig. 1 Map showing the geographic location of New Caledonia and the South West lagoon study area (*inset*)

Table 1 Habitat description and surface cover of major Sargassaceae species for all beds surveyed

Sargassum bed designation	S01	S02	S03	S04	A	B	C	D	E	F	G
Estimated surface area (km^2)	0.007	0.023	0.302	0.027	0.019	0.138	0.239	2.138	4.882	00.252	0.999
Number of stations surveyed	1	1	1	1	1	1	2	6	5	1	1
Habitat description[a]											
Reef flat	X	–	X	–		–	–	–	–	–	–
Lagoon floor	–	X	–	X	X	X	X	X	X	X	X
Sandy rocky floor	–	X	–	X	X	X	X	X	X	X	X
Rubble	X	–	X	–	–	–	–	–	–	–	–
Maerl	X	–	X	–	–	–	X	–	–	–	–
Depth (m)[b]	2.5	10	4.5	2.5	7	10	7.5	9	12	11	10
Number of species[b]	6	4	7	3	3	6	3	4	5	4	5
Mean surface cover[b] (R^x)[c]											
Sargassum spinuligerum[d]	9.3	13.9	7.2	32.6	51.6	3	12.7	12.2	14.9	9.1	26.4
Sargassum decurrens	<1	17.3	<1	<1	0	24.3	17	8	6.7	21.4	5.3
Sargassum howeanum	38.3	0	26	0	0	0	0	0	0	0	0
Sargassum crassifolium	<1	0	<1	0	<1	<1	0	0	<1	<1	<1
Sargassum polycystum	<1	0	<1	0	0	<1	0	<1	0	0	<1
Hormophysa cuneiformis	0	9	6.8	0	0	2.4	5.2	4.5	2.8	3.3	0
Cystoseira trinodis	<1	<1	<1	<1	<1	<1	0	<1	<1	0	<1
Total	47.6	40.2	40	32.6	51.6	29.3	34.9	24.7	24.4	33.8	31.7

[a] For beds S01–S04, the mean annual surface cover is given; for beds A–G, mean surface cover was assessed during campaigns in January and June 2006
[b] When several stations were surveyed within a bed, the mean is given
[c] Surface cover of species x
[d] Includes *S. spinuligerum* var. *crispata*

Receptacles were sectioned transversally using a freezing microtome and observed with a light microscope (Olympus BH2). Morphological taxonomic ambiguities were improved using nuclear *ITS-2* (intergenic transcribed spacer 2) and chloroplastic *rbc*LS (rubisCo spacer) genetic markers as described in Stiger et al. (2000, 2003) and Phillips et al. (2005). Finally, morphotypes were identified at species level using the available *Sargassum* taxonomic regional literature (C. Agardh 1820; J. Agardh 1848, 1889; Grunow 1916a, b; May 1953; Womersley and Bailey 1970; Womersley 1987; Garrigue and Tsuda 1988; Noro and Abbott 1994) and the study of type specimens and/or specimens from the type locality whenever possible.

Surface cover and biomass estimation

To estimate the surface cover and spatial and temporal variability in biomass, two types of sampling were carried out, each requiring a different level of sampling effort.

Sampling 1 was a time-consuming, destructive sampling that focused on four beds deemed representative of habitats and species variability from among the 11 beds. Beds S01 to S04 (see Table 1) were surveyed by SCUBA every 3 months from August 2005 to January 2007 using the same protocol (Fig. 2). Surface cover per species (R^x) was measured for each station along four parallel 10 m transects using the Line Intersect method (English et al. 1997). The floristic content from four random 0.25 m^2 quadrats was then collected, sorted into species, oven-dried at 60°C for 48 h and weighed to measure mean biomass per species and per station (W^x).

Sampling 1 provided the temporal and spatial variation of the species listed for each bed. Seasonal variation significance was tested using Statistica software (Statsoft, Maisons-Alfort, France). Variances were homogeneous (Barthell test, $P>0.05$) but data were not distributed normally, even when transformed. Thus, a non-parametric Kruskal-Wallis test was used followed by a post-hoc Mann Whitney U test.

Linear relationships between surface cover and biomass, derived from transect and quadrat measurements, were estimated for each species separately and considering all Sargassaceae species together. They were constrained to intersect the origin. Several relationships could then be computed as follows:

- considering all surveys, and all four beds, the relationship for each species x is:

$$W^x = a^x \times R^x \tag{1}$$

where a^x is the slope of the linear relationship between the biomass per unit of surface area, W^x, and the surface cover, R^x, for a given species x in all four beds at all seasons.

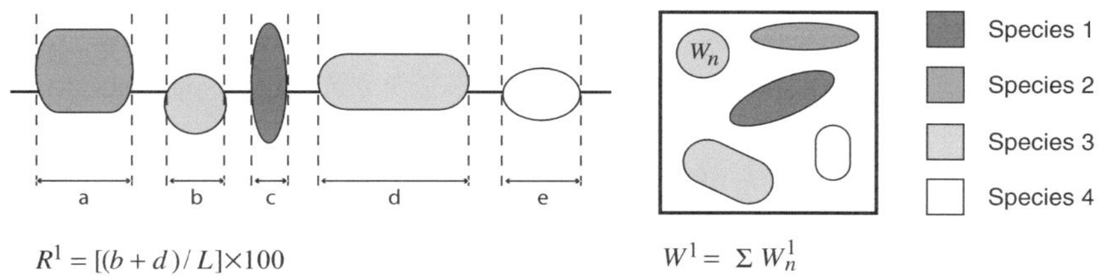

Fig. 2 Transect and quadrat measurement methods. Surface cover of species 1 (denoted R^1) is the sum of transect intercepts measured for this species divided by the total transect length (L=10 m). Thus ,the total surface cover at a station is the sum of each species surface cover at this station. The biomass of species 1 (W^1) is the sum of W_n^1 (g dry matter) measured in 1 m^2 using a quadrat

- considering all surveys, all four beds, and considering all Sargassaceae species ($\sum x$), the unique relationship applicable to all species is:

  $$W^{\sum x} = a^{\sum x} \times R^{\sum x} \qquad (2)$$

  where $a^{\sum x}$ is the slope of the linear relationship between the biomass per unit of surface area, $W^{\sum x}$, and the surface cover, $R^{\sum x}$ of all Sargassaceae species in all four beds at all seasons
- considering all surveys for each of the four beds and for all species, bed-scale annual mean biomass per unit of surface area is: W_{S01}^{mean}, W_{S02}^{mean}, W_{S03}^{mean}, W_{S04}^{mean} and $W_{S01-S04}^{mean}$ is the average over the 4 beds.

Sampling 2, which was a less time-consuming method, was conducted over the remaining seven algal beds (beds A–G, see Table 1). Only non-destructive surface cover per species was measured along four parallel 10 m transects per station. A total of 17 stations were sampled for the seven beds between January and June 2006 (Table 1). Mean surface cover was calculated for each species x, and also taking into account all Sargassaceae species ($\sum x$) in each bed α, thus R_α^x and $R_\alpha^{\sum x}$ (in %), respectively.

Four different methods to estimate the biomass for each of the 11 *Sargassum* beds were designed from Samplings 1 and 2. All methods required at least the bed surface area estimated from satellite imagery (S, in km^2) and the annual surface cover and biomass measured over the four representative beds. Decreasing levels of complementary data requirements for each method were then considered as follow, (1) bed-specific species surface cover and quadrat-based biomass measurements (from Sampling 1), (2) bed-specific surface cover for each species (from Sampling 2), (3) bed-specific surface cover for all Sargassaceae species pooled together (from Sampling 2), and (4) surface area of the bed from satellite imagery and the annual mean biomass, i.e. the minimum requirements as stated above.

Thus the total biomass of bed α could be estimated in four different ways:

First,

$$W_{1,\alpha} = S_\alpha \times W_\alpha^{mean} \qquad (3, \text{method1})$$

where $W_{1,\alpha}$ is the total biomass of bed α [in tons of (DM)], S_α is the remote sensing estimated surface of bed α (in km^2) and W_α^{mean} the mean biomass measured in the field with destructive quadrat sampling (in g m^{-2} DM). These means have been computed for beds S01–S04 where quadrats were sampled. Method 1, which required more field data than subsequent methods, is considered here as the reference method.

Second,

$$W_{2,\alpha} = S_\alpha \times \sum_x \left(a^x \times R_\alpha^x\right) \qquad (4, \text{method2})$$

where $W_{2,\alpha}$ is the total biomass of bed α (in tons of DM), S_α is the remote-sensing-estimated surface of bed α (in km^2), a^x is the slope of the linear regression for species x (from Eq. 1) and R_α^x is the mean surface cover measured for species x in bed α (from Sampling 2).

Third,

$$W_{3,\alpha} = S_\alpha \times a^{\sum x} \times R_\alpha^{\sum x} \qquad (5, \text{method3})$$

where $W_{3,\alpha}$ is the total Sargassaceae biomass of bed α (in tons of DM), S_α is the remote sensing estimated surface of bed α (in km^2), $a^{\sum x}$ is the slope of the linear regression for all Sargassaceae species (from Eq. 2) and $R_\alpha^{\sum x}$ is the mean surface cover measured for all Sargassaceae species in bed α (from Sampling 2).

Fourth,

$$W_{4,\alpha} = S_\alpha \times W_{S01-S04}^{mean} \qquad (6, \text{method4})$$

where S_α is the remote-sensing-estimated surface of bed α (in km^2), and thus the only bed-specific information. It is assumed here that the annual mean biomass $W_{S01-S04}^{mean}$ is close to the mean annual biomass of any bed found in the studied area.

Sensitivity analysis on biomass estimates

To compute absolute errors on biomass estimates at bed-scale, estimation of errors of mean biomass estimation and errors of estimated beds surface areas would be necessary. However, errors on bed surface areas could not be fully quantified here. For this, we would need to estimate the uncertainty due to the image resolution and due to errors in bed delineation and misclassification of each pixel, in the form of an error matrix (see Andréfouët et al. 2004). Nevertheless, it was possible to compute the relative differences between methods, taking $W_{1,\alpha}$ (method 1) as the reference. To assess the differences between methods, we computed $\Delta_{i/j}$ as follows:

$$\Delta_{i/j} = \frac{\left(W_{i,\alpha}^{estimated} - W_{j,\alpha}^{estimated}\right)}{W_{j,\alpha}^{estimated}} \times 100$$

where $\Delta_{i/j}$ is the relative difference between biomasses estimated with methods i and j (in %). The estimation differences $\Delta_{i/1}$ could be calculated only for beds S01–S04 as method 1 was applied only to those beds. $\Delta_{i/j}$ ($j \neq 1$) were calculated for all 11 beds (i.e. S01–S04 and A–G).

Correction of patchiness in *Sargassum* bed extent estimates

It might be possible to account for intra-bed patchiness (small bare holes within the bed), which is not detectable at the 30 m resolution of the Landsat images, by using 2.4-m resolution Quickbird (QB) images and comparing the extents computed from both images, if the images were concurrent. If this is not the case, we have to assume that *Sargassum* beds are stable in extent, cover and biomass across time. Here, we used a QB image from 25 December 2004, i.e. a different date from all of the Landsat acquisition dates from our archive. The QB image covered three of the previously targeted *Sargassum* beds and 15% of the Landsat study area. For the inter-sensor comparison, a Landsat image from 3 August 2002 was used. Under the assumptions above, a correction coefficient k was computed using:

$$k = \sum \frac{S_{QB}}{S_{L7}}$$

where S is the surface estimated from either L7 or QB images in square kilometres. The correction factor k was applied to all bed surfaces estimated from the selected L7 image.

Results

Mapping and surface areas

Image contrast enhancement of Landsat data was efficient to detect bed boundaries even with different water qualities because (1) beds were shallow (between 2.5 and 12 m), (2) dark *Sargassum* beds were surrounded by bright soft and hard bottoms, and (3) *Sargassum* beds boundaries were sharp at the resolution of the Landsat 7 image (30 m). However, in some areas, boundaries could be fuzzier due to adjacent seagrass beds, or because of larger depth gradient within the *Sargassum* bed. Both factors can blur the *Sargassum* bed radiometric signal.

Image analysis shows virtually no variation of the *Sargassum* bed extents between August, September and October 2002. Extents are comparable to those observed in the November 1999 and September 2000 images. However, while the bed boundaries do not vary, the relative contrast with the surrounding habitats changes. This could be interpreted as variation in the algal surface cover over time within beds of stable extent. We conclude that bed boundaries are quite stable at the inter-annual scale, but that changes in surface cover are detectable. Therefore, the surface areas estimated here are valid all year round and can be used in the different biomass formulations (Eqs. 3–6).

The factor k used to correct patchiness between sensors at different resolutions was equal to 0.658 and was applied to correct surfaces estimated with the reference Landsat 7 image. The individual surface areas of the 11 targeted beds were estimated; the total surface occupied by *Sargassum* beds in the study area was 9.03 km^2 (Table 2, Fig. 3).

Species identification

Eight species belonging to the Sargassaceae were identified, six of which were part of the genus *Sargassum*. They belonged to the subgenera *Phyllotrichia* [*S. decurrens* (Brown ex. Turner) C. Agardh] and *Sargassum* section *Malacocarpicae* (*S. spinuligerum* Sonder, *S. spinuligerum* Sonder var. *crispata* Grunow and *S. howeanum* Lucas), section *Zygocarpicae* (*S. polycystum* C. Agardh) and section *Acanthocarpicae* (*S.* cf. *crassifolium* J. Agardh). Two other species were identified as *Hormophysa cuneiformis* (J.F. Gmelin) P.C. Silva and *Cystoseira trinodis* (Forsskål) C. Agardh.

Lobophora variegata (Lamouroux) Womersley ex Oliveira (Phaeophyceae, Dictyotales) was also found to be abundant in several beds; however, it belongs to a different functional–form group and was considered separately. Several other algal species belonging to various families were observed but not considered further due to their rarity.

Sargassum bed typology

The major characteristics of the beds surveyed are listed in Table 1. The eight Sargassaceae species and the whole range of habitat diversity were represented in beds S01–S04. Among the 11 considered beds, 2 occurred on shallow

Table 2 Estimated surface area for each bed surveyed, mean biomass and standard deviations measured for beds S01–S04, estimated biomass and estimation differences for all beds. *S* Surface area, W^{mean} mean biomass, W_i biomass estimated with method i, $\Delta_{i/j}$ estimation of difference between method i and method j

Bed designation	S (km^2)	W^{mean} ($g\ m^{-2}$)	W_1 (t)	W_2 (t)	W_3 (t)	W_4 (t)	$\Delta_{2/1}$ (%)	$\Delta_{3/1}$ (%)	$\Delta_{4/1}$ (%)	$\Delta_{3/2}$ (%)	$\Delta_{4/2}$ (%)	$\Delta_{4/3}$ (%)
S01	0.007	456±128	3.2±0.9	3.4	3.7	3.3	6.2	15.6	3.1	8.8	−2.94	−10.8
S02	0.023	390.1±123.7	9.0±2.8	10.9	9.9	10.6	21.1	10.0	17.8	−9.2	−2.75	7.1
S03	0.302	429±102.9	129.6±31	132.5	129.6	138.8	2.2	0.0	7.1	−2.2	4.75	7.1
S04	0.027	566±122.6	15.3±3.3	11.7	9.5	12.4	−23.5	−37.9	−19.0	−18.8	5.98	30.5
A	0.019	–	–	12.9	10.5	8.7	–	–	–	−18.6	−32.56	−17.1
B	0.138	–	–	40.0	43.6	63.8	–	–	–	9.0	59.50	46.3
C	0.239	–	–	95.1	89.7	110.1	–	–	–	−5.7	15.77	22.7
D	2.138	–	–	644.9	567.6	984.3	–	–	–	−12.0	52.63	73.4
E	4.882	–	–	1461.9	1280.2	2247.2	–	–	–	−12.4	53.72	75.5
F	0.252	–	–	90.2	91.4	115.8	–	–	–	1.3	28.38	26.7
G	0.999	–	–	395.5	340.3	459.8	–	–	–	−14.0	16.26	35.1
Mean for S01–S04	0.089±0.142	460.3±119.3	41±9.5	39.6±62	38.2±61	41.3±66	1.5±18.6	−3.1±24.1	2.3±15.4	−5.3±11.6	1.3±4.8	8.5±17
Mean for small beds[a]	0.019±0.009	–	–	9.7±4.3	8.4±3.2	8.7±3.9	–	–	–	−9.4±13	−8.1±16.8	2.4±21.4
Mean for medium beds[b]	0.386±0.348	–	–	150.7±281	139±117	177.7±161	–	–	–	−2.3±8.5	24.9±21.1	27.6±14.6
Mean for large beds[c]	3.510±1.940	–	–	1,053.4±578	924±504	1,615.7±893	–	–	–	−12.2±0.3	53.2±0.8	74.5±1.5
Mean for all beds	0.821±1.491	–	–	263±445	234±388	378±686	–	–	–	−6.7±9.9	18.1±28.4	27±30.3
Total	9.027	–	–	2,899.0	2,576.0	4,154.8	–	–	–	−73.7	198.7	296.6
Minimum	0.007	–	–	3.4	3.7	3.3	–	–	–	−18.8	−32.6	−17.1
Upper value	4.882	–	–	1461.9	1280.2	2247.2	–	–	–	9.0	59.5	75.5

[a] Surface area between 0 and 0.1 km^2: beds S01, S02, S04 and A
[b] Surface area between 0.1 and 1 km^2: beds S03, B, C, F and G
[c] Surface area greater than 1 km^2: beds D and E

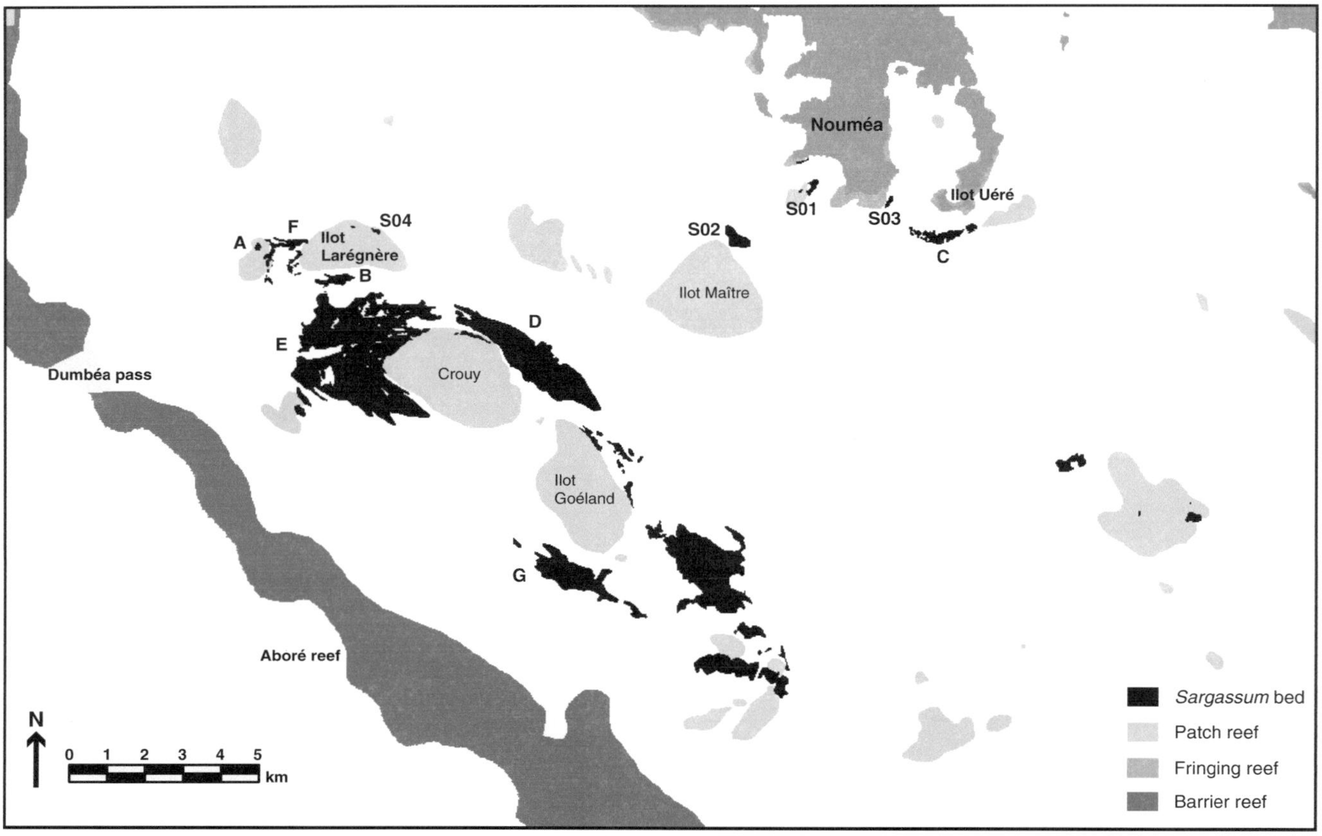

Fig. 3 Distribution of *Sargassum* beds in the South West lagoon of New Caledonia

reef flats less than 4.5 m deep on rubble and/or maerl substrata (S01 and S03). All others occupied the lagoon floor with sandy rocky bottoms and ranged from 2.5 to 12 m deep. As an overall trend, *S. howeanum* was dominant on shallow rubble reef flats whereas *S. spinuligerum* was found to be abundant on shallow-to-deep sandy rocky lagoon floors.

The total surface cover of Sargassaceae ranged from 24.4% in bed E to 51.6% in bed A, with an overall mean of 35.5%. *Sargassum spinuligerum* (including *S. spinuligerum* var. *crispata*), *S. decurrens*, *S. howeanum* and *Hormophysa cuneiformis* were the major species components of the beds, while *S. crassifolium*, *S. polycystum* and *Cystoseira trinodis* were less abundant, with specific surface cover of less than 1%. Sargassaceae species diversity ranged between three and seven species per bed and *S. spinuligerum* occurred at all stations (Table 1).

Seasonal in situ biomass measurements

The mean biomass and standard deviation per station for the different species are reported in Table 3 for the four beds (S01–S04) where destructive quadrat sampling was conducted. Seasonal variations per bed and species are represented in Fig. 4. No significant differences were found except between *S. howeanum* and *S. spinuligerum* during winter 2005 and autumn 2006 ($P<0.05$) within bed S01. However, some trends could be detected. Specifically, bed S01 was dominated by *S. howeanum* throughout the year, except during spring 2006 where *S. spinuligerum* was dominant. At bed S02, the total biomass was shared in different proportions according to season by *S. spinuligerum*, *S. decurrens* and *H. cuneiformis*. Within bed S03, *S. howeanum* dominated the biomass during the winters of 2005 and 2006, spring 2006 and summer 2007, whereas *H. cuneiformis* was more important during summer 2005 and autumn 2006. Bed S04 was dominated by *S. spinuligerum* all year round.

The highest specific biomass was measured for *S. spinuligerum* in bed S04 during summer 2006 (686 g m^{-2} DM), whereas the lowest was measured for the same species in bed S02 during autumn 2006 (6 g m^{-2} DM). As an overall trend, *S. spinuligerum* biomass was higher during warm periods (spring and summer) and lower mostly during cool periods (autumn and winter). *Sargassum howeanum* biomass was maximum during autumn 2006

Table 3 Mean seasonal biomass (in g m^{-2}) per species and per beds surveyed (S01–S04) and standard deviations. *W05* Winter 2005, *S06* Summer 2006, *A06* Autumn 2006, *W06* Winter 2006, *Sp06* Spring 2006, *S07* Summer 2007

Beds	Species/seasons	W05	S06	A06	W06	Sp06	S07
S01	*S. howeanum*	556±148.9	270±185.1	563±58.9	307±85.3	188±131.2	170±140
	S. spinuligerum	9±18	110±77.1	31±54.3	69±92.6	231±119.2	68±89.2
S02	*S. decurrens*	222±164.4	132±92.4	163±32.3	137±48.7	268±117.8	73±71.5
	S. spinuligerum	55±50.1	162±152.6	6±12	0	87±77.7	285±132.2
	H. cuneiformis	0	134±128.5	41±82	452±410.2	137±240.5	65±81.2
S03	*S. decurrens*	0	0	16±18.7	0	0	0
	S. howeanum	249±168.3	145±28.3	124±72.6	281±118.4	222±12.4	344±212.4
	S. spinuligerum	29±19.9	121±111.4	78±44.1	101±103.3	19±38	86±106
	H. cuneiformis	47±47.7	185±125.1	175±84.6	28±32.9	0	38±76
S04	*S. spinuligerum*	313±194.4	686±269.2	679±41.4	603±211	423±131.7	618±201.7

and minimum during summer 2007 at S01, whereas the tendency was inverted at S03 with the highest values observed in summer 2007 and the lowest in autumn 2006. *Sargassum decurrens* was well represented in S02 compared to the three other stations. Its highest and lowest biomasses were measured in spring 2006 and in summer 2007, respectively. *Hormophysa cuneiformis* was conspicuous at S02 and S03. During winter 2006 its highest biomass was recorded for S02 whereas it was the lowest biomass period at S03.

The above results highlight the year-long turnover of different species within each bed, and the large inter-bed variations that we observed in terms of species composition and biomass.

These specific seasonal trends disappeared when the total Sargassaceae biomass per station and for the two contrasting periods of the year (namely warm and cool) were considered as outlined in Table 4. Mean biomass differences between stations were large but were not significant between the warm and the cool periods, or within periods.

Biomass estimation and sensitivity analysis at bed scale

All linear regressions showed a good (r^2>0.70) and significant (P<0.005) correlation between biomass and surface cover. Linear regressions per species (Eq. 1) and for all Sargassaceae species (Eq. 2) including slope coefficients are

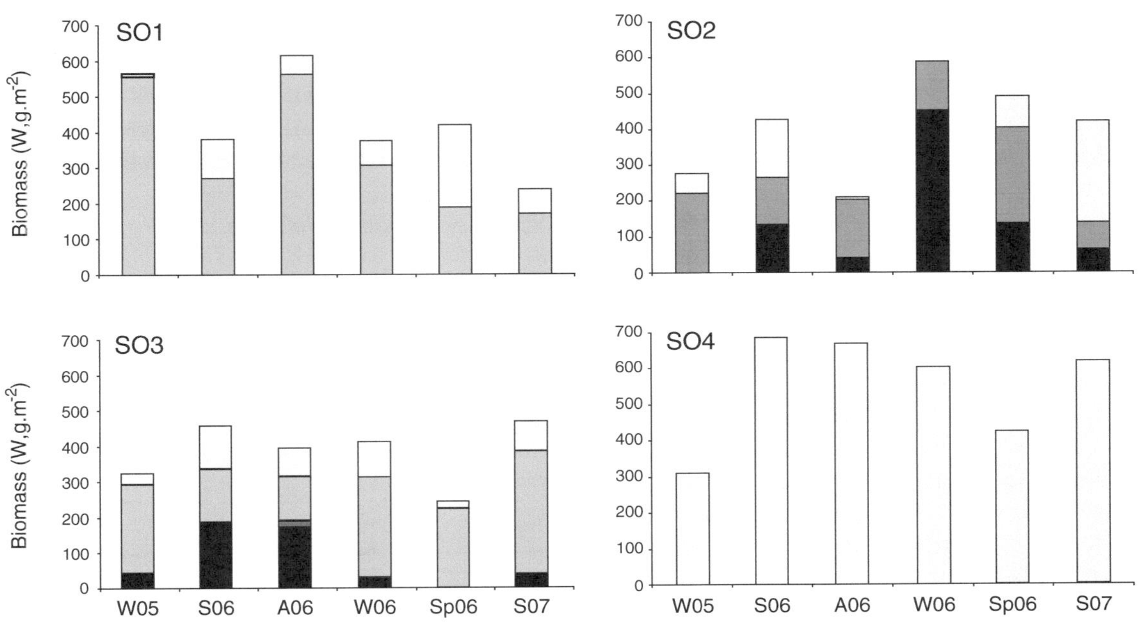

Fig. 4 Mean biomass [in g m^{-2} dry matter (DM)] per species at stations S01–S04 for August 2005 and at intervals of 3 months from January 2006 to January 2007 (*W* winter, *Sp* spring, *S* summer, *A* autumn). White *S. spinuligerum*, light grey *S. howeanum*, dark grey *S. decurrens*, black *H. cuneiformis*

Table 4 Mean biomass and standard deviation (g m^{-2} DM) during warm and cool periods for each of the stations surveyed seasonally (S01–S04)

Station	Warm period[a]	Cool period[b]
S01	345.7±95.3	502.3±109.4
S02	447.7±38.5	358.7±202.3
S03	386.7±126.4	376.0±44.9
S04	575.7±136.5	531.7±193

[a] From October to March
[b] From April to September

detailed in Table 5. Overall annual means per bed (W_{S01}^{mean}, W_{S02}^{mean}, W_{S03}^{mean}, W_{S04}^{mean}) and the average for the four beds ($W_{S01-S04}^{mean}$) are shown Table 2.

The estimated biomasses W_1 (beds S01–S04) and W_2, W_3 and W_4 (all 11 beds) are provided in Table 2. The total biomass for the 11 beds could not be computed with method 1 because of the data requirements for this estimation method. However, the total biomass estimated with methods 2, 3 and 4 were 2,899 t DM, 2,576 t DM and 4,154.8 t DM, respectively.

Relative biomass differences $\Delta_{i/1}$ for beds S01–S04 and series $\Delta_{3/2}$, $\Delta_{4/2}$ and $\Delta_{4/3}$ for all beds are summarised in Table 2. The mean difference estimation between method 2 and method 1 ($\Delta_{2/1}^{mean}$) was 1.5±18.6%. However, bed specific $\Delta_{2/1}$ shows large variations between beds. Estimated biomasses $W_{2,S01}$ and $W_{2,S03}$ were slightly overestimated compared to $W_{1,S01}$ and $W_{1,S03}$ (6.2 and 2.2% respectively), whereas $W_{2,S02}$ was overestimated by 21.1% and $W_{2,S04}$ was underestimated by 23.5%. Nevertheless, we note that W_2 values were contained within the range of uncertainty (standard deviation) of W_1, with the exception of $W_{2,S04}$.

The mean difference estimation between method 3 and method 1 ($\Delta_{2/1}^{mean}$) was −3.1±24.1%. Individually, $W_{3,S01}$ and $W_{3,S02}$ were overestimated, $W_{3,S04}$ was severely underestimated, and $W_{3,S03}$ and $W_{1,S03}$ were similar ($\Delta_{3/1}$=0). The mean difference between method 4 and method 1 ($\Delta_{4/1}^{mean}$) was 2.3±15.4%. The same pattern as for $\Delta_{2/1}^{mean}$ was observed with W_4 values within the range of W_1 standard deviations, except $W_{4,S02}$.

When considering only beds S01–S04, $\Delta_{3/2}^{mean}$, $\Delta_{4/2}^{mean}$, and $\Delta_{4/3}^{mean}$ were all quite low. This was expected, since the parameterisation of all methods came from beds S01–S04, which were deemed representative of the South West Lagoon bed diversity. When considering all 11 beds together, the mean differences $\Delta_{3/2}^{mean}$, $\Delta_{4/2}^{mean}$, and $\Delta_{4/3}^{mean}$showed that, on average, method 3 provided slightly lower results than method 2 ($\Delta_{3/2}^{mean} = -6.7 \pm 9.9$), whereas both methods 3 and 4 provided higher estimates than method 2 ($\Delta_{4/2}^{mean} = 18.1 \pm 28.4\%$, and $\Delta_{4/3}^{mean} = 27 \pm 30.3\%$) (Table 2).

Discussion

Diversity and typology of the *Sargassum* beds

Sargassum beds in the South West Lagoon of New Caledonia are multi-species assemblages, including six species of *Sargassum* and other members of Sargassaceae. These mixed beds represent less than 40% of the New Caledonian diversity of the genus, which currently comprises 15 valid species belonging to tropical, subtropical and temperate waters (L. Mattio, unpublished data). This genus is well represented compared to others regional flora; e.g. 18 species reported from Western Australia (Womersley 1987), 3 from French Polynesia (N'Yeurt and Payri 2007), 4 from the Solomon Islands (Womersley and Bailey 1970), 5 from the Hawaiian Islands (Abbott and Huisman 2004) and 6 from the Marianna Islands (Lobban and Tsuda 2003). The six species are never found all together in the same beds. Species assemblages vary in composition and density between location and within location as well as between depths, making grouping species into general environmental categories difficult. Nevertheless, two main types of *Sargassum* beds can be proposed based on geomorphology, substratum, depth, species diversity and surface cover. The first type, dominated by *S. howeanum*—a short and compact species with a high biomass-to-size ratio and high surface cover—occurs in shallow reef flats with coral rubble. This species is often mixed with *S. spinuligerum* occurring in lesser abundance (surface cover and biomass). The second type, dominated by *S. spinuligerum* associated with the less abundant *S. decurrens* and *H. cuneiformis*, occurs in the deeper lagoon floor down to 12 m depth, with large areas of hard stable substrata (sandy rocky bottom). In both types of *Sargassum* beds, a single species dominates a mixed assemblage. A similar pattern was previously reported by Kendrick (1993) in Australian *Sargassum* beds.

Spatial and seasonal variation

Our study identified large differences in species diversity between beds, reflecting variation in habitat as discussed

Table 5 Slope (*a*) of linear regressions between biomass and surface cover per species and for all major Sargassaceae with correlation coefficient r^2 and *P*-value

Taxa	*a*	r^2	*P*-value
S. spinuligerum	13.25	0.75	<0.05
S. howeanum	9.15	0.93	<0.05
S. decurrens	8.70	0.89	<0.05
H. cuneiformis	15.63	0.70	<0.05
Sargassaceae	10.87	0.79	<0.05

above; however, there is no clear spatial distribution of species. The variation of the mean species surface cover is the only evidence of depth influence, except for *S. howeanum*, which seems to be restricted to the intertidal zone. *Sargassum spinuligerum* and *S. decurrens* are the only species recorded at all stations sampled, regardless of depth. However, for both species, abundances are generally lower in the intertidal zone compared to the subtidal zone. Hence, dominance shifts from *S. howeanum* in shallow beds to a *S. spinuligerum-decurrens* assemblage in deep subtidal beds. Similar spatial distribution patterns have been described in Australia by Kendrick (1993), and in the Philippines by Ang (1986) but further investigations are needed to understand the factors that control species abundance within each bed.

Regarding seasonality, all the species persist throughout the year as macroscopic thalli, but all display temporal variation. Abundance peaks can occur during warm periods (*S. spinuligerum* either in deep or shallow beds), or in cooler periods (*S. decurrens* and *S. howeanum*). Despite these specific trends, there is no clear overall seasonal pattern. *Sargassum* seasonality is well documented, for example in Guam (Tsuda 1971), the Philippines (Hurtado and Ragaza 1999), and in French Polynesia (Payri 1987; Stiger and Payri 1999). All these studies reported highest biomass for *Sargassum* during cool periods. Conversely, Murase et al. (2000) reported the highest *S. macrocarpum* productivity during summer in Japan, and Kilar and Norris (1988) did not detect any clear seasonal pattern in *S. filipendula* from Caribbean Panama.

Extent of *Sargassum* beds

To estimate their extent and distribution, *Sargassum* beds in the South West Lagoon, were characterised using 30 m resolution Landsat 7 images, and the coarseness of the observations was corrected using a unique *k* factor derived from a QB image (4 m resolution). The accuracy of each bed surface estimation remains unknown at this stage, even if field verification and acoustic data (not shown here, but see Payri et al. 2007) have confirmed a good agreement between methods and their respective limitations. In our study, despite possible contiguous seagrass habitats and depth gradients, identification of the boundaries of *Sargassum* beds was fairly straightforward due to their size, density, and contrast with nearby bright soft- and hard-bottom habitats. Furthermore, the comparison between different Landsat images across time shows a high stability of the boundaries. The image contrast differences, which we interpreted as changes in surface cover, are qualitatively consistent with the related biomass variation trends observed in situ at bed scale. Even if in situ measurements of total biomass at bed-scale and across time are not significantly different, there is nevertheless a non-negligible difference, which could explain the optical variations seen on the Landsat time-series.

Observations on bed boundary stability and biomass variation show that the assumptions necessary for using *k* are only partially verified. Bed outer boundaries can be considered as stable, but surface cover varies. This warrants further investigation of the patch dynamics inside algal beds to assess how images from different sensors can be combined to improve biomass estimates.

Sensitivity of biomass estimation at bed scale

The challenge met by our study was to set up effective yet simple methods to predict the abundance of *Sargassum* beds in the South West Lagoon of New Caledonia. The total surface area of the lagoon occupied by *Sargassum* beds could be estimated with simple optical remote sensing techniques. Measuring surface cover and biomass directly from space has proved possible in a variety of studies (mostly for seagrass). We have not yet attempted this approach, but used field data to provide a first scaled estimate. Four methods were used to estimate the total biomass of the different beds. Three of the methods can be carried out without destructive sampling to any bed once a generic cover-biomass relationship has been established (Methods 2–4). The estimation differences ($\Delta_{i/j}$), computed over the 11 beds considered in this study, were lower for methods 2 and 3 (measuring bed-specific surface cover at species or assemblage level) than for method 4 (using a mean annual value). The lower agreement achieved with method 4 is due to the absence of bed-specific heterogeneity parameters in Eq. 6 (see Materials and methods) compared to methods 2 and 3 (Eqs. 4 and 5, see Materials and methods). Furthermore, the errors due to smoothing bed heterogeneities increase logically as the beds get larger. Thus, for large beds (surface areas >1 km^2 n=2, see Table 2) the overestimation of method 4 vs methods 2 and 3 is larger than for medium-sized beds (surface area between 0.1 and 1 km^2, n=5) or small beds (surface area <0.1 km^2).

The relative differences achieved between methods suggest that methods 2 and 3 are acceptable, when considering method 1 (measuring bed-specific surface cover and quadrat-based biomass per species) as the reference. Method 4 is not recommended, especially for large beds, but it could be used for a gross preliminary rapid estimation. Method 2 requires specific diversity and surface cover and avoids destructive and time-consuming sampling of quadrats. Method 3 was designed on the same principle as method 2 but used only total Sargassaceae surface cover, thus avoiding both quadrat sampling and species identification. This method appears as a good compromise, representing a rapid, inexpensive and simple way of estimating biomass over large areas with acceptable difference compared to the reference method 1.

This study improved our knowledge on the extent, diversity and biomass of *Sargassum* beds in New Caledonia. It clarified the spatial and temporal variations of the targeted populations, and represents a baseline for further monitoring studies. We concluded that biomass estimation method 3, involving the measurement of bed-specific surface cover at assemblage-level, is an effective method with which to monitor *Sargassum* bed dynamics.

Acknowledgements This research is part of the PhD projects of L.M. and G.D. carried out at the Institut de Recherche pour le Développement (IRD) of Nouméa, New Caledonia, with financial support from the ZoNéCo program and from the Province Sud of New Caledonia (Prix d'encouragement à la recherche de la Province Sud). A NASA Interdisciplinary grant NNG04GO90G to S.A. provided extra support for this study. Satellite images were provided by the Millennium Mapping Project (NASA grant NAG5–10908 to SA) and by ZoNéCo. The authors thank the IRD divers' team and boat pilots for their valuable assistance in the field. They also sincerely thank Dr. C. Chevillon for his work with the RoxAnn system and Dr. V. Stiger for her efficient assistance with molecular techniques. Finally, we thank Dr. ADR N'Yeurt for improving the English of the manuscript and the two anonymous reviewers for their in depth valuable comments.

References

Abbott IA, Huisman JM (2004) Marine green and brown algae of the Hawaiian Islands. Bishop Museum Press, Honolulu

Agardh CA (1820) Species algarum rite cognitae, cum synonymis, differentiis specificis et descriptionibus succinctis, vol 1(1). Berling, Lund

Agardh JG (1848) Species genera et ordines algarum, seu descriptiones succinctae specierum, generum et ordinum, quibus algarum regnum constituitur. In: Algas fucoideas complectens, vol 1. Gleerup CWK, Lund

Agardh JG (1889) Species Sargassorum Australiae descriptae et dispositae. Kongl Svenska Vetenskapsakademiens Handlingar 23:1–133

Andréfouët S, Kramer P, Torres-Pulliza D et al (2003) Multi-sites evaluation of IKONOS data for classification of tropical coral reef environments. Remote Sensing Environ 88:128–143

Andréfouët S, Zubia M, Payri C (2004) Mapping and biomass estimation of the invasive brown algae *Turbinaria ornata* (Turner) J Agardh and *Sargassum mangarevense* (Grunow) Setchell on heterogeneous Tahitian coral reefs using 4-meter resolution IKONOS satellite data. Coral Reefs 23:26–38

Ang PO Jr (1986) Analysis of the vegetation structure of a *Sargassum* community on the Phillipines. Mar Ecol Prog Ser 28:9–19

Bellwood DR, Hughes TP, Hoey AS (2006) Sleeping functional group drives coral-reef recovery. Curr Biol 16:2434–2439

Belsher T, Mouchot MC (1992) Use of satellite imagery in management of giant kelp resources, Morbihan Gulf, Kerguelen archipelago. Oceanol Acta 15:297–307

Chase M, Hills H (1991) Silica gel: an ideal material for field preservation of leaf samples for DNA studies. Taxon 40:215–220

Chevillon C (2005) Caractérisation des types de fonds et habitats benthiques par système de discrimination acoustique dans le lagon sud-ouest de la Nouvelle-Calédonie. Rapport final d'opération ZoNéCo 2000–2004, IRD, Nouméa

Critchley AT, Farnham WF, Yoshida T, Norton TA (1990) A bibliography of the invasive alga *Sargassum muticum* (Yendo) Fensholt (Fucales, Sargassaceae). Bot Mar 33:551–562

Davis TA, Volesky B, Vieira VHSF (2000) *Sargassum* seaweed as biosorbent for heavy metals. Water Res 34:4270–4278

English S, Wilkinson C, Baker V (1997) Survey manual for tropical marine resources, 2nd edn. Australian Institute of Marine Science, Townsville

Garrigue C, Tsuda RT (1988) Catalog of the Benthic Algae from New Caledonia. Contrib Univ Guam Mar Lab 268:53–70

Godoy EAS, Coutinho R (2002) Can artificial beds of plastic mimics compensate for seasonal absence of natural beds of *Sargassum furcatum*. J Mar Sci 59:S111–S115

Grunow A (1915) Additamenta ad cognitionem Sargassorum. Verh Zool-Bot Ges Wien 65:329–448

Grunow A (1916a) Additamenta ad cognitionem Sargassorum. Verh Zool-Bot Ges Wien 66:1–48

Grunow A (1916b) Additamenta ad cognitionem Sargassorum. Verh Zool-Bot Ges Wien 66:136–185

Guiry MD, Nic Dhonncha E (2007) AlgaeBase version 30 World-wide electronic publication. National University of Ireland, Galway http://wwwalgaebaseorg searched on 2 October 2007

Hurtado A, Ragaza A (1999) *Sargassum* studies in Currimao, Ilocos Norte, Northern Philippines, I: Seasonal variation in the biomass of *Sargassum carpophyllum* J. Agardh, *Sargassum ilicifolium* (Turner) C. Agardh and *Sargassum siliquosum* J. Agardh (Phaeophyta, Sargassaceae). Bot Mar 42:321–325

Jacquet S, Delesalle B, Torréton JP, Blanchot J (2006) Response of phytoplankton communities to increased anthropogenic influences (southwestern lagoon, New Caledonia). Mar Ecol Prog Ser 320:65–78

Kendrick GA (1993) *Sargassum* beds at Rottnest Island: species composition and abundance. In: Wells FE, Walker DI, Kirkman H, Lethbridge R (eds) Proceedings of the fifth International Marine Biological Workshop Western Australian Museum, Perth, 1993

Kilar JA, Norris JN (1988) Composition, export, and import of drift vegetation on a tropical, plant-dominated, fringing-reef platform (Caribbean Panama). Coral Reefs 7:93–103

Labrosse P, Fichez R, Farman R, Adams T (2000) New Caledonia. In: Sheppard C (ed) Seas at the millenium: an environmental evaluation, Vol 2. Elsevier Science, Amsterdam, pp 723–736

Lobban CS, Tsuda RT (2003) Revised checklist of benthic marine macroalgae and seagrasses of Guam and Micronesia. Micronesica 35–36:54–99

May V (1953) Some marine algae from New Caledonia collected by Mrs Catala. Contrib NSW Nat Herb 2(1):38–66

McClanahan TR, Nugues M, Mwachireya S (1994) Fish and sea urchin herbivory and competition in Kenyan coral reef lagoons: the role of reef management. J Exp Mar Biol Ecol 184(2):237–254

McCook LJ (1999) Macroalgae, nutrients and phase shifts on coral reefs: scientific issues and management consequences for the Great Barrier Reef. Coral Reefs 18:357–367

Moseley LM, Aalbersberg WGL (2003) Nutrient levels in sea and river water along the 'Coral Coast' of Viti Levu, Fiji. S Pac J Nat Sci 21:35–40

Mukai H (1971) The phytal animal on the thalli of *Sargassum serratifolium* in the *Sargassum* region, with reference to their seasonal fluctuations. Mar Biol 8:170–182

Mumby P, Green J, Edwards AJ (1997) Measurement of seagrass standing crop using satellite and digital airborne remote sensing. Mar Ecol Prog Ser 159:51–60

Mumby PJ, Green EP, Clark CD (1998) Digital analysis of multispectral airborne imagery of coral reefs. Coral Reefs 17:59–69

Murase N, Kito H, Mizukami Y, Maegawa M (2000) Productivity of *Sargassum macrocapum* (Phaeophta, Fucales) in Fukawa Bay, Sea of Japan. Fish Sci 6:270–277

N'Yeurt ARD, Payri CE (2007) Marine algal flora of French polynesia, I: Phaeophceae (Ochrophyta, brown algae). Cryptogamie Algol 27:111–152

Nizamuddin M (1962) Classification and distribution of the Fucales. Bot Mar 4:191–203

Noro T, Abbott IA (1994) A tentative identification of some specimens of *Sargassum* from the tropical South Pacific. In: Abbott IA (ed) Taxonomy of economic seaweeds, vol 4. California Sea Grant College Program, pp 33–34

Okamura K (1932) The distribution of marine algae in Pacific waters. Rec Oceanogr Works Jpn 4:30–150

Ornellas AB, Coutinho R (1998) Spatial and temporal patterns of distribution and abundance of a tropical fish assemblage in a seasonal *Sargassum* bed, Cabo Frio Island, Brazil. J Fish Biol 53 (A):198–208

Payri CE (1987) Zonation and seasonal variation of the commonest algae on Tiahura reef (Moorea Island, French Polynesia). Bot Mar 30:141–149

Payri C, Andréfouët S, Mattio L et al (2007) Typologie des algueraies à sargasses et leurs cartographies par méthodes optiques et acoustiques. Rapports de convention, IRD, Nouméa

Phillips N (1995) Biogeography of *Sargassum* (Phaeophyta) in the Pacific basin. In: Abbott IA (ed) Taxonomy of economic seaweeds, vol 5. California Sea Grant College Program, pp 107–144

Phillips N, Smith C, Morden CW (2005) Testing systematic concepts of *Sargassum* (Fucales, Phaeophyceae) using portion of the *rbc*LS operon. Phycol Res 53:1–10

Riegl B, Moyer R, Morris L, Virnstein RW, Purkis SJ (2005) Distribution and seasonal biomass of drift macroalgae in the Indian River Lagoon (Florida, USA) estimated with acoustic seafloor classification (QTCView, Echoplus). J Exp Mar Biol Ecol 326:89–104

Rossier O, Kulbicki M (2000) A comparison of fish assemblages from two types of algal beds and coral reefs in the south-west lagoon of New Caledonia. Cybium 24:3–26

Schaffelke B (1999) Particulate organic matter as an alternative nutrient source for tropical *Sargassum* species (Fucales, Pheaophyceae). J Phycol 35:1150–1157

Sheng PX, Ting YP, Chen JP, Hong L (2004) Sorbtion of lead, copper, cadmium, zinc and nickel by marine algal biomass: characterization of biosorptive capacity and investigation of mechanisms. J Colloid Interface Sci 275:131–141

Stiger V, Payri CE (1999) Spatial and seasonal variations in the biological characteristics of two invasive brown algae, *Turbinaria ornata* (Turner) J. Agardh and *Sargassum mangarevense* (Grunow) Setchell (Sargassaceae, Fucales) spreading on the reefs of Tahiti (French Polynesia). Bot Mar 42:295–306

Stiger V, Horiguchi T, Yoshida T, Coleman AW, Masuda M (2000) Phylogenetic relationships of *Sargassum* (Sargassaceae, Phaeophyceae) with reference to a taxonomic revision of the section *Phyllocystae* based on *ITS*-2 nrDNA sequences. Phycol Res 48:251–60

Stiger V, Horiguchi T, Yoshida T, Coleman AW, Masuda M (2003) Phylogenetic relationships inferred from *ITS*-2 nrDNA comparisons within the genus *Sargassum* (Fucales, Phaeophyceae) from the Pacific basin, with an emphasis on the taxonomic subdivision of the genus. Phycol Res 51:1–10

Stiger V, Deslandes E, Payri CE (2004) Phenolic contents of two brown algae, *Turbinaria ornata* and *Sargassum mangarevense* on Tahiti (French Polynesia): interspecific, ontogenic and spatio-temporal variations. Bot Mar 47:402–409

Tanaka MO, Leite FPP (2003) Spatial scaling in the distribution of macrofauna associated with *Sargassum stenophyllum* (Martens) Martius: analysis of faunal group, gammarid life habits, and assemblage structure. J Exp Mar Biol Ecol 293:1–22

Tenorio RRB, Le Borgne R, Rodier M, Neveux J (2005) The impact of terrigenous inputs on the bay of Ouinné (New Caledonia) phytoplankton communities: aspectrofluorometric and microscopic approach. Estuarine Coastal Shelf Sci 64:531–545

Tsuda RT (1971) Morphological, zonational, and seasonal studies of two species of *Sargassum* on the reefs of Guam. In: Proc 7th Int Seaweed Symp, University of Tokyo Press, Japan, pp 40–44

White WH, Harborne AR, Sotheran IS, Walton R, Foster-Smith RL (2003) Using an acoustic ground discrimination system to map coral reef benthic classes. Int J Remote Sensing 24:2641–2660

Womersley HBS (1987) The marine benthic flora of southern Australia Part II. South Aust Gov Print Div, Adelaide pp 418–446

Womersley HBS, Bailey A (1970) Marine algae of the Solomon islands. Phil Trans R Soc B 259:257–352

Yoshida T (1989) Systematics of *Sargassum* (Fucales, Phaeophyceae). Korean J Phycol 4:107–10

Zubia M, Payri CE, Deslandes E, Guezennec J (2003) Chemical composition of attached and drift brown algae, *Sargassum mangarevense* and *Turbinaria ornata* (Phaeophyta: Fucales), from Tahiti (French Polynesia). Bot Mar 46:562–571

Sargassum muticum (Yendo) Fensholt in Ireland: an invasive species on the move

Stefan Kraan

Originally published in the Journal of Applied Phycology, Vol 20, No 5, 375–382.
DOI: 10.1007/s10811-007-9208-1

Abstract The invasive marine macroalga *Sargassum muticum* was recorded from Kilmore Quay, Co. Waterford, for the first time in the Republic of Ireland in 2001. It was found at several other places in that year and in subsequent years. An intensive survey was carried out from April 2003 to November 2003 to map the geographical distribution and spread of this invasive brown macroalga, and data on distribution were collected in subsequent years through 2006. This study indicated that *S. muticum* most probably arrived in the early or mid 1990s and has spread all around the Irish coastline colonising Co. Donegal for the first time in 2006. The results indicate that spreading is facilitated by boating and perhaps via shellfish transport. A rough rate of spread of 2–3 km year^{-1} has been calculated within one bay and for the Irish coastline of about 54 km year^{-1}. Observations showed that *S. muticum* has been found growing in seagrass beds and in rock pools, which might have serious consequences for the biodiversity in rock pools and for the protective status of seagrass habitats. It is recommended that monitoring of *S. muticum* should continue, especially in areas of high amenity value and economic importance in order to observe possible effects on local flora and fauna, aquaculture, the seaweed industry and tourism.

Keywords Geographical distribution · Ireland · Point introduction · Rock pools · *Sargassum muticum* · Seagrass beds · Spreading rate · Transport · Vector

S. Kraan (✉)
Irish Seaweed Centre, Martin Ryan Institute,
National University of Ireland, Galway,
Galway, Ireland
e-mail: Stefan.kraan@nuigalway.ie

Introduction

Sargassum muticum (Yendo) Fensholt was recorded for the first time in the Republic of Ireland at Kilmore Quay in August 2001 (Fahy 2001). Kilmore Quay is a popular yachting harbour attracting yachts from France and the United Kingdom. It was subsequently found at Rath Strand, Kenmare River, Co. Kerry (Winder 2002); Cashel Bay, Co. Galway (Loughnane and Stengel 2002); and Drum Cliff Bay, Co. Sligo (Callagher, personal communication, 2002; Fig. 1, see names marked with a black asterisk). It is important to note that these plants were attached, indicating the presence of established populations.

Sargassum muticum was endemic to Japan prior to 1944, when it was introduced to Pacific Canada (Ribera and Boudouresque 1995). In the last 20–40 years, the geographical range of this species has spread dramatically. It was discovered for the first time in Europe in 1973 off the Isle of Wight in Britain (Farnham et al. 1973; Farnham and Jones 1974) and was then found in the Mediterranean by 1980 (Ribera and Boudouresque 1995). Moreover, it was found in the Isles of Scilly, along the entire English Channel coast (Hiscock and Moore 1986) and the east coast north to Suffolk; however, the Norfolk population appears to be no longer extant (Critchley et al. 1983). Attached specimens were found in Strangford Lough, Northern Ireland, in 1995 (see Fig. 1 marked with ©) and in north Cornwall in 1991 (Boaden 1995). Elsewhere in Europe it is known from the Mediterranean and along the North Sea and Atlantic coasts of Portugal, Spain, France, Belgium, the Netherlands, Germany, Denmark, southern Norway and Sweden (Critchley et al. 1983; Kornmann and Sahling 1994; Rueness 1989). In Denmark the first specimen of *S. muticum* was found in 1984 in Nissum Bredning, Limfjorden (Christensen 1984). Since then it

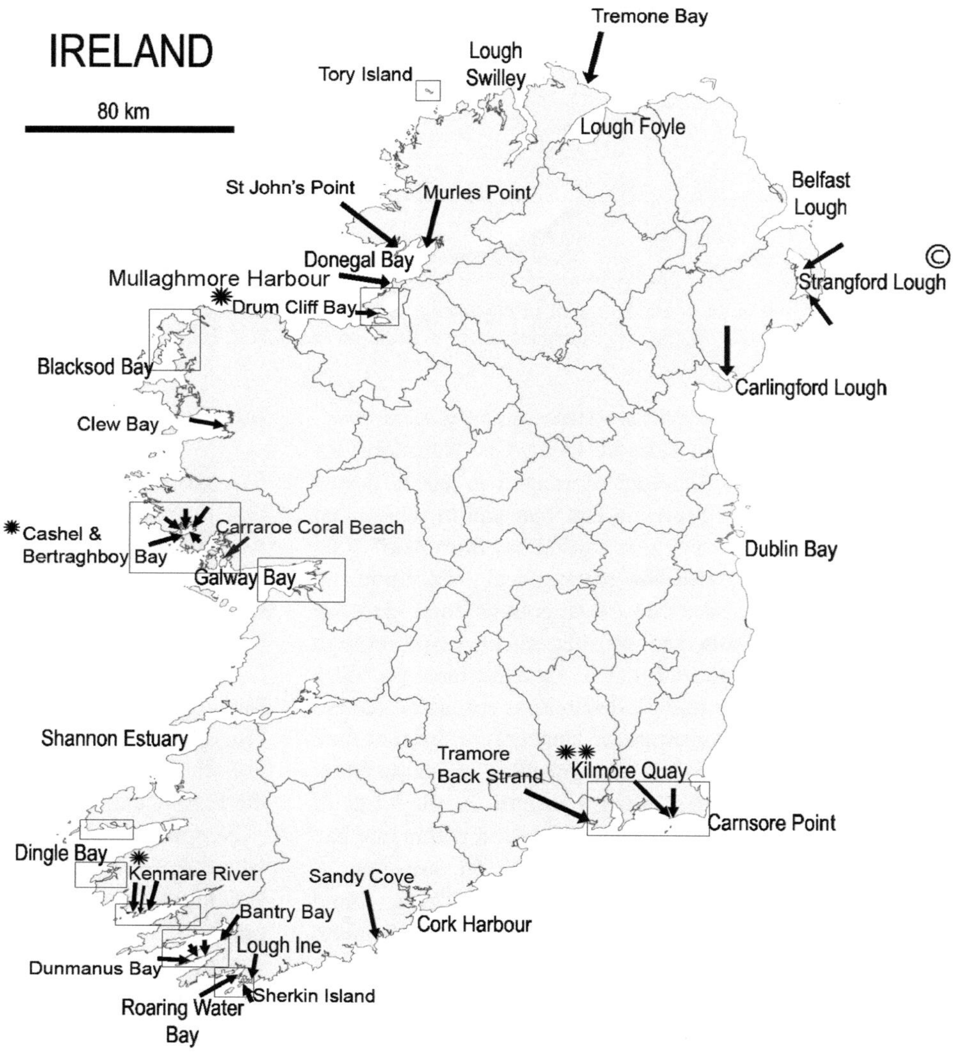

Fig. 1 Records of *Sargassum muticum* in Ireland since its discovery in Strangford Lough, Northern Ireland, in 1995 (marked with ©) and subsequently for the first time in the Republic of Ireland in Kilmore Quay in 2001 (marked with two black asterisks) and other places later that year (marked with one black asterisk). Records have been collected up to May 2007. Boxed areas were surveyed in 2003. Arrows indicate established beds or attached plants of *S. muticum*

has spread throughout the fjord, where it is now the dominant macroalga. Remarkably, it was estimated that *S. muticum* at the current rate of spreading would arrive in Irish coastal waters during the mid 1980s. However, it took more than 15 years to establish itself in Irish waters.

It is not clear by what mechanism *S. muticum* was introduced into Europe; however, the most likely vectors were out-growing of imported oysters or boating activities (Farnham and Jones 1974; Ribera and Boudouresque 1995). According to Critchley et al. (1990), it was an associated unintentional introduction with commercial introductions of oysters from the Canadian state of British Columbia or Japan to France. Spreading from northern France is presumed to have occurred by natural means. Propagules or fragments may be transported in ballast water, on ships' hulls and by rafting or floating of entire plants or detached fragments. Marginal dispersal (up to 30 miles) is most likely to occur by the latter method (Farnham et al. 1981). The spread along the English south coast was calculated at about 30 km year^{-1} and along the northwest American coast at an average rate of about 60 km year^{-1}, mostly by drifting of fertile adult fragments (Farnham et al. 1981). In Denmark a colonisation rate of 15 km year^{-1} was found in Limfjorden indicating that the colonisation rate is strongly dependent on factors such as currents, prevailing winds and geography (Stæhr et al. 2000). *Sargassum muticum* can outcompete local species due to its fast growth rate and can reproduce within the first year of life. It has a diplontic life cycle and, being monoecious, can fertilize itself (Yendo 1907; Jensen 1974). Fertilized embryos remain attached to the receptacle until they develop tiny, adhesive rhizoids. At this point they are released and recruit to the substratum in close proximity to the parent plant (Deysher and Norton 1982).

Reasons for successful invasions

Sargassum muticum possesses characteristics of both r- and K-selected species, such as a high growth rate of up to 4 cm day^{-1} (r: Norton 1976; Rueness 1989; Arenas et al. 1995), large propagule production (r: Norton 1976; Rueness 1989), high thallus differentiation (K: Jensen 1974) and long lifespan (K: Critchley 1981). Individuals of *S. muticum* have the highest biomass in July-August (Wernberg et al. 2001). Fertile receptacles, which are cast off during the summer months, are able to float and can survive for up to 3 months (Farnham et al. 1981). A single branch is capable of producing a large progeny. Adult plants are tolerant of a wide range of salinities and temperatures (−1 to 25°C), nevertheless the fertilization process and small germlings seem to be sensitive to low salinities (Steen 2004). The combination of these factors is thought to be the reason for its successful invasion of American and European waters (Hales and Fletcher 1989; Rueness 1989).

Effects on the environment and commercial interests

According to Farnham and Jones (1974) *S. muticum* can be a nuisance, both economically and ecologically. It causes the physical displacement of native species by over-growing and shading underlying species. It has the potential drastically to alter the ecosystem of shallow coastal waters and rock pools by out-competing other species and reducing the overall (seaweed) biodiversity (Critchley et al. 1986). There has been documented replacement of *Saccharina latissima* (Linnaeus) C.E. Lane, C. Mayes, Druehl & G.W. Saunders and *Zostera marina* Linnaeus at Grandcamp in Normandy on the French Atlantic coast (Givernaud et al. 1991), impacts on other algae and overall biodiversity at the American Pacific coast (Britton-Simmons 2004) and decrease in cover of several indigenous species belonging to the genera *Laminaria* J.V. Lamouroux, *Fucus* Linnaeus, and *Codium* Stackhouse indicating competitive displacement through competition for hard substrate and light (Stæhr et al. 2000).

In Britain, *S. muticum* has been found growing in rock pools and channels and in eelgrass beds (Critchley et al. 1986; Raines et al. 1992). Apparently, *Halidrys siliquosa* (Linnaeus) Lyngbye can be displaced by *S. muticum* as the dominant species (Raines et al. 1992; Stæhr et al. 2000). Withers et al. (1975) reported a rich epiphytic community associated with *Sargassum* collected from the east Solent, United Kingdom, suggesting that native epiphytic species are not particularly affected. *Sargassum muticum* has been responsible for fouling fishing gear and boat propellers (causing boating accidents), as well as blocking intake pipes (aquaculture) and has been reported to interfere with recreational use of waterways, particularly when it becomes detached from holdfasts and floats off forming large masses (Farnham 1980; Critchley et al. 1986). It is also a fouling organism on oyster beds and a nuisance to commercial fishermen, fouling their nets and 'stealing' colonised oysters by rafting (Critchley et al. 1986).

Eradication

The eradication of this species in British waters has been attempted but has failed. In common with almost all other marine invaders, *S. muticum* cannot be eradicated, although it can be controlled by seasonal harvesting in problem areas such as shellfish grounds and shallow channels frequented by small motor vessels. Removing *Sargassum* by hand is extremely time-consuming and needs to be repeated, probably indefinitely (Farnham 1980). Removal by trawling, cutting and suction has also been tried. Chemical methods using herbicide have been tried but failed due to lack of selectivity and the large doses needed. Small germlings are consumed by molluscs and amphipods but this seems to have no restrictive effect on *S. muticum*. Whatever method is used, the alga always quickly regrows, and effective methods for its permanent removal have not been found, although the preferred method is cutting and suction (Farnham et al. 1981; Critchley et al. 1986; Schaffelke et al. 2006).

The recent arrival of *S. muticum* on shores in the Republic of Ireland has initiated an investigation into the current distribution of this brown invasive macroalga in Ireland. The aim of this investigation is to determine the present status, distribution and invasive capability of *S. muticum* on the north, west and south coasts of Ireland.

Materials and methods

Sargassum survey areas and methodology

Based on records of *S. muticum* in 2001 and 2002, Cashel Bay, Co. Galway; Killmore Quay, Co. Waterford; Kenmare River, Co. Kerry; and Drum Cliff Bay, Co. Sligo, and areas in close proximity to these bays were chosen for surveying and are marked with square boxes in Fig. 1. Surveying took place by foot in the lower intertidal zone at low water, and along drift line at high water, by kayak at various stages of the tide spanning intertidal to subtidal, and by snorkelling or SCUBA diving at various stages of the tide spanning intertidal to subtidal. Names of surveyors, dates and places were recorded together with a site description and other general information such as bottom substrata and marine flora. Drift plants of *S. muticum* were noted, and attached plants were recorded using a GPS (Magellan, Meridian Marine) together with Map Send BlueNav Charts (Thales

Table 1 Bays and inlets where *S. muticum* was found during the survey and in the subsequent years 2004–2006 together with their GPS coordinates

Place	County	GPS	Bed or individual plants	Density per m²	Survey method	Activity	Reference
Inner Cashel Bay	Co. Galway	N53°25′01″ W9°48′67″ to N53°24′87″ W9°48′17″	B	3	K	LF, F, Y	Irish Seaweed Centre 2003
Middle Cashel Bay		N53°24′97″ W9°49′72″ to N53°24′81″ W9°48′97″	B	3–5	K, S	LF, F, Y	Irish Seaweed Centre 2003
Outer Cashel Bay		N53°23′82″ W9°50′68″	15 over 300 m	< 0.1	OF	LF, F, Y	Irish Seaweed Centre 2003
		N53°24′06″ W9°51′71″	40 over 100 m	< 0.1			
		N53°23′25″ W9°51′88″ to N53°23′55″ W9°52′73″	25 over 900 m	< 0.1			
Blackhaven Bay		N53°23′81″ W9°53′37″	B	0.2	K	LF, F	Irish Seaweed Centre 2003
		N53°23′70″ W9°53′17″	B	3			
		N53°23′28″ W9°53′63″	B	1			
Roundstone Bay		N53°23′31″ W9°54′32″	FI	-	OF	Y, LF F	Irish Seaweed Centre 2003
Illaunaknock east of Glinsk Pier		N53°23′04″ W9°48′67″	B	5	S	LF, F	Irish Seaweed Centre 2003
Leitir Árd Quay, Inishtreh Aound		N53°21′92″ W9°53′48″	B	3	S	LF	Irish Seaweed Centre 2003
Carraroe coral beach		N53°14′92″ W9°37′80′	FI	5	OF	RP	Stefan Kraan, pers. obs., 2007
Reen Point Loughaun Pier, Bantry Bay	Co. Cork	N51°38′75″ W9°34′97″	B	6–8	S	A	Irish Seaweed Centre 2003
Farranamagh, Kilcrohan, Dunmannus Bay		N51°34′65″ W9°41′20″	B	3	D	LF, Y	P. Whelan, pers. obs., Botany Dept. University College Cork, 2002
Kitchen Cove Dunmannus Bay		N51°35′73″ W9°38′25″	FI	-	S	LF, F	Irish Seaweed Centre 2003
Carriglea Rock, Dunmannus Bay		N51°36′62″ W9°35′29″	RP	7	OF	-	Irish Seaweed Centre 2003
Sherkin Island, Roaring Water Bay		N51°28′60″ W9°24′80″	MI	-	OF	LF, Y	Marie Mahon, pers. comm., 2004
White Hall, Roaring Water Bay		N51°31′15″ W9°25′60″	FI	-	K	A	Irish Seaweed Centre 2003
Lough Hyne		N51°29′80″ W9°17′80″	FI	-	-	Y	Simkanin 2004
Sandy Cove, Kinsale Harbour		N51°40′54″ W8°31′33″	B	3	OF	Y	Irish Seaweed Centre 2005
Rath Strand, Caher Daniel	Co. Kerry	N51°45′30″ W10°06′09″	RP	60	K	Y	Irish Seaweed Centre 2003
Rath Slip, Caher Daniel		N51°45′90″ W10°04′78″	B	1–2	K	Y, LF	Irish Seaweed Centre 2003
West Cove		N51°46′10″ W10°03′20″	B	3	K	Y, LF	Irish Seaweed Centre 2003
Rosses Point, Yachting club	Co. Sligo	N54°18′36″ W8°34′52″	FI	-	S	Y	Irish Seaweed Centre 2003
Mullaghmore Harbour, Donegal Bay		N54°27′85″ W8°26′55″ to N54°27′70″ W8°26′45″	B	1–5	OF	Y, LF	D. Minchin, Marine Organism Investigations, 2005
Ross Laher, Kilmeena, Clew Bay	Co. Mayo	N53°50′80″ W09°34′20″	MI	On oysters and trestle	OF	A	Padraig Gannon, Croagh Patrick Seafoods, 2003
Killmore Quay, Forlorn Point	Co. Waterford	N52°10′29″ W06°35′47″ to N52°10′63″ W06°33′56″	B	3–5	OF	Y, LF, F	Irish Seaweed Centre 2003

Calingford Marina	Co. Louth	N54°03'11" W06°11'38" to N54°03'12" W06°11'40"	FI	-	OF	Y	D. Minchin, Marine Organism Investigations, 2005
East Bay, St John's Point	Co. Donegal	N54°32'92" W08°27'51"	B	3	S	LF	Julia Nunn, Ulster Museum and Botanic Gardens, Belfast, 2006
Murles Point	Co. Donegal	N54°34'12" W08°17'64"	MI	-	RP	LF	Julia Nunn, Ulster Museum and Botanic Gardens, Belfast, 2006
Tremone Bay, Inishowen Peninsula		N55°16'32" W07°04'19"	RP	-	S	F, LF	Julia Nunn, Ulster Museum and Botanic Gardens, Belfast, 2006

A Aquaculture, *B* beds, *D* dredging, *F* fisheries, *FI* few individuals, *MI* many individuals but not beds, *K* kayak, *S* snorkelling/diving, *Y* yachting, *LF* lobster fisheries, *OF* on foot, *RP* rock pool

Navigation, UK, 2002) and software (version 4.0). The site and *S. muticum* plants (drift and attached) were photographed using a digital camera.

Wind and current profiles around Ireland

Prevailing wind data and seawater current profiles were obtained from the Irish Marine Institute (Marine Institute 2006), showing a southwesterly wind as the prevailing wind 28% of the time and a northwesterly wind 23% of the time. The water circulation in this area is dominated by the poleward-flowing slope current. This persists throughout the year north of Porcupine Bank and is stronger in the summer. South of the bank the current is present in the winter months, but breaks down in the summer, when flow becomes complex. There is also a weaker current flowing north from Brittany and splitting east and west along the Irish coast. In general there is a current flowing clockwise around Ireland.

Biomass and extent of the beds at Kilmore Quay, Co. Waterford

A sub-sample of plants were measured for length (in cm) and wet weight (in grams) where large beds were established, using a measure tape and a hanging balance respectively. A total of 29 attached plants were measured at Cashel Bay, Connamara, Co. Galway (N53°25.106′ W9°48.732′) on the 7th of August 2003; 5 attached plants at Rath Slip, Rath Strand, Kenmare River, (N51°45.432′ W10°05.728′) on the 12th and 13th of August 2003; and 8 attached plants at Cloonile Bay, Connamara, Co. Galway (N53°23.699′ W9° 53.174′) on the 8th of August 2003. A quadrat was used to determine the number of plants per square meter. With the average wet weight per plant and number of plants per square meter, a standing stock for a given area could be calculated for August. A study by Wernberg et al. (2001) showed that individuals of *S. muticum* had a significantly higher biomass in July-August than at any other time of the year. One area with an extensive large bed was at Kilmore Quay. Biomass for August was calculated for Kilmore Quay using the surface area of the bed, the number of plants per square meter and the average wet weight of the plant measurements.

Results

Distribution and length and wet weight measurements of *S. muticum* in Ireland

Areas with attached *S. muticum* found during the survey in 2003 and in the subsequent years 2004, 2005 and 2006 are shown by arrows in Fig. 1. Additional data are presented in Table 1. The overall distribution map of *S. muticum* in Ireland (Fig. 1) shows that the species has spread all around the island of Ireland (excluding the sandy beaches at the east coast between Carlingford Lough and Carnsore Point, as these beaches offer no substrata for seaweeds to grow on), with the latest records from Lough Hyne and Sandy Cove, Co. Cork; St John's Point and Tramone Bay, Co. Donegal; and Carraroe Coral Beach, Co. Galway (Fig. 1). Once established, the alga starts spreading from the original point of introduction. Average length and weight measurement were taken of *S. muticum* beds in three bays in August 2003 (Table 2). The smallest plant was 50 cm and the tallest plant 470 cm. The lightest plant was 75 g and the heaviest was 2,300 g wet weight.

Biomass estimation for the bed at Kilmore Quay

The area from the eastern harbour wall of the marina at Kilmore Quay exposed a large scattered *S. muticum* bed with 3 plants/m^2. Plants of up to 4 m were commonly found. The most easterly *S. muticum* plants were found at N52°10′63″ W6°33′56″ and were attached on rocks within a seagrass bed. The bed at Kilmore Quay had a calculated surface area of 234,600 m^2 containing approximately 700,000 plants. The standing stock of the Kilmore Quay *S. muticum* bed for the month of August using the average weight measurements (Table 2) of 696±501 g per plant is estimated to be 487 t wet weight with a minimum of 137 t and a maximum of 838 t. This large, well-established population indicates that it has probably been present since the early 1990s and most probably has acted as the source for further distribution around the Irish west coast via boating.

Table 2 Average length (cm) and wet weight (g) with standard deviations (SD) of *S. muticum* plants sampled at Cashel Bay, Cloonile Bay and Kenmare River in August 2003

Sample area and date		*n*	Average	SD
Cashel Bay, Connamara, Co. Galway; 7 August 2003	Length	29	168.2	111.3
	Wet weight	29	383.1	321.5
Cloonile Bay, Connamara, Co. Galway; 8 August 2003	Length	8	235	71.5
	Wet weight	8	715.6	467.1
Rath Slip, Rath Strand and Westcove, Kenmare River, Co. Kerry; 12–13 August 2003	Length	5	162	168.6
	Wet weight	5	760	892.8

Rate of spreading

One bay was suited for calculating the rate of spreading, Cashel Bay, Connamara, Co. Galway; however, the following assumptions had to be made: (1) records from 2001 represented a point introduction, and (2) the place of introduction was only the place where the alga first was recorded and that spreading originated from the point of introduction. In Cashel Bay, the record dates from 2001 and large fertile plants were found indicating that these plants have arrived at least 1 year before in 2000. Assuming that the introduction at Cashel Bay was a point introduction by boat or yacht it is possible to calculate the rate of spread. The distance from the boathouse (first place of discovery in 2001) in middle Cashel Bay to Leitir Árd (survey observation of 2003) is approximately 7.5 km indicating a rate of spreading of 2–3 km $year^{-1}$. A rate of spreading around the coast has been calculated as well using the record from Cashel Bay from 2001 and the arrival in the east bay of St John's Point, Co. Donegal, in 2006 (which in previous surveys of 2003 and 2004 had no *S. muticum* established). The distance between Cashel Bay and St John's Point is about 270 km by sea, giving a rate of spreading of about 54 km $year^{-1}$.

Discussion

The records and survey data of this study show that *S. muticum* has spread all around the Republic of Ireland within 5 years of its discovery at Kilmore Quay in 2001. The extent of the beds, e.g., Farramanagh, Bantry Bay and Kilmore Quay, strongly suggests that these beds were established at least several years ago, most probably in the early 1990s. This would agree with the discovery of *S. muticum* in Strangford Lough, Northern Ireland, in 1995. It is therefore difficult to establish a precise introduction date and a rate of spreading as records showed up simultaneously at the west and south coast in 2001. Comparison with spreading rates of *S. muticum* found in other countries shows that in Ireland the spreading rate along the coastline is similar to that of the northwest American coastline and almost twice as fast compared to the English coast (Farnham et al. 1981). The spreading rate calculated for Bertraghboy Bay, Co. Galway, is five times slower compared to the Limfjorden, Denmark, indicating that the colonisation rate is strongly dependent on factors such as currents, prevailing winds and geography (Stæhr et al. 2000).

Of some concern is the establishment of *S. muticum* in seagrass beds at Kilmore Quay. *Sargassum muticum* is known to replace certain native macroalgae and is able to replace seagrass beds (Givernaud et al. 1991; Wernberg et al.

2001; Britton-Simmons 2004). Seagrass beds are protected as part of a specific habitat under Habitats Directive 1992. Moreover, seagrass beds are often part of a habitat that occurs in a number of areas proposed as Special Areas of Conservation (SACs) under the EC Habitats Directive. Recent initiatives arising from the EU Habitats Directive and the Convention on Biological Diversity have led to seagrass habitats being specifically targeted for conservation and restoration and include provisions requiring measures to prevent the introduction of, or to control the spread of, non-native species, especially those that threaten native or protected species (Wynne et al. 1995).

Once *S. muticum* has established itself in an area, it can affect the surrounding biodiversity. In Denmark the invasion of *S. muticum* started in 1984 and has changed the macroalgal community structure to a less complex system with fewer species overall and fewer species of intermediate dominance (Stæhr et al. 2000; Wernberg et al. 2001).

Britton-Simmons (2004) concluded from his experimental field data from San Juan Islands, Washington, that *S. muticum* competes with native canopy and under-storey algae and has a negative, indirect effect on the native herbivore sea urchin *S. droebachiensis*. Moreover, *S. muticum* reduced native algal biodiversity and decreased the growth rate of the native kelp *Laminaria bongardiana* Postels & Ruprecht. *Sargassum muticum* has been present in this area since the 1950s.

In Portugal and Spain, the most southerly distribution limit of *S. muticum* in Europe, sheltered tide pools form a specific habitat which *S. muticum* is able to colonise. Recently colonised rock pools appear to be completely dominated by *S. muticum*, out-competing other macroalgal species (Engelen et al. 2003; Arenas et al. 2003).

Sargassum muticum is a relatively new arrival in Ireland and possible effects on the biodiversity and surrounding fauna and flora may be difficult to establish, although it has been observed in seagrass beds and rock pools. The first record at Rath Strand in 2001 described a few plants in a rock pool. This survey revealed 65 plants in the same rock pool. A continuous monitoring program may reveal the possible effects this may have. It is possible that *Halidrys siliquosa*, a common species of intertidal rock pools, could be out-competed and might disappear from rock pools altogether as seen in Portugal (Engelen et al. 2003; Arenas et al. 2003). While this is speculative it indicates that proper studies should be conducted on the effect of *S. muticum* in rock pool colonisation and seagrass bed invasions. In general, non-native species may displace native organisms by preying on them or out-competing them for resources such as food, space, nutrients and light. In some cases this has led to the elimination of indigenous species from certain areas (Britton-Simmons 2004).

In Cashel Bay, it appears that *S. muticum* occupies a niche which is not used by other macroalgae. It grows below the *Fucus serratus* Linnaeus zone on small rocks on a muddy bottom, sometimes with *Saccharina latissima*. The area *S. muticum* occupies is most probably too deep for other intertidal algae to occupy and too sheltered for *Laminaria* species like *L. digitata* (Hudson) J. V. Lamouroux and *L. hyperborea* (Gunnerus) Foslie to settle.

The invasion of *S. muticum* in Ireland may perhaps change the intertidal flora make-up and may have implications for seaweed harvesters in rural areas, affecting their traditional way of life and their heritage. Moreover, it might have an effect on seaweed processors that harvest *Ascophyllum nodosum* (Linnaeus) Le Jolis for the horticulture, animal feed, alginates and biotechnology industries, as *S. muticum* may be harvested with the commercial harvested alga thus lowering the value of the processed end-product.

Sargassum muticum was often found in conjunction with mooring areas, anchorage sites and pontoons, indicating that the most likely vector for distribution in Irish waters is by boat and not oyster aquaculture, although it is often difficult to pinpoint a pathway for a specific introduction (Schaffelke et al. 2006). It is possible that yachts arriving from colonised areas within Ireland and from the United Kingdom and France may be responsible for transporting this species to new areas with established marinas or anchorage sites by carrying *S. muticum* on their hull, rudder, fishing gear or anchor chain (for an extensive review on pathways, see Schaffelke et al. 2006). An example is the marina in Mullaghmore Harbour, Co. Sligo. This marina was surveyed in August 2003, and no plants of *S. muticum* were encountered. The harbour was surveyed again in August 2005, and large beds were encountered. This indicates that, in areas free of *S. muticum*, fragments may drift or be brought in by boat and establish beds within 2 years.

At present *Sargassum muticum* has established itself all around the Irish coast, and it is recommended that monitoring continue in order to observe possible effects on local flora and fauna. Specifically seagrass beds and rock pool communities are of interest as in other studies it has been shown that effects take place in these specific habitats.

Acknowledgements The author is very grateful to Anita Gallagher of the Botany Department, NUI, Galway, and Aurore Raveneau of the Agricultural college, ENESAD, Dijon, France, for their assistance with fieldwork; to Astrid Werner, Jim Morrissey and Robert Wilkes for help with the surveys; and particulary to Lin Baldock, Clare Heardman, Marie Mahon Padraig Whelan and Padraig Gannon for additional records of *S. muticum*. Special thanks to Dan Minchin of Marine Organism Investigations and Julia Nunn of the Ulster Museum and Botanic Gardens, Belfast, for records of 2005 and 2006. This project was partially funded by the Heritage Council under the Wildlife Grant Scheme 2003 and received support from NUI Galway's Millennium Fund.

References

Arenas F, Fernadéz C, Rico JM, Fernandéz E, Haya D (1995) Growth and reproductive strategies of *Sargassum muticum* (Yendo) Fensholt and *Cystoseira nodicaulis* (Whit.). Roberts Sci Mar 59 (1):1–8

Arenas F, Sanchez I, Fernandez C, Hawkins SJ (2003) The impact of invasive species on European shores: the case of *Sargassum muticum* in intertidal rock pools. In: Maggs C (ed) Programme and book of abstracts. Third European Phycological Congress. Taylor & Francis Group, Belfast, Northern Ireland, p 49

Boaden PJS (1995) The invasive seaweed *Sargassum muticum* (Yendo) Fensholt in Strangford Lough, Northern Ireland. Ir Nat J 25:111–113

Britton-Simmons KH (2004) Direct and indirect effects of the introduced alga *Sargassum muticum* on benthic, subtidal communities of Washington State, USA. Mar Ecol Prog Ser 277:61–78

Christensen T (1984) Sargassotang, en ny algeslægt i Danmark. Urt 4:99–104

Critchley A (1981) Age determination of *Sargassum muticum* (Yendo) Fensholt. Br Phycol J 16:134

Critchley AT, Farnham WF, Morell SL (1983) A chronology of new European sites of attachment for the invasive brown alga *Sargassum muticum* 1973–1981. J Mar Biol Assoc UK 63: 799–811

Critchley AT, Farnham WF, Morrell SL (1986) An account of the attempted control of an introduced marine alga*Sargassum muticum*, in southern England. Biol Cons 35:313–332

Critchley AT, Farnham WF, Yoshida T, Norton TA (1990) A bibliography of the invasive alga *Sargassum muticum* (Yendo) Fensholt (Fucales; Sargassaceae). Bot Mar 33:551–562

Deysher LE, Norton TA (1982) Dispersal and colonisation in *Sargassum muticum* (Yendo) Fensholt. J Exp Mar Biol Ecol 56:179–195

Engelen A, Santos R, Alves C (2003) Demography of the alien species *Sargassum muticum* at its southern European distribution limit. In: Maggs C (ed) Programme and book of abstracts. Third European Phycological Congress. Taylor & Francis Group, Belfast, Northern Ireland, p 48

Fahy D (2001) Harmful seaweed found on Wicklow coast. Irish Times, 20 August. http://www.ireland.com/newspaper/science/2001/0920/archive.01092000088.html. Cited 1 July 2005

Farnham WF (1980) Studies on aliens in the marine flora of southern England. In: Price JH, Irvine DEG, Farnham WF (eds) The shore environment, vol. 2. Ecosystems. Academic, London, pp 875–914

Farnham WF, Jones EBG (1974) The eradication of the seaweed *Sargassum muticum* from Britain. Biol Cons 6:57–58

Farnham WF, Fletcher RL, Irvine LM (1973) Attached *Sargassum* found in Britain. Nature 243:231–232

Farnham WF, Murfin C, Critchley A, Morrell S (1981) Distribution and control of the brown alga *Sargassum muticum*. In: Levring T (ed) Proceedings of the Xth International Seaweed Symposium. Goteborg, Sweden. Academic Press, Walter de Gruyter, Berlin, 11–15 August 1980, pp 277–282

Givernaud T, Cosson J, Givernaud-Mouradi A (1991) Etude des populations de *Sargassum muticum* (Yendo) Fensholt sur les côtes de Basse-Normandie (France). In: Elliott M, Ducrotoy JP (eds) Estuaries and coasts: spatial and temporal intercomparisons. ECSA 19th symposium. Olsen & Olsen, Fredesburg, Denmark, Caen, September 1989, pp 129–132

Hales JM, Fletcher RL (1989) Studies on the recently introduced brown alga *Sargassum muticum* (Yendo) Fensholt. IV. Effect of temperature, irradiance and salinity on germling growth. Bot Mar 32:167–176

Hiscock K, Moore J (1986) Surveys of harbours, rias and estuaries in southern Britain: Plymouth area including the Yealm. CSD report no. 752. Natural England

Jensen JB (1974) Morphological studies in Cystoseiraceae and Sargassaceae (Phaeophyceae) with special reference to apical organization. University of California Publications in Botany, vol. 68. University of California, p 61

Kornmann P, Sahling PH (1994) Marine algae of Helgoland: second supplement. Helgol Meeresunters 48(4):365–406

Loughnane C, Stengel DB (2002) Attached *Sargassum muticum* (Yendo) Fensholt found on the west coast of Ireland. Ir Nat J 27 (2):70–72

Marine Institute (2006) Weather buoy data. http://www.marine.ie. Cited 5 June 2007

Norton TA (1976) Why is *Sargassum muticum* so invasive? Br Phycol J 11:199

Raines P, Nunny R, Cleator B (1992) Isles of Scilly expedition 1992. Coral Cay Conservation Sub-Aqua Club

Ribera MA, Boudouresque CF (1995) Introduced marine plants, with special reference to macroalgae: mechanisms and impact. Progr Phycol Res 11:187–268

Rueness J (1989) *Sargassum muticum* and other introduced Japanese macroalgae: biological pollution of European coasts. Mar Poll Bull 20:173–176

Schaffelke B, JE Smith, CL Hewitt (2006) Introduced macroalgae – a growing concern. J Appl Phycol 18:529–541

Simkanin C (2004) The invasive seaweed *Sargassum muticum* (Yendo) Fensholt in Lough Hyne Marine Nature Reserve, Co. Cork. Ir Nat J 27:481–482

Stæhr PA, Pedersen MF, Thomsen MS, Wernberg T, Krause-Jensen D (2000) Invasion of *Sargassum muticum* in Limfjorden (Denmark) and its possible impact on the indigenous macroalgal community. Mar Ecol Progr Ser 207:79–88

Steen H (2004) Effects of reduced salinity on reproduction and germling development in *Sargassum muticum* (Phaeophyceae, Fucales). Eur J Phycol 39:293–299

Wernberg T, Thomsen MS, Staehr PA, Pedersen MF (2001) Comparative phenology of *Sargassum muticum* and *Halidrys siliquosa* (Phaeophyceae: Fucales) in Limfjorden, Denmark. Bot Mar 44:31–39

Winder F (2002) *Sargassum muticum* in Kenmare River, Co. Kerry. Ir Nat J 27:85

Withers RG, Farnham WF, Lewey S, Jephson NA, Haythorn RM, Gray PWG (1975) The epibiota of *Sargassum muticum* in British waters. Mar Biol 31:79–81

Wynne G, Avery M, Campbell L, Gubbay S, Hawkswell S, Juniper T, King M, Newberry P, Smart J, Steel C, Stones T, Stubbs A, Taylor J, Tydeman C, Wynde R (1995) Biodiversity challenge, 2nd edn. Royal Society for the protection of Birds, RSPB, Sandy, UK

Yendo K (1907) The Fucaceae of Japan. J Coll Sci Tokyo Imperial Univ 21(12):1–174

Effect of seawater temperature on the productivity of *Laminaria japonica* in the Uwa Sea, southern Japan

Satoshi Suzuki · Kazushige Furuya · Tadashi Kawai · Ichiro Takeuchi

Originally published in the Journal of Applied Phycology, Vol 20, No 5, 383–394.
DOI: 10.1007/s10811-007-9283-3

Abstract Recent studies on global climate change report that increase in seawater temperature leads to coastal ecosystem change, including coral bleaching in the tropic. In order to assess the effect of increased seawater temperature on a temperate coastal ecosystem, we studied the inter-annual variation in productivity of *Laminaria japonica* using long-term oceanographic observations for the Uwa Sea, southern Japan. The annual productivity estimates for *L. japonica* were 2.7±2.5 (mean±SD) kg wet wt. m^{-1} (length of rope) (2003/2004), 1.0±0.6 kg wet wt. m^{-1} (2004/2005) and 12.1±12.5 kg wet wt. m^{-1} (2005/2006). Our previous study using the same methodology at the same locality reported that the productivity was estimated for the 2001/2002 (33.3±15.2 kg wet wt. m^{-1}) and 2002/2003 (34.0±8.7 kg wet wt. m^{-1}) seasons. Productivity in 2003/2004 and 2004/2005 was significantly lower than in years 2001/2002, 2002/2003 and 2005/2006. A comparison of oceanographic conditions among the 5 years revealed the presence of threshold seawater temperature effects. When the average seawater temperature during the first 45 days of each experiment exceeded 15.5°C, productivity was reduced to about 10 % of that in cooler years. Moreover the analysis of growth and erosion rates indicates that when the seawater temperature was over 17.5°C, erosion rate exceeded growth rate. Thus, an increase of seawater temperature of just 1°C during winter drastically reduces the productivity of *L. japonica* in the Uwa Sea.

Keywords Global warming · *Laminaria japonica* · Productivity · Seawater temperature · Uwa Sea

S. Suzuki · I. Takeuchi (✉)
Department of Life Environment Conservation,
Faculty of Agriculture, Ehime University,
3-5-7 Tarumi, Matsuyama,
Ehime 790-8566, Japan
e-mail: takeuchi@agr.ehime-u.ac.jp

K. Furuya
Maruwa Ltd., 4263 Yusu,
Uwajima, Ehime 798-0103, Japan

T. Kawai
Hokkaido Nuclear Energy Environmental Research Center,
261-1 Miyaoka, Kyowa,
Hokkaido 045-0123, Japan

Introduction

Global climate change affects physico-chemical conditions of the oceans (e.g., seawater temperature, sea level and salinity), and therefore it significantly impacts marine ecosystems (e.g., Short and Neckles 1999). Coral bleaching, i.e., loss of most of the symbiotic zooxanthellae normally found within coral tissue, has occurred with increasing frequency in the last two decades in reefs throughout tropical to subtropical regions (Coles and Brown 2003; Hoegh-Guldberg and Salvat 1995; Jokiel and Brown 2004; Jones et al. 1998; Obura 2005; Sabatés et al. 2006). It is well known that when mean summer maximum temperature increases by only 1°C above the norm, coral bleaching occurs (Hoegh-Guldberg and Salvat 1995; Jokiel and Brown 2004).

For the seaweeds and seagrasses in temperate regions, decreases in standing biomass as a result of increasing water temperatures have been reported in the last decade. For example, there is increasing environmental stress on *Zostera marina* at the southern end of its distribution range on the east coast of the U.S. where eelgrass in North Carolina and Chesapeake Bay grows well during the cooler months but dies back during warmer summer periods (Short and Neckles 1999). A similar decrease in standing biomass of *Laminaria religiosa* has occurred along the Japan Sea coast of Hokkaido (Tsuda et al. 2006), and *Ecklonia* spp. biomass in Tosa Bay, Kochi (Hiraoka et al. 2005) has declined with increasing seawater temperature.

We have studied growth and productivity of the introduced kelp *Laminaria japonica* off Yusu in Uwajima Bay, Uwa Sea, which is located on the west coast of the Shikoku Island, southwestern Japan (Suzuki et al. 2006). The annual productivity of *L. japonica* per meter of rope on which the thalli were grown was equivalent to the wild productivity on 1.1–2.2 m^2 of substratum in the native habitat off Hokkaido. Thus, *L. japonica* is highly adaptable, as expressed by its ability to maintain high levels of productivity when grown in warmer waters south of its natural range. However, the seawater temperature of the Uwa Sea coast (where we conducted the *L. japonica* growth experiments) increased significantly from 1981 to 2004. The rate of increase in the winter months (December to February) has been especially remarkable (Suzuki and Takeuchi 2007).

Accordingly, we conducted a field study to clarify the relationship between environmental conditions and inter-annual variation in the productivity of *Laminaria japonica* transplanted to the Uwa Sea.

Materials and methods

Study site and period The experiment was conducted off Yusu in Uwajima Bay, Uwa Sea, which is located on the west coast of the Shikoku Island, southwestern Japan (Fig. 1; 33° 11′ N, 132° 28′ E). The experimental facility for this study was a pearl aquaculture raft. The *Laminaria japonica* growth experiment was conducted from December 2003 to August 2004 (2003/2004), from December 2004 to August 2005 (2004/2005) and from December 2005 to August 2006 (2005/2006).

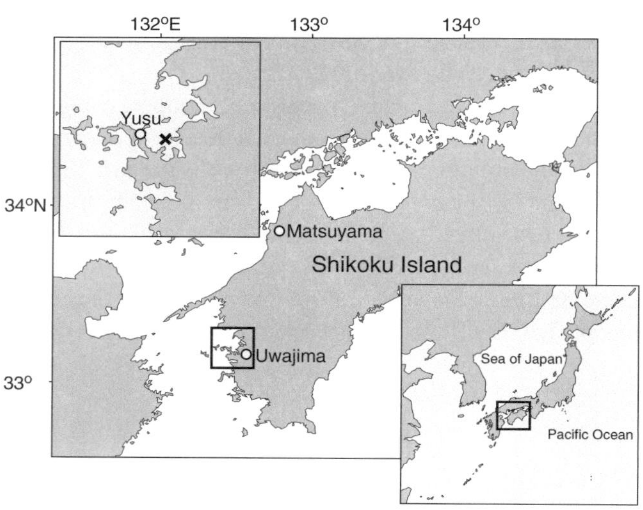

Fig. 1 Map showing the location of the study site

Table 1 Relationship between total length and wet weight

Study period	Parameters		
2003/2004	$y=0.0018x^{2.2}$	$r^2=0.92$	n=38
2004/2005	$y=0.0007x^{2.2}$	$r^2=0.92$	n=13
2005/2006	$y=0.0015x^{2.1}$	$r^2=0.98$	n=105

x; total length (cm) y; wet weight (g)

Experimental method The experimental period and method follow those of our previous study (Suzuki et al. 2006). A seed rope with zoospores attached was introduced into the study area from Aomori Prefecture, which is within the native area of *Laminaria japonica*. On 20 Dec 2003, 19 Dec 2004 and 21 Dec 2005, the seed rope was cut into 5 cm lengths and these were inserted at 15 cm intervals along a 20 m long culture rope and immediately suspended horizontally at a depth of about 1–3 m.

Fifty 30 cm sections were labeled on each culture rope. The number of individuals of *L. japonica* was counted, and their total length measured in six randomly chosen rope sections. In the period 11 Jan to 8 Feb 2004 counts were made in only three rope sections. Growth and erosion rates were measured by the hole punching method in other rope sections chosen at random. Holes were punched in the blades (8 to 15 kelp individuals in each experiment) 20 cm above the base of the holdfast on each measurement date, and the distances from the holdfast to the previously punched holes and total thallus length were recorded. Hole punching was started on 21/22 Feb in each experiment. The measurements were made at about 15 days intervals throughout the study period.

Rope sections not used for the above measurement were used for sampling 1–30 intact *Laminaria japonica* individuals (including the holdfasts). In the laboratory, periphyton was removed, and total length and wet weight of thalli were measured.

The water temperature during the study period was measured at a depth of 5 m using an electronic thermometer attached to a chlorophyll meter (COMPACT-CL; Alec Electronics Co., Ltd., Kobe, Japan). On each sampling occasion, a vertical salinity profile was measured using a STD (AST200-PK; Alec Electronics Co., Ltd.); we calculated an average of the salinities from 1–3 m water depth. Photon irradiance was measured at a depth of 3 m using a light intensity recorder (MDS-MkV/L; Alec Electronics Co., Ltd.). The light collector was cleaned on each sampling visit to prevent underestimation caused by fouling. We used data collected in the five days after set

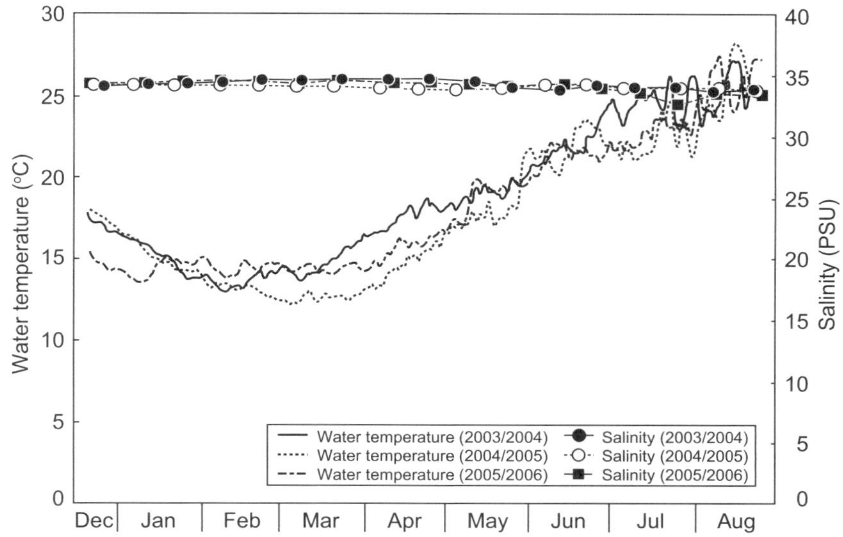

Fig. 2 Changes in seawater temperature at a depth of 5 m in 2003/2004, 2004/2005 and 2005/2006, and the salinity average over depths of 1–3 m in 2003/2004, 2004/2005 and 2005/2006. Left axis indicates seawater temperature (°C) and right axis indicates salinity (PSU)

up or cleaning. Water transparency was measured with a secchi disk. Water samples collected for measurement of dissolved nutrients at the study site were taken at 1, 3 and 5 m depth using a Van-Dorn bottle. Samples were frozen until analysis. The nutrients (ammonium-N, nitrate-N, nitrite-N and phosphate-P) were analyzed with a Technicon AutoAnalyzer II (Tarrytown, NY, USA).

Productivity of the Laminaria japonica Productivity (P: g wet wt. d^{-1}) of individual *L. japonica* thalli between time 0 and time 1 (time 0: the last measurement, time 1: the present measurement) was calculated from the mean growth rate (cm d^{-1}) using the relationship between total length (TL: cm) and wet weight (WW: g) (Table 1) according to Suzuki et al. (2006). Productivity at the population level was obtained from individual productivity rate and density, and was set to 0 when all of the individuals used in the growth measurements were washed away.

Results

Environmental conditions During study periods, the seawater temperature decreased from December, and reached minima of 12.9°C (9 Feb 2003/2004), 12.2°C (4 Mar 2004/2005) and 13.5°C (8 Jan 2005/2006) (Fig. 2). The temperature then increased, and reached maxima of 27.4°C (14 Aug 2003/2004), 28.3°C (15 Aug 2004/2005) and 27.3°C (23 Aug 2005/2006). The salinity ranged from 32.6 to 34.6 PSU throughout the experiment period (Fig. 2).

The daytime photon irradiance at a depth of 3 m was low during winter (478.6–540.9 μmol photons m^{-2} s^{-1} in

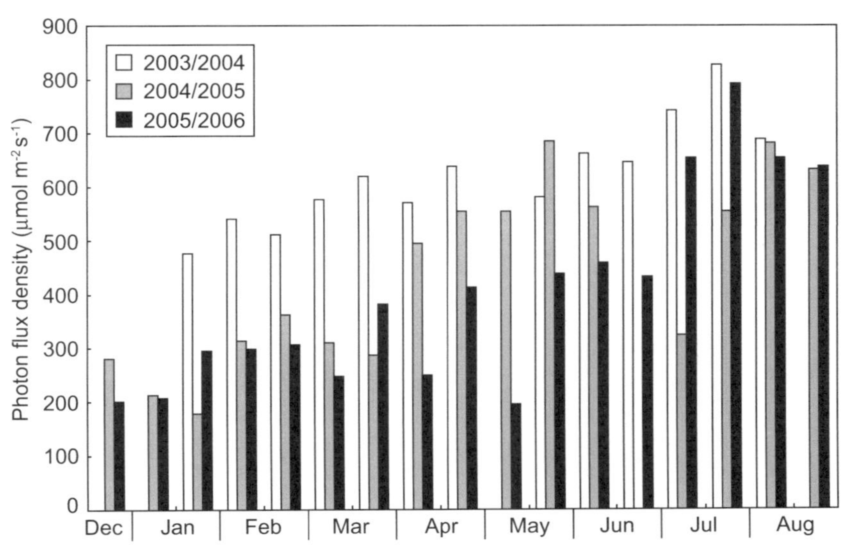

Fig. 3 Seasonal changes in daytime photon flux irradiance at the depth of 3 m in 2003/2004, 2004/2005 and 2005/2006

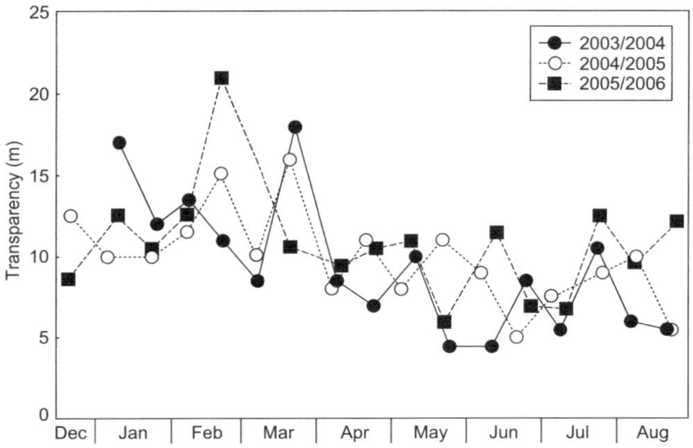

Fig. 4 Seasonal changes in seawater transparency in 2003/2004, 2004/2005 and 2005/2006

2003/2004, 177.5–362.1 μmol m^{-2} s^{-1} in 2004/2005 and 200.3–305.7 μmol photons m^{-2} s^{-1} in 2005/2006) (Fig. 3). Values increased through to summer (645.8–826.4 μmol photons m^{-2} s^{-1} in 2003/2004, 324.3–683.4 μmol photons m^{-2} s^{-1} in 2004/2005 and 430.6–794.0 μmol photons m^{-2} s^{-1} in 2005/2006) (Fig. 3). Transparency ranged from 4.5 to 21.0 m throughout the experiment period (Fig. 4).

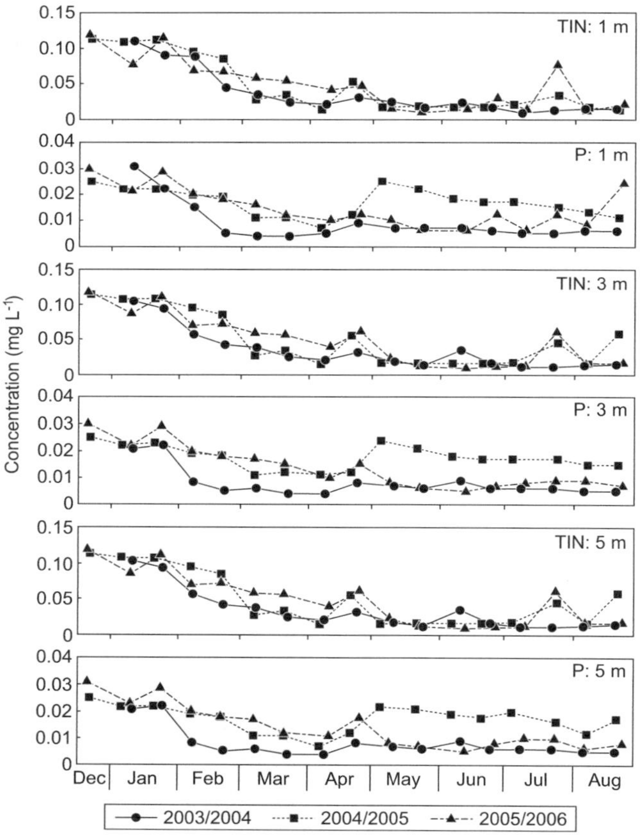

Fig. 5 Seasonal changes in total inorganic nitrogen (TIN) and phosphate-phosphorus (P) at 1, 3 and 5 m depth in 2003/2004, 2004/2005 and 2005/2006

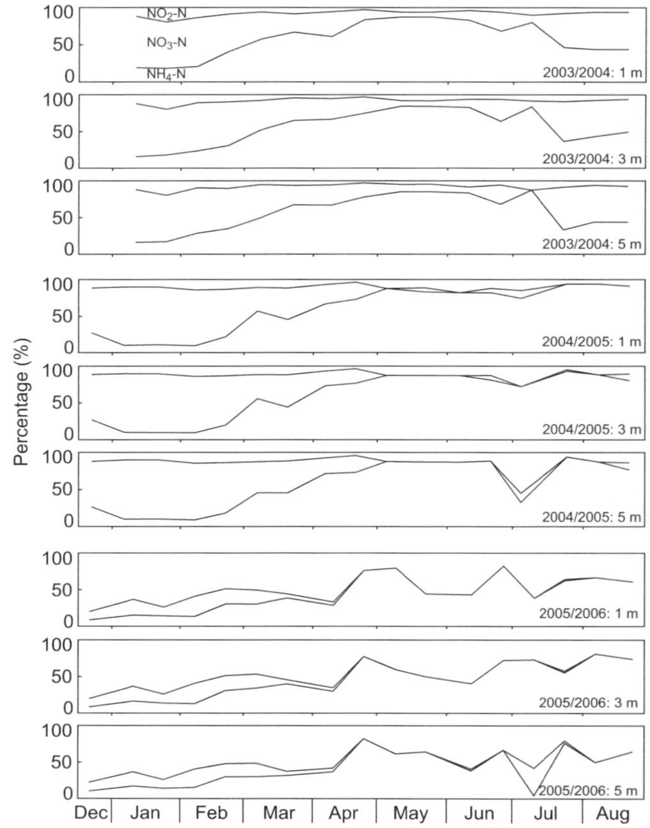

Fig. 6 Seasonal changes in relative abundance of inorganic nitrogen compounds (TIN: total inorganic nitrogen) at 1, 3 and 5 m depth in 2003/2004, 2004/2005 and 2005/2006

Seasonal changes in dissolved total inorganic nitrogen (TIN) and phosphate-phosphorus (phosphate-P) are shown in Fig. 5. In all the experiments, the TIN concentration was high in winter at all depths (0.044–0.119 mg L^{-1}). The TIN concentration began to decrease in March, and reached its minimum from May to August at all depths. The composition of TIN is shown in Fig. 6. Nitrite (NO_2-N) or

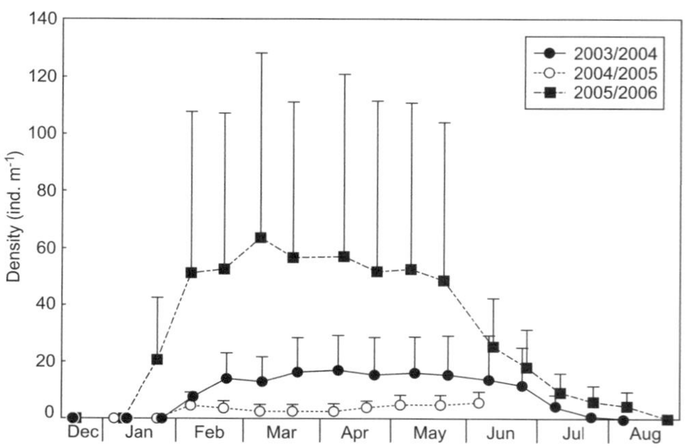

Fig. 7 Seasonal changes in density of *Laminaria japonica* in 2003/2004, 2004/2005 and 2005/2006. Means±SD (n=6 except 3 from 11 Jan to 8 Feb 2004)

nitrate (NO_3-N) accounted for 60–90 % of the TIN content during the high TIN period. A similar seasonal pattern was found for phosphate-P concentration except in 2004/2005 (Fig. 5) when the concentration was also high in summer. The concentration of phosphate-P in winter was 0.005–0.031 mg L^{-1}.

Population dynamics From the time of germination of zoospore in January, the density of *Laminaria japonica* increased (Fig. 7). In 2003/2004, the density reached a maximum of 17.2±12.4 ind. m^{-1} rope (mean±SD, here and in following text) on 8 Apr, and stayed near 15 ind. m^{-1} until 25 Jun. In 2004/2005, the density reached a maximum of 5.6±4.0 ind. m^{-1} on 7 Jun, but the culture experiment was accidentally terminated afterwards. In 2005/2006, the density reached a maximum of 63.9±64.6 ind. m^{-1} on 7 Mar, and stayed near 55 ind. m^{-1} until 23 May. After July (2003/2004) or June (2005/2006), the density decreased gradually to zero in August.

In 2003/2004, the distribution ranged widely across size classes, due to new germination and growth from 8 Feb to 9 Apr (Fig. 8a). Total length measurements ranged from 0 to 200 cm on 9 Apr. After late April, no new germination was observed, and total length ranged from 20 to 200 cm. After early May, the range of the distribution across size classes became more limited. In 2004/2005, the size class structure was changed by growth from 7 Feb to 7 Apr (Fig. 8b). After late April, new germination was observed again, and the size class structure showed two peaks between 10 and 50 cm and between 160 and 260 cm. This pattern of size class structure in 2003/2004 was found again in 2005/2006 (Fig. 8c). In 2003/2004 and 2005/2006, there were no marked peaks in size class structure. Individual thalli were scattered across size classes.

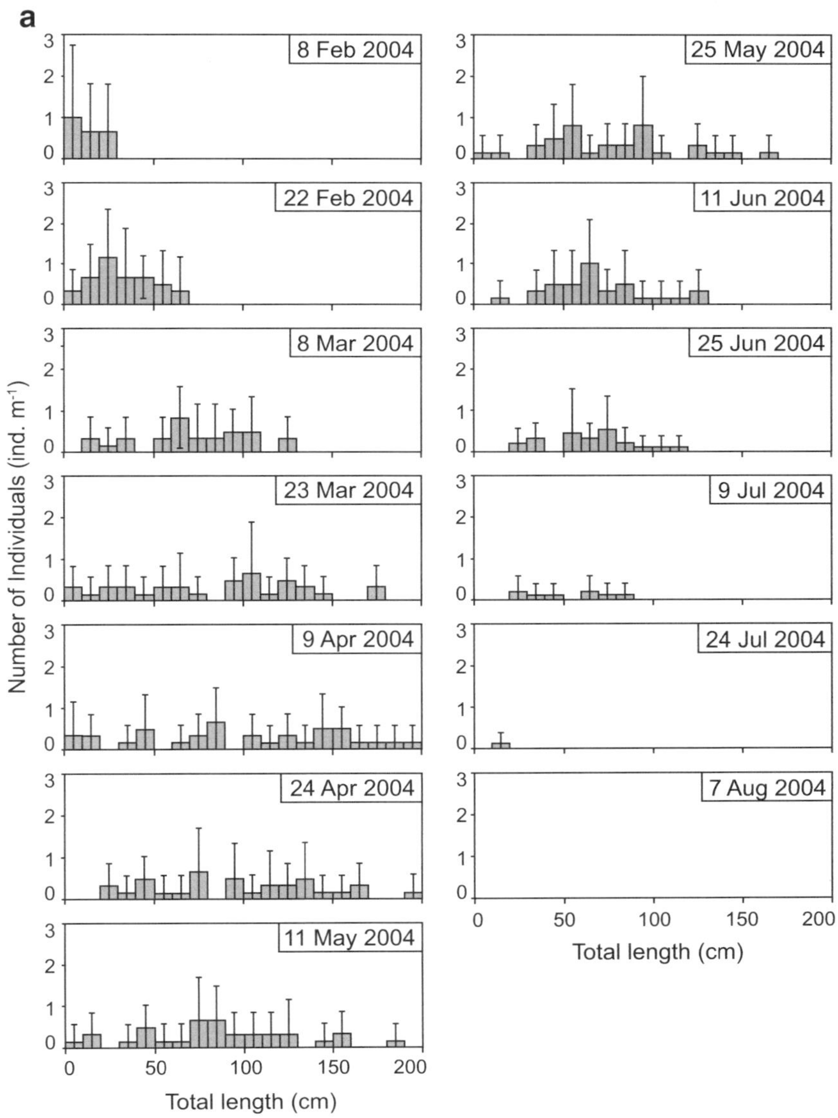

Fig. 8 **a** Seasonal fluctuations in the size class distribution of *Laminaria japonica* in 2003/2004. Means±SD (n=6 except 3 from 11 Jan to 8 Feb 2004). **b** Seasonal fluctuations in the size class distribution of *Laminaria japonica* in 2004/2005. Means±SD (n=6). **c** Seasonal fluctuations in the size class distribution of *Laminaria japonica* in 2005/2006. Means±SD (n=6)

Fig. 8 (continued)

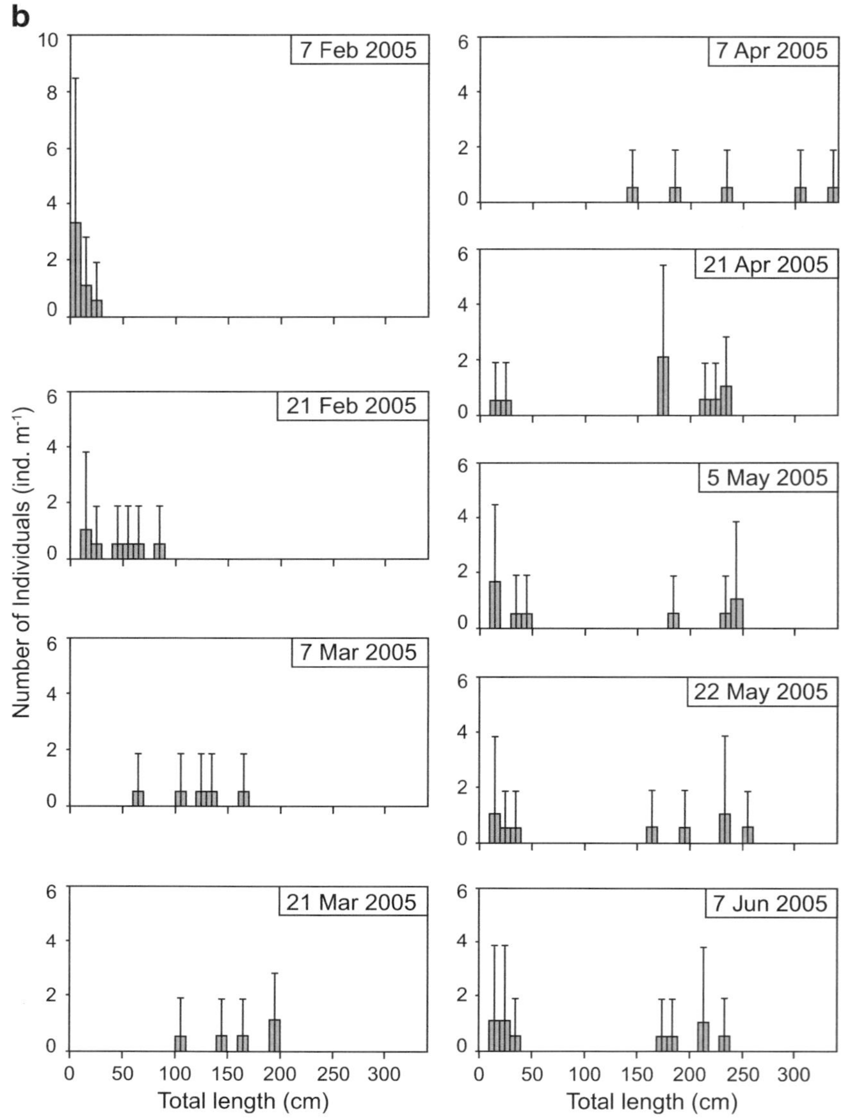

The mean total length increase followed a sigmoid curve from 0 cm in late December up to 99.6±55.7 cm (191 cm in maximum) on 9 Apr 2003/2004, up to 240.0±81.5 (340 cm in maximum) on 7 Apr 2004/2005 and up to 148.0±68.2 cm (293 cm in maximum) on 25 Apr 2005/2006 (Fig. 9). After April, mean total length decreased steadily until it reached zero in late August.

Seasonal changes in wet weight were obtained from the total length measurements by using the regression equations in Table 1. Mean wet weight reached 62.5± 56.6 g ind^{-1} on 9 Apr 2003/2004, 135.3±92.9 g ind^{-1} on 7 Apr 2004/2005 and 67.3±55.9 g ind^{-1} on 25 Apr 2005/2006 (Fig. 10).

Biomass Biomass increase followed a sigmoid curve from zero on January up to 1076.9±613.8 g wet wt. m^{-1} on 9 Apr 2003/2004, up to 376.0±585.8 g wet wt. m^{-1} on 7 Apr 2004/2005 and up to 3514.2±4920.5 g wet wt. m^{-1} on 25 Apr 2005/2006 (Fig. 11). Afterwards, biomass decreased steadily, reaching zero in August.

Growth and erosion rates The rate of growth indicated by the distal movement of punched holes was maximal from late February to early March in all studies, with decreases during subsequent months (Fig. 12). The algae elongated not only in the basal part of the blade (the meristematic zone), but also grew in the middle part of the blade, as judged from extended distance of two punched holes on the middle part of the blade.

The growth rates were high from February through April (0.9 to 3.2 cm d^{-1} in 2003/2004, 2.1 to 4.6 cm d^{-1} in 2004/2005 and 1.7 to 4.2 cm d^{-1} in 2005/2006), especially in

Fig. 8 (continued)

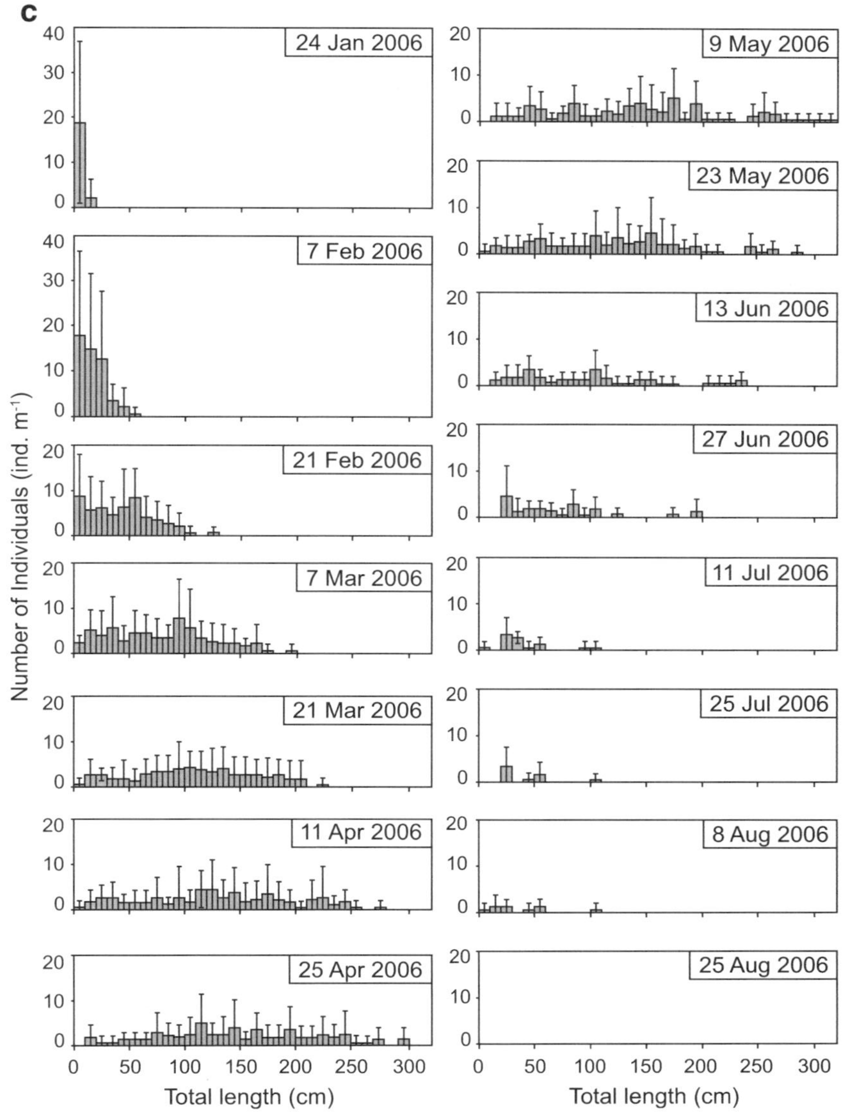

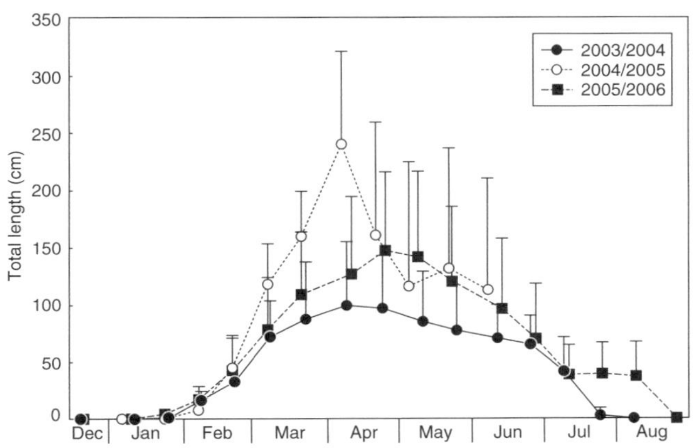

Fig. 9 Seasonal changes in length of *Laminaria japonica* in 2003/2004, 2004/2005 and 2005/2006. Means±SD (n=1–31 in 2003/2004, 5–10 in 2004/2005 and 9–115 in 2005/2006)

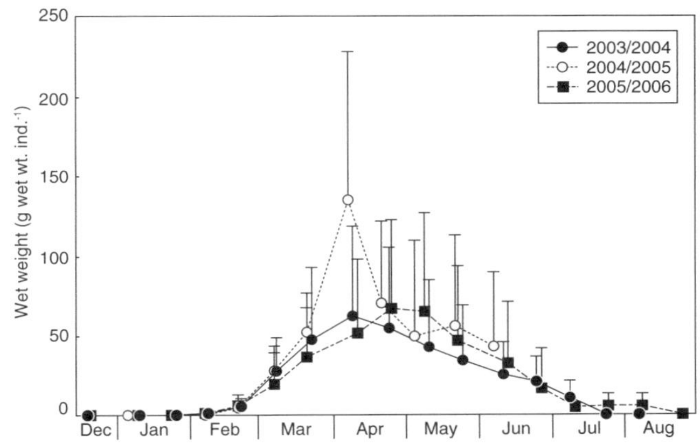

Fig. 10 Seasonal changes in wet weight of *Laminaria japonica* in 2003/2004, 2004/2005 and 2005/2006. Means±SD (n=1–31 in 2003/2004, 5–10 in 2004/2005 and 9–115 in 2005/2006)

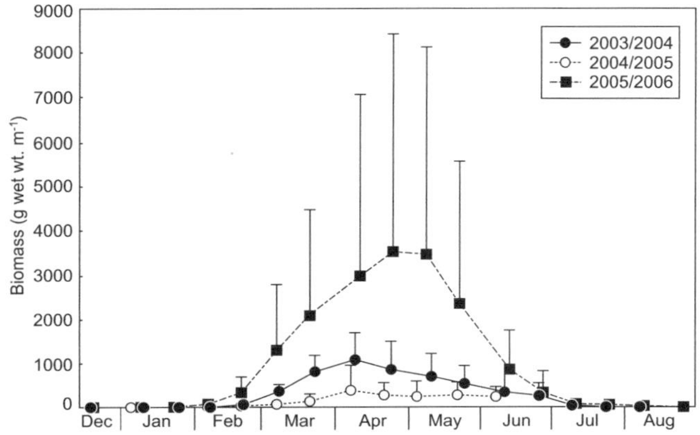

Fig. 11 Seasonal changes in biomass of *Laminaria japonica* per meter of cultivation rope in 2003/2004, 2004/2005 and 2005/2006. Means±SD (n=6 except 3 from 11 Jan to 8 Feb 2004)

comparison with those after May (0 to 0.4 cm d^{-1} in 2003/2004, 0.4 to 1.3 cm d^{-1} in 2004/2005 and 0.1 to 1.2 cm d^{-1} in 2005/2006) (Fig. 13a).

The mean erosion rates ranged from 0.5±0.2 to 1.9 cm d^{-1} in 2003/2004, from 0.1±0.2 to 2.2±0.6 cm d^{-1} in 2004/2005 and from 0.1±0.1 to 4.1±2.6 cm d^{-1} in 2005/2006 (Fig. 13a). Erosion rates exceeded growth rates from early May 2003/2004 and 2005/2006, and from late April 2004/2005.

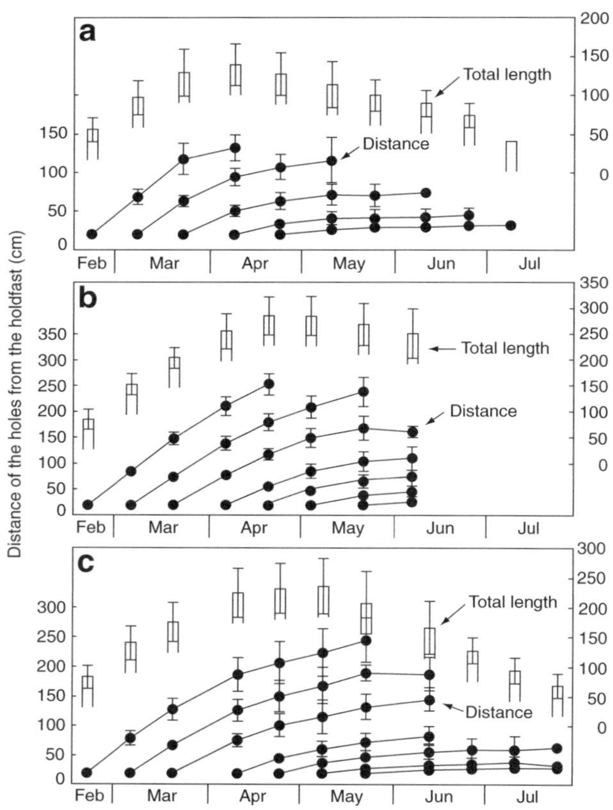

Fig. 12 Elongation of *Laminaria japonica* thalli (bar) and displacement of punched holes on the blade (line) in 2003/2004 (**a**), 2004/2005 (**b**) and 2005/2006 (**c**). Means±SD (n=1–15 in 2003/2004, 6–8 in 2004/2005 and 3–18 in 2005/2006)

Productivity and erosion rates The mean productivity rates (individual level) increased to a maximum of 3.5±1.7 g wet wt. d^{-1} from 8–23 Mar in 2003/2004, to 4.0±0.9 g wet wt. d^{-1} from 21 Mar to 7 Apr in 2004/2005 and to 3.1±1.2 g wet wt. d^{-1} from 21 Mar to 11 Apr in 2005/2006 (Fig. 13b). Thereafter in each year, the mean productivity rates decreased. Mean erosion rates were low in February and March when productivity rates were high (Fig. 13b). The mean erosion rates reached a maximum of 2.4±1.3 g wet wt. from 9–24 April in 2003/2004, of 2.9±1.0 g wet wt. from 5–21 May in 2004/2005 and 3.7±4.4 g wet wt. from 9–23 May in 2005/2006.

In 2003/2004, the mean productivity (population levels) increased steadily during the period from December to March, and reached a maximum of 885.0±652.7 g (n=6) wet wt. m^{-1} per 15 days from 8–23 Mar (approximately 60 g wet wt. m^{-1} d^{-1}) (Fig. 14a). Thereafter, mean

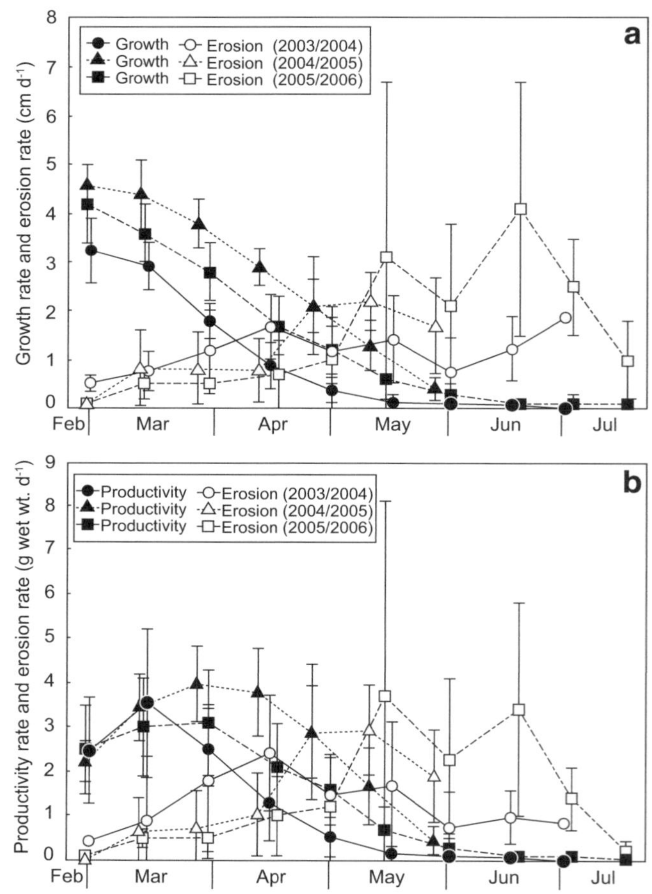

Fig. 13 Seasonal changes in growth and erosion rates (**a**), and in individual productivity and erosion rates (**b**) of *Laminaria japonica*. Means±SD (n=1–15 in 2003/2004, 6–8 in 2004/2005 and 3–18 in 2005/2006)

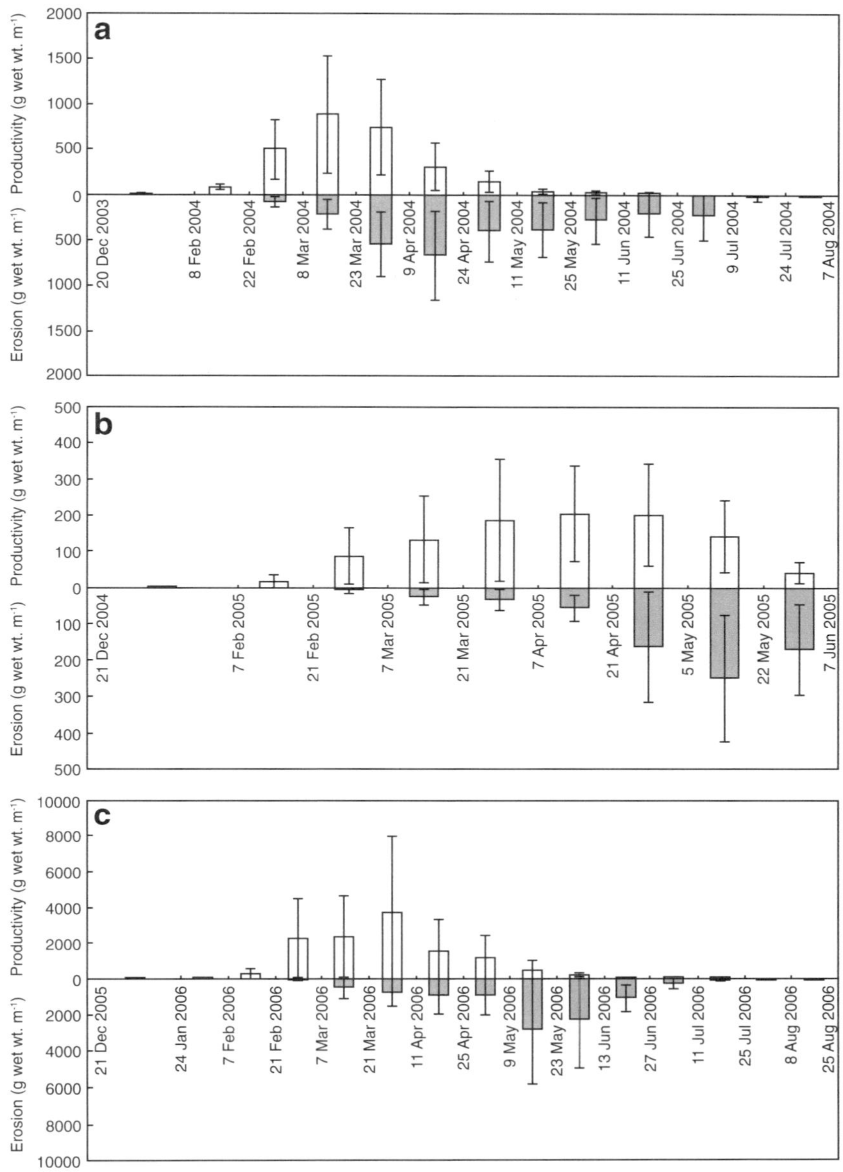

Fig. 14 Seasonal changes in population productivity and erosion of *Laminaria japonica* in 2003/2004 (**a**), 2004/2005 (**b**) and 2005/2006 (**c**). Means±SD (n=6 except 3 from 20 Dec 2003 to 8 Feb 2004)

productivity decreased into July, with no net productivity from 25 Jun to 9 Jul. The annual productivity was estimated at 2.7±2.5 kg wet wt. m^{-1}. The mean erosion rate increased during the period from February into April, and reached a maximum of 668.8±485.6 g wet wt. m^{-1} per 15 days from 9–24 Apr. In 2004/2005, the mean productivity increased steadily during the period from December to April, and reached a maximum of 205.8±132.8 g (n=6) wet wt. m^{-1} per 14 days from 7–21 Apr (approximately 15 g wet wt. m^{-1} d^{-1}) (Fig. 14b). Thereafter, the mean productivity decreased to only 40.9±29.7 g wet wt. m^{-1} per 16 days from 22 May to 7 Jun. Hence, annual productivity was estimated to be 1.0±0.6 kg wet wt. m^{-1}, assuming that the productivity after 7 Jun can be ignored (see Discussion). The mean erosion rate increased during the period from February into May, and reached a maximum of 249.1± 174.0 g wet wt. m^{-1} per 17 days from 5–22 May. In 2005/2006, mean productivity increased steadily during the period from December to April, and reached a maximum of 3761.2±4192.6 g (n=6) wet wt. m^{-1} per 21 days from 21 Mar to 11 Apr (approximately 180 g wet wt. m^{-1} d^{-1}) (Fig. 14c). Thereafter, mean productivity decreased into July, falling to only 2.6±2.4 g wet wt. m^{-1} per 14 days from 11–25 Jul. Therefore, the annual productivity was estimated to be 12.1±12.5 kg wet wt. m^{-1}. The mean erosion rate increased during the period from February to May, and reached a maximum of 2832.6±2996.9 g wet wt. m^{-1} per 14 days from 9–23 May.

Discussion

The experiments reported here for a single site in southern Japan, in combination with our earlier work (Suzuki et al. 2006) reveal relatively high inter-annual variation in kelp productivity. In present study, the annual productivity of *Laminaria japonica* introduced into Uwa Sea was estimated at 2.7±2.5 kg wet wt. m^{-1} (2003/2004), 1.0±0.6 kg wet wt. m^{-1} (2004/2005) and 12.1±12.5 kg wet wt. m^{-1} (2005/2006). The annual productivity in 2003/2004 and 2004/2005 were significantly lower than that of 2005/2006 (Mann-Whitney U-test; $p<0.05$). Annual productivity in 2001/2002 and 2002/2003 were estimated at 33.3±15.2 kg wet wt. m^{-1} and 34.0±8.7 kg wet wt. m^{-1}, respectively (Suzuki et al. 2006). While there was no significant difference in production among 2005/2006, 2001/2002 and 2002/2003, production in 2003/2004 and 2004/2005 was significantly lower than in 2001/2002 and 2002/2003 (Mann-Whitney U-test; $p<0.05$).

Why was annual productivity of *Laminaria japonica* in 2003/2004 and 2004/2005 much lower than in the other study years? Among several possible factors, e.g, light saturation, nutrients, seawater temperature, the lack of a sufficiently cold period in the winters of 2003/2004 and 2004/2005 (especially from December to January) may have had important effects on the development of kelp sporophyte directly, or encouraged grazing activity of herbivorous fishes.

The zoospore released from the sporophyte produces a young sporophyte (about 5 mm) via the gametophyte stage. This process requires about 45 days (Ohno 1987; Kawashima 2004). In the present study, the mean quantum irradiances in the daytime for the 45 days after the experiments were started were about 480 μmol m^{-2} s^{-1} in 2003/2004, about 220 μmol m^{-2} s^{-1} in 2004/2005 and about 230 μmol m^{-2} s^{-1} in 2005/2006 (Fig 3). The meteorological data for Uwajima provided by the Japan Meteorological Agency indicate that no significant differences existed among the five study years in the duration of sunshine for the relevant 45 day periods. There was also no significant difference in water transparency for the relevant period in the five years (present study and Suzuki et al. 2006). Moreover, the light saturation point of *L. japonica* is 200 μmol m^{-2} s^{-1} (Lüning 1981). Thus, the above data clearly indicate that quantum irradiance was not the inhibitory factor responsible for growth differences among years.

The mean TIN and phosphate-P concentrations at 1–5 m depth during this period were 0.099 and 0.023 mg L^{-1} for 2003/2004, 0.110 and 0.023 mg L^{-1} for 2004/2005 and 0.105 and 0.027 mg L^{-1} for 2005/2006 (Fig. 5). Among the years, there was very little difference in nutrients concentrations. In addition, Kawabata et al. (1998) reported that the TIN and phosphate-P concentrations at Yusu changed little from 1990 to 1997, and concentrations were similar to present studies. Accordingly, we suggest that nutrient concentrations were similar in the years 2001/2002 through 2005/2006. Moreover, the TIN and phosphate-P concentrations in our study area are about 80 % of the values for Osatsube, Hokkaido coast within the native range of *L. japonica* (Maita et al. 1991). Accordingly, nutrient concentrations are likely not limiting the growth of *L. japonica* off Yusu.

Kinoshita (1947) reported that zoospores of *Laminaria japonica* are unable to form gametophytes when temperatures exceed 14.2°C, and the optimum range is 0.8–9°C. Okada and Sanbonsuga (1980) reported a range of 3–18°C for sporophyte formation, with an optimum of 8–13°C. In addition, Kirihara et al. (2003) reported that individuals of *L. japonica* off the coast of Shiriyazaki, Shimokita Peninsula, Aomori Pref. occurred at high density (>50 ind. m^{-2}) in years when the seawater temperature in January was <8°C, while density decreased to <30 ind. m^{-2} in years when the January temperature was >8°C. Thus, the upper limit of seawater temperature for young sporophyte formation is <13°C. These studies indicate that the aquaculture of *L. japonica* in the Uwa Sea is undertaken close to the upper thermal limit of *L. japonica*. Accordingly, we examined the relations between mean temperature and annual productivity (Fig. 15). The mean temperatures for the 45 days after the experiments were started in the Uwa Sea were 15.5°C (range: 13.7–17.8°C) in 2003/2004, 15.5°C (13.2–18.0°C) in 2004/2005 and 14.5°C (13.5–15.4°C) in 2005/2006. In 2001/2002 and 2002/2003, the mean temperatures for the 45 days were 15.2°C (13.9–16.9°C) and 15.0°C (13.2–16.8°C) (Suzuki et al. 2006). The mean water temperature was <15.5°C in 2001/2002, 2002/2003 and 2005/2006 when significantly higher annual

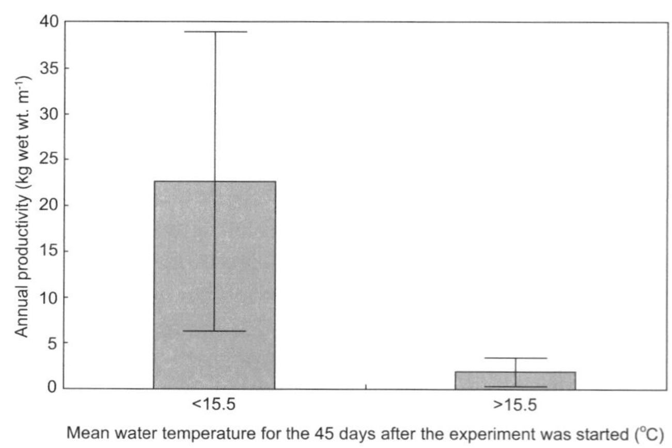

Fig. 15 Relationship between mean temperature for the 45 after the experiment began and annual productivity from 2001/2002 to 2005/2006

productivity occurred (22.6±16.3 kg wet wt. m^{-1}), while it was >15.5°C in 2003/2004 and 2004/2005 when there was significantly lower annual productivity (1.9±1.6 kg wet wt. m^{-1}). The annual productivity per meter of rope at Yusu when the mean temperature for the first 45 experimental days was <15.5°C was approximately equivalent to the productivity on 2.7–4.5 m^2 of substratum off the Hokkaido coast, but equivalent to productivity on only 0.2–0.4 m^2 of Hokkaido coast substratum when the mean temperature was >15.5°C (Suzuki et al. 2006). This comparison also indicates that the optimal seawater temperature for young sporophyte generation off Yusu was <15.5°C.

Kelp beds (*Ecklonia cava*, *E. kurome* and *Eisenia arborea*) along the western coast of Kyushu to Shizuoka Prefecture, southern Japan have undergone destruction through by herbivorous fishes (*Calotomus japonicus*, *Prionurus scalprum* and *Siganus fuscescens*) since the late 1990s (Hasegawa et al. 2003; Kiyomoto et al. 2000; Nakayama and Arai 1999). These fish species were previously endemic to the Okinawa Islands off the southern coast of Japan. It is possible that fish herbivory during years with warmer winter temperatures also contributed to the decrease in annual net kelp productivity at our experimental site.

Kitadai and Kadowaki (2003) reported that kelp erosion was caused by a high water temperature. Analyses of the relationship between "growth and erosion rates" and "productivity and erosion rates" versus "seawater temperature" from 2001/2002 to 2005/2006 indicate that growth and erosion rates were similar at approximately 17.5°C (Fig. 16). At higher temperatures, erosion rates exceeded growth rates. Hence, *Laminaria japonica* can achieve a net gain in biomass at temperatures up to 17.5°C in our study area. Seawater temperatures reach 17.5°C in April–May. Thus, the culture period of *L. japonica* in the Uwa Sea is restricted to 4–5 months.

Global seawater temperature has risen 0.5°C/100 yr (Japan Meteorological Agency 2006). Since the mid-1990s, in particular, there have frequently been years with record high water temperatures (IPCC 2001), and this tendency might continue in the future. In the Uwa Sea, a significant trend of annual mean seawater temperature increase was found at 0.5, 2.0 and 5.0 m depths between 1982 and 2004

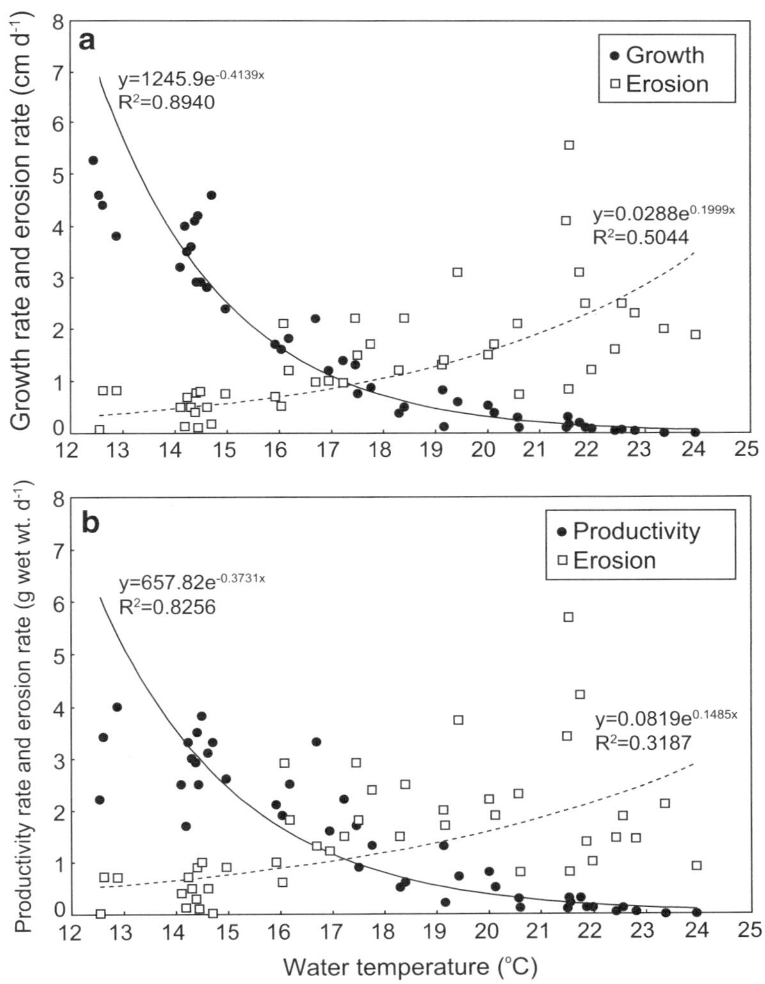

Fig. 16 Relationships between "growth and erosion rates" and "seawater temperature" (**a**) and between "productivity and erosion rates" and "seawater temperature" (**b**) from 2001/2002 to 2005/2006

(Suzuki and Takeuchi 2007). The increases occurred mainly in winter (from December to February). If the increasing seawater temperature trend in winter continues, annual productivity of *L. japonica* off the coast of the Uwa Sea may continue to decrease through decreased formation of young sporophytes. Furthermore, the seawater temperature in April and May have also increased, so that erosion rates will exceed growth rates earlier in the year (cf. Fig. 16). Accordingly, the growth period of *L. japonica* in the Uwa Sea will be shortened and productivity will continue to decrease.

Acknowledgements The present study was partially supported by the special research found from Faculty of Agriculture, Ehime University.

References

Coles SL, Brown BE (2003) Coral bleaching-capacity for acclimatization and adaptation. Adv Mar Biol 46:183–223

Hasegawa M, Koizumi K, Konagaya T, Noda M (2003) Grazing of herbivorous fish as a continuous factor of Isoyake off the coast of Hainan, Shizuoka Prefecture. Bull Shizuoka Pref Fish Exp Stn 38:19–25

Hiraoka M, Ura Y, Haraguchi H (2005) Relationship between seaweed beds and seawater temperature in the Tosa Bay. Aquabiology 27 (5):485–493

Hoegh-Guldberg O, Salvat B (1995) Periodic mass-bleaching and elevated sea temperatures: bleaching of outer reef slope communities in Moorea, French Polynesia. Mar Ecol Prog Ser 121: 181–190

IPCC (2001) Climate change 2001: the scientific basis. Cambridge University Press, Cambridge, p 881

Japan Meteorological Agency (2006) Long-term variation in surface seawater temperature. http://www.data.kishou.go.jp/kaiyou/shindan/a_1/glb_warm/glb_warm.html

Jokiel PL, Brown EK (2004) Global warming, regional trends and inshore environmental conditions influence coral bleaching in Hawaii. Glob Chan Biol 10:1627–1641

Jones RJ, Hoegh-Guldberg O, Larkum AWD, Schreiber U (1998) Temperature-induced bleaching of corals begins with impairment of the CO_2 fixation mechanism in zooxanthellae. Plant Cell Env 21:1219–1230

Kawabata Z, Shiba S, Sotomaru Y, Kusumoto D, Ohgaki N, Nakashige E, Udaka N, Kuranishi M, Shimizu N, Yamamoto C (1998) Water quality monitoring in Usu Bay. In: Report on environmental survey of aquaculture ground in Uwajima Bay 1984–1998. Usu Fishery Cooperative Union, Ehime, Japan, pp 603–644

Kawashima S (2004) Laminariales. In: Ohno M (ed) Biology and technology of economic seaweeds. Uchida Rokakuho Publishing Co., Ltd., Tokyo, Japan, p 575

Kinoshita T (1947) Research on enhancement of *Laminaria* and *Undaria*. Hoppo Press, Sapporo, Japan, p 79

Kirihara S, Nakamura T, Notoya M (2003) Effect of water temperature on the growth of *Laminaria japonica* (Laminariales, Phaeophyceae) at the coast of Siriyazaki, Shimokita Peninsula, Japan. Suisanzoshoku 51(3):273–280

Kitadai Y, Kadowaki S (2003) The growth process and N, P uptake rates of *Laminaria japonica* cultured in coastal fish farms. Suisanzoshoku 51(1):15–23

Kiyomoto S, Yoshimura T, Arai S, Kiriyama T, Fujii A, Yotsui T (2000) Recovery of *Ecklonia kurome* after the occurrence of blade disappearing phenomenon at Nomozaki, Nagasaki Prefecture. Bull Seikai Natl Fish Res Inst 78:57–65

Lüning K (1981) Light. In: Lobban CS, Wynne MJ (eds) The biology of seaweeds. Univ. of Calif. Press, CA, pp 326–355

Maita Y, Mizuta H, Yanada M (1991) Nutrient environment in natural and cultivated grounds of *Laminaria japonica*. Bull Fac Fish Hokkaido Univ 42(3):98–106

Nakayama Y, Arai S (1999) Grazing of the brown alga *Ecklonia cava* by three herbivorous fishes on the coast of Nakagi, South Izu, central Japan. Jpn J Phycol 47:105–112

Obura DO (2005) Resilience and climate change: lessons from coral reefs and bleaching in the Western Indian Ocean. Est Coastal Shelf Sci 63:353–372

Ohno M (1987) Present seaweeds cultivation. In Tokuda H, Ohno M, Ogawa H (eds) The resources and cultivation of seaweeds. Midori Syobo Co., Ltd., Tokyo, Japan, p 354

Okada Y, Sanbonsuga Y (1980) Effects of temperature on the growth and maturation of female gametophytes of Laminariaceous plants. Bull Hokkaido Reg Fish Res Lab 45:51–56

Sabatés A, Martín P, Lloret J, Raya V (2006) Sea warming and fish distribution: the case of the small pelagic fish, *Sardinella aurita*, in the western Mediterranean. Glob Chan Biol 12:2209–2219

Short FT, Neckles HA (1999) The effects of global climate change on seagrasses. Aquatic Botany 63:169–196

Suzuki S, Takeuchi I (2007) Inter-annual variation in seawater temperature off the Uwa Sea coast, Ehime Prefecture. La mer 45:35–46

Suzuki S, Furuya K, Takeuchi I (2006) Growth and annual production of the brown alga *Laminaria japonica* (Phaeophyta, Laminariales) introduced into the Uwa Sea in southern Japan. J Exp Mar Biol Ecol 339:15–29

Tsuda F, Agatsuma Y, Taniguchi K (2006) History and current state of "Isoyake" phenomenon along the Japan Sea coast in Hokkaido, Japan. Kaiyo Monthly 38(3):210–215

Eisenia bicyclis bed coverage off Oshika Peninsula, Japan, in relation to sporophyte survival and *Strongylocentrotus nudus* abundance

Daisuke Muraoka

Originally published in the Journal of Applied Phycology, Vol 20, No 5, 395–401.
DOI: 10.1007/s10811-008-9337-1

Abstract It has recently been reported that the *Eisenia bicyclis* beds on the coast of Tomarihama in the Oshika Peninsula in Miyagi Prefecture (northern part of Japan) have been progressively receding from the offshore side of the beds every year. In this study, to determine the factors causing the change in the distribution of *E. bicyclis* beds as well as the extent of the change, an ecological survey of *E. bicyclis* and grazing animals was conducted. A 150 m-long fixed line was set up in the rocky coastal area on the Oshika Peninsula in Miyagi Prefecture, and the number of *E. bicyclis* individuals inhabiting a 1-m width was then measured. In addition, the numbers of individuals of sea urchins (*Strongylocentrotus nudus*) and abalone (*Haliotis discus hannai*) were measured. Although the *E. bicyclis* bed extended to the 64-m point on the fixed line at the time of the survey in August 1999, by July 2001, its lower limit was at the 53-m point; thus, a retreat of the outer edge of the bed of 11 m was observed in 2 years. The density of adult *E. bicyclis* sporophytes in the bed was reduced from 1.08 to 1.80 individuals (ind.) m^{-2}, respectively. Juvenile sporophytes were also observed, but most of those that had settled near the lower limit of the bed failed to survive to become adult sporophytes. There were few sea urchins, the main grazing animal on the site, inside the *E. bicyclis* bed (3.5 ind. m^{-2} in October 2001) while there were many near the bed lower limit and on the deep-site side (12.9 ind. m^{-2} in October 2001). These results indicate that the high grazing pressure of sea urchins reduced the survival of juvenile sporophytes near the bed lower limit, thus impeding the recruitment of juvenile sporophytes to replace adult sporophytes that died due to senescence, storm action and/or grazing.

Keywords Grazing · *Eisenia bicyclis* · Phaeophyta · Sea urchins · *Strongylocentrotus nudus* · Degeneration · Kelp beds · Monitoring

D. Muraoka (✉)
Tohoku National Fisheries Research Institute,
3-27-5, Shinhama-cho,
Shiogama, Miyagi 985-0001, Japan
e-mail: dmuraoka@affrc.go.jp

Introduction

Kelp beds play an important role as one of the primary producers in temperate coastal ecosystems; they provide a place for marine animals to spawn and develop, and supply food to algae-grazing species such as sea urchins and abalone (e.g., Dayton 1985; Steneck et al. 2002). In particular, *Eisenia* and *Ecklonia* beds, which account for approximately 20% of the kelp beds extending along Japan's coastline (Nature Conservation Bureau 1994), have been recognized as being very important kelp beds for the fishing industry because of their high productivity and their high food value for sedentary algae-grazing animals. Thus, numerous projects to develop kelp beds have been carried out to preserve these rocky coastal resources (Steneck et al. 2002). Furthermore, loss or decline of kelp beds has been reported on a worldwide scale. The main reasons have been considered to be overgrazing, mainly by sedentary algae-grazing sea urchins (Lawrence 1975; Elner and Vadas 1990; Sivertsen 2006), and grazing by herbivorous fish (Harris et al. 1984). It has been reported that, on the coasts of Japan, 1,687 ha of *Eisenia* and *Ecklonia* beds disappeared in the 13 years after 1978 (Nature Conservation Bureau 1994).

Recently, a gradual degeneration and shrinkage from the offshore side has been observed in multiple *Eisenia bicyclis* beds off the coast of Minami-Sanriku (northern Japan; Taniguchi 1996). Because changes in kelp beds directly affects the harvest of economically high-valued inshore resources in this marine area, especially *Strongylocentrotus nudus* and *Haliotis discus hannai*, which feed on seaweeds as their main food sources, the degeneration of *E. bicyclis* beds has significant consequences to both the ecosystem and the local fishing community dependent on these resources. In this paper, in order to explore the changes in *E. bicyclis* beds, and the causes of such changes, the rocky coastal area of Tomarihama in the Oshika Peninsula in Miyagi Prefecture (northern part of Japan) was selected as the seaweed test site, and an ecological survey was conducted on *E. bicyclis* and the main algae-grazing species in that area, *S. nudus* and *H. discus hannai*. Based on the findings, the relationship between the two was examined in terms of the extent of coverage of the *E. bicyclis* bed.

Materials and methods

The rocky coastal area of Tomarihama in the Oshika Peninsula, Miyagi Prefecture (northern Japan) was set up as the test sea area (Fig. 1). At the start of the survey, an *E. bicyclis* bed occurred in the depth range of 2–5 m. *Sargassum yezoense* beds were also seen to a depth of around 2 m. From 5 to 10 m depth crustose coralline algae dominated.

In 1999, dan buoys were put on the starting point (0 m; inside the *E. bicyclis* bed) and the terminal (150 m; inside of coralline algae flats area) sea bottom (Fig. 1). From 1999 through 2001, a 150-m long measuring rope (fixed line) was connected between the 0 m point and the 150 m point

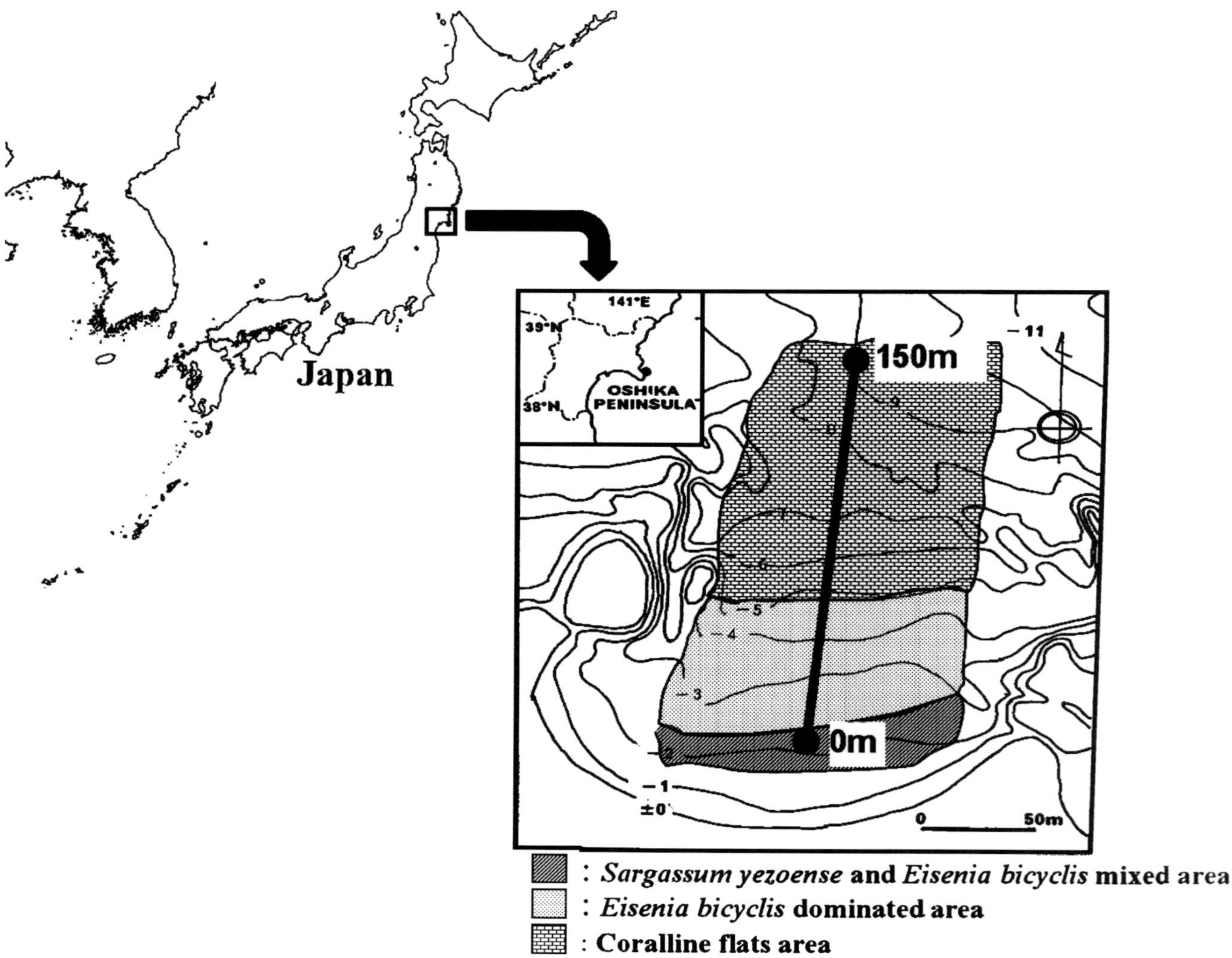

Fig. 1 Tomarihama, the survey sea area and the location of the fixed line

on each investigation. The surveys conducted are described below.

Water temperature survey

In December 1999, an automatic Temperature Recorder, MDS-T (Alec Electronics, Kobe, Japan), was set up inside the *E. bicyclis* beds near the fixed line (3 m in depth), and the water temperature was then measured for approximately 2 years until January 2002.

Eisenia bicyclis ecological survey

In August 1999, July 2000, and February, May, and July 2001, SCUBA diving surveys were conducted, and the number and location of the emergence of all *E. bicyclis* juveniles (less than 1 year old; not yet having dichotomous branching) and adults (1 year old or older; having dichotomous branching) sporophytes growing within a 1-m width (i.e., 0.5 m either side) of the 150 m-long fixed line. Juvenile and adult sporophytes individuals were distinguished by the method of Taniguchi and Kato (1984). At the same time, arm length (the length between top of the stipe and base of the frond) of all *E. bicyclis* individuals was measured, and used to estimate the age according to the method developed by Taniguchi and Kato (1984).

Eisenia bicyclis maturity survey

In June, August, September, October, and December 1999; April, June, July, August, October, and December of 2000; and February, May, and July 2001, at a location of 1m or greater away from the fixed line, six individuals or more of *E. bicyclis* that were deemed 2 years old or older based on the arm length (>5.65 cm; Taniguchi and Kato 1984) were sampled. Zoospores were released according to the method of Nakahara (1984). The formation of zoosporangial sori was then verified for each individual. The individuals with zoosporangial sori were then taken back to the laboratory and, after being preserved chilled overnight, the zoosporangial sori part was soaked in sea water under bright lighting to verify the release of zoospores.

Ecological survey of *S. nudus* and *H. discus hannai*

In October 1999, 2000 and 2001, all *S. nudus* and *H. discus hannai* that had grown within a 1 m-width of the 150 m-long fixed line, were sampled by SCUBA diving. Samples were carried up from the bottom of the sea to the boat. On the boat, the location of every *S. nudus* and *H. discus hannai* was recorded by using the fixed line to pinpoint their exact position. The test length was also recorded. Furthermore, in the October 2001 survey, the total weight of *S. nudus* and *H. discus hannai* was measured at intervals of 10 m on the fixed line. After measurement, every *S. nudus* and *H. discus hannai* was returned to its original location.

Results

Water temperature survey

Figure 2 shows the fluctuation in water temperature near the seaweed test site in 2000 and 2001. In 2000, the maximum and minimum temperatures were 23.9°C in August and 6.7°C in March, respectively, and in 2001 the corresponding temperatures were 22.4°C in August and 4.3°C in March, respectively.

Eisenia bicyclis ecological survey

Figure 3 shows the distribution of the density of *E. bicyclis* individuals (juvenile and adult sporophytes) for each 10 m interval of the line for each survey, and the lower limit of *Eisenia bicyclis* adult sporophytes. The densities of adult *E. bicyclis* in bed were 1.08 (July 2000) to 1.80 (February 2001) individuals (ind.) m^{-2}. At the start of the survey (August 1999), the *E. bicyclis* bed extended to 64 m on the fixed line. Although the lower limit of adult sporophytes receded to 61 m in July 2000, the settlement of the juvenile sporophytes could be seen up to 80 m. However, the juvenile sporophytes that settled at 60–80 m had all disappeared by February 2001; the lower limit of adult

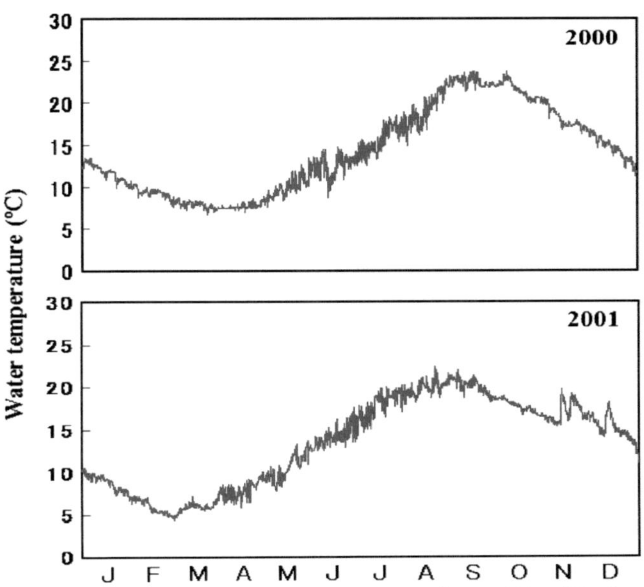

Fig. 2 Water temperature fluctuation in Tomarihama, Miyagi Prefecture, Japan (in 2000 and 2001; surveyed at 3 m depth)

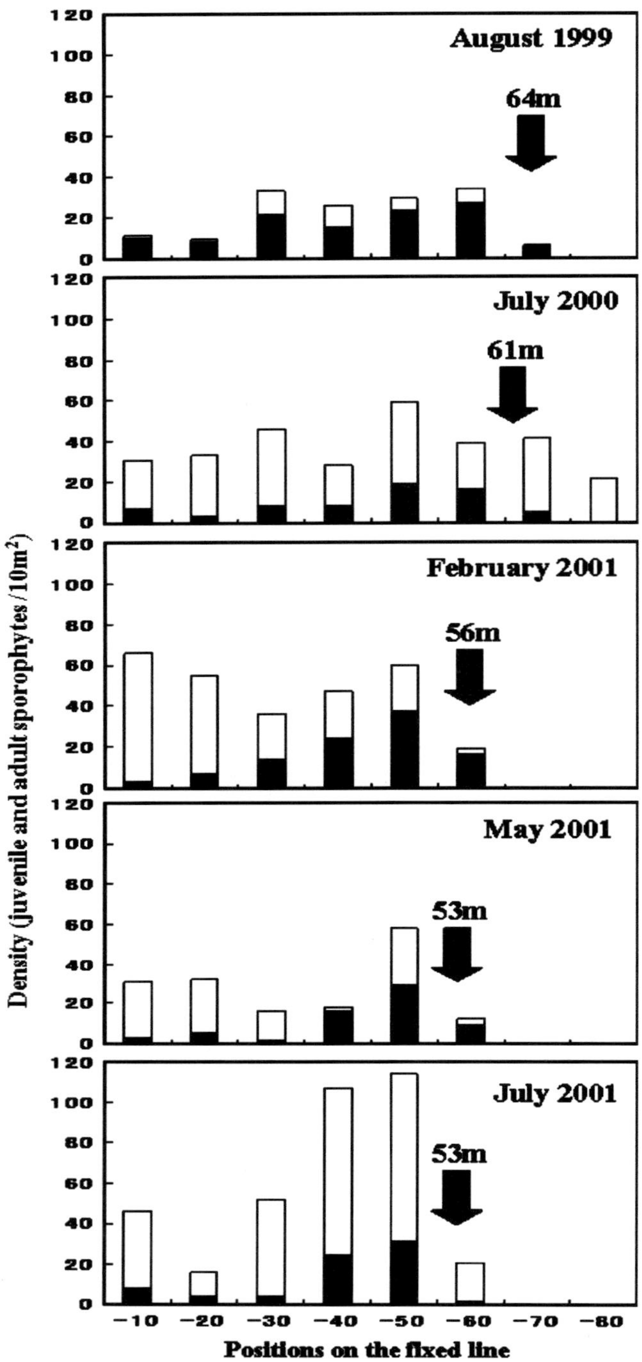

Fig. 3 Density of individuals in 10 m intervals within a 1m-width of the 150 m-long fixed line (in Tomarihama). ■ *Eisenia bicyclis* adult sporophytes, □ juvenile sporophytes; *arrows* position of lower limit of *E. bicyclis* adult sporophytes

sporophytes was 53 m in the May 2001 survey. In the July 2001 survey, the settlement of a large number of juvenile sporophytes (283 ind.) could be observed at 0–60 m on the fixed line. Yet, no settlement could be confirmed at 60–80 m. Also, only a few isolated macrophytes could be observed deeper than the *E. bicyclis* bed lower limit, thus forming a coralline algae flat.

Eisenia bicyclis maturity survey

The maturity survey (the percentage of individuals having zoosporangial sori) of *E. bicyclis* 2 years old or older between June 1999, and July 2001, showed 100% maturity in September, October, and December 1999 and in October and December 2000 (Fig. 4). Although data were not collected in all months, these results suggest that the maturity period is mostly from autumn to winter. Furthermore, in the zoospore release test, the release of zoospores from the zoosporangial sori was verified in all survey months in which the formation of zoosporangial sori was observed.

Strongylocentrotus nudus and *H. discus hannai* ecological survey

Figure 5 shows the density distribution of *S. nudus* and *H. discus hannai* in October in 1999, 2000 and 2001. Of the two species at the survey site, *S. nudus* accounted for 91.1% (1999), 97.3% (2000) and 96.8% (2001) of the entire amount in terms of numbers of individuals, revealing that the main algae-grazing animal in the survey sea area was *S. nudus*. The total number of *S. nudus* different in each survey year, with 448 individuals in 1999, 829 individuals in 2000, and 1,400 individuals in 2001. Moreover, in terms of the distribution on the fixed line, the distribution of the two species differed between years, with a boundary around the 50–60 m site, which was the lower limit of the *E. bicyclis* bed. For this reason, whereas *H. discus hannai* was distributed with relatively high frequency on the shallow side (inside the *E. bicyclis* bed), most *S. nudus* were found on the deeper side (inside the coralline algae flat; Fig. 5). For example, the density of *S. nudus* inside the *E. bicyclis* bed was 3.5 ind. m^{-2} in October 2001,

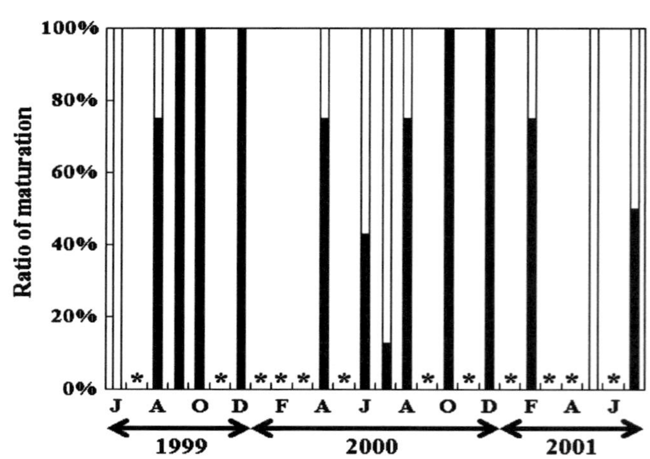

Fig. 4 Ratio of maturation in each survey of *E. bicyclis* (2 years old or older) that grow in Tomarihama, the survey sea area (■: mature, □ immature). *Asterisks* Months with no data

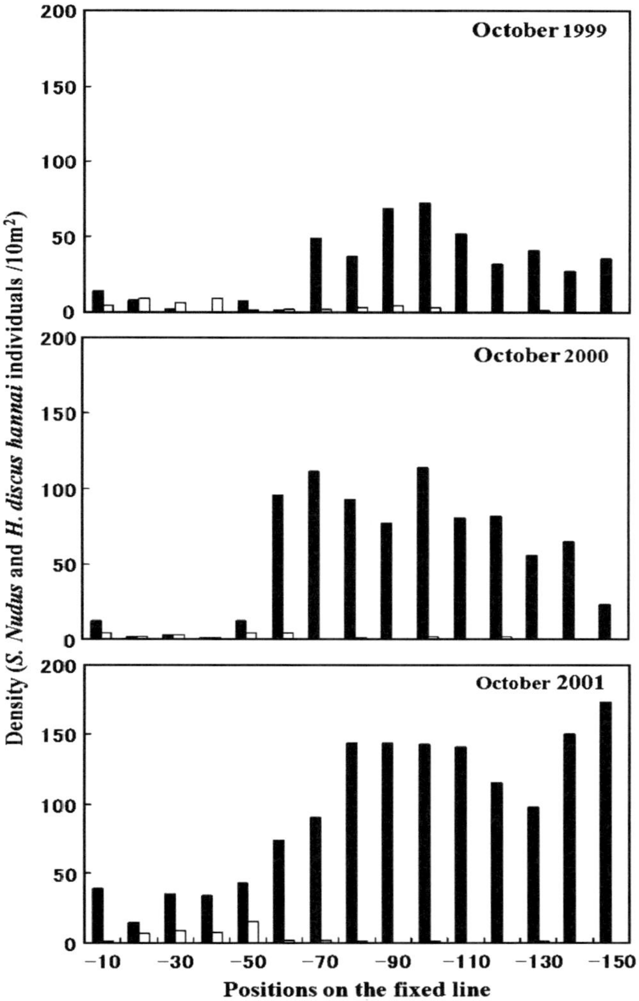

Fig. 5 Density of individuals of *Strongylocentrotus nudus* (■) and *Haliotis discus hannai* (□) at 10 m intervals within a 1 m-width of the 150 m-long fixed line (in Tomarihama)

whereas the density inside the coralline algae flat was 12.9 ind. m^{-2}.

As a result of surveying the test diameter of *S. nudus*, it was found that, although there was dispersion in the number of individuals by survey year, a large number of small individuals with a test diameter of less than 4 cm were growing in the coralline algae flat that spread in areas deeper than *E. bicyclis* bed (Fig. 6). In October 2001, a relatively larger number of *S. nudus* individuals were observed within the *E. bicyclis* bed. Also, 86.7% of the individuals within the *E. bicyclis* bed were large individuals with 4 cm or greater test diameter. On the other hand, in the coralline algae flat, 66.9% of *S. nudus* individuals were less than 4 cm in test diameter.

In the 2001 weight measurement, the total weight of *S. nudus* corresponded to 95.6% of the entire weight (Fig. 7). In *S. nudus*, the biomass of *S. nudus* inside the *E. bicyclis* bed was 0.06 kg m^{-2}. On the other hand, the biomass inside the coralline algae flat was 0.40 kg m^{-2}. From the weight measurement survey of *S. nudus* conducted on the line in October 2001, the biomass at the 60–70 m site was 7.2 kg, the maximum value, which was followed by 5.0 kg, a high value, at the 70–80 m site (Fig. 7).

Discussion

Sebens (1985) indicated that colonies of erect algae can be changed into coralline algae flats by grazing sea urchins. The survey in this study revealed that in the 2 years from August 1999 through July 2001, the lower limit of the *Eisenia bicyclis* bed (adult sporophytes) had receded 11 m towards the coastal side (Fig. 3). There is also data indicating that, as of 1981 in Tomarihama, *E. bicyclis* beds existed to 8 m in depth (Taniguchi et al. 1987). In the test area, 8 m depth is situated at approximately 120 m on the

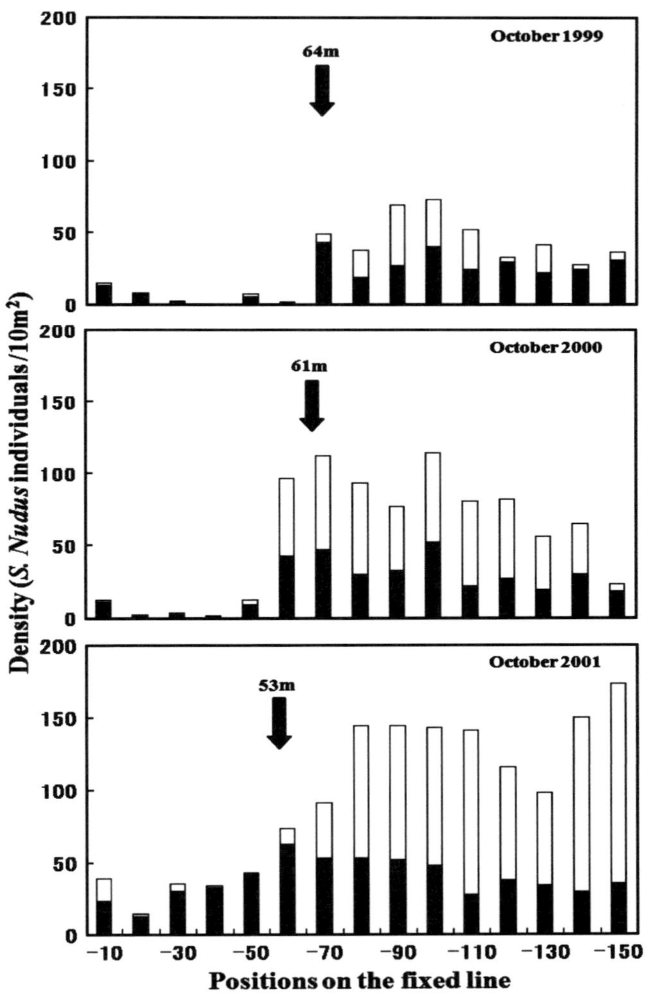

Fig. 6 Density of individuals of *S. nudus* (■ ≥4 cm in diameter; □ <4 cm in diameter) at 10 m intervals within a 1 m-width of the 150 m-long fixed line (in Tomarihama); *arrows* positions of lower limit of *E. bicyclis* adult sporophytes in each year

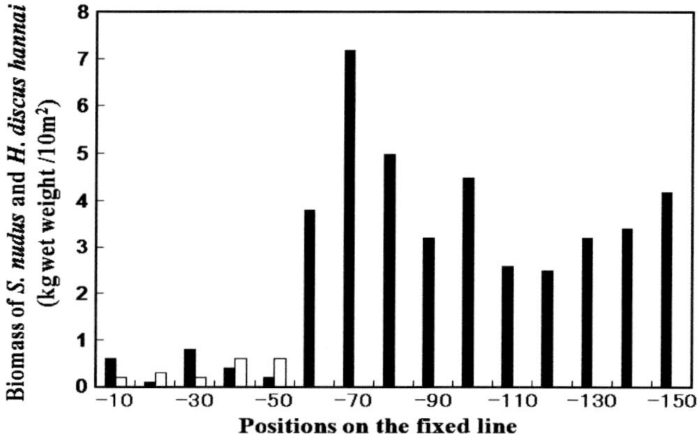

Fig. 7 Biomass of *S. nudus* (■) and *H. discus hannai* (□) at 10 m intervals within a 1 m-width of the 150 m-long fixed line in October 2001 (in Tomarihama)

fixed line. If this is the case, the bed has retreated approximately 67 m in 20 years. Not just in the case of *E. bicyclis*, the disappearance of a kelp bed occurs because, after the existing individuals die of old age, disappear during storms and/or are grazed, recruitment of the next generation of individuals does not occur for some reason. The reason could be that there was no supply of spores from the parent seaweed in the first place, or that for some reason the juvenile sporophytes that had been generated did not survive to become adult sporophytes. The highest water temperature in the test sea area in the survey period was 23.9°C in September 2000. Kurashima et al. (1996) found that the optimal water temperature for photosynthesis by *E. bicyclis* was 25°C (using filtered seawater). This suggests that *E. bicyclis* in the test sea area were not adversely affected by high water temperatures and did not suffer maturation impediments due to extreme temperature. In fact, the results of the survey findings verified that adult sporophytes composing the *E. bicyclis* bed formed zoosporangial sori mainly in autumn and winter (Fig. 4), and zoospore release was also verified in a laboratory test. Thus, it became clear that the cause of the receding of the lower limit of the *E. bicyclis* bed in the test sea area was because the juvenile sporophytes do not survive to become adult sporophytes. This is supported by the fact that growth of the *E. bicyclis* juvenile sporophytes at the 60–80 m site, observed in July 2000, had completely disappeared by February 2001 (Fig. 3). *S. nudus*, the main algae-grazing species in the test sea area, occurred mainly from the lower limit of the *E. bicyclis* bed to the coralline algae flat. Kawamata (1998) suggested that the sweeping action of kelp fronds in turbulent flow was an effective mechanism to deter urchin invasion. This might be one of the reasons why most *S. nudus* were not found inside the *E. bicyclis* bed but on the coralline algae flat. The results of weight measurement showed the maximum biomass at 60–80 m, where the juvenile sporophytes of *E. bicyclis* had disappeared (Fig. 7).

Based on these findings, it can be inferred that the receding of the lower limit of the *E. bicyclis* bed in the test sea area has occurred in the following way: after *E. bicyclis* adult sporophytes growing near the lower limit of the *E. bicyclis* bed individuals died of old age, disappeared during storms and/or were grazed, the settled juvenile sporophytes of *E. bicyclis*, which typically would mature as adult sporophytes, were consumed by the high grazing pressure of *S. nudus* and thus did not survive to become adult sporophytes. Agatsuma et al. (1997) removed sea urchins at a coralline algae flat on the Hokkaido coast of the Sea of Japan and confirmed that seaweeds would settle there if the urchin grazing pressure was removed. Vega et al. (2005) reported that an abrupt change in the spatial-temporal patterns of abundance of the black sea urchin *Tetrapygus niger*, the most conspicuous benthic grazer in northern Chile, produced local extinctions and compression of bathymetric distribution of *Macrocystis integrifolia* and *Lessonia trabeculata* (Laminariales, Phaeophyceae). On the seaward face of Port Erin (Isle of Man, United Kingdom) breakwater large algae do not occur on the lowest 3 m. The density of sea urchin (*Echinus esculentus*) was high but, after removal of sea urchin, kelp (*Laminaria hyperborea*) and other algae occurred in the lower area (Jones and Kain 1967). In California, following mass mortality by disease of a localized population of sea urchins (*Strongylocentrotus franciscanus*), a kelp forest resulted from the rapid seaward expansion of four species of brown algae (Pearse and Hines 1979). These reports show that the grazing pressure of sea urchins directly affects the colony formation of kelp beds.

Most of the *S. nudus* living on the coralline algae flats were small-sized individuals, while large individuals were dominant within the *E. bicyclis* bed. This finding is consistent with the behavioral pattern of *S. nudus*, which is believed to settle on a coralline algae flat area due to the effect of the settlement inducer dibromomethane (Taniguchi et al. 1994), and then move into kelp beds in order to find food to grow (Agatsuma and Kawai 1997; Sano et al. 1998).

Is it possible for once-degenerated *E. bicyclis* beds to recover, and the distribution area to expand once again? Taniguchi and Kito (1988) reported that, in a year when the Kurile current was strong and cold, the developmental density of the juvenile sporophytes increased and, as a result, *E. bicyclis* beds expanded into deeper places. In northern Chile, during the 1997–1998 El Niño, the spatial-temporal patterns of abundance of the kelp assemblage were maintained by the continuity of coastal upwelling, which buffered and moderated superficial warming of the sea and depletion of nutrients. In this context, localities associated with coastal upwelling areas could function as "sources" of reproductive propagules after passage of El Niño, thus increasing kelp recolonization rates in "sink" localities that have suffered local kelp extinctions (Vega et al. 2005). In this test sea area, observations were further continued after 2001, but the expansion of *E. bicyclis* beds has not been observed. Therefore, in order to understand the mechanism of *E. bicyclis* bed fluctuations more clearly, longer-term observations (include oceanographic events) are necessary.

Acknowledgments I would like to thank Dr. K. Saitoh, Dr. H. Takami, Mr. M. Fukuda, Dr. H. Tamaki, Ms. K. Inomata and Mr. J.W. Renshaw for their support and assistance during the course of this study. This article was issued as FRA-TNFRI-B104.

References

Agatsuma Y, Kawai T (1997) Seasonal migration of the sea urchin *Strongylocentrotus nudus* in Oshoro Bay of southwestern Hokkaido, Japan (in Japanese). Nippon Suisan Gakkaishi 63: 557–562

Agatsuma Y, Matsuyama K, Nakata A, Kawai T, Nishikawa N (1997) Marine algal succession on coralline flats after removal of sea urchins in Suttsu Bay on the Japan Sea coast of Hokkaido, Japan (in Japanese). Nippon Suisan Gakkaishi 63:672–680

Dayton PK (1985) Ecology of kelp communities. Annu Rev Ecol Syst 16:215–245

Elner RW, Vadas RL (1990) Inference in ecology: the sea urchin phenomenon in the northwestern Atlantic. Am Nat 136:108–125

Harris LG, Ebeling AW, Laur DR, Rowley RJ (1984) Community recovery after storm damage: a case of facilitation in primary succession. Science 224:1336–1338

Jones NS, Kain JM (1967) Subtidal algal colonization following the removal of *Echinus*. Helgoländer Wiss Meeresunters 15:460–466

Kawamata S (1998) Effect of wave-induced oscillatory flow on grazing by a subtidal sea urchin *Strongylocentrotus nudus* (A. Agassiz). J Exp Mar Biol Ecol 224:31–48

Kurashima A, Yokohama Y, Aruga Y (1996) Physiological characteristics of *Eisenia bicyclis* Setchell and *Ecklonia cava* Kjellman (Phaeophyta) (in Japanese). Jpn J Phycol 44:87–94

Lawrence JM (1975) On the relationships between marine plants and sea urchins. Oceanogr Mar Biol Annu Rev 13:213–286

Nakahara H (1984) Alternation of generations of some brown algae in unialgal and axenic cultures. Sci Pap Inst Alg Res Hokkaido Univ 7:77–194

Nature Conservation Bureau (1994) The report of the marine biotic environment survey on the 4th national survey on the natural environment, vol 2. Algal and sea-grass beds (in Japanese)

Pearse JS, Hines AH (1979) Expansion of a central California kelp forest following the mass mortality of sea urchins. Mar Biol 51:83–91

Sano M, Omori M, Taniguchi K, Seki T, Sasaki R (1998) Distribution of the sea urchin *Strongylocentrotus nudus* in relation to marine algal zonation in the rocky coastal area of the Oshika Peninsula, Northern Japan. Benth Res 53:79–87

Sebens KP (1985) The ecology of the rocky subtidal zone. Am Sci 73:548–557

Sivertsen K (2006) Overgrazing of kelp beds along the coast of Norway. J Appl Phycol 18:599–610

Steneck RS, Graham MH, Bourque BJ, Corbett D, Erlandson JM, Estes JA, Tegner MJ (2002) Kelp forests ecosystems: biodiversity, stability, resilience and future. Environ Cons 29:436–459

Taniguchi K (1996) Primary succession of marine algal communities in the sublittoral zone off Oshika Peninsula, Japan (in Japanese). Nippon Suisan Gakkaishi 62:765–771

Taniguchi K, Kato F (1984) On age and growth of *Eisenia bicyclis* (Kjellman) Setchell (Phaeophyceae, Laminariales) (in Japanese). Bull Tohoku Reg Fish Res Lab 46:15–19

Taniguchi K, Kito H (1988) Age composition in the population of *Eisenia bicyclis* (Laminariaceae; Phaeophyta) (in Japanese). Nippon Suisan Gakkaishi 54:1583–1588

Taniguchi K, Kurata K, Maruzoi T, Suzuki M (1994) Dibromomethane, a chemical inducer of larval settlement and metamorphosis of the sea urchin *Strongylocentrotus nudus*. Fish Sci 60:795–796

Taniguchi K, Sato Y, Osada Y, Suenaga H (1987) On the structure of the *Eisenia bicyclis* population on the coast of Oshika Peninsula in northern Honshu, Japan (in Japanese). Bull Tohoku Reg Fish Res Lab 49:103–109

Vega JMA, Vasquez JA, Buschmann AH (2005) Population biology of the subtidal kelps *Macrocystis integrifolia* and *Lessonia trabeculata* (Laminariales, Phaeophyceae) in an upwelling ecosystem of northern Chile: interannual variability and El Niño 1997–1998. Rev Chil Hist Nat 78:33–50

Algal communities at Gouqi Island in the Zhoushan archipelago, China

S. Y. Zhang · Lei Wang · W. D. Wang

Originally published in the Journal of Applied Phycology, Vol 20, No 5, 403–411.
DOI: 10.1007/s10811-008-9338-0

Abstract The kelp bed of Gouqi Island lies in the east of the East China Sea, in the Shengsi Archipelago. It is a key component of the island-reef ecosystem. This study evaluated kelp beds on intertidal and subtidal zones from 2004 to 2006. We evaluated seasonal variations in the community structure of macroalgae, and their relationship with water temperature, nitrogen and phytoplankton. The Gouqi Island kelp bed consists mainly of species of *Sargassum*. *Sargassum horneri* was the dominant species in subtidal zones, representing 90% of the total biomass. The dominant species in intertidal zones were *S. fusiforme*, *S. horneri*, *S. thunbergii* and *Undaria pinnatifida*. Fifty-four phytoplankton species were found in the Gouqi Island kelp bed, of which 51 species were found inside the kelp beds, 43 species outside the kelp beds, and 40 species were present both inside and outside the kelp beds. Diatoms were dominant both inside and outside the kelp beds. Except in winter, *Skeletonema costatum* was dominant, and its abundance in autumn was over 98% of the total abundance. Water temperature was found to be the primary factor influencing the growth of *S. horneri* in the Gouqi Island kelp bed. *S. horneri* grew slowly from November to March, and rapidly from March to June. The upper temperature limit for growth of *S. horneri* was about 18°C. Higher water temperatures can result in a decline in the kelp bed. Nitrogen seemed to be a limiting factor for macroalgal growth. This was particularly true for *S. horneri* and phytoplankton. Since nitrate is the primary nutrient for *S. horneri*, its absorption by *S. horneri* resulted in seasonal changes of nitrate in the ecosystem.

Keywords Gouqi Island · Kelp beds · Nutrition · Phytoplankton · Water temperature

S. Y. Zhang (✉) · L. Wang
Shanghai Fisheries University,
334 Jungong Rd.,
Yangpu D.C., Shanghai 200090, China
e-mail: syzhang@shfu.edu.cn

W. D. Wang
Marine Fisheries Research Institute of Zhejiang Province,
25 Xiaoxihu Lane,
Putuo D.C., Zhoushan City, Zhejiang Province 316100, China

Introduction

Kelp beds are a common habitat in coastal inshore ecosystems. Macroalgae provide primary productivity to biological communities and support the food web in the kelp beds. Macroalgae offer ideal habitats for many plants and animals in the coastal ecosystem. These beds provide spawning and feeding grounds, and refuges for many fish species as well as micro and macro aquatic organisms (Komatsu and Tatsukawa 1999). After macroalgae die, their thalli are decomposed by bacteria and microorganisms in the water. This process increases the level of nutrients in the system, leading to increased productivity of the phyto- and zoo-plankton that provide the key food for juvenile fish inhabiting the kelp bed.

Gouqi Island lies in the east of the East China Sea, composing, together with Shengshan, Huaniao, Lvhua, Sijiao and Huanglong Islands, the Shengsi Archipelago (Fig. 1). It has a subtropical climate and semi-diurnal tides; the average surface water temperatures range from 17 to 19°C. The western area of the Island is influenced by the Changjiang and Qiangtang Rivers, and the eastern open waters are controlled by the Taiwan Warm Current and Zhebei Upwelling. Salinity therefore has large horizontal gradients from the east to the west coast, ranging between

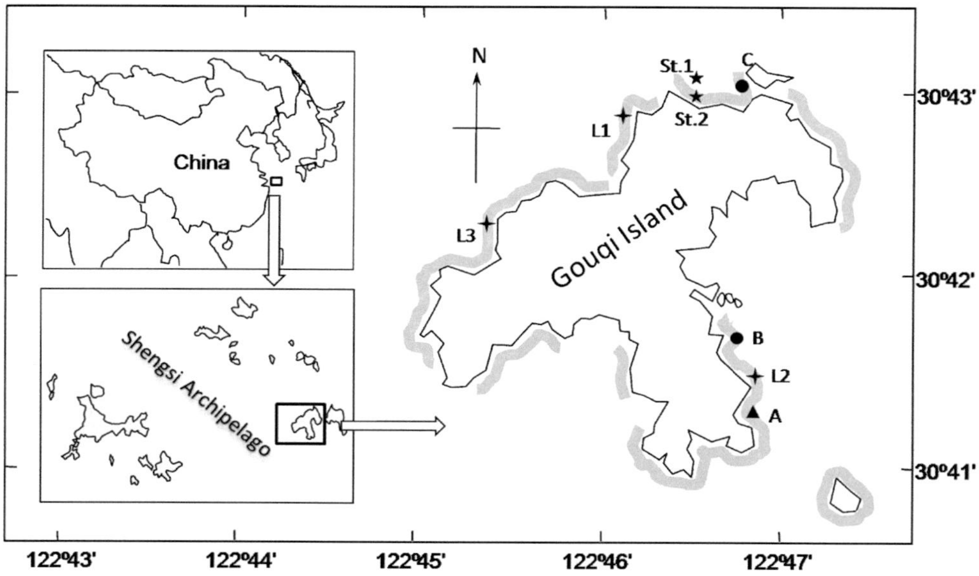

Fig. 1 Map of Gouqi Island, with distribution of kelp beds and sampling sites

12.02 and 29.10‰. The lowest salinity occurs in July and October on the West coast. Because of mixing of different water systems and the existence of multitudinous islands, primary productivity is high and fisheries resources are rich in the waters of the Shensi Archipelago. In addition to traditional fishing, such as longline, gillnet, stow net and small dragnet, there is also cage aquaculture in the Lvhua area, mussel raft cultivation farms in the Gouqi area, and artificial reefs in the centre of the area. Macroalgae are widely distributed in the intertidal and subtidal zones of the Gouqi, Huania and Lvhua Islands, which are on the east and north of the Shensi Archipelago. A large number of juvenile fish, shrimps and crabs inhabit the kelp beds (Zhang et al. 2007).

To understand the community structure and function of the island kelp bed ecosystem in the Shengsi Archipelago, we conducted a series of surveys from 2004 to 2006 to evaluate the Gouqi kelp beds on the east of the Shensi Archipelago. We evaluated seasonal variations in the community structure and biomass of macroalgae and abundance of phytoplankton to improve our understanding of the ecosystem dynamics of kelp beds.

Materials and methods

Gouqi Island has an area of about 6×10^6 m^2, and lies about 85,000 m from the Chinese continent. It is influenced greatly by the open ocean. Salinity is always higher than 25‰ (Ocean Reconnaissance and Design Academy in Zhoushan 2004). Tidal amplitude is 4–5 m. Gouqi Island has a coastline of more than 22.5×10^3 m (Statistic Bureau in Zhoushan City 2001). With the exception of a 5,000 m sand beach on the east side of the island, the coast is rocky. Macroalgae can be found in the intertidal and subtidal zones. Mass mussel raft culture activities, at a distance of only 100 m away from the land, are present on the southwest, northwest and northeast of the island.

We chose independent sampling stations in the intertidal and subtidal zones, sampled the seaweed quantitatively, and collected water samples within and outside of the kelp beds to measure nutrients in the water.

Intertidal zone sampling

Macroalgae were sampled from February to June 2004 in three sample areas (A1, A2 and A3) near station A on the southeast of the Gouqi island intertidal zone. A1 was located in the middle of the intertidal zone; A2 and A3 were located at the underside of the intertidal zone, each having an area of 1×1 m. Macroalgae were identified by species, and biomass and size of individual samples were measured every month during the spring survey. At the same time, photos showing the coverage of macroalgae on the rocky floor in the three sampling areas were taken. As a supplementary investigation in the early stages of the study, we sampled macroalgae entirely in two sampling areas of 0.5×0.5 m each at stations B and C on the northeast of Gouqi Island in August and June 2006.

Subtidal zone sampling

In order to determine the dominant species and the biomass distribution of the Gouqi kelp bed, we chose three sites lying at the northeast (L1), southeast (L2) and northwest (L3) of subtidal zones during the blooming period of the kelp bed (20–22 May 2004; Fig. 1). The following procedure was used in the survey: (1) the breadth of the kelp bed was measured by tying one end of a rope to the shore. A diver then swam straight out from the land with the other end of the rope until no macroalgae were observed, and then the distance was measured; (2) three 0.5 × 0.5 m quadrats were selected at the three sites at equal distance on each transect for sampling macroalgae, photos were taken, and the depths were measured; and (3) macroalgae samples were collected from rocks with shovels and knives and taken to shore for measuring and identification. In order to identify correlations among seasonal variation of macroalgae species richness and biomass, nutrients and plankton, four sampling trips were made: August and November 2005, and March and June 2006. The sites for the survey were on the St.1 and St.2 transects, for which the macroalgae and environment were similar to those for the L1 sites. Macroalgae were sampled by divers in a 0.5 × 0.5 m quadrat. Phytoplankton samples were collected at the same time with a type-III plankton net [JF62 (0.077), JP80(0.077)], and zooplankton samples were also collected with the type-I_ plankton net [CQ14(0.505), JP9.4 (0.507)].The samples were preserved with formaldehyde solution (5%), and taken to the laboratory for further analyses.

Laboratory work

We estimated the proportion of areas covered by macroalgae using the photos taken. We divided the 1 × 1 m sample area into 25 sub-areas of 0.2 × 0.2 m, estimated the cover for each sub-area, and calculated the average to represent the cover for the sampling area.

Phytoplankton species were identified using a Nikon Eclipse 80i microscope. Phytoplankton abundance was determined by concentrating the samples to 0.1 L, from which three 1×10^{-4} L sub samples were taken randomly for counting of phytoplankton cells under the microscope, and the average was calculated. Finally, based on the water depth in the survey sites and the diameter of the plankton net, sampling water volume was calculated, and the average value of abundance of phytoplankton in the sampling sites was estimated. We also identified macroalgae species, and measured their wet weight and height. Nutrient concentrations (nitrate, nitrite, ammoniac nitrogen, phosphate and silicate) were measured for the sampling sites (The Ocean Monitor Norm of China, CSBTS 1998).

Results

Three phyla and 20 species of macroalgae were identified in the intertidal and subtidal zones around the Gouqi Island from 2004 to 2006.

Intertidal zone

Three phyla, 16 genera and 19 species of macroalgae were identified in the intertidal zone. Brown algae accounted for 2 genera, red algae for 11 genera, and green algae for 3 genera. The dominant species of the intertidal zone in winter and spring appeared to be *Sargassum thunbergii* and *Undaria pinnatifida*. *Sargassum thunbergii*, *S. horneri* and *S. fusiforme* biomass accounted for 55.2% of the total samples. In August 2005 and June 2006, the dominant species in the other two sample areas of the intertidal zone were found to be *S. thunbergii* and *Chondria crassicaulis*, respectively. The species identified and biomass calculated are summarised in Table 1.

Subtidal zone

Three phyla, 5 genera and 7 species of macroalgae were identified in the subtidal zone, with brown algae being the primary group (Table 1). *Sargassum horneri* was the dominant species in spring 2004 with an average height of 0.8 m, accounting for 85.8% of the biomass. *Sargassum thunbergii* and *U. pinnatifida* accounted for 1.7% and

Table 1 The species and proportion of total biomass of macroalgae in the intertidal zone in summer and autumn, and species in the subtidal zones of sites L1, L2, and L3

Species	August 2005	June 2006	Area
Phaeophyta			
Sargassum horneri [a]	–	–	Intertidal and subtidal
Sargassum fusiforme	–	< 0.01	Intertidal
Sargassum thunbergii	0.21	0.52	Intertidal and subtidal
Undaria pinnatifida[a]	–	–	Intertidal and subtidal
Sargassum vachellianum	–	–	Subtidal
Rhodophyta			
Chondria crassicaulis	0.54	0.21	Intertidal
Pachymenia carnosa	<0.01	–	Intertidal
Corallina officinalis	0.09	0.02	Intertidal and subtidal
Grateloupia filicina	0.09	–	Intertidal
Ceramium kondoi	–	< 0.01	Intertidal
Gelidium divaricatum[a]	–	–	Intertidal
Gelidium amansii [a]	–	–	Intertidal
Chondrus ocellatus	–	0 .04	Intertidal
Laurencia composita	–	< 0.01	Intertidal
Symphyocladia pennata	–	< 0.01	Intertidal
Champia parvula	–	< 0.01	Intertidal
Gigartina intermedia	–	< 0.01	Intertidal and subtidal
Chlorophyta			
Codium fragile	0.05	–	Intertidal
Bryopsis plumosa [a]	–	–	Intertidal
Ulva lactuca	0.03	0.18	Intertidal and subtidal

[a] Macroalgae found in the intertidal zone in summer and winter, but not included in the total biomass

12.6% of biomass, respectively. A small quantity of *S. vachellianum*, *Corallina officinalis*, *Gigartina intermedia* and *Ulva lactuca* were found in the subtidal zone from 2005 to 2006.

Changes in the biomass of *S. horneri* are shown in Table 2. The highest average height of the species, recorded in June, was 0.75 m.

Phytoplankton community structure. The primary phytoplankton species within and outside of the kelp beds were diatoms and dinoflagellates, with the diatoms accounting for over 90% of the total abundance. The phytoplankton samples collected in the experiment area belonged to 4 phyla, 31 genera and 55 species. There were 23 genera and 41 species of diatoms, 5 genera and 10 species of dinoflagellates (Table 3). There were 51 species of phytoplankton inside the kelp beds, 44 species outside the kelp beds, and 42 species both inside and outside the kelp beds. *Coscinodiscus spinosus*, *Chaetoceros denticulatus*, and *Thalassiosira nordenskioldi* were found only inside the kelp beds. *Hemiaulus hauckii*, *Bacterastrum hyalinum* and *Eucampia zoodiacus* were found only outside the kelp beds.

Between May 2005 and March 2006, the highest abundance of phytoplankton was recorded in August. The abundance of phytoplankton within the kelp beds was 15.63×10^6 cells m^{-3}, and was twice that outside the kelp beds. The density of the phytoplankton in the kelp beds was $0.43–15.63 \times 10^6$ cells m^{-3}, with a mean of 6.80×10^6 cells m^{-3}. The density of the phytoplankton outside the kelp beds was $0.35–8.19 \times 10^6$ cells m^{-3}, with a mean of 5.29×10^6 cells m^{-3}. The abundance of phytoplankton in the kelp beds was higher than that outside the kelp beds except in May.

The species diversity of phytoplankton was the highest in August. There were 37 and 35 species inside and outside the kelp beds, respectively. The largest difference seen between inside and outside the kelp beds was 19 species, and occurred in November, when there were 36 species inside the kelp beds and 17 species outside the kelp beds. Except in May, the species diversity of phytoplankton in the kelp beds was higher than that outside the kelp beds.

Seasonal changes in nitrogen

Figure 2 shows the annual N concentration changes in the Gouqi Island kelp bed from May 2005 to March 2006. The largest difference in total N (nitrate, nitrite, ammoniac nitrogen) between inside and outside the kelp bed was seen in May. Nitrate concentration in the kelp bed was much lower, with only 1.3% of total nitrogen. Outside the kelp bed, ammoniac nitrogen was highest, then nitrate, and then nitrite. In August, total nitrogen was lowest, both inside and outside the kelp bed. In November and March, the difference in total nitrogen between inside and outside the kelp bed was the least. The component proportion of nitrogen during the period was as follows: NO_3^- was the primary component with increasing trend, followed by ammoniac N with reducing trend, and lastly NO_2^- with increasing trend.

Discussion

Factors affecting macroalgae

The dominant species in the Gouqi Island kelp bed are *Sargassum* spp. *Sargassum horneri* comprised over 90% of the total biomass and is the primary food of sea urchins, holothurians and abalone, which often affect the growth of macroalgae. The degeneration phenomenon of the kelp bed as 'Isoyake' has in the past been considered to result from the deterioration of the aquatic environment. However, recent Japanese studies implicate grazing damage from phytophagous animals (Watanuki 2006). We found that the number of sea urchins was limited in the kelp bed, especially on rocks inhabited by *S. horneri*. Sea urchins such as *Hemicentrotus pulcherrimus* sporadically appeared in the area where there was low macroalgal growth,

Table 2 The length and density of *Sargassum horneri* at stations St.1 and St.2 from 2005 to 2006

	Station	16 August 2005	26 November 2005	27 March 2006	11 June 2006
Length range (m)	St.1	0.01–0.09	0.09–0.60	0.08–0.99	0.21–1.74
	St.2	0.01–0.09	0.10–0.56	0.10–0.73	0.25–0.92
Average length (m)	St.1	0.056	0.270	0.305	0.879
	St.2	0.044	0.295	0.372	0.614
Number of plants	St.1	631	90	52	25
	St.2	830	130	64	24
Plant density (individuals m^{-2})	St.1	2524	360	208	100
	St.2	3320	520	256	96
Biomass (kg m^{-2})	St.1	0.692	1.505	1.456	4.420
	St.2	0.540	3.185	2.384	1.300

Table 3 The species, abundance [(B) $\times 10^5$ cells m^{-3}] and relative abundance (B%) of phytoplankton within (*In*) and outside (*Out*) the kelp beds in each season

Species	26 May 2005				16 August 2005				26 November 2005				30 March 2006			
	In		Out		In		Out		In		Out		In		Out	
	B	B%	B	B%	B	B%	B	B%	B	B%	B	B%	B	B%	B	B%
Bacillariophyta																
Skeletonema costatum	7.55	0.53	30.84	0.47	69.01	0.44	41.07	0.50	95.46	0.98	59.64	0.99	0.96	0.23	1.45	0.42
Ditylum brightwelli	0.04	<0.01	0.19	<0.01	1.01	0.01	0.54	0.01	–	–	–	–	0.10	0.02	0.04	0.01
Ditylum sol	–	–	–	–	0.20	<0.01	0.08	<0.01	–	–	–	–	0.03	0.01	–	–
Cyclotella comta	–	–	–	–	0.02	<0.01	0.02	<0.01	<0.01	<0.01	–	–	0.03	0.01	–	–
Coscinodiscus radiatus	–	–	–	–	0.26	<0.01	0.41	0.01	–	–	–	–	0.11	0.03	0.03	0.01
Coscinodiscus concinnus	–	–	–	–	–	–	–	–	–	–	–	–	0.01	<0.01	0.01	<0.01
Coscinodiscus jonesianus	0.04	<0.01	0.39	0.01	0.37	<0.01	0.16	<0.01	0.07	<0.01	0.01	0.00	0.41	0.10	0.04	0.01
Coscinodiscus nitidus	–	–	–	–	–	–	–	–	–	–	–	–	0.15	0.04	0.02	0.01
Coscinodiscus spinosus	–	–	–	–	–	–	–	–	–	–	–	–	0.17	0.04	–	–
Coscinodiscus argus	0.01	<0.01	0.19	<0.01	–	–	–	–	–	–	–	–	–	–	–	–
Coscinodiscus oculusiridis	–	–	–	–	0.40	<0.01	0.05	<0.01	0.02	<0.01	<0.01	<0.01	–	–	–	–
Chaetoceros pseudocurvisetus	0.12	0.01	3.69	0.06	17.34	0.11	8.90	0.11	0.06	<0.01	0.05	<0.01	0.77	0.19	0.82	0.24
Chaetoceros affinis	–	–	–	–	22.50	0.14	10.95	0.13	0.14	<0.01	–	–	0.05	0.01	0.32	0.09
Chaetoceros lorenzianus	–	–	–	–	4.63	0.03	2.11	0.03	0.19	<0.01	0.02	<0.01	0.18	0.04	0.21	0.06
Chaetoceros denticulatus	–	–	–	–	–	–	–	–	0.08	<0.01	–	–	–	–	–	–
Chaetoceros atlanticus	0.22	0.02	0.19	<0.01	7.21	0.05	2.98	0.04	0.07	<0.01	0.04	<0.01	0.06	0.02	0.35	0.10
Thalassiothrix frauenfeldii	0.07	0.01	–	–	1.60	0.01	0.63	0.01	0.10	<0.01	0.01	<0.01	–	–	–	–
Nitzschia pungens	0.92	0.06	6.60	0.10	12.33	0.08	5.19	0.06	0.07	<0.01	0.03	<0.01	0.10	0.02	0.05	0.02
Nitzschia debilis	2.24	0.16	12.03	0.18	7.12	0.05	2.88	0.04	0.04	<0.01	–	–	0.03	0.01	0.01	<0.01
Triceratium favus	–	–	–	–	0.01	<0.01	0.11	<0.01	–	–	–	–	–	–	–	–
Eucampia zoodiacus	–	–	1.55	0.02	–	–	–	–	–	–	–	–	–	–	–	–
Biddulphia sinensis	–	–	–	–	0.12	<0.01	0.02	<0.01	0.02	<0.01	0.03	<0.01	0.13	0.03	0.01	<0.01
Biddulphia pulchella	–	–	–	–	0.01	<0.01	–	–	0.01	<0.01	<0.01	<0.01	0.11	0.03	0.05	0.01
Aulacoseira granulata	2.48	0.17	–	–	4.47	0.03	1.87	0.02	0.23	<0.01	0.14	<0.01	0.18	0.04	–	–
Melosira sulcata	0.32	0.02	8.15	0.12	–	–	–	–	–	–	–	–	0.09	0.02	–	–
Fragilaria islandica	–	–	–	–	–	–	–	–	–	–	–	–	0.20	0.05	–	–
Navicula tenera	–	–	–	–	–	–	–	–	–	–	–	–	–	–	0.01	<0.01
Thalassiosira nordenskioldi	0.04	<0.01	–	–	–	–	–	–	–	–	–	–	–	–	–	–
Thalassionema nitzschioides	–	–	0.58	0.01	1.79	0.01	0.74	0.01	0.02	<0.01	0.02	<0.01	–	–	–	–
Bacterastrum hyalinum	–	–	0.19	<0.01	–	–	–	–	–	–	–	–	–	–	–	–
Hemiaulus hauckii	–	–	0.39	0.01	–	–	–	–	–	–	–	–	–	–	–	–
Hemiaulus sinensis	–	–	–	–	0.02	<0.01	0.01	<0.01	0.01	<0.01	<0.01	<0.01	–	–	–	–
Rhizosolenia calcaravis	–	–	–	–	1.93	0.01	0.18	<0.01	0.01	<0.01	–	–	–	–	–	–
Rhizosolenia stolterfothii	–	–	–	–	0.70	<0.01	0.42	0.01	–	–	–	–	–	–	–	–
Rhizosolenia styliformis	–	–	0.19	<0.01	0.37	<0.01	0.13	<0.01	0.01	<0.01	–	–	–	–	–	–
Rhizosolenia robusta	–	–	–	–	0.10	<0.01	0.03	<0.01	0.01	<0.01	–	–	–	–	–	–
Leptocylindrus danicus	–	–	–	–	2.20	0.01	1.20	0.01	–	–	–	–	–	–	–	–
Corethron hystrix	–	–	–	–	0.42	<0.01	0.32	<0.01	0.01	<0.01	–	–	–	–	–	–
Bellerochea malleus	–	–	–	–	0.03	<0.01	0.02	<0.01	0.04	<0.01	–	–	–	–	–	–
Pleurosigma affine	–	–	–	–	0.02	<0.01	0.02	<0.01	0.01	<0.01	–	–	–	–	–	–
Arachnoidiscus ehrenbergii	–	–	–	–	–	–	–	–	0.01	<0.01	–	–	–	–	–	–
Hyalodiscus stelliger	–	–	–	–	–	–	–	–	0.02	<0.01	–	–	–	–	–	–
Pyrrophyta																
Noctiluca scientillans	0.07	<0.01	0.19	<0.01	0.04	<0.01	0.05	<0.01	0.01	<0.01	0.01	<0.01	0.01	<0.01	0.02	0.01
Ceratiun macroceros	–	–	–	–	–	–	–	–	0.01	<0.01	–	–	–	–	–	–
Ceratium breve	0.01	<0.01	0.19	<0.01	0.02	<0.01	0.13	<0.01	0.01	<0.01	–	–	–	–	–	–
Ceratium furca	–	–	–	–	0.01	<0.01	–	–	0.01	<0.01	–	–	–	–	–	–
Ceratium fusus	0.01	<0.01	0.39	0.01	0.17	<0.01	0.10	<0.01	0.18	<0.01	0.04	<0.01	0.03	0.01	0.01	<0.01
Ceratium tripos	0.03	<0.01	0.19	<0.01	0.04	<0.01	0.08	<0.01	0.01	<0.01	<0.01	<0.01	0.04	0.01	0.01	<0.01

Table 3 (continued)

Species	26 May 2005				16 August 2005				26 November 2005				30 March 2006			
	In		Out		In		Out		In		Out		In		Out	
	B	B%	B	B%	B	B%	B	B%	B	B%	B	B%	B	B%	B	B%
Ceratium furca Dujardin	–	–	–	–	0.02	<0.01	0.12	<0.01	0.01	<0.01	–	–	–	–	–	–
Peridinium depressum	–	–	–	–	0.15	<0.01	0.26	<0.01	0.02	<0.01	0.01	<0.01	–	–	–	–
Pyrocystis fusiformis	–	–	–	–	0.01	<0.01	0.02	<0.01	0.01	<0.01	–	–	–	–	–	–
Pyrocystis nuctiluca	–	–	–	–	0.02	<0.01	0.09	<0.01	0.01	<0.01	–	–	–	–	–	–
Xanthophyta																
Halosphaera viridis	–	–	–	–	–	–	–	–	0.01	<0.01	–	–	0.03	0.01	–	–
Chlorophyta																
Pandorina morum	–	–	–	–	–	–	–	–	–	–	–	–	0.20	0.05	–	–
Pediastrum sp.	–	–	–	–	–	–	–	–	–	–	–	–	0.01	<0.01	–	–

suggesting that the existence of sea urchins affected the growth of macroalgae to a certain extent. In areas where sea urchins appeared more frequently, their density usually was 2–4 individuals (ind.) m^{-2}. Sea urchins tend to move in aggregations, and the direction of urchin movement is found to be either random or weakly directional (Duggan and Miller 2001). Sea urchin movement rate decreases as the macrophyte density increases (Konar and Estes 2003). This may result in a low frequency of urchins being spotted in large areas unless the location of a large aggregation happened to be found. The average length of *S. horneri* in the kelp bed was 0.6–0.8 m in May and June, with the highest recorded length being 3.02 m. It has been shown that urchins present at a density of the order of 100 urchins m^{-2} move at a rate of up to 6 m day^{-1}, reducing above-ground seagrass to less than 2% of its initial biomass (Maci'a and Lirman 1999). The density of urchins in the Gouqi Island kelp bed was much lower than this.

Changes in water temperature were the primary factor affecting the growth of *S. horneri* in the Gouqi Island kelp bed. This study suggests that the growth period of *S. horneri* is 1 year. Algal spores usually accrete on the seabed in May–June, growing until they mature, fall off and die in May–July the following year (Table 4).

The average water temperature in August 2005 was 24.99°C, the average length of *S. horneri* was 0.05 m, and its density reached up to 2,922 ind. m^{-2}, indicating the beginning of a new life cycle of the species. Their growth rate was restricted by the unfavourable water temperature. When the water temperature fell to 17.55°C in November, the length was 0.28 m, and average daily growth rate was 2.25×10^{-3} m. At the same time, due to growth competition, the density reduced rapidly to 440 ind. m^{-2}, a reduction of 84.9%. Along with the unfavourable water temperature in autumn–winter, the macroalga grew much more slowly. Until March 2006, the water temperature was 11.96°C; the average length of *S. horneri* was 0.34 m and the average daily growth rate was only 0.49×10^{-3} m, a reduction by 47.3%, and the density reduced to 232 ind. m^{-2}. When the water temperature increased, *S. horneri* entered a fast-growing period and began to mature. In June 2006, when the water temperature was 20.76°C, the average length was 0.75 m and the average daily growth rate was 5.32×10^{-3} m, but the density was reduced by 57.8%, to 98 ind. m^{-2}. During this period, many floating seaweeds appeared on the surface of the sea, showing that *S. horneri* parts from the seabed, and the kelp bed began to wane in June 2006.

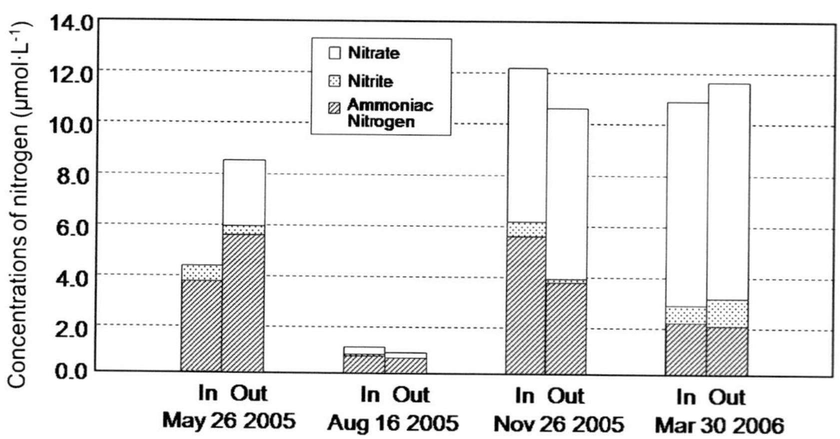

Fig. 2 Concentrations of nitrate, nitrite and ammoniac nitrogen in four separate months at Gouqi Island

Table 4 Five day average water temperature around the investigation day, and the length and density of *S. horneri* on the investigation day in the kelp bed

	21 May 2004	16 August 2005	26 November 2005	27 March 2006	11 June 2006	23 September 2006	17 May 2007
Five day average water temperature(°C)	18.08	24.99	17.55	11.96	20.76	24.81	19.97
Average length (cm)	80	5	28	34	75	8	87
Density (individuals m^{-2})	23	2,922	440	232	98	190	124

Sargassum horneri experienced the lowest 5 day average water temperature around 10 February during the slow growing period from November to March. The investigation was delayed because of foul weather. Thus we cannot calculate the growth rate of seaweed in the declining and rising period of water temperature from November to March, separately. Water temperature decreased and increased by 9.14°C (17.55–8.41°C) and 3.55°C (8.41–11.96°C), respectively. Comparing the decrease of 7.44°C (24.99–17.55°C) from August to November with an increase of 8.8°C (11.96–20.76°C), the magnitude of the decrease was higher. Considering sunlight and lower water clarity in winter, the growth rate would not be higher than in summer–autumn and spring–summer. Thus, we deem that the period from November to March was the slow growing period for *S. horneri*, which is consistent with what has been observed on the coast of Europe (Shen and Shi 2002).

Similar to other kelp beds, the growth period of Gouqi Island kelp beds occurred in May, but this might vary from year to year with annual fluctuations in water temperature. During the kelp bed survey on 21 May 2004, *S. horneri* was in bloom. A few floating seaweeds were found on the sea surface and the water temperature of this month ranged from 15.79 to20.12°C. On 17 May 2007, most *S. horneri* had decomposed and a mass of floating seaweeds was visible; the water temperature was 17.22–23.77°C. The 5-day average water temperatures during the period of surveys and average water temperature from 1 May to 19 May 2004 were 1.89°C (ranging from 19.97 to18.08°C) and 1.67°C (ranging 18.7 to17.03°C) lower than the average water temperature in 2007, showing that the kelp bed in 2007 waned earlier than in 2004.

Based on water temperature and biomass data from March to June during 2004–2006, the water temperature ranged from 16 to 22°C for the corresponding period for which the average length of *S. horneri* was 0.7 m in the Gouqi Island kelp bed. The 5-day average water temperature in May 2004 was 18.08°C. Some floating seaweeds could be seen. Considering the seaweed's appearance and the water temperature of 19.97°C at the same time in 2007, we conclude that the upper limit of temperature for *S. horneri* growth is about 18°C. If the water temperature is higher than that, the kelp bed would begin to decline. This situation was very different from the water temperature 20°C (Liu et al. 2001) observed in the kelp bed inhabiting the other *Sargassum* community. One reason for this difference may be increased wave action due to the typhoon after May, which picked up deciduous seaweeds from the seabed.

Factors affecting phytoplankton

The zooplankton species collected in the area in March, May, August and November 2005 represented 26 species in total, including *Calanus sinicus*, *Sagitta enflata* and *Diphyes chamissonis*. There were 22 species in the kelp bed, and 19 species outside the kelp bed. The highest biomass appeared when their biomass in the kelp bed was 403.36×10^{-3} g m^{-3}, and 131.76×10^{-3} g m^{-3} outside the kelp bed. The lowest was in March, 127.40×10^{-3} g m^{-3} and 18.78×10^{-3} g m^{-3} inside and outside the kelp beds, respectively. Several investigations indicated that both individual density and total biomass of zooplankton in the Gouqi Island kelp bed were greater inside than that outside of the kelp bed. The seasonal changes in zooplankton biomass and phytoplankton abundance were similar to the trend outside the kelp bed. At the same time, phytoplankton abundance in the kelp bed was 1.421×10^{6} cells m^{-3}, which was the only time when it was lower than outside the bed. Obviously, the impacts of zooplankton on phytoplankton in the kelp bed cannot be ignored.

According to an investigation of plankton in the East China Sea (121°–123°30′E, 27°–32°N) in spring 2005 (Sun et al. 2007), *C. sinicus* was the most important mid-size zooplankton in the area, and the rate of grazing phytoplankton can be described by the following equation: $\mathrm{Gp} = 807.429\mathrm{P}_0/(4.093 + \mathrm{P}_0 + 0.007\mathrm{P}_0^2)$, where P_0 is the standing stock of phytoplankton. Because we used a type-I_ plankton net [CQ14(0.505), JP9.4(0.507)] to collect zooplankton (mesh $=0.505\times10^{-3}$ m), the zooplankton collected are all mid-sized. Calculated from the equation above, the daily grazing phytoplankton speed of zooplankton (173.16 ind. m^{-3}) was only 69.16 ind. L^{-1} inside the kelp bed in May 2005, which was less than 0.003% of phytoplankton abundance inside the kelp bed. Thus, it was impossible for mid-sized zooplankton to cause the decline of phytoplankton abundance inside the kelp bed in May. Considering the work of Sun et al. (2007) and Calbet and Landry (1999, 2004), we speculate that micro-zooplankton may lead to the decline of phytoplankton abundance inside the kelp bed in May. Except for the lower zooplankton biomass and grazing pressure, the large increase in phytoplankton inside the kelp bed in August is also connected to the concentration and types of nutrients available, as discussed in the correlations between phytoplankton and *S. horneri*.

Mussel cultivation could be another primary factor influencing phytoplankton abundance in the Gouqi Island kelp bed and surrounding areas. The area around Gouqi Island is one of the main mussel cultivation areas in China. During the mid-1990s, mussel cultivation began to increase, and it has since become one of the key industries in the area. Phytoplankton account for 15% of food for the mussel, resulting in an increase in water transparency in the area. For example, transparency was 4 m on 11 June 2006, with 1.94 m inside the kelp bed, and 2.42 m outside the kelp bed. According to an experiment by Chang and Wu (2007), with a water depth of 20–30 m and a cultivation density of 3.0–4.5 kg m^{-2}, *Mytilus coruscus* larvae can graze 7.75×10^7 cells L^{-1} phytoplankton per day. The weight of *M. coruscus* is 1.5 times that of *Mytilus edulis*, which is being cultured in the area at the same size. The quantity of food that mussels graze is determined mainly by their size, i.e. 5% of their weight (Li et al. 1999). We calculated that 2.75×10^7 kg *M. edulis* could graze 7.57×10^7 cells L^{-1} phytoplankton in a 5×10^6 m^2 area of mussel cultivation in 2005. The investigation in May showed that the average density of phytoplankton was 4.18×10^3 cells L^{-1}, which was lower than the former estimation by four orders of magnitude. Thus, we conclude that mussel cultivation can maintain significant grazing pressure on phytoplankton in this area. The shortest distance between outside the kelp bed and the nearest mussel cultivation farm is about 100 m. The phytoplankton in this area was more severely suppressed compared with that in the 200 m between the *inside* of the kelp bed and the mussel cultivation farm. Therefore the average phytoplankton abundance (5.29×10^6 cells m^{-3}) outside the kelp bed was lower than the inside (6.80×10^6 cells m^{-3}) for the whole year. In addition, according to Lu and Zhang (1996), the average phytoplankton abundance at Gouqi Island in autumn 1990 was 7.44×10^7 cells m^{-3}, which is much greater than 7.86×10^6 cells m^{-3} in autumn 2005. The reason may be that the scale of mussel cultivation was smaller in the 1990s, at 1.5×10^6 m^2, i.e. less than one-third of the present scale.

Correlation between phytoplankton, *S. horneri* and nutrition

Both phytoplankton and macroalgae take up N, P and Si. When the concentration of nutrient salts is sufficient for the growth of phytoplankton, the phytoplankton in the sea take up N, P and Si according to the Redfield ratio N:P:Si= 16:1:1 (Redfield 1958). Since the brown seaweed *S. horneri* were the supporting species, due to the limited absorption of silicate (Pu and Wu 2000), we considered only the effects of N and P. The ratios of N:P in the kelp bed in May, August, November and March from 2005 to 2006 were 7.17, 0.95, 7.29 and 7.67, respectively, which were all smaller than the Redfield ratio of 16. The corresponding of total nitrogen concentrations were 4.39, 1.07, 12.51 and 11.13 μmol L^{-1}, and those of phosphate were 0.61, 1.13, 1.72 and1.45 μmol L^{-1}, respectively. This is greater than the lowest thresholds of 1 μmol L^{-1} and 0.1 μmol L^{-1} required for phytoplankton to grow. The range of phosphate was 0.613–1.710 μmol L^{-1}, with an average of 1.226 μmol L^{-1}. The range of total nitrogen was 12.500–1.071 μmol L^{-1}, with an average of 4.500 μmol L^{-1}. Therefore, we consider nitrogen as the limiting nutrient for macroalgae such as *S. horneri*.

Some studies have indicated that, under co-occurrence of ammoniac nitrogen and nitrate, the former is first absorbed by phytoplankton, limiting the absorption of nitrate (Wheeler and Kokkinakis 1990). When ammoniac nitrogen is present at >1 μmol L^{-1}, its uptake is reduced or absent (Liu and Dong 2001). Concentrations of ammoniac nitrogen in the kelp bed in May, August, November and March from 2005 to 2006 were 3.7, 0.7, 5.6 and 2.1 μmol L^{-1}, respectively. The decline in nitrate concentration should not be considered to be due primarily to absorption by phytoplankton. As described previously (Fig. 2a), during the blooming period of *S. horneri*, nitrate was reduced greatly to 0.057 μmol L^{-1}, accounting for 1.3% of total nitrogen in the kelp bed. At the same time, nitrate accounted for 31.1% of total nitrogen outside the kelp bed—a difference of almost 24 fold. During the period of slow growth from November to March, the average concentration of nitrate was 62.4% of the total nitrogen in the kelp bed, close to 68.9% measured outside the kelp bed at the same time. Therefore, in the Gouqi Island kelp bed, because of the growth of *S. horneri*, the absorption of nitrate could be the primary driving force for seasonal changes in the plant community, with the growth of *S. horneri* causing fluctuations in the nitrate concentration, which thus affect the total N concentration and proportions of nitrate, nitrite and ammoniac nitrogen, as well as the phytoplankton community in the kelp bed.

Diatoms, which have a requirement for Si, accounted for over 98% of the total phytoplankton abundance. However, the concentrations of silicate in the kelp bed in May, August, November and March from 2005 to 2006 were 5.97, 2.18, 3.39 and 4.25 μmol L^{-1}, with the ratios of N:P:Si being 0.74:0.1:1, 0.49:0.52:1, 3.69:0.51:1 and 2.62:0.34:1, respectively. With reference to the Redfield ratio and the lowest threshold 2 μmol L^{-1} of Si for phytoplankton growth (Justic et al. 1995a, b), we infer that

Si was not the limiting factor for the growth of phytoplankton, but rather the necessary condition required to sustain high diatom abundance. Thus, both phytoplankton and macroalgae are nitrogen restricted in the Gouqi Island kelp bed.

The N-limitation of the phytoplankton affected the seasonal changes in phytoplankton abundance as a result of seasonal changes in the various species of N. The concentration of ammoniac nitrogen in the kelp bed in August 2005 was 0.7 (<1) μmol L^{-1}. Under these conditions, both ammoniac nitrogen and nitrate are nutrient sources for the growth of phytoplankton. In August, warm water temperatures caused the mass reproduction of phytoplankton, reaching 15.63×10^6 cells m^{-3}, the highest value in the whole year. Meanwhile, phytoplankton growth reduced the total N to 0.015 mg L^{-1}, the lowest value in the whole year. Similar results were observed outside the kelp bed. Phytoplankton abundance reached a highest concentration of 8.19×10^6 cells m^{-3}, and total nitrogen was lowest at 0.012 mg L^{-1}. That the lower abundance was inside the kelp bed compared to outside reflects the impacts of the kelp bed environment. Because the macroalgae size was least in August, the difference was small. This may be due to the shorter distance to the mussel cultivation farm and higher grazing pressure. In May and November 2005 and March 2006, when ammoniac nitrogen was >1 μmol L^{-1}, phytoplankton absorbed ammoniac nitrogen. Abundance is also affected by other factors, such as water temperature, sunlight, transparency and grazing.

In the future we plan to sample more frequently to evaluate the life history process of *S. horneri* and the relationships between *S. horneri* and environmental factors such as water temperature. We also plan to do experiments to improve our understanding of the N-uptake process of *S. horneri*, which will facilitate the development of a growth model of *S. horneri*. This will greatly improve our understanding of the ecosystem dynamics of kelp beds.

Acknowledgements We would like to thank Dr. T. Komatsu and Dr. K. Tatsukawa of Ocean Graduate School, Tokyo University for their help in developing the experiments, and from Dr. Y Chen of the University of Maine for comments and editing of an early version of the manuscript. Dr. J.P. Jiao participated in the species identification and account of phytoplankton. The project is supported by Shanghai Leading Academic Discipline Project (No.T1101) and National 863 Program (No.2006AA100303) by Ministry of Science & Technology of P. R. China and National Natural Science Foundation of China (No.30471332).

References

Calbet A, Landry MR (1999) Mesozooplankton influences on the microbial food web: direct and indirect trophic interactions in the oligotrophic open ocean. Limnol Oceanogr 44:1370–1380

Calbet A, Landry MR (2004) Phytoplankton growth, microzooplankton grazing, and carbon cycling in marine systems. Limnol Oceanogr 49:51–57

Chang KM, Wu JF (2007) Study on artificial propagation of mussel *Mytilus coruscus*. South China Fish Sci 3(3):26–30

Duggan RE, Miller RJ (2001) External and internal tags for the green sea urchin. J Exp Mar Biol Ecol 258:115–122

Justic D, Rabalais NN, Turner RE, Dortch Q (1995a) Changes in nutrient structure of river-dominated coastal waters: stoichiometric nutrient balance and its consequences. Limnol Oceanogr 40:339–356

Justic D, Rabalais NN, Turner RE (1995b) Stoichiometric nutrient balance and origin of coastal eutrophication. Mar Pollut Bull 30:41–46

Komatsu T, Tatsukawa K (1999) The kelp beds and aquatic product resource. Ocean, No.17, Japan

Konar B, Estes JA (2003) The stability of boundary regions between kelp beds and deforested areas. Ecology 84:174–185

Li H, He H, Jin QZ (1999) Trophic niches and potential yields of three economic bivalves in Daya Bay. Trop Oceanol 18(14):53–60

Liu JW, Dong SL (2001) Nutrient metabolism and the major nutrient uptake kinetics of seaweeds. Plant Physiol C 37:325–330

Liu JW, Dong SL, Ma S (2001) Effects of temperature and salinity on growth of *G. tenuistipitata* var *liui*, *U. pertusa*, *G. filicina* and NH_4-N uptake of *G. tenuistipitata* var *liui*. Acta Oceanolo Sin 23(2):109–116

Lu DD, Zhang ZD (1996) Studies on phytoplankton and red tide causative species in the sea area near Ma'an Islands of Zhejiang. Donghai Mar Sci 14(1):44–51

Maci'a S, Lirman D (1999) Destruction of Florida Bay seagrasses by a grazing front of sea urchins. Bull Mar Sci 65:593–601

Ocean Reconnaissance and Design Academy in Zhoushan (2004) Aquiculture arrangement and layout in Shengsi, China. http://www.zj-ssyb.com.cn/news.asp?ID=8. Cited 20 Nov 2006

Pu XM, Wu YL (2000) Review of nutrient limitation of phytoplankton. Mar Sci 24(2):27–30

Redfield AC (1958) The biological control of chemical factors in the environment. Am Sci J 46:205–222

Shen GY, Shi BZ (2002) Marine ecology. Science Publishing, Beijing

CSBTS (State Bureau of Quality and Technical Supervision of China) (1998) Seawater analysis. In: The specification for marine monitoring part 4. CSBTS, Beijing, China

Statistic Bureau in Zhoushan City (2001) Zhoushan Statistical Almanac. Chinese Statistic, Beijing

Sun J, Song SQ, Wang D, Xu ZL (2007) Estimating *Calanus sinicus* grazing on phytoplankton and microzooplankton. Acta Ecol Sin 27(8):3302–3315

Watanuki A (2006) Restoring kelp beds to healthy and strong from the Isoyake. News Letter, No.151, Ocean Policy Research Foundation. http://www.sof.or.jp/ocean/newsletter/151/a01.php. Cited 2 Aug 2007

Wheeler K, Kokkinakis D (1990) Ammonium recycling limits nitrate use in the oceanic subarctic Pacific. Limnol Oceanogr 35:1267–1278

Zhang SY, Wang ZH, Lin J (2007) Variation of fisheries resources in summer and autumn in seaweed beds of Gouqi Island. Mar Fish Res 28(1):45–52

A field test of porous carbonated blocks used as artificial reef in seaweed beds of *Ecklonia cava*

Kumi Oyamada · Mario Tsukidate · Keiji Watanabe · Tatsuhito Takahashi · Tsuneo Isoo · Toshinobu Terawaki

Originally published in the Journal of Applied Phycology, Vol 20, No 5, 413–418.
DOI: 10.1007/s10811-008-9332-6

Abstract Marine forests are the main primary producer in coastal waters, supplying food to fish and shellfish as well as providing their spawning and growing sites. It is important to conserve marine forest in order to protect coastal marine environments. A "Marine Block" comprised of steelmaking slag particles combined by $CaCO_3$ has several merits: (1) solidification of CO_2, (2) stability and safety, (3) porous materials covered with $CaCO_3$, (4) harmony with the seabed environment, (5) recycling of iron by-products, and (6) the ability to be mass produced in coastal ironworks. The objectives of this study were to observe the algal succession, the growth of *Ecklonia cava* and other attached organisms on Marine Blocks and concrete blocks compared to natural seaweed beds. In November 2001, five 1 m^3 Marine Blocks and five 1 m^3 concrete blocks were installed close to *E. cava* beds around the coastal frontage of Jogashima at the mouth of Tokyo Bay. Scuba-diving observations were continued until April 2006. In the field, a large number of seaweed and animal species were observed on the Marine Blocks, and it was found that the speed of succession, the plant length and the wet weight of *E. cava* were greater on Marine Blocks than on concrete blocks. The main advantages of Marine Blocks are considered to be the recycling of steelmaking slag, the absorption and solidification of atmospheric CO_2, and the conservation of coastal environments.

Keywords *Ecklonia cava* · Artificial reef · CO_2 · Marine block · Marine forest · Porous carbonated blocks · Seaweed bed

K. Oyamada (✉)
JFE R&D Corporation,
1-1, Minamiwatarida-cho, Kawasaki-ku,
Kawasaki, Kanagawa 210-0855, Japan
e-mail: k-oyamada@jfe-rd.co.jp

M. Tsukidate
Marine Biological Research Corporation,
Yokohama, Kanagawa 221-0834, Japan

K. Watanabe
JFE Steel Corporation,
Chiyoda-ku,
Tokyo 100-0011, Japan

T. Takahashi · T. Isoo
JFE Refractories Corporation,
Ako, Hyogo 678-0232, Japan

T. Terawaki
Fisheries Research Agency,
Kamisu, Ibaragi 220-6115, Japan

Introduction

Seaweed beds play an important role in biological production, water quality purification, and sediment stabilization, etc. The various functions of a seaweed bed combine to work in a mixed manner. The following seaweed bed functions are known: (1) basic production, (2) conservation of the detritus food chain and primary consumers, (3) provision of spawning and farming sites, (4) provision of feeding and hiding places, (5) environmental stabilization, and (6) supply of drifting seaweed (Marine Ecology Research Institute 1991). Therefore, it is essential to conserve and cultivate marine forests in order to protect coastal marine environments (Harrold and Pearse 1987). Furthermore, the preservation and development of seaweed beds with such functions contributes to the increase in catches of useful seafood (Watanuki and Yamamoto 1990). The worldwide demand for seaweed—a useful resource for ingredients and processed food—is growing (Critchley and Ohno 1998).

However, due to coastal development or coastal environment degradation, vegetation is decreasing (Environmental Agency of Japan 1994), and it is now imperative that we recover seaweed vegetation (Ohno 1985; Nakatsu 2005). Common methods for recovering seaweed beds are transplanting algae, supplying seeds, placing artificial reefs, removing herbivorous fishes and animals, and making breakwaters (Kawasaki 1992). In particular, artificial reefs have often been used to recover and develop seaweed beds (Largo and Ohno 1993; Choi et al. 2002; Yoshida et al. 2006). The materials conventionally used as artificial reef are mostly fieldstones and concrete blocks, and their efficiency has been affirmed (Terawaki et al. 2001; Watanuki 2003). Successful "Management-Free Techniques" for restoration of *Eisenia* and *Ecklonia* beds have been reported along the central Pacific coast of Japan (Terawaki et al. 2001); in this latter study, artificial reefs were made of concrete blocks.

As an alternative to concrete, large Marine Blocks, $1 m^3$ in size, have been developed using only steelmaking slag and CO_2 without applying cement (Isoo et al. 1999, 2000a, b). This porous material, which is covered with $CaCO_3$, is completely novel in this field.

Many large plants of the annual brown alga *Sargassum* showed healthy growth on small test pieces (10 cm×5 cm) of Marine Blocks in the Seto Inland Sea area (Isoo et al. 2000a, b). The present study was designed to observe the growth of *Ecklonia cava* on Marine Blocks and ordinary concrete blocks, and to survey the formation of seaweed beds on Marine Blocks. The laminariaceous brown algae of *Ecklonia cava* and *Eisenia bicyclis* form a community called "marine forest" in coastal waters at 2–25 m depth. Marine forests play an important role as nursery and breeding spaces for commercial fish, shellfish, and other animals such as abalone and sea urchin (Terawaki et al. 2001), while having high productivity (Yokohama et al. 1987). Therefore, it is important to study the growth of *Ecklonia cava* on a Marine Block in order to evaluate the suitability of this new material as artificial reef.

Materials and methods

Calcium oxide (CaO), which accounts for 40–60% of slag, increases its alkalinity, and steelmaking slag disintegrates easily in the presence of water vapor and CO_2 because of an increase in its volume. However, when the CaO is carbonated and stabilized as calcium carbonate ($CaCO_3$), the slag can be used in various ways. The carbonation reaction in the slag is essentially a reaction between the CaO and CO_2 contained in the slag ($CaO + CO_2 \rightarrow CaCO_3$). The manufacturing process of Marine Blocks is as follows: the martial slag is placed in an airtight mold with an adequate amount of water; CO_2 is injected into the slag from the bottom; Fig. 1 shows a large solid carbonated block (1 m×1 m×1 m). Marine Blocks have the following properties: (1) solidification of CO_2 in exhaust gas emitted during steel manufacturing, (2) they are stable and safe, (3) the material is porous, (4) the calcium carbonate coating is in harmony with the seabed environment, (5) they permit high-level recycling of steelmaking slag, and (6) they offer convenience of material transportation because many Japanese steel plants are located in coastal marine areas. For these reasons, it is efficient and effective to use Marine Blocks as artificial reefs in order to develop seaweed beds. The microstructure of a Marine Block (Kawashima et al. 1999) is quite different from that of a concrete block. A concrete block has a fine structure, while a Marine Block is a porous body having countless continuous pores several hundreds of micrometers in size. Accordingly, there are some pores whose size is several hundreds of micrometers on the surface. The density of Marine Block is $2.4 t m^{-3}$, i.e., almost the same as that of a concrete block.

We studied two types of artificial-reef blocks placed at a depth of 7 m around the costal frontage of Jogashima, Kanagawa, Japan, on 15 November 2001. We replaced five concrete blocks with five cubic Marine Blocks with a size of 1×1×1 m. A concrete block is an appropriate reference substrate for evaluating artificial substrates for seaweed bed development, because there are already many case studies on concrete blocks in the literature (Watanuki et al. 1987; Terawaki 1988; Terawaki et al. 2000). The area consists mainly of rocky ledges, and there is a natural marine forest of *Ecklonia cava*. The laminariaceous brown algae *Ecklonia cava* is large perennial brown alga, whose lifetime is 3–4 years. The blocks were placed on relatively flat areas of sea floor scattered across this terrain. Scuba-diving observations and measurements, to count the number of *Ecklonia cava* and *Sargassum* plants and other attached organisms attached to each block, were continued until April 2006. In particular, we measured the number and length of all *Ecklonia cava* individuals on the upper surface (1×1 m) of each of the five blocks, and identified the

Fig. 1 Marine block (1 m×1 m×1 m)

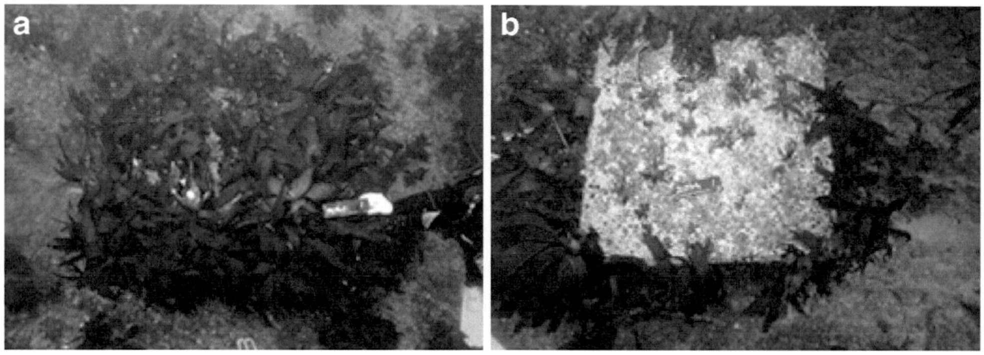

Fig. 2 **a,b** Photos of top surfaces in October 2002. **a** Marine Block. The surface is covered with *Ecklonia cava*. **b** Concrete Block. *Ecklonia cava* is attached to the edge of the block

major seaweeds and attached organisms. The observation frequency was set to be once every 2–5 months. The sampling procedure conducted in July 2005 was as follows: all *Ecklonia cava* individuals, with their holdfasts, on the upper surfaces of two Marine Blocks and two concrete blocks were removed by a diver using a scraper so as not to damage the rhizoids. The sampled *Ecklonia cava* was brought to a laboratory at JFE R&D Corporation, and the quantity at each age was counted, and the sample was weighed. The same procedures were repeated for control groups sampled at two natural reef areas of 1 m×1 m. The age of *Ecklonia cava* was determined by counting the number of rhizoid tiers, because a new rhizoid tier of adult *Ecklonia cava* and its related species *Eisenia bicyclis* is formed at the lower portion of the stem annually in winter at this marine location (Terawaki and Goto 1986; Terawaki et al. 1991). The age of *Ecklonia cava* in a near sea area was estimated by counting the number of rhizoid tiers (Kimura and Nanba 1995). Thus, in this study, we surveyed, observed, and recorded the formation process of rhizoids of *Ecklonia cava* on Marine Blocks and natural reefs, and then confirmed that the age of *Ecklonia cava* can be estimated by counting the number of rhizoid tiers.

Results

Succession in algal vegetation

During the initiation phase, diatoms were dominant on both types of blocks. Two months later, *Ectocarpus* spp. proliferates, and a change in dominant species was observed. Three months later, the growth of *Ectocarpus* spp. declined on Marine Blocks but was still dominant on concrete blocks. Succession to *Ectocarpus* spp. and *Colpomenia sinuosa* was dominant on both types of blocks. From the 5th month onwards, they were replaced by *Ecklonia cava*. When *Ecklonia cava* became dominant, crustose coralline algae began to cover the surface. From the 11th month onwards, a community of *Ecklonia cava*-crustose coralline algae had developed. In addition, large brown algae other than *Eckloniacava*, i.e., *Sargassum* including *Sargassum horneri*, *Sargassum ringgoldianum* and *Sargassum yamamotoi*, began to appear on both types of blocks.

Changes in the average number of *Ecklonia cava* individuals

The number of *Ecklonia cava* attached to Marine Blocks exceeded 70 individuals m^{-2} in April 2002 and decreased to 40 individuals m^{-2} in July 2002. This remained constant for a while before decreasing again to 20 individuals m^{-2} in September 2003, 36 months after the installation of the blocks. Then, zero-age *Ecklonia cava* appeared on Marine Blocks and the number again increased to 70 individuals m^{-2} in January 2004 and February 2005. On the other hand, the number of *Ecklonia cava* attached to concrete blocks was 3–4 individuals m^{-2} until the end of the 1st year (Fig. 2), and then increased to 30 individuals m^{-2}. This remained constant with little change and increased to 80 individuals m^{-2}, i.e., exceeding that of *Ecklonia cava* on Marine Blocks, in March 2005, 40 months after the installation. Numbers then decreased and became almost the same as that of *Ecklonia cava* on Marine Blocks (Fig. 3). On two adjacent natural reefs (1 m^2 of one reef area) investigated as a control group, there was a stable marine forest of *Ecklonia cava*, and the number of *E. cava* measured at the time of collection of *E. cava* ranged from 9 to 11 individuals m^{-2}. This number is

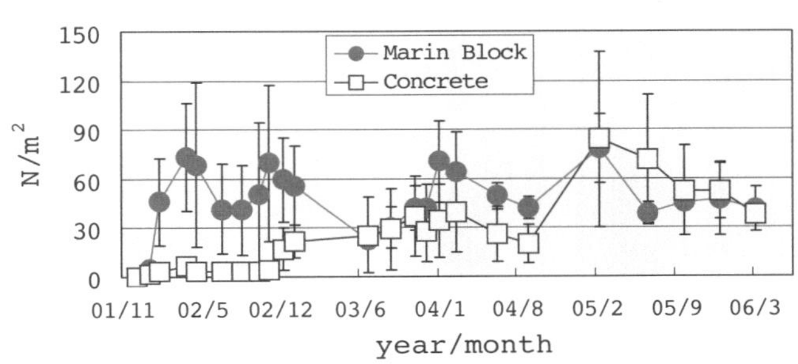

Fig. 3 Changes in average numbers of individuals of *Ecklonia cava* plants (n=5, *bar* SD)

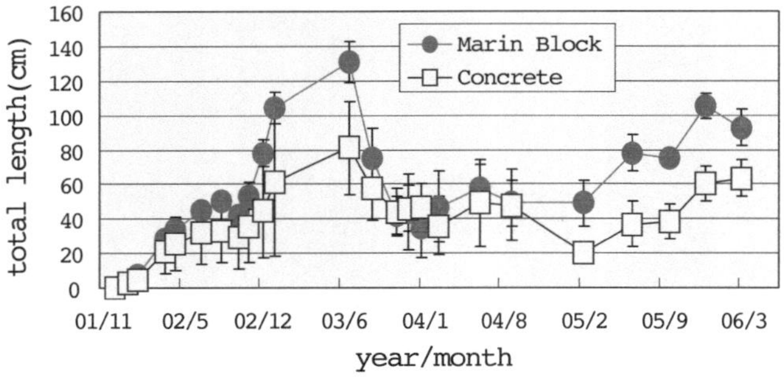

Fig. 4 Change of average plant length of *Ecklonia cava* (*n*=5, *bar* SD)

near to the growth density of the natural reef around Miura Peninsula and Izu Peninsula on the Pacific Coast (Terawaki 1991).

Average length of *Ecklonia cava*

After 29 months (July 2003), the average length of *Ecklonia cava* was 130 cm on Marine Blocks and 80 cm on concrete blocks. Length then decreased on both types of blocks until January 2004 and then increased again (Fig. 4). For the first 3 years from the start of the experiment, there was no significant difference in the length of *Ecklonia cava* between the two block types, excluding the period in which the length reached a maximum in July 2003. However, from the 4th year of the experiment, i.e., from February 2005, the length of *Ecklonia cava* on Marine Blocks was greater than that on concrete blocks. A t-test showed significant differences ($P<0.05$) in July 2003, March 2005, October 2005, and January 2006; and significant differences ($P<0.01$) in July 2005. On two adjacent natural reefs, the length of *Ecklonia cava* measured at the time of collection in July 2005 was 104 cm.

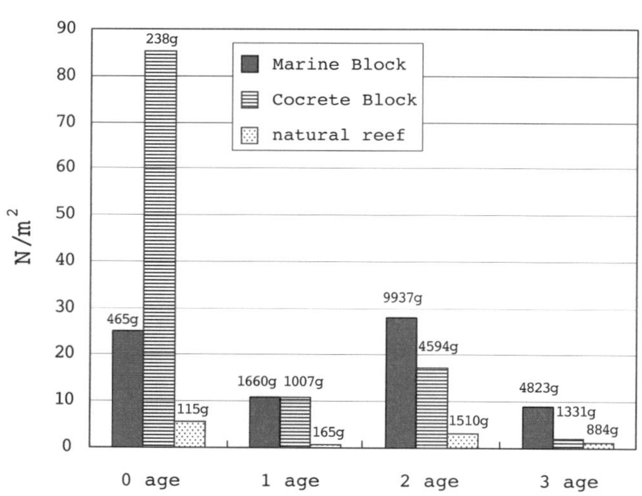

Fig. 5 Age composition and wet weight of *Ecklonia cava* in July 2005 (after 4 years)

Ecklonia cava after 4 years

The number of individuals was greater on concrete blocks than on Marine Blocks. While multi-year *Ecklonia cava* was dominant on Marine Blocks, zero-age *E. cava* was dominant on concrete blocks. The wet weight on Marine Blocks was twice that on concrete blocks. On Marine Blocks, there were 25 age 0 individuals, 11 at age 1, 28 age 2, and 9 age 3. The age composition was similar to that on natural reefs. On concrete blocks, there were 85 age 0 individuals, 11 age 1, 17 age 2, and 2 at age 3 (Fig. 5). The average weight of *Eckloina cava* was 16,885 g m^{-2} on Marine Blocks, 7,170 g m^{-2} on concrete blocks, and 2,674 g m^{-2} on natural reefs.

Composition of attached species

Kimoto's degree of similarity (1976) was calculated from the dominancy of species, population of species, and total population of full species. This degree is 1 if two communities are identical, and 0 if two communities are completely different. From the results of Kimoto's degree of similarity, it was found that the difference in block material significantly influences the formation of biologic communities until 6 months from the installation of a block, and Marine Blocks and concrete blocks were different. In addition, Marine Blocks and concrete blocks had low similarities with the control group until 11 months after block installation. However, 20 months after block installation, the degree of similarity between Marine Blocks and the control group was as high as 0.79. Meanwhile, the degree of similarity between concrete blocks and the control group was as low as 0.5. The attached organisms at the time of collection of *Ecklonia cava* are listed in

Table 1 Composition of attached organism species in July 2005 (after 4 years)

		Marine Block (2 blocks)	Concrete block (2 blocks)	Natural reef 2 districts (1 district = 1 m^2)
Average number of species /m^2	Total	31.4	26.0	41.0
	Algae	10.4	10.4	14.0
	Animal	21.4	15.6	27.0
Dominant species or biological taxon on each block and natural reef		*Ecklonia cava*	*E. cava*	*E. cava*
		Crustose coralline algae	Crustose coralline algae	Crustose coralline algae
		Sargassum yamamotoi	*S. yamamotoi*	Hydrozoa
		Hydrozoa	Hydrozoa	-

Table 1. With regard to seaweed, *Sargassum yamamoto* was rarely observed on natural rock reefs, while it was the second most abundant seaweed after *E. cava* on both types of block, especially on concrete blocks. The number of animal species was smaller on both types of blocks than in the control group. The primary attached invertebrates were Hydrozoa, and followed by *Ergalatax contractus*.

Discussion

In this study, the succession in the algal vegetation on blocks was similar to that in the vegetation in the surrounding natural marine forests of *Ecklonia cava*. In the succession, diatoms first covered blocks, and then small annual algae appeared, and ultimately *E. cava*, a large perennial alga, became dominant and formed stable vegetation on nearby natural rock reefs. Such processes are normally seen in natural seaweed vegetation (Terawaki et al. 1991; Choi et al. 2002), and so it can be concluded that both Marine Blocks and concrete blocks can form artificial reefs having the same functions as natural rocky reefs. Further study showed that the number and weight of *E. cava* on natural reefs was smaller than that on blocks, but *E. cava* on natural reefs was longer. It can be inferred that the marine forest of *E. cava* on the blocks had not yet reached its climax, or that a difference in size or shape of the substratum influences the number, weight and length of *E. cava*. When the length of *E. cava* on the artificial reefs is longer, the growth plant density shrinks adversely on artificial reefs, according to a report by Terawaki (1991).

It is still necessary to study why the succession process on Marine Blocks was more rapid and the amount and size of *E. cava* on Marine Blocks was larger. As of now, it can be inferred that Marine Blocks are favorable to seaweed because a Marine Block is a porous body having countless pores several hundreds of micrometers in diameter on its surface. A rugged surface generates local turbulence, and a propagule of a seaweed is more likely to settle (Furukawa et al. 1994), the adhesion of seaweed is improved, and it is hard to remove by external force such as a surge (Watanuki 2003). What kind of harmful compounds elute from concrete is not known. There have been no studies on phenomena occurring on concrete surface before seaweed are firmly established (Watanuki 2003). The calcium carbonate covering Marine Blocks is extremely chemically stable (Isoo et al. 1999, 2000a, b). In order to investigate the mechanism of attachment of *E. cava*, scanning electron microscope observations will be conducted on the holdfasts of the algae as well as on the cross-sectional surface of Marine Blocks, and the biological affinity of algae to $CaCO_3$ will also be investigated.

Marine Blocks contribute to the high-level recycling of steelmaking slag and to the solidification of CO_2 from exhaust gas. When Marine Blocks are used as artificial reefs to develop seaweed beds and forests, ecosystems in the newly created seaweed beds, as well as the primary productivity of the algal communities on Marine Blocks, are also considered to contribute to marine and coastal conservation and the increase of fishery resources and, accordingly, also to the development of technologies for global environmental protection.

Acknowledgments We would like to extend special thanks to Dr. Daisuke Fujita, Assistant Professor of Tokyo University of Marine Science and Technology Association for the study of the seaweed bed of the Waterfront Vitalization and Environment Research Center (WAVE), and Kanagawa Prefectural Fisheries Technology Center; all of them gave us useful instructions. We also appreciate the cooperation of Jyogashima Fisheries Cooperative Association. In addition, part of this study was conducted with a grant provided by the Program for Promoting the Development of Practical Use of New Energy and Industrial Technology Development Organization (NEDO) R&D Results.

References

Choi CG, Takeuchi Y, Terawaki T (2002) Ecology of seaweed beds on two types of artificial reef. J Appl Phycol 14:343–349

Critchley AT, Ohno M (eds) (1998) Seaweed resources of the world. Kanagawa International Fisheries Training Center, JICA, Tokyo

Environmental Agency of Japan (1994) The 4th basic survey of natural environment conservation. The report of survey in coastal organism environment. Seaweed bed. Marine Park Center, Tokyo

Furukawa K, Murozen I, Hosokawa Y (1994) Examination about the speed distribution around the salient for biofouling promotion to the harbor structure. Port Harbour Res Inst Rep 33:3–25

Harrold C, Pearse JP (1987) The ecological role of echinoderms in kelp forests. In: Jangoux M, Lawrence JM (eds) Echinoderm studies, vol 2. Balkema, Rotterdam, pp 137–233

Isoo T, Takahashi T, Numata N, Nishi K (1999) Manufacturing process of large carbonated steelmaking slag blocks. CAMP-ISIJ 13:142

Isoo T, Takahashi T, Okamoto N, Furukawa K (2000a) Development of large steelmaking slag blocks using a new carbonation process. Adv Cement Res 12:97–101

Isoo T, Takahashi T, Okada M (2000b) Evaluation of carbonated solids as seaweed substratum. Nippon Suisan Gakkaishi 66:647–650

Kawasaki Y (1992) The present conditions of the research and development of making marine forest technology. In: Proceedings of association of Japanese assessment seminar. Japan Association of Environment Assessment, Tokyo

Kawashima T, Watanabe K, Isoo T, Takashi T (1999) New carbonation mechanism for steelmaking slag block production. CAMP ISIJ 12:829

Kimoto S (1976) A method of study of animal crowd - variety and kind composition. Kyoritsu, Tokyo, pp 54–94

Kimura S, Nanba T (1995) Comprehensive undertaking for improving the production environment of Isone Fishery; the marine product increase laboratory in Wakayama prefecture, Rep. No27:85–89

Largo DB, Ohno M (1993) Constructing an artificial seaweed bed. In: Ohno M, Chitchley AT (eds) Seaweed cultivation and marine ranching. JICA, Tokyo, pp 113–130

Marine Ecology Research Institute (1991) Existing information on the structures and functions of seaweed beds. Marine Ecology Research Institute, Tokyo

Ohno M (1985) Marine forest—its ecology and constructing technology. Kaiyo Kagaku 17:706–713

Nakatsu T (2005) Promotion of seaweed bed creation in fisheries infrastructure improvement. Fish Eng 42:185–187

Terawaki T (1988) Preliminary study for creation of kelp forest. II On relation between surface form of artificial algal substratum and colonization of young *Ecklonia cava* plant. Abiko Research Laboratory, CRIEPI, Rep. No.U88037:1–26

Terawaki T (1991) Development of techniques for marine macrophyte bed creation in sandy sea beds. IV. Growth of *Eisenia, Ecklonia* plants on concrete blocks setted in sandy sea beds. Abiko Research Laboratory, CRIEPI, Rep. U91024:1–31

Terawaki T, Goto H (1986) Preliminary study for creation of kelp forest. Seasonal changes of lamina and growth of root of *Eisenia bicyclis* in Odawa Bay, Miura Peninsula. Abiko Research Laboratory, CRIEPI, Rep. No.U88037:1–23

Terawaki T, Kawasaki Y, Honda M et al (1991) Verification of technologies for kelp forest creation on sandy sea beds. II Ecology and growing characteristics of *Eisenia bicyclis* and *E. cava* at western sea coast of Miura peninsula, central Japan. Abiko Research Laboratory, CRIEPI, Rep. U91022:1–69

Terawaki T, Yoshida G, Yoshikawa K, Arai S, Murase N (2000) "Management-free techniques" for the restoration of *Sargassum* beds using subtidal, concrete structure on sandy substratum along the coast of the western Seto Inland Sea, Japan. Environ Sci 7:165–175

Terawaki T, Hasegawa H, Arai S (2001) Management-free techniques for restoration of *Eisenia* and *Ecklonia* beds along the central Pacific coast of Japan. J Appl Phycol 13:13–17

Watanuki K (2003) The design of the artificial reef. In: Notoya M (ed) Seaweed of marine forest and its development technology. Seizando-shoten, Tokyo, pp 251–253

Watanuki A, Yamamoto H (1990) Settlement of seaweed on coastal structures. Hydrobiologia 204/205:275–280

Watanuki K, Yamamoto S, Arai S (1987) Effects of surface roughness on settlement of *Ecklonia stolonifera* OKAMURA (Phaophyta, Laminaliales) in the early stage of growth. Aquacult Sci 35:69–75

Yokohama Y, Tanaka J, Chihara M (1987) Productivity of the *Ecklonia cava* community in a bay of Izu Peninsula on the Pacific coast of Japan. Bot Mag 100:129–141

Yoshida G, Yoshikawa K, Arai S, Terawaki T (2006) Seaweed colonization on the experimental substrata set *Sargassum horneri* community. Fish Eng 42:267–273

Seaweed of the littoral zone at Cove Island in Long Island Sound: annual variation and impact of environmental factors

A. Pedersen · G. Kraemer · C. Yarish

Originally published in the Journal of Applied Phycology, Vol 20, No 5, 419–432.
DOI: 10.1007/s10811-008-9316-6 © Springer Science + Business Media B.V. 2008

Abstract A site in the western part of Long Island Sound was monitored from January 2000 to May 2002. The littoral was divided into five different zones from the supra-littoral fringe (A) to the infra-littoral fringe (E). The midshore was dominated by *Fucus vesiculosus* L. and the sublittoral fringe by *Chondrus crispus* Stackh. There was a significant change in community structure over the years and the predominant change occurred between 2001 and 2002. The alternation in community structure was caused by an increase in abundance of species like *Porphyra suborbiculata* Kjellm., *Porphyra leucosticta* Type A and C (Neefus et al. 2000), *Ceramium virgatum* Roth, and *Codium fragile* subsp. *tomentosoides* (van Goor)Silva and a decrease in abundance in *Fucus vesiculosus*, *Blidingia minima* (Nägeli ex Kütz.) Kylin and *Ulva lactuca* L. The changes in community structure coincided with the change in environmental conditions. Air temperature as well as surface seawater temperature (depth <2 m) were the most important factors of those analyzed. Temperature seems to be the bottom-up force regulating the community structure.

A. Pedersen (✉)
Norwegian Institute of Water Research,
Gaustadalléen 21,
NO-0349 Oslo, Norway
e-mail: are.pedersen@niva.no

G. Kraemer
Division of Natural Science, Purchase College,
State University of New York,
Purchase, NY 10577, USA

C. Yarish
Department of Ecology and Evolutionary Biology,
University of Connecticut at Stamford,
Stamford, CT 06901, USA

Keywords Community structure · New England · Temperature · Salinity · Nutrients · *Porphyra* spp

Introduction

Population and communities in the littoral zone can be complex and are influenced by various biotic and abiotic factors like climate, nutrient availability, predation, grazing, competition, symbionts, parasites, substrate characteristics, exposure and tidal variation. Species living in the littoral zone are exposed to an extreme environment.

Littoral communities have been extensively studied over centuries. Several studies of the community structure in the intertidal have focused on the coast from Rhode Island and up to Maine (Bertness and Leonard 1976; Leonard 2000; Lubchenco 1980, 1983; Mathieson et al. 1976, 1981a, b; Mathieson and Penniman 1986; Menge 1976, 1991; Petraitis 1987). However, very few studies have been published on the littoral assemblages in Long Island Sound (LIS), even though sites in the vicinity of Dominion Nuclear Power Station (Niantic, CT) have been thoroughly monitored for over 20 years (Keser et al. 2003, 2005).

The main factors regulating the littoral community structure in New England are exposure and top-down regulations such as predation and grazing pressure (Hunter and Price 1992). The main predators are *Nucella lapillus* L. and *Asterias* spp., while periwinkles are the main grazers (mainly *Littorina littorea* L.) (Dudgeon et al. 1999; Lubchenco 1983; Menge 1976, 1978a, b, 1983; Petraitis 1987). The intertidal zone is dominated by fucoids (*Fucus vesiculosus* and *Ascophyllum nodosum* (L) Le Jol.), and in the infra-littoral fringe Irish moss (*Chondrus crispus*) is the most abundant. In highly exposed areas, the mussel

(*Mytilus edulis* L.) outcompetes *Chondrus crispus* as predation pressures are reduced.

Studies have recently focused on the importance of bottom-up regulation. These factors, i.e. variation in nutrient and other environmental conditions, have been associated with changes in algal cover, usually in larger spatial scales (10–100 km) (Menge 2000). Other bottom-up processes like recruitment, congestion for space, nutrient and food abundance (POC, PON and chlorophyll-a) have also been shown to have a bottom-up effect on the community structure (Bertness et al. 1999a, b; Menge et al. 1999; Menge 2000). However, a recent study demonstrated that a bottom-up process interact with top-down forces in the regulation of community structure locally in the littoral zone in New England (Leonard et al. 1998).

Apart from fucoids and Irish moss which are the key algal species in the two main zones on these shores, several species of *Porphyra* are also frequently found in the littoral zone along New England's shores (Sears 1998). However, the genus *Porphyra* contains several species that are difficult to identify (Brodie et al. 1998; Broom et al. 2002; Lindstrom and Cole 1993; Neefus et al. 2000, 2002; Nelson et al. 2001) since their morphologies are very similar (Nelson et al. 2000). As several *Porphyra* are of great importance in aquaculture, especially in Korea, China and Japan (Oohusa 1992; Tseng and Fei 1987), domestic species have gained attention in recent years as potentially valuable resources, and several studies have recently been initiated (Carmona et al. 2001; Chopin et al. 2001a, b, c; Levine 1998; Yarish et al. 1997, 1998, 1999). As a part of this new interest in *Porphyra* in New England, this study aimed to gain information on the spatial and temporal occurrence of *Porphyra* species in LIS. However, the main goal of this study was to test the importance of climatic factors as bottom-up forces in regulating the community structure of the shore line.

Materials and methods

A site at Cove Island, Stamford, CT (41°2.644′N, 73° 30.133′W) (Fig. 1) was surveyed 23 times over a period of 2 years, from February 2000 to March 2002. The mid-littoral zone was divided into five vertical zones, from supra-littoral fringe (A) to the infra-littoral fringe (E) (classification of vertical zonation according to Hiscock and Mitchell 1980). The procedure of dividing the phytal zone into five different zones was based on biological as well as physical differences between the zones. The upper zone A represented the supra-littoral zone and extended into the upper part of the balanoid/fucoid belt in the mid-littoral zone. The extensive fucoid belt, present at the time, was divided into two zones, B and C. The two zones were

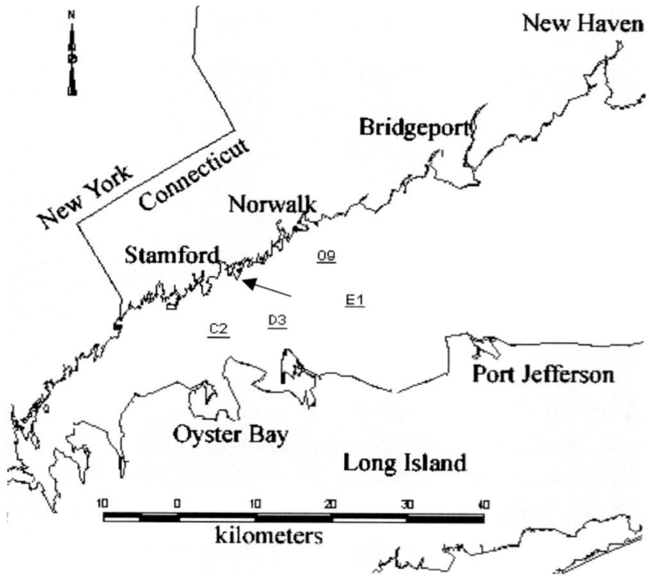

Fig. 1 NOAA's hydrographical stations in Long Island Sound, used to calculate average values for environmental variables (modified after NOAA's station map). *Arrow* points at Cove Island

thought to response differently to prolonged periods of abnormal environmental conditions. The two lower zones had a smaller angle of inclination than the above fucoid zones. They were separated into two zones. Zone D, just above the infra-littoral fringe, was flat with a few fucoids but periodically with high abundance of *Codium fragile.* The flats were occasionally covered with sediments and were daily exposed to air due to tidal fluctuations as opposed to the lowest zone (E) within the infra-littoral fringe which was only exposed to air a few days every month. The horizontal elevation from the Mean Lower Low Water (MLLW) was measured by laser level. MLLW was set to 0 m depth and the zones represented the following depth intervals (cm above MLLW) A=270–220 cm, B=220–150 cm, C=150–50 cm, D=50–0 cm and E=0 to −40 cm.

The dominating species in the lowest zone was also different from the above. It was also assumed that zone D would be more exposed to sedimentation than the above fucoid belt. The species occurring within each zone were recorded by placing a frame (0.5 × 0.5 m) subdivided into smaller (10×10 cm) squares randomly within the zones during each registration. Each sub-square represented 4% cover and enabled us to add up better the percentage cover of each species. Five replicate frames were recorded within each zone. All species were registered as percentage cover at low tide. The registrations were based on non-destructive sampling. However, small pieces of those species hard to identify in the field were collected either within or outside the frames for further identification in the laboratory.

Low taxonomic resolution reflected environmental pollution gradients even more clearly than did higher taxonomic levels (Gray et al. 1990; Olsgard et al. 1998; Olsgard and Somerfield 2000), especially when trying to relate community structure to environmental data (Clarke and Ainsworth 1993; Terlizzi et al. 2002). Hence, running multivariate analysis of the community structure of Cove Island included grouping of species into higher taxa. This was done to compensate for the existence of species difficult to distinguish and identify in the field. It has been shown that grouping species into higher taxa does not alter the main outcome and can sometimes improve the results of multivariate analysis of community structure (Clarke 1993; Gray et al. 1990; Lasiak 2003; Olsgard et al. 1997; Terlizzi et al. 2003; Warwick 1988).

Small inconspicuous species (<1–3 mm or microscopic) were not included. Species of similar morphology difficult to differentiate in the field, such as species from the genera *Ectocarpus* and *Pilaiella* as well as species within the *Ceramium*, *Polysiphonia, Neosiphonia*, *Enteromorpha* and *Cladophora* genera, were grouped for statistical analysis. Sedimentation was measured on a semi-quantitatively scale from 0 to 4, where 0=no obvious sedimentation, 1=ca. 1 mm, 2=ca. 3 mm, 3=ca. 4 mm and 4 severe sedimentation of > 5 mm.

Data for temperature, nitrate and salinity from LIS were obtained from National Oceanic and Atmospheric Administration (NOAA). These were obtained from NOAA's hydrographical stations 09, C2, D3 and E1 (Fig. 1) and averaged across a 1 month period prior to the dates when community structure were registered. Monthly values for nitrate (μM), salinity (ppt) and temperature (°C) were averaged for each season (i.e., winter, spring, summer and fall). Winter included the months December to March, spring from April to May, summer from June to September and fall from October to December. Missing values for temperature in February and May of 2001, nitrate + nitrite and salinity for May 2001, were interpolated based on the previous and the following months. Differences among environmental data were tested with two-sample *t* tests or paired *t* tests where applicable. To test the influence of climatic factors on the community structure, a dataset was obtained from the National Climate Data Center at NOAA for Stamford, CT, including: DPNT (departure from normal monthly temperature), DT00 (number of days with minimum air temperature less than or equal to –17.8°C), DT32 (number of days with minimum air temperature less than or equal to 0°C), DT90 (number of days with maximum air temperature greater or equal to 32°C), DX32 (number of days with maximum air temperature less than or equal to 0°C), EMNT (extreme minimum air temperature for the month), EMXT (extreme maximum air temperature for the month), EMXP (extreme maximum daily precipitation in the month), MMNT (monthly mean minimum air temperature), MMXT (monthly mean maximum air temperature) and MNTM (monthly mean air temperature).

Multidimensional scaling (MDS) was used to analyze quantitative data on species and taxa abundances over time. MDS is an ordination method and will present temporal differences in community structure in a graphical plot. The longer the distance between two samples, the greater is the difference in community structure between the samples. Due to the high number of quadrats recorded, the five replicate quadrats within each zone were averaged prior to the multivariate analysis. This was done to reduce number of permutation and make the computations possible to execute. Bray–Curtis similarity index was used in calculating the species similarity matrix. Prior to analysis, the species data were transformed by a fourth root transformation to make the datasets more similar to a normal distribution. The similarity matrices were input for MDS analysis. The same similarity matrices were used for hierarchical agglomerative clustering using group average linkage to cross-check the results obtained via MDS analysis, when the stress factor approached 0.2 (Clarke 1993). One-way ANOSIM permutation test (which is a simulated ANOVA, PRIMER ver. 5) was used to test differences between the species composition at different years, zones and seasons. SIMPER (Similarity percentage procedure) was used on the abundance data matrices to test the different species contribution to the Bray–Curtis dissimilarities between years, zones and seasons, i.e. the analysis ranks and quantify the importance of each species which causes the differences among the samples in the MDS plot. Similarity matrices of the environmental data, i.e. temperature, salinity and nitrate concentration in 2 m water depth in LIS and climate data from Stamford, were based on normalized Euclidean distance as opposed to a Bray–Curtis similarity matrices for the biological datasets. Principal component analysis (PCA) of environmental data from LIS and climate data from Stamford were plotted in two dimensions. To compare the environmental data and climate datasets against biological datasets, BIO-ENV (PRIMER ver.5) was used. These procedures compare the environmental matrices (environmental data matrix from LIS and climate data matrix from Stamford as well as a combination of them) against the biota matrices within each zone. BIO-ENV calculates a measure of agreement between the two dissimilarity matrices by rank correlation of the matching elements in the two matrices. The coefficient of agreement used was a Spearman rank correlation coefficient ρ which ranges from –1 complete discordance between ranked values) and +1 (complete concordance).

Results

Environmental conditions

The average temperature (at 2 m depth) in Long Island Sound off the study site, varied from 0.5°C in January to 23°C in August/September (Fig. 2). In 2002, the average temperature (T) in the coldest winter months was twice as high as in 2000 and 2001, 4.5°C as opposed to 1.8°C and 1.7°C, respectively. No difference was detected between summer temperatures in 2000 and 2001.

Nitrate (=NO_3+NO_2) concentration (N) in the upper water column varied dramatically over the year with low N in the summer months and high in the winter months. Figure 2 shows that N dropped dramatically from February to March in 2000 and from January to February in 2001. However, in 2002, high N was observed into May.

The salinity (S) at 2 m depth in the inner part of LIS varied between 25.4 and 29.1 psu (Fig. 2). S was lower during the spring/summer months than during winter presumably due to runoff from land during spring. S during the summer months in 2000 and 2001 varied from 25.4 to 26.8 psu, while the average S in the winter months from November 2000 to March 2001 (27.5 psu) was significantly lower than in the same period during the 2001/2002 winter (28.9, $p<0.001$) (Fig. 2).

Principal Component Analysis (PCA) of the environmental data placed the samples as shown in Fig. 3. The data analyzed are monthly average values for S, N and T concentration from the inner part of LIS. It shows that the conditions follow a cyclic pattern and that conditions varied more in late winter/early spring than in late fall/early winter. Conditions in January and February 2002 were quite different from the previous 2 years. PC 1 explained more than 50% of the variation (Fig. 3; Table 1) and was mainly represented by differences in N at the different sampling occasions (0.701 in Table 2). The relative importance of T represented almost all of the 34.6% variation explained by PC2 (0.967) (Tables 1 and 2). S was the least influential environmental factor of the three (Fig. 3; Tables 1 and 2).

Climate data from National Climate Data Center at NOAA also shows that the climate in January in 2000 and 2001 as well as December in 2000 was different from the same periods in 2002. The constellation of stations in Fig. 4 A, B was mainly due to the mean monthly minimum and maximum temperatures (MMNT, MMXT), monthly mean temperature (MNTM) and extreme temperatures in winter and summer months (EMNT, EMXT) (Table 4). Departure from normal monthly temperature (DPNT) gave the best discrimination along PC2 where it explained 0.62 of the variation.

The annual variation in climate over the monitored period showed that 2002 was a warmer year than the previous two years (Fig. 2). No days during 2002 had maximum temperature below 0°C whereas 2000 and 2001 had 13 and 5 days, respectively. The winter temperature in surface waters (>2 m) was much higher during 2002 than the previous two years (Fig. 2). This was due to an air temperature that was higher than normal from August 2001 until April 2002. During this period the temperatures were on average 2.4°C above the normal (based on 30 years variation, NOAA-datasets). The previous winter temperatures were not different from the 30-year range. DPNT (departure from normal monthly temperature) was the factor that best described the variation over PC2 as well as other temperature factors as DT00 and DX32 (Tables 3 and 4).

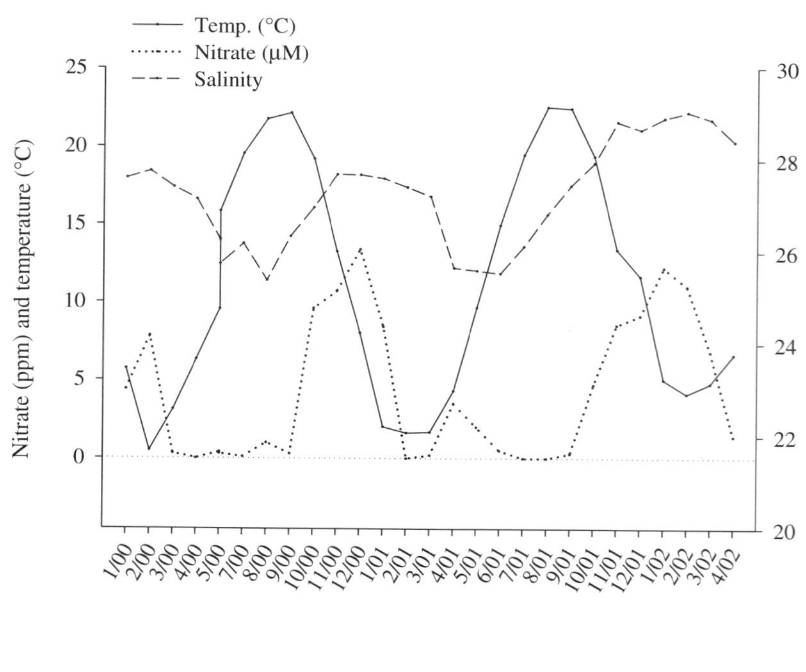

Fig. 2 Average nitrate concentrations, temperature and salinity at 4 stations (NOAA) in LIS from January 2000 to April 2002. (Note that values for June 2000 are missing.)

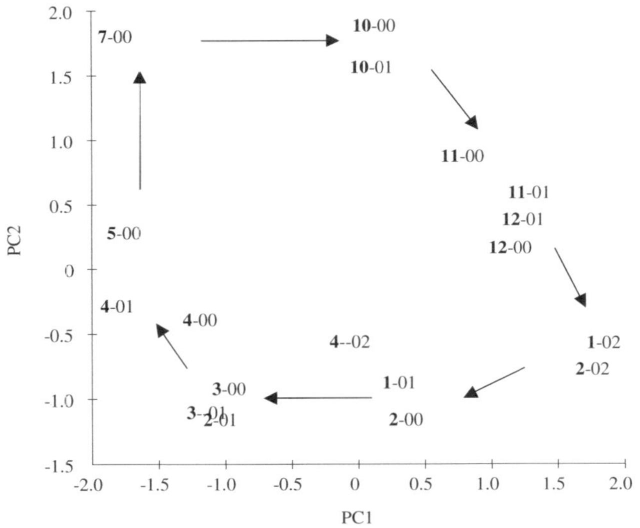

Fig. 3 PCA of the environmental conditions in the inner part of LIS during the period February 2000 to April 2002. The *arrows* indicate succession in environmental condition during the years. The labels on the x-axis are in month-year format

Biological succession at Cove Island, LIS

The intertidal zone at Cove Island was dominated by *Fucus vesiculosus* with a total average annual cover of 34%, followed by *Chondrus crispus* (19%) and *Semibalanus* sp. (11%) (Table 5). The occurrence varied within the five different zones and, although *F. vesiculosus* was found in all five zones, it was most pronounced in the intermediate zones B (68%), C (55%) and D (37%) (Table 5; Fig. 5). *Fucus vesiculosus* is a perennial, but the percentage cover varied both seasonally and annually. During 2000 and until May 2001, the average cover (5 replicates) within each zone were fairly stable. However, during summer 2001, the percentage cover dropped significantly in all zones (two-sample *t* test: $p<0.05$,; Fig. 5a).

Chondrus crispus dominated in the sublittoral fringe, zone E, and the percentage cover was stable over the years. There was a drop in cover in spring 2000 which coincided with an increased cover of *Ulva lactuca* in the same zone (Fig. 5d). *Ulva lactuca* was epiphytic on *C. crispus* and concealing it. Hence, *C. crispus* coverage was underestimated.

Green algae like *Ulothrix* spp. and *Urospora* spp. (combined in the analysis) showed the same pattern as *Ulva lactuca*, with a significantly ($p<0.01$) higher occurrence in winter/spring of year 2000 than in 2001 and 2002. *Codium fragile* subsp. *tomentosoides*, introduced in LIS in 1956 (cited in (Sears 1998), showed opposite trends than the other green algae by having a significant higher percentage cover (in zone D) in 2002 than the previous years ($p<0.01$; Fig. 5c). Another green alga that occurred early spring, sometimes in very high percentage cover, was *Blidingia minima*. However, *B. minima* showed high occurrence only in 2001 in the upper two zones, A and B (Fig. 5b).

Table 1 Eigenvalues and explained variation on 3 PC-axis in Fig. 3

PC	Eigenvalues	Variation (%)	Cumulative variation (%)
1	1.55	51.8	51.8
2	1.04	34.6	86.5
3	0.41	13.5	100.0

Table 2 Eigenvectors representing the relative importance of the three variables *N* (nitrate + nitrite), *T* (temperature) and *S* (salinity) in describing the variation along the different PC-axis

Variable	PC1	PC2	PC3
N	0.710	0.150	-0.688
T	0.038	0.967	0.250
S	0.703	-0.203	0.682

Among the red alga, *Ceramium virgatum* varied over the period with low percentage cover in 2000 and a slight increase in 2001. In 2002, the increase in percentage cover of *C. virgatum* was significantly higher than the previous years ($p<0.01$; Fig. 5e) similar to *Codium fragile* subsp. *tomentosoides* and *Neosiphonia harveyi* (Bailey) Kim, Choi, Guiry & Saunders. Three different species of *Porphyra* spp. were found at the site. They occurred most frequently in winter/spring, but their occurrence varied significantly over the years. *Porphyra suborbiculata*, which is easily identified by peripheral 2–4 cell teeth along the edge of the thallus in young sporophytes, was first observed in October and increased with a peak in late March. The teeth did, however, disappear later in spring and on older individuals. The round to oval thallus varied in thickness between 23–30 μm, but was usually found to be about 25 μm. Older sporophytes formed numerous endosporangia (Figs. 5–10 in Nelson et al. 1998) in March/April and these individuals were still found in late June, but not in August. The overall percentage cover of *P. suborbiculata* increased significantly from 2000 to 2001 ($p<0.05$) and slightly from 2001 to 2002 (n.s.; Fig. 5f). However, the increase was different within each zone and most pronounced in zones A and C. *Porphyra leucosticta* Thur. is a difficult taxa to identify and there may be five or more cryptic species behind this designation (Neefus et al. 2000). The types found at Cove Island resembled Type A and Type C in Neefus et al. (2000). *Porphyra leucosticta* Type A occurs epiphytic on *C. crispus* in the period January–May in the lower intertidal to shallow subtidal in LIS (zones D and E; Fig. 5g). The Type A strain has a lanceolate blade with ruffled margins and were between 1–5 cm wide and 5–12 cm long. The thallus thickness ranged from 18–24 μm.

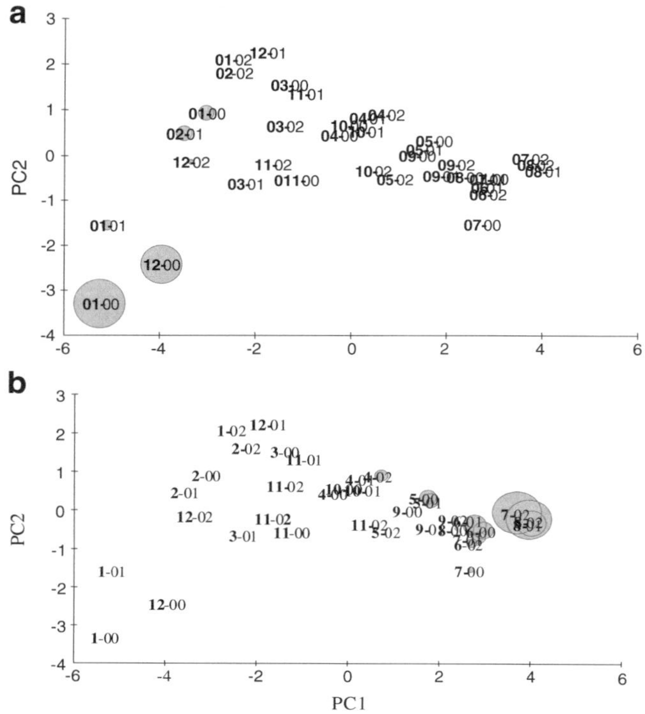

Fig. 4 PCA of the climate data from Stamford, CT, during 2000, 2001 and 2002. The labels on the x-axis are in month-year format. **a** Values for number of days with maximum temperatures below 0°C are superimposed on plot. **b** Values for number of days with maximum temperature above 32°C are superimposed on the plot

Porphyra leucosticta Type A increased over the years from 0.5% cover in zone D in 2000 to 5.5% cover in 2002 (n.s.). Type C of *P. leucosticta* was found epiphytic on different algae or epilithic in the mid-littoral zone from November to May (Fig. 5h). It was mainly ovate with ruffled edges of 3–8 cm wide and 5–10(–15) cm long. The thallus thickness varied between 30–40 μm (21–50 μm) in thickness. It was found in all zones at Cove Island from A to E as opposed to Type A found exclusively in zones D and E. In general, all types of *Porphyra* spp found at the site increased in percentage coverage from 2000 to 2002 as did *Ceramium virgatum, Neosiphonia harveyi* and *C. fragile* subsp *tomentosoides*.

Table 3 Eigenvalues and explained variation on 5 PC-axis in Fig. 4

PC	Eigenvalues	Variation (%)	Cumulative variation (%)
1	6.86	62.4	62.4
2	1.35	12.2	74.6
3	1.08	9.8	84.4
4	0.73	6.6	91.0
5	0.49	4.4	95.5

Multivariate analysis of community structure

Average percentage cover for all species was compared for all sampling dates. The community structure shows that some zones form more distinct group than others (A+B+C and D and E) (Fig. 6). Zones B and C show less separation than others and are also indicated with the lowest pairwise *r*-value between zones (B and C in Table 6). There are significant differences between all zones (one-way ANOSIM, $p<0.001$; Table 6). The community structure of upper three zones (A, B, C) are more alike as opposed to the well separated two lower zones (Fig. 6; Table 6). In zones A–D, the occurrence of *F. vesiculosus* contributed most to the zones' characteristics (25–34%) as did *C. crispus* in zone E (39%). The species that contributed most in separating the zones are listed in Table 7. Other important species in separating the different zones were *Blidingia minima, Mytilus edulis* L. and balanoids which were abundant in the upper zones, *Enteromorpha spp* in the intermediate zones and *U. lactuca*, *Ceramium virgatum, Polysiphonia stricta* (Dwil.) Grev. and *Codium fragile* subsp. *tomentosiodes* in the lower two zones. All contributed from 4 to 15% in separating the different zones (Table 7).

To check the effect of redundancy in the dataset, MDS, SIMPER and ANOSIM were performed on the same dataset excluding *F. vesiculosus*, which was the dominating species in the upper four zones. The MDS plot did not change dramatically and still showed distinct zones significantly different from each other ($p<0.001$). However, excluding *F. vesiculosus* resulted in even less separation of zones A, B and C (smaller *r*-values). Several species had different abundances over the period monitored. Figure 7 shows the total community structure as in Fig. 6, but with years superimposed on the samples. *Fucus vesiculosus*,

Table 4 Eigenvectors representing the relative importance of the 11 variables (coefficients in the linear combinations of variables making up PC's)

Variable	PC1	PC2	PC3	PC4	PC5
DPNT	0.009	0.620	0.424	0.612	−0.106
EMNT	0.371	−0.086	0.023	−0.017	0.031
EMXP	0.152	−0.371	−0.446	0.764	0.112
EMXT	0.364	0.003	0.112	0.040	−0.058
MMNT	0.377	−0.089	0.022	−0.034	−0.056
MMXT	0.376	−0.064	0.077	−0.044	−0.135
MNTM	0.378	−0.076	0.051	−0.039	−0.098
DT00	−0.186	−0.444	0.496	0.117	−0.601
DT90	0.252	−0.085	0.510	−0.025	0.644
DT32	−0.363	0.015	0.028	0.087	0.316
DX32	−0.224	−0.499	0.306	0.117	0.256

See Materials and methods for abbreviations

Table 5 Average (plus minimum and maximum) percentage cover of species and taxa within the zones used in the analysis of community structure at Cove Island, LIS during January 2000 to May 2002

Species	Zones (depth)					
	Sum all 5 zones (0–310 cm)	A (0–50 cm)	B (50–120 cm)	C (120–220 cm)	D (220–270 cm)	E (270–310 cm)
	Av. (min–max)	Av. (min–max)	Av. (min–max)	Av. (min–max)	Av. (min–max)	Av. (min–max)
Algae						
Agardhiella subulata	0 (0–0)	0 (0–0)	0 (0–0)	0 (0–0)	0 (0–0)	0 (0–0)
Ascophyllum nodosum	1 (0–30)	0 (0–0)	0 (0–0)	0 (0–6)	2 (0–30)	0 (0–1)
Erythrocladia irregularis	0 (0–1)	0 (0–0)	0 (0–0)	0 (0–0)	0 (0–1)	0 (0–0)
Audouniella daviesii	0 (0–1)	0 (0–0)	0 (0–0)	0 (0–0)	0 (0–1)	0 (0–1)
Bangia atropurpurea	1 (0–15)	2 (0–15)	1 (0–15)	0 (0–0)	0 (0–0)	0 (0–0)
Blidingia minima	5 (0–63)	14 (0–63)	7 (0–63)	3 (0–10)	0 (0–1)	0 (0–0)
Bryopsis plumose	0 (0–2)	0 (0–0)	0 (0–0)	0 (0–0)	0 (0–2)	0 (0–2)
Ceramium virgatum	2 (0–35)	0 (0–0)	0 (0–0)	0 (0–1)	5 (0–35)	2 (0–16)
Chondrus crispus	19 (0–98)	0 (0–0)	0 (0–0)	0 (0–2)	17 (0–39)	78 (0–98)
Chorda filum	0 (0–2)	0 (0–0)	0 (0–0)	0 (0–2)	0 (0–0)	0 (0–0)
Cladophora sp.	0 (0–3)	0 (0–0)	0 (0–0)	0 (0–1)	0 (0–3)	0 (0–1)
Codium fragile subsp. *tomentosoides*	3 (0–43)	0 (0–0)	0 (0–0)	0 (0–0)	13 (0–43)	1 (0–3)
Dasya baillouviana	0 (0–1)	0 (0–0)	0 (0–0)	0 (0–0)	0 (0–0)	0 (0–1)
Desmarestia viridis	0 (0–1)	0 (0–0)	0 (0–0)	0 (0–0)	0 (0–0)	0 (0–1)
Colonial Diatoms	2 (0–62)	0 (0–1)	0 (0–0)	1 (0–5)	7 (0–62)	2 (0–22)
Ectocarpales indet.	2 (0–48)	0 (0–0)	1 (0–8)	1 (0–4)	3 (0–23)	5 (0–48)
Elachista fucicola	1 (0–6)	0 (0–0)	1 (0–4)	1 (0–5)	1 (0–6)	0 (0–0)
Enteromorpha linza	3 (0–40)	2 (0–40)	3 (0–28)	8 (0–37)	3 (0–10)	0 (0–3)
Enteromorpha prolifera	2 (0–30)	1 (0–14)	3 (0–30)	6 (0–27)	2 (0–13)	0 (0–1)
Fucus vesiculosus	34 (0–90)	8 (0–23)	68 (0–86)	55 (0–90)	37 (0–71)	1 (0–10)
Hildenbrandia rubra	1 (0–10)	0 (0–2)	0 (0–2)	2 (0–10)	1 (0–5)	0 (0–0)
Laminaria juv.	0 (0–1)	0 (0–0)	0 (0–0)	0 (0–0)	0 (0–0)	0 (0–1)
Laminaria saccharina	0 (0–1)	0 (0–0)	0 (0–0)	0 (0–0)	0 (0–1)	0 (0–0)
Petalonia fascia	0 (0–2)	0 (0–0)	0 (0–0)	0 (0–0)	0 (0–2)	0 (0–1)
Polysiphonia fucoids	0 (0–1)	0 (0–0)	0 (0–0)	0 (0–0)	0 (0–1)	0 (0–1)
Neosiphonia harveyi	1 (0–11)	0 (0–0)	0 (0–0)	0 (0–0)	1 (0–10)	2 (0–11)
Polysiphonia nigrescens	0 (0–1)	0 (0–0)	0 (0–0)	0 (0–0)	0 (0–0)	0 (0–1)
Polysiphonia urceolata	0 (0–4)	0 (0–0)	0 (0–0)	0 (0–1)	1 (0–2)	1 (0–4)
Porphyra suborbiculata	1 (0–8)	2 (0–8)	1 (0–7)	1 (0–7)	0 (0–0)	0 (0–0)
Porphyra leucosticta A	0 (0–5)	0 (0–0)	0 (0–0)	0 (0–0)	0 (0–2)	1 (0–5)
Porphyra leucosticta C	1 (0–10)	0 (0–3)	1 (0–10)	1 (0–5)	1 (0–8)	0 (0–2)
Scytosiphon lomentaria	0 (0–2)	0 (0–0)	0 (0–2)	0 (0–0)	0 (0–0)	0 (0–1)
Sphacelaria sp.	0 (0–6)	0 (0–0)	0 (0–0)	0 (0–0)	1 (0–6)	0 (0–1)
Sphacelaria plumosa	0 (0–1)	0 (0–0)	0 (0–0)	0 (0–0)	0 (0–1)	0 (0–1)
Spongomorpha pallida	0 (0–1)	0 (0–0)	0 (0–0)	0 (0–0)	0 (0–0)	0 (0–1)
Ulothrix/Urospora spp.	3 (0–66)	11 (0–66)	2 (0–7)	3 (0–11)	1 (0–8)	0 (0–0)
Ulva lactuca	3 (0–47)	0 (0–0)	0 (0–0)	2 (0–9)	6 (0–18)	9 (0–47)
Fauna						
Alcyonidium undet.	0 (0–7)	0 (0–0)	0 (0–0)	0 (0–0)	1 (0–7)	2 (0–6)
Balanoids undet.	11 (0–97)	18 (0–97)	20 (1–43)	19 (0–50)	0 (0–2)	0 (0–0)
Bryozoa undet.	0 (0–1)	0 (0–0)	0 (0–0)	0 (0–0)	0 (0–1)	0 (0–1)
Crepidula fornicate	0 (0–2)	0 (0–0)	0 (0–0)	0 (0–0)	0 (0–2)	0 (0–0)
Littorina obtusata	0 (0–0)	0 (0–0)	0 (0–0)	0 (0–0)	0 (0–0)	0 (0–0)
Littorina spp.	0 (0–0)	0 (0–0)	0 (0–0)	0 (0–0)	0 (0–0)	0 (0–0)
Mytilus edulis	1 (0–10)	0 (0–3)	4 (0–10)	1 (0–7)	0 (0–0)	0 (0–0)
Ostrea sp.	0 (0–2)	0 (0–1)	1 (0–2)	0 (0–2)	0 (0–0)	0 (0–0)
Porifera undet.	0 (0–8)	0 (0–0)	0 (0–0)	0 (0–0)	0 (0–8)	0 (0–2)
Sediment: unclassified	3 (0-	0 (0–0)	0 (0–2)	2 (0–14)	10 (0–51)	1 (0–5)

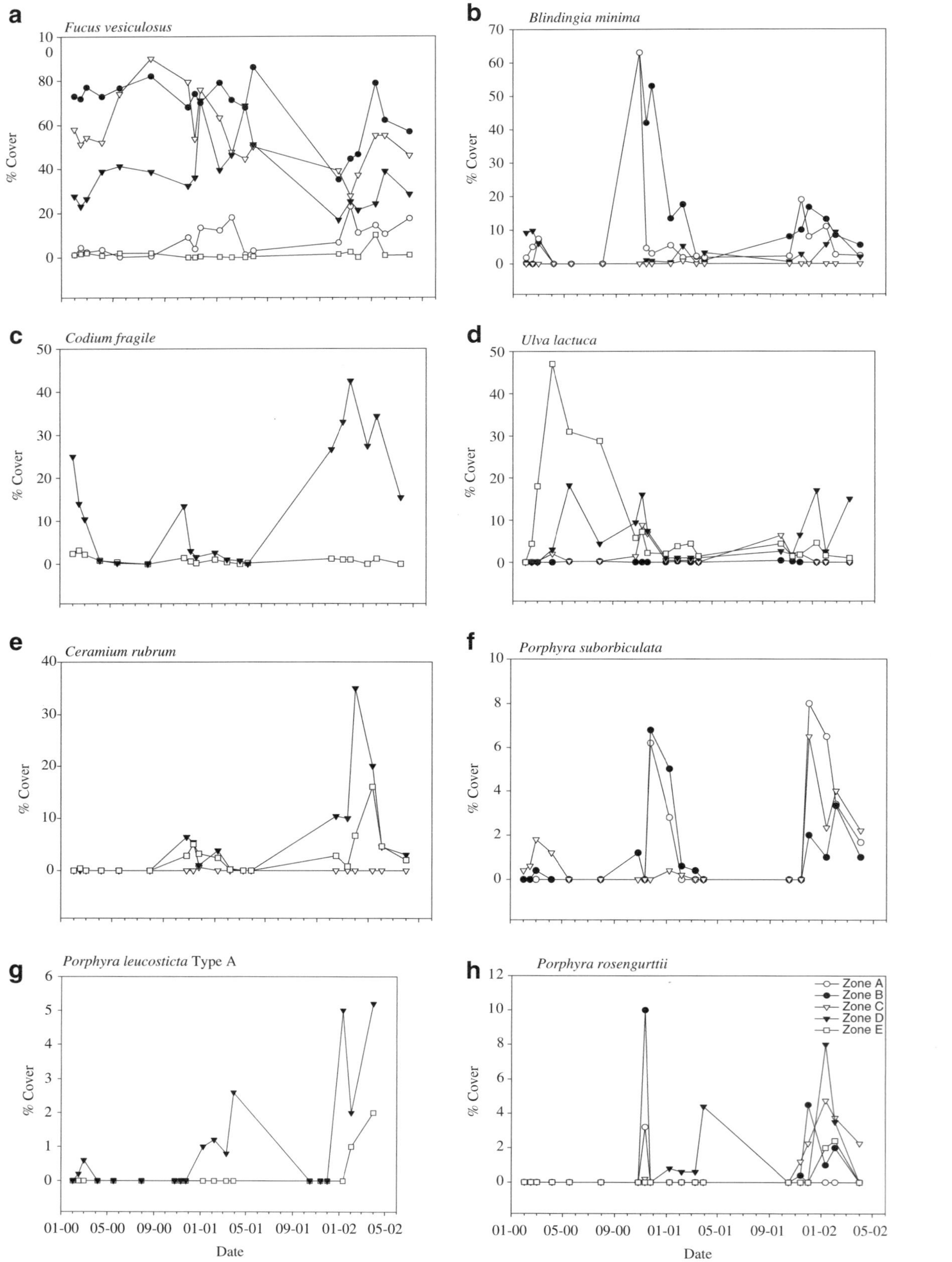

Fig. 5 Difference in percentage cover of 8 important species in the littoral zones (**a–h**) at Cove Island, LIS. Note different scale of percentage cover axis. (Note that data between June and September 2001 are missing.)

Ulothrix/Urospora and *U. lactuca* decreased from January 2000 to May 2002, whereas other species like the *Porphyra* spp., *C. virgatum, N. harveyi* and *C. fragile* subsp. *tomentosiodes* increased in abundance over the 2 years. Two-way crossed simulated ANOVA's (ANOSIM) were used to test for differences between years (averaged across all zones) and a two-way nested ANOSIM tested for differences between seasons (averaged across zone groups) as well as among zones across seasons groups. Both analysis showed that all years were significantly different from each other (global r=0.23, p<0.001) and that all zones were significant different (Table 8). Differences between seasons were even more distinct from each other than years (global r=0.32, p<0.001) and the difference between zones became even more evident (Table 9).

Correlation between biological and environmental/climate data

BIO-ENV (for explanation see Materials and methods) between N, S and T at 2 m depth in the inner part of LIS, and the observed community structure at the respective sampling occasions, resulted in an average correlation coefficient among all depth interval of r=0.37. Even though it is a positive correlation, the coefficient is not significant as the numbers of variables are only three. Salinity and nitrogen concentrations gave the best correlation coefficient between the two matrices.

Best correlation coefficients were obtained when running BIO-ENV on the combined matrices for environmental data from LIS and climate data from Stamford, against the biological data matrix. The Spearman rank correlation coefficient was significant at the 5% level for zones A, B and C (r=0.54 when ρ=0,;Table 13 in Pearson and Hartley 1966) but not for D and E. The average correlation coefficient (r) for all zones was 0.39, but this is not significant.

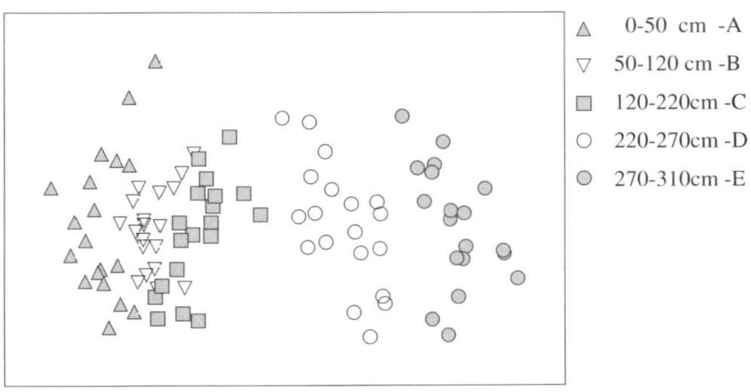

Fig. 6 Multidimensional Scaling of the community structure at Cove Island from January 2000 to May 2002. The *symbols* represent the different zones from upper littoral fringe down to the sublittoral fringe. 270 cm is Mean Lower Level Water

Table 6 Result of a simulated one-way ANOVA (ANOSIM) on the community structure with zones (A–E) as factors (ref. Fig. 6)

Groups	Pairwise tests	
	R statistic	Significance level %
A, B	0.418	0.1
A, C	0.477	0.1
A, D	0.969	0.1
A, E	0.999	0.1
B, C	0.233	0.1
B, D	0.949	0.1
B, E	1.000	0.1
C, D	0.755	0.1
C, E	0.995	0.1
D, E	0.531	0.1

Global R=0.712 (p<0.001)
Number of permutations: 999 (random sample from a large number)
Number of permuted statistics greater than or equal to Global R: 0

Discussion

Community characteristics

The most abundant species occurring in the zones, from 50 to 220 cm (B, C, D) above the MLLW at Cove Island, was bladder wrack *Fucus vesiculosus* with its peak in zone B, closely followed by zone C. In the sublittoral fringe, *Chondrus crispus* was the main species with an overall average cover of 78%. Such a zonal pattern was also found earlier in the intertidal zone along the New England coast (Lubchenco and Menge 1978; Lubchenco 1980; Menge 1976). Other studies found that, at more exposed coasts, barnacles and the mussel *Mytilus edulis* were more dominant (Lubchenco and Menge 1978; Menge 1976).

By recording percentage cover at low tide by placing a frame randomly on top of the algae cover underestimates the abundance or percentage cover of many species. Underestimation is especially the case for understory flora and fauna, as well as for thin filamentous and membranous algae. In the fucoid belt, periwinkles and small predators, as well as other species of algae, might have been hidden under the fucoids. It was, however, easier to detect such species in the *Chondrus* belt at MLLW due to the upright shape of the Irish moss. *Porphyra* spp. were especially problematic to estimate at low tide. Thin membranous species will collapse and percentage cover will be significantly reduced. Hence, the maximum percent cover of *Porphyra* spp. and herbivores like periwinkles, were underestimated at low tide, but the consistency in the method made registrations applicable in the analysis.

The *Porphyra* species occurring at Cove Island varied in abundance within the zones, and also their occurrences peaked at different times during fall to spring. Thallus thickness of *P. leucosticta* Type C, which occurred in all

Table 7 Average abundance (percentage cover) and contribution of the 7 most important species/taxa groups to the distribution of samples within each zone in the MDS analysis (Fig. 6)

Species	Av. abund.	Av. abund.	Contrib. %	Cum. %	Species	Av. abund.	Av. abund.	Contrib. %	Cum. %
	Zones A and B: average dissimilarity=50.21					Zones C and D: average dissimilarity=63.12			
	Zone A	Zone B				Zone C	Zone D		
Fucus vesiculosus	8.18	68.03	12.46	12.46	*Chondrus crispus*	0.19	17.24	8.81	8.81
Mytilus edulis	0.27	4.5	9.96	22.42	Balanoids	18.99	0.29	7.91	16.72
Balanoids	18.16	19.61	9.17	31.6	*Codium fragile*	0	13.26	7.6	24.32
Ulothrix/Urospora	10.98	1.93	8.99	40.59	*Blidingia minima*	2.95	0.04	4.48	28.8
Blidingia minima	13.75	7.44	8.57	49.16	*Enteromorpha linza*	7.67	2.89	4.47	33.27
Enteromorpha linza	2.4	2.62	6.95	56.11	*Ceramium rubrum*	0.03	5.26	4.25	37.52
Porphyra suborbiculata	1.57	1.14	6.21	62.32	*Hildenbrandia rubra*	2.45	0.59	4.21	41.73
	Zones A and C: average dissimilarity=56.66					Zones A and E: average dissimilarity=91.39			
	Zone A	Zone C				Zone A	Zone E		
Fucus vesiculosus	8.18	55.42	9.69	9.69	*Chondrus crispus*	0	77.61	15.15	15.15
Enteromorpha linza	2.4	7.67	8.52	18.22	*Ulva lactuca*	0	8.99	7.66	22.81
Blidingia minima	13.75	2.95	8.15	26.37	Balanoids	18.16	0	7.32	30.12
Balanoids	18.16	18.99	7.87	34.24	*Blidingia minima*	13.75	0	7.2	37.33
Ulothrix/Urospora	10.98	2.79	7.69	41.93	*Ulothrix/Urospora*	10.98	0.06	5.69	43.02
Hildenbrandia rubra	0.29	2.45	6.35	48.28	*Fucus vesiculosus*	8.18	1.38	4.32	47.33
Ulva lactuca	0	1.61	6.18	54.46	*Polysiphonia urceolata*	0	1.44	4.23	51.56
	Zones B and C: average dissimilarity=42.89					Zones B and E: average dissimilarity=89.34			
	Zone B	Zone C				Zone B	Zone E		
Enteromorpha linza	2.62	7.67	8.36	8.36	*Chondrus crispus*	0.01	77.61	12.46	12.46
Mytilus edulis	4.5	1.23	7.72	16.08	*Fucus vesiculosus*	68.03	1.38	9	21.46
Blidingia minima	7.44	2.95	7.02	23.1	Balanoids	19.61	0	8.36	29.82
Hildenbrandia rubra	0.42	2.45	6.54	29.64	*Ulva lactuca*	0.06	8.99	5.63	35.44
Ulothrix/Urospora	1.93	2.79	6.44	36.08	*Mytilus edulis*	4.5	0	5.62	41.07
Ostrea sp	0.67	0.21	6.35	42.43	*Blidingia minima*	7.44	0	5.34	46.4
Enteromorpha prolifera	3.03	5.75	6.35	48.78	*Polysiphonia urceolata*	0	1.44	3.53	49.94
	Zones A and D: average dissimilarity=79.44					Zones C and E: average dissimilarity=82.81			
	Zone A	Zone D				Zone C	Zone E		
Chondrus crispus	0	17.24	9.17	9.17	*Chondrus crispus*	0.19	77.61	12.67	12.67
Codium fragile	0	13.26	7.3	16.47	*Fucus vesiculosus*	55.42	1.38	8.88	21.55
Blidingia minima	13.75	0.04	6.82	23.29	Balanoids	18.99	0	8.17	29.72
Balanoids	18.16	0.29	6.09	29.38	*Enteromorpha linza*	7.67	0.34	4.68	34.41
Ulva lactuca	0	5.67	5.96	35.34	*Hildenbrandia rubra*	2.45	0	4.11	38.51
Ulothrix/Urospora	10.98	1.19	4.82	40.16	*Blidingia minima*	2.95	0	4.01	42.52
Enteromorpha linza	2.4	2.89	4.68	44.84	*Ulva lactuca*	1.61	8.99	3.96	46.49
	Zones B and D: average dissimilarity=70.2					Zones D and E: average dissimilarity=57.79			
	Zone B	Zone D				Zone D	Zone E		
Chondrus crispus	0.01	17.24	8.56	8.56	*Fucus vesiculosus*	36.53	1.38	9.58	9.58
Balanoids	19.61	0.29	7.86	16.43	*Chondrus crispus*	17.24	77.61	6.31	15.89
Codium fragile	0	13.26	6.95	23.38	*Codium fragile*	13.26	0.91	5.33	21.22
Mytilus edulis	4.5	0	6.17	29.55	*Enteromorpha linza*	2.89	0.34	5.18	26.4
Blidingia minima	7.44	0.04	5.71	35.26	*Ceramium rubrum*	5.26	2.47	4.81	31.21
Ulva lactuca	0.06	5.67	5.11	40.37	Ectocarpaceae	2.68	5.17	4.8	36.01
Ceramium rubrum	0	5.26	3.9	44.27	*Ulva lactuca*	5.67	8.99	4.13	40.14

zones from A to E, seemed to increase the higher it was found in the mid-littoral. The same pattern has been found for species of *Porphyra* occurring in China (personal communication). This might be due to the specimens different exposure to desiccation. Thick cell walls might preserve water better during period of desiccation than thin specimens, hence the gradient in thickness. The thinnest *Porphyra* sp was *P. leucosticta* Type A which was found in

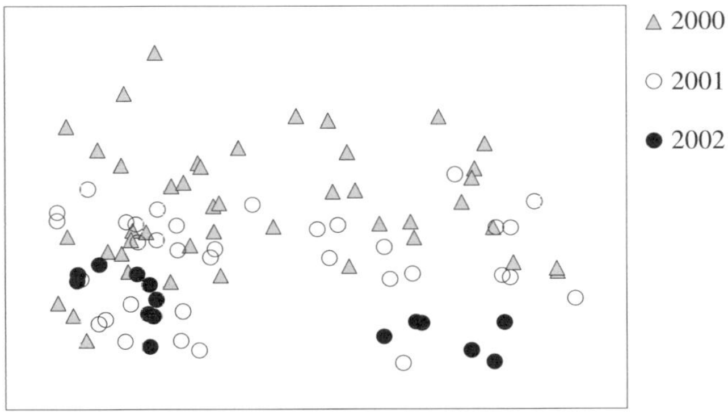

Fig. 7 Multidimensional Scalingof the community structure at Cove Island from January 2000 to May 2002 with years as overlay

the infra-littoral fringe and thereby less exposed to desiccation.

At Cove Island, *C. cripsus* was the dominating species around MLLW and beyond (−40cm). A top-down regulation was suggested to promote establishment of *C. crispus* in sheltered areas in New England; predators like starfish (*Asteria forbesi* and *Asterias vulgaris*), dogwinkles (*Nucella lapillus*) and periwinkles (*Littorina littorea*) have been shown to prevent establishment of the competitive dominant *Mytilus edulis* in mid- and lower mid-littoral zones (Lubchenco and Menge 1978; Menge 2000; Petraitis 1983, 1987). Periwinkles (*Littorina littorea, L. obtusata* L. and small *Littorina* sp.) were observed within the frames among *F. vesiculosus* at Cove Island, but their abundance was low (average < 0.4%). A top-down regulation suggested by Lubchenco (1983) explains to some extent the main zonation pattern at Cove Island. A mid-littoral zone consisting of more or less small rocks and pebbles next to the study site (2–3 m) was dominated by periwinkles. Very little vegetation was found here and the periwinkles most probably prevented all recruitment of alga by heavy grazing, hence supporting a potential impact at our study site.

Table 8 Result of a two-way crossed ANOVA (ANOSIM) on the community structure between years and between zones with years as samples

Groups	Pairwise tests	
	R statistic	Significance level %
A – B	0.405	0.1
A – C	0.633	0.1
A – D	0.978	0.1
A – E	0.998	0.1
B – C	0.351	0.1
B – D	0.952	0.1
B – E	1	0.1
C – D	0.786	0.1
C – E	0.993	0.1
D – E	0.559	0.1

Tests for differences between Year groups (averaged across all Zone groups): Global test; sample statistic (Global R): 0.226 ($p<0.001$)
Tests for differences between Zone groups (averaged across all Year groups): Global test; sample statistic (Global R): 0.743 ($p<0.001$)

Table 9 Result of two-way nested ANOVA (ANOSIM) on community structure between seasons and between zones with season groups as samples. (A-E are the different zones in the intertidal)

Groups	Pairwise tests	
	R statistic	Significance level %
A – B	0.854	2.9
A – C	0.844	2.9
A – D	1.000	2.9
A – E	1.000	2.9
B – C	0.927	2.9
B – D	1.000	2.9
B – E	1.000	2.9
C – D	1.000	2.9
C – E	1.000	2.9
D – E	0.969	2.9

Global R=0.324, $p<0.001$ for tests between season groups (averaged across zone groups)
Global R=0.942, $p<0.001$ for test between zone groups (using season groups as samples)

Climate and environmental data

The environmental factors salinity, temperature and nitrate-nitrite concentration in the surface water (<2 m depth) clearly showed a natural seasonal pattern (Fig. 2), and when run in a PCA the seasonal pattern resulted in a cyclic pattern (Fig. 3). The reduction in nutrients during spring of the respective years indicates that the spring bloom started earlier in 2001 than the other years, and in 2002 the seawater was still not nitrate-depleted in April. The bacterial remineralization of organic matter to nitrate and nitrite started in September–October. The PCA showed that January and February 2002 did separate from the same periods the previous years (Fig. 3), and do to some extent coincide with the configuration in the PCA for climate data (Fig. 4). The position of December 2001, February 2002 and especially January 2002 were placed farthest off on PC2 with respect to these months in 2000 and 2001, showing the climate influence on the surface water environment in LIS (Fig. 4).

Multivariate analysis of community structure

The MDS plot of the biological data (Fig. 6) showed that the community structure formed distinct zones, and they were placed in the MDS plot according to the order in which they formed the littoral zone from the upper zone A to the lowest zone E. The zones were significantly different from each other (Table 6). As shown in Table 8, *F. vesiculosus* dominated in the middle zones B, C and D, and *C. crispus* was irrefutably the dominant species in the sublittoral fringe (zone E). These patterns coincide with the structure found in the littoral shores of protected and semi-protected shores in New England (Lubchenco 1980, 1983; Menge 1976).

The community structure from January to April was different in 2002 than in the two previous years. This corresponds with differences in environmental data from LIS (Fig. 3) and climate data from Stamford (Fig. 4), where the winter months were plotted far from each other. High temperature in LIS and in the air during the winter 2001/2002 coincided with the changes in species abundances during same period. High temperature significantly promoted the occurrence of *Codium fragile, Porphyra* spp. and other red alga like *Ceramium virgatum* ($p<0.01$) at the sacrifice of *Blidingia minima, Fucus vesiculosus* and *Ulva lactuca.*

Statistically comparing biological and environmental data

BIO-ENV was used to test the concordance between environmental (including climate) data and biological data. One might expect that EMNT (Extreme minimum air temperature of the month) would be more important in zones A and B than further down on the shoreline, as A and B are the zones mostly exposed to air. The water temperature will reflect air temperatures, but at a much slower and delayed response and less fluctuating patterns. This was also the case in our findings.

Our analysis includes a limited numbers of environmental factors. PAR (photosynthetically active radiation) is an important environmental factor for structuring communities and algal growth, but it has not been recorded here. Hence, one cannot conclude that the combination of environmental factors resulting in a significant concordance with the biological matrices in this paper are the only factors determining the community structure. However, correlations found here between community structure and environmental factors are indications that the factors are important in regulating the community structure at our site at Cove Island.

In general, the community structure at Cove Island was similar to other shorelines described for medium to sheltered New England shorelines, with a dominating zone of *F. vesiculosus* in mid-littoral zone and a luxurious *C. cripus* belt in the upper infra-littoral zone (Bertness and Leonard 1976; Leonard 2000; Lubchenco 1980, 1983; Mathieson et al. 1976, 1981a, b; Mathieson and Penniman 1986; Menge 1976, 1991; Petraitis 1987). Three *Porphyra* spp were recorded and they all occurred in the fall to spring period, although with spatial and temporal differences in peak abundances. The different species showed a conspicuous variation among the years and all seem to increase during the unusually warm winter of 2001/2002.

Significant annual variation in different environmental factor was reflected in differences in community structure between 2001 and 2002 as documented in this paper. Of the environmental factors tested, temperature was shown to be the most important factor. The most prominent responses to increase in temperature were increases in several Rhodophytes and decline in the Fucoids population. A decline in *Ascophyllum nodosum* populations as a response to an increase in water temperature (Keser et al. 2005) for the eastern LIS, coincides with the results in this paper. The unusual high temperature during winter/spring 2002 turned back to normal temperatures in fall 2002. We have no data on the community structure in late 2002 and winter 2003, but one would expect the community structure during fall and winter 2002/2003 to oscillate back to similar assemblages occurring in 2000–2001. The community structure at Cove Island is most probably regulated both by top-down forces like grazing by periwinkles as suggested by several authors (Dudgeon et al. 1999; Lubchenco 1983; Menge 1976, 1978a, b, 1983; Petraitis 1987) but also as shown in this paper, from bottom-up regulating forces (Bertness et al. 1999a, b; Menge et al. 1999; Menge 2000) like air and seawater surface temperatures (depth <2 m) in LIS.

References

Bertness MD, Leonard GH (1976) The role of positive interactions in communities: Lessons from intertidal habitats. Ecology 78:1976–1989

Bertness MD, Leonard GH, Levine JM, Bruno JF (1999a) Climate-driven interactions among rocky intertidal organisms caught between a rock and a hot place. Oecologia 120:446–450

Bertness MD, Leonard GH, Levine JM, Schmidt PR, Ingraham AO (1999b) Testing the relative contribution of positive and negative interactions in rocky intertidal communities. Ecology 80:2711–2726

Brodie J, Hayes PK, Barker GL, Irvine LM, Bartsch I (1998) A reappraisal of Porphyra and Bangia (Bangiophyceae, Rhodophyta) in the Northeast Atlantic based on the rbcL-rbcS intergenic spacer. J Phycol 34:1069–1074

Broom JE, Nelson WA, Yarish C, Jones WA, Rosas RA, Rosas LEA (2002) A reassessment of the taxonomic status of *Porphyra suborbiculata*, *Porphyra carolinensis* and *Porphyra lilliputiana* (Bangiales, Rhodophyta) based on molecular and morphological data. Eur J Phycol 37:227–235

Carmona R, Kraemer GP, Zertuche JA, Chanes L, Chopin T, Neefus C, Yarish C (2001) Exploring *Porphyra* species for use as nitrogen scrubbers in integrated aquaculture. J Phycol 37:10–10

Chopin T, Buschmann AH, Halling C, Troell M, Kautsky N, Neori A, Kraemer GP, Zertuche-Gonzalez JA, Yarish C, Neefus C (2001a) Integrating seaweeds into marine aquaculture systems: A key toward sustainability. J Phycol 37:975–986

Chopin T, Yarish C, Neefus C, Kraemer G, Zertuche-Gonzalez J, Belyea E, Carmona R (2001b) *Aquaculture from a different angle: the seaweed perspective, and the rationale for promoting integrated aquaculture*. Cape Cod Press, Falmouth USA

Chopin T, Yarish C, Neefus C, Kraemer GP, Belyea E, Carmona R, Saunders GW, Bates C, Page F, Dowd M (2001c) Underutilized tools: seaweeds as bioremediation and diversification tools and bioindicators for integrated aquaculture and coastal management. J Phycol 37:12–12

Clarke KR (1993) Non-parametric multivariate analysis of changes in community structure. Aust J Ecol 18:117–143

Clarke KR, Ainsworth M (1993) A method of linking multivariate community structure to environmental variables. Mar Ecol Prog Ser 92:205–219

Dudgeon SR, Steneck RS, Davison IR, Vadas RL (1999) Coexistence of similar species in a space-limited intertidal zone. Ecol Monogr 69:331–352

Gray JS, Clarke KR, Warwick RM, Hobbs G (1990) Detection of initial effects of pollution on marine benthos: An example from the Ekofisk and Eldfisk oilfields, North Sea. Mar Ecol Prog Ser 66:285–299

Hiscock K, Mitchell R (1980) The discription and classification of sublittoral epibenthic ecosystems. In Farnham WF, Irvine DEG, Price JH (eds) The shore environment, Vol. 2. Ecosystems. Academic Press, London, 323–370

Hunter MD, Price PW (1992) Playing chutes and ladders: heterogeneity and the relative roles of bottom-up and top-down forces in natural communities. Ecology 73:724–732

Keser M, Swenarton JT, Vozarik JM, Foertch JF (2003) Decline in eelgrass (Zostera marina L.) in Long Island Sound near Millstone Point, Connecticut (USA) unrelated to thermal input. J Sea Res 49:11–26

Keser M, Swenarton JT, Foertch JF (2005) Effects of thermal input and climate change on growth of *Ascophyllum nodosum* (Fucales, Phaeophycea) in eastern Long Island Sound (USA). J Sea Res 54:211–220

Lasiak T (2003) Influence of taxonomic resolution, biological attributes and data transformations on multivariate comparisons of rocky macrofaunal assemblages. Mar Ecol Prog Ser 250:29–34

Leonard GH (2000) Latitudinal variation in species interactions: A test in the New England rocky intertidal zone. Ecology 81:1015–1030

Leonard GH, Levine JM, Schmidt PR, Bertness MD (1998) Flow-driven variation in intertidal community structure in a Maine estuary. Ecology 79:1395–1411

Levine IA (1998) Commercial cultivation of *Porphyra* (nori) in the United States. World Aquac 29:37–47

Lindstrom SC, Cole KM (1993) The systematics of *Porphyra*: Character evolution in closely related species. Hydrobiologia 260–261:151–157

Lubchenco J (1980) Algal zonation in the New England rocky intertidal community: an experimental analysis. Ecology 61:333–344

Lubchenco J (1983) Littorina and Fucus: Effects of herbivores, substratum heterogeneity, and plant escapes during succession. Ecology 64:1116–1123

Lubchenco J, Menge BA (1978) Community development and persistence in a low rocky intertidal zone. Ecol Monogr:67–94

Mathieson AC, Penniman CA (1986) Species composition and seasonality of New England seaweeds along an open coastal-estuarine gradient. Bot Mar 29:161–176

Mathieson AC, Shipman JW, O'Shea JR, Hasevlat RC (1976) Seasonal growth and reproduction of estuarine fucoid algae in New England. J Exp Mar Biol Ecol 25:273–284

Mathieson AC, Hehre EJ, Reynolds NB (1981a) Investigations of New England marine algae I: A floristic and descriptive ecological study of the marine algae at Jaffrey Point, New Hampshire, U.S.A. Bot Mar 24:521–532

Mathieson AC, Reynolds NB, Hehre EJ (1981b) Investigations of New England marine algae II: The species composition, distribution and zonation of seaweeds in the Great Bay Estuary System and the adjacent open coast of New Hampshire. Bot Mar 24:533–546

Menge BA (1976) Organization of the New England rocky intertidal community: Role of predation, competition and environmental heterogeneity. Ecol Monogr 46:355–393

Menge BA (1978a) Predation intensity in a rocky intertidal community. Relation between predator foraging activity and environment harshness. Oecologia 34:1–16

Menge BA (1978b) Predation intensity in a rocky intertidal community. Effect of an algal canopy, wave action and desiccation on predator feeding rates. Oecologia 34:17–35

Menge BA (1983) Components of predation intensity in the low zone of the New England rocky interdital region. Oecologia 58:141–155

Menge BA (1991) Generalizing from experiments: Is predation strong or weak in the New England rocky intertidal? Oecologia 88:1–8

Menge BA (2000) Top-down and bottom-up community regulation in marine rocky intertidal habitats. J Exp Mar Biol Ecol: 257–289

Menge BA, Daley BA, Lubchenco J, Sanford E, Dahlhoff E, Halpin PM, Hudson G, Burnaford JL (1999) Top-down and bottom-up regulation of New Zealand rocky intertidal communities. Ecol Monogr 69:297–330

Neefus CD, Mathieson AC, Yarish C, Klein A, West A, Teasdale B, Hehre EJ (2000) Five cryptic species of *Porphyra* from the Northwest Atlantic. J Phycol 36:suppl

Neefus CD, Mathieson AC, Klein AS, Teasdale B, Bray T, Yarish C (2002) *Porphyra birdiae* sp. nov. (Bangiales, Rhodophyta): A new species from the Northwest Atlantic. Algae 17:203–216

Nelson WA, Knight GA, Hawkes MW (1998) *Porphyra lilliputiana* sp. nov. (Bangiales, Rhodophyta): A diminutive New Zealand endemic with novel reproductive biology. Phycol Res 46:57–61

Nelson WA, Broom JE, Farr TJ (2000) Confusing convergent morphologies: Diversity and difficulties in New Zealand Erythropeltidales and Bangiales. J Phycol 36

Nelson WA, Broom JE, Farr TJ (2001) Four new species of *Porphyra* (Bangiales, Rhodophyta) from the New Zealand region described using traditional characters and 18S rDNA sequence data. Cryptogamie: Algol 22:263–284

Olsgard F, Somerfield PJ (2000) Surrogates in marine benthic investigations – which taxonomic unit to target? J Aquat Ecosyst Stress Recovery 7:25–42

Olsgard F, Somerfield PJ, Carr MR (1997) Relationships between taxonomic resolution and data transformations in analyses of a macrobenthic community along an established pollution gradient. Mar Ecol Prog Ser 149:1–3

Olsgard F, Somerfield PJ, Carr MR (1998) Relationships between taxonomic resolution, macrobenthic community patterns and disturbance. Mar Ecol 172:25–36

Oohusa T (1992) Recent trends in nori products and markets in Asia. J Appl Phycol 5:155–159

Pearson ES, Hartley HO (1966) Biometrika Tables for Statisticians. Cambridge

Petraitis PS (1983) Grazing patterns of the periwinkle and their effect on sessile intertidal organisms. Ecology 64:522–533

Petraitis PS (1987) Factors organizing rocky intertidal communities of New England: Herbivory and predation in sheltered bays. J Exp Mar Biol Ecol 109:117–136

Sears JR (1998) *NEAS Keys to the Benthic Marine Algae of the Northeastern Coast of North America from Long Island Sound to the Strait of Belle Isle*. Available from the UMass Dartmouth Campus Book Store, Dartmouth, MA

Terlizzi A, Fraschetti S, Guidetti P, Boero F (2002) The effects of sewage discharge on shallow hard substrate sessile assemblages. Mar Pollut Bull 44:544–550

Terlizzi A, Bevilacqua S, Fraschetti S, Boero F (2003) Taxonomic sufficiency and the increasing insufficiency of taxonomic expertise. Mar Pollut Bull 46:556–561

Tseng CK, Fei XG (1987) Economic aspects of seaweed cultivation: Macroalgal commercialization in the Orient. Hydrobiol 151–152:167–172

Warwick RM (1988) Effects on community structure of a pollutant gradient – summary. Mar Ecol Prog Ser 46:207–211

Yarish C, Frankenstein G, Sperr AE, Fei XC, Mathieson AC, Levine I (1997) Domestication of *Porphyra* (=nori) for north-east America. J Shellf Res 16:296–297

Yarish C, Wilkes R, Chopin T, Fei XG, Mathieson AC, Klein AS, Neefus CD, Mitman GG, Levine I (1998) Domestication of indigenous *Porphyra* (nori) species for commercial cultivation in Northeast America. World Aquac 29:26–30

Yarish C, Chopin T, Wilkes R, Mathieson AC, Fei XG, Lu S, Parsons GJ (1999) Domestication of nori for northeast America: The Asian experience. Bull Aquac Assoc Can 99–1:11–17

Gametophyte-sporophyte coalescence in populations of the intertidal carrageenophyte *Mazzaella laminarioides* (Rhodophyta)

César Vera · Paulina Lobos · Héctor Romo

Originally published in the Journal of Applied Phycology, Vol 20, No 5, 433–437.
DOI: 10.1007/s10811-007-9238-8

Abstract *Mazzaella laminarioides* has consistently been reported as a typical coalescent/clump species with a triphasic life history of the *Polysiphonia*-type in which the haploid gametophyte is the predominant phase with respect to the diploid sporophyte. Preliminary observations of intertidal populations revealed that, in some instances, cystocarpic and tetrasporic fronds emerged from the same clump (G-T clumps), implying a coalescent process of haploid and diploid thalli by fusion of their corresponding adjacent basal holdfasts. Population surveys at three sites in Coliumo Bay, central Chile, were carried out to characterize frond demography as well as to asses the frequency of gametophyte-tetrasporophyte (G-T) coalescence. Visual and resorcinol methods were employed to determine the phases of the fronds collected over central transects of 15 randomly sampled clumps. Coalescence of G-T clumps was infrequent, with gametophytes dominating over tetrasporophyte thalli.

Keywords Frond demography · Gametophyte-sporophyte fusion · Gigartinaceae · Intertidal seaweed

Abbreviations

G-T	Gametophyte - Tetrasporophyte
G-G	Gametophyte - Gametophyte
T-T	Tetrasporophyte - Tetrasporophyte
G-T clumps	Gametophyte - Tetrasporophyte clumps
EBM	Marine Biological Station

C. Vera · P. Lobos · H. Romo (✉)
Departamento de Oceanografía, Universidad de Concepción,
Casilla 160-C,
Concepción, Chile
e-mail: hromo@udec.cl

Introduction

Coalescence is a common process in which adjacent germlings fuse together; this occurs in some Rhodophyceae species as well as in other vegetable and animal organisms. The fusion is even more absolute given the connection between secondary pit connections of cells from the two adjacent germlings (Tveter-Gallagher and Mathieson 1980). This fusion results in a clumpy individual composed of more than one genet, with many fronds derived from the two original neighboring spores emerging from a common attachment holdfast. More than 30 species of red algae belonging to seven orders have been identified as having discoidal development followed by the formation of a basal crust and coalescence when two co-specific spores germinate (or two germlings grow) close enough to one another that cells of both germlings come into contact and establish secondary pit connections between them (Santelices et al. 1999). Although coalescence in seaweeds with basal disks from which erect shoots emerge was first documented by Rosenvinge (1931), intensive research on structural interconnections between different coalescent red algae, as well as the physiological and evolutional implications of such complex organisms, was only done after the Maggs and Cheney (1990) review of coalescence in several species and that of Martínez and Santelices (1992), which referred specifically to the coalescence of *Mazzaella laminarioides* (Bory de Saint Vincent). See Santelices et al. (1999) and Santelices (2004) for extensive recent reviews on coalescent macroalgae.

Mazzaella laminarioides (Bory de Saint Vincent) Fredericq is an endemic intertidal Chilean species distributed along the continental shore of Chile from Bosque Fray Jorge National Park (30° 34′ S; 61° 06′W; Vásquez and Vega 2004) to the Diego Ramírez Islands (56° 30′S; 68° 43′ W;

Mansilla and Navarro 2003). Its wide distribution indicates a great thermal tolerance; whereas the mean sea surface temperature in the northern locations fluctuates between 12°C and 17°C (winter-summer), at the Subantarctic limits of the species, it ranges from 4°C to 9°C (winter-summer), with nearly freezing air temperatures during winter low tides.

As all Gigartinaceae species, *M. laminarioides* exhibits a "*Polysiphonia* type" life history, with free-living tetrasporophyte and gametophyte phases and carposporophytes that always live in the cystocarp of the fertilized female gametophyte. Coalescence between *M. laminarioides* germlings has been demonstrated by Santelices et al. (1999, 2004) and in vitro contact between intra and inter-specific haploid and diploid sporelings was reported by Santelices et al. (2003b).

Studies of *M. laminarioides* (as *Iridaea*) at Pelancura (33°35′ S; 71°38′W; Hannach and Santelices 1985) and at Matanza (33° 56′S, 71° 53′W; Santelices and Norambuena 1987) revealed a clear gametophytic dominance in the northern populations. Thornber and Gaines (2003, 2004) also found haploid dominance at 10 sites between Punta Talca (30° 57′ S; 71° 35′W) and Pichilemu (34° 24′ S; 71°59′W). Contrasting with these findings, Westermeier et al. (1987) pointed out a lack of clear dominance between both phases at the more southern site of Mehuin (39°24′S; 73′14′W).

Coalescence among nearby clumps have been studied in the field at Caleta Maitencillo (32° 30′S; 71° 29′W) and Topocalma (34° 05′S; 71° 58′W), and the results suggested that coalescence between *M. laminarioides* clumps could be a frequent process (Santelices et al. 2003a, b, 2004). Visual observations done in Coliumo Bay (H. Romo, personal observations), central Chile, showed a demographic pattern of haploid gametophytic dominance in the *M. laminarioides* population and a likely low frequency of coalescence among gametophytic and tetrasporophytic clumps (biphasic clumps). This is the first report of the frequency of coalescence between gametophytic and sporophyte thalli in the field.

Material and methods

Samples were collected at three sites along the inner shore of Coliumo Bay (36° 32′ S; 72° 56′W) in spring 2004. The sites (Fig. 1) were rocky areas at: i) Litril, with rocky platforms adjacent to an area of boulders in the innermost part of the bay; ii) Marine Biological Station (EBM) at the Universidad de Concepción, where rocky platforms are surrounded by sandy beaches, and iii) Caleta Villarrica, a beach with boulders and cobblestones (with *M. laminarioides* on top) intermingled with gravel on the east coast of the bay (Fig. 1).

Individual clumps of *M. laminarioides* were sampled haphazardly after careful visual examination to be sure of the spatial integrity and continuity of the basal disks. Twenty seven clumps were sampled at Litril and Caleta Villarrica and 30 at EBM, totaling 84 clumps for the three sites. Clump diameter of basal disks varied between 1 and 13 cm and the number of fronds sampled from each clump varied between 7 and 63 depending on the size of the clumps. Each clump was sampled by means of one transect, consisting in a 20 cm straight steel wire, placed trough the fronds, and crossing the center of the main diameter of the clump (Fig. 2). All fronds in contact with one side of the wire were detached in order; from one end of the clump to the other (Fig. 2). The fronds were strung across an iron pin in the same order that they were detached (Fig. 3). The determination of the karyological phase (haploid gametophyte or diploid tetrasporophyte) of the fronds from each clump was carried out by the resorcinol method (Garbary and De Wreede 1988). Thus, the order of the frond sequence analyzed from each pin coincided with the sampled frond sequence of each clump in the field. In the paper, clumps composed exclusively of gametophytes are identified as G-G, those consisting only of tetrasporophyte fronds as T-T, and biphasic clumps (sequences of gametophytes followed by tetrasporophyte fronds or vice-versa) as G-T (Fig. 3). In this paper, and based on its "*Polysiphonia* type" life history, we have assumed that gametophytes (resorcinol + fronds) are haploids and tetrasporophytes (resorcinol - fronds) are diploids notwithstanding nobody have elucidate the karyotype of both phases in this species.

Results

The karyological phase composition displayed by the fronds differed to some degree at the three sites. Whereas the T-T

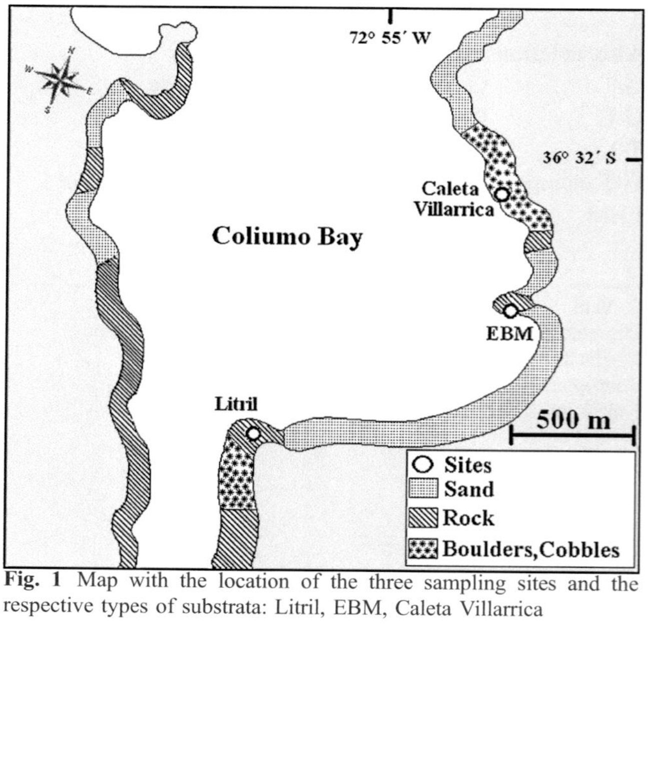

Fig. 1 Map with the location of the three sampling sites and the respective types of substrata: Litril, EBM, Caleta Villarrica

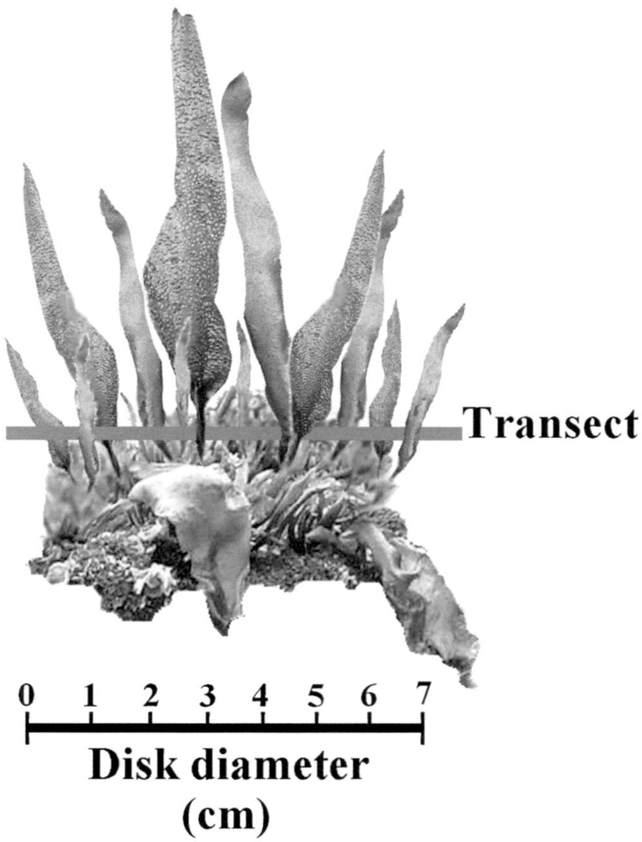

Fig. 2 Disposition of the fronds of *Mazzaella laminarioides* along transect on the basal crust

sequence was low but surprisingly identical for clumps at all sites (Table 1), the G-G sequence varied from 14 clumps at Litril to 21 at EBM. The G-T sequence was similar at Litril and EBM, but not observed at all at Villarrica (Table 1). The frond composition sequence of the pooled clumps for all three sites was 60.7% G-G, 21.4% T-T, and 17.9% G-T (Table 2), for a ratio of 6:2:2 (χ^2=0.286; 0.90<P<0.75).

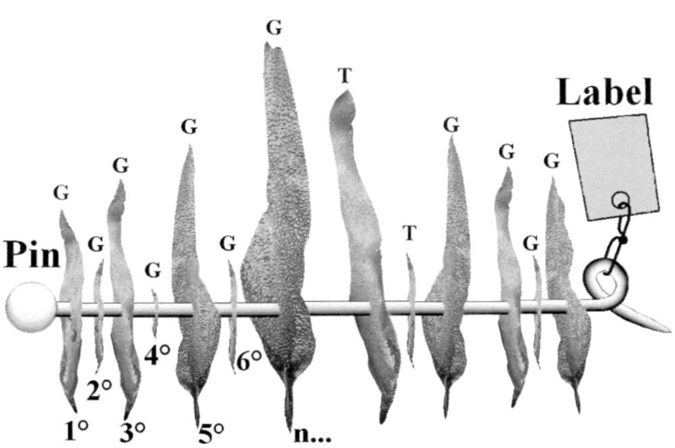

Fig. 3 Disposition of the fronds of *Mazzaella laminarioides* obtained from the basal disk in the pin

Table 1 Number of clumps and percentage of gametophytic (G-G), tetrasporophytic (T-T), and gametophytic-tetrasporophytic (G-T) frond sequences in clumps of *Mazzaella laminarioides* in Coliumo Bay

Type of clumps	Litril	EBM	C. Villarrica	Total	%
G-G	14	16	21	51	60.7
T-T	6	6	6	18	21.4
G-T	7	8	0	15	17.9
Total clumps	27	30	27	84	

Clumps with biphasic sequences were found only at Litril and EBM, and most of these samples had a pattern of the type: G-G-...-G-T-T-...-T or, inversely, T-T-...-T-G-G-...-G, depending on the end of the clump diameter where we began the sampling. Only one sample from EBM and two from Litril had mixed biphasic sequences of the G-G-T-G-...-G-T-T-T type, that is, at least one tetrasporophyte (or gametophyte) frond was found among a series of fronds of the opposite phase. Clumps having both mature cystocarpic and tetrasporic fronds were not frequent during the study period (spring 2004); only one such clump was observed at Litril and two at EBM (3.6% of all 84 clumps analyzed).

The analysis of the proportions of gametophytic and tetrasporophytic fronds showed a general strong dominance (71.5%) of haploids (Table 2); this did not differ significantly from the 76% reported by Thornber and Gaines (2004) for northern sites (χ^2=0.666; 0.75>P>0.50). In the three stands, the G: T ratios were always greater than 1 (Table 2), reaching 3.66 at Caleta Villarrica and 2.50 for the pooled data from all three sites.

Discussion

This research confirmed the preliminary assumption about natural field coalescence between karyologically different individuals of *M. laminarioides*, which was based on preliminary field observations of clumps that exhibit both cystocarpic and tetrasporic fronds (Fig. 4).

Of the 84 clumps analyzed, the G-T frond sequences reached 17.9%, similar to the 21.4% observed for T-T sequences; the statistical similarity between these two

Table 2 Number of fronds of each phase and G: T ratio of *Mazzaella laminarioides* clumps in the sampled transects in Coliumo Bay

Type of fronds	Litril	EBM	C. Villarrica	Total	%
Gametophyte	208	306	315	829	71.5
Tetrasporophyte	111	134	86	331	28.5
Total fronds	319	440	401	1160	100
G : T ratio	1.87	2.28	3.66	2.50	

Fig. 4 Coalescent clump in the field showing a mature cistocarpic frond (G) and a tetrasporic (T) one

sequences suggest a very close relationship between G-T coalescence and the amount of diploid tetrasporophytic fronds at nearby sites. However, no reasonable explanation was found for the absence of G-T clumps at Caleta Villarrica. The G-T frequency at this site could have been so low that it was not detected by the sampling, although there were no obvious causal factors - biological or environmental - explaining the G-T absence. The most visible attributes of the shore at Caleta Villarrica are its unstable, rocky substrate and the very small basal disk of clumps growing on top of the boulders (up to 5 cm diameter in our sampling). The site is composed mainly of boulders and stones intermingled with coarse sand and gravel. Periodically, strong winter storms (3–4 per year) create surges between the boulders. The possible interference on coalescence by frequent collisions by stones on boulders during storms could be producing more frequent disruption of basal disks. This fact could be the explanation for the absence or low frequency of G-T sequences and the small size of the disks, but this study does not have available data to sustain this hypothesis. The harvest pressure is rather similar for the three sites because the population of hand harvesters that operate during spring low tides is the same and no great differential pressure seems to exist between sites. A survey of the interaction between substrate stability/instability and basal disk size at sites with and without unstable rocky substrata could be one way to help decipher this result. The G-T coalescence pattern between karyologically different *M. laminarioides* partners mostly showed fronds with G-G-G-G-...-T-T-T type arrangements and, less frequently, G-G-G-T-G-...T-T. In the first case, coalescence was the product of two karyologically different spores (or two karyologically different group of spores) that initially germinated and grew as two spatially segregated groups. The growth was such that, after the fusion of their basal disks, one or several tetraspores developed G-G-G-G... sequence of fronds in one side of the crust, whereas the other side of the disk was the result of the germination of one or several carpospores giving a...T-T-T sequence of fronds. In this case a natural bifacial chimera with two different karyologies occurs. In the first case, clumps composed of disorderly G-T frond sequences (*e.g.* G-G-G-**T**-G-...T-T) constituted a multifacial chimera, also with fronds of different karyologies, but generated by groups of carpospores and tetraspores that germinated spatially intermingled.

This is a general case of chimerism in *M. laminarioides* but a more subtle one was reported by Santelices (2004, see Fig. 5 of cited paper) using PCR RAPD analysis in a *M. laminarioides* gametophyte clump. In that study, three genetically different kinds of fronds were detected when eight haploid fronds were sampled from the same clump. The same chimerical process could be occurring in both T-T and G-T clumps.

On the other hand, the low observed frequency of biphasic clumps reflected an inequality of phases such that, at least in its northern distribution, the species is always strongly gametophyte dominant (Santelices and Norambuena 1987; Thornber and Gaines 2003, 2004). Variations in the proportion of phases may be found in the more southern populations, as was observed at Mehuin, Valdivia Province (39° 24′S; 73° 14′W), approximately 300 km south of Coliumo Bay, where Westermeier et al. (1987) found no differences between phase proportions. Moreover, the species' demography is absolutely unknown along the 1350 km from Mehuin to the Diego Ramírez Islands (56° 30′S; 68° 43′), the site reported as the southern distributional limit (Mansilla and Navarro 2003). In order to determine the homogeneity or heterogeneity of the population along the species' distributional range, further field studies on the proportion of phases and intra and interphase frequency of coalescence are required in the Subantarctic areas. Here, the species lives in cold environmental conditions that come close to freezing during winter low tides, receiving higher UV radiation and suffering extreme winter-summer photoperiods.

No available data on the temporal variation of G-T coalescence exists, but it can be assumed as a hypothesis that due the perennial quality of the basal disks no great changes could be at least expected in small temporal scale.

As an overall conclusion, in these kinds of clumpy algae, the concept of a thallus as an individual composed of an encrusting holdfast with many fronds of the same origin is clearly misleading and can not be applied to coalescent species because a clump is not always a genetically homogeneous individual.

Acknowledgments This research was carried out with funds provided by the Facultad de Ciencias Naturales y Oceanográficas and the Dirección de Investigación, Universidad de Concepción. We thank Mr. Carlos Jara for improving our original figures and to Danielle Barriga for her kind corrections on syntaxes and grammar and two anonymous reviewers for their helpful comments.

References

Garbary DJ, De Wreede RE (1988) Life history phases in natural populations of Gigartinaceae (Rhoodophyta): quantification using resorcinol. In: Lobban CS, Chapman DJ, Kremer BP (eds) Experimental phycology. A laboratory manual. Cambridge University Press, Cambridge, pp 174–178

Hannach G, Santelices B (1985) Ecological differences between the isomorphic reproductive phases of two species of *Iridaea*. Mar Ecol Prog Ser 22:291–303

Maggs CA, Cheney DP (1990) Competition studies of marine macroalgae in laboratory culture. J Phycol 26:18–24

Mansilla A, Navarro N (2003) Contribución a la flora ficológica de Islas Diego Ramírez. In: Mansilla A, Werlinger C, Navarro N (eds) Macroalgas en ambientes subantárticos. Ediciones Universidad de Magallanes, Punta Arenas, pp 85–89

Martínez E, Santelices B (1992) Size hierarchy and −2/3 "Power law" relationship in a coalescent seaweed. J Phycol 28:259–264

Rosenvinge LK (1931) The marine algae of Denmark, contributions to their natural history. Part IV: Rhodophyceae IV: (Gigartinales, Rhodymeniales, Nemastomatales). K Dan Vidensk Selsk Skr 7,4:489–628

Santelices B (2004) A comparison of ecological responses among aclonal (unitary), clonal and coalescing macroalgae. J Exp Mar Biol Ecol 300:31–44

Santelices B, Norambuena R (1987) A harvesting method for *Iridaea laminarioides* in central Chile. Hydrobiologia 151/152:329–333

Santelices B, Correa JA, Aedo D, Flores V, Hormazábal M, Sánchez P (1999) Convergent biological processes in coalescing Rhodophyta. J Phycol 35:1127–1149

Santelices B, Aedo D, Hormazábal M, Flores V (2003a) Field testing of inter-and intraspecific coalescence among mid-intertidal red algae. Mar Ecol Progr Ser 250:91–103

Santelices B, Correa JA, Hormazábal M, Flores V (2003b) Contact responses between spores and sporelings, karyological phases and cystocarps of coalescing Rhodophyta. Mar Biol 141:381–392

Santelices B, Hormazábal M, Correa J, Flores V (2004) The fate of overgrown germlings in coalescing Rhodophyta. Phycologia 43:346–352

Thornber CS, Gaines SD (2003) Spatial and temporal variation of haploids and diploids in populations of four congeners of the marine alga *Mazzaella*. Mar Ecol Progr Ser 258:65–77

Thornber CS, Gaines SD (2004) Population demographics in species with biphasic cycles. Ecology 85:1661–1674

Tveter-Gallagher E, Mathieson AC (1980) Sporeling coalescence in *Chondrus crispus* (Rhodophyceae). J Phycol 12:110–118

Vásquez JA, Vega JMA (2004) Ecosistemas marinos costeros del Parque Nacional Fray Jorge. In: Squeo FA, Gutiérrez JR, Hernández IR (eds) Historia Natural del Parque Nacional Bosque Fray Jorge. Ediciones Universidad de la Serena, La Serena, Capitulo 13:235–252

Westermeier R, Rivera P, Chacana M, Gómez I (1987) Biological basis for the management of *Iridaea laminarioides* Bory in southern Chile. Hydrobiologia 151/152:313–328

Growth and survival performance of the gametophyte of *Gigartina skottsbergii* (Rhodophyta, Gigartinales) under defined nutrient conditions in laboratory culture

A. Mansilla · M. Palacios · N. P. Navarro · M. Avila

Originally published in the Journal of Applied Phycology, Vol 20, No 5, 439–446.
DOI: 10.1007/s10811-007-9279-z

Abstract *Gigartina skottsbergii* is a commercially important carrageenan producer that has been suffering severe extraction pressure in Chile's Magellan Region and Cape Horn Archipelago since 1998. In order to create baseline information for its cultivation and repopulation, we studied the effects of agricultural fertilizers on growth of *G. skottsbergii* early developmental stages. The culture media utilized were: a) seawater + Bayfoland, b) seawater + Superphosphate, c) seawater + Urea, d) seawater + Provasoli and e) seawater as a control. The culture conditions were: a) 12L:12D photoperiod; b) temperature 8±1°C and c) irradiance at 45 μmol photons m^{-2} s^{-1}. After 60 days, higher relative growth rates between treatments were observed; the treatments that included Bayfoland and Provasoli showed greater growth (382±55 and 378±50 μm, respectively,) compared to Superphosphate (88±16 μm), control (78±10 μm) and Urea (70±11 μm) treatments, after 81 days. The Urea treatment and the control had inhibitory effects on *G. skottsbergii* germlings growth and survival, as evidenced by progressive loss of pigmentation and death after 60 days. These results showed that Bayfoland was an excellent alternative to develop cultures.

A. Mansilla (✉) · M. Palacios · N. P. Navarro
Departamento de Ciencias y Recursos Naturales,
Facultad de Ciencias, Universidad de Magallanes,
casilla 113-D,
Punta Arenas, Chile
e-mail: andres.mansilla@umag.cl

M. Avila
Instituto de Ciencia y Tecnología, Universidad Arturo Prat,
Puerto Montt, Chile
e-mail: marcela.avila@unap.cl

Keywords *Gigartina skottsbergii* · Growth · Nutrient requirements

Introduction

Over the last several years, there has been a global increase in the demand for red algae in the carrageenan and agar industries (Bixler 1996; Zemke-White and Ohno 1999; Buschmann et al. 2001; McNeill et al. 2003; Buschmann et al. 2004). About 20% of the world demand for carrageenan has been supplied by extractive industries in Chile (Avila et al. 2001a, b). However, these local carrageenan industries depend entirely on the harvesting of natural beds of *Gigartina skottsbergii* (Buschmann et al. 2001), and other carrageenan producing red algae such as *Sarcothalia crispata* and *Mazzaella laminarioides*; all are harvested as a single group, known as "lugas" (Marin et al. 2002; Avila et al. 2001a, b). Since, at the global level, the cultivation of commercially important algae represents nearly 22% of all red algae used (Zemke-White and Ohno 1999; McNeill et al. 2003), culture techniques and repopulation methods are an important alternative for the recolonization of natural algae beds to ensure the supply of primary material for the hydrocolloid industry.

Due to its high level of carrageenan content (>70% of dry weight) (Piriz and Cerezo 1991), *Gigartina skottsbergii* is considered to be the most important carrageenan producing red alga commercially in Chile. Export volumes of carrageenan have increased from 26 tons in 1989 to 3,772 ton in 2005, totally US$ 29.5 million in value. The Magellan Region contributes approximately 9% of the total biomass produced in Chile and has over the last five years become the third most important port of landing for the resource (Zamorano, Pers. comm.). There exists a great internal demand and significantly increasing extraction of the species, which constitutes an alternative resource for artisenal fisheries, especially during the closed seasons for sea urchin and king crab and during outbreaks of red tide, a characteristic phenomenon of the region (Avila et al. 2001b). This situation has provoked an over-exploitation and in some cases the collapse of natural populations (Avila et al. 1999; Buschmann et al. 2004; Romo et al. 2006). The growing demand for this resource and the apparent deterioration of natural populations makes evident the importance of developing a production technology to allow the acceleration of some aspects of the developmental process, such as growth (Avila et al. 1999; Romo et al. 2006). In addition, the recently declared Cape Horn Biosphere Reserve has as a primary objective the establishment of research and demonstration projects that focus on the sustainable use of biodiversity for the betterment of social wellbeing in the archipelago (Rozzi et al. 2006).

Several endeavours on the part of researchers from diverse institutions, financed through a variety of sources, have allowed the generation of background information on the ecology, biology, physiology, and cultivation of *G. skottsbergii* in its distribution south of 40°S (Westermeier et al. 1999; Buschmann et al. 1999, 2001; Avila et al. 1999, 2003; Romo et al. 2001, 2006; Buschmann et al. 2004; Faugeron et al. 2004; Martínez et al. 2005). However, no prior studies exist on the effects of agricultural fertilizers on the development of *G. skottsbergii* in controlled systems. The objective of this study, therefore, was to determine the effect of traditional agricultural fertilizers on the growth and survival rates of the initial developmental stages of *G. skottsbergii* under laboratory conditions, which renders production more efficient, lowers costs associated with repopulation, and develops techniques valuable for the sustainable use of this resource.

Material and methods

Mature tetrasporic fronds of *G. skottsbergii* (Fig. 1) were collected by scuba diving from natural populations at Santa Ana Point (53°18'11.4"S; 70°23'30, 5"W), 70 km southwest from the city of Punta Arenas on the Strait of Magellan (Fig. 2). The material was transported in seawater inside hermetic containers to the laboratory at the Center for Aquaculture and Marine Resources, Department of Natural Sciences and Resources, University of Magallanes (UMAG). Once in the laboratory the material was washed with filtered, sterilized seawater to eliminate epiphytes and epibionts that could affect the subsequent cultures.

Tetraspore release

Liberation of tetraspores was induced by stress through dehydration of the fronds at ambient temperature for approximately 45 min, after which the fronds were

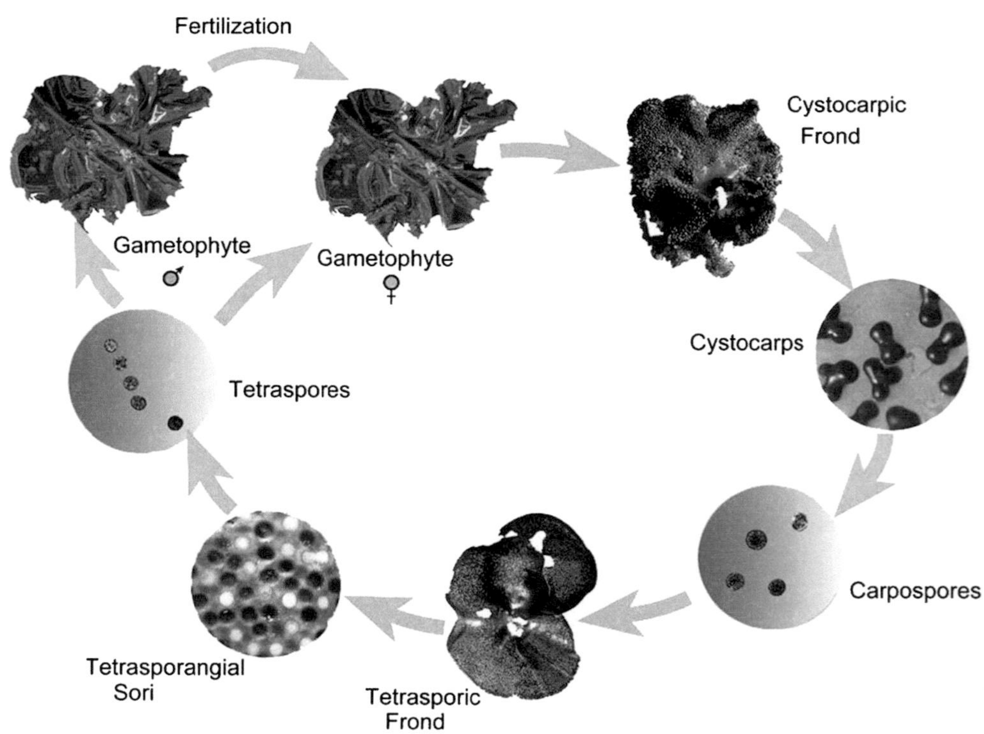

Fig. 1 The life history of *Gigartina skottsbergii*. (Mod. Romo et al. 2004)

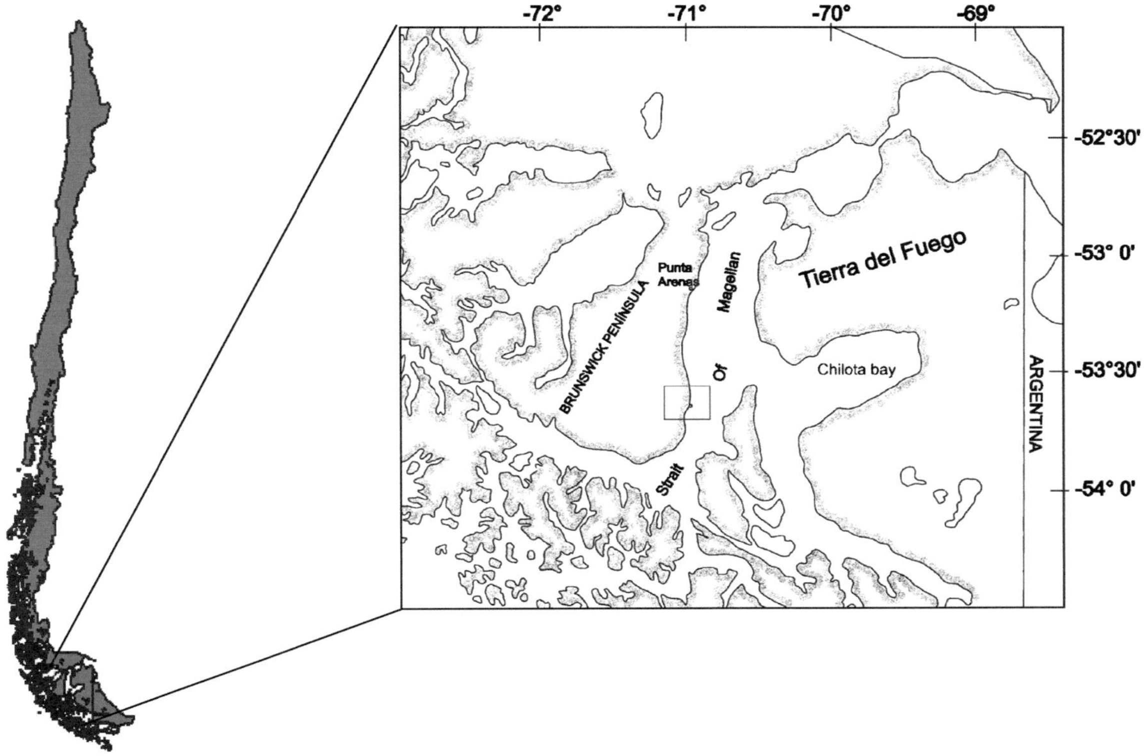

Fig. 2 Geographic location of collection site of *Gigartina skottsbergii* at Santa Ana Point, Strait of Magellan, Chile

rehydrated in sterilized and filtered (0.2 μm) seawater. Once the tetraspores were released, the fronds were removed from the water and the resulting suspension was filtered through a sieve in order to catch the remaining tissues and impurities, obtaining a homogenous density of approximately 2,000 tetraspores mL^{-1} to use as an inoculation solution on glass microscope slides.

Cultivation methods used

Five treatments were performed: three commercial fertilizers (Bayfoland 250SL, Superphosphate and Urea, Table 1), a laboratory medium (Provasoli) and a control (seawater without additives). Superphosphate and Urea were used as phosphate and nitrogen sources, respectively. Both are sold in their solid state, and for this reason it was necessary to prepare stock solutions. The concentrations of each nutrient were adjusted to the corresponding values of nitrogen and phosphorus as has been used in *Gracilaria chilensis* (Alveal et al. 1991): 0.06 g $L^{-1}$of sodium nitrate (10 mg L^{-1} of nitrogen) and 0.01 g L^{-1} of sodium phosphate (3 mg L^{-1} of phosphorus). Bayfoland is a liquid commercial foliar fertilizer and was used at a concentration of 0.1 ml L^{-1}, corresponding to 1/3 of the required minimum concentration for the cultivation of initial growth stages. The quantities of different agricultural fertilizers required to achieve the indicated concentrations of nitrogen and phosphorus are shown in Table 2.

The cultures were maintained inside a temperature-controlled room at 8±1°C and 45 μmol photons m^{-2} s^{-1} PAR (provided by Philips TLT 20W/54 daylight fluorescent tubes) at 12:12h light-dark cycle. The culture media for each treatment was renewed weekly.

Table 1 Commercial entities, chemical composition and components of the nutrient supply used in experimental treatments

Fertilizers	Commercial entities	Chemical composition	Components
Superphosphate (45% $P_2 0_5$)	Jarditec	$Ca(H_2PO_4)_2$	Calcium Phosphate, Phosphoric acid
Urea (45% Nitrogen)	Jarditec	$(NH_2)_2CO$	Diamid Carbonic acid
Bayfolan	Bayer	N:11%, P_2O_5:8%, K_2O.6% (Soluble concentrate)	Mineral elements (N, P_2O_5, K_2O, Fe, Mn, B, Cu, Zn, Ni, Co, Mo, Cl, Na, S.)

Table 2 Concentration of fertilizer additives in culture media used in experimental treatments

Fertilizer (treatments)	Preparation of stock solution	Concentration used
Superphosphate	7 g in 500 mL distilled water	5 $mL.L^{-1}$
Urea	Romo and Paula (1995) without Tri-phosphate	5 $mL.L^{-1}$
Bayfoland	Bayer	0.1 $mL.L^{-1}$
Provasoli	4.3 g in 500 mL distilled water	20 $mL.L^{-1}$

Experimental design

Slides inoculated with tetraspores were placed in 250 mL capacity Petri dishes. Fifteen experimental units (dishes) were prepared, distributed in the 5 treatments of 3 replicates each. The effects of the distinct nutrient solutions on the development of *G. skottsbergii* tetraspores were evaluated through the mortality and growth of 45 individuals of each experimental unit. Weekly and during the 81 days of cultivation, 45 individuals were photographed using a microscope integrated imaging system connected to an image analyzing software (Imagen Pro-Plus version 4.1), providing data on the increase in diameter of discs of *G. skottsbergii*.

Growth rate and data analyses

Increase of basal diameter was used as the parameter for estimating growth rates. The relative growth rate of *G. skottsbergii* discs was estimated using the formula proposed by Hansen (1980):

$$RGR(\%) = \left(100 \times \ln\left(t_f/t_0\right)\right)/T$$

Where:

RGR (%)	relative growth rate expressed in daily percentage
t_0	initial diameter of disc (μm)
t_f	final diameter of disc (μm)
T	time in days.

Subsequently, the values for the growth rates and mortality percentages were transformed by applying arcsine (Zar 1999) to achieve normal distributions and then subjected to an analysis of variance analysis (ANOVA), using the software program Statistica 5.0. When significant differences were found, a Newman-Keul test was applied *a posteriori* (Zar 1999). All conclusions were based on a 95% confidence level ($p<0.05$).

Results

Growth rates

Gametophytes of *G. skottsbergii* cultivated with different nutrient solutions showed different responses in growth during their initial developmental stages. At the end of day 81, the greatest increase in disc diameter was obtained with Bayfoland (381.87±54.99 μm) and Provasoli (377.88± 49.92 μm). Growth was less in the treatments with

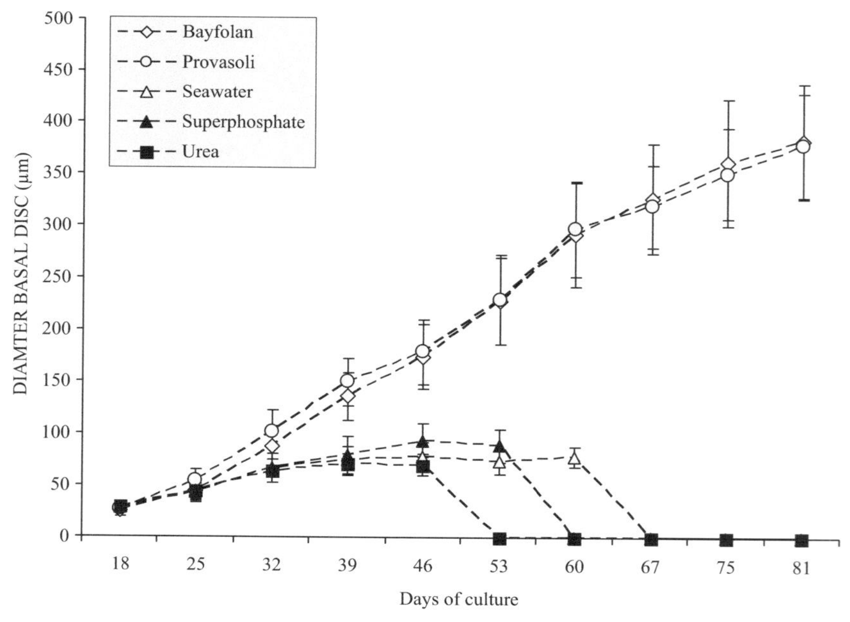

Fig. 3 Diameter increase (μm ± SD) of *G. skottsbergii* discs during 81 days of culture using different nutrient solutions

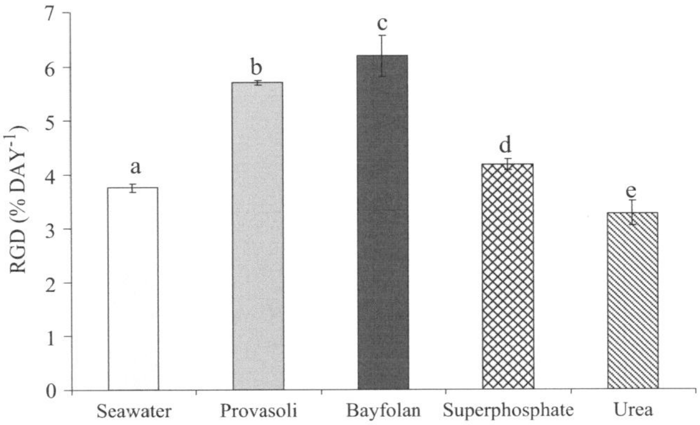

Fig. 4 Daily growth rate (% day^{-1}±SD) of *G. skottsbergii* discs during the first 35 days of the study in different treatments. Different letters refer to significant differences between mean values with *Newman-Keuls* post-hoc test ($p<0.05$)

Superphosphate (88.36±16.27 µm), in the control solution (77.65±9.93 µm) and with Urea (70.06±10.88 µm) (Fig. 3).

Therefore the highest growth rates were observed in the gametophytes cultivated with Bayfoland (6.19±0.38% day^{-1}) and Provasoli (5.70±0.035% day^{-1}), whereas the lowest growth rates were observed in the control and Urea treatments (3.76±0.07% day^{-1} and 3.27±0.23% day^{-1}, respectively) (Fig. 4). The differences between all the treatments utilized were statistically significant ($p=0.000$) (Table 3).

The *a posteriori* test applied to the interaction between the different treatments used and the time of cultivation, considering weekly periods, showed that during the first week, the gametophytes cultured with Provasoli had a higher growth rate than those under the other nutrient regimes (Fig. 5). During the second week, significant differences were observed in growth under Provasoli and Superphosphate treatments ($p=0.021$) and Urea ($p=0.026$) (Fig. 5, Table 4).

In the third week, the differences observed between the treatments were only in Bayfoland v/s the other cultures (control, $p=0.043$; Superphosphate, $p=0.028$; Urea, $p=0.000$) (Fig. 5, Table 4). During the fourth week, the Provasoli and Bayfoland treatments showed significant differences in growth rate related to the other treatments (Fig. 5, Table 4).

Finally, beginning in the fifth week of the study, differences in growth rates became evident, mainly between the treatments Provasoli and Bayfoland compared to control, Superphosphate and Urea treatments (Fig. 5, Table 4). The *a posteriori* test indicated that the differences derived mostly from the control (least growth) versus Bayfoland ($p=0.033$) and Provasoli ($p=0.021$), Urea versus Provasoli ($p=0.011$) and Bayfoland ($p=0.016$), which showed the greatest growth at the end of the study (Table 4). The variance test used on the weekly growth rates showed that the nutrient solutions used, the time under cultivation, and the interaction between them, acted in a synergistic manner on the growth of the gametophytes.

A more detailed analysis of the behavior of the growth rates of different treatments (control, Provasoli, Bayfoland, Superphosphate, and Urea) during the five weeks of the study indicated that, although the growth rates of gametophytes cultivated with Bayfoland and Provasoli did not show significantly greater growth during the first 7 days of cultivation, they began in the second and third weeks to show better developmental responses in relation to the other treatments, recording rates of greater than 8% daily growth as compared with growth rates that did not surpass 6% in the other treatments. ANOVA carried out on the growth results of the Provasoli and Bayfoland treatments, the solutions that noticeably promoted the growth of *G. skottsbergii* gametophytes, showed no statistically significant difference ($p=0.431$) between the growth rates of these two treatments (Table 3).

Survival

Beginning with day 39 and day 46, respectively, mortality was observed in the Urea (45%) and control treatments (11%). At 67 days of cultivation, all of the gametophytes cultured under the control treatment, Superphosphate, and Urea had died. The treatments using Provasoli and Bayfoland had disc survival rates of 100% on the 81st (final) day of the experiment ($p=0.000$) (Table 3). The *a posteriori* test indicated that the observed differences were between the treatments of Bayfoland and Provasoli ($p=1.00$) and cultures enriched with Superphosphate ($p=0.000$) and Urea ($p=0.000$) (Table 5).

Table 3 Analysis of variance (ANOVA) for the weekly growth rate, mortality and different enrichment solutions of *G. skottsbergii* gametophytes, cultured with different treatments, during the first 35 days. Treatments: Provasoli (Provasoli), Bayfoland (Bayfolan). (*) Indicates significant differences ($p<0.05$)

	SS Effect	df	MS Effect	SS Err	df Err	MS Err	F	p
RGD (% day^{-1})	19.23	4	4.81	0.43	10	0.04	111.37	0.000*
Mortality (%)	8700.27	4	2175.07	81.05	10	8.10	268.37	0.000*
Provasoli/ Bayfolan	0.001	1	0.001	0.005	4	0.001	0.765	0.431

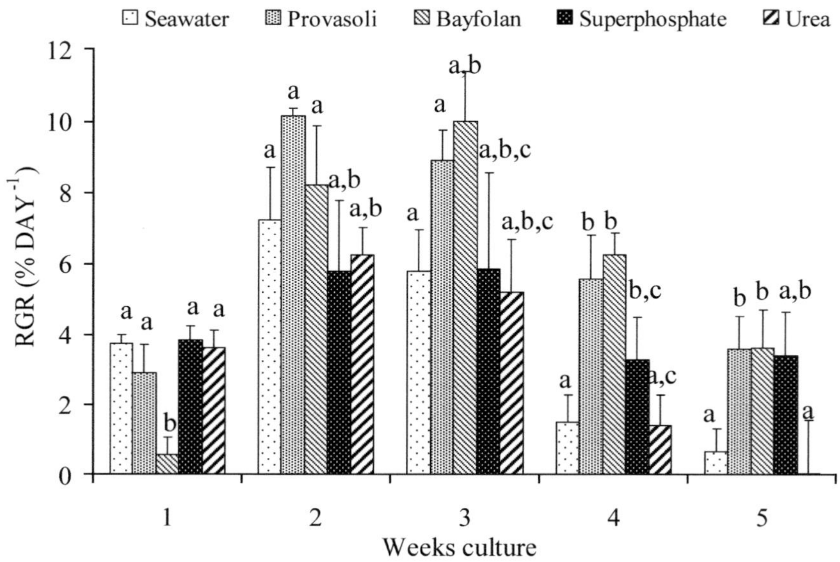

Fig. 5 Growth rates (% day^{-1}±SD) of *G. skottsbergii* disks during the first five weeks of culture (n=3). (*) Indicates significant differences (p<0.05)

Discussion

Physiological explanations for differential response to cultivation media

There is a tendency to think of all algae as having similar nutritional requirements, but the heterogeneity of seaweeds makes such generalization difficult (Lobban and Harrison 1994). A wide variety of physiological and biological factors, such as inter-seaweed variability, nutritional history, type of tissue, life history stage, age, the surface area to volume ratio of the thallus, and morphology (Harrison et al. 1986; Harrison and Hurd 2001), will influence growth and uptake of nutrients.

Nitrogen and phosphorus are two nutrients that limit seaweeds growth and yields in most natural environments (Harrison and Hurd 2001). Nitrate and ammonia are important environmentally available nitrogen sources for the growth of marine algae. After these ions are absorbed from the environment, they are generally incorporated into amino acids and proteins, or used in the synthesis of pigments such as phycoerithrin, which is considered to be a store of available nitrogen (Lapointe 1985; Ryther et al. 1981). However, in many marine algae, nitrogen in the form of ammonia is frequently absorbed and assimilated more rapidly than nitrate and Urea (De Boer 1981). At the same time, it has also been found that Urea is an excellent source of nitrogen for many marine algae (Nars et al. 1968; Probyn and Chapman 1982; Thomas et al. 1985). However, experiments carried out by Buschmann et al. (2004), showed that *G. skottsbergii* achieves most of its growth rate with the addition of ammonia, rather than nitrate.

Table 4 A Newman- Keuls *a posteriori* test of the weekly growth rates of *G. skottsbergii* gametophytes, cultured in different enrichment solutions, during the first 81 days. Treatments: Seawater (S), Provasoli (P), Bayfoland (B), Superphosphate (S), Urea (U). (*) Indicates significant differences (p<0.05)

Treatments/ weeks	1	2	3	4	5
Seawater x Provasoli	0.154	0.065	0.094	0.001*	0.021*
Seawater x Bayfolan	0.005*	0.400	0.043*	0.000*	0.033*
Seawater x Superphosphate	0.836	0.444	0.953	0.049	0.081
Seawater x Urea	0.155	0.400	0.676	0.899	0.489
Provasoli x Bayfolan	0.001	0.120	0.441	0.422	0.967
Provasoli x Superphosphate	0.233	0.021	0.045	0.015*	0.212
Provasoli x Urea	0.022	0.026	0.078	0.001*	0.011*
Provasoli x Superphosphate	0.000*	0.204	0.028*	0.009*	0.390
Provasoli x Urea	0.000*	0.233	0.032*	0.000*	0.016*
Superphosphate x Urea	0.099	0.706	0.877	0.091	0.057

Table 5 Newman-Keuls *a posteriori* test for mortality (%) of *G. skottsbergii* gametophytes, cultured in different enrichment solutions, during the first 5 weeks. Treatment: Seawater (S), Bayfoland (B), Superphosphate (S), Urea (U). (*) Indicates significant differences (p<0.05)

Treatments	Provasoli	Bayfolan	Superphosphate	Urea
Seawater	0.000*	0.000*	0.281	0.002*
Provasoli	—	1.000	0.000*	0.000*
Bayfolan	—	—	0.000*	0.000*
Superphosphate	—	—	—	0.005*

In our experiments, *G. skottsbergii* germlings grew more rapidly when cultivated in a solution of Bayfoland 250 SL and Provasoli, while growth rates in Urea and Superphosphate were significantly lower. These differences may be due to the chemical composition of each of the nutrient solutions used. A comparison of the solutions of Urea and Superphosphate with those of Bayfoland and Provasoli reveals that the latter solutions contain a broad variety of macro and micronutrients, namely all those required for the development of plants. In fact, the Provasoli solution was formulated specifically for the enrichment of media for cultivation of seaweeds and likely for this reason produced favorable growth in comparison with Urea and Superphosphate. On the other hand, Bayfoland has been formulated for the growth and development of terrestrial plants, and contains macro (N, P and K) and micronutrients (Fe, Cu, Mo, Ni, Mn and Bo among others), thus preventing nutritional problems in plants and making it also an effective medium for the cultivation of *G. skottsbergii*.

For their part, Urea and Superphosphate provide N (46%) and P (48%). Urea is an excellent N source for some seaweeds (e.g., kelp), but other seaweeds show reduced growth on Urea (De Boer 1981; Navarro-Angulo and Robledo 1999). While it is true that P and N are considered essential limiting elements for growth of algae (Lobban and Harrison 1994), in this study, their addition via Urea and Superphosphate did not increase growth rates in the *G. skottsbergii* germlings, possibly due to the absence of other elements which are present in Bayfoland and Provasoli.

This decrease in growth in two treatments, even in the presence of limiting nutrient, could be due to saturation of the uptake capacity for N and P. Some seaweeds do not show such saturation at high environmental nutrient concentration (as described by the Michaelis-Menten Equation), and their uptake rate increases linearly with high nutrient additions (Harrison et al. 1986; Taylor et al. 1998). However, here *G. skottsbergii* may need other elements and nutrients or have difficulty absorbing and assimilating Urea and Superphosphate. Nitrate is not as easily assimilated as ammonia, and there is a waste of energy in its capture and assimilation (Hurd 2000). Also, since our experiments were done without aeration, it must be considered that water motion is extremely important for seaweeds in natural conditions, since it determines the thickness of velocity and diffusion boundary layers (DBL's) around the thallus and hence the movement of ions and gases to and from the thallus surface (Hurd 2000). Nonetheless, an adequate nutrient supply is important in the early stages (germlings) of *G. skottsbergii* development, as it determines future growth and reproduction processes.

Acknowledgements We gratefully acknowledge Dr. Christopher Anderson for valuable suggestions and improvements of this manuscript. The manuscript also has been improved with the suggestions of two anonymous reviewers. This work was financed in part by Project PRN 00251 (University of Magallanes) to AM and Project Fondef D03I1164 and Project 010201010085 (University Arturo Prat) to MA.

References

Avila M, Candia A, Núñez M, Romo H (1999) Reproductive biology of *Gigartina skottsbergii* (Gigartinaceae, Rhodophyta) from Chile. Hydrobiologia 398/399:149–157

Avila M, Pavez H, Candia A, San Martín R, Cáceres J (2001a) Effects of temperature and Irradiance on the germination and growth of *Sarcothalia crispata* in southern Chile. In: Chapman ARO, Anderson RJ, Vreeland VJ, Davison I (eds) Proceedings of the 17th International Seaweed Symposium. Oxford Univ Press, pp 27–33

Avila M, Candia A, Romo H, Pavez H, Torrijos C (2001b) Exploitation and cultivation of *Gigartina skottsbergii* in southern Chile. In: Chapman ARO, Anderson RJ, Vreeland VJ, Davison I (eds) Proc 17th International Seaweed Symposium. Oxford Univ Press, pp 137–143

Avila M, Candia A, Romo H, Pavez H, Torrijos C (2003) Exploitation and cultivation of *Gigartina skottsbergii* in southern Chile. In: Chapman ARO, Anderson RJ, Vreeland VJ, Davison IR (eds) Proc 17th International Seaweed Symposium, pp. 137–143

Bixler JH (1996) Recent developments in manufacturing and marketing carrageenan. Hydrobiology 362/327:35–57

Buschmann AH, Correa JA, Westermeier R (1999) Recent advances in the understanding of the biological basis for *Gigartina skottsbergii* Rhodophyta cultivation in Chile. Hydrobiologia 398/399: 427–434

Buschmann AH, Correa J, Westermeier R, Hernandez M, Norambuena R (2001) Red algal farming in Chile: a review. Aquaculture 194:203–220

Buschmann AH, Varela D, Cifuentes M, Hernandez Gonzalez MC, Henriquez L, Westermeier R, Correa JA (2004) Experimental indoor cultivation of the carrageenophytic red alga *Gigartina skottsbergii*. Aquaculture 241:357–370

Conway HL, Harrison P, Davis CO (1976) Marine diatoms grown in chemostats under silicate or ammonium limitation. II. Transient response of *Skeletonema costatum* to a single addition of the limiting nutrients. Mar Biol 35:187–199

De Boer JA (1981). Nutrients. In: Lobban CS, Wynne MJ (eds) The Biology of seaweeds. Oxford Blackwell Scientific, pp 356–391

Faugeron S, Martinez EA, Correa J, Cardenas L, Viard F, Destombe C, Valero M (2004) Reduced genetic diversity and increased population differentiation in peripheral and overharvested populations of *Gigartina skottsbergii* (Rhodophyta, Gigartinales) in southern Chile. J Phycol 40:454–462

Hansen JE (1980) Physiological considerations in the mariculture of red algae. In: Abbott IA, Foster MS, Eklund LF (eds) Pacific Seaweed Aquaculture. Publications of the California Sea Grant. Coll. Program, La Jolla, pp 80–91

Harrison PJ, Hurd CL (2001) Nutrient physiology of seaweeds: application of concepts to aquaculture. Cah Biol Mar 42:71–82

Harrison PJ, Druehl LD, Lloyd KE, Thompson PA (1986) Nitrogen uptake kinetics in three year-classes of *Laminaria groenlandica* (Laminariales: Phaeophyta). Mar Biol 93:29–35

Hein M, Pedersen MF, Sand-Jensen K (1995) Size-dependent nitrogen uptake in micro and macroalgae. Mar ecol Prog ser 118:247–253

Hurd CL (2000) Water motion, marine macroalgal physiology and production. J Phycol 36:453–472

Lapointe B (1985) Strategies for pulse nutrient supply to *Gracilaria* cultures in the Florida Keys: interactions between concentration and frequency of nutrient pulses. J Exp Mar Biol Ecol 93:211–221

Lobban CS, Harrison PJ (1994) Seaweed Ecology and Physiology. Cambridge Univ Press, Cambridge, p 366

Marin SL, Westermeier R, Melipillan J (2002) Simulation of alternative management strategies for red algae, luga roja, (*Gigartina skottsbergii* Setchell and Gardner) in southern Chile. Ecol Mod 154:121–133

Martínez EA, Correa J, Faugeron S, Mansilla A, Avila M, Camus P (2005) Demographic and genetic survey of the red seaweed *Gigartina skottsbergii* Setchell et Gardner (Rhodophyta, Gigartinaceae) along its distribution range in the southern Pacific. Cienc Tecnol Mar 28(1):53–63

McNeill SE, Page M, Falshaw R (2003) Field trials to optimize the timing and frequency of pruning for cultivation of a New Zealand carragenophyte, *Gigartina atropurpurea*. J Appl Phycol 15:391–405

Navarro-Angulo L, Robledo D (1999) Effects of nitrogen source, N: P ratio and N-pulse concentration and frequency on the growth of *Gracilaria cornea* (Gracilariales, Rhodophyta) in culture. Hydrobiologia 398/399:315–320

Nars AH, Bekheet I, Ibrahim RK (1968) The effects of different nitrogen and carbon source on amino acids synthesis in *Ulva, Dictyota* and *Pterocladia*. Hydrobiologia 31:7–16

Pedersen MF (1994) Transient ammonium uptake in the macroalgae *Ulva lactuca* (Chlorophyta): Regulation, and the consequences for choice of measuring technique. J Phycol 30:980–986

Piriz M, Cerezo A (1991) Seasonal variation of carrageenans in tetrasporic, cystocarpic and "sterile" stages of *Gigartina skottsbergii* S. et G. (Rhodophyta, Gigartinales). Hydrobiology 226:65–69

Probyn TA, Chapman ARO (1982) Nitrogen uptake characteristics of *Chordaria flagelliformis* (Phaeophyta) in batch mode and continuous mode experiments. Mar Biol 71:129–133

Romo H, Paula E (1995) Métodos experimentales para el cultivo de *Porphyra*. In: Alveal K, Ferrario M, Oliveira y E, Sar E (eds) Manual de Métodos Ficológicos. Universidad de Concepción, Editorial Anibal S.A., Chile, pp 551–562

Romo H, Avila M, Candia A (2001) Manual de Técnicas de Cultivo y Repoblación de "Luga Roja" (*Gigartina skottsbergii*). Proyecto FONDEF D97I1064. Universidad de Concepción-IFOP, Chile, p 32

Romo H, Avila M, Candia A (2004) Manual de Técnicas de Cultivo y Repoblación de "Luga Roja" (Gigartina skottsbergii). Proyecto FONDEF D97I1064 y DOOI1109. Universidad de Concepción-IFOP, Chile, p 32

Romo H, Avila M, Nuñez M, Perez R, Candia A, Aroca G (2006) Culture of *Gigartina skottsbergii* (Rhodophyta) in southern Chile. A pilot scale approach. J Appl Phycol 18:307–314

Rozzi R, Massardo F, Anderson C, Berghoefer A, Mansilla A, Mansilla M, Plana J, Berghoefer U, Barros E, Araya P (2006) La RB Cabo de Hornos. Ediciones de la Universidad de Magallanes, Punta Arenas, Chile, p 260

Ryther JH, Corwin N, DeBusk TA, Williams TD (1981) Nitrogen uptake and storage by the red alga *Gracilaria tikvahiae* (McLachlan, 1979). Aquaculture 26:107–115

Taylor RB, Peek JT, Rees TAV (1998) Scaling of ammonium uptake by seaweeds to surface area: volume ratio: Geographical variation and the role of uptake by passive diffusion. Mar Ecol Prog Ser 169:143–148

Thomas TE, Harrison PJ, Taylor EB (1985) Nitrogen uptake and growth of the germlings and mature thallus of *Fucus distichus*. Mar Biol 84:267–272

Vallentinus I (1984) Comparisons of nutrients uptake rates for Baltic macroalgae with different thallus morphologies. Mar Biol 80:215–225

Westermeier R, Aguilar A, Sigel J, Quintanilla J, Morales J (1999) Biological basis for the management of *Gigartina skottsbergii* Gigartinales, Rhodophyta in southern Chile. Hydrobiologia 398/399:137–147

Zar JH (1999) Biostatistical analysis, 4th edn. Prentice- Hall, Inc, New Jersey, p 663

Zemke-White WL, Ohno M (1999) World seaweed utilization: An end-of-century summary. J Appl Phycol 11:369–376

UVB effects on early developmental stages of commercially important macroalgae in southern Chile

N. P. Navarro · A. Mansilla · M. Palacios

Originally published in the Journal of Applied Phycology, Vol 20, No 5, 447–456.
DOI: 10.1007/s10811-007-9276-2

Abstract High levels of ultraviolet-B radiation (UVB) could represent a danger to seaweeds by affecting their physiological processes and development. The aim of this work was to study the effects of UVB radiation on early developmental stages of commercially and ecologically important marine algal species in southern Chile, considering spores survival and embryos growth. Spores of *Mazzaella laminarioides*, *Gigartina skottsbergii*, *Sarcothalia crispata* and embryos of *G. skottsbergii* and *Macrocystis pyrifera* were submitted to treatments of a) photosynthetically active radiation (PAR: Control), b) PAR+UVA (PA) and c) PAR+UVA+UVB (PAB). UV radiation did not affect spore survival of *M. laminarioides S. crispata* and *G. skottsbergii* (P=0.55, P=0.6 & P=0.25 respectively), but did provoke differences in the growth rate of *G. skottsbergii* embryos (P=0.00). Differences in survival and growth of *M. pyrifera* embryos were also observed (P=0.001 & P= 0.007, respectively). Differences in growth of *M. pyrifera* embryos were observed only in the first five days, whereas changes in survival persisted until the end of the experiment. Additionally, UVB provoked morphological alteration in *M. pyrifera* embryos, as evidenced by progressive curling. These results suggest that the initial stages of the subtidal algae species *G. skottsbergii* and *M. pyrifera* cultivated in laboratory conditions were sensitive to UVA and UVB radiation, and their recruitment and development could be affected as well in natural conditions found in southern South America, where the ozone layer has thinned more than in other parts of the planet.

N. P. Navarro (✉) · A. Mansilla · M. Palacios
Facultad de Ciencias, Universidad de Magallanes,
Casilla 113-D, Avenida Bulnes,
01855 Punta Arenas, Chile
e-mail: nelso.navarro@umag.cl

Keywords Commercial seaweeds · Spores · UVB impact

Introduction

One of the most recognised atmospheric changes during the last few decades has been the thinning of the stratospheric ozone layer (Kirchhoff et al. 1997). This phenomenon has resulted in increasing levels of ultraviolet radiation B (UVBR: 280–320 nm) (Seckmeyer and Mckenzie 1992), reaching not only the Antarctic region, but also the southernmost part of South America (Chile and Argentina) (Kirchhoff et al. 1997; Bianciotto et al. 2003). During spring, when the area of the ozone hole increases and the stratospheric vortex elongates, the southern tip of South America is often under the influence of the ozone hole (Diaz et al. 2006).

Since ultraviolet radiation also penetrates the water column (Smith et al. 1992; Figueroa 2002), marine organisms are exposed to its harmful effects as well. Benthic macroalgae, in contrast to phytoplankton, are fixed and restricted to their growth sites, thus lacking the ability to avoid solar radiation, especially during low tide (Franklin and Forster 1997).

Sensitivity and tolerance of certain species to UVB radiation depends on their capacity to prevent and repair damage (van De Poll et al. 2001). When the levels of ultraviolet radiation increase, as they have in recent years, algal defense mechanisms may be insufficient, causing irreversible cellular damage. This sensitivity to UVB radiation might also vary according to developmental stages of species (Dring et al. 1996).

Ultraviolet radiation has been found to affect marine macroalgae in several ways, including effects on photosynthesis, DNA lesions, nitrogen metabolism and growth, as shown in numerous publications (reviewed by Franklin and Forster 1997; Xue et al. 2005). However, most of these studies have been focused primarily on the adult stages in the life history of macroalgae (Cordi et al. 2001).

Research evaluating the effects of UVB radiation on juvenile stages and spores is not frequent in literature, but early growth forms are of great importance since reproductive structures and the first stages of development are essential for recruitment, especially for those species of great ecological and economic importance, as their recovery over time depends almost entirely on reproductive ability. Considering the smaller size and greater structural simplicity of microscopic life stages, any kind of stress that affects the biology of a species may exert more evident effects on their early developmental stages, as has been shown for zoospores and germlings of various brown algae (Dring et al. 1996; Wiencke et al. 2000, 2006; Altamirano et al. 2002), unicells of Ulvales (Cordi et al. 2001) and for early life stages of Gigartinales (Roleda et al. 2004a).

Knowledge of the sensitivity of these ontogenetic stages is crucial, since recruitment of species depends on the survival of these stages, and conclusions obtained with macroscopic stages should not be extrapolated to microscopic ones (Altamirano et al. 2002). However, information regarding macroalgae, especially commercially important species, is still limited. The aim of this work was to evaluate the effects of UVB radiation on early developmental stages of commercially and ecologically important species in southern Chile, considering spore survival of *Mazzaella laminarioides*, *Gigartina skottsbergii*, *Sarcothalia crispata* and embryo growth of *G. skottsbergii* and *Macrocystis pyrifera*.

In Chile, *Sarcothalia crispata* (Bory) Leister, *Gigartina skottsbergii* Setchell & Gardner, and *Mazzaella laminarioides* (Bory) Fredericq are considered important carragenan-producing red alga. For this reason, all are harvested as a single group to obtain this important phycocolloid, employed in the food and pharmaceutical industry (Marín et al. 2002). *Macrocystis pyrifera* (Linnaeus) C. Agardh is an economically and ecologically important brown algae used as food in abalone farming and as an alginate resource. In the Magellan Region is also an ecosystem engineer that serves as a refuge and reproductive site for a great diversity of organisms such as sea urchins, king crabs, snow crabs and scallops.

Material and methods

Reproductive fronds of *Mazzaella* and *Sarcothalia* were collected by hand from intertidal zones, while *Gigartina* and *Macrocystis* were collected by SCUBA diving in the subtidal zone of Punta Santa Ana (53 37′ S; 70 59′ W), Strait of Magellan (Chile) in autumn 2005 and transported to the Center for Aquiculture and Subantarctic Marine Resources Laboratory at the Department of Natural Sciences and Resources, University of Magallanes, where unialgal cultures were established as described by Oliveira et al. (1995). All cultures were maintained with Provasoli enriched seawater (20 mL L^{-1}; 31 psu salinity), in a temperature-controlled room at 9±1°C with artificial irradiation of 55 µmol photons m^2 s^{-1} PAR (provided Philips TLT 20W/54 daylight fluorescent tubes) on a 12–12 h light-dark cycle.

Segments (2 cm^2) from different reproductive fronds were cut and put in Petri dishes with sterile filtered seawater. These were kept in a culture chamber until the release of spores. Three hundred spores from differents species were cultivated in triplicate in plastic Petri dishes. In order to obtain embryos of *Gigartina*, tetraspores were cultivated. To obtain young sporophytes (embryos) of *Macrocystis*, gametophytes were obtained from zoospores in the laboratory. Afterwards, gametophytes were cultivated for 2 weeks to reach 3 mm long young sporophytes. Afterwards, the sporophytes were unfastened and collected by Pasteur pipettes and put in plastic Petri dishes immediately.

Experimental light conditions The UV radiation was supplemented with UV-A (provided by UV-A-340 fluorescent tubes, Q-Panel Lab Products, Cleveland, USA) and UV-B (TL20W/12RS, Philips, Netherlands). Petri dishes were covered with different radiation filters in order to set three light/UV conditions: PAR only ($\lambda \geq 400$ nm, achieved by Ultraphan URUV, Digefra, Munich, Germany), PAR+UV-A ($\lambda \geq 320$ nm, Folanorm SF-AS, Folex, Cologne, Germany) and PAR+UV-A+UV-B ($\lambda \geq 295$ nm, Ultraphan URT, Digefra, Munich, Germany). Radiation conditions in the experimental treatments are summarised in Table 1.

PAR was kept low and constant during the entire experimental period and the algae were exposed to a constant UVB and UVA irradiance for only 3 h a day (from 11.00 to 14.00 h). Under these conditions, the spores of red macroalgae were exposed for 4 subsequent days to calculate survival, while *Macrocystis* and *Gigartina* embryos were exposed for 15 days.

Survival and growth measurements From each Petri dish that contained red macroalgae spores, survival was calculated by percentage alive after two and four days. In each sample about 300 spores were counted and the percentage of live spores was determined using a light microscope. The live cells were easily distinguishable (pigmented and dividing cells) from the dead cells (no pigment and sometimes not settled).

In order to calculate the survival of *Macrocystis* and *Gigartina* embryos, survival after 15 days was determined. Thirty 0.1 µm diameter and 1 week old gametophytes were

Table 1 Irradiance treatments applied to test survival and growth rates of spores and embryos of commercially important macroalgae in the laboratory

	PAR (μmol photons $m^{-2}.s^{-1}$)	UVA ($W.m^{-2}$)	UVB ($W.m^{-2}$) unweighted	UVB ($W.m^{-2}$) weighted
P (control)	55			
PA	55	0.9		
PAB	55	0.9	0.2	0.003

For weighted irradiance, the action spectra for DNA damage (Setlow 1974) was applied
P = PAR photosynthetic active radiation, PA=PAR+UVA, PAB=PAR+UVA+UVB

obtained from tetraspores of *G. skottsbergii*, which were cultivated in Petri dishes in triplicate for growth rate and survival determination. Diameter was recorded for a period of 15 days. Relative growth rates (RGR) were estimated from the following equation:

$$RGR(\%\ per\ day) = \left((\ln D_f - \ln D_i)*t^{-1}\right)*100$$

(Lignel and Pedersén 1989), where D_i is the initial and D_f is the final diameter of the young tetrasporophytes after t days of culture under the different treatments. The diameter of discoid embryos was measured in triplicate, using photographs, and analyzed by optical microscopy software Image Pro Plus 4.0.

Recovery After the experiment with UV radiation (PA and PAB), embryos of *Gigartina*, which had been cultivated under this radiation, were transferred to the control treatment (P) for 14 days, to observe if there was recovery of the growth rate after radiation.

Data analysis Arcsine transformation was applied to percentage data of spores and embryo survival. Data were compared by one-way (UV irradiance treatment) analysis of

Table 2 One-way (UV irradiance treatment) analysis of variance (ANOVA) and *Newman-Keuls* test for spores survival of *Mazzaella laminarioides*, *Sarcothalia crispata* and *Gigartina skottsbergii*, and growth rate of *Macrocystis pyrifera* and *G. skottsbergii* embryos exposed to different conditions (P = PAR photosynthetic active radiation (control), PA=PAR+UVA, PAB=PAR+UVA+UVB)

		P-values				
	df	*F*		P × PA	P × PAB	PA × PAB
M. laminarioides carpospore survival at 2 days	2	1.80	0.243 ns			
M. laminarioides carpospores survival at 4 days	2	1.56	0.284 ns			
M. laminarioides tetraspores survival at 2 days	2	0.64	0.557 ns			
M. laminarioides tetraspores survival at 4 days	2	0.63	0.561 ns			
S. crispata tetraspores survival at 2 days	2	4.99	0.052 ?	0.079 ns	0.048 *	0.363 ns
S. crispata tetraspores survival at 4 days	2	0.55	0.602 ns			
G. skottsbergii tetraspores survival at 3 days	2	0.93	0.445 ns			
G. skottsbergii tetraspores survival at 6 days	2	1.74	0.253 ns			
G. skottsbergii embryos growth after 15 days	2	315.9	0.000 *	0.000 *	0.000 *	0.000 *
G. skottsbergii embryos growth between the beginning and the 3th day	2	5.28	0.048 *	0.056 ns	0.039 *	0.739 ns
G. skottsbergii embryos growth between the 3th and the 7th day	2	9.39	0.014 *	0.012 *	0.137 ns	0.042 *
G. skottsbergii embryos growth between the 7th and the 9th day	2	11.80	0.008 *	0.032 *	0.007 *	0.086 ns
G. skottsbergii embryos growth between the 9th and the 15th day	2	65.82	0.000 *	0.002 *	0.000*	0.001*
G. skottsbergii embryos recovery after14 days	2	3.41	0.102 ns			
G. skottsbergii embryos recovery between the beginning and the 7th day	2	0.181	0.839 ns			
G. skottsbergii embryos recovery between the 7th and the 14th day	2	5.37	0.046 *	0.22 ns	0.11 ns	0.039 *
M. pyrifera embryos growth after 9 days	2	8.80	0.008 *	0.748 ns	0.007 *	0.011 *
M. pyrifera embryos growth between the beginning and the 5th day	2	56.14	0.000 *	0.276 ns	0.000 *	0.000 *
M. pyrifera embryos growth between the 5th and the 9th day	2	0.66	0.542 ns			
M. pyrifera embryos survival after 16 days	2	13.11	0.006 *	0.082 ns	0.006 *	0.024 *
M. pyrifera embryos survival between the beginning and the 5th day	2	17.27	0.003 *	0.077 ns	0.003 *	0.011 *
M. pyrifera embryos survival between the 5th and the 9th day	2	84.10	0.001 *		0.001 *	
M. pyrifera embryos survival between the 9th and the 16th day	1	60.62	0.000 *		0.000 *	

* = significant (p<0.05); ns = not significant.

variance (ANOVA). *Newman-Keuls* test was used to establish statistical differences. In all cases the RGR mean value of each Petri dish was considered as a replicate, thus for each treatment the number of replicate was three. All statistical tests were performed in accordance with Zar (1999).

Results

Spore survival In general, spore survival of red macroalgae studied was not affect by UV radiation after the first 4 days of UV exposure (Table 2). A higher percentage of survival (>80%) was observed in tetraspores and carpospores of *Mazzaella* exposed to all treatments tested (Figs. 1 and 2). ANOVA of carpospores and tetraspores percentage survival of *Mazzaella* showed no significant effects among all tested treatments (P=0.561 & P=0.284, respectively).

For *Sarcothalia* a higher percentage of survival was observed in tetraspores exposed to the control conditions (92±4%), when compared with PA (73±6%) and PAB (64±17%) treatments at two days (Fig. 3). However, there were no differences among percentage survival of tetraspores exposed to all treatments (P=0.052). Nonetheless, this p-value is at the limit of accepting hypothesis of the existence of significant differences in the survival percentages observed in the used treatments. In fact, the Newman-Keuls test showed differences between control and PAB treatment (P=0.048).

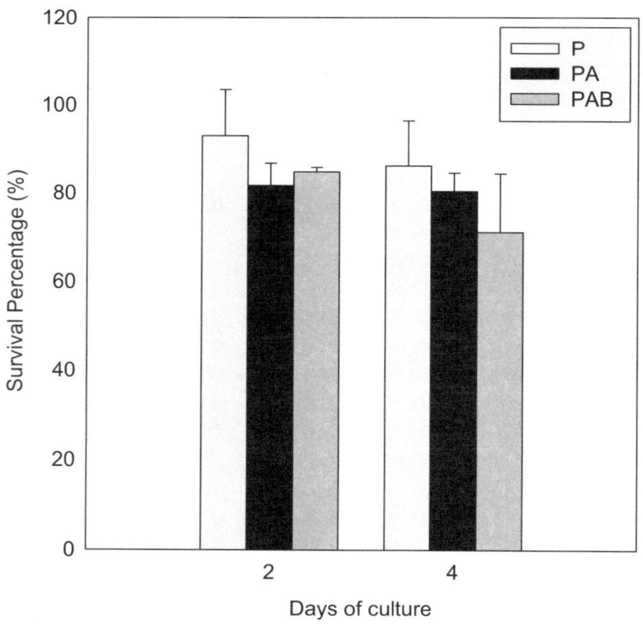

Fig. 1 Mean survival percentage of tetraspores of *Mazzaella laminarioides* exposed to different conditions (P=PAR photosynthetic active radiation (control), PA=PAR+UVA, PAB=PAR+UVA+UVB) after 2 and 4 days. Data are expressed as mean values ± S.D. (n=3). Letters on the graph show the result of the *Newman-Keuls* test (P<0.05); different letters refer to significant differences between mean values

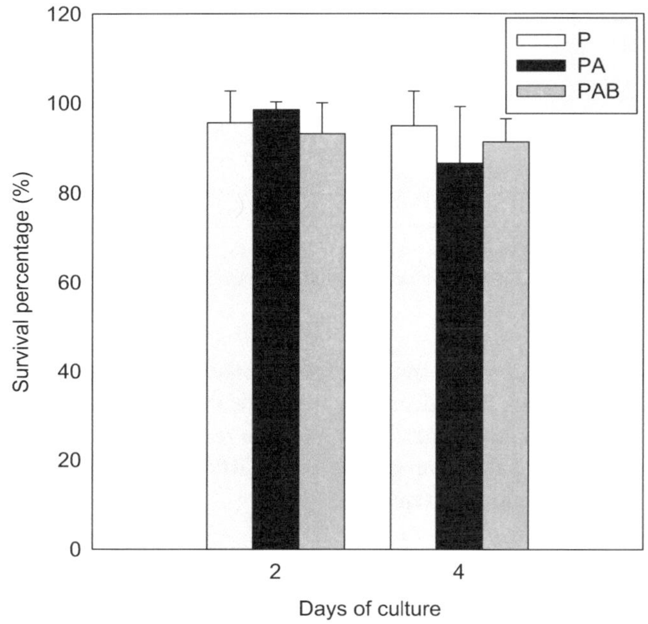

Fig. 2 Mean survival percentage (± S.D.) of carpospores of *Mazzaella laminarioides* exposed to different conditions (P=PAR photosynthetic active radiation (control), PA=PAR+UVA, PAB=PAR+UVA+UVB) after 2 and 4 days. ANOVA results are as in Fig. 1

In the case of *Gigartina*, UV radiation (UVA and UVB) did not lead to differences in tetraspore survival (P=0.445 at three days and P=0.253 at six days). However, spore survival was lower in all treatments (Fig. 4).

Growth of embryos

Gigartina

Higher RGRs were observed in *Gigartina* embryos exposed to the control conditions when compared with those cultivated at PA and PAB (P=0.000: Fig. 5). Detailed analysis of the relative growth rates at all data collection dates (days 3 to 15) showed that there was daily variation (Table 2: Fig. 6). Higher RGRs were observed in embryos cultivated in control conditions, when compared with those exposed to UVR treatments at 3, 9 and 15 days, except at day 7, where the higher RGRs were similar between control and PAB treatment, but different to PA treatment.

Those embryos of *Gigartina* which were transferred from PA and PAB to P treatment showed a similar growth rate to those of the control treatment after 14 days (Fig. 7), indicating that the effect of ultraviolet radiation on the growth rates of *Gigartina* embryos could be reversible.

Macrocystis

The analysis of the growth rate calculated from the length of the *Macrocystis* embryos over the total period of the

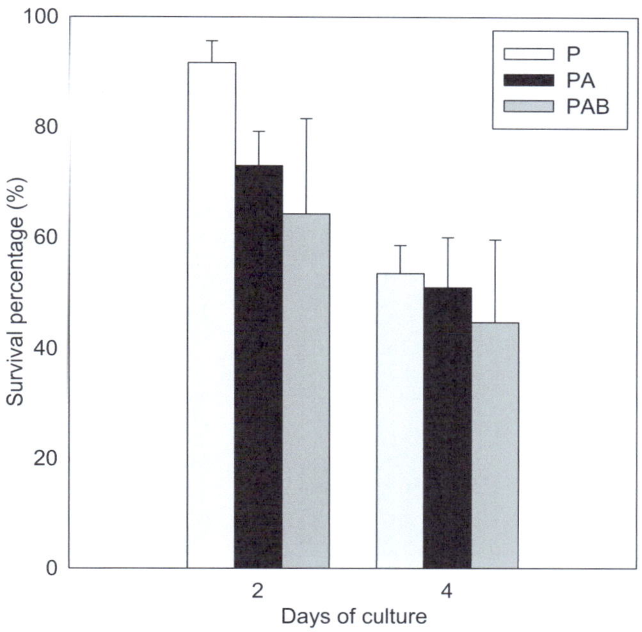

Fig. 3 Mean survival percentage (± S.D.) of tetraspores of *Sarcothalia crispata* exposed to different conditions (P=PAR photosynthetic active radiation (control), PA=PAR+UVA, PAB=PAR+UVA+UVB) after 2 and 4 days. ANOVA results are as in Fig. 1

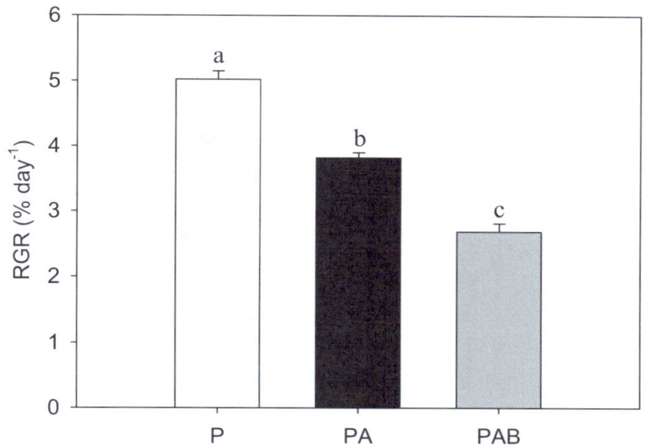

Fig. 5 Mean relative growth rate (RGR; % day-1) of *Gigartina skottsbergii* embryos exposed to different conditions (P=PAR (control), PA=PAR+UVA, PAB=PAR+UVA+UVB) after 15 days. Data are expressed as mean values ± S.D. (n=4); different letters refer to significant differences between mean values with *Newman-Keuls* post-hoc test (P<0.05)

experiment (9 days) showed differences between the treatments (P=0.007). The *post-hoc* test indicated that the differences were between the P treatment (control) and the PAB, with lower growth being observed under this last treatment (Fig. 8). A more detailed analysis showed that the major differences in the growth rates were observed in the first 5 days of cultivation (P=0.000), whereas in the period between the 5th and the 9th day there were no significant differences (P=0.542) (Fig. 9).

Regarding survival, it was found that embryos cultivated under the PAB treatment showed a survival rate lower than

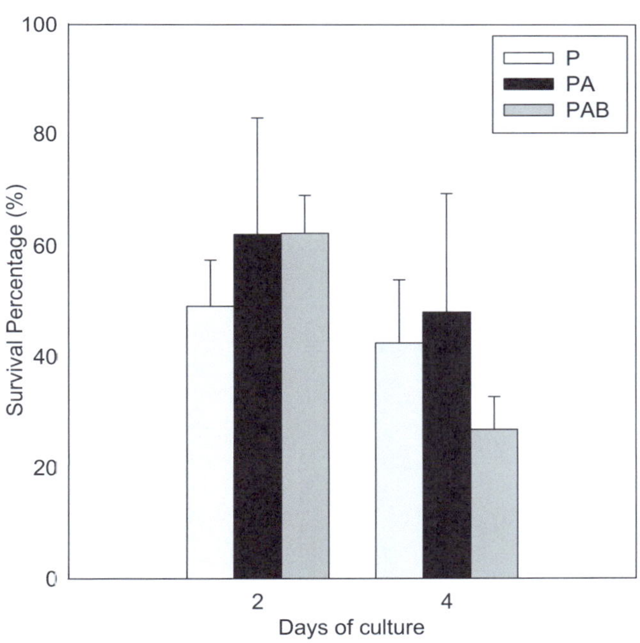

Fig. 4 Survival percentage of tetraspores of *Gigartina skottsbergii* exposed to different conditions (P=PAR Photosynthetic Active Radiation (control), PA=PAR+UVA, PAB=PAR+UVA+UVB) after 3 and 6 days. Data are expressed as mean values ± S.D. (n=3). Letters on the graph show the result of the Newman-Keuls test (P<0.05); different letters refer to significant differences between mean values

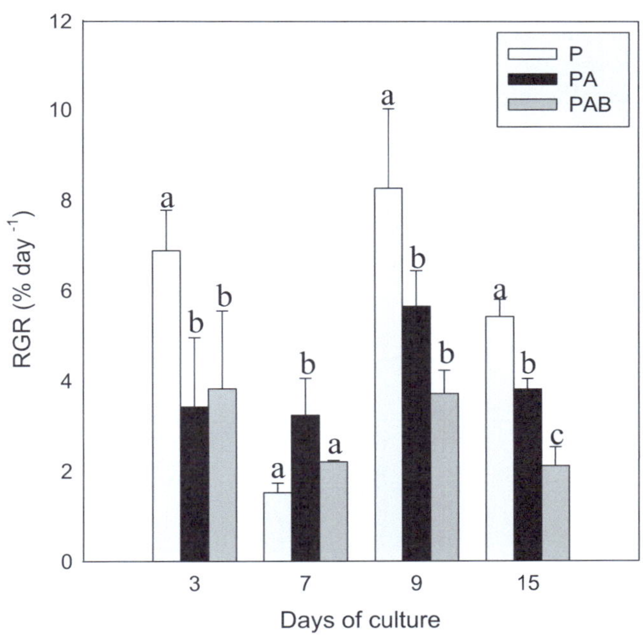

Fig. 6 Mean relative growth rate (RGR; % day-1) of *Gigartina skottsbergii* embryos exposed to different conditions (P=PAR (control), PA=PAR+UVA, PAB=PAR+UVA+UVB) of all data collection dates (days 3 to 15). Data are expressed as mean values ± S.D. (n=4); different letters refer to significant differences between mean values with *Newman-Keuls* post-hoc test (P<0.05)

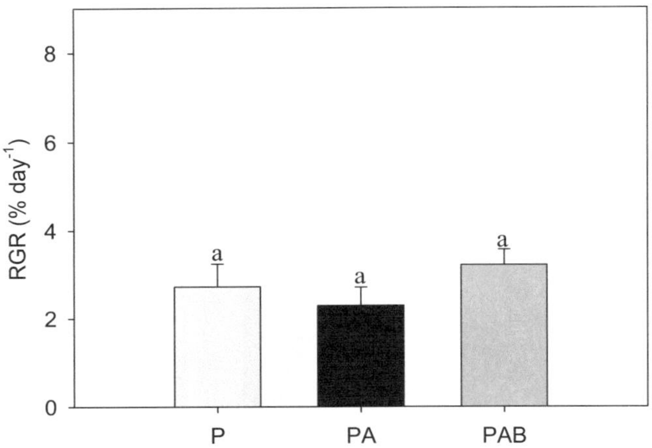

Fig. 7 Relative growth rate (RGR; % day-1) of *Gigartina skottsbergii* embryos after 14 days of recovery period under only PAR. Insert shows daily growth rate. Data are expressed as mean values ± S.D. (n=4); different letters refer to significant differences between mean values with *Newman-Keuls* post-hoc test (P<0.05)

20%, whereas the survival of those cultivated under the control condition (P) was higher than 70% (Fig. 10).

After the evaluation of the growth and survival rates, some embryos (sporophytes) of *Macrocystis* continued to be cultivated with the purpose of determining the effect of UVB radiation on their morphology. The observations performed showed that UVB damages the morphological development of the sporophytes (Fig. 11), with samples maintained in the treatment with ultraviolet B (PAB), from the 9th day of cultivation, showing changes in their morphology. These changes appeared as sporophyte curling, alterations at the edges of the thalli and/or tissue necrosis (Fig. 11); alterations that caused a decrease in the

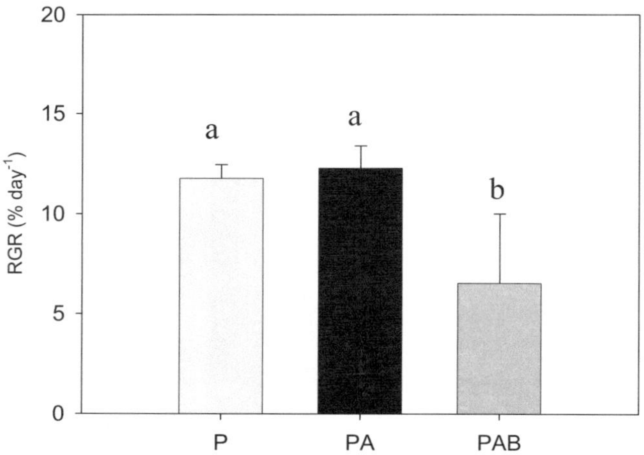

Fig. 8 Mean relative growth rate (RGR; % day-1) of *Macrocystis pyrifera* embryos exposed to different conditions (P=PAR Photosynthetic Active Radiation (control), PA=PAR+UVA, PAB=PAR+UVA+UVB) after 9 days. Data are expressed as mean values ± S.D. (n=4); different letters refer to significant differences between mean values with *Newman-Keuls* post-hoc test (P<0.05)

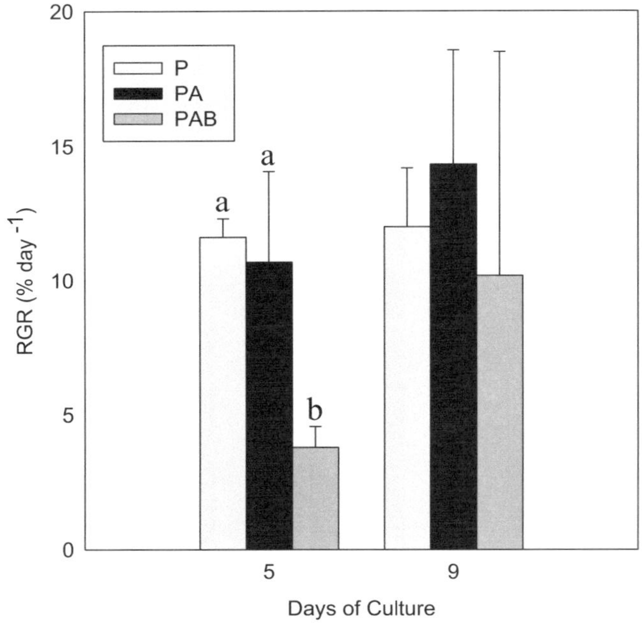

Fig. 9 Detailed analysis of the mean relative growth rates (RGR; % day^{-1}) of *Macrocystis pyrifera* embryos of all data collection dates (days 5 to 9) under different UV conditions (P=PAR Photosynthetic Active Radiation (control), PA=PAR+UVA, PAB=PAR+UVA+UVB). Data are expressed as mean values ± S.D. (n=4); different letters refer to significant differences between mean values with *Newman-Keuls* post-hoc test (P<0.05)

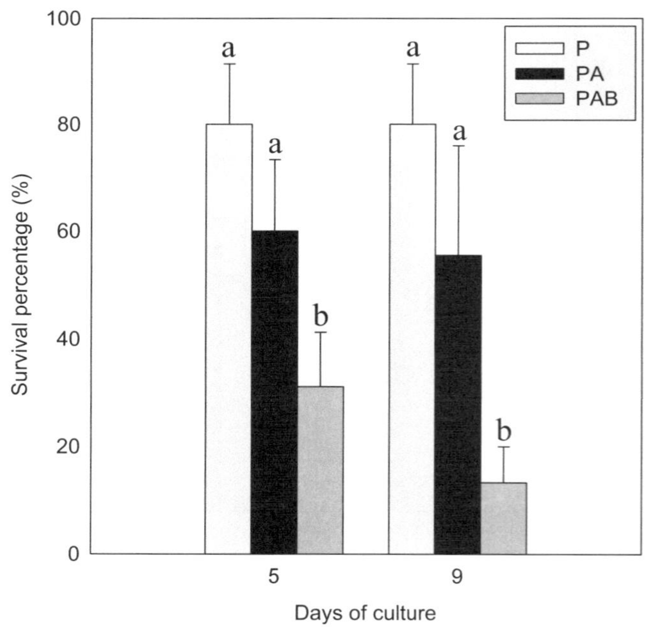

Fig. 10 Mean survival percentage (± S.D.) of *Macrocystis pyrifera* embryos exposed to different conditions (P=PAR Photosynthetic Active Radiation (control), PA=PAR+UVA, PAB=PAR+UVA+UVB) after 5 and 9 days. Data are expressed as mean values ± S.D. (n=4)

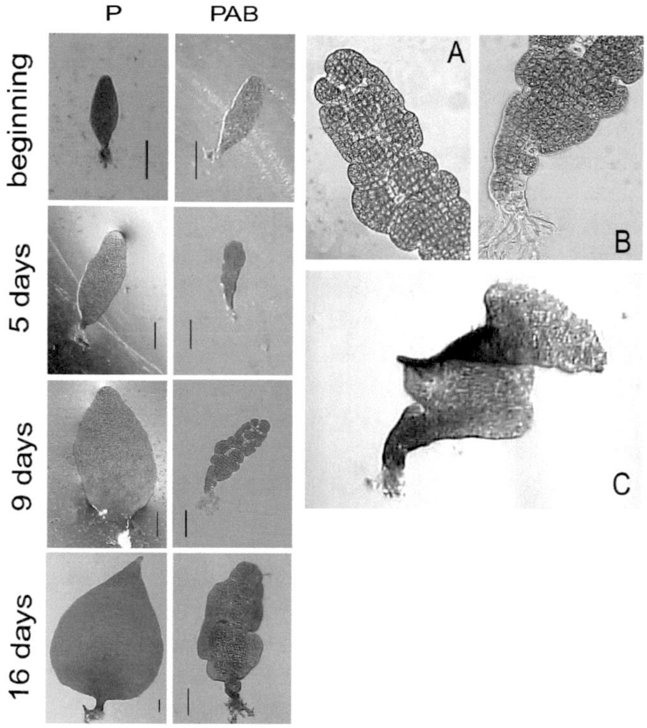

Fig. 11 Mean morphological responses (± S.D.) of *Macrocystis pyrifrea* embryos after 16 days under different condition (P=PAR photosynthetic active radiation (control) and PAB=PAR+UVA+UVB). Letters A, B and C show details of *M. pyrifera* embryos exposed to UVB radiation at 9 days

growth (P=0.000) (Fig. 12). This morphological alteration was observed mainly in embryos cultivated under the PAB treatment.

Discussion

In natural ecosystems, seaweeds are influenced not only by PAR, but also by ultraviolet radiation (UVA and UVB) (Makarov and Voskoboinikov 2001). These radiations may or may not be harmful for the development of particular species. It has been shown previously that there was decreased sensitivity to UV irradiation with increasing age (Dring et al. 1996) and with high visible light stress (Lüning 1980; Hanelt et al. 1997). In this regard, our results showed that, in general, there were no negative effects on the survival of spores, but negative effects were indeed observed in embryo growth of *Gigartina* and the sporophytic embryos of *Macrocystis*.

These data seem contrary to the evidence presented by the above-mentioned authors; however, our data should be analyzed separately, considering the bathymetric distribution of these species and their evolutionary history. Of the species studied, *Mazzaella* is an abundant species within the upper eulittoral to the upper sublittoral zone; *Sarcothalia* inhabits the lower eulittoral and upper sublittoral, whereas *Gigartina* and *Macrocystis* are found in sublittoral environments. In their natural environment, eulittoral (intertidal) species are more exposed to solar radiation and therefore are more likely to be adapted to UV exposure. On the other hand, sublittoral macroalgae are less exposed and may not be adapted to enhanced levels of this radiation. In fact, van De Poll et al. (2001) also reported that littoral species had a greater tolerance to UVB radiation than subtidal taxa.

In our experiments, spores (tetraspores and carpospores) of *Mazzaella* showed a high survival rate in the treatments used, which would indicate the acclimation of spores through the existence of defense mechanisms against UV, or that levels of UVB used are not harmful for the survival of the spores. In the case of the survival of *Sarcothalia* tetraspores, differences were observed among the treatments with a higher survival in the control treatment and a lower level in the treatment with UVB. However, these differences, besides being small, did not remain over time, since on the sixth day survival was similar among the treatments. Similar results have been reported in the other two Gigartinales species that co-inhabit the rocky littoral zone of North Atlantic coasts. Carpospores of *Mastocarpus stellatus* were found to be more tolerant of the deleterious effect of UVR compared with *Chondrus crispus* carpospores (Roleda et al. 2004a). Therefore, differences in the sensitivity between *Mastocarpus* and *Chondrus* carpospores to high light and UVB radiation reflect the differences in their zonation.

While *Mastocarpus* recruits at higher shore levels, *Chondrus* would limit recruitment at lower shore levels (Roleda et al. 2004a). In this way, it is possible that differences in the sensitivity between *Mazzaella* and *Sarcothalia* spore survival to UVB radiation could play a

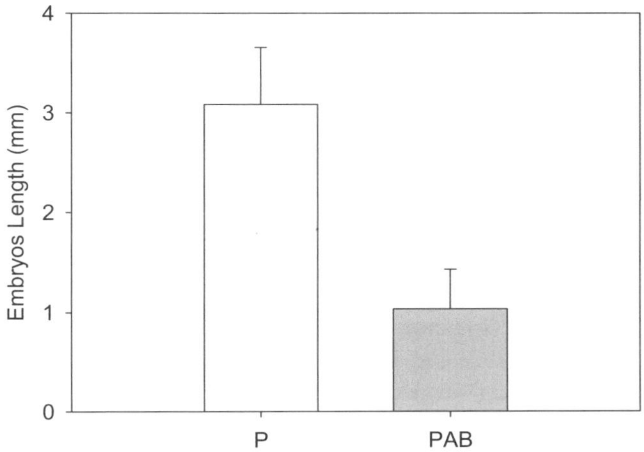

Fig. 12 Mean size measurements (± S.D.) of *Macrocystis pyrifera* embryos exposed to different conditions (P=PAR (control) and PAB=PAR+UVA+UVB) after 16 days. ANOVA values are as Fig. 1

role in the establishment of the zoning pattern at southern South American shorelines, at least for *Mazzaella* and *Sarcothalia*.

Sarcothalia, like *Mazzaella*, seems to have defense mechanisms against UV. However these mechanisms would not act immediately, but in a gradual acclimation process, since the *Sarcothalia* is not directly exposed to UV, with the exception of those days of spring tide. One of the most important physiochemical acclimation mechanisms against biologically harmful UV radiation involves the biosynthesis and accumulation of UV-screening substances. The most common photoprotective sunscreens in Antarctic macroalgae are the mycosporine-like amino acids (MAAs), which have been reported for species of *Mazzaella* and *Sarcothalia* in Chile (Huovinen et al. 2004).

Because *Gigartina* inhabits the sublittoral zone and often is protected from UV by the shadowing of big brown algae, mainly *Macrocystis*, we expected that its spores would be sensitive to UV radiation, however UVA and UVB radiation did not lead to differences in tetraspore survival when compared with the control treatment. On the other hand, the existence of a low survival of tetraspores in the treatments used, could be due to the use of PAR, whose intensity in the experiment was higher than that suggested by Romo et al. (2001) for the cultivation of this species in its first stages of development. However, the possibility of the existence of mycosporine-like amino acid compounds can not be dismissed. Research into this phenomenon could provide evidence of the synthesis of mycosporine-like amino acids in spores.

Embryo growth The growth of *Gigartina embryos* was negatively affected by UV, being significantly more inhibited by the UVB. These results are in agreement with those reported Mansilla et al. (2006), which showed that embryos of *Gigartina* cultivated under the different intensities of UVB had lower growth rates in comparison with those cultivated without UVB, referring to subtidal algae, and therefore more sensitive to UVB. In this study, the embryos were also treated with UVA, unlike Mansilla et al. (2006). UVA also inhibited the growth of embryos of *Gigartina*, even though high levels of UVA and PAR have been reported to increase the induction of mechanisms for repairing damage and protection against UVB and to increase tolerance to UVB (Takayanagi et al. 1994; van De Poll et al. 2001). From our results, it is difficult to assume this. However, it can be concluded that the effects caused by UVA and UVB radiation do not remain over time, since on the exclusion of UVA and UVB during cultivation allowed the recovery of the growth rate. This would indicate that the embryos of *Gigartina* have mechanisms to avoid harm, changing the energy initially destined for growth, towards a process of protection.

In the case of *Macrocystis*, inhibiting effects of UVB were also detected in the growth of sporophytic embryos. In addition, morphological changes were observed. The interaction of these effects would have significant effects on the survival of those embryos treated with UVB. On the other hand, those surviving embryos have the capacity to acclimate to UVB, since the differences in the growth rate among the treatments were only maintained for the first 5 days and disappeared on the ninth day of cultivation.

The mechanisms that could be involved in the acclimation of embryos to UVB could be a) the establishment of physical barriers that protect the tissues (or part of them) as morphological changes, or b) the induction or synthesis of phlorotannin compounds, which have a photoprotective function in brown algae (Schoenwaelder 2002; Henry and Van Alstyne 2004). An induction of phlorotannins due to UVB and UVA radiation has been described for *Macrocystis integrifolia* (Swanson and Druehl 2002) and in zoospores of five species of Laminariales (Wiencke et al. 2004).

Phlorotannins occur in cells in vesicles known as physodes and are also deposited in cell walls (Schoenwaelder and Clayton 1998, 2000). These compounds play a role in the determination of depth zonation of brown algae (Bischof et al. 2006). Recently it was demonstrated that zoospores of Arctic *Alaria esculenta* and *Saccorhiza dermatodea* are less sensitive to UVR or have a better capacity to recover from UVR-induced stress than zoospores of species from deeper water, *Laminaria digitata*, *L. saccharina* and *L. solidungula* (Wiencke et al. 2004).

Although the existence of phlorotannins has not been shown in adults or the early development stages of *Macrocystis pyrifera*, evidence does exist that to believe that it does occur in them, or there is at least some other mechanism that protects them from UV radiation. First, although *Macrocystis* is a sublittoral alga, its apical meristem sections reach the water's surface, thus being exposted to solar radiation (PAR and UV). Consequently, it is essential to guarantee protection of the delicate meristematic cells to UV radiation by means of photoprotective compounds. Second, zoospores and embyos of *Macrocystis* preferentially grow a few centimeters from parental plants, as demonstrated by the zoospores of *Sargassum spinuligerum* (Kendrick and Walker 1991). For this reason, zoospores and embryos of *Macrocystis* would be more sensitive to UV radiation than the adults, which could be attenuating the UVB radiation that arrives in the sublittoral zone. However, in our experiments some embryos had the capacity to acclimate to UVB. This suggests the existence of a protection and acclimation system. Also, kelp phlorotannin exudates from macroalgal sources have been proposed to reduce the impact of UVB radiation by forming UV refugia for kelp zoospores within the water column (Swanson and Druehl 2002). Additionally, it has been suggested that kelp spores and embryos are pre-adapted to the UV conditions of the parent plant (Swanson and Druehl 2000). Genetic adaptations to PAR light have been previously demonstrated within the Laminariales (Gerard 1988), and may also exist for potent environmental stresses such as UV (Swanson and Druehl 2000).

On the other hand, embryos of *Macrocystis* show morphological changes, seen as deformation in the thallus and the loss of cells, possibly from necrosis. These changes have been reported for *Laminaria ochroleuca* (Roleda et al. 2004b) and *L. solidungula* (Michler et al. 2002), and the red algae *Iridaea cordata* (Navarro 2004). In this last case, the deformation was manifest as curving of the thallus, diminishing in the area of exposure to UVB, and thus being a protective factor against radiation.

Implications Our results provide firm evidence of UVB radiation effects on the development of *Macrocystis*, *Gigartina*, *Sarcothalia* and *Mazzaella*, all of which are of great commercial and interest in southern Chile. Such effects have a potential to interfere with natural populations of these macroalgal species, as well.

In nature, the early developmental stages of macroalgae species could grow with the protection of nearby adults. Adult individuals could be attenuating the UVB radiation to the young stages, reducing radiation levels which could cause irreparable harm. The occurrence of microhabitats could favour the survival of spores and juveniles stages (Brawley and Johnson 1991). This protection has been suggested for other species (Ang 1991; Brawley and Johnson 1991; Johnson and Brawley 1998) and could attenuate the UVB effect, preventing the irreparable harm caused by high levels. Nonetheless, if the adults are extracted because of their economic importance, the result would be an increase of UVB levels reaching the intertidal and subtidal zone. In this scenario, the effects on growth and morphology may not be reversible, and the defense mechanisms might be insufficient, causing irreparable cellular damage and unbalances in early developmental stages.

The importance of kelp forests of *Macrocystis* is being increasingly recognised in southern Chile for both its role as a keystone species and ecosystem engineer. In addition *Macrocystis* has high commercial value and is under increasing pressure for industrial exploitation. Within kelp forests we also find many other species of algae, including *Gigartina* (Zamorano and Westermeier 1996), which would be affected by the future extraction of *Macrocystis*. Consequently, for both ecological and economic reasons, there is a great necessity to understand the role of UVB on natural populations of algae and their long-term recruitment systems. In addition, complementary studies regarding the effects of UVB on the carrageen and algine acid content of these species will allow researchers to better understand the real effects that this kind of radiation has on commercial species of macroalgae in Chile. A final aspect that should be included in future work is the need to incorporate an outlook of long-term recuperation of stratospheric ozone levels, which are currently projected to not reach historic levels until 2050 (WMO/UNEP 2002; Valkama et al. 2003).

Acknowledgement We gratefully acknowledge Susana Ovando, Clare Brown and Dr. Christopher Anderson for valuable suggestions and improvement of this manuscript. This work was financed in part by Project Mineduc Acuicultura PR 334 (Chilean Ministry of Education), and by Project 26600 (University of Magallanes).

References

Altamirano M, Flores-Moya A, Figueroa F (2002) Effects of the radiation and temperature on growth of germling of three species of *Fucus* (Phaeophyta). Aquat Bot 1606:1–12

Ang PO Jr (1991) Natural dynamics of a *Fucus distichus* (Phaeophyta, Fucales) population: reproduction and recruitment. Mar Ecol Prog Ser 78:71–85

Bianciotto OA, Pinedo LB, San Roman NA, Blessio AY, Collantes MB (2003) The effects of natural UV-B radiation on a perennial *Salicornia* salt-marsh in Bahía San Sebastián, Tierra del Fuego, Argentina: a 3-year field study. J Photochem Photobiol B Biol 70:177–185

Bischof K, Gómez I, Molis M, Hanelt D, Karsten U, Lüder U, Roleda MY, Zacher K, Wiencke C (2006) Ultraviolet radiation shapes seaweed communities. Rev Environmental Sci Biotechnol 5:141–166

Brawley SH, Johnson LE (1991) Survival of fucoid embryos in the intertidal zone depends upon developmental stage and microhabitat. J Phycol 27:179–186

Cordi B, Donkin ME, Peloquin J, Price DN, Depledge MH (2001) The influence of UV–B radiation on the reproductive cells of the intertidal macroalga, *Enteromorpha intestinalis*. Aquat Toxicol 56:1–11

Diaz S, Camilio C, Deferrari G, Fuenzalida H, Armstrong R, Booth C, Paladini A, Cabrera S, Casiccia C, Lovengreen C, Pedroni J, Rosales A, Zagarese H, Vernet M (2006) Symposium-in-Print: UV effects on aquatic and coastal ecosystems ozone and UV radiation over southern south America: climatology and anomalies. Photochem Photobiol 82:834–843

Dring M, Makarov V, Schoschina E, Lorenz M, Lüning K (1996) Influence of ultraviolet-radiation on chlorophyll fluorescence and growth in different life-history stages of three species of *Laminaria* (Phaeophyta). Mar Biol 126:183–191

Figueroa FL (2002) Bio-optical characteristics of Gerlache and Bransfield Strait waters during an Antarctic summer cruise. Deep Sea Res II 49:675–691

Franklin L, Forster R (1997) The changing irradiance environment: consequences for marine macrophyte physiology, productivity and ecology. Eur J Phycol 32:207–237

Gerard VA (1988) Ecotypic differentiation in light-related traits of the kelp *Laminaria saccharina*. Mar Biol 97:25–36

Hanelt D, Wincke C, Karsten U, Nultsch W (1997) Photoinhibition and recovery after high light stress in different developmental and life-history stages of *Laminaria saccharina* (Phaeophyta). J Phycol 33:387–395

Henry EB, Van Alstyne KL (2004) Effects of UV radiation on growth and phlorotannins in *Fucus gardneri* (Phaeophyceae) juveniles and embryos. J Phycol 40:527–533

Huovinen P, Gómez I, Figueroa F, Ulloa N, Morales V, Lovengreen C (2004) Ultraviolet-absorbing mycosporine-like amino acids in red macroalgae from Chile. Bot Mar 47:21–29

Johnson LE, Brawley SH (1998) Dispersal and recruitment of canopy-forming intertidal alga: the relative roles of propagule availability in post-settlement processes. Oecologia 117:517–526

Kendrick GA, Walker D (1991) Dispersal distances for propagules of *Sargassum spinuligerum* (Sargassaceae, Phaeophyta) measured directly by vital staining and venturi suction sampling. Mar Ecol Prog Ser 79:133–138

Kirchhoff VWJH, Zamorano F, Casiccia C (1997) UVB enhancements at Punta Arenas, Chile. J Photochem Photobiol B Biol 38:174–177

Lignel A, Pedersén M (1989) Agar composition as a function of morphology and growth rate. Studies on some morphological strains of *Gracilaria secundata* and *Gracilaria verrucosa* (Rhodophyta). Bot Mar 32:219–227

Lüning K (1980) Critical levels of light and temperature regulating the gametogenesis of three *Laminaria* species (Phaeophyceae). J Phycol 16:1–15

Makarov MV, Voskoboinikov GM (2001) The influence of ultraviolet-B radiation on spores release and growrh of kelp *Laminaria saccharina*. Bot Mar 44:89–94

Mansilla A, Werlinger C, Palacios M, Navarro NP, Cuadra P (2006) Effects of UVB radiation on the initial stages of growth of *Gigartina skottsbergii*, *Sarcothalia crispata* and *Mazzaella laminarioides* (Gigartinales, Rhodophyta). J Appl Phycol 18:451–459

Marín SL, Westermeier R, Melipillan J (2002) Simulation of alternative management strategies for red algae, luga roja, (*Gigartina skottsbergii* Setchell and Gardner) in southern Chile. Ecol Mod 154:121–133

Michler T, Aguilera J, Hanelt D, Bischof K, Wiencke C (2002) Long-term effects of ultraviolet radiation on growth and photosynthetic performance of polar and cold-temperature macroalgae. Mar Biol 140:1117–1127

Navarro N (2004) Efeitos da radiação UVB em *Iridaea cordata* (Gigartinales, Rhodophyta). Dissertação Universidade de São Paulo

Oliveira EC, De Paula EJ, Plastino EM, Petti R (1995) Metodologías para el cultivo no axenico de macroalgas marinas *in vitro*. In: Alveal K, Ferrario M, Oliveira E, Sar E (eds) Manual de métodos ficológicos. Universidad de Concepción, Chile, pp 429–447

Roleda M, van de Poll W, Hanelt D, Wiencke C (2004a) PAR and UVBR effects on photosynthesis, viability, growth and DNA in different life stages of two coexisting Gigartinales: implications for recruitment and zonation pattern. Mar Ecol Prog Ser 281:37–50

Roleda M, Hanelt D, Krabs G, Wiencke C (2004b) Morphology, growth, photosynthesis and pigment in *Laminaria ochroleuca* (Laminariales, Phaeophyta) under ultraviolet radiation. Phycologia 43(5):603–613

Romo H, Avila M, Nuñez M (2001) Manual de técnicas de cultivo y repoblación de "Luga Roja" (*Gigartina skottsbergii*) Proyecto FONDEF D97I1064. IFOP- Universidad de Concepción, Chile

Schoenwaelder M (2002) The occurrence and cellular significance of physodes in brown algae. Phycologia 41:125–139

Schoenwaelder M, Clayton M (1998) Secretion of phenolic substances into the zygote wall and cell plate in embryos of *Hormosira* and *Acrocarpia* (Fucales, Phaeophyceae). J Phycol 34:969–980

Schoenwaelder M, Clayton M (2000) Physode formation in embryos of *Phyllospora comosa* and *Hormosira banksii* (Phaeophyceae). Phycologia 39:1–9

Seckmeyer G, Mckenzie RL (1992) Elevated ultraviolet radiation in New Zealand (45 S) contrasted with Germany (45 N). Nature 359:135–137

Setlow RB (1974) The wavelengths in sunlight effective in producing skin cancer: a theoretical analysis. Proc Nat Acad Sci USA 71:3363–3366

Smith RC, Prézelin BB, Baker KS, Bidigare R, Boucher N, Coley T, Karentz D, Macintyre S, Matlick H, Menzies D, Ondrusek M, Wan Z, Waters J (1992) Ozone depletion: ultraviolet radiation and phytoplankton biology in Antarctic waters. Science 255:952–959

Swanson A, Druehl L (2000) Differential meiospore size and tolerance of ultraviolet light stress within and among kelp species along a depth gradient. Mar Biol 136:657–664

Swanson A, Druehl L (2002) Induction, exudation and the UV protective role of kelp phlorotannins. Aquat Bot 73:241–253

Takayanagi S, Trunk J, Sutherland JC, Sutherland BM (1994) Alfalfa grown outdoor are more resistant to UV-induced DNA damage than plants grown in a UV-free environmental chamber. Photochem Photobiol 60:363–367

Valkama E, Kivimaenpaa M, Hartikainen H, Wulff A (2003) The combined effects of enhanced UV-B radiation and selenium on growth, chlorophyll fluorescence and ultrastructure in strawberry (*Fragaria* x *ananassa*) and barley (*Hordeum vulgare*) treated in the field. Agric For Meteorol 120:267–278

van De Poll W, Eggert A, Buma A, Breeman A (2001) Effects of UV-induced DNA damage and photoinhibition on growth of temperature marine red macrophytes: habitat - related differences in UV-B tolerance. J Phycol 37:30–37

Wiencke C, Gómez I, Pakker H, Flores-Moya A, Altamirano M, Hanelt D, Bischof K, Figueroa FL (2000) Impact of UV-radiation on viability, photosynthetic characteristics and DNA of brown algal zoospores: implications for depth zonation. Mar Ecol Prog Ser 197:217–229

Wiencke C, Clayton M, Schoenwaelder M (2004) Sensitivity and acclimation to UV radiation of zoospores from five species of Laminariales from the Arctic. Mar Biol 145:31–39

Wiencke C, Roleda M, Gruber A, Clayton MN, Bischof K (2006) Susceptibility of zoospores to UV radiation determines upper depth distribution limit of Arctic kelps: evidence through field experiments. J Ecol 94:455–463

Xue L, Zhang Y, Zhang T, An L, Wang X (2005) Effects of enhanced ultraviolet-B radiation on algae and cianobacteria. Crit Rev Microbiol 31:79–89

WMO/UNEP (2002) Scientific assessment of ozone depletion: 2002. Available at http://www.wmo.ch/web/arep/reports/o3_assess_rep_2002_front_page.html

Zamorano J, Westermeier R (1996) Phenology of *Gigartina skottsbergii* (Rhodophyta, Gigartinales) in Ancud Bay, southern Chile. Hydrobiologia 326/327:253–259

Zar JH (1999) Biostatistical analysis, 4th edn. Prentice-Hall, Upper Saddle River, NJ

Can macroalgae recover, 13 months after the 2004 Tsunami?: a case study at Talibong Island, Trang Province, Thailand

Anchana Prathep · Jaruwan Mayakun · Piyalap Tantiprapas · Anuchit Darakrai

Originally published in the Journal of Applied Phycology, Vol 20, No 5, 457–464.
DOI: 10.1007/s10811-007-9287-z © Springer Science + Business Media B.V. 2007

Abstract The tsunami that hit the coast of Thailand along the Andaman Sea on 26 December 2004 caused great damage to human life, property and coastal resources. Here we report the effect of the tsunami on the seaweed community, and its recovery, at Talibong Island, one of the affected areas in Trang province. We made surveys at three sites around the island from April 2004 to January 2006. Fifteen 0.5 m × 0.5 m permanent plots were set up on the west coast of the island to monitor changes in the seaweed community. Eighteen species were found. *Sargassum stolonifolium* and *Laurencia composita* were the most abundant, covering 90% and 39%, respectively, of the rocky substrate. Thirteen species varied among sites and seasons. Eleven species, however, were strongly affected by the tsunami. *L. composita* and *Padina sanctae-crucis*, for example, were initially washed up onto the shore by the strong wave action, which clearly resulted in a decrease in percentage cover. Also, many permanent plots were covered by sediment causing anoxic conditions, an indirect impact of the tsunami. A difference in species composition revealed a change in overall diversity. Particular morphologies and life forms were more sensitive to the extremes of the tsunami than others. Therefore, the recovery ability of populations of the affected species varied. We did not find any seasonal pattern to the recovery of seaweed after the tsunami within the 13 months of this study.

Keywords Community · Diversity · Macroalgae · Recovery · Thailand · Tsunami

A. Prathep (✉) · J. Mayakun · P. Tantiprapas · A. Darakrai
Seaweed and Seagrass Research Unit,
Centre for Biodiversity of Peninsular Thailand,
Department of Biology, Faculty of Science,
Prince of Songkla University,
Hat Yai, Songkhla 90112, Thailand
e-mail: anchana.p@psu.ac.th

Introduction

The tsunami of 26 December 2004, a natural disaster in the Indian Ocean, caused severe damage to humankind as well as to marine life in coastal areas. Considerable effort has been made towards recovery. Studies that contribute to understanding the effects of the tsunami on marine organisms include those on microbial life in both seawater and coastal sediments (Ramesh et al. 2006), the coral reef community (Chavanich et al. 2005; Hughes et al. 2005; Kumaraguru et al. 2005; Phongpaichit et al. 2006), mangrove ecosystems (Dahdouh-Guebas et al. 2005; Danielsen et al. 2005; Vermaat and Thampanya 2006) and on some littorinid mollusks (Sanpanich et al. 2006). However, only a few preliminary studies on the seagrass and seaweed community have been reported (Mantri 2005, 2006; Prathep and Tantiprapas 2006).

The 2004 tsunami hit six provinces on the Andaman Sea coast in Southern Thailand, including Talibong island in Trang province where we had made a preliminary survey in May 2003. Later, starting in April 2004, a bimonthly monitoring of the macroalgal community was carried out at Talibong. The island, which is a protected area with no hunting or fishing allowed, has an area of around 4,000 ha and is the biggest island in Trang Province. Talibong Island has recently been designated as Ramsar site No. 1182 with a connection to the Had Chao Mai Marine National Park and the Trang River Estuaries. These are interconnected wetland ecosystems including riverine, estuarine, and coastal wetlands, mangroves, sandy beach and rocky marine shores, mud flats, coral reefs and seagrass beds. The island supports a broad diversity of marine organisms, including migrating seabirds and an important endangered species of dugong (*Dugong dugon*). Talibong is also known to have the healthiest and most richly diverse seagrass ecosystem in Thailand: 8 of the 12 known species reported in Thailand are found there (Changsang and Poovachiranon 1994; Poovachiranon and Changsang 1994).

Having a contemporary dataset before the event allows us to clearly assess the effect of tsunami or other disturbances on individual species and communities. For example, disturbances such as occurred with *Caulerpa brachypus* forma *parvifolia* in Florida in the Unites States could be evaluated (Lapointe et al. 2006). Also, the recovery and resilience of the community can be monitored and better understood. The results of this study allow us to report (1) the before and after tsunami diversity and community structure of seaweeds at Talibong island, and (2) the recovery ability of the seaweed community.

Materials and methods

Monitoring seaweed biodiversity and percentage cover began under the Biodiversity Study of Seaweed in South East Asia. Talibong was chosen as a monitoring site for the Andaman Sea, Thailand. The preliminary survey around the island carried out from 2–5 May 2003, showed a high diversity and abundance of macroalgae. Fifteen permanent plots, 0.5 m × 0.5 m, were randomly set up at three different sites along the west coast, with five plots at each site. The estimates of percentage cover of seaweed were made following Saito and Atobe (1970). Sites 1 (7° 16.86′N, 99°22.68′E) and 2 (7°15.69′N, 99° 22.027′E) have exposed sandy and rocky substrates, while Site 3 (7° 12.551′N, 99° 23.534′E) has sheltered rocky substrates (Fig. 1). Each plot was marked using GPS (GARMIN GPS 76); fluorescent tape was used to mark the corner of each plot. The monitoring study started in April 2004. In a sudden twist, these sites became important for the study of biodiversity and community structure immediately preceding the December 2004 Indian Ocean tsunami that hit the Andaman Coast where the study site is located. Diversity and percentage cover of macroalgae at each plot were estimated and compared at the sites from April 2004 to January 2006.

Talibong Island is influenced by the Southwest Monsoon from May–October, which represents the rainy season, and by the Northeast Monsoon from November–April, the dry season. Field collections were made in April, September and November 2004 (before the tsunami) and January, March, May, August, October 2005 and January 2006. These collections bridge the differences in seasons, sites and percentage cover as well as the effects of tsunami on the seaweed community. To assess the effects of the tsunami, the datasets of only four collections were used:

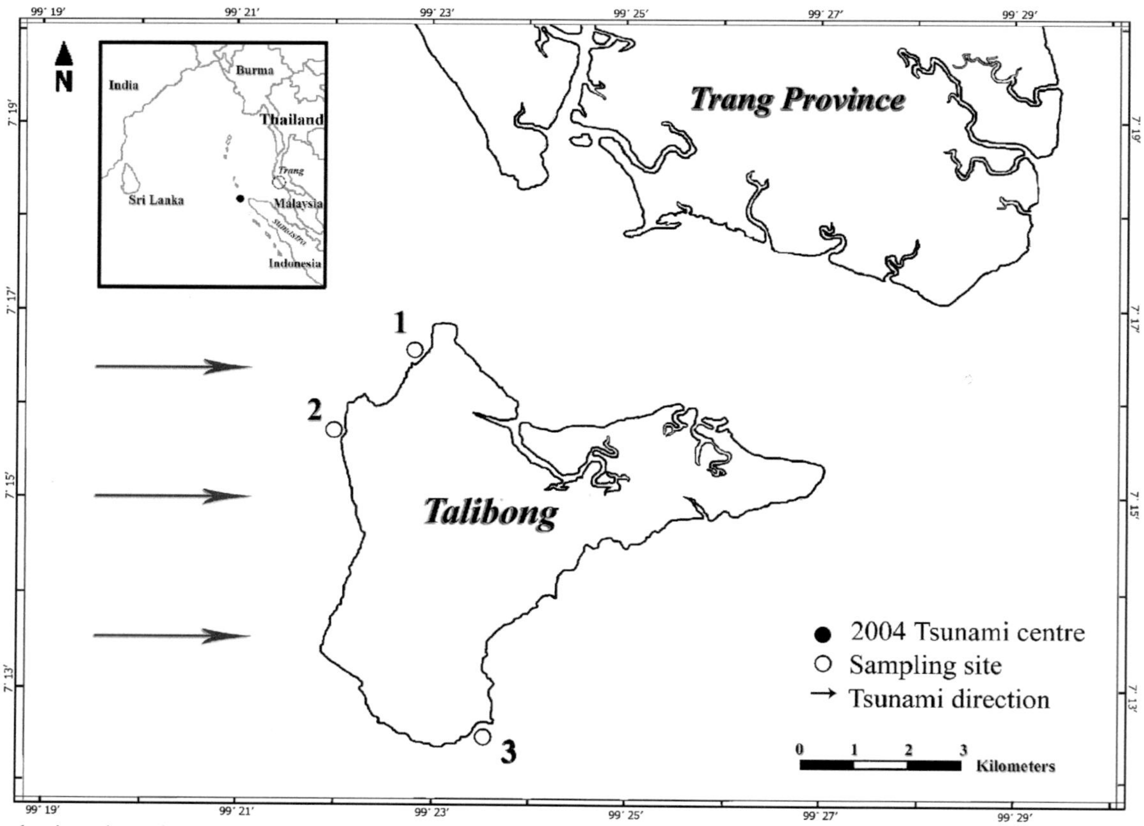

Fig. 1 Study sites along the coast of Talibong island, Trang Province, Andaman Sea, Southern Thailand, which were affected by the 2004 tsunami

Table 1 List of species and their distribution

Taxa	Site 1									Site 2									Site 3								
	Apr 04	Sep 04	Nov 04	Jan 05	Mar 05	May 05	Aug 05	Oct 05	Jan 06	Apr 04	Sep 04	Nov 04	Ja 05	Mar 05	May 05	Aug 05	Oct 05	Jan 06	Apr 04	Sep 04	Nov 04	Jan 05	Mar 05	May 05	Aug 05	Oct 05	Jan 06
Phylum Cyanobacteria																											
Lyngbya majuscula (Dillwyn) Harvey ex Gomont			X			X						X			X									X			
Symploca sp.			X								X	X															
Phylum Chlorophyta																											
Caulerpa racemosa (Forsskål) J. Agardh	X							X	X	X																	
C. taxifolia (Vahl) C. Agardh	X	X																	X								
Dictyospheria carvernosa (Forsskål) Børgesen	X																										
Valonia pachynema (Martens) Børgesen	X																										
Division Ochrophyta																											
Dictyota dichotoma (Hudson) Lamouroux	X																										
D. cervinoides Kützing								X	X								X	X	X							X	X
Padina sanctae-crucis Børgesen	X	X	X		X	X	X	X	X			X			X		X	X	X	X	X				X	X	X
Sargassum stolonifolium Phang et Yoshida							X					X	X						X	X	X	X	X	X	X	X	X
Turbinaria ornata (Turner) J. Agardh	X		X			X													X								
Phylum Rhodophyta																											
Filamentous red alga	X																										
Galaxuara obtusata (Ellis et Solander) Lamouroux																			X	X	X				X	X	X
Gelidiella acerosa (Forsskål) Feldmann and Hamel	X				X		X	X	X	X		X													X		
Gelidiopsis sp.			X																								
Gelidium sp.	X	X	X	X	X						X																
Jania sp.												X															
Laurencia composita Yamada	X		X		X						X	X	X	X	X	X	X	X	X	X	X						
Number of species	11	3	7	1	4	3	3	4	4	2	3	7	2	1	3	1	3	3	7	4	4	1	1	2	4	4	4

X; Observed

April 2004 represented the first field collection, November 2004 represented macroalgae community before the tsunami and January 2005 represented the macroalgae community 5 days after the tsunami hit; and January 2006 represented the recovery of macroalgae, a year after the tsunami catastrophe.

Statistical analysis

Two-way analyses of variance (ANOVA) were employed to test whether there were differences in percentage cover of macroalgae between (1) sites and seasons and (2) sites and tsunami (comparing April 2004, November 2004, January 2005 and January 2006). The graphical outputs show only the dominant species at each site and focus on the effects of the tsunami and species recovery abilities. Prior to analysis, homogeneity of variance was tested using Cochran's test and, where necessary, data were transformed. When significant differences between means were found, a Tukey's multiple comparison was applied (Zar 1984). Statistical analyses were carried out using STATISTICA version 6.0.

Results

There were variations in the diversity and community structure of seaweed between sites and seasons. A total of 18 species were found in this study. The highest diversity was found in April 2004 at Site 1, with 11 species of macroalgae (Table 1). *Laurencia composita*, *Lyngbya majuscula*, *Padina sanctae-crucis* and *Gelidella acerosa* were found at all three sites. *L. composita* and *P. sanctae-crucis* were the most common species. Five days after the tsunami, 1 January 2005, only four species remained. One year later at the beginning of January 2006, seven species were found. Thirteen months after the pre-tsunami study in April 2004, only 50% of the original flora was present. This suggests that the recovery of seaweed diversity was slow. Thirteen species showed significant differences in percentage cover with respect to sites, seasons and their interactions (Table 2), while 11 species showed significant differences in percentage cover with respect to sites, before/after the tsunami and their interactions (Table 2). *P. sanctae-crucis*, *L. composita* and *S. stolonifolium* were dominant and had the highest percentage cover at sites 1, 2 and 3, respectively. The percentage cover of *P. sanctae-crucis* fluctuated significantly throughout the study. The highest percentage cover was 31±10.77% (mean±S.E.) in May 2005. However, the population had disappeared from all sites in the immediate post-tsunami study in January 2005 (Fig. 2a). The fan-shaped form of *P. sanctae-crucis* recovered very slowly and did not reach the same abundance as before the tsunami. The percentage cover of *L. composita*, the dominant red alga, decreased sharply. Only 3±2% was left in January 2005 compared to 34±9.72% in November 2004, a month prior the tsunami catastrophe (Fig. 2b). We also observed these seaweeds as well as the seagrass, *Syringodium isoetifolium* (Aschers.) Dandy, floating and washed up on the shore in January 2005. The recovery of both *P. sanctae-crucis* and *L. composita*, continues to be very slow and they have not, even in 2007, reached pre-tsunami levels.

Although study Site 3 was not directly exposed to the impact of the tsunami, the seaweed community there was also affected. The percentage cover of *S. stolonifolium* dropped by 26%, from 83±3% in November 2004 to 57±2% (Fig. 2c). Shorter fronds were observed and many *S. stolonifolium* specimens were washed up on the shore. However, recovery was relatively fast and the percentage cover had already reached 76±5.57% by January 2006. The increase in percentage cover of all three species came from old populations, which were not removed during the tsunami. No new recruitment was observed.

There was considerable sedimentation, especially at site 2. We observed greater than 15 cm sediment covering the permanent plots. After 5 days, the sediment surface was covered by large patches of microbial organisms, turning the sediment surface white (Fig. 3a). Under the sediment, many benthic organisms, such as corals, molluscs, sponges and macroalgae were still present (Fig. 3b). These organisms died off and black anoxic conditions were observed. The strength of the tsunami wave differed from site to site. Site 3 was less affected since it does not directly face the direction from which the tsunami wave came, while Site 1 was less affected because of the more gradual slope of the beach.

Discussion

The variations in diversity of seaweed at Talibong Island seem to be the result of differences in substrates and degree of wave exposure. The complex nature of substrates is known to influence the diversity of algae and marine benthos (Dean and Connell 1987). The greater diversity in Site 1 might be due to the variety of substrate as well as having been exposed to less active waves during the rainy season. The difference in substrates also influenced the species composition at each site. *Caulerpa* spp. was dominant at Site 1, on the sandy substrate that allows stolons and rhizoids to grow into the sand. *S. stolonifolium* was dominant on the rocky substrate of Site 3, where the discoid holdfast is adapted to attach.

In general, after the monsoon season, macroalgal abundance drops, perhaps due to the high wave energy,

Table 2 Summary of significant results (*F*-ratio) from two-way ANOVA test for effects of (1) different months on three different sites, and (2) tsunami event on three different sites

	Season			Tsunami event		
	Site	Month	Site* month	Site	Before/after the tsunami	Site before/after the tsunami
Caulerpa taxifolia	1.93^{ns}	1.23^{ns}	1.72^{ns}	3.22*	1.4^{ns}	1.13^{ns}
C. racemosa	9.86***	10.79***	9.93***	9.86**	10.79**	9.92***
Dictyosphaeria carvernosa	2.0^{ns}	2.39**	2.55**	0.96^{ns}	0.97^{ns}	0.98^{ns}
Dictyota dichotoma	4.32*	4.47***	4.46***	4.29**	4.44**	4.42**
D. cervinoides	2.15^{ns}	6.9***	5.87***	2.96^{ns}	6.42***	7.59***
Filamentous red algae	1.1.3^{ns}	1.13^{ns}	1.13^{ns}	1.12^{ns}	1.12^{ns}	1.12^{ns}
Galaxuala obtusata	1.35^{ns}	0.77^{ns}	1.23^{ns}	0.97^{ns}	0.75^{ns}	1.31^{ns}
Gelideila acerosa	2.20^{ns}	5.31***	1.51^{ns}	0.29^{ns}	12.12***	1.21^{ns}
Gelidiopsis sp.	15.53***	15.67***	15.69***	15.51***	15.54***	15.62***
Gelidium sp.	1.54^{ns}	3.38***	2.50**	1.76^{ns}	0.70^{ns}	0.71^{ns}
Jania sp.	14.47***	14.57***	14.59***	14.42***	14.47***	14.52***
Laurencia composita	15.24***	6.03***	8.57***	86.78***	8.4**	13.19***
Lyngbya sp.	43.57***	35.71***	23.15***	79.85***	93.07***	80.51***
Padina sanctae-crucis	28.12***	10.07***	7.38***	13.79***	23.84***	9.28***
Sargassum stolonifolium	600.96***	2.56*	56.24***	599.39***	9.99***	5.85***
Symploca sp.	2.30^{ns}	3.12**	3.54***	3.50*	3.51*	3.52**
Turbinaria ornata	5.66**	6.88***	8.57***	1.08^{ns}	0.83^{ns}	0.92^{ns}
Valonia pachynema	0.97^{ns}	0.98^{ns}	0.98^{ns}	0.97^{ns}	0.97^{ns}	0.97^{ns}

*$P<0.05$, ** $P<0.1$, *** $P<0.001$; ns not significant; seaweed species not presented showed no significant relationship

which stirs up sediments and covers small benthic algae such as *G. acerosa*. Fragments of *L. composita*, a fragile red alga, broken by the strong wave action were found floating. The typical seasonal pattern in the area has greater diversity and abundance in early wet season (Prathep 2005). The post-tsunami macroalgal community is similar to that of the post-monsoon period, but the magnitude of the tsunami wave energy was much stronger. Also, post-tsunami recovery is slower. The magnitude of the wave is comparable to those occurring during hurricanes in Australia, the United States and the Caribbean, which damage marine communities (Fourqurean and Rutten 2004; Preen et al. 1995; Lapointe et al. 2006). The diversity and abundance of macroalgae at all three monitoring sites at Talibong Island decreased either directly from the physical impact or indirectly from sediments.

The strong wave action removed not only the macroalgae but also the substrates. This, however, varied in degree of impact. The direct impact of the severe wave energy on shallow nearshore habitats (including coral reef ecosystems, seagrasses and mangroves) was extensive, although dependent on the amount of wave energy these ecosystems are normally exposed to. Organisms already adapted to high wave energy, as seen on temperate shores (Lewis 1964; Stephenson and Stephenson 1972) and in intertidal tropics shores (Prathep 2005; Prathep et al. 2007), were less affected than sheltered habitat organisms. On the other hand, shallow bays typically protected from high wave action, such as Site 3, suffered less extensive damage. However, *S. stolonifolium* at the latter site declined greatly after the tsunami but fully recovered after 1 year.

Damage by wave energy is species specific. Some species, such as *L. composita* and *P. sanctae-crucis*, are delicate and did not withstand the turbulent high energy environments. Neither species has recovered yet. Growth from solid, deeply anchored rhizome or rhizoid systems, combined with flexible fronds as seen in both seaweeds and seagrasses, is adapted to resist perturbation by hurricanes or storms (Cruz-Palacios and van Tussenbroek 2005). However, since the magnitude of the tsunami wave energy was much higher than the usual storms, many seagrasses were also uprooted. The thin root-rhizome of *Syringondium isoetifolium* was more vulnerable than the deeply rooted *Enhalus acriodes*. *Caulerpa brachypus* forma *parvifolia* (Harvey) Cribb, with its small stolon and rhizoid system, was also extensively removed by hurricanes in Florida (Lapointe et al. 2006).

Although wave forces and currents are known to influence recruitment, dispersal, community structure and dislodgement of algae and marine benthic organisms (Ballantine 1961; Lewis 1964; Norton 1991; Denny 1995), the impacts on macroalgae have only recently been studied (Thomsen and Wernberg 2005). Thus, more studies are still needed to understand what causes the dislodgment of macroalgae. During catastrophic events such as tsunami or hurricanes, measurements of current and wave energy

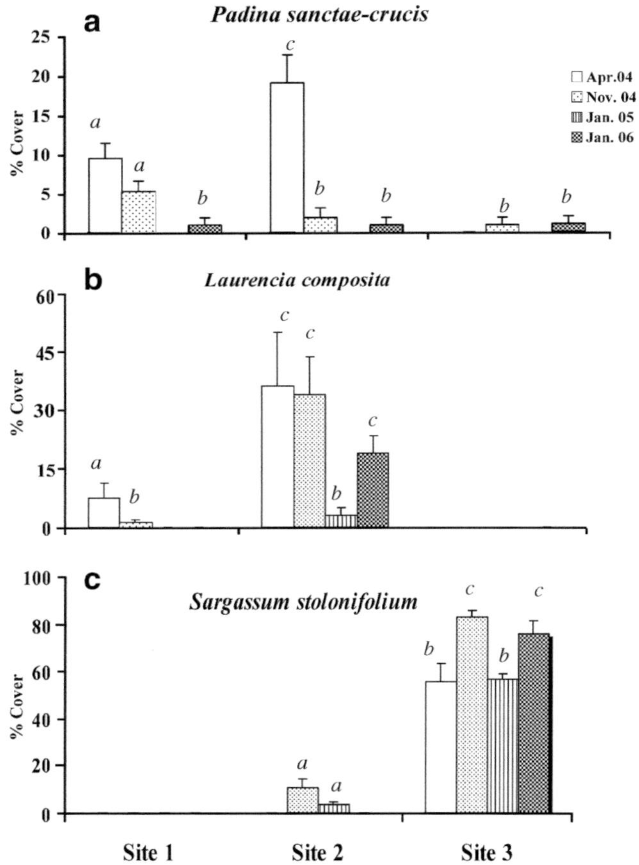

Fig. 2 Variations in percentage cover of some dominant macroalgae. **a** *Padina sanctae-crucis*, **b** *Laurencia composita*, **c** *Sargassum stolonifolium*. *Bars* Standard error ($n=5$). Letters *a–c* indicate significant group of means found with the Tukey HSD tests

were not reported; such information would provide us with a better understanding of the impacts on the benthic community, including its dislodgement and recovery.

The indirect impact of the tsunami was sedimentation. The water became more turbid with sediment, which was deposited, covering the macroalgae and corals, within 5 days after the tsunami. Sediment 15 cm deep was deposited on the permanent plots at Site 2. Although sediment dynamics (removal and deposition) are seasonal and play important roles in the marine community (Prathep et al. 2003), this much greater deposition had a critical influence on all marine communities. Such sediment deposition and removal are characteristic of hurricanes (Ballantine 1984; Glynn et al. 1964). Experiments on hurricane-like disturbances have shown that seagrasses and seaweeds buried under 10 cm sediment were more likely to suffer permanent damage than those buried under only 5 cm or without any sediment (Cruz-Palacios and van Tussenbroek 2005). However, since the macroalgal community at Site 2 was buried under about 15 cm, many organisms were quickly subjected to anoxic conditions and died only 3–5 days after the tsunami. Also, recovery from such burial conditions seems to be very slow.

The resilience of coastal marine organisms and their ability to recover has been extensive discussed (Connell 1997; Nyström et al. 2000; Bengtsoon et al. 2003; Turner et al. 2003; Hughes et al. 2005; Pascual and Guichard 2005; Gallopín 2006; Woodroffe 2007). The recovery of the macroalgal community after 13 months is very slow and varied from site to site and species to species. The high sedimentation in Site 2 has totally changed the substrate to a soft bottom, thus limiting the recruitment and succession of those macroalgae previously adapted to a solid substrate. Once a species is removed, the process of successful recovery is complex and dependent on factors such as the degree and time of removal (Sousa 1984a, 1985), available

Fig. 3 Greater sediment causing anoxic conditions at the study sites 5 days after the tsunami. **a** Large area covered by microbes, making the sediment surface white; **b** various small benthic organisms die off after being buried, turning the sediment underneath black

space (Sousa 1984b; Farrell 1989) and dispersal mechanisms (Keough 1984).

Abundance was affected by whether or not some parts of plants were left, in which case the recovery was faster such as with *L. composita* and *S. stolonifolium*. However, recovery is unlikely to establish the same community as was present before the tsunami since some species had still not recovered a year after the tsunami. A small scale disturbance, e.g. such as that from a boat anchor as experimentally tested in seagrass beds, showed that some associated seaweeds would recover after 4–5 months, but some species might take longer than 13 months (Creed and Amado Filho 1999). However, the high magnitude of the 2004 tsunami wave energy caused greater damage and it is taking much longer for the community to become reestablished. Further investigations and monitoring will allow us to understand the acute effects of the disturbance on the fragile macroalgae community, which is less well known than seagrasses, coral and mangroves. Such monitoring will also provide a better understanding of various ecological theories related to natural disturbances, recruitment and succession.

Acknowledgements We are very grateful for the support of the members of the Seaweeds and Seagrass Research Unit, Centre of Biodiversity in Southern Peninsular Thailand. Thanks also to Professor Larry B. Liddle for manuscript improvement. Knowledge Network Institute of Thailand, Unocal Songkhla and Prince of Songkla University, Thailand provided funding; and Japanese Society for the Promotion of Science and Professor Hisao Ogawa provided support for the study.

References

Ballantine WJ (1961) A biologically-defined exposure scale for the comparative description of rocky shore. Field Stud 1:1–19

Ballantine DL (1984) Hurricane-induced mass mortalities to a tropical subtidal algal community and subsequent recoveries. Mar Ecol Prog Ser 20:75–83

Bengtsoon J, Angelstam P, Elmqvist T, Emanuelsson U, Folke C, Ihse M, Moberg F, Nyström M (2003) Reserves, resilience and dynamic landscapes. Ambio 32(6):389–396

Changsang H, Poovachiranon S (1994) The distribution and species composition of seagrass beds along the Andaman sea coast of Thialand. Phuket Mar Biol Cent Res Bull 59:43–52

Chavanich S, Siripong A, Sojisuporn P, Menasveta P (2005) Impact of Tsunami on the seafloor and corals in Thailand. Coral reefs 24:535

Connell JH (1997) Disturbance and recovery of coral assemblages. Coral Reefs 16(Suppl.):S101–S113

Creed JC, Amado Filho GM (1999) Disturbance and recovery of the macroflora of a seagrass (*Halodule wrightii* Ascherson) meadow in the Abrolhos Marine National Park, Brazil: an experimental evaluation of anchor damage. J Exp Mar Biol Ecol 235:285–306

Cruz-Palacios V, van Tussenbroek BI (2005) Simulation of hurricane-like disturbance on a Caribbean seagrass bed. J Exp Mar Biol Ecol 324:44–60

Dahdouh-Guebas F, Jayatissa LP, Di Nitto D, Bosire JO, Lo Seen D, Koedam N (2005) How effective were mangroves as a defence against the recent tsunami? Curr Biol 15:R443–R447

Danielsen F, SØrensen MK, Olwig MF, Selvam V, Parish F, Burgess ND, Hiraishi T, Karunagaran VM, Rasmussen MS, Hansen LB, Quarto A, Suryadiputra N (2005) The Asian tsunami: a protective role coastal vegetation. Science 310:643

Dean RL, Connell JH (1987) Marine invertebrates in an algal succession. II. Test of hypotheses to explain changes in diversity with succession. J Exp Mar Biol Ecol 109:217–247

Denny M (1995) Predicting physical disturbance—mechanistic approaches to the study of survivorship on wave-swept shores. Ecol Monogr 65:372–418

Farrell TM (1989) Succession in a rocky intertidal community: the importance of disturbance size and position within a disturbed patch. J Exp Mar Biol Ecol 128:57–73

Fourqurean JW, Rutten LM (2004) The impact of Hurricane Georges on soft-bottom back reef communities: site and species-specific effects in south Florida seagrass beds. Bull Mar Sci 75:239–257

Gallopín GC (2006) Linkages between vulnerability, resilience, and adaptive capacity. Global Environ Change 16:293–303

Glynn PW, Almodóvar LR, Gonzales JG (1964) Effects of Hurricane Edith on marine life in la Parguera, Puerto Rico. Carib J Sci 4:335–345

Hughes TP, Bellwood DR, Folke C, Steneck RS, Wilson JE (2005) New paradigms for supporting the resilience of marine ecosystems. Trends Ecol Evol 20(7):380–386

Keough MJ (1984) Effects of patch size on the abundance of sessile marine invertebrates. Ecology 65:423–437

Kumaraguru AK, Jayakumar K, Jerald Wilson J, Ramakritinan CM (2005) Imapact of the tsunami of 26 December 2004 on the coral reef environment of Gulf of Mannar and Palk Bay in the Southeast coast of India. Curr Sci 89(10):1729–1741

Lapointe BE, Bedford BJ, Brumberger R (2006) Hurricanes Frances and Jeanne Remove Blooms of the Invasice Green Alga *Cualerpa brachypus* forma *parvifolia* (Hrvey) Cribb from coral reefs off northern Palm Beach County, Florida. Estuar Coast 29 (6A):966–971

Lewis JR (1964) The ecology of rocky shores. English University Press, London

Mantri VA (2005) Changes in local intertidal seaweed habitats in the Andaman andNicobar islands after 26 December 2004 tsunami. Curr Sci 89(7):1071–1072

Mantri VA (2006) Seaweed floristic studies along tsunami affected Indian coasts: a litmus test scenario after 26th December 2004. J Earth Syst Sci 115(3):371–378

Norton TA (1991) Conflict in constraints on the form of intertidal algae. Br Phycol J 126:203–218

Nyström MC, Folke C, Moberg F (2000) Coral reef disturbance and resilience in a human-dominated environment. Trends Ecol Evol 15:413–417

Pascual M, Guichard F (2005) Criticality and disturbance in spatial ecological systems. Trends Ecol Evol 20(2):88–95

Phongpaichit S, Preedana S, Rungjindama N, Sakayaroj J, Benzies C, Chuapat J, Plathong S (2006) Aspergillosis of the gorgonian sea fan *Annella* sp. after the 2004 tsunami at Mu Ko Similan National Park, Andaman Sea, Thailand. Coral Reefs 25:296

Poovachiranon S, Changsang H (1994) Community structure and biomass of seagrass beds in th Andaman sea. I Mangrove associated seagrass beds. Phuket Mar Biol Cent Res Bull 59:53–64

Prathep A (2005) Spatial and temporal variations in diversity and percentage cover of macroalgae at Sirinart Marine National Park, Phuket Province, Thailand. Sci Asia 31:225–233

Prathep A, Tantiprapas A (2006) Preliminary report on the diversity and community structure of macroaglae before and after the 2004 tsunami at Talibong island, Trang province, Thialand. Coast Mar Sci 30(1):189–195

Prathep A, Marrs RH, Norton TA (2003) Spatial and temporal variations in sediment accumulation in an algal turf and their impact on associated fauna. Mar Biol 142:381–490

Prathep A, Wichachucherd B, Thongroy P (2007) Spatial and temporal variation in density and thallus morphology of *Turbinaria ornata* in Thailand. Aquat Bot 86:132–138

Preen AR, Long WJ, Coles RG (1995) Flood and cyclone related loss, and partial recovery of more than 1000 km^2 of seagrass in Harvey Bay, Queenland, Australia. Aquat Bot 52:3–17

Ramesh S, Jayaprakashvel M, Mathivanan N (2006) Micorbial status in seawater and coastal sediments during pre-and post-tsunami periods in the Bay of Bengal, India. Mar Ecol 27:198–203

Saito Y, Atobe S (1970) Phytosociological study of intertidal marine algae I. Usujiri Benten-Jima, Hokkaido. Bull Fac Fish Hokkaido Univ 21(2):37–69

Sanpanich K, Wells FE, Chitramvong Y (2006) Effects of the 26 December 2004 tsunami on littorinid mollusks near Phuket, Thailand. J Molluscan Stud 72:311–325

Sousa WP (1984a) The role of disturbance in natural communities. A Rev Ecol Syst 15:353–391

Sousa WP (1984b) Intertidal mosaics: patch size, propagule availability, and spatially variable patterns of succession. Ecology 65:1918–1935

Sousa WP (1985) Disturbance and patch dynamics on rocky intertidal shores. In: Pickett STA, White PS (eds). The Ecology of natural disturbance and patch dynamics. Academic, Orlando, pp 101–124

Stephenson TA, Stephenson A (1972) Life between the Tide-Marks on a Rocky Shore. Freeman, San Francisco

Thomsen MS, Wernberg T (2005) Miniview: What affects the forces required to break or dislodge macroalgae? Eur J Phycol 40 (2):139–148

Turner MG, Collins SL, Lugo AL, Magnuson JJ, Rupp TS, Swanson FJ (2003). Disturbance dynamics and ecological response: the contribution of long-term ecological research. BioScience 53 (1):46–56

Vermaat JE, Thampanya U (2006) Mangrove mitigate tsunami damage: a further response. Est Coast Shelf Sci 69:1–3

Woodroffe CD (2007) The natural resilience of coastal systems: primary concepts. In: McFadden L, Penning-Rowsell E, Nicholls RJ (eds) Managing Coastal Vulnerability. Elsevier, Amsterdam, pp 45–60

Zar JH (1984) Biostatistical analysis. Prentice-Hall, London

Differing patterns of *hsp70* gene expression in invasive and native kelp species: evidence for acclimation-induced variation

Sarah K. Henkel · Gretchen E. Hofmann

Originally published in the Journal of Applied Phycology, Vol 20, No 5, 465–474.
DOI: 10.1007/s10811-007-9275-3

Abstract Temperature is one of the primary factors determining the geographic boundaries of seaweeds. Thus, investigations of how seaweeds cope with temperature stress and what affects their ability for range expansion are particularly important when studying invasive species. In physiological ecology, an established index of thermotolerance is the up-regulation of heat shock genes and subsequent synthesis of heat shock proteins (Hsps). The goal of this study was to examine the up-regulation of the *hsp70* gene to assess physiological tolerances of the Eastern Pacific invasive kelp, *Undaria pinnatifida*, as compared to a potentially competing native kelp, *Egregia menziesii*. Individuals of both species were collected from six known *Undaria* invasion sites on the west coast of North America and held in the laboratory for 1–2 weeks for acclimation at 10°C. Samples were then heat shocked at 7 temperatures for 1 h. RNA was extracted, reverse transcribed, and amplified in quantitative PCR reactions to determine relative amounts of *hsp70* transcript. Results indicate that the native *Egregia* may be locally adapted to different thermal regimes across latitude, while the invasive *Undaria* populations exhibit similar expression profiles across latitude but differ by habitat.

Keywords *Egregia* · Heat shock · Real time PCR · Seaweed · Temperature · *Undaria*

S. K. Henkel (✉) · G. E. Hofmann
Department of Ecology, Evolution, and Marine Biology,
University of California,
Santa Barbara, CA 93106, USA
e-mail: henkel@lifesci.ucsb.edu

Introduction

The frequency of non-native species introductions in the marine environment and the magnitude of their impacts appear to be increasing (Harris and Tyrrell 2001). Consequently, biological invasions are becoming of great interest to scientists, economists, and, increasingly, the general public. When introduced species establish themselves in a new location and become invasive, they can change the community structure of that habitat by altering community composition (Battershill et al. 1998; Britton-Simmons 2004; Casas et al. 2004). Amongst invasives, large-sized algae (macrophytes) can have especially strong potential impacts by changing light availability (Britton-Simmons 2004; Sanchez et al. 2005), nutrient cycling (Torres et al. 2004), and food availability for herbivores (Britton-Simmons 2004). Additionally, as habitat forming organisms, invasive macrophytes can alter the species composition of an area resulting in economic damage to fishing, aquaculture, and SCUBA-based tourism (reviewed in Ribera and Boudouresque 1995). These losses may be due to the decline of economically valuable fishing species, interfering with the harvest of native kelps and/or changing underwater landscapes (Verlaque 1994).

It has been hypothesized that introduced species that become invasive may occupy a wider niche than in their native region, and their morphological, reproductive, and genetic characters may be different (Nei et al. 1975; Barton 1989; Voisin et al. 2005). Furthermore, it has been shown that exotics are sometimes able to outperform natives in their environments, indicating that the native species are not optimally adapted to their environments (reviewed in Sax et al. 2007). The question arises: do invaders have some selective or ecological advantage in their new habitat? The goal of this study was to explore whether a physiological character, thermotolerance, differed in invasive populations of the Eastern Pacific kelp, *Undaria pinnatifida* (Harvey) Suringar, compared to a native kelp species, *Egregia menziesii* (Turner) Areschoug.

In North America, *U. pinnatifida* (native to southeast Asia) was first identified in California in 2000 and has been found in intertidal and subtidal habitats, especially harbors, from Monterey Bay, California, to Baja California, México, (Silva et al. 2002; Aguilar-Rosas et al. 2004). *U. pinnatifida* is listed on the Global Invasive Species Database as one of the "100 worst invasive alien species" by the International Union for Conservation of Nature and Natural Resources. *U. pinnatifida* is especially worrisome as an introduced macrophyte because invasive populations seem to be more robust than populations in its native biogeographic range. For example, in France, *U. pinnatifida* grows to twice the average size that it does in its native Japan (Perez et al. 1988), and New Zealand and California populations have been shown to exhibit a more vigorous reproductive cycle than native *U. pinnatifida* populations in Japan, presumably based on differences in water temperatures (Hay and Villouta 1993; Thornber et al. 2004).

Much research has been conducted at the organismal level on invasive species in terms of their environmental tolerances (Peters and Breeman 1992; Viejo et al. 1995; Gacia et al. 1996; Gillespie et al. 1997), ability to colonize new areas (Forrest et al. 2000; Arrontes 2002), and competition versus native species (Arrontes 2002; Valentine and Johnson 2004). However, little research to date has investigated molecular-level, physiological traits that may enable seaweeds to invade new habitats. This study employed molecular techniques to identify physiological characters related to thermotolerance that may facilitate the invasion of *U. pinnatifida*. Investigations of thermotolerance in *U. pinnatifida* may be especially interesting as it can survive and reproduce in temperatures ranging from 0 to 30°C (Floc'h et al. 1991; Peters and Breeman 1992), and is the first macrophyte known to have successfully crossed the tropics from northern temperate waters to invade a temperate region in the southern hemisphere (Hay and Luckens 1987; Hay and Villouta 1993), indicating it may possess a range of temperature tolerances not previously recorded in kelp species. As temperature has long been thought to be one of the primary factors determining the geographic boundaries of seaweeds (Lüning et al. 1987; Breeman 1988; Lüning and Freshwater 1988), this study would elucidate a possible mechanism for how seaweeds cope with temperature stress, and, furthermore, what factors might limit their ability to expand their range.

Although there are many approaches to assessing an organism's response to temperature stress, in physiological ecology the up-regulation of heat shock genes (termed the heat shock response) is often used as an established index of thermotolerance (Feder and Hofmann 1999). In this process, rapidly synthesized molecular chaperones called heat shock proteins (Hsps) act to protect cellular proteins from thermal denaturation. The presence of Hsps and their induction by environmental stress has been shown to be a central tolerance mechanism in many organisms (Feder and Hofmann 1999). Investigations of the heat shock response in non-mobile organisms such as seaweeds are especially interesting because they are unable to escape from unfavorable conditions; thus they must make appropriate physiological and biochemical changes in order to deal with heat stress. Heat shock studies have been carried out with intertidal invertebrates (e.g., Hofmann 1999; Halpin et al. 2002; Tomanek 2002; Hamdoun et al. 2003; Sagarin and Somero 2006) and in terrestrial plants (e.g., Howarth 1991; Downs et al. 1998; Hong and Vierling 2000; Manitasevic et al. 2007). However, the heat shock response has been looked at in few species of macroalgae to date (Vayda and Yuan 1994; Lewis et al. 1998; Ireland et al. 2004; Li and Brawley 2004). Physiological comparisons in response to thermal stress between the native species and a competitive invader may help determine if thermal microhabitats are influential in facilitating invasions. This is especially important in benthic species such as kelps where community composition is often controlled by the order in which species colonize a habitat; thus, rising sea temperatures may give invasive species from warmer climates the competitive edge (Stachowicz et al. 2002).

We examined the up-regulation of the *hsp70* gene as a means to assess the temperature tolerances of the Eastern Pacific invasive kelp, *U. pinnatifida*, as compared to a potentially competing native kelp, *E. menziesii*. We hypothesized that the invasive *U. pinnatifida* would have a greater range of thermotolerance (have higher induction and peak expression temperatures of the *hsp70* gene) than the native *E. menziesii*. Knowledge of the potential for *U. pinnatifida* to expand beyond its current level of invasion in California is important for assessing where to focus prevention efforts as well as methods for controlling transmission.

Materials and methods

Undaria pinnatifida was collected from six locations where it has invaded the west coast of North America: Monterey Bay Harbor (MB), Santa Barbara Harbor (SBA), Los Angeles Harbor (LAX), Catalina Island (CAT), Coronado Island in San Diego Bay (SDO), and Todos Santos Island (TSI) in Baja California, Mexico (Fig. 1). Individuals from harbor populations were collected from the docks or pilings and were found growing at or just below the zero tide line. Catalina Island and Todos Santos Island populations were

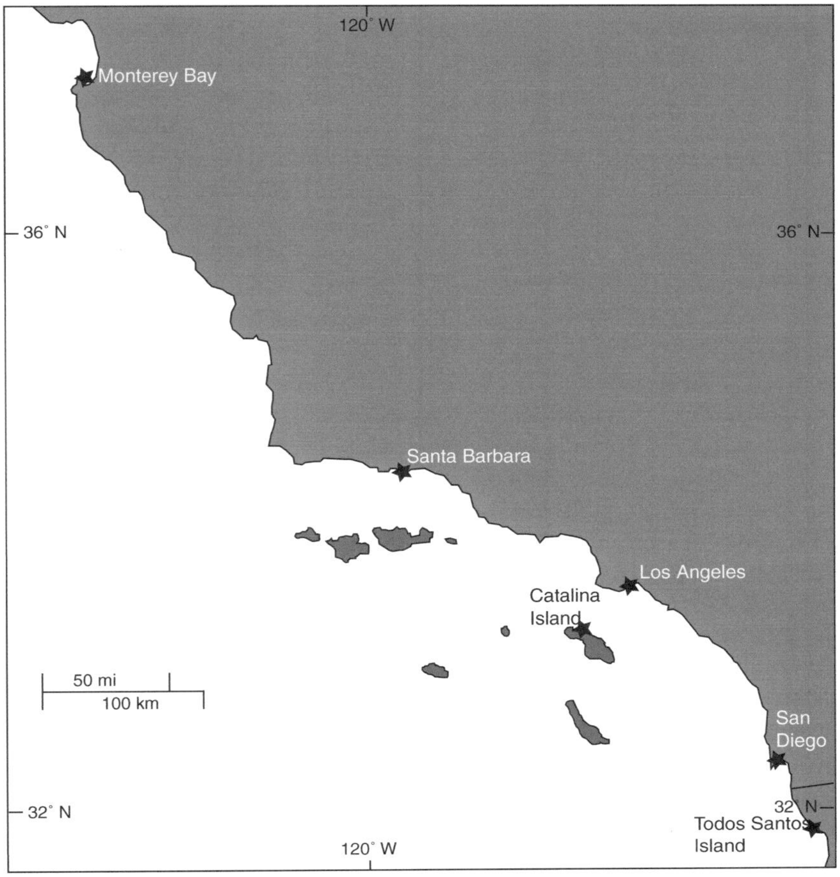

Fig. 1 *Undaria pinnatifida* and *Egregia menziesii* collection sites

subtidal, and individuals were collected from depths of 10–12 m and 7–8 m, respectively. The native kelp, *Egregia menziesii*, was collected at intertidal sites near *Undaria* sites. All samples were collected March–April 2006. Temperature data for mainland harbor sites were obtained from SCOOS (Southern California Ocean Observing System; http://www.sccoos.org/) and NOAA (http://www.ndbc.noaa.gov/, http://www.nodc.noaa.gov/). Benthic temperatures for Catalina Island were not available due to a lost logger. Benthic temperatures for Todos Santos Island were obtained from Centro de Investigacion Cientifica y de Educacion Superior de Ensenada (CICESE). A summary of collection sites and conditions is listed in Table 1.

Collected samples were brought back fresh from the field and acclimated by holding in recirculating seawater tanks at 10°C (the lowest temperature seen by any population; Santa Barbara Harbor, 10.2°C) for 5–14 days so that all individuals would have "unstressed" starting

Table 1 Location and habitat type of collection for *Undaria pinnatifida* and *Egregia menziesii*

Undaria pinnatifida		*Egregia menziesii*	
Collection site	Habitat type (temperature)	Collection site	Habitat type
Monterey Bay Harbor	Harbor (12.9–13.3°C)	Monastery Beach, Carmel	Intertidal
Santa Barbara Harbor	Harbor (12.7–14.0°C)	Ledbetter Beach, Santa Barbara	Intertidal
Cabrillo Harbor (Los Angeles)	Harbor (13.1–13.9°C)	Cabrillo Beach, Los Angeles	Intertidal
Long Point, Catalina Island	Benthic: 10–12 m (N/A)	Fisherman Cove, Catalina Island	Intertidal
Coronodo Island (San Diego)	Harbor (16.3–17.8°C)	La Jolla Shores, San Diego	Intertidal
Todos Santos Island	Benthic: 7–8 m (11.1–13.8°C)	Todos Santos Island	Intertidal

Temperatures reported are the average daily minimum and maximum recorded in April 2006 in the harbors and near the *Undaria* collection site at Todos Santos Island. Coronodo Island harbor temperatures are from Navy Pier in San Diego Bay, approx 1.5 km N of the collection site on Coronado Island conditions. After acclimation, 150–200 mg blade tissue from each specimen was placed in a 20-ml scintillation vial filled with filtered seawater and placed in an aluminum gradient heat block at approximately 12, 17, 22, 26, 31, 33, and 36°C for 1 h. After thermal incubation, samples were removed from heat, rinsed in 1X PBS, blotted dry, flash frozen in liquid N_2 and stored at −80°C until extraction.

For analysis of *hsp70* gene induction, total RNA was extracted from heat shocked samples following a modified protocol from Chang et al. (1993). Samples were ground to a powder using plastic pestles in liquid N_2; 600 µl of extraction buffer (pre-warmed at 65°C: 2% CTAB (w/v); 100 mM Tris-HCl (pH 8.0); 25 mM EDTA; 2 M NaCl; 2% PVP (w/v); 0.05% spermidine (w/v); 2% ß-mercaptoethanol (v/v)) was added to the powdered tissue and the sample was homogenized using the pestle for 30 s. Tubes were centrifuged at 12,000 *g* for 15 min at 10°C, and supernatant was subjected to two extractions with equal volume of 24:1 chloroform:isoamyl. RNA was precipitated from the final supernatant overnight at 4°C using 1/4 volume of 10 M LiCl. RNA was collected by centrifugation at 12,000 *g* for 20 min at 10°C. The pellet was then washed twice with cold (−20°C) 70% EtOH in DEPC-H_2O using 10 min 12,000 *g* spins at 10°C, dried under vacuum, and resuspended in DEPC-treated water. Samples were then DNase treated using the *Turbo-DNA free* kit (Ambion) and stored at −80°C. Total RNA (600 ng) was transcribed into cDNA using Stratascript cDNA synthesis kit (Stratagene) with oligo dT priming.

Partial sequences of *hsp70* were obtained for *Egregia* and *Undaria*. Degenerate primers were designed to obtain a central fragment of the gene based on conserved regions of *Ascophyllum nodosum hsp70* (accession no. X90948), *Guillardia theta hsp70* (accession no. X90991), and *Cyanophora paradoxa hsp70* (accession no. X90949). Polymerase chain reactions (PCRs) were carried out using 30 cycles of 45 s at 94°C, 45 s at 56°C (*Egregia*) or 50°C (*Undaria*), and 1 min at 72°C on cDNA isolated as described above. PCR products were ligated into a T-vector (pGEM-T easy; Promega; Fisher Scientific, Hampton, N.H., USA), and transformed into heat shock competent *Escherichia coli* (Promega; Fisher Scientific). Colonies were grown on ampicillin Luria-Bertani (LB) agar plates. Several colonies containing the ligated PCR product were selected and plasmids were isolated from liquid culture using GenElute Plasmid Miniprep kit (Sigma-Aldrich, St. Louis, Mo., USA) and sequenced on a CEQ 8000 (Beckman-Coulter, Fullerton, Calif., USA) using the DCTS Quick Start Kit and M13 primers. Consensus sequences were submitted to GenBank (*hsp70 Egregia menziesii* accession no. EF537648; *hsp70 Undaria pinnatifida* accession no. EF537647).

Beacon Designer (Bio-Rad, Hercules, Calif., USA) was used to design primers for qPCR based on conserved regions of *hsp70* for *Egregia* and *Undaria* and on published sequences of *ribulose* for *Egregia* (accession no. AF318970) and *Undaria* (accession no. AF318966). Sequences were as follows: *kelphsp70*-forward (CAGGTG TACGAGGGTGAG), *kelphsp70*-reverse (GTTAAGGAT ACCGTTGCG); *Egregia ribulose*-forward (ATGT CAAAAGGATGGGCTGT), *Egregia ribulose*-reverse (CAGGTTCGTTTGAAGGTCGT), *Undaria ribulose*-forward (CACAAGGATGTTTTTCGTTTTT), *Undaria ribulose*-reverse (GAAGACCCCATAATTCCCAAT).

Quantitative PCR (qPCR) was conducted by amplifying cDNA synthesis products (2 µL) with primers for *hsp70* and *ribulose* in 20 µL SYBR green supermix (Bio-Rad) reactions (run in duplicate). Primer concentrations were empirically determined based on lowest C_t values and highest efficiencies. A single cDNA standard was run on each plate to allow comparison across plates. Four 10-fold serial dilutions of the standard sample from each PCR plate were used as a standard curve to calculate PCR efficiency. Fluorescence threshold values were set at levels to maximize PCR efficiency, and melt curve analysis was performed to confirm that only a single product was amplified. To estimate for genomic DNA (gDNA) contamination, non-transcribed RNA from representative samples was also used as template for qPCR. Amplification showed that gDNA contamination constituted no more than 1/1,000 of the amplification observed with transcribed cDNA samples. Reported relative mRNA levels were calculated in the following fashion: (1) *hsp70* and *ribulose* C_t values were first normalized to that of the corresponding product of the cDNA standard from each plate, then (2) resultant *hsp70* values were normalized to the resultant *ribulose* values for the each sample. The *ribulose* gene was deemed to be an appropriate internal control because the *ribulose* mRNA levels did not differ significantly across heat incubations ($p=0.171$ for *Egregia* and $p=0.310$ for *Undaria*).

Each data set was examined for variance homogeneity using Cochran's test and transformed if necessary to meet assumptions of ANOVA using $y=x^{-0.2}$, chosen according to the method of Box and Cox (1964). Because ANOVA is robust to non-normality (Underwood 1997), departures were not considered reason to reject parametric procedures. Two-way ANOVA models were employed to analyze the effects of the fixed factors temperature and site and temperature and habitat on the transformed *hsp70* expression data. Where differences were significant, a Student–Newman–Keuls' (SNK) multiple comparison test was used to facilitate interpretation of ANOVA results. During the heat shock experiments, all kelp blades held at 36°C turned green and enough RNA to conduct qPCR was rarely recovered from these samples, likely due to cell autolysis. Because of the few data points for the 36°C incubation, these data were not included in the analyses.

Results

We measured levels of transcript of the inducible *hsp70* gene in harbor and benthic subtidal populations of the invasive kelp, *Undaria pinnatifida*, and in intertidal populations of the native kelp, *Egregia menziesii*, to determine if they responded differently to heat stress. Following 5–14 days of acclimation at 10°C, intertidal *Egregia* and subtidal *Undaria* populations differed in the thermal set points of the heat shock response (Table 2). Specifically, *Egregia* maintained population specific responses to heat stress that varied with latitude, while the invasive *Undaria* exhibited differences in the thermal set points and magnitude of expression of *hsp70* on the basis of habitat depth.

For *Egregia*, temperatures of induction (T_{on}) were usually equivalent to peak expression (T_{max}) of *hsp70* within a population (Table 2) and correlated negatively with the latitude of collection (T_{max} vs latitude; R^2=0.833, p= 0.005) (Fig. 2a). The general pattern of these set points was Baja California, México > southern California > Monterey Bay. Specifically, Todos Santos Island had both induction and peak expression at 33°C, T_{on} and T_{max} were 26°C for Santa Barbara, Catalina Island, and San Diego, Los Angeles had a T_{on} of 17°C followed by a T_{max} of 22°C, while the most northern population (Monterey Bay) had induction and peak expression at 17°C. The observed differences in the pattern of *hsp70* expression among populations were statistically significant (2-way ANOVA, F=19.661, p<0.001). Notably, the Los Angeles samples showed a response to temperature that was significantly different from all other populations, with much lower expression at all temperatures. Differences in *hsp70* expression among temperatures were not significant (2-way ANOVA, F=0.851, p=0.518), and there was no significant interaction between temperature and population (2-way ANOVA, F=1.171, p=0.293).

In contrast to the gradient of responses observed in *Egregia*, *Undaria* populations exhibited expression profiles more similar to one another (Fig. 2b), with T_{on} of 17°C or 22°C, T_{max} at 26°C for most populations (Table 2), and T_{off} between 26°C and 31°C. For *Undaria*, differences in *hsp70* expression both among populations (F=20.530, p<0.001) and across various temperatures (F=11.249, p<0.001) were statistically significant. Letters above the bars on Fig. 2b indicate significant subsets of temperature means. Again, there was no interaction between temperature and population (F=1.114, p=0.343).

The T_{max} for *Undaria* did not correlate with the latitude of collection (T_{max} vs latitude; R^2=0.011, p=0.699) as it did in *Egregia*. However, we did observe smaller scale differences among the populations. While the T_{on} of the *Undaria* population from Coronado Island in San Diego Bay was 22°C, like the other harbor populations, there was not a large amount of *hsp70* induction until 26°C. Notably, the water temperatures in San Diego Bay were considerably warmer than the other harbors during the study period (Table 1). Additionally, the T_{on} of the two benthic, subtidal populations was at a lower temperature (17°C) than the T_{on} of the four harbor populations (22°C) (Table 2). Since the range of temperatures experienced by most harbor populations of *Undaria* were similar during the study period and, in general, warmer than temperatures at depth where benthic *Undaria* were collected (Table 1), we grouped the results from harbor collected individuals and compared them to benthic individuals, as a group (Fig. 3). Two-way ANOVA indicated that both temperature (F=4.834, p<0.001) and habitat (F=8.495, p=0.004) were significant. SNK post-hoc analysis indicated that the difference between habitats was significant only at the17°C incubation temperature (p= 0.009), and not at 12°C (p=0.074) nor higher temperatures (p=0.123–0.861). The interaction between temperature and habitat was not significant (F=0.855, p=0.514).

Table 2 Temperatures of *hsp70* induction (T_{on}) and peak expression (T_{max}) for *Egregia menziesii* and *Undaria pinnatifida*. Site abbreviations: *MB* Monterey Bay, *SBA* Santa Barbara, *LAX* Los Angeles, *CAT* Catalina Island, *SDO* San Diego, *TSI* Todos Santos Island in Baja California, Mexico

	MB	SBA	LAX	CAT	SDO	TSI
Undaria						
T_{on} (°C)	22	22	22	17	22	17
T_{max} (°C)	26	26	22	17	26	26
Egregia						
T_{on} (°C)	17	26	17	26	26	33
T_{max} (°C)	17	26	22	26	26	33

Discussion

This is the first study to quantify the heat shock response in kelp. While direct comparison of the two species was not possible due to the use of species-specific primer sets and collection from different habitats, we believe that all populations exhibited responses reflective of their thermal history: *Egregia menziesii* varied along a latitudinal gradient while *Undaria pinnatifida* varied with depth. Thus, this study provides evidence for habitat specific gene expression responses to changing environmental conditions in two kelp species.

We found evidence for local acclimation along a latitudinal gradient in the thermal set-points of *hsp70* mRNA expression in *Egregia*, suggesting that kelp living in more northern, cooler waters maintain lower T_{on} and T_{max} than southern populations. The population specific response of *Egregia* is similar to that of many other

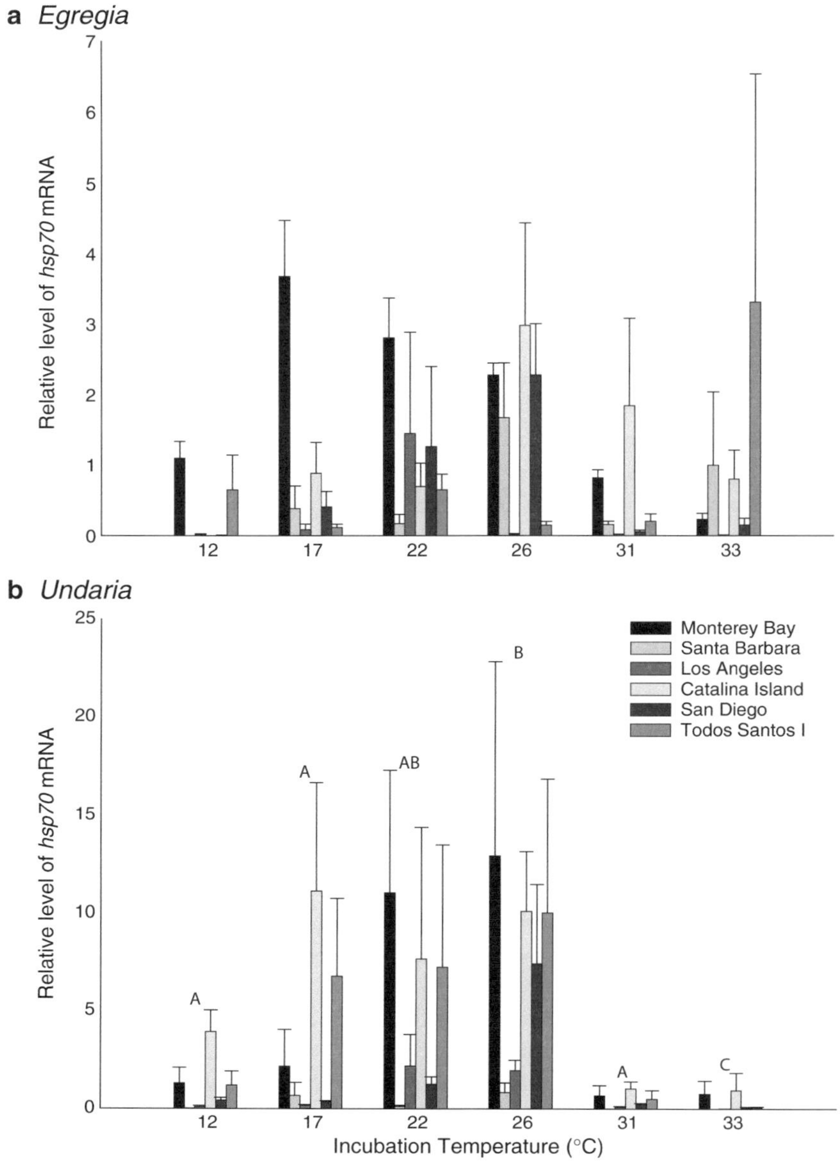

Fig. 2 Relative levels of hsp70 expression (±1 SE) from six populations at seven heat shock temperatures. Values are relative to a common standard for each species. **a** *Egregia menziesii*, *n*= 2–4. **b** *Undaria pinnatifida*, *n*= 4, except SBA 31 and 33°C and CAT all temperatures, *n*=3. *Letters* above the *bars* represent statistically significant subsets of means based on Student–Newman–Keuls' test results

organisms previously reported where intertidal invertebrates (Hofmann and Somero 1995; Tomanek and Somero 1999, 2000) and terrestrial plants (Kimpel and Key 1985; Barua et al. 2003) from warmer environments induce heat shock proteins at a higher temperature than those from colder environments. While we found differences in temperature set points along a latitudinal gradient, we did not see differences in the magnitude of *hsp70* expression among *Egregia* populations from different thermal regimes as has been observed in some studies (Osovitz and Hofmann 2005; Fangue et al. 2006).

The response to thermal stress exhibited by *Undaria* also suggests local acclimation. However, in contrast to the latitudinal differences observed in *Egregia*, differences in *Undaria* responses were observed along a vertical gradient. Harbor populations live near the surface and experience daily and seasonal heating due to solar radiation; these populations had a higher T_{on} and less overall expression of *hsp70* mRNA than benthic populations. The benthic populations sampled in this study were collected between 7 and 12 m and experienced lower temperatures, despite being at lower latitudes. The heat shock response of benthic individuals turned on at a lower temperature and was more robust than the harbor populations. The habitat specific heat shock response observed in the harbor versus benthic populations of *Undaria* is consistent with a number of studies in which differences were seen in the same species between different habitats and between subtidal versus

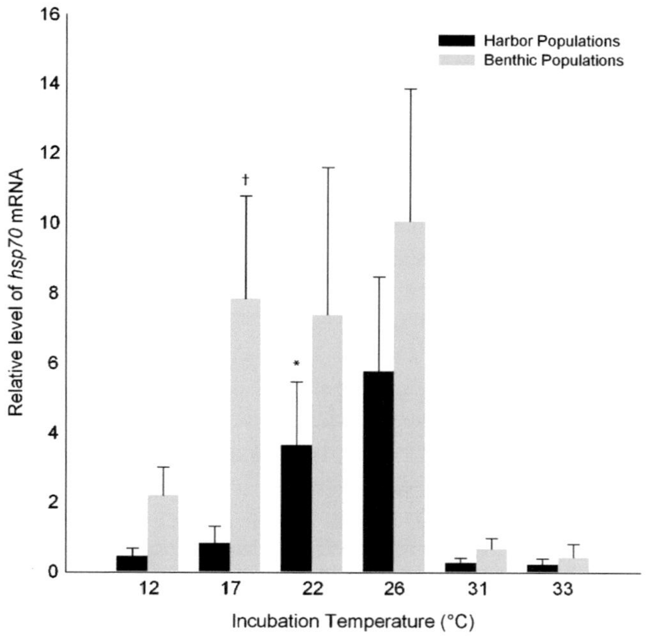

Fig. 3 Relative levels of *hsp70* expression (±1 SE) in *Undaria pinnatifida* grouped by habitat. n=15–16 for harbor specimens, n=7 for benthic specimens. † indicates T_{on} for benthic populations, * indicates T_{on} for harbor populations

intertidal invertebrate species (Sharp et al. 1994; Helmuth and Hofmann 2001; Halpin et al. 2002; Hamdoun et al. 2003).

In many cases, *hsp70* expression may be up-regulated at a lower temperature or in greater amounts in organisms that are from colder habitats than in those from warmer habitats (Hochachka and Somero 2002). This is often because organisms from warm environments maintain higher standing stocks of Hsps in their cells (Barua et al. 2003). For example, the magnitude of *hsp70* induction is greater in northern killifish, *Fundulus heteroclitus*, than in southern killifish (Fangue et al. 2006), and the mussel *Mytilus trossulus* has a lower T_{on} in winter than in summer (Buckley et al. 2001). The heat shock response of purple sea urchins, *Stongylocentrotus purpuratus*, exhibits both these trends: urchins collected from colder, Oregon waters have both a lower T_{max} and greater *hsp70* expression than those from warmer, southern California waters (Osovitz and Hofmann 2005).

Similar trends of local acclimation have been observed at the microhabitat level in protein expression. *Mytilus californianus* from warmer, horizontal substrata in California exhibit higher standing stocks of inducible Hsp72 protein than mussels attached to cooler, north-facing surfaces of rocks on the same shore (Helmuth and Hofmann 2001). Similarly, *Cellana ornata* limpets collected from open flat areas have higher amounts of Hsp70 than those collected from crevices, and *M. californianus* and *Lottia digitalis* from higher in the intertidal in Oregon have more Hsp72 than those from the mid and low-zones (Halpin et al. 2002). The relationship between standing stock of heat shock proteins and induction of more protein synthesis has been exhibited in the Pacific oyster, *Crassostrea gigas*, where levels of Hsc70 were as much as 100-fold greater in summer than in winter and the temperature of Hsp69 induction was lower in winter than in summer (Hamdoun et al. 2003). Habitat specific Hsp70 expression has also been observed in terrestrial plants where *Iris pumila* growing in open habitats exhibit more Hsp70s than samples collected from shaded areas (Manitašević et al. 2007).

One reason induction thresholds may shift is that Hsps are costly to make (Heckathorn et al. 1996; Feder and Hofmann 1999). Thus, local acclimation may prevent over expression of Hsps, effectively conserving cellular resources and energy. Furthermore, over expression may be correlated with reduced metabolic condition (Viant et al. 2003), may alter the specific activity of important enzymes (Krebs and Holbrook 2001), and may even be deleterious to growth and development (Krebs and Feder 1997), making a correctly tuned heat shock response vital. Many organisms have been shown to have a plastic thermotolerance response (Barua et al. 2003) in order to be maximally efficient in making the right amount of Hsps under changing environmental conditions.

Because the heat shock response can vary according to growth conditions, it is not surprising that intertidal *Egregia* exhibited site-specific differences on the basis of latitude while subtidal populations of *Undaria* did not. Intertidal temperatures along the coast can vary greatly, ranging from 7.5°C to 25.5°C during the course of this study (data not shown), while the temperatures experienced in the four harbors sampled differed much less (absolute range: 10.6–18.7°C). These intertidal temperatures include water temperatures during immersion and air temperatures during emersion. However, it has been well established that air temperatures alone are not an accurate measure of the body temperature of an intertidal organism during emersion and that factors such as solar radiation and wind have much more to do with determining an organism's body temperature (Helmuth 2002). Thus, intertidal *Egregia* likely experience, on a daily basis, a range of stressful abiotic environmental conditions, making it even more essential for *Egregia* to be locally acclimated to the variable conditions experienced along the latitudinal gradient of the west coast of North America.

From a mechanistic standpoint, inducible Hsp expression in all organisms is regulated by heat shock transcription factors (HSFs), which are complexed with Hsp70 in the cytoplasm and dissociate from the protein during stressful conditions when chaperoning is necessary (Wu 1995; Shi et al. 1998). This dissociation frees the HSF to translocate to the nucleus, bind to the heat shock element (HSE), a domain of the promoter region of heat shock genes, and activate transcription of more Hsps (Pelham 1982; Westwood et al. 1991). Observed differences in *hsp70* induction and expression in different kelp populations in this study were likely due to changes in heat shock factor binding as changes in HSF1 binding activity have been shown to parallel the induction pattern of *hsp70* mRNA (Lund et al. 2006). Furthermore, the kinetics of HSF1 DNA-binding activity vary as a function of acclimation temperature in fish (Buckley and Hofmann 2002, 2004; Lund et al. 2006).

Summary

The results of our investigation measuring levels of *hsp70* gene expression in invasive *Undaria pinnatifida* and native *Egregia menziesii* support the findings of a number of studies which indicate that the pattern of the heat shock response can be set by acclimation to different environmental conditions (Barua et al. 2003). Both species appear to be acclimated to the thermal regimes of their specific habitats: *Egregia* varies with latitude, and *Undaria* varies with depth. An alternative explanation for the absence of latitudinal variation in the heat shock response in *Undaria* may be the lack of genetic diversity among populations. Studies of the genetic diversity of *Undaria* in California and in Baja California, Mexico, indicate that the Monterey, Santa Barbara, and Los Angeles populations are the same haplotype, which is different from the individuals collected from Mexico (Voisin et al. 2005; Uwai et al. 2006). While the Catalina Island population was not sampled in those studies, it is likely it is the same haplotype as the Los Angeles population, owing to the great amount of vessel traffic between Los Angeles Harbor and Catalina Island.

Given the economic and biological impact of invasive macrophytes, studies that assess the physiological tolerances of these species are becoming more relevant and important. Although this study was not able to predict the competitive success of *Undaria* over *Egregia* or make predictions of further range expansion in *Undaria*, it has shown that investigations of heat shock gene expression in kelps are possible, and provides baseline data for continued studies in differential responses to environmental stress between native and invasive kelp species. Further studies including more native species and in complementary habitats are underway and will hopefully be able to isolate species-specific responses that are related to invasiveness.

Future studies comparing native and invasive macrophytes should be accompanied by measurements of photosynthetic performance in order to fully characterize differential temperature tolerances. Additionally, studies using both gene expression and photosynthetic performance measures on kelp gametophytes (alternate, microscopic life-history stage) may be most significant for assessing the potential spread and establishment of introduced kelp species. Studies have shown that *Undaria* gametophytes have high survival and recovery rates at temperatures 28.0–33.5°C, indicating they could survive passage through the tropics attached to ship hulls or in ballast water (Peters and Breeman 1992), making them the more likely vector for introduction.

Invasive species that are not competitive dominants have been shown to proliferate when native species are in decline due to climate change, increased storms, anthropogenic effects, and changes in food webs (Ambrose and Nelson 1982; Walker and Kendrick 1998; Valentine and Johnson 2004), and high temperature stress has led to the decline of native kelps in southern California (Seymour et al. 1989). If global temperatures rise and disturbance of the coastal zone continues to increase, information on how species will respond at both physiological and ecological levels to changes in temperature and other stressors such as pollution will become important not only for predicting invasions but for conservation of native species.

Acknowledgements This study was funded primarily by an EPA STAR grant S.K.H. Additional funds were obtained from a PISCO grant to G.E.H. The authors wish to acknowledge the following individuals who made this study possible: Steve Lonhart of the Monterey Bay National Marine Sanctuary for identification of *Undaria* locations in Monterey Harbor, and Harbor Master Steve Pryor for access to Monterey Harbor; Marla Ranelletti for identification of *Undaria* sites in Santa Barbara Harbor; Rachel Woodfield for identification of *Undaria* sites in Los Angeles Harbor; Erin Maloney of Moss Landing Marine Laboratories for identification of *Undaria* sites in San Diego Bay; Kathy Ann Miller of the University of California Berkeley for *Undaria* collection from Catalina Island; and Eugenio Carpizo of UABC and Elizabeth Hoaglund of UCSB for assistance with *Undaria* collection from Todos Santos Island in Baja California, Mexico. We appreciate temperature data for Todos Santos Island shared by Julio Palleiro of CICESE. This is contribution number 243 from PISCO, the Partnership for Interdisciplinary Studies of Coastal Oceans funded primarily by the Gordon and Betty Moore Foundation and David and Lucile Packard Foundation.

References

Aguilar-Rosas R, Aguilar-Rosas LE, Avila-Serrano G, Marcos-Ramírez R (2004) First record of *Undaria pinnatifida* (Harvey) Suringar (Laminariales, Phaeophyta) on the Pacific coast of Mexico. Bot Mar 47:255–258

Ambrose RF, Nelson BV (1982) Inhibition of giant kelp recruitment by an introduced brown alga. Bot Mar 25:265–267

Arrontes J (2002) Mechanisms of range expansion in the intertidal brown alga *Fucus serratus* in northern Spain. Mar Biol 141:1059–1067

Barton NH (1989) Founder effect speciation. In: Otte D, Endler JA (eds) Speciation and its consequences. Sinauer, Sunderland, Mass., pp 229–256

Barua D, Downs CA, Heckathorn SA (2003) Variation in chloroplast small heat-shock protein function is a major determinant of variation in thermotolerance of photosynthetic electron transport among ecotypes of *Chenopodium album*. Funct Plant Biol 30:1071–1079

Battershill C, Miller K, Cole R (1998) The Understorey of Marine Invasions. Seafood NZ 6:31–33
Box GEP, Cox DR (1964) An analysis of transformations. J R Stat Soc B 26:211–252
Breeman AM (1988) Relative importance of temperature and other factors in determining geographic boundaries of seaweeds: experimental and phenological evidence. Helgo Meer 42:199–241
Britton-Simmons KH (2004) Direct and indirect effects of the introduced alga *Sargassum muticum* on benthic, subtidal communities of Washington State, USA. Mar Ecol Progr Ser 277:61–78
Buckley BA, Hofmann GE (2002) Thermal acclimation changes DNA-binding activity of heat shock factor 1 (HSF1) in the goby *Gillichthys mirabilis*: implications for plasticity in the heat shock response in natural populations. J Exp Bio 205:3231–3240
Buckley BA, Hofmann GE (2004) Seasonal patterns and in vitro kinetics of HSF1 activation and Hsp70 mRNA production in the goby, *Gillichthys mirabilis*. Physiol Biochem Zool 77:570–581
Buckley BA, Owen ME, Hofmann GE (2001) Adjusting the thermostat: the threshold induction temperature for the heat-shock response in intertidal mussels (genus *Mytilus*) changes as a function of thermal history. J Exp Bio 204:3571–3579
Casas G, Scrosati R, Piriz ML (2004) The invasive kelp *Undaria pinnatifida* (Phaeophyceae, Laminariales) reduces native seaweed diversity in Nuevo Gulf (Patagonia, Argentina). Biol Inv 6:411–416
Chang S, Puryear J, Cairney J (1993) A simple and efficient method for isolating RNA from pine trees. Plant Mol Biol Rep 11:113–116
Downs CA, Heckthorn SA, Bryan JK, Coleman JS (1998) The methionine-rich low-molecular-weight chloroplast heat-shock protein: evolutionary conservation and accumulation in relation to thermotolerance. Am J Bot 85:175–183
Fangue NA, Hofmeister M, Schulte PM (2006) Intraspecific variation in thermal tolerance and heat shock protein gene expression in common killifish, *Fundulus heteroclitus*. J Exp Biol 209:2859–2872
Feder ME, Hofmann GE (1999) Heat-shock proteins, molecular chaperones, and the stress response: evolutionary and ecological physiology. Annu Rev Physiol 61:243–282
Floc'h JY, Pajot R, Wallentinus I (1991) The Japanese brown alga *Undaria pinnatifida* on the coast of France and its possible establishment in European waters. J Conseil 47:379–390
Forrest BM, Brown SN, Taylor MD, Hurd CL, Hay CH (2000) The role of natural dispersal mechanisms in the spread of *Undaria pinnatifida* (Laminariales, Phaeophyceae). Phycologia 39:547–553
Gacia E, Rodriguez-Prieto C, Delgado O, Ballesteros E (1996) Seasonal light and temperature responses of *Caulerpa taxifolia* from the northwestern Mediterranean. Aquat Bot 53:215–225
Gillespie RD, Meinesz A, Critchley AT (1997) Growth responses of *Caulerpa taxifolia* (Ulvophyceae, Chlorophyta) from the South African aquarist trade. A potential invasive of South African coastal waters. S Afr J Bot 63:480–483
Halpin PM, Sorte CJ, Hofmann GE, Menge BA (2002) Patterns of variation in levels of Hsp70 in natural rocky shore populations from microscales to mesoscales. Integr Comp Biol 42:815–824
Hamdoun AM, Cheney DP, Cherr GN (2003) Phenotypic plasticity of HSP70 and HSP70 gene expression in the Pacific Oyster (*Crassostrea gigas*): Implications for thermal limits and induction of thermal tolerance. Biol Bull 205:160–169
Harris LG, Tyrrell MC (2001) Changing community states in the Gulf of Maine: synergism between invaders, overfishing and climate change. Biol Inv 3:9–21
Hay CH, Luckens PA (1987) The Asian Kelp *Undaria pinnatifida* (Phaeophyta, Laminariales) found in a New Zealand harbor. NZ J Bot 25:329–332
Hay CH, Villouta E (1993) Seasonality of the adventive Asian kelp *Undaria pinnatifida* in New Zealand. Bot Mar 36:461–476
Heckathorn SA, Poeller GJ, Coleman JS, Hallberg RL (1996) Nitrogen availability alters patterns of accumulation of heat stress-induced proteins in plants. Oecologia 105:413–418
Helmuth B (2002) How do we measure the environment? Linking intertidal thermal physiology and ecology through biophysics. Int Comp Biol 42:837–845
Helmuth BST, Hofmann GE (2001) Microhabitats, thermal heterogeneity, and patterns of physiological stress in the rocky intertidal zone. Biol Bull 201:374–384
Hochachka PW, Somero GN (2002) Biochemical Adaptation. Oxford University Press, New York
Hofmann GE (1999) Ecologically relevant variation in induction and function of heat shock proteins in marine organisms. Am Zool 39:889–900
Hofmann GE, Somero GN (1995) Evidence for protein damage at environmental temperatures-Seasonal changes in levels of ubiquitin conjugates and Hsp70 in the intertidal mussel *Mytilus trossulus*. J Exp Biol 198:1509–1518
Hong SW, Vierling E (2000) Mutants of *Arabidopsis thaliana* defective in the acquisition of tolerance to high temperature stress. Proc Natl Acad Sci USA 97:4392–4397
Howarth CJ (1991) Molecular responses of plants to an increased incidence of heat-shock. Plant Cell Environ 14:831–841
Ireland HE, Harding SJ, Bonwick GA, Jones M, Smith CJ, Williams JHH (2004) Evaluation of heat shock protein 70 as a biomarker of environmental stress in *Fucus serratus* and *Lemna minor*. Biomarkers 9:139–155
Kimpel JA, Key JL (1985) Heat-shock in plants. Trends Biochem Sci 10:353–357
Krebs RA, Feder ME (1997) Deleterious consequences of Hsp70 overexpression in *Drosophila melanogaster* larve. Cell Stress Chaperon 2:60–71
Krebs RA, Holbrook SH (2001) Reduced enzyme activity following Hsp70 overexpression in *Drosophila melanogaster*. Biochem Genet 39:73–82
Lewis S, May S, Donkin ME, Depledge MH (1998) The influence of copper and heatshock on the physiology and cellular stress response of *Enteromorpha intestinalis*. Mar Environ Res 46:421–424
Li R, Brawley SH (2004) Improved survival under heat stress in intertidal embryos (*Fucus* spp.) simultaneously exposed to hypersalinity and the effect of parental thermal history. Mar Biol 144:205–213
Lund SG, Ruberte MR, Hofmann GE (2006) Turning up the heat: The effects of thermal acclimation on the kinetics of *hsp70* gene expression in the eurythermal goby, *Gillichthys mirabilis*. Comp Biochem Physiol A 143:435–446
Lüning K, Freshwater W (1988) Temperature tolerance of northeast Pacific marine algae. J Phycol 24:310–315
Lüning K, Guiry MD, Masuda M (1987) Upper temperature tolerance of North Atlantic and North Pacific geographical isolates of *Chondrus crispus* (Rhodophyta). Helgo Meer 41:297–306
Manitašević S, Dunderski J, Matic G, Tucic B (2007) Seasonal variation in heat shock proteins Hsp70 and Hsp90 expression in an exposed and a shaded habitat of *Iris pumila*. Plant Cell Environ 30:1–11
Nei M, Maruyama T, Chakraborty R (1975) The bottleneck effect and genetic variability in populations. Evolution 29:1–10
Osovitz CO, Hofmann GE (2005) Thermal history-dependent expression of the *hsp70* gene in purple sea urchins: Biogeographic patterns and the effect of temperature acclimation. J Exp Mar Biol Ecol 327:134–143
Pelham HR (1982) A regulatory upstream promoter element in the *Drosophila* hsp70 heat-shock gene. Cell 30:517–528
Perez R, Durand P, Kaas R, Barbaroux O, Barbier V, Vinot C, Bourgeay-Causse M, Leclercq M, Moigne JY (1988) *Undaria pinnatifida* on the French coasts. Cultivation method; biochemical

composition of the sporophyte and the gametophyte. In: Staedler T, Mollion J, Verdus MC, Karamanos Y, Morvan H, Chiristiaen D (eds) Algal biotechnology. Elsevier, London, pp 315–328

Peters AF, Breeman AM (1992) Temperature responses of disjunct temperate brown-algae indicate long-distance dispersal of microthalli across the tropics. J Phycol 28:428–438

Ribera MA, Boudouresque CF (1995) Introduced marine plants with special reference to macroalgae: mechanisms and impact. In: Round FE, Chapman DJ (ed) Progress in phycological research. Biopress, Bristol, pp 217–268

Sagarin RD, Somero GN (2006) Complex patterns of expression of heat-shock protein 70 across the southern biogeographical ranges of the intertidal mussel *Mytilus californianus* and snail *Nucella ostrina*. J Biogeogr 33:622–630

Sanchez I, Fernandez C, Arrontes J (2005) Long-term changes in the structure of intertidal assemblages after invasion by *Sargassum muticum* (Pheophyta). J Phycol 41:942–949

Sax DF, Stachowicz JJ, Brown JH, Bruno JF, Dawson MN, Gaines SD, Grosberg RK, Hastings A, Holt RD, Mayfield MM et al (2007) Ecological and evolutionary insights from species invasions. TREE 22:465–471

Seymour RJ, Tegner MJ, Dayton PK, Parnell PE (1989) Storm wave induced mortality of giant kelp, *Macrocystis pyrifera*, in southern California. Estuar Coast Shelf Sci 28:277–292

Sharp VA, Miller D, Bythell JC, Brown BE (1994) Expression of low molecular weight HSP70 related polypeptides from the symbiotic sea anemone *Anemonia viridis* Forskall in response to heatshock. J Exp Mar Biol Ecol 179:179–193

Shi Y, Mosser DD, Morimoto RI (1998) Molecular chaperones as HSF1-specific transcriptional repressors. Genes Dev 12:654–666

Silva PC, Woodfield RA, Cohen AN, Harris LH, Goddard JHR (2002) First report of the Asian kelp *Undaria pinnatifida* in the northeastern Pacific Ocean. Biol Inv 4:333–338

Stachowicz JJ, Terwin JR, Whitlatch RB, Osman RW (2002) Linking climate change and biological invasions: Ocean warming facilitates nonindigenous species invasions. Proc Natl Acad Sci USA 99:15479–15500

Thornber CS, Kinlan BP, Graham MH, Stachowicz JJ (2004) Population ecology of the invasive kelp *Undaria pinnatifida* in California: environmental and biological controls on demography. Mar Ecol Prog Ser 268:69–80

Tomanek L (2002) The heat-shock response: Its variation, regulation and ecological importance in intertidal gastropods (genus *Tegula*). Int Comp Biol 42:797–807

Tomanek L, Somero GN (1999) Evolutionary and acclimation-induced variation in the heat-shock responses of congeneric marine snails (genus *Tegula*) from different thermal habitats: Implications for limits of thermotolerance and biogeography. J Exp Biol 202:2925–2936

Tomanek L, Somero GN (2000) Time course and magnitude of synthesis of heat-shock proteins in congeneric marine snails (Genus *Tegula*) from different tidal heights. Physol Biochem Zool 73:249–256

Torres AI, Gil MN, Esteves JL (2004) Nutrient uptake rates by the alien alga *Undaria pinnatifida* (Pheophyta) (Nuevo Gulf, Patagonia, Argentina) when exposed to diluted sewage effluent. Hydrobiologia 520:1–6

Underwood AJ (1997) Ecological Experiments: Their Logical Design and Interpretation using Analysis of Variance. Cambridge University Press, Cambridge

Uwai S, Nelson W, Neill K, Wang WD, Aguilar-Rosas LE, Boo SM, Kitayama T, Kawai H (2006) Genetic diversity in *Undaria pinnatifida* (Laminariales, Phaeophyceae) deduced from mitochondria genes - origins and succession of introduced populations. Phycologia 45:687–695

Valentine JP, Johnson CR (2004) Establishment of the introduced kelp *Undaria pinnatifida* following dieback of the native macroalga *Phyllospora comosa* in Tasmania, Australia. Mar Freshw Res 55:223–230

Vayda ME, Yuan M-L (1994) The heat shock response of an antarctic alga is evident at 5°C. Plant Mol Biol 24:229–233

Verlaque M (1994) Inventaire des plantes introduites en Méditerraneé: origine et répercussions sur l'environnement et les activités humaines. Oceanol Acta 17:1–23

Viant MR, Werner I, Rosenblum ES, Gantner AS, Tjeerdema RS, Johnson ML (2003) Correlation between heat-shock protein induction and reduced metabolic condition in juvenile steelhead trout (*Oncorhynchus mykiss*) chronically exposed to elevated temperature. Fish Physiol Biochem 29:159–171

Viejo RM, Arrontes J, Andrew NL (1995) An experimental evaluation of the effect of wave action on the distribution of *Sargassum muticum* in northern Spain. Bot Mar 38:437–441

Voisin M, Engel CR, Viard F (2005) Differential shuffling of native genetic diversity across introduced regions in a brown alga: aquaculture vs. maritime traffic effects. Proc Natl Acad Sci USA 102:5432–5437

Walker DI, Kendrick GA (1998) Threats to macroalgal diversity: marine habitat destruction and fragmentation, pollution, and introduced species. Bot Mar 41:105–112

Westwood JT, Clos J, Wu C (1991) Stress-induced oligomerization and chromosomal relocalization of heat-shock factor. Nature 353:822–827

Wu C (1995) Heat shock transcription factors: structure and regulation. Annu Rev Cell Dev Biol 11:441–469

Early development of germlings of *Sargassum thunbergii* (Fucales, Phaeophyta) under laboratory conditions

Ziguo Zhao · Fengjuan Zhao · Jianting Yao · Jingmei Lu · Put. O. Ang Jr. · Delin Duan

Originally published in the Journal of Applied Phycology, Vol 20, No 5, 475–481.
DOI: 10.1007/s10811-008-9311-y

Abstract Morphology and culture studies on germlings of *Sargassum thunbergii* (Mertens et Roth) Kuntze were carried out under controlled laboratory conditions. Growth characteristics of these germlings grown under different temperatures (from 10 to 25°C), irradiances (from 9 to 88 μmol photons m^{-2} s^{-1}), and under blue and white light conditions are described. The development of embryonic germlings follows the classic "8 nuclei 1 egg" type described for Sargassaceae. Fertilized eggs spent 5–6 h developing into multicellular germlings with abundant rhizoids after fertilization. Under conditions of 20°C, 44 μmol photons m^{-2} s^{-1} and photoperiod of 12 h, young germlings with one or two leaflets reached 2–3 mm in length after 8 weeks. Temperature variations (10, 15, 20, 25°C) under 88 μmol photons m^{-2} s^{-1} significantly influenced the growth rate within the first week, although this effect became less obvious after 8 weeks, especially at 15 and 20°C. Variation in germling growth was highly significant under different irradiances (9, 18, 44, 88 μmol photons m^{-2} s^{-1}) at 25°C. Low temperature (10°C) reduced germling growth. Growth of germlings cultured under blue light was lower than in white light. Optimal growth of these germlings occurred at 25°C and 44 μmol photons m^{-2} s^{-1}.

Z. Zhao · J. Lu
School of Life Sciences, Northeast Normal University,
Changchun 130024, China

Z. Zhao · F. Zhao · J. Yao · D. Duan (✉)
Institute of Oceanology, Chinese Academy of Sciences,
Qingdao 266071, China
e-mail: dlduan@ms.qdio.ac.cn

P. O. Ang Jr.
The Chinese University of Hong Kong,
Shantin, N. T., Hong Kong SAR, China

Keywords *Sargassum thunbergii* · Germling · Growth and development · Temperature · Irradiance

Introduction

Sargassum thunbergii (Mertens et Roth) Kuntze is an important brown algae species forming extensive beds in the littoral regions along the coast of China, Japan and Korea. These beds are spawning, nursery and feeding ground for fish, shellfish and other marine organisms (Tsukidate 1984). In addition, *S. thunbergii* is also of great economic importance. It is widely used in alginate production (Tseng 1983), the extraction of natural bioactive products (Zhang et al. 1994; Wei et al. 2006; Lin et al. 2006), and biosorption of heavy metal ions (Padilha et al. 2005). It is also used as a source of feed for holothurian aquaculture in China. Increasing demand for *S. thunbergii* as raw material, especially in the rapidly developing aquaculture industries, has resulted in the depletion of natural populations along the northern coast of China. Hence, mariculture and artificial breeding of *S. thunbergii* have become crucial. Recently, artificial raft cultivation using holdfasts for regeneration of young germlings, or using wild juvenile plants as the seedling source, has been attempted in Weihai, Shandong Province of China (Zou et al. 2005; Yuan et al. 2006). However, the need for a large quantity of seedlings becomes a bottleneck for successful large scale farming. Therefore, it is necessary to establish a stable, efficient, and practical seedling production system to fit the breeding needs, and to guarantee the sustainable utilization of the wild *S. thunbergii* resource.

Previous trials to produce seedlings from zygotes in related species of *Sargassum*, such as *S. fusiforme* (Harv.) Setchell and *S. fulvellum* (Turn.) C. Agardh, have been documented (Ruan and Xu 2001; Pang et al. 2005, 2006; Hwang et al. 2006). Many studies have examined the biological processes of receptacle formation, egg release and fertilization (Sun et al. 1996; Ruan and Xu 2001), the optimal irradiance and temperature needed for the development of juvenile sporophytes obtained from zygotes (Hwang et al. 1997), and the tumble culture of receptacles and synchronized expulsion of eggs and sperm under controlled conditions (Pang et al. 2005).

Successful development of a new cultivation method for many seaweeds relies on the effective control of their reproduction and, hence, knowledge about their life history. For *S. thunbergii*, however, most studies have been concentrated on its ecology and phenology (Nakamura et al. 1971; Umezaki 1974; Wang 2003), phlorotannin content (Yan et al. 1996), or the extraction of bioactive products (Shi and Xu 1997). Knowledge of its reproduction and developmental characteristics of *S. thunbergii* is very limited as only preliminary investigations have been reported (Aral et al. 1985; Zheng and Chen 1993).

In this study, we cultured zygote-derived germlings of *S. thunbergii* under laboratory conditions with the aim of understanding the early developmental characteristics. We investigated the effects of different temperatures and irradiances on the growth of the early developmental stages of germlings. This information will be useful for the artificial seedling production of *S. thunbergii*.

Materials and methods

Specimens of *Sargassum thunbergii* with fertile receptacles were collected on 26 July 2006, in an intertidal zone of the Huiquan Bay (36°04′N, 120°21′E), Qingdao. The plants were brushed to eliminate surface epiphytes and washed thoroughly with sterile seawater. The distal parts of branchlets with receptacles, 6–8 cm in length, were excised and transferred to beakers containing 1,000 ml sterile seawater. Each beaker contained 3–4 branchlets from different individuals. The pre-culture conditions were 20°C, 40 μmol photons m^{-2} s^{-1} and a 12L:12D (light:dark cycle) until eggs or young germlings were released.

Treatment of germlings used for growth study

Shortly after the eggs or germlings were released, the algal plants were removed from the beaker. The remaining seawater suspension containing eggs or germlings was collected, combined, and distributed proportionally to about 20 petri dishes (60 mm × 15 mm) containing sterile seawater. The eggs or germlings in suspension were stored in a 10°C culture chamber at an irradiance of 18 μmol photons m^{-2} s^{-1} for 3 days to inhibit the growth of other algal species. Approximately 100 germlings were cultured in each petri dish, with a settlement density of 5~10 germlings per microscopic field (magnification: 10 × 10). Finally the petri dishes were transferred to culture chambers with different test conditions.

Effects of temperature and irradiance on growth

The petri dishes containing germlings were cultured in four light- and temperature-controlled incubators (Jiangnan, China) with the following factorial experimental design. Two petri dishes per culture condition were allocated. For temperature treatment, the germlings were cultured under four temperature conditions: 10, 15, 20, 25°C at an irradiance of 88 μmol photons m^{-2} s^{-1}. For irradiance treatment, the germlings were cultured under four irradiances of 9, 18, 44, and 88 μmol photons m^{-2} s^{-1} with the temperature kept at 25°C. An additional experiment was also carried out to examine the effect of light quality on germling growth. Blue LEDs (light-emitting diode) with maximum wavelength of 470 nm and four white 8-W fluorescent lamps (Philips, Shanghai, China) were used as the light sources. Both were then adjusted to the same irradiance of 73 μmol photons m^{-2} s^{-1}. All treatments were kept under a 12L:12D light:dark cycle. Sterilized filtered seawater was used as the culture medium and changed every 3 days. The whole experiment lasted 8 weeks.

Growth of germlings was measured weekly in terms of their length extension, excluding rhizoid, under a dissecting microscope. For each culture condition, 50 germlings from each petri dish were measured.

Morphological observation of germlings during the early development

The surface structures of fertile receptacles and the detailed early developmental process of germlings were observed respectively under a dissecting microscope and a light microscope and recorded with a digital camera. The germlings in two petri dishes incubated separately under conditions of 20°C and 44 μmol photons m^{-2} s^{-1} were selected as the main samples for morphological observation.

Data analysis

One-way ANOVA was performed followed by the Tukey's multiple comparison test to compare the length variations among the germlings grown under different temperature and irradiance treatments. Student's *t* test was used to compare the growth of germlings under white and blue light treatments. All statistical analyses were carried out using SPSS with the level of significance set at $p<0.01$. The parametric assumptions of the statistical models were evaluated and data transformed if necessarily to fit these assumptions. All data are expressed as mean ± SE.

Results

Morphological changes of germlings during the early developmental stage

One week after the algae were cultured under the pre-treatment experimental conditions, eggs were released from the conceptacles of receptacles in all female branchlets (Fig. 1a). This took place often in early morning. The eggs were fertilized within 24 h. The unfertilized eggs and young germlings tended to adhere to the surface of the receptacle for 10–24 h until shed naturally. A few of the germlings took up to 2–3 days before shedding and, at this stage, rhizoids had differentiated and attached onto the surface of the receptacle (Fig. 1b).

Freshly released unfertilized eggs had multiple nuclei scattered in the cytoplasm (Fig. 1f), which then changed into a large nucleus in the center after fertilization (Fig. 1c). The first cell division 1–2 h after fertilization was transverse across the longer axis of the zygote, dividing it into a two-cell germling with polarity, being obtuse at one end and cuspid at the another (Fig. 1d). The second cell division was transverse in the cuspid part of the germling (Fig. 1e), dividing it unequally into a larger body cell and a smaller rhizoid cell. Through subsequent cell divisions within 5–6 h, a multicellular germling was formed, characterized by multiple bundles of rhizoids (Fig. 1f,g).

After 24 h, the length of the rhizoids was almost equal to that of the germling (Fig. 2a). During development, the number of rhizoids was variable, but usually 8–10 were formed in the end (Fig. 2b). As well as this, numerous filaments composed of a row of long-columnar cells (Fig. 2c) developed on the distal parts of the germlings cultured for 2 weeks (Fig. 2c,e,f). The length of the germlings and rhizoids increased (Fig. 2d,e,f), and branchlets differentiated from some germlings after 3 weeks (Fig. 2f). These germlings attained a length of about 1.5 mm after 5 weeks (Fig. 2f). Some morphological differences, such as germling length and the number of rhizoids, were observed for germlings cultured under blue and white light conditions for 5 weeks (Fig. 2h,i).

Effects of temperature on germling growth

Temperature significantly affected growth (increase in length) of young germlings under 88 μmol photons m^{-2} s^{-1} irradiance as determined 1 week and 8 weeks after the start of the culture (ANOVA: $F=122.28$, $p<0.01$ and $F=108.32$, $p<0.01$, respectively) (Fig. 3). Germling growth increased with increasing temperature from 10 to 25°C. The maximum growth of 1,083 μm (1 week) and 2,840 μm (8 weeks) were both recorded under 25°C. In the first week, significant differences in germling growth were detected at all temperatures (Tukey's test, $p<0.01$), with the germling growth increasing linearly between 15 and 25°C. After 8 weeks, there was no significant difference in germling growth in culture under 15 and 20°C (Tukey's test, $p<0.01$).

Effects of irradiance on germling growth

Effects of irradiance on the growth of germlings at 25°C were evaluated at the same time as the temperature effects (Fig. 4). Germling growth occurred under all irradiances from 9 to 88 μmol photons m^{-2} s^{-1}. Significant difference in germling growth among treatments was already detected in the first week of the experiment (ANOVA: $F=43.25$, $p<0.01$). The difference was even more obvious and significant after 8 weeks (ANOVA: $F=195.45$, $p<0.01$). Howev-

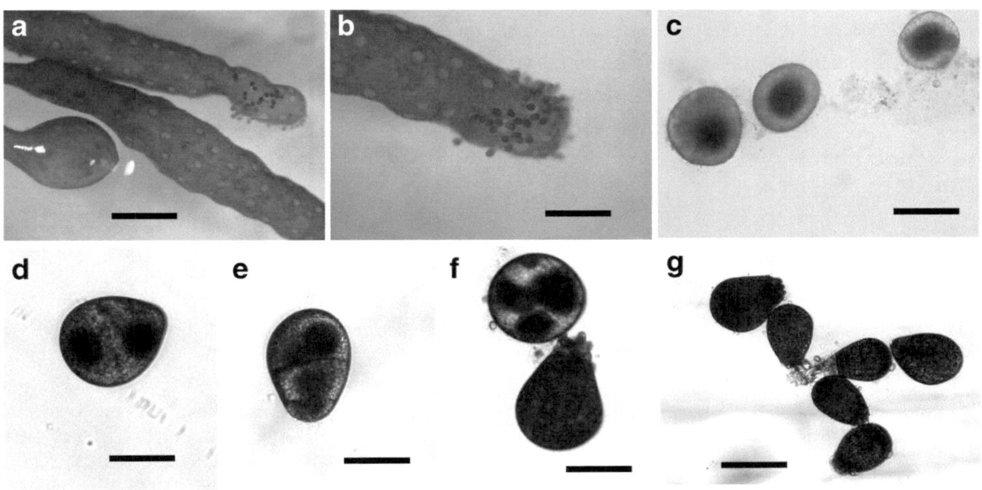

Fig. 1 Observation on early development of *Sargassum thunbergii*. **a** Profile of fertile female receptacle, *bar*= 1,000 μm; **b** eggs or germlings released from conceptacles on the receptacle surface, *bar*= 600 μm; **c** three fertilized eggs released to culture medium, *bar*=135 μm; **d** first cell division of fertilized eggs, *bar*=100 μm; **e** second cell division of fertilized eggs, *bar*=100 μm; **f** unfertilized eggs (with multiple nuclei) and developing germlings, bar= 100 μm; **g** naturally shed germlings, showing the abundance of rhizoids, *bar*=165 μm

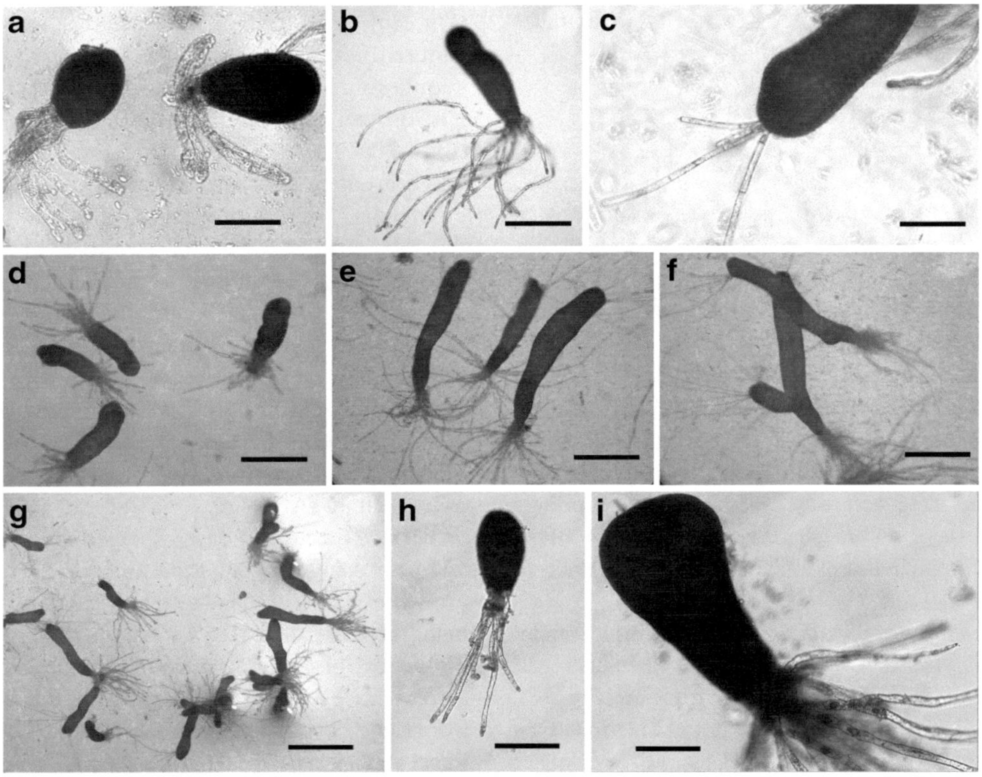

Fig. 2 Germlings of *Sargassum thunbergii*. **a** Young germlings with rhizoids on 3rd day after being released, *bar*=165 μm; **b** young germling cultured for 1 week under 20°C and 40 μmol photons m^{-2} s^{-1}, showing the number of rhizoids, *bar*=150 μm; **c** detail of the distal part of germling, showing the structure of the filaments, *bar*=150 μm; **d** young germlings cultured for 3 weeks, *bar*=720 μm; **e** developing germlings after 3 weeks, showing filaments and the increase in length, *bar*=720 μm; **f** branch differentiated from the developing germlings cultured for 4 weeks, *bar*=720 μm; **g** young germlings 1.5–2 mm in length cultured for 5 weeks, *bar*=51,050 μm; **h** young germling cultured at 10°C, under blue light, and irradiance of 73 μmol photons m^{-2} s^{-1} for 5 weeks, *bar*=250 μm; **i** young germling cultured at 10°C under white light, and irradiance of 73 μmol photons m^{-2} s^{-1} for 5 weeks, *bar*=150 μm

er, the increase in germling length did not follow the increase of irradiance (Fig. 4). In the first week, the highest mean germling growth length of 1,083±322 μm was recorded under 88 μmol photons m^{-2} s^{-1}. Significant differences in germling growth were detected under all pairs of irradiance levels (Tukey's test, $p<0.01$). After 8 weeks, the highest mean germling growth length of 3,065±427 μm was recorded under 44 μmol photons m^{-2} s^{-1}. The growth of germlings under low irradiance of 9 and 18 μmol photons m^{-2} s^{-1} became not significantly different (Tukey's test, $p>0.05$), and those under the high irradiance of 88 μmol photons m^{-2} s^{-1}even decreased significantly (Fig. 4).

Effects of light quality on germling growth

Effects of blue and white light on germling growth were examined separately 1, 5 and 8 weeks after the start of the culture (Fig. 5). Blue light significantly inhibited the growth of germlings when compared with white light (Student's *t*- test performed separately on all three groups of data: $p<0.01$; Fig. 2h,i). After 8 weeks, the means of germling lengths under blue and white light were 436±63 μm and 11,08±95 μm, respectively. In addition, there were no significant differences between the lengths of the germlings grown under blue light for 5 and 8 weeks,

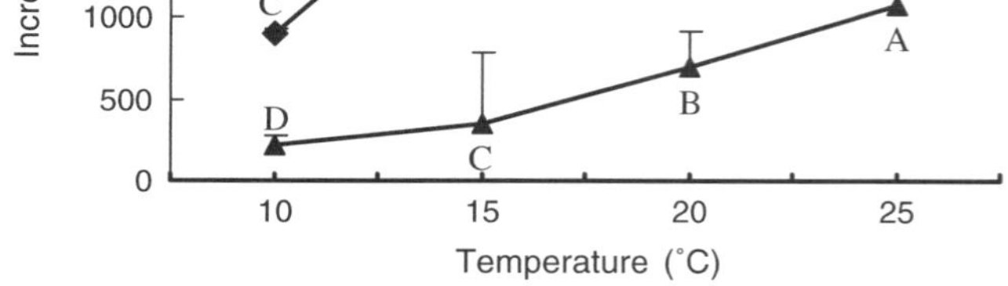

Fig. 3 The growth of *Sargassum thunbergii* germlings after 1 (▲) and 8 (■) weeks at different temperatures. Values are means ± SE. Statistical significances (ANOVA, Tukey) are indicated by a capital *letter* ($P<0.01$)

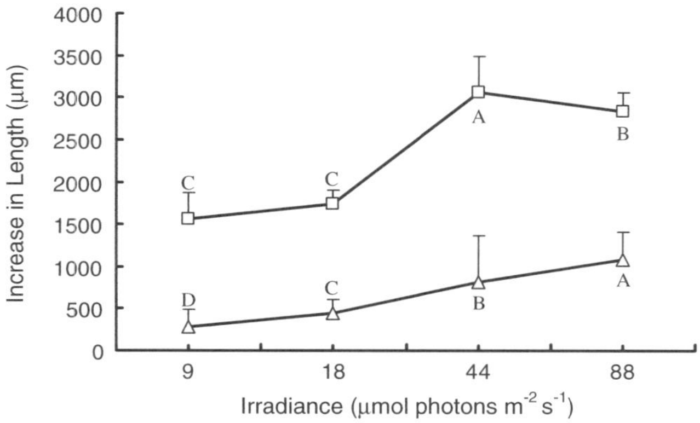

Fig. 4 The growth of *Sargassum thunbergii* germlings after 1 (△) and 8 (□) weeks under different irradiance levels. Values are means ± SE. Statistical significances (ANOVA, Tukey) are indicated by a capital letter ($P<0.01$)

indicating the low growth rate of germlings of about 1 µm day^{-1} under blue light.

Discussion

Comparative studies on early development of embryonic germlings of 22 species of Fucaceae have been previously studied, and six different developmental types of germlings were proposed (Inoh 1949). The early development process of embryonic germlings of *S. thunbergii* observed in the present study belongs to the classic "8 nuclei 1egg" type described by Inoh (1949). Only one of the eight nuclei would complete the fertilization with sperm and form a central nucleus. In the meanwhile, the other seven nuclei would gradually disappear. The early mode of cell division to form rhizoid cells was similar to that of *S. crassifolium* J. Agardh and *S. fusiforme* (Harv.) Setchell. Branchlets were differentiated and the young germlings obtained under indoor conditions eventually reached 3 mm in length, suggesting the potential of setting up an artificial zygote-derived seedling nursery. In addition, it was observed that filaments arose from the distal part of the germlings during the early development under the higher temperature regimes of 20 and 25°C. It is suspected that these filaments might be a callus-like structure since the number of filaments increased when the germlings were physically damaged.

To date, the relationships between growth rate of *Sargassum* germlings and some essential physical factors such as temperature have been investigated only in *S. echinocarpum* J. Agardh, *S. obtusifolium* J. Agardh, *S. oligocystum* Montagne, *S. polyphyllum* J. Agardh (De Wreede 1976, 1978), and *S. muticum* (Yendo) Fensholt (e.g., Norton 1977; Hales and Fletcher 1989). Steen (2004) explored the effects of reduced salinity on reproduction and germling development of *S. muticum* from Verdens Ende (Norway), and found that low salinity could be a physiological barrier that restricted the expansion of this alga into brackish habitats. Our study provides information on the effects of temperature and irradiance on the growth of germlings of *S. thunbergii,* and these factors could explain the distribution of this species along the coast of northern west Pacific as well as the optimal conditions for culturing this species.

The germlings of *S. thunbergii* exhibited a broad tolerance to temperature and irradiance conditions during the 8 weeks of culture, with growth recorded from 10°C to 25°C, and 9 to 88 µmol photons m^{-2} s^{-1}. Optimal growth occurred at 25°C and 44 µmol photons m^{-2} s^{-1}. A low temperature of 10°C and high irradiance of 88 µmol photons m^{-2} s^{-1} markedly reduced growth. Alhough temperature variations (10, 15, 20, 25°C) had a significant effect the growth of germlings within the first week of culture, this effect diminished in the later weeks, especially at higher temperatures of 15 and 20°C. This indicates that, while *S. thunbergii* germlings have a wide tolerance to a broad spectrum of temperature, they grew better at higher temperatures. Similar observations have been made for the vegetative branchlets of this species (Haraguchi et al. 2005).

Sargassum thunbergii is found in the intertidal from central Japan to southeastern China. The adult plants are therefore exposed naturally to a wide temperature range of < 5°C in winter (around Japan Sea) and around 25°C in summer (along the coast of Fujian Province Dongshan Island, southeastern China). It is therefore not surprising to find similar tolerance of its germlings to a wide temperature range. However, this species reproduces in summer, in time

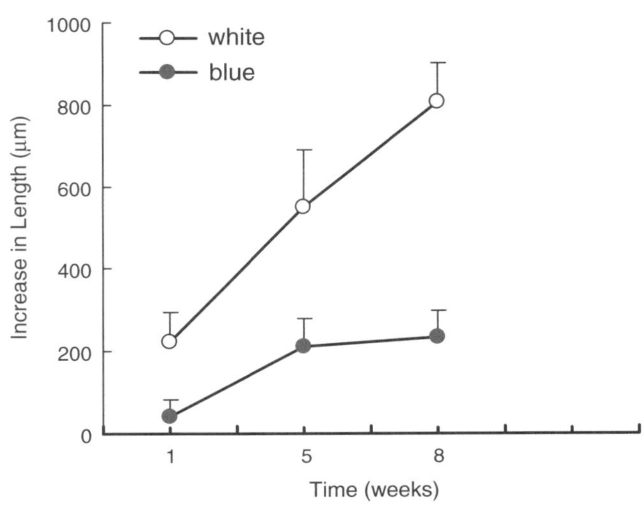

Fig. 5 Effect of white and blue light on the growth of germlings cultured under 10°C, 73 µmol photons m^{-2} s^{-1}. Values are means ± SE

for its germlings to grow better at higher temperatures. Such timing of reproduction to coincide with a better growth condition for its germlings has also been observed in other *Sargassum* species (De Wreede 1978).

The germlings of *S. thunbergii* grew best under 44 μmol photons m^{-2} s^{-1}. This is comparable to that reported by Hales and Fletcher (1989) for *S. muticum*. Given that *S. thunbergii* is an intertidal species, the lower growth of its germlings under the highest irradiance of 88 μmol photons m^{-2} s^{-1} was surprising. Individuals of this species tend to grow in clumps, as is true for many other *Sargassum* species. The environment under the canopy tends to have the advantage of lower desiccation, reduced grazing, etc. This self-shaded environment, hence with a reduced level of irradiance, must provide a better place for the germlings to grow.

Effects of light quality on the growth and development of brown algae have been illustrated previously, especially for *Laminaria* gametophytes (Lüning and Dring 1972, 1975; Lüning and Neushul 1978; Shi et al. 2005). Comparable studies have not been made for *Sargassum* spp. The low growth of *S. thunbergii* germlings under blue light shown in this present study indicates the importance of light quality to germling growth. Although the physiological mechanism involved is not clear, the effects of light quality on algal growth might be species-specific.

Our present study suggests the potential of using zygote-derived germlings as a source of seedlings for future development of a seedling nursery. Laboratory conditions for the optimal growth of these germlings could be further fine-tuned. A larger scale of culture may also be used to further test the feasibility of using this approach of growing germlings to address the shortage of seed materials for out-growing of *S. thunbergii* in the field.

Acknowledgements The authors gratefully acknowledge NSFC (40618001), N_CUHK438/06 and the Shandong Agriculture Seed-stock Breeding Project for financial assistance, and the anonymous reviewers for their constructive comments. Thanks are also kindly given to Mr. J.D. Liu for his assistance in the field sampling.

References

Aral A, Aral S, Miura A (1985) Growth and maturation of *Sargassum thunbergii* (Mertens ex Roth) O. Kuntze (Phaeophyta, Fucales) at Kominto, Chiba Prefecturs. Jpn J Phycol 33:160–166

De Wreede RE (1976) The phenology of three species of *Sargassum* (Sargassaceae, Phaeophyta) in Hawaii. Phycologia 15:175–183

De Wreede RE (1978) Growth in varying culture conditions of embryos of three Hawaiian species of *Sargassum* (Phaeophyta, Sargassaceae). Phycologia 17:23–31

Hales JM, Fletcher RL (1989) Studies on the recently introduced brown alga *Sargassum muticum* (Yendo) Fensholt. IV. The effect of temperature, irradiance and salinity on germling growth. Bot Mar 32:167–176

Haraguchi H, Murase N, Mizukami Y, Noda M, Yoshida G, Terawaki T (2005) The optimal and maximum critical temperatures of nine species of the Sargassaceae in the coastal waters of Yamaguchi Prefecture, Japan. Jpn J Phycol 53:7–13 (In Japanese with English abstract)

Hwang EK, Cho YC, Sohn CH (1997) Culture conditions on the early growth of *Hizikia fusiformis* (Phaeophyta). J Aquac 10:190–211 (in Korean with English abstract)

Hwang EK, Park CS, Baek JM (2006) Artificial seed production and cultivation of the edible brown alga, *Sargassum fulvellum* (Turner) C. Agardh: Developing a new species for seaweed cultivation in Korea. J Appl Phycol 18:251–257

Inoh S (1949) The development of seaweeds. Take Press, Tokyo (in Japanese with English abstract)

Lin C, Yu SG, Guo DS, Wei YX, Ai GH (2006) The antibacterial activity of phlorotannins from *Sargassum thunbergii* O. Kuntze. Mar Sci 30:94–97 (in Chinese with English abstract)

Lüning K, Dring MJ (1972) Reproduction induced by blue light in female gametophytes of *Laminaria saccharina*. Planta 104:252–256

Lüning K, Dring MJ (1975) Reproduction, growth and photosynthesis of gametophytes of *Laminaria saccharina* growth in blue and red light. Mar Biol 29:195–200

Lüning K, Neushul M (1978) Light and temperature demands for growth and reproduction of Laminarian gametophyte in southern and central California. Mar Biol 45:297–309

Nakamura H, Tatewakl YM, Nakahara H, Saito S (1971) The seasonal variation of standing crops of *Sargassum thunbergii*. Interim Report of the Kuroshio Littoral region Research Group for JIBP PM. pp 15–17

Norton TA (1977) Ecological experiments with *Sargassum muticum*. J Mar Biol Assoc UK 57:33–43

Padilha FP, Franca FP, Costa ACA (2005) The use of waste biomass of *Sargassum* sp. for the biosorption of copper from simulated semiconductor effluents. Bioresour Technol 96:1511–1577

Pang SJ, Chen LT, Zhuang DG, Fei XG, Sun JZ (2005) Cultivation of the brown alga *Hizikia fusiformis* (Harvey) Okamura: enhanced seedling production in tumbled culture. Aquaculture 245:321–329

Pang SJ, Gao SQ, Sun JZ (2006) Cultivation of the brown alga *Hizikia fusiformis* (Harvey) Okamura: controlled fertilization and early development of seedlings in raceway tanks in ambient light and temperature. J Appl Phycol 18:723–731

Ruan JH, Xu LG (2001) A preliminary study on the reproductive and development biology of *Sargassum fusiforme* (Harvs) Setch. J Zhejiang Univ (Science Edn) 28:315–320 (in Chinese with English abstract)

Shi XR, Xu ZH (1997) Antimicrobial activities of several seaweeds in Qingdao coast. J Mar Drugs 6:16–19

Shi CJ, Kataoka H, Duan DL (2005) Effects of blue light on gametophyte development of *Laminaria japonica* (Laminariales, Phaeophyta). Chin J Oceanol Limnol 23:323–329

Steen H (2004) Effects of reduced salinity on reproduction and germling development in *Sargassum muticum* (Phaeophyceae, Fucales). Eur J Phycol 39:293–299

Sun JZ, Fang JZ, Zhu ZF (1996) A preliminary study on the reproductive biology of *Sargassum fursiforme*. J Zhejiang College Fish 15:243–249 (in Chinese with English abstract)

Tseng CK (1983) Common seaweeds of China. Science Press, Beijing (in Chinese)

Tsukidate J (1984) Studies on the regenerative ability of the brown algae, *Sargassum muticum* (Yendo) Fensholt and *Sargassum tortile* C. Agardh. Hydrobiologia 116/117:393–397

Umezaki I (1974) Ecological studies of *Sargassum thunbergii* (Mertens) G. Kuntze in Maizuru Bay. Japan Sea Bot Mag Tokyo 87:285–292 (in Japanese with English abstract)

Wang WD (2003) Investigation and study on *Sargassum* and *Hizikia* in Zhejiang Province. J Shanghai Fish Univ 12:227–232 (in Chinese with English abstract)

Wei YX, Li J, Zhao AY, Wang CY, Yu SG, Cong PJ (2006) Preliminary studies on the preparation of polysaccharide from *Sargassum thunbergii* Kuntze (PST) and its anticoagulation activity. Chin J Mar Drugs 25:41–44 (in Chinese with English abstract)

Yan XJ, Zhou TC, Lou QX, Wang XJ (1996) Seasonal changes of content of phlorotannins. Mar Sci 20:39–42 (in Chinese with English abstract)

Yuan YD, Zhang SH, Sun AF, Liu HY (2006) Trial culture of *Sargassum thunbergii* with long line rafts. Transact Oceanol Limnol 2:125–128 (in Chinese with English abstract)

Zhang EX, Yu LJ, Xiao X (1994) Studies on physiological activity and biochemistry of alcohol extracts from *Sargassum thunbergii*. Chin J Mar Drugs 3:1–10 (in Chinese with English abstract)

Zheng Y, Chen ZH (1993) The seasonal growth and reproduction of *Sargassum thunbergii* (Phaeophyta) in Pingtan Islang, Fujian Province. J Fujian Norm Univ 9:81–85 (in Chinese with English abstract)

Zou JX, Li YQ, Liu YX, Zhang TW, Wang YM (2005) Studies on biological features and raft culture of *Sargassum thunbergii*. Shangdong Fish 22:25–28 (in Chinese)

Seaweeds preferred by herbivorous fishes

G. Tolentino-Pablico · N. Bailly · R. Froese · C. Elloran

Originally published in the Journal of Applied Phycology, Vol 20, No 5, 483–488.
DOI: 10.1007/s10811-007-9290-4 © Springer Science + Business Media B.V. 2007

Abstract Marine macrobenthic algae (or seaweeds), epiphytic microalgae, and other aquatic plants constitute the main food items of marine herbivorous fishes. About 5% of all fish species are herbivorous; only 30% of these are marine, most of them living in coral reefs. An analysis was performed on all the seaweeds that formed part of the natural diet of these fishes, based on information contained in FishBase (http://www.fishbase.org). The results showed that many coral-reef-associated marine herbivorous fishes, such as the families Blennidae, Kyphosidae and Siganidae, fed selectively on filamentous and turf fleshy seaweeds, which they prefer over calcareous coralline and encrusting species. In particular, Chlorophyceae of the genera *Cladophora*, *Enteromorpha* and *Ulva* were preferred by *Scartichthys viridis* (Blennidae), *Girella* spp. (Kyphosidae), *Sarpa salpa* (Sparidae), and Phaeophyceae in the genera *Sargassum* and *Dictyota* were preferred by *Kyphosus* spp. (Kyphosidae) and *Siganus* spp. (Siganidae). A web-based tool was developed to provide information on plants (algae, seagrasses, terrestrial plants and fruits) preferred as food by herbivorous fishes (http://www.incofish.org/herbitool.php). The tool is intended to assist aquaculturists, conservationists and ecosystem-based fisheries managers.

Keywords Aquaculture · Diet composition · FishBase · Food items · Herbivores tool

G. Tolentino-Pablico (✉) · N. Bailly · C. Elloran
Philippine Office / Natural Resource Management,
WorldFish Center,
Khush Hall, IRRI College, Los Baños,
Laguna 4031, Philippines
e-mail: g.pablico@cgiar.org

N. Bailly
e-mail: n.bailly@cgiar.org

C. Elloran
e-mail: c.elloran@cgiar.org

R. Froese
Leibniz-Institut für Meereswissenschaften IfM-GEOMAR,
Düsternbrooker Weg 20,
24105 Kiel, Germany
e-mail: rfroese@ifm-geomar.de

Introduction

Filamentous and turf seaweeds, Bacillariophyta, Dinophyceae, Cyanobacteria, associated meiofauna and detritus are food resources for grazing herbivorous fishes (Ojeda and Muñoz 1999; Crossman et al. 2001; Choat et al. 2002) and form a complex assemblage called the epilithic algal matrix (EAM) (Wilson et al. 2003). The bulk of the diet consists mainly of turf and filamentous algae, and macrophytes, but a small amount of animal materials may be ingested (Horn et al. 1982); other food items ingested are detritus or "unidentified organic matter". Wilson et al. (2003) have even suggested that fishes assimilate more detritus than algae and should be considered as detritivores.

The susceptibility of algae to grazing by herbivorous fishes depends on algal morphology (Hay 1984; Lewis 1985; Hay 1997). Most herbivorous fishes prefer fleshy seaweeds over calcareous coralline and encrusting seaweeds (Montgomery and Gerking 1980; Horn et al. 1982; Ojeda and Muñoz 1999). This preference appears to have developed from the temporal and spatial variation in food availability (Horn et al. 1982, 1986; Andrew and Jones 1990; Jones and Andrew 1990); the presence or absence of secondary metabolites produced as chemical defenses against herbivory (Lewis 1985; Steinberg 1986; Duffy and Paul 1992; Cronin and Hay 1996; Cronin et al. 1997; Hay 1997; Cetrulo and Hay 2000; Erickson et al. 2006); and by nutritional qualities in terms of energy, micro-nutrient, calorie and protein content (Montgomery and Gerking 1980; Edwards and Horn 1982; Neighbors and Horn 1991; Pillans et al. 2004).

The objectives of this paper are (1) to identify the dominant seaweed groups among the Florideophyceae, Chlorophyceae and Phaeophyceae that are ingested by herbivorous fish, and (2) to develop a web-based tool for identifying the food preferences of herbivorous fishes by country, FAO area or ecosystem. This would also provide knowledge on the distribution and diversity of herbivorous fishes in different localities and ecosystems around the world. The eventual goal is to provide a tool for natural resource managers involved in aquaculture, fisheries, and conservation of marine herbivorous fishes.

Materials and methods

Seaweeds as food

The name "fish" refers to non-Tetrapoda Vertebrata species. Larvae were not included in the study. "Herbivorous fishes" or "herbivores" refer to adults and juveniles of fish species with trophic levels between 2.0 to 2.2. Trophic levels were estimated (1) from diet composition data, covering the whole range of food items consumed by a given species at a given locality and season, and (2) from food items studies, if no information on diet was available (Pauly and Sa-a 2000).

In FishBase, food items of fishes were classified into 58 categories, defined as functional groups of taxa that shared the same consumers and resources within a food web (Sa-a et al. 2000). The food items and diet data were drawn from peer-reviewed publications, theses and dissertations that contained lists of seaweed species that have been found in the stomach or otherwise known to be ingested by a given fish.

An analysis was performed on the seaweeds that formed part of the natural diet and stomach contents of herbivorous fishes, based on information contained in FishBase. Seaweeds consumed by a given species were grouped by phylum and by family; the taxonomy and validity of algal names were verified using AlgaeBase (Guiry and Guiry 2007; http://www.algaebase.org).

Herbivorous fish tool

The web-based tool available in the Incofish website was developed to retrieve information on herbivorous fishes from FishBase (Fig. 1). The tool was developed using open source resources such as PHP Hypertext Preprocessor (PHP), Hypertext Markup Language (HTML), Cascading Style Sheet (CSS) and MySQL for the backend database.

Results

Seaweeds as food

Herbivorous fishes accounted for only 5% (1,660/29,500) of the total number of fish species known globally. Of this quantity, 89% are from tropical waters (1,441/1,660) while 3% are temperate species (49/1,660).

Seventy percent, 28% and 2% are freshwater, marine, and brackishwater species, respectively (Froese and Pauly 2007).

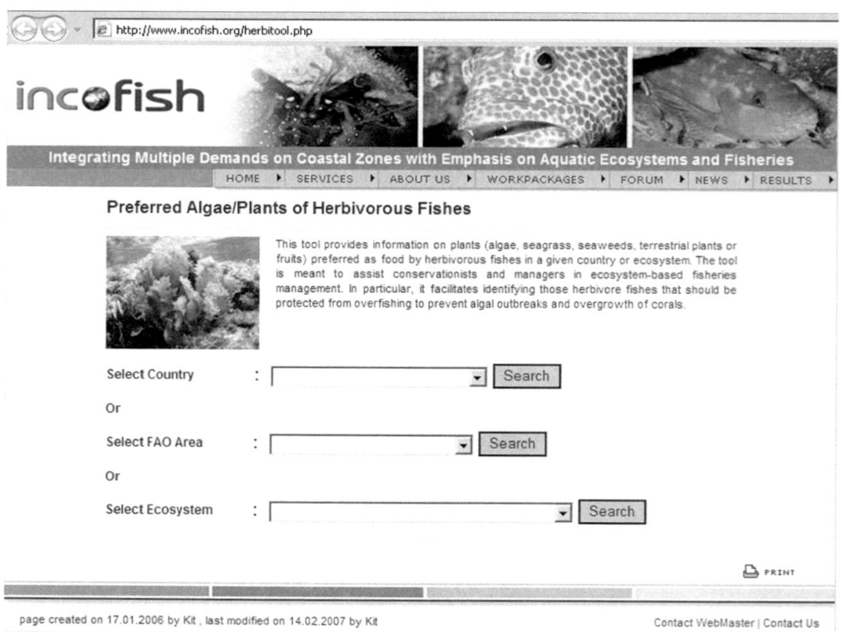

Fig. 1 The search page of the herbivorous fish tool, http://www.incofish.org/herbitool.php

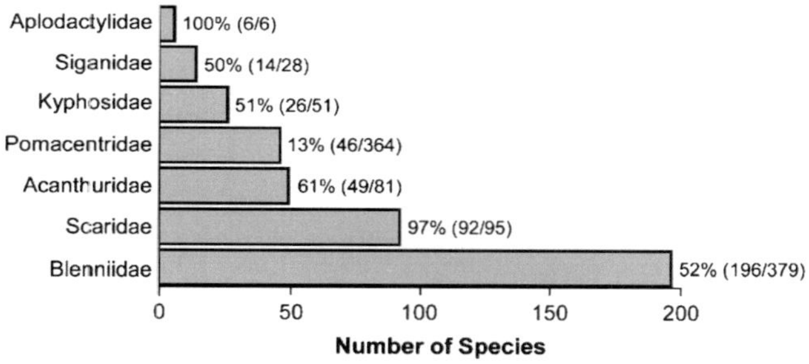

Fig. 2 Number of species and percentage of marine herbivorous fishes per family

The marine herbivores were mainly in coral-reef-associated areas while others were demersal or benthopelagic.

Families with the most representatives of herbivorous species are shown in Fig. 2. The following families had the most representatives of marine herbivores: Blenniidae (combtooth blennies), Scaridae (parrotfishes), Acanthuridae (surgeonfishes), Pomacentridae (damselfishes), Kyphosidae (sea chubs), Siganidae (rabbitfishes) and Aplodactylidae (marblefishes). These families represent the greatest number of species of marine herbivorous fishes in coral reefs, with Acanthuridae and Scaridae being the most abundant.

The total numbers of all seaweed species consumed by fishes reported in FishBase are shown in Table 1. The numbers exclude any unidentified seaweeds species recorded. Class Florideophyceae was the most often reported component of herbivore diets, followed by the class Phaeophyceae, dominated by families Dictyotaceae and Sargassaceae, and class Chlorophyceae, recorded mainly from the family Ulvaceae.

The seaweed higher taxa most often ingested by herviborous fishes are shown in Table 2. In class Florideophyceae, the genera *Ceramium*, *Corallina*, *Jania*, *Gelidium*, *Laurencia*, and *Polysiphonia* were ingested mainly by *Sarpa*, *Scartichthys*, *Hermosilla*, *Girella* spp. and *Siganus* spp.; in Phaeophyceae, the genera *Dictyota* and *Sargassum* were ingested mainly by *Acanthurus*, *Naso*, *Kyphosus*, *Holacanthus*, *Amphiprion* and *Sarpa* spp. In addition, the fleshy and filamentous members of the Chlorophyceae, such as *Ulva*, *Enteromorpha* and *Cladophora* were also ingested by some of these fishes.

A comparison of the number of species of each class Florideophyceae, Phaeophyceae and Chlorophyceae ingested by marine herbivorous fishes is shown in Fig. 3. Kyphosidae, with mostly benthopelagic species, consumed the highest number of Florideophyceae. Among the other families, where species were mostly coral-reef-associated, only Acanthuridae ingested a higher number of Chlorophyceae. Marine benthic algae were preferred mainly by members of the acanthurids, blenniids, kyphosids and siganids. These herbivores were nearly always reef-associated species, except for families Kyphosidae and Blenniidae, which were benthopelagic and demersal, respectively.

Table 1 Total numbers of seaweed families and species ingested by herbivorous fishes recorded in FishBase

	Number of families	Number of species
Florideophyceae	33	145
Phaeophyceae	20	81
Chlorophyceae	14	33

Table 2 Dominant seaweeds consumed by herbivorous fishes

Frequently ingested seaweeds		Herbivorous fish
Florideophyceae:		
Ceramiaceae	*Ceramium*	*Sarpa salpa*
Corallinaceae	*Corallina*	*Scartichthys* spp.
	Jania	*Hermosilla azurea*
Gelidiaceae	*Gelidium*	*Girella* spp.
Rhodomelaceae	*Laurencia*	*Siganus* spp.
	Polysiphonia	
Phaeophyceae:		
Sargassaceae	*Sargassum*	*Acanthurus* spp.
		Naso spp.
		Kyphosus spp.
Dictyotaceae	*Dictyota*	*Holacanthus passer*
		Amphiprion spp.
		Sarpa salpa
Chlorophyceae:		
Ulvaceae	*Ulva,*	*Girella* spp.
	Enteromorpha	*Sarpa salpa*
Cladophoraceae	*Cladophora*	*Scartichthys viridis*
		Liza vaigiensis

Herbivorous fish tool

This tool provides information on plants (algae, seagrasses, terrestrial plants or fruits) preferred as food by adults and juveniles of herbivorous fishes. Lists of herbivorous fishes are created if searched by country, FAO area or ecosystem. It provides the fish common names, maximum length, trophic level, general aquaculture use, climate zone and diet. Selecting a species opens its species summary page in FishBase, which provides information on its taxonomy, distribution, biological and environmental characteristics,

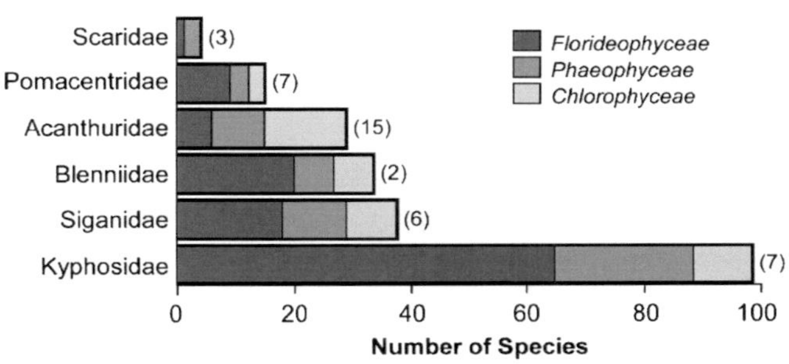

Fig. 3 Number of seaweed species consumed by fishes recorded in FishBase. Numbers of herbivores examined are in parentheses

and other topics and links. Selecting the trophic level of a given species opens a report showing the list of its preferred algae/plants, and identifies the algal food group or its taxonomic classification. The algal scientific names are linked to AlgaeBase. Selecting the "Diet" button of herbivorous fish will show up the species' diet composition, which is based on studies in percent weight or volume reported for a sample. The unavailability of information on food items and diet composition of some herbivores illustrates the research gaps.

Discussion

Seaweeds as food

The analysis of food items and diets from published references indicates the expected diversity of feeding modes in herbivorous fishes through ontogeny. The larval and post-larval stages of most herbivorous fishes are planktivorous, progressively becoming omnivorous during their juvenile stages, and then herbivorous upon reaching maturity (Benavides et al. 1994). A shift from animal to plant diet is related to an increase in the capability of assimilating the walls of seaweed and plant cells as the length of the digestive tract increases during ontogeny (Horn 1989; Clements and Choat 1993; Benavides et al. 1994). Many herbivorous fishes have the low stomach pH levels and specialized guts required for the digestion of plant materials (Horn and Messer 1992).

Higher temperatures facilitate plant material digestion when cooler temperatures may not allow herbivorous fishes to process enough food material to meet their metabolic demands (Floeter et al. 2005). Hence, temperature-related feeding and digestive processes are most likely the principal inducers of geographical distribution patterns. Accordingly, as shown from studies compiled in FishBase, these fishes are more abundant at lower than higher latitudes, especially in coral reef areas, where they present high species diversity and biomass. The herbivorous fishes are thus the major component of the coral reef ecosystems balancing seaweed production, whereas sea-urchins, gastropods and crustaceans perform this function in temperate waters. Moreover, fish herbivory often induces adaptations of seaweed morphology, structure, and biochemistry to resist grazing (Hay 1997).

Knowledge on herbivorous fishes and their food items is thus important in the design and management of marine protected areas, e.g., avoid seaweed overgrowth and outbreaks.

Montgomery and Gerking (1980) found that Chlorophyceae were superior to Phaeophyceae, and Phaeophyceae superior to Florideophyceae in terms of caloric and nutrient contents. Some studies have shown the importance of toughness and calcification as feeding deterrents. Moreover, Paul and Hay (1986) found that many tough calcified seaweeds produce unusual secondary metabolites that could act as chemical defenses and thus lead to low susceptibility to herbivory. Accordingly, in FishBase, most of the dominant Chlorohyceae and Phaeophyceae ingested by herbivorous fishes possess fleshy filamentous and leathery foliose thalli possessing alleged morphological and/or secondary compounds that are highly susceptible to fish grazing (Lewis 1985). Hay (1984) also reported that some Phaeophyceae that produce polyphenolics or phenolic acids, were consumed at high intermediate rates, suggesting that these secondary compounds were not effective deterrents for some herbivorous fishes. In addition, turf filamentous and branching Florideophyceae, such as *Ceramium*, *Polysiphonia* and *Gelidium*, were the major food items of many reef-associated species. For instance, the diet of *Scartichthys viridis* individuals less than 13 cm total length consisted of the foliose Chlorophyceae, *Ulva* and *Enteromorpha*, but when it reached over 22 cm total length, the tough Florideophyceae *Gelidium* comprised the bulk of the gut contents (Muñoz and Ojeda 2000). Ochavillo et al. (1992) also reported that turf algae were the major component of the diet of Scaridae. In addition, large fishes such as *Hermosilla*, *Girella* and *Sarpa* frequently ingest crustose and coralline seaweeds such as *Corallina*, *Jania* and *Gelidiella*.

Crossman et al. (2001) have demonstrated that detritus from EAM is high in protein amino acids and of potential food value for grazing fishes. Diet composition studies of some species of acanthurids and scarids recorded as herbivorous in FishBase showed that at least 50% of the diet was composed of detritus. Many of the coral reef fishes selectively feed on detritus and this represents a major dietary constituent (Wilson et al. 2003). Gut segments of detritivorous scarids and acanthurids had significantly higher total amino acid concentrations than planktivores and omnivores, with algivores being the lowest (Crossman et al. 2005). Nevertheless, the relative length of the gut of the species with trophic levels between 2.0 and 2.2 should be explored since it might indicate the importance of plant material assimilation (Horn and Messer 1992).

Herbivorous fish tool

The herbivorous fish tool was developed to provide information on biological data, taxonomy, distribution and dominant algal food. This decision support tool is intended to assist fish farming, ecosystem-based fishery management, and conservation domains.

A compilation of the different algal/plant food items of a fish species can be used to identify food preferences in herbivorous fishes for which detailed diet composition data are not available, and in preliminary estimates of trophic levels (Sa-a et al. 2000). It facilitates identification of those herbivorous fishes that should be protected from overfishing to prevent algal outbreaks in coral reefs and other tropical areas. Knowledge of the diet of herbivorous fish in terms of species and nutrient content is important to improve their aquaculture by delivering the most efficient food.

A next step could be to create a "Detritivores tool" and link it to the "Herbivores tool" when the corresponding herbivory/detritivory part of the diet is elucidated.

Summary and recommendations

Seaweeds have been shown to be a source of minerals and nutrients in fish diets and the main food for many herbivorous fishes. Some seaweed characteristics have important direct and indirect effects on seaweed–herbivorous fish interactions. The presence of chemical deterrents, as well as the morphological and nutritional characteristics of some seaweed species may prevent intense feeding by herbivorous fishes. Defended seaweeds become safe sites for small and less mobile invertebrates, and habitats for larvae and juveniles fishes. On the other hand, if the seaweeds are not removed by herbivorous fishes, they could become competitively dominant organisms and may eventually destroy the coral reefs. The exploitation of these areas must be well-managed.

Developing herbivorous fish aquaculture in tropical developing countries is a pro-poor alternative to carnivorous fish aquaculture, which usually requires fish in the feeds. This means more pressure on the already globally depleted fish stocks, and a protein-rich diet is usually the most expensive component of aquaculture. This could hamper further growth of farms, especially in poor areas, for subsistence food. The less costly approach is to provide herbivorous fishes with algal supplements and natural periphyton in ponds (Keshavanath et al. 2002). The availability and abundant year-end supply of seaweeds are economical for aquaculture for a maximal production/cost ratio. An integrated intensive aquaculture approach, such as fish or shrimp culture associated with shellfish and/or seaweeds, is currently under development and shows economical and environmental sustainability (Neori et al. 2004).

The herbivorous fish tool will be further enhanced and developed, in particular to better identify herbivorous fish–seaweed species pairings for aquaculture usage. However, food items and diet information for many herbivorous fishes are quite lacking in the literature. These research gaps must be filled in the coming years by further studies on species that lack any record. The herbivorous tool provides a list of these species. Moreover, detritivory might be more important than herbivory in some fishes known to be herbivores, which could impact aquaculture practices such as facilitating sedimentation to create EAM.

Acknowledgements The authors profoundly thank Dr. Masao Ohno, who supported the senior author to attend the 19th International Seaweed Symposium and Lotta Jarnmark, Incofish Project WP1 for support. We acknowledge the FishBase team for the data encoded and assistance provided on web programming of the herbivorous fish tool, and an anonymous reviewer for comments and suggestions. This is WorldFish Center Contribution No. 1846.

References

Andrew NL, Jones GP (1990) Patch formation by herbivorous fish in a temperate Australian kelp forest. Oecologia 85:57–68

Benavides AG, Cancino JM, Ojeda FP (1994) Ontogenetic change in the diet of *Aplodactylus punctatus* (Pisces: Aplodactylidae): an ecophysiological explanation. Mar Biol 118:1–5

Cetrulo GL, Hay ME (2000) Activated chemical defenses in tropical versus temperate seaweeds. Mar Ecol Prog Ser 207: 243–253

Choat JH, Clements KD, Robbins WD (2002) The trophic status of herbivorous fishes on coral reefs. I. Diet analyses. Mar Biol 140:613–623

Clements KD, Choat JH (1993) Influence of season, ontogeny and tide on the diet of the temperate marine herbivorous fish *Odax pullus* (Odacidae). Mar Biol 117:213–220

Cronin G, Hay ME (1996) Within-plant variation in seaweed palatability and chemical defenses: optimal defense theory versus the growth-differentiation balance hypothesis. Oecologia 105:361–368

Cronin G, Valerie JP, Hay ME, Fenical W (1997) Are tropical herbivores more resistant than temperate herbivores to seaweed chemical defenses? diterpenoid metabolites from *Dictyota acutiloba* as feeding deterrents for tropical versus temperate fishes and urchins. J Chem Ecol 23:289–302

Crossman DJ, Choat JH, Clements KD, Hardy T, McConochie J (2001) Detritus as food for grazing fishes on coral reefs. Limnol Oceanogr 47:1596–1605

Crossman DJ, Choat JH, Clements KD (2005) Nutritional ecology of nominally herbivorous fishes on coral reefs. Mar Ecol Prog Ser 296:129–142

Duffy JE, Paul VJ (1992) Prey nutritional quality and the effectiveness of chemical defenses against tropical reef fishes. Oecologia 90:333–339

Edwards TW, Horn MH (1982) Assimilation efficiency of a temperate-zone intertidal fish (*Cebidichthys violaceus*) fed diets of macroalgae. Mar Biol 67:247–253

Erickson AA, Paul VJ, Van Alstyne KL, Kwiatkowski LM (2006) Palatability of macroalgae that use different types of chemical defenses. J Chem Ecol 32:1883–1895

Floeter SR, Behrens MD, Ferreira CEL, Paddack MJ, Horn MH (2005) Geographical gradients of marine herbivorous fishes: patterns and processes. Mar Biol 147:1435–1447

Froese R, Pauly D (eds) (2007) FishBase. World wide web electronic publication. http://www.fishbase.org. Cited February 2007

Guiry MD, Guiry GM (2007) AlgaeBase version 4.2. World-wide electronic publication, National University of Ireland, Galway. http://www.algaebase.org. Cited 14 March 2007

Hay ME (1984) Predictable spatial escapes from herbivory: how do these affect the evolution of herbivore resistance in tropical marine communities? Oecologia 64:396–407

Hay ME (1997) The ecology and evolution of seaweed-herbivore interactions on coral reefs. Coral Reefs 16:68–76

Horn M (1989) Biology of marine herbivorous fishes. Oceanogr Mar Biol Annu Rev 27:167–272

Horn MH, Messer KS (1992) Fish guts as chemical reactors: a model of the alimentary canals of marine herbivorous fishes. Mar Biol 113:527–535

Horn MH, Murray SN, Edwards TW (1982) Dietary selectivity in the field and food preferences in the laboratory for two herbivorous fishes (*Cebidichthys violaceus* and *Xiphister mucosus*) from a temperate intertidal zone. Mar Biol 67:237–246

Horn MH, Neighbors MA, Murray SN (1986) Herbivore response to a seasonally fluctuating food supply: growth potential of two temperate intertidal fishes based on the protein and energy assimilated from their macroalgal diets. J Exp Mar Biol Ecol 103:217–234

Jones GP, Andrew NL (1990) Herbivory and patch dynamics on rocky reefs in temperate Australasia: the roles of fish and sea urchins. Aust J Ecol 15:505–520

Keshavanath P, Gangadhar B, Ramesh TJ, van Dam AA, Beveridge MCM, Verdegem MCJ (2002) The effect of periphyton and supplemental feeding on the production of the indigenous carps *Tor khudree* and *Labeo fimbriatus*. Aquaculture 213:207–218

Lewis SM (1985) Herbivory on coral reefs: algal susceptibility to herbivorous fishes. Oecologia 65:370–375

Montgomery WL, Gerking SD (1980) Marine macroalgae as foods for fishes: an evaluation of potential food quality. Env Biol Fish 5:143–153

Muñoz AA, Ojeda FP (2000) Ontogenetic changes in the diet of the herbivorous *Scartichthys viridis* in a rocky intertidal zone in central Chile. J Fish Biol 56:986–998

Neighbors MA, Horn MH (1991) Nutritional quality of macrophytes eaten and not eaten by two temperate-zone herbivorous fishes: a multivariate comparison. Mar Biol 108:471–476

Neori A, Chopin T, Troell M, Buschmann AH, Kraemer GP, Halling C, Shpigel M, Yarish C (2004) Integrated aquaculture: rationale, evolution and state of the art emphasizing seaweed biofiltration in modern mariculture. Aquaculture 231:361–391

Ochavillo DG, Dixon PI, Aliño PM (1992) The daily food ration of parrotfishes in the fringing reefs of Bolinao, Pangasinan, Northwestern Philippines. Proceedings 7th International Coral Reef Symposium. Guam 2:927–933

Ojeda FP, Muñoz AA (1999) Feeding selectivity of the herbivorous fish *Scartichthys viridis*: effects on macroalgal community structure in a temperate rocky intertidal coastal zone. Mar Ecol Prog Ser 184:219–229

Paul VJ, Hay ME (1986) Seaweed susceptibility to herbivory: chemical and morphological correlates. Mar Ecol Prog Ser 33:255–264

Pauly D, Sa-a P (2000). Estimating trophic levels from individual food items. In: Froese R, Pauly D (eds) FishBase 2000: concepts, design and data sources. ICLARM, Los Baños, Laguna, Philippines, p 185

Pillans RD, Franklin CE, Tibbetts IR (2004) Food choice in *Siganus fuscescens*: influence of macrophyte nutrient content and availability. J Fish Biol 64:297–309

Sa-a P, Palomares ML, Pauly D (2000) The FOOD ITEMS table. In: Froese R, Pauly D (eds) FishBase 2000: concepts, design and data sources. ICLARM, Los Baños, Laguna, Philippines, pp 182–188

Steinberg PD (1986) Chemical defenses and the susceptibility of tropical marine brown algae to herbivores. Oecologia 69:628–630

Wilson SK, Bellwood DR, Choat JH, Furnas MJ (2003) Detritus in the epilithic algal matrix and its use by coral reef fishes. Oceanogr Mar Biol Annu Rev 41:279–309

Making the links: towards a global taxonomy for the red algal genus *Porphyra* (Bangiales, Rhodophyta)

Juliet Brodie · Agnes Mols Mortensen · María Eliana Ramirez · Stephen Russell · Barbara Rinkel

Originally published in the Journal of Applied Phycology, Vol 20, No 5, 489–499.
DOI: 10.1007/s10811-008-9315-7

Abstract Species assigned to *Porphyra* sensu lato (Bangiales) make up one of the largest groups of red algae in the world. While the Bangiales are monophyletic, *Porphyra* is not, but the number of genera remains unresolved and a consensus needs to be reached on how to proceed. Here, it is proposed that a global taxonomy for the group would enable resolution and consensus. Using examples from our work on *Porphyra* in both the northern and southern hemispheres, including the North Atlantic, Mediterranean and Pacific, notably Chile, we have addressed the following questions: i) how many species are there, ii) how many species are there in different parts of the world, and iii) how do these species relate to those in different parts of the world? A cumulative total of species described in the North Atlantic and Mediterranean since 1800 indicates that there are still species to be added to the flora for these regions. Determination of intraspecific diversity or diversity within species complexes was examined using the mitochondrial *cox1*. The results split the species into two distinct groups, revealed differences between the British and Faroese floras and indicated a new species in the Faroes. Detection of species diversity in different geographical areas of the world and their relationship with the North Atlantic and southeast Pacific (Chile) was studied using the plastid Rubisco spacer and partial rDNA SSU data. While the Rubisco spacer was limited for global comparisons, it was useful for species identification adding five more species to the Atlantic flora. The partial rDNA SSU data, while its shortness made it unsuitable for generic interpretation, revealed the spread of taxa from different geographical areas throughout the tree. We have concluded that to achieve a global taxonomy standard molecular markers should be used and we propose that the *rbc*L and complete rDNA SSU would be suitable. We have also concluded that while a few regions of the world have been well studied, before a consensus on generic boundaries can be achieved detailed taxonomic studies are needed in many other parts of the world including circumpolar Arctic, mid and southwest Atlantic and southeast Asia.

Keywords Chile · *cox1* · Mediterranean · North Atlantic · *Porphyra* · rDNA SSU · Rhodophyta · Rubisco spacer · Taxonomy

J. Brodie (✉) · S. Russell · B. Rinkel
Department of Botany, The Natural History Museum,
Cromwell Road,
London SW7 5BD, UK
e-mail: J.Brodie@nhm.ac.uk

A. Mols Mortensen
Department of Phycology, Institute of Biology,
University of Copenhagen,
Øster Farimagsgade 2D,
DK-1353 Copenhagen K, Denmark

M. E. Ramirez
Museo Nacional de Historia Natural,
Santiago, Chile

Introduction

The genus *Porphyra* sensu lato (*Porphyra* and *Bangia*) is one of the largest genera of red algae (Brodie and Zuccarello 2007), and although the exact number of species is unknown, there are almost certainly well over 150. While the Bangiales is monophyletic (Saunders and Hommersand 2004; Milstein and de Oliveira 2005), the genus is polyphyletic (Oliveira et al. 1995; Müller et al. 1998; Nelson et al. 2006) although the exact number of genera is currently unresolved. Five genera are currently recognised (*Porphyra*, *Bangia*, *Dione*, *Pseudobangia*, *Minerva*) although

Nelson et al. (2006) speculated, based on the analysis of the 18S rDNA for 127 samples, that there would be at least ten genera. Species occur in the intertidal and shallow subtidal throughout the world. Although they are particularly abundant in boreal and cold temperate waters, they are also found in the Arctic and Antarctic and warm temperate and tropical regions (Kapraun and Lemus 1987), with evidence that some species in the tropics occur during the rainy season (Sahoo et al. 2006; Fujiyoshi et al. 2007). Fossil evidence based on diagnostic cell division suggests that bangiacean red algae occurred at least 1.2 billion years ago, with *Bangiomorpha pubescens* Butterfield representing the oldest taxonomically resolved eukaryote (Butterfield 2000). The remarkable diversity of life histories exhibited by the species (e.g. Notoya 1997; Brodie and Irvine 2003) and an array of life history strategies (Holmes and Brodie 2004) may reflect this long evolutionary history. Species of *Porphyra* are economically important, and have traditionally been eaten in many parts of the world, notably as nori in Japan, laver in Wales and luche in Chile. This has also led to cultural differences in different parts of the world in the way that the species are prepared and consumed, ranging from subsistence foods to haute cuisine. The success of the nori industry, owes much to the discovery by Kathleen Drew of the *Porphyra* life cycle (Drew 1949, 1954) where the blade phase, which is conspicuous on the shore, produces spores which grow into shells to produce a shell-boring filamentous or conchocelis phase which is sometimes visible as pink staining on shells. The diversity of species, global spread and antiquity of the genus suggest a remarkably successful group of organisms with the ability of both the blade and conchocelis phases to survive environmental changes over millions of years.

Species of *Porphyra* are often notoriously difficult to identify, so despite at least 250 years of study, the complete number of species remains unknown. However, major advances in recent years with the application of molecular techniques, the ability to use these methods on type or historic material plus the sharing of knowledge from different parts of the world, means that the goal of achieving a complete taxonomy for *Porphyra* is timely and a real possibility.

This paper focuses on the species of *Porphyra* in the North Atlantic and their possible relationships with other parts of the world. *Porphyra* is very common in the North Atlantic and while there is evidence of speciation in the region (Robba et al. 2006), and at least one species, *P. dioica*, appears to be endemic to the northeastern part, the genus almost certainly originally evolved in the Pacific, and it has also been proposed that the origin for the modern day Bangiales is centred upon New Zealand (Broom et al. 2004). Given the apparent antiquity of the genus, there are different routes and different times by which species could potentially have arrived since the formation of the North Atlantic, e.g. via the Tethys Sea, the Panama Isthmus, the Arctic Seaways, or from the South Atlantic (see Brodie and Nielsen 2005).

To achieve a global taxonomy, we need to resolve the following questions:

1. How many species are there?
2. How many species are there in different parts of the world?
3. How do these species relate to those in different parts of the world?
4. How many genera are there?

The aim of this paper is to use examples, notably from the North Atlantic but also from the Pacific coast of Chile to put forward a possible approach to achieve a global taxonomy, concentrating on highlighting some of the major problems which require resolution.

Materials and methods

Cumulative total of species in the North Atlantic and Mediterranean

In order to predict whether there are still species to be discovered in the North Atlantic and Mediterranean, a cumulative total of species was derived by adding up the number of species described before 1800, and within 50 year time intervals since that time until 2000 and then after that date. Species were only included if there was enough evidence that they were considered distinct.

Molecular analysis

Sequence data were analysed for mitochondrial (*cox1*), plastid (Rubisco spacer), and nuclear (partial rDNA SSU) regions. The *cox1* sequence data were used to compare differences within species in Great Britain and the Faroes. Rubisco spacer sequences were mainly generated from specimens collected in the North Atlantic and were used for species identification primarily in this region. The partial rDNA SSU was used to compare specimens from the North Atlantic with those from the Pacific, including Japan, New Zealand, Australia and South Africa, based on the phylogenetic tree for the Bangiales (see Nelson et al. 2006) and from our collections from Chile. In addition to our own data, sequences were selected based on the genera and clades recognised by Nelson et al. (2006).

Specimens used for molecular analysis are given in Table 1. DNA extraction, PCR amplification and sequencing for *cox1* and Rubisco spacer sequences generated in this study followed methods given in Robba et al. (2006).

Table 1 Taxa from which sequence data were obtained with their location and GenBank accession numbers

Taxa	Code	Location	GenBank accession no.			Reference
			*rbc*L-*rbc*S spacer	partial rDNA SSU	*cox1*	
Bangia atropurpurea	NH	Atlantic US	-	AF043353	-	Müller et al. (1998)
B. atropurpurea	CA-At	Atlantic Canada	-	AF043353	-	Müller et al. (1998)
B. fuscopurpurea	JB72	Wales	AJ010785	-	-	Brodie et al. (1998)
B. fuscopurpurea	IR	Ireland	-	AF175534	-	Müller et al. (2005)
B. fuscopurpurea	JB212	England	-	-	DQ442886	Robba et al. (2006)
B. fuscopurpurea	JB238	England	-	-	DQ442886	Robba et al. (2006)
B. fuscopurpurea	225	England	-	-	DQ442887	Robba et al. (2006)
Bangia sp.	MA	Atlantic US	-	AB013179	-	Müller et al. (2005)
Bangia sp.	BOHNZ	New Zealand	-	AY909600	-	Nelson and Broom (2005)
Bangia sp.	BMWNZ	New Zealand	-	AY184344	-	Broom et al. (2004)
'*Dione arcuata*'	NZ	New Zealand	-	AY184338	-	Broom et al. (2004)
Minerva aenigmata	NZ	New Zealand	-	AY184347	-	Broom et al. (2004)
P. amplissima	JB91	Scotland	AJ010780	-	-	Brodie et al. (1998)
P. '*amplissima*'	CA-At	Atlantic Canada	-	L36048	-	Oliveira et al. (1995)
P. birdiae	10A	Faroes	AY458657	-	-	Brodie and Nielsen (2005)
P. cinnamomea	NZ	New Zealand	-	AH008010	-	Broom et al. (1999)
P. coleana	NZ	New Zealand	-	AY136423	-	Broom et al. (1999) (as PAP52)
P. columbina?	CH7	Chile	-	AM943385	-	This paper
P. cuneiformis	CH3	Chile	-	AM943386	-	This paper
P. dentata	JP	Japan	-	AB013183	-	Kunimoto et al. (1999)
P. dioica	P163	Scotland	AJ010779	-	-	Brodie et al. (1998)
P. dioica	JB261	England	-	AM943387	-	This paper
P. dioica	JB347	England	-	-	DQ191340	Robba et al. (2006)
P. dioica	JB137	Wales	-	-	DQ442888	Robba et al. (2006)
P. dioica	JB246	Scotland	-	-	DQ442889	Robba et al. (2006)
P. dioica	JB314	Scotland	-	-	DQ442889	Robba et al. (2006)
P. drachii	Pdr	Wales	AJ010788	-	-	Brodie et al. (1998)
P. gardneri	MX	Mexico	-	DQ084423	-	Nelson et al. (2006)
P. haitanensis	JP	Japan	-	AB013181	-	Kunimoto et al. (1999)
P. katadae	JP	Japan	-	AB013184	-	Kunimoto et al. (1999)
P. kunthiana?	CH49	Chile	AM943010	AM943388	-	This paper
P. kunthiana?	CH54	Chile	-	AM943389	-	This paper
P. lanceolata	MX	Mexico	-	AY909594	-	Nelson and Broom (2005)
P. '*leucosticta*'	JB402	England	AM943011	AM943390	-	This paper
P. leucosticta	Pgu	Channel Islands	AJ010789	-	-	Brodie et al. (1998)
P. leucosticta	AF175557	Atlantic Canada	-	AF175557	-	Müller et al. (2001a)
P. leucosticta	GE	Helgoland	-	AF342746	-	Müller et al. (2001b)
P. leucosticta	Ca-At	Atlantic Canada	-	L26199	-	Ragan et al. (1994)
P. leucosticta	AMM06SF1240	Faroes	-	-	AM943398	This paper
P. leucosticta	JB322	England	-	-	DQ191338	Robba et al. (2006)
P. leucosticta	JBLR21	England	-	-	DQ191337	Robba et al. (2006)
P. leucosticta	JB340	England	-	-	DQ442890	Robba et al. (2006)
P. linearis	125	Wales	AJ010781	AM943391	-	Brodie et al. (1998); this paper
P. lucasii	AU-ln	Australia	-	AY139685	-	Farr et al. (2003)
P. miniata	Psam	Faroes	AJ010786	-	-	Brodie et al. (1998)
P. miniata	CA-At	Atlantic Canada	-	AF175540	-	Müller et al. (2005)
P. olivii	TES23	Greece	DQ834623	-	-	Brodie et al. (2007)
P. olivii	TES19	Greece	-	DQ834632	-	This paper
P. perforata	MX	Mexico	-	AY909592	-	Nelson and Broom (2005)
P. pseudolinearis	P2	Japan	AJ010787	-	-	Brodie et al. (1998)
P. pseudolinearis	JP	Japan	-	AB013185	-	Kunimoto et al. (1999)
P. purpurea	P127	England	AJ010776	-	-	Brodie et al. (1998)
P. purpurea	JB257	England	-	AM943392	-	This paper

Table 1 (continued)

Taxa	Code	Location	GenBank accession no.			Reference
			*rbc*L-*rbc*S spacer	partial rDNA SSU	*cox1*	
P. purpurea	CA-At	Atlantic Canada	-	L26201	-	Ragan et al. (1994)
P. purpurea	(GenBank)		-	-	NC0020007	GenBank
P. rosengurttii	MAL24	Spain	DQ834627	-	-	Brodie et al. (2007)
P. rosengurttii	MAL26	Spain	-	DQ834638	-	This paper
P. rosengurttii	AMM06SF1308	Faroes	-	-	AM943399	This paper
P. rosengurttii	JB286	England	-	-	DQ191335	Robba et al. (2006)
P. rakiura	NZ	New Zealand	-	AF136425	-	Broom et al. (1999) (as RAK49)
P. spiralis	BR	Brazil	-	L26177	-	Ragan et al. (1994)
P. suborbiculata	NZ	New Zealand	-	AF136424	-	Broom et al. (1999)
P. thulaea Holotype	PthH	Greenland	AM943012	-	-	This paper
P. thulaea	564557	Iceland	AM943013*[1]	AM943393	-	This paper
P. umbilicalis	P126	England	AJ010782	-	-	Brodie et al. (1998)
P. umbilicalis	JB258	England	-	AM943394	-	This paper
P. umbilicalis	AMM06SF1246	Faroes	-	-	AM943400	This paper
P. umbilicalis	AMM06SF1144	Faroes	-	-	AM943401	This paper
*P. umbilicalis**[2]	JB178	Scotland	-	-	DQ442891	Robba et al. (2006)
P. umbilicalis	JB329	Northern Ireland	-	-	DQ191334	Robba et al. (2006)
P. umbilicalis	AMM06SF1351	Faroes	-	-	AM943402	This paper
P. umbilicalis	JB342	England	-	-	DQ191336	Robba et al. (2006)
P. umbilicalis	JB334	England	-	-	DQ191333	Robba et al. (2006)
P. umbilicalis	JB324	England	-	-	DQ442892	Robba et al. (2006)
P. umbilicalis	JB357	England	-	-	DQ191332	Robba et al. (2006)
P. umbilicalis	JB352	England	-	-	DQ191332	Robba et al. (2006)
P. umbilicalis	AMM06SF4241	Faroes	-	-	AM943403	This paper
P. virididentata	NZ	New Zealand	-	AF136421	-	Broom et al. (1999)
P. yezoensis	Pye	Japan	AJ010783	-	-	Brodie et al. (1998)
P. yezoensis	JP	Japan	-	AB013177	-	Kunimoto et al. (1999)
P. yezoensis sensu Kornmann	Pmu	Helgoland	AJ010778	-	-	Brodie et al. (1998)
Porphyra sp.	PO2	Faroes	AJ634466	-	-	Brodie and Nielsen (2005)
Porphyra sp.	AZ23	Azores	AM943014	-	-	This paper
Porphyra sp.	RNO2	Denmark	AM943016	-	-	This paper
Porphyra sp	AZ21	Azores	AM943015	-	-	This paper
Porphyra sp.	JB403	England	-	AM943395	-	This paper
Porphyra sp.	SYJP	Japan	-	AB013182	-	Kunimoto et al. (1999)
Porphyra sp.	CH5	Chile	-	AM943396	-	This paper
Porphyra sp.	CH4	Chile	-	AM943397	-	This paper
Porphyra sp.	ZGR SA	South Africa	-	AY292631	-	Jones et al. (2004)
Porphyra sp.	ZEKSA	South Africa	-	AY292630	-	Jones et al. (2004)
Porphyra sp.	GRB368	New Zealand	-	AF136420	-	Broom et al. (1999)
Porphyra sp.	CHLCH	Chile	-	AY913952	-	Nelson and Broom (2005)
Porphyra sp.	Wal	Wales	-	AF175554	-	Müller et al. (2005)
Porphyra sp.	AMM06SF2347	Faores	-	-	AM943404	This paper
Pseudobangia kaycoleia	-	Virgin Islands	-	AF043364	-	Müller et al. (1998)
Chondrus crispus	-	England	-	-	DQ191341	Robba et al. (2006)
Corallina officinalis	-	England	-	-	DQ191342	Robba et al. (2006)
Erythrocladia sp.	-		-	L26188	-	Ragan et al. (1994)
Erythrotrichia carnea	-		-	L26189	-	Ragan et al. (1994)
Gracilaria gracilis	-	England	-	-	DQ191347	Robba et al. (2006)
Cyanidium caldarium	(GenBank)		-	-	Z48930	GenBank

*[1] sequence not used in analysis
*[2] proposed neotype of *Porphyra umbilicalis* (Brodie et al. in press)

Sequence data obtained for the Xs region of rDNA SSU followed the methods given in Broom et al. (1999) as described in Brodie et al. (2007). All other sequences used in this paper were obtained from GenBank.

Data analysis

Sequence data obtained were assembled in Lasergene (DNA*Star Inc., 2005) and aligned with BioEdit Sequence Alignment Editor v6.0.5 (Hall 2004), with a final alignment by eye. GenBank sequences (Table 1) were then included in the analysis. The sequences were analysed using PAUP* 4.0 (Swofford 2002). Thirty taxa of 533 bp for *cox1*, 22 taxa consisting of 99 bp (22 bp of the 3′ end of *rbc*L and the spacer which is 77 bp) were used for the Rubisco spacer analysis and 54 taxa of 501 base pairs (bp) for the rDNA SSU analysis. No exclusions were made within the sequences. Parsimony analysis was undertaken using a full heuristic search with tree bisection reconnection branch-swapping (TBR), MULTrees option on, collapse branching on if maximum length equalled zero for 1000 random-additional sequence replicates. For the *cox1* data, which generated 1 tree, a phylogram was generated, rooted against the outgroups *Chondrus crispus*, *Corallina officinalis*, *Gracilaria gracilis* and *Cyanidium caldarium* and bootstrap analysis undertaken using 20 repetitions with 1000 additional repetitions. For the Rubisco spacer data, where parsimony analysis generated 62 trees, an unrooted strict consensus tree was produced. Bootstrap analysis was then undertaken for clade support using 100 repetitions with 1000 additional repetitions. For partial rDNA SSU parsimony analysis generated 200 trees. The strict consensus tree was rooted against the outgroups *Erythrocladia* sp. and *Erythrotrichia carnea*. Bootstrap analysis undertaken for clade support used 10 repetitions with 1000 additional repetitions. AS-IS maximum likelihood analysis was completed for the three datasets using default settings for the HKY85 model. On completion of the analysis, bootstrap analysis was undertaken for 100 repetitions with 1000 additional repetitions for both the *cox1* and Rubisco datasets and 10 repetitions with 1000 additional repetitions for the rDNA SSU. Bayesian analysis was run in MrBayes v3.1.2. The model parameters (1set + prset) were selected using MrModelTest2.2 (Nylander 2005).

Results and discussion

Determination of species diversity: Example: North Atlantic and Mediterranean species

The results of the number of species reported from the North Atlantic and Mediterranean (Table 2) show that the first half of the 18th century and the second half of the 20th century were periods of time when more species were described than at others. The evidence from the first seven years of the 21st century also suggests that there are still more species to be added to the list. In addition to those listed in Table 2, we know that there are other species that will be added to the list (Neefus 2007; Brodie, pers. obs.) However, not all these species will necessarily be new to science. *Porphyra carolinensis* is a synonym of *P. suborbiculata* Kjellman, a cosmopolitan species (Milstein and de Oliveira 2005) and is possibly an introduced species in the North Atlantic (Brodie et al. 2007). *Porphyra yezoensis* which was being grown by an aquaculture company in Maine, USA as a source of nori (Levine 1998), escaped and was confirmed on the New Hampshire coast by West et al. (2005). It is probable that further species will be introduced by direct anthropogenic activities but with the warming of the Arctic waters, we might predict that more species will find their way into the North Atlantic and some species will possibly move in the opposite direction. Some species listed may represent species complexes, notably *P. leucosticta*, a name which has been misapplied to a number of different species in the North Atlantic (Neefus 2007).

Determination of intraspecific diversity or diversity within species-complexes: Example: *cox1* as a tool to resolve intraspecific diversity in Great Britain and Faroes specimens

The *cox1* data (Fig. 1) are useful for species identification whilst suggesting differences within species. As previously reported (Robba et al. 2006), variation within species is between 0 and 4 bp. While the number of species included and the geographical range is more restricted in the dataset than for the other two molecular markers used in this study, the results provide an excellent preliminary comparison between Great Britain and the Faroes which have similar *Porphyra* floras but with southerly and northerly elements respectively (see Brodie and Nielsen 2005). Here, *Porphyra* sensu lato is split into two groupings separated by strong bootstrap support. One of these contains species which have been considered as the 'Atlantic Group' (Brodie et al. 1998): *P. dioica*, *P. purpurea*, *P. umbilicalis* and *Bangia fuscopurpurea*. *Porphyra dioica* and *P. umbilicalis* both show slight variation within the species and there is a suggestion of geographical variation. In both species, there are northerly and southerly groupings, but whether this will hold true with the inclusion of more samples remains to be tested. The sample of *P. umbilicalis* from Scotland which groups with specimens from Northern Ireland and the Faroes has been selected as the neotype for the species (Brodie et al. in press). It is well established that *Porphyra*

Table 2 Cumulative total with time for species described or reported in the North Atlantic and Mediterranean

Species	Time period					
	<1800	1800–1849	1850–1899	1900–1949	1950–1999	>2000
P. atropurpurea (Olivi) De Toni	+					
P. purpurea (Roth) C. Agardh		+				
P. linearis Greville		+				
P. umbilicalis (Linnaeus) Kützing		+				
P. miniata (C. Agardh) C. Agardh		+				
P. leucosticta Thuret in Le Jolis			+			
P. amplissima (Kjellman) Setchell & Hus ex Hus				+		
P. suborbiculata (as *carolinensis*) J. Coll & J. Cox					+	
P. rosengurttii J. Coll & J. Cox					+	
P. drachii J. Feldmann					+	
P. thulaea I.M. Munda & P.M. Pedersen					+	
P. yezoensis sensu Kornmann					+	
P. dioica Brodie & L. Irvine					+	
P. yezoensis Ueda						+
P. birdiae Neefus & Mathieson						+
P. olivii Orfanidis, Neefus & Bray						+
Total	1	4	1	1	6	3
Cumulative total	1	5	6	7	13	16

and *Bangia* are not separate taxonomic entities (Müller et al. 2005) and the diversity observed in *Bangia* was already noted in Robba et al. (2006). The other group contains the '*P. leucosticta*' complex (Brodie et al. 2007), including *P. rosengurttii* which is recognised for the first time in the Faroes flora. The data also reveal an unidentified species which groups with the *P. leucosticta* complex. That there were almost certainly unidentified species in the Faroese flora was predicted by Brodie and Nielsen (2005), who found several species that they were unable to name based on morphological observations. The presence of *P. rosengurttii* in the Faroese flora extends the known distribution of this species in the North Atlantic and Mediterranean.

These results indicate the value of such comparisons between floras from different regions and are vital to resolve the global taxonomy of the genus. This was demonstrated in Brodie et al. (2007) where two species thought to be confined to the Mediterranean were much more widespread in the North Atlantic following comparison of data between research groups in the region. One species of the species which was about to be described as new from Spain (Mediterranean) was established as *P. rosengurttii* (described from North Carolina). The other, *P. olivii*, while considered new in Greece, had concurrently been recognised as undescribed in the northeastern US. By joining up research groups and sharing data, unnecessary descriptions of species were avoided and a much better understanding of the taxonomy, ecology and distribution were able to be made.

All the Faroese samples were collected from one fjord which was sampled at several different times during a year (Mortensen et al. 2007) and will be the subject of a separate paper. This work demonstrates the value of such detailed studies and comprehensive collecting over a small area at different times of the year in really understanding the *Porphyra* flora of a region, as species can easily be overlooked or misidentified. Furthermore, species may be ephemeral or have restricted times of the year when they occur.

Detection of species diversity in different geographical areas of the world and their relationship: Example: North Atlantic and SE-Pacific (Chilean) *Porphyra* flora

While the Rubisco spacer data are unsuitable for phylogenetic analysis, the unrooted tree (Fig. 2) suggests three groupings. The first (top group in Fig. 2), separated by strong bootstrap support (98), has species from the Azores (mid Atlantic) to the Faroes (northeastern Atlantic) and contains members of the 'Atlantic Group' (Brodie et al. 1998), notably *P. dioica*, *P. linearis*, *P. purpurea* and *P. umbilicalis*. *Porphyra dioica* appears to be confined to the northeastern Atlantic (Brodie and Irvine 2003) and has not been reported from the eastern US although we have looked for it. *Porphyra linearis* and *P. umbilicalis* occur in the northeastern and northwestern Atlantic although the names have been misapplied both in this region and elsewhere (Brodie pers obs.). *Porphyra purpurea* is common in the North Atlantic and occurs in the northeast-

Fig. 1 Phylogram of *cox1* for *Porphyra* species from Great Britain and the Faroes. Numbers from left to right represent: i) parsimony bootstrap values, ii) maximum likelihood bootstrap values and iii) Bayesian posterior probabilities; zero values are where statistical support was ≤50

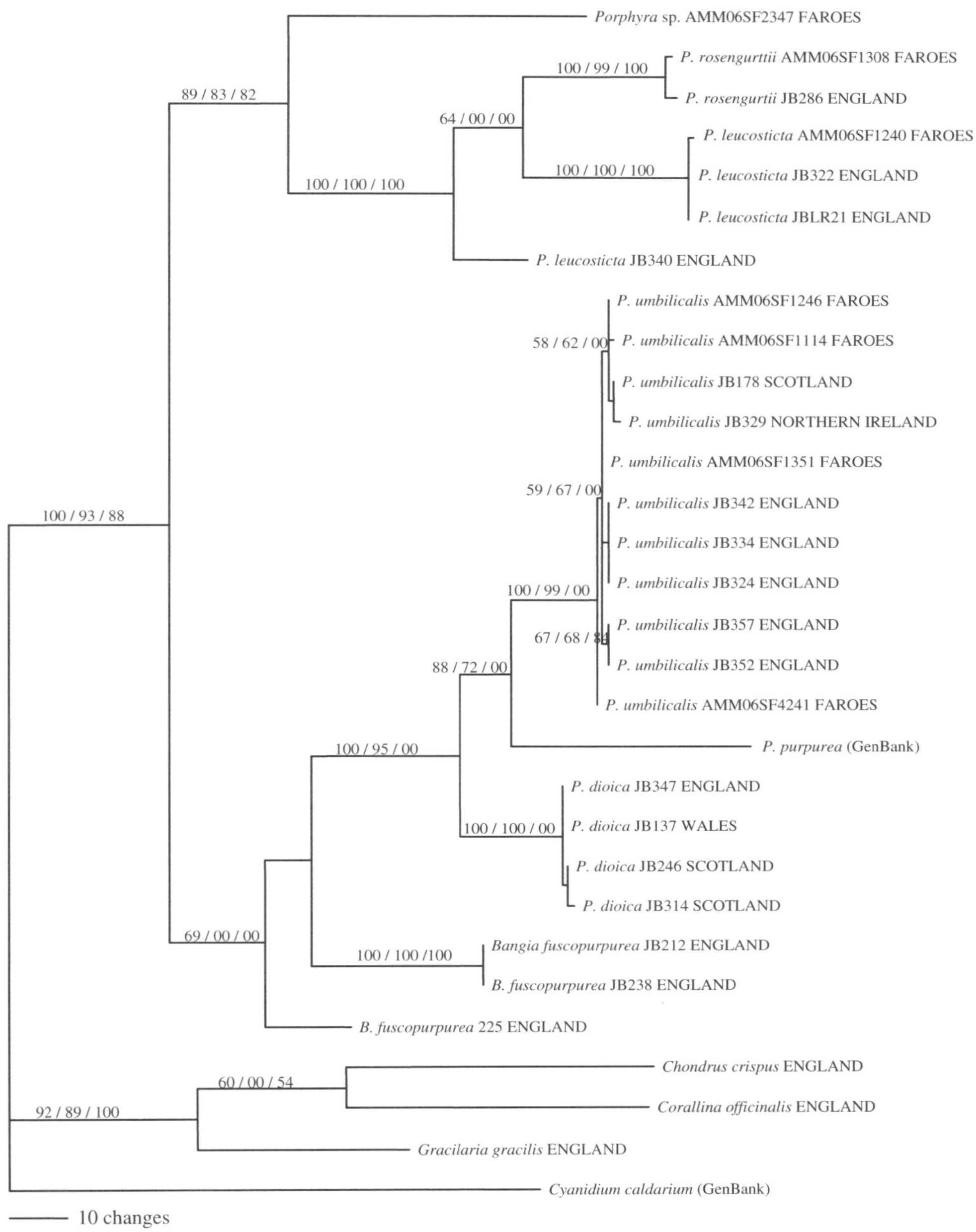

ern Pacific coast of the US (as *P. rediviva* Stiller and Waaland; Bray et al. 2006; see also Lindstrom and Fredericq 2003; Lindstrom 2008), although we need to know more about its distribution in this region.

The inclusion of one sample of *Porphyra* from Chile, and the closeness to species in the group points to the need for work to continue to establish the relationships between floras in different parts of the world. Further evidence of a possible link between the Chilean flora and those in the North Atlantic can be found between *P. amplissima* (North Atlantic) and *P. cuneiformis* (southern Chile) where the Rubisco spacer region (not shown in this paper) is identical (Brodie, unpubl. data).

The second group (middle of the unrooted tree) contains the holotype of *P. thulaea* from Greenland. The sequence for *P. thulaea* from Iceland (564557) is identical to that of the holotype. The species has only been confirmed from Greenland and Iceland and appears to be confined to these more northerly regions, although the sequence from *Porphyra* sp. RNO2 from Denmark only differs by an extra bp and all three are close to *P. pseudolinearis* from Japan. However, a single bp in the Rubisco spacer suggests that these are distinct. *Porphyra dioica* was separated from *P. purpuea* by 1 bp (in conjunction with morphological and ecological characters; Brodie et al. 1998) and this is bourne out for at least *P. thulaea* and *P. pseudolinearis* from the

Fig. 2 Unrooted strict consensus tree of the Rubisco spacer (including the 3′ end of *rbc*L) for *Porphyra* species. Numbers from left to right represent: i) parsimony bootstrap values, ii) maximum likelihood bootstrap values and iii) Bayesian posterior probabilities; zero values are where statistical support was ≤50

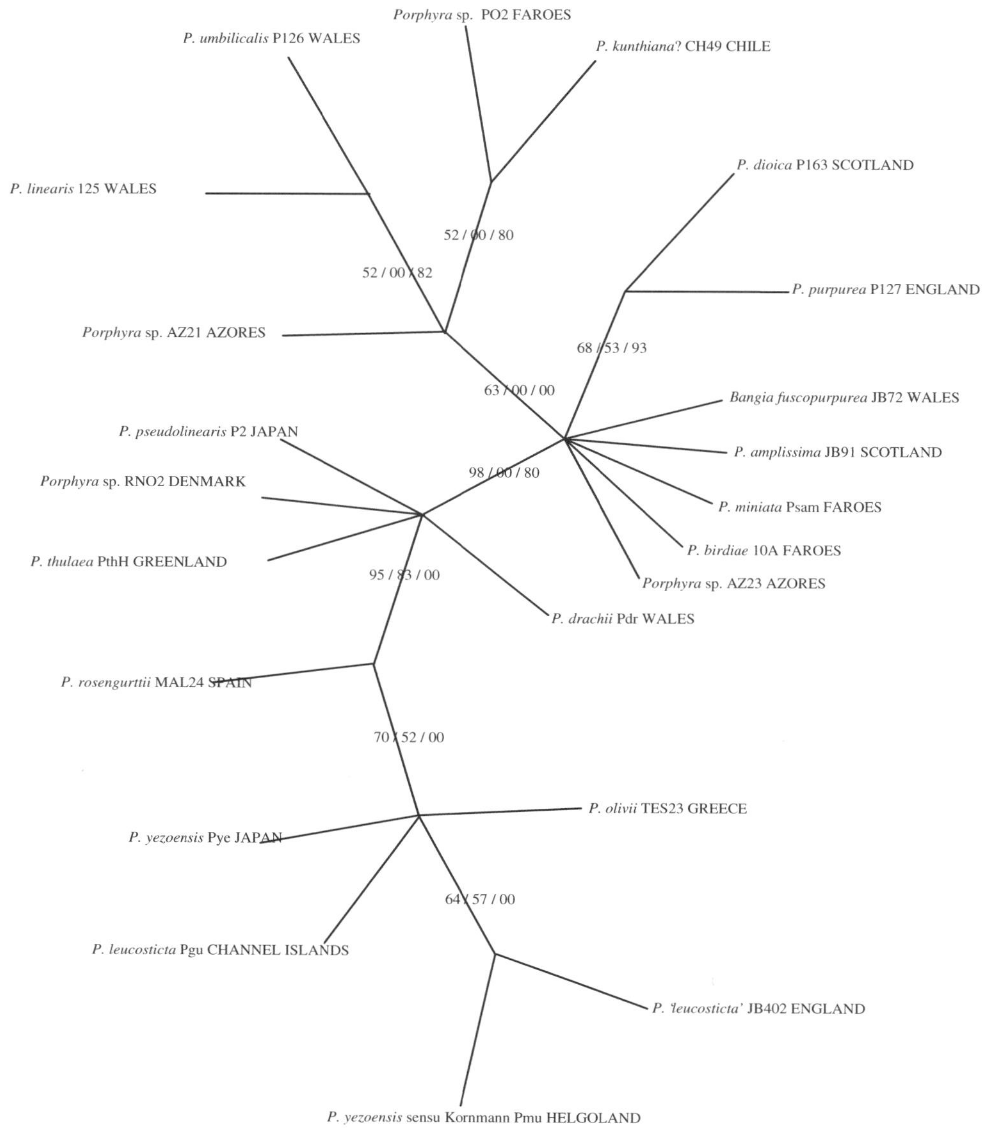

partial rDNA SSU data (Fig. 3) where they are clearly separated.

Porphyra drachii also occurs with the second group, although given the distinctive, *P. leucosticta*-like male sori, might have been predicted to come out with the third group, separated by high bootstrap support (95), which contains species of the *P. leucosticta* complex (Brodie et al. 2007). It includes *P. rosengurttii* and *P. olivii* which occur in the North Atlantic and Mediterranean (Brodie et al. 2007), other species under the name *P. leucosticta* (Neefus 2007) and closely related *P. yezoensis* from Japan. *Porphyra yezoensis* sensu Kornmann, recorded from Helgoland, may represent another species in the '*leucosticta*' complex.

There are five species to add to the North Atlantic flora based on these results alone. Preliminary data indicate that in addition to the two unidentified species from the Azores, there are probably at least three more unidentified species from this region (Brodie, pers. obs.). There are still several species in the *P. leucosticta* complex that remain unresolved.

The discovery of a possible Atlantic group within the *cox1* and Rubisco spacer data can be compared with Milstein and de Oliveira (2005), who studied the molecular phylogeny of the Bangiales based on the complete rDNA SSU with the emphasis on Brazilian *Porphyra* species. Here they recovered two monophyletic sister groups one of which (G1) contained *P. umbilicalis*, *P. purpurea* and *B. atropurpurea*, species of our Atlantic group, but also species from New Zealand, South Africa and Australia. Their second group (G2) contained *P. leucosticta* and all the species they included in their analysis from Brazil as well as Pacific representatives from Japan and New Zealand.

While the Rubisco spacer has proved valuable as an initial identification tool for *Porphyra* species and as a pointer for further investigation, at only 77 bp long, its use is almost certainly limited for global comparisons. A more useful plastid marker that would give much the same information as well as being informative phylogenetically would be the *rbc*L (e.g. Klein et al. 2003; Lindstrom and Fredericq 2003; Lindstrom 2008) and it is proposed that this should be one of the molecular markers used in determining a global taxonomy.

For a much more comprehensive analysis of a Bangiales rDNA SSU dataset and discussion of generic concepts, refer to Nelson et al. (2006). In our study the partial rDNA SSU dataset represents just under a third of the gene in a region that is useful for species identification in *Porphyra*. The results (Fig. 3) show several groupings, although these are not clearly resolved, and the relatively short length of the sequences for this gene and the limited number of taxa included in the tree make the data unsuitable for any generic interpretation. However, notable in the data, is the spread of

Table 3 Details of parsimony analysis for the three molecular markers

Marker	Rubisco spacer	Partial rDNA SSU	*cox1*
No. of of taxa	22	54	30
No. of bp included in analysis	99	492	533
No. of trees generated by parsimony analysis	62	54*	1
No. of parsimony informative characters	20	142	194
Tree length	46	630	698
CI	0.826	0.529	0.593
RI	0.924	0.710	0.762

*saved no more than one tree ≥ a score of 10 each replication.

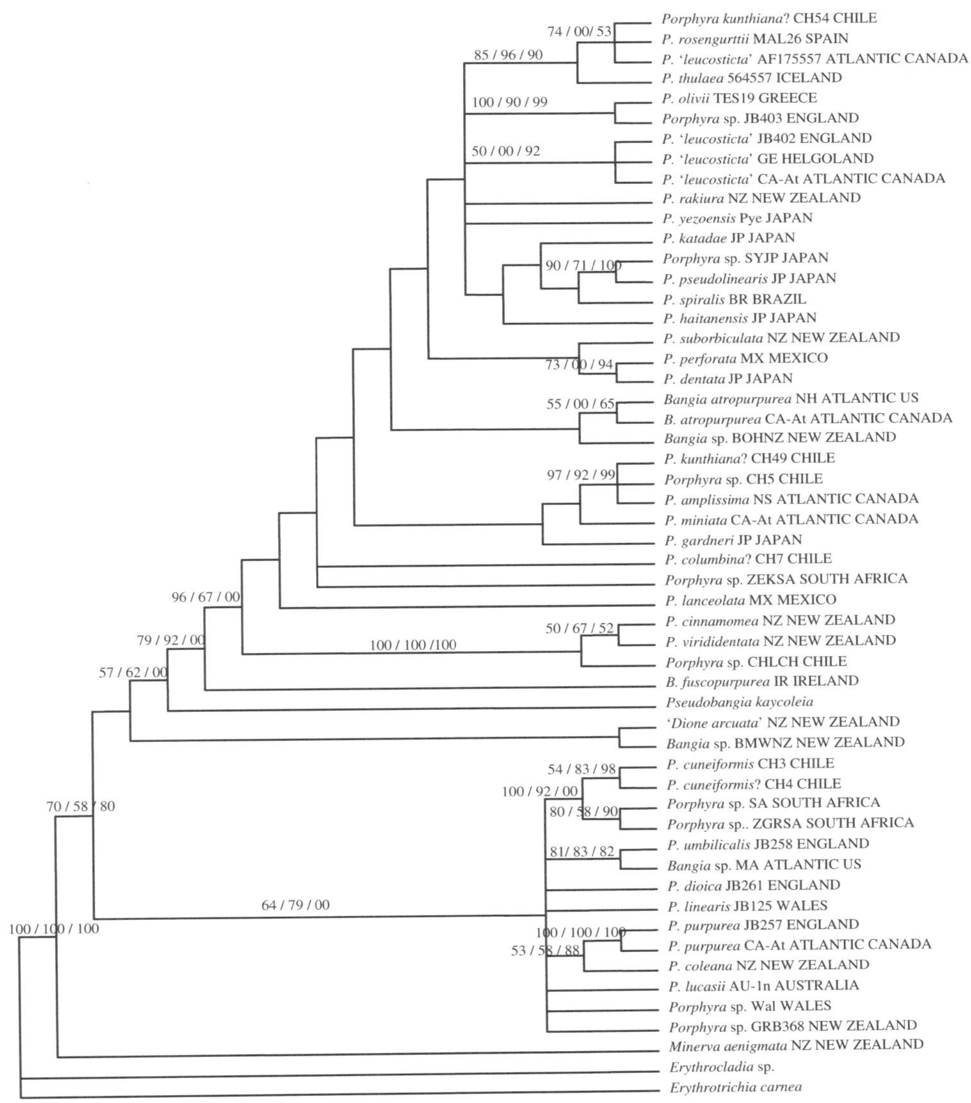

Fig. 3 Strict consensus tree of partial rDNA SSU for *Porphyra* species. Numbers from left to right represent: i) parsimony bootstrap values, ii) maximum likelihood bootstrap values and iii) Bayesian posterior probabilities; zero values are where statistical support was ≤50

taxa from different geographical areas throughout the tree, which highlights the need for more comprehensive sampling in targeted areas of the world where the taxonomy of *Porphyra* has been under worked to be certain of determining reliable generic concepts. The taxa included in the data set from Chile indicate that the *Porphyra* flora of this region is rich and that there are a considerable number of unidentified species. Species from this area occur throughout the tree, notably one of the taxa identified as *P. kunthiana*? (CH54) grouping with *P. rosengurttii* (MAL26) from the Mediterranean and a species identified as *P. leucosticta* (AF175557) from Eastern Canada, and another species named *P. kunthiana*? (CH49) close to *Porphyra* sp. (CH5) from Chile and a sample identified as *P. amplissima* from Nova Scotia. Specimens identified as *P. cuneiformis* appear to be closely related to an unidentified South African species. Additional data (not shown here) also suggest a close link between the Chilean and South African *Porphyra* flora. Names have almost certainly been misapplied in the region, notably *P. columbina* whose presence in the Chilean flora remains to be confirmed. Further detailed work is required on the *Porphyra* flora of Chile to sort out the names and study the relationships of these species, notably with South Africa, Japan and the North Atlantic.

The sequence '*Dione arcuata*' AY184338 from Nelson et al. (2006) does not match the type sequence (Nelson et al. 2005) but groups with a species of *Bangia* (BMWNZ).

Parsimony analysis of the three molecular markers showed that *cox1* (Table 3) had the highest number of parsimoniously informative characters and just one tree. The large number of trees and lowest number of parsimony informative characters for the Rubisco spacer reflects the short length of the sequences. The partial rDNA SSU represents approximately a third of the gene, excluding introns, and for a more complete analysis see Nelson et al. (2006). Nevertheless, all three markers provide useful information in our understanding of the number of species of *Porphyra*, relationships both between and within species in different parts of the world.

Resolving the genera involved in worldwide Porphyra sensu lato

Before determining the taxonomy at the generic level, it is crucial to include as many species as possible in the analysis. It is also necessary to decide which molecular markers to use for this work. The Rubisco spacer has been shown as unsuitable for this work, but the *rbc*L is suitable along with the whole rDNA SSU. While the *cox1* data may be unsuitable for generic determination, they are valuable for species discrimination and intra-specific variation.

It is evident that we do not yet know the number of species of *Porphyra* in the world, or how many species occur in different parts of the world. While there is evidence that different regions have distinct floras, it is important to consider that species may be much more widely distributed than recognised and may be going under different names in different regions. Conversely species may be going under the same name but be different species. A lot more work needs to be done to establish how species in different parts of the world relate to each other. While detailed molecular studies of the Bangiales have been undertaken in some parts of the world, notably parts of the northeastern Atlantic (Brodie et al. 1998; Klein et al. 2003), the northeastern Pacific (Lindstrom and Fredericq 2003; Lindstrom 2008), South Africa (Jones et al. 2004) and New Zealand (Nelson et al. 2006), many other regions of the world remain to be studied. It has also been noted previously (Brodie and Nielsen 2005) that northern areas of the North Atlantic and Arctic require investigation and there is evidence (Brodie, pers. obs.) of additional unidentified species in these regions.

Here, the results indicate that more detailed taxonomic studies are required for the Azores and Chile but many other regions need such study, e.g. circumpolar Arctic, southwest Atlantic and southeast Asia. While not the focus of this paper, in addition to the molecular work, detailed morphological descriptions need to be made of all the species along with consideration of ecological data.

Acknowledgements We thank Notoya Masahiro for inviting JB to speak at the 19th International Seaweed Symposium in the symposium 'Regional distribution and variation of the genus *Porphyra*: its utilization and application'. We thank Ruth Nielsen and Ian Tittley for specimens. We are extremely grateful to Chris Neefus for generating the Bayesian probabilities, to Wendy Nelson and Judy Broom for molecular advice and Linda Irvine for helpful discussions. We also acknowledge the helpful suggestions of the two referees. JB was funded by The Natural History Museum Enhancement fund to collect in Chile and by the Royal Society to travel to the symposium in Japan.

References

Bray TL, Neefus CD, Mathieson AC (2006) Morphological and molecular variability of *Porphyra purpurea* (Roth) C. Agardh (Rhodophyta, Bangiales) from the Northwest Atlantic. Nova Hedwig 82:1–22

Brodie JA, Irvine LM (2003) Seaweeds of the British Isles volume 1 part 3B Bangiophycidae. Intercept, Hampshire

Brodie J, Nielsen R (2005) The diversity of the Bangiophycidae (Rhodophyta) of the Faroes in the context of the northern Atlantic. Biofar Proceedings Fróðskaparrit (Annales Societatis Scientarum Færensis), pp 52–61

Brodie J, Zuccarello GC (2007) Systematics of the species-rich algae: red algal classification, phylogeny and speciation. In: Hodkinson TR, Parnell J, Waldren S (eds) The taxonomy and systematics of large and species-rich taxa: building and using the Tree of Life. Systematics Association Series, CRC, Boca Raton, FL, pp 317–330

Brodie J, Hayes PK, Barker GL, Irvine LM (1996) Molecular and morphological characters distinguishing two *Porphyra* species (Rhodophyta: Bangiophycidae). Eur J Phycol 31:303–308

Brodie J, Hayes PK, Barker GL, Irvine LM, Bartsch I (1998) A reappraisal of *Porphyra* and *Bangia* (Bangiales, Rhodophyta) in the northeast Atlantic based on the *rbc*L-*rbc*S intergenic spacer. J Phycol 34:1069–1074

Brodie J, Bartsch I, Neefus C, Orfinidis S, Bray T, Mathieson A (2007) New insights into the cryptic diversity of the North Atlantic-Mediterranean '*Porphyra leucosticta*' complex: *P. olivii* sp. nov. and *P. rosengurttii* (Bangiales, Rhodophyta). Eur J Phycol 42:3–28

Brodie JA, Irvine LM, Neefus CD, Russell SJ (in press) *Ulva umbilicalis* L. and *Porphyra umbilicalis* Kütz. (Rhodophyta, Bangiaceae): a molecular and morphological redescription of the species, with a typification update. Taxon.

Broom JE, Jones WA, Hill DF, Knight GA, Nelson WA (1999) Species recognition in New Zealand *Porphyra* using 18S rDNA sequencing. J Appl Phycol 11:421–428

Broom JES, Farr TJ, Nelson WA (2004) Phylogeny of the *Bangia* (Rhodophyta) flora of New Zealand suggests a southern origin for *Porphra* and *Bangia*. Mol Phylogenet Evol 31:1197–1207

Butterfield NJ (2000) *Bangiomorpha pubescens* n. gen., n. sp.: implications for the evolution of sex, multicellularity and the Mesopreterozoic- Neoproterozoic radiation of eukaryotes. Paleobiology 26:386–404

Drew KM (1949) *Conchocelis*-phase in the life history of *Porphyra umbilicalis* (L.) Kütz. Nature 164:748–749

Drew KM (1954) Life history of *Porphyra*. Nature 173:1243

Farr TJ, Nelson WA, Broom JES (2003) A challenge to the taxonomy of *Porphyra* in Australia: the New Zealand red alga *Porphyra rakiura* (Bangiaales, Rhodophyta) occurs in sourthern Australia, and is distinct from *P. lucasii*. Aus J Syst Bot 16:569–575

Fujiyoshi E, Kikuchi N, Ogawa H (2007) *Porphyra* in southeast Asia. XIXth International Seaweed Symposium, Kobe, Japan p 60 [abstract]

Hall TA (2004) BioEdit: a user-friendly biological sequence alignment editor and analysis program for Windows 95/98/NT. Nucl Acids Symp Ser 41:95–98

Holmes MJ, Brodie J (2004) Morphology, phenology and life history of *Porphyra dioica* (Bangiales, Rhodophyta) in the field and in laboratory culture. Phycologia 43:176–188

Jones WA, Griffin NJ, Jones DT, Nelson WA, Farr TJ, Broom JE (2004) Phylogenetic diversity in South African *Porphyra* (Bangiales, Rhodophyta) determined by nuclear SSU sequence analysis. Eur J Phycol 39:197–211

Kapraun DF, Lemus AJ (1987) Field and culture studies of *Porphyra spiralis* var. *amplifolia* Oliveira Filho et Coll (Bangiales, Rhodophyta) from Isla de Margarita, Venezuela. Bot Mar 30:483–490

Klein AS, Mathieson AC, Neefus CD, Cain DF, Taylor HA, West AL, Hehre EJ, Brodie J, Yarish C, Teasdale B, Wallace AL (2003) Identification of north-western Atlantic *Porphyra* (Bangiaceae, Bangiales) based on sequence variation in nuclear SSU and *rbc*L genes. Phycologia 42:109–122

Kunimoto M, Kito H, Kaminishi Y, Mizukami Y, Murase N (1999) Molecular divergence of the SSU rRNA gene and internal transcribed spacer 1 in *Porphyra yezoensis* (Rhodophyta). J Appl Phycol 11:211–216

Levine I (1998) Commercial cultivation of *Porphyra* (nori) in the United States. World Aquac 29:37–47

Lindstrom SC (2008) Cryptic diversity, biogeography and genetic variation in northeast Pacific species of *Porphyra sensu lato* (Bangiales, Rhodophyta). J Appl Phycol DOI 10.1007/s10811-008-9313-9

Lindstrom SC, Fredericq S (2003) *rbc*L gene sequences reveal relationships among north-east Pacific species of *Porphyra* (Bangiales, Rhodophyta) and a new species, *P. aestivalis*. Phycol Res 51:211–224

Milstein D, de Oliveira MC (2005) Molecular phylogeny of Bangiales (Rhodophyta) based on small subunit rDNA sequencing: emphasis on Brazilian *Porphyra* species. Phycologia 44:212–221

Mortensen AM, Brodie J, Nielsen R, Pedersen PM (2007) A study of *Porphyra* (Rhodophyta) species and their communities in the Faroes Islands: an ecological, morphological and molecular approach. Seaweed Symposium, Kobe, Japan, p 174 [abstract]

Müller KM, Sheath RG, Vis ML, Crease TJ, Cole KM (1998) Biogeography and systematics of *Bangia* (Bangiales, Rhodophyta) based on the Rubisco spacer, *rbc*L gene and 18S rRNA gene sequences and morphometric analysis. 1. North America. Phycologica 37:195–207

Müller KM, Cannone JJ, Gutell RR, Sheath RG (2001a) A structural and phylogenetic analysis of the group ICI introns in the order Bangiales (Rhodophyta). Mol Biol Evol 18:1654–1667

Müller KM, Oliveira MC, Sheath RG, Bhattacharya D (2001b) Ribosomal DNA phylogeny of the Bangiophycidae (Rhodophyta) and the origin of secondary plastids. Am J Bot 88:1390–1400

Müller KM, Cannone JJ, Sheath RG (2005) A molecular phylogenetic analysis of the Bangiales (Rhodophyta) and description of a new genus, *Pseudobangia kaycoleia*. Phycologia 45:146–155

Neefus CD (2007) Untangling the *Porphyra leucosticta* complex. International Seaweed Symposium, Kobe, Japan, pp 121–122 [abstract]

Nelson WA, Broom JE (2005) Contributions of molecular biology to understanding systematics and phylogeny in the order Bangiales. Nat Hist Res Spec Issue 8:1–12

Nelson WA, Farr TJ, Broom JES (2006) Phylogenetic relationships and generic concepts in the red order Bangiales: challenges ahead. Phycologia 45:249–259

Notoya M (1997) Diversity of life history in the genus *Porphyra*. Nat Hist Res Spec Issue 3:47–56

Nylander JAA (2005) MrModeltest v2.2. Program distributed by the author. Evolutionary Biology Centre, Uppsala University

Oliveira MC, Kurniawan J, Bird CJ, Rice EL, Murphy CA, Singh RK, Gutell RR, Ragan MA (1995) A preliminary investigation of the order Bangiales (Bangiophycidae, Rhodophyta) based on sequences of nuclear small-subunit ribosomal RNA genes. Phycol Res 43:71–79

Ragan M, Bird CJ, Rice EL, Gutell R, Murphy CA, Singh RK (1994) A molecular phylogeny of the marine red alae (Rhodophyta) based on the nuclear small-subunit rRNA gene. Proc Nat Acad Sci U S A 91:7276–7280

Robba L, Russell SJ, Barker GL, Brodie J (2006) Assessing the use of the mitochondrial *cox1* marker for use in DNA barcoding of red algae (Rhodophyta). Am J Bot 93:1101–1108

Sahoo D, Baweja P, Kushwah N (2006) Developmental studies in *Porphyra vietnamensis*: a high temperature resistant species from the Indian coast. J Appl Phycol 18:279–286

Saunders GW, Hommersand MH (2004) Assessing red algal supraordinal diversity and taxonomy in the context of contemporary systematic data. Am J Bot 91:1494–1507

Swofford DL (2002) PAUP*: Phylogenetic Analysis Using. Parsimony (* and other methods). Version 4.0. Sinauer Associates, Sunderland, MA

West AL, Mathieson AC, Klein AS, Neefus CD, Bray TL (2005) Molecular ecological studies of New England species of *Porphyra* (Rhodophyta, Bangiales). Nova Hedwigia 80:1–24

Cryptic diversity, biogeography and genetic variation in Northeast Pacific species of *Porphyra* sensu lato (Bangiales, Rhodophyta)

Sandra C. Lindstrom

Originally published in the Journal of Applied Phycology, Vol 20, No 5, 501–512.
DOI: 10.1007/s10811-008-9313-9 © Springer Science + Business Media B.V. 2008

Abstract We sequenced the chloroplast rubisco large subunit (*rbc*L) gene in 236 samples of *Porphyra* sensu lato from the northeast Pacific. Comparisons of sequences within the study area as well as comparisons with published sequences revealed up to five cryptic species among the 22 named species: a species closely related to *Porphyra abbottiae*, a species previously identified as *P. pseudolinearis*, a species closely related to *P. pseudolanceolata* and previously identified as that species, a previously unknown species from the eastern Aleutian Islands, and a species closely related to *P. schizophylla* and previously identified as that species. All of these previously unrecognized species had high bootstrap values separating them from the other species. In addition, our wide geographic sampling allowed us to extend, curtail or clarify the geographic ranges of a number of the species. We also provide published sequences for *P. gardneri* and *P. smithii* for the first time. We compared amount of sequence divergence within species grouped on the basis of sexuality (monoecious, sectored into separate male and female "halves", or dioecious), habitat (high, mid, or low intertidal/subtidal), and seasonality (winter, spring, or summer) using Tukey's HSD *t* test, but we observed no significant differences between species grouped in this manner. Different species showed different levels of genetic variation in the *rbc*L gene apparently unrelated to these traits. Also, we observed no differences in the patterns of genetic variation in a species based on whether the specimens were collected from outside or from within the region covered by ice during Pleistocene glaciations.

S. C. Lindstrom (✉)
Department of Botany, #3529 – 6270 University Blvd,
University of British Columbia,
Vancouver, BC, Canada V6T 1Z4
e-mail: sandracl@interchange.ubc.ca

Keywords Alaska · Aleutian Islands · Genetic diversity · Geographic range · *rbc*L gene · Seaweed

Introduction

Porphyra is an important genus commercially as well as culturally. Although species are not currently being cultivated in the northeast Pacific, they have been, and there may be interest in doing so again (Lindstrom 1989; Merrill 1990; Lindstrom et al. 2008). Among indigenous people of the region, *Porphyra* remains an important food and cultural article (Betts 1991; Turner 2003).

In any organismal study, the correct identification of species is necessary for proper attribution of results to particular taxa. Two species being incorrectly identified as one, or a species being misidentified as another, may result in confusion or misinformation about the taxa in the literature. Although *Porphyra* is one of the best-studied genera in the northeast Pacific from a systematics perspective (it has been monographed by Hus 1902; Krishnamurthy 1972; Conway et al. 1975; Garbary et al. 1980), more species continue to be added to the flora (Lindstrom and Cole 1990a, 1992; Stiller and Waaland 1996; Lindstrom and Fredericq 2003). Efforts are continuing to insure the correct identification of species in the region, to determine their geographic distributions, and to document other features that may be helpful in identifying species for commercial, cultural or other purposes.

The aim of the present study was to confirm the identity and distribution of the species of *Porphyra* occurring along the northeast Pacific coast from northern California to the Aleutian Islands, Alaska. This paper expands the earlier study of Lindstrom and Fredericq (2003) by reporting sequences of the *rbc*L gene from more than 230 specimens of *Porphyra* collected in this region. Because multiple samples of most species were sequenced, we also provide a preliminary comparison of population differentiation within species occurring in the region.

Materials and methods

Specimens of *Porphyra* spp were collected as indicated in Table 1. The number of specimens collected from various areas is shown graphically in Fig. 1. Specimens were damp dried in the field and desiccated in silica gel for later DNA extraction. For larger specimens, part of the sample was used to make a herbarium voucher; for smaller specimens, a separate individual was used for the voucher.

DNA was extracted in the laboratory using the CTAB mini-extraction procedure, as described by Lindstrom and Fredericq (2003). The *rbc*L gene was amplified using primer pairs F57-R753, F57-R1150_Psc, F57-RrbcSst or F753-RrbcsSt (primers as listed in Hommersand et al. 1994, except R1150_Psc, which was designed specifically to amplify *Porphyra schizophylla* Hollenberg DNA and which works well for other species of the genus: 5′-GCATTTGGCCAGCATGAATACC-3′). Approximately 1,300 to 1,350 base pairs (bp) of the 1,467 bp-long gene (~89–92%) were amplified for most specimens. The amplification protocol was as described by Lindstrom and Fredericq (2003). PCR products were cleaned using the Qiaquick PCR purification kit (Qiagen, Canada). Sequencing was done by the Nucleic Acid Protein Service Unit, University of British Columbia, using ABI Applied Biosystems Big Dye Terminator V.3.1 cycle sequencing kit and protocol.

Newly obtained sequences (Table 1) together with sequences from our earlier work (Lindstrom and Fredericq 2003) were aligned using BioEdit (Tom Hall, Ibis Therapeutics, Carlsbad, Calif., USA). Maximum parsimony (MP) and maximum likelihood (ML) analyses were carried out using PAUP* 4.0b10 (Swofford 2002) as implemented by Lindstrom and Fredericq (2003). The appropriate model of evolution for likelihood analysis was determined using the Modeltest program of Posada and Crandall (1998). Bootstrap proportions were determined based on 20,000 replicates for MP and 100 replicates for ML. *Palmaria palmata* and *Halosaccion glandiforme* were used as outgroups in all analyses (Lindstrom and Fredericq 2003). Levels of genetic variation for each species (number of variable sites/number of samples) were compared between species grouped on the basis of sexuality (monoecious, sectored into separate male and female "halves", or dioecious), habitat (high, mid, or low intertidal/subtidal), and seasonality (winter, spring, or summer) using Tukey's HSD *t* test (http://faculty.vassar.edu/lowry/hsd.html). Finally, the geographic distribution of genetic divergence in each species was documented by comparing the sequence of each specimen to the outgroup for that species (i.e., its closest relative; Fig. 2).

Results

Specimen-replete analyses produced cladograms of similar topology to that in Fig. 2. In addition to named species, we observed sufficient genetic divergence from sequences of those species to suggest the occurrence of up to five probably undescribed species:

1. Unknown #1. Specimens from a number of sites varied from *Porphyra abbottiae* at 8 of 1,350 nucleotide positions (0.6% sequence divergence). These specimens were collected from the Victoria, British Columbia, area in the eastern Strait of Juan de Fuca, and from the Kenai Peninsula and the eastern Aleutian Islands, Alaska. Fixed differences, including at collection sites in the Victoria area where the species co-occurred, suggested distinct species. The species (Unknown #1 and *P. abbottiae*) were difficult to distinguish from each other morphologically. However, specimens of the undescribed species from the Victoria area occurred epiphytically, in contrast to *P. abbottiae*, which has never been reported to be epiphytic.
2. Unknown #2. Specimens we had previously identified as *P. pseudolinearis* (Lindstrom and Fredericq 2003) occurred on a separate subclade from those from the northwest Pacific, the type locality of *P. pseudolinearis*. The northeast Pacific specimens differed from the northwest Pacific specimens by 2.0% (GenBank AB11856) and 2.3% (GenBank AB11852) in sequence divergence.
3. Unknown #3. Specimens from Whiffin Spit, along the northern Strait of Juan de Fuca, British Columbia, and from Sitka Sound, Southeast Alaska, varied from other *P. pseudolanceolata* by 1% sequence divergence. These two sites were distant from each other and occurred within the distribution range of true *P. pseudolanceolata*, which we recorded from Oregon to the Aleutian Islands. The fixed nucleotide differences, including among specimens from Whiffin Spit where the species co-occurred, suggested distinct species. However, the species were morphologically indistinguishable.
4. Unknown #4. A species from Akutan Harbor in the Aleutian Islands diverged from its closest relative, *P. aestivalis*, by 7.6% sequence divergence. No species with a similar *rbc*L sequence has been published, and further comparisons with other northerly species are

Table 1 Sources of specimens of *Porphyra* for which the *rbc*L gene was sequenced during this study

Extract no.	Collection site, date and geographic area (in parentheses)	Herbarium voucher no.	GenBank no.
Porphyra abbottiae V. Krishnamurthy			
P130	Baker Beach, CA, 25 May 2002 (NCA)	UBC A85586	EU223021
P218	Peehan, AK, 12 May 1995 (SEA)	ALAJ 742, 814, 1432	EU223022
P220	Nanwallek Reef, AK, 30 Jun 2003 (CI)	SCL 11159, 11165	EU223023
P224	Harling Pt, BC, 25 Apr 2005 (SJF)	SCL 12567	EU223024
P236	Clover Pt, BC, 25 Apr 2005 (SJF)	SCL 12563	EU223025
P386	Tonquin Park, BC, 18 May 2007 (BB)	UBC A85703	EU223026
P394	Harling Pt, BC, 5 May 2007 (SJF)	UBC A85660	EU223027
P396	Harling Pt, BC, 5 May 2007 (SJF)	UBC A85660	EU223028
P412	Chugach Bay, AK, 9 Jul 2002 (KEN)	SCL 10218	EU223029
P423	Baker Beach, CA, 25 May 2002 (NCA)	SCL 11401	EU223030
P425	Baker Beach, CA, 25 May 2002 (NCA)	SCL 11402	EU223031
Porphyra aestivalis S.C. Lindstrom *et* Fredericq			
P99	South Head, AK, 8 Jul 2002 (CI)	SCL 9984	EU223032
P166	Captains Bay, AK, 4 Aug 2004 (EAL)	SCL 13256	EU223033
P214	John Brown's Beach, AK, 19 Aug 2005 (SEA)	SCL 12765	EU223034
P216	John Brown's Beach, AK, 19 Aug 2005 (SEA)	SCL 12767	EU223035
P263	Helianthus Cove, AK, 5 Jul 2006, Leg. M. Lindeberg (EAL)	ALEUT06_089	EU223036
P 378	Akutan Harbor, AK, 30 Jul 2004 (EAL)	SCL 11467	EU223037
Porphyra brumalis Mumford			
P304	Stanley Park, BC, 19 Feb 1999 (SOG)	UBC A84426	EU223038
Porphyra conwayae (S.C. Lindstrom et K.M. Cole) S.C. Lindstrom et Fredericq			
P80	French Beach, BC, 20 Apr 1999 (SJF)	SCL 13257	EU223039
P131	Baker Beach, CA, 25 May 2002 (NCA)	UBC A85587	EU223040
P389	Tonquin Park, BC, 17 May 2007 (BB)	UBC A85691	EU223041
P390	Chesterman Beach, BC, 17 May 2007 (BB)	UBC A85690	EU223042
P391	Chesterman Beach, BC, 17 May 2007 (BB)	SCL 13208	EU223043
P430	French Beach, BC, 12 Mar 2007 (SJF)	SCL 13109	EU223044
P431	French Beach, BC, 12 Mar 2007 (SJF)	SCL 13109	EU223045
P441	Chesterman Beach, BC, 17 May 2007 (BB)	SCL 13204	EU223046
P445	Chesterman Beach, BC, 17 May 2007 (BB)	SCL 13206	EU223047
Porphyra cuneiformis (Setchell et Hus) V. Krishnamurthy			
P306	Kaflia Bay, AK, 15 Jun 2003, Leg. D. Lees (KAT)	SCL 10491	EU223048
P307	Russian Harbor, AK, 23 Jun 2005 (KOD)	SCL 12437	EU223049
P308	Point Brown, AK, 13 Jun 2006 (SEA)	SCL 12898	EU223050
P309	Humpback Bay, Unalaska I., AK, 7 Jun 2005 (EAL)	SCL 13258	EU223051
P311	Alimuda Bay, Unalaska I., AK, 2 Jun 2005 (EAL)	UBC A85184	EU223052
P312	Iniskin Bay, AK, 28 Aug 2004 (CI)	SCL 11666	EU223053
P314	Shakun Islets, AK, 16 Jun 2003 (KAT)	SCL 10810B	EU223054
P361	Kake, AK, 21 Apr 2007 (SEA)	UBC A85669	EU223055
Porphyra fallax S.C. Lindstrom et K.M. Cole			
P172	Clover Pt, BC, 27 Apr 2002 (SJF)	SCL 13259	EU223056
P191	Harling Pt, BC, 25 Apr 2005 (SJF)	UBC A85153	EU223057
P204	Akun I., AK, 31 Jul 2005 (EAL)	SCL 12637	EU223058
P206	Akun I., AK, 31 Jul 2005 (EAL)	SCL 12634	EU223059
P207	Akun I., AK, 31 Jul 2005 (EAL)	SCL 12641	EU223060
P208	Akun I., AK, 31 Jul 2005 (EAL)	SCL 12640	EU223061
P219	Hallo Bay, AK, 17 Jun 2003 (KAT)	SCL 10629	EU223062
P221	Katmai Bay, AK, 12 Jun 2003 (KAT)	SCL 10326	EU223063
P225	Akutan Bay, AK, 31 Jul 2004 (EAL)	UBC A85649	EU223064
P271	Outside Beach, AK, 6 Jun 1996, Leg. M. Lindeberg (CI)	ALAJ 844, 845	EU223065
P274	Gull I., Kachemak Bay, AK, 6 Jun 1996 (CI)	SCL 8867	EU223066
P356	Lucy I., BC, 18 Apr 2007 (NBC)	UBC A85668	EU223067
P357	Lucy I., BC, 18 Apr 2007 (NBC)	SCL 13135	EU223068
P359	Lucy I., BC, 18 Apr 2007 (NBC)	SCL 13260	EU223069
P368	Humpback Bay, Unalaska I., AK, 6 Jun 2005 (EAL)	UBC A85652	EU223070
P371	Volcano Bay, Unalaska I., AK, 9 Jun 2005 (EAL)	UBC A85152	EU223071
P372	Akun I., AK, 31 Jul 2005 (EAL)	SCL 12639	EU223072
P402	Hallo Bay, AK, 17 Jun 2003 (KAT)	SCL 10649	EU223073
P404	Shakun Islets, AK, 16 Jun 2003 (KAT)	SCL 10809	EU223074
P405	Shakun Islets, AK, 16 Jun 2003 (KAT)	SCL 10805	EU223075
P407	Kiukpalik I., AK, 16 Jun 2003 (KAT)	SCL 10774	EU223076

Table 1 (continued)

Extract no.	Collection site, date and geographic area (in parentheses)	Herbarium voucher no.	GenBank no.
P408	Kiukpalik I., AK, 16 Jun 2003 (KAT)	SCL 10773	EU223077
P409	Kiukpalik I., AK, 16 Jun 2003 (KAT)	SCL 10771	EU223078
P436	Pasagshak Bay, AK, 19 May 2005 (KOD)	SCL 13261	EU223079
PW2	Driftwood Bay, Unalaska I., AK, 8 Jul 2006, Leg. M. Lindeberg (EAL)	UBC A85407	EU223080
PW9	Skan Bay, Unalaska I., AK, 7 Jun 2005 (EAL)	UBC A85150	EU223081
PW10	Cannery Bay, Unalaska I., AK, 5 Jun 2005 (EAL)	SCL 12197	EU223082
PW24	Pumicestone Bay, AK, 3 Jun 2005 (EAL)	SCL 12166	EU223083
PW28	Shakun Islets, AK, 16 Jun 2003 (KAT)	SCL 10806	EU223084
Porphyra fucicola V. Krishnamurthy			
P38	Sunshine Cove, AK, 16 Apr 1996 (SEA)	ALAJ 746	EU223085
P141	Kiukpalik I., AK, 16 Jun 2003 (KAT)	SCL 10766	EU223086
P203	Akun I., AK, 31 Jul 2005 (EAL)	SCL 12620	EU223087
P213	Japonski I., AK, 19 Aug 2005 (SEA)	SCL 12715	EU223088
P244	Peacock Point, AK, 14 Jul 2006, Leg. M. Lindeberg (EAL)	ALEUT06_243	EU223089
P328	nr Foul Bay, AK, 25 May 2005 (KOD)	SCL 12017	EU223090
P369	Skan Bay, AK, 7 Jun 2005 (EAL)	SCL 12238	EU223091
P383	Akutan Harbor, AK, 30 Jul 2004 (EAL)	UBC A85662	EU223092
P399	Alimuda Bay, AK, 2 Jun 2005 (EAL)	UBC A85676	EU223093
P416	Chugach Bay, AK, 9 Jul 2002 (KEN)	SCL 10260	EU223093
Porphyra gardneri (G.M. Smith et Hollenberg) M.W. Hawkes			
P129	Granite Passage, AK, 13 Jul 2002, Leg. M. Lindeberg (KEN)	SCL 10203	EU223095
P326	Ushagat I., AK, 25 Aug 2006, Leg. S. Duncan (KEN)	SCL 13105	EU223096
Porphyra kanakaensis Mumford			
P105	Baker Beach, CA, 25 May 2002 (NCA)	SCL 11405	EU223097
P132	Baker Beach, CA, 25 May 2002 (NCA)	UBC A85588	EU223098
P222	Olympic Pen., WA, 31 May 2003 (SJF)	SCL 10932	EU223099
P322	Baker Beach, CA, 25 May 2002 (NCA)	SCL 11412	EU223100
Porphyra kurogii S.C. Lindstrom			
P145	Katmai Bay, AK, 12 Jun 2003 (KAT)	SCL 10327	EU223101
P281	Udamak Cove, AK, 9 Jun 2005 (EAL)	UBC A85154	EU223102
P291	Gaada'x, AK, 12 May 1995 (SEA)	ALAJ 807, 813	EU223103
P292	Outside Beach, AK, 6 Jun 1996 (CI)	ALAJ 846	EU223104
P293	Sandy Beach, Sitka, AK, 20 Apr 2000 (SEA)	SCL 9935	EU223105
P300	Seldovia Pt, AK, 4 Jun 1996 (CI)	SCL 8814	EU223106
P301	Udamak Cove, AK, 9 Jun 2005 (EAL)	SCL 12275	EU223107
P302	Portage Bay, Unalaska I., AK, 5 Jun 2005 (EAL)	UBC A85671	EU223108
P384	Akutan Harbor, AK, 30 Jul 2004 (EAL)	UBC A85663	EU223109
Porphyra mumfordii S.C. Lindstrom et K.M. Cole			
P160	Chile, Leg. M. Avila	SCL 13262	EU223110
P175	Rosario Beach, WA, 5 Mar 2005 (SJF)	SCL 12543	EU223111
P178	Rosario Beach, WA, 5 Mar 2005 (SJF)	SCL 12543	EU223112
P234	Devils Punchbowl, OR, 8 Jan 1993, Leg. E. Henry (OR)	UBC A85651	EU223113
P438	Orlebar Pt*, BC, 3 Feb 2002 (SOG)	SCL 13265	EU223114
Porphyra nereocystis C.L. Anderson			
P262	Umnak I., AK, 18 Jul 2006, Leg. M. Lindeberg (EAL)	UBC A85403	EU223115
P320	Passage I., AK, 30 Jun 2003, Leg. M. Lindeberg (CI)	SCL 11215	EU223116
P325	Ushagat I., AK, 25 Aug 2006 (KEN)	SCL 13101, 13102	EU223117
Porphyra occidentalis Setchell et Hus			
P317	Cape Palmerston, BC, 17 May 2003 (NVI)	SCL 10842	EU223118
P319	Olympic Peninsula, WA, 31 May 2003 (SJF)	SCL 10935	EU223119
Porphyra papenfussii V. Krishnamurthy			
P158	Seldovia Pt, AK, 1 Jul 2003, Leg. S. Saupe (CI)	SCL 11140	EU223120
P268	Telegraph Cove, BC, 18 May 2003 (NVI)	SCL 10874	EU223121
P360	Ridley I., BC, 17 Apr 2007 (NBC)	SCL 13266	EU223122
P433	Perevalnie Passage, AK, 23 May 2005 (KOD)	SCL 11879	EU223123
Porphyra perforata J. Agardh			
P133	Baker Beach, CA, 25 May 2002 (NCA)	SCL 11415	EU223124
P134	Baker Beach, CA, 25 May 2002 (NCA)	UBC A85590	EU223125
P142	Birch Bay, WA, 21 Jun 2003 (SOG)	SCL 13267	EU223126
P232	Crescent City Light, CA, 24 May 2002, Leg. P. Gabrielson (NCA)	SCL 13268	EU223127
P237	Clover Pt., BC, 25 Apr 2005 (SJF)	UBC A85583	EU223128
P264	Three Saints Bay, AK, 20 Jun 2005 (KOD)	SCL 12337	EU223129

Table 1 (continued)

Extract no.	Collection site, date and geographic area (in parentheses)	Herbarium voucher no.	GenBank no.
P266	Campbell River, BC, 16 May 2003 (SOG)	SCL 13269	EU223130
P267	Brady's Beach, BC, 8 Aug 2002 (BB)	SCL 13269	EU223131
P269	Afognak I,, AK, 22 Aug 2006 (KOD)	SCL 13057	EU223132
P295	Humboldt Bay, CA, 2 Apr 1994 (NCA)	SCL 13270	EU223133
P298	Cape Arago, OR, 16 Mar 1996 (OR)	SCL 9161	EU223134
P388	Tofino, BC, 18 May 2007 (BB)	UBC A85680	EU223135
P422	Seppings I., BC, 9 Aug 2002 (BB)	SCL 10400	EU223136
P446	Tonquin Park, BC, 18 May 2007 (BB)	SCL 13213	EU223137
Porphyra pseudolanceolata V. Krishnamurthy			
P143	Amalik Bay, AK, 13 Jun 2003 (KAT)	SCL 10343	EU223141
P149	Akutan Pt, AK, 30 Jul 2004 (EAL)	UBC A85646	EU223142
P179	Rosario Beach, WA, 5 Mar 2005 (SJF)	SCL 11576	EU223143
P181	Rosario Beach, WA, 5 Mar 2005 (SJF)	SCL 11576	EU223144
P193	Harling Pt, BC, 5 Feb 2005 (SJF)	SCL 12538	EU223145
P195	Izhut Bay, AK, 21 May 2005 (KOD)	SCL 11813	EU223146
P196	Big Bay, Shuyak I., AK, 24 May 2005 (KOD)	SCL 11970	EU223147
P197	Sedanka Pt, AK, 3 Jun 2005 (EAL)	UBC A85156	EU223148
P198	Humpback Bay, Unalaska I., AK, 6 Jun 2005 (EAL)	UBC A85157	EU223149
P199	Cape Sitkinak, AK, 24 Jun 2005 (KOD)	SCL 12485	EU223150
P227	Skan Bay, AK, 7 Jun 2005 (EAL)	UBC A85158	EU223151
P231	Deadman Bay*, WA, 23 Jan 2002 (SJF)	SCL 13273	EU223152
P233	French Beach, BC, 20 Apr 1999 (SJF)	SCL 13274	EU223153
P239	Botany Bay, BC, 3 Mar 2006 (BB)	SCL 12864	EU223154
P242	Fishboat Bay, BC, 6 Jan 2006 (SJF)	SCL 12860	EU223155
P242V	Marine Gardens, OR, 8 Jan 1993 (OR)	SCL 13275	EU223156
P253	Perevalnie Passage, AK, 23 May 2005 (KOD)	SCL 11885	EU223157
P336-P339	Whiffin Spit, BC, 12 Mar 2007 (SJF)	UBC A85575	EU223158
P346	Harling Point, BC, 12 Mar 2007 (SJF)	UBC A85659	EU223159
P348	Lucy I., BC, 18 Apr 2007 (NBC)	UBC A85664	EU223160
P349	Lucy I., BC, 18 Apr 2007 (NBC)	UBC A85664	EU223161
P350	Dundas I., BC, 18 Apr 2007 (NBC)	SCL 13136	EU223162
P351	Dundas I., BC, 18 Apr 2007 (NBC)	SCL 13136	EU223163
P373	Lucy I., BC, 18 Apr 2007 (NBC)	SCL 13125B	EU223164
P411	Sedanka Pt, AK, 3 Jun 2005 (EAL)	SCL 12137	EU223165
P418	Big Bay, Shuyak I., AK, 24 May 2005 (KOD)	SCL 11973	EU223166
P421	Perevalnie Passage, AK, 23 May 2005 (KOD)	SCL 11881	EU223167
Porphyra rediviva Stiller et Waaland			
P265	Sitka Natl Historical Park, AK, 31 Jul 2003 (SEA)	SCL 11353	EU223181
Porphyra schizophylla Hollenberg			
P98	Carmel*, CA, 27 May 2002, Leg. P. Gabrielson (CCA)	SCL 13282	EU223182
P159	Mendocino Headland, CA, 26 May 2002, Leg. P. Gabrielson (NCA)	NCU	EU223183
Porphyra smithii Hollenberg et I.A. Abbott			
P108	Nudibranch Pt, BC, 10 Aug 2002 (BB)	UBC A85585	EU223224
Porphyra tasa (Yendo) Ueda			
P3	Kiska I., AK, Leg. W. Schofield (WAL)	SCL 13288	EU223225
P185	Spray Cape, AK, 4 Jun 2005 (EAL)	SCL 12178	EU223226
P245	Tigalda Bay, AK, 2 Jul 2006, Leg. M. Lindeberg (EAL)	ALEUT06_028	EU223227
P258	Carlisle I., AK, 15 Jul 2006, Leg. M. Lindeberg (CAL)	ALEUT06_261	EU223228
Porphyra torta V. Krishnamurthy			
P71	Whiffin Spit, BC, 3 Feb 2002 (SJF)	SCL 13286	EU223229
P194	Harling Pt, BC, 5 Feb 2005 (SJF)	SCL 12541	EU223230
P250	Botany Bay, BC, 3 Mar 2006 (BB)	SCL 12865	EU223231
P251	Skan Bay, AK, 7 Jun 2005 (EAL)	UBC A85183	EU223232
P252	Pumicestone Bay, AK, 3 Jun 2005 (EAL)	UBC A85653	EU223233
P254	Pirate Cove, AK, 12 Apr 1997 (SEA)	ALAJ 877	EU223234
P330	Big Bay, Shuyak I., AK, 24 May 2005 (KOD)	SCL 11968, 11969	EU223235
P439	Vista Mar*, San Juan I., WA, 22 Jan 2002 (SJF)	SCL 13289	EU223236
Porphyra variegata (Kjellman) Kjellman			
P259	Kagamil I., AK, 16 Jul 2006, Leg. M. Lindeberg (CAL)	ALEUT06_308	EU223237
P313	Akutan Harbor, AK, 30 Jul 2004 (EAL)	UBC A85589	EU223238
P318	Akutan Harbor, AK, 30 Jul 2004 (EAL)	SCL 11502	EU223239

Table 1 (continued)

Extract no.	Collection site, date and geographic area (in parentheses)	Herbarium voucher no.	GenBank no.
Porphyra unknown #1 (near *P. abbottiae*)			
P366	Alimuda Bay, AK, 2 Jun 2005 (EAL)	UBC A85679	EU223011
P367	Alimuda Bay, AK, 2 Jun 2005 (EAL)	UBC A85678	EU223012
P395	Harling Pt, BC, 5 May 2007 (SJF)	UBC A85661	EU223013
P397	Harling Pt, BC, 5 May 2007 (SJF)	UBC A85661	EU223014
P410	Spray Cape, AK, 4 Jun 2005 (EAL)	UBC A85674	EU223015
P414	Chugach Bay, AK, 9 Jul 2002 (KEN)	SCL 10221	EU223016
P415	Chugach Bay, AK, 9 Jul 2002 (KEN)	SCL 10240	EU223017
PW1	Peacock Pt, AK, 14 Jul 2006, Leg. M. Lindeberg (EAL)	UBC A85424	EU223018
PW4	Akun I., AK, 5 Jul 2006, Leg. M. Lindeberg (EAL)	UBC A85414	EU223019
PW25	Sedanka Pt, AK 3 Jun 2005 (EAL)	SCL 12123	EU223020
Unknown #2 ("*Porphyra pseudolinearis*" sensu Lindstrom and Fredericq 2003)			
P120	South Head, AK, 8 Jul 2002 (CI)	SCL 10003	EU223168
P144	Katmai Bay, AK, 12 Jun 2003 (KAT)	SCL 10328	EU223169
P260	Carlisle I., AK, 15 Jul 2006, Leg. M. Lindeberg (CAL)	UBC A85402	EU223170
P261	Sarannaya Bay, Russia, 9 Aug 2006, Leg. O. Selivanova	NCU	EU223171
P289	Bridget Cove, AK, 29 Apr 2006 (SEA)	SCL 12890	EU223172
P364	Captains Bay, AK, 4 Aug 2004 (EAL)	UBC A85650	EU223173
P370	Volcano Bay, AK, 9 Jun 2005 (EAL)	UBC A85162	EU223174
P377	Akutan Harbor, AK, 30 Jul 2004 (EAL)	SCL 11468	EU223175
P398	Alimuda Bay, AK, 2 Jun 2005 (EAL)	UBC A85677	EU223176
P400	Skan Bay, AK, 7 Jun 2005 (EAL)	SCL 12231	EU223177
P401	Volcano Bay, AK, 9 Jun 2005 (EAL)	UBC A85161	EU223178
P420	Pasagshak Bay, AK, 19 May 2005 (KOD)	SCL 11797	EU223179
P442	Driftwood Bay, AK, 8 Jul 2006, Leg. M. Lindeberg (EAL)	UBC A85742	EU223180
Unknown #3 ("*Porphyra pseudolanceolata*" sensu Lindstrom and Fredericq 2003)			
P200	Whiffin Spit, BC, 3 Feb 2002 (SJF)	SCL 13271	EU223138
P228	Whiffin Spit, BC, 3 Feb 2002 (SJF)	SCL 13271	EU223139
P235	John Brown's Beach, AK, 5 Apr 1996 (SEA)	ALAJ 1430	EU223140
Unknown #4			
P148	Akutan Harbor, AK, 1 Aug 2004 (EAL)	SCL 11613	EU223240
P381	Akutan Harbor, AK, 30 Jul 2004 (EAL)	SCL 11553	EU223241
P382	Akutan Harbor, AK, 30 Jul 2004 (EAL)	SCL 11552	EU223242
P427	Akutan Harbor, AK, 31 Jul 2004 (EAL)	SCL 11581	EU223243
Unknown #5 ("*Porphyra schizophylla*" sensu Lindstrom and Fredericq 2003)			
P92	Creyke Point, BC, 28 Apr 2002 (SJF)	SCL 13276	EU223184
P93	Nudibranch Pt, BC, 10 Aug 2002 (BB)	SCL 13277	EU223185
P96	Sombrio Beach, BC, 30 Apr 2002 (SJF)	SCL 13278	EU223186
P107	Baker Beach, CA, 25 May 2002 (NCA)	SCL 13279	EU223187
P109	Morning Cove, AK, 12 Jul 2002 (KEN)	SCL 10157	EU223188
P126	near McArthur Pass, AK, 12 Jul 2002 (KEN)	SCL 13280	EU223189
P127	Fishboat Bay, BC, 29 Apr 2002 (SJF)	SCL 13281	EU223190
P136	Sombrio Beach, BC, 30 Apr 2002 (SJF)	SCL 13278	EU223191
P138	Botany Bay, BC, 30 Apr 2002 (BB)	SCL 13283	EU223192
P147	Halibut Pt, AK, 2 Aug 2003 (SEA)	SCL 11367	EU223193
P150	Akutan Pt, AK, 30 Jul 2004 (EAL)	UBC A85647	EU223194
P157	Akutan Pt, AK, 30 Jul 2004(EAL)	UBC A85647	EU223195
P164	Clover Pt, BC, 27 Apr 2002 (SJF)	SCL 13284	EU223196
P168	Yugnat Rocks, AK, 16 Jun 2003, Leg. P. Thuringer (KAT)	SCL 10619	EU223197
P169	Akutan Pt, AK, 30 Jul 2004 (EAL)	UBC A85647	EU223198
P171	Seppings Is., BC, 18 Apr 2003 (BB)	UBC A85165	EU223199
P173	Amalik Bay, AK, 13 Jun 2003 (KAT)	SCL 10347	EU223200
P174	Cape Palmerston, BC, 17 May 2003 (NVI)	UBC A85654	EU223201
P180	Port Dick, AK, 10 Jul 2002 (KEN)	SCL 13285	EU223202
P183	Pumicestone Bay, AK, 3 Jun 2005 (EAL)	UBC A85169	EU223203
P184	Humpback Bay, Unalaska I., AK, 6 Jun 2005 (EAL)	UBC A85171	EU223204
P187	Shark Point, AK, 27 Jun 2005 (KOD)	UBC A85714	EU223205
P188	near Big Bay, Shuyak I., AK, 24 May 2005 (KOD)	SCL 11971	EU223206
P189	Alitak Bay, AK, 21 Jun 2005 (KOD)	SCL 12380	EU223207
P192	Harling Pt, BC, 25 Apr 2005 (SJF)	UBC A85176, A85177, A85178	EU223208
P202	Akun I., AK, 31 Jul 2005 (EAL)	SCL 12611	EU223209
P211	John Brown's Beach, AK, 19 Aug 2005 (SEA)	SCL 12768	EU223210

Table 1 (continued)

Extract no.	Collection site, date and geographic area (in parentheses)	Herbarium voucher no.	GenBank no.
P212	Japonski I., AK, 19 Aug 2005 (SEA)	SCL 12732	EU223211
P217	Harling Pt, BC, 25 Apr 2005 (SJF)	UBC A85176, A85177, A85178	EU223212
P223	Sayward, BC, 18 May 2003 (NVI)	SCL 10881	EU223213
P239V	Creyke Pt, BC, 28 Apr 2002 (SJF)	SCL 13276	EU223214
P240V	Nudibranch Pt, BC, 19 Apr 2003 (BB)	UBC A85166	EU223215
P241V	Spray Cape, AK, 4 Jun 2005 (EAL)	SCL 13287	EU223216
P352	Chatham Sound, BC, 18 Apr 2007 (NBC)	UBC A85667	EU223217
P353	Dundas I., BC, 19 Apr 2007 (NBC)	UBC A85670	EU223218
P354	Dundas I., BC, 19 Apr 2007 (NBC)	UBC A85670	EU223219
P355	Dundas I., BC, 19 Apr 2007 (NBC)	UBC A85670	EU223220
P387	Tonquin Park, BC, 18 May 2007 (BB)	UBC A85702	EU223221
P392	Harling Pt, BC, 5 May 2007 (SJF)	UBC A85567	EU223222
P393	Harling Pt, BC, 5 May 2007 (SJF)	SCL 13169	EU223223

Collection site abbreviations: Canada: *BC* British Columbia; USA: *AK* Alaska, *CA* California, *OR* Oregon, *WA* Washington. Geographic area: see Table 3 for abbreviations. Herbarium voucher abbreviations: codes from the Index Herbariorum website: http://sciweb.nybg.org/science2/IndexHerbariorum.asp except SCL, which is the personal herbarium of the author. *A collection from the type locality of the species. All collections were made by the author unless noted otherwise.

required prior to describing this species as new to science.

5. Unknown #5. Specimens of *P. schizophylla* showed major genetic divergence along its distribution range from central California to the Aleutian Islands, with one major break occurring at Cape Mendocino in northern California and a second between southern Vancouver Island and northern Vancouver Island. The specimens from Cape Mendocino and Carmel, California, differed from more northerly specimens by a minimum divergence of 1.4% and 1.6%, respectively, suggesting that they represent true *P. schizophylla*, which has Carmel as its type locality, and the more northerly specimens represent a distinct species. Among the specimens collected from north of Cape Mendocino, those from northern Vancouver Island northward showed only two fixed differences (0.1%) from their more southerly conspecifics, suggesting population differentiation rather than additional speciation.

In addition to the species described above and those published previously (Lindstrom and Fredericq 2003), we obtained for the first time DNA sequences for *Porphyra gardneri* (G.M. Smith et Hollenberg) M.W. Hawkes and *P. smithii* Hollenberg et I.A. Abbott. Both appear in the terminal clade of Fig. 2 (clade 5 of Lindstrom and Fredericq). *Porphyra gardneri* occurred

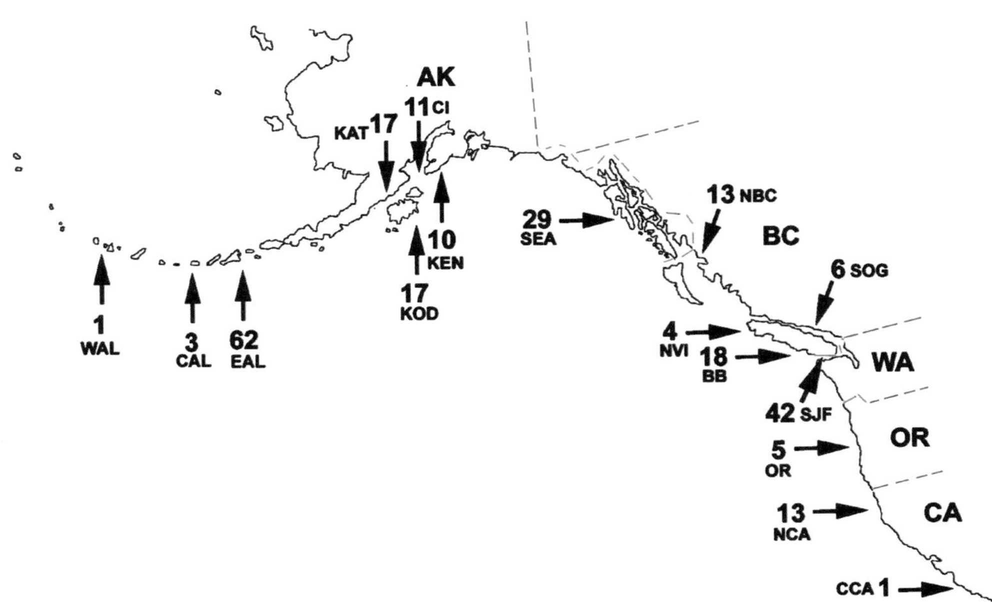

Fig. 1 Collection areas for *Porphyra* in the northeast Pacific and numbers of specimens collected in each area. See Table 3 for area abbreviations

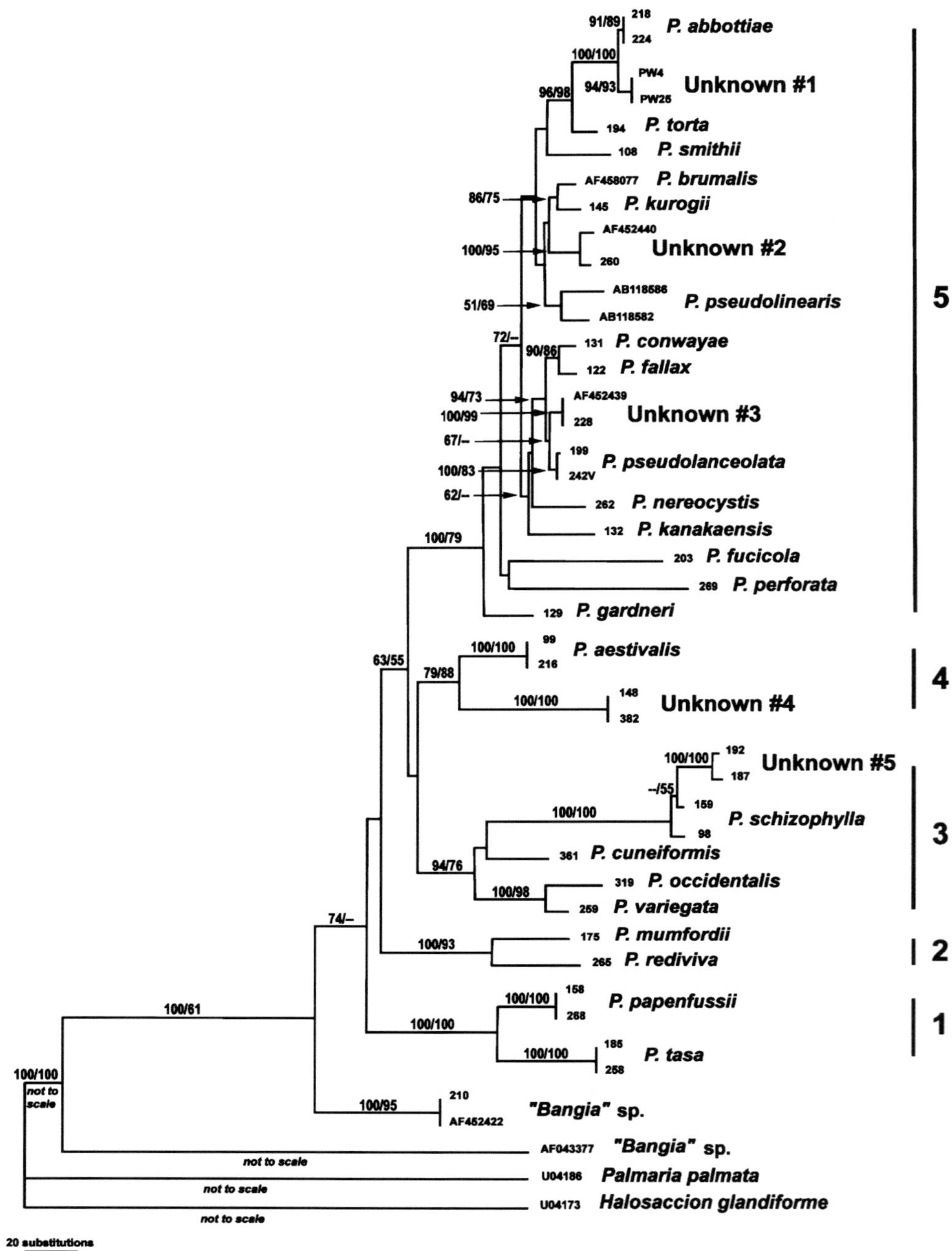

Fig. 2 Tree produced by maximum likelihood analysis using the GTR + G + I model of evolution; −lnL=9056.4622. Bootstrap values are for maximum parsimony (*left*) and maximum likelihood (*right*). Clades are numbered as discussed in the text

on a branch above *P. perforata* J. Agardh (with poor bootstrap support) but below *P. fucicola* V. Krishnamurthy (with no bootstrap support) in MP, but basal to both *P. fucicola* and *P. perforata* in ML (with no bootstrap support), as shown in Fig. 2. *Porphyra smithii* occurred basal to the subclade containing *P. abbottiae*, Unknown #1, and *P. torta* in both the MP and ML analyses but without bootstrap support in either analysis. The variability of the position of *P. gardneri* in different analyses and the failure of both *P. gardneri* and *P. smithii* to form close relationships with other species indicate the distinctiveness of these two species.

In addition to records of "new" species, this survey extended some geographic ranges and curtailed others (Table 2). For example, the geographic range of *P. abbottiae* was extended south to northern California, but its northern range was curtailed in the western Gulf of Alaska in the Kodiak archipelago. Specimens west of the Gulf of Alaska that resembled *P. abbottiae* were all Unknown #1. Also, *P. perforata* and *P. papenfussii* had

Table 2 Revised geographic ranges of northeast Pacific species of *Porphyra* as confirmed by this study

	Previous geographic range	Confirmed geographic distribution
P. abbottiae	WA to Aleutian Is., AK; Northwest Pacific	Northern CA to Kodiak Is., AK
P. aestivalis	Southeast AK to Kodiak Is., AK	Southeast AK to Aleutian Is., AK
P. fallax	OR to Southeast Alaska	OR to Aleutian Is., AK
P. fucicola	OR to Kodiak Is., AK[1]	OR to Aleutian Is., AK
P. kanakaensis	Northern CA to northern BC	Northern CA to eastern GOA[2]
P. kurogii	Southern BC to Japan	Southeast AK to Aleutian Is., AK
P. mumfordii	OR to southern BC	Northern CA to southern BC; Chile
P. papenfussii	WA to northern BC	WA to Kodiak Is., AK
P. perforata	Baja California to Aleutian Is., AK; Okhotsk Sea; Kamchatka	CA to Kodiak Is., AK
P. pseudolanceolata	WA to Aleutian Is., AK	OR to Aleutian Is., AK
P. rediviva	Northern CA to northern WA[3]	Northern CA to Northern WA; Southeast AK
P. schizophylla	Central CA to Aleutian Is., AK	Central CA to Cape Mendocino, CA
P. tasa	Aleutian Is., AK, to Kurile Is.[4]	Aleutian Is., AK, to Kurile Is.
P. variegata	WA to Aleutian Is., AK; Northwest Pacific	Northern GOA to Aleutian Is., AK; Northwest Pacific
P. unknown #1	Previously included in *P. abbottiae*	Southern BC to Aleutian Is., AK
P. unknown #2	Previously included in *P. pseudolinearis*	Southeast AK to Russia
P. unknown #3	Previously included in *P. pseudolanceolata*	Whiffin Spit, BC; Sitka Sd., AK
P. unknown #4	Previously unknown	Akutan Harbor, AK
P. unknown #5	Previously included in *P. schizophylla*	North of Cape Mendocino, CA, to Aleutian Is., AK

Species for which we did not alter the published ranges as reported in Scagel et al. (1993) include *P. brumalis*, *P. conwayae*, *P. cuneiformis*, *P. gardneri*, *P. nereocystis*, *P. occidentalis*, *P.* "*pseudolinearis*", *P. smithii*, and *P. torta*; these species are not listed. Previous geographic ranges are taken from Scagel et al. (1993) unless noted otherwise. *GOA* Gulf of Alaska. For other geographic abbreviations, see Table 1.

[1] Lindstrom and Cole (1992)

[2] ALA 1598, ALA 1618, collected outside LaChaussee Spit, Lituya Bay, AK, 16 Aug 1971, Leg. G. J. Mueller (herbarium abbreviations from Index Herbariorum-see Table 1).

[3] Stiller and Waaland (1996)

[4] Wynne (1972)

their northern range limited to the Kodiak archipelago. Specimens that were responsible for the previous distribution records of *P. perforata* in the Aleutian Islands and eastern Russia are probably *P. fallax*, a species with which it has been frequently confused (Lindstrom and Cole 1990a, b); the distribution record of *P. fallax* is extended westward to the Aleutian Islands in the present study. Although it had been suggested that *P. papenfussii* from the eastern Pacific might be conspecific with *P. tasa* (Yendo) Ueda from the western Pacific (Scagel et al. 1993), our results show that the eastern Pacific taxon is clearly distinct (4.7% sequence divergence) from the Aleutian species, which we have identified as *P. tasa*.

We also examined the sequence data for patterns of genetic divergence (Table 3). We observed no significant differences in levels of divergence between groups of species based on sexuality (monoecious with sexes intermixed, monoecious with sexes separated on different "halves" of the thallus, or dioecious), vertical elevation (high, mid or low intertidal/subtidal) or seasonality (winter, spring, or summer) using Tukey's HSD t test. Also, we observed no differences in the patterns of genetic variation in a species based on whether the specimens were collected from outside or from within the region covered by ice during Pleistocene glaciations.

Discussion

The results of the present analysis conform to those of our earlier study of *Porphyra* in the northeast Pacific (Lindstrom and Fredericq 2003). Both studies reveal five moderately to strongly supported clades of *Porphyra* (Fig. 2). The present study adds species to four of those five clades. *Porphyra tasa* joins *P. papenfussii* in clade 1; the species are clearly distinct despite similar habitats, morphologies, and earlier suggestions that they might be conspecific (Scagel et al. 1993). The sequence data for multiple specimens of *P. schizophylla* (clade 3) suggest that northern populations should be considered a separate species. The sequence for an unknown species from Akutan Harbor, Aleutian Islands, Alaska, is only distantly related to *P. aestivalis* in clade 4.

Clade 5, which was the most speciose clade in our earlier study, contains the other species not previously sequenced (*Porphyra gardneri*, *P. smithii*) or described (unknowns #1–3). *Porphyra gardneri*, which like *P. nereocystis* is an epiphyte of kelps, is not closely related to any other species in this clade. *Porphyra smithii*, which like *P. kanakaensis* has two chloroplasts per cell, also shows no close affinities. Among the "unknown" species in clade 5, two have been previously published under other names. Unknown #2 has

Table 3 Geographic distribution of genetic divergence in species of *Porphyra* in this study, levels of genetic variation, species traits, and outgroups

Species	No. of samples	No. of variable sites	CCA	NCA	OR	SJF	BB	SOG	NVI	NBC	SEA	KEN	KOD	CI	KAT	EAL	CAL	WAL	Var	Sex	Hab	Sea	Otgr
			|←Not glaciated→|			|←――――――――――――― Glaciated												――――――――――――――→|					
P. abbottiae	12	1	–	0	–	0	0	–	–	–	1	0	–	0	–	–	–	–	0.08	MM	MI	SP	U#1
P. aestivalis	8	0	–	–	–	–	–	–	–	–	0	–	–	0	–	0	–	–	0.00	MS	MI	SU	U#4
P. brumalis	2	0	–	–	–	–	–	0	–	–	–	–	–	–	–	–	–	–	0.00	MS	HI	WI	Pku
P. conwayae	10	3	–	1	2	2	1	–	–	–	–	–	–	–	–	–	–	–	0.30	DI	MI	SP	Pfa
P. cuneiformis	9	2	–	–	–	–	–	–	–	–	1	–	1	1	1	1	–	–	0.22	MM	LO	SP	Pva
P. fallax	30	3	–	–	–	1	–	0	–	–	–	–	1	1	1	3	–	–	0.10	MM	MI	SP	Ppse
P. fucicola	11	1	–	–	–	–	–	–	–	–	0	0	0	–	0	1	–	–	0.09	MM	MI	SP	Ppe
P. gardneri	2	0	–	–	–	–	–	–	–	–	–	0	–	–	–	–	–	–	0.00	MM	LO	SU	Ppe
P. kanakaensis	5	6	–	2	–	4	–	–	–	–	–	–	–	–	–	–	–	–	1.20	MM	MI	SP	Pne
P. kurogii	10	1	–	–	–	–	–	–	–	–	0	–	–	–	0	1	–	–	0.10	MS	MI	SP	Pbr
P. mumfordii	6	0	–	–	0	0	–	–	–	–	–	–	–	–	–	–	–	–	0.00	MS	HI	WI	Pre
P. nereocystis	4	3	–	–	–	–	–	–	–	–	2	–	–	1	–	0	–	–	0.75	MM	LO	SU	Pka
P. occidentalis	3	1	–	–	–	1	–	–	0	–	–	–	–	–	–	–	–	–	0.33	DI	LO	SU	Pva
P. papenfussii	5	0	–	–	–	–	–	–	0	0	0	–	0	0	–	–	–	–	0.00	DI	MI	SU	Pta
P. perforata	15	3	–	3	0	2	0	2	–	–	3	–	3	–	–	–	–	–	0.20	MM	MI	WI	Pfu
P. pseudolanceolata	30	3	–	–	1	2	2	–	–	2	–	–	3	–	2	2	–	–	0.10	DI	HI	WI	U#3, Pne
P. rediviva	2	1	–	–	–	1	–	–	–	–	1	–	–	–	–	–	–	–	0.50	MS	HILO	WISU	Pmu
P. schizophylla	2	8	4	4	–	–	–	–	–	–	–	–	–	–	–	–	–	–	4.00	DI	HI	SP	Pcu
P. smithii	1	–	–	–	–	–	0	–	–	–	–	–	–	–	–	–	–	–	0.00	MM	MI	SU	Pt
P. tasa	4	1	–	–	–	–	–	–	–	–	–	–	–	–	–	0	–	1	0.25	DI	MI	SP	Ppa
P. torta	9	5	–	–	–	4	1	–	–	–	1	–	1	–	–	1	–	–	0.56	MM	HI	WI	Pab
P. variegata	4	10	–	–	–	–	–	–	–	–	–	–	–	4	–	4	5	–	2.50	MS	LO	SP	Poc
Unknown #1	10	1	–	–	–	0	–	–	–	–	–	0	–	–	–	1	–	–	0.10	MM	MI	SU	Pab, Pto
Unknown #2	13	14	–	–	–	–	–	–	–	–	9	–	4	4	4	10	4	–	1.08	DI	LO	SP	Pbr, Pku
Unknown #3	4	0	–	–	–	0	–	–	–	–	0	–	–	–	–	–	–	–	0.00	DI	HI	WI	Ppse
Unknown #4	4	0	–	–	–	–	–	–	–	–	–	–	–	–	–	0	–	–	0.00	DI	HI	WI	Pae
Unknown #5	41	11	–	2	–	7	2	–	2	2	4	3	4	–	4	2	–	–	0.27	DI	HI	SP	Psc

Abbreviations: *CCA* central California, *NCA* northern California, *OR* Oregon, *SJF* Strait of Juan de Fuca (southern BC, northern Washington, including San Juan Is.), *BB* Botany Bay and Barkley Sound, *SOG* Strait of Georgia, *NVI* northern Vancouver I., *NBC* northern British Columbia, *SEA* Southeast Alaska, *KEN* Kenai Peninsula, *KOD* Kodiak archipelago, *CI* Cook Inlet, *KAT* Katmai National Park (northeastern Alaska Peninsula), *EAL* eastern Aleutian Is., *CAL* central Aleutian Is., *WAL* western Aleutian Is. *Var* number of variable sites/number of samples, *Sex* sexuality (*MM* monoecious and mixed, *MS* monoecious and sectored, *DI* dioecious), *Hab* habitat (*HI* high intertidal, *MI* mid intertidal, *LO* low intertidal/subtidal); *Sea* season (*WI* winter, *SP* spring, *SU* summer), *Otgr* outgroup (abbreviations: first letter of genus name + first two letters of species epithet)

Numbers under geographic areas refer to the number of nucleotide positions at which specimens from that area varied from their conspecifics using the outgroup method to determine the polarity of change. - means no sample of that species was sequenced from that area. *CCA*, *NCA* and *OR* were not glaciated during Pleistocene glaciations whereas the other areas were.

been called *P. pseudolinearis* (Lindstrom and Fredericq 2003), based on its morphological similarities to that species and the fact that a more or less continuous distribution across the northern rim of the Pacific between its type locality in Korea and Alaska has been assumed (Scagel et al. 1993). Differences in sequences between western Pacific and Alaska specimens (this study) and in the behavior of the conchocelis in culture (Lindstrom et al. 2008) suggest that the eastern and western North Pacific populations represent distinct species. Unknown #3 had been identified previously as *P. pseudolanceolata* by Lindstrom and Fredericq (2003), and it is called by that name in Lindstrom et al. (2008). The two species share the same habitat and morphology.

Unknown #1 has undoubtedly been confused with *P. abbottiae* previously based on the recorded distribution of that species from Russia to northern Washington (Scagel et al. 1993). All of our specimens tentatively identified as *P. abbottiae* from west of Kodiak Island turned out to be Unknown #1. Like the other unknown species in clade 5, Unknown #1 shares the same habitat and morphology as the species with which it has been confused except that it has been found as an epiphyte, a habitat not reported for *P. abbottiae* (Conway et al. 1975). We were unable to confirm the distribution of *P. abbottiae* further west than Kodiak Island, Alaska, although we extended its southern distribution to northern California. Cannon (1989) had previously reported on the species in northern California, but we had been unable to verify its occurrence there because of lack of voucher specimens (Gabrielson et al. 2004).

Besides *Porphyra abbottiae*, two other species appear to reach their western limit in the Kodiak archipelago: *P. papenfussii* and *P. perforata*. The Kodiak archipelago was one of the areas of Alaska where a relatively high percentage of species of seaweeds reach their northern (western) distribution limit (Lindstrom 2006). Ranges have been adjusted for other species based on the present data, including a new record of *P. rediviva* in Southeast Alaska, where it was found in the low intertidal in summer in contrast to its occurrence in the high intertidal in winter at its type locality in northern Washington (Stiller and Waaland 1996), and new records of *P. aestivalis*, *P. fallax*, and *P. fucicola* in the Aleutian Islands. Previously, Wynne (1972) had recorded *P. purpurea*, *P. umbilicalis*, *P. yezoensis*, *P. perforata*, *P. ochotensis*, *P. pseudolinearis*, *P. tasa*, *P. variegata* and *P. amplissima* from Amchitka Island in the central Aleutian Islands. The identity of most of these species needs to be verified in light of our current understanding of these species.

Comparisons of features such as sexuality, seasonality and habitat with genetic diversity failed to reveal any patterns, as had been previously observed using isozymes (Sosa and Lindstrom 1999), in which significant differences were observed between measures of population differentiation and both habitat and mating system (sexuality). This may be due to a different system being used (a chloroplast gene versus isozymes), a different measure (percent of variable sites per specimen versus mean heterozygosity and percent polymorphic loci), different taxon sampling or sampling intensity, or other unknown factors. Still, some species showed little or no genetic variation despite geographic distances of up to several thousands of kilometers whereas other species showed notably more variation over much shorter distances (Table 3). High levels of genetic variation were not limited to areas outside those covered by glaciers during the Pleistocene Ice Ages, namely Oregon and California. Although there was genetic variation in most of the species collected outside the glaciated region, there was genetic variation in some species in all areas along the coast, including areas that were heavily glaciated during the Pleistocene (northern Washington, British Columbia and Alaska).

This present survey of northeast Pacific *Porphyra* confirms that this region is rich in species of this ancient form-genus, and it continues the discovery of cryptic species diversity. However, further work is required to adequately characterize these newly discovered cryptic species so that they can be readily identified by workers in other disciplines. Moreover, widely different levels of genetic diversity within species along the northeast Pacific coast suggest that this is also an area of research requiring more intensive investigation.

Acknowledgements The support of the Natural Sciences and Engineering Research Council of Canada is gratefully acknowledged. I appreciated the help of the following students in the laboratory: Vida Omidvar, Tina Chen, Alexis Kwong and Penny White. Many of the collections were made while participating in field trips at the invitation of Jon Houghton or Susan Saupe—to both I am extremely grateful. I am also grateful to Mandy Lindeberg, Paul Gabrielson, Max Hommersand and others for providing specimens for this work. Frank Shaughnessy generously introduced me to Baker Beach, and Scott Pilcher provided transportation to remote locations in northern British Columbia.

References

Betts M (1991) Seven criteria worksheet for findings on: Customary and traditional uses of seaweed and kelp in southeast Alaska. Report to the Alaska Board of Fisheries. Division of Subsistence, Region I, Alaska Department of Fish & Game, Juneau, Alaska

Cannon MI (1989) Cell division patterns and diurnal cycle of *Porphyra abbottae* (Rhodophyta, Bangiales) carpogonial and carpospore development. J Phycol 25:612–615

Conway E, Mumford TF Jr, Scagel RF (1975) The genus *Porphyra* in British Columbia and Washington. Syesis 8:185–244

Gabrielson PW, Widdowson TW, Lindstrom SC (2004) Keys to the Seaweeds and Seagrasses of Oregon and California North of

Point Conception. Phycological Contribution No. 6, PhycoID, Hillsborough, North Carolina, USA

Garbary DJ, Hansen GI, Scagel RF (1980) The marine algae of British Columbia and northern Washington: Division Rhodophyta (red algae), Class Bangiophyceae. Syesis 13:137–195

Hommersand MH, Fredericq S, Freshwater DW (1994) Phylogenetic systematics and biogeography of the Gigartinaceae (Gigartinales, Rhodophyta) based on sequence analysis of *rbc*L. Bot Mar 37:193–203

Hus HTA (1902) An account of the species of *Porphyra* found on the Pacific coast of North America. Proc Calif Acad Sci, Ser 3, Botany 2:173–241

Krishnamurthy V (1972) A revision of the species of the algal genus *Porphyra* occurring on the Pacific coast of North America. Pac Sci 26:24–49

Lindstrom S (1989) Nori aquaculture makes waves in British Columbia. Bull Aquacult Assoc Can 89–2:36–40

Lindstrom SC (2006) Biogeography of Alaskan seaweeds. J Appl Phycol 18:637–641

Lindstrom SC, Cole KM (1990a) *Porphyra fallax*, a new species of Rhodophyta from British Columbia and northern Washington. Jpn J Phycol 38:371–376

Lindstrom SC, Cole KM (1990b) An evaluation of species relationships in the *Porphyra perforata* complex (Bangiales, Rhodophyta) using starch gel electrophoresis. Proc Int Seaweed Symp 13:179–183

Lindstrom SC, Cole KM (1992) A revision of the species of *Porphyra* (Rhodophyta: Bangiales) occurring in British Columbia and adjacent waters. Can J Bot 70:2066–2075

Lindstrom SC, Fredericq S (2003) *rbc*L gene sequences reveal relationships among north-east Pacific species of *Porphyra* (Bangiales, Rhodophyta) and a new species, *P. aestivalis*. Phycol Res 51:211–224

Lindstrom SC, Conitz JM, Hall S, Stekoll MS (2008) Induction of conchospore release: ecotypic variation in northeast Pacific species of *Porphyra*. J Appl Phycol DOI 10.1007/s10811-007-9261-9

Merrill J (1990) The current status of sea vegetable farming. Fish Farm News 3(9):40, 50

Posada D, Crandall KA (1998) Modeltest: testing the model of DNA substitution. Bioinformatics 14:817–818

Scagel RF, Gabrielson PW, Garbary DJ et al (1993) A Synopsis of the Benthic Marine Algae of British Columbia, Southeast Alaska, Washington and Oregon. University of British Columbia, Vancouver, Phycological Contribution No. 3, p 532

Sosa PA, Lindstrom SC (1999) Isozymes in macroalgae (seaweeds): genetic differentiation, genetic variability and applications in systematics. Eur J Phycol 34:427–442

Stiller JW, Waaland JR (1996) *Porphyra rediviva* sp. nov. (Rhodophyta): a new species from northeast Pacific salt marshes. J Phycol 32:323–332

Swofford DL (2002) Phylogenetic Analysis Using Parsimony (* and Other Methods) Version 4.0b10. Sinauer, Sunderland, Mass.

Turner NJ (2003) The ethnobotany of edible seaweed (*Porphyra abbottae* and related species; Rhodophyta: Bangiales) and its use by First Nations on the Pacific coast of Canada. Can J Bot 81:283–293

Wynne MJ (1972) The genus *Porphyra* at Amchitka Island, Aleutians. Proc Int Seaweed Symp 7:100–104

Taxonomy and distribution of *Sargassum* (Phaeophyceae) in the Gulf of Thailand

T. Noiraksar · T. Ajisaka

Originally published in the Journal of Applied Phycology, Vol 20, No 5, 513–527.
DOI: 10.1007/s10811-008-9327-3 © Springer Science + Business Media B.V. 2008

Abstract Ten species of *Sargassum* (Sargassaceae, Phaeophyceae) were found along the Gulf of Thailand. Morphological characteristics of *Sargassum baccularia* (Mertens) C.A. Agardh, *S. binderi* Sonder, *S. cinereum* J.G. Agardh, *S.crassifolium* J.G. Agardh, *S. longifructum* Tseng et Lu, *S. oligocystum* Montagne, *S. polycystum* C.A. Agardh, *S. siliquosum* J.G. Agardh, *S. swartzii* (Turner) C.A. Agardh and one unidentified species were examined and are described in detail. The most common species were *S. polycystum* distributed widely in almost all the study sites, *S. crassifolium* restricted to Prachuap Khirikhan Province, *S. longifructum* restricted to Chumphon Province, *S. siliquosum* restricted to Surat Thani Province and one unidentified species restricted to Songkhla Province. Three species (*S. cinereum, S. longifructum* and *S. swartzii*) are new records for the algal flora of Thailand. Five species (*S. baccularia, S. cinereum, S. longifructum, S. polycystum* and the unidentified species) belong to the section Zygocarpicae (J.G. Agardh) Setchell.

Keywords Phaeophyceae · *Sargassum* · Taxonomy · Distribution · The Gulf of Thailand

T. Noiraksar (✉)
Institute of Marine Science, Burapha University,
Bangsaen, 20131 Chon Buri, Thailand
e-mail: sargassum2005@yahoo.com

T. Ajisaka
Hall of Global Environmental Research,
Graduate School of Global Environmental Studies,
Kyoto University,
Kyoto, 606-8502, Japan

Introduction

Sargassum C. Agardh (Sargassaceae, Fucales), a very large brown algal genus with nearly 400 species, is widely distributed in warm and temperate waters, especially in the Indo-west Pacific region and Australia (Tseng et al. 1985). Thailand is a tropical country in south-east Asia with a coastline of approximately 2,650 km, consisting of 1,880 km along the Gulf of Thailand (Pacific Ocean) and 770 km along the Andaman Sea (Indian Ocean); both coasts have a diverse seaweed flora. Thai species of *Sargassum* were first recorded by Reinbold in "Flora of Koh Chang" from the specimens collected by Schmidt (1900–1916) during the Danish Expedition to Siam; *S. polycystum* was reported from Koh Kahdat, Trat Province situated on the east coast of the Gulf of Thailand. Lewmanomont (1988) reported *S. polycystum* from coral reefs along the Gulf of Thailand, and *S. crassifolium* and *S. polycystum* from the Andaman Sea. Egerod (1974) collected *S. grevillei* from Koh Ra, Ranong Province, the Andaman Sea on the Fifth Thai–Danish Expedition of 1966. Nateewathana et al. (1981) and Aungtonya and Liao (2002) compiled and updated the extensive checklist, adding *S. crassifolium*, *S. granuliferum* and *S. siliquosum* from the Reference Collection of the Phuket Marine Biological Center (PMBC). Ajisaka and Lewmanomont (2004) reported *S. stolonifolium* from the Andaman Sea. Noiraksar et al. (2006) collected *S. baccularia*, *S. binderi*, *S. oligocystum* and *S. polycystum* from the east coast of the Gulf of Thailand. While there are numerous reports, there are very few detailed studies of *Sargassum* species in the Gulf of Thailand. The object of the present study is to present detailed descriptions and distribution of *Sargassum* species along the Gulf of Thailand.

Materials and methods

The specimens were collected from the following nine provinces along the Gulf of Thailand: Chon Buri, Rayong, Chanthaburi, Trat (the east coast); Prachuap Khirikhan, Chumphon, Surat Thani, Nakhon Si Thammarat and Songkhla (the west coast) (Fig. 1) between 2001 and 2006. Whole thalli (with holdfasts) were collected during low tide. Specimens from the deeper areas were collected by SCUBA. Some of the collected specimens were fixed in 4% formaldehyde-seawater, and the remainder was dried on herbarium sheets. Important parts of the thallus were drawn, and sections of receptacles were prepared and stained with 1% aniline blue intensified with 1% HCl and mounted on glass slides using 50% glucose syrup (Karo® Syrup, Corn Products™). Identification of specimens were based on the following taxonomic references: C.A. Agardh (1824); J.G. Agardh (1848, 1889); Reinbold (1913); Grunow (1915); Yamada (1925, 1942); Setchell (1935); Pham (1967); Chou and Chiang (1981); Tseng (1983); Yoshida (1988); Trono (1992, 1997); Tseng and Lu (1992, 1995a, b, 1997); Chiang et al. (1992); Noro et al. (1994); Lewmanomont and Ogawa (1995); Phang et al. (1995); Ajisaka et al. (1995a, b, 1997, 1999) and Ajisaka (2002). The specimens from this study are deposited at the Bangsaen Institute of Marine Science (BIMS), Burapha University and Kasetsart University Museum of Fisheries (KUMF).

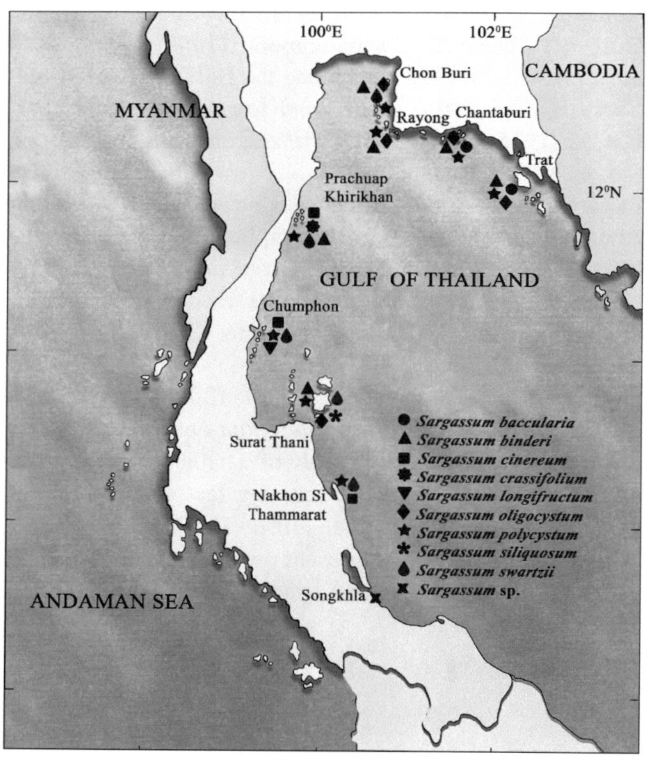

Fig. 1 Study area and distribution of collecting sites for *Sargassum* along the Gulf of Thailand

Results

Key to the species of *Sargassum* from the Gulf of Thailand

1. Secondary holdfasts present transformed from stolons; main branches terete, with warts or spines............... ....*S. polycystum*
1. Secondary holdfasts lacking; main branches terete to flattened, without spines...2
   2. Primary branches terete; plants dioecious..............3
   2. Primary branches compressed or flattened; plant monoecious..7
3. Leaves large, up to 5.0–6.5 cm long..........................4
3. Leaves small less than 4.5 cm long...........................6
   4. Leaves membranous..............................*S. cinereum*
   4. Leaves thick, or slightly so....................................5
5. Male receptacles short and terete.............*S. siliquosum*
5. Male receptacles long and terete or slightly compressed and sometimes triquetrous at the apices, with dentate margins..................................*Sargassum* sp.1
   6. Receptacles longer than 20 mm.......*S. longifructum*
   6. Receptacles shorter than 20 mm..........*S. baccularia*
7. Receptacles cymosely arranged................................8
7. Receptacles racemosely arranged...............................9
   8. Leaves thick, double serrulate at apices......... ....*S. crassifolium*
   8. Leaves lanceolate to linear lanceolate, margins dentate to slightly entire..........................*S. swartzii*
9. Receptacles flattened, often twisted with dentate magins, vesicle stalks flattened......................*S. binderi*
9. Receptacles compressed, not twisted, with apical spines, vesicle stalks terete.....................*S. oligocystum*

Description of the species

Sargassum baccularia (Mertens) C.A. Agardh (Fig. 2)
Tseng and Lu 1992, p. 16, Fig. 17; Trono 1992, p. 46, Fig. 93; 1997, p. 127; Ajisaka et al. 1999, p. 25, Fig. 2

Holdfast discoid, up to 17 mm in diameter. Stem terete, warty, up to 3 mm in diameter and 1.5 cm long, producing 3–5 spirally arranged primary branches. Primary branches terete, smooth, up to 200 cm long and 2 mm in diameter; leaves large, lanceolate, simple, with asymmetrical bases, up to 45 mm long and 13 mm wide, with rounded to acute apices, margins dentate with small teeth, midribs distinct near apices, cryptostomata small and scattered to in rows on both sides of the midrib. Secondary branches spirally arranged, terete, smooth, up to 32 cm long and 9 cm in branching interval; leaves lanceolate to linear, simple, with asymmetrical bases, up to 40 mm long and 10 mm wide, with rounded apices, margins dentate with small teeth, midribs distinct near apices, small cryptostomata scattered or in rows on both sides of the midrib (Fig. 2b). Vesicles spherical, sometimes elliptical, up to 4.5 mm long, to 4 mm wide and 3 mm thick, entire at the apices, sometimes mucronate, stalks terete, usually shorter than the vesicles (Fig. 2c).

Plants dioecious. Male receptacles long, terete, up to 10 mm long and 2 mm wide, with smooth surface, simple to once to twice furcate (Fig. 2d,e). Female receptacles triquetrous, up to 10 mm long and 2 mm wide, with a warty surface, sometimes bearing spines near the apex, simple to furcate (Fig. 2f,g). Male and female receptacles arranged in a raceme, pseudozygocarpic, vesiculate.

Remarks: *Sargassum baccularia* and *S. polycystum* share some morphological characters especially in young plants. We consider that the most important characters to separate the two species are the forms of the vesicles and female receptacles. In *S. baccularia,* the vesicles are spherical to elliptical, and female receptacles are triquetrous, while the vesicles of *S. polycystum* are spherical to obovate, and the female receptacles are terete to slightly compressed.

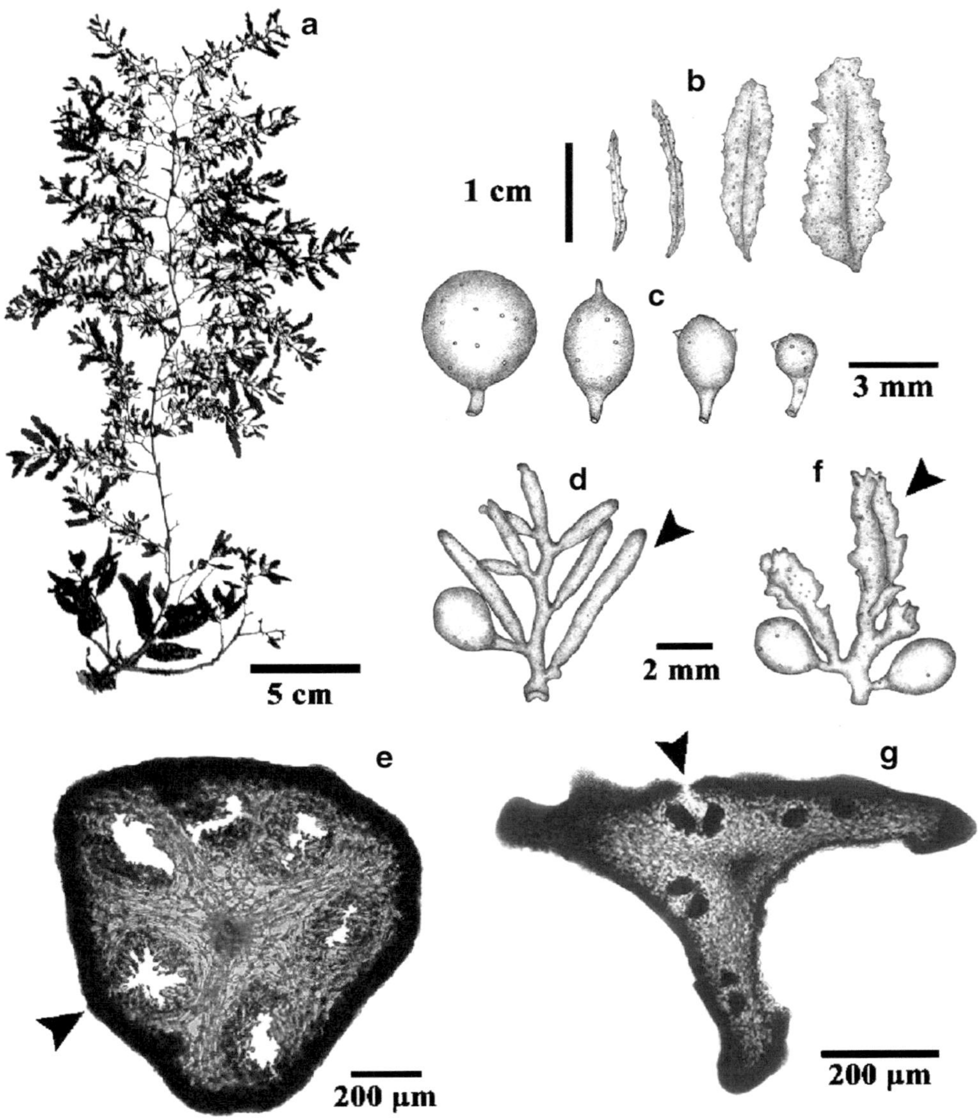

Fig. 2 *Sargassum baccularia* (Mertens) C.A. Agardh. **a** Habit, **b** leaves, **c** vesicles, **d** male receptacles (*arrowhead*), **e** transverse section of male receptacle showing male conceptacles (*arrowhead*), **f** female receptacles (*arrowhead*), **g** transverse section of female receptacle showing female conceptacles (*arrowhead*)

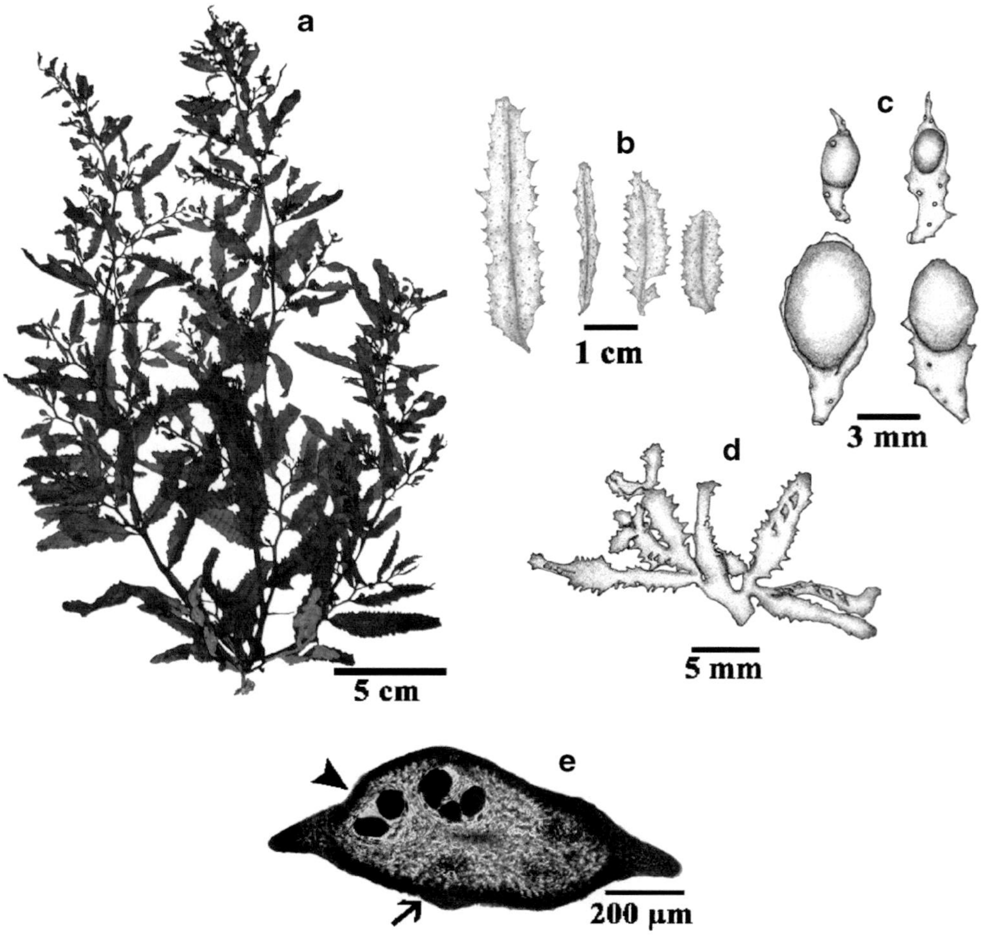

Fig. 3 *Sargassum binderi* Sonder. **a** Habit, **b** leaves, **c** vesicles, **d** androgynous receptacles, **e** transverse section of receptacles showing male conceptacles (*arrow*) and female conceptacles (*arrowhead*)

Sargassum binderi Sonder (Fig. 3)
J.G. Agardh 1889, p. 87, Figs. 26–2; Yamada 1925, p. 245; Chou and Chiang 1981, p. 143; Yoshida 1988, p.10; Trono 1992, p. 46, Figs. 1–4, 109; 1997, p. 128, Fig. 86; Tseng and Lu 1995b, p. 96, Fig. 3; Ajisaka et al. 1999, p. 28, Fig. 3

Holdfast discoid, up to 12 mm in diameter. Stem terete, smooth to warty, up to 3 mm in diameter, 1 cm long, producing 6–8 spirally arranged primary branches. Primary branches flattened to compressed, smooth, up to 46 cm long and 5 mm wide; leaves large, lanceolate to slender lanceolate, simple, with asymmetrical bases, up to 77 mm long and 16 mm wide, with rounded to slightly acute apices, margins entire to dentate with small teeth, midribs distinct near apices or vanishing midway, small cryptostomata scattered. Secondary branches distichously arranged, slightly compressed, smooth, up to 40 cm long and 3.8 cm in branching interval; leaves lanceolate to linear, simple with asymmetrical to cuneate bases, up to 64 mm long and 15 mm wide, with rounded to acute apices, margins dentate with sharp teeth, midribs vanishing near apices, small cryptostomata scattered, sometimes arranged in rows on both sides of the midrib (Fig. 3b). Vesicles spherical to elliptical, up to 10 mm long, to 6 mm wide and 5 mm thick, often mucronate at the apices, sometimes entire, stalks flattened, usually slightly longer than the vesicles (Fig. 3c).

Plants monoecious. Receptacles androgynous, flattened, often twisted, up to 18 mm long and 2 mm wide, sharply dentate at the margin, simple to furcate, racemosely arranged (Fig. 3d,e).

Remarks: *Sargassum binderi* and *S. oligocystum* overlap in many morphological characters, particularly when they are immature. Specimens of *S. binderi* from Thailand are closely related to the specimens from China and Malaysia (Tseng and Lu 1995b; Ajisaka et al. 1999) in the form of their flattened primary branches; spherical to elliptical vesicles which are often mucronate at their apices or sometimes entire, with flattened stalks, and flattened, twisted receptacles, with a sharply dentate margin.

Sargassum cinereum J.G. Agardh (Fig. 4)
Tseng and Lu, 1988, p. 33, Figs. 5, 33; Ajisaka 1999, p. 29, Figs. 3–5, 7

Holdfast discoid, up to 16 mm in diameter. Stem terete, smooth, up to 3 mm in diameter, 10 mm long, producing 4–

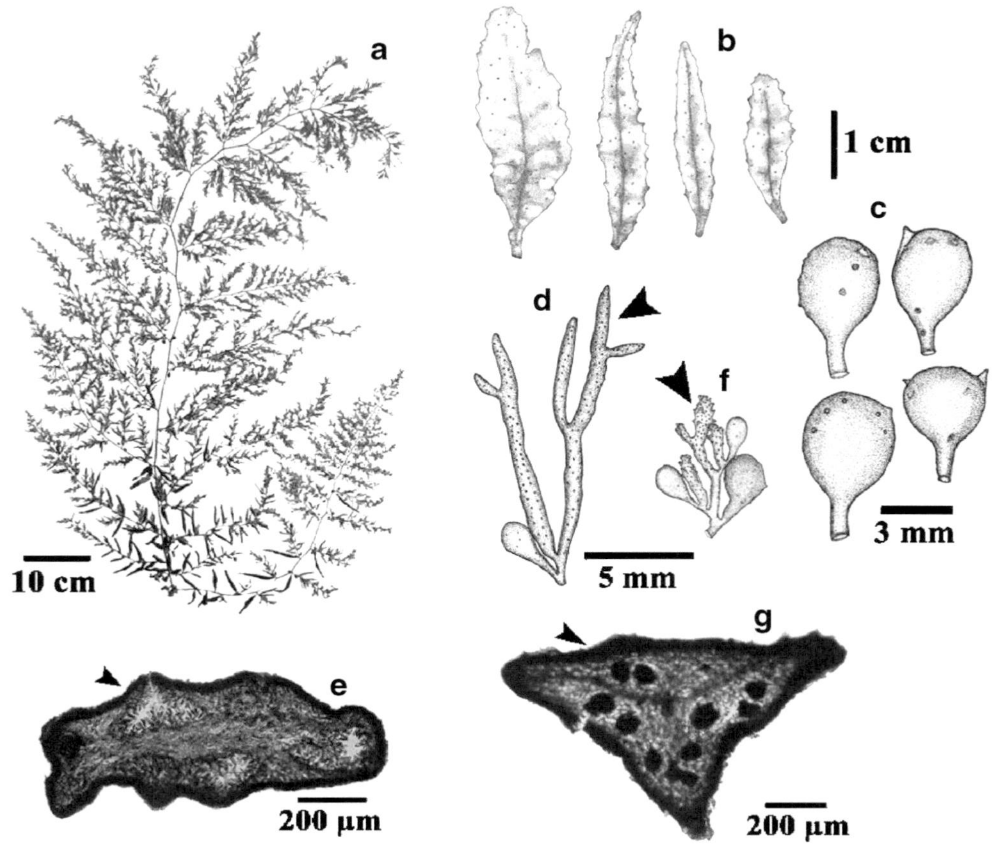

Fig. 4 *Sargassum cinereum* J.G. Agardh. **a** Habit, **b** leaves, **c** vesicles, **d** Male receptacles (*arrowhead*), **e** transverse section of male receptacle showing male conceptacles (*arrowhead*), **f** female receptacles (*arrowhead*), **g** transverse section of female receptacle showing female conceptacles (*arrowhead*)

8 spirally arranged primary branches. Primary branches terete to subterete, smooth, up to 125 cm long and 3 mm in diameter. Leaves membranous, lanceolate to linear lanceolate, simple, with asymmetrical to cuneate bases, up to 65 mm long and 14.7 mm wide, with rounded apices; margins dentate with small teeth, midribs distinct near apices, small cryptostomata scattered to one row or disposed two rows. Secondary branches spirally arranged, terete, smooth, up to 38 cm long and 6 cm in branching interval; leaves lanceolate to linear, simple, with asymmetrical to cuneate bases, up to 48 mm long and 8.9 mm wide, with rounded or sometimes acute apices, margins dentate with small teeth, midribs distinct near apices, small cryptostomata in one row to disposed two rows (Fig. 4b). Vesicles spherical, sometimes obovoid to elliptical, simple, up to 9 mm long, 4.6 mm wide and 4.1 mm thick, entire at the apices, sometimes provided with appendages on one or both sides; stalks terete, sometimes compressed, usually shorter than the vesicles (Fig. 4c).

Plants dioecious. Male receptacles long, terete, up to 18 mm long and 1.1 mm wide, simple to 1–3 times furcate (Fig. 4d,e). Female receptacles normally short compressed and triquetrous, up to 8.3 mm long and 1.5 mm wide, with a warty surface, bearing spines at the apex, simple to once furcate. Male and female receptacles arranged in a raceme, pseudozygocarpic, vesiculate (Fig. 4f,g).

Remark: *Sargassum cinereum* is included in the species group Cinerea Setchell in subsection Pseudozygocarpicae Setchell (Ajisaka et al. 1995b). Thai specimens of *S. cinereum* closely resemble to the specimens from China (Tseng and Lu 1988) and Malaysia (Ajisaka et al. 1999) in the form of their leaves, which are membranous, lanceolate to linear lanceolate, with asymmetrical to cuneate bases and dentate margins; spherical to elliptical vesicles with entire apices, sometimes provided with an appendage, terete stalks; and female receptacles that are short and triquetrous, with a warty surface and bearing spines at the apex. The morphology of male receptacles of Malaysian specimens resembles Thai specimens in having lengths up to 15–18 mm, but Chinese specimens are only 8–10 mm long. This species is a new record for Thailand.

Sargassum crassifolium J.G. Agardh (Fig. 5)
Trono 1992, p. 50, Figs. 9–11, 112; 1997, p. 131, Figs. 88A-88B; Tseng and Lu, 1997, p. 23, Fig. 14; Ajisaka et al. 1997, p. 34, Fig. 12

Holdfast discoid, up to 17 mm in diameter. Stem terete, smooth, up to 4.6 mm in diameter, 5.3 mm long, producing 4–8 spirally arranged primary branches. Primary branches with slightly compressed in lower parts and slightly terete in upper parts, smooth, up to 29 cm long and 3.8 mm in

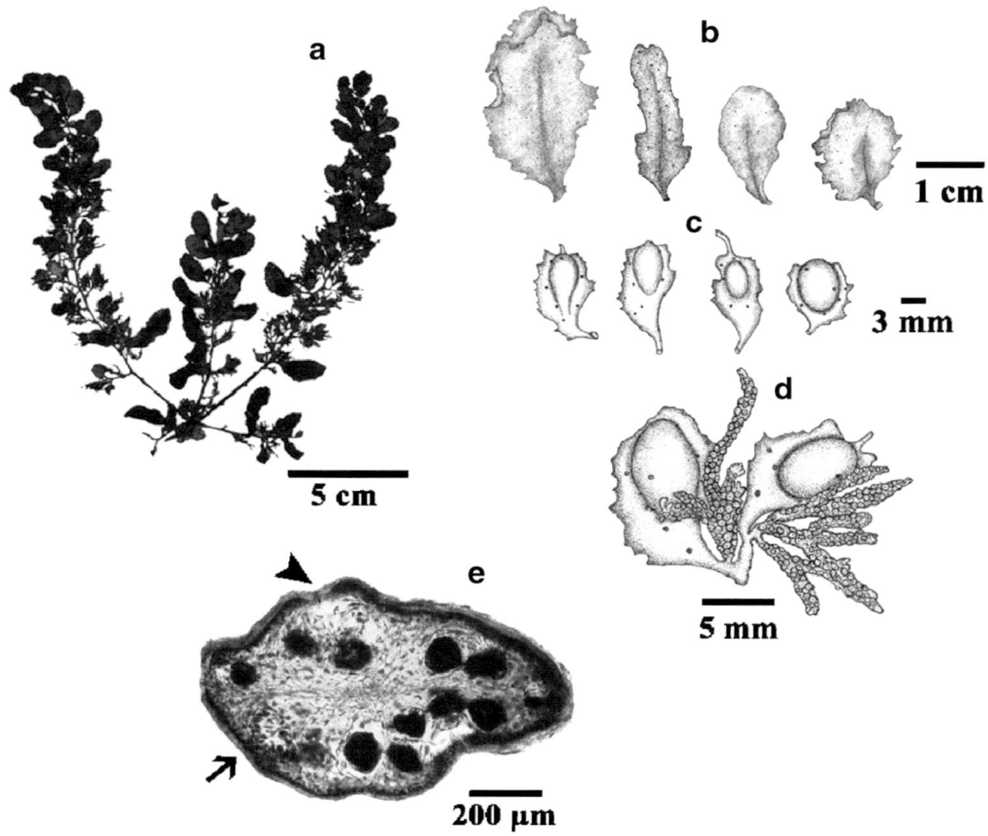

Fig. 5 *Sargassum crassifolium* J.G. Agardh. **a** Habit, **b** leaves, **c** vesicles, **d** androgynous receptacles, **e** transverse section of receptacles showing male conceptacles (*arrow*) and female conceptacles (*arrowhead*)

diameter. Leaves thick and vertically expanded, large elliptical, elliptical-oval to oblong, simple, with asymmetrical to cuneate bases, up to 32 mm long and 16 mm wide, usually with rounded and double serrulate apices, margins dentate, midribs vanishing midway to distinct near apices, small cryptostomata scattered. Secondary branches alternately arranged, terete, smooth, up to 4.7 cm long and 1.5 cm in branching interval; leaves elliptical-oval, obovate, oblong to lanceolate, simple, with asymmetrical to cuneate bases, up to 19 mm long and 11.3 mm wide, with rounded and double serrulate apices, margins undulate to dentate with small teeth, midribs distinct near apices, small cryptostomata scattered to one row (Fig. 5b). Vesicles spherical, elliptical, ovate to obovoid, simple, up to 8.8 mm long, 7.7 mm wide and 6.4 mm thick, entire or with a pointed end or a short leafy crown at the apices, phyllocysts provided with marginal appendages or wings: stalks with lower part terete and upper part compressed to foliaceous, usually shorter than the vesicles or sometimes longer or equal to the vesicles (Fig. 5c).

Plants monoecious. Receptacles terete in lower part and slightly compressed in upper part, up to 18 mm long and 1.8 mm wide, simple to furcated up to four times, with a warty surface, bearing spines at the apex, arranged in a cymose to compoundly cymose, pseudozygocarpic, vesiculate (Fig. 5d,e).

Remarks: *Sargassum crassifolium* has the important distinguishing characters of very thick, vertically expanded, elliptical to oblong hard leaves with blunt apices, sometimes with double serrulate margins; elliptical vesicles, which are entire or with apices pointed or with a short leafy crown leaf, phyllocysts with marginal appendages or wings, stalks with terete lower part and compressed or foliaceous upper part; receptacles with terete lower part and slightly compressed upper part, bearing spines at the apex, arranged in a cymose to compoundly cymose (Trono 1992, 1997; Tseng and Lu 1997; Ajisaka et al. 1997).

Sargassum longifructum Tseng et Lu (Fig. 6)
Tseng and Lu 1988, p. 43, Figs. 10, 25, 26; Ajisaka et al. 1995a, p. 45, Figs. 6–15; Ajisaka et al. 1995b, p. 23, Figs. 11–15

Holdfast discoid, up to 15 mm in diameter. Stem terete, smooth, up to 3.1 mm in diameter, 4.2 mm long, producing 4–9 spirally arranged primary branches. Primary branches terete to subterete, smooth, up to 78 cm long and 3 mm in diameter. Leaves linear-elliptical to linear-lanceolate, simple, with slightly asymmetrical to cuneate bases, up to 32 mm long and 10.9 mm wide, with rounded to acute apices, margins entire to dentate with small teeth, midribs distinct near apices, small cryptostomata scattered. Secondary

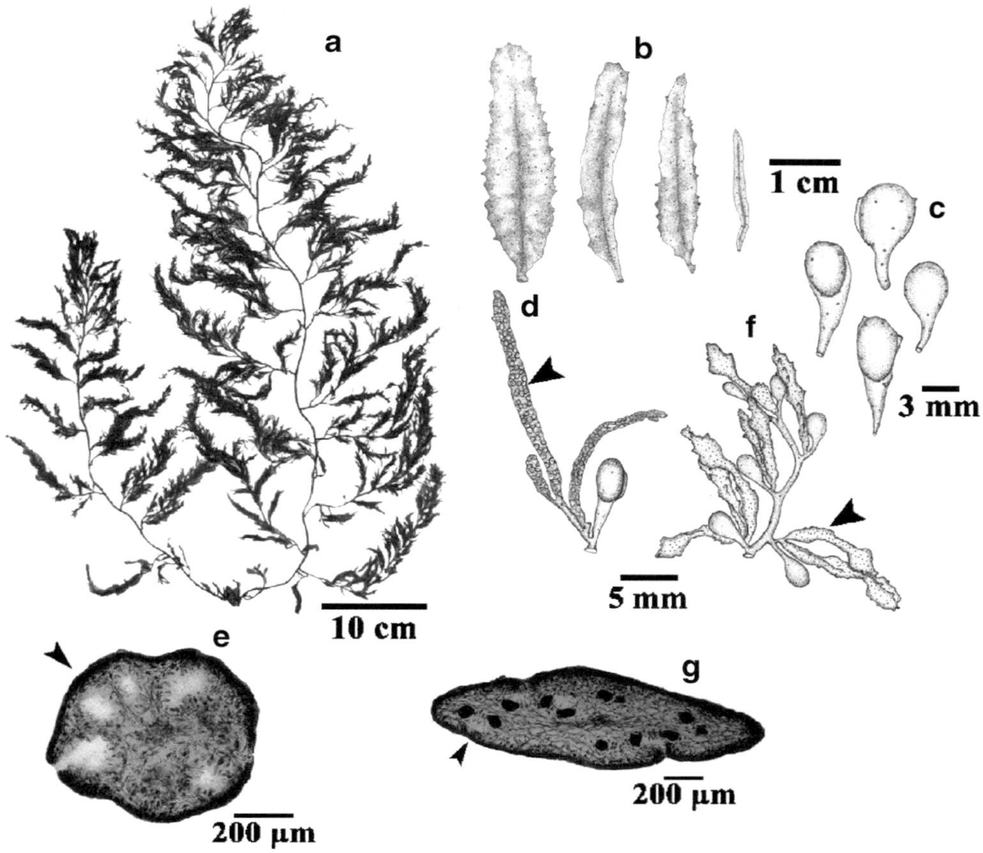

Fig. 6 *Sargassum longifructum* Tseng et Lu. **a** Habit, **b** leaves, **c** vesicles, **d** male receptacles (*arrowhead*), **e** transverse section of male receptacle showing male conceptacles (*arrowhead*), **f** female receptacles (*arrowhead*), **g** transverse section of female receptacle showing female conceptacles (*arrowhead*)

branches spirally arranged, terete, smooth, up to 26 cm long and 5 cm in branching interval; leaves linear-elliptical, linear-lanceolate to linear, simple, with slightly asymmetrical to cuneate bases, up to 31 mm long and 7.6 mm wide, with acute or sometimes rounded apices, margins entire to dentate with small teeth, midribs distinct near apices, small cryptostomata scattered to one row (Fig. 6b). Vesicles slightly spherical, obovoid or elliptical, simple, up to 5.7 mm long, 4.7 mm wide and 3.7 mm thick, entire at the apices, sometimes provided with appendages on one or both sides, stalks terete or sometimes compressed, usually shorter than the vesicles (Fig. 6c).

Plants dioecious. Male receptacles long, terete to slightly compressed and with a few spines at the apex, up to 26 mm long and 1.6 mm wide, simple to forked one to two times (Fig. 6d,e). Female receptacles flattened and sometimes triquetrous, up to 27 mm long and 3.7 mm wide, margins dentate with spines, sometimes twisted, simple to once to twice furcate. Male and female receptacles arranged in a raceme, pseudozygocarpic, vesiculate and leaf (Fig. 6f,g).

Remarks: *Sargassum longifructum* from China, Japan, Vietnam and Thailand display large morphological variations in leaves, vesicles, and receptacles. Leaves varied from elliptical to linear. Vesicles were entire or with winged or crowned apices. Male receptacles were terete to compressed, sometimes with spines at the apices. Both male and female receptacles were holozygocarpic or pseudozygocarpic (Tseng and Lu 1988; Ajisaka et al. 1995a, b). This species is difficult to distinguish from *S. cinereum* when the plants are immature, as features of receptacles are important for identification. *Sargassum longifructum* has male receptacles up to 26 mm long and 1.6 mm wide; female receptacles flattened and sometimes triquetrous, up to 27 mm long and 3.7 mm wide. In contrast, *S. cinereum* has male receptacles up to 18 mm long and 1.1 mm wide; female receptacles normally short compressed and triquetrous, up to 8.3 mm long and 1.5 mm wide. This species is a new record for Thailand.

Sargassum oligocystum Montagne (Fig. 7)
Reinbold 1913, p. 156; Grunow 1915, p. 383; Pham 1967, p. 295; Tseng 1983, p. 234, pl. 118, Fig. 1; Noro et al. 1994, p. 27, Fig. 3; Tseng and Lu 1995b, p. 100, Fig. 10; Lewmanomont and Ogawa 1995, p. 83; Ajisaka et al. 1999, p. 34, Fig. 8

Holdfast discoid, up to 12 mm in diameter. Stem terete, smooth, up to 4 mm diameter, 12 mm long, producing 6–8 spirally arranged primary branches. Primary branches

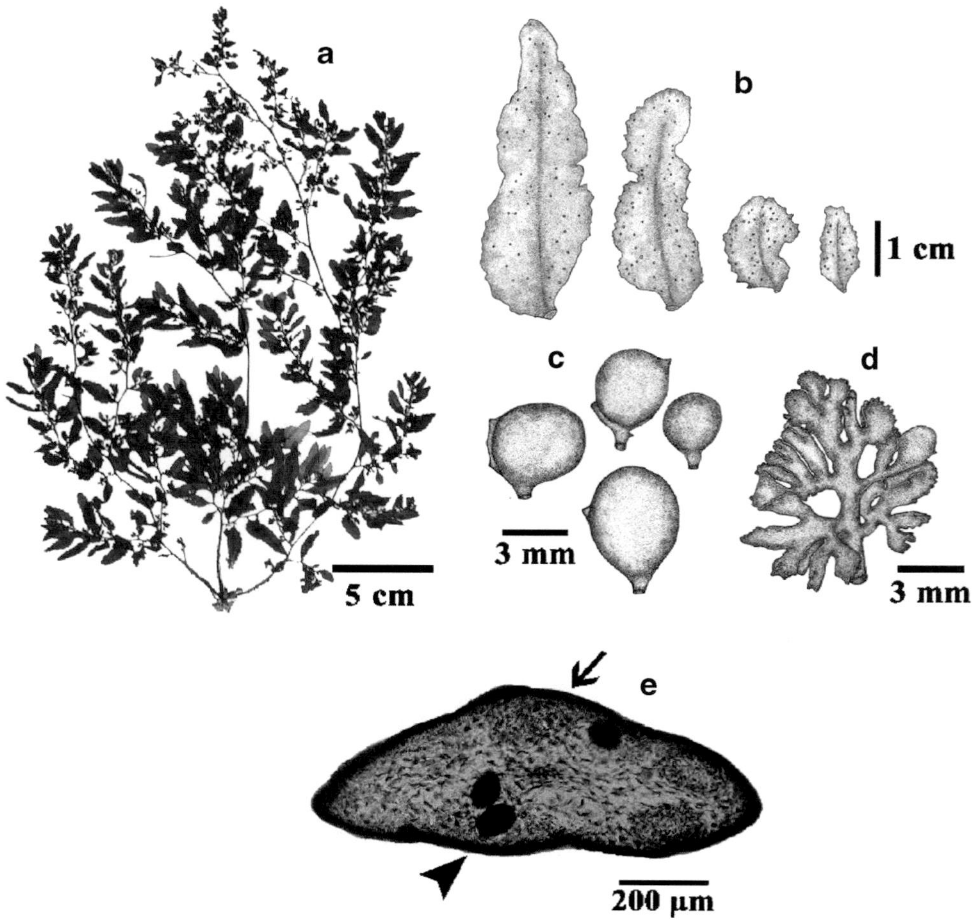

Fig. 7 *Sargassum oligocystum* Montagne. **a** Habit, **b** leaves, **c** vesicles, **d** androgynous receptacles, **e** transverse section of receptacles showing male conceptacles (*arrow*) and female conceptacles (*arrowhead*)

flattened to compressed, smooth, up to 49 cm long and 4 mm wide; leaves lanceolate to spatulate, simple, with asymmetrical bases, up to 54 mm long and 18 mm wide, with rounded apices, margins entire to dentate with small teeth, midribs vanishing to distinct near apices, small cryptostomata scattered. Secondary branches distichously arranged, terete to slightly compressed, smooth, up to 19 cm long and 5.3 cm in branching interval; leaves lanceolate to spatulate, simple, with asymmetrical bases, up to 34 mm long and 10 mm wide, with rounded apices, margins entire to dentate with small teeth, midribs vanishing near apices, small cryptostomata scattered (Fig. 7b). Vesicles spherical to elliptical, up to 4 mm long, 4 mm wide and 3 mm thick, entire at the apices, sometimes provided with appendages, stalks terete, usually shorter than the vesicles (Fig. 7c).

Plants monoecious. Receptacles androgynous, slightly compressed, up to 9 mm long and 3 mm wide, warty or with a few marginal spines, simple to furcate two or three times, racemosely arranged, pseudozygocarpic, vesiculate (Fig. 7d,e).

Remarks: *Sargassum oligocystum* from Thailand closely resembles the specimens from Malaysia (Ajisaka et al. 1999) in the form of their compressed primary branches; spherical to elliptical vesicles with entire apices, sometimes with small appendages on one side or both sides, terete stalks; slightly compressed receptacles that are warty or have a few spines at the margin. Our specimens differ from those of China (Tseng and Lu 1995b) and the Philippines (Trono 1992; Noro et al. 1994) in having flattened primary branches, and plants are dioecious.

Sargassum polycystum C.A. Agardh (Fig. 8)
Chiang et al. 1992, p. 36, Figs. 1–12; Trono 1992, p. 63; 1997, p. 147, Fig. 96; Ajisaka et al. 1995b, p. 34, Figs. 19–20; Lewmanomont and Ogawa 1995, p. 84; Ajisaka et al. 1999, p. 36, Fig. 6

Holdfast discoid, up to 13 mm in diameter. Stem terete, warty, up to 3 mm in diameter and 4 mm long, bearing up to 5 stolons and 7–9 spirally arranged primary branches. Stolons terete to slightly compressed at their proximal parts, up to 12 cm long and 2 mm wide. Primary branches usually muricate with prolifically branched spines (Fig. 8h), transformed into a stolon and secondary holdfasts (Fig. 8i), up to 200 cm long and 2 mm in diameter; leaves elliptical, lanceolate to linear, simple, with asymmetrical to cuneate

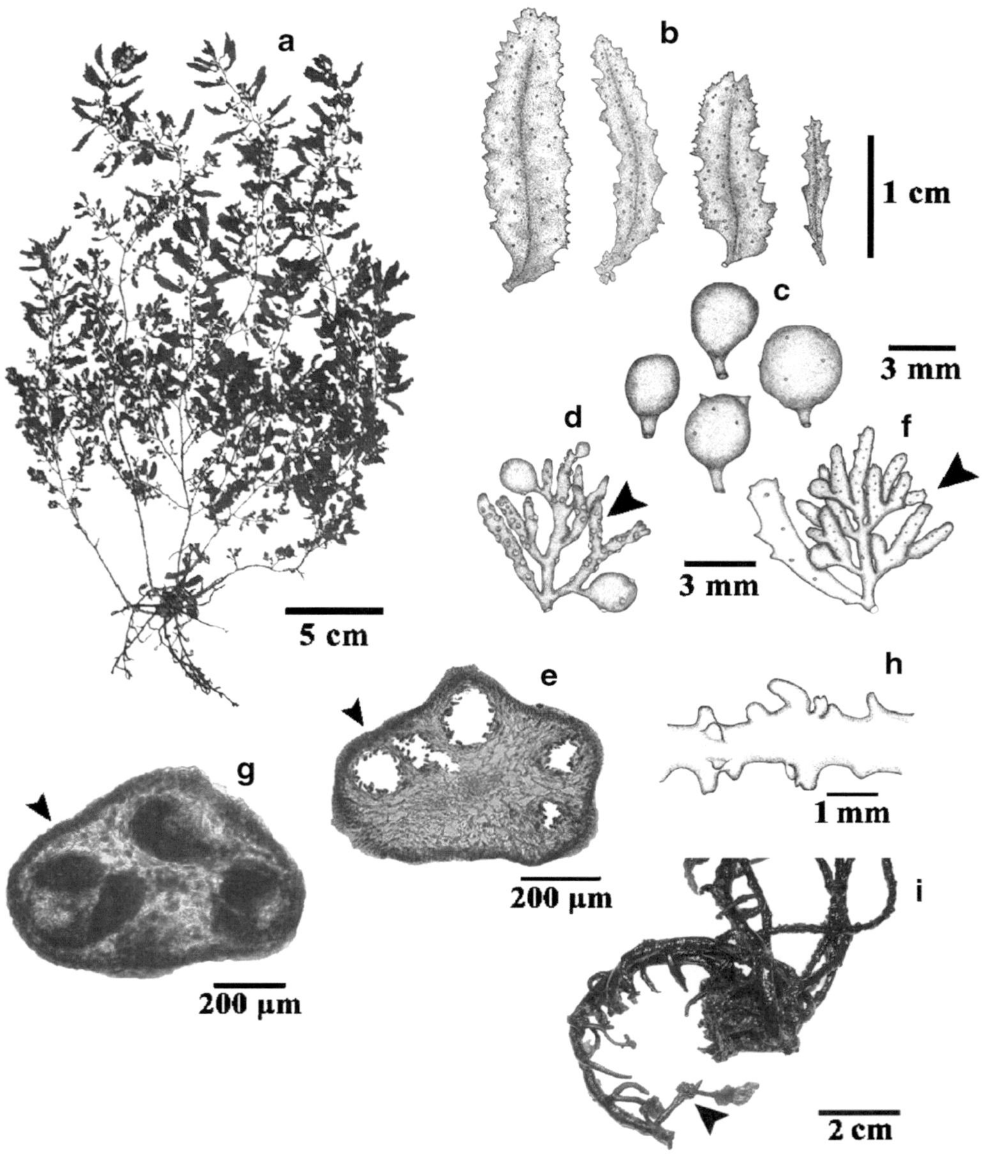

Fig. 8 *Sargassum polycystum* C.A. Agardh. **a** Habit, **b** leaves, **c** vesicles, **d** male receptacles (*arrowhead*), **e** transverse section of male receptacle showing male conceptacles (*arrowhead*), **f** female receptacles (*arrowhead*), **g** transverse section of female receptacle showing female conceptacles (*arrowhead*), **h** spines on primary branch, **i** secondary holdfast (*arrowhead*)

bases, up to 45 mm long and 13 mm wide, with rounded to acute apices, margins dentate with coarse teeth, midribs vanishing to distinct near apices, cryptostomata small and scattered to in rows on both sides of the midrib. Secondary branches spirally arranged, terete, crowded with spines, up to 60 cm long and 16 cm in branching interval; leaves linear-lanceolate to spatulate, simple, asymmetrical with cuneate bases, up to 30 mm long and 7 mm wide, with rounded to acute apices, margins dentate with coarse teeth, midribs vanishing to distinct near apices, small cryptostomata scattered to arranged in rows on both sides of the midrib (Fig. 8b). Vesicles spherical to obovate, to 8 mm long, 6 mm wide and 5 mm thick, entire at the apices or with ear-like wings on both sides of the vesicles, stalks terete, usually shorter than the vesicles (Fig. 8c).

Plants dioecious. Male receptacles long, terete, up to 15 mm long and 1 mm wide, with a warty surface, simple to once furcate (Fig. 8d,e). Female receptacles terete to slightly compressed, up to 4 mm long and 1 mm wide, with a warty surface, simple to once furcate (Fig. 8f,g). Male and female receptacles arranged in a raceme, halozygocarpic and vesiculate.

Remarks: *Sargassum polycystum* is different from S. *baccularia* in having secondary holdfasts transformed from the stolon and heavily muricate on main branches (Chiang et al. 1992; Ajisaka et al. 1995b, 1999). From our observations, some specimens of *S. polycystum* have no secondary holdfasts and some have scattered or absent spines on primary branches, which make them similar to S. *baccularia* when growing together these two species are not easily distinguished.

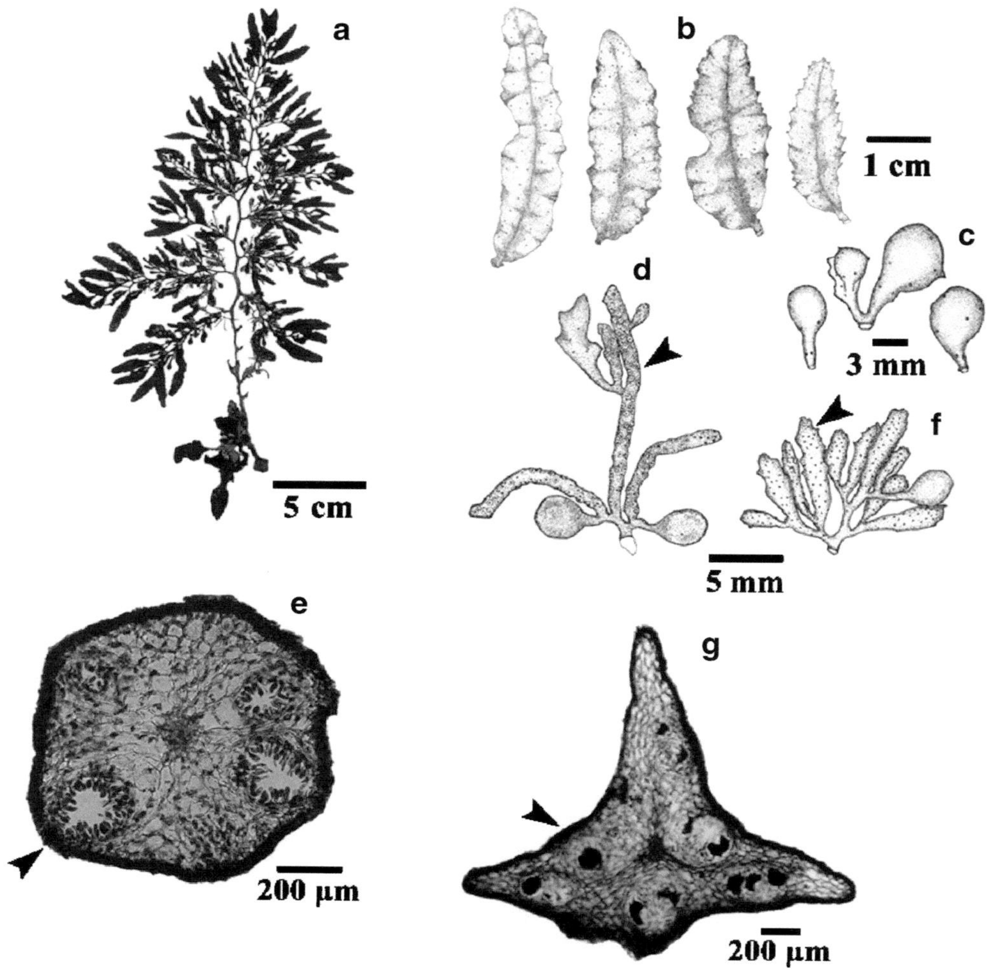

Fig. 9 *Sargassum siliquosum* J.G. Agardh. **a** Habit, **b** leaves, **c** vesicles, **d** male receptacles (*arrowhead*), **e** transverse section of male receptacle showing male conceptacles (*arrowhead*), **f** female receptacles (*arrowhead*), **g** transverse section of female receptacle showing female conceptacles (*arrowhead*)

Sargassum siliquosum J.G. Agardh (Fig. 9)
Tseng and Lu 1992, p. 20 Figs. 4, 20; Trono 1992, p. 65 Fig. 97; Phang et al. 1995, p. 55, Figs. 1–8; Trono 1997, p. 149

Holdfast discoid, up to 16 mm in diameter. Stem terete, smooth, sometimes warty, up to 5.3 mm in diameter and 16 mm long, producing 5–9 spirally arranged primary branches. Primary branches terete, smooth, up to 80 cm long and 3.9 mm in diameter; leaves broad lanceolate, thick, simple, with asymmetrical to cuneate bases, up to 55 mm long and 18 mm wide, with rounded blunt apices, margins entire to dentate with small teeth and wavy leaf edge, sometimes unequal leaf blades on both sides of the midrib, midribs distinct near apices, cryptostomata small and scattered. Secondary branches alternately arranged, terete, smooth, up to 37 cm long and 3.9 cm in branching interval; leaves lanceolate to oblong, simple, asymmetrical with cuneate bases, up to 39 mm long and 12.6 mm wide, with rounded apices, margins dentate with small teeth, midribs distinct near apices, cryptostomata small and scattered (Fig. 9b). Vesicles sub-spherical to obovoid, simple, sometimes in pairs, up to 7.4 mm long, 5.8 mm wide and 5.6 mm thick, entire at the apices, sometimes mucronate, stalks terete, sometimes compressed, usually shorter than the vesicles, sometimes as long as the vesicles (Fig. 9c).

Plants dioecious. Male receptacles long, terete, sometimes slightly compressed and triquetrous in the upper part, with small spines at the apices and margins, terete in the lower part, up to 20 mm long and 1 mm wide, simple to once to twice furcate (Fig. 9d,e). Female receptacles triquetrous, up to 11 mm long and 2 mm wide, with a warty surface, bearing spines at the apex, simple to once furcate. Male and female receptacles arranged in a raceme, pseudozygocarpic with vesicles and leaves sometimes halozygocarpic (Fig. 9f,g).

Remark: *Sargassum siliquosum* occurred only in Surat Thani Province. Thai specimens closely resemble specimens from China, Malaysia and Singapore (Tseng and Lu 1992; Phang et al. 1995) in important features such as terete primary branches; lanceolate leaves with asymmaetrical bases and rounded apices, vesicles that are solitary or paired, obovoid to subspherical, entire at the apices with

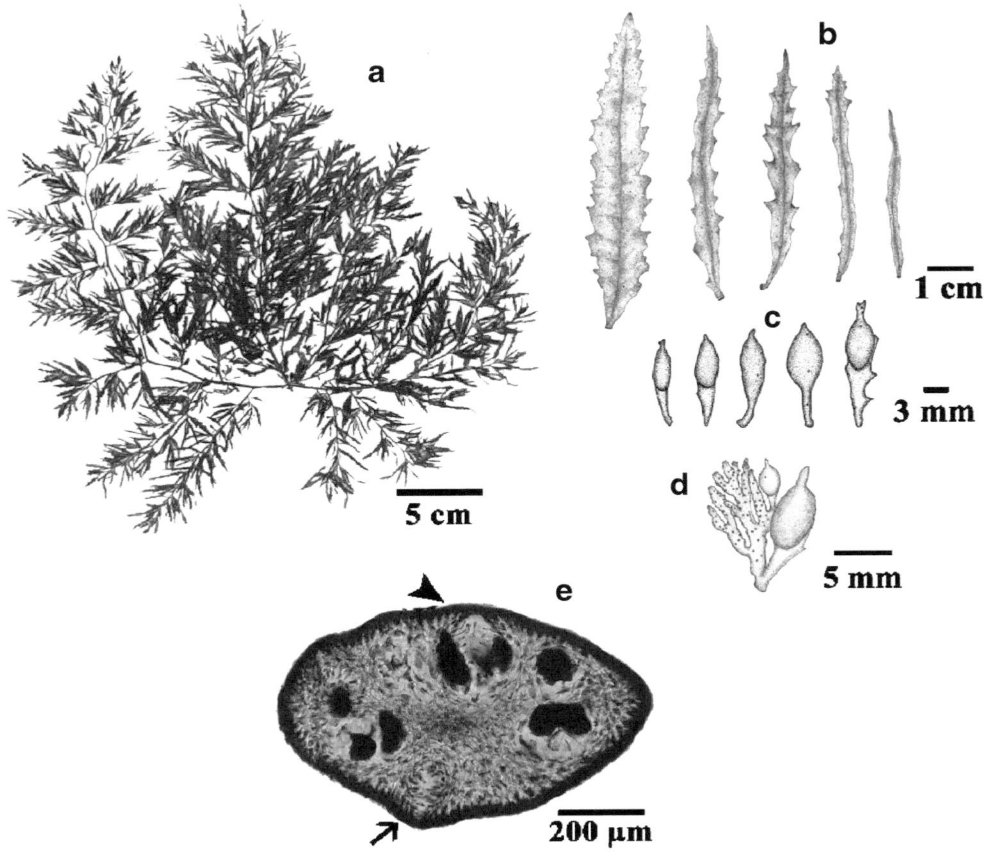

Fig. 10 *Sargassum swartzii* (Turner) C.A. Agardh. **a** Habit, **b** leaves, **c** vesicles, **d** androgynous receptacles, **e** transverse section of receptacles showing male conceptacles (*arrow*) and female conceptacles (*arrowhead*)

terete stalks, and receptacles arranged in a raceme. The Thai specimens slightly differ from those from the Philippines (Trono 1992, 1997) in having terete to slightly compressed primary and secondary branches; leaves of plants with more variation, and female receptacles are twisted.

Sargassum swartzii (Turner) C.A. Agardh (Fig. 10)
C.A. Agardh 1824, p. 296; J.G. Agardh 1848, p. 328; 1889, p. 85, pl. 26.1; Reinbold 1913, p. 157; Grunow 1915, p. 381; Setchell 1935, p. 3, pl. 1; Yamada 1942, p. 25, Fig. 1; Pham 1967, p. 293, Fig. 13; Tseng 1983, p. 238, pl. 120, Fig. 1; Noro et al. 1994, p. 27, Figs. 4, 9; Tseng and Lu 1995a, p. 79, Fig. 10; Ajisaka 2002, p. 82, Figs. 12–15

Holdfast discoid, up to 17 mm in diameter. Stem terete, smooth, up to 3.9 mm in diameter and 7.3 mm long, producing 4–8 spirally arranged primary branches. Primary branches compressed, smooth, up to 140 cm long and 5.9 mm wide. Leaves elongated lanceolate to linear lanceolate, simple, with asymmetrical bases, up to 80 mm long and 11 mm wide, usually with acute (or sometimes obtuse) at apices, margins dentate with small teeth, midribs distinct near apices, cryptostomata small and scattered to one row. Secondary branches distichously arranged, compressed, smooth, up to 63 cm long and 8.4 cm in branching interval; leaves linear lanceolate, simple, with asymmetrical to cuneate bases, up to 74 mm long and 9 mm wide, with acute apices, margins slightly entire to dentate with small teeth, midribs vanishing to distinct near apices, small cryptostomata arranged in one row (Fig. 10b). Vesicles elliptical, simple, up to 10 mm long, 6.4 mm wide and 5.3 mm thick, apex pointed or with a short leafy crown, margins entire, stalks terete to slightly compressed or terete in the lower part and compressed in the upper part, usually shorter than the vesicles or sometimes longer (Fig. 10c).

Plants monoecious. Receptacles slightly terete, up to 16 mm long and 2.1 mm wide, simple to furcate 2–5 times, bearing spines at the apex, arranged in a cymose to compound cymose, pseudozygocarpic, vesiculate (Fig. 10d,e).

Remarks: Thai specimens of *S. swartzii* closely resemble those from China (Tseng and Lu 1995a, b) and Malaysia (Ajisaka 2002) in having leaves that are lanceolate to linear lanceolate and acute at the apices, vesicles that are elliptical, pointed at the apex, with entire margins and terete stalks, and compressed receptacles. Thai specimens distinctly differ from those of Australia (Noro et al. 1994) which have linear leaves; subspherical vesicles, and terete receptacles. *Sargassum swartzii* shows some resemblance to *S. binderi* and *S. oligocystum*, but can be distinguished

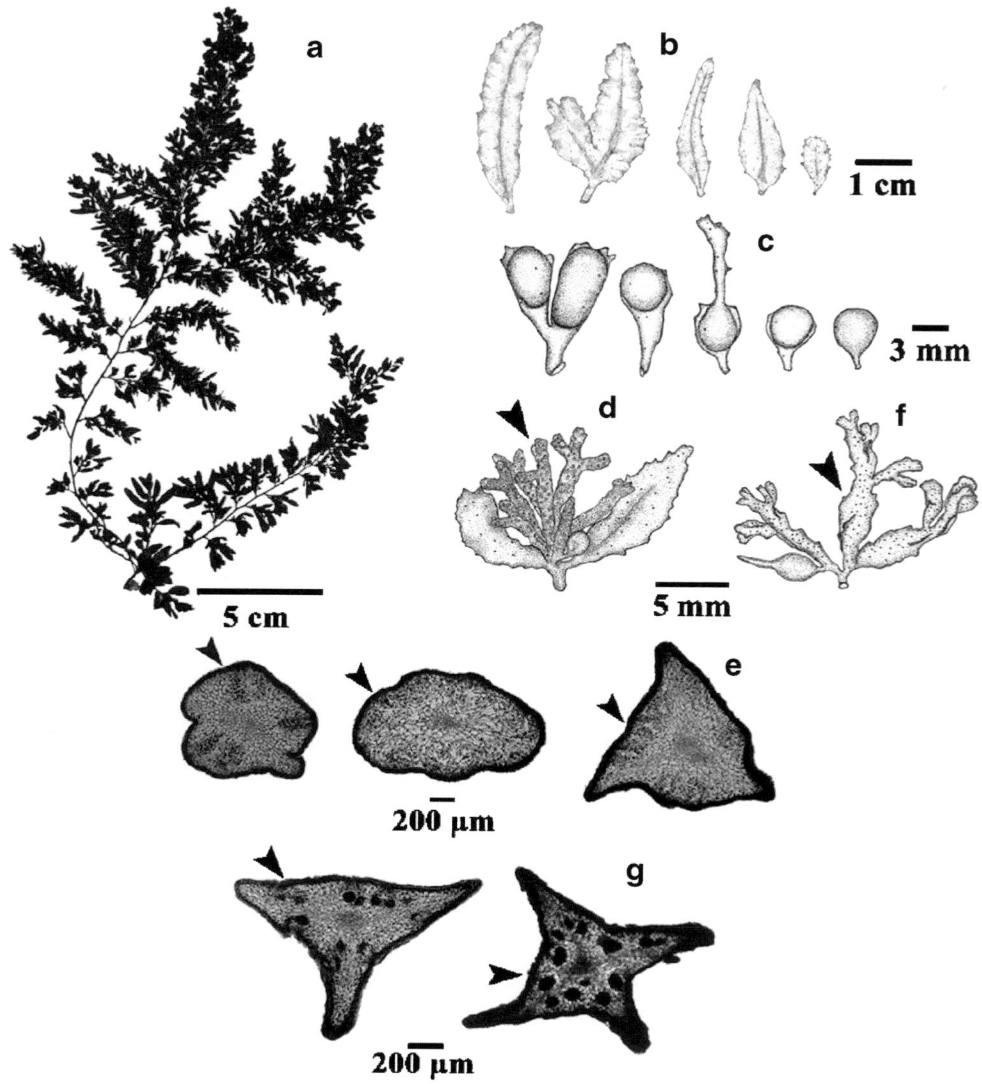

Fig. 11 *Sargassum* sp. **a** Habit, **b** leaves, **c** vesicles, **d** male receptacles (*arrowhead*), **e** transverse section of male receptacle showing male conceptacles (*arrowhead*), **f** female receptacles (*arrowhead*), **g** transverse section of female receptacle showing female conceptacles (*arrowhead*)

from those species by their slender leaves with shallow dentate to entire margins; small vesicles with terete stalks and entire margins; and receptacles arranged in a cymose. This species is also a new record for Thailand.

Sargassum sp. (Fig. 11)

Holdfast discoid, up to 13 mm in diameter. Stem terete, smooth, sometimes warty, up to 2.9 mm in diameter, 7.2 mm long, producing 3–14 spirally arranged primary branches. Primary branches terete, smooth, up to 54 cm long and 2.2 mm in diameter; leaves rather short, lanceolate to oblanceolate, simple, sometimes furcate with asymmetrical bases, up to 54 mm long and 10 mm wide, with rounded apices, margins dentate with small teeth, midribs distinct near apices, small cryptostomata scattered and arranged in one or two rows. Secondary branches alternately arranged (or in immature plants arranged alternately in the lower part and spirally in the upper part), terete, smooth, up to 18 cm long and 2.4 cm in branching interval; leaves lanceolate, linear lanceolate to oblanceolate, simple, sometimes furcate with asymmetrical bases, up to 29 mm long and 7 mm wide, with rounded to slightly rounded or acute apices, margins dentate with small teeth, midribs distinct near apices, cryptostomata small and arranged in one or two rows (Fig. 11b). Vesicles spherical, sometimes obovoid or elliptical, simple, sometimes in pairs, up to 6.7 mm long, 4.3 mm wide and 4.3 mm thick, entire at the apices, sometimes mucronate or with a leafy crown at the apex, up to 5.8 mm long and 1.24 mm wide, stalks terete, sometimes compressed, usually shorter than the vesicles, sometimes longer (Fig. 11c).

Plants dioecious. Male receptacles long, terete, sometimes slightly compressed and triquetrous at the upper part, with small spines at the apices and margins, and terete in the lower part, up to 14 mm long and 1.5 mm wide, simple

Table 1 Comparison of ten species of Thai *Sargassum* in the Gulf of Thailand

Character	Species									
	S. baccularia	*S. binderi*	*S. cinereum*	*S. crassifolium*	*S. longifructum*	*S. oligocystum*	*S. polycystum*	*S. siliquosum*	*S. swartzii*	*Sargassum* sp.
Holdfast	Discoid	Discoid	Discoid	Discoid	Discoid	Discoid	Discoid	Discoid	Discoid	Discoid
Stem	Terete, warty	Terete, smooth to warty	Terete, smooth	Terete, smooth	Terete, smooth	Terete, smooth	Terete, warty	Terete, smooth	Terete, smooth	Terete, smooth to warty
Primary branch	Terete, smooth	Flattened to compressed, smooth	Terete to subterete, smooth	Compressed to terete, smooth	Terete to subterete, smooth	Flattened to compressed, smooth	Terete, muricate with prolifically branched spines, transformed into stolon and secondary holdfast	Terete, smooth	Compressed, smooth	Terete, smooth
Secondary branch	Terete, smooth	Slightly compressed, smooth	Terete, smooth	Terete, smooth	Terete, smooth	Terete to slightly compressed, smooth	Terete, crowded with spines	Terete, smooth	Compressed, smooth	Terete, smooth
Leaves										
Primary leaves	Lanceolate to linear, asymmetrical bases, rounded apices, margins dentate, midribs distinct near apices, cryptostomata scattered or in rows on both sides of the midrib	Lanceolate to slender lanceolate, asymmetrical bases, rounded to slightly acute apices, margins entire to dentate, midribs distinct near apices or vanishing midway, cryptostomata scattered	Membranous, lanceolate to linear lanceolate, asymmetrical to cuneate bases, rounded apices, margins dentate, midribs distinct near apices, cryptostomata scattered to one or two rows	Thick, elliptical to oblong, asymmetrical to cuneate bases, rounded and double serrulate apices, margins dentate, midribs vanishing midway to distinct near apices, cryptostomata scattered	Linear-elliptical to linear-lanceolate, asymmetrical to cuneate bases, rounded to acute apices, margins entire to dentate, midribs distinct near apices, cryptostomata scattered	Lanceolate to spatulate, asymmetrical bases, rounded apices, margins entire to dentate, midribs distinct, vanishing near apices, cryptostomata scattered	Elliptical, lanceolate to linear, asymmetrical to cuneate bases, with rounded to acute apices, margins dentate, midribs distinct, vanishing near apices, cryptostomata scattered to arranged in rows on both sides of the midrib	Thick, lanceolate, asymmetrical to cuneate bases, rounded blunt apices, margins entire to dentate, wavy leaf edge, sometimes unequal leaf blades on both sides of the midrib, midribs distinct near apices, cryptostomata scattered	Elongated lanceolate to linear-lanceolate, asymmetrical bases, acute or sometimes obtuse apices, margins dentate, midribs distinct near apices, cryptostomata scattered to in one row	Lanceolate, oblanceolate or sometimes furcated, asymmetrical bases, rounded apices, margins dentate, midribs distinct near apices, cryptostomata scattered to in one or two rows
Secondary leaves	Lanceolate to linear, asymmetrical bases, rounded apices, margins dentate, midribs distinct near apices, cryptostomata scattered or in rows on both sides of the midrib	Lanceolate to linear, asymmetrical to cuneate bases, rounded to acute apices, margins dentate, midribs vanishing near apices, cryptostomata scattered, sometimes arranged in rows on both sides of the midrib	Lanceolate to linear, asymmetrical to cuneate bases, rounded or sometimes acute apices, margins dentate, midribs distinct near apices, cryptostomata in one to two rows	Elliptical-oval, lanceolate to oblong, asymmetrical to cuneate bases, rounded and double serrulate apices, margins undulate to dentate, midribs distinct near apices, cryptostomata scattered to one row	Linear-elliptical to linear, asymmetrical to cuneate bases, acute or sometimes rounded apices, margins dentate to entire, midribs distinct near apices, cryptostomata scattered to one row	Lanceolate to spatulate, asymmetrical bases, rounded apices, margins entire to dentate, midribs vanishing near apices, cryptostomata scattered	Linear-lanceolate to spatulate, asymmetrical to acunate bases, rounded to acute apices, margins dentate, midribs distinct, vanishing near apices, cryptostomata scattered to arranged in rows on both sides of the midrib	Lanceolate to oblong, asymmetrical to cuneate bases, rounded apices, margins dentate, midribs distinct near apices, cryptostomata scattered	Linear lanceolate, asymmetrical to cuneate bases, acute apices, margins dentate to entire, midribs distinct near apices, cryptostomata arranged in one row	Lanceolate, linear lanceolate to oblanceolate, or sometimes furcate, asymmetrical bases, rounded to acute apices, margins dentate, midribs distinct near apices, cryptostomata arranged in one or two rows

Vesicles	Spherical to elliptical	Spherical to elliptical	Spherical, obovoid to elliptical	Spherical, elliptical, ovate to obovoid	Slightly spherical, obovoid or elliptical	Spherical to elliptical	Spherical to obovate	Sub-spherical to obovoid, simple to sometimes in pair	Elliptical	Spherical ,obovoid or elliptical
Receptacle										
Monoecious		Flattened, often twisted		Terete to slightly compressed		Slightly compressed, warty to few spines at the margin, simple to furcate two or three times			Slightly terete, small spines at the apices and margins	
Dioecious										
Male	Long, terete, simple to once to twice furcate		Long, terete		Long, terete to slightly compressed		Long, terete, warty surface, simple to once furcate	Long, terete, sometimes slightly compressed and triquetrous		Long, terete, sometimes slightly compressed and triquetrous at the upper part
Female	Triquetrous, simple to furcate		Short compressed and triquetrous		Flattened and sometimes triquetrous		Terete to slightly compressed, warty surface, simple to once furcate	Triquetrous, simple to once furcate		Short triquetrous

to furcate once to twice (Fig. 11d,e). Female receptacles generally short triquetrous, up to 7 mm long and 2 mm wide, with a warty surface, bearing spines at the apex, simple to furcate three times. Male and female receptacles arranged in a raceme, pseudozygocarpic, vesiculate, sometimes halozygocarpic (Figure 11f,g).

Remarks: *Sargassum* sp. was found only in Songkhla Province. This unidentified species has many unique important characters that make identification difficult, including the the form of their leaves found at primary and secondary branches, which are rather short and lanceolate to furcated. Male receptacles are terete, sometimes slightly compressed and triquetrous at the upper part. Female receptacles of *Sargassum* sp. closely resemble those of *S. baccularia*. In *Sargassum* sp. female receptacles are short, triquetrous, up 7 mm long and 2 mm wide, while those of *S. baccularia* are triquetrous, up 10 mm long and 2 mm wide. Male and female receptacles of *Sargassum* sp. are arranged in a raceme, and are zygocarpic.

Discussion

In this study, ten species of *Sargassum* are reported from the Gulf of Thailand, three of which are new records for Thailand. Five species belong to the section Zygocarpicae (J.G.. Agardh) Setchell. The most common species was *S. polycystum* distributed widely in almost all of the study sites. *Sargassum crassifolium*, *S. siliquosum* and *S. swartzii* were the few found with zygocarpic receptacles. However, this is minor characteristic for identification. Morphological characters are summarized for these ten species of Thai *Sargassum* in Table 1.

Acknowledgements We are deeply indebted to Professor Khanjanapaj Lewmanomont for reading and improving the manuscript. Our thanks go to the Institute of Marine Science, Burapha University; the Faculty of Fisheries, Kasetsart University and the Department of Natural Resources, Graduate School of Global Environmental Studies, Kyoto University. This study was supported by the JSPS Multilateral Core University Program on Coastal Marine Science, Japan.

References

Agardh CA (1824) Systema algarum. Berling, Lund

Agardh JG (1848) Species genera et ordines algarum 1. Species, genera of ordines fucodearum. Gleerup, Lund

Agardh JG (1889) Species Sargassorum austliae descriptae et dispositate. Kgl Svenska Vet-Acad Handl 23:1–133

Ajisaka T (2002) *Sargassum* specimens from Singapore and Malaysia in the herbarium of the Bishop Museum. In Abbott IA, McDermid K (eds) Taxonomy of Economic Seaweeds with Reference to Some Pacific Species vol. 8, California Sea Grant College System: 77–88

Ajisaka T, Lewmanomont K (2004) Variations in the basal system and stolons of *Sargassum stolonifolium* in the Andaman Sea. In Abbott IA, McDermid K (eds) Taxonomy of Economic Seaweeds with Reference to Some Pacific and Other Locations vol. 9, Hawaii Sea Grant College Program: 57–72

Ajisaka T, Nang HQ, Dinh NH (1995a) Studies of two zygocarpic species of *Sargassum* (subgenus *Sargassum*) from Vietnam. In Abbott IA (ed) Taxonomy of Economic Seaweeds with Reference to Some Pacific Species vol. 5, California Sea Grant College System: 45–53

Ajisaka T, Noro T, Yoshida T (1995b) Zygocarpic *Sargassum* species (Subgenus Sargassum) from Japan. In Abbott IA (ed) Taxonomy of Economic Seaweeds with Reference to Some Pacific Species vol. 5, California Sea Grant College System: 11–44

Ajisaka T, Nang HQ, Dinh NH, Lu B, Ang Put Jr, Pang SM, Noro T, Yoshida T (1997) Taxonomic and nomenclatural study of *Sargassum duplicatum* Bory and related species. In Abbott IA (ed) Taxonomy of Economic Seaweeds with Reference to Some Pacific Species vol. 6, California Sea Grant College System: 27–36

Ajisaka T, Phang SM, Yoshida T (1999) Preliminary report of *Sargassum* species collected from Malaysian coasts. In Abbott IA (ed) Taxonomy of Economic Seaweeds with Reference to Some Pacific Species vol. 7, California Sea Grant College System: 23–41

Aungtonya C, Liao LM (2002) Marine flora (algae and seagrasses) in the reference collection of the Phuket Marine Biological Center, Thailand. Phuket Mar Biol Cent Res Bull 64:65–80

Chiang Y-M, Yoshida T, Ajisaka T, Trono GC Jr, Tseng CK, Lu B (1992) Distribution and variation in *Sargassum polycystum* C.A. Agardh (Fucales, Phaeophyta). In Abbott IA (ed) Taxonomy of Economic Seaweeds with Reference to Some Pacific and Western Atlantic Species vol. 3, California Sea Grant College: 35–42

Chou HN, Chiang YM (1981) The *Sargassum* of Taiwan. Acta Oceanogr Taiwanica Sci Rep Natl Taiwan Univ 12:132–149

Egerod L (1974) Report of the marine algae collected on the fifth Thai-Danish expedition of 1966. Bot Mar 17:130–157

Grunow A (1915) Additamena ad cognitionem Sargassorum. KK Zool-Bot Ges Verhandl Vienna 66:1–48 136–185

Lewmanomont K (1988) Marine algae of coral reefs of Thailand. Thai Fisheries Gazette 41(6):561–568

Lewmanomont K, Ogawa H (1995) Common seaweeds and seagrasses of Thailand. Faculty of Fisheries, Kasetsart University

Nateewathana A, Tantichodok P, Busarawich S, Sirivejabandhu R (1981) Marine organisms in the reference collection. Phuket Mar Biol Cent Res Bull 28:43–82

Noiraksar T, Ajisaka T, Kaewsuralikhit (2006) Species of *Sargassum* in the east coast of the Gulf of Thailand. ScienceAsia 32(Suppl 1):99–106

Noro T, Ajisaka T, Yoshida T (1994) Species of *Sargassum* subgenus *Sargassum* (Fucales) with compressed primary branches. In Abbott IA (ed) Taxonomy of Economic Seaweeds with Reference to Some Pacific Species, vol 4, California Sea Grant College: 23–31

Pham HH (1967) Contribution a l'atude des algues littorales d'Vietnam 1: le genre *Sargassum*. Ann Fac Sci Saigon, pp 259–332

Phang SM, Noro T, Yoshida T (1995) Variation in *Sargassum siliquosum* J. Agardh (Fucales, Phaeophyta) found in Malaysia and Singapore. In Abbott IA (ed) Taxonomy of Economic Seaweeds with Reference to Some Pacific Species, vol 5, California Sea Grant College System: 55–66

Reinbold T (1913) *Sargassum*. In: Weber-van Bosse A (ed) Myxophyceae, Chlorophyceae, Phaeophyceae. Siboga Expeditie, Vol. 59a. Brill, Leiden, pp 150–174

Schmidt J (1900–1916) Flora of Koh Chang. Contributions to the knowledge of the vegetation in the Gulf of Siam. Bianco Luno, Copenhagen

Setchell WA (1935) Hong Kong Seaweed. IV. Hong Kong Nat Suppl 4:1–24, 1–17

Trono GC Jr (1992) The Genus *Sargassum* in the Philippines. In Abbott IA (ed) Taxonomy of Economic Seaweeds with Reference to Some Pacific and Western Atlantic Species vol. 3, California Sea Grant College: 43–94

Trono GC Jr (1997) Field Guide and Atlas of the Seaweed Resources of the Philippines. Bookmark, Philippines

Tseng CK (1983) Common Seaweeds of China. Science Press, Beijing

Tseng CK, Lu B (1988) Studies on Chinese species of Zygocarpic *Sargassum*. In Abbott IA (ed) Taxonomy of Economic Seaweeds with Reference to Some Pacific and Caribbean Species vol. 2, California Sea Grant College Program: 23–54

Tseng CK, Lu B (1992) Studies on the Malacocarpic *Sargassum* of China: II. Racemosae J. Agardh. In Abbott IA (ed) Taxonomy of Economic Seaweeds with Reference to Some Pacific and Western Atlantic Species vol. 3, California Sea Grant College: 11–34

Tseng CK, Lu B (1995a) Studies on the Glomerulate *Sargassum* of China: II. The species group *Swartzia*. In Abbott IA (ed) Taxonomy of Economic Seaweeds with Reference to Some Pacific Species vol. 5, California Sea Grant College System: 75–92

Tseng CK, Lu B (1995b) Studies on the Glomerulate *Sargassum* of China: III.The species group *Binderia*. In Abbott IA(ed) Taxonomy of Economic Seaweeds with Reference to Some Pacific Species, Vol 5, California Sea Grant College System: 93–106

Tseng CK, Lu B (1997) Studies on the Glomerulate *Sargassum* of China: IV. series of Platycarpae. In Abbott IA (ed) Taxonomy of Economic Seaweeds with Reference to Some Pacific Species vol. 6, California Sea Grant College System: 9–25

Tseng CK, Yoshida T, Chiang YM (1985) East Asiatic species of *Sargassum* subgenus *Bactrophycus* J. Agardh (Sargassaceae, Fucales), with keys to the section and species. In Abbott IA, Norris JN (eds) Taxonomy of Economic Seaweeds with Reference to Some Pacific and Caribbean Species California Sea Grant College Program: 1–14

Yamada Y (1925) Studien über die Meeresalgen von der Insel Formosa. 2. Phaeophyceae, Bot Mag Tokyo 39:239–254

Yamada Y (1942) Notes on *Sargassum* from southern parts of Japan. I and II. J Jpn Bot 18:369–381 503–519

Yoshida T (1988) Japanese and Taiwanese species of *Sargassum* subgenus *Sargassum*. In Abbott IA (ed) Taxonomy of Economic Seaweeds with Reference to Some Pacific and Caribbean Species, Vol 2, California Sea Grant College Program: 5–21

Phylogeography of the genus *Ulva* (Ulvophyceae, Chlorophyta), with special reference to the Japanese freshwater and brackish taxa

Satoshi Shimada · Naoko Yokoyama · Shogo Arai · Masanori Hiraoka

Originally published in the Journal of Applied Phycology, Vol 20, No 5, 529–539.
DOI: 10.1007/s10811-007-9296-y

Abstract The nuclear-encoded ITS and associated 5.8S rDNA regions were sequenced for 72 specimens of *Ulva* collected from 44 rivers across Japan, including *U. prolifera* Müller from the Shimanto River, Kochi Prefecture, as well as 26 samples originally identified as *U. linza* L. from 20 coastal marine areas. Sequence data revealed that the samples fall into six distinct clades: the *U. flexuosa* Wulfen clade (2 samples), the *Ulva linza-procera-prolifera* (LPP) complex clade (75 samples), *Ulva* sp. 1 clade (3 samples), *Ulva* sp. 2 clade (7 samples), *Ulva* sp. 3 clade (4 samples) and *Ulva* sp. 4 clade (7 samples). The LPP complex contained a mixture of 26 samples collected from seashores and 49 samples obtained from rivers, including *U. prolifera* from the Shimanto River, and GenBank data for *U. linza* and *U. procera* Ahlner. The samples of the LPP complex differed by only 0–7 substitutions (0–1.149%). Subsequent phylogeographic analyses of the LPP complex based on the 5S rDNA spacer region revealed the presence of two further groupings: a group including 22 strictly marine littoral *U. linza* samples and a *U. prolifera* group composed of a mixture of 4 marine samples and all 49 river samples. The monophyly of all river samples indicates that adaptation to low salinity might have occurred only once in the evolutionary history of the LPP complex.

Keywords Taxonomy · River · Freshwater · Japan · Phylogeography · *Ulva*

S. Shimada (✉)
Creative Research Initiative Sousei, Hokkaido University,
Sapporo 060-0810, Japan
e-mail: sshimada@sci.hokudai.ac.jp

N. Yokoyama
Department of Natural History Sciences, Faculty of Science,
Hokkaido University,
Sapporo 060-0810, Japan

S. Arai
Marine Algal Research Co., Ltd.,
Shingu-cho, Kasuya-gun,
Fukuoka 811-0114, Japan

M. Hiraoka
Usa Marine Biological Institute, Kochi University,
Usa-cho, Tosa,
Kochi 785-0167, Japan

Introduction

The green algal genus *Ulva*, including species previously placed in the genus *Enteromorpha* (Hayden et al. 2003), is well known for its wide distribution in marine, freshwater and brackish environments throughout the world (Canter-Lund and Lund 1995; Martins et al. 1999; McAvoy and Klug 2005; Reed and Russell 1979; Shimada et al. 2007; van den Hoek et al. 1995). Some 100 species are now included in the genus (Guiry and Guiry 2007), of which only 18 species are currently recognized in Japan (Yoshida et al. 2005).

Ulva prolifera Müller can be observed growing from the estuary to 7 km upstream in the Shimanto River in Kochi Prefecture (Shikoku Island, Japan) (Ohno and Takahashi 1988). This species was originally described from Nebbelund, Lolland Island, Denmark (Müller 1778), and first reported in Japan by Nakura (1921) based on materials collected from brackish waters in Nabeta, Kaibe, Aichi Prefecture. It is one of the most expensive edible Japanese algae and costs about 8,000–10,000 yen kg^{-1} (dry weight) (Ohno 2004). Although many studies have been reported for the green algal genus *Ulva* in relation to fisheries (Ohno and Takahashi 1988; Hiraoka et al. 2003;

Pandey and Ohno 1985; Dan et al. 2002, 2003), the taxonomic diversity of its representatives growing in freshwater and brackish rivers across Japan still remains poorly understood.

The aim of this study was thus to elucidate the species diversity of fresh and brackish water Japanese *Ulva*, and clarify their phylogenetic and phylogeographic relationships.

Materials and Methods

Materials used in this study were gathered from rivers and marine littoral habitats and are listed in Table 1 while the locations shown in Fig. 1. Voucher herbarium specimens are deposited in the Herbarium of the Graduate School of Science, Hokkaido University, Sapporo (SAP).

Total DNA was extracted by using the Dneasy Plant Mini Kit (QIAGEN, Valencia, Calif., USA) according to the specifications of the manufacturer. The regions selected for PCR amplification and automated sequencing were the nuclear-encoded ITS (Internal Transcribed Spacers) region including the 5.8S rDNA (Shimada et al. 2003), and the 5S rDNA spacer region (Yotsukura et al. 2006). The following primer pairs were used: ITS1 (5′-TTTGTACACAC CGCCCG-3′)-ITS2 (5′-GAATTCTGCAATTCACA-3′) and ITS3 (5′-CTCTCAACAACGGATATCT-3′)-ITS4 (5′-TCCTCCGCTTATTGATATGC-3′) for the ITS (Coat et al. 1998; Malta et al. 1999); 5S-F(5′- GGTTGGGCAGGAT TAGTA -3′)- 5S-R (5′- AGGCTTAAGTTGCGAGTT -3′) for the 5S rDNA spacer. PCR amplification was run on a thermocycler (Program Temp Control System, Astec, Fukuoka, Japan) and the profile of the reactions consisted of one initial denaturation of 1 min at 94°C followed by 35 cycles of denaturation of 45 sec at 94°C, primers annealing of 45 sec at 50°C and extension of 60 sec at 72°C, terminated by a final hold at 4°C. The presence of the PCR-amplified products was visualized under UV light after agarose gel electrophoresis and staining with ethidium bromide. PCR-amplified products of ITS regions were cleaned using the QIAquick PCR Purification Kit (QIAGEN). On the other hand, PCR using the primer pair 5S-F - 5S-R to locate the 5S rDNA tandem arrays amplified multiple DNA fragments. The shortest DNA fragment from each PCR of the 5S rDNA spacer region was excised from the gels on a transilluminator with a razor blade and the DNA was collected using QIAquick Gel Extraction Kit (QIAGEN. DNA sequencing was performed as described by Uchimura et al. (2006), using the same primers that were used for PCR. When the ITS and 5S rDNA spacer fragments could not be directly sequenced, PCR products were cloned as described in Shimada et al. (2003) before determining the nucleotide sequences.

The generated sequences were compiled and aligned manually using DNASIS (Version 2.0, Hitachi Software Engineering, Yokohama, Japan). Sequences of 33 samples of Ulvacean species were downloaded from GenBank and included in the alignment of the ITS sequences. *Umbraulva amamiensis* (Tanaka) Bae et Lee (AB097640) and *Umbraulva olivascens* (Dangeard) Bae et Lee (AJ234322) were used as outgroups for the ITS tree following Shimada et al. (2003).

Phylogenetic analyses were performed using the maximum likelihood (ML) algorithm available in the computer program PAUP* 4.0 b10 (Swofford 2002). Identical sequences were excluded from the alignment. The program MODELTEST version 3.7 (Posada and Crandall 1998) was used to find the model of sequence evolution that best fit each dataset by a hierarchical likelihood ratio test (α=0.01). When the best sequence evolution model was determined, ML tree searches were performed using the estimated model parameters with the following options: starting tree option = obtained by stepwise addition, and branch swapping algorithm = TBR. Bootstrap values based on 100 re-samplings in ML, 2,000 re-samplings in maximum parsimony (MP) and neighbor-joining (NJ) of the dataset were calculated (TBR, full heuristic search option) (Felsenstein 1985).

A statistical parsimony network (SPN) tree for the 31 haplotypes of the 5S rDNA spacer region were inferred using TCS versus 1.18 (Clement et al. 2000). Each gap was treated as missing data in the construction of a SPN tree.

Results

ITS analysis

The phylogenetic tree obtained with the ML method is presented in Fig. 2. Likelihood settings from the best-fit model (GTR + I + G) were selected by a hierarchical likelihood ratio test in the program MODELTEST version 3.7: assumed nucleotide frequencies A=0.14400, C= 0.37800, G=0.32330, and T=0.15470; substitution-rate matrix with AC=1.226900, AG=2.180000, AT=2.801000, CG=1.401000, CT=3.666100 and GT=1; proportion of invariable sites=0.3781; gamma distribution with shape parameter=0.5223. Based on these settings, a heuristic search was performed with the TBR branch swapping option (-ln L=4599.99070) after 10,369 rearrangements (Fig. 2).

The 98 samples were resolved into six clades. The largest clade contained all 26 seashore samples and 49 river samples, including *U. prolifera* from the Shimanto River, and GenBank data for *U. linza* and *U. procera*. It was referred to as *Ulva linza-procera-prolifera* (LPP) complex and received 82% bootstrap support in ML, 94% in MP and 100% in NJ analyses (Fig. 2). There were only 0–7 substitutions (0–1.149%) in the LPP complex. Two samples (E29, C263) had sequences identical to *U. flexuosa* Wulfen

Table 1 Specimens of *Ulva* used in this study

Species	Sample	Locality (population no.): collection date, (PSU), site	SAP	Accession no. (identical sample no.)		Haplotypes
				ITS	5S spacer	
U. linza	C189	Jeju Island, Korea, 2000/2/9, (no data), Sea	102935	AB299439	AB299441	-
	C503	Carraroe.co.Galway.Ireland, 2004/9/4, (no data), Sea	102936	AB298633	AB298669	-
	C513	Clogherhead.co.Louth.Ireland, 2004/6/30, (no data), Sea	102937	AB298634	(C584)	-
	C469	Dolinsk, Sakhalin, Russia, 2003/6/30, (no data), Sea	102938	(C503)	AB298684	-
	C37	Bokoi, Muroran, Hokkaido (58), 2000/6/13, (no data), Sea	95060	(C503)	c1:AB298670 c2:AB298671	-
	C48	Oshoro, Otaru, Hokkaido (69), 2000/6/25, (no data), Sea	102940	(C503)	AB298672	-
	C631	Oshoro, Otaru, Hokkaido (69), 1994/3/1, (no data), Sea	102941	(C503)	AB298673	-
	C632	Oshoro, Otaru, Hokkaido (69), 1994/3/1, (no data), Sea	102942	(C503)	(C48)	-
	C528	Higasishizunai, Shinhidaka, Hokkaido (54), 2005/8/7, (no data), Sea	102943	(C503)	(C48)	-
	C529	Higasishizunai, Shinhidaka, Hokkaido (54), 2005/8/7, (no data), Sea	102943	(C503)	AB298674	-
	NY112	Bannaguro, Ishikari, Hokkaido (55), 2004/8/19, (no data), Sea	100132	(C503)	c1: C631 c2:AB298676 c3:AB298675	-
	C187	Shimoda, Shimoda, Shizuoka (56), 1998/3/28, (no data), Sea	102952	AB298636	AB298678	-
	C80	Tanabe, Wakayama, Wakayama (57), 2000/9/11, (no data), Sea	102953	(C189)	AB298677	-
	C584	Seaside of Daishouji R., Kaga, Ishikawa (60), 2005/11/21, (30), Sea	102960	(C503)	AB298685	-
	C484	Mimatsu, Ohi, Fukui (61), 2005/3/29, (no data), Sea	102961	(C503)	AB298686	-
	C181	Shishigawa, Togitsu, Nagasaki (62), 1993/12/24, (no data), Sea	102966	AB298635	AB298679	-
	C191	Usuki, Usuki, Ooita (63), 2001/1/23, (no data), Sea	102967	(C503)	(C181)	-
	C595	Hirakawa, Kagoshima, Kagoshima (64), 2005/11/24, (no data), Sea	102972	(C503)	AB298680	-
	C596	Hirakawa, Kagoshima, Kagoshima (64), 2005/11/24, (no data), Sea	102973	(C503)	AB298681	-
	C245	Kubotsu, Tosashimizu, Kochi (66), 2004/2/8, (no data), Sea	102982	(C503)	AB298682	-
	KM2	Port of Murotsu, Muroto, Kochi (67), 2003/3/4, (no data), Sea	102983	AB299440	AB298683	-
	TM2	Mugi, Mugi, Tokushima (68), 2003/2/19, (no data), Sea	102984	(C503)	(C245)	-
U. flexuosa	E29	Yoshino R., Tokushima, Tokushima (70), 2002/5/22, (no data), 2km upstream	103008	AB097644	-	-
	C263	Ohmizaki, Matsue, Shimane (71), 2001/8/1, (no data), salted lake	103006	AB097644	-	-
U. prolifera	NY105	Nobusha River, Rumoi, Hokkaido (1), 2004/8/19, (no data), 100m upstream	100130	AB298308	AB298639	N
	NY106	Nobusha R., Rumoi, Hokkaido (2), 2004/8/19, (no data), 50m upstream	100130	(NY105)	AB298640	Q
	NY128	Furubira R., Yoichi, Hokkaido (3), 2004/9/15, (no data), 10m upstream	100142	(NY105)	(NY105)	N
	C455	Sekiguchi R., Yamada, Iwate (4), 2005/5/2, (2), 10m upstream	102944	AB298316	AB298639	N
	C458	Natori R., Natori, Miyagi (5), 2005/5/3, (2), 20m upstream	102945	AB298314	AB298646	O
	C460	Ohta R., Minami-Sohma, Fukushima (6), 2005/5/3, (0), 10m upstream	102946	(C455)	AB298645	X
	C461	Juhoh R., Hitachi, Ibaraki (7), 2005/5/3, (1), 10m upstream	102947	(C458)	(E1)	A
	C462	Tone R., Kamisu, Ibaraki (8), 2005/5/4, (3), 2.5km upstream	102948	(C455)	(C458)	O
	C463	Hamano R., Chiba, Chiba (9), 2005/5/4, (8), 950m upstream	102949	(C455)	c1:AB298647 c2:AB298648	c1: H c2: 1

Table 1 (continued)

Species	Sample	Locality (population no.): collection date, (PSU), site	SAP	Accession no. (identical sample no.)		Haplotypes
				ITS	5S spacer	
	C444	Tagoshi R., Zushi, Kanagawa (10), 2005/3/20, (3), 700m upstream	102950	(C455)	(E1)	A
	C445	Tagoshi R., Zushi, Kanagawa (11), 2005/3/20, (9), 500m upstream	102951	(C455)	(E1)	A
	C561	Koyoshi R., Yurihonjo, Akita (12), 2005/11/19, (8), 10m upstream	102954	(NY105)	(NY105)	N
	C562	Koyoshi R., Yurihonjo, Akita (12), 2005/11/19, (8), 10m upstream	102954	(NY105)	AB298641	R
	C563	Toyo R., Sakata, Yamagata (13), 2005/11/19, (4), 20m upstream	102955	AB298317	AB298649	M
	C565	Sanze R., Tsuruoka, Yamagata (52), 2005/11/19, (no data), 10m upstream	102956	AB298315	AB298654	Z
	C577	Nadachi R., Nadachi, Niigata (14), 2005/11/20, (0), 10m upstream	102957	(NY105)	(NY105)	N
	C578	Fishing Port of Kurobe, Toyama (59), 2005/11/20, (no data), Sea	102958	(NY105)	AB298642	S
	C629	Fishing Port of Kurobe, Toyama (59), 2005/11/20, (no data), Sea	102958	(NY105)	(NY105)	N
	C630	Fishing Port of Kurobe, Toyama (59), 2005/11/20, (no data), Sea	102958	(NY105)	(NY105)	N
	C581	Daishouji R., Kaga, Ishikawa (15), 2005/11/21, (7), 100m upstream	102959	(NY105)	AB298662	P
	C587	Yura R., Maizuru, Kyoto (16), 2005/11/21, (7), 1km upstream	102962	(NY105)	AB298643	T
	C588	Yura R., Maizuru, Kyoto (16), 2005/11/21, (7), 1km upstream	102962	(NY105)	(NY105)	N
	C267	Honjo, Matsue, Shimane (17), 2000/11/24, (no data), salted lake	102963	AB298309	(NY105)	N
	C268	Honjo, Matsue, Shimane (17), 2000/11/24, (no data), salted lake	102963	(NY105)	AB298663	U
	C269	Honjo, Matsue, Shimane (17), 2000/11/24, (no data), salted lake	102963	(NY105)	AB298664	V
	C270	Irie, Matsue, Shimane (18), 2000/11/24, (no data), salted lake	102963	AB298310	AB298655	Y
	C613	Mumomi R., Fukuoka, Fukuoka (19), 2005/12/11, (2), 2.7km upstream	102964	AB298312	AB298661	Aa
	C614	Mumomi R., Fukuoka, Fukuoka (19), 2005/12/11, (2), 2.7km upstream	102965	(C613)	(C565)	Z
	C392	Yanyuu R., Ohshima, Kagoshima (22), 2005/3/15, (2), 70m upstream	102974	(C458)	AB298644	W
	C393	Ura R., Ohshima, Kagoshima (23), 2005/3/15, (1), 90m upstream	102975	(C458)	AB298650	K
	C403	Nakakiku R., Ohshima, Kagoshima (24), 2005/3/17, (0), 200m upstream	102976	(C458)	(E1)	A
	C404	Yuwan R., Ohshima, Kagoshima (25), 2005/3/18, (0), 200m upstream	102977	(C455)	AB298651	J
	C399	Naze Port, Amami, Kagoshima (65), 2005/3/16, (35), Sea	102978	(C503)	(E1)	A
	C342	Koume R., Nishinoomote, Kagoshima (26), 2004/12/19, (0), 1km upstream	102979	(C455)	(C270)	Y
	C350	Saikyou R., Nishinoomote, Kagoshima (27), 2004/12/18, (3), 90m upstream	102980	AB298311	(E1)	A
	C352	Minato R., Nishinoomote, Kagoshima (28), 2004/12/18, (2), 30m upstream	102981	(C455)	AB298652	L
	C106	Shimanto R., Shimanto, Kochi (29), 2001/2/26, (4.7), 7km upstream	102985	(C455)	c1: (C270) c2:AB298656	c1: Y c2: Ab
	E21	Shimanto R., Shimanto, Kochi (29), 2001/2/29, (no data), 7km upstream	95087	AB298320	c1:AB298658 c2:AB298660	c1: Ad c2: Ae
	C110	Shimanto R., Shimanto, Kochi (30), 2001/2/26, (21.9), 500m upstream	102986	AB298313	(E1)	A
	C107	Shimanto R., Shimanto, Kochi (53), 2001/2/26, (34.4), 10m upstream	102987	AB298319	c1:AB298637 c2:AB298657	c1: F c2: Ac
	E1	Hiwasa R., Kaibe, Tokushima (31), 2000/2/29, (no data), 200m upstream	95080	(C455)	AB298665	A
	E8	Hiwasa R., Kaibe, Tokushima (32), 2000/4/20, (no data), 100m upstream	95082	c1: (C455) c2: (C107)	c1: (C565) c2: (C270)	c1: Z c2: Y

	E2	Niyodo R., Tosa, Kochi (33), 2000/2/7, (no data), 100m upstream	95081	(C455)	c1: (E1) c2:AB298666	c1: A c2: B
	E3	Niyodo R., Tosa, Kochi (33), 2000/2/7, (no data), 100m upstream	95081	(C455)	(E1)	A
	E6	Niyodo R., Tosa, Kochi (33), 2000/2/7, (no data), 100m upstream	95081	(C107)	c1:AB298638 c2: (E1)	c1: G c2: A
	KM7	Murotsu R., Muroto, Kochi (34), 1999/3/3, (no data), 80m upstream	102988	(C455)	AB298653	C
	E11	Yoshino R., Tokushima, Tokushima (35), 2001/5/23, (no data), 12km upstream	95083	(C455)	c1:(E1) c2:AB298667	c1: A c2: D
	E19	Yoshino R., Tokushima, Tokushima (35), 2001/5/23, (no data), 12km upstream	95084	(C455)	c1: (C613) c2:(E1)	c1: Aa c2: A
	E20	Yoshino R., Tokushima, Tokushima (35), 2000/5/7, (no data), 12km upstream	95086	(C455)	c1:AB298668 c2:(E1)	c1: E c2: A
	C555	Shio R., Motobu, Okinawa (44), 2005/10/4, (20), 80m upstream	102998	AB298318	(E1)	A
	C617	Shio R., Motobu, Okinawa (44), 2005/10/4, (20), 80m upstream	102999	AB298321	(E1)	A
	C618	Shio R., Motobu, Okinawa (44), 2005/10/4, (20), 80m upstream	102999	(C455)	(E1)	A
	C619	Shio R., Motobu, Okinawa (44), 2005/10/4, (20), 80m upstream	102999	(C617)	(E1)	A
	C381	Okumura R., Chichijima, Tokyo (51), 2005/3/3, (4), 250m upstream	103005	(NY105)	(C581)	P
Ulva sp. 1	TT005	Ara R., Onna, Okinawa (37), 2005/5/10, (3), 80m upstream	102990	AB298454	-	-
	TT006	Ara R., Onna, Okinawa (37), 2005/5/10, (3), 80m upstream	102991	AB298455	-	-
	TT013	Fukuchi R.,, Higashi, Okinawa (38), 2005/5/10, (0), 400m upstream	102992	AB298456	-	-
Ulva sp. 2	E16	Yoshino R., Tokushima, Tokushima (36), 2000/7/18, (no data), 7km upstream	102989	AB298457	-	-
	C380	Okumura R., Chichijima, Tokyo (50), 2005/3/3, (5), 300m upstream	103004	(E16)	-	-
	C266	Honjo, Matsue, Shimane (17), 2000/11/24, (no data), salted lake	103007	(E16)	-	-
	NY140	Todoroki R., Ishigaki, Okinawa (45), 2004/9/30, (3), 210m upstream	103000	AB298459	-	-
	NY138	Hirakubo R., Ishigaki, Okinawa (46), 2004/9/30, (20), 20m upstream	103001	AB298458	-	-
	C591	Yahusa R., Ichikikushikino, Kagoshima (20), 2005/11/24, (no data), 500m upstream	102968	AB298460	-	-
	C593	Yojirougahama, Kagoshima, Kagoshima (21), 2005/11/24, (no data), 700m upstream	102971	AB298461	-	-
Ulva sp. 3	TT001	Hohtoku R., Tomigusuku, Okinawa (39), 2005/5/9, (10), 1.7km upstream	102993	AB298462	-	-
	TT002	Yuhhi R., Naijou, Okinawa (40), 2005/5/9, (15), 700m upstream	102994	AB298463	-	-
	TT012	KiseR., Nago, Okinawa (41), 2005/5/10, (7), 400m upstream	102995	(TT001)	-	-
	TT017	Ufuguimu R., Motobu, Okinawa (42), 2005/5/10, (0), 200m upstream	102996	(TT001)	-	-
Ulva sp. 4	NY131	Small stream, Ohdo, Itoman, Okinawa (43), 2004/9/28, (0), 10m upstream	102997	AB298464	-	-
	NY132	Small stream, Ohdo, Itoman, Okinawa (43), 2004/9/28, (0), 10m upstream	102997	(NY131)	-	-
	NY133	Small stream, Ohdo, Itoman, Okinawa (43), 2004/9/28, (0), 10m upstream	102997	(NY131)	-	-
	NY145	Small stream, Ohdo, Itoman, Okinawa (43), 2004/9/28, (0), 10m upstream	102997	(NY131)	-	-
	C378	Yatsuse R., Chichijima, Tokyo (47), 2005/3/3, (0), 1km upstream	103002	AB298465	-	-
	C386	Tama R., Hahajima, Tokyo (48), 2005/3/7, (35), 90m upstream	103003	(C378)	-	-
	C387	Tama R., Hahajima, Tokyo (49), 2005/3/7, (3), 180m upstream	103004	(C378)	-	-

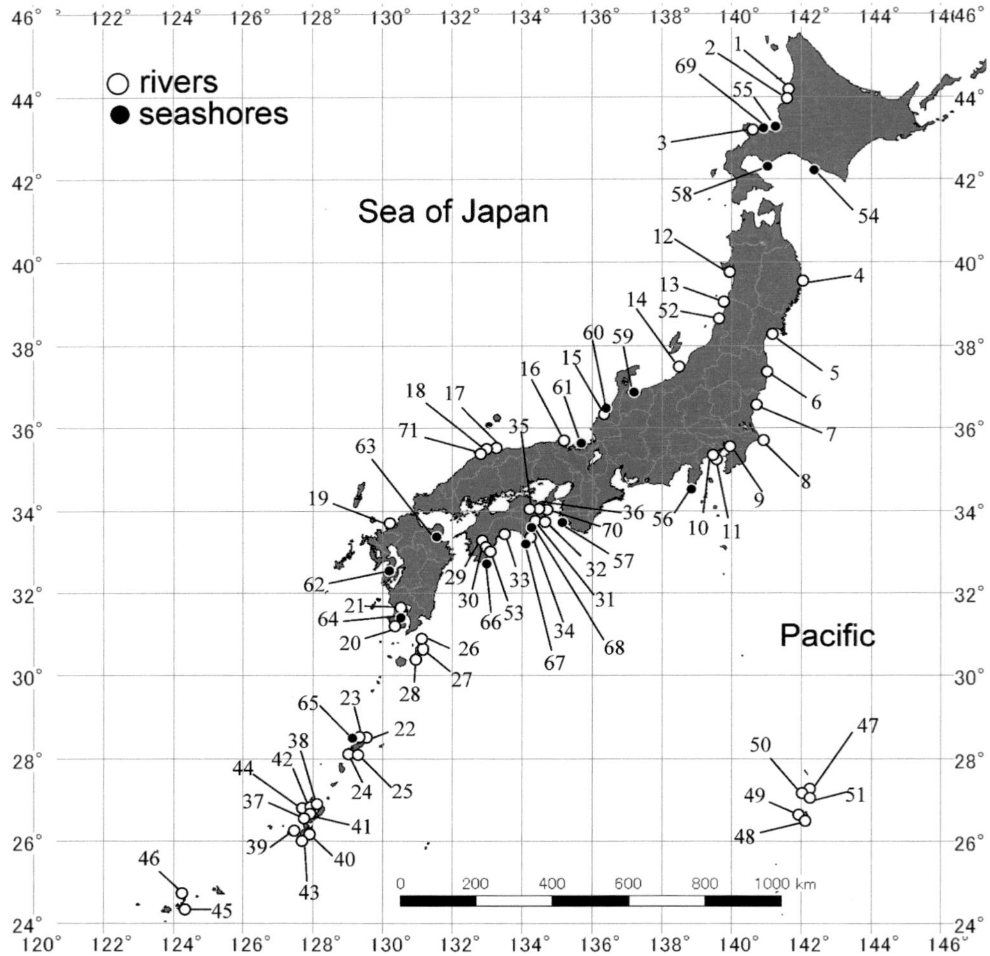

Fig. 1 Map of sampling locations of *Ulva* in Japan

and formed a distinct lineage. The remaining 4 clades did not include GenBank samples with which they could be identified and were thus provisionally referred to as *Ulva* sp. 1 (three samples), *Ulva* sp. 2 (seven samples), *Ulva* sp. 3 (four samples), and *Ulva* sp. 4 (seven samples), respectively. They clustered with high (99–100%) bootstrap values in all analyses. The *Ulva* sp. 1 clade and *Ulva* sp. 3 clade consisted of samples from Okinawa. The *Ulva* sp. 2 clade was widespread in the moderately warm regions of Okinawa, Kyushu, Shikoku and Ogasawara Islands. Specimens of the *Ulva* sp. 4 clade originated from the Islands of Okinawa and Ogasawara.

The 5S rDNA spacer analysis of the LPP complex

The unrooted phylogenetic tree obtained with ML analysis of this region (389 bp) is presented in Fig. 3. For the ML method, likelihood settings from the best-fit model (HKY85+G) were selected by a hierarchical likelihood ratio test in the program MODELTEST version 3.7: assumed nucleotide frequencies A=0.16370, C=0.42320, G=0.23210, and T=0.18100; gamma distribution with shape parameter= 0.7872. Based on these settings, a heuristic search was performed with the TBR branch swapping option (-ln L= 1213.55640) after 388,72 rearrangements (Fig. 3).

Seventy-five samples of the LPP complex were divided into two groupings: the *U. linza* group including only marine littoral representatives (22 samples) and the *U. prolifera* group composed of a mixture of both marine (4 samples) and all river representatives (49 samples). These two groupings were supported by 100% bootstrap values in all analyses. Four marine samples contained in the *U. prolifera* group did not form a monophyletic clade. Pairwise distances in the LPP complex varied between 0 and 29 bp (0–11.417%) and were higher in the *U. linza* group [0–15 bp (0–5.769%)] than in the *U. prolifera* group [0–10 bp (0–2.717%)].

Statistical parsimony network (SPN) tree of the *U. prolifera* group

The distribution of 31 haplotypes among the *U. prolifera* group is shown in Fig. 4a, and the SPN tree is presented in Fig. 4b. Samples possessing haplotype A (filled red circle)

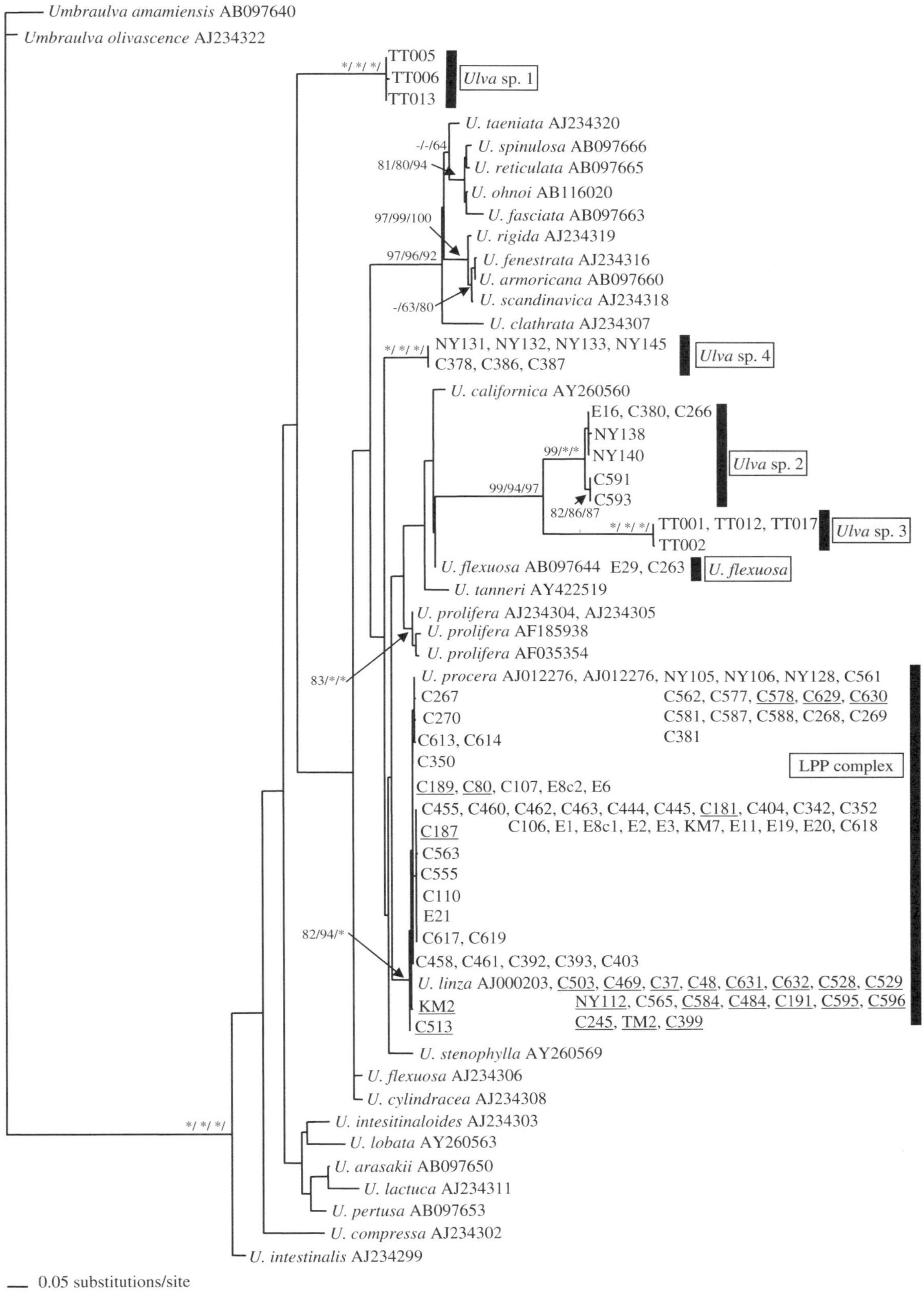

Fig. 2 Phylogenetic tree of the maximum likelihood (ML) analysis inferred from the nuclear encoded ITS regions including 5.8S rDNA of *Ulva*. *Umbraulva amamiensis* and *U. olivascens* were used as outgroups. Numerals at internal nodes are bootstrap values >50% for 100 replicates in ML, 2,000 replicates in maximum parsimony (MP) and neighbor-joining (NJ) analyses (ML/MP/NJ). The *underlined* sample numbers were collected from seashores. * denotes 100% bootstrap values

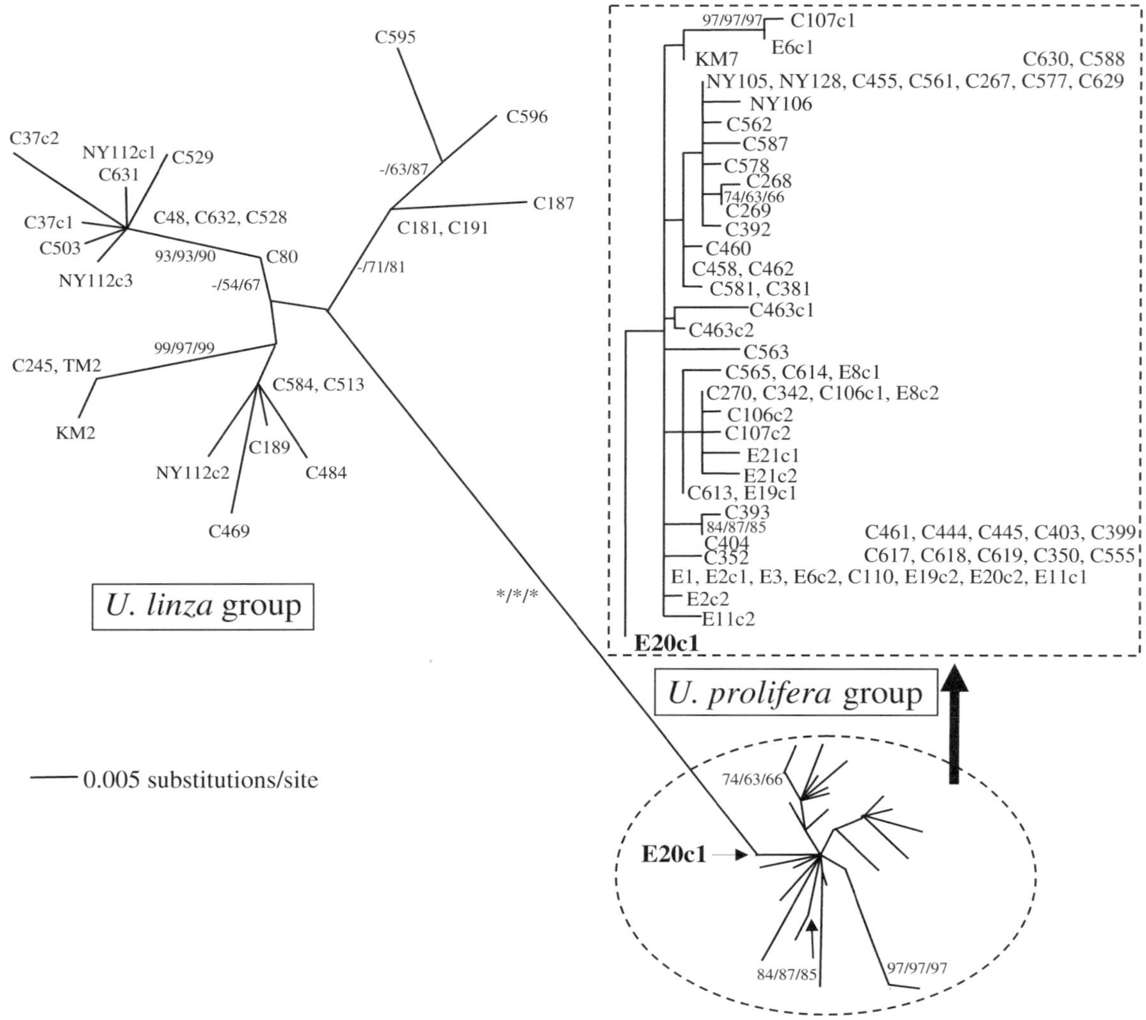

Fig. 3 Unrooted ML tree of the 5S rDNA spacer region of the LPP complex. Numerals at internal nodes are bootstrap values >50% for 100 replicates in ML, 2,000 replicates in maximum parsimony (MP) and neighbor-joining (NJ) analyses (ML/MP/NJ). The *Ulva prolifera* group is focused on the upper side. * denotes 100% bootstrap values

were widely distributed on the Pacific coast of Japan (Fig. 4a). In the SPN tree, haplotype A connected with 11 other haplotypes (Fig. 4b). Two lineages (blue and green) were also related to haplotype A (Fig. 4b). The blue lineage was differentiated into 11 haplotypes (Fig. 4b). Haplotype N (filled blue circle) extended from the Pacific coast of the Tohoku region to the Sea of Japan (Fig. 4a). The green lineage was differentiated into 7 haplotypes (Fig. 4b), distributed along both the Pacific and Japan Sea Coasts (Fig. 4a).

Discussion

We investigated the species diversity of the genus *Ulva* growing in rivers, using nuclear encoded ITS regions including 5.8S rDNA. Six groupings, of which four groups including unidentified species, were detected. The four unidentified species-clades await further investigation to clarify their taxonomic status. The largest group of Japanese *Ulva* growing in rivers consisted of a combination of three species (*U. linza, U. procera* and *prolifera*) and was referred to as the LPP complex. The LPP complex included *U. prolifera* collected from Japanese rivers. However, European *U. prolifera* from the Ythan Estuary (AJ234304, AJ234305) and Portaferry (AF035354, AF185938) were separated from the LPP complex in the ITS tree. Therefore, further studies including materials from the type locality of *U. prolifera* are clearly needed to clarify its taxonomic identity. A further conclusion emerging from this study is that, in the Japanese warm regions, the diversity of *Ulva* species growing in rivers may be higher than currently appreciated. Members of the LPP complex were distributed all around Japan (from Okinawa in the south to Hokkaido in the north); the fact that the monthly seawater temperature average at Oshoro (near Yoichi town

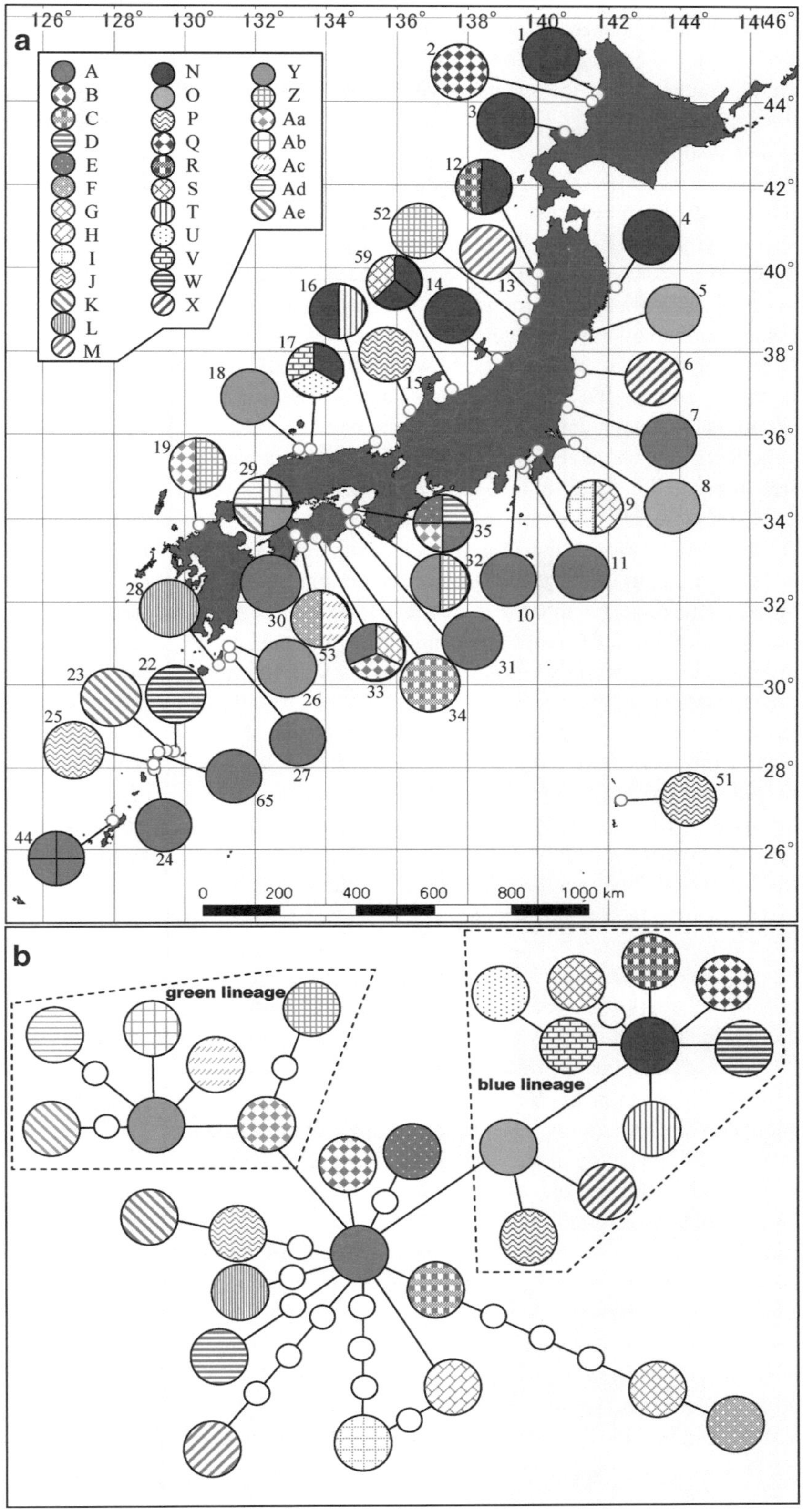

Fig. 4 Geographical distributions of DNA haplotypes of sequences of the 5S rDNA spacer region, and the SPN tree of the *Ulva prolifera* group in the LPP complex. Combinations of *color* and *pattern* represent different haplotypes. **a** Geographical distributions; **b** SPN tree. Each *line* connecting haplotypes corresponds to one base mutation

in Hokkaido) has been reported to vary between 3 and 10°C from winter to spring seasons (Shimada and Masuda 2003) clearly suggests that the species involved also tolerate low temperatures.

Detailed phylogenetic analyses of the LPP complex, using a more resolved DNA marker, the 5S rDNA spacer, has already been reported to be useful for inferring inter-species phylogeny of the Phaeophyceae (Yotsukura et al. 2002; Yotsukura et al. 2006). It was about 10 times more variable than the ITS region and thus allowed a better understanding of the phylogenetic relationships within the LPP complex. The LPP complex was further subdivided into two groups: the *U. linza* group and the *U. prolifera* group. This result indicates that these two entities are genetically separated. However, although they were identified here as *U. linza* and *U. prolifera,* additional investigations including crossing experiments are still needed for further confirmation. All 49 samples originating from rivers were clustered in the *U. prolifera* group with high bootstrap values. This suggests that the adaptation to low salinity for members of the LPP complex might have occurred only once in the evolutionary history of this group, as was also demonstrated by Sherwood et al. (2002) for the European freshwater species of *Hildenbrandia*. The fact that the *U. linza* group was more variable than the *U. prolifera* group in the sequences of the 5S rDNA spacer region also suggests that the *U. prolifera* group was recently separated from the marine *U. linza* group. In the 5S rDNA spacer tree, although four marine samples preliminarily identified as *U. linza* were contained in the *U. prolifera* group, they did not form a monophyletic clade. This may be explained by an initial adaptation of these marine samples to low salinity in rivers followed by a return and readaptation to the marine environment.

Although we did not find *U. intestinalis* L. in Japanese rivers, *U. intestinalis* has been reported to have an adaptation to low salinity (Kamer and Fong 2000; Martins et al. 1999; McAvoy and Klug 2005; Reed and Russell 1979). Blomster et al. (1998) clearly showed, using molecular phylogenetic analyses, that only *U. intestinalis* samples, not *U. compressa* L., were found close to freshwater outlets. Pringle (1986) collected *U. intestinalis* from brackish water and was able to culture it in >2 PSU seawater. Despite its demonstrated ability to regulate turgor pressure, *U. intestinalis* died after exposure to 0 PSU for as little as 6 days (Martins et al. 1999). Edwards et al. (1988) and Ritchie and Larkum (1985) showed that *U. intestinalis* could not persist under 0 PSU conditions for extended periods (3–5 weeks). Kamer and Fong (2000) suggested that *U. intestinalis* can tolerate short-term fluctuations in salinity but can be negatively impacted by persistent salinity reductions. Ohno et al. (1999) reported that *U. prolifera* in the Shimanto River, Kochi Prefecture, can survive under 0.01 to 18 PSU conditions. Correspondingly, the populations of *U. prolifera* in the Shimanto River disappear above 8 km upstream (Hiraoka and Shimada 2004). Therefore, we conclude that *U. intestinalis* and *U. prolifera* are probably adapted to brackish waters, but not to freshwater environments.

The SPN tree (Fig. 4a) and the distribution of haplotypes of the 5S rDNA spacer sequences around Japanese coasts (Fig. 4b) suggest the possibility that: (1) the first divergence in the *U. prolifera* group arose around the Pacific coast of Japan; (2) haplotype A first invaded Okinawa and Kanto and this was followed by a secondary differentiation into two lineages;(3) a blue lineage derived from the haplotype A in Kanto radiated to the northern part of Honshu before invading the Sea of Japan through the Tsugaru channel; and (4) the green lineage originated around Shikoku and Kyushu and then expanded into the Sea of Japan from Kyushu to Tohoku. Expansion from Kyushu to Hokkaido in the Sea of Japan was speculated for *Laurencia nipponica* (Masuda et al. 1992) and *Chondrus nipponicus* (Brodie et al. 1991). Our results suggest the possibility of a bidirectional expansion of the *U. prolifera* group into the Sea of Japan: one through the Tsugaru channel and the other from Kyushu. This represents the first report of such a bidirectional expansion into the Sea of Japan, which was closed during the last ice-age (some 70,000 to 30,000 years ago) because the sea level subsided by 120 m (Horikoshi et al. 1996). The Tsugaru channel was reopened some 20,000 years ago, and seawater entered the Sea of Japan from the Tsugaru channel (Horikoshi et al. 1996). It is hypothesized that the blue lineage of the *U. prolifera* group entered the Sea of Japan at that time through the Tsugaru channel. As for the expansion of the green lineage of *U. prolifera* to the Sea of Japan from Kyushu, it is hypothesized that it was associated with the flow of the Tsushima current through the Tsushima channel some 8,000 years ago.

Acknowledgements We thank Dr. A. Kurihara of Hokkaido University and Dr. M. Iima of Nagasaki University, who generously collected some of the materials. We are deeply indebted to Dr. E.J. Faye of Port and Airport Research Institute, Yokosuka, for helpful discussions. The help of Professor C.A. Maggs of Queen's University, Belfast, with stimulating discussion, constructive criticism and valuable suggestions on an early version of the manuscript was greatly appreciated. This research was partially supported by the Ministry of Education, Science, Sports and Culture, Grant-in-Aid for Encouragement of Young Scientists (A), 17687007, 2005–2007.

References

Blomster J, Maggs CA, Stanhope MJ (1998) Molecular and morphological analysis of *Enteromorpha intestinalis* and *E. compressa* (Chlorophyta) in the British Isles. J Phycol 34:319–340

Brodie J, Guiry MD, Masuda M (1991) Life history and morphology of *Chondrus nipponicus* (Gigartinales, Rhodophyta) from Japan. Br Phycol J 26:33–50

Canter-Lund H, Lund JWG (1995) Freshwater Algae. Their microscopic world explored. Biopress, Bristol

Clement M, Posada D, Crandall KA (2000) TCS: a computer program to estimate gene genealogies. Mol Ecol 9:1657–1659

Coat G, Dion P, Noailles MC, De Reviers B, Fontaine JM, Bergaer-Perrot Y, Loiseaux-De Goër S (1998) *Ulva armoricana* (Ulvales, Chlorophyta) from the coasts of Brittany (France). II. Nuclear rDNA ITS sequences analysis. Eur J Phycol 33:81–86

Dan A, Hirosawa A, Makino K, Ohno M, Critchley AT (2002) Observations on the effect of salinity and photon fluency rate on the induction of sporulation and rhizoid formation in the green alga, *Enteromorpha prolifera* (Müller) J. Agardh (Chlorophyta, Ulvales). Fish Sci 68:1182–1188

Dan A, Hirosawa A, Makino K, Ohno M, Notoya M (2003) Low temperature storage of young *Enteromorpha prolifera* (Ulvacea; Chlorophyta). Suisanzoushoku 51:7–14

Edwards DM, Reed RH, Stewart WDP (1988) Osmoacclimation in *Enteromorpha intestinalis*: long-term effects of osmotic stress on organic solute accumulation. Mar Biol 98:467–476

Guiry MD, Guiry GM (2007) AlgaeBase version 4.2. World-wide electronic publication, National University of Ireland, Galway. http://www.algaebase.org; searched on 14 March 2007

Felsenstein J (1985) Confidence limits on phylogenies: an approach using the bootstrap. Evolution 39:783–791

Hayden HS, Blomster J, Maggs CA, Silva PC, Stanhope MJ, Waaland JR (2003) Linnaeus was light all along: *Ulva* and *Enteromorpha* are not distinct genera. Eur J Phycol 38:277–294

Hiraoka M, Dan A, Shimada S, Hagihira M, Migita M, Ohno M (2003) Different life histories of *Enteromorpha prolifera* (Ulvales, Chlorophyta) from four rivers on Shikoku Island. Phycologia 42:275–284

Hiraoka M, Shimada S (2004) Biology of a special green laver, *Ulva prolifera* from the Shimanto River. Aquabiology 26:508–515

Horikoshi M, Nagata Y, Sato T, Handa N (1996) Nature in Japan: Sea around Japanese archipelago. Iwanami-shoten, Tokyo

Kamer K, Fong P (2000) A fluctuating salinity regime mitigates the negative effects of reduced salinity on the estuarine macroalga, *Enteromorpha intestinalis* (L.) Link. J Exp Mar Biol Ecol 254:53–69

Malta EJ, Draisma SGA, Kamermans P (1999) Free-floating *Ulva* in the southwest Netherlands: species or morphotypes? A morphological, molecular and ecological comparison. Eur J Phycol 34:443–454

Martins I, Oliveria JM, Flindt MR, Marques JC (1999) The effect of salinity on the growth rate of the macroalgae *Enteromorpha intestinalis* (Chlorophyta) in the Mondego estuary (west Portugal). Acta Oecol 20:259–265

Masuda M, Abe T, Saito Y (1992) The conspecificity of *Laurencia yendoi* Yamada and *L. nipponica* Yamada (Ceramiales, Rhodophyta). Jpn J Phycol 40:125–133

McAvoy KM, Klug JL (2005) Positive and negative effects of riverine input on the estuarine green alga *Ulva intestinalis* (syn. *Enteromorpha intestinalis*)(Linnaeus). Hydobiologia 545:1–9

Müller OF (1778) Icones plantarum. Florae danicae, vol. 5, fasc 13. . Copenhagen

Nakura G (1921) Report of seaweeds, vol 6. Rep Aichi Fish Res Inst 24:1–18

Ohno M (ed) (2004) Biology and technology of economic seaweeds. Uchida-Roukakuho, Tokyo

Ohno M, Mizutani Y, Taino, Takahashi I (1999) Ecology of the edible green alga *Enteromorpha prolifera* in Shimanto River, southern Japan. Bull Mar Sci Fish Kochi Univ 19:27–35

Ohno M, Takahashi I (1988) The horizontal and vertical distribution of the food alga *Enteromorpha prolifera* in Shimanto River, southern Japan. Rep Usa Mar Biol Inst Kochi Univ 10:45–54

Pandey RS, Ohno M (1985) An ecological study of cultivated *Enteromorpha*. Rep Usa Mar Biol Inst Kochi Univ 7:21–31

Posada D, Crandall KA (1998) Modeltest: testing the model of DNA substitution. Bioinformation 14:817–818

Pringle JD (1986) Swarmer release and distribution of life-cycle phases of *Enteromorpha intestinalis* (Chlorophyta) in relation to environmental factors. J Exp Mar Biol Ecol 100:97–112

Reed RH, Russell G (1979) Adaptation to salinity stress in populations of *Enteromorpha intestinalis* (L.) Link. Estuar Coast Mar Sci 8:251–258

Ritchie RJ, Larkum AWD (1985) Potassium transport in *Enteromorpha intestinalis*: II. Effects of medium composition and metabolic inhibitors. J Exp Bot 36:394–412

Sherwood AR, Shea TB, Sheath RG (2002) European freshwater *Hildenbrandia* (Hildenbrandiales, Rhodophyta) has not been derived from multiple invasions from marine habitats. Phycologia 41:87–95

Shimada S, Masuda M (2003) Reassessment of the taxonomic status of *Gelidium subfastigiatum* (Gelidiales, Rhodophyta). Phycol Res 51:271–278

Shimada S, Hiraoka M, Nabata S, Iima M, Masuda M (2003) Molecular phylogenetic analyses of the Japanese *Ulva* and *Enteromorpha* (Ulvales, Ulvophyceae), with special reference to the free-floating *Ulva*. Phycol Res 51:99–108

Shimada S, Yokoyama N, Masuda M (2007) Genus *Ulva* (Ulvophyceae, Chlorophyta) in Hokkaido, Japan. Jpn J Bot 82 (in press)

Swofford DL (2002) PAUP*: Phylogenetic Analysis Using Parsimony (*and other methods), Version 4. Sinauer, Sunderland, Mass.

Uchimura T, Faye EJ, Shimada S, Arai S, Inoue T, Nakamura Y (2006) Are-evaluation of the taxonomic status of *Halophila euphlebia* Makino (Hydrocharitaceae) based on morphological features and ITS sequence data. Bot Mar 49:111–121

van den Hoek C, Mann DG, Jahns HM (1995) Algae. An Introduction to Phycology. Cambridge University Press, Cambridge

Yoshida Y, Shimada S, Yoshinaga K, Nakajima Y (2005) Checklist of marinealgae of Japan. Jpn J Phycol 53:179–228

Yotsukura N, Kawai T, Kawashima S, Ebata H, Ichimura T (2006) Nucleotide sequence diversity of the 5S rDNA spacer in the simple blade kelp genera *Laminaria*, *Cymathaere* and *Kjellmaniella* (Laminariales, Phaeophyceae) from northern Japan. Phycol Res 54:269–279

Yotsukura N, Kawai T, Motomura T, Ichimura T (2002) Tandem 5S ribosomal RNA genes and spacer region sequences of three Japanese *Laminaria* species. J Appl Phycol 14:233–239

Observations on the division characterization of diploid nuclear in *Porphyra* (Bangiales, Rhodophyta)

Wei Zhou · Jianyi Zhu · Songdong Shen · Shan Lu · Jinfeng Wang · Jianrong Xu · Pu Xu

Originally published in the Journal of Applied Phycology, Vol 20, No 5, 541–549.
DOI: 10.1007/s10811-007-9235-y

Abstract Nuclear divisions of carpospores, conchocelis and conchospores of *Porphyra yezoensis*, *P. haitanensis*, *P. katadai* var. *hemiphylla* and *P. oligospermatangia* from China were investigated. The observations showed diploid chromosome numbers of 2n=6 for *P. yezoensis* and *P. oligospermatangia*, and 2n=10 for *P. haitanensis* and *P. katadai* var. *hemiphylla*. For all four species, somatic pairing of chromosome sets was observed in late prophase. Sister chromosomes separated at anaphase as mitosis took place in carpospores, conchocelis filamentous cells, conchosporangial branch cells and sporangial cells (conchospore formation). Chromosome configurations of tetrad and ring-shaped in conchospore germination were observed, demonstrating the occurrence of meiosis. The characteristics of diploid nuclear division in 2n=6 species are the same as those of 2n=10 species. The influence of somatic pairing on nuclear division of diploid cells in *Porphyra* was discussed.

Keywords *Porphyra* · Conchocelis · Conchospores · Somatic pairing · Meiosis

W. Zhou · J. Zhu (✉) · J. Wang · J. Xu · P. Xu
Department of Biology, Changshu Institute of Technology, Changshu 215500, China
e-mail: zhujianyint@sina.com.cn

W. Zhou · S. Shen
School of Life Sciences, Soochow University, Suzhou 215123, China

S. Lu
School of Life Sciences, Nanjing University, Nanjing 210093, China

Introduction

In 1949, Drew proved that *Conchocelis rosea* was the germinated carpospores which were released from the leafy (haploid) *Porphyra* plants (Drew 1949). This established the diploid phase (namely conchocelis) of *Porphyra* life history. However, there have been disputed results on the observation of chromosome morphology of different species ever since, especially on the occurrence of meiosis.

Although sexual reproduction has been conclusively demonstrated in *P. gardneri* (Hawkes 1978) and implied by evidence from cytological studies in numerous other species (Yabu and Tokida 1963; Migita 1967; Giraud and Magné 1968; Kito 1967, 1968, 1974, 1978; Yabu 1969a, b, 1971, 1972, 1978; Hawkes 1977, 1978; Tseng and Sun 1989; Mumford and Cole 1977; Krishnamurthy 1984; Maand Miura 1984; Burzycki and Waaland 1987; Kapraun and Freshwater 1987), haploid nuclei were also reported at the conchocelis stage of several species. A haploid nucleus with n=5 was reported throughout the life history of *P. purpurea* (Krishnamurthy 1959), while n=4 for *P. carolineusis* (Freshwater and Kapraun 1986). Conway and Cole (1973) also described haploid conchocelis of *P. papenfussii*. In a study on 15 *Porphyra* species from the west coast of the North America, Mumford and Cole (1977) reported 3 diploid conchocelis and 4 haploid ones. In China, the presence of both haploid and diploid conchocelis also has been observed (Sun and Tseng 1987).

The disputed reports of *Porphyra* conchocelis ploidy were closely related to another controversy about the site of meiosis. In 1955, Tseng and Chang reported meiosis at the first division of the fertilized carpogonia in *P. tenera*, which concluded that carpospores must be haploid (Tseng and Chang 1955). However, some other observations (Magné, 1952; Yabu and Tokida 1963; Yabu 1969a, b, 1971, 1972;

Kito 1967, 1968, 1974; Mumford and Cole 1977; Hawkes 1977, 1978) demonstrated that the carpospores were produced without meiosis and had diploid chromosomes, although none of these observations could suggest when meiosis might occur. In between, Migita (1967), Giraud and Magné (1968) and Kito (1974, 1978) reported that the divisions leading to the formation of conchospores in the conchosporangia were meiosis. However, these reports were based exclusively on the visualiazation of synapsis at diplonema and diakinesis and no supporting evidence was presented that demonstrated a change in ploidy immediately following the presumed meiotic division (Burzycki and Waaland 1987). The observations which Migita, Giraud and Kito interpreted as prophase of meiosis could also be examples of somatic pairing (Burzycki and Waaland 1987; Tseng and Sun 1989). Wang et al. (2006) reported that prophase of meiosis I began during the formation of conchosporangial branches, while metaphase I, anaphase I and telophase I took place during the maturation of conchosporangial branches, and Meiosis II occurred at the first nuclear division during conchospore germination in *P. yezoensis*.

With the support of the genetic analysis (Miura and Kunifuji 1980; Ohme et al. 1986; Ohme and Miura 1988; Mitman and van der Meer 1994), Ma and Miura (1984), Burzycki and Waaland (1987) and Tseng and Sun (1989) agreed that meiosis occurred during the germination of conchospores. This conclusion was accepted by numerous researchers.

With the historical disagreements on the development of the conchocelis stage of *Porphyra* it is possible that the important characteristics of nuclear divisions have not been touched on. In this study we try to decipher it by studying the nuclear morphology of *Porphyra* species with different diploid chromosome numbers.

Materials and methods

The leafy thalli of *P. yezoensis* were collected from the field of Nantong, Jiangsu Province, China and those of *P. haitanensis* were from the field of Pintan, Fujian Province, China. *P. oligospermatangia* and *P. katadai* var. *hemiphylla* thalli were wild type collected from Qingdao, Shandong Province, China.

Methods

Collection and fixing of carpospores

Mature thalli of *P. yezoensis*, *P. haitanensis*, *P. katadai* var. *hemiphylla* and *P. oligospermatangia* were dried in the shade and then immersed in sterile seawater while being agitated to accelerate carpospore release. Slides were laid in the blotting silk filtered spore-water for carpospore attachment. The culture condition for carpospore germination was 20~22°C, 60 μmol photons m^{-2} s^{-1}, 12L:12D in boil-sterilized seawater. For sampling, the slides with spores were fixed in Carnoy's solution (3:1 ethanol:glacial acetic acid) at 2~3 h post sunset for more than 12h.

Collection and fixing of conchocelis

For conchocelis development, carpospores were inoculated into shells at about 350 cells cm^{-2}. The culture conditions varied with different developmental stages. It was 18~22°C, 50~60 μmol photons m^{-2} s^{-1}, 14L:10D at the conchocelis filamentous stage, 25~26°C, 50~60 μmol photons m^{-2} s^{-1}, 12L:12D at early conchosporangial branch stage, and 23~24°C, 50 μmol photons m^{-2} s^{-1}, 10L:14D for late conchosporangial branch stage for *P. yezoensis*, *P. katadai* var. *hemiphylla* and *P. oligospermatangia*. For *P. haitanensis*, the temperature was 2°C higher throughout. Provasoli enriched seawater (PES, Provasoli 1966) was used as the culture medium. At each stage, the shells were treated with Pereny's solution (0.4% HNO_3: 28% ethanol: 0.15% CrO_3) and then the conchocelis was fixed at 3~5 h post sunset.

Collection and fixing of conchospores

At late conchosporangial branch stage, the shell boring conchocelis was cultured with running seawater. Large amounts of conchospores were released when the culture temperature was switched to 5°C lower for *P. yezoensis* and *P. haitanensis* for 7~10 days, and 10°C lower for *P. katadai* var. *hemiphylla* and *P. oligospermatangia* for 10~15 days.

The spores were collected on slides as for carpospores, and cultured at 18~20°C, 70~80 μmol photons m^{-2} s^{-1}, 12L:12D, with boiling-sterilized seawater for *P. yezoensis*, *P. katadai* var. *hemiphylla* and *P. oligospermatangia*, while the temperature was 3°C higher for *P. haitanensis*. The spores were fixed at 2h post sunset on the first day or 3h before sunset on the second day.

All the fixed materials were stained with aceto-iron-haematoxylin-chloral hydrate of Wittmann (1965) or iron alum-haematoxylin. Stained slides were observed and photographed under oil immersion on Nikon E800 microscope.

Observations

Carpospore germination

Carpospores were cell wall-less and amoeboid when released. During germination, the spores first gave rise to germination tubes.

In the prophase of germination, chromatic threads condensed gradually and the phenomenon of homologous chromosomes pairing appeared later on, which were observed as 3 chromosomes for *P. yezoensis* (Fig. 1). In metaphase, chromosomes lay in the equatorial plane while homologous chromosomes were parallel with or approached one another. Then, sister chromosomes separated into two sets, one of which moved into germination tube. At telophase, cell wall formed in the basal part of germination tube.

Nuclear division of conchocelis filamentous cells

Conchocelis filaments developed from carpospores, of which cells and nuclei were both threadlike. Pit-connections exist between cells (Fig. 2a).

At late prophase of nuclear division in conchocelis filamentous cells of *P. yezoensis*, chromosomes aligned themselves to the longitude axis of the cells. Homologous chromosomes lay closely together and only 3 of them could be observed (Fig. 2b,c). Figure 2b–e indicates different views of the nuclear division in conchocelis filamentous cells from late prophase to early metaphase. At this time, paired chromosomes generally formed a Y- or V-shape (Fig. 2e), and spread apart a little along with the division. At metaphase, chromosomes lay at the equatorial plane along the longitude axis of the filamentous cells. Homologous chromosomes corresponded to each other, showing the presence of 6 chromosomes (Fig. 2f–h). At late metaphase, sister chromosomes started to separate (Fig. 2i,j) so that 7–12 chromosomes could be seen in different views (Fig. 2k). At anaphase, sister chromosomes separated into two sets and moved to the two poles along the longitudinal axis of the filamentous cells where 12 chromosomes could be observed clearly (Fig. 2l). Finally, a new cell wall was produced horizontally at telophase.

In *P. haitanensis*, homologous chromosomes also paired up at late prophase of the nuclear division in conchocelis filamentous cells and 5 chromosomes could be seen (see Fig. 7a–c). At metaphase, the distance between paired chromosomes increased while 10 chromosomes lay at the equatorial plane (Fig. 7d,e). Figure 7d shows 10 chromosomes corresponding to each other vertically to the longitudional axis of the cell. In most conditions, chromosomes lined up along the longitudional axis of the cell. Although the chromosome number was hard to count, 5–10 chromosomes could generally be found. In Fig. 7e, 10 chromosomes can be distinguished clearly.

The conchocelis filamentous cells of *P. katadai* var. *hemiphylla* were 2n=10, in which nuclear divisions were similar to that of *P. haitanensis*. Separation of sister chromosomes could be observed at anaphase (see Fig. 13a,b).

Nuclear division of conchosporangial branch cells

Along with the development of conchocelis, filamentous threadlike cells gradually become condensed, forming conchosporangial branches. Early conchosporangial branch cells are still long, and the ratio of length to width gradually decreases with development. Ramose cells developed from interjacent cells are more mature. Late conchosporangial branch cells were nearly square with pit-connections in between (Fig. 3a).

At late prophase of early conchosporangial branch cells of *P. yezoensis*, homologous chromosomes paired up closely into 3 chromosomes (Fig. 3b). The distance between chromosomes in these cells was a little longer than that in conchocelis filamentous cells. From late prophase to metaphase, paired chromosomes gradually moved apart, forming a reticulate configuration (Fig. 3c,d), while the number of chromosomes was counted as 4–6. At metaphase, 6 chromosomes lay at the equatorial plane and homologous ones were still parallel to one another (Fig. 3e,f). In some early conchosporangial branch cells, the widths were so short that 6 chromosomes still lined up along the longitudional axis of cells, just like the metaphase of conchocelis filamentous cells (Fig. 3f). At late metaphase, sister chromosomes separated (Figs. 3g,h) and then moved to the two poles.

The nuclear divisions of late conchosporangial branch cells were mainly the same as the earlier ones. Homologous chromosomes also matched up at late prophase. In metaphase, 6 chromosomes moved to the equatorial plane, forming two lines while the space between chromosomes increased compared to early conchosporangial branch cells. The arrangement patterns of chromosomes were different according to cell types (Fig. 4a–h). In anaphase, sister chromosomes separated and a new wall appeared horizontally at telophase.

In *P. haitanensis*, nuclear divisions of early conchosporangial branch cells were similar to those of *P. yezoensis*. At late prophase, homologous chromosomes connected tightly into 5 sets and the 10 chromosomes lay at the equatorial plane at metaphase (see Fig. 8a). Then sister chromosomes separated at late metaphase to anaphase (see Fig. 8b). In nuclear divisions of late conchosporangial branch cells, homologous chromosomes also paired up closely at late prophase to early metaphase (see Fig. 9a,b). The nuclear division of this phase in *P. katadai* var. *hemiphylla* was the same as that of *P. haitanensis*, and at metaphase of late conchosporangial branch cells, 10 chromosomes presented in the equatorial plane (see Fig. 14).

Nuclear division of sporangial cells

Sporangial cells developed from conchosporangial branch cells, in which the cell divisions were called conchospore

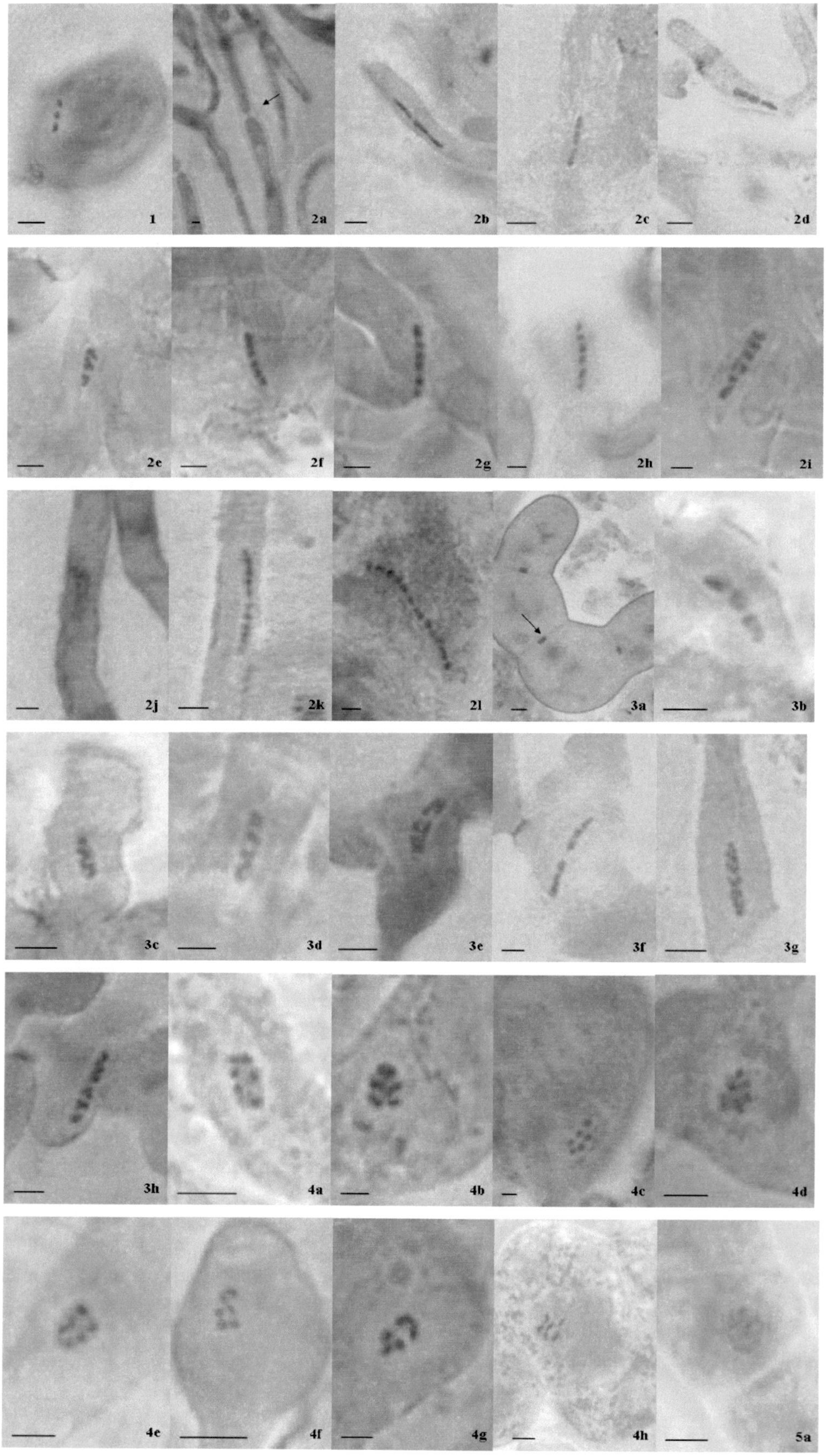
1
2a
2b
2c
2d
2e
2f
2g
2h
2i
2j
2k
2l
3a
3b
3c
3d
3e
3f
3g
3h
4a
4b
4c
4d
4e
4f
4g
4h
5a

◀ **Fig. 1** Late prophase of carpospore germination in *P. yezoensis*, homologous chromosomes paired up, with 3 chromosomes observed. Scale bar = 2.5μm

Fig. 2 a Pit-connection of conchocelis filamentous cells in *P. yezoensis*. **b–d** Late prophase of nuclear division of conchocelis filamentous cell in *P. yezoensis*, homologous chromosomes paired up to 3 sets. **e** Late prophase to early metaphase of nuclear division of conchocelis filamentous cell in *P. yezoensis*, homologous chromosomes showing a Y or V shape. **f–h** Metaphase of nuclear division of conchocelis filamentous cell in *P. yezoensis*, 6 chromosomes lying at the equatorial plane. **i,j** Late metaphase of nuclear division of conchocelis filamentous cell in *P. yezoensis*, sister chromosomes separated. **k** Anaphase of nuclear division of conchocelis filamentous cell in *P. yezoensis*, 7~12 chromosomes can be seen. **l** Anaphase of nuclear division of conchocelis filamentous cell in *P. yezoensis*, showing 12 chromosomes. Scale bars = 2.5μm

Fig. 3 a Pit-connection of conchosporangial branch cells in *P. yezoensis*. **b–d** Late prophase to early metaphase of nuclear division of early conchosporangial branch cell in *P. yezoensis*, homologous chromosomes paired up, and reticulate structure shown in Fig. 3c and d. **e,f** Metaphase of nuclear division of early conchosporangial branch cell in *P. yezoensis*, 6 chromosomes lying at the equatorial plane. **g,h)** Late metaphase to early anaphase of early conchosporangial branch cell in *P. yezoensis*, sister chromosomes separated. Scale bars = 2.5μm

Fig. 4 a–h Metaphase of nuclear division of late conchosporangial branch cell in *P. yezoensis*, the arrangement patterns of chromosomes different according to cell types. Scale bars = 2.5μm

Fig. 5 a Prophase of nuclear division of sporangial cell in *P. yezoensis*, showing chromatin of little grains shape. **b,c)** Prophase of nuclear division of sporangial cell in *P. yezoensis*, chromosomes gradually coiled, homologous chromosomes paired up closely. **d** Metaphase of nuclear division of sporangial cell in *P. yezoensis*, 6 chromosomes lying at the equatorial plane. **e** Late metaphase to early anaphase of sporangial cell in *P. yezoensis*, sister chromosomes separated. **f** Metaphase of second nuclear division of sporangial cell in *P. yezoensis*, showing 6 chromosomes. Scale bars = 2.5μm

formation. Sporangial cells generally divided only once, producing two conchospores. Second division sometimes occurs, forming 3–4 conchospores. Sporangial cells have a cell wall around them, but newly formed cells have neither cell wall nor pit-connections.

At the onset of sporangial cell nuclear division in *P. yezoensis*, the nucleoplasm was filled with granules (Fig. 5a), and thin coiled chromatic threads appeared subsequently. At prophase, homologous chromosomes paired up (Fig. 5b,c). During metaphase, chromosomes were condensed and short laid at the equatorial plane, while the distance between paired chromosomes was further than that in prophase (Fig. 5d). Then sister chromosomes separated and moved to two poles (Fig. 5e). Later diaphragm formed horizontally at telophase. 6 chromosomes could be clearly observed at metaphase in the sporangial cells taking second division (Fig. 5f).

The nuclear division of sporangial cells in *P. haitanensis* is similar to that in *P. yezoensis*. At prophase, homologous chromosomes paired up closely into 5 sets (see Fig. 10a). From late prophase, the distance between paired chromosomes increased while diploid chromosomes could be distinguished. Ten chromosomes lay at the equatorial plane at metaphase (Fig. 10b), and sister chromosomes separated at anaphase.

Conchospore germination

At prophase of conchospore germination of *P. yezoensis*, homologous chromosomes paired up more tightly while only 3 chromosomes could be seen (Fig. 6a,b). It is difficult to trace the entire course of matching. At metaphase, tetrad configurations could sometimes be distinguished, while in most cases 3 ring-shaped chromosomes could be observed, demonstrating the separation of synapsis chromosomes (Fig. 6c,d). During anaphase, homologous chromosomes separated into two sets (Fig. 6e–g) and moved to the two poles, respectively, when the spindle fibrils could be clearly observed (Fig. 6h,i).

For the second division during conchospore germination, the upper one of the two cells of the germling usually divided first. 3 chromosomes could be observed at metaphase (Fig. 6j), and sister chromosomes separated at anaphase (Fig. 6k).

Relating to the above sections, see Figs. 7, 8, 9, and 10.

The metaphase of conchospore germination in *P. oligospermatangia* (see Fig. 12) also showed the same morphology of 3 ring-shaped chromosomes. Synapsis chromosomes presented at late prophase of conchospore germination of *P. haitanensis* (Fig. 11a–b). 5 ringed chromosomes showing homologous chromosomes began to separate at metaphase (Fig. 11c). In *P. katadai* var. *hemiphylla*, late prophase and metaphase were the same as those of *P. haitanensis* (see Fig. 15a,b), and homologous chromosomes separated and moved apart at anaphase (Fig. 15c,d).

Relating to the above sections, see Figs. 11, 12, 13, 14, and 15.

Discussion

Characterization of nuclear division stages of conchocelis-phase of *Porphyra*

The carpospore germination of *P. yezoensis* was described by Migita (1967) and Kito (1978) in detail. It was suggested that the nuclear division at this stage was the same as mitosis of higher plants. Their results were supported by Sun and Tseng (1987). In our observations reported here, carpospore germination was indeed mitosis, but homologous chromosomes paired at prophase and metaphase.

At the stage of conchocelis filamentous cells of *P. yezoensis*, Migita (1967) and Kito (1978) found that the

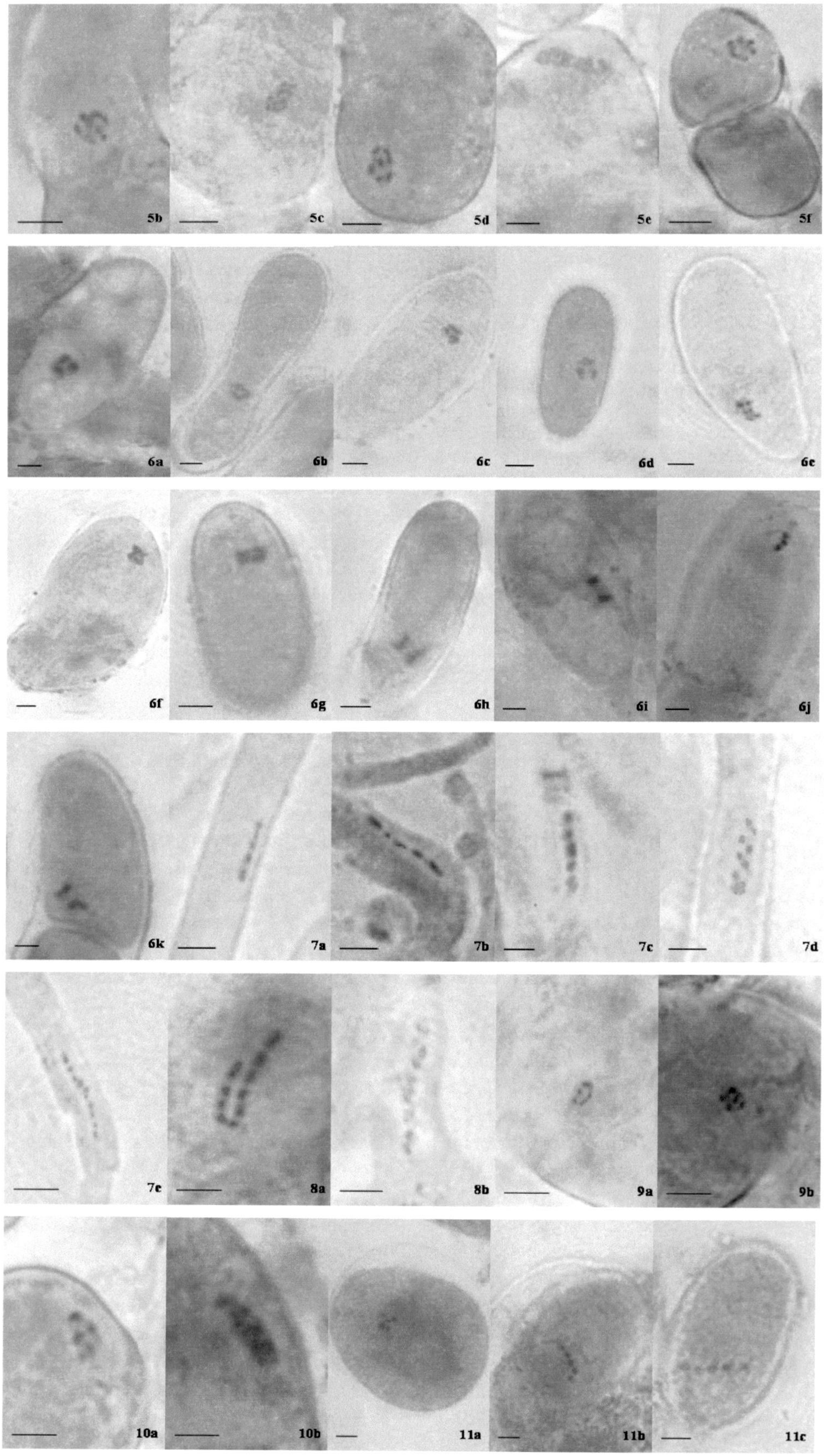
5b
5c
5d
5e
5f
6a
6b
6c
6d
6e
6f
6g
6h
6i
6j
6k
7a
7b
7c
7d
7e
8a
8b
9a
9b
10a
10b
11a
11b
11c

◀ **Fig. 6 a,b** Prophase of conchospore germination in *P. yezoensis*, homologous chromosomes paired up more tightly, only 3=chromosomes can be seen. **c,d** 3-ringed structure at metaphase of conchospore germination in *P. yezoensis*. **e,f** Late metaphase of conchospore germination in *P. yezoensis*, homologous chromosomes separated. **g** Early anaphase of conchospore germination in *P. yezoensis*, two sets began to move apart. **h,i** Anaphase of conchospores germination in *P. yezoensis*, two sets moved apart, spindle fibers could be seen clearly. **j** Metaphase of upper cell in the individual germinated from conchospore in *P. yezoensis*, showing 3 chromosomes. **k** Anaphase of upper cell in the individual germinated from conchospore in *P. yezoensis*, sister chromosomes separated. Scale bars = 2.5μm

Fig. 7 a–c) Late prophase of nuclear division of conchocelis filamentous cell in *P. haitanensis*, 5 chromosomes could be seen. **d–e)** Metaphase of nuclear division of conchocelis filamentous cell in *P. haitanensis*, 10 chromosomes lying at the equatorial plane. Scale bars = 2.5μm

Fig. 8 a Metaphase of nuclear division of early conchosporangial branch cell in *P. haitanensis*, 10 chromosomes lying at the equatorial plane. **b** Late metaphase to early anaphase of early conchosporangial branch cell in *P. haitanensis*, sister chromosomes separated. Scale bars = 2.5μm

Fig. 9 a,b) Late prophase of nuclear division of late conchosporangial branch cell in *P. haitanensis*, homologous chromosomes paired in 5 sets. Scale bars = 2.5μm

Fig. 10 a Late prophase of nuclear division of sporangial cell in *P. haitanensis*, homologous chromosomes paired in 5 sets. **b** Metaphase of nuclear division of sporangial cell in *P. haitanensis*, 10 chromosomes lying at the equatorial plane and parallel to one another. Scale bars = 2.5μm

Fig. 11 a,b Late prophase of conchospore germination in *P. haitanensis*, chromosomes paired in 5 sets. **c** Metaphase of conchospore germination in *P. haitanensis*, 5 ringed structure appeared. Scale bars = 2.5μm

process of nuclear division could not be traced clearly while only 5 chromosomes could be observed at metaphase. Sun and Tseng (1987) described nuclear division of *P. tenera*, *P. yezoensis* and *P. katadai* var. *hemiphylla* at this stage as mitosis, and homologous chromosomes laid closely with crosses at one point or several points, while some others paired completely in late prophase and early metaphase. These reports were consistent with the result of *P. yezoensis, P. oligospermatangia*, *P. haitanensis* and *P. katadai* var. *hemiphylla* in our study.

Although, at this stage, Krishnamurthy (1959) reported 5 chromosomes at late prophase of *P. purpurea*, and Freshwater and Kapraun (1986) also reported 4 chromosomes in *P. carolineusis*, which suggested haploid nuclear. In 1978, Kito proved 10 chromosomes in conchocelis filamentous cells of *P. purpurea* by cytological research (Kito 1978). In addition, diploid conchocelis of *P. purpurea* was also affirmed by Mitman and van der Meer (1994) through cytological and genetic research. Based on more observations, it seems that the influence of homologous chromosomes pairing in the division might be neglected. Homologous chromosomes pairing in diploid stage of *Porphyra* made it hard to count the chromosomes number before metaphase.

At the conchosporangial branch stage of *P. yezoensis*, Migita (1967) classified it into immature and mature types. The divisions of immature conchosporangial cells were generally mitosis. At the beginning of nuclear division in mature conchosporangial cells, 6 chromosomes appeared in pairs and 2 chromosomes of each pair associated very intimately. The phase was considered as the diplotene stage. In diakinesis, 3 bivalent chromosomes were observed, and then separated into two sets which began to move apart toward the opposite poles of the cell. Mature conchosporangial cells reported by Migita (1967) were actually sporangial cells, and the observation is consistent with our study reported here. It could be inferred from numerous observations that the characteristics of leptotene and the following course of nuclear division in diploid cells of *Porphyra* were difficult to distinguish. The diplotene mentioned by Migita (1967) was similar to metaphase of sporangial cells and the diakinesis was identical with the somatic paring of homologous chromosomes. Therefore, the recognition of meiosis was influenced by the wrong estimation of stages in nuclear division.

Kito (1978) similarly classified conchosporangial branches into immature and mature types. He described how homologous chromosomes paired into 3 bivalent chromosomes, considered to be diakinesis in mature conchosporangial branch cells. From our observations, homologue pairing in conchosporangial branch cells is not related to meiosis, but is consistent with somatic pairing at late prophase in mitosis.

Wang et al. (2006) reported that meiosis I began during the formation of conchosporangial branches and continued through the maturity, and meiosis II occurred at the first nuclear division of conchospore germination. According to our observations, there were a series of mitoses during the formation and maturation of conchosporangial branches. As important evidence, the diakinesis described by Wang is the same as the results of earlier Japanese researchers. In their observations, the universality of somatic pairing in conchocelis phase had apparently not been noticed.

Characterization of somatic pairing in mitosis of conchocelis phase in *Porphyra*

Cole et al. (1983) reported that somatic pairing of the two chromosome sets showed in prophase preparations of most diploid cells in *Bangia*, which was also reported in conchosporangial branch cells of *P. torta* by Burzycki and Waaland (1987) and conchocelis filamentous cells in *P. tenera*, *P. yezoensis* and *P. katadai* var. *hemiphylla* by Tseng and Sun (1989). Even earlier, Tseng and Chang (1955) described the homologous chromosomes synapsis

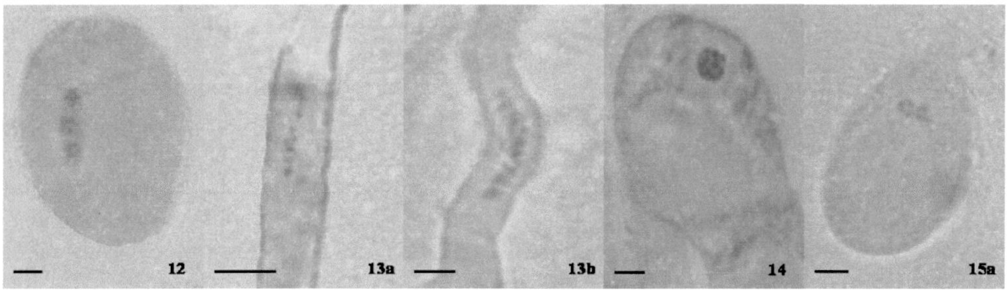

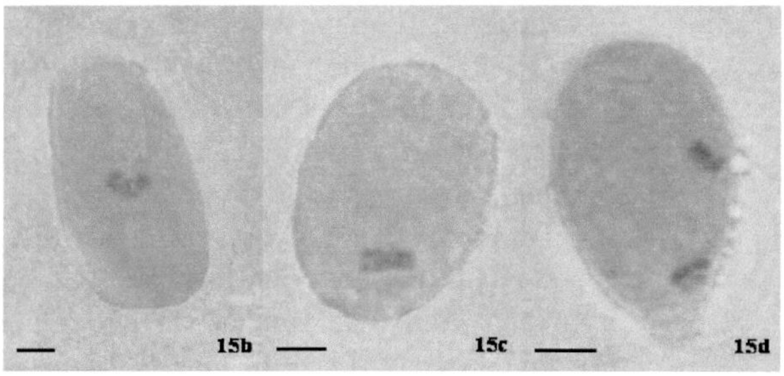

Fig. 12 Metaphase of conchospore germination in *P. oligospermatangia*, showing 3 tetrad configurations. Scale bar = 2.5μm

Fig. 13 **a** Metaphase of nuclear division of conchocelis filamentous cell in *P. katadai* var. *hemiphylla*, 10 chromosomes lying at the equatorial plane. **b** Late metaphase to early anaphase of conchocelis filamentous cell in *P. katadai* var. *hemiphylla*, sister chromosomes separated. Scale bars = 2.5μm

Fig. 14 Metaphase of nuclear division of late conchosporangial branch cell in *P. katadai* var. *hemiphylla*, 10 chromosomes lying at the equatorial plane. Scale bar = 2.5μm

Fig. 15 **a** Late prophase of conchospore germination in *P. katadai* var. *hemiphylla*, homologous chromosomes paired up to 5 sets. b Metaphase of conchospore germination in *P. katadai* var. *hemiphylla*, showing 5 ringed chromosomes configurations. **c** Late metaphase of conchospore germination in *P. katadai* var. *hemiphylla*, homologous chromosomes separated. **d** Anaphase of conchospore germination in *P. katadai* var. *hemiphylla*, two sets moved apart. Scale bars = 2.5μm

and chiasma in nuclear division of fertilized carpogonia in *P. tenera*. According to our results, homologous chromosomes paired at prophase and metaphase in diploid cells and the phenomenon of somatic pairing was the important characteristic of conchocelis-phase in *Porphyra*. The figures show that the morphology is similar to homologue pairing in the prophase of meiosis also presented at late prophase of cells at different stages of conchocelis-phase. Some figures show much the same phenomenon with the typical diakinesis stage. This might help to explain how the wrong estimation of meiosis came about.

Characterization of meiosis in *Porphyra*

In conchospore germination of *P. yezoensis*, Ma and Miura (1984) observed homologous chromosomes association at late prophase, 3 ring-shaped chromosomes presented at metaphase, and finally homologous chromosomes separated. Burzycki and Waaland (1987) reported that chromosomes still contained two chromatids at anaphase of conchospore germination in *P. torta*. Tseng and Sun (1989) described the meiosis of *P. tenera* and *P. yezoensis*. In all these reports, the course of meiosis and its cytological characteristics had been demonstrated for those 2n=6 species. With *P. purpurea* of 2n=10, Mitman and van der Meer (1994) suggested that meiosis took place at the germination of conchospores as well. But the cytological characteristics of meiosis had not been adequately described.

In our work, we reveal a tendency for closer association of homologous chromosomes at the prophase of conchospore germination, compared to other diploid cells, which Ma and Miura (1984) described as diplonema and diakinesis. A more important characteristic is that the ring-shaped chromosomes were observed at metaphase of conchospore germination in all the four species, which was obviously different from nuclear divisions of other cells, and this could be confirmed by comparing conchospore germination with adjacent divisions. The ring-shaped chromosomes at metaphase of conchospore germination demonstrated the movement of synapsis chromosomes, which suggested that chiasma in non-sister chromosomes at the separation of homologous chromosomes. The characteristic of meiosis in the species of 2n=6 is the same as the species of 2n=10.

Acknowledgements This research was supported by the National Natural Science Foundation of China (NSFC, 40576074).

References

Burzycki GM, Waaland JR (1987) On the position of meiosis in the life history of *Porphyra torta* (Rhodophyta). Bot Mar 30:5–10

Cole KM, Hymes BJ, Sheath RG (1983) Karyotypes and reproductive seasonality of the genus *Bangia* (Rhodophyta) in British Columbia, Canada. J Phycol 19:136–145

Conway E, Cole K (1973) Observations on an unusual form of reproduction in *Porphyra* (Rhodophyceae, Bangiales). Phycologia 12:213–225

Drew KM (1949) Conchocelis phase in the life history of *Porphyra umbilicalis* (L.) Kutz. Nature 164:748–749

Freshwater DW, Kapraun DF (1986) Field, culture and cytological studies of *Porphyra carolinensis* Coll et Cox (Bangiales, Rhodophyta) from North Carolina. Jap J Phycol 34:251–262

Giraud A, Magné F (1968) La place de la méiose dans le cycle de dévelopment de *Porphyra umbilicalis*. C. R. Acad Sci Paris 267:586–588

Hawkes MW (1977) A field, culture and cytological study of *Porphyra gardneri* (Smith and Hollenberg) comb. nov. (Bangiales, Rhodophyta). Phycologia 16:457–469

Hawkes MW (1978) Sexual reproduction in *Porphyra gardneri* (Smith et Hollenberg) Hawkes (Bangiales, Rhodophyta). Phycologia 17:329–353

Kapraun DF, Freshwater DW (1987) Karyological studies of five species of *Porphyra* (Bangiales, Rhodophyta) from the North Atlantic and Mediterranean. Phycologia 26:82–87

Kito H (1967) Cytological studies of several species of *Porphyra*. Mitosis in carpospore-germlings of *Porphyra yezoensis*. Bull Fac Fish Hokkaido Univ 18:201–207

Kito H (1968) Cytological studies of several species of *Porphyra*. Chromosome number of *Porphyra tenera* Kjellm. Bull. Tohoku Reg. Fish Res Lab 28:137–140

Kito H (1974) Cytological observations on the conchocelis phase in three species of *Porphyra*. Bull. Tohoku Reg. Fish Res Lab 33:101–117

Kito H (1978) Cytological studies on genus *Porphyra*. Bull. Tohoku Reg. Fish Res Lab 39:29–84

Krishnamurthy V (1959) Cytological investigations on *Porphyra umbilicalis* (L.) kuetz. var. *laciniata* (Lightf.) J. Ag. Annls Bot N S 23:147–176

Krishnamurthy V (1984) Chromosome numbers in *Porphyra* C. Ag. Phykos 23:185–190

Ma JH, Miura A (1984) Observations of the nuclear division in the conchospores and their germlings in *Porphyra yezoensis* Ueda. Jap J Phycol 32:373–378

Migita S (1967) Cytological studies on *Porphyra yezoensis* Ueda. Bull Fac Fish Nagasaki Univ 24:55–64

Mitman GG, van der Meer JP (1994) Meiosis, blade development and sex determination in *Porphyra purpurea* (Rhodophyta). J Phycol 30:147–159

Miura A, Kunifuji Y (1980) Genetic analysis of the pigmentation types in the seaweed Susabi nori (*Porphyra yezoensis*). Iden 34 (9):14–20

Mumford J, Cole K (1977) Chromosome numbers for fifteen species in the genus *Porphyra* (Bangiales, Rhodophyta) from the west coast of North America. Phycologia 16:373–377

Ohme M, Miura A (1988) Tetrad analysis in conchospore germlings of *Porphyra yezoensis* (Rhodophyta, Bangiales). Plant Sci 57:135–140

Ohme M, Kunifuji Y, Miura A (1986) Cross experiments of the color mutants in *Porphyra yezoensis* Ueda. Jap J Phycol 34: 101–106

Provasoli L (1966) Media and prospects for the cultivation of marine algae. In: Watanabe A, Hattori A (eds) Cultures and Collections of Algae. Proc. U. S.-Japan Conf. Hakone, Sept. 1966. Jap Soc Plant Physiol pp 63–75

Sun AS, Tseng CK (1987) Cytological studies of *Porphyra*. Nuclear division of conchocelis-phase. Chinese Sci Bull 32: 707–710

Tseng CK, Chang TJ (1955) Studies on *Porphyra*. Sexual reproduction of *Porphyra*. Acta Bot Sin 4:153–166

Tseng CK, Sun AS (1989) Studies on the alternation of the nuclear phases and chromosome numbers in the life history of some species of *Porphyra* from China. Bot Mar 32:1–8

Wang J, Dai JX, Zhang YT (2006) Nuclear division of the vegetative cells, conchosporangial cells and conchospores of *Porphyra yezoensis* (Bangiales, Rhodophyta). Phycological Res 54: 204–210

Wittmann W (1965) Aceto-iron-haematoxylin-chloral hydrate for chromosome staining. Stain Technol 40:161–164

Yabu H (1969a) Observation on chromosomes in some species of *Porphyra*. Bull Fac Fish Hokkaido Univ 19:239–243

Yabu H (1969b) Mitosis in *Porphyra tenera* Kjellm. Bull Fac Fish Hokkaido Univ 20:1–7

Yabu H (1971) Observation on chromosomes in some species of *Porphyra*. Bull Fac Fish Hokkaido Univ 21:253–258

Yabu H (1972) Observation on chromosomes in some species of *Porphyra*. Bull Fac Fish Hokkaido Univ 22:261–266

Yabu H (1978) Chromosome numbers in species of *Porphyra* from Nova Scotia, Canada. Jap J Phycol 26:97–104

Yabu H, Tokida J (1963) Mitosis in *Porphyra*. Bull Fac Fish Hokkaido Univ 14:131–136

MIX
Papier aus verantwortungsvollen Quellen
Paper from responsible sources
FSC® C105338

If you have any concerns about our products,
you can contact us on
ProductSafety@springernature.com

In case Publisher is established outside the EU,
the EU authorized representative is:
Springer Nature Customer Service Center GmbH
Europaplatz 3, 69115 Heidelberg, Germany

Printed by Libri Plureos GmbH
in Hamburg, Germany